FEW BODY DYNAMICS

FEW BODY DYNAMICS

Proceedings of the VII International Conference on
Few Body Problems in Nuclear and Particle Physics
(Delhi, December 29, 1975 - January 3, 1976)

Sponsored by the Indian National Science Academy, University Grants Commission, and I.U.P.A.P. - U.N.E.S.C.O. Also supported by Kothari Scientific Research Institute. Hosted by the University of Delhi

Edited by:

ASOKE N. MITRA
IVO SLAUS
V.S. BHASIN
V.K. GUPTA

N·H
P C

1976

NORTH-HOLLAND PUBLISHING COMPANY
AMSTERDAM · NEW YORK · OXFORD

North-Holland ISBN for this Volume: 0 7204 04819

Published by:

NORTH-HOLLAND PUBLISHING COMPANY
AMSTERDAM, NEW YORK, OXFORD

Distributors for the U.S.A. and Canada:

ELSEVIER/NORTH-HOLLAND INC.
52 VANDERBILT AVENUE
NEW YORK, N.Y. 10017

Library of Congress Cataloging in Publication Data

International Conference on Few Body Problems in Nuclear and Particle Physics, 7th, University of Delhi, 1975-1976.
Few body dynamics.

Includes index.
1. Nuclear physics--Congresses. 2. Particles (Nuclear physics)--Congresses. 3. Problem of few bodies--Congresses. I. Mitra, Asoke Nath, 1929- II. Indian National Science Academy. III. India (Republic). University Grants Commission. IV. International Union of Pure and Applied Physics. V. Title.
QC770.I4977 1975 539.7 76-17270
ISBN 0-7204-0481-9

PRINTED IN THE NETHERLANDS

INTERNATIONAL ADVISORY COMMITTEE

1. M.G.K. Menon (India) - Chairman
2. H.A. Bethe (U.S.A.)
3. R. Bondelid (U.S.A.)
4. G.E. Brown (Denmark)
5. H. Bruckmann (W. Germany)
6. H.E. Conzett (U.S.A.)
7. L.M. Delves (U.K.)
8. L.D. Faddeev (U.S.S.R.)
9. S.S. Hanna (U.S.A.)
10. P.K. Iyenger (India)
11. V. Komarov (U.S.S.R.)
12. F. Levin (U.S.A.)
13. I. Lovas (Hungary)
14. I.E. McCarthy (Australia)
15. M.J. Moravcsik (U.S.A.)
16. V.I. Neudatchin (U.S.S.R.)
17. H.P. Noyes (U.S.A.)
18. P. Radvanyi (France)
19. R. Ramanna (India)
20. J.R. Richardson (Canada)
21. L. Rosen (U.S.A.)
22. W. Sandhas (W. Germany)
23. T. Sasakawa (Japan)
24. M. Simonius (Switzerland)
25. Yu. A. Simonov (U.S.S.R.)
26. I. Slaus (Yugoslavia)
27. R.J. Slobodrian (Canada)
28. J.A. Tjon (Netherlands)
29. L. van Hove (CERN, Switzerland)
30. A.N. Mitra (India) Secretary and Convener

LOCAL ORGANIZING COMMITTEE

1. F.C. Auluck
2. V.S. Bhasin
3. V.K. Gupta
4. S.K. Monga
5. S.C. Pancholi
6. K.K. Singh (Co-opted)
7. A.N. Mitra (Chairman)

NATIONAL COMMITTEE

1. M.G.K. Menon (Chairman)
2. F.C. Auluck
3. V.S. Bhasin
4. S.N. Biswas
5. A.S. Divatia
6. H.S. Hans
7. P.K. Iyenger
8. M.Z.R. Khan
9. D.N. Kundu
10. D.S. Kothari
11. M.K. Mehta
12. B.D. Nag Chaudhuri
13. M.K. Pal
14. S.P. Pandya
15. R. Ramanna
16. B.M. Udgaonkar
17. A.R. Verma
18. Y. Waghmare
19. C.S. Warke
20. A.N. Mitra (Secretary and Convener)

RECEPTION COMMITTEE

1. J.D. Anand
2. (Miss) Asha Lal
3. F.C. Auluck
4. V.S. Bhasin
5. M. Choudhary
6. P.N. Dheer (Co-opted)
7. S.L. Gupta (Joint Secretary)
8. V.K. Gupta
9. D.L. Katyal
10. R.S. Kaushal
11. L.S. Kothari (Chairman)
12. K. Mahalingam
13. M.M. Malhotra
14. A.N. Mitra
15. S.K. Monga
16. N.K. Nayyar
17. S.C. Pancholi
18. D.N. Parashar
19. C.S.G.K. Setty
20. R.K. Shivpuri
21. K.K. Singh
22. G.P. Srivastava -(Secretary)
23. M.P. Srivastava (Co-Opted)

EDITORIAL PREFACE

The history of two-body conferences dates back to the London Conference on Few Nucleon Problems and Nuclear Forces in 1959, which was in a way the first organized occasion to focus attention on the exciting possibilities of studies with few body systems. After the London Conference, the next organized discussion took place in Brela, Yugoslavia, 1967. The subsequent conferences in this series were held in Birmingham (1969), Budapest (1971), Los Angeles (1972) and Quebec (1974). There was an increasing realization that few body systems, apart from conventional manifestations, have a much wider role to play as connecting links - bridges - between different physical systems and even between different areas of physics - from atomic to subnuclear. Some of the obvious bridges are: nuclear structure - nuclear forces; particle physics - nuclear physics; atomic and molecular physics - nuclear physics; and few body problems - nuclear reactions. This bridge concept was envisaged as a theme for the Delhi Conference, extending the traditional domains of few body research from eV regions (relevant to electron systems) at one end, and to GeV regions (relevant to mesons and resonances) at the other.

Consequently, the format of the conference was adjusted to suit this purpose, viz. organization of parallel sessions and a workshop, with invited speakers for specialized topics. This was to be followed by Plenary sessions with perspective talks by invited speakers on the status of the field and its future, while summaries of deliberations in parallel session were to be provided by rapporteurs. However, in areas such as few electron systems, and mesons and resonances which had been included essentially for the first time in these conferences so that a small number of contributed papers was available, the rapporteurs were encouraged to give a broader review of the current status of research in these respective fields.

The organization of parallel sessions was entrusted to discussion leaders who were assisted by scientific secretaries to keep track of the trend of discussion and to bring the salient features to the attention of rapporteurs.

Through this device, it was felt that the discussions at parallel session would be effectively reflected in the reports of the rapporteurs.

The conference was organized in a three-tier fashion; the International Advisory Committee, the National Committee and the Local Organizing Committee. The Convener of the Conference was a bridge between these Committees, under the aegis of the Indian National Science Academy. The editors were entrusted to ensure rapid publication of the proceedings on the Los Angeles Conference pattern. Therefore it was decided to employ the camera-ready procedure and to request all contributors except rapporteurs to have their material ready by the time of the conference. In the same vein only those discussions in plenary sessions which were received in a written form from the participants concerned by the end of the conference were incorporated in these Proceedings.

In addition to more than 150 participants, there were about 50 observers at this Conference. Besides invited papers there were about 100 contributed papers, after an initial screening by the local members of the editorial committee. Most of these papers, which were available in a camera-ready form have been included in these Proceedings.

The conference was opened by Prof. M.G.K. Menon, F.R.S., Chairman of the Electronics Commission and Scientific Adviser to the Defence Minister. It was presided over by Prof. R.C. Mehrotra, the Vice-Chancellor of the University of Delhi, which acted as the host for the conference. The Indian National Science Academy, the main sponsor of the Conference, was represented at the inauguration by its immediate past President, Prof. D.S. Kothari; and the University Grants Commission, another cosponsor, by its Secretary Mr. R.K. Chhabra. A grant from the Kothari Scientific Research Institute helped the publication of the abstracts and other literature to be received by the participants at the beginning of the Conference. The Department of Physics and Astrophysics where the sessions were held was represented by its Chairman, Prof. L.S. Kothari. The sponsorship of this Conference by I.U.P.A.P. was acknowledged by the Convener of the Conference, who is a member of the Indian National Committee for I.U.P.A.P.

The Conference concluded with thanks to the participants on behalf of the Indian National Science Academy by its Foreign Secretary, Prof. B.D. Nagchaudhuri, and with thanks to the Organizers on behalf of the participants by Prof. S.S. Hanna, Stanford University.

The editors express their gratitude to Drs. B.V. Subbarayappa and A. Natesaiyer of the Indian National Science Academy and to Mr. M.M. Malhotra, I.N.S.A.'s Special Officer for this Conference, for their continuous service to ensure a smooth conduct of the Conference. Mr. K. Mahalingam and Mr. J.P. Joshi from the Department of Physics and Astrophysics, Delhi University have also helped greatly in this respect. The editors acknowledge the help received from Messers Shiv Kumar Sharma, S.K. Luthra, M.L. Balwani, L.R. Varma, R. Saini and Miss Madhu in connection with the Proceedings. Finally thanks are due to the North Holland Publishing Company, and Dr. P. Bolman in particular, for facilitating rapid publication of these Proceedings, with full understanding of the difficulties of the Editors.

Delhi, Jan. 8, 1976

Asoke N. Mitra
Ivo Slaus
V.S. Bhasin
V.K. Gupta

PROGRAMME

Monday, 29th December 1975

OPENING ADDRESS:

M.G.K. Menon, Chairman Electronic Commission, and Scientific Adviser to Defence Minister

Chairman: R.C. Mehrotra, Vice Chancellor, University of Delhi

PLENARY SESSION 1:

VEC at Calcutta (invited talk)

A.S. Divatia, VEC, Calcutta

Chairman: D.N. Kundu, Saha Institute, Calcutta

PARALLEL SESSIONS:

I. DYNAMIC EQUATIONS AND APPROXIMATION METHODS

Discussion Leader : H.P. Noyes, SLAC

Scientific Secretaries: V.S. Bhasin, K.K. Datta, A.V. Lagu

II. COMPUTATION TECHNIQUES AND FEW BODY BOUND STATES

Discussion Leader : L.P. Kok, Groningen-Weizmann

Scientific Secretaries: V.K. Gupta, H. Jacob, V.S. Mathur

III. WORKSHOP ON EXPERIMENTAL TECHNIQUES

Discussion Leader : R.O. Bondelid, N.R.L., Washington

Scientific Secretaries: R. Allas, J. Lambert, M.K. Mehta, S.C. Pancholi

Reception by the President, Indian National Science Academy (Dr. B.P. Pal)

Tuesday, 30th December, 1975

PLENARY SESSION 2:

Inclusive Reactions (invited talk)

R. Rajaraman, University of Delhi

Chairman: F. Levin, Brown Univ.

N-N Interaction (invited talk)

R. Vinh Mau, Univ. Paris

Chairman: M.J. Moravcsik, Univ. of Oregon

PARALLEL SESSIONS:

IV. BREAK-UP REACTIONS AND POLARIZATION

Discussion Leader: R.J. Slobodrian, Univ. Laval

Scientific Secretaries: D.P. Goyal, G.K. Mehta, R.K. Shivpuri

V. FEW ELECTRON SYSTEMS

Discussion Leader: I.E. McCarthy, Flinders University

Scientific Secretaries: Y.S.T. Rao, M.P. Srivastava

VI. PHOTON AND ELECTRON PROBES ON FEW BODY SYSTEMS

Discussion Leader: M.K. Sundaresan, Carleton University

Scientific Secretaries: R. Majumdar, R. Nath

VII. FEW BODY REACTIONS WITH MESONS AND RESONANCES

Discussion Leader: A.N. Mitra, University of Delhi

Scientific Secretaries: G. Bhamathi, K.C. Tripathy, M.K. Srivastava, C.S. Warke

VIII.NUCLEAR REACTIONS AND SCATTERING - FEW BODY ASPECTS

Discussion Leader: M.K. Banerjee, University of Maryland

Scientific Secretaries: S.K. Monga, S. Mukerjee (Saha Inst.)

Reception by the Vice-Chancellor, Univ. of Delhi (Prof.R.C. Mehrotra)

Wednesday, 31st December,1975

PLENARY SESSION 3:

Relativistic Aspects in Few Body Systems (invited talk)

F. Gross, College of William and Mary, Virginia

Chairman: M.I. Haftel, N.R.L., Washington

Coulomb Problem and Separation of Electromagnetic Effects (invited talk)

P.U. Sauer, Techn.Univ., Hannover

Chairman: I. Lovas, CRIP, Budapest

PLENARY SESSION 4:

Preliminary Results of the Research on Few Body Problems at TRIUMF (invited talk)

J.R. Richardson, TRIUMF

Status of the C.P. Anderson Meson Physics Facility (LAMPF) (invited talk)

M. Jain, L. Rosen, LAMPF

Chairman: R. Ramanna, BARC, Bombay

PLENARY SESSION 5:

Dynamical Equation and Approximation Methods (invited talk)

W. Sandhas, Univ. Bonn

Chairman: E. Schmid, Univ. Tubingen

Rapporteur's talk

Y.E. Kim, Purdue Univ.

Chairman: M. Fabre dela Ripelle, Orsay

PLANEARY SESSION 6:

Three Body Forces in Nuclei (invited talk)

B.H.J. McKellar, Melbourne Univ.

Chairman: S.P. Pandya, PRL, Ahmedabad

CONFERENCE BANQUET:

Speakers: Ivo Slaus and M.G.K. Menon

Friday, 2nd January, 1976

PLENARY SESSION 7:

Breakup Processes - a Bridge, (invited talk)

Ivo Slaus, Institute "R. Boskovic"

Rapporteur's talk

H. Pugh, National Science Foundation, Washington DC

Chairman: S.S. Hanna, Stanford University

PLENARY SESSION 8:

Few Electron Systems (invited talk)

E. Gerjouy, Univ. of Pittsburgh

Chairman: P.K. Iyenger, BARC, Bombay

Rapporteur's talk:

L. Spruch, New York Univ.

Chairman: D. Kouri, Univ. of Houston

PLENARY SESSION 9:

Computation Techniques - Rapporteur's talk

J.A. Tjon, Inst. Theor. Fys., Utrecht

Chairman: V. Komarov, Moscow State Univ.

Bound States - Rapporteur's talk

T. Sasakava

Chairman: S. Oryu, Tohoku Univ. Sci. Univ. of Tokyo.

<u>Reception by the President of India</u>

<u>Concert</u>

Saturday, 3rd January, 1976

PLENARY SESSION 10

Quark Physics (invited talk)

R.H. Dalitz, Oxford University

Resonances - Rapporteur's talk

G. Shaw, U.C. Irvine

Chairman: V. Singh, TIFR, Bombay

PLENARY SESSION 11:

Symmetries (invited talk)

E.M. Henley, Univ. of Washington

Chairman: H. Zingl , Univ. of Graz

Photon and Lepton Probes on Few Body Systems - Rapporteur's talk

M.K. Sundaresan, Carleton Univ., Ottawa

Chairman: K. Bleuler, Universitat Bonn

PLENARY SESSION 12:

Polarisation - Rapporteurs' talk

H.E. Conzett, LRL, Berkeley and
R.J. Slobodrian, Laval Univ., Quebec

Chairman: W. Gruebler, ETH, Zurich

PLENARY SESSION 13:

Meson Reactions with Few Body Systems - Rapporteur's talk

I.R. Afnan, Flinders University

Chairman: I.J.R. Aitchison, Oxford Univ.

PLENARY SESSION 14:

Nuclear Reactions and Scattering - Few Body Aspects - Rapporteur's talk

F. Levin, Brown University

Chairman: V. Vanzani, Padova Univ.

CLOSING CEREMONY:

B.D. NagChaudhuri, J.N.U., New Delhi

S.S. Hanna, Stanford University.

*****

TABLE OF CONTENTS

3.2. Dynamical Equations and Approximation Methods

3.3. Few Body Bound States and Computational Techniques

3.4. Scattering , Breakup Reactions, Polarization

3.5. Few Body Reactions with Mesons and Resonances

3.6. Symmetries, Photon and Lepton Probes on Two Body Systems

***** *****

PART 1

STATUS AND RESULTS FROM EXPERIMENTAL FACILITIES

VARIABLE ENERGY CYCLOTRON AT CALCUTTA

A.S. Divatia
Variable Energy Cyclotron Project
Bhabha Atomic Research Centre
Calcutta-700064, India

INTRODUCTION

The Bhabha Atomic Research Centre is building a 224 cm Variable Energy Cyclotron (VEC) at Calcutta [1,2]. It is based on the design of Berkeley 88" and Texas A&M Cyclotrons and it can accelerate various ions to a maximum energy of 130 Q^2/A MeV and protons to a maximum energy of 70 MeV, the limitation on proton energy being due to the frequency limitation. This machine will be the first of its kind in the country, and will be a leading nuclear research facility. It will be used by scientists from various research institutions and universities. In addition to the equipment for nuclear physics experiments, facilities will be provided for studies in solid state physics including radiation damage, radio-chemistry, bio-medicine etc.; isotope production facilities have also been planned. A cyclotron of this type will undoubtedly be a major asset for the study of few body problems in nuclear physics, as is evident from the programme of this Conference.

Let us first have a look at the progress made in various systems of the main machine and then discuss the beam transport system, research facilities and utilisation of VEC. Table 1 gives the design parameters of the machine and some properties of the beam.

Table 1

DESIGN PARAMETERS - VEC CALCUTTA

BEAM		HILL GAP	19.05 cm
MAXIMUM ENERGY	130 Q^2/A MeV	VALLEY GAP	29.97 cm
INTERNAL BEAM CURRENT	1mA	MAX. HILL FIELD	21.1kG
EXTERNAL BEAM CURRENT	100 μA	MAX. VALLEY FIELD	14.1kG
ENERGY RESOLUTION		ACCELERATION SYSTEM	
a) UNANALYSED BEAM	0.5% (FWHM)	DEE	1;180°
b) ANALYSED BEAM(1mm Slit)	0.024%(FWHM)	FREQUENCY RANGE	5.5-18.0 MHz
MAGNET		DEE VOLTAGE	70 kV (MAX)
POLE DIAMETER	224 cm	EXTRACTION SYSTEM	
SPIRAL SECTORS	3	DEFLECTOR VOLTAGE	120 kV(MAX)

SYSTEMS OF THE CYCLOTRON

1. Magnet

The 262 tonne main magnet frame has been fabricated at the Heavy Engineering Corporation, Ranchi and it has been assembled at the VEC Laboratory at Calcutta. Fig.1 shows the assembled magnet including the main coils. The yokes and yoke legs have been made from cast steel and the pole pieces from forged steel. Ultrasonic testing has been done to check for porosity. The composition of the steel is given in Table 2. The machining accuracy required is high and the accuracy has been tested by doing precise magnet gap measurements. The gap has been measured using an optical tilting level as well as an internal dial gauge. It has been found that the deviations at various points of the gap from the specified value are within 250 microns which is the required tolerance over most of the pole piece area. It is observed that only near the outer edge of the pole piece the deviation exceeds 250 microns. This implies that the magnetic field configuration is likely to be satisfactory.

Fig. 1 : Cyclotron Magnet

Table 2

Comparison of Specified and achieved steel composition

	Specified Composition	Achieved Composition (Typical)
Carbon	0.12% max.	0.11%
Silicon	0.35% max.	0.25%
Manganese	0.50% max.	0.30%
Phosphorous	0.04% max.	0.011%
Sulphur	0.05% max.	0.023%
Iron	Balance	Balance

The main coils were fabricated at the Bharat Heavy Electricals, Bhopal. These coils are made from square copper conductor, with a central hole for water cooling and they are designed for a total of 600,000 ampere-turns. Fig. 2 shows a completed pancake at the factory. As already seen earlier, the main coils have been installed by now.

There are total 17 trim coil pairs and 15 valley coil pairs, there being 5 coils per valley. These coils are made from a single copper conductor with a central hole and they are epoxy potted. Contribution of a trim coil to the main field will be about 160 gauss and a valley coil pair can generate 20-25 gauss first harmonic at the desired points.

Fig. 2 : Main Coil Pancake

The power supplies for all these coils have been fabricated at the VEC Laboratory and they have been installed in the pit area just below the cyclotron vault. The main magnet power supply has been designed to give a maximum current of 2800 A at 170 volts for the coil resistance of 55 milliohms. Its stability has been checked to be better than 1 part in 10^4. The current is continuously adjustable from 10% to 100% of the rated output value. The design is based on a 12 phase, SCR controlled system; magnetic transductors are used as current sensors and a temperature compensated zener diode kept in a temperature controlled oven is used for reference. All the power supplies have been tested using simulated resistance loads.

Fig. 3 : First Harmonic Measurement Apparatus

Preliminary Phase-I magnetic field measurements have been carried out with about 200A in the main coils. It was found that at 200A first harmonic amplitude was about 1 gauss at the radii of 25 cm and 100 cm without the ion source plug in position. At the intermediate radii it falls upto 0.4 gauss. Fig. 3 shows the first harmonic measurement apparatus. A difference signal of two identical search coils placed 120° apart and at the same radius is plotted from 0° to 360°. Fourier analysis of this difference curve gives the desired first harmonic amplitude. Phase-I measurements of the magnetic field are about to begin. Fig. 4 shows schematically the Phase-II of the magnetic field measurements. The programme consists of measuring the field at various points on the median plane at about 30 main current levels and contributions of trim and valley coils at 5-6 principal main current levels. A computer controlled rapid mapper system will be used for data collection.

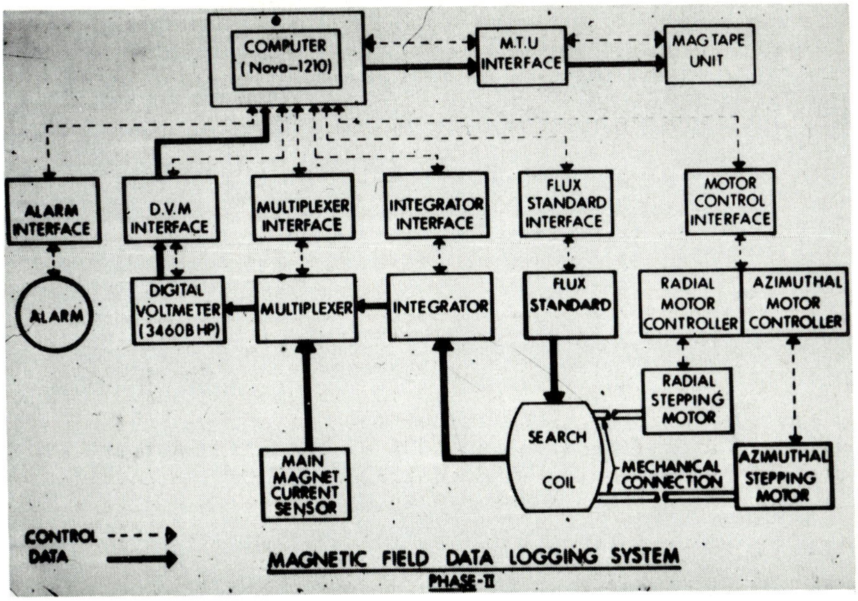

Fig. 4

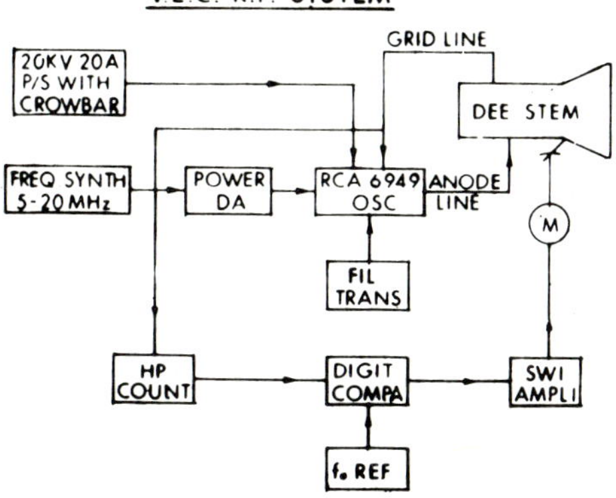

Fig. 5

2. RF System

Fig. 5 shows a schematic diagram of the RF system. The system utilizes a 400 kW RCA 6949 oscillator tube. Frequency variation is done by movable panels(Fig. 6) which alongwith dee, dee stem, drive system, resonator tank and other mechanical systems of the RF system are ready. The RF panels have been slightly redesigned to increase the maximum resonant frequency from 16.5 MHz to 18.0 MHz, thus pushing the maximum achievable energy for protons to 70 MeV. Fig. 7 shows the copper dee and dee stem. The power supply for the RF oscillator tube is a 20 kV, 20A, d.c. power supply.

Fig. 6 : RF Panels

Fig. 7 : Dee and Dee-stem

3. Injection & Extraction System

Fabrication of the PIG ion-source, which will be located in the central hole of the lower pole piece and the lower yoke, has been completed alongwith its drive system. It can deliver beam currents of the order of 5 mA for p, d and α-particles.

The electrostatic deflector assembly consists of two sets of electrodes. Two inner electrodes are joined together and kept at ground. Two outer electrodes, which can be moved to change the channel width for optimum extraction, are insulated and given a maximum high potential of 120 kV d.c. Fabrication of the deflector has been completed and the deflector power supply is ready.

4. Control System

Control console has been fabricated and installed alongwith a closed circuit television system for viewing the beam and various components when the machine is operating.

5. Vacuum System

Components of the vacuum system such as two 89 cm diffusion pumps, pneumatically operated gate valves and freon cooled chevron baffles are ready and they have been fully tested. Fig. 8 shows one diffusion pump. All the components have been made by the Technical Physics Division, BARC, Trombay.

All the other auxiliary facilities like low conductivity cooling water for various coils and other components, compressed air, softened water etc. are ready. Fig.9 shows a view of the low conductivity water (LCW) cooling system.

Fig. 8 : 89 cm Diffusion Pump

Fig. 9 : LCW Cooling System

With all these systems ready and assembly of the cyclotron in progress, we except to get the internal beam by the middle of 1976 and the external beam 6-8 months thereafter.

6. Beam Transport System

The characteristics of the extracted beam are expected to be 2.5 x 32 mm-mrad radially and 11.4 x 7 mm-mrad axially. Dispersion in the fringe field will be of the order of 50 x $\Delta E/E$ mm, i.e. about 2.5 mm. So a conservative estimate of the radial extent of the cyclotron source is about 5 mm. As mentioned earlier, the machine will deliver a maximum of about 100 μA of beam current which has to be transported to various target stations.

As shown in Fig. 10 there will be nine channels in four experimental caves for the experiments. Three channels will be used for experiments requiring low energy resolution (0.5% FWHM or 300 keV for 60 MeV protons). Almost 100% of the extracted beam will be available in these channels. In the rest of the six channels 1-2 μA beam for experiments with very high energy resolution (0.024% FWHM or about 15 keV for 60 MeV protons will be available). However, in these channels about 40% of the extracted beam can be made available with low resolution of about 0.5% by adjusting the parameters of various beam transport magnets so that the analysing magnet operates in a high transmission mode. Major bending and analysis of the extracted beam is accomplished with the help of one 159.50° analysing magnet and two switching magnets - the first switches the beam to low resolution channels placed at 0°, +20° and +40° and the second switches the analysed beam to six high resolution channels placed at -50°, -40°, -15°, 0°, +35° and +45°.

V.E.C. LABORATORY — EXPERIMENTAL AREA

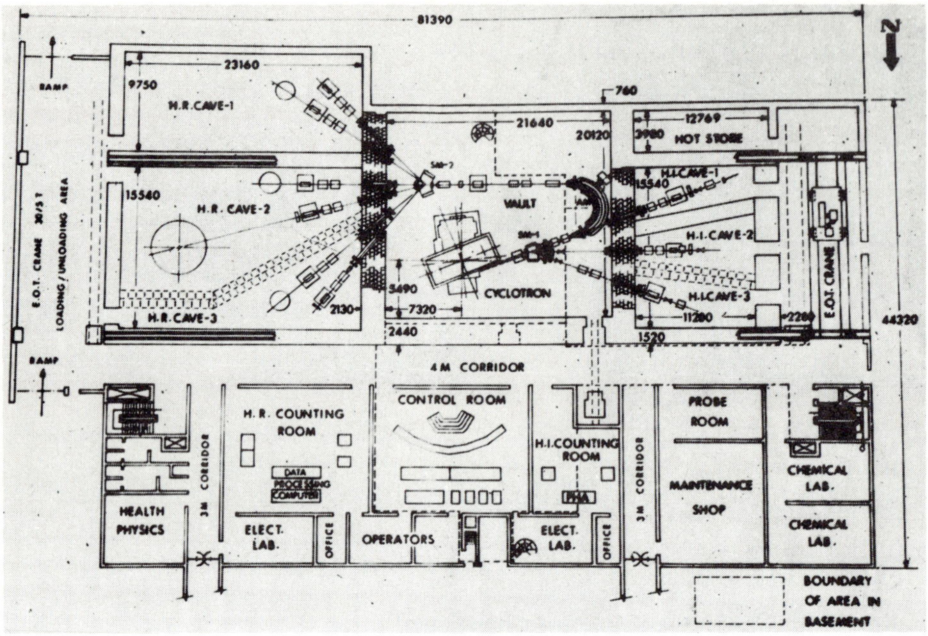

Fig. 10

The first switching magnet has been designed to bend 65 MeV deuteron beam to a maximum angle of 50°. Fabrication work of this magnet is 90% completed. Its frame has been made from low (0.07%) carbon steel castings and coils are made of hollow aluminium conductor with a square cross section. Its 150 V, 300 A SCR controlled power supply has also been fabricated. We plan to use this magnet for energy analysis to some extent. Calculations show that about $1\,\mu A$ of beam with an energy resolution of about 0.1% can be obtained provided $\theta'_{0}{}^{2}$ aberration is eliminated. A method has been developed to design shims to minimize this aberration.

The analysing magnet which is similar to the one at Texas A&M Cyclotron is $n = 1/2$ double focussing C-shaped magnet with normal entry and exit. Its mean bending radius is 213.36 cm and the mean gap is 6.35 cm. Both the entry and exit edges are concave each with a radius of curvature 795.0 cm for eliminating θ_0^2 aberration. Source and image distances are each 201.9 cm and magnification in both the planes is unity. Maximum size of the dispersed radial image is about 40 mm. The magnet will be operated at a maximum field of 9 kG with 0.001% stability. A 160 V, 500 A SCR controlled power supply with 0.001% regulation by series transistors has been fabricated for this magnet. Castings and forgings for the magnet frame are ready at the Heavy Engineering Corporation, Ranchi. Coil winding work is in progress at the Bharat Heavy Electricals, Bhopal.

All the quadrupole doublets are identical. Physical length of each element of a typical doublet is 20.3 cm and the aperture is 10.20 cm. Maximum field will be about 9.80 kG and maximum field gradient about 1.93 kG/cm. At the maximum coil excitation current of 300 A each element will have a focal strength of 0.86 assuming effective length to be about 25 cm. Fig. 11 shows a prototype quadrupole magnet fabricated at the VEC Workshops. Hyperbolic cross section of the pole pieces has been approximated by a circle of radius 1.234 x radius of the aperture. This leaves the useful region of the constant gradient to about 50% of the aperture.

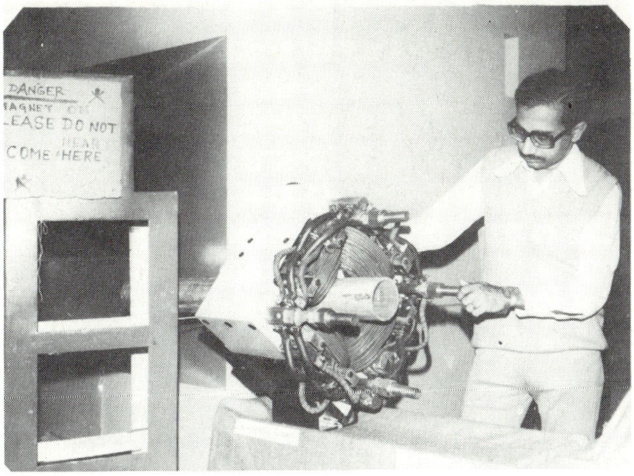

Fig. 11 : Quadrupole Magnet

Fabrication of prototype of many other beam transport components has been completed at the VEC Workshops. Fig. 12 shows a beam steering magnet, and Fig. 13 shows a Faraday cup.

Fig. 12: Beam Steering Magnet

Fig. 13 : Faraday Cup

RESEARCH FACILITIES

For effective utilization of VEC in pure and applied sciences, sophisticated research facilities have been planned. The research facilities proposed have been outlined in Table 3.

Table 3

FACILITIES

1. SCATTERING CHAMBER
2. ELECTRONICS FACILITY
3. ON-LINE COMPUTER
4. DETECTOR FACILITY
5. TARGET PREPARATION FACILITY
6. ION SOURCES
7. MAGNETIC SPECTROMETER
8. NEUTRON PRODUCTION FACILITY
9. CRYOGENIC SYSTEM
10. ON-LINE MASS SEPARATOR FACILITY
11. COMPUTER SYSTEMS
12. RADIATION DAMAGE AND NON DESTRUCTIVE ANALYSIS OF MATERIALS
13. CHEMISTRY, RADIO CHEMISTRY AND RADIO ACTIVE ISOTOPE PRODUCTION
14. BIOLOGICAL STUDIES
15. CANCER THERAPY FACILITY

1. Scattering Chamber

Ordinary and cryogenic scattering chambers using solid state detectors have been planned. Fabrication of a 915 mm diameter scattering chamber is underway at the Central Workshops, BARC, Trombay. Arrangements for changing and reorienting the target with respect to the beam, flexibility of changing the detector angles and turbo-molecular pump for evacuating the chamber are some of the features of these chambers.

2. Electronic Support Facility

A centralized electronic support facility will be set up at the VEC Laboratory to maintain and repair the electronic systems. An additional activity of this facility will be the design and development of standardized modular systems such as the USAEC-NIM system and the very recent CAMAC system in order to facilitate efficient processing of data generated by the experimental programmes.

3. Computer Facility

A Unichannel-15 on line computer system for data acquisition and on line data processing obtained through the International Atomic Energy Agency, has already been installed. It has got two types of memories viz. a core memory of 8K words and a main memory of 16K words with a word length of 18 bits. The peripheral equipment includes a CRT display, a 1.2 M words magnetic disc unit, a 9 track tape unit, a paper tape reader/punch unit and a dual DEC tape unit. This computer will be linked with transducers to work as a real time processor and as a buffer to a large computer. A large R-1050 computer, having a memory of 1 million byte and word length 36 bits, has been planned. Table 4 gives the R-1050 computer specifications and Fig. 14 shows the computer configuration.

4. Detector Facility

To meet the heavy demand for semi-conductor detectors for various experiments with the VEC, work on setting up of a semi-conductor detector fabrication and reactivation facility has been started. E type Si(Li) detector with active depth 20 mm and 5 mm, E type Ge(Li) detectors with active depth 10 mm and dE/dX type totally depleted Si surface barrier detectors with depth 10 μm to 500 μm and partially depleted Si diffused detectors with depth 500 μm to 1000 μm will be produced with this facility.

Table 4

R-1050 COMPUTER SPECIFICATIONS

1.	R-1050 MEMORY	: 1 MILLION BYTES
2.	9 TRACK TAPE UNITS	: 6 UNITS
3.	29 M DISC UNITS	: 4
4.	CARD READERS (1200 CPM)	: 2
5.	CARD PUNCH (25 CPM)	: 1
6.	LINE PRINTER(650-890 LPM)	: 3
7.	CONSOLE	: 2
8.	CALCOMP PLOTTER	: 1
9.	KEY PUNCHES	: 8
10.	CARD REPRODUCER	: 1

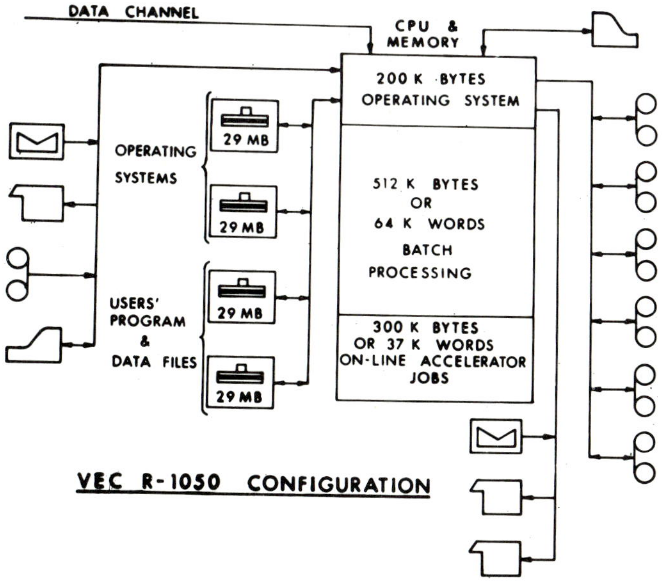

Fig. 14

5. Target Facility

The annual requirement of targets of various enriched isotopic materials has been estimated to be about 100. For the preparation of these targets a facility comprising the Van Arkel-de-Boer process for purification, vacuum evaporation, precision rolling mill, metal press etc. is being set up at the VEC Laboratory.

6. Heavy Ion Sources & Polarised Ion Sources

Work has been started for modifying a standard internal source which will be used for heavy ion acceleration with the VEC. Work has also begun on the fabrication of a source similary to the Berkeley internal PIG source with 3 kW arc power. These sources will enable the acceleration of ions like $^{7}Li^{2+}$, $^{12}C^{3+}$, $^{14}N^{5+}$, $^{14}N^{6+}$, $^{16}O^{3+}$, $^{16}O^{4+}$, $^{20}Ne^{5+}$, $^{56}Fe^{9+}$, $^{56}Fe^{10}$ etc. Development of a "Resonance" electron beam ion source (REBIS) is also underway at this Laboratory for obtaining highly stripped heavy ions. A new technique has been employed which will increase the effective well life time even at higher pressures so that the trapped ion in the potential well undergoes many impacts to achieve high charge states without getting neutralized. Work on a polarized ion source has also been planned.

7. Magnetic Spectrometer

It is proposed to instal a Berkeley type high resolution magnetic spectrometer. Specifications and design details for this spectrometer have been finalised. Designed for a maximum energy of 90 MeV for tritons, this spectrometer has an energy range of 30% of the maximum and a maximum resolving power of 10,000. Very large vertical acceptance of the order of 100 mrad, ability to record extreme backward and forward angles from -125° to +180°, active focal plane detector, ability to adjust the orientation of the focal planes and a sextupole magnet to minimize θ_o^2 aberration are some of its important features. MWPC and SWPC detectors will also be developed alongwith the spectrometer at the VEC Laboratory.

8. Neutron Production Facility

Facility will be set up for obtaining nearly monoenergetic high energy neutrons, using the $^{7}Li(p,n)^{7}Be$ reaction. Polarized neutrons with energy of 20-40 MeV will also be available with 25-40 percent polarization which will be obtained by the $^{2}H(d,n)^{3}He$ reaction.

VEC UTILIZATION

In general, high and variable energy, along with the high energy resolution and precision, expected from the VEC, will enable investigations of many interesting phenomena in nuclear physics. As a result, enormous interest has been shown in utilizing the VEC for various experiments for studies in nuclear spectroscopy, reaction mechanism, nuclear structure, fission etc. A number of specific experiments were suggested with the initial alpha beam, at a VEC utilization seminar arranged during December 1974, at the Nuclear Physics and Solid State Physics Symposium, held at Bombay [3]. Some of the studies proposed are as follows.

A study of gamma decay of high spin states, produced by the (α, xn) reaction, at incident alpha energies around 30 MeV, which has been planned, yields information on the rotational bands of nuclei, which leads us to the phenomena of back-bending. The (α, xn) reactions also open up avenues for in-beam nuclear spectroscopic studies where spin and parity determination using internal conversion electron angular distribution and gamma ray linear polarization, life-time determination using internal conversion electrons, magnetic moment measurements using Perturbed Angular Distribution (PAD) techniques, and electric quadrupole moments of isomeric states can be studied.

Reaction mechanism studies using elastic scattering of alpha particles and other heavy ions have been planned. Through optical model analysis, they can lead to detailed information about the nucleon-nuclear interaction, e.g. it has been found that an l-dependent imaginary potential is needed to explain the presence or absence of structure observed in elastic scattering excitation functions. Another interesting aspect to be studied is the 'quasimolecular state' or 'intermediate state' manifested through scattering of alpha particles. A measurement of total reaction cross sections for protons and alpha particles which is also envisaged, will throw light on the imaginary part of the optical potential, and its variation with incident energy and ground state nuclear configuration (clustering) of the target nucleus.

Study of alpha particle induced fission of bismuth and lead isotopes, say, at about 40 MeV incident energy, with refernece to the fragement energy mass and angular correlations have been proposed. They provide a means of investigating the mechanism of mass division in the fission process, which is still the least understood aspect of the fission process.

Current problems in nuclear physics, that can be studied with a cyclotron such as the VEC have been highlighted in a recent review by C. Mayer-Böricke [4], where studies involving scattering, pre-compound and compound processes, high spin states and giant resonances have been discussed.

Of particular relevance at this conference is the role that can be played by the VEC in the study of few body problems. Reactions involving three nucleons, such as D(p,pn)P, D(p, 2p)n, D(n, 2n)P, P(d,2p)n are of great importance in understanding the basic nucleon-nucleon interaction, and experiments providing sensitive information such as spin observables, inelastic scattering, Wolfenstein parameters in elastic N-d scattering and break up cross sections are still needed for evaulating sophisticated models as discussed by I. Slaus[5]. A study of the ^{3}He(p,2p)d and ^{3}He(p, 2p)pn reactions can provide information about the important ground state wave functions of the ^{3}He nucleus, and (α, 2α) reactions can lead us to understand the $\alpha-\alpha$ interaction, clustering and off-mass shell effects.

It can thus be seen that the Variable Energy Cyclotron at Calcutta offers unique and exciting opportunities to the physicists for seeking a basic understanding of the nuclear forces.

References

1. C. Ambasankaran and D.Y. Phadke, IEEE Tran. Nucl. Sc., NS-20, No. 3, (June 1973) 236.

2. C. Ambasankaran et. al., Proceedings of the VIIth International Conference on Cyclotrons and their Applications, Zürich, 1975 (Birkhauser Verlag, Basel) p. 84.

3. Proceedings of the Nuclear Physics and Solid State Physics Symposium, Bombay, Vol. 17 (1974).

4. C. Mayer-Böricke, Proceedings of the VIIth International Conference on Cyclotrons and their Applications, Zürich, 1975 (Birkhauser Verlag, Basel) p. 481.

5. I. Slaus, Few Particle Problems in the Nuclear Interaction, (North Holland, Amsterdam, 1972) p. 272.

PRELIMINARY RESULTS OF THE RESEARCH ON FEW BODY PROBLEMS AT TRIUMF

J. Reginald Richardson*
TRIUMF, Vancouver, B.C., Canada

SUMMARY

TRIUMF is the meson factory at Vancouver, Canada. The first 500 MeV extracted beam of protons was obtained in December, 1974 and in the past year the experimental program using this beam has gradually begun to get under way and produce some preliminary results. Many of the experiments under way on few body problems make use of the variable energy feature (180-525 MeV). Among those experiments which have produced results is that on the elastic scattering of protons by ^{4}He (200-500 MeV) over the angular range 3-15° (lab). Nucleon-nucleon scattering is represented by programs on the p-p and n-p interaction making use of the variable energy polarized proton beam. Some results have also been obtained on the reaction $p + p \rightarrow \pi^+ + d$ at various energies and angles and this experiment will also eventually include the use of the polarized beam. Preliminary experiments are being performed on other few-body reactions, including $\pi^- + p \rightarrow \gamma + n$, $\pi^- + p \rightarrow \pi^\circ + n$, $\pi^- + {}^3$He x-rays and $\pi^+ \rightarrow e^+ + \nu$.

INTRODUCTION

The year 1975 at TRIUMF has been devoted to three main objectives.

1. Investigating and improving the properties of the cyclotron beam in the energy range 180-525 MeV and the characteristics of the three external beam lines.
2. Improving the reliability of the major components of the facility.
3. Starting some research programs in intermediate energy nuclear science.

My purpose in this paper is to describe briefly some of the research that is under way relevant to the topic "Few Body Problems in Nuclear and Particle Physics". Since objectives 1 and 2 above have necessarily received higher priority, most of the research programs that I will describe have not yet obtained definitive results. However, there are two experiments which have been completed.

The H^- cyclotron at TRIUMF[1] delivers two simultaneous and independent external proton beams of differing energies, one into the proton hall, the other into the meson hall. Because of lack of shielding the proton beams have been held down to 0.5 μA with 100% duty factor, although 50 μA at 1% duty factor has been accelerated as a test. The design current for the proton beam is 100 μA. The energy resolution of the proton beams, FWHM is less than 1 MeV in the energy range from 200 to 525 MeV. This will be improved further by improving the radial emittance of the cyclotron.

A polarized H^- beam of 250 nA has been passed through the injection system of the cyclotron. In accordance with our previous experience with our unpolarized source, we will expect at least 30 nA of 80% polarized H^- ions to be accelerated to 500 MeV and this should be achieved in January, 1976. Here also the energy range will be 180-525 MeV.

The following facilities are in operating condition:

a. Variable energy neutron beam from p + d (200-500 MeV)
b. Gas jet system for study of exotic nuclei and time-of-flight mass identification facility
c. Scattering chamber for heavy fragments of nuclei
d. Large scattering stand for light nuclei
e. Target and pion spectrometer for (p,π) reaction on protons, deuterons, etc.
f. Stopping π and μ channel

---

* On leave from UCLA

g. Special channel for studies with muon spin precession in chemistry and solid state physics
h. Two large NaI crystals for rare processes involving the emission of high energy gamma rays or electrons or neutral pions
i. Biomedical channel for development of radiotherapy with negative pions

THE EXPERIMENTAL PROGRAM RELEVANT TO FEW BODY PROBLEMS

1. Elastic Scattering — $^4He(p,p)^4He$

One of the first experiments to collect usable data at TRIUMF was that on small angle elastic scattering. In particular, some of the data on the $^4He(p,p)^4He$ elastic scattering has undergone preliminary analysis.[2] The experiment was set up on a scattering stand on a low intensity beam line and the diagram of the apparatus for the small angle scattering is shown in Figure 1. The whole chamber was filled with helium gas and coincidences were required between the protons as recorded by the multi-wire proportional chamber (MWPC) and plastic scintillator on the one hand and the knock-on alpha particle as recorded in the solid state detector on the other hand. The relative time-of-flight between the proton scintillator pulse and the alpha particle was recorded and gave a clean separation between rf groups corresponding to real and accidental coincidences. However in practice, the scatter diagram of the proton angle from the wire chamber pulses versus the pulse height from the alpha particle gave an adequate separation. Note that the apparatus was symmetric in the horizontal plane.

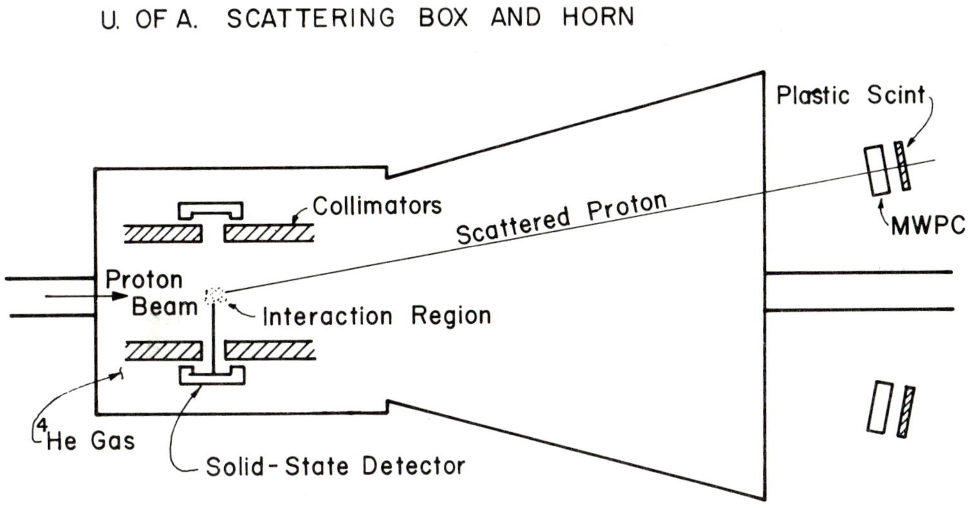

Figure 1: Apparatus for elastic p - ^4He at small angles. Coincidences are required between the proton (angle measured) and alpha particle (pulse height measured).

The advantages of this technique can be listed as follows:

a. The effective angular resolution is limited only by the energy loss of the alpha particle across the width of the beam. This gives $\Delta\theta \lesssim 0.1°$.
b. The effective solid angle and target length is (almost) independent of beam parameters.
c. The density of the target gas is easily determined.

The lower limit of 3° on the laboratory angle of measurement is imposed both by the disappearance of the alpha particle signal into the noise and the MWPC running into the proton beam pipe.

The results of the scattering cross section vs. laboratory angle are shown in Figure 2 for incident proton energies of 500 MeV and 350 MeV. Results at 200 MeV have not yet been analyzed. Measurements were taken every 0.2° on the average and a selection of some 4 million events is shown in the figure. The results (including a rough survey of the results at 200 MeV) indicate that the nuclear-coulomb interference is destructive at the lower energies.

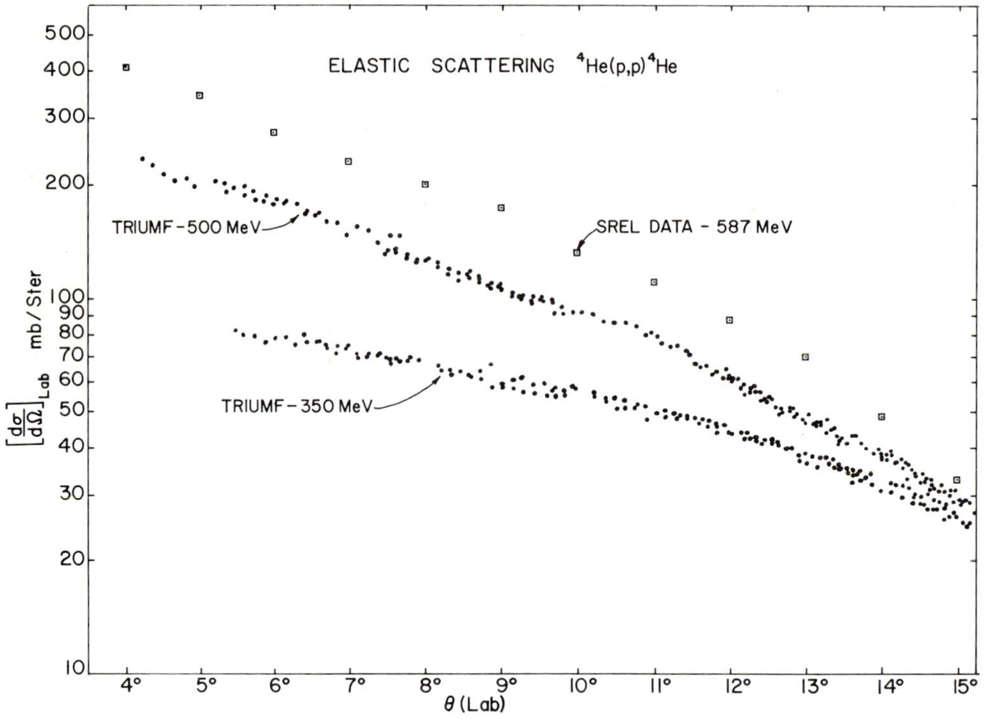

Figure 2: Preliminary results on the angular distribution of the elastic scattering cross section for p - ^{4}He in the laboratory system. Recent normalization requires the values in the graphs to be divided by the factor 1.03 at 500 MeV and 0.875 at 350 MeV.

2. <u>The Nucleon-nucleon Elastic Scattering Amplitudes</u>
   (BASQUE group) Bedford College, London; AERE, Harwell; University of Surrey; Queen Mary College, London; University of British Columbia; University of Victoria.

The objective of this series of experiments (Bugg) is a precise determination of the pp and np elastic scattering amplitudes in the energy range 200-520 MeV by means of measurements of Wolfenstein parameters, polarization and $d\sigma/d\Omega$. The Wolfenstein parameters and polarization will be measured by scattering polarized beams of protons or neutrons from liquid hydrogen and analyzing the polarization of scattered protons and neutrons in a polarimeter. This polarimeter consists of a carbon plate (3 or 6 cm thick, according to energy) placed centrally in an array of 12 multi-wire proportional chambers. The front six chambers, defining the proton incident on the polarimeter, have active areas of 50 cm^2; the back six, defining the scattered proton, have active areas of 100 cm^2. The efficiency for a useful scatter in 6 cm of carbon is 5%, with an analyzing power which varies with energy and scattering angle, but is typically 35%. The array has an efficiency >99%, and is capable of recording 250 events/sec.

The first stage of the experiment consists of calibrating this polarimeter using protons of known polarization from pp elastic scattering. The spin of the protons is precessed alternately ±90° using a superconducting solenoid. This calibration has been completed at two energies with statistics of 5 x 10^5 scatters at each energy. The preliminary result, shown in Figure 3, at 380 MeV gives an analyzing power somewhat higher than that observed by Aebischer et al[3] but in close agreement with the older results of Birge and Fowler.[4] This calibration will be done at 40 MeV steps.

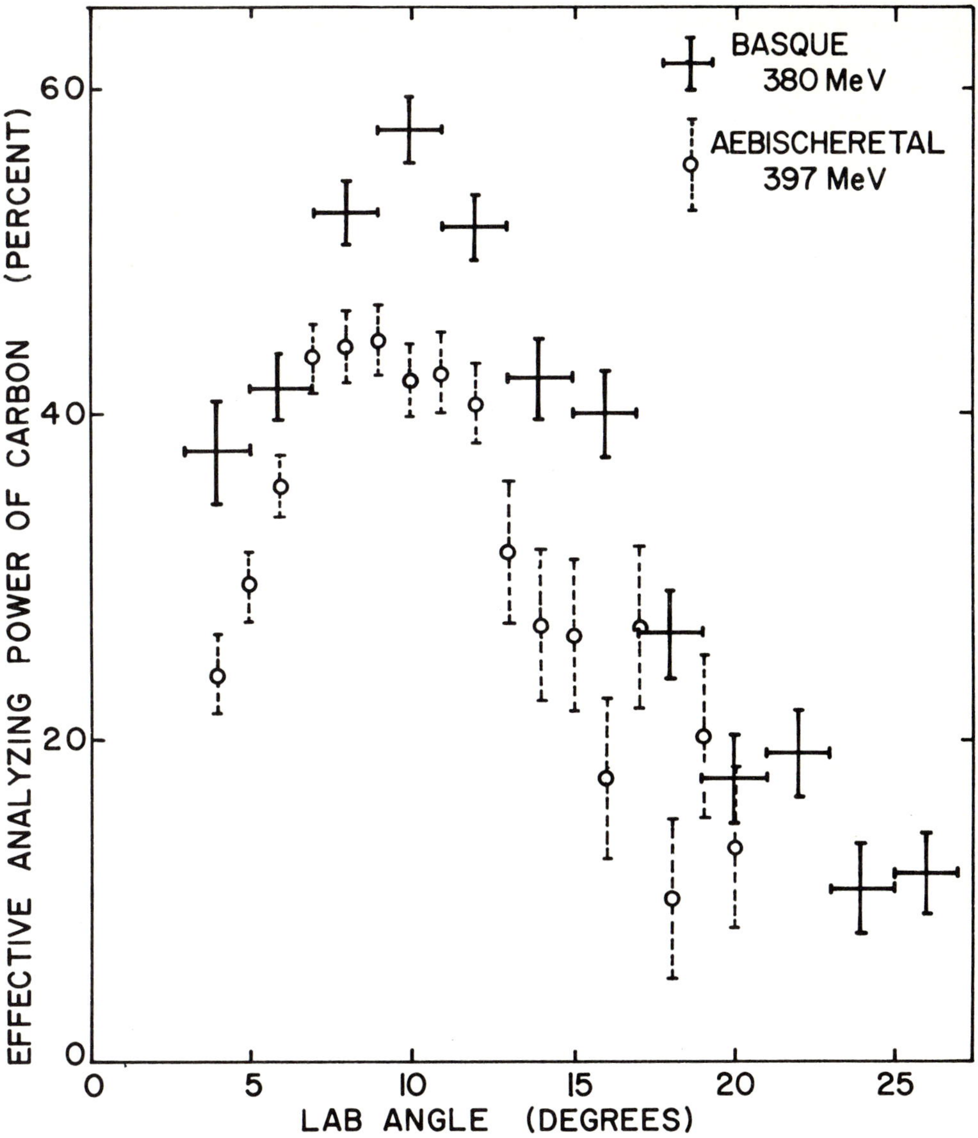

Figure 3: Preliminary results on the analyzing power of carbon at a proton energy of 380 MeV. The values given in reference 3 are shown for comparison.

3. Nucleon-nucleon Bremsstrahlung

A. Experiment

Although proton-proton bremsstrahlung has received much attention from both theorists and experimentalists, it is now clear that most of the experiments were not capable of giving new information about the off-shell characteristics of the nucleon-nucleon force. More recent experiments[5] show large discrepancies from potential model calculations. The discrepancies are typically in regions of phase space where the photon energy is a sizeable fraction of the maximum allowed by the kinematics. The TRIUMF cyclotron has two features which will allow us to explore these "far out" regions of phase space more extensively.

It is planned to carry out a series of experiments to study proton-proton bremsstrahlung in the Harvard geometry at very small and equal or differing angles and over a range of bombarding energies. The apparatus to measure $pp\gamma$ at a proton laboratory scattering angle as small as 10° has been constructed and is currently being calibrated. The measurement will utilize a gaseous hydrogen target and two proton detector telescopes composed of plastic scintillators and multiwire proportional counters. Initial measurements will be made at an incident energy of 200 MeV.

The key factors which should make these small angle measurements possible are the low emittance and large duty factor of the TRIUMF beam. The measurements on the small angle ^{4}He scattering described above utilized proton detectors quite similar to those which will be used for $pp\gamma$. This experiment showed that with a thin target it is possible to routinely operate plastic scintillators and multiwire counters at angles as small as 3°.

B. Theory

Previous theoretical analyses of nucleon-nucleon bremsstrahlung have usually been based on low energy potential models, and thus are not particularly appropriate for analyzing medium energy experiments, especially those at the upper end of the energy range accessible to TRIUMF. In consequence, additional calculations are necessary.

Harold W. Fearing[6], at TRIUMF, has taken as his objective a theoretical analysis of the bremsstrahlung process oriented towards finding the experimental geometry and energy which is most likely to be sensitive to the off-energy-shell matrix elements in the nucleon-nucleon interaction. As a first step he has made calculations for the medium energy range using a Low theorem approach. This theorem expresses the total bremsstrahlung amplitude in powers of the photon momentum k

$$M = A/k + B + Ck + \cdots \qquad [1]$$

where A and B depend on the elastic nucleon-nucleon phase shifts and are independent of off-energy-shell effects. Fearing has separately calculated the leading two terms of $|M|^2$ (this separation has not previously been done) so as to provide some information on the region of convergence of the Low expansion. This approach has the distinct advantages of being relativistically invariant, gauge invariant, and simple and of utilizing the maximum amount of on-shell information. It however does not provide any direct prediction of the third and higher order terms which contain off-shell information, as would be provided by a potential or other model calculation.

So far these calculations have been used primarily to explore the kinematic region to find out where the Low terms might be small or where the second term is relatively large compared to the first. Presumably it is these regions where results will be most sensitive to the interesting third order terms.

Indications are that the planned TRIUMF experiment will explore kinematic regions which at 200 MeV will be further, and at 450 MeV significantly further, from the mass shell than previous experiments. Furthermore in the most sensitive region of the experiment both first and second terms in the Low expansion are small and the second term is relatively large, roughly 20-30% of the first. Thus one might expect reasonable sensitivity to the off-shell terms.

For the future, these calculations will be extended, perhaps to include some simple models. Also a similar analysis will be made to examine the various polarization configurations which might be measured utilizing the TRIUMF polarized beam to see if there are particular correlations which are especially sensitive to off-mass shell effects.

4. The $pp \to d\pi$ Reaction for Protons of 325 and 375 MeV[7]

One of the initial experiments under way, which makes use of the unique feature of the continuously variable proton energy of TRIUMF concerns the reaction $pp \to d\pi$. This reaction is dominated, in the "intermediate" energy region, by isobar formation characterized by the production of p-wave (or higher angular momentum) pions. Non-isobar mechanisms such as those described by the soft pion theories would thus be expected to be most evident near threshold where s-wave pion production is significant.

Pions, over an energy range of 15 to 100 MeV, are detected by a broad range Browne-Buechner magnetic spectrograph. The spectrograph, which subtends an angle of 0.85° at the target, can be rotated over the angular range of 35-145°, with a solid angle of about 2 msr. The energy resolution of the spectrograph as determined by the size of the scintillation counters constituting the 24 element hodoscope on the focal plane is 2% of the central energy (i.e. 1 MeV for 50 MeV pions), a value consistent with the energy spread of the proton beam and the spot size at the target. There are long scintillation counters placed above the hodoscope which provide the coincidence constraints required for the definition of pion "events", as well as dE/dx and some coarse range information. The region within the magnet and hodoscope assembly was designed to operate at a pressure of less than 10^{-4} T in order to enable momentum calibration of the spectrograph by means of alpha sources.

Although use of a LH_2 target is planned, measurements to date have employed a CH_2 target of 1/16 inch thickness. A typical measurement of the $pp \to d\pi$ pion line is shown in Figure 4, where the event coincidence rate is plotted as a function of magnetic field. The position of the line affords a convenient determination of the incident proton energy, which can be obtained by this means to an accuracy of ±0.5 MeV.

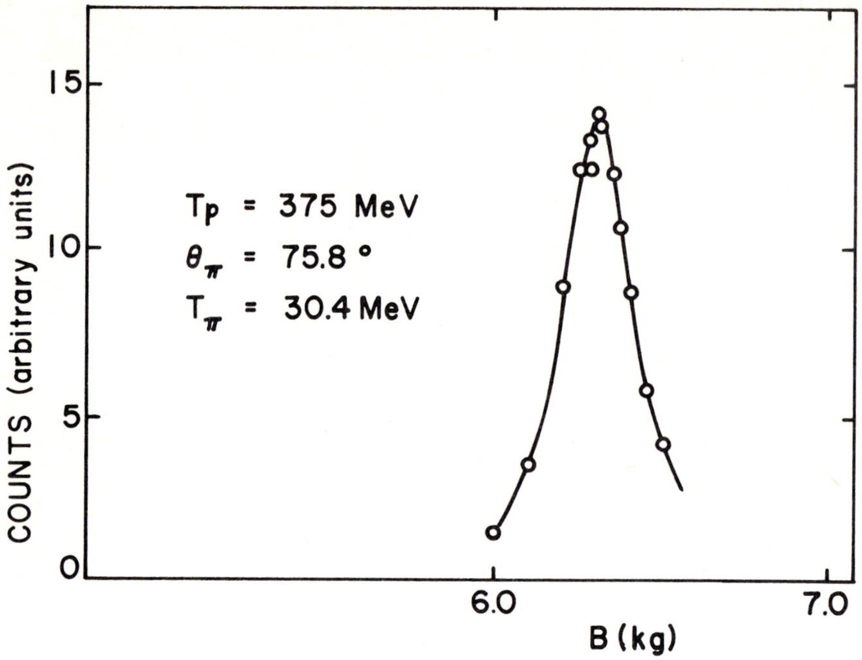

Figure 4: The coincidence rate of pions from pp → dπ as a function of the magnetic field in the Browne-Buechner spectrograph. The energy of the pions gives a good check on the proton energy.

Preliminary angular distribution measurements have now been made at proton energies of 375 and 325 MeV by recording such pion lines at a number of angular positions with the results normalized to the count rate obtained in a scintillation counter telescope fixed at 90° to the beam. The results were fitted to a distribution of the form

$$d\sigma/d\Omega \propto A + \cos^2\theta \text{ (in the cm system)}. \qquad [2]$$

The value of A obtained from these measurements together with those of previous measurements are shown on Figure 5. The solid lines shown on this figure illustrate the expected dependence according to the phenomenological theory of Gell-Mann and Watson as fitted to total pp → dπ cross section data. More measurements are planned at lower energy (310 MeV) as well as measurements of the azimuthal asymmetry of pion production when the TRIUMF polarized proton beam becomes available.

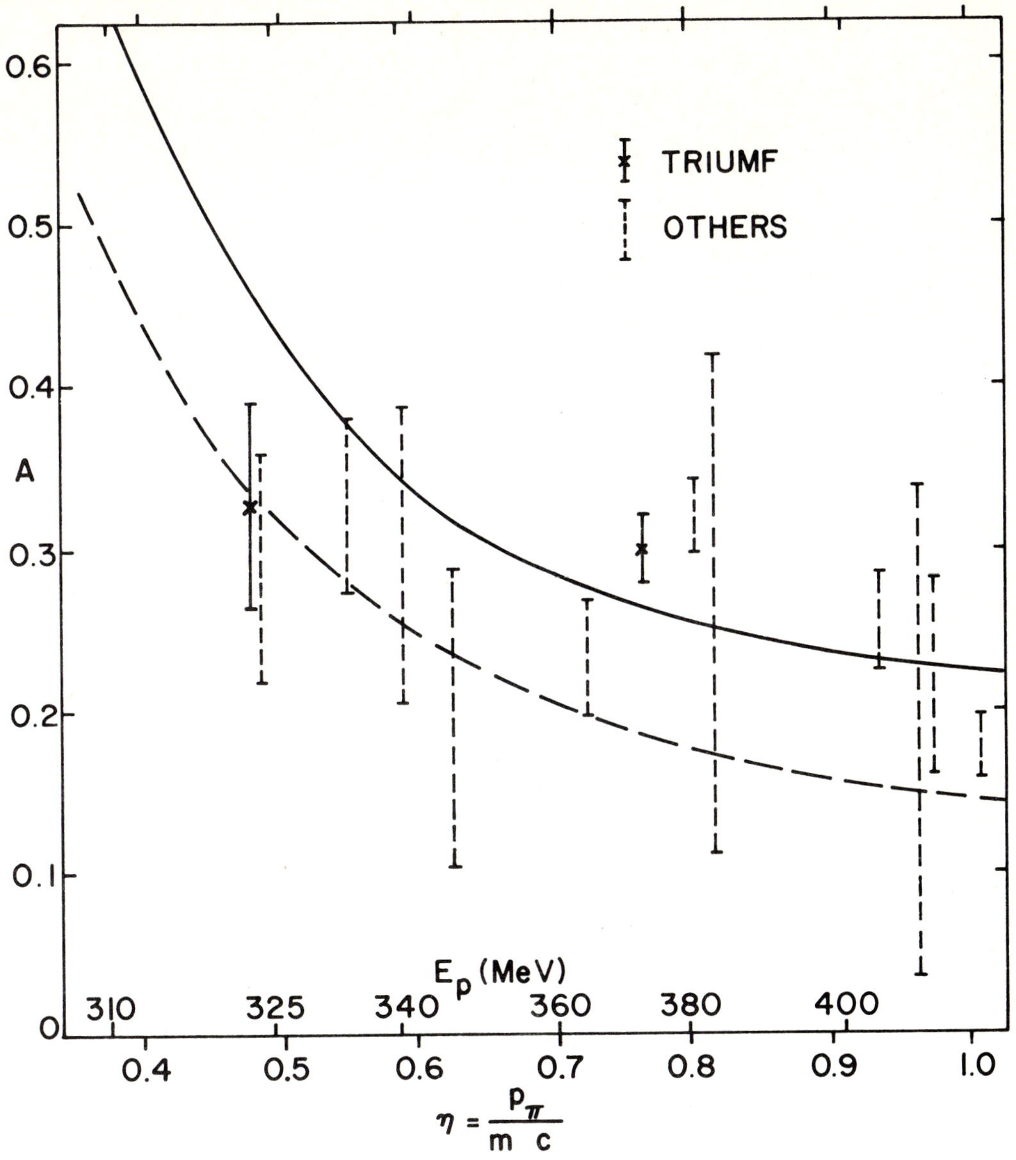

Figure 5: *The parameter A in* $\frac{d\sigma}{d\Omega} \sim A + \cos^2\theta$ *for the reaction* $pp \to d\pi$. *The curves are the results of phenomenoligical calculations due to Measday (solid curve) and Crawford and Stevenson (broken curve). Measurements at 310 MeV incident proton energy are in progress.*

5. Few-Body Problems Involving Pions and the Detection of γ-rays or Electrons [8]

This group of experiments makes use of one or two large NaI scintillators of which the larger (TINA) has a diameter of 46 cm and a length of 51 cm. Figure 6 shows the response of the large scintillator to a 96 MeV/c positron beam in the stopping π,μ channel. A large part of the width $\Delta E/E = 4.4\%$ is due to energy spread in the positron beam. The initial measurements will start with a check of the Panofsky ratio in liquid hydrogen

$$\frac{\omega(\pi^- + p \rightarrow \pi^\circ + n)}{\omega(\pi^- + p \rightarrow \gamma + n)} = 1.53 \pm 0.02 \text{ (present value)} \qquad [3]$$

and proceed to lowering of upper limits for the reactions (at rest):

$$\pi^- + d \rightarrow \pi^\circ + n + n \text{ (present branching ratio} < 0.2\%) \qquad [4]$$

$$\pi^- + {}^6Li \rightarrow \pi^\circ + {}^6He \text{ (present branching ratio} < 0.4\%) \qquad [5]$$

Also an attempt will be made to improve the measured value of the branching ratio

$$\frac{\pi^+ \rightarrow e^+ + \nu}{\pi^+ \rightarrow \mu + \nu} \text{ from its present value of } (1.24 \pm 0.03) \times 10^{-4}$$

The intensity of the proton beams will be increased in 1976 and at that time the pion fluxes will be large enough to measure the following reactions.

$$\begin{matrix} \pi^- + p \rightarrow \pi^\circ + n \\ \pi^- + p \rightarrow \gamma + n \end{matrix} \quad \text{for } 20 \text{ MeV} < E_\pi < 65 \text{ MeV}$$

Also both NaI scintillators will be used in coincidence to investigate the radiative decay mode $\pi^+ \rightarrow e^+ + \bar{\nu}_e + \gamma$.

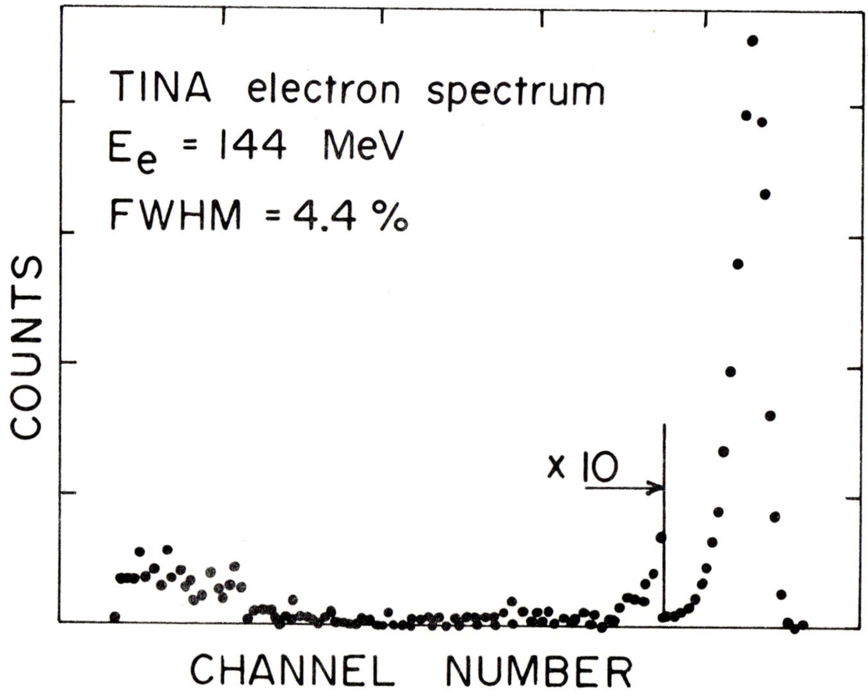

Figure 6: The pulse height spectrum of the large NaI scintillator from a 144 MeV beam of positrons in the stopping π,μ channel. The spread of 4.4% in pulse height is primarily due to the energy spread of the positrons.

6. X-ray Transitions in Pionic ^3He [9]

The absorption of π^- by ^3He will usually occur from the 1s state. Both the shift in the experimentally observed energy of the 2p - 1s line, as compared with the value calculated from the Klein-Gordon equation, and the broadening of the line, will provide useful information about the strong interaction. Figure 7 shows recent results obtained at CERN[10,11] and elsewhere on the 1s level shifts for light nuclei. A comparison with the theoretical calculations shown in the figure indicates that the Ericson potential, as presently used, is apparently not satisfactory for light nuclei. A measurement of the shift in the case of the ^3He transition should at least determine the sign of the isovector term in the interaction.

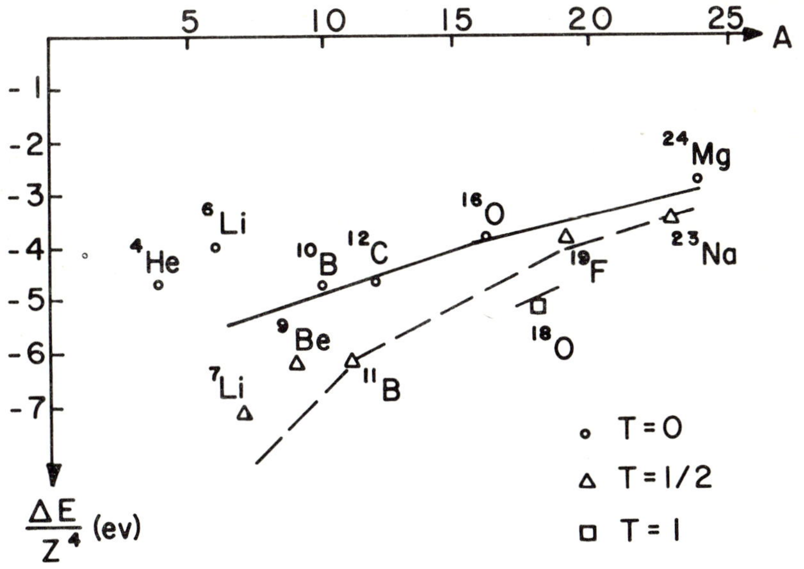

Figure 7: The 1s strong interaction shift for elements Z < 13. The lines are theoretical calculations of Krell and Ericson.

The experimental apparatus is shown in Figure 8 and consists of a target of liquid ^3He and a Si(Li) detector for the x-rays. Preliminary measurements are in progress.

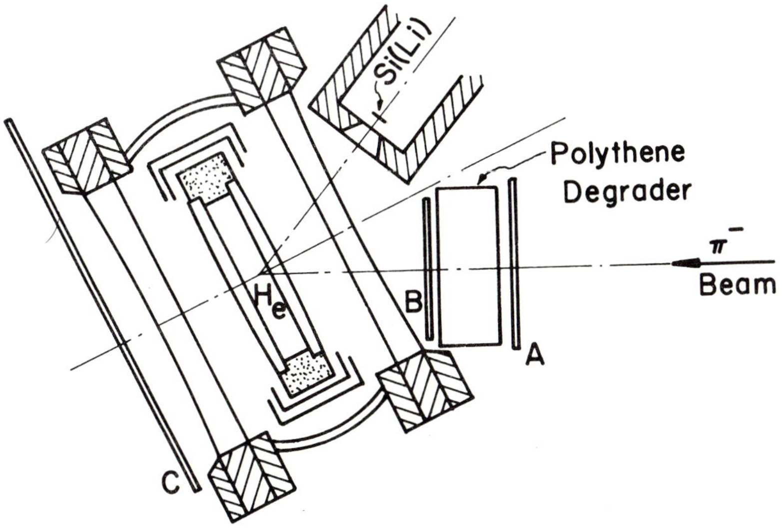

Figure 8: The experimental arrangement for the measurement of the 2p-1s x-rays in ^3He. The pions are stopped in liquid He and the x-rays measured with a Si(Li) detector.

7. Pion Production by Protons[12]

Although not directly relevant to few body problems, the first completed experiment at TRIUMF is important in evaluating the sources of high flux pion beams for the study of these problems. A pion range telescope was used with the variable energy feature of TRIUMF to measure the cross sections for the production of π^+ at proton energies from 400 to 500 MeV from targets of carbon and copper. The pion energy spectrum from 20 to 100 MeV was investigated at production angles of 60, 100 and 150°. A feature of the pion range telescope used in the experiment to reduce the background was the requirement of a $\pi^+ \to \mu^+$ decay in the events used. A small selection of the results are shown in Figure 9 for the cross section for the production of 32 MeV π^+. The results of this experiment are a factor of three smaller than the rather mysterious cross sections reported more than a decade ago for carbon at 450 MeV but are consistent with recent measurements at SREL and Berkeley.

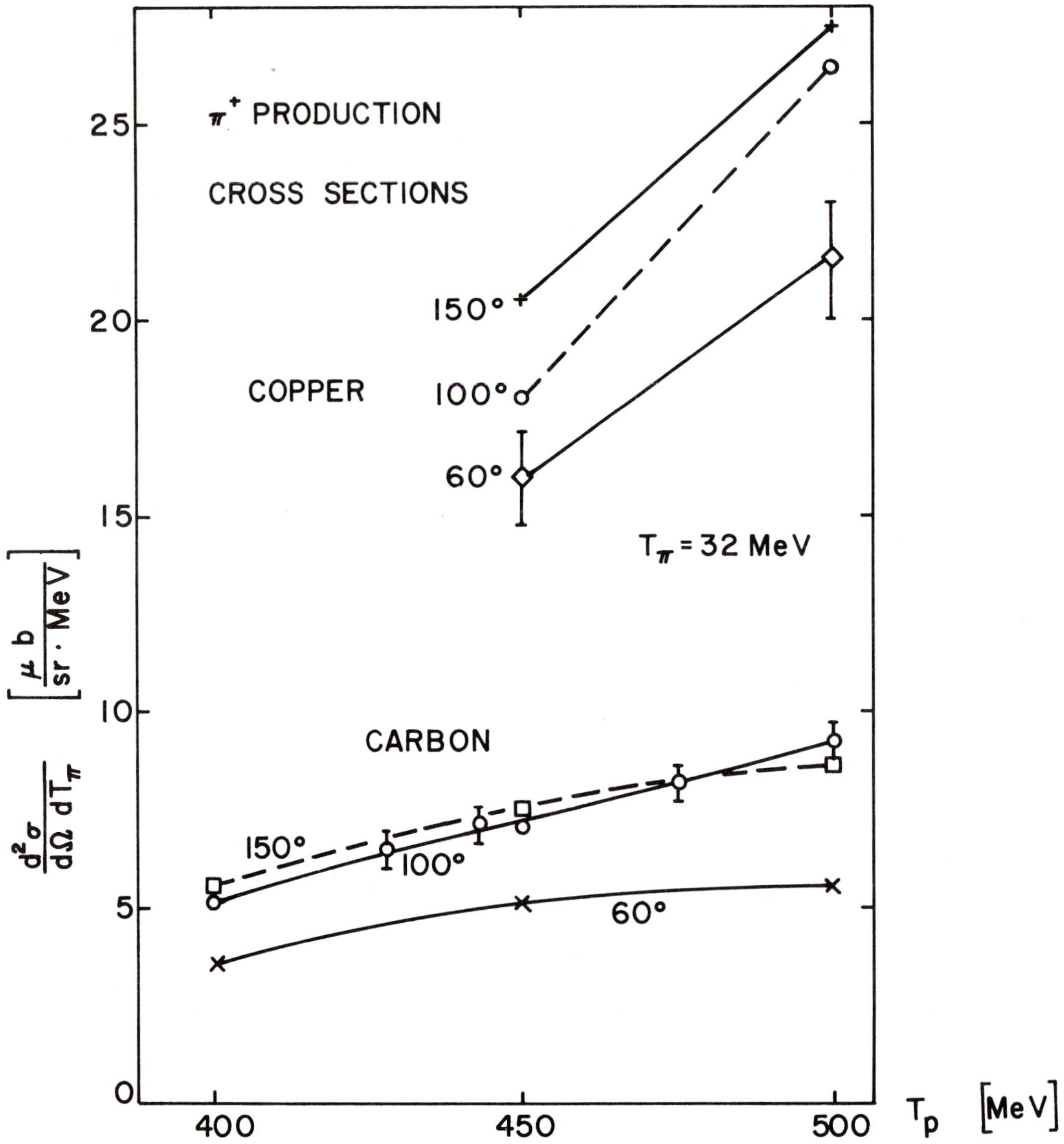

Figure 9: The differential cross section for the production of 32 MeV π^+ from carbon and copper at various proton energies and production angles. Typical errors are shown.

CONCLUSION

Because of the short period since the start of its operation, only two experiments have been completed at TRIUMF, but a considerable number are under way and many of these involve interesting and important few-body problems. Its unique variable energy feature makes TRIUMF particularly valuable in the study of many of the few-body problems in nuclear and particle physics. Strong research programs are under way in the two-nucleon systems and in the three, four and five nucleon interactions. The variable energy feature of the facility, together with the good energy resolution of the proton beam, can be used to produce a π^+ beam of variable energy and very high luminosity through the reaction $pp \to \pi d$. Also intense beams of pions and muons will be useful in the few-body studies involving leptons and the weak interaction.

In addition to those individuals mentioned in the references, credit should be given to E.W. Blackmore, M.K. Craddock, G. Dutto, C.J. Kost, G.H. Mackenzie, P.W. Schmor and M. Zach for important contributions to the productive operation of the beams required for experiments.

REFERENCES

[1] J.R. Richardson, The Status of TRIUMF, Proc. of the 7th International Conf. on Cyclotrons and their Applications, Birkhäuser Verlag, (1975)

[2] J.M. Cameron, R.H. McCamis, C.A. Miller, G.A. Moss, B. Murdoch, J.G. Rogers, G. Roy, A.W. Stetz, W.T.H. van Oers, private communication

[3] D. Aebischer, B. Favier, G. Greeniaus, R. Hess, H. Junod, C. Lechanoine, J.C. Nickles, D. Rapin and D. Werren, Nucl. Instr. & Meth. 124, 49 (1975)

[4] R.W. Birge and W.B. Fowler, Phys. Rev. Letters 5, 254 (1960)

[5] A. Willis, V. Comparat, R. Frascaria, N. Marty, M. Morlet, and N. Willis, Phys. Rev. Lett. 28, 1063 (1972); L.G. Greeniaus, J.V. Jovanovich, R. Kerchner, T.W. Millar, C.A. Smith, and K.F. Suen, Phys. Rev. Lett. 35, 696 (1975)

[6] H.W. Fearing, private communication

[7] T. Masterson, P. Walden, D. Ottewell, E. Auld, R. Johnson, G. Jones, private communication

[8] D.F. Measday, M.D. Hasinoff, M. Salomon, J.-M. Poutissou, TRIUMF Experiment Proposal #9, A study of the reaction of $\pi^- + p \to \gamma + n$ at pion kinetic energies from 20-200 MeV

[9] G.R. Mason, M. Krell, J.S. Vincent, private communication

[10] G. Backenstoss et al, Nucl. Phys. A232, 519 (1974)

[11] G. Backenstoss et al, Nucl. Phys. B66, 125 (1973)

[12] D.A. Bryman, L.P. Robertson, R.A. Olin, J.S. Vincent, E. Mathie, G.A. Beer, G.R. Mason, R.R. Johnson, J.B. Warren, private communication

STATUS OF THE CLINTON P. ANDERSON MESON PHYSICS FACILITY (LAMPF)†

Mahavir Jain*
Texas A&M University
College Station, TX, 77843, U.S.A.

Louis Rosen
University of California, Los Alamos Scientific Laboratory
Los Alamos, NM, 87545, U.S.A.

INTRODUCTION

The history of the LAMPF is the history of the ingenious and creative people who invented this machine and the several hundred dedicated and competent scientists, engineers, and medical doctors who put this machine and experimental areas together and are now performing beautiful experiments. In this short presentation it would be impossible to give credits or to even mention all the research. We start by a short historical interlude (Livingston [1]).

The growing awareness of the importance of research in meson physics and the necessity to maintain a high level of accomplishment in nuclear science were the basic contributing factors which led a small group of physicists at Los Alamos Scientific Laboratory to formulate the initial plans and proposals for the construction of LAMPF in 1962. From this evolved the concept of the "meson factory," which is a very high-intensity proton accelerator of medium energy and which produces intense secondary and tertiary radiations - pions, muons, neutrons and neutrinos. Such a research facility was expected to provide new tools for the study of the fundamental interactions and to open new fields of basic research and practical applications. To achieve the above capability, a proton linear accelerator with the design characteristics shown in Fig. 1 was proposed, even though the highest energy proton linac operating at that time was only 70 MeV. The main departure was high external beam intensity, 10 000 times higher than previously achieved. The formal proposal to build such a large nuclear science facility at Los Alamos with an estimated construction budget of $55 million was submitted to the Atomic Energy Commission in 1964. It was authorized in 1966, but major construction funds were not released until October 1968. Since then, in the short span of four years, the accelerator was completed and on June 9, 1972, the first 800 MeV protons were accelerated at

Beam Energy

800 MeV, Continuously Variable

Beam Intensity

Average Current: 1 mA
Extraction Efficiency: 100%
Pulse length: 500 μsec - 1000 μsec
Repetition Rate: 120 cps

Beam Quality

Energy Spread: $\pm$ 0.4%
Beam Area in Transverse Phase Space: π mrad-cm
Macro Duty Factor: 6% - 12%
rf microstructure: 0.25 ns pulses separated by 5 ns

Fig. 1. Design parameters of LAMPF.

†Work performed under the auspices of U.S. Energy Research and Development Administration.

*Currently at the Los Alamos Scientific Laboratory, Los Alamos, NM, 87545, U.S.A. Travel funds received under N.S.F. Grant No. OIP76-06014.

low intensity (Rosen [2]). On completion, the facility was named the Clinton P. Anderson Meson Physics Facility in honor of the Senator who did so much to advance U.S. science and technology.

LAMPF is located at Los Alamos, amidst the scenic grandeur of the Sangre de Cristo and Jemez mountains. From its inception LAMPF was planned to be an open, multi-faceted, multidisciplinary facility that would bridge the intellectual and technological gap between nuclear and subnuclear physics and would address practical applications as well. The breadth and depth of the experimental program is visualized in Fig. 2. About 300 proposals already received reflect interest in every area depicted here. It has already operated efficiently as a multiple-use facility, by simultaneously providing beam, on the average, to six experiments. During 1974 the full energy beam of average intensity up to 14 μA was supplied to 65 experiments involving 400 scientists from 56 institutions and many foreign countries. It is truly an international research facility as was promised.

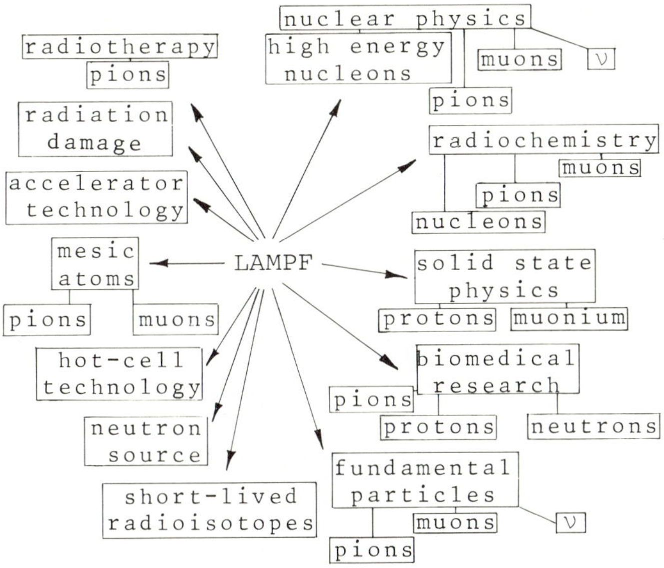

Fig. 2. Scientific purposes of LAMPF.

During 1974 the mode of operation was acceleration of H^- beam at 5-6% duty factor. During 1975, the facility was shut down to allow time for realignment of the linac as well as installation of more shielding and many new experiments. In September 1975 both H^+ and H^- were accelerated simultaneously, and average beam currents of 100 μA H^+ and 3 μA H^- were achieved at 800 MeV. During 1976 average beam currents up to 100 μA will be supplied in experimental areas. Full design energy and intensity beam was also achieved at a very low duty factor and for a short time. At this time the higher intensity is limited by the desire to keep the radioactivity at low levels in the machine and by the amount of beam experimenters can effectively use. A concentrated effort is under way to achieve reliable facility operation and to reduce beam spill along the machine. Ninety-nine percent transmission has already been achieved at 100 μA. It is anticipated that LAMPF will achieve full design intensity of 1 mA in 1977. Although LAMPF is the most complex and most powerful of the meson factories, it achieved full energy on schedule, both time-wise and budget-wise. Furthermore, it has already provided more than 60 000 μAh of beam for experiments.

DESCRIPTION OF THE ACCELERATOR

The linear accelerator was selected mainly to minimize beam losses at high intensities and to have the flexibility of energy variability and good beam quality. LAMPF is a three-stage machine comprising a 750-keV Cockcroft-Walton (C-W), a 100-MeV drift-tube accelerator, and a new concept in wave-guides (Nagle [3]). These are shown in Fig. 3 and are discussed below.

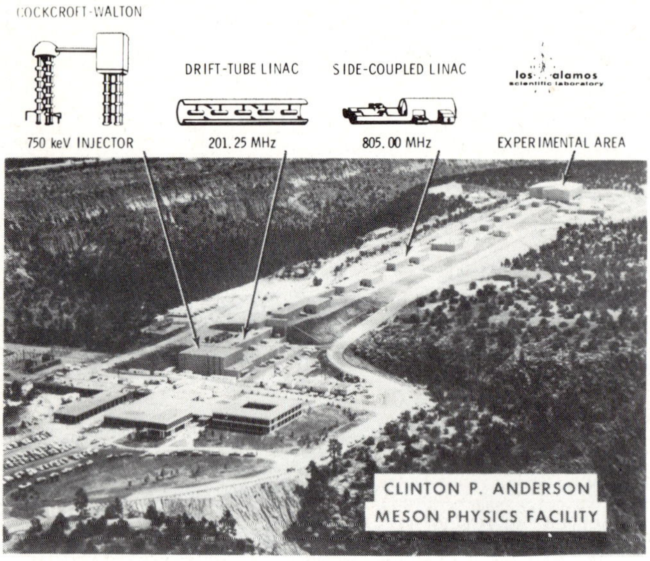

Fig. 3. Accelerator layout.

Two C-W injectors of 750-keV energy provide simultaneous injection of H^+ and H^- beams into the drift-tube linac. A third C-W injector will be used later for polarized beams. A duoplasmatron ion source with "Pierce"-type extraction electrode is used for protons. The H^- ion source is of the charge-transfer type. The beams proceed through separate transport systems to a buncher which forms the input to the drift-tube linac. Simultaneous acceleration of H^+ and H^- is achieved by appropriate longitudinal separation in that H^+ and H^- beams are accelerated on opposite rf phases.

The beams then proceed through the 201.25-MHz linac. It uses four successive copper-lined tanks with the drift tubes mounted along their axes to accelerate the 750-keV beam to 5.4, 41, 72.7, and 100 MeV, respectively. Radial beam focusing is provided by quadrupoles mounted inside the drift tubes. The novel feature of the drift-tube linac is an assembly of copper rods opposite each drift tube. These form resonant circuits with the drift tubes and effectively convert the system to a $\pi/2$ mode of operation from the usual 2π mode. The phase and amplitude of the rf fields in the linac are adjusted by systematically varying the rf phase and amplitude and observing the amount of current accelerated to the proper energy in each module. The empirical curves are compared to analytical solutions and adjustments are made until satisfactory agreement is obtained (Jameson [4]). The beams pass through the transition region, where the relative phase of H^- beam is shifted to form suitable input to the 805-MHz linac.

The 805-MHz linac is a side-coupled cavity system operating in $\pi/2$ mode, which was developed at LAMPF (Knapp [5]). The $\pi/2$ mode is the most stable one for linac operation and is most tolerant of alignment, machining, and frequency errors. It is also very efficient (> 30%) for converting rf power into beam power at 6% duty factor (Hagerman [6]). In a side-coupled cavity chain, the segments off the beam line are resonant cavities which couple fields between adjacent accelerating cavities. Various beam-handling devices are mounted in the linac. At the end of the linac the beams enter a switchyard where they are separated, focused, and diverted to the various experimental areas.

A unique feature of LAMPF is a centralized on-line computer control for the operation of the linac and highly sophisticated beam diagnostics procedures (Butler [7]). The computer is interfaced with the accelerator through 64 modules. The interface between a machine operator and the computer is via a control console, equipped with display graphics, track ball, and function buttons, which provide for automatic and manual modes of operation of any selected part of the accelerator.

EXPERIMENTAL AREAS

Figure 4 illustrates the LAMPF switchyard and experimental areas. Line X provides an achromatic H^- beam to experimental areas B and C by partial stripping of H^-

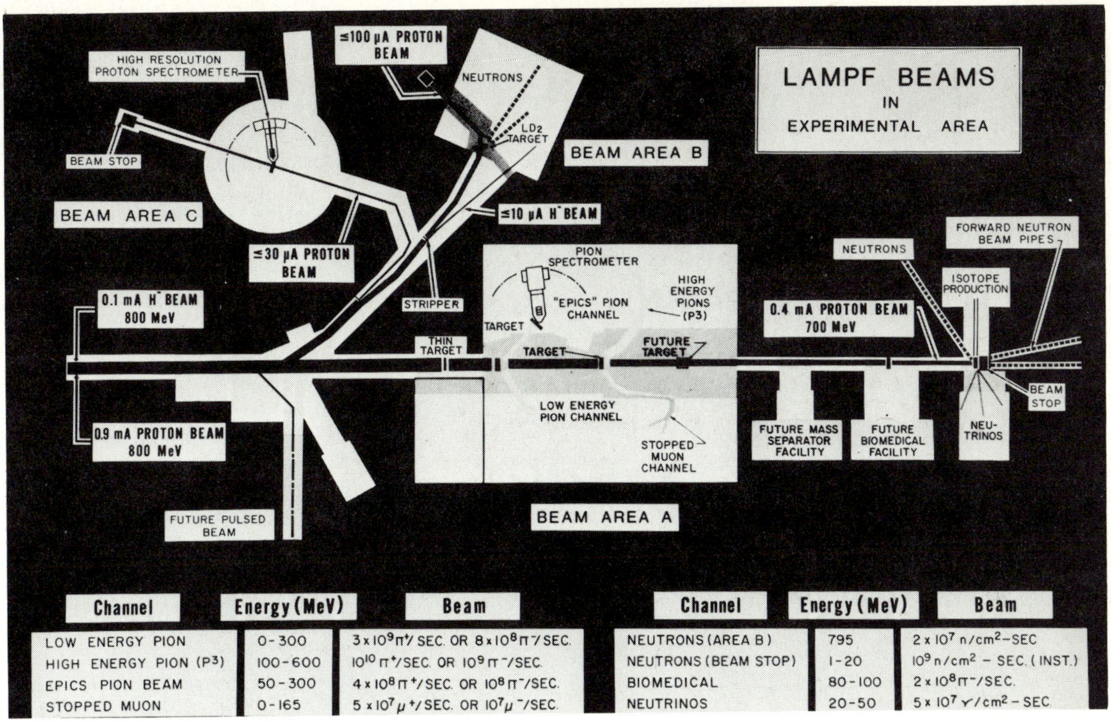

Fig. 4. LAMPF switchyard and experimental areas.

beam. Line A provides the high intensity proton beam in the main experimental area A. The transport optics utilizes quadrupole triplets to focus the transmitted beam onto successive production targets (usually graphite). Various pion and muon channels in area A are described below (Howard [8]).

The "Energetic Pion Channel and Spectrometers" (EPICS) is designed to provide a high-energy intense pion beam with good energy and angular resolution (< 100 keV and ~ 10 mr). It accepts pions in the energy range of 100-300 MeV. The channel with a 700-MeV/c spectrometer is capable of a momentum resolution of 3×10^{-4} FWHM with beam momentum spread of 2%.

The "Low Energy Pion Channel" (LEP) is designed to provide general purpose $\pi^{\pm}$ beams from 20 to 300 MeV. Design features include good momentum resolution (up to 0.05%), a small achromatic beam spot for stopped pion experiments. and high-intensity beams at moderate resolution.

The "High-Energy Pion Channel" (P^3) is a very flexible general purpose channel for the highest energy pions. The channel features high intensity and moderate resolution.

The "Stopped Muon Channel" (SMC) is a low-energy, high-intensity facility, primarily for stopped-particle experiments. A simplified view of the channel divides it into a pion collection and analyzing portion, a pion decay and muon collection portion, and a pion rejection and muon momentum analysis portion. The momentum range of the channel is from 50 to 250 MeV/c. Contaminations due to e, π, p, and n are 2%, < 0.1%, << 0.1%, and ~ 10^4/s, respectively.

The "Biomedical Facility" was designed for the dosimetry, radiobiology, and radiotherapy studies to make a definitive clinical trial of the use of negative pions in radiation therapy. The channel is shown in Fig. 5. It transports a restricted energy range of pions, through bending magnets and quadrupoles, to the patient. The variation of wedge thickness is adjusted to give almost monoenergetic pions, which stop at approximately the same depth. The thickness of the range shifter can be varied with time to vary the stopping π^- distances. The patient is translated slowly in a horizontal plane to achieve the desired volume of irradiation. With 0.5-mA protons and a 5-cm carbon target, the overall dose rate in a 1000-cm^3 volume is > 20 rad/min.

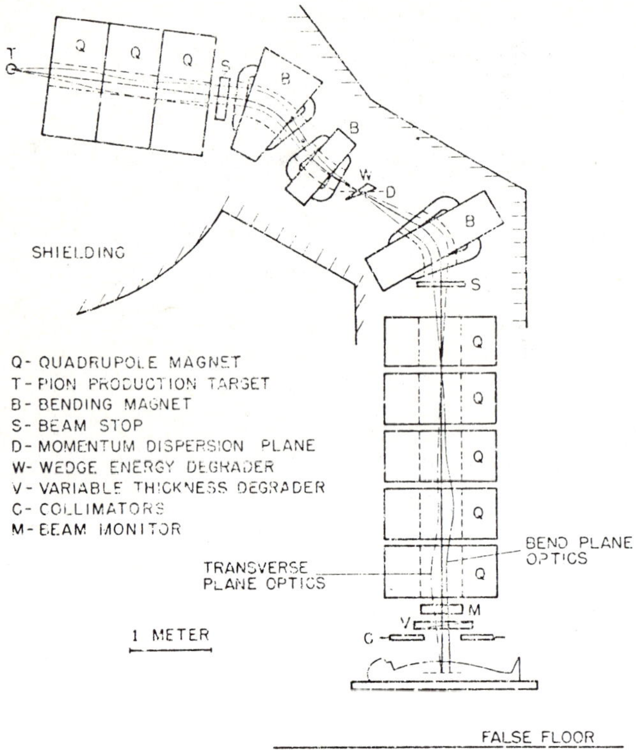

Fig. 5. LAMPF biomedical pion channel.

The main beam stop area, A-6, is a multi-use experimental facility. It has provision for the production of radioisotopes, study of radiation effects, neutron time-of-flight experiments, and various neutrino experiments. The neutrino facility has 5.8 m of steel shielding. The main neutrino energy is 35 MeV and the maximum energy is 53 MeV. The other facilities are the "Thin Target Area" for on-line nuclear studies and "Pulsed Neutron National Defense Facility," which is under construction.

The "Nucleon Physics Laboratory" (NPL), located at the end of line B, provides high-quality neutron beams and a low-intensity proton beam (EPB) at 300 to 800 MeV. The neutron beams are formed by collimation of neutrons, emerging from a liquid deuterium target placed in the proton beam. The transmitted proton beam is focused and used for nuclear chemistry irradiations.

The EPB is intended to be very flexible and can accommodate a wide range of experiments. Usually a pencil beam of 3 mm diam of low intensity is provided. At 800 MeV the energy spread of the beam is 3.5 MeV.

Experimental area C is the "High-Resolution Spectrometer" (HRS). The optics are designed to give a momentum resolution of the order of 10^{-5}, an angular resolution of $\pm 10^{-3}$ rad for a momentum acceptance of 2%, and a solid angle acceptance of 3.6 msr. With these characteristics, the HRS covers a large range of experiments.

RESULTS

We shall briefly comment on a few of the experimental results from LAMPF. The choice of selected topics is rather arbitrary, dictated mainly by time limitations and availability of final experimental results. At energies of several hundred MeV, the basic N-N interaction is not well determined, especially in I=0 states

(Simmons [9]). Also the low energy π-N phase shifts are not well known. Many experimental groups have performed elastic and inelastic scattering of pions and nucleons from protons and various polarization experiments are projected for 1976-77. Out of such studies will come an improved understanding of the basic π-N and N-N interactions which will provide a basis for future understanding of pion and nucleon interactions with nuclei.

Neutron experiments were performed by utilizing the 0° neutrons from p + d → n + 2p reaction. The neutron beam is dominated by a high energy peak (FWHM ≲ 13 MeV) and almost isolated from the relatively small background of lower energy neutrons (Bjork [10]). The n-p charge exchange scattering was measured at 647 and 800 MeV. Figure 6 shows the differential cross section measurements at 647 MeV, which resolve existing discrepancies among previous measurements around this energy (Evans [11]). Other neutron experiments (n + p → π + x, p + p → n + x, and n + d → n + d) will be discussed elsewhere in this conference.

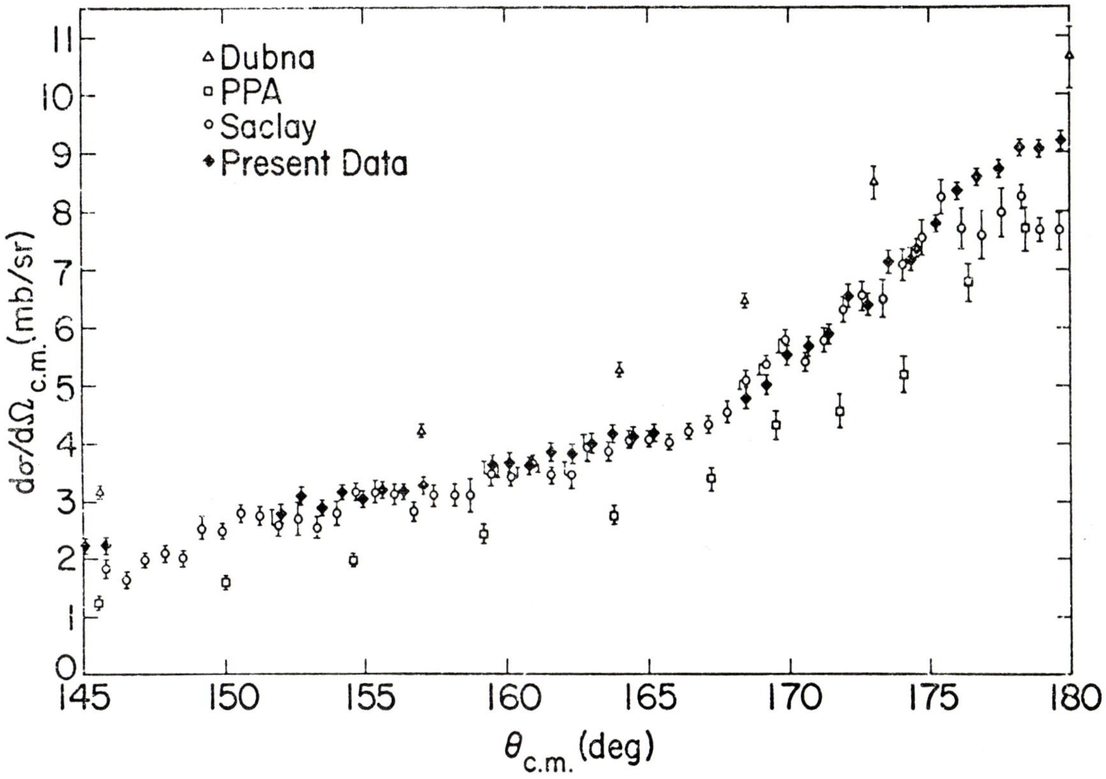

Fig. 6. Differential cross section data for n-p charge exchange scattering near 640 MeV; Dubna (630 MeV, Ref. 12); PPA (640 MeV, Ref. 13); Saclay (645 MeV, Ref. 14); LAMPF data (647 MeV, Ref. 11).

Differential cross sections for p-p elastic scattering have been measured at 647 and 800 MeV to an accuracy of 3% (Willard [15]). New phase shift analysis using these data should be able to improve some phase parameters.

The elastic scattering of pions from protons (Frank [16]) and other nuclei has been studied. The experiments give information on the π-N interaction and various optical model analyses. Figure 7 shows the π^{+}-^{40}Ca elastic scattering data at 204 MeV (Phillips [17]). The real and imaginary parts of the forward elastic amplitude were determined through coulomb-nuclear interference. Figure 8 shows elastic scattering of π^{+} from ^{12}C at 50 MeV and indicates the inadequacy of current optical

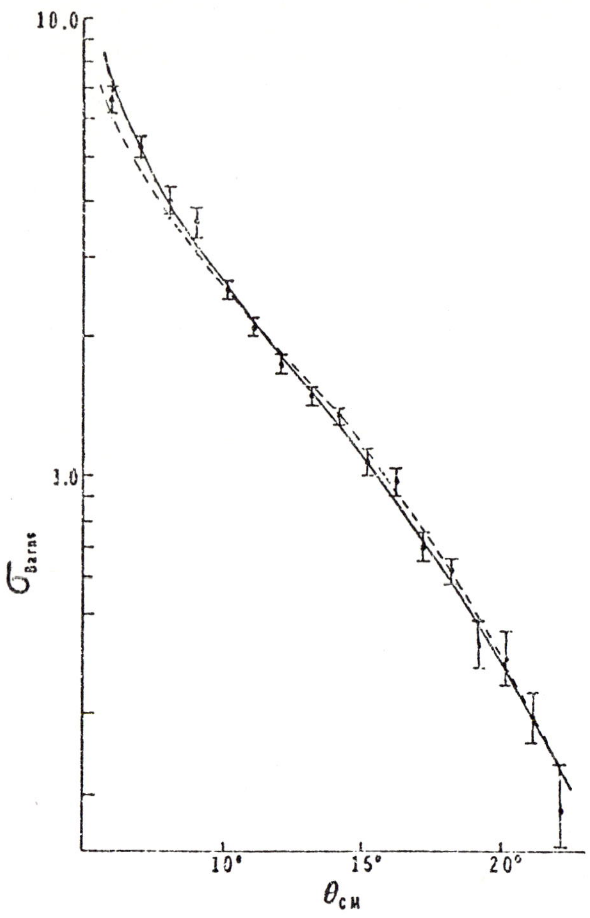

Fig. 7. Elastic differential cross section for $\pi^{\pm}$ ^{40}C at 204 MeV. The solid curve is the best fit of the diffraction model (Ref. 18), while the dashed curve is the best optical model fit (Ref. 19).

model analyses (Amann [20]). It may be pointed out that the theoretical fits do not take account of nuclear Fermi motion and recoil, which gives mixing of π-N partial waves (Hess [23]).

Data on total cross section for $\pi^{\pm}$ from various isotopes of C and Ca were taken at 90 MeV (Boyd [24]). Figure 9 shows the π^{+}-^{4}He partial cross section as a function of detector size and extrapolation to zero solid angle giving σ_{tot} = 186 ± 8 mb. In addition, a new method of dealing with coulomb-nuclear interference was developed. Using this method, the difference between rms radii of neutron and proton distributions, Δ, can be deduced from a comparison of $\pi^{\pm}$ scattering from different isotopes (Cooper [25]). For ^{13}C a preliminary value of Δ is 0.25 ± 0.15 f.

The direct lepton production in high-energy hadron-hadron collisions has been observed recently in many experiments (reviewed by Lederman [26]). The results give the ratio of lepton to pion production ~ 10^{-4}, almost independent of transverse momentum, c.m. energy, target nucleus, and production angle. These results created a flurry of excitement and many theoretical models were proposed. Two possibilities which are still viable are an undiscovered low-mass particle or a continuum of low-mass virtual photons. The experiment was started at LAMPF to test the former or to put a lower limit for the threshold energy for direct lepton production. The preliminary results seem to rule out the former possibility, and it is hoped that the results will permit a definitive statement in the realm of $\ell/\pi \approx 10^{-5}$ (Mischke [27]).

The three-body problem has received attention via a study of d(p,2p)n reaction at 800 MeV for different configurations of detector angles. Figure 10 shows the proton spectra corresponding to the kinematic situation which enhances neutron-proton final stage interaction (Phillips [28]). Using FSI theories fit to the spectra was obtained by including the singlet and triplet contribution to n-p scattering plus a constant term. The ratio of the triplet to singlet contribution was allowed to vary and was found to be larger than 3:1 ratio predicted by spin statistics. Data have also been taken for quasi-free p-p and p-n scattering. For angles corresponding to p-p elastic scattering, the data are in good agreement with the Hulthén deuteron wave function up to a spectator momentum of 180 MeV/c (Phillips [28]). Another experiment to look at the 3N systems is via π^{-} capture on ^{3}H and ^{3}He. The data from $^{3}H(\pi^{-}, \gamma)3n$ reactions are in final stages of reduction and would give information on the 3n system (Crowe [29]).

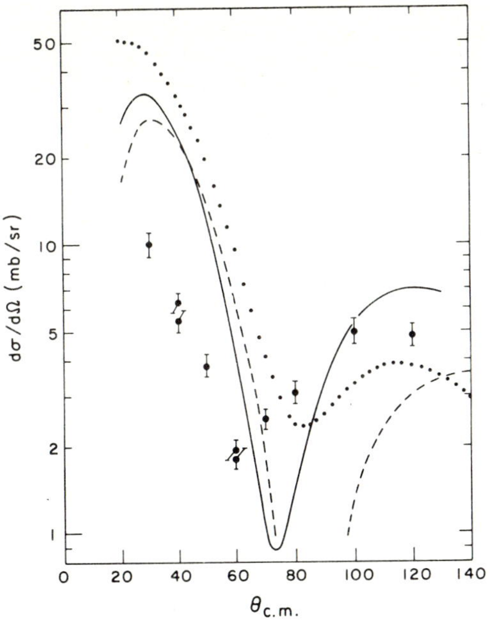

Fig. 8. Elastic scattering data for $\pi^{\pm}$ ^{12}C at 50 MeV. Theoretical fits are from separable potential (Ref. 21) solid line, local laplacian potential (Refs. 22, 23) dashed line, and the Kisslinger potential (Refs. 22, 23) dotted line.

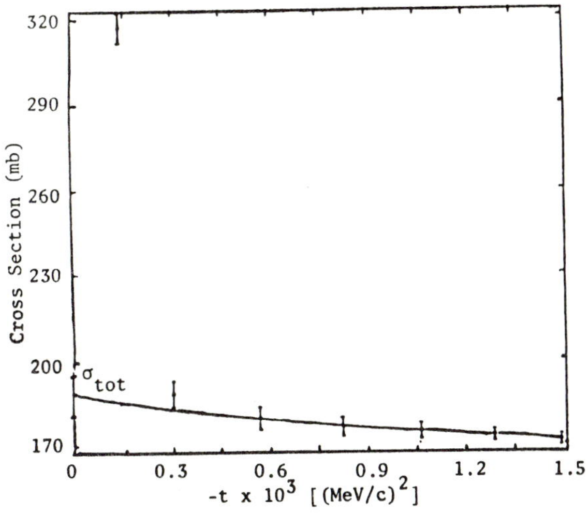

Fig. 9. Corrected partial cross sections for π^+ on ^4He at 90 MeV. The curve is a least square fit with statistical correlations taken into account.

The angular distributions for π^{+} + d → p + p reaction were obtained at 40, 50 and 60 MeV (Preedom [30]). Due to the large momentum mismatch between the entrance and exit channels the reaction is sensitive to the high momentum components of the deuteron. The calculations also show sensitivity to π-N p-wave off-shell parameter and require a 4% D-state admixture to fit the data at 50 MeV as shown in Fig. 11.

Various theoretical models for the excitation function of pion single charge exchange predict either a minima or maxima in the region of (3,3) resonance. The (π^+, π^o) excitation functions, on light nuclei in the energy range of 70 to 250 MeV, measured by activation techniques are shown in Fig. 12 (Shamai [32]). The data show neither a marked maximum or minimum at this resonance and provide a test for various theoretical calculations.

In the muon channel, the ground state muonium hyperfine structure interval has been measured to an accuracy of 0.36 ppm (Casperson [33]). The agreement with theory provides a major confirmation of conventional muon electrodynamics. Assuming that the QED corrections are known, the ratio of the muon mass to the electron mass was deduced to an accuracy of 2 ppm. Another experiment derived the quadrupole and hexadecapole moment of ^{165}Ho from the hfs of the 3d muonic states. The resulting values are Qo = 7.44 ± 0.07b and πo = 0.52 ± 0.10 b^2 (Powers [34]). Within the framework of deformed Fermi charge distribution and rotational model, an indication for the angular variation in the skin thickness was found. From extensive muonic x-ray measurements, charge radii and isotope shifts for 16 isotopes in mass 60 region were deduced (Shera [35]). These results coupled with electron scattering and Hartree-Fock calculations have considerably improved the understanding of nuclear charge distributions in these nuclei (Rinker [36]).

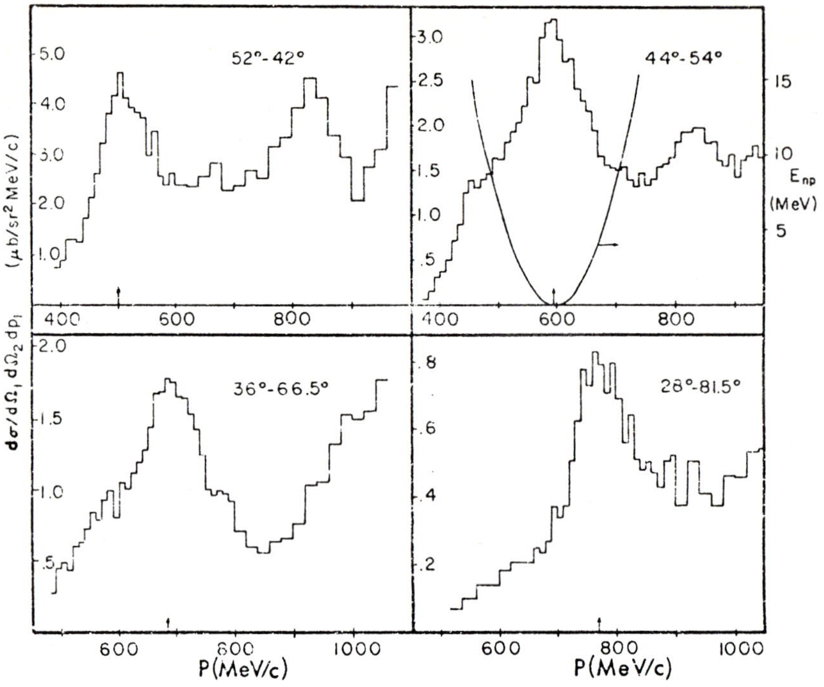

Fig. 10. Proton momentum spectra, in D(p,2p)n reaction at 800 MeV, showing final state interactions.

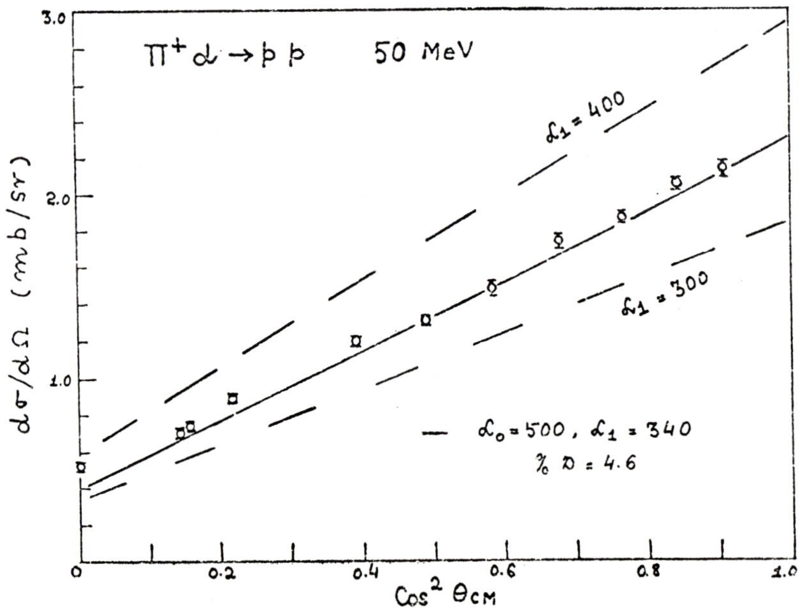

Fig. 11. Pion-deuteron absorption at 50 MeV [29] as calculated by Goplen [31]. α_0 and α_1 are the momentum cut-off parameters for s- and p-waves, respectively.

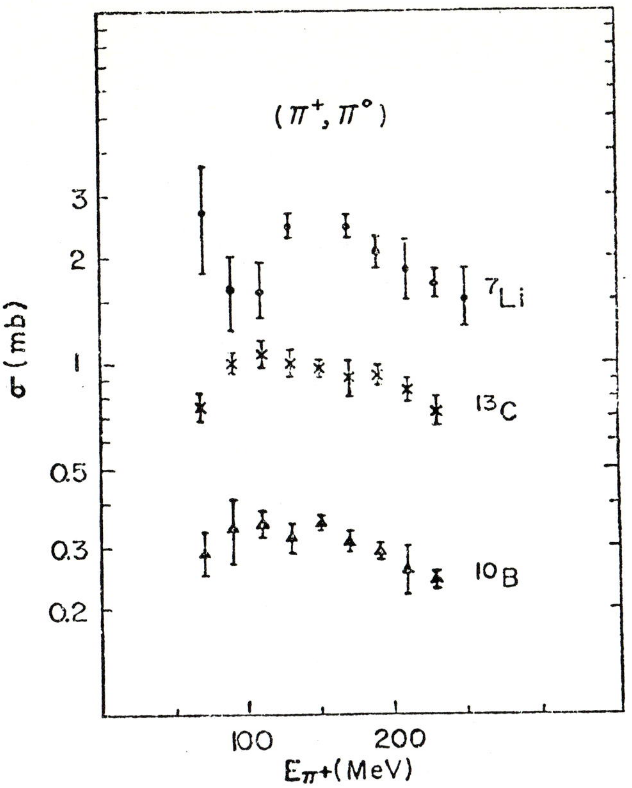

Fig. 12. Activation cross section for the (π^+, π^0) reaction on ^{7}Li, ^{10}B, and ^{13}C.

The spallation of uranium by protons was studied with dE/dx and time-of-flight techniques. Figure 13 shows the sodium mass spectra with a mass resolution of 1.5% (Butler [37]). Also a new time pickoff with a time resolution of 200 ps FWHM has been developed (Bowman [38]). These techniques would greatly improve the detection and separation of heavy ion fragments and would be valuable for astrophysics, biophysics and isotope hunting.

PRACTICAL APPLICATIONS

We are very enthusiastic about using LAMPF beams and technology for practical applications and have already enjoyed rather dramatic success. The most exciting development is the application of negative pions to the treatment of cancer. The main reason for using π^- is the production of a large amount of energy at the end of its range by the formation of "stars." The ionizing particles in such "stars" are primarily protons and alpha particles with typical ranges of a few microns in tissue, thereby leading to local deposition of high LET radiation (Rosen [39]). Quite often the dose to the tumor is limited by the radiation damage to the tissues outside the tumor. The negative pions should help to overcome this limitation and the anoxic problem. In addition, the pions can be tailored to the shape of tumor volume by magnetic fields and to the depth by variation of energy. Many biological specimens were irradiated in an exploratory-controlled program of treating subcutaneous tumors using Co γ rays and π^- (Kligerman [40]). Preliminary results indicate superiority of π^- irradiations.

A temperature measuring and recording instrument was developed to determine the feasibility of mass screening for breast cancer. It is presently undergoing tests. Another contribution of LAMPF technology to medicine is the construction of almost all new megavolt x-ray therapy units utilizing our side-coupled cavity accelerator design (Knapp [5]). The selling price of these clinical units exceeds the construction cost of LAMPF. The residual proton beam at the beam stop will become a major source of radioisotopes to be used for diagnosis and treatment in medicine and other fields. Recently, chemists separated the first pure sample of ^{82}Sr, which is the parent of ^{82}Rb (Grant [41]). The latter appears ideally suited for blood flow studies.

The radiation damage facility will permit exposure of materials under controlled conditions of temperature and mechanical stress to high-energy neutrons. The muonic x rays have proved to be a very useful nondestructive elemental analysis tool.

More recently, one of us (L. Rosen) has suggested that the primary proton beam may provide a powerful means for simulation of fast neutron radiation damage phenomena, and thereby to facilitate the search for materials which will be adequately immune to radiation damage in breeder reactors.

In conclusion, we expect that LAMPF will become an important resource for learning and practicing the art and science of interdisciplinary problem-solving.

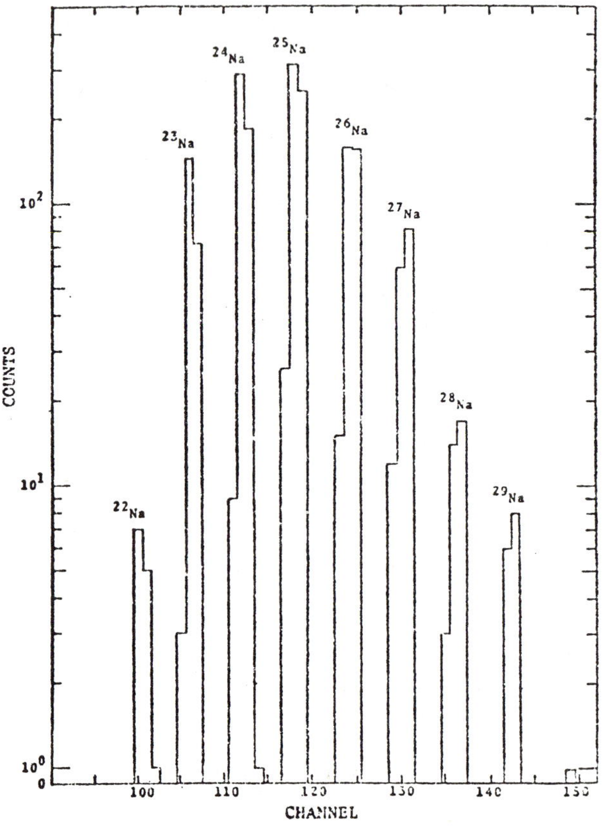

Fig. 13. Sodium mass spectrum (Flight path = 4.3 m) from 800 MeV protons on uranium.

ACKNOWLEDGEMENT

We thank various experimental groups for permission to show their unpublished results. One of us (M. Jain) acknowledges helpful discussions with J. C. Allred, J. Amann, E. P. Chamberlin, D. C. Hagerman, A. T. Hess, R. A. Jameson, R. E. Mischke, G. J. Rinker, Jr., and especially, J. E. Simmons.

REFERENCES[a]

[1] M. S. Livingston, "Origins and History of the Los Alamos Meson Physics Facility," Los Alamos Report, LA-5000 (1972).
[2] L. Rosen, Proc. Nat. Acad. Sci. U.S.A., 70 (1973) 603.
[3] D. E. Nagle, E. A. Knapp and B. C. Knapp, Rev. Sci. Instr. 38 (1967) 1583.
[4] R. A. Jameson, T. F. Turner, and N. W. Kindsay, IEEE Trans. Nucl. Sci. NS-12 (1965) 138.
[5] E. A. Knapp, B. C. Knapp and J. M. Potter, Rev. Sci. Instr. 39 (1968) 979.
[6] D. C. Hagerman, IEEE Trans. Nucl. Sci. NS-20 (1973).
[7] H. Butler, IEEE Trans. Nucl. Sci. 1 (1971) 336.
[8] H. H. Howard, "LAMPF USERS HANDBOOK," Los Alamos Report, MP-DO-1-UHB (1974). It describes details of beam channels, LAMPF organization, and procedure for submitting proposals.
[9] J. E. Simmons, Proc. High Energy Phys. and Nucl. Structure Conf., Santa Fe (1975) 103.
[10] C. Bjork et al., Proc. Int'l. Conf. on Few Body Problems in Nucl. and Part. Phys., Quebec (1974) 443.
[11] M. L. Evans et al., Submitted for publication to Phys. Rev. Lett.
[12] N. C. Amaglobeli and Yu. M. Kazarinov, Soviet Phys.-JETP 10 (1960) 1125.
[13] P. F. Shepard, T. J. Devlin, R. E. Mischke and J. Solomon, Phys. Rev. D10 (1974) 2735.
[14] G. Bizard et al., Nucl. Phys. B85 (1975) 14.
[15] H. B. Willard et al., High Energy Phys. and Nucl. Structure Conf. (Contributed papers), Santa Fe (1975) 252.
[16] J. S. Frank et al., ibid, p. 39.
[17] G. C. Phillips et al., to be published.
[18] F. Binon et al., Nucl. Phys. B33 (1971) 42.
[19] E. V. Hungerford, Nucl. Instr. & Meth., III (1973) 509.
[20] J. F. Amann et al., Phys. Rev. Lett. 35 (1975) 426.
[21] R. Silbar and M. Sternheim, Ann. Rev. Nucl. Sci. 24 (1975) 249.
[22] D. Koltun, Adv. Nucl. Phys. 3, (1969) 71.
[23] A. T. Hess, private communication.
[24] T. J. Boyd et al., Los Alamos Report, LA-5959-PR (1975) 43.
[25] M. D. Cooper and M. Johnson, accepted for publication in Nucl. Phys. A, and private communication.
[26] L. M. Lederman, Int'l. Sym. on Lepton and Photon Int., Stanford (1975).
[27] R. E. Mischke et al., private communication.
[28] G. C. Phillips et al., Los Alamos Report, LA-6109-PR (1975) 71.
[29] K. M. Crowe et al., Los Alamos Report LA-5959-PR (1975) 49.
[30] B. M. Preedom et al., submitted for publication to Phys. Rev. and private communication.
[31] B. Goplen, W. R. Gibbs and E. L. Lomon, Phys. Rev. Lett. 32 (1974) 1012.
[32] Y. Shamai et al., submitted for publication to Phys. Rev. Lett.
[33] D. E. Casperson et al., High Energy Phys. & Nucl. Structure Conf. (Contributed papers), Santa Fe (1975) 145.
[34] R. J. Powers et al., Phys. Rev. Lett. 34 (1975) 492.
[35] E. B. Shera et al., High Energy Phys. & Nucl. Structure Conf. (Contributed papers), Santa Fe (1975) 147.
[36] G. A. Rinker, Jr. and J. W. Negle, ibid, p. 148 and private communication.
[37] G. W. Butler, et al., private communication.
[38] J. D. Bowman et al., private communication.
[39] L. Rosen, Nucl. Appl. 5 (1968) 375.
[40] M. M. Kligerman, Proc. Int'l Particle Radiation Therapy Workshop, Key Biscayne, FL (1975).
[41] P. M. Grant, B. R. Erdal, and H. A. O'Brien, Jr., J. of Nucl. Medicine 16 (1975) 300.

[a] An abbreviated form is used.

PART 2

INVITED AND CONTRIBUTED PAPERS AT PARALLEL SESSIONS

AN OBE-MODEL INCLUDING THE N-Δ-INTERACTION, TWO-NUCLEON AND NUCLEAR MATTER RESULTS

by K. B l e u l e r, K. H o l i n d e, R. M a c h l e i d t
Institut für Theoretische Kernphysik der Universität Bonn
Nussallee 14-16, D-53 B O N N, W.-Germany

Recently[1] we have proposed a new one-boson-exchange potential (HM2) in which the phenomenological cutoff of dipole type used so far (e.g. HM1[2]) has been replaced by a so-called eikonal form factor, which arises from a simplified summation of multiple vector meson exchange processes. This scheme was used so far only in high energy physics, e.g. in the vector dominance model of electron-proton scattering; we show, however, that in this way also a good description of the nucleon-nucleon scattering phase parameters (χ^2/datum=2.77) and the deuteron data (E_b=2.2246 Mev, Q=o.2864 fm^2, P_D=4.32%) is obtained, whereby the vector meson coupling constants to be used in this new fit are in fact close to the values obtained from high energy physics, very much in contrast to the values obtained in the former case (see ref.[2]). In the same time rough estimates suggest nearly correct values for the binding energies of the triton and O^{16}, whereas a calculation of nuclear matter properties within the framework of first-order Brueckner theory yields a saturation energy of as much as -23.5 Mev at a Fermi momentum of k_F=1.77 fm^{-1}, i.e. a considerable overbinding as compared to the empirical value of about -16 Mev. (This fact can be traced back to a rather small new value of the deuteron D-state admixture (4.32%) as compared to 5.75% for HM1, which in turn yields -11.8 Mev at k_F=1.48 fm^{-1} for nuclear matter). Such a large overbinding is, however, compensated by a new type of corrections which are related to the introduction of the inner excitation of the nucleons (Δ-resonance state) during the scattering process: In the presence of other nucleons within nuclear matter the Pauli principle gives now rise to a characteristic change of binding energies. Such an inner excitation of the nucleon is in fact, physically speaking, inevitable; it replaces in the same time part of the action of the "unphysical" σ-boson, which had to be used so far. As a first step, we consider only the π-exchange at the N-Δ-vertex wherein a strong cutoff (of dipole type) is needed in order to simulate the compensating contribution from the N-Δ-ρ-vertex. Our results depend essentially on the corresponding cutoff parameter Λ; as an example we present here our results for Λ=55o Mev, which amounts to a replacement of about 1/3 of the whole σ-contribution. (The rest should be replaced by the 2π-contributions which have still to be included explicitly in our calculations).

We now add such a Δ-contribution to HM2 and make a new fit to all partial waves. Again, we achieve a good description of all N-N-scattering phase shifts (χ^2/datum=3.8) and of the deuteron data (E=2.2246 Mev). On the other hand, a standard first-order Brueckner--Bethe calculation for nuclear matter yields a saturation energy of only -1o.7 Mev at $k_F=1.38\ fm^{-1}$ which means that the many-body (Pauli- and dispersion-) corrections connected with the introduction of the Δ-resonance correspond to a shift of as much as 13 Mev. It is realized, however, that in view of the very strong density-dependence of these corrections the satisfactory value of the binding energy of the triton is practically not altered.

In order to compare our result with calculations from other sources we list in some detail for $k_F=1.4\ fm^{-1}$ the various corrections to the binding energy of nuclear matter due to the introduction of the Δ-excitation(see table below). We specify these corrections according to various partial waves (L-values), to the effect of the Pauli principle P_Δ and to the so-called dispersion effect D_Δ . In addition $D_{\Delta\Delta}$ represents the dispersion effect arising from the Δ-excitation of both nucleons, in which case there is no Pauli effect. In the first row of our table we give all these values in Mev, whereas the second row indicates the density-(k_F)-dependence of these contributions through the power n of k_F. Our results show that the corrections for higher angular momenta ($L \geq 1$) and the corrections $D_{\Delta\Delta}$ which were disregarded so far[3], play an essential role. The discrepancy of the value P_Δ ($^1S_0$) as compared to ref.[3] (our value is about a factor 5 smaller) might be due to certain approximations in that paper. In order to avoid our strong dipole cutoff which had to be used so far in the N-Δ-π-vertex we are just making further calculations in which also the N-Δ-ρ-vertex is included.

Table:

	$P_\Delta(^1S_o)$	$D_\Delta(^1S_o)$	$P_\Delta(L \geq 1)$	$D_\Delta(L \geq 1)$	$D_{\Delta\Delta}$	total
Mev	o.56	1.57	1.25	1.51	1.96	6.85
n	4.o	3.5	5.3	5.5	4.3	4.5

References

1) K. Holinde and R. Machleidt, OBEP and Eikonal Form Factor, Nucl.Phys., in press.
2) K. Holinde and R. Machleidt, Nucl.Phys. A247 (1975) 495
3) A.M.Green and J.A. Niskanen, Nucl.Phys. A249 (1975) 493

COMPLETENESS RELATIONS IN SCATTERING THEORY FOR NON-HERMITIAN POTENTIALS

Y. E. Kim† and A. Tubis*
Department of Physics, Purdue University
W. Lafayette, Indiana 47907, U. S. A.

McKay[1] and McKellar (MM) have recently discussed a number of interesting aspects of nonrelativistic scattering theory for non-hermitian (energy-dependent and/or complex) potentials. If $|\Psi_{\vec{k}}^{(\pm)}\rangle$, the eigenstates of a non-hermitian two-body Hamiltonian, are assumed to be complete, it is a straightforward procedure to construct an off-energy-shell continuation of the two-body T-matrix, which may be used in Faddeev calculations[2]. MM have not given a general proof of the completeness property, but have instead derived necessary and sufficient conditions on the $|\Psi_{\vec{k}}^{(\pm)}\rangle$ for completeness to hold.

We give in this paper a simple perturbative proof of the completeness of the $|\Psi_{\vec{k}}^{(\pm)}\rangle$. In order to keep our presentation as simple as possible, we follow MM in limiting the discussion to potentials which are non-singular in coordinate space (no hard cores, etc.), fall off faster than 1/r for large interparticle separations, and give no bound states. The second restriction allows us to use the usual form of formal scattering theory[3]. The analysis is easily generalized to the case of bound states and singular core interactions.

Assume that the two-body center-of-mass Hamiltonian H is

$$H = H_o + V, \quad V \neq V^\dagger \tag{1}$$

where

$$H_o = \frac{p^2}{2\mu} + V_o, \quad V_o^\dagger = V_o \quad . \tag{2}$$

$|\varphi_{\vec{k}}^{(\pm)}\rangle$, the outgoing-(incoming-) wave eigenstates of H_o, satisfy

$$|\varphi_{\vec{k}}^{(\pm)}\rangle = |\vec{k}\rangle + \frac{1}{E_k \pm i\epsilon - \frac{p^2}{2\mu}} V_o |\varphi_{\vec{k}}^{(\pm)}\rangle = |\vec{k}\rangle + \frac{1}{E_k \pm i\epsilon - H_o} V_o |\vec{k}\rangle, \tag{3}$$

where $\epsilon \to 0^+$ and $E_k = k^2/2\mu$. These eigenstates of the hermitian Hamiltonian H_o satisfy the usual orthonormality and completeness relations.

$$\langle \varphi_{\vec{k}'}^{(\pm)} | \varphi_{\vec{k}'}^{(\pm)} \rangle = \langle \vec{k}' | \vec{k} \rangle = \delta(\vec{k} - \vec{k}') \tag{4}$$

$$\int d\vec{k} |\varphi_{\vec{k}}^{(\pm)}\rangle \langle \varphi_{\vec{k}}^{(\pm)}| = 1 \ . \tag{5}$$

$|\Psi_{\vec{k}}^{(\pm)}\rangle$, the eigenstates of the full Hamiltonian satisfy

$$|\psi_{\vec{k}}^{(\pm)}\rangle = |\varphi_{\vec{k}}^{(\pm)}\rangle + \frac{1}{E_k \pm i\epsilon - H_o} V |\psi_{\vec{k}}^{(\pm)}\rangle = |\varphi_{\vec{k}}^{(\pm)}\rangle + \frac{1}{E_k \pm i\epsilon - H} V |\varphi_{\vec{k}}^{(\pm)}\rangle, \tag{6}$$

but, because V is non-hermitian, these states do not satisfy relations similar to (4) and (5). However, we may give a formal construction of a set of states $|\tilde{\psi}_{\vec{k}}^{(\pm)}\rangle$ which are bi-orthogonal to the $|\psi_{\vec{k}}^{(\pm)}\rangle$.

$$|\tilde{\psi}_{\vec{k}}^{(\pm)}\rangle = |\varphi_{\vec{k}}^{(\pm)}\rangle + \frac{1}{E_k \pm i\epsilon - H_o} V^\dagger |\tilde{\psi}_{\vec{k}}^{(\pm)}\rangle = |\varphi_{\vec{k}}^{(\pm)}\rangle + \frac{1}{E_k \pm i\epsilon - H^\dagger} V^\dagger |\varphi_{\vec{k}}^{(\pm)}\rangle \tag{7}$$

The orthonormality property

$$\langle \tilde{\Psi}_{\vec{k}'}^{(\pm)} | \Psi_{\vec{k}}^{(\pm)} \rangle = \delta(\vec{k}' - \vec{k}) \tag{8}$$

may be easily demonstrated using (6) and (7).

The completeness properties of the $|\psi_{\vec{k}}^{(\pm)}\rangle$ would follow provided we can show that

$$\int d^3k |\tilde{\psi}_{\vec{k}}^{(\pm)}\rangle \langle\psi_{\vec{k}}^{(\pm)}| = \int d^3k |\psi_{\vec{k}}^{(\pm)}\rangle \langle\tilde{\psi}_{\vec{k}}^{(\pm)}| = 1 . \tag{9}$$

(9) may be verified to any order in V by substituting (6), (7), and the expansion

$$\frac{1}{E_k \pm i\epsilon - H^\dagger} = \frac{1}{E_k \pm i\epsilon - H_o} + \frac{1}{E_k \pm i\epsilon - H_o} V^\dagger \frac{1}{E_k \pm i\epsilon - H_o} + \ldots, \tag{10}$$

and then examining matrix elements of the left-side of (9) with respect to the complete sets $|\varphi_{\vec{k}}^{(\pm)}\rangle$. It should be noted that the steps in the proof of (9) are algebraically equivalent to those in the proof of the completeness of the $|\psi_{\vec{k}}^{(\pm)}\rangle$ for the case, $V^\dagger = V$, $H^\dagger = H$. Since the completeness of the $|\psi_{\vec{k}}^{(\pm)}\rangle$ is known to be generally true in this case, the result (9) should be highly plausible even without a detailed proof.

References

[1] C. McKay and B. H. J. McKellar, "Scattering Theory for Energy Dependent Potentials", preprint, Service de Physique Theorique, CEN - Saclay, and School of Physics, University of Melbourne (1975), to be published in Nuclear Physics.

[2] Y.E. Kim, C.M. McKay and B.H.J. McKellar, "The Three-Body Problem with Energy Dependent Potentials", preprint, Research Institute for Theoretical Physics, University of Helsinki (1975), and contribution to this conference.

[3] See e.g., M.L. Goldberger and K.M. Watson, Collision Theory, John Wiley and Sons, New York (1964), Chapter 5.

† Supported in part by the U.S. National Science Foundation.

* Supported in part by the U.S. Energy Research and Development Administration.

AMBIGUITIES IN PHASE-SHIFT ANALYSIS

L.P. Kok

Weizmann Institute of Science, Rehovot, Israel, and
Institute for Theoretical Physics, Groningen, The Netherlands

One can distinguish two steps in the process of obtaining information on the basic dynamics in a scattering process from the experimentally measured quantities. Roughly the first step is the determination, starting from angular distributions, of all on-shell information, and the second step is going off-shell. Both steps in general are highly non-unique. I want to discuss the first step in its form of phase-shift analysis (p.s.a.).

The basic problem features already in simple two-body elastic spinless chargeless scattering at fixed energy. Ambiguities in p.s.a. can originate from various sources, and obviously errors in measurement play a role. Confining ourselves to data taken with infinite precision at all physical angles $-1 \leq z = \cos\theta \leq 1$ we are still left with two classes of discrete and continuum ambiguities. The distinction is a practical one rather than a matter of principle.

Often polynomial amplitudes truncated at order L in z are used in p.s.a. If only one channel is open (all elasticities $\eta_\ell = 1$), can different amplitudes of this type give the same prediction for observable angular distributions (apart from "trivially different", $\delta_\ell \rightarrow \pm\delta_\ell + n\pi$)? This question was answered yes by Crichton[1] in 1966 for L=2, and understood some years later[2]. All the analogous ambiguities for L=3, L=4 have been given[3,4]. De Roo proves their existence for arbitrary L by providing an example[5]. The corresponding angular distributions have a specific shape, which for L=2,3,4 may be unlikely to occur in nature. For high L the number of possible forms of $\sigma(z) = k^2\, d\sigma/d\Omega$ however probably is very high. This point deserves further investigation. The basic reason for the discrete ambiguities is that if the amplitude has a complex zero at $z=z_0$, $F(z) = (z-z_0)\,G(z)$, then the different amplitude $F'(z) = (z-z_0^*)\,G(z)$ will fit the experimental data equally well. The fact that both F and F' are to satisfy elastic unitarity imposes a severe constraint on the possible corresponding angular distributions.

Above the first inelastic threshold unitarity requires $0 \leq \eta_\ell \leq 1$ instead of $\eta_\ell = 1$. This restriction being much less severe makes that for many physically measured angular distributions the discrete ambiguity exists[6,7,8,9]. Apart from the mechanism of conjugation of zeros a new source of ambiguity now is revealed, amplitudes have a rotational degree of freedom $\exp(i\phi)$ for values of ϕ consistent with the unitarity limitations. This is the most trivial example of the continuum ambiguity, the basic reason for which is simply that the inelastic channels have not been subject to measurement.

In order to allow a full systematic investigation into the continuum ambiguity Atkinson e.a.[10] consider F(z) and $\sigma(z) = F(z)\ F^*(z^*)$ continued into the complex z-plane where it is analytic within the small Martin ellipse, until we hit the first cross-channel singularity. There the partial wave expansion of F(z) should diverge, whereas until now we considered only a finite number of partial waves. Relaxing this condition we admit an infinite number of partial waves, however still requiring analyticity within the small Martin ellipse. This then guarantees that the partial waves F_ℓ decrease with ℓ exponentially. Thus we find the continuum ambiguity by changing the inelastic contributions continuously (the η_ℓ). The method presented in[10] is an iterative method for generating a new set of phase shifts $\{\eta_\ell, \delta_\ell\}$, given an initial set, such that angular distributions are unchanged. This particular Newton iteration scheme guarantees solutions which are unitary, and analytic within the small Martin ellipse. We may vary freely all but a finite number N of the η_ℓ's and find new δ_ℓ's, and the remaining N η_ℓ's. The latter have to be constrained in order to assure analyticity of the amplitudes at each stage of the iteration process. Every zero of the dispersive part of F(z), $D(z) = \frac{1}{2}(F(z)-F^*(z^*))$, within a suitably chosen (or the Martin) ellipse in the complex z-plane leads to one constrained elasticity. We thus generate new sets of $\{\eta_\ell, \delta_\ell\}$ which can be reached via continuous trajectories from the original set. The unitarity inequalities $0 \leqq \eta_\ell \leqq 1$ set the limit to the size of the continuum ambiguity. If the total cross-section is known experimentally we just have to add one more constrained η_ℓ.

Practical application of this theory followed in[11] for $\alpha\alpha$-scattering data[12] with Coulomb effects taken out. Both theoretically and practically this program now has been extended.

1. For spin 0 - spin $\frac{1}{2}$ scattering, with exactly known $d\sigma/d\Omega$ and polarization P(z), with or without σ_{tot} kept fixed, use has been made of the Barrelet formalism[7] to explore the continuum ambiguity[13]. Also in spin $\frac{1}{2}$ - spin $\frac{1}{2}$ scattering[14] with 6 independent amplitudes (5 for identical particles) the continuum ambiguity exists in the inelastic region, knowing $d\sigma/d\Omega$, P, (P_t), D, D_t and C_{nn}. In higher spin cases the same picture will prevail.

2. The case of N two-body channels, with $\frac{1}{2}N(N+1)$ measurable cross sections, has been looked at[15]. For 2 channels the measurement of $\sigma_{12}(z) = \sigma_{21}(z)$ reduces the continuum ambiguity in the p.s.a. of $\sigma_{11}(z)$, but still a continuum ambiguity remains. To resolve this also $\sigma_{22}(z)$ has to be measured. What this seems to implicate is that if the second channel is a three-body channel an analysis free of continuum ambiguity would require knowledge of three-body scattering processes $A+B+C \rightarrow A+B+C$.

3. We can allow $\sigma(z)$ to vary, and investigate how the phase shifts vary as a consequence of this variation. To ensure analyticity we can constrain as before

some of the η_ℓ, but now also, or instead, we may constrain some of the partial wave projections of $\sigma(z)$. Thus we may follow the phase-shifts as a function of the variation of $\sigma(z)$ with energy, or with experimental error. This has been carried out in the neighbourhood of Crichton-like cross-sections, and for $\alpha\alpha$-scattering as a function of energy[9,16]. The ultimate goal of our program here is to set up a formalism which probes changes in angular distributions and elasticities which leave only the quantity χ^2 unchanged at a certain value. Whether this is feasible in practical examples is not yet clear.

There are many suggestions how to resolve the ambiguity, or at least to reduce its extent.

1. Continuation in the energy variable[17]. Inherent instability of numerical analytic continuation here is the main problem. Certain solutions may become non-unitary with increasing energy. In practice it helps for discrete ambiguities. A shortest path algorithm in general selects one answer, whether it gives the physical amplitudes is however uncertain.

2. Coulomb interference if measured with infinite precision kills the continuum ambiguity. At high energy, or with non-forward measurements only,it does not help. As soon as the Coulomb amplitude is approximated by a pole the effect is similar to the constraint of σ_{tot} kept fixed. In practical cases at low energy the ambiguity has to be explored starting from χ^2 as a function of the phase shifts, allowing small variations in $\sigma(z)$.

3. Use knowledge on the inelastic processes. Easiest to handle is the constraint that σ_{tot} is kept fixed, and experimentalists are requested to measure this quantity where possible! Inclusion of further inelastic scattering information of course helps too.

4. For instance in spin 0 - spin $\frac{1}{2}$ scattering different solutions may lead to different $A(\theta)$ and $R(\theta)$, measurement of which may reject one or more solutions.

5. Most other methods for the removal of ambiguities are theoretical in nature[9]. If one has a model to fix the high partial waves (Born or Regge Ansatz) no ambiguities exists apart from accidental (discrete) ones. Fixed t dispersion relations may distinguish between solutions. In all these cases the result of the p.s.a. is only as believable as the model.

To conclude we list processes considered so far. $\alpha\alpha$-scattering[8,9,11] was investigated from 17.4 to 60.0 MeV c.m. Considerable discrete and continuum ambiguity was found neglecting Coulomb effects. Also ^{3}He-α scattering was looked at, from 28 to 45 MeV c.m. Here the ambiguity is much worse, and a systematical exploration seems impossible. Much more unique turned out to be the N-α p.s.a.[18] in the $D_{3/2}$ resonance region, where that F_ℓ can vary only a few hundredths in all directions in the Argand circle. Various analyses[19] of a p+^{12}C have been looked at, from 9 to 18 MeV. The general trends in these p.s.a. seem significant.

As an example considering the Wienhard p.s.a.[19] we "conjugated" one Barrelet zero at energies 10.35, 10.375, 10.4, 10.425, two at 10.450, one at 10.5 and 10.525, and one at 10.7, 10.75, 10.8 MeV, to obtain new amplitudes with a smoother energy dependence. Even smoother dependence then can be obtained using the continuum ambiguity. In passing we remark that of the $2^6=64$ possible ways to conjugate Barrelet zeros at each of the 25 energies considered some 50% of all newly generated amplitudes proved to be unitary! Neglecting Coulomb effects in this case is hardly justified, to be contrasted with πN and KN scattering. For πN the full exploration[20] of the continuum ambiguity including spin and isospin is near completion showing considerable ambiguity corridors at higher energies, confirming results in[21].

We found in practice that if much continuum ambiguity is present there is much discrete ambiguity, and vice versa, and the smaller the η_ℓ are, the more ambiguity there is. A p.s.a. with many small η_ℓ therefore tends to be unreliable. A different way to quickly estimate the amount of ambiguity is the relatively easy determination of the number N of zeros of the dispersive part of the amplitude close to the physical region. If N is almost as large as the number of serious partial waves there is little ambiguity[9,11].

I am grateful to the authors in ref. 19, who made many data available, and to Mees de Roo and my colleagues in Groningen for their contribution to this work.

1. J.H. Crichton, Nuovo Cim. 45A (1966) 256.
2. A. Martin, Nuovo Cim. 59A (1969) 131, D. Atkinson, P.W. Johnson, N. Metha, M. de Roo, Nucl. Phys. B55 (1973) 125.
3. F.A. Berends, S.N.M. Ruysenaars, Nucl. Phys. B56 (1973) 507, 525.
4. H. Cornille, J.M. Drouffe, Nuovo Cim. 20A (1974) 401.
5. M. de Roo, private comm.
6. A. Gersten, Nucl. Phys. B12 (1969) 537; A219 (1974) 317.
7. E. Barrelet, Nuovo Cim. 45A (1972) 331.
8. A.C. Heemskerk, L.P. Kok, M. de Roo, Nucl. Phys. A244 (1975) 15.
9. M. de Roo, thesis, Groningen (1974) unpubl.
10. D. Atkinson, G. Mahoux, F.J. Yndurâin, Nucl. Phys. B54 (1973) 263.
11. D. Atkinson, P.W. Johnson, L.P. Kok, M. de Roo, Nucl. Phys. B77 (1974) 109.
12. P. Darriulat, thesis, Univ. of Paris (1965), unpubl.
13. D. Atkinson, G. Mahoux, F.J. Yndurâin, Nucl. Phys. B66 (1973) 429.
14. M. Manolessou-Grammaticou, Nucl. Phys. B98 (1975) 298.
15. D. Atkinson, G. Mahoux, F.J. Yndurâin, Nucl. Phys. (1975) 521, F. Hoekman, Groningen report 102, 1973.
16. D. Atkinson, M. Kaekebeke, M. de Roo, J. Math. Phys. 16 (1975) 685.
17. H. Burkhardt, A. Martin, Preprint Ref. TH 2017 CERN (May 1975).
18. B. Hoop, H.H. Barschall, Nucl. Phys. 83 (1966) 65.
19. K. Wienhard, G. Clausnitzer, G. Hartmann, Z. Physik 256 (1972) 457, G. Hartmann, Thesis, Erlangen (1968) unpubl., H.O. Meyer, G.R. Plattner, Nucl. Phys. A199 (1973) 413, H.O. Meyer, W. Weitkamp, private comm.
20. D. Atkinson, A.C. Heemskerk, D. Swierstra, Groningen preprint (Dec. 1975).
21. F.A. Berends, Leiden preprint (June 1973).

PHYSICAL PROPERTIES INVARIANT FOR PHASE EQUIVALENT POTENTIALS

C. S. WARKE
Tata Institute of Fundamental Research, Colaba, Bombay 400005, India

Let there be two potentials V and V' which are phase shift equivalent, such that the corresponding Hamiltonians $H = P^2/2 + V$ and $H' = P^2/2 + V'$ have the same bound state energies. These two restrictions require that the corresponding wave functions ψ_i and ψ_i' are asymptotically equal for all the eigenvalues of H and H'. As the spectrum of H and H' is the same, the complete set of wave functions ψ_i' are related to ψ_i by an unitary transformation U, $\psi_i' = U\psi_i$. The asymptotic equality of ψ_i and ψ_i' requires that

$$\lim_{r\to\infty} \psi_i'(r) = \lim_{r\to\infty} \int_0^\infty (r|U|r')\psi_i(r')dr' = \lim_{r\to\infty} \psi_i(r)$$

Hence

$$(r|U|r') \to \delta(r-r') \text{ as } r\to\infty. \tag{1}$$

The phase equivalent potentials V and V' have equal Fredholm determinants $D^+(k)$ of the outgoing solutions ψ_k^+ and $\psi_k'^+$. In the following G_k^+ ($\mathcal{G}_k^+$) is the non-interacting (interacting) $\ell = 0$ partial wave Green's function. According to the above statements V and V' are related by a unitary transformation U. The Fredholm determinant

$$\begin{aligned} D'^+(k) &= \mathrm{Det}[1-G_k^+V'] = \mathrm{Det}[1 - G_k^+(U^+HU-H_0)] \\ &= \mathrm{Det}[1 - G_k^+V - G_k^+(U^+HU-H)] \\ &= \mathrm{Det}[1-G_k^+V]\, \mathrm{Det}[1-(1-G_k^+V)^{-1}G_k^+(U^+HU-H)] \\ &= \mathrm{Det}[1-G_k^+V]\, \mathrm{Det}[1-\mathcal{G}_k^+(U^+HU-H)] \\ &= \mathrm{Det}[1-G_k^+V] = D^+(k). \end{aligned} \tag{2}$$

As the second determinant in Eq. (2) can easily be shown to be unity, when it is evaluated in the basis in which $\mathcal{G}_k^+$ is diagonal and use is made of the invariance of the spectrum of H with respect to a transformation U.

The low energy three body scattering amplitude for three free particles is the same for phase equivalent potentials. Amado (1) have shown that the connected scattering amplitude for three free particles can be written in the form $f = AE^{-1} + BE^{-\frac{1}{2}} + c\log E + 0(1)$, A, B, C are completely determined by kinematics and the two-body scattering amplitude at low energy, which is the same for all the phase equivalent potentials.

The full off-shell t-matrix for $s \to -E_b$ is the same for all the phase equivalent potentials which are obtained from the unitary transformation U such that $\psi_b' = U\psi_b = \psi_b$. From the reference Mongan (2)

$$\langle k|t(s)|p\rangle \to \langle k|V|\psi_b\rangle\langle\psi_b|V|p\rangle/(s+E_b) \quad as \quad s \to -E_b,$$

$$= (E_b - E_k)(E_b - E_p)\langle k|\psi_b\rangle\langle\psi_b|p\rangle. \tag{3}$$

$$\text{and } \langle k|t'(s)|p\rangle \to \langle k|V'|\psi_b'\rangle\langle\psi_b'|V'|p\rangle/(s+E_b)$$

$$= (E_b - E_k)(E_b - E_p)\langle k|\psi_b'\rangle\langle\psi_b'|p\rangle. \tag{4}$$

From Eqs. (3) and (4) and with $\psi_b' = \psi_b$, we have

$$\lim_{s\to -E_b} \langle k|t(s)|p\rangle = \lim_{s\to -E_b} \langle k|t'(s)|p\rangle. \tag{5}$$

The phase shift and $|D^+(k)|$ sum rules are the same for all phase equivalent potentials. It follows from the fact that for these potentials Fredholm determinants are identical. The sum rules are derived using only $D^+(k)$, and its asymptotic expansion by Puff (3) and Warke (3). The coefficients of the asymptotic expansion of $D^+(k)$ are also the same for phase equivalent potentials. For example, the volume integral of optical potential of a nucleon-nucleus scattering is the same for any two potentials which fit the same scattering data.

The off-shell generalization of the elastic unitarity relation remains the same for all the phase equivalent potentials. As follows from the reference Newton (4), it means that

$$\langle k|t|p\rangle - \langle p|t|k\rangle^* - \int_0^\infty dk' \langle k|t|k'\rangle\langle p|t|k'\rangle^* \left[\frac{1}{p^2-k'^2+i\varepsilon} - \frac{1}{k^2-k'^2-i\varepsilon}\right]$$

$$+ (p^2-k^2)\sum_n \langle k|\psi_n\rangle\langle\psi_n|p\rangle$$

$$= \langle k|t'|p\rangle - \langle p|t'|k\rangle^* - \int_0^\infty dk' \langle k|t|k'\rangle\langle p|t|k'\rangle^* \left[\frac{1}{p^2-k'^2+i\varepsilon} - \frac{1}{k^2-k'^2-i\varepsilon}\right]$$

$$+ (p^2-k^2)\sum_n \langle k|\psi_n'\rangle\langle\psi_n'|p\rangle. \tag{6}$$

for the t and t' matrices of two phase equivalent potentials V and V' respectively. The half off-shell transition matrix $\langle k|t|p\rangle = \langle k|t(p^2+i\varepsilon)|p\rangle$. From the definition of the half off-shell t-matrix and the use of Eq. (1), it is not difficult to show that

$$\langle k|t'|p\rangle = \langle k|t|p\rangle + (p^2-k^2)\langle k|U-1|\psi_p^+\rangle,$$

and

$$\langle k|t|p\rangle = \langle k|t'|p\rangle + (p^2-k^2)\langle k|U^+-1|\psi_p'^+\rangle. \tag{7}$$

From the use of Eq. (7), the left hand side (L. H. S.) of Eq. (16) becomes

$$\text{L.H.S.} = \langle k|t|p\rangle - \langle p|t|k\rangle^* + (p^2-k^2)\sum_n \langle k|\psi_n\rangle\langle\psi_n|p\rangle$$
$$- \int_0^\infty dk'\left[\frac{1}{p^2-k'^2+i\varepsilon} - \frac{1}{k^2-k'^2-i\varepsilon}\right]\Big\{\langle k|t'|k'\rangle\langle p|t'|k'\rangle^*$$
$$+ \langle k|U^+-1|\psi_{k'}'^{+}\rangle(k'^2-k^2)\langle k|t'|k'\rangle^*$$
$$+ \langle k|t'|k'\rangle(k'^2-p^2)\langle p|U^+-1|\psi_{k'}'^{+}\rangle^*$$
$$+ \langle k|U^+-1|\psi_{k'}'^{+}\rangle\langle p|U^+-1|\psi_{k'}'^{+}\rangle(k'^2-k^2)(k'^2-p^2).$$

Using the transformation properties of the wave functions and Eq. (7), the above equation can be reduced to a required form

$$= \langle k|t'|p\rangle - \langle p|t'|k\rangle^* - \int_0^\infty dk'\langle k|t'|k'\rangle\langle p|t'|k'\rangle^*\left[\frac{1}{p^2-k'^2+i\varepsilon}\right.$$
$$\left.- \frac{1}{k^2-k'^2-i\varepsilon}\right] + (p^2-k^2)\sum_n \langle k|\psi_n'\rangle\langle\psi_n'|p\rangle.$$

One knows that the full off-shell transition matrix is determined from the half off-shell t-matrix. We prove that given the half off-shell t-matrices of two phase equivalent potentials V and V', the unitary transformation U connecting V and V', can be obtained by solving a inhomogeneous linear integral equation. Let us denote $U - 1 = u$, for the sake of simplifying the notation. Adding and subtracting Eqs. (7), we obtain

$$(\varepsilon_p-\varepsilon_k)\left[\langle k|u|\psi_p^+\rangle + \langle k|u^+|\psi_p'^{+}\rangle\right] = 0$$

$$\langle k|t'-t|p\rangle = \tfrac{1}{2}(\varepsilon_p-\varepsilon_k)\left[\langle k|u|\psi_p^+\rangle - \langle k|u^+|\psi_p'^{+}\rangle\right]. \tag{8}$$

The asymptotic boundary condition in Eq. (1) is consistent with the assumption that $U^+ = U$. In the following U is assumed to be hermitian. Using now the momentum space representation of the scattering solution, one obtains from Eq. (8)

$$\langle k|\delta\psi|p\rangle = \int_0^\infty dk'\,\langle k|u|k'\rangle\langle k'|K|p\rangle. \tag{9}$$

In Eq. (9), we have defined $\langle k|\delta\psi|p\rangle = \langle k|t-t'|p\rangle/(\varepsilon_p-\varepsilon_k+i\varepsilon)$ and

$$\langle k|K|p\rangle = \int_0^\infty dk'\langle k|\tfrac{1}{2}(t+t')|k'\rangle\langle k'|\delta\psi|p\rangle/(k'^2-k^2+i\varepsilon). \tag{10}$$

The required unitary transformation can be determined from Eq. (9) using the known half off-shell t and t' matrices.

References:

(1) R. D. Amado and Morton H. Ruben, Phys. Rev. Letts. 25 (1970), 194.

(2) T. R. Mongan, Phys. Rev. 184 (1969), 1888.

(3) R. D. Puff, Phys. Rev. A11 (1975), 154; C. S. Warke, To appear in Pramana, and other references quoted there.

(4) R. G. Newton, Scattering Theory of Waves and Particles (McGraw Hill, N. Y. 1966) p. 189.

AN INTERPRETATION OF THE TABAKIN RANK-ONE POTENTIAL

M.J. Englefield
Department of Mathematics, Monash University,
Clayton, Victoria 3168, Australia.

A realistic representation of the S-state nucleon-nucleon interaction by a separable potential usually assumes a rank of at least two, with a repulsive and an attractive component, in order to obtain the negative phase shifts observed when $E > 300$ MeV. However Tabakin (1) obtained a rank-one potential with a phase shift that changed sign. This was of interest because the Fadeev-Mitra equations for the 3-body problem are most easily solved for a rank-one potential. Tabakin's potential has unusual properties: (i) the phase shift decreases to $-\pi$ as the energy increases (or, equivalently, the zero-energy phase shift is π); (ii) a bound state exists at the energy (240 MeV) where the phase decreases through zero; (iii) the scattering wave functions have an extra node near $r = .45$ fm, the position varying little with energy; (iv) when the interaction was used by Alessandrini (2) and by Beam (3) in the 3-body problem, a binding energy of more than 300 MeV was obtained.

The purpose of this paper is to point out that these properties can be understood in terms of the orthogonality condition model of the interaction between composite bodies, and to present an attempt to calculate the 3-body energy in the light of this interpretation. This model was originally introduced by Saito (4) as a simple approximation to the resonating-group treatment of the α-α interaction. It consists of an attractive potential, plus conditions of orthogonality to redundant states (in the α-α relative coordinate), which are excluded by the Pauli principle between the constitutent nucleons. (An angular momentum independent α-α interaction proposed by Rahman (5) at the 1972 conference could perhaps be interpreted in this way.) The repulsive features of the α-α interaction, e.g. negative phase shift (or rather decrease of phase shift through multiples of π), are reproduced by the orthogonality conditions, which also eliminate spurious bound states of the attractive potential, and induce in the scattering wave functions extra nodes at positions which hardly vary with energy. Proposals to apply this model to the nucleon-nucleon interaction have assumed a local potential and one redundant state giving one orthogonality condition. According to Englefield (6), suitable local potentials usually have a spurious (negative energy) bound state which to is removed by the orthogonality condition. In the simplest model, due to Neudatchin (7), the redundant state is identified with the spurious bound state of the local potential, and then the orthogonality condition is automatically satisfied by the scattering wave functions of the two-body problem.

Now there is no reason why the attractive potential should not be separable rather than local, and then there is nothing against a spurious bound state having positive energy rather than negative. It is proposed here to interpret the Tabakin rank-one potential in this way. Properties (i) - (iii) above then become expected properties of the system. (For the α-α interaction, separable representations proposed by Kukulin (8) are constructed to ensure just these features).

If this interpretation is correct, then orthogonality conditions must be used when solving the 3-body problem. Calculations using such constraints include the treatment of ^{6}Li as the 3-body system αnp by Wackman (9), and the treatments of ^{12}C = 3α by Smirnov (10) and by Horiuchi (11). The latter calculations use harmonic oscillator wave functions. With the Tabakin potential, it is easiest to work in momentum space. Since the momentum space wave function of the positive energy bound state (the redundant state) is not short-ranged, and presumably not well approximated by a Gaussian function, the exact projector off this state has been diagonalized numerically. Using the 4 symmetric S-states with 0 and 2 oscillator quanta gives just one state exactly satisfying the orthogonality conditions; using 7 states with 0, 2 and 4 quanta gives three states satisfying the conditions. The energy matrix with respect to these three states was then diagonalized. Since the exact projector is used, the oscillator parameter can be varied. Even so, in this approximation the 3-body system is unbound. Without the orthogonality conditions, using 3 states gives -410 MeV (taking the singlet interaction between all pairs).

There are two comparable calculations which suggest that more calculation is necessary to determine whether the 3-body system is bound. When Jackson (11) used a similar low-level approximation, the Reid potential failed to bind the 3-body system. Perhaps the better comparison is with the excited state which exists in the absence of the orthogonality conditions, as this wave function must also satisfy an orthogonality condition, namely orthogonality to the 3-body ground state. Using 7 states without the orthogonality condition, and the triplet interaction between all pairs, gives energies (in MeV) -215 and +131, compared to the exact values -237 and -11 obtained by Beam (3). This suggests poor convergence for any state having to satisfy any orthogonality condition. However, it is doubtful whether calculating higher approximations is worthwhile, because it is clear that the original reason for using the Tabakin rank-one potential - a simple 3-body calculation - is invalid.

REFERENCES

1. F. Tabakin, Phys. Rev. 174 (1968) 1208.
2. V.A. Alessandrini and C.A. Garcia Canal, Nucl. Phys. A133 (1969) 590.
3. J.E. Beam, Phys. Lett. 30B (1969) 67.
4. S. Saito, Prog. Theor. Phys. 40 (1968) 893; 41 (1969) 705.
5. M. Rahman, S. Ali and S.A. Afzal. Proc. of Int. Conf. on Few Particle Problems, Los Angeles, 1972 (North-Holland).
6. M.J. Englefield and H.S.M. Shoukry, Prog. Theor. Phys. 52 (1974) 1554.
7. V.G. Neudatchin, I.T. Obukovskii and Yu. F. Smirnov, Phys. Lett. 43B (1973) 13.
8. V.I. Kukulin and V.G. Neudatchin, Nucl. Phys. A157 (1970) 609.
9. P.H. Wackman and N. Austern, Nucl. Phys. 30 (1962) 529.
10. Yu. F. Smirnov, I.T. Obukhovsky, Yu. M. Tchuvil'sky and V.G. Neudatchin, Nucl. Phys. A235 (1974) 289.
11. H. Horiuchi, Prog. Theor. Phys. 51 (1974) 1266; 53 (1975) 447.
12. A.D. Jackson, A. Lande and P.U. Sauer, Nucl. Phys. A156 (1970) 1.

A POSSIBLE PHENOMENOLOGICAL FORM FOR THE TWO BODY LOCAL POTENTIAL

A.V.Lagu

Department of Physics, Banaras Hindu University,
Varanasi-221005, India.

There now exist several realistic two nucleon interactions, some purely phenomenological and others derived from more fundamental considerations. In all cases, however, the short range part is treated phenomenologically. A common feature is present in most of these potentials viz. that in the short range the dependence of the potential on the nucleon separation is invariably represented through Yukawa, Gaussian forms or their variants alongwith a core or momentum dependence. It is rightfully argued that since much ambiguity exists both experimentally and theorywise in this region, it is not very meaningful to specify other complicated forms. Interestingly, however, conclusions are often drawn regarding the dependence of the three and the many body properties on the short range part of the potential. Two such examples are (i) insensitivity of the three body parameters to the short range behaviour of the interaction (1) and (ii) that realistic N-N potentials do not give the right binding energy and density for nuclear matter (2). We feel that at this stage it may be premature to draw these inferences because the likelihood exists that these conclusions are really model dependent in the sense that the model for the short range is usually the same. It may be that a more fundamental description of the N-N interaction used in conjunction with a modified wave equation would actually prove both (i) and (ii) to be correct, but at the present level of sophistication it is not yet done. In the absence of such an approach we cannot ignore the possibility that both (i) and (ii) may be subject to revision if entirely different short range model is used. We, therefore, feel that a case does exist for constructing a new phenomenological potential with entirely new short and intermediate range feature.

We have, therefore, studied the basic form $V(r)=r^{-mr}=\exp(-mr \log r)$ and its variants for possible use as the short range part of the interaction. This form has a singularity at the origin. For that matter the Yukawa form also has a singularity at the origin but it is necessarily retained there because of its derivation from more fundamental considerations. In the present case no such compulsion exists and hence we can do away with the singularity altogether by modifying the form to $V(r)=\exp(-mr \log (a+r))$ where a is positive. This enables us to overcome any analyticity problems. Clearly this form has all the desired features of a nuclear potential and can be used for both short and intermediate regions. We, however, feel that the OPEP should be retained for the long range part i.e. after 2.5 fm. since it is very well established by now. If we compare this form with the usual Yukawa form viz. $\exp(-\mu r)/r$, we obtain $\mu=(mr \log(a+r)-\log r)/r$ which could be identified with the inverse range and can be seen to be different than the usual concept of range and thus emphasises the different feature of this form. To account for the change in sign of the phase shifts, we can introduce a multiplier (r-c) where now the core could be said to be a super soft core. We do not feel that any more manipulations would be required to fit the data and the parameters would be only m,a,c and an overall multiplier V_0 (generally negative) in each eigenchannel. We are currently undertaking such a parameter search and are hopeful that with only four parameters in each channel, a realistic data fitting could be done with a reasonable chi-square per data.

It may be mentioned that the present form is also interesting because it exhibits most of the desirable features of a molecular potential and with suitable modifications can be used there. It can also be used as an effective interaction.

We feel that three and many body calculations done with this new form would clearly establish as to what extent these properties are dependent on the short range part of the interaction. Another potential worth more investigation in the spirit of our arguments is by Neudatchin et al (3) which also has some basic features sufficiently different from the usual ones. A potential derived from more fundamental considerations and incorporating a slightly unusual phenomenological short range is due to the Paris group (4). Three and many body calculations with these also would help in clarifying points already mentioned. We realise that no remarkable and new results have been presented here and the whole note is more argumentative and suggestive in nature than concrete; we, nevertheless, feel that it is worth the trouble to do more calculations before firm conclusions are drawn about the dependence of three and many body properties on the short range behaviour of the nuclear force.

After completing this draft we received two preprints by Agarwal(5) in which essentially similar form for the potential is suggested and parameters given. This potential looks very similar to ours but differs in two ways viz. that it is singular at the origin and that OPEP is not incorporated for the long range part. This potential appears to have as good a fit to the two body data as the Reid Soft Core in all respects. We, however, feel that it would be preferable to do away with the singularity and also to use the OPEP for the long range part unlike Agarwal.

Financial assistance from the CSIR, New Delhi is gratefully acknowledged.

REFERENCES

1. I.R.Afnan and J.M.Read, Phys.Rev. C12 (1975) 293.
2. F.Coester, S.Cohen, B.Day and C.M.Vincent, Phys.Rev.C1(1970)769.
3. V.Neudatchin et al, Phys.Rev. C11(1975)128.
4. M.Lacombe et al, Few Body Problems in Nuclear and Particle Physics, Ed. R.J.Slobodrian et al, Les Presses de l'Universite Laval, Quebec (1975) 120.
5. B.K.Agarwal, ICTP Preprint IC/75/16(1975).
B.K.Agarwal and K.J.Narain, ICTP Preprint IC/75/17(1975).

TWO BODY BOUND STATE AND PARAMETRIZATION OF HALF-OFF-SHELL t-MATRIX

A.V. Lagu and C. Maheshwari

Department of Physics, Banaras Hindu University,
Varanasi-221005, India.

We give here a method of parametrizing the half shell function when the eigenchannel has a bound state. Earlier, Baranger et al (BGMS) (1) had given a method when there was no bound state and Sauer (2) had shown how field theoretic constraints could be incorporated. Haftel (3) extended the method of BGMS to include a bound state also and Mathur et al (4) have shown how the constraint of range can be imposed in such a case. Other approaches in this direction are due to Amado (5) and Picker et al (6). The present method is numerically simpler than that by Haftel and reduces to Amado's as a special case but has the same limitations as Picker's viz. using the Low equation may not be possible. Interestingly, however, the bound state plays an active role from the very beginning. The idea is to break the half shell t-matrix element $\langle k'|t(k^2+io)|k\rangle$ into two parts t' and t" with the stipulation that the information about the bound state should completely determine the term $\langle k'|t'(k+io)|k\rangle$ while nothing is stipulated about t". There is clearly no loss of logic in making such a stipulation. Writing t' in terms of a real part $\varphi'(k,k')$ and a phase part $\delta'(k)$ and t" in terms of $\varphi''(k,k')$ and $\delta''(k,k')$, we get

$$\varphi(k,k')\exp(i\delta(k)) = \varphi'(k,k')\exp(i\delta'(k)) + \varphi''(k,k')\exp(i\,\delta''(k,k')) \qquad (1)$$

where $\delta(k)$ are known experimental phase shifts, φ' and δ' are known since they are derived from the bound state information and φ is the part which is to be parametrized. The structure of this equation is such that if Φ is assumed to be two dimensional vector with angle δ in complex plane, then equation (1) can be written as

$$\Phi = \Phi' + \Phi'' \qquad (2)$$

with similar meanings attached to other terms. In equation (1), either of φ'' or δ'' can be taken to be arbitrary. The reason why δ' is taken to be a function of k alone while δ'' of both k and k' would become clear shortly. For the determination of φ' and δ' we recall the work of Lovelace, Fuda and Harms (see ref. 7) on the UPA and UPE where it was shown that a function $|\Psi(-B)\rangle$ can be constructed from the bound state wavefunction $|B\rangle$. Subsequently it was shown (8) that φ' and δ' can be expressed completely in terms of $|\Psi(-B)\rangle$ and hence δ' is a function of k alone and is a sort of UPA phase shift. No such meaning can be attached to δ'' and hence it is a function of both k and k' and more so because otherwise it would lead to an untenable situation. Taking φ' and δ' to be UPA things has a sound basis because such t' can be looked upon as the UPA of the unknown t-matrix and is a good approximation to it in a fair energy range. Whatever uncertainty now left is, therefore, transferred to t" i.e. to φ'' or δ''. It is not mandatory to use the UPA only for determining φ' and δ' - any other method can also be used. Using equation (2) now, it is easy to obtain $\varphi(k,k')$ which is

$$\varphi(k,k') = \varphi'(k,k')\cos(\delta(k)-\delta'(k)) \pm (\varphi''^2(k,k') - \varphi'^2\sin^2(\delta(k)-\delta'(k)))^{1/2} \qquad (3)$$

with φ'' subject to on-shell condition which is simple to derive. $\varphi(k,k')$ can alternatively be written in terms of $\delta''(k,k')$ also. Since no notice has been taken about the completeness of states in this method, using subtracted Low equation is doubtful. The unitarity of the s-matrix is, however, inherently built in.

If we take φ' to be Amado's φ_m and put $\delta = \delta'$, we get $\varphi(k,k') = \varphi'(k,k') \pm \varphi''(k,k')$ which is very similar to Amado's expression with his $(k^2-k'^2)\Delta(k,k')$ replaced by our $\varphi''(k,k')$. Interestingly, however, whereas there was no condition on Δ, φ'' is subject to an on-shell condition which is such that there is no need of the multiplier $(k^2-k'^2)$ because φ'' is zero on-shell when $\delta = \delta'$. The present method, therefore, appears more general than Amado's and clearly is easier to use. It is also interesting to observe certain similar features between equation (3) and Haftel's equation (17).

Despite a little digression, it is interesting to note that t" can be looked upon as the sum total of rest of the terms in a UPE if φ' and δ' are taken as UPA things. Now since φ'' is arbitrary, in the terminology of the UPE, an arbitrary choice of φ'' might amount to the choice of an arbitrary set of functions $|\Psi_n(-B)\rangle$ subject to the condition $\langle\Psi_n(-B)|G_0(-B)|\ \Psi_m(-B)\rangle = -\delta_{nm}$ coupled with the relation $|B\rangle = G_0(-B)\ |\Psi(-B)\rangle$ when n=1, where G_0 is the free resolvent operator. Choosing such a set of functions is also a logical way of parametrizing the half shell function, however, it appears cumbersome.

To test the method, we assumed the 'experimental' quantities $\delta(k)$, $|B\rangle$ and B to be those given by a hard core square well potential. Taking different choices of φ'' subject to on-shell constraint, φ was constructed and compared with exact φ obtained earlier from the potential. It was found that the method works quite well and is simple to use. We were also able to explain the error curves obtained earlier (8). We have, however, not given these details here because our choice of φ'' may turn out to be untenable after the imposition of field theoretic constraints on which we are presently working.

Financial assistance from the CSIR, New Delhi and the DAE, Govt. of India is gratefully acknowledged.

REFERENCES

1. M.Baranger, B.Giraud, S.K.Mukhopadhyay and P.U.Sauer, Nucl.Phys. A138(1969)1.
2. P.U.Sauer, Ann.Phys. (N.Y.) 80(1973)242.
3. M.I.Haftel, Phys.Rev.Lett. 25(1970)120.
4. V.S.Mathur, A.V.Lagu and C.Maheshwari, Contribution S-VM/4 to this Conference.
5. R.D.Amado, Phys.Rev. C2(1970)2439.
6. H.S.Picker, E.F.Redish and G.J.Stephenson, Phys.Rev.C4(1971)287.
7. M.K.Srivastava and D.W.L.Sprung, Advances in Nucl.Phys.Vol.8, Ed. E.Vogt and M.Baranger (1974).
8. A.V.Lagu, C.Maheshwari and V.S.Mathur, Phys.Rev.C11(1975)1443.

THE PHYSICAL BASIS OF THE UNITARY POLE EXPANSION - SOME REMARKS

A.N.Mantri and A.V.Lagu

Department of Physics, Banaras Hindu University,
Varanasi-221005,India.

The unitary pole expansion (UPE) due to Harms (1) makes use of a separable expansion to the two nucleon t-matrix which is given by the relation

$$t^M_{UPE}(s) = \sum_{n,m}^{M} |\psi_n(-B)\rangle \Delta^M_{n,m}(s) \langle\psi_m(-B)| \ , \quad (1)$$

where the two-body potential V admits a bound state at -B, and the form factors $|\psi_n(-B)\rangle$ are the eigenfunctions of the kernel of the Lippman-Schwinger (LS) equation. It is customary to define a function $|\chi_n(-B)\rangle$ in terms of $|\psi_n(-B)\rangle$ which is a solution of the Schrodinger like bound state equation for the modified potential $\lambda_n(-B)V$ with the eigenvalue -B.

As is well known (2,3) the convergence of the UPE investigated for various potentials is very fast in the sense that taking more terms in the expansion (1) hardly improves over the results obtained using only the first term (M=1) in equation (1). This one term approximation to the UPE is known (4) as the unitary pole approximation (UPA) and is nothing but the simple deuteron problem where the potential V(λ_n for n=1 is unity) binds the system with the binding energy B. We took up the problem of examining the physical basis of the higher order terms in the UPE and the results are presented herein.

Confining ourselves to s-wave channel and defining

$$\langle \vec{r} | \chi_{n,l=0}(-B)\rangle = \phi_n(r)/r \quad (2)$$

it is seen that the wavefunction $\phi_n(r)$ satisfies the Schrodinger equation for the potential $\lambda_n(-B)V$. For numerical calculations we have taken examples of (i) Herzfeld potential which is an attractive square well potential with an infinitely repulsive core and (ii) an exponential potential with an infinitely repulsive core. The eigenvalues $\lambda_n(-B)$ of the kernel of the LS equation are easily obtained by solving the appropriate equations in each case. We find that $\lambda_n(-B)$ for $n\neq1$ are all positive, greater than unity and increase very fast with n. This means that for $n\neq1$, a larger depth of the potential is needed to give the same binding energy and since no reduction in the range is made this leads to compression of the wavefunction. As a result the wavefunctions $\phi_n(r)$, besides the zero at hard core radius, start crossing the axis at other points and contain exactly (n-1) nodes and are therefore unphysical quantities for $n\neq1$ as far as the deuteron system description is concerned. The deuteron does not have any excited states and in fact the functions $\phi_n(r)$ for $n\neq1$ may be looked upon as "forbidden". We summarise the situation in the accompanying table where we give the potential parameters, eigenvalues λ_n and the corresponding nodal properties of the wavefunction.

Having shown the unphysical character of the wavefunctions $\phi_n(r)$ for $n\neq1$ the next significant question one faces is then why the UPE appears to behave so well despite the unphysical character of these terms? We set out to resolve this apparent anomalous situation and our findings showed that the unphysical nature of the higher order wavefunctions does not show up for low momentum values only,because in the t-matrix, the terms containing their effect (which appear in the form of product of form factors, see eqn.1) nearly cancel out (5). Now the product of form factors changes its sign around certain k values (which can be connected with the position of nodes in the corresponding wavefunctions) and therefore above these k values the said cancellation will not take place and the t-matrices for UPA and UPE will differ. For the Herzfeld potential, as an example, the UPA and the UPE (with two terms) results for the matrix $\langle k|t(s)|k'\rangle$ start differing for k,k' values greater than 0.8 fm^{-1} (the product of the form factors $\langle k|\Psi_1\rangle$ and $\langle k|\Psi_2\rangle$ is seen to change sign around this k value) which is seen to be connected with the occurrence of a node in the wavefunction $\chi_2(r,-B)$ at r ~ 1.29 fm.

Herzfeld potential	Exp. potential with hard core
$V = +\infty,\ r < a$ $= -V_0,\ a \leqslant r < b$ $= 0,\ r \geqslant b$	$V = +\infty,\ r < a$ $= -V_0 \exp(-b(r-a)),\ r \geqslant a$
$V_0 = 1.54\ fm^{-2}$ $a = 0.4$ fm $b = 1.737$ fm $B = \kappa^2 = 0.0104\ fm^{-2}$	$V_0 = 6.2735\ fm^{-2}$ $a = 0.4$ fm $b = 2.0831\ fm^{-1}$ $B = \kappa^2 = 0$
$\phi_n(r) = \sin\sqrt{\lambda_n V_0 - \kappa^2}(r-a),\ a \leqslant r < b$ $= 0$ at $r \leqslant a$	$\phi_n(r) = J_0((2\sqrt{V_0\lambda_n}/b)\exp(\frac{-b(r-a)}{2})),\ r \geqslant a$ $= 0$ at $r \leqslant a$
$n = 1,\ \lambda_1 = 1$, no node	$n = 1,\ \lambda_1 = 1$, no node
$n = 2,\ \lambda_2 = 8.17$, one node $\phi_2(r) = 0$ at $r = 1.29$ fm	$n = 2,\ \lambda_2 = 5.27$, one node $\phi_2(r) = 0$ at $r = 1.2$ fm
$n=3,\ \lambda_3=22.51$, two nodes $\phi_3(r)=0$ at $r=0.93$ & 1.47 fm	$n=3,\ \lambda_3=12.95$, two nodes $\phi_3(r)=0$ at $r=0.83$ & 1.6 fm

Financial assistance from the UGC, New Delhi, the CSIR, New Delhi and the DAE, Govt. of India is gratefully acknowledged.

REFERENCES

1. E.Harms, Phys.Rev. C1(1970)1667.
2. M.K.Srivastava and D.W.L.Sprung, Advances in Nucl.Phys., Ed. E.Vogt and M.Baranger, 8(1974).
3. J.S.Levinger, Springer Tracts in Modern Physics 71(1974)88.
4. M.G.Fuda, Nucl.Phys. A116(1968)83.
5. Similar conclusions have been arrived at through a different procedure by B.L.G.Bakker and W.Sandhas, Few Body Problems in Nuclear and Particle Physics, Ed. R.J.Slobodrian et al, Les Presses de l'Universite Laval, Quebec(1975)509.

PARAMETRIZATION OF THE HALF SHELL t-MATRIX WHEN THE TWO BODY BOUND STATE IS DELINEATED

V.S.Mathur, A.V.Lagu and C.Maheshwari
Physics Department, Banaras Hindu University, Varanasi-221005, India.

The problem of parametrizing the half-shell function $\phi(k,k')$, using the s-wave phase shifts $\delta(k)$ and the bound state wavefunction $\langle r|B\rangle$ as input data, was first considered by Haftel (1). His method required one to choose a 'model' potential V_M, which admitted the same bound state as the actual potential and introduce a function $\phi_W(k,k')$ by

$$\phi_W(k,k') = \langle \chi^\circ_{k'} | V-V_M | \psi^\circ_k \rangle \tag{1}$$

where $|\psi^\circ_k\rangle$ is the scattering state for the actual potential V while $|\chi^\circ_k\rangle$ is that for the model potential V_M, the notation being used are those of Baranger et al.(2). Then, if $\phi_W(k,k')$ could be determined, it would be possible to determine the required half shell function in terms of the half-shell function $\phi_M(k,k')$ and the phase shifts $\delta_M(k)$ of the model potential, according to the following equation:

$$\phi(k,k') = \mathrm{Cos}(\delta(k)-\delta_M(k))\,\phi_M(k,k')+\mathrm{Cos}\,\delta_M(k')\phi_W(k,k') + (k'^2-k^2)\,P\int_0^\infty dq\,\phi_M(q,k')\phi_W(k,q)(q^2-k^2)^{-1}(q^2-k'^2)^{-1}. \tag{2}$$

As regards the determination of $\phi_W(k,k')$ itself, Haftel suggested that its symmetric part $\sigma_W(k,k')$ be chosen arbitrarily, the only guideline being its prescribed on shell behaviour viz.,

$$\sigma_W(k,k') = -(2k/\pi)\,\mathrm{Sin}(\delta(k)-\delta_M(k)) \tag{3}$$

and $\phi_W(k,k')$ be obtained therefrom, using the method of Baranger et al. (2).

We use a separable potential V_{sep}, constructed (3) from the given bound state wavefunction $\langle r|B\rangle$ viz., $V_{sep} = -|\psi(-B)\rangle\langle\psi(-B)|$

$$\text{where} \quad \langle k|\psi(-B)\rangle = -(k^2+B)\langle k|B\rangle \tag{4}$$

to be the model potential envisaged by Haftel (1). This choice is justified because such a separable potential would naturally admit the delineated bound state and has the advantage that, being UPA to the exact (unknown) potential it is expected to reproduce the half-shell function to a considerable extent (4,5) leaving not much to be bridged by arbitrary correcting terms. To obtain guidelines for the (otherwise arbitrary) choice of the symmetric function $\sigma_W(k,k')$ for the case where $V_M=V_{sep}$ we have derived an analytic expression for $\phi_W(k,k')$ viz.

$$\phi_W(k,k') = -(2k'/\pi)\,\mathrm{Cos}\,\delta_M(k')\,\mathrm{Sin}\,\delta(k) + (2k/\pi)\,\mathrm{Cos}\,\delta(k)\,\mathrm{Sin}\,\delta_M(k') + (k^2-k'^2)\int_0^\infty v^M_{k'}(r)(\psi^\circ_k(r)-v^\circ_k(r))dr, \tag{5}$$

where $v^\circ_k(r)$ is the phase shifted free wavefunction for the actual potential given by $v^\circ_k(r) = kr\,j_0(kr)\mathrm{Cos}\,\delta(k)-kr\,\eta_0(kr)\mathrm{Sin}\,\delta(k)$, and $v^M_k(r)$ (with a similar definition) is that for the 'model' potential. We find that eq.(5) reproduces the on-shell behaviour as per eq.(3). The only unknown on its right hand side is the 'free scattering defect' $\psi^\circ_k(r)-v^\circ_k(r)$ which may be suitably parametrised(6) e.g. by

$$\psi^\circ_k(r)-v^\circ_k(r) = -(2/\pi)^{1/2}\,\mathrm{sin}\,\delta(k)\,\exp(-\beta r). \tag{6}$$

We have done a couple of test calculations wherein we assume that the 'experimental' data viz. phase shifts and bound state are the ones given by (i) square well and (ii) Herzfeld potentials. Then, forgetting about the potential we use the 'data' to construct the separable potential Vsep (3) and obtain the quantities $\delta_M(k)$ and $\phi_M(k,k')$ for the model potential as in ref.(5). Thereafter we calculate Haftel's function $\phi_w(k,k')$ from eq.(5) parametrizing the 'free scattering defect' as in eq.(6). The values of β in eq.(6) are chosen to be 0.7 and 1.0 fm^{-1}, respectively, for the two test cases which incidentally are of the same order as the inverse range of the corresponding potentials. Finally $\phi(k,k')$ is calculated using eq.(2) and is compared with the half shell function calculated directly from the potential (5). For an alternative check, we have also calculated $\phi_w(k,k')$ back from the exact $\phi(k,k')$, $\phi_M(k,k')$ and $\delta_M(k)$ using eq.(14) of Haftel (1), and compared the results of this back calculation with the values obtained from eqs. (5) and (6). We find that for low values of k and k' the above comparison indicates very good agreement which is expected since, for this case, the first two terms of the right hand side of eq.(5)(called the Sauer analogous terms) dominate over the third term, the correction integral. For large k and k', however, the values of both Sauer analogous term and the correction integral are large, comparable and of opposite signs. So the uncertainty in the choice of the free scattering defect would show up quite significantly in $\phi_w(k,k')$, hence in $\phi(k,k')$, when k and k' are large. This feature is shown also by the equal value contours for $\sigma_w(k,k')$ drawn on the basis of eq.(5) and (6) as well as from the back calculations (Figs.1 and 2). We also find trivially that if the exact expression for the free scattering defect is put into eq.(5) (we know these in the test cases where the potentials are known) the agreement of $\phi(k,k')$ calculated according to the above scheme is almost perfect with exact $\phi(k,k')$.

Equal value contours of $\sigma_w(k,k')$: obtained by parametrizing $\psi_k^o(r)-V_k^o(r)$ for β=0.7 ----: Zero-value, Resolution: 01

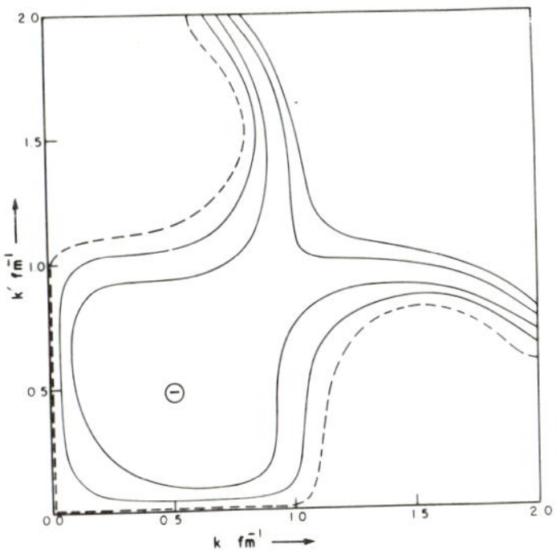

fig. 1

Equal value contours of $\sigma_w(k,k')$: obtained by back calc. ----: Zero-value, Resolution: 0 (Square well pot)

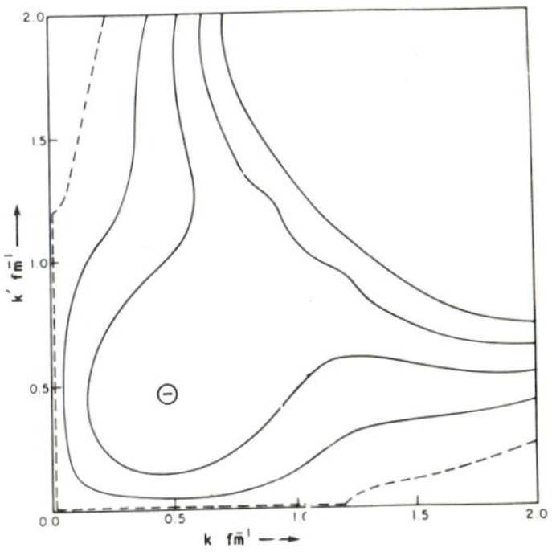

fig. 2

REFERENCES

1. M.I.Haftel; Phys.Rev.Lett. 25(1970)120.
2. M.Baranger, B.Giraud, S.K.Mukhopadhyaya and P.U.Sauer; Nucl.Phys. A138(1969)1.
3. M.G.Fuda; Nucl.Phys. A116(1968)83.
4. J.S.Levinger and J.O'Donoghue; Preprint, Renssaeler Polytech. Inst. (1972).
5. A.V.Lagu, C.Maheshwari and V.S.Mathur; Phys.Rev. C11(1975)1443.
6. P.U.Sauer; Ann.Phys. (N.Y.) 80(1973)242.

A SIMPLE METHOD OF CALCULATING SCATTERING PARAMETERS FOR SEPARABLE POTENTIALS WITH COULOMB INTERACTION

H. Kumpf

Zentralinstitut für Kernforschung der Akademie der Wissenschaften der DDR, Rossendorf, DDR 8051 Dresden, PSF 19

Up to now separable potentials are widely used for solving the Faddeev equations of the three nucleon problem in order to extract information about the off-shell behaviour of the nuclear interaction. Besides it is now well established that scattering parameters too depend on the interaction off-shell, if the Coulomb force is taken into account 1,2). Realistic potentials to be used in three nucleon calculations should now simultaneously fit the experimental p-p parameters if the electrostatic force is included and n-n parameters if that force is not added. In 3) simple expressions for the low energy scattering parameters of charged particles interacting by separable potentials are derived. They are suitable also for potentials given in a numerical form as gained e.g. by solving the inverse scattering problem. If the interaction is of rank one:

$$V(p,p') = -g(p)\,g(p') \quad ,$$

the formfactor in configuration space is:

$$G(r) = \sqrt{8}\int_0^\infty dp\,\sin(pr)\,g(p) \quad .$$

The scattering length a with Coulomb may then be expressed by

$$a = R_c J_I/(2J_I J_K + 2J_{IK} - 1)$$

$$J_I = \int_0^\infty dr\, r^{1/2}\, I_1(2(r/R_c)^{1/2})\,G(r)$$

$$J_K = \int_0^\infty dr\, r^{1/2}\, K_1(2(r/R_c)^{1/2})\,G(r)$$

$$J_{IK} = \int_0^\infty dr\, r^{1/2} G(r) \int_0^r dr'\, r'^{1/2}\, G(r') \cdot$$

$$\left[I_1(2(r'/R_c)^{1/2})\,K_1(2(r/R_c)^{1/2}) - I_1(2(r/R_c)^{1/2})\,K_1(2(r'/R_c)^{1/2})\right] \quad .$$

Here R_c is the Coulomb radius of the proton, I_1 and K_1 are the usual Bessel functions of imaginary argument. In 3) formulae are compiled expressing the other scattering parameters by quadratures involving Bessel functions I_n and K_n. Generalizations for rank N potentials are only slightly more complicated. The derived relations were applied to those separable potentials having been used in 4) for deuteron break-up calculations. Results are shown in the table.

name	a_{nuc}	r_{nuc}
Y	-18.8	2.92
E	-17.3	2.95
M	-18.1	2.88
L,Q,S	a_{coul} not existing	

For the forms named L,Q and S the effective range expansion does not exist at all if the Coulomb force is included, because the condition that the formfactor in configuration space should decrease at least as $\exp(-r^{1/2})$ is not fulfilled in these cases. For the other forms the table contains the pure nuclear scattering length a_{nuc} and effective range r_{nuc} in fm if the potentials are adjusted such as to fit the corresponding p-p values $a_{coul} = -7.823$ fm, $r_{coul} = 2.794$ fm. For the E and M cases a_{nuc} agrees fairly well with the measured neutron-neutron scattering length $a_{nn} = -16.6$ $\pm$ 0.5 fm, especially if one considers some minor corrections. In the deuteron break-up study 5) all six forms describe the general features of experimental data but none of them is able to reproduce all the spectra in details.

References

1) H.Kumpf, Yad. Fiz. 17 (1973) 1156

2) P.U.Sauer, Phys. Rev. C7 (1973) 943

3) H.Kumpf, Report ZfK-289, 1975

4) J.W.Bruinsma, W.Ebenhöh, J.H.Stuivenberg, R.van Wageningen, Nucl.Phys. A228 (1974) 52

5) B.Kühn, H.Kumpf, J.Mösner, W.Neubert, G.Schmidt, Nucl.Phys. A247 (1975) 21

ANALYTICITY CONSTRAINTS ON THE NUCLEON-NUCLEON T-MATRIX

S. K. Mukhopadhyay*
Theoretical Physics Section
Tata Institute of Fundamental Research, Bombay 5, India

P. U. Sauer
Theoretical Physics Institute
TU Hannover, 3 Hannover, West Germany

The two-nucleon transition matrix T is defined in terms of the nuclear potential V,

$$T(\omega) = V + V(\omega - K)^{-1} T(\omega) \tag{1}$$

K is the kinetic energy operator of relative motion in units of $\hbar^2/M$. T, not V, is close to experiment and to the effective interaction of nuclear structure calculation. A parametrization of T in momentum space, which avoids the cumbersome detour to a potential, but is consistent with our complete knowledge of the nuclear force, has, therefore, great practical and theoretical advantages. A method for doing so was suggested in ref. 1 for an uncoupled partial wave channel without a bound state.

The objective of this paper is to consider the constraints resulting from the analyticity properties of the T-matrix. Following ref. 2, one can assume some analyticity properties in the complex k_1^2 plane of the half-shell T-matrix $\langle k_2 | T(k_1^2 + i0) | k_1 \rangle$ in an uncoupled partial wave channel without a bound state. One should remember, however, that the analyticity properties assumed have been verified rigorously only in the case of potential scattering.

Assuming that $\langle k_2 | T(k_1^2 + i0) | k_1 \rangle$ has a right-hand unitarity cut and that it falls off faster than a constant as $k_1^2 \to \infty$ (i.e., excluding singular core-type interactions), we can write down a dispersion representation for the $^1S_0$ case, for example, subtracting at the on-shell point

$$\langle k_2 | T(k_1^2 + i0) | k_1 \rangle = \langle k_2 | T(k_2^2 + i0) | k_2 \rangle + \frac{k_1^2 - k_2^2}{\pi} \int^{\infty} dk^2 \frac{\operatorname{Im} \langle k_2 | T(k^2 + i0) | k \rangle}{(k^2 - k_1^2)(k^2 - k_2^2)}$$
$$+ \frac{k_1^2 - k_2^2}{2\pi i} \int dk^2 \frac{\operatorname{disc} \langle k_2 | T(k^2 + i0) | k \rangle}{(k^2 - k_1^2)(k^2 - k_2^2)} \tag{2}$$

The first term on the right-hand side is the on-shell T-matrix; the second term is the contribution from the unitarity cut, the third term is the contribution from cuts other than the unitarity cut. The normalization is such that

$$\langle k | T(k^2 + i0) | k \rangle = -\exp[i\eta(k)] \sin \eta(k)/k \tag{3}$$

$\eta(k)$ being the phase shift at the center of mass energy $\hbar^2 k^2/M$. Unitarity implies

$$\operatorname{Im} \langle k_2 | T(k_1^2 + i0) | k_1 \rangle = -k_1 \langle k_2 | T(k_1^2 + i0) | k_1 \rangle \langle k_1 | T(k_1^2 + i0) | k_1 \rangle^* \tag{4}$$

Noting that the function, $\exp[u(k_1)]$, where

$$u(k_1) = \frac{k_1^2 - k_2^2}{2\pi i} \int \frac{\eta(k)\, dk^2}{(k^2 - k_1^2 - i0)(k^2 - k_2^2 - i0)} \tag{5}$$

*Present Address: VEC Project, BARC, Trombay, Bombay 85.

has the same phase as $\langle k_2 | T(k_1^2 + i0) | k_1 \rangle$ on the unitarity cut, we can write

$$\langle k_2 | T(k_1^2 + i0) | k_1 \rangle = \langle k_2 | T(k_2^2 + i0) | k_2 \rangle \exp[u(k_1)] + \frac{k_1^2 - k_2^2}{2\pi i} \exp[u(k_1)] \int dk^2 \frac{\exp[-u(k)] \operatorname{disc} \langle k_2 | T(k^2 + i0) | k \rangle}{(k^2 - k_1^2)(k^2 - k_2^2)} \tag{6}$$

In eqn. (6) the discontinuity across the cuts disc $\langle k_2 | T(k^2 + i0) | k \rangle$ can be calculated in terms of the masses and coupling constants of the mesons exchanged.

However, unless the phase shifts and the approximations for calculating the discontinuities are compatible unitarity will be violated. On the other hand, the labor involved in making them mutually consistent will be comparable to that involved in fitting a potential to the phase shifts. We, therefore, propose to use experimental phase shifts in computing the function $u(k)$. Moreover, in calculating the disc T we consider only those mesons whose masses and coupling constants are known with reasonable certainty and that too upto second Born term. $u(k)$ and disc T calculated according to the ansatz outlined above are introduced into the right-hand side of eqn. (6). This will give us a half-shell T. Moreover, we take the symmetric part

$$\sigma(k_1 k_2) = \left(\frac{k_1 k_2}{\pi}\right) \Big[\langle k_1 | T(k_2^2 + i0) | k_2 \rangle \exp[-i\eta(k_2)] + \langle k_2 | T(k_1^2 + i0) | k_1 \rangle \exp[-i\eta(k_1)] \Big] \tag{7}$$

and then follow the procedure of ref. 1 in generating the half-shell T and later the full T. The full T generated in this fashion is unitary. Besides, it incorporates the dynamical constraints on the T-matrix within the framework of meson theory. Furthermore, the once-subtracted form, as noted in ref. 2 , suppresses the contributions from very high energies. We also incorporate quite naturally the constraints due to the short-range nature of the nucleon-nucleon interaction in momentum space.

The authors acknowledge discussions with S.M. Roy.

REFERENCES

1. M. Baranger, B. Giraud, S.K. Mukhopadhyay and P.U. Sauer Nucl. Phys. A138 (1969) 1
2. M.J. Reiner, Phys. Rev. Lett. 32 (1974) 236

A FEW APPLICATIONS OF COULOMBIAN ASYMPTOTIC STATES

H. van Haeringen

Natuurkundig Laboratorium der Vrije Universiteit,

Amsterdam, The Netherlands

In this note we report a few applications of our asymptotic states $|k\infty\rangle$ in partial wave (p.w.) space for $\ell = 0$ and for $\ell = 1$. The important equations are (2) and (7). We use the notation of previous papers [1,2]. From one of the expressions for $\langle\vec{p}|\vec{k}\infty\rangle$ (see [2]), we derive that the p.w. projected asymptotic states can be defined by

$$\langle p|k\infty\rangle \equiv \langle p|k\ell\infty\rangle = \lim_{\varepsilon\downarrow 0} (2ik)^{i\gamma}\Gamma(1+i\gamma) \frac{\varepsilon}{\pi pk} \frac{(p-k+i\varepsilon)^{-i\gamma}}{(p-k)^2+\varepsilon^2}. \qquad (1)$$

Notice that this expression does not depend on the value of ℓ. The p.w. Coulomb scattering state in co-ordinate representation is

$$\langle r|k\ell+\rangle = (2/\pi)^{\frac{1}{2}}\exp(i\sigma_\ell+\tfrac{1}{2}i\pi\ell)\, F_\ell(kr)/(kr),$$

where F_ℓ is the well-known (real-valued) regular Coulomb wave function. In order to derive, later on, Lippmann-Schwinger and related equations, we wish to prove that the following equation

$$\langle p|T_{c,\ell}(k^2)|k\infty\rangle = \langle p|V_{c,\ell}|k\ell+\rangle \qquad (2)$$

is valid for all $\ell = 0,1,2,\ldots$. The r.h.s. is known to be (cf. Dolinskiĭ and Mukhamedzhanov [3]),

$$\langle p|V_{c,\ell}|k\ell+\rangle =$$

$$\frac{e^{-\frac{1}{2}\pi\gamma}}{-i\pi p}\Gamma(1+i\gamma) \lim_{\varepsilon\downarrow 0}\left[(p-k-i\varepsilon)^{i\gamma}(p+k-i\varepsilon)^{-i\gamma} - \text{c.c.}\right] \quad \text{if } \ell = 0, \qquad (3)$$

$$\frac{e^{-\frac{1}{2}\pi\gamma}}{-i\pi p}\frac{\Gamma(1+i\gamma)}{1-i\gamma} \lim_{\varepsilon\downarrow 0}\left[\left(\frac{p^2+k^2+\varepsilon^2}{2pk}+i\gamma\right)\frac{(p-k-i\varepsilon)^{i\gamma}}{(p+k-i\varepsilon)^{i\gamma}} - \text{c.c.}\right] \quad \text{if } \ell = 1. \qquad (4)$$

In [1], eqs.(26)-(30), we obtained $\langle p|T_{c,\ell=0}(k^2)|p'\rangle$ in closed form. Upon application of the asymptotic state of eq.(1) we find

$$\langle p|T_{c,\ell=0}(k^2)|k\infty\rangle = i/(\pi p)\,\Gamma(1+i\gamma) \lim_{\varepsilon\downarrow 0}\{(ia)^{i\gamma}-(i/a)^{i\gamma}\}, \qquad (5)$$

with $a \equiv (p-k-i\varepsilon)/(p+k+i\varepsilon)$. The r.h.s.'s of eqs.(3) and (5) turn out to be equal, which confirms eq.(2) for $\ell = 0$. We have also obtained $\langle p|T_{c,\ell=1}(k^2)|p'\rangle$ in closed form. This is a relatively complicated function which will be given elsewhere. Applying $|k\infty\rangle$ to this function we find that eq.(2) is also valid for $\ell = 1$.

For Coulomb plus short-range potentials we can use <u>exactly</u> the same asymptotic states as we use for the pure Coulomb potential. This can be proved in general. From the theoretical point of view, it is always interesting and stimulating to have explicit closed formulae. Such formulae can be obtained in the present case if we choose the short-range potential V_s to be separable with rational form factors (cf. [1], section 8). The simplest specimen of the class of rational separable potentials is the Yamaguchi-type which has been used very often. It is given by

$$V_{s,\ell} = |g_\ell\rangle \lambda_\ell \langle g_\ell| ,$$

where the form factors g_ℓ are defined by

$$\langle p|g_\ell\rangle = (2/\pi)^{\frac{1}{2}} p^\ell (p^2+\beta^2)^{-\ell-1}, \quad \beta > 0, \; \ell = 0, 1, \ldots . \qquad (6)$$

Apart from the simplicity argument, there are also physical motivations for the use of this type of form factor, see Cattapan, Pisent and Vanzani [4]. The T matrix for the total potential $V_c + V_s$ is given by $T_c + T_{cs}$ where T_{cs} is separable with the so-called Coulomb-modified form factor $|g^c(k^2)\rangle$, see e.g.[1], sections 5 and 6, in particular eq.(56). Now we expect that the following equality holds,

$$\langle k\infty-|g_\ell^c(k^2)\rangle = \langle k\ell-|g_\ell\rangle . \qquad (7)$$

The r.h.s. can be evaluated in co-ordinate representation, yielding the closed formula (cf.[4], eq.(3,7)),

$$\langle k\ell-|g_\ell\rangle = \langle g_\ell|k\ell+\rangle = e^{-\frac{1}{2}\pi\gamma} \frac{\Gamma(\ell+1+i\gamma)}{\Gamma(\ell+1)} \frac{(2/\pi)^{\frac{1}{2}} k^\ell}{(\beta^2+k^2)^{\ell+1}} \left(\frac{\beta-ik}{\beta+ik}\right)^{i\gamma} . \qquad (8)$$

In eq.(66) of [1] we have obtained $\langle p|g_{\ell=0}^c(k^2)\rangle$ in closed form and from that expression we find that eq.(7) is valid for $\ell = 0$. Further we have derived $\langle p|g_{\ell=1}^c(k^2)\rangle$ from the above mentioned closed formula for $\langle p|T_{c,\ell=1}(k^2)|p'\rangle$. Applying the asymptotic state given in eq.(1), we find that eq.(7) is also valid for $\ell = 1$.

Finally we note the interesting but seemingly little noticed fact that the p.w. Coulomb scattering states in momentum representation are relatively simple objects possessing nice properties. For example, we derive eq.(8) for $\ell = 0$ in momentum representation by means of a simple contour integration as follows.
We have (cf.[3])

$$\langle p|k\ell+\rangle_{\ell=0} = e^{-\frac{1}{2}\pi\gamma}\Gamma(1+i\gamma)$$
$$\times \frac{1}{i\pi} \lim_{\varepsilon\downarrow 0}\left[\frac{1/p}{(p-i\varepsilon)^2-k^2}\frac{(p-k-i\varepsilon)^{i\gamma}}{(p+k-i\varepsilon)^{i\gamma}} - \frac{1/p}{(p+i\varepsilon)^2-k^2}\frac{(p+k+i\varepsilon)^{i\gamma}}{(p-k+i\varepsilon)^{i\gamma}}\right]. \qquad (9)$$

Because the second term in brackets is just equal to the first term with p replaced by -p, eq.(9) is even in p, and so we get, utilizing eq.(6),

$$\langle g_\ell|k\ell+\rangle_{\ell=0} = (2/\pi)^{\frac{1}{2}}\, e^{-\frac{1}{2}\pi\gamma}\Gamma(1+i\gamma)$$
$$\times \frac{1}{i\pi}\lim_{\varepsilon\downarrow 0}\int_{-\infty}^{\infty}\frac{p\,dp}{p^2+\beta^2}\,\frac{1}{(p-i\varepsilon)^2-k^2}\,\frac{(p-k-i\varepsilon)^{i\gamma}}{(p+k-i\varepsilon)^{i\gamma}}\,. \qquad (10)$$

Since the only singularity of the integrand in the lower p-half-plane is a first-order pole at $p = -i\beta$, we close the contour by a large semi-circle and confirm thus eq.(8) for $\ell = 0$.

This investigation forms a part of the research program of the Foundation for Fundamental Research of Matter (FOM) which is financially supported by the Netherlands Organization for Pure Scientific Research (ZWO).

References:

1. H. van Haeringen and R. van Wageningen, J.Math.Phys.16(1975)1441
2. H. van Haeringen, Coulombian Asymptotic States, preprint (1975)
3. É.I.Dolinskiĭ and A.M.Mukhamedzhanov, Sov.J.Nucl.Phys.,3(1966)180
4. G.Cattapan,G.Pisent and V.Vanzani, Nucl.Phys.A241(1975)204

NUCLEAR THREE-BODY FORCES ARISING FROM TRIPLE-BOSON COUPLINGS

T. Ueda, T. Sawada and S. Takagi
Faculty of Engineering Science, Osaka University, Toyonaka, Japan

Three-nucleon potentials (TNP) due to triple-boson couplings and the 3-3 resonance are studied and their effects on the properties of the nuclear matter [1,2] are investigated. Three-nucleon (N) forces due to the two-π exchange have been investigated by many people [3-5]. They may be important in the long range part ($r \gtrsim 1.4$ fm), however, the inner part may be considerably influenced by other mechanisms due to heavy bosons.

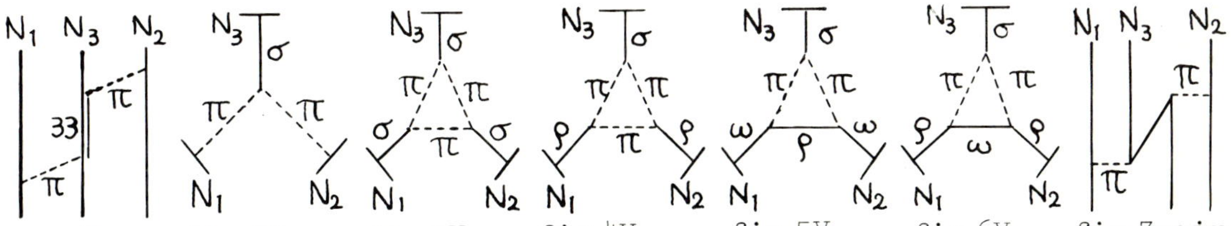

fig.1V_{33} fig.2$V_{\sigma\pi\pi}$ fig.3$V_{\sigma\sigma\sigma}$ fig.4$V_{\sigma\rho\rho}$ fig.5$V_{\sigma\omega}$ fig.6$V_{\sigma\rho}$ fig.7pair

The one-boson exchange potentials (OBEP), incorporating π, σ($I=0$, $J^P = 0^+$), ρ and ω, have been very successful in describing the two-N interaction [6,7]. From the similar point of view we can depict six important diagrams shown by figs. 1-6 for the three-N interaction. Then we obtain TNP from the corresponding Feynman amplitudes in the non-relativistic approximation and in reduction to the effective two-N potential with variables of the third N integrated over:

$$\frac{1}{4}\mathop{\mathrm{Tr}}_{\vec{\tau}_3}\mathop{\mathrm{Tr}}_{\vec{\sigma}_3}\rho\int d\vec{r}_3\,[\eta(\vec{r}_1-\vec{r}_3)\eta(\vec{r}_2-\vec{r}_3)]^2\,V(\vec{r}_1,\vec{r}_2,\vec{r}_3;\vec{\sigma}_1,\vec{\sigma}_2,\vec{\sigma}_3;\vec{\tau}_1,\vec{\tau}_2,\vec{\tau}_3), \qquad (1)$$

where V is the Fourier transform of the amplitude, ρ the density and η the correlation function.

The correlation function is parametrized as $\eta(r) = 1 - \exp[-K^2r^2]$, where K is determined by the excitation parameter: $\kappa_0 = \rho\int d\vec{r}(1-\eta)^2$. The coupling constants, $g_{\rho NN}$, $f_{\rho NN}$ and $g_{\omega NN}$, appearing in the amplitudes are known from the N-N scattering [7]; the σ mass, m_σ, and the product of coupling constants, $g_{\sigma\pi\pi}\cdot g_{\sigma NN}$, are known from the π-N scattering [8]; the coupling constants, $g_{\omega\rho\pi}$ and $g_{\rho\pi\pi}$, are determined by decay widths of $\omega\rightarrow\pi^0\gamma$ and $\rho\rightarrow\pi\pi$, respectively. The pion propagators in the amplitudes for figs. 1, 2 and 3 are modified as $1/[(k^2+\mu^2)(k^2+\Lambda_\pi^2)]$, where the second factor represents the virtuality correction and μ is the π mass. Λ_π is taken to be 420 MeV, consistent with $pp\rightarrow\pi^+pn$ [9]. The three diagrams of figs.4, 5 and 6 involve logarithmic divergences. These are removed by the cutoff function, $1/(k^2+\Lambda^2)$, introduced into the divergent integrals, where Λ is tentatively chosen to be 1 GeV.

Thus we obtain TNP, $V_{\sigma\pi\pi}$, $V_{\sigma\sigma\sigma}$, $V_{\sigma\rho\rho}$, $V_{\sigma\omega}$ and $V_{\sigma\rho}$, corresponding to figs. 2,3,4,5 and 6, respectively. First, we summarize the qualitative features of our TNP.

(1) $V_{\sigma\pi\pi}$, $V_{\sigma\sigma\sigma}$, $V_{\sigma\rho\rho}$ and $V_{\sigma\omega}$ have similar spin-isospin structure to U_π, U_σ, U_ρ and U_ω, respectively, where U_B represents the two-N B-exchange potential.

(2) TNP have longer ranges than the corresponding two-N potentials. In place of the Yukawa function, $e^{-\mu r}/r$ for the two-N potentials, TNP have an exponential function, for example, $e^{-m_\sigma r}$ for $V_{\sigma\sigma\sigma}$.

(3) $V_{\sigma\pi\pi}$ is considerably large. In some literatures [3,5] this has

been ignored for the reason that $V_{\sigma\pi\pi}$ is canceled by V(pair)(fig.7). In nuclear matter, however, V(pair) may be largely suppressed because the exclusion principle works for the initial and final third nucleons which must exist simultaneously for the interaction. Thus the cancellation of $V_{\sigma\pi\pi}$ by V(pair) is questionable.

(4) $V_{\sigma\sigma\sigma}$ has a considerable attractive central potential.

(5) The three-N tensor potential tends to strengthen the two-N tensor potential.

(6) The three-N LS potentials are very small, compared with the LS potentials in the two-N states.

Table I. The binding energy per nucleon, BE/A, the potential energy per nucleon, PE/A, in each state and the excitation parameter κ are shown. PE/A for the spin-triplet state is a sum over the three states with a (2j + 1) weighting. U = OBEP(UGI), $\overline{33} = V_{33}$, $\sigma\pi\pi = V_{\sigma\pi\pi}$ etc. all = $\overline{33} + \sigma\pi\pi + \sigma\sigma\sigma + \sigma\rho\rho + \sigma\omega + \sigma\rho$.

Model	U	U+all	U+$\overline{33}$	U+σππ	U+σσσ	U+σρρ	U+σω	U+all	U+all
k_F (fm^{-1})	1.4	1.4	1.4	1.4	1.4	1.4	1.4	1.4	1.7
κ_0		.0895	.0895	.0895	.0895	.0895	.0895	0.0	.117
PE/A(MeV) $^1S_0$	-17.6	-12.4	-16.7	-5.8	-19.0	-17.4	-17.5	-13.4	-9.7
$^3S_1$	-18.1	-16.9	-19.4	-11.6	-18.8	-18.5	-18.1	-18.8	-14.0
$^1P_1$	4.9	3.3	5.4	0.9	4.6	5.0	4.9	3.4	1.4
^3P	-1.4	-6.2	-1.1	-6.6	-4.5	-0.8	-1.1	-7.4	-19.2
$^1D_2$	-3.0	-4.6	-3.5	-4.4	-3.4	-3.0	-3.0	-4.9	-10.0
^3D	-3.3	-5.3	-3.9	-5.6	-3.7	-3.4	-3.3	-5.7	-12.1
$^1F_3$	0.9	2.3	1.1	3.4	0.8	0.9	0.9	2.4	5.9
^3F	0.5	1.4	0.7	2.7	0.1	0.5	0.5	1.5	2.2
$\ell \geq 4$	-0.6	-2.8	-0.8	-4.8	-0.7	-0.6	-0.6	-2.8	-10.2
BE/A(MeV)	13.3	16.9	13.7	7.4	20.1	13.1	12.8	21.4	29.8
κ	.0895	.106	.0996	.130	.0812	.0890	.0902	.105	.142

Using the same computational method as in reference 1, we investigate the effects of our TNP on the properties of the nuclear matter. The Bethe-Goldstone (BG) equation is solved for the effective two-N interaction, U + V, where U is the OBEP of Ueda and Green (UG-I)[7], and V is TNP. The results are shown in Table I. Observing the results we remark the followings.

(1) The V_{33} contribution to the binding energy per nucleon, BE/A, is 0.4 MeV attraction. This is very small, compared with those of the other authors. For example, 3.5 MeV attraction is predicted by Tanaka et al, who solved also the BG equation. Main reason for the difference between ours and theirs seems to be that we take the virtuality correction (or a pionic form factor) into account but they do not [3].

(2) The state by state contributions of $V_{\sigma\pi\pi}$ are quite large. To the total BE/A, however, considerable cancellation occurs among contributions from various states. The first-order transition due to V(pair) between states in the Fermi sea is forbidden by the exclusion principle. However, off-shell transitions between occupied states and unoccupied states due to V(pair) are possible. Then, one may expect some cancellation between V(pair) and $V_{\sigma\pi\pi}$. Thus in the off-shell transitions we may somewhat overestimate the effects of $V_{\sigma\pi\pi}$.

(3) The $V_{\sigma\sigma\sigma}$ contribution is appreciable in each state and in the total binding energy. The primary contribution is due to the attractive central potential.

(4) $V_{\sigma\rho\rho}$, $V_{\sigma\omega}$ and $V_{\sigma\rho}$ make rather small contributions in BE/A.

(5) The large correlation effect on BE/A is observed in comparison

of the two cases of κ_0:0.0895 (UG-I value) and 0.0 (no correlation). (6) Contribution of TNP to BE/A increases with density. This is primarily due to the increase of the density multiplied in eq. (1). Consequently, we find it hard to reproduce the saturation at the normal density. This may indicate the necessity to include three-N correlations due to both two- and three- N potentials.

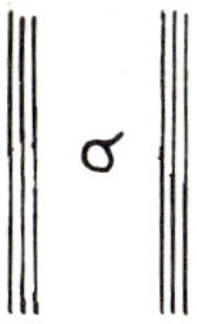

Fig. 8

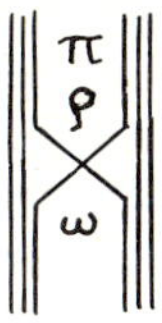

Fig. 9

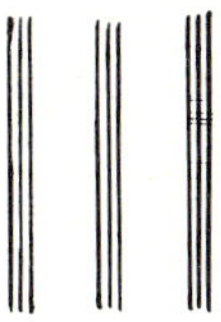
Fig.10

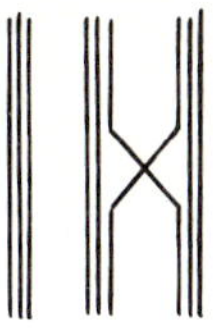
Fig.11

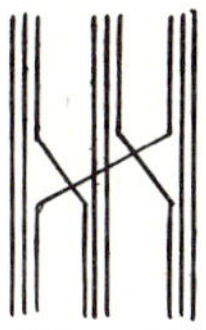
Fig.12

Fig.13

Fig.14

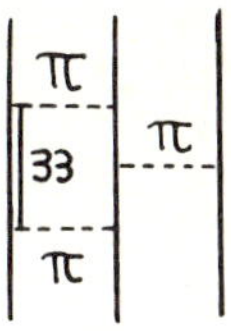

Fig.15

Our final comment is on interpretation of the six three-N interactions calculated, in terms of quark physics. In the OBEP for the two-N interaction the σ exchange can be interpreted as no quark (q) exchange shown by the quark diagram of fig. 8, and the π, ρ and ω exchange as the exchange of one quark-antiquark ($\bar{q}$) pair (fig.9) [10]. Similarly for three-N interactions we can depict no q exchange (fig. 10), one-$q\bar{q}$ exchange (fig.11), three-q exchange (fig.12) and two-$q\bar{q}$ exchange (fig.13). Then the σσσ coupling (fig.3) represents no q exchange (fig.10). A sum of the σππ(fig.2), σρρ(figs.4 and 6) and σωω(fig.5) couplings represents the one-$q\bar{q}$ exchange (fig.11). The 3-3 resonance contribution (fig.1) represents the three-q exchange (fig.12). A part of the two-$q\bar{q}$ exchange (fig.13) corresponds to the N-$\bar{N}$ pair term (fig.7) and the two-π exchange in retardation (fig.14). We note that the three-π exchange mechanism shown by fig.15 is topologically equivalent to fig.11 and thus conjectured to be dual with a part of the interactions of figs. 2, 4, 5 and 6. Three-N interactions due to ωρπ and ρππ couplings belong to fig.12 and may be dual with a part of the 3-3 resonance contribution. However, those two interactions vanish in the present approximations.

In conclusion we remark that the contributions from $V_{\sigma\pi\pi}$ and $V_{\sigma\sigma\sigma}$ are noteworthy for the three-nucleon potential.

References

[1] C.W. Wong and T. Sawada. Ann. Phys. 72 (1972) 107.

[2] K.Takada, S.Takagi and W.Watari,Prog.Theor.Phys,38(1967)144.

[3] T.Kasahara, Y.Akaishi and H.Tanaka, Prog.Theor.Phys.Suppl.No.56 (1974)96; M.Sato, Y.Akaishi and H.Tanaka, ib.No.56(1974)76.

[4] J. Fujita and H. Miyazawa, Prog. Theor. Phys. 17 (1957) 360; D.W.E. Blatt and B.H.J. McKellar, Phys. Rev. C11 (1975) 614.

[5] B.A. Loiseau and Y. Nogami, Nucl. Phys. B2 (1967) 470.

[6] S.Ogawa, S.Sawada, T.Ueda, W.Watari and M.Yonezawa, Prog.Theor. Phys.Suppl.No.39(1967)140. T.Ueda,Proceedings of the International Symposium on Nuclear Many-Body problem (Rome 1972).

[7] T. Ueda and A.E.S. Green, Phys. Rev. 174 (1968) 1304.

[8] N. Hiroshige and T. Tsujimura, Lett. Nuovo Cim. 11 (1974) 330; prp. OCU-11.

[9] T. Ueda, Prog. Theor. Phys. 29 (1963) 829.

[10] T. Matsuoka, K. Ninomiya and S. Sawada, Prog. Theor. Phys. 42 (1969) 56.

CONSISTENT CALCULATION OF $\alpha-\alpha$ INTERACTION BY THE RESONATING GROUP METHOD

S. A. Afzal and S. Ali, Atomic Energy Centre, Dacca, Bangladesh.

In almost all previous RGM calculations of the $\alpha-\alpha$ interaction[1-4] using simplified N-N forces, an inconsistency prevailed namely that the α-particle wave function parameter which enters in the calculation of the kernel was fixed from the r.m.s. radius of the α-particle and not from variational calculation of the α-particle energy E_α as should have been the case. Some calculations were of course done by Niem, Heiss and Hackenbroich[5] who, employing a two-channel approximation and an N-N force of the soft core type used fragment wave functions determined by a variational procedure starting from the same assumptions about the internucleon forces as made in the scattering calculations. However, two points need be noted about their calculations. Firstly their approximation for treating the short range correlation effects are rather crude. Secondly their calculated α-particle binding energy (21.5 MeV) is about 6.8 MeV smaller than the experimentally observed value. Since it is known that the kernel $K(\underline{r}, \underline{r}')$ of the RGM depends explicitly on the total energy E, such a discrepancy even in single channel scattering, can have a sizeable effect on the phase shifts. Thus the results of Niem et al have been regarded as fortuitous. The purpose of the present paper is to report some RGM calculations for the $\alpha-\alpha$ interaction, performed for the sake of completeness of earlier calculations with an N-N potential frequently used in the past but with a consistent variational procedure for the alpha-particle energy. Thus our calculations used

$$V_{ij}(r_{ij}) = -V_0 (w + mP_x + bP_\sigma - hP_\tau) \exp(-\beta r_{ij}^2) \qquad (1)$$

which is expressed in the usual form $V_{ij} = y\, V_{Serber} + (1-y)\, V_{Rosenfeld}$ and an alpha particle wave function

$$\phi_\alpha(12,34) \sim \exp\Big(-\gamma \sum_{i<j=1}^{4} r_{ij}^2\Big) \qquad (2)$$

Table I shows the values of the deuteron-binding energy E_d and the triplet and singlet scattering parameters a_t, r_t, a_s, r_s for a potential set A to E[6] of form (1). The variational energies E_α and the corresponding values of γ and also the r.m.s. radius of the α-particle are shown for each case in tables I & II

TABLE-I

(n-p parameters)

Potential	V_0(MeV)	β(fm^{-2})	a_t(fm)	r_t(fm)	E_d(MeV)	w+m-b-h (=x)	a_s(fm)	r_s(fm)
A	-72.98	0.460	5.36	1.70	2.223	0.630	-14.66	2.29
B	-45.00	0.266	6.09	2.16	1.850	0.600	-22.83	3.01
C	-48.50	0.279	5.63	2.07	2.234	0.586	-23.50	2.93
D	-51.39	0.300	5.60	2.01	2.224	0.598	-23.92	2.81
E	-62.20	0.379	5.45	1.84	2.230	0.630	-23.73	2.50
Expt. values[9]			5.43 ± 0.004	1.76 ± 0.005	2.226 ± 0.002		-23.715 ± 0.013	2.66 ± 0.09

TABLE-II

(Minimum energy and r.m.s. radius of α-particle for different N-N potentials)

Potential	γ (fm^{-2})	E_α (MeV)	$\langle r^2\rangle_\alpha^{1/2}$ (fm)	y
A	0.1291	-39.726	1.043	0.67
B	0.0786	-26.534	1.337	0.70
C	0.0841	-29.544	1.293	0.72
D	0.0901	-31.090	1.249	0.65
E	0.1106	-36.584	1.127	0.65
Expt. value		-28.2	1.44	

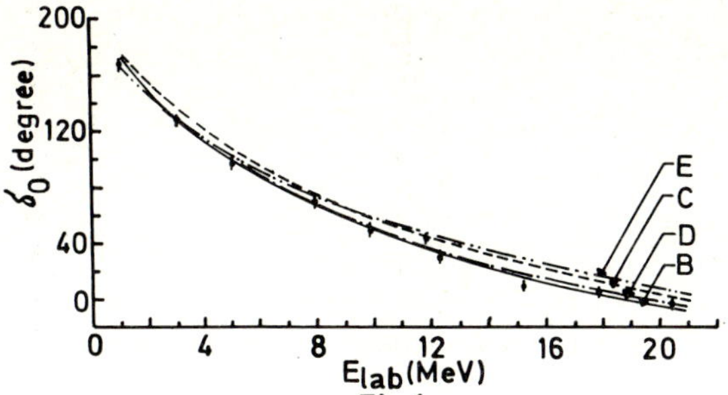

Fig.1.

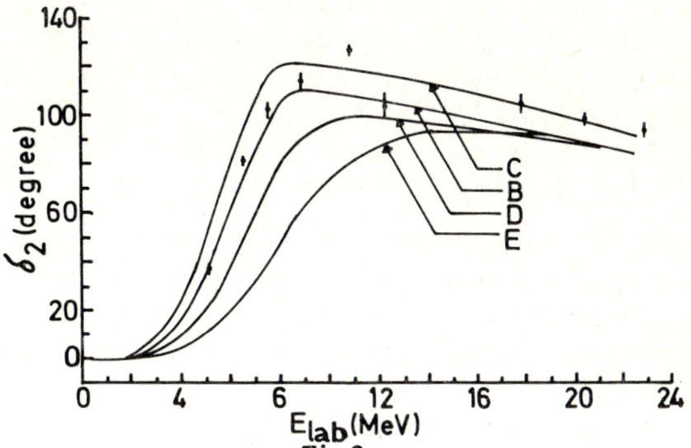

Fig.2.

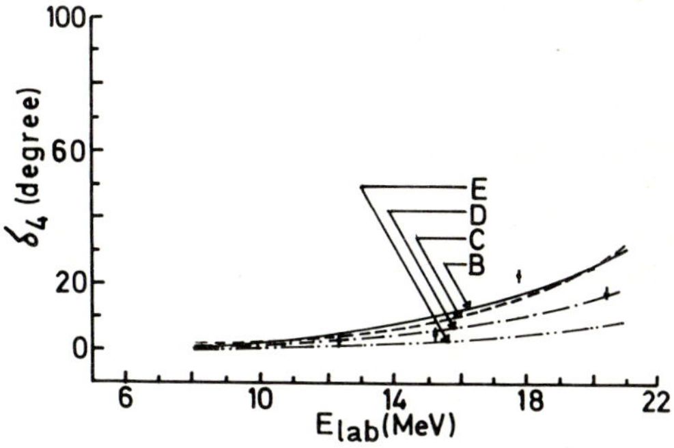

Fig.3.

Figs.1,2 & 3 show the calculated S, D and G phase shifts respectively together with the experimental points[1]. A combined study of the tables and the figures reveals that potential D or one intermediate between C and D but closer to D would seem to represent the α-α data not too badly without sacrificing the N-N and α-particle properties. Potential D has a Serber component of $y=0.65$ and is very similar to the Biel[7] potential (70% Serber + 30% Rosenfeld). Also as seen from table II, earlier authors using potential A would have obtained $y=0.67$ (rather than 0.94) if calculations were done in the true variational spirit. However, in building up cluster-cluster interaction within the framework of RGM, a unique value of y with a near-Serber force does not seem to emerge. Moreover earlier calculations indicate[8] that the direct component of the force could be substantially reduced with admission of distortion effects. In the present calculations however all distortion effects may be assumed to have been taken into account only in an effective way through using y as an adjustable parameter.

With the simple potential used here, it is not reasonable to expect more than what has been achieved here as a reasonable compromise between the low-energy N-N scattering parameters, the α-particle binding energy and its r.m.s. radius and the α-α scattering data. Our results indicate that even if the N-N forces are non-saturating, one can still fix the α-particle wave function variationally without there being appreciable collapse. In fact the r.m.s. radius of the α-particle determined by the variational parameters is found to be only 10-15% smaller than the experimentally observed value.

REFERENCES

(1) S. A. Afzal, A. A. Z. Ahmad and S. Ali, Rev. Mod. Phys. 41 (1969)247 for earlier work
(2) D. R. Thompson, I. Reichstein, W. Mclure and Y. C. Tang, Phys. Rev. 185(1969)1351
(3) I. Reichstein and Y. C. Tang Nucl. Phys. A139(1969)144
(4) R. E. Brown and Y. C. Tang, Nucl. Phys. A170(1971)225
(5) Le-Chi-Niem, P. Heiss and H. H. Hackenbroich, Z. Physik 244(1971)346
(6) S. A. Afzal and S. Ali, Nucl. Phys. A157(1970)363
(7) S. J. Biel, A. C. Butcher and J. M. McNamee, Proc. Int. Conf. on Nuclear Forces and the Few-Nucleon problem, London, 1960(Pargamon Press, 1960) p641
(8) A. Herzenberg and A. S. Roberts, Nucl. Phys. 3(1957)314
(9) J. C. Davis and H. H. Barchell, Phys. Lett. 27B (1968)636.

TENSOR CONTRIBUTION TO NUCLEAR MATTER IN THE FIRST ORDER PERTUBATION THEORY

A.K. DEKA and P. MAHANTA
Department of Physics, Dibrugarh University, Dibrugarh-786 004, Assam, India.

The tensor forces are necessary for a fit to the two-body phase shift data and are important for their role in producing the saturation in the nuclear binding energy. The widely used nucleon-nucleon potentials in the nuclear calculations are usually defined parity wise for the zero and unit spin states of the two-particle system. Due to the applicability of the identity.

$$\sum_{\mathcal{J}} (2\mathcal{J}+1)\langle l s \mathcal{J} | S_{12} | l s \mathcal{J}\rangle = 0$$

the tensor contribution to the first order infinite nuclear matter binding energy vanishes for such potentials. A non-vanishing contribution can, however, be expected from the potentials whose parameters depend on the total angular momentum $\mathcal{J}$. A suitable potential for this purpose is that due to Ulehla et al[1]. having radial dependence

$$V_{ST}^{l\mathcal{J}}(r) = \sum_{n} \left(a_n + b_n \mathcal{J}(\mathcal{J}+1)\right) e^{-nX}/X$$

where $X = \mu r$, $\mu = .707\,fm^{-1}$

We used this finite charge-dependent potential to calculate the tensor contribution to the infinite nuclear matter binding energy in the first order perturbation theory. The integrals needed for the calculations are

$$\int_0^\infty \int_0^{k_F} j_L^2(kr)\left(2k^2 - 2k^3/k_F + k^5/k_F^3\right) \tilde{V}_{ST}^{l\mathcal{J}}(r)\, r^2 dr\, dk$$

with $$\tilde{V}_{ST}^{l\mathcal{J}} = \sum_{\mathcal{J}} (2\mathcal{J}+1) V_{ST}^{l\mathcal{J}} \langle l s \mathcal{J} | S_{12} | l s \mathcal{J}\rangle$$

where $V_{ST}^{l\mathcal{J}}$ expresses the radial dependence for a definite spin S, isospin T, total angular momentum $\mathcal{J}$. The integration over r can be obtained in the analytical form leaving the integration over k. We considered only the P and D waves and the results for these at $k_F = 1.4\ fm^{-1}$ are shown in table 1.

Table 1

Tensor contribution to potential energy per particle

$^3P_0$	$^3P_1$	$^3P_2$	$^3D_3$	$^3D_2$	$^3D_1$
-5.320	7.336	-2.016	1.296	-3.954	2.511

The contribution from the P waves are cancelled out, the $^3P_1$ contribution being equal and opposite to the $^3P_2$ and $^3P_0$ contributions. In ^3D state there is a net contribution of -.14MeV. This contribution is very small compared to the large +ve potential energy generated by finite realistic potential in the first order perturbation theory. The small contribution is in conformity with the good fit of the phase-shift data by the potentials whose forms do not explicitly depend on the total angular momentum. In view of this small value of the tensor force for this $\mathcal{J}$-dependent class of potential, the concept of the vanishing contribution of tensor force in first order perturbation theory remains approximately valid.

REFERENCE

1) I. Ulehla 1972 The nuclear many body problem (editrice compositori international physics series) edited by F. Calogero and C. Ciofi Degli atti, Vol.1. p. 145.

PAIRING-PLUS-SURFACE-TENSOR-INTERACTION AND STRUCTURE OF ^{51}V

D. Banerjee
Department of Physics, University College of Science, Calcutta-9, India.

Banerjee [1] and Richert have proposed a pairing-plus-surface-tensor interaction (PSTI) to be used as the effective two-body interaction in shell-model calculations. The form of the interaction is

$$V_{12} = -A q_{12} + B_T \left(\frac{r^2}{R_0^2}\right) \delta(r_1 - R_0)\, \delta(r_2 - R_0)\, S_{12} \qquad (1)$$

where
$$S_{12} = \vec{\sigma}_1 \cdot \hat{r}\ \vec{\sigma}_2 \cdot \hat{r} - \frac{1}{3} \vec{\sigma}_1 \cdot \vec{\sigma}_2 \qquad (2)$$

and the operator q_{12} is defined by its antisymmetrized matrix elements

$$\langle j_a j_b JM | q_{12} | j_c j_d JM \rangle = [(2j_a+1)(2j_c+1)]^{\frac{1}{2}} \delta_{j_a j_b} \delta_{j_c j_d} \delta_{J0} \qquad (3)$$

SECTION 1

In this section, the basic difference between PSTI and the realistic forces, namely the Kuo-Brown [2] interaction will be pointed out. One finds from a systematic analysis of the empirical sets of two-body matrix elements that it is necessary to have average repulsion for $f_{7/2}\, p_{3/2}$ matrix elements for good reproduction of energy levels. In particular, the $(f_{7/2}\, p_{3/2})_{J=5}$ matrix element should be highly repulsive. For PSTI, the $f_{7/2}\, p_{3/2}$ matrix elements for J=2,3,4 and 5 are respectively +0.92, +0.52, -0.03 and -0.72 Mev. The corresponding bare matrix elements of Kuo-Brown interaction are +0.63, +0.21, +0.22 and +0.12 MeV. These values are changed respectively to +0.86, +0.03, +0.05 and -0.15 MeV when core-renormalization is taken into consideration. It is clear that for Kuo-Brown interaction the averagenvalue is highly attractive. Here positive indicates attraction. For PSTI the average value is slightly repulsive and for J = 5 state the matrix element is highly repulsive. The J = 5 state is unique in L-S coupling. It is L = 4 and S = 1 i.e. a pure spin-triplet. Since it is T = 1, the only way to get this matrix element highly repulsive is to have triplet repulsion. The realistic forces, being Serber-like gives attraction for J = 5 state. The core-renormalisation makes it slightly repulsive.

SECTION 2

The results of shell-model calculation of ^{51}V is shown in Fig. 1. For comparison the levels determined by Kuo-Brown interaction in a larger vector space [3] is shown with caption A. Our result is good. We got better density of states than that reported previously. Our wave functions give better result for spectroscope factors also.

REFERENCES

1. D. Banerjee and J. Richert, Nuovo Cimento Lett. 3 (1972) 30.
2. T.T.S. Kuo and G.E. Brown, Nucl. Phys. A114)1968) 241.
3. M.L. Rustgi, R.P. Singh, B. Barman Roy, R. Raj and C.C. Fu, Phys. Rev. C3 (1971) 2238.

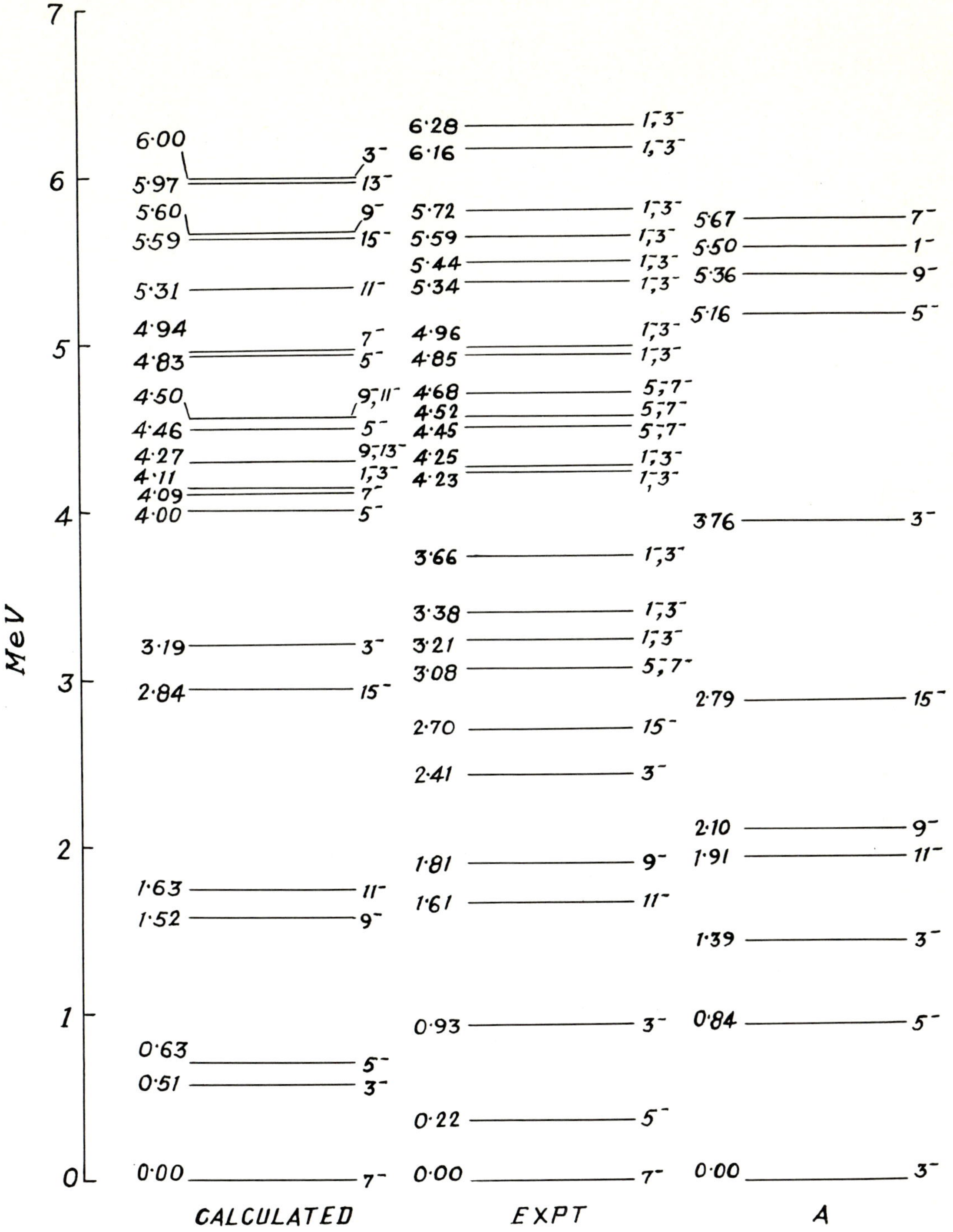
MeV
CALCULATED
EXPT
A
0·00 7⁻
0·51 3⁻
0·63 5⁻
1·52 9⁻
1·63 11⁻
2·84 15⁻
3·19 3⁻
4·00 5⁻
4·09 7⁻
4·11 1⁻,3⁻
4·27 9⁻,13⁻
4·46 5⁻
4·50 9⁻,11⁻
4·83 5⁻
4·94 7⁻
5·31 11⁻
5·59 15⁻
5·60 9⁻
5·97 13⁻
6·00 3⁻
0·00 7⁻
0·22 5⁻
0·93 3⁻
1·61 11⁻
1·81 9⁻
2·41 3⁻
2·70 15⁻
3·08 5⁻,7⁻
3·21 1⁻,3⁻
3·38 1⁻,3⁻
3·66 1⁻,3⁻
4·23 1⁻,3⁻
4·25 1⁻,3⁻
4·45 5⁻,7⁻
4·52 5⁻,7⁻
4·68 5⁻,7⁻
4·85 1⁻,3⁻
4·96 1⁻,3⁻
5·34 1⁻,3⁻
5·44 1⁻,3⁻
5·59 1⁻,3⁻
5·72 1⁻,3⁻
6·16 1⁻,3⁻
6·28 1⁻,3⁻
0·00 3⁻
0·84 5⁻
1·39 3⁻
1·91 11⁻
2·10 9⁻
2·79 15⁻
3·76 3⁻
5·16 5⁻
5·36 9⁻
5·50 1⁻
5·67 7⁻

LOCAL STRONG AND COULOMB POTENTIALS IN THE THREE-NUCLEON SYSTEM

E.O. Alt
Institut für Physik,
Universität Mainz,
West Germany

1. INTRODUCTION

Attempts to use local potentials in three-nucleon calculations with the Faddeev equations are impeded by the fact that for increasing energies contributions from higher and higher subsystem angular momentum states become important which quickly make the system of coupled equations unwieldy. However, if long range interactions like the Coulomb potential were added such a procedure would not be useful at all. In fact, several approaches exist which deal with the problems arising from the infinite range of the latter. In the work of Noble [1] and Bencze [2], the Faddeev equations are modified so that the Coulomb potential occurs only in the form of the analytically known Coulomb Green's functions which replace the free Green's functions appearing for short range interactions. Drastic approximations had to be made by Adya [3] in an attempt to use these equations for practical calculations. On the other hand, employing screened Coulomb potentials Veselova [4] reformulated the Faddeev equations in such a way that the resulting equations have, for energies below the break-up threshold, kernels which are well behaved in the limit of zero screening.

Another approach which is applicable for local interactions is the quasiparticle method of Alt et al. [5] in which the locality of the potential can be accounted for in a perturbative manner. In this way it was possible for Alt et al. [6] to estimate the influence of the two-nucleon interaction in higher angular momentum states on three-nucleon observables. Also the Coulomb interaction can be implemented without great difficulties thus making calculations of proton-deuteron scattering feasible (for all energies), as will be discussed.

2. THE QUASIPARTICLE APPROACH

It is well known that the Faddeev equations for the three-body operators $U_{\beta\alpha}$,

$$U_{\beta\alpha} = \bar{\delta}_{\beta\alpha} G_0^{-1} + \sum_\gamma \bar{\delta}_{\beta\gamma} T_\gamma G_0 U_{\gamma\alpha} , \tag{2.1}$$

with $G_0(z) = (z-H_0)^{-1}$ and $\bar{\delta}_{\beta\alpha} = (1-\delta_{\beta\alpha})$, reduce to much simpler equations when the two-body interactions V_γ, and with them the corresponding transition operators T_γ, are taken to be separable. However, a similar reduction in complexity can be achieved also for local potentials, as was shown by Alt et al. [5].

For this purpose the local potentials V_γ are split into a separable part and a remainder

$$V_\gamma = V_\gamma^s + V_\gamma' = \sum_r |\chi_{\gamma r}\rangle \lambda_{\gamma r} \langle \chi_{\gamma r}| + V_\gamma' . \tag{2.2}$$

Defining T_γ' to be the amplitude to be calculated from a Lippman-Schwinger (L.S.) equation with potential V_γ', a decomposition for T_γ similar to (2.2) is obtained

$$T_\gamma = T_\gamma^s + T_\gamma' = \sum_{rs} |\gamma r\rangle \Delta_{\gamma,rs} \langle \gamma s| + T_\gamma' . \tag{2.3}$$

Hereby, form factors $|\gamma r\rangle = (1+T_\gamma' G_0)|\chi_{\gamma r}\rangle$ have been introduced, and a matrix Δ_γ whose elements are defined as

$$(\Delta_\gamma^{-1})_{rs} = \delta_{rs}\lambda_{\gamma r}^{-1} - \langle\chi_{\gamma r}|G_0|\gamma s\rangle \ . \tag{2.4}$$

Note that the separable term V_γ^s has to be chosen such that the remainder V_γ' is too weak to support any bound state or resonance.

When the splitting (2.3) is inserted into the Faddeev equations (2.1), effective two-body multichannel L.S. equations can be derived which can be written in concise matrix notation as

$$\mathcal{T} = \mathcal{V} + \mathcal{V}\mathcal{G}_0\mathcal{T} \ . \tag{2.5}$$

Those elements $\mathcal{T}_{\beta n,\alpha m}$ in which the indices m and n characterize two-body bound states, give, when taken between plane wave states corresponding to the free motion of the incoming and outgoing two particles, on-shell just the desired physical bound-state scattering amplitudes. The effective potential $\mathcal{V}$ is defined as

$$\mathcal{V}_{\beta n,\alpha m} = \langle\beta n|G_0 \ U_{\beta\alpha}' \ G_0|\alpha m\rangle \ , \tag{2.6}$$

whereas the effective Green's function $\mathcal{G}_0$ equals the quantity Δ_γ of eq. (2.4). The three-body operators $U_{\beta\alpha}'$ appearing in eq. (2.6) fulfill Faddeev equations (2.1) but with the full two-particle transition operators T_γ replaced by the remainder T_γ' .

This latter fact can be exploited to construct an approximation scheme for $\mathcal{V}$. Namely, in the case that T_γ' has been made sufficiently small, $U_{\beta\alpha}'$ can be expanded in a multiple scattering series in T_γ'

$$U_{\beta\alpha}' = \bar{\delta}_{\beta\alpha}G_0^{-1} + \sum_\gamma \bar{\delta}_{\beta\gamma}\bar{\delta}_{\gamma\alpha}T_\gamma' + \sum_{\gamma\delta} \bar{\delta}_{\beta\gamma}\bar{\delta}_{\gamma\delta}\bar{\delta}_{\delta\alpha}T_\gamma' G_0 T_\delta' + \ldots \tag{2.7}$$

And when this expansion is introduced into the definition (2.6) it gives rise to the so-called quasi-Born series, the lowest order term of which, $\mathcal{V}^{(0)}$, is just the particle exchange contribution obtained for separable interactions. The next term, $\mathcal{V}^{(1)}$, contains the first order corrections due to the residual interaction V_γ'. These two terms are represented in graphical form in Fig. 1.

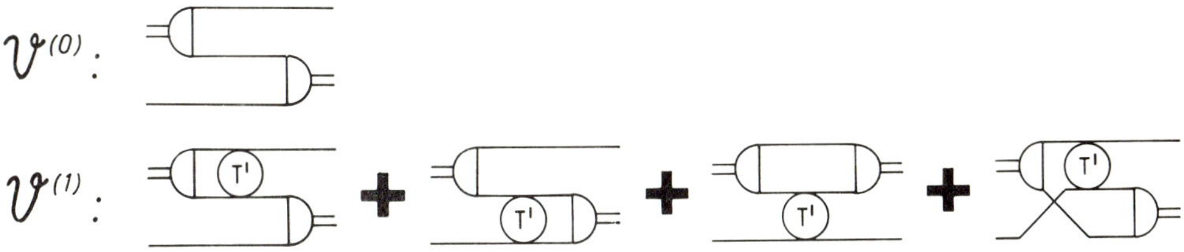

Fig. 1: Graphical representation of the two lowest order terms of the quasi-Born expansion of the effective potential $\mathcal{V}$.

The important point to note is that the effective potential includes via T_γ' the effects of all subsystem partial waves. In particular, if the separable terms V_γ^s act only in one partial wave it follows from (2.2) that for all other partial waves V_γ' is identical to the original potential V_γ. The same holds true for the amplitudes.

3. APPLICATIONS

To begin with I mention the investigation of the nnp system performed by Alt et al. [6]. The local potentials considered were those of Malfliet et al. [7] containing a soft core. Their pole approximations, constructed by Bakker [8], were used as separable parts. By approximating the effective potential by the terms displayed in Fig.

1 the influence of the two-nucleon interaction in higher partial waves on various three-nucleon observables could be estimated. The result was that the triton binding energy is increased by approximately 1/4 MeV, and correspondingly the doublet scattering length is decreased by approximately 1/4 fm. Low-energy elastic nd scattering cross sections exhibit only very small changes.

As a second application consider the approach of Alt et al. [9] for calculating elastic pd scattering, with emphasis on the correct inclusion of the Coulomb repulsion between the two protons. For simplicity the strong interaction between the nucleons is assumed to be purely separable. Then the Coulomb interaction is the only non-separable part in eq. (2.2). This implies that for the np system the scattering amplitude (2.3) is also purely separable, whereas in the pp system it is just the usual sum of the pure Coulomb amplitude plus a separable, Coulomb-modified strong amplitude.

Due to this particularly simple form of the subsystem transition operators the multiple scattering expansion of $U'_{\beta\alpha}$ reduces exactly to the first two terms written down in eq. (2.7) with γ denoting the pp subsystem. (This multiple scattering series would not terminate if the strong interactions were taken to be local.) Inserting this result into eq. (2.6) provides us with a closed and exact expression for the effective potential. Its graphical representation is the one displayed in Fig. 1, with T' being the Coulomb amplitude T_c.

The separable potentials are chosen to act in S-waves only but charge independence of the nuclear forces is not assumed. Symmetrization between the two protons then leads for total spin S = 3/2 to a single equation, and for S = 1/2 to a set of three equations one of which, however, decouples. For the calculation of three-nucleon bound state properties the latter equations can be used as they stand. This has been done in an isospin formalism by Alessandrini et al. [10] and by Zankel et al. [11] in their attempt to calculate the binding energy difference between the triton and Helium-3. Timm [12] has used these equations to estimate the difference between the pd and the nd quartet scattering lengths. However, for energies in the scattering region difficulties can arise from the long range of the Coulomb potential. Since they are the same in the doublet and the quartet channel we detail our approach by means of the uncoupled equation for the quartet channel amplitude $\mathcal{T}_{dd}^{(3/2)}$ where the index d implies that the np subsystem forms a deuteron (the index 3/2 will be omitted in the following). For simplicity we make the approximation of replacing the Coulomb amplitude T_c by V_c. This should be justified for a repulsive interaction, in particular at higher energies. We now proceed similarly to Veselova [4]. We first replace the Coulomb potential by a Yukawa potential $V_\mu(r) = e^2 \exp(-\mu r)/r$ (i.e. we use exponential screening). To indicate this all quantities will be given an index μ. Consider then the effective potential. Looking at Fig. 1 it is apparent that only in the third graph of $\mathcal{V}^{(1)}$ singularities can show up for $\mu \to 0$. The analytical expression corresponding to this graph which we call $\mathcal{U}_\mu^{(1)}$ is (whenever possible, explicit energy dependence will be suppressed)

$$\mathcal{U}^{(1)}_{\mu,dd}(\vec{q}\,',\vec{q}) = V_\mu(\vec{q}\,'-\vec{q})F_d(\vec{q}\,',\vec{q};q_e) \qquad (3.1)$$

where V_μ is the two-particle Yukawa potential, and the smooth function F_d is related to the deuteron form factor. The on-shell momentum q_e is determined by the three-body energy E and the deuteron binding energy B_d as $E = 3q_e^2/4M + B_d$ (M is the nucleon mass). Thus we expect from the term (3.1) singular contributions in the zero screening limit which are closely connected with those appearing in the genuine two-particle Coulomb problem.

To see this in more detail we investigate the singularity structure of the kernel of eq. (2.5), $\mathcal{K}_{dd}^{(\mu)} = \mathcal{V}_{dd}^{(\mu)} \mathcal{G}_{o,d}$. It is well known that the propagator $\mathcal{G}_{o,d}(q^2)$ has a pole for $q^2 = q_e^2$

$$\mathcal{G}_{o,d}^{-1}(q^2) = \frac{3}{4M}(q_e^2+i\varepsilon-q^2)\hat{F}_d(q,q_e) \tag{3.2}$$

which gives rise to the bound-state scattering cut. The quantity $\hat{F}_d(q,q_e)$ equals, for $q = q_e$, the normalization integral of the deuteron wave function. Thus by separating out the contribution from $\mathcal{V}_\mu^{(1)}$, the kernel can be decomposed as follows

$$\mathcal{K}_{dd}^{(\mu)}(\vec{q}\,',\vec{q}) = \frac{e^2}{2\pi^2} \cdot \frac{F_d(\vec{q}\,',\vec{q};q_e)/\hat{F}_d(q,q_e)}{[(\vec{q}\,'-\vec{q})^2+\mu^2][\frac{3}{4M}(q_e^2+i\varepsilon-q^2)]} + \tilde{\mathcal{K}}_{dd}^{(\mu)}(\vec{q}\,',\vec{q}), \tag{3.3}$$

where $\tilde{\mathcal{K}}^{(\mu)}$ has no worse singularities when $\mu \to 0$ than those which are already present for short range two-particle interactions. We can simplify the first term by rewriting F_d as

$$F_d(\vec{q}\,',\vec{q};q_e) = \{\hat{F}_d(q',q_e)\hat{F}_d(q,q_e)\}^{1/2}+\{\hat{F}_d(q',q)-[\hat{F}_d(q',q_e)\hat{F}_d(q,q_e)]^{1/2}\}$$
$$+ \{F_d(\vec{q}\,',\vec{q};q_e)-\hat{F}_d(q',q)\} . \tag{3.4}$$

The second bracket vanishes for $q = q' = q_e$ while the third one vanishes for $\vec{q} = \vec{q}\,'$ since it can be shown that $F_d(\vec{q},\vec{q};q_e) = \hat{F}_d(q,q)$. Thus, when this splitting is introduced in eq. (3.3), the latter two terms give nonsingular contributions and can, therefore, be added to the kernel $\tilde{\mathcal{K}}^{(\mu)}$. The singular part which we denote by $\mathcal{K}_{\mu,dd} = \mathcal{V}_{\mu,dd}\mathcal{G}_{o,d}$ is completely determined by the first term of eq. (3.4) and equals, apart from the numerator, the kernel of the two-particle L.S. equation for a Yukawa potential. We thus end up with the following equation (in operator notation)

$$\mathcal{T}_{dd}^{(\mu)} = (\mathcal{V}_{\mu,dd}+\tilde{\mathcal{V}}_{dd}^{(\mu)}) + (\mathcal{V}_{\mu,dd}+\tilde{\mathcal{V}}_{dd}^{(\mu)})\mathcal{G}_{o,d}\mathcal{T}_{dd}^{(\mu)} \tag{3.5}$$

where $\tilde{\mathcal{V}}_{dd}^{(\mu)}$ is defined as $\tilde{\mathcal{K}}_{dd}^{(\mu)} = \tilde{\mathcal{V}}_{dd}^{(\mu)}\mathcal{G}_{o,d}$ and contains all the terms of the effective potential which are well behaved in the zero screening limit.

Now we are in the position to apply the well-known Faddeev trick [13], namely to explicitly invert the singular part of the kernel. Shifting $\mathcal{V}_{\mu,dd}\mathcal{G}_{o,d}$ to the left hand side of eq. (3.5), and multiplying the resulting equation by $(1-\mathcal{V}_{\mu,dd}\mathcal{G}_{o,d})^{-1}$, we obtain

$$\mathcal{T}_{dd}^{(\mu)} = \mathcal{T}_{\mu,dd} + (1+\mathcal{T}_{\mu,dd}\mathcal{G}_{o,d})\tilde{\mathcal{V}}_{dd}^{(\mu)}(1+\mathcal{G}_{o,d}\mathcal{T}_{dd}^{(\mu)}) \tag{3.6}$$

with $\mathcal{T}_{\mu,dd}$ defined as

$$\mathcal{T}_{\mu,dd} = (1-\mathcal{V}_{\mu,dd}\mathcal{G}_{o,d})^{-1}\mathcal{V}_{\mu,dd} \Rightarrow \mathcal{T}_{\mu,dd} = \mathcal{V}_{\mu,dd}+\mathcal{V}_{\mu,dd}\mathcal{G}_{o,d}\mathcal{T}_{\mu,dd}. \tag{3.7}$$

From the above discussion of the kernel $\mathcal{K}_\mu$ of eq. (3.7) it follows immediately that $\mathcal{T}_{\mu,dd}(\vec{q}\,',\vec{q})$ can be constructed explicitly from the two-body amplitude $T_\mu(\vec{q}\,',\vec{q})$ describing the scattering of a charged particle of mass M off another charged particle of mass 2M via an exponentially screened Coulomb potential

$$\mathcal{T}_{\mu,dd}(\vec{q}\,',\vec{q}) = \{\hat{F}_d(q',q_e)\hat{F}_d(q,q_e)\}^{1/2}\, T_\mu(\vec{q}\,',\vec{q}) , \tag{3.8}$$

the smooth function multiplying T_μ being equal to one for on-shell values of the momenta $q = q' = q_e$. The second term on the r.h.s. of eq. (3.6) is the Coulomb-modified strong amplitude, usually denoted by $\mathcal{T}_{sc}^{(\mu)}$, for which all relations known to hold in the two-body system, can be derived here, too. Eq. (3.6), therefore, looks the same as the one obtained for a genuine two-body problem with short range and exponentially screened Coulomb interactions. Thus we can take over the well-known result that, after renormalization, the on-shell amplitudes $\mathcal{T}^{(\mu)}(\vec{q}\,',\vec{q})$, $\mathcal{T}_\mu(\vec{q}\,',\vec{q})$ and $\mathcal{T}_{sc}{}^{(\mu)}(\vec{q}\,',\vec{q})$ tend in the limit $\mu\to 0$ to their unscreened counterparts.

The coupled equations for the doublet channel can be treated in complete analogy leading to the same type of equations (3.6)-(3.8) since only the elements $\mathcal{T}_{dn}^{(1/2)}$(n = d or s) require special care.

The equations discussed above are being employed for a numerical investigation of the ppn system by Alt et al. [9]. In a first crude calculation we have used a simple charge-independent Yamaguchi interaction with the parameters of Aaron et al. [14]. The binding energy of the triton comes out as -10.99 MeV, and the one of Helium-3 as -10.19 MeV, leading to a binding energy difference of 0.8 MeV. Such a large value is known to arise from the inadequacy of a simple Yamaguchi potential without repulsion. Preliminary results have also been obtained for the quartet effective range parameters and S-wave scattering phase shifts. In these calculations the screening radius was varied from 30 fm up to 100 fm. We obtain a=13.3 fm and r_o= 1.9 fm as compared to the experimental values of 11.88 fm and 2.63 fm, respectively, given by Arvieux [15]. In Table 1 the Coulomb-modified strong S-wave phase shifts δ_{sc} are presented for a few energies and compared with the results of a phase shift analysis of Arvieux [15]. The deviations at higher energies might be due to numerical inaccuracies. For comparison we also show the nd phase shifts obtained by switching off the Coulomb interaction.

E_{Lab}	1.MeV	3.MeV	5.MeV
δ_{sc} (radians)	2.50	2.04	1.82
Arvieux [15]	2.503	2.020	1.755
δ_s (radians)	2.33	1.90	1.71

Table 1: Coulomb-modified nuclear quartet S-wave scattering phase shifts δ_{sc} for pd scattering, as compared to the experimental ones of Arvieux [15]. δ_s are the corresponding phase shifts for nd scattering. (At 5.MeV δ is the real part of the phase shift.)

REFERENCES

[1] J.V. Noble, Phys. Rev. 161 (1967) 945
[2] G. Bencze, Nucl. Phys. A196 (1972) 135
[3] S. Adya, Phys. Rev. 166 (1968) 991; ibid. 177 (1969) 1406
[4] A.M. Veselova, Teor. Mat. Fiz. 3 (1970) 326; preprint Kiev,1973
[5] E.O. Alt, P. Grassberger and W. Sandhas, Nucl. Phys. B2 (1967) 167
[6] E.O. Alt and W. Sandhas, in Few-Body Problems in Nuclear and Particle Physics, edited by R.J. Slobodrian, B. Cujec and K. Ramavataram. Quebec: Les Presses de l'Universitê Laval 1975.
[7] R.A. Malfliet and J.A. Tjon, Nucl. Phys. A127 (1969) 161
[8] B.L.G. Bakker, Z. Physik A122 (1975) 335
[9] E.O. Alt, H. Zankel and H. Ziegelmann: to be published
[10] V.A. Alessandrini, C.A. Garcia and H. Fanchiotti, Phys. Rev. 170 (1968) 935
[11] H. Zankel, C. Fayard and G.H. Lamot, Nuovo Cim. Lett. 12 (1975) 221
[12] W. Timm, Diplomarbeit, University of Münster, 1975: unpublished
[13] L.D. Faddeev, Mathematical Aspects of the Three-Body Problem in the Quantum Scattering Theory (Israel Program for Scientific Translations, Jerusalem, 1966)
[14] R. Aaron, R.D. Amado and Y.Y. Yam, Phys. Rev. 140 (1965) B1291
[15] J. Arvieux, Nucl. Phys. A221 (1974) 253

DYNAMICAL EQUATIONS FOR THE N-PARTICLE TRANSIT OPERATORS

Gy. Bencze

Central Research Institute for Physics
1525 Budapest 114. P.O.Box 49, Hungary

1. Introduction

In the years following Faddeev's pioneering work on the three-body problem it has become dear what are the requirements that have to be satisfied by any reasonable N-particle scattering theory based on integral equations. As a consequence, by now there exist several variants of N-particle equations which differ in the classification of the N-particle amplitudes, i.e. the number of coupled equations and/or the connectivity structure of their kernel. It is interesting to note that among the existing formulations one finds the two extremes as far as the number of coupled equations is concerned. The Yakubovskii theory [1] and its modifications [2-4] are based on the most detailed classification of the N-particle amplitudes so that the number of coupled equations is the possible maximum. Due to the presence of various disconnected terms N-2 iterations are needed to make the Yakubovskii kernel connected. On the other hand in the approach proposed by Weinberg [5] a single connected kernel equation is constructed for the resolvent of the N-particle Hamiltonian. Unfortunately, as is well known, the Weinberg equation may have spurious bound state solutions.

The most physical approach to N-particle scattering is to derive equations for the transit operators. In such a way the quantities needed for the physical description of the scattering process are directly obtained. In the following the advantages of such an approach will be briefly discussed.

2. Integral Equations for the N-Particle Transit Operators

Integral equations for the N-particle transit operators have been first obtained by Bencze [6] . The same equations were also derived by Redish [7] using different methods. The coupling in these equations is "minimal", i.e. only transit operators connecting two-cluster channels are coupled together. Thus for N distinguishable particles there are $2^{N-1}-1$ coupled equations. The kernel becomes connected after a single iteration and its separable expansion reduces the equations to a set of effective two-body equations in a complete analogy with the three-body case. By choosing suitable off-shell continuations for the transit operators the BR-equations can be written in two equivalent forms

$$U^{L}_{\mu\nu} = V^{\mu}_{\nu} + \sum_{\sigma\neq\mu} K_{\mu\sigma} G_0 U^{L}_{\sigma\nu} \quad , \qquad (1)$$

$$U^{AGS}_{\mu\nu} = (1-\delta_{\mu\nu})(z-H_0-V_{\mu\nu}) + \sum_{\sigma\neq\mu} K_{\mu\sigma} G_0 U^{AGS}_{\sigma\nu} \quad , \qquad (2)$$

where μ, ν and σ denote two-cluster partitions. As it is indicated by the superscripts these equations are in fact the generalisations of the Lovelace and AGS three-body equations, respectively, for the case of an arbitrary number of particles.

Recently another approach to obtaining coupled equations for the transit operators has been proposed by Kouri and Levin [8] and Tobocman [9]. (hereafter KLT). The most distinctive feature of the KLT-approach is the introduction of a channel coupling array which makes it possible to couple together channels in various ways provided a connected kernel results after a finite number of iterations. In the case of minimal coupling the KLT kernel becomes connected after $2^{N-1}-2$ iterations.

Due to the different structure of the kernel the BR and KLT equations are seemingly unrelated. However, in a recent paper by L'Huillier [10] et al. it has been shown that the BR-equations can also be derived by the KLT-method by using a special coupling scheme. Thus in fact the two approaches are equivalent. It is important to stress that both formulations are free of the spurious homogeneous solutions as well. Since these questions will be discussed in other talks there is no need to go into details here.

The problem of the scattering of identical particles can also be most conveniently studied in terms of physical quantities like transit operators. The first explicit treatment of the scattering of N identical particles has been given by Bencze and Redish [11] using the BR-equations. A similar method has been developed recently by Tobocman [12] in the frame of the KLT-approach. Such a general treatment of identical particles does not exist so far for any other variant of N-particle equations.

3. Channel Decoupling in the N-Particle Equations

Let a and b denote two arbitrary partitions channels of the N-particle system and let us define the transit operator connecting them as

$$U_{ba} = V^b + V^b G V^a \tag{3}$$

By making use of the resolvent relation

$$G = G_c + G_c V^c G \tag{4}$$

with arbitrary $c \neq b$ and $G_c = (z - H_0 - V_c)^{-1}$, (3) can be rewritten in the form

$$U_{ba} = V^b G_c G_a^{-1} + V^b G_c U_{ca} \quad , \tag{5}$$

therefore two different transit operators are connected by a linear operator relation. It is clear that by means of relation (5) it is possible to eliminate all but one transit operator in the coupled equations so that channel decoupling results. It is important, of course, that the kernel of the decoupled equation should be connected. As an example let us take the BR-equations

$$U_{\mu\nu} = I_{\mu\nu} + \sum_{\sigma\neq\mu} K_{\mu\sigma} G_0 U_{\sigma\nu} \quad , \tag{6}$$

where the inhomogeneous term $I_{\mu\nu}$ corresponds to the off-shell continuation (3) of the transit operators. According to (5) one has

$$U_{\sigma\nu} = V^\sigma G_\mu G_\nu^{-1} + V^\sigma G_\mu U_{\mu\nu} \quad , \tag{7}$$

which when substituted into (6) yields the decoupled equation .

$$U_{\mu\nu} = I_{\mu\nu} + X_\mu G_\mu G_\nu^{-1} + X_\mu G_\mu U_{\mu\nu} \quad , \tag{8}$$

where the shorthand notation $X_\mu = \sum_{\sigma\neq\mu} K_{\mu\sigma} G_0 V^\sigma$ has been introduced. From the definition of $K_{\mu\sigma}$ follows [6] that X_μ and consequently the kernel of eq.(8) is completely connected. It can be shown that eq.(8) has no spurious solutions either. The inhomogeneous term in (8) can be simplified by suitable off-shell transformations. In particular it can be achieved that satisfies the Lippmann-Schwinger-type equation

$$U_{\mu\mu} = X_\mu + X_\mu G_\mu U_{\mu\mu} \tag{9}$$

so that X_μ can be regarded as a generalised "optical potential" operator. It has to be noted that in the KLT equations a certain choice of the channel coupling array can also produce decoupling by iterations. From the above considerations it follows, however, that decoupling is possible for any choice of the coupling scheme and an L-S-type equation (8) can always be obtained. The decoupled equation (8) always implies a single connected kernel equation for the resolvent operator due to the relation

$$G = G_\mu + G_\mu U_{\mu\nu} G_\nu \tag{10}$$

Thus starting with coupled equations for the transit operators Weinberg's programme can be carried out by the decoupling procedure and depending on the coupling scheme various equations can be obtained for the resolvent G. These equations, however, do not suffer from the "Federbush-disease". It is interesting to note that the equation proposed by Avishai [13] can be recovered from the KLT-equations by decoupling.

4. Conclusions

The above considerations have shown that the integral equations for the transit operators have numerous advantages over other formulations of N-particle scattering. In particular it is always possible to decouple channels so that the solution of the N-particle problem is reduced to that of a single equation. The price one has to pay for this is, of course, the more complicated structure of the kernel. However, the considerable freedom in the decoupling procedure and in the coupling scheme makes these equations very suitable for practical applications.

References

[1] O.A.Yakubovskii, Yad.Fiz. 5(1967) 1312 [Sov.J.Nucl.Phys. 5(1967) 937]

[2] V.Vanzani, Nuovo Cim. 2A (1971) 525

[3] W.Sandhas, The N-body problem, Univ. of Bonn preprint PIB-2-157(1974)

[4] B.R.Karlson and E.M.Zeiger, Phys.Rev. D9 (1974) 761

[5] S.Weinberg, Phys.Rev. 133(1964)B232
[6] Gy.Bencze, Nucl.Phys. A210(1973)568
[7] E.F.Redish, Nucl.Phys. A225(1974)16
[8] D.J.Kouri and F.S.Levin, Phys.Lett 50B(1974)421
[9] W.Tobocman, Phys.Rev. C9(1974)2466
[10] M.L'Huillier, E.F.Redish and P.C.Tandy, preprint 1975.
[11] Gy.Bencze and E.F.Redish, Nucl.Phys. A238(1975)240
[12] W.Tobocman, Phys.Rev. C12(1975)1146
[13] Y.Avishai, Nucl.Phys. A161(1971)621

GENERALIZED FADDEEV THEORY OF NUCLEAR REACTIONS

Edward F. Redish
University of Maryland, College Park, MD 20742 USA

Essentially all the information we possess about nuclei comes from a nuclear scattering or reaction of some kind. For many years, reaction theory in nuclear physics has been thought of as a way to make a better DWBA; however, as we study more reactions in greater detail, the simplified picture of nuclear reaction mechanisms has receded until we have reached the point where the reaction mechanism itself is often as interesting and revealing of the many-body aspects of nuclear processes as is the structure of the bound state.

In order to have a complete many-body theory of the nucleus, we therefore require a complete many-body theory of scattering. We would like to develop a rigorous, exact formalism which can stand as a framework for the development of approximation schemes and for studies of the convergence properties of these schemes. When we deal with the general theory of scattering, bound state methods can be generalized up to the point that we need to include the effects of rearrangement and breakup. We then encounter three formal difficulties:

1. In order to maintain approximate unitarity, processes must be summed to all orders. A convenient way to do this is with integral equations. If we permit more than two continuum clusters, we immediately run into kernels which cannot possibly be Fredholm [1]. *(The Dangerous Delta Difficulty)*

2. Because the continuum states are infinitely degenerate, we must be very careful to specify boundary conditions. An outgoing state in one arrangement can masquerade as an incoming state in another arrangement [2]. *(The Foldy-Tobocman Problem)*

3. There is no single unperturbed Hamiltonian which can be used to choose a suitable starting basis. The different arrangements each have a different unperturbed Hamiltonian. *(The Non-Orthogonality Problem)*

Conventional nuclear reaction theories are unable to address these difficulties. Therefore, the kinds of reaction mechanisms considered have necessarily been limited. These problems have been solved for the case $N = 3$ by the formalism of Faddeev [3]. We have developed a formalism which overcomes these limitations in general and can provide a basis for a complete theory of nuclear reactions.

In principle, there are two ways of attacking the general N-body problem. First, we can proceed by induction. We begin by solving the two-body problem. The solution is used as input for solving the three-body problem; then, we use the output of the solution to the three-body problem as input for solving the four-body problem, etc. This approach has been investigated by a number of groups [4] and is an important part of learning how to handle the many-body problem. It does not help us directly in understanding complicated nuclear processes such as Uranium-Uranium scattering. We may consider a second approach, namely, the imbedding of n-cluster reaction models in the N-body problem where $n \ll N$. This is what is done in conventional reaction theory. The DWBA is equivalent to two-body scattering plus first order perturbation theory. In this case $n = 2$. Our investigations permit us to generalize this method to the case $n > 2$.

The Faddeev method has been highly successful for the study of nuclear systems with $N = 3$ but in the case $N > 3$ it has been useful for only a few special cases (e.g., $d + \alpha$ and $^{12}C \to 3\alpha$). Even in those nuclear reactions where three-body model studies indicate that two-body processes should not suffice, two-body models are still used almost exclusively. Part of the reason for this is that it is not obvious how to produce a realistic three-body model for a many-body reaction. There are a number of difficulties to be considered. (1) In the nuclear case, all the particles are identical. How can we superpose many three-body problems without introducing serious overcounting? How can exchange effects be included? (2) How can the complex structure of the bound states be introduced into the model? Where do the spectroscopic factors go? (3) What does one do about the existence of three-body channels which differ from each other by rearrangements? Can these be included without overcounting? (4) What about Coulomb effects? They are known to be important, and even to dominate some aspects of nuclear reactions. The inclusion of the Coulomb force in the Faddeev theory does not yet have a practical solution and we will not consider it at this time.

We begin by developing generalized Faddeev equations for the N-body problem. We have done this both for the scattering operators [5] and for generalized Faddeev wave functions [6]. Both the operators and the wave functions are labelled by partition indices a, b, c, ... specifying the way the N particles are divided into clusters. The transition operators are given by

$$T^{ab} = V^a + V^a G V^b. \tag{1}$$

We use upper indices, as in V^a, to indicate a residual interaction and lower indices to indicate an internal interaction ($V_b = H - H_b$). Both upper and lower indices may appear on the single potential like V^a_b. This indicates the sum of those pair interactions which are both external to partition a and internal to partition b. We restrict ourselves to problems initiated by two incoming clusters and specify a two-cluster partition by using Greek indices α, β, γ, ... The N-cluster partition is represented by 0. We represent the number of clusters in the partition a by n_a. If we wish to specify the number of clusters in a partition a and $n_a = j$ we will write a_j instead of simply a.

The crucial step in the method is to distribute the residual interaction over other partitions. The way this happens can be seen in a particular case by studying Table I below. In the four-body problem the residual interaction for the partition a = (1)(234) is equal to $V_{12} + V_{13} + V_{14}$. Table I indicates which of these three pair interactions appears internally in the other partitions, b.

TABLE I

	II'			II"			III					
b	(2)(134)	(3)(124)	(4)(123)	(12)(34)	(13)(24)	(14)(23)	(1)(2)(34)	(1)(3)(24)	(1)(4)(23)	(2)(3)(14)	(2)(4)(13)	(3)(4)(12)
V_{12}	0	1	1	1	0	0	0	0	0	0	0	1
V_{13}	1	0	1	0	1	0	0	0	0	0	1	0
V_{14}	1	1	0	0	0	1	0	0	0	1	0	0

Table of V^a_b for N = 4, a = (1)(234). The elements of the table indicate whether a pair potential external to V^a is also internal to partition b. If it is, the table element is 1, if not, 0. The group of partitions II' are the two-cluster partitions of the 1 + 3 type; II' are the two-cluster partitions of the 2 + 2 type; III are the three-cluster partitions. Note that $\sum_{b\in \mathrm{II}} V^a_b - 2 \sum_{b\in \mathrm{III}} V^a_b = V^a$ as claimed in Eq. (2).

Each of the pair interactions in V^a appears exactly twice among the partitions II', once among II", and once among the partitions III. In general, one can show that any residual interaction V^a may be distributed over <u>all</u> partitions by an equation of the form $V^a = \sum_c K_{ac} V^a_c$ where K_{ac} is a numerical coefficient matrix. Inserting this into Eq. (1) we obtain T as a sum of terms. In the c-th term of this sum we use the resolvent equation $G = G_c + G_c V^c G$. The use of a different resolvent equation for each term in the decomposition of the residual interaction leads to coupled equations for the different T operators. The optimal choice for the coefficient matrix is

$$V^a = \sum_b (-1)^{n_b} (n_b - 1)! \; V_b^a \tag{2}$$

where the sum runs over all possible partitions b. In this case (and only in this case) all the disconnected terms in the kernel can be shown to vanish identically [7] leaving

$$T^{a\beta} = V_\beta^a + \sum_\gamma W_\gamma^a \, G_0 \, T^{\gamma\beta}. \tag{3}$$

(The Born term has been simplified by an on-shell transformation.) The kernel operator W_γ^a is the sum of all perturbation graphs with connectivity γ ending with an interaction external to a. This is the operator we must end up with in the kernel if we want a true connected kernel formalism, i.e., one which becomes completely connected upon first iteration. If the first iterated kernel is connected then the kernel may be squared without squaring any delta functions. The operator W_γ is the only operator which is "almost completely connected". Addition of one more potential interaction (one in V^γ) will connect it completely. This operator plays a crucial role in correlating the particles unambiguously to get the proper outgoing waves in the various rearrangement channels [8]. It also has the nice property that, because it correlates all the particles of each cluster of the partition γ, it restricts the coordinates in the multi-dimensional configuration space to a two-cluster tube (loosely).

Since we plan to truncate, it may not be necessary to distribute the residual interaction over all the partitions as in Eq. (2). We can see that in the case of the example given in Table I, V^a may be distributed over partitions II', II", III, or over any combination of those groups. Many relations of the type of Eq. (2) exist. Some lead to partially connected formalisms which may be useful for certain truncations. This degree of freedom is now under study [7].

The equations (3) have been symmetrized using standard techniques for the case of identical particles and, as in the three-body case, the number of equations reduces dramatically (from $2^{N-1} - 1$ to [N/2]). The burden of the symmetry is carried by the kernel and the driving term [9].

We must determine a method for expanding this formalism in a manner which permits the introduction of physical insights and the extraction of calculable models. For this we invert the *Yakubovskii cluster-expansion* [10]. This expansion expresses the full T matrix (internal to a given partition a) in terms of parts having specified connectivity. Essentially it states that the T matrix for a partition a is equal to the sum of all graphs with connectivities obtainable by decomposing the clusters of the partition a. Explicitly, it takes the form

$$T_a = \sum_{a \supseteq b} W_b. \tag{4}$$

This set of equations relates the set of operators T_a and the set of operators W_a. The equations may be inverted by Gaussian elimination to get the W's as a function of the T's. Doing this and employing the relations $T_a G_0 = V_a G_a$ yields the equation [11]

$$W_\gamma^a G_0 = \sum_{\gamma \supseteq b} N(\gamma,b) \, V_b^a \, G_b = V_\gamma^a \, G_\gamma - \sum_{\gamma \supset b_3} V_{b_3}^a \, G_{b_3} + \ldots \tag{5}$$

We refer to this as the *anti-cluster expansion* for W. Use of this expansion allows the introduction of appropriate complete sets at the Green functions. The complete set of physically allowable channels including rearrangement now appears naturally in the equations without the need for introducing overcomplete sets (as is required in some other methods). Truncations based on data or physical intuition can now be made.

We want a formalism which reduces in a simple and direct manner to the usual approximations of nuclear reaction theory. These approximations have been tested and proved successful in numerous cases. We don't wish to supplant these methods but rather to extend them in a manner which permits a loosening of their more binding constraints. We have therefore proved that our formalism (Eq. (3)) with the anti-cluster expansion (Eq. (5)) has the following properties:

1. It reduces directly to the DWBA for nuclear rearrangement reactions [11].

2. It reduces to the DWIA for nucleon-nucleus elastic and inelastic scattering [12].

In both of these cases the formalism provides extensions of the usual prescriptions which are interesting and may by of practical value.

In the first case (rearrangement reactions) our formalism provides an unambiguous multistep DWBA series without non-orthogonality corrections [11]. The prescription for the transition operators which we obtain agrees with that obtained by Udagawa, et al. [13] by transforming away the usual non-orthogonality term. The kernel generating our multistep DWBA series is of the form $(E^{+} - K - U)^{-1}V$ where U is an optical potential and V a transition operator. The convergence properties of the multistep DWBA series can possibly be understood by studying this operator. Furthermore, the first natural correction to the transition operator to the DWBA for deuteron stripping is the Johnson-Soper correction which has proved so valuable in extending the DWBA to higher energy regimes [14].

In the second case (elastic and inelastic scattering) it provides a three-body generalization of the KMT two-body prescription for the optical potential matrix. It provides a particular way of knitting together the N simultaneous three-body problems implicit in the reaction mechanism.

We can also approach the N-body problem through the Schrödinger many-body wave function, rather than through the scattering operator. In this case we begin by decomposing the wave function into parts corresponding to the two-cluster partitions of the N-body problem. We obtain the following representation for the wave function [6]

$$\psi_{\beta} = \sum_{\gamma} \psi_{\beta}^{(\gamma)} \tag{6}$$

with the components given by

$$\psi_{\beta}^{(\gamma)} = \delta_{\gamma\beta}\, \phi_{\beta} + G_{0}\, W_{\gamma}^{0}\, G_{0}\, T^{\gamma\beta}\, \phi_{\beta} \tag{7}$$

We can also show [8] that the W operator appearing in the above expression serves the purpose of correlating the N outgoing particles into γ-clusters. Specifically, the connectivity of the expression for the scattered wave determines the kind of outgoing waves permitted. The wave function component $\psi_{\beta}^{(\gamma)}$ only has outgoing waves of the type γ, that is, either two bound clusters of the partition γ or direct breakup states of this partition ($\gamma \supset b_{k}$). There is no rearrangement present.

The wave function components can be shown to satisfy generalized Faddeev equations of the form

$$(E - H_{0} - \vartheta_{\gamma})\, \psi_{\beta}^{(\gamma)} = \vartheta_{\gamma} \sum_{\alpha \neq \gamma} \psi_{\beta}^{(\alpha)} \tag{8}$$

where the effective interaction is given by

$$\vartheta_{\gamma} = W_{\gamma}^{0}\, G_{0}\, G_{\gamma}^{-1} = W_{\gamma}^{0}(1 - G_{0}\, V_{\gamma}). \tag{9}$$

The anti-cluster expansion for W can now be used in Eq. (9) to permit the construction of few-body models. It is not yet clear whether a model obtained by using the wave function formalism and truncating is equivalent to those obtained by using the T-operator formalism and truncating. This point is now under investigation.

These formalisms provide possible starting points for a general N-body theory of scattering. The T-operator formalism (which we have investigated more fully) is known to have the following properties:

1. The kernel of the integral equation obtained is fully connected. Furthermore, the equations can be truncated to n-body processes with n < N and equations with connected kernel are obtained for any n. *The dangerous delta difficulty is therefore solved in a convenient way.*

2. The folklore in this field states that one needs at least as many integral equations as there are two-cluster partitions to specify the boundary conditions uniquely. This is precisely the number we have. We do not yet have a firm proof, but *we believe the Foldy-Tobocman problem does not arise if these equations are used.*

3. The approach is dispersive rather than perturbative. As a result, all physical processes are summed over in intermediate states in a natural way. There is no need to introduce overcomplete sets and *the non-orthogonality problem does not arise.*

Similarly, the difficulties with extracting few-body models for nuclear reaction mechanisms are solved in this formalism. Identity of particles can easily be included without complicating the formalism. The bound states of the subsystems go in as initial, final, and intermediate states and therefore are only used in calculating matrix elements for the kernel and driving terms of the integral equations. Any nuclear model for the bound state may therefore be used without changing the formalism, and spectroscopic factors enter naturally. Different three-body arrangements are permitted, and there is no overcounting or non-orthogonality correction required.

The utility of the formalism of course depends on its providing improved fits to data. Calculations using this formalism in the scattering of light nuclei are now being prepared.

References

[1] S. Weinberg, Phys. Rev. 133 (1964) B232.
[2] L. Foldy and W. Tobocman, Phys. Rev. 105 (1957) 1099.
[3] L. D. Faddeev, JETP (U.S.S.R.) 39 (1960) 1459 (Eng. trans. Sov. Phys. JETP 12 (1961) 1014.
[4] J. Tjon, Phys. Lett. 56B (1975) 217; S. K. Adhikari and I. Sloan, Phys. Rev. C12 (1975) 1152.
[5] Gy. Bencze, Nucl. Phys. A210 (1973) 568; E. F. Redish, Nucl. Phys. A225 (1974) 16.
[6] M. L'Huillier, E. F. Redish, and P. C. Tandy, in preparation.
[7] P. Benoist-Gueutal, M. L'Huillier, E. F. Redish, and P. C. Tandy, in preparation; M. L'Huillier, et al., U. of Md. Tech. Rept. 76-068.
[8] P. C. Tandy, E. F. Redish, and M. L'Huillier, in preparation.
[9] Gy. Bencze and E. F. Redish, Nucl. Phys. A238 (1975) 240.
[10] O. A. Yakubovskii, Yad. Fiz. 5 (1966) 1312.
[11] E. F. Redish, Phys. Rev. C10 (1974) 67.
[12] E. F. Redish, Nucl. Phys. A225 (1974) 82.
[13] T. Udagawa, H. H. Wolter, and W. R. Coker, Phys. Rev. Lett. 31 (1973) 1507.
[14] R. C. Johnson and P. Soper, Phys. Rev. C1 (1970) 976.

THREE-AND FOUR-BODY EQUATIONS WITH HALF-OFF-SHELL INPUT+

Bengt R. Karlsson
Institute of Theoretical Physics, Fack, S-40220 Goteborg, Sweden
and
Enrique M. Zeiger
Stanford Linear Accelerator Center, Stanford University, Stanford, Ca 94305,USA.

1. THE THREE-BODY CASE

The Faddeev reformulation of the three-body Schrodinger equation amounts to converting this equation into a set of multiple-scattering equations in which the basic (pair) interaction is represented by full subsystem transition amplitudes. In general, these amplitudes have to describe a two-body scattering in the presence of a third particle. Therefore, they necessarily enter the Faddeev equations off-the-energy-shell, in the conventional formulation even completely-pff-shell, i.e. on the form $t(\vec{q},\vec{q}\,'; E-\tilde{p}^2-io)$ where $\tilde{q}^2$, $\tilde{q}\,'^2$ and $E-\tilde{p}^2$ are all different (E is the energy of the three-body system and $E-\tilde{p}^2$ is the energy of the two-body subsystem).

As was recently shown [1], the Faddeev equations can naturally be converted to a form where only half-off-shell two-body transition ampl itudes (and vertex functions) appear. This is done as follows. In order to make explicit the singularity structure of the kernel of the Faddeev equations

$$|\Psi_\beta\rangle = \delta_{\beta\alpha}|\vec{p}_\alpha^{(0)}\varphi_\varkappa^\alpha\rangle - G_\beta V_\beta \sum_{\gamma\neq\beta}|\Psi_\gamma\rangle \tag{1}$$

we employ a complete set of eigenstates of the channel hamiltonian $H=\tilde{p}^2+\tilde{q}^2+V_\beta$, $\{|\vec{p}_\beta\varphi_\varkappa\rangle, |\vec{p}_\beta\,\Psi^-_{\vec{q}_\beta}\rangle\}$. For the wave function components we thereby find the representation $(\tilde{p}_\alpha^{(0)2}-\varkappa_\alpha^2=E)$

$$\langle \vec{p}_\beta\vec{q}_\beta|\Psi_\beta\rangle = \delta_{\beta\alpha}\,\delta(\vec{p}_\alpha-\vec{p}_\alpha^{(0)})\langle\vec{q}_\alpha|\varphi_\varkappa^\alpha\rangle - \frac{\langle\vec{q}_\beta|\varphi_\varkappa^\beta\rangle}{\tilde{p}_\beta^2-\varkappa_\beta^2-E-io}\,\mathcal{H}_{\beta\alpha}(\vec{p}_\beta,\vec{p}_\alpha^{(0)};E+io)$$

$$-\int\langle\vec{q}_\beta|\Psi^-_{\vec{q}\,'_\beta}\rangle\frac{d\vec{q}\,'_\beta}{\tilde{p}_\beta^2+\tilde{q}_\beta'^2-E-io}\,\mathcal{E}_{\beta\alpha}(\vec{p}_\beta,\vec{q}\,'_\beta;\vec{p}_\alpha^{(0)};E+io) \tag{2}$$

in terms of the (singularity-free) physical transition amplitudes for elastic/rearrangement ($\mathcal{H}_{\beta\alpha}$) and breakup ($\Sigma_\beta\,\mathcal{E}_{\beta\alpha}$) processes. The equations for these amplitudes are directly obtained from Eq.(1), and we get (in a shorthand notation)

$$\mathcal{H}_{\beta\alpha} = \mathcal{V}_{\beta\alpha} - \sum_{\gamma\neq\beta}\mathcal{V}_{\beta\gamma}\frac{1}{\tilde{p}_\gamma^2-\varkappa_\gamma^2-E-io}\mathcal{H}_{\gamma\alpha} - \sum_{\gamma\neq\beta}\mathcal{W}_{\beta\gamma}\frac{1}{\tilde{p}_\gamma^2+\tilde{q}_\gamma'^2-E-io}\mathcal{E}_{\gamma\alpha} \tag{3}$$

with a similar equation for $\mathcal{E}_{\beta\alpha}$. The "potentials" of these equations, $\mathcal{V}_{\beta\alpha}$, $\mathcal{W}_{\beta\alpha}$ etc. contain the two-body input only through half-off-shell transition amplitudes $t(\vec{q},\vec{q}\,'; \tilde{q}^2+io)$ and bound state vertex functions*. Therefore, at no point in this formulation of the three-body theory there is need neither for completely-off-shell two-body transition amplitudes nor for two-body potentials. Moreover, the "potentials" of Eq.(3) are all independent of the three-body energy E and they can in fact easily be made real after angular momentum decomposition.

---

* Recall that for instance $\langle\Psi^-_{\vec{q}\,'_\beta}|V_\beta|\vec{q}_\beta\rangle = t_\beta(\vec{q}\,'_\beta,\vec{q}_\beta;\tilde{q}_\beta'^2+io)$

\+ Presented as invited talk by B.K.

This reformulation of the three-body equations also provides a new starting point for computational procedures. For instance, Bollé [2] has decoupled the Eqns. (3) in such a way that

$$\mathcal{H}_{\beta\alpha} = \mathcal{H}^{(0)}_{\beta\alpha} + \sum_{\gamma} \mathcal{H}^{(0)}_{\beta\gamma} \frac{1}{\tilde{p}_\gamma^2 - \varkappa_\gamma^2 - E - i0} \mathcal{H}_{\gamma\alpha}$$

$$\mathcal{E}_{\beta\alpha} = \mathcal{E}^{(0)}_{\beta\alpha} + \sum_{\gamma} \mathcal{E}^{(0)}_{\beta\gamma} \frac{1}{\tilde{p}_\gamma^2 - \varkappa_\gamma^2 - E - i0} \mathcal{H}_{\gamma\alpha} \tag{4}$$

(5)

where $\mathcal{H}^{(0)}_{\beta\alpha}$ and $\mathcal{E}^{(0)}_{\beta\alpha}$ satisfy equations with $(\tilde{p}^2+\tilde{q}^2-E-i0)^{-1}$-type green functions. An obvious scheme is now to use approximate expressions for $\mathcal{H}^{(0)}_{\beta\alpha}$ and $\mathcal{E}^{(0)}_{\beta\alpha}$ and solve the one-dimensional (after angular momentum decomposition) Eqn.(4) for $\mathcal{H}_{\beta\alpha}$ exactly. $\mathcal{E}_{\beta\alpha}$ Can subsequently be obtained from Eqn.(5) through quadrature.

It should finally be mentioned that the present formulation has been extended to the singular core interaction case by Kim and Tubis [3].

2. THE FOUR-BODY CASE

The four-body Faddeev-Yakubovski (FY) equations can also be turned into a set of equations that only require half-off-shell input [4], and this section contains a pre-view of the results obtained. To simplify the presentation we here only consider pair interactions without two-body bound states but with one three-body bound state in each channel σ of the type (123)(4).

To make explicit the singularity structure of the kernel of the FY equations (we use the notation of ref. [5])

$$|\psi^\sigma_\beta\rangle = \delta^{\sigma\tau}|\vec{r}\,\phi^{(\sigma)}_\beta\rangle - G_0 \sum_{\gamma<\sigma} K^\sigma_{\beta\gamma} \sum_{\varrho\supset\gamma} \bar{\delta}^{\tau\varrho}\, |\psi^\varrho_\gamma\rangle \tag{6}$$

we employ the complete set of eigenstates of the channel hamiltonian $H^\sigma = \tilde{r}^2 + \tilde{p}^2 + \tilde{q}^2 + \sum_{\gamma<\sigma} V_\gamma$, $\{|\vec{r}\,\phi^{(\sigma)}_{\varkappa_\sigma}\rangle, |\vec{r}\,\psi^{-(\sigma)}_{\vec{p}\vec{q}}\rangle\}$ (for σ = (123)(4), ϕ and ψ^- are bound and continuum three-body states; for σ = (12)(34), ϕ is absent and ψ^- should properly carry different momentum labels). After separation of the singularities of G_0 and $K^\sigma_{\beta\gamma}$, we get for the kernel of the FY equations

$$\begin{aligned} G_0 K^\sigma_{\beta\gamma} = {} & \int |\vec{r}\,\phi^{(\sigma)}_{\beta;\varkappa_\sigma}\rangle \frac{d\vec{r}}{\tilde{r}^2 - \varkappa_\sigma^2 - E - i0} \langle \vec{r}\,\phi^{(\sigma)}_{\varkappa_\sigma}|\bar{V}^{(\sigma)}_\gamma \\ & + \int |\vec{r}\,\psi^{-(\sigma)}_{\beta;\vec{p}\vec{q}}\rangle \frac{d\vec{r}\,d\vec{p}\,d\vec{q}}{\tilde{r}^2+\tilde{p}^2+\tilde{q}^2 - E - i0} \langle \vec{r}\,\psi^{-(\sigma)}_{\vec{p}\vec{q}}|\bar{V}^{(\sigma)}_\gamma \\ & + G_0(E+i0)\Big\{V_\beta \bar{\delta}_{\beta\gamma} - \int |\vec{r}\,\phi^{(\sigma)}_{\beta,\varkappa_\sigma}\rangle d\vec{r} \langle \vec{r}\,\phi^{(\sigma)}_{\varkappa_\sigma}|\bar{V}^{(\sigma)}_\gamma - \int |\vec{r}\,\psi^{-(\sigma)}_{\beta;\vec{p}\vec{q}}\rangle d\vec{r}\,d\vec{p}\,d\vec{q} \langle \vec{r}\,\psi^{-(\sigma)}_{\vec{p}\vec{q}}|\bar{V}^{(\sigma)}_\gamma\Big\} \end{aligned} \tag{7}$$

where $\bar{V}^{(\sigma)}_\gamma = \sum_{\gamma'<\sigma} \bar{\delta}_{\gamma\gamma'} V_{\gamma'}$, and where β-components of the three-body statevectors now appear. In analogy with Eqn.(2), $|\psi^\sigma_\beta\rangle$ has a representation on the form

$$\langle \vec r \vec p \vec q | \Psi^{\sigma}_{\beta} \rangle = \delta^{\sigma\tau}\delta(\vec r - \vec r^{(\omega)}) \langle \vec p \vec q | \phi^{(\sigma)}_{\beta,\varkappa_\sigma}\rangle$$

$$- \frac{\langle \vec p \vec q | \phi^{\sigma}_{\beta,\varkappa_\sigma}\rangle}{\tilde r^2 + \varkappa_\sigma^2 - E - i0}\, \mathcal{H}^{\sigma\tau}(\vec r, \vec r^{(\omega)}; E+i0)$$

$$- \int \langle \vec p \vec q | \Psi^{-(\sigma)}_{\beta,\vec p' \vec q'}\rangle \frac{d\vec p'\, d\vec q'}{\tilde r^2 + \tilde p'^2 + \tilde q'^2 - E - i0}\, \mathcal{E}^{\sigma\tau}(\vec r \vec p' \vec q'; \vec r^{(\omega)}; E+i0)$$

$$- \frac{1}{\tilde r^2 + \tilde p^2 + \tilde q^2 - E - i0}\, \mathcal{J}^{\sigma\tau}_{\beta}(\vec r \vec p \vec q; \vec r^{(\omega)}; E+i0) \qquad (8)$$

and it can be shown that $\mathcal{H}^{\sigma\tau}$ and $\sum_\sigma \mathcal{E}^{\sigma\tau}$ are indeed the physical four-body elastic/rearrangement and breakup transition amplitudes. The remaining amplitude $\mathcal{J}^{\sigma\tau}_{\beta}$ vanishes identically upon summation over $\beta \subset \sigma$ and therefore never contributes to the full wavefunction or to the physical amplitudes.

The equations for the amplitudes are directly obtained from the FY equation (6)

$$\mathcal{H}^{\sigma\tau} = \mathcal{V}^{\sigma\tau} - \sum_{\rho\neq\sigma} \mathcal{V}^{\sigma\rho} \frac{1}{\tilde r^2 - \varkappa_\rho^2 - E - i0}\, \mathcal{H}^{\rho\tau}$$

$$- \sum_{\rho\neq\sigma} \mathcal{W}^{\sigma\rho} \frac{1}{\tilde r^2 + \tilde p^2 + \tilde q^2 - E - i0}\, \mathcal{E}^{\rho\tau} - \sum_{\rho\neq\sigma}\sum_{\gamma\subset\rho} \mathcal{W}^{\sigma\rho}_{\gamma} \frac{1}{\tilde r^2 + \tilde p^2 + \tilde q^2 - E - i0}\, \mathcal{J}^{\rho\tau}_{\gamma} \qquad (9)$$

with similar equations for $\mathcal{E}^{\sigma\tau}$ and $\mathcal{J}^{\sigma\tau}_{\beta}$. It can be shown that the "potentials" $\mathcal{V}^{\sigma\tau}, \mathcal{W}^{\sigma\tau}, \mathcal{W}^{\sigma\tau}_{\gamma}$ etc. of Eqn.(9) are independent of the four-body energy E, and only require two-and three-body half-off-shell transition amplitudes and vertex functions as input. Eqns.(9) therefore are acceptable four-body analogues of the three-body Eqns.(3) and are our main result.

Since the kernel of the Faddeev-Yakubovski equations is only two-body connected there is need for a detailed index structure (σ and β indices) to make sure that the kernel becomes connected after a few (2) iterations. It is a remarkable feature of Eqn.(9) that this detailed indexing only shows up in the non-physical amplitude $\mathcal{J}^{\sigma\tau}_{\beta}$. The implications of this fact are at the moment somewhat unclear.

We finally note that it would be interesting to apply an analysis similar to the one presented here to other four-body equations, in particular to those of Sloan [6].

To summarize, we have converted the four-body Faddeev-Yakubovski equations into a set of equations which only require half-off-shell two-and three-body inputs. The method used is applicable also to other four-body equations.

REFERENCES

1. B.R. Karlsson and E.M. Zeiger, Phys. Rev. D11 (1975) 939.
2. D. Bollé, to be published.
3. Y.E. Kim and A. Tubis, Phys. Rev. D11 (1975) 947.
4. B.R. Karlsson and E.M. Zeiger, to be published.
5. B.R. Karlsson and E.M. Zeiger, Phys. Rev. D10 (1974) 1291.
6. I.H. Sloan, Phys. Rev. C6 (1972) 1945.

THEORY OF THE THREE-BODY SEPARABLE EXPANSION AMPLITUDE

Shinsho Oryu

Department of Physics, Faculty of Science and Technology,
Science University of Tokyo, Noda-shi, Chiba 278 JAPAN

In the momentum space calculation for the Faddeev equation, the construction of two-body amplitudes which are separable is essential if we are to be able to calculate the three-body system exactly[1]. Furthermore, the investigation of the four-body problems using the Faddeev formalism requires the separable amplitudes of the two-body and the three-body systems. Systematic research for obtaining the separable two-body amplitudes from a realistic potential, however, has just begun[2,3]. Also, we have no general method for obtaining the three-body separable amplitudes from the three-body Faddeev equation. Noyes and Kowalski[4] have proposed a method to make a two-body separable amplitude with a form factor which depends systematically upon the potential and the energy. Alt et. al.[5] have analyzed the four-nucleon system by using a separable potential model which might be an application of the Noyes and Kowalski model to the two-nucleon and the three-nucleon systems. Noyes-Kowalski model, however, has a defect in that it has an unphysical divergent pole when the two-body phase shift becomes zero[6]. Recently, we have shown that the pole can be canceled out by an increase in rank [2]. In this paper, we will propose a method for obtaining the separable amplitudes of the three-body equations as well as the two-body amplitudes.

First of all we will start from the standard partial wave Amado cquation with the Yamaguchi potential[7], for the three-identical particle system:

$$X_\ell(q',q;E)=B_\ell(q',q;E)+\int_0^\infty B_\ell(q',q'';E)\tau(z'')X_\ell(q'',q;E)dq''. \quad (1)$$

It is well known that it can be solved numerically by the contour deformation method, i.e. the set of values $\{q''\}$ are changed to the complex values $\{\bar{q}''\}$ so that we have the secondary Amado equation[8]

$$X_\ell(\bar{q}'',q;E)=B_\ell(\bar{q}'',q;E)+\int_C B_\ell(\bar{q}'',\bar{q}''';E)\tau(\bar{z}''')X_\ell(\bar{q}''',q;E)d\bar{q}''' \quad (2)$$

where the complex values $\{\bar{q}''\} = \{\bar{q}'''\}$ on the contour C. The integral kernel of this equation has some poles which appear from the two-body propagator and logarithmic cut of the function $B_\ell(\bar{q}'',\bar{q}''';E)$. The region of those singularities was shown by Cahill-Sloan[8] and Ebenhöh[9]. The solution of eq.(2) is restricted to the complex values of $\{\bar{q}''\}$ and $\{q\}$ on the contour and is obtained by the same method as the two-body generalized separable expansion method[2],

$$X_\ell(\bar{q}'',q;E)=\sum_{I=1}^{N}\sum_{J=1}^{N}\frac{\Theta_{IJ}(E)}{\Theta(E)}\Phi_\ell(\bar{q}'',k_J;E)\Phi_\ell(k_I,q;E)+\Gamma_\ell^{(N+1)}(\bar{q}'',q;E). \quad (3)$$

The first terms on the right hand side are separable amplitudes of rank N and the second is a nonseparable term which satisfies the following integral equation,

$$\Gamma_\ell^{(N+1)}(\bar{q}'',q;E)=B_\ell^{(N+1)}(\bar{q}'',q;E)+\int_C B_\ell^{(N+1)}(\bar{q}'',\bar{q}''';E)\tau(\bar{z}''')\Gamma_\ell^{(N+1)}(\bar{q}''',q;E)d\bar{q}''' \quad (4)$$

where the inhomogeneous term and the kernel consist of an error function which will be mentioned below. Function $\Phi_\ell(\bar{q}'',k_J;E)$ and $\Phi_\ell(k_I,q;E)$ are the three-body form factors which are given by modification of the "three-body potential" $B_\ell(x',x;E)$. The modification is given by the following integral equations,

$$\Phi_\ell(\bar{q}'',k_J;E)=B_\ell(\bar{q}'',k_J;E)+\int_C B_\ell^{(N+1)}(\bar{q}'',\bar{q}''';E)\tau(\bar{z}''')\Phi_\ell(\bar{q}''',k_J;E)d\bar{q}''' \quad (5)$$

and

$$\Phi_\ell(k_I,q;E)=B_\ell(k_I,q;E)+\int_C \Phi_\ell(k_I,\bar{q}''';E)\tau(\bar{z}''')B_\ell^{(N+1)}(\bar{q}''',q;E)d\bar{q}'''. \quad (6)$$

We can think of those formalisms as a distortion of $B_\ell(x',x;E)$ by the error integral kernel:

$$B_\ell^{(N+1)}(x',x;E)\tau(x,E)=\Big[B_\ell(x',x;E) - \sum_{I=1}^{N}\sum_{J=1}^{N}\frac{\Delta_{IJ}(E)}{\Delta(E)}B_\ell(x',k_J;E)B_\ell(k_I,x;E)\Big]\tau(x,E) \quad (7)$$

where $\Delta(E)=\det[B_\ell(k_I,k_J;E)]$ and $\Delta_{IJ}(E)$ is IJ co-factor of $\Delta(E)$. The k_1, k_2, ... k_N are the parameters which should be chosen to minimize the norm of the error integral kernel. In particular, the points of the poles on the original kernel in eq.(2) must be chosen as the parameters k_1, k_2, ... k_n and the other parameters can be chosen at N-n points near the contour C (see Figs. 1 and 2). Since, the kernel of eq.(5) has no singularity on the contour C, we can easily solve eq.(5). The form factor on the real values {q'} can be given by translation of the simple integral

$$\Phi_\ell(q',k_J;E)=B_\ell(q',k_J;E)+\int_C B_\ell^{(N+1)}(q',\bar{q}'';E)\tau(\bar{z}'')\Phi_\ell(\bar{q}'',k_J;E)d\bar{q}''. \quad (8)$$

In general, eq.(8) shows the translation of {$\bar{q}''$} on the contour into an arbitrary {q'} which may be complex. Analyticity of $\Phi_\ell(q',k_J;E)$ is given by this equation. In the same manner, $\Phi_\ell(k_I,q;E)$ is obtained from eq.(6) and the conjugate integral equation of eq.(5). $\Theta(E)$ is a determinant which consists of N x N matrix elements $A^\ell_{IJ}(E)$. They are defined by the following equation,

$$A^\ell_{IJ}(E)=B_\ell(k_I,k_J;E)-\int_D B_\ell(k_I,\hat{q}'';E)\tau(\hat{z}'')\Phi_\ell(\hat{q}'',k_J;E)d\hat{q}''. \quad (9)$$

If we choose k_I (I=1,2, ... N) at the values which are shown in Figs. 1 and 2, the contour D can be taken on the real axis from zero to infinity except for the poles of $\tau(\hat{q}'',E)$. Therefore, the integration of the second term of eq.(9) is obvious. The numerator $\Theta_{IJ}(E)$ is the IJ co-factor of $\Theta(E)$.

Now, by using eqs.(6) and (8), it is easily found that the separable formalism of eq.(3) can be redefined on the real axis of q' by the analytic continuation of the form factors. It is important that the analytic continuation becomes possible only when the parameters k_I (I=1,2, ... N) are chosen by our method. Therefore, the direct definition of the separable solution of eq.(1) can be obtained:

$$X_\ell(q',q;E)=\sum_{I=1}^{N}\sum_{J=1}^{N}\frac{\Theta_{IJ}(E)}{\Theta(E)}\Phi_\ell(q',k_J;E)\Phi_\ell(k_I,q;E)+\Gamma_\ell^{(N+1)}(q',q;E) \quad (10)$$

where the form factors are given by eqs.(5), (6) and (8). The error function $\Gamma_\ell^{(N+1)}(q',q;E)$ is defined by

$$\Gamma_\ell^{(N+1)}(q',q;E)=B_\ell^{(N+1)}(q',q;E)+\int_C B_\ell^{(N+1)}(q',\bar{q}'';E)\tau(\bar{q}'',E)\Gamma_\ell^{(N+1)}(\bar{q}'',q;E)d\bar{q}''. \quad (11)$$

It is easily shown that $\Gamma_\ell^{(N+1)}$ is exactly zero in the cases of on-energy shell and half on-energy shell. Therefore the solution of eq.(1) is given exactly by the separable form in the cases of on- and half on-energy shells. As for the analysis of experimental data for the three-body problem, the remainder of the amplitude is only the separable terms, since the initial and final momenta q and q' are on-energy shell. In the three-body subsystem of the four-body problem, the effects of $\Gamma_\ell^{(N+1)}$ are negligible.

Second, we apply this method to the three-nucleon scattering problem at the low energies. According to many authors we take five channels, that is, channel 1=[n_1,(n_2p) singlet], channel 2=[n_1,(n_2p) triplet], channel 3=[p,(n_1n_2) singlet], channel 4=[n_2,(n_1p) singlet] and channel 5=[n_2,(n_1p) triplet] for α, β, δ and σ which indicate not only the particle channels but also the state channels. Then the above theory can be easily generalized into the coupled channels case and the error function of eq.(10) is exactly zero. Therefore, a transition matrix for the process from an α-channel to a β-channel is given by the separable form:

$$X_\ell(q',q;\beta,\alpha;E)= \sum_{\delta,\sigma}^{5}\sum_{I,J}^{N} \frac{\Theta_{IJ}^{\delta\sigma}(E)}{\Theta(E)}\Phi_\ell(q',k_J;\beta,\sigma;E)\Phi_\ell(k_I,q;\delta,\alpha;E). \quad (12)$$

The three-body form factor $\Phi_\ell(q',k_J;\beta,\sigma;E)$ is given as follows,

$$\Phi_\ell(q',k_J;\beta,\sigma;E)=B_\ell(q',k_J;\beta,\sigma;E)+\sum_\gamma\int_C B_\ell^{(N+1)}(q',\bar{q}'';\gamma,\sigma;E) \times\tau_\gamma(\bar{q}'',E)\Phi_\ell(\bar{q}'',k_J;\gamma,\sigma;E)d\bar{q}'' \quad (13)$$

and the form factor $\Phi_\ell(k_I,q;\delta,\alpha;E)$ is obtained by the conjugate equation of eq.(13). The denominator function $\Theta(E)$ is given by 5N×5N-determinant which consists of the elements $A_\ell(k_I,k_J;\delta,\sigma;E)$. The elements satisfy the following equation,

$$A_\ell(k_I,k_J;\delta,\sigma;E)=B_\ell(k_I,k_J;\delta,\sigma;E)-\sum_\gamma\int_D B_\ell(k_I,\hat{q}'';\delta,\gamma;E) \times\tau(\hat{q}'',E)\Phi_\ell(\hat{q}'',k_J;\gamma,\sigma;E)d\hat{q}''. \quad (14)$$

Then the rearrangement amplitude from channel α to β is given by

$$X(\mathbf{q}'_\beta,\mathbf{q}_\alpha;\beta,\alpha;E)= \sum_{\ell=0}^{\infty}(2\ell+1)P_\ell(\cos\theta_{\beta\alpha})X_\ell(q',q;\beta,\alpha;E) \quad (15)$$

and the break-up amplitude is obtained as follows,

$$Y(\mathbf{q}'_\beta,\mathbf{p}'_\beta;\mathbf{q}_\alpha;\beta,\alpha;E)=g_\beta(p'_\beta)\tau_\beta(z'_\beta)X(\mathbf{q}'_\beta,\mathbf{q}_\alpha;\beta,\alpha;E) \quad (16)$$

where $\mathbf{q}'_\beta$ is the momentum of the spectator particle of the final channel β and $\mathbf{q}_\alpha$ is the momentum of the incident particle of the channel α in the three-body center of mass system. $\mathbf{p}'_\alpha$ is the relative momentum between the correlating particles of the final channel β. $\theta_{\beta\alpha}$ is an angle defined by $\cos\theta_{\beta\alpha}=\mathbf{q}_\beta\mathbf{q}_\alpha/|\mathbf{q}_\beta\mathbf{q}_\alpha|$, and $z'_\beta=E-3q'^2_\beta/4$. Furthermore we can approximate eq.(13) by the first term, if we take the rank to be sufficiently large to suit our needs, because the values which come from the second term of eq.(13) is less than the value of the first term. Nevertheless, the second term of eq.(13) should be calculated by using two kinds of contours for the cases $0\leq q'^2\leq E$ and $E<q'^2\leq 4E/3$. By these approximations, eq.(12) can be written as follows,

$$X_\ell(q',q;\beta,\alpha;E)\simeq\sum_{\sigma=1}^{5}\sum_{J=1}^{N}C_\ell^\alpha(\sigma,J;E)B_\ell(q',k_J;\beta,\sigma;E) \quad (17)$$

and eq.(14) is given by

$$A_\ell(k_I,k_J;\delta,\sigma;E)\simeq B_\ell(k_I,k_J;\delta,\sigma;E)-\sum_\gamma\int_D B_\ell(k_I,\hat{q}'';\delta,\gamma;E) \times\tau(\hat{q}'',E)B_\ell(\hat{q}'',k_J;\gamma,\sigma;E)d\hat{q}'' \quad (18)$$

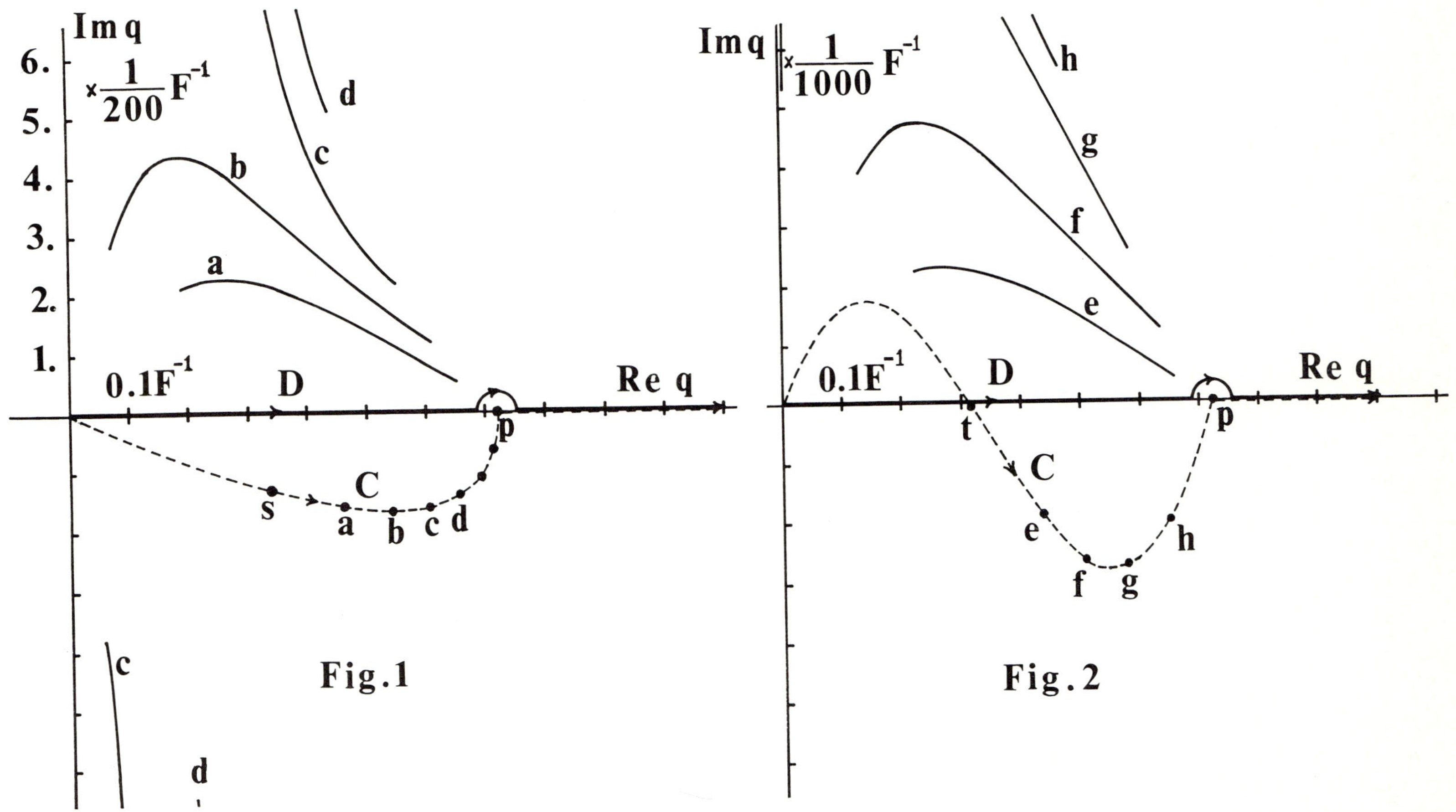

Fig.1 An integral contour D along the real axis and the locations of cuts(a, b, c and d) corresponding to the parameters k_I (I=a, b, c and d) on a quadratic contour C are shown. p is a location of a n-p triplet pole and s is a limit of parameters which are allowed on the contour C.

Fig.2 An integral contour D along the real axis and the locations of cuts(e, f, g and h) corresponding to the parameters k_I (I=e, f, g and h) on a cubic contour C are shown. p is a location of a n-p triplet pole and t is a limit of parameters which are allowed on the contour C.

where eq.(18) consists of only the first Born and the second Born terms of the original integral equation(1). Then C_ℓ^α of eq.(17) can be redefined by using eq.(18). Therefore the break-up and the rearrangement amplitudes become quite simple forms which are given by a kind of superposition of the first Born and the second Born terms in which the analyticities are already well known.

Finally, it should be noted that equations similar to ours have been proposed by Kharchenko et. al.[10] based on the Bateman's Method. However, our final equations differ in several respects from theirs because our equation has many singularities in it's integral kernel and our choice of parameters for separating the amplitudes is unique. Also numerical results for positive energies have not yet been obtained in their framework. We would like to stress that our present results should be understood mainly as a first step for the general calculations of the few-body problems in which we expect some encouraging checks of the applicability of our new method.

References:

[1] L. D. Faddeev, Zh. Eksp. i Teor. Fiz. 39 (1960) 1459 [English transl: Soviet Phys. JETP 12 (1961),1014].
R. Amado, Phys. Rev. 132 (1963) 485.
C. Lovelace, Phys. Rev. 135 (1964) B1225.

[2] S. Oryu, Prog. Theor. Phys. (Kyoto) 52 (1974) 550.
S. Oryu, T. Ishihara and S. Shioyama, Prog. Theor. Phys.(Kyoto) 51 (1974) 1626.

[3] S. K. Adhikari, Phys. Rev. C10 (1974) 1623.

[4] H. P. Noyes, Phys. Rev. Letters 15 (1965) 538.
K. L. Kowalski, Phys. Rev. Letters 15 (1965) 798.

[5] E. O. Alt, P. Grassberger and W. Sandhas, Phys. Rev. C1 (1970)85.

[6] T. A. Osborn, Nucl. Phys. A138 (1969) 305.

[7] Y. Yamaguchi, Phys. Rev. 95 (1954) 1628.

[8] R. T. Cahill and I. H. Sloan, Nucl. Phys. A165 (1971) 161.

[9] W. Ebenhöh, Nucl. Phys. A191 (1972) 97.

[10] V. F. Kharchenko, S. A. Strozhenko and V. E. Kuzmichev, Nucl. Phys. A188 (1972) 609.
H. Bateman, Proc. Roy. Soc. A100 (1921) 441; Messenger Math. 37 (1908) 179.

A NEW DYNAMICS OF NN SCATTERING*

D. D. Brayshaw
Stanford Linear Accelerator Center
Stanford University, Stanford, California 94305 USA

The meson theory of nuclear forces provides a good qualitative, and often quantitative, description of NN scattering, but nevertheless suffers from some major defects. Thus, by adjusting certain πN and $\pi\pi$ amplitudes (including a fictitious "σ") one can construct a 2π exchange potential which, together with OBE terms corresponding to π, ρ, and ω exchange, can produce a reasonably good fit to the phase shifts (with certain exceptions such as the $^3D_1$). See, for example, A. D. Jackson [1]. However, after 25 years of effort the form of the (crucial) 2π contribution is still somewhat ambiguous, and one cannot with any certainty predict the effect of still higher order diagrams. Furthermore, effective "potentials" derived from this theory come equipped with a sizeable number of adjustable "coupling constants" and "regularization parameters", which are not (for the most part) independently measurable. This introduces a latitude in the description which is highly unsatisfactory for a basic theory.

Moreover, this approach obscures our understanding of related phenomena, such as π production and absorption, and mesonic corrections to electromagnetic (EM) form factors. The reason is simply that the mesonic degrees of freedom are lost in constructing the effective "potential". The resulting predicament is well illustrated by attempts to marry field theory and nonrelativistic wave functions in calculating exchange corrections to the deuteron form factor, and the rather embarassing comparison to recent data at large q^2 reported by R. G. Arnold [2]. A similar situation could well arise in the near future when accurate data on π-N and π-nucleus scattering become available. My objective in this talk is to discuss an alternative approach suggested by recent developments in hadron scattering at high energies.

Historically, strong interaction field theory was constructed in imitation of the EM interaction, going back to the Yukawa postulate in 1934 that the force is mediated via the exchange of a massive analogue of the photon. The discovery of the π in 1947 was of course a major triumph, and the subsequent success of QED suggested an appropriate formalism. Physically, the concept of a quantized field is strongly related to the idea of point particles, in that the Wick argument $R < \hbar c/\Delta E$, relating the range R to the energy fluctuation ΔE, permits the exchange of an arbitrarily large number n of π's ($\Delta E = n\mu c^2$) providing that R can be arbitrarily small. This implies that the NN system is intrinsically a many-body system at small spacial separations, and a second-quantized field is a natural way of building in the essentially infinite degrees of freedom. On the other hand, if the nucleons had some finite intrinsic size (not due to the pion field), the number of π's would be finite and the field concept inappropriate.

A sizeable body of evidence has accumulated in the last several years which indicates that this is in fact the case. Thus, a great variety of direct experiments suggest that hadrons are in fact composite (e.g., made up of quarks), in addition to the indirect evidence of unitary symmetry. In fact, both EM and weak probes (deep inelastic electron scattering at SLAC and neutrino experiments at CERN) have given considerable support to the Gell-Mann-Zweig quark model. The relevance to nuclear physics has been noted by Neudatchin and coworkers [3], who observe that the concept of composite nucleons provides a simple explanation of the repulsive core seen empirically in NN (and other dihadron) scattering. The physics is quite simple; given that the constituents obey some exclusion principle, two clusters such as NN will resist interpenetration. The effect is thus to keep hadrons apart (they have a characteristic "size"). As I noted at Laval, this interpretation also provides a ready explanation of the approximate constancy of the logarithmic derivative (LD) of the NN wave function at $r_b = .7$ fm, and the relation of r_b to the hard core radius r_c. This empirical fact has led to a very successful phenomenology as developed by Feshbach, Lomon and collaborators [4].

*Work supported by the National Science Foundation, and the U. S. Energy Research and Development Administration.

This suggests a rather different picture of dihadron scattering, in which the primary effects arise from the core properties, but are obscured at low energies by the exchange of (at most a few) light mesons. Phenomenologically, the core-core behavior can be well represented by an effective radius and a constant LD, which can be determined empirically from NN data at $T_L \gtrsim 200$ MeV. Thus, to the lowest level of approximation, one takes

$$\begin{aligned}\psi_\ell(r) &= 0 \ , \qquad\qquad r < r_b \\ &= j_\ell(\kappa r) + ie^{i\delta_\ell} \sin \delta_\ell \, h_\ell(\kappa r) \ , \quad r > r_b \ ;\end{aligned} \tag{1}$$

with δ_ℓ determined by the boundary condition (BC) $(\psi_\ell'/\psi_\ell)_{r_b} = \lambda_\ell^c$. By suitably adjusting r_b and λ_ℓ^c one can fit δ_ℓ to the data at energies approaching the π-production threshold. However, the low energy behavior will be incorrect, and one will in general need to introduce an energy-dependent LD $\lambda_\ell(\kappa^2)$ of the form

$$\lambda_\ell(\kappa^2) = \lambda_\ell^c + \sum_i \frac{r_{\ell,i}}{\kappa^2 - \beta_{\ell,i}^2} \tag{2}$$

in order to obtain realistic phases. Even if this is done, of course, $\psi_\ell(r)$ will not be realistic except asymptotically.

It is clear that what this description lacks is the effect of meson exchanges at distances $r > r_b$. The approach of Feshbach and Lomon is to add meson theoretic potentials to represent 1- and 2-π exchange (since $r_b \simeq (2\mu)^{-1}$, this is presumably sufficient). This leads to many of the same problems noted above, since the exchange potentials cannot be unambiguously constructed and the pionic coordinates are lost. However, there is an alternate way to proceed. One may instead treat this as a coupled channel problem, in which the (virtual) pionic channels are taken explicitly into account. This means that at the next level of approximation one regards NN scattering as a special case of NNπ scattering in which the pion is only present in intermediate states. In practice, this requires that one employ a fully covariant description of NNπ scattering as a three-body problem, extract the part corresponding to NN initial and final states, and analytically continue this amplitude to energies below the threshold for actual pion production. Technically, one just identifies t_{NN} as the residue of the double pole arising in $T_{NN\pi}$ from a P_{11} πN combination at the N mass in both the initial and final states. This is formally equivalent to regarding the nucleon as an Nπ bound state, and the prescription is identical with that used in extracting Nd scattering from 3N $\rightarrow$ 3N. Actually, the same prescription arises in field theory, or S-matrix theory; the difference here is that $T_{NN\pi}$ is to be calculated on the basis of a three-body scattering theory, and not according to some set of field-theoretic diagrams.

In order to play this game one clearly requires a fully covariant three-body formalism capable of dealing with two-particle "interactions" characterized by boundary conditions. Fortunately, such a theory may be derived as an unambiguous generalization of the corresponding nonrelativistic formalism, as I have recently shown [5]. Despite the fact that the "interaction" is nonseparable, the corresponding equations reduce to one-dimensional form in a partial-wave decomposition, and hence are readily amenable to numerical solution. Preliminary applications to πd scattering and the ω 3π system were reported at Laval; more recent results include an analysis of the A_1 state of three pions [6].

The obvious first approximation is to use the nucleon as an s-wave spectator of the πN state which contains the nucleon pole (P_{11}), and the pion as a p-wave spectator of the appropriate NN s-wave ($^1S_0$ to drive the $^3S_1$ calculation, and $^3S_1$ to drive the $^1S_0$ calculation). After antisymmetrization in the nucleon variables, the equation takes the form

$$X_i(q_i') = \bar{K}_{i2}(q_i') + \sum_{j=1}^{2} \int_0^{Q_j} dq_j \, q_j^2 \, K_{ij}(q_i', q_j) \, X_j(q_j) \tag{3}$$

Here X_1 and X_2 represent series of pairwise rescatterings initiated by a πN pair at the nucleon pole; $X_1(X_2)$ corresponds to a final NN (πN) scattering. The variable q_j is the three-momentum of the spectator particle in the c.m. frame of the pair (j=1 corresponds to a spectator pion, j=2 to a nucleon). The Lorentz frame used to describe each pair configuration is uniquely specified by requiring that the pair remain in its own c.m. system when the spectator recedes to infinite distance. This reduces to the usual definition nonrelativistically, but introduces important kinematic effects in eq. (3). For three particles of mass m_α, m_β, and m_γ treated as free outside the region excluded by the cores, and using a real spectator momentum $q \geq 0$, the c.m. energy for the $\beta\gamma$ pair is

$$\left(m_\beta^2 + \kappa^2\right)^{1/2} + \left(m_\gamma^2 + \kappa^2\right)^{1/2} = \left(s + m_\alpha^2 q^2/M_\alpha^2\right)^{1/2} - \left(m_\alpha^2 + m_\alpha^2 q^2/M_\alpha^2\right)^{1/2}, \qquad (4)$$

with $M_\alpha^{-1} = m_\alpha^{-1} + (m_\beta + m_\gamma)^{-1}$, and $s = P^2$, the invariant four-momentum squared of the three-particle system. The upper limit on this energy, and hence on the energy where we need the two-body input for our equation, is achieved at $q^2=0$, while the lower limit, implied by the fact that eq. (4) can be satisfied only for $\kappa^2 \geq -\min\left(m_\beta^2, m_\gamma^2\right)$, fixes an upper limit $q=Q_\alpha$ (infinite only if $m_\beta = m_\gamma$).

Since the c.m. energy of the $\beta\gamma$ pair is bounded by $\sqrt{s} - m_\alpha$, any three-body treatment of the NN system requires two-body input always a pion mass below the two-body output to be computed. Thus, in order to calculate NN scattering near elastic threshold ($\sqrt{s} \simeq 2M$), we require only NN input for $-M^2 \leq \kappa^2_{NN} \leq -M_\mu (1-\mu/4M)$, and πN input in the narrow range $-\mu^2 \leq \kappa^2_{\pi N} \leq -\mu^2(1-\mu^2/4M^2)$; the immediate vicinity of the nucleon pole. As noted above, at any level of approximation one can partially account for neglected channels by employing energy-dependent LD's obtained from NN and πN scattering data and analytically continued to the required region (since λ_ℓ must be meromorphic, the extrapolation is essentially unique). Having obtained such fits for the NN system, it turns out that λ_ℓ has essentially achieved its asymptotic value (λ_ℓ^c) in the kinematic region required for a threshold calculation ($\kappa^2_{NN} \leq -M_\mu$). The NN input thus consists of the constant LD parameters taken to represent the core-core (quark) structure, and obtained empirically from the high energy NN phase shifts. Using $r_b = .7$ fm, we take $\lambda_0^c = 0.30$ for the $^1S_0$, and $\lambda_0^c = 1.8$ for the $^3S_1$ (from the $^3S_1 - {}^3D_1$ coupled channel fit of Feshbach and Lomon [7]).

The input for the P_{11} amplitude presents more of a problem, since the nucleon pole is only a pion mass below πN threshold and, in contrast to the NN situation, we are most sensitive to data up to about a pion mass above threshold, where they are poorly known. We know the position of the pole, and its residue can be inferred from the requirement that our three-body formulation yields the correct OPE singularity. The simplest fit (one pole term in eq. (2)) thus requires only a single parameter (in addition to the core radius $r_{\pi N}$), and it is possible to obtain quite reasonable fits to the P_{11} phase shift obtained by J. R. Carter [8] for $r_{\pi N} \simeq .2$ fm (approximately $\hbar c/M$). However, in view of the uncertainties in this phase we simultaneously consider several alternative fits, and also compare values based on $G^2 = 14.6$ and 15.3.

At this level of approximation all parameters are thus determined, and we may apply eq. (3) to calculate the low energy properties of the NN s-waves, and in particular the existence of bound states. The results of the calculation are given in Table I. We see that in spite of uncertainties engendered by the P_{11} amplitude, the most significant features of the NN s-waves—namely, two bound states close to zero in units of the π mass and split by approximately 2 MeV —are stably reproduced. Considering the simplicity of the model at this level and its close connection to empirical results found in quite different experiments, this close agreement with experiment ($\epsilon_d = 2.2$, $\epsilon_0 = -.07$) is quite remarkable [9].

Table I

$r_{\pi N}$ (fm)	G^2	ϵ_d (MeV)	ϵ_0 (MeV)
0.180	14.6	3.26	1.41
0.186	14.6	3.14	1.34
0.196	15.3	2.96	1.10
0.198	15.3	3.02	1.17
0.220	15.3	2.59	0.73

In order to go beyond these results one must work a bit harder. Clearly, there is no point in computing phase shifts until the scattering lengths are brought into agreement with experiment. By introducing a phenomenological term of range $(2\mu)^{-1}$ and adjusting its size to produce a singlet scattering length $a_s = -24.4$ fm, it is possible to produce an excellent prediction for the $^1S_0$ phase up to $T_L = 50$ MeV (i.e., the effective range and shape parameter are generated automatically). However, if a similar adjustment is made to produce a correct value for ϵ_d in the $^3S_1$ state the fit is not nearly as good ($a_t \simeq 4.6$ fm instead of 5.4 fm). The reason is that while coupling to neglected channels may indeed be represented by such a term, it is virtually impossible to guess energy-dependence to sufficient accuracy. This problem is much more acute in the triplet channel, which properly must be treated as a coupled $^3S_1 - ^3D_1$ system by including the nucleon as a d-wave spectator of the P_{11} πN state.

Unfortunately, the code used to produce these results was not sufficiently general to permit an investigation of additional three-body channels. Quite recently, a new covariant three-body code was completed which is suitable for this purpose. In order to complete a calculation in time for this conference, I have concentrated on the simpler $^1S_0$ state. Although a number of possible contributions were assayed (S_{11}, S_{31}, P_{31}), the only significant channel turned out to be P_{33} coupled to a d-wave nucleon spectator; this is suppressed by the d-wave character at low energy, but becomes important as one nears the π-production threshold. This is shown in fig. 1, in which the solid curve corresponds to the two-channel

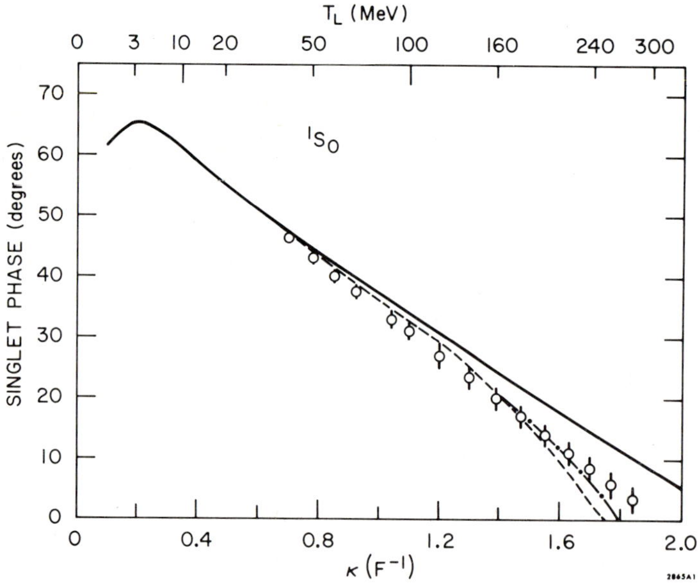

Fig. 1. Calculated $^1S_0$ phase (solid curve) for two channel model; experimental points from ref. [10]. Effects due to the P_{33} coupling (dashed curve) and energy-dependence in the $^3S_1$ (dashed-dot) are also shown.

model discussed above, and the dashed shows the effect of including the P_{33} (all are adjusted to produce $a_s = -24.4$). It is clear that this channel contributes a significant repulsion which brings the computed curve into good agreement with the experimental points of M. MacGregor [10] for $T_L < 200$ MeV (this role for the $\Delta(1236)$ was also noted in a simpler calculation by P. Haapakoski [11]). It should be emphasized that the curves shown are not a "fit", but an unadjusted theoretical prediction based on data from other experiments. It would almost certainly be possible to produce a high precision fit with minor adjustments of the parameters, but our purpose at this stage is merely to explore the general consequences of our dynamical picture. Fine details will require more effort in pinning down accurate input parameters, as well as including more channels. This becomes apparent in the figure as T_L approaches the π threshold. For example, the energy-dependence of the $^3S_1$ LD parameter begins to show up, as shown by comparing the dashed to the dashed-dot curve (which includes this effect). Nevertheless, it seems fair to conclude that such an embodiment of quark dynamics may provide an attractive alternative to field theoretic models.

REFERENCES

[1] A. D. Jackson, D. O. Riska and B. Verwest, Nucl. Phys. A249 (1975) 397.
[2] R. G. Arnold et al., Phys. Rev. Lett. 35 (1975) 776.
[3] V. G. Neudatchin, V. I. Kukulin, V. L. Korotkich and V. P. Korennoy, Phys. Lett. 34B (1971) 581.
[4] H. Feshbach and E. Lomon, Phys. Rev. Lett. 6 (1961) 635.
[5] D. D. Brayshaw, Phys. Rev. D 11 (1975) 2583.
[6] D. D. Brayshaw, Stanford Linear Accelerator Center SLAC-PUB-1677 (November 1975).
[7] E. L. Lomon and H. Feshbach, Ann. Phys. (N.Y.) 48 (1968) 94.
[8] J. R. Carter, D. V. Bugg, and A. A. Carter, Nucl. Phys. B58 (1973) 378.
[9] For more detail see D. D. Brayshaw and H. Pierre Noyes, Phys. Rev. Lett. 34 (1975) 1582.
[10] M. MacGregor, R. Arndt, and R. Wright, Phys. Rev. 182 (1969) 1714.
[11] P. Haapakoski, Phys. Lett. 48B (1974) 307.

ZERO RANGE COVARIANT THREE BODY EQUATIONS*

H. Pierre Noyes

Stanford Linear Accelerator Center
Stanford University, Stanford, California 94305

The fixed past-uncertain future program for a particulate quantum mechanics [1] maintains covariance by using only free particle wave functions. All change is to be generated by the appearance and disappearance of finite mass particles due to the Wick-Yukawa mechanism. This program is Democritean [2] in that it gives material significance only to particles and the void. In a deeper sense it may be characterized as Epicurean since the random least motion or "swerve" of Epicurus [3] is hauntingly reminiscent of quantum mechanical fluctuations. In order to make a dynamical theory out of this kinematic phenomenology [2], the first step is to obtain three particle dynamical equations which contain only two particle observables as input. Attempts to do so in the non-relativistic context [4, 5] have not been particularly successful. The technical problems have been solved [6, 7], but the resulting ansatz is by no means unique.

Fortunately Brayshaw has recently succeeded in articulating a closely related covariant three body phenomenology for hadrons starting from the singular core model [8]. The common feature is that, outside the two particle cores, the hadrons are free particles. But physically his model assumes that inside the cores there are "quark" degrees of freedom which to a surprisingly good approximation communicate with the free hadrons only through a boundary condition on the wave function at the core, which can be determined empirically by hadron-hadron scattering. Philosophically this empirical element introduces dimensional parameters which do not necessarily refer to free particle motion. If these parameters, which are now vaguely associated with phenomena described by "quarks" or "partons", can be given objective and quantitative significance, Brayshaw's model may prove to be a very fruitful basis for further research. However, we still believe it is worth pursuing the aim of creating a physics in which the only dimensional parameters are $\hbar$, c, and some finite stable particle mass [9]. To pursue this idea, all we need do is take the zero range limit of Brayshaw's model, which he has recently shown is well defined [10]. The difference between the two theories will then contain only finite modifications of the driving terms and kernels of the integral equations. A minimal accomplishment of our approach will then be to isolate the "quark" dynamics of Brayshaw's model from the empirical two-particle scattering input.

Rather than simply taking the zero range limit in Brayshaw's theory, we present here a direct derivation of the three body equations using a method developed by us in the non-relativistic context [11]. The method used is to apply the two-body scattering boundary condition at zero range to each of the pairs in the three body configuration space wave function. Essentially the same derivation is given by Brayshaw [8] applying the boundary condition at finite range, but we feel that his more abstract treatment tends to conceal the essential simplicity of the approach. Since he gives the general result for a finite number of angular momentum states, and also proves the on-shell unitarity of the resulting three body amplitude, we confine ourselves here to the case of three spinless particles with zero total and relative angular momentum.

The basic physical insight which allowed Brayshaw to make a simple generalization from the non-relativistic three body equations to the relativistic case was that we should use for each pair its own zero momentum coordinate system. Then, when the third (spectator) particle is removed to infinity, we are left with an (asymptotically, except in the zero range case) known two body scattering wave function. In the non-relativistic case, if we formulate the problem in the three body zero momentum system, Galilean invariance insures that the only effect of using these three different coordinate systems, other than a simple linear transformation between the relative momenta, is absorbed in the Galilean invariant phase factor exp $i(\underline{K}\cdot\underline{X} - Et)$, and hence can be ignored. In the relativistic case,

*Work supported by the U. S. Energy Research and Development Administration.

the Lorentz transformations to these three coordinate systems introduce additional kinematic factors, which we will exhibit explicitly below.

We start by writing the wave function in the zero momentum frame in terms of the coordinates of the α pair using the usual non-relativistic definitions: $\underline{p}_\alpha$ is the relative momentum of the pair with conjugate coordinate $\underline{x} = \underline{r}_\beta - \underline{r}_\gamma$, and $\underline{q}_\alpha$ the momentum of the spectator relative to that pair with conjugate coordinate $\underline{y}$. The incident plane wave has momenta $\underline{q}'_\alpha$, and $\underline{p}'_\alpha = \underline{k}_\alpha(q'_\alpha, s)$ where the on-shell value k_α is defined relativistically by

$$\left(k_\alpha^2 + m_\beta^2\right)^{1/2} + \left(k_\alpha^2 + m_\gamma^2\right)^{1/2} = \left(s + \left(\frac{Mq'_\alpha}{M-m_\alpha}\right)^2\right)^{1/2} - \left(m_\alpha^2 + \left(\frac{Mq'_\alpha}{M-m_\alpha}\right)^2\right)^{1/2} \quad (1)$$

$$M = \sum_\alpha m_\alpha \; ; \qquad s^{1/2} = M + W \; ; \qquad \mu_\alpha^{-1} = m_\beta^{-1} + m_\gamma^{-1} \; ; \qquad n_\alpha^{-1} = (M - m_\alpha)^{-1} + m_\alpha^{-1}$$

$$k_\alpha^2 \underset{v^2 \ll c^2}{\longrightarrow} 2\mu_\alpha\left(W - q_\alpha^2/2n_\alpha\right)$$

Here s is the invariant four-momentum squared. Note that only in this particular coordinate system does $\underline{x}$ correspond to the difference between the positions of m_β and m_γ. To calculate cross sections we must introduce the actual particle coordinates and make appropriate Lorentz transformations.

With this kinematic definition, we can now write down a three body wave function in terms of three T matrices (normalized in the single scattering approximation, to the two body amplitudes $\tau = e^{i\delta} \sin \delta/k$), and project out the zero angular momentum part. This is

$$xy\,\psi^{(\alpha)}(x,y) \equiv U^{(\alpha)}(x,y) = \frac{\sin k_\alpha(q'_\alpha, s)}{k_\alpha(q'_\alpha, s)}\,\frac{\sin q'_\alpha y}{q'_\alpha} + \int_0^\infty q\,dq \sin qy\, T_\alpha(q, q'_\alpha; s)\, e^{ik_\alpha(q_\alpha, s)y}$$

$$+ \sum_{\beta \neq \alpha} \frac{2}{\pi} \int_0^\infty q''^2\, dq''\, T_\beta(q'', q'_\beta; s) \int_0^\infty \frac{p''^2\, dp''}{p''^2 - k_\beta^2(q'', s)} \int_{-1}^{1} \frac{\sin P_{\alpha\beta} x}{P_{\alpha\beta}}\,\frac{\sin Q_{\alpha\beta} y}{Q_{\alpha\beta}}\, d\xi \quad (2)$$

Here $\xi = \hat{p}'' \cdot \hat{q}''$, and the only difference between this and the non-relativistic three body wave function (in the zero range theory) comes from the Lorentz transformation which relates $P_{\alpha\beta}$ and $Q_{\alpha\beta}$ in the α coordinate system to their values in the β system where T_β and the variables of integration are defined. These equations are

$$\left(m_\beta^2 + P_{\alpha\beta}^2\right)^{1/2} = \frac{m_\beta^2 + \epsilon_{\beta\beta}\epsilon_{\beta\gamma} \mp Mp''q''\xi/(M-m_\beta)}{\left[m_\gamma^2 + m_\beta^2 + 2\epsilon_{\beta\beta}\epsilon_{\beta\gamma} \mp 2Mp''q''\xi/(M-m_\beta)\right]^{1/2}}$$

$$\left(m_\beta^2 + Q_{\alpha\beta}^2\right)^{1/2} = \frac{\epsilon_{\beta\alpha}(\epsilon_{\beta\beta} + \epsilon_{\beta\gamma}) + p''^2 \pm Mp''q''\xi/(M-m_\beta)}{\left[m_\gamma^2 + m_\beta^2 + 2\epsilon_{\beta\beta}\epsilon_{\beta\gamma} \mp 2Mp''q''\xi/(M-m_\beta)\right]^{1/2}} \quad (3)$$

where ($\pm$) correspond to (α, β) as (cyclic, anticyclic) and

$$\epsilon_{\beta\alpha} = \left(m_\alpha^2 + p''^2\right)^{1/2};\; \epsilon_{\beta\gamma} = \left(m_\gamma^2 + p''^2\right)^{1/2};\; \epsilon_{\beta\beta} = \left(m_\beta^2 + (Mq''/(M-m_\beta))^2\right)^{1/2} \quad (4)$$

Since we have three ways to write the total wave function $U^{(\alpha)}$ depending on which value of α we choose, we can apply the boundary condition on the pairwise scattering separately to these three representations. All the spectator does is provide the additional momentum that allows the energy uncertainty principle to take the whole system off-shell, without affecting the on-shell scattering of the α pair. Hence we can define a one-variable function $U_q^{(\alpha)}(x)$ by

$$U_q^{(\alpha)}(x) = \frac{2}{\pi}\int_0^\infty \sin qy\ U^{(\alpha)}(x,y)\ dy \tag{5}$$

and require it to satisfy the zero range boundary condition

$$\lim_{x\to 0}\left[\frac{dU_q^{(\alpha)}(x)}{dx}\right]/U_q^{(\alpha)}(x) = \tau^{-1}(k^2) + ik = k\ \mathrm{ctn}\,\delta \tag{6}$$

We see immediately that the only term in U_q to survive at x=0 is the one containing T_α, and since the derivative of this term brings down the requisite ik, this boundary condition insures that T_α is proportional to $\tau(k^2) = e^{i\delta}\sin\delta/k$. We therefore take this factor out explicitly, and separate off the single scattering term by defining

$$T_\alpha(q,q';s) = \tau_\alpha(k^2(q,s))\left[\frac{\delta(q-q')}{qq'} + Z_\alpha(q,q';s)\right] \tag{7}$$

With this definition inserted in Eq. (2), our zero range boundary condition (Eq. (6)) therefore leads immediately to the three coupled zero range equations

$$Z_\alpha(q,q'_\alpha;s) = \sum_{\beta\neq\alpha}\frac{2}{\pi q}\int_0^\infty dq''\,K_{\alpha\beta}(q,q'';s)\left[Z_\beta(q'',q'_\beta;s) + \frac{\delta(q''-q'_\beta)}{q''q'_\beta}\right] \tag{8}$$

where

$$K_{\alpha\beta} = \tau_\beta\left(k_\beta^2(q'',s)\right)\int_0^\infty \frac{p''^2q''^2dp''}{p''^2-k_\beta^2(q'',s)-i0^+}\int_{-1}^1 d\xi\,\frac{\delta(q-Q_{\alpha\beta}(p'',q'',\xi))}{Q_{\alpha\beta}(p'',q'',\xi)} \tag{9}$$

The kinematics of Eq. (1) restrict the singularity in the kernel in the unequal mass case to finite values of q'', since $k_\beta^2(q'',s)$ can never be more negative than $-\min(m_\gamma^2, m_\alpha^2)$; hence in discussing convergence we need only consider the case of three equal masses where q'' can run up to infinity. Just as in the non-relativistic case, the δ-function gives a θ function which restricts the range of the p'' integration. The upper limit $p_+(q'',q)$ is that root of

$$2p_+E_+ + p_+E'' - \frac{3}{2}q''E_+ = \frac{3}{2}q\sqrt{2m^2 + 2E_+E'' + 3p_+q''} \tag{10}$$

$$E_+^2 = p_+^2 + m^2; E'' = \frac{9}{4}q''^2 + m^2$$

which reduces to $q''/2+q$ in the non-relativistic limit. For $mE'' > 9q^2/2+m^2$, the lower limit is that root of

$$2p_-E_- + p_-E'' - \frac{3}{2}q''E_- = -\frac{3}{2}q\sqrt{2m^2 + 2E_-E'' + 3p_-q''} \tag{11}$$

which reduces to $q''/2 - q$, while for $mE'' < 9Q^2/2+m^2$, it is that root of

$$2p_-E_- + p_-E'' + \frac{3}{2}q''E_- = \frac{3}{2}q\sqrt{2m^2 + 2E_-E'' + 3p_-q''} \tag{12}$$

which reduces to $q'' - q/2$. The second change compared to the non-relativistic case is a kinematic factor, which reduces to unity in the non-relativistic case, and is

$$3\left[F(p'', q'', q) + \frac{1}{4}E_q^2 - E_q F^{1/2}\right]^{3/2} / 2E_q\left[F - \frac{1}{2}E_q F^{1/2}\right] \tag{13}$$

where

$$E_q^2 = \frac{9}{4}q^2 + m^2;\quad E_p^2 = p''^2 + m^2;\quad F = E_p E'' + E_p^2 + E_q^2/4$$

In the non-relativistic case the $p_\pm$ limits gives a logarithmic term that for q, q'' both large gives a $qq''/(q^2+q''^2)^2$ convergence factor, and $\tau(k^2(q'', s))$ goes like $1/q''^2$, either factor by itself making the integral equation convergent. It may be that the relativistic $p_\pm$ limits still provide sufficient convergence, but we have not yet been able to show this, and so far have to rely on convergence provided by τ. The difficulty here is that as q'' goes to infinity k^2 goes to the finite value $-m^2$, and we can only guarantee convergence by requiring that $\tau(-m^2) = 0$. This is unpleasant, as we have no physical argument for making this requirement.

For our overall program, the most interesting case is the bound state of three pions in the I=1, $J^P=0^-$ state. Brayshaw discovered that, using an empirical fit to π-π I=0 s-wave phase shifts with a core radius $r_c = \hbar/4m_\pi c$, there is only one such bound state, and that it lies close to the mass of the pion [8]. By requiring it to be exactly at the mass of the pion, he found that the integral equation develops a singularity, and that consistency requires that, for this core radius, the s-wave scattering length $-a_0^0 \approx 0.23\ \hbar/m_\pi c$, in not too bad disagreement with other estimates. Our aim is to see if we can achieve a comparable result in a zero range theory.

In order to make the crudest possible model for the pion bootstrap, we assume that the π-π phase shift rises to $\pi/2$ at 2π production threshold ($k^2 = 3m^2$) and then stays there, making the amplitude purely imaginary beyond that point. That at high energy the internal structure of the pion (or any other hadron) forces the wave function to have a zero derivative at the boundary condition radius is consistent with Neudatchin's ideas about internal structure, as Brayshaw has already point out [12]. We simply let this radius go to zero. This simple model in its crudest form is then

$$\tau_{\pi\pi}(k^2) = A/\left(\sqrt{1 - k^2/3m^2} - ik\right),\qquad 1/\sqrt{3} > Am = -a_0^0 m > 0$$

or (14)

$$k^2 < 3m^2:\qquad k \operatorname{ctn}\delta = \sqrt{1 - k^2/3m^2}/A$$

$$k^2 > 3m^2:\qquad \delta = \pi/2;\ \eta = \frac{\sqrt{3}\,mkA - \sqrt{k^2 - 3m^2}}{\sqrt{3}\,mkA + \sqrt{k^2 - 3m^2}}$$

Inserting this into our zero range equation, and requiring this to have a bound state at $s = m^2$ then gives an integral equation defining a_0^0 in terms only of the pion mass (and $\hbar$ and c). If this integral equation does not converge, we will have to make a subtraction at $k^2 = -m^2$ to make $\tau(-m^2)$ vanish, but we still achieve our first crude pion bootstrap. Whether this crude model will give a quantitative result comparable to Brayshaw's, or a value of a_0^0 that is unreasonable is still uncertain.

REFERENCES

[1] H. Pierre Noyes, Found. of Phys. 5 (1975) 37; see also SLAC-PUB-1351.

[2] H. Pierre Noyes, Found. of Phys. (in press); see also SLAC-PUB-1404.

[3] Elizabeth Asmis, "The Epicurean Theory of Free Will and Its Origins in Aristotle," Ph.D. Thesis, Cornell (1974).

[4] H. Pierre Noyes, "Three Body Forces" in Few Particle Problems, I. Slaus et al., eds. (North-Holland, Amsterdam, 1972), p. 122.
[5] H. Pierre Noyes, Czech. J. Phys. B24 (1974) 1205.
[6] H. Pierre Noyes, "Comments on the Minimal Three Body Equations," submitted to Phys. Rev. Letters; see also SLAC-PUB-1676.
[7] H. Pierre Noyes, "A Separable Dispersion-Theoretic Amplitude for Three Body Problems," submitted to Czech. J. Phys.; see also SLAC-PUB-1695.
[8] D. D. Brayshaw, Phys. Rev. D 11 (1975) 2583.
[9] Pierre Noyes, "Non-Locality in Particle Physics" in Revisionary Philosophy and Science, Rupert Sheldrake and Dorothy Emmet, eds. (Macmillan, London, in press); see also SLAC-PUB-1405 (rev. Nov 1975).
[10] D. D. Brayshaw, private communication.
[11] H. Pierre Noyes, "Zero Range Three Body Equations," SLAC-PUB-1519 (1974), unpublished.
[12] D. D. Brayshaw, Phys. Rev. D 10 (1974) 2827.

THE THREE BODY PROBLEM WITH ENERGY DEPENDENT POTENTIALS

Y.E. Kim*
Department of Physics, Purdue University,
W. Lafayette, Indiana, 47907 U.S.A.
C.M. McKay and B.H.J. McKellar
School of Physics, University of Melbourne,
Parkville, Vic., Australia

Recent investigations of the nucleon-nucleon potential [1] have emphasized that the energy dependence of the potential cannot be ignored. This makes it important to generalize the formulation of the many body problem, and in particular the three body problem, to handle such energy dependent potentials. Such a generalization is presented in this paper. We anticipate that this generalization will also be useful in extending three body models of nuclear reactions [2] to include energy dependent optical model potentials.

Writing the Faddeev equations [3] for the three body problem,

$$T^{(i)}(s) = t_i(s) + \sum_{j\neq i} t_i(s)G_o(s)\, T^{(j)}(s), \tag{1}$$

in which $T = \sum_i T^{(i)}$ is the three body T matrix, t_i is a two body matrix in the 3 particle space, and $G_o(s)$ is the propagator of the free three body system, we see that the two body t matrices $t_i(s)$ are required for a range of energies below the elastic threshold, where the energy dependence of the potential is usually not defined. This demonstrates the need for a generalization of these equations to include energy dependent potentials.

The key to our generalization is the study of the two body problem with energy dependent potentials made by two of us [4]. The complete set of eigenstates of the Hamiltonian, Ψ_a, defined by

$$H_{12}(E_a)\Psi_a = (T_1 + T_2 + V_{12}(E_a))\psi_a = E_a\Psi_a \tag{2}$$

do not form an orthonormal set [5]. We introduce the complementary biorthogonal set $\{\tilde{\Psi}_a\}$, such that

$$\langle\tilde{\Psi}_a|\Psi_b\rangle = \delta_{ab}$$

and

$$\sum_a |\Psi_a\rangle\langle\tilde{\Psi}_a| = \sum_a |\tilde{\Psi}_a\rangle\langle\Psi_a| = 1. \tag{3}$$

We can then define an energy independent but non Hermitian Hamiltonian operator

$$\mathcal{H}_{12} = \sum_a E_a|\Psi_a\rangle\langle\tilde{\Psi}_a| \tag{4}$$

which has eigenstates Ψ_a with the eigenvalues E_a. The generalization to three body space is straightforward. We simply consider

$$\mathcal{H}^{(3)} = T_3 + \mathcal{H}_{12} \tag{5}$$

as a Hamiltonian operator of the three body system, whose eigenstates are $\Psi_a^{(3)} = \Psi_a(1,2)\; X(3)$ with the eigenvalues $E_a + E_3 = E_a^{(3)}$. Using the corresponding Green's function

$$\mathcal{G}^{(3)}(z) = \frac{1}{\mathcal{H}^{(3)} - z} = \sum_a \frac{1}{E_a^{(3)} - z}|\Psi_a^{(3)}\rangle\langle\tilde{\Psi}_a^{(3)}|\,, \tag{6}$$

we can define the two body T matrix in the three body space off the energy shell as

$$t_3(z) = (H_o - z) + (H_o - z)\,\mathcal{G}^{(3)}(z)(H_o - z). \tag{7}$$

Since eq. (7) does not contain E_{12} it provides a unique continuation of the two body t matrix to any energy, whether or not we have a definition of $V_{12}(E_{12})$ at that energy. This $t_i(z)$ can now be used in the Faddeev equation (1) to achieve our aim--a unique consistent formulation of the three body problem with energy dependent potentials.

With this as the starting point, one can readily generalize other three body equations, for example those of Karlsson and Zeiger [6]. The generalized Karlsson-Zeiger equations (which are too lengthy to incorporate in this brief report) retain their desirable properties of being free from primary singularities and requiring only half-shell t matrices as input. However even this does not avoid the necessity of constructing the biorthogonal set of wave functions. They appear explicitly for the bound states and are implicit in the half off shell two body t matrices

$$\langle \vec{q}_\beta | t^{(-)}(q_\beta^2) | \vec{q}_\beta^{(1)} \rangle = \langle \tilde{\Psi}^{(-)}_{q_\beta} | (\mathcal{H}_\beta - T_1 - T_2) | \vec{q}_\beta^{(1)} \rangle. \quad (8)$$

References

1. G.N. Epstein and B.H.J. McKellar, Phys. Rev. D10 (1974)1005; K. Erkelenz, Phys. Reports 13 (1974) 191; A.D. Jackson, D.O. Riska and B. Verwest, Nucl. Phys. A249 (1975) 397; M. Lacombe, B. Loiseau, J-M Richard, R. Vinh Mau and R. de Toureil, Orsay preprint (1975).
2. P.E. Shanley, Ann. Phys. (N.Y.) 44 (1966) 363; K. King and B.H.J. McKellar, Phys. Rev. Letters 30 (1973) 562; Phys. Rev. C9 (1974) 1309.
3. For a recent review of three body equations and the three nucleon problem, see Y.E. Kim and A. Tubis, Ann. Rev. Nucl. Sci. 24 (1974) 69.
4. C. McKay and B.H.J. McKellar, to be published.
5. A direct perturbative proof of the completeness of Ψ_a for a wide class of non-hermitian potentials is given by Y.E. Kim and A. Tubis, to be published; see also Y.E. Kim and A. Tubis, contribution to this conference.
6. B.R. Karlsson and E.M. Zeiger, Phys. Rev. D11 (1975) 939; see also Y.E. Kim and A. Tubis, Phys. Rev. D11 (1975) 947.

* Supported in part by the U.S. National Science Foundation.

GENERALIZED SEPARABLE EXPANSION METHOD OF THE TWO-BODY AND THE THREE-BODY SCATTERING AMPLITUDES

Shinsho Oryu and Toshihide Ishihara

Department of Physics, Faculty of Science and Technology, Science University of Tokyo, Noda-shi, Chiba 278 JAPAN

We propose a systematic method for obtaining new N-rank separable amplitudes of the two-body and the three-body equations. First of all, we will start from the Amado equation which is modified from the three-body Faddeev equation by using the two-body Yamaguchi potential for the nucleon-nucleon interaction[1]. It is well known that the Amado equation can be integrated on the real axis because the kernel has a logarithmic cut on the real axis. However, we found a separable three-body form factor which is regular on the real axis except for the cut. Therefore we can give a separable formalism for a solution of the partial wave Amado equation as follows

$$X_{\beta\alpha}(q',q;E)=B_{\beta\alpha}(q',q;E)+\sum_{\delta}\int_0^\infty B_{\beta\delta}(q',q'';E)\tau_\delta(z'')X_{\delta\alpha}(q'',q;E)dq'' \quad (1)$$

$$=\sum_{\delta,\sigma}\sum_{I,J}^{N}\frac{\Theta_{IJ}(E)}{\Theta(E)}\Phi_{\beta\sigma}(q',k_J;E)\Phi_{\delta\alpha}(k_I,q;E)+\Gamma^{(N+1)}_{\beta\alpha}(q',q;E) \quad (2)$$

where Φ is a three-body form factor which satisfies the following integral equation:

$$\Phi_{\gamma\sigma}(\bar{q}'',k_J;E)=B_{\gamma\sigma}(\bar{q}'',k_J;E)+\sum_{\delta}\int_C B^{(N+1)}_{\gamma\delta}(\bar{q}'',\bar{q}''';E)\tau_\delta(\bar{z}''')\Phi_{\delta\sigma}(\bar{q}''',k_J;E)d\bar{q}''' \quad (3)$$

where $\{\bar{q}''\}$ and $\{\bar{q}'''\}$ are defined only on the contours C proposed by Ebenhöh[2]. In the kernek of eq.(3), $B^{(N+1)}$ is an (N×Channel-numbers +1)×(N×Channel-numbers+1) determinant which consists of $B_{\delta\sigma}(k_I,k_J;E)$, $B_{\delta\sigma}(k_I,q;E)$ and $B_{\delta\sigma}(q',k_J;E)$. These elements are ordered so as to minimize the norm of the integral kernel[3]. Φ on the real axis can be obtained by analytic continuation using the following equation:

$$\Phi_{\beta\sigma}(q',k_J;E)=B_{\beta\sigma}(q',k_J;E)+\sum_{\gamma}\int_C B_{\beta\gamma}(q',\bar{q}'';E)\tau_\gamma(\bar{z}'')\Phi_{\gamma\sigma}(\bar{q}'',k_J;E)d\bar{q}'' \quad (4)$$

Next, Θ is defined by $\det\{A_{\delta\sigma}(k_I,k_J;E)\}$ in which the elements are given as follows, (Θ_{IJ} is the IJ co-factor of Θ):

$$A_{\delta\sigma}(k_I,k_J;E)=B_{\delta\sigma}(k_I,k_J;E)-\sum_{\gamma}\int_D B_{\delta\gamma}(k_I,\hat{q}'';E)\tau_\gamma(\hat{z}'')\Phi_{\gamma\sigma}(\hat{q}'',k_J;E)d\hat{q}'' \quad (5)$$

where integral contour D can be defined on the real axis except for some poles of $\tau(z'')$ on the real axis for which some parameters of $\{k_I\}$ should be defined at poles of $\tau_\sigma(z'')$ ($\sigma=1,2,\ldots n;\ n\leqslant N$), and the others should be given in the region where $\mathrm{Re}(3k_I^2/4)>$(Break-up threshold:B) or at least in a limited region so as to minimize the kernel $B^{(N+1)}_{\delta\sigma}(E)\tau_\sigma(z'')$. It is easily shown that the lowest two-body bound pole (P in Figs. 1 and 2) or $k_1=2\sqrt{(E+\alpha^2)/3}$ in the three-nucleon system is a fixed point as was shown by Ebenhöh[2]. If there are some other poles like resonance poles or double poles (which appear in the case where $2\,|r_0|\gtrsim|a|$ which sandwich the real axis in the three-body elastic region, we should check and take them as the parameters. $\Gamma^{(N+1)}_{\beta\alpha}$ is a non-separable error function of the solution but is exactly zero in the cases of on-energy and half on-energy shells.

Finally, we can show that the Amado equation (1) is similar to the two-body Lippmann-Schwinger(L-S) equation. The L-S equation, however, has a simple analytic form as compared with the three-body case. In this case, the integral contour can be allowed on the real axis because we have no singularity on this axis except for a single pole which can be removed by our method. Then the partial wave two-body amplitude is given by the similar form to eq.(2) in which the channel subscripts are negrected[3]. Whenever we have positive energy E, the first fixed point(or the first parameter) is $k_1=k=\sqrt{E}$ and the second point k_2 is parameter depending upon the potential and the energy and which can be obtained by the χ^2-method. The third point k_3 is a parameter which depends upon the potential, energy and the second parameter. Therefore, we can find the N-1 parameters successively by using the χ^2-method. Some examples and parameters are shown below:

[Yukawa] e^{-x}/x, $\mu=0.714F^{-1}$; $k_2=3k$, $k_3=6k$, $k_4=1.5k$, $k_5=10k$

[Exponential] e^{-x}, $\mu=1.41\ F^{-1}$; $k_2=3.25k$, $k_3=5.69k$

[Gaussian] e^{-x^2}, $\mu=0.68\ F^{-1}$; $k_2=1.98k$, $k_3=3.25k$

[Reid Soft-Core(T=1)] $(-he^{-x}-1650.6e^{-4x}+6484.2e^{-7x})/x$, h=10.463MeV, $\mu=0.7\ F^{-1}$; $k_2=1.56k$, $k_3=24.2k$,

where E=20MeV and S-wave potentials in which the ranges are $\mu=x/r$. For the Yukawa potential, we can illustrate the relation between the second parameter k_2, energies and the partial wave parameters ℓ in Fig.3, where $k_2=6.5k\{(n_2-1)/(31-n_2)\}$ and n_1 corresponds to $k_1=k=\sqrt{E}$.

References

[1] R. D. Amado, Phys. Rev. 132 (1963) 485.
[2] W. Ebenhöh, Nucl. Phys. A191 (1972) 97.
[3] S. Oryu, Prog. Theor. Phys. 52 (1974) 550.

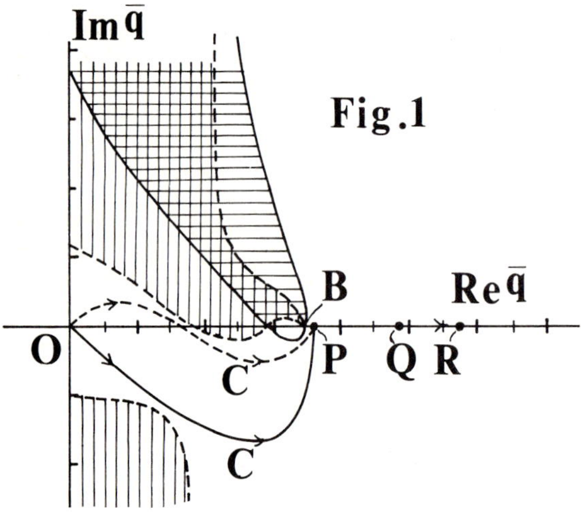

Fig.1

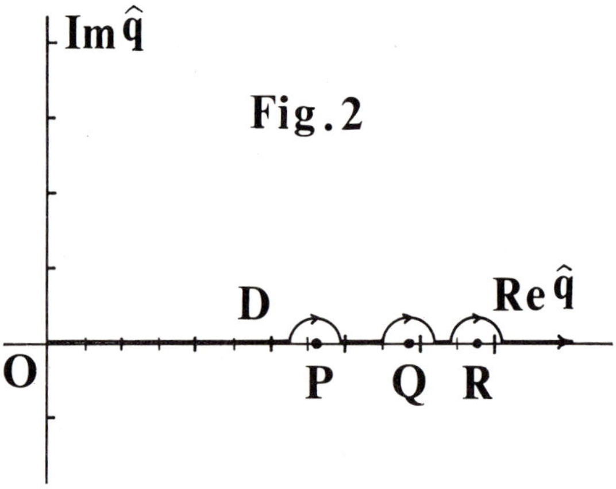

Fig.2

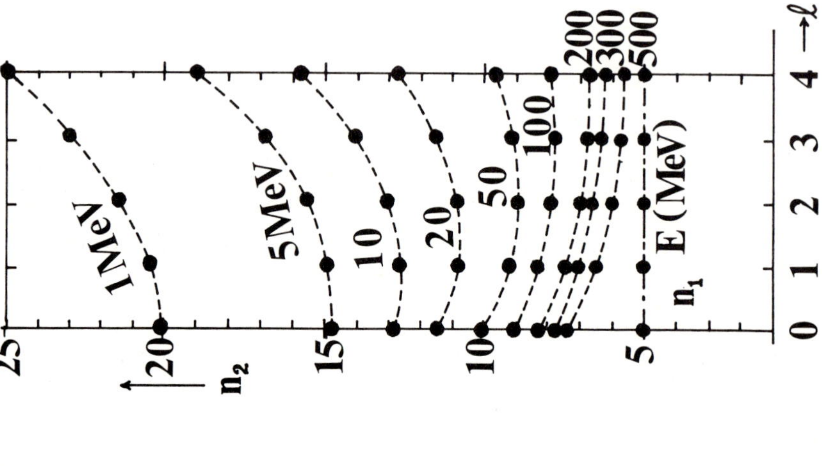

Fig.3

HOW DIFFERENT IS THE RESONATING-GROUP METHOD FROM THE INTEGRAL-EQUATION APPROACH TO FEW-PARTICLE SCATTERING ?

J. Schwager
Institut für Theoretische Physik der Universität Tübingen
D-74 Tübingen, Germany

In the last decade, a quantum mechanical scattering theory for few-particle systems has been developed which is based on exact integral equations. Of course, for more than three particles there is no hope for solving these equations directly on the computer. Besides, there is no interest in a complete solution as a function of 3N variables. It is important, therefore, to reduce the equations to effective two-particle equations. This can be achieved by approximating the kernels by pole terms which represent the dominant subsystem properties.

Such a procedure leads into the neighbourhood of conventional reaction theories as far as they are designed for a quantitative description of reaction phenomena from a microscopic point of view. There are essentially two theories of this kind: The shell-model reaction theory, which is well suited for the description of the scattering of a single particle from a heavy nucleus, and the cluster-model reaction theory or resonating-group method, which has been applied successfully in the calculation of scattering data of composite projectiles from light nuclei.

Unfortunately, little attention has been paid as yet to the relation between the integral-equation approach and the customary reaction theories. An attempt to throw some light on this connection was recently made by the author [1]. In the hope that such a comparision may be helpful for a further development of either approach, an outline of some essential points will be given in this contribution.

In conventional reaction theories much of the difficulty associated with a general solution of the many-particle scattering problem is circumvented by excluding channels in which the system is broken into more than two fragments. In this case, the correct boundary conditions to be imposed on the solution of the Schrödinger equation are intuitively clear and are easily met if this equation is solved in a linear space of functions which is restricted by these constraints right from the beginning of the calculation. Accordingly, in the resonating-group method a trial wave function is chosen of the following form:

$$\Psi_i = \sum_f \mathcal{A}[\varphi_f \cdot \chi_{fi}(\vec{R}_f)] + \sum_n a_{ni} D_n . \tag{1}$$

The sum over f extends over all possible and distinguishable two-cluster final states, while the antisymmetrizer $\mathcal{A}$ implies a sum over the indistinguishable ones. φ_f is a product of the bound-state wave functions of the clusters in channel f and $\vec{R}_f$ is their c.m. distance. The D_n are fixed (and correctly antisymmetrized) compound structures (distortion functions), the range of which is limited to the reaction region. The relative motion functions χ_{fi} must have the asymptotic behaviour

$$\chi_{fi}(\vec{R}_f) \xrightarrow[R_f \to \infty]{} \delta_{fi}\, e^{i\vec{q}_i \vec{R}_f} + T_{fi}(\vec{q}_f, \vec{q}_i)\, \frac{e^{iq_f R_f}}{R_f} . \tag{2}$$

These functions and the coefficients a_{ni} are determined by the set of equations which follows from the projection of $(H-E)\Psi = 0$ onto the basis states φ_f and D_n. If the variation of χ is allowed to be completely free, we obtain coupled effective two-body Schrödinger equations for these functions with nonlocal potentials arising from the exchange terms as well as from a formal elimination of the parameters a_{ni}. If, alternatively, χ is expanded into a discrete set of functions (Hulthén-Kohn procedure), one gets a system of linear algebraic equations for the coefficients (including T_{fi} and a_{ni}). In any case, this procedure amounts to a variational determination of the scattering amplitudes T_{fi}, which means that their error will be of second order in the error of the wave function. In addition, unitarity is satisfied if the K matrix formalism (standing waves) is used instead of the T matrix.

The connection with the integral equation approach can be established if the latter is also formulated in terms of wave functions. Except for the neglect of breakup channels, the splitting (1) of the wave function can be shown [1] to be identical with the Faddeev-Yakubovsky decomposition [2] in the asymptotic region when partial sums are taken over all components belonging to the same two-cluster partition. Each component can be defined by the application of a string of products of subsystem Green's functions and interactions on Ψ.

To avoid unnecessary complications, we will consider in the following the three-particle problem with only one bound state φ_α in each pair α. The Faddeev decomposition reads

$$\Psi_\gamma = \sum_{\alpha=1}^{3} \Psi_\gamma^{(\alpha)} \qquad \text{with} \qquad \Psi_\gamma^{(\alpha)} = G_0 V_\alpha \Psi_\gamma . \tag{3}$$

Below the breakup threshold, we have from (3) outside the reaction region (in an obvious notation)

$$\Psi_\gamma \xrightarrow[R\to\infty]{} \sum_{\alpha=1}^{3} \varphi_\alpha(\vec{r}_\alpha)\, \chi_{\alpha\gamma}(\vec{R}_\alpha) , \tag{4}$$

which is identical with the asymptotic form of (1) for the case considered here. (The same holds for indistinguishable particles.)

The Faddeev equations can be written as

$$\Psi_\gamma^{(\alpha)} = \delta_{\alpha\gamma}\Phi_\gamma + G_\alpha V_\alpha \sum_{\beta\neq\alpha} \Psi_\gamma^{(\beta)} , \tag{5}$$

where Φ_γ is φ_γ times a plane wave in $\vec{R}_\gamma$. Approximation of the subsystem Green's functions G_α by their pole terms (projectors on the two-body bound states) then implies that Ψ has the product form (4) in the whole configuration space, which coincides with the resonating-group ansatz in the "no-distortion approximation". In spite of this fact, it makes a great difference whether the Faddeev equations or the Schrödinger equation are solved in such a restricted function space. The difference does not arise from the fact that the equations (5) are integral equations (they are easily converted into a differential form), but from the fact that a restrictive assumption is made for each individual Faddeev component, and not only for the sum. The difference is obvious in the case of identical particles. The set (5) then reduces to only one independent equation which is not symmetric with respect to particle exchange, while the Schrödinger equation is. In contrast to the latter, the Faddeev equations, when solved in pole approximation, do not yield variational estimates for the scattering parameters, and test calculations on a three-boson system show [1] that much poorer results are obtained. (Note that for this model the resonating-group calculation with a clever choice of distortion functions gives practically the exact elastic phase shifts for energies not too high above the breakup threshold.)

However, the usual and successful method of pole approximation consists in replacing the subsystem transition operators t_α in equations of the form

$$W_{\alpha\gamma} = \delta_{\alpha\gamma} V_\alpha \Phi_\gamma + t_\alpha G_0 \sum_{\beta \neq \alpha} W_{\beta\gamma} \qquad (6)$$

by their pole terms. Here, $W_{\alpha\gamma} = V_\alpha \Psi_\gamma$ is a quantity (like the transition amplitude) in which the wave function is cut off by a short-range interaction. This leads a bit further away from the resonating-group method. But it again means that one assumes a product form,

$$W_{\alpha\gamma} \simeq V_\alpha(\vec{r}_\alpha)\, \varphi_\alpha(\vec{r}_\alpha)\, \chi_{\alpha\gamma}(\vec{R}_\alpha) , \qquad (7)$$

and that the eqs. (6) are practically solved in a function space restricted by the requirement (7). This, however, is a weaker restriction than the previous one, because the wave function itself is given by

$$\Psi_\gamma = G_0 \sum_{\alpha=1}^{3} W_{\alpha\gamma} . \qquad (8)$$

In particular, therefore, three free particles are not excluded as intermediate or final states. (In the resonating-group method, they can only occur as intermediate states through the distortion functions.) The price one has to pay for this is the occurence of rather complicated singularities in the kernels of the integral equations (at least above the breakup threshold) since the free propagator G_0 in (6) is treated exactly.

The most successful approximation is obtained by the inclusion of a special energy dependent factor in the pole term of t_α. In this way, one gets a solution which is exact for a corresponding separable potential and, therefore, satisfies (off-shell) unitarity ("Unitary Pole Approximation").

Procedures have been proposed [3,4] of how to extend the pole approximation methods to more than three particles, as well as to include correction terms. According to the problem under consideration, different formulations may be useful. However, in cases which imply a competition with the resonating-group method, the following questions might be relevant: Can unitarity be preserved ? Can one profit from a variational procedure ? What configurations are possible as intermediate or final states ? Are properties of the compound system important, besides the subsystem properties ?

On the other hand, with the background provided by the exact integral equations, it seems worth while to attack the breakup problem in the framework of the resonating-group method from both a practical and a fundamental point of view.

References:

1. J. Schwager, preprint (1975).

2. L.D. Faddeev, Soviet Phys. JETP 12 (1961) 1014;
O.A. Yakubovsky, Soviet J. Nucl. Phys. 5 (1967) 937.

3. P. Grassberger and W. Sandhas, Nucl. Phys. B2 (1967) 181;
E.O. Alt, P. Grassberger, and W. Sandhas, JINR Report E4-6688, Dubna (1972); W. Sandhas, Czech. J. Phys. B25 (1975) 251.

4. Gy. Bencze, Nucl. Phys. A210 (1973) 568;
E.F. Redish, Nucl. Phys. A235 (1974) 82;
Gy. Bencze and E.F. Redish, Nucl. Phys. A238 (1975) 240.

FOUR-BODY EQUATION

Tatuya Sasakawa
Department of Physics, Tohoku University, 980 Sendai, Japan

1) Equivalence of Yakubovski equation and six coupled equations

The wave function of the four-body system is expressed referring to the final state interactions as

$$\Psi = \Psi_{12} + \Psi_{13} + \Psi_{14} + \Psi_{23} + \Psi_{24} + \Psi_{34} \quad . \tag{1}$$

Yakubovski decomposed Ψ_{12} as

$$\Psi_{12} = \Psi_{12}^{123} + \Psi_{12}^{124} + \Psi_{12}^{12,34} \tag{2}$$

$$\Psi_{12}^{123} = G_o t_{12}(\Psi_{13} + \Psi_{23}), \quad \Psi_{12}^{12,34} = G_o t_{12} \Psi_{34} \quad . \tag{3}$$

From these equations one obtains the Yakubovski[1] equation for Ψ_{12}^{123}, Ψ_{13}^{123} and Ψ_{23}^{123}

$$\Psi_{12}^{123} - G_o t_{12} \Psi_{13}^{123} - G_o t_{12} \Psi_{23}^{123} = G_o t_{12} (\Psi_{13}^{134} + \Psi_{13}^{13,24} + \Psi_{23}^{234}$$

$$+ \Psi_{23}^{23,14}) \quad \text{and cyclic permutations of 123.} \tag{4}$$

Karlson-Zeiger[2] have shown the equivalence of the Alt-Grassberger-Sandhas equation[3] to the Yakubovski equation.

Now, we note that

$$\Psi_{13}^{134} = G_o t_{13} (\Psi_{14} + \Psi_{34}), \Psi_{13}^{13,24} = G_o t_{13} \Psi_{24} \quad . \tag{5}$$

Then Eq.(4) is alternatinely expressed as

$$\Psi_{12}^{123} - G_o t_{12} \Psi_{13}^{123} - G_o t_{12} \Psi_{23}^{123} = G_o t_{12} G_o (t_{13} + t_{23})(\Psi_{14} + \Psi_{24} + \Psi_{34}). \tag{6}$$

Clearly, Eq.(6) is a natural extention of the three-body Faddeev equation. From these equations, we obtain

$$\Psi_{12}^{123} = G_o W_{12}^{123} (\Psi_{14} + \Psi_{24} + \Psi_{34}) \tag{7}$$

where W_{12}^{123} is fully connected with respect to 123. The suffix 12 shows the final interaction. Using Eq.(2), we obtain the equation

$$\Psi_{12} = G_o W_{12}^{123} (\Psi_{14} + \Psi_{24} + \Psi_{34}) + G_o W_{12}^{124} (\Psi_{13} + \Psi_{23} + \Psi_{34})$$

$$+ G_o W_{12}^{12,34} (\Psi_{13} + \Psi_{23} + \Psi_{14} + \Psi_{24}) . \tag{8}$$

This is the equation found by Mitra et. al.[4] and some other authors.[5)-8)] Since the use in actual problems of the Yakubovski equation and Eq.(8) is rather inconvenient, we present new equation which consists of seven coupled equations.

2) Seven coupled equations

We define ψ^{123} and $\psi^{12,34}$ by

$$\psi^{123} = \psi^{123}_{12} + \psi^{123}_{13} + \psi^{123}_{23} \quad , \psi^{12,34} = \psi^{12,34}_{12} + \psi^{12,34}_{34} \quad . \tag{9}$$

Then Eq.(1) is alternatively expressed as

$$\psi = \psi^{123} + \psi^{124} + \psi^{134} + \psi^{234} + \psi^{12,34} + \psi^{13,24} + \psi^{23,14} \quad . \tag{10}$$

Equating Eq.(1) and Eq.(10), we have

$$\psi_{14} + \psi_{24} + \psi_{34} = \psi^{124} + \psi^{134} + \psi^{234} + \psi^{12,34} + \psi^{13,24} + \psi^{23,14} - (\psi_{12} + \psi_{13} + \psi_{23} - \psi^{123}) \quad . \tag{11}$$

On the other hand, we obtain from Eqs.(5) and (9), another equation

$$\psi_{12} + \psi_{13} + \psi_{23} - \psi^{123} = G_o(t_{12} + t_{13} + t_{23})(\psi_{14} + \psi_{24} + \psi_{34}) . \tag{12}$$

Eliminating the left hand side of Eq.(12), the right hand side of Eq.(11) is expressed as

$$\psi_{14} + \psi_{24} + \psi_{34} = 1/[1 + G_o(t_{12} + t_{13} + t_{23})] . (\psi^{124} + \psi^{134} + \psi^{234} + \psi^{12,34} + \psi^{13,24} + \psi^{23,14}) \quad . \tag{13}$$

Thus, putting

$$F^{123} = W^{123}/[1 + G_o(t_{12} + t_{13} + t_{23})] , \tag{14}$$

we obtain the equation for ψ^{123} as

$$\psi^{123} = G_oF^{123}(\psi^{124} + \psi^{134} + \psi^{234} + \psi^{12,34} + \psi^{13,24} + \psi^{23,14}) . \tag{15}$$

Similarly, we obtain the equation

$$\psi^{12,34} = G_oF^{12,34}(\psi^{123} + \psi^{124} + \psi^{134} + \psi^{234} + \psi^{14,23} + \psi^{24,13}) , \tag{16}$$

where

$$F^{12,34} = W^{12,34}/[1 + G_o(t_{12} + t_{34})] .$$

Here F^{123} ($F^{12,34}$) is connected with respect to 123 (12 and 34 separately).

We can show that by Eqs.(15) and (16) we can treat the four-body problem involving any bound substructures.

1) O.A.Yakubovski, Sov.J.Nucle.Phys.5(1967),937.
2) E.O.Alt,P.Grassberger and W.Sandhas, in Few Particle Problems in the Nuclear Interactions,ed:I.Slaus et. al.(North-Holland, Amsterdam,1972) p.299.
3) B.R.Karlson and E.M.Zeiger,SLAC-PUB-1346(1973).
4) A.N.Mitra,J.Gillespie and N.Panchapakesan,Phys.Rev.140B(1965),1336
5) L.Rosenberg,Phys.Rev.140B(1965),217.
6) Y.Takahasi and N.Mishima,Prog.Theor.Phys.(Kyoto)34(1965),498,ibid. 35(1966)440.
7) V.A.Alessandrini,J.Math.Phys.7(1966)215.
8) T.Sasakawa,Phys.Rev.158(1967)1249.

K-MATRIX FORMALISM FOR THE FOUR-BODY PROBLEM AND BOUND STATE SCATTERING THEORY

V.K. Sharma
Department of Physics, Dharma Samaj College, Aligarh 202001, India

Recently some attempts have been made to obtain Faddeev-type equations for N particles. We mention, in particular, the formulation by Yakubovsky [1] and its two variants proposed by Faddeev [2] and Sandhas [4] and later on Karlsson and Zeiger [3] for four-body problem. Among these Karlsson-Zeiger (KZ) approach seems to be interesting , since it shares the attractive characteristic features of Faddeev and Lovelace-type formalism.

In the present paper, our objectives is to construct a K-matrix formalism for four-body scattering problem and to discuss wave function formalism for bound state scattering within the framework of KZ theory. Here we have used AGS-type [4] approach, which is based on a scheme for writing down three-body relations as matrix versions of two-body relations.

Let us consider that the T-operator for four-body system in a matrix form satisfy Lippmann-Schwinger (LS) equation

$$T = V + VG_0T \; . \qquad (1)$$

Similar to two-body problem, the four-body Green function G_0 is divided into two parts [5]

$$G_0 = \mathcal{G}_0 + ig_0 \; , \qquad (2)$$

where $\mathcal{G}_0 = \mathcal{P}(z-H_0)^{-1}$ and $g_0 = -\pi\delta(z-H_0)$. (3)

The splitting (2) allows us to write (1) into the form

$$T = K + Kig_0T = K + Tig_0K \; , \qquad (4)$$

where K-operator satisfies the LS-type equation

$$K = V + V\mathcal{G}_0K \; . \qquad (5)$$

The unitary condition is given by

$$T - T^{\dagger} = 2T^{\dagger} ig_0 T \; . \qquad (6)$$

We introduce following set of matrix notations for KZ operators:

$$\begin{aligned} &T = \{\tilde{T}^{\mu\nu}\} = \{\tilde{T}^{\mu\nu}_{\alpha\beta}(z)\}, \quad K = \{\tilde{K}^{\mu\nu}\} = \{\tilde{K}^{\mu\nu}_{\alpha\beta}(z)\}, \\ &V = \{\bar{\delta}^{\mu\nu}\tilde{T}^{\mu}\} = \{\bar{\delta}^{\mu\nu}\tilde{T}^{\mu}_{\alpha\beta}\}, \quad V = \{\bar{\delta}^{\mu\nu}\tilde{K}^{\mu}\} = \{\bar{\delta}^{\mu\nu}\tilde{K}^{\mu}_{\alpha\beta}\}, \\ &\Psi = \{\Psi^{\mu\nu}\} = \{\Psi^{\mu\nu}_{\alpha\beta}\}, \quad G_0 = \{\delta^{\mu\nu}G_0\} = \{\delta^{\mu\nu}\delta_{\alpha\beta}G_0(z)\}, \\ &\Phi = \{\Phi^{\mu}\delta^{\mu\nu}\} = \{\Phi^{\mu}_{\alpha}\delta^{\mu\nu}\delta_{\alpha\beta}\}, \quad \mathcal{G}_0 = \{\delta^{\mu\nu}\mathcal{G}_0\} = \{\delta^{\mu\nu}\delta_{\alpha\beta}\mathcal{G}_0(z)\} \; . \end{aligned} \qquad (7)$$

Using the above matrix notations, one can write LS equations(1) as Faddeev-type matrix equations

$$\tilde{T}^{\mu\nu} = \bar{\delta}^{\mu\nu}\tilde{T}^{\mu} + \sum_{\sigma}\tilde{T}^{\mu}\bar{\delta}^{\mu\sigma}G_0\tilde{T}^{\sigma\nu}, \qquad (8a)$$

$$\tilde{K}^{\mu\nu} = \bar{\delta}^{\mu\nu}\tilde{K}^{\mu} + \sum_{\sigma}\tilde{K}^{\mu}\bar{\delta}^{\mu\sigma}\mathcal{G}_0\tilde{K}^{\sigma\nu}; \qquad (8b)$$

or in more detailed,

$$\tilde{T}^{\mu\nu}_{\alpha\beta}(z) = \bar{\delta}^{\mu\nu}\tilde{T}^{\mu}_{\alpha\beta}(z) + \sum_{\gamma\subset\mu}\sum_{\sigma\supset\gamma}\tilde{T}^{\mu}_{\alpha\gamma}(z)\bar{\delta}^{\mu\sigma}G_0(z)\tilde{T}^{\sigma\nu}_{\gamma\beta}(z), \qquad (9a)$$

$$\tilde{K}^{\mu\nu}_{\alpha\beta}(z) = \bar{\delta}^{\mu\nu}\tilde{K}^{\mu}_{\alpha\beta}(z) + \sum_{\gamma\subset\mu}\sum_{\sigma\supset\gamma}\tilde{K}^{\mu}_{\alpha\gamma}(z)\bar{\delta}^{\mu\sigma}\mathcal{G}_0(z)\tilde{K}^{\sigma\nu}_{\gamma\beta}(z). \qquad (9b)$$

Similarly, the Heitler equations for various scattering states are written as,

$$\tilde{T}^{\mu\nu}_{\alpha\beta}(z) = \tilde{K}^{\mu\nu}_{\alpha\beta}(z) + \sum_{\sigma\supset\gamma}\sum_{\gamma\subset\sigma} \tilde{K}^{\mu\sigma}_{\alpha\gamma}\, i\,g(z)\, \tilde{T}^{\sigma\nu}_{\gamma\beta}(z), \tag{10a}$$

$$\tilde{T}^{\mu\nu}_{\alpha\beta}(z) = \tilde{K}^{\mu\nu}_{\alpha\beta}(z) + \sum_{\sigma\supset\gamma}\sum_{\gamma\subset\sigma} \tilde{T}^{\mu\sigma}_{\alpha\gamma}\, i\,g_0(z)\, \tilde{K}^{\sigma\nu}_{\gamma\beta}(z). \tag{10b}$$

The Heitler eqs.(10) are on-shell in nature, and are used to obtain on-shell unitary relations as

$$\tilde{T}^{\mu\nu}_{\alpha\beta}(z) - \tilde{T}^{\mu\nu}_{\alpha\beta}(z)^{\dagger} = 2\sum_{\sigma}\sum_{\gamma} \tilde{T}^{\mu\sigma\dagger}_{\alpha\gamma}\, i\,g_0(z)\, \tilde{T}^{\sigma\nu}_{\gamma\beta}(z). \tag{11}$$

Let us add two comments on practical utilization of the K-matrix formalism:
(i) Using angular momentum decomposition, the on-shell Heitler equations of the form (10) can be reduced to one-dimensional integral equations as given by Cahill [6].
(ii) The off-shell values of the T-matrix can be obtained using Heitler damping equations and unitarity relations following the method of Kouri and Levin [7].

Now, using above matrix notations, we can write the integral equations for bound state scattering wave-functions in T matrix as

$$|\Psi^{\mu\nu}_{\alpha\beta}\rangle = |\Phi^{\mu}_{\alpha}\rangle\,\delta^{\mu\nu}\delta_{\alpha\beta} + G_0(z)\sum_{\gamma\subset\mu}\sum_{\sigma\supset\gamma} \bar{\delta}^{\mu\sigma}\,\tilde{T}^{\mu}_{\alpha\gamma}(z)\,|\Psi^{\sigma\nu}_{\gamma\beta}\rangle, \tag{12a}$$

$$|\Psi^{\mu\nu}_{\alpha\beta}\rangle = |\Phi^{\mu}_{\alpha}\rangle\,\delta^{\mu\nu}\delta_{\alpha\beta} + G_0(z)\sum_{\gamma\subset\nu} \tilde{T}^{\mu\nu}_{\alpha\gamma}(z)\,|\Phi^{\nu}_{\gamma}\rangle. \tag{12b}$$

The bound state transition amplitude is given by

$$M^{\mu\nu} = \langle\Phi^{\mu}\bar{V}^{\mu}\Psi^{\nu}\rangle = \langle\Phi^{\mu}V^{\nu}\bar{\delta}^{\mu\nu}\Phi^{\nu}\rangle + \sum_{\sigma}\langle\Phi^{\mu}\tilde{T}^{\mu\sigma}\bar{\delta}^{\nu\sigma}\Phi^{\nu}\rangle. \tag{13}$$

This is further rewritten in terms of vertex functions ($\langle\Phi^{\mu}_{\alpha}V^{\mu}\Psi\rangle = \phi_{\mu}$, the vertex function of four-body subsystem similar to Lovelace three-body form factor) as,

$$M^{\mu\nu}_{\alpha\beta}(z) = \phi_{\mu}G_0(z)\phi_{\nu}\bar{\delta}^{\mu\nu} + \phi_{\mu}G_0(z)\sum_{\sigma\supset\beta}\tilde{T}^{\mu\sigma}_{\alpha\beta}(z)\bar{\delta}^{\nu\sigma}G_0(z)\phi_{\nu}. \tag{14}$$

Similarly, we proceed to discuss the bound state scattering in K-matrix formalism. The transition amplitude for bound state scattering is expressed in terms of KZ K-matrix as,

$$R^{\mu\nu}_{\alpha\beta}(z) = \langle\Phi^{\mu}_{\alpha}V^{\mu}G_0(z)V^{\nu}\Phi^{\nu}_{\beta}\rangle(1-\delta^{\mu\nu}) + \sum_{\sigma}\langle\Phi^{\mu}_{\alpha}V^{\mu}G_0(z)\tilde{K}^{\mu\sigma}_{\alpha\beta}(z)G_0(z)V^{\nu}\Phi^{\nu}_{\beta}\rangle. \tag{15}$$

It is important to note that the integral eqs. (14) and (15) for amplitude of bound state scattering can be used to obtain the physical scattering amplitudes on on-energy-shell. This wave-functions formalism for four-body problem suggests that the Lovelace-type bound state scattering amplitudes can be obtained in terms of vertex functions of the four-body subsystem.

REFERENCES

1. O.A. Yakubovskii, Sov. J. Nucl. Phys. 5 (1967) 937.
2. L.D. Faddeev, in Three Body Problem in Nuclear and Particle Physics, ed.: J.S.C. McKee and P.M. Rolph (North Holland, Amsterdam, 1970) p. 154.
3. B.R. Karlson and E.M. Zeiger, Phys. Rev. D9 (1974) 1761.
4. E.O. Alt, P. Grassberger, and W. Sandhas, in Few Particle Problems in the Nuclear Interaction, ed.: I. Slaus (North Holland, Amsterdam, 1972).
5. T.A. Osborn, Ann. Phys. (N.Y.) 58 (1970) 417.
6. R.T. Cahill, Nucl. Phys. A. 194 (1972) 599.
7. D.J. Kouri and F.S. Levin, Phys. Rev. C10 (1974) 2096.

INTEGRAL EQUATION FOR THREE- AND FOUR-NUCLEON PROBLEM IN RESONATING GROUP APPROXIMATION

V.K. Sharma
Department of Physics, Dharma Samaj College, Aligarh,202001, India

After the pioneering work of Faddeev on the formulation and solution of the three-body problem, several attempts have been made to obtain Faddeev-type equations for N particles. Among these, the Yakubovsky equations are the best, as their homogeneous set of equations are free from the nonphysical solutions [1]. The Faddeev-Yakubovsky (FY) equations for four-body problem have been presented by Faddeev [2].

In the three-body problem, the kernels of the Faddeev equations contain one-delta function which can be removed by using nonlocal separable potential to obtain "equivalent two-body" problem as in Mitra's formalism [3]. This formalism provides the concept of Wheeler's resonating group substructures[4]. Thus this method with nonlocal separable potential can be extended to more than three-body problem. However, in contrast to three-body problem, the kernels of the FY four-body equations contain two delta functions. These delta functions disappear after taking two iterations of the kernels. Recently, some techniques have been developed [5,6], for solving the FY equations with separable pair interactions, leading to a set of two-dimensional integral equations. However, in our opinion by using the resonating group structure of the wave functions and nonlocal separable potential, one can remove these delta functions to obtain a set of one-dimensional coupled integral equations of Fredholm type.

According to Wheeler's resonating group method [4,7] a completely antisymmetric total wave function of four-nucleon system may be written as,

$$\Psi^{ST} = \sum_{(ijk,\ell)} \sum_{\substack{t_{ij}\sigma_{ijk}\tau_{ijk} \\ \pi}} \psi_{t_{ij}}^{\sigma_{ijk}\tau_{ijk}}(\vec{k}_{ij}, \vec{p}_{ij,k})\, H_{\pi}^{ST}(\vec{q}_{ijk,\ell})\, \Big| \xi^{S\lambda_S T\lambda_T}_{t_{ij}\sigma_{ijk}\tau_{ijk}} \Big\rangle + \sum_{(ij,k\ell)} \sum_{\substack{t_{ij}t_{k\ell} \\ \pi}} \phi_{t_{ij}}(\vec{k}_{ij})\, \phi_{t_{k\ell}}(\vec{k}_{k\ell})\, F_{\pi}^{ST}(\vec{s}_{ij,k\ell})\, \Big| \xi^{S\lambda_S T\lambda_T}_{t_{ij}\, t_{k\ell}} \Big\rangle, \tag{1}$$

where $\psi_t^{\sigma\tau}$ and ϕ_t are the space part of the normalized wave functions for three- and two-nucleon bound states respectively; H_{π}^{ST} and F_{π}^{ST} are the scattering functions of the corresponding arguments; $|\xi^{S\lambda_S T\lambda_T}_{t\sigma\tau}\rangle$ and $|\xi^{S\lambda_S T\lambda_T}_{t_\alpha t_{\bar\alpha}}\rangle$ are the antisymmetric spin-isospin states, with conserving quantum numbers S,T of four nucleon systems. Substituting the form (1) for Ψ^{ST} into FY equations [5] following the resonating group method [7], we get a set of integral equations with nonlocal separable potential as:

$$H_{\pi}^{ST}(\vec{q}) = H_{0\pi}^{ST}(\vec{q}) + \left(\frac{3}{2\sqrt{2}}\right)^3 \sum_{t_\beta\sigma_b\tau_b} \int d\vec{q}\,' \langle t_\alpha\sigma_a\tau_a | K_{HH}^{t_\alpha t_\beta\sigma_b\tau_b} | t_\beta\sigma_b\tau_b\rangle H_{\pi}^{ST}(\vec{q}\,') + (-)^{\pi}\left(\sqrt{\tfrac{3}{2}}\right)^3 \sum_{t_\beta t_{\bar\beta}} \int d\vec{q}\,' \langle t_\alpha\sigma_a\tau_a | K_{HF}^{t_\alpha t_\beta\sigma_a\tau_a} | t_\beta t_{\bar\beta}\rangle F_{\pi}^{ST}(\vec{q}\,'), \tag{2a}$$

$$F_{\pi}^{ST}(\vec{q}) = F_{0\pi}^{ST}(\vec{q}) + \left(\sqrt{\tfrac{3}{2}}\right)^3 \sum_{t_\beta\sigma_b\tau_b} \int d\vec{q}\,' \langle t_\alpha t_{\bar\alpha} | K_{FH}^{t_\alpha t_{\bar\alpha} t_\beta t_{\bar\beta}} | t_\beta\sigma_b\tau_b\rangle H_{\pi}^{ST}(\vec{q}\,'); \tag{2b}$$

where

$$\langle t_\alpha \sigma_a \tau_a | K_{HH}^{t_\alpha t_\beta \sigma_b \tau_b} | t_\beta \sigma_b \tau_b \rangle = C_{t_\beta \sigma_b \tau_b}^{t_\alpha \sigma_a \tau_a} \Big[\int d\vec{P} \frac{U_{t_\alpha t_\beta}^{\sigma_a \tau_a}(\vec{P}, \vec{Q}_1; Q_1^2 - b) F_\pi^{\sigma_a \tau_a}(\vec{P}, B) F_\pi^{\sigma_b \tau_b}(\vec{Q}_2, B)}{(z + b - P^2 - q^2)} +$$

$$+ \sum_{t_\gamma} \iint d\vec{P} d\vec{K}' \frac{U_{t_\alpha t_\gamma}^{\sigma_a \tau_a}(\vec{K}', \vec{Q}_1; Q_1^2 - b) F_\pi^{\sigma_a \tau_a}(\vec{P}, B) F_\pi^{\sigma_b \tau_b}(Q_2, B)}{(z + b - P^2 - q^2)} \tau_{t_\gamma}(z - q^2 - K'^2) X_{t_\gamma t_\beta}^{\sigma_a \tau_a}(\vec{P}, \vec{K}'; z - q^2) \Big], \quad (3a)$$

$$\langle t_\alpha \sigma_a \tau_a | K_{HF}^{t_\alpha t_\beta \sigma_a \tau_a} | t_\beta t_{\bar\beta} \rangle = D_{t_\beta t_{\bar\beta}}^{t_\alpha \sigma_a \tau_a} \Big[\int d\vec{P} \frac{U_{t_\alpha t_\beta}^{\sigma_a \tau_a}(\vec{P}, \vec{R}_1; R_1^2 - b) F_\pi^{\sigma_a \tau_a}(\vec{P}, B) \phi_{t_{\bar\beta}}(\vec{R}_2)}{(z + b - P^2 - q^2)} +$$

$$+ \sum_{t_\gamma} \iint d\vec{P} d\vec{K}' \frac{U_{t_\alpha t_\gamma}^{\sigma_a \tau_a}(\vec{K}', \vec{R}_1; R_1^2 - b) \phi_{t_{\bar\beta}}(\vec{R}_2) F_\pi^{\sigma_a \tau_a}(\vec{P}, B)}{(z + b - P^2 - q^2)} \tau_{t_\gamma}(z - q^2 - K'^2) X_{t_\gamma t_\beta}^{\sigma_a \tau_a}(\vec{P}, \vec{K}'; z - q^2) \Big], \quad (3b)$$

and

$$\langle t_\alpha t_{\bar\alpha} | K_{FH}^{t_\alpha t_{\bar\alpha} t_\beta t_{\bar\beta}} | t_\beta \sigma_b \tau_b \rangle = 2 B_{t_\beta \sigma_b \tau_b}^{t_\alpha t_{\bar\alpha} ST} \lambda_{t_\beta}^{-1} \Big[\frac{\phi_{t_{\bar\beta}}(\vec{R}_1) F_\pi^{\sigma_b \tau_b}(\vec{S}_2, B)}{(S_1^2 + q^2 - z - b)} +$$

$$+ \int d\vec{x} \frac{\phi_{t_{\bar\alpha}}(\vec{x}) F_\pi^{\sigma_b \tau_b}(\vec{R}_2, B)}{(x^2 + q^2 - z - b)} Y_{t_\alpha t_{\bar\alpha} t_\beta t_{\bar\beta}}^{\pm}(\vec{x}, \vec{S}_1; z - q^2) \Big]. \quad (3c)$$

Here following notations are used

$$\vec{Q}_1 = \frac{3\vec{q}' + \vec{q}}{2\sqrt{2}}, \quad \vec{Q}_2 = \frac{3\vec{q} + \vec{q}'}{2\sqrt{2}}, \quad \vec{R}_1 = \sqrt{3}\frac{\vec{q}' - \vec{q}}{\sqrt{2}}, \quad \vec{R}_2 = \sqrt{3}\frac{\vec{q} - \vec{q}'}{\sqrt{2}}, \quad \vec{S}_1 = \sqrt{3}\frac{\vec{q}' + \vec{q}}{\sqrt{2}}, \quad \vec{S}_2 = -\sqrt{3}\frac{\vec{q} - \vec{q}'}{\sqrt{2}}, \quad (4)$$

with $B_{t_\beta \sigma_b \tau_b}^{t_\alpha t_{\bar\alpha} ST}$, $C_{t_\beta \sigma_b \tau_b}^{t_\alpha \sigma_a \tau_a}$ and $D_{t_\beta t_{\bar\beta}}^{t_\alpha \sigma_a \tau_a}$ as the spin-isospin recoupling coefficients whose values are obtained by using the explicit mattix representations of the symmetric groups S_4 [6]. The functions $X_{t_\alpha t_\beta}^{\sigma_a \tau_a}$ and $Y_{t_\alpha t_{\bar\alpha} t_\beta t_{\bar\beta}}^{\pm}$ of four-nucleon subsystems (i.e. 3+1, and 2+2 type) satisfy the eqs.,

$$X_{t_\alpha t_\beta}^{\sigma_a \tau_a}(\vec{P}, \vec{P}'; z) = U_{t_\alpha t_\beta}^{\sigma_a \tau_a}(\vec{P}, \vec{P}'; z) + \sum_{t_\gamma} \int d\vec{P}'' U_{t_\alpha t_\gamma}^{\sigma_a \tau_a}(\vec{P}, \vec{P}''; z) \tau_{t_\gamma}(z - P''^2) X_{t_\gamma t_\beta}^{\sigma_a \tau_a}(\vec{P}'', \vec{P}'; z), \quad (5a)$$

$$Y_{t_\alpha t_{\bar\alpha} t_\beta t_{\bar\beta}}^{\pm}(\vec{x}, \vec{x}'; z) = (-)^\pi W_{t_\alpha t_{\bar\alpha} t_\beta t_{\bar\beta}}(\vec{x}, \vec{x}'; z) + (-)^{\pi + S + T} \sum_{t_\gamma t_{\bar\gamma}} \int d\vec{x}'' W_{t_\alpha t_{\bar\alpha} t_\gamma t_{\bar\gamma}}(\vec{x}, \vec{x}''; z)$$

$$\times \tau_{t_\gamma}(z - x''^2) Y_{t_\gamma t_{\bar\gamma} t_\beta t_{\bar\beta}}^{\pm}(\vec{x}'', \vec{x}'; z); \quad (5b)$$

with

$$U_{t_\alpha t_\beta}^{\sigma_a \tau_a}(\vec{P}, \vec{P}'; z) = 2\left(\tfrac{2}{\sqrt{3}}\right)^3 \frac{g\left(\frac{1}{\sqrt{3}}\vec{P} + \frac{2}{\sqrt{3}}\vec{P}'\right) g\left(\frac{2}{\sqrt{3}}\vec{P} + \frac{1}{\sqrt{3}}\vec{P}'\right)}{z - 4/3(P^2 + P \cdot P' + P'^2)} \Lambda_{t_\alpha t_\beta}^{\sigma_a \tau_a}, \quad (6a)$$

$$W_{t_\alpha t_{\bar\alpha} t_\beta t_{\bar\beta}}(\vec{x}, \vec{x}'; z) = \frac{g_{t_\alpha}(\vec{x}) g_{t_\beta}(\vec{x}')}{z - x^2 - x'^2} \delta_{t_\alpha t_{\bar\beta}} \delta_{t_{\bar\alpha} t_\beta}. \quad (6b)$$

Here $F_\pi^{\sigma\tau}$ is called as the Spectator Function of three-nucleon system satisfies the integral equation (which can .be obtained using above method),

$$F_{\pi}^{\sigma_a \tau_a}(\vec{P}, z) = F_{o\pi}^{\sigma_a \tau_a}(\vec{P}, z) + \sum_{t_\alpha t_\beta} \int d\vec{P}' \frac{U_{t_\alpha t_\beta}^{\sigma_a \tau_a}(\vec{P}, \vec{P}'; P'^2 - b)}{(z - P^2 + b)} F_{\pi}^{\sigma_a \tau_a}(\vec{P}') . \quad (7)$$

For $n+He^3$ (or $p+H^3$) scattering; S=1, T=0, t=0, σ =3/2,1/2, τ =1/2, $F_{o\pi}^{10}$=0. $H_{o\pi}^{10}$ = $(2\pi)^3 \delta(\vec{q}-\vec{q}_0)$ and $z = q_0^2 - B + i0$, where B is binding energy of He^3 (or H^3). Similarly, for d+d scattering; S=2, T=0, t=0, σ =3/2, τ =1/2, $F_{o\pi}^{20} = (2\pi)^3 (\vec{q}-\vec{q}_0)$, $H_{o\pi}^{20} = 0$ and $z = q^2 - 2b + i0$, where b is the binding energy of deuteron. The homogeneous set of integral equations corresponding to eqs.(2) with $F_{o\pi}^{ST} = H_{o\pi}^{ST} = 0$, $z = -\varepsilon + i0$ ε is the total energy, describes the four-nucleon bound states.

The existence of very similar equations (7) for three-nucleon problem obtained by Mitra [3], Sitenko and Kharchenko [8] suggests that our method could be applied for 4- and N-nucleon systems, since the resonating group method which has been applied successfully in the calculation of scattering data of light nuclei is closely related to the method of pole approximation by which the exact 3-body integral equations can be reduced to effective two-particle (without spin-isospin) equations [9]. With respect to the 4-nucleon equations obtained by other authors [5,6] our equations have the following advantages. Firstly, they are superior than the pole approximation because the effect of other channels is also taken into account. Secondly, their inhomogeneous terms have a transparent physical meaning in terms of scattering mechanisms, i.e. inhomogeneous terms are plane waves. And, thirdly, the equations are one-dimensional with the kernels containing functions of two- and three-nucleon bound and scattering states. The numerical solution of these equations for four-body systems are in progress.

The author is grateful to Professor A.N. Mitra for his stimulating interest in the subject of this work.

REFERENCES

1. O.A. Yakubovsky, Sov. J. Nucl. Phys. 5 (1967) 937.
2. L.D. Faddeev, in Three Body Problem in Nuclear and Particle Physics,ed.: J.S.C. McKee and P.M. Rolph (North Holland, Amsterdam, 1970) p. 154.
3. A.N. Mitra, Nucl. Phys. 32 (1962) 529.
4. J.A. Wheeler, Phys. Rev, 52 (1937) 1083.
5. V.F. Kharchenkov and V.E. Kuzmichev, Nucl. Phys. A183 (1972) 606.
6. I.M. Narodetsky and I.L. Grach, Preprint ITEP-955 (1972) Moscow.
7. V.K. Sharma, Nuovo Cimento A23 (1974) 679.
8, A.G. Sitenko and V.F. Kharchenko, Nucl. Phys. 49 (1963) 15.
9. J. Schwager, contribution to this Conference.

ON A TIME-INDEPENDENT THEORY OF MULTICHANNEL

QUANTUM MECHANICAL SCATTERING

Colston Chandler

Department of Physics and Astronomy The University of New Mexico
Albuquerque, New Mexico, 87131 U.S.A.

A.G. Gibson

Department of Mathematics and Statistics The University of New Mexico
Albuquerque, New Mexico, 87131, U.S.A.

In the continuing search for computationally viable theories of few body scattering a number of variants of the Faddeev equations and their many body generalizations have been proposed [1-7]. The general method seems to be to develop equations for the scattering operators in the full N-body Hilbert space. Problems such as uniqueness of solution and connectedness of the kernels of the resulting integral equations are dispatched in this full Hilbert space. Finally, at the end of the calculation, the projections onto the asymptotic scattering states are made.

In this paper we report on a study in which the projections onto the asymptotic states are made at the beginning and kept throughout the calculation. We adopted this approach in the belief that it is physically more natural and in the hope that it would lead to simpler equations for the transition operators. This hope is not yet completely realized but we are optimistic.

The mathematical vehicle for our study is the two-Hilbert-space form of scattering theory [8]. In this theory the asymptotic channel subare $\mathcal{H}_\alpha \equiv P_\alpha \mathcal{H}$, where α denotes the channel, P_α denotes a projection operator, and $\mathcal{H}$ is the full N-body Hilbert space. Channel subspaces are combined into a direct sum Hilbert space $\mathcal{H}' \equiv \oplus \mathcal{H}_\alpha$ and a bounded linear operator $J:\mathcal{H}'\to\mathcal{H}$ is defined by $J\Phi = \oplus\phi_\alpha$. The scattering operator $S:\mathcal{H}'\to\mathcal{H}'$ is then given in the case of short range interactions by the formula

$$S-1 = \underset{\varepsilon\to 0}{\text{w-lim}}\ (-2\pi i)\int_\lambda\int_\mu dE'_\lambda\ \delta_\varepsilon(\lambda-\mu)T(\frac{\lambda+\mu+i\varepsilon}{2})dE'_\mu. \tag{1}$$

Here E'_λ is the spectral family of the "free" Hamiltonian H', $H'\Phi \equiv \oplus H_\alpha\phi_\alpha$, where the H_α are the channel Hamiltonians. The function $\delta_\varepsilon(x) \equiv (\varepsilon/\pi)(x^2+\varepsilon^2)^{-1}$ enforces energy conservation in the limit $\varepsilon\to 0$. The operator $T(z)$ can be any of several operators, for example the Alt-Grassberger-Sandhas [3] operator

$$T(z) \equiv (z-H')\{J^*(z-H)^{-1}J-(z-H')^{-1}\}(z-H'). \tag{2}$$

The derivation of equations of the Lippmann-Schwinger type for the operator $T(z)$ is relatively straightforward. The resolvent equation

$$J^*(z-H)^{-1} = (z-H')^{-1}J^* + (z-H')^{-1}V^*(z-H)^{-1}, \tag{3}$$

where $V^* = J^*H-H'J^*$, is combined with Eq. (2) to obtain

$$T(z) = (J^*J-1)(z-H') + V^*(z-H)^{-1}J(z-H'). \tag{4}$$

The problem now is to insert a J^* to the left of $(z-H)^{-1}$ in Eq. (4). One way of doing this is to insert $1=(JJ^*)^{-1}JJ^*$ to the left of $(z-H)^{-1}$ in Eq. (4). By referring to Eq. (2), one obtains the final equation

$$T(z) = Z(z) + V^*(JJ^*)^{-1}J(z-H')^{-1}T(z), \tag{5}$$

where $Z(z) = (J^*J-1)(z-H') + V^*(JJ^*)^{-1}J$.

That Eq. (5) has a unique solution for Im $z \neq 0$ is easily shown. To do this we must prove that the homogeneous equation

$$X = [J^*H - H'J^*](JJ^*)^{-1}J(z-H')^{-1}X \tag{6}$$

has only the trivial solution. Trivial manipulations of Eq. (6) lead to the equation

$$(z-H')\Pi Y = J^*(H-z)(JJ^*)^{-1}JY, \tag{7}$$

where $\Pi = 1 - J^*(JJ^*)^{-1}J$ and $Y = (z-H')^{-1}X$. We note that Π is a projection operator and that $\Pi J^* = 0$. It then follows that

$$(z- \Pi H'\Pi)\Pi Y = 0. \tag{8}$$

Since Π is self-adjoint, then $(z-\Pi H'\Pi)$is invertible and Eq. (8) implies that $\Pi Y = 0$. Equation (7) then reads

$$0 = J^*(H-z)(JJ^*)^{-1}JY, \tag{9}$$

from which it follows that $JY = J(z-H')^{-1}X = 0$. Substitution of this result into Eq. (6) yields the desired result $X = 0$.

We have asked if the foregoing theory can be generalized to include theories of the type proposed by Kouri and Levin [6] and by Tobocman [7]. This can apparently be done by introducing a nonsingular operator $W:\mathcal{H}'\to\mathcal{H}'$. The representation $1 = (JWJ^*)^{-1}JWJ^*$ is then inserted to the left of $(z-H)^{-1}$ in Eq. (4) and the calculation pursued as before. The proof of uniqueness fails, however, because the operator $\Pi = 1 - J^*(JWJ^*)^{-1}JW$ now appearing in Eq. (8) is not now self-adjoint for $W\neq 1$. Since an important ingredient of their theory is a choice of W different from the identity, an alternative proof is necessary if their approach is to be developed within our formalism. We are presently working on such an alternative proof.

About the problem of connectedness we can at present say only the most obvious things. We write out the kernel of Eq. (5) channel by channel

$$(V^*(JJ^*)^{-1}J)_{\beta\alpha} = P_\beta(H-H_\beta)(JJ^*)^{-1}P_\alpha. \tag{10}$$

For the three body problem the kernel operator is connected if β is not the free particle channel; the equation does not have to be iterated to obtain the connectedness. If β is the free particle channel, however, the kernel is not connected, thus presenting a problem which is not yet resolved, but upon which we have made interesting progress.

The only other theory we know that incorporates the channel projections in the basic equations is that of Karlsson and Zeiger [5] and Osborn and Kowalski [4]. To put the essentials of their theory in a two-Hilbert-space form we must use a different direct sum Hilbert space, $H'' = \oplus'\{P_\alpha H \oplus (1-P_\alpha)H\}$. The symbol $\oplus'$ here denotes a direct sum over all channels except the one in which all the particles are free. The "free" Hamiltonian is defined by $H''\Phi = \oplus'\{H_\alpha\phi_\alpha \oplus H_\alpha\psi_\alpha\}$ where $\Phi = \oplus'\{\phi_\alpha \oplus \psi_\alpha\}$. The operator $J'':H''\to H$ analogous to J is defined by $J''\Phi = \Sigma'\ (\phi_\alpha + \psi_\alpha)$, and a transition operator T' is defined in a manner analogous to Eq. (2). An analog of Eq. (4) is then derived, with V* replaced by $J''^*H - H''J''^*$. We now observe that for any ψ in H the β - channel component of $(J''^*H - H''J''^*)\psi$ is given by

$$\{(J''^*H - H''J''^*)\psi\}_\beta = \{J''^*(J''H'' - H_0 J'')W_\beta J''^*\psi\}_\beta \ , \tag{11}$$

where H_0 is the Hamiltonian for the free particle channel and W_β is a (singular) matrix with elements $(W_\beta)_{\gamma\delta} = \delta_{\gamma\delta}(1-\delta_{\beta\gamma})$. The operator J''^* is thus inserted into Eq. (10) and the calculation can proceed as before. The transition operators with the free particle channel as initial or final space are then constructed from the parts of T' that have the spaces $(1-P_\alpha)H$ as initial or final space.

The equations of Karlsson and Zeiger, Osborn and Kowalski are more complicated in their two-Hilbert-space form than those we have advocated. The benefit of this additional complexity is that the connectedness problems of the kernel are solved. The drawback is that the way to extend the theory to systems of more than three particles is not clear, a drawback we hope to circumvent with our techniques.

<u>References</u>

[1] L.D. Faddeev, <u>Mathematical Aspects of the Three-Body Problem in the Quantum Scattering Theory</u> (Israel Program for Scientific Translations, Jerusalem, 1965)

[2] O.A. Yakubovskii, Sov. J. Nucl. Phys. 5 (1967) 937

[3] E.O. Alt, P. Grassberger and W. Sandhas, Nucl. Phys. B2 (1967) 167 and P. Grassberger and W. Sandhas, Nucl. Phys. B2 (1967)181

[4] T.A. Osborn and K.L. Kowalski, Ann. Phys. (N.Y.) 68 (1971) 361

[5] B.R. Karlsson and E.M. Zeiger, Phys. Rev. D11 (1975) 939

[6] D.J. Kouri and F.S. Levin, Phys. Rev. C11 (1975) 352

[7] W. Tobocman, Phys. Rev. C9 (1974) 2466; Phys. Rev. C11 (1975)43

[8] C. Chandler and A.G. Gibson, J. Math. Phys. 14 (1973) 1328

EFFECT IN LEVEL SPECTRUM OF THREE RESONANTLY INTERACTING PARTICLES*

Vitaly Efimov
Leningrad Nuclear Physics Institute, Gatchina, Leningrad District, 188350, USSR

Introduction

At the Conference on Few Particle Problems in Nuclear Interaction (Los Angeles, 1972) Amado [1] has discussed in considerable detail the interesting effect in level spectrum of three particles, interacting via two-body resonant forces. It turns out that under certain conditions the series of three-particle levels which have rather peculiar features is produced; number of these levels can be large and even infinite. After the Los Angeles Conference there have been further investigations of this effect. Because of attention this problem attracts it seems useful to discuss physical patternof the effect, as it is seen today, five years after the first publication [2].

Properties of a physical system are usually calculated starting from given interparticle forces. But there are certain cases when the levels appear as a direct result of the existence of some other levels. The effect under consideration is an example of such a relation between two- and three-body levels. It is a rather rigid relation: it does not depend on actual form of two-body forces which are only assumed to be of short range; the effect exists whatever the forces are, if only they give rise to certain two-body levels.

From the theoretical point of view, somewhat unexpected conclusions follow as a result of the effect study. Firstly, it turns out that the short range of forces does not prevent formation of large number of spatially extended loosely bound states. This was usually believed to occur only for particles, interacting via long-range forces. Secondly, there proves to be possible for the number of levels to decrease very rapidly with the growth of two-body forces. This is also in contrast with ordinary beliefs.

From the experimental point of view, the effect is interesting mostly as a phenomenon which can exist in any quantum three-body system with short-range forces. Therefore it may be found in nuclei, elementary particles, molecules, etc.

Essence of the effect

We shall describe the effect through the example of a simple three-body system - three identical spinless neutral particles. Every pair of these particles is supposed to form a bound or virtual state with zero angular momentum and small energy e_0 (the precise formulation of what energies are considered to be small will be given later on). This level produces a series of three-particle levels numbering

$$N = -(2\pi)^{-1} \ln[|e_0| m r_0^2] \tag{1}$$

where m is the particle mass and r_0 is the range of the two-body forces ($\hbar=1$). This is our basic statement. Notice that at $e_0=0$ the number of the levels (1) is infinite.

* This paper was scheduled as an invited paper, but the author was not present at the Conference. This paper has been marginally edited - Editors.

Note the following properties of these levels all of which are symmetric under permutations and have $J^P = 0^+$: (i) their binding energies are small; (ii) spatial sizes are large (more than r_0). Levels with higher number (N) are bound less tightly. In fact the binding goes to zero and the size goes to infinity as N grows; (iii) there is a scaling in the spectrum.

Formula (1) appears to describe a somewhat unusual situation. To give a feeling, let us choose two-body forces so as to form a virtual level with small energy e_0 and then let the forces grow. Indeed e_0 first decreases, passes through zero and eventually grows corresponding to the energy of a bound state. According to formula (1), under these circumstances the number of three-particle levels at first increases (approaching infinity at $e_0 = 0$) and then drops sharply. Thus we are faced with a decrease of the number of levels as the strength of the force increases.

How does one reconcile this situation with the evident assertion that the binding energy of every level must increase with the strength of the force? The answer lies in the dependence of three-body energy E on the strength g of the two-body force. The latter has a critical value g_0 at which the two-body energy e_0 passes through zero, and above which (i.e., $g > g_0$) the three-body continuum sets in (See figure in ref. [2]). As Amado and Noble [3] have picturesquely pointed out, the threshold "overruns and devours" the levels one after another.

For a more quantitative description, let us specify energies which have been considered above to be small. Two-body levels were assumed to be shallow, i.e. $e_0 \ll 1/mr_0^2$. For example, if two-body forces can be represented by attractive potentials, the energy level has to be much less than the well depth. The three-body energy levels were assumed to satisfy the same condition. Three particles may also have deep states. They are not influenced by the two-body level. Three identical particles are known to have no more than several such states, if the forces are strong enough to form the first shallow two-body level. Producing the states (1), this level may be said to drastically change the spectrum of three particles. It is also worth stressing that the actual form and origin of forces is absolutely unessential for the above description of the effect which thus may be called a universal three-body law.

If a three-particle system is more complex, e.g., if particles have spins, charges, different masses, etc., then the pattern of the effect also becomes more complicated. We will discuss this matter below. However, the basic content of the effect remains essentially the same, viz, certain levels of two particles produce a series of three-body levels with the above characteristic features.

Now we give a qualitative explanation of the effect. It is founded on the fact that a shallow two-body 0^+-level produces an effective long-range attraction of three particles. As a result of this attraction many levels are formed with small energies, in the well known case of a particle moving in a long-range field.

How does long-range attraction of three particles arise? We explain the mechanism at first for a bound two-body level, whose radius is

$$a \sim 1/\sqrt{m|e_0|} \quad . \tag{2}$$

With a third particle approaching the bound pair up to distance a, an exchange becomes possible. One particle of the pair passes to the third particle and forms just thesame bound pair. This exchange process is allowed to be repeated many times. It is very important that the exchange three-body interaction produced in that way has a large radius a which grows without limit as $e_0 \to 0$.

To understand why this interaction may be attractive, consider the well known example of exchange forces. Let the masses of two particles be large; the third particle is assumed to have a bound state with each of the heavy ones. This is the simplest model of molecular bond. Interaction between heavy particles arises as a result of light particle exchange. If the latter particle is in a symmetric state, this interaction is known to be attractive; in an anti-symmetric state a repulsion is produced. This example suggests that an attraction may be generated by exchange, if the configuration of three particles is selected properly. Thus to obtain attraction in a system of three identical particles one has to select symmetric states. The form of exchange attraction is

$$U(R) \approx -\frac{1}{mR^2}, \tag{3}$$

where $R^2 = \frac{2}{3}(r_{12}^2 + r_{23}^2 + r_{31}^2)$ and $\vec{r}_{ik}$ are the relative coordinates of particles. At $R \sim a$ this attraction goes to zero, since the exchange interaction radius is equal to a. On the other hand there occurs a saturation at $R \sim r_0$. When all three particles approach distances $r_{ik} \sim r_0$ the forces of various pairs overlap and create attraction of about three times the well depth. In this case the interaction (3) remains at the value $\sim 1/mr_0^2$ for $R \lesssim r_0$.

The simple form (3) is caused by the fact that at $r_0 \ll R \ll a$ the characteristic distances between particles also satisfy the condition $r_0 \ll \bar{r}_{ik} \ll a$. Therefore, in first approximation one may put $r_0 = 0$ and $a = \infty$. As the result, all dimensional parameters which specify two-body forces disappear. In this situation the form of the interaction (3) seems to be quite natural. It is the simplest symmetric quantity of appropriate dimension, that can be constructed from relative coordinates of particles.

Until now we were considering the case of bound two-body levels. If the level is virtual, there is little change. Now the exchange occurs between the virtual states of pairs with interaction radius $(m/e_0)^{-1/2}$ as before. This is a characteristic distance of correlations between particles in pairs. The virtual and bound state cases can be both treated on an equal footing, if a two-body scattering length is used. This quantity appears to be just the radius of the pair correlations. For a bound pair the scattering length is equal to the pair size a, Eq.(2). In the case of a virtual pair, the scattering length is negative but its magnitude is $|a|$. Taking into account these properties, we can say that the radius of the "exchange" three-particle interaction (3) is $|a|$.

It is now easy to explain the effect. "Radial" motion of three particles (i.e. the motion changing the coordinate R) takes place in the attractive long-range field, Eq.(3), bounded by $R = r_0$ and $R = |a|$. The number of levels in such a field is known to be equal to $\sim \ln \frac{|a|}{r_0}$, which are just the levels (1).

For further discussion it is convenient to rewrite formula (1) in terms of the scattering length:

$$N \approx \frac{1}{\pi} \ln \frac{|a|}{r_0} \tag{4}$$

As long as the forces are not strong enough to form two-body shallow levels, there is no three-body long-range attraction ($|a| \lesssim r_0$). When the virtual shallow level is formed, the long-range tail (radius $\sim |a|$) increases with the strength of the force and produces a number $\sim \ln \frac{|a|}{r_0}$ of three-body states. At $a = \infty$ an infinite number of such states is produced. Then the shallow two-body level becomes bound and the radius of the long-range interaction gradually decreases, thus leading to a sharp decrease in the number of three-body levels. When the bound level is no longer shallow, the long-range tail disappears.

Now the features (i)-(iii) of our three-particle levels can be understood rather simply. Indeed, while moving in a long-range field, particles are mostly at large relative distances. Hence the spatial size of their states may be very large. Binding energies of such states are certainly very small. Besides, it appears to be natural for attraction to manifest itself first of all in states with highest possible symmetry and without centrifugal repulsion. For three scalar particles such states have positive parity. Thus, the quantum numbers 0^+_{sym} of the levels are favoured. As for the effect in states with other quantum numbers, we will discuss this matter later on. Last property of the spectrum, the scaling, emerges because the interaction $1/mR^2$ does not contain any dimensional parameters which could fix the scale of energy or length.

Proper mathematical treatments have been worked out to justify the description and explanation of the effect given above. Various approaches have been used. In ref. [2,4] the boundary condition method has been applied. It is known [5] that the shallow two-body levels may be taken into account by means of boundary conditions on three-particle wave function

$$\frac{\partial(r_{ik}\Psi)}{\partial r_{ik}} = -\frac{1}{a_{ik}} r_{ik}\Psi, \quad r_{ik} \to 0 \qquad (5)$$

where a_{ik} is the scattering length of the pair ik. Collision of each pair being possible, there are three such conditions. Besides pair collisions, a triple collision may happen. A proper boundary condition has been formulated for this case. Thus the mathematical problem reduces to solution of the free three-particle Schrodinger equation with the boundary conditions given. As the result, the interaction (3) is obtained and the above pattern of the effect is reproduced.

Another approach is that of the Faddeev equation [3]. Here the shallow level is taken into account as a pole of two-body scattering amplitude, the latter being a part of the Faddeev kernel. This pole term is shown to dominate the kernel at small momenta of three particles i.e. in states with large spatial size). The Faddeev equation with such a kernel gives $\sim \ln|a|$ bound states. The statement that there is an infinite number of levels at $e_0 = 0$ has been proved [6] as a strict mathematical theorem.

A third approach based on numerical calculations of three-particle states has also been employed.with two-body separable forces. Calculated trajectories of several levels as functions of force strength [7] for three identical spinless particles reveal the same general features described above.

To summarize, the pattern of the effect appears today to be rather well founded.

Influence of various factors

We divide the factors into three groups. In the first group there are those factors, which change the quantum numbers of states. Here are the results of analysis of these factors' influence.

I(i): Centrifugal forces. Angular momentum of all levels arising as the result of the effect in a system of three spinless particles is equal to zero. This is because the centrifugal barrier, being of long-range ($1/R^2$) form, overpowers the attraction (3) even for L = 1. The only exception is a molecular system of three particles where a rotational spectrum is formed (see discussion at the end of this section).

I(ii): Parity. For our levels there is relation between angular momentum and parity: $P = (-1)^L$. Hence, the quantum numbers of the levels are 0^+, with the exception of molecular system which may have $1^-, 2^+, \ldots$ etc. states.

I(iii): Symmetry. Our states have the highest symmetry which is possible in a given three-body system. For example, they are the symmetric ones for the case of three identical particles, and the Σ_g^+-states for a molecular system with two identical nuclei.

I(iv): Addition of particles. A general question may be asked whether a shallow n-particle level ($n > 2$) can produce the level series for an (n+1)-particle system in the same manner as above. The answer is negative [8] since there is no long-range attraction which is necessary for the effect's appearance. This is caused in turn by properties of a shallow n-particle level whose size unlike a two-particle level, does not grow without limit as the binding disappears.

II(i): Three-body forces. They have no influence on theeffect. The reason is that the three-body forces change interaction of three particles only at $R \lesssim r_0$ and bear no relation to the long-range attraction which is responsible for the effect.

II(ii): Shallow two-body levels with non-zero momentum. There is no effect in this case. Owing to the centrifugal barrier the size of the two-body state does not increase as $e_0 \to 0$. Hence, the long-range attraction cannot arise.

II(iii): Zero-momentum shallow two-body levels formed by short-range forces which have a barrier with low penetrability. This case is interesting because of a strong difference between the radius of forces and the usual effective range. It is the effective range that the formulas (1) and (4) imply [9].

II(iv): Excited shallow two-body levels. Here we have in mind the following situation. Let a shallow zero-momentum level be an excited state of two particles. Our three-body levels produced would in this case be unstable, channels of decay into pair in lower states and free particle being open, The three-body levels become resonances which in some cases are rather strongly pronounced.

II(v): Only two pairs have shallow levels. If one pair has no shallow level, the strength of the attraction (3) weakens.

The third group includes the factors concerning the parameters of the particles.

III(i): Charges. Coulomb forces being long-range strongly influence the effect. At large distances they should dominate over the attraction (3) because of their 1/R-law. Hence, the number of the levels produced by the attraction (3) is now finite in any case. For example, the estimate of the maximum number of such levels in a system of three identical particles is

$$N_{max} \approx \frac{1}{\pi} \ln \frac{a_c}{r_0} ,$$

where a_c is the Bohr radius (cf. (4)): $a_c = 1/mq^2$. As the particle's charge q grows, the radius of the region where the attraction (3) is effective decreases. At $q \sim (mr_0)^{-1/2}$ this radius becomes as small as r_0, and the effect disappears.

III(ii): Spins. The effect is influenced by the particle spin in two ways. On the one hand, the strength of the attraction (3) weakens because of possible non-existence of two-body shallow levels at some values of pair spins (there may be even no attraction at all for some sets of shallow levels). For this reason, the number of our three-particle levels is reduced in comparison to the spinless case. On the other hand, a given three-body spin can be composed of individual particle spins in several ways, so that in some cases a larger number of levels result [10]. The first tendency seems to prevail in nature. So the effect may be said to be less pronounced for particles with spin.

III(iii): Masses. Influence of particles' mass ratio is significant. As the ratio changes, so does the chracter of particle motion, and that leaves an imprint on the effect's pattern. There are three regimes of motion which gradually turn one into another, as the mass ratio changes, viz, identical particle regime, heavy center regime and molecular regime. The first regime has been described above in detail. Heavy center regime comes when: (a) masses of two particles are of the same order of magnitude, the third particle (the center) being much heavier; (b) there is no shallow level formed by two light particles. For example, let the light particles be noninteracting with each other Were the mass of the center infinite, this case would be that of two noninteracting particles in external field. There is nothing like our effect in the level spectrum of this simple system. But for any finite mass of the center the effect does exist. The number of levels is given by a formula similar to (4), but with a small coefficient before the logarithm:

$$N \sim \frac{m_L}{m_H} \ln \frac{|a|}{\varepsilon_0}$$

where m_L and m_H are the masses of the light particles and the center respectively.

This regime is unfaourable in view of extremely shallow two-body levels required to produce even several of our three particle levels.

On the other hand, in a molecular regime the number of our levels is considerable under rather weak restrictions. The molecular regime comes when (a) two particles are much heavier than the third, (b) the light particle forms shallow levels with both heavy ones. Here the interaction (3) is created by exchange of the light particle and represents the energy of a molecular term. Applying the molecular classification, the 0^+-levels (4) may be named as vibrational ones because they result from adiabatical motion of heavy particles in the molecular term field (3). The heavier the particles the more dense the vibrational spectrum is. This corresponds to growth of coefficient before logarithm in a formula similar to (4), as the m_H/m_L grows. For example, in the case of identical particles the number of our levels is

$$N \sim \left(\frac{m_H}{m_L}\right)^{1/2} \ln \frac{|a|}{\varepsilon_0}$$

Diatomic molecule formed in such way has somewhat unusual structure: nuclei of the molecule are moving within a large region, radius of which is determined by two-body scattering lengths.

Such a molecule also has rotational spectrum. It should be interpreted as the result of the effect in L^P-states other than 0^+. Series of the $1^-, 2^+, \ldots$-levels arise. The highest possible momentum is

$$L \sim \left(\frac{m_H}{m_L}\right)^{1/2}.$$

In conclusion, the effect takes place under a rather wide range of conditions. Influence of different factors is very significant, since many new features in the effect's pattern arise. However, the essence of the effect remains the same.

Effect in nuclei, molecules and continuous media

The effect was mentioned to manifest itself in any quantum three-body system, provided the proper conditions are realized. Consider a system of three nuclei. According to the general idea, one ought to choose the pair of nuclei which has shallow levels with zero angular momentum, and then search for our levels in a system of three such nuclei.

Coulomb forces between nuclei, which were said in the preceeding section to have a negative influence on the effect, restrict the mass region of its possible existence. An estimate gives the boundary of this region to be $A \lesssim 20$. It hardly can be expected to find very shallow two-nuclei levels, since their formation is in a certain sense a matter of chance. Assuming the two-nuclei scattering lengths to exceed the radius of internuclear forces by no more than ten times, we find from a formula of the kind of (4) that at most only one level of our structure is produced.

When we are predicting only one level instead of a level series, doubts may arise as to whether this level is indeed formed through our mechanism. Therefore it seems useful to collect here the level's properties that follow from the theory described:

i) Level is located near the threshold of the decay into three nuclei. The word "near" means here "no more than several MeV".

ii) Angular momentum and parity are 0^+.

iii) Radius of the state exceeds that of a usual nuclear state; this feature affects, e.g., the charge form factor.

iv) The state has a cluster structure. This is a complex of three well separated nuclei. They are just those very nuclei whose decay thresholds are near. Owing to this feature, the reduced three-particle width of the level is large.

v) Pairs of the clusters, while free, must have shallow levels with zero angular momentum. In these two-body states the clusters are also well separated. Hence, the fifth feature can be formulated in following way: in the two-nuclei subsystems there should exist the zero-momentum cluster states located near two-body threshold.

To give the representation of a concrete situation, consider the nucleus ^{12}C. Here the second excited level has the excitation energy 7.65 MeV and quantum numbers 0^+. We hypothesize it to be made up from three α-particles by means of our mechanism. What are arguments in favour? First of all, this level is located near the 3α-threshold (the distance is 0.4 MeV). Its quantum numbers are proper. Unfortunately, there is as yet no definite experimental information on the radius of this state. The level has large α-width (of the order of Wigner limit). The unusual structure of this level may be inferred in an indirect way by means of the observation that the shell model calculations give for it an excitation energy which is much larger than 7.65 MeV. As for the required shallow level in the system of two α-particles, this is the ground state of nucleus ^{8}Be. This state is believed to be well clusterized into two α-particles. Thus, our hypothesis on the structure of the 7.65 MeV level in ^{12}C seems to have some plausibility.

Besides a system of three nuclei, such levels may be expected to exist in systems of two neutrons + nucleus and Λ^0-hyperon + two nuclei.

However, a situation that is most favourable for the effect's manifestation does not appear to be realized in nuclear physics. We have in mind two heavy particles and a light one, with at least two particles being neutral. It occurs in atomic systems. A simple example of that kind is the negative ion of a diatomic molecule, treated as a three-body system: electron + two neutral atoms. The scattering lengths of the electron on these atoms being large, the level series near the threshold of the decay into two atoms and an electron (or an atom and an atomic ion) would be observed.

The low energy electron-atom scattering data are rather scanty [11]. As for experimental study of negative molecular ion spectra, this is yet at an initial state [12]. So in order to find the effect in molecular ions, more experimental data are needed. Besides the molecular ions, the effect may be observed in other atomic systems, for instance, in three- and many-atomic molecules.

For all cases considered above, it is not possible to menage a shallow level formation. We would like to note a case, which though somewhat speculative, is free from this defect. Consider the particles placed into a medium whose parameters (say, density) may be varied continuously. If the shallow levels of two particles were formed under variation of the medium parameters, three-particle complexes would arise through our mechanism. Their level spectrum changes as the medium parameters change. Realization of this possibility would be apparently the most comprehensive manifestation of the effect.

Thus, the effect is expected to be observed in many quantum systems. It should be stressed that there are yet no systematic investigations along these lines, since the main attention was focused in previous years on the study of the "physics" of the effect and the development of proper mathematical treatment. Further investigations of experimental spectra are likely to produce interesting results.

Conclusion

The main conclusion of this report is that the physical pattern of the effect is well understood today. However, for a more complete understanding of the phenomena arising in three-particle systems under the influence of two-body levels, it would be desirable that certain questions get their answers. Firstly, it is interesting to elucidate how the two-body levels influence the continuous spectrum of three particles (particle-pair elastic and inelastic scattering, three-particle resonances) at low energies [13,14]. Secondly, how does the effect influence the analytic structure of three-body scattering amplitude? Besides studying the manifestation of these phenomena in nature, it is important to develop a practical technique of calculations for arbitrary masses, spins, scattering lengths and other characteristics.

Task of experimental study should be, in our opinion, systematic measurements of scattering lengths and other parameters to find shallow two-body levels, and the search and investigation of the three-particle phenomena related to these levels.

The effect may be a prototype or starting point to find other forms of relationship between properties of various quantum systems.

REFERENCES

1. R.D. Amado, in: Few Particle Problems in Nuclear Interaction, ed. I. Slaus, S.A. Moszkowski, R.P. Haddock and W.T.H. van Oers (Amsterdam - London, North Holland; New York, Elsevier,1972) p. 254.
2. V. Effimov, Phys. Lett. 33B (1970) 563; Yad. Fiz. 12 (1970) 1080.
3. R.D. Amado and J.V. Noble, Phys. Rev. D5 (1972) 1992.
4. V. Efimov, Nucl. Phys. A210 (1973) 157.
5. L.H. Thomas, Phys. Rev. 47 (1935) 903;
   G.V. Skornyakov and K.A. Ter-Martirosian, Zh. Exp. Teor. Fiz. 31 (1956) 775.
6. D.R. Yafaev, Matematicheskii sbornik 94 (1974) 567.
7. A.T. Stelbovics and L.R. Dodd, Phys. Lett. 39B (1972) 450.
8. R.D. Amado and F.C. Greenwood, Phys. Rev. D7 (1973) 2517.
9. A.M. Dyugaev, Yad. Fiz. 18 (1973) 787.
10. A. Bulgac and V. Efimov, Yad. Fiz. 22 (1975) 296.
11. B. Bederson and L.J. Kieffer, Rev. Mod. Phys. 43 (1971) 601.
12. G.J. Schulz, Rev. Mod. Phys. 45 (1973) 423.
13. S.K. Adhikari and R.D. Amado, Phys. Rev. C6 (1972) 1484.
14. D.D. Brayshaw and R.F. Peierls, Phys. Rev. 177 (1969) 2539;
    I.J.R. Aitchison and D. Krupa, Nucl. Phys. A182 (1972) 449.

TRINUCLEON PROPERTIES WITH ONE-BOSON-EXCHANGE POTENTIALS

R.A. Brandenburg and P.U. Sauer
Technische Universität Hannover

R. Machleidt
Universität Bonn
Bonn, W.-Germany

The properties of the bound three-nucleon systems, ^3H and ^3He, have been the subject of considerable interest in the past several years [1] . The theoretical formulations of the problem are now well established, and the numerical methods for solving the equations are such that the wave functions for these nuclei can be calculated using the best existing nucleon-nucleon potentials. Until recently, the purely phenomenological potentials (such as the Hamada-Johnston [2] and the Reid soft-core [3] (RSC) potentials) gave the best fits to the two-nucleon scattering and bound state data, and most of the work on the three nucleon systems has assumed one of these potentials as the starting point of the calculation. There now exists, however, equally good fits to the two-nucleon data by one-boson-exchange (OBE) potentials [4] . As the OBE potentials have a stronger theoretical basis, it is of interest to compare their predictions of three-nucleon bound state properties to those of phenomenological potentials.

Presented here are the results of calculations with two OBE potentials, HM1 [5] and EHM' [6] , and the RSC potential. Implicit in the calculations is the assumption of charge independence in the nucleon-nucleon interaction and the neglect of the Coulomb potential, so that the wave functions of ^3H and ^3He are identical. The wave functions are obtained from a solution of Faddeev equations, in which only the $^1S_0$ and $^3S_1$- $^3D_1$ partial wave components of the potentials are assumed to be nonzero.

The two OBE potentials are fit to the same two-nucleon data except in the $^1S_0$ partial wave. There HM1 fits the experimental n-p scattering phase shifts while EHM' fits the Coulomb-subtracted p-p phase shifts of RSC. An additional feature of the two OBE potentials is their similarity in the $^3S_1$- $^3D_1$ partial waves. In Table 1 the low-energy effective-range parameters and the deuteron properties of HM1, EHM', and RSC are tabulated.

The physical observables of the bound trinucleon systems which are discussed here are the triton binding energy (E_T) and the ^3He charge form factor (CFF). The experimental value for E_t is 8.48 MeV, while the measured CFF has a diffraction minimum at $Q^2_{MIN} \sim 11.8$ fm^{-2}. The calculated values, listed in the order (RSC, EHM', HM1), are E_T = (7.0, 7.2, 7.5) MeV and Q^2_{MIN} = (13.9, 15.3, 15.5) fm^{-2} (the uncertainty in these values due to numerical approximations is estimated to be 0.1 MeV in E_T and 0.3 fm^{-2} in Q^2_{MIN}). Fig. 1 shows the CFF from all three potentials as well as the experimental data [7, 8]. While all three potentials give satisfactory results for the CFF for $Q^2 \lesssim 7.0$ fm^{-2}, they fail in the high momentum-transfer region where the predicted diffraction minima are too far out and the secondary maxima too low. The OBE potentials are in worse agreement with experiment in this region than RSC. This is not unexpected since work with phase-shift-equivalent potentials [9-11] indicates a correlation between increasing E_T and Q^2_{MIN}, which is the effect seen here since both OBE potentials give more binding than RSC.

The authors would like to acknowledge many helpful discussions with Dr. K. Holinde.

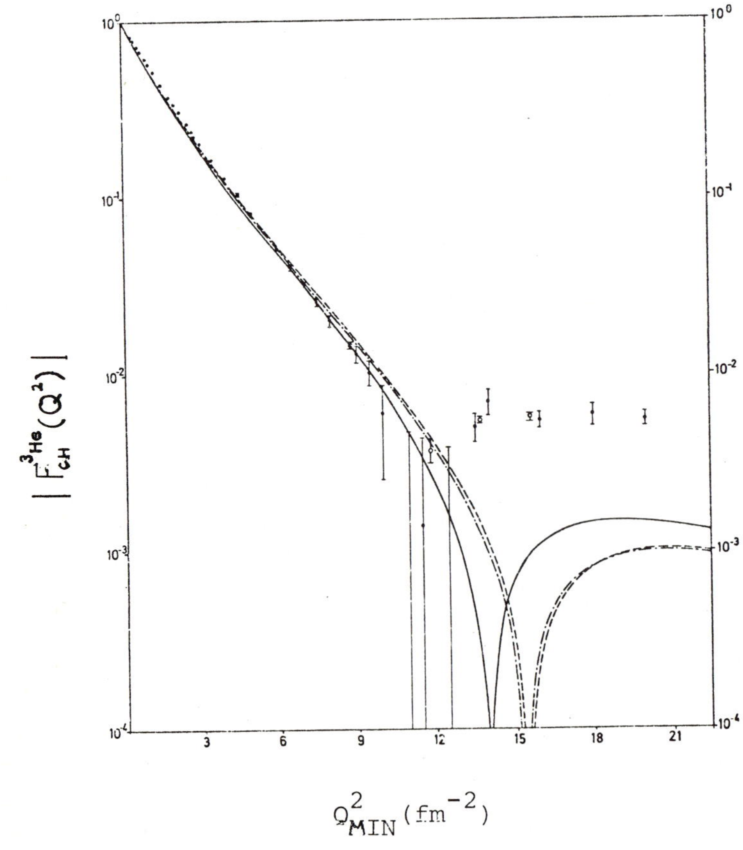

Figure 1

The ^{3}He charge form factors of RSC, EHM', and HM1 are represented by the solid, dot-dashed, and dashed curves, respectively. The solid data points are from ref. 7 and the open circles are data from ref. 8.

Table 1

The effective range parameters and deuteron properties predicted by the potentials RSC, EHM' and HM1 .

	HM1	EHM	RSC
a_s (fm)	-23.69	-17.17	-17.17
r_s (fm)	2.68	2.77	2.80
a_t (fm)	5.50	5.50	5.38
r_t (fm)	1.86	1.86	1.72
E_D (MeV)	-2.224	-2.2246	-2.2246
P_D (%)	5.75	5.71	6.47
Q_D (fm^2)	0.284	0.284	0.280

1. For a review of the three-nucleon problem, see Y.E. Kim and A. Tubis, Annu. Rev. of Nucl. 24, (1974) 69.
2. T. Hamada and I.D. Johnston, Nucl. Phys. 34, (1962) 382.
3. R.V. Reid, Ann. Phys. (N.Y.) 50, (1968) 411.
4. K. Erkelenz, Phys. Reports 13C, (1974) 191.
5. K. Holinde and R. Machleidt, Nucl. Phys. A247, (1975) 495.
6. The potential EHM' is an unpublished version of OBEB (II) of K. Erkelenz, K. Holinde, and R. Machleidt, Phys. Lett. 49B (1974) 209.
7. J.S. McCarthy, I. Sick, R.R. Whitney and M.R. Yearian, Phys. Rev. Lett. 25, (1970) 884.
8. M. Bernheim, D. Blum, W. McGill, R. Riscalla, C. Trail, T. Stovall, and D. Vinciguerra, Lett. Nuovo Cimento 5, (1972) 431.
9. E.P. Harper, Y.E. Kim and A. Tubis, Phys. Rev. C6, (1972) 1601.
10. M.I. Haftel, Phys. Rev. C7, (1973) 80.
11. P.U. Sauer and J.A. Tjon, Nucl. Phys. A216, (1973) 541.

THE TRINUCLEON SYSTEM BOUND BY A ONE-BOSON-EXCHANGE FORCE*

E. Harper
Physics Department
Indiana University, Bloomington, Indiana, USA.

In this paper we investigate the bound three-nucleon system when the N-N interaction is given by the ONE-BOSON-EXCHANGE potential (OBEP) of Holinde and co-workers.(1) Two distinct versions of the potential are used. These differ in their description of the short range force. In the first case (A) we use a phenomenological monopole vertex function in order to secure convergence of the unitarizing two-body Lippmann-Schwinger equation. In case (B) the more realistic vertex function obtained by summing in an eikonal approximation the crossed ladder diagrams and vertex contributions(2) (irreducible graphs) is used as a convergence generating form factor. Holinde et al.(1) have fitted the parameters of both these descriptions by obtaining quantitative fits to the N-N scattering and bound state data. In case (A) the exchanged bosons have the (J^π, I) quantum numbers $(0^-,1)\pi$, $(0^-,0)\eta$, $(0^+,1)\delta$, $(0^+,0)\sigma$, $(1^-,1)\rho$, $(1^-,0)\omega$, $(1^-,0)\varphi$ while in case (B) the φ exchange was not considered. In case (A) the effect of the monopole vertex function on the scalar and pseudoscalar propagators which occur in the OBEP, $V^\mu_{OBE}(\vec{p},\vec{q})$ is summarized in the following equation

$$\frac{\Lambda_1^2}{\Lambda_1^2-t}\left[\frac{1}{\mu^2-t}\right]\frac{\Lambda_2^2}{\Lambda_1^2-t} = \frac{\Lambda_1^2\Lambda_2^2}{(\Lambda_1^2-\mu^2)(\Lambda_2^2-\mu^2)}\left[\frac{1}{\mu^2-t} + \frac{\mu^2-\Lambda_1^2}{\Lambda_1^2-\Lambda_2^2}\frac{1}{\Lambda_2^2-t} - \frac{\mu^2-\Lambda_2^2}{\Lambda_1^2-\Lambda_2^2}\frac{1}{\Lambda_1^2-t}\right] \quad (1)$$

while the vector propagators require an additional Pauli-Villars renormalization of the propagator and assume the form

$$\frac{\Lambda_1^2}{\Lambda_1^2-t}\left[\frac{1}{\mu^2-t} - \frac{1}{\Lambda_0^2-t}\right]\frac{\Lambda_2^2}{\Lambda_2^2-t} = \Lambda_1^2\Lambda_2^2\left[\frac{1}{(\Lambda_1^2-\mu^2)(\Lambda_2^2-\mu^2)}\frac{1}{\mu^2-t} - \frac{1}{(\Lambda_1^2-\Lambda_0^2)(\Lambda_2^2-\Lambda_0^2)}\frac{1}{\Lambda_0^2-t} + \right.$$

$$\left.\frac{\mu^2-\Lambda_0^2}{\Lambda_1^2-\Lambda_2^2}\left\{\frac{1}{(\Lambda_1^2-\mu^2)(\Lambda_1^2-\Lambda_0^2)}\frac{1}{\Lambda_1^2-t} - \frac{1}{(\Lambda_2^2-\mu^2)(\Lambda_2^2-\Lambda_0^2)}\frac{1}{\Lambda_2^2-t}\right\}\right] \quad (2)$$

where μ and Λ_0 are the appropriate meson and renormalizing masses and Λ_1, Λ_2 differ from each other by a small amount and

$$t = (E_p - E_q)^2 - (\vec{p} - \vec{q})^2, \quad E_p = \sqrt{\vec{p}^2 + m^2}, \quad m = \text{nucleon mass.} \quad (3)$$

In case (B) it has been shown(2) that the effect of the procedure mentioned already is to multiply the "bare" OBEP by the factor

$$\text{Exp } 2i\,[\chi(t) + \chi(u) - \chi(u^2) - \chi(4m^2 - \mu^2 - s)]$$

where $u = (E_p - E_q)^2 - (\vec{p} + \vec{q})^2$ and $\sqrt{s}$ is the two-nucleon total energy. (4)

The function $i\chi(t)$ can be readily evaluated and is found to be

$$-i\chi(t) = \begin{cases} \dfrac{2\gamma(2m^2-t)}{\sqrt{t(4m^2-t)}}\tan^{-1}\sqrt{\dfrac{t}{4m^2-t}} & 0 \le t < 4m^2 \\[2ex] \dfrac{2\gamma(2m^2-t)}{\sqrt{t(t-4m^2)}}\ln\left[\sqrt{\dfrac{-t}{4m^2}} + \sqrt{1 - \dfrac{t}{4m^2}}\right] & t \le 0 \end{cases} \quad (5)$$

The parameter γ was fitted by Holinde et al(1) and found to have the value 1·25.

It is to be noticed that in case (A) the vertex function has not been normalized to the value unity at the point $t = \mu^2$ while the vertex function of case (B) has been so normalized. It is this normalization that introduces the explicit energy dependence into the potential.

The two-nucleon t-matrix satisfying the minimal requirement of relativity[1] that of relativistic unitarity, was obtained by solving a Lippmann-Schwinger equation and the results used to solve the Faddeev equations to find the trinucleon binding energy and wave function from which the He^3 charge form factor was obtained. Table I gives the potential parameters for case (A) and (B) and Table II gives the trinucleon results for both cases.

For case (B) it was found that if the natural interpretation for the two-body energy (that is the total energy in the three-body system at which two nucleons are interacting) is used then the potential becomes very weak at the energies necessary for the solution of the bound trinucleon. Thus for

$$S = (P-k)^2 = (3m - E_3 - \sqrt{4/3\, \vec{q}^2 + m^2})^2 - 4/3\, \vec{q}^2 \qquad (6)$$

E_3 = trinucleon binding energy, $\vec{q}$ = momentum of spectator nucleon
k = 4-momentum of spectator nucleon, P = Total 4-momentum in trinucleon C.M
we get a binding energy $E_3 \sim 3\text{-}4$ MeV.

The reason for the ambiguity in the interpretation of the parameter lies in the fact that s was introduced [1,3] on the basis of the relation

$$s + t + u = 4m^2 \qquad (7)$$

which is true only for on-energy-shell scattering if s is to be interpreted as the square of the total two-body C.M energy. For off-shell scattering s cannot be so interpreted. In this case (7) gives

$$s = 4E_pE_q \text{ (for on-mass-shell nucleons)} \qquad (8)$$

If this identification is made at sub-threshold two-body energies then the eikonal form factor does not damp the potential sufficiently. It should be remembered that normalizing the form factor so that the residue of the potential at $t = \mu^2$ is to be interpreted as the empirical coupling constant is necessary for the on-shell Born term only. The continuation off-shell is ambiguous. The two plausible interpretations $\sqrt{s_1} = 2m$ and $\sqrt{s_2} = 2m-2/3E_3$ are made. The fact that E_3 differs by so much for these two cases is an indication that the normalization used in fitting N-N scattering data gives rise to an inadequate sub-threshold description of the force between nucleons. It is possible to omit the normalization entirely but in this case the fitted parameters could not be compared with coupling constants obtained from sources other than N-N scattering. In the light of the difficulties encountered this may be the best line of approach.

Table I

Meson masses and coupling constants used in the potentials. Upper numbers are case (A), lower numbers case (B). All masses in MeV.

Parameter \ Meson	η	π	σ	δ	ω	ρ	φ
g^2	6.0	13.0	4.63	4.74	14.0	1.5	7.0
	2.0	14.0	5.68	0.71	10.0	0.5	-
f/g	-	-	-	-	0.0	3.5	0.0
	-	-	-	-	0.0	6.2	-
μ	548.5	138.5	500.0	960.0	782.8	763.0	1020.0
	548.5	138.0	520.0	960.0	782.8	711.0	-

$\Lambda_1 = 1945,\ \Lambda_2 = 1955,\ \Lambda_0 = 1250,\ \gamma = 1.25$

Table II

Calculated trinucleon results. (R is the ratio of the strength of the diffraction maximum in the He^3 charge form factor to the experimental value)

Potential \ Parameter	E_3(MeV)	$\langle r_{ch} \rangle$	Q^2_{min}	Q^2_{max}	R
A	6.75	1.70	≈16.8	≈22.8	8.8
B_1	6.77	1.97	~14.0	~18.4	7.2
B_2	7.48	1.98	≈13.7	≈18.2	7.6
Rsc (4)	6.98	1.96	~13.9	≈20.0	3.5
Expt	8.49	1.88	≈11.8	≈18.0	1.0

References

1. K. Holinde, K. Erkelenz, and R. Alzetta, Nucl. Phys. A194 (1972) 161.
K. Holinde, (Private Communication)
2. R. Woloshyn and A. Jackson, Nucl. Phys. A185 (1972) 131.
M. Islam, Nuovo Cimento 5A (1971) 315.
3. A. D. Jackson, D. O. Riska, and B. Verwest, (to be published).
4. E. P. Harper, U. E. Kim and A. Tubis, Phys. Rev. Lett. 28 (1972) 1533.
R. A. Brandenburg, Y. E. Kim and A. Tubis, Phys. Rev. C, 12 (1975) 1368.

* Work supported in part by NSF under grant No. 48-308-27.

THREE-BODY UNITARY TRANSFORMATIONS, THREE-BODY FORCES AND TRINUCLEON BOUND STATE PROPERTIES

Michael I. Haftel
Naval Research Laboratory
Washington, D. C. 20375 USA

Investigations in the three-nucleon (3N) bound state problem have shown that observables predicted purely from the nonrelativistic Schroedinger (or Faddeev) equations with reasonable two-body forces cannot predict all of the observed static and electromagnetic properties. Some have hypothesized that 3N forces are necessary while others cite the need for meson-exchange current corrections in the 3N form factors. In another contribution to this conference the author[1] has shown that by including three-body forces by a unitary transformation method, and by including meson-exchange currents, one can indeed fit the triton binding energy (E_T) and the charge form factors of ^3He, ^3H. This paper addresses the possible interaction dependence of the ^3He Coulomb energy (ΔE_c) and the interplay (if any) between likely 3N forces and possible charge asymmetry in the N-N interaction. The nuclear interactions considered, and the unitary transformations method of producing 3N forces, appear in ref.[1]. Table 1 gives E_T and ΔE_c for both a point proton and a proton charge distribution given by the experimental proton charge form factor. With the point charge ΔE_c scales roughly with E_T, at least for the two-body potentials. However, with the extended charge distribution values range only from .58 to .67 MeV. Figure 1 shows $F_{ch}(q^2)$ (^3He, ^3H) for each potential with exchange current [1], [2] corrections. Potentials S and GRPA give the best fits to the experimental $F_{ch}(q^2)$ and predict $\Delta E_c \approx .60-.63$ MeV. Other models give somewhat poorer fits and, significantly, give predictions of ΔE_c outside the range of .60-.65 MeV. Calculations with other potentials, including unitary transformation cases, also confirm that the electron scattering data favors $\Delta E_c \approx .60-.65$ MeV. These results agree with those of Friar and Fabre de la Ripell[3] even though they did not take exchange currents into account. The value of ΔE_c is largely determined from $F_{ch}(q^2)$ whether there is an explicit 3N force or not.

TABLE I

Triton Binding Energies and Coulomb Energies of ^3He

Potential	E_T a) (MeV)	ΔE_c(pt. charge) (MeV)	ΔE_c(ext. charge) (MeV)
Reid (R)	6.34 (7.0)	.608	.580
De Tourreil-Sprung A (S)	7.02 (7.64)	.640	.604
Graz Preliminary (GRP)	7.94 (8.5)	.735	.668
GRPA b)	7.94 (8.5)	.670	.625

a) See ref. [1]   b) GRPA has a 3N force described in ref.[1]

In conclusion, ΔE_c is fairly insensitive to the interaction employed to obtain the 3N wave function. With the further constraint of the e^- - ^3He and e^- - ^3H data, the present calculations confirm previous estimates of ΔE_c. The discrepancy between ΔE_c and the observed ^3H - ^3He energy difference ($\Delta E \approx .76$ MeV) indicates some charge-symmetry-breaking (CSB) in the N-N interaction. Given the sensitivity of 3N binding energies[4] to a_{pp} and a_{nn}, if the CSB were attributable to changes in the scattering length alone we would have $a_{pp} - a_{nn} \approx -5$ fm. However, estimates of CSB from elementary considerations[5] give $a_{pp} - a_{nn} \approx 1$ fm. Alternatively, given the sensitivity of 3N binding energies[4] to r_{nn}, r_{pp}, a CSB of $r_{nn} - r_{pp} \approx .1$ fm could account for the Coulomb anomaly in ^3H, ^3He. This explanation seems more likely and therefore a value of $r_{nn} \approx 2.75 - 2.8$ fm seems indicated.

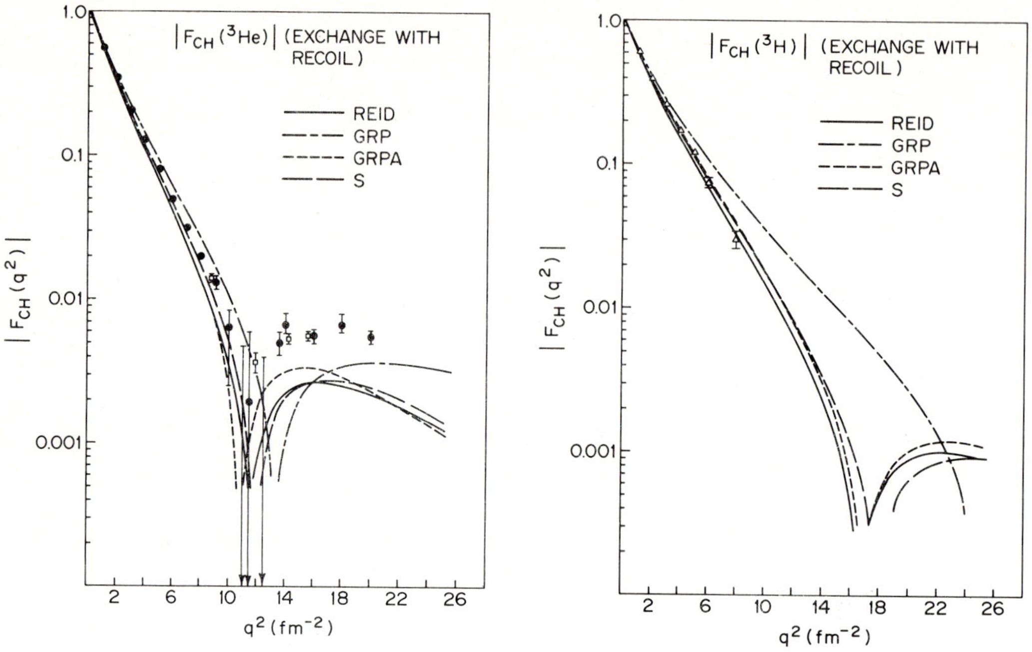

(a) (b)

Figure 1. Trinucleon charge from factors with exchange current corrections.

REFERENCES

[1] M. I. Haftel, this conference.

[2] W. M. Kloet and J. A. Tjon, Few Body Problems in Nuclear and Particle Physics (R. J. Slobodrian, B. Cujec and K. Ramavataram, Les Presses de L'Université Laval, 1975) 523.

[3] J. L. Friar, Nucl. Phys. A156, (1970) 43; M. Fabre de la Ripell, Fizika 4, (1971) 1.

[4] B. F. Gibson and G. J. Stephenson, Phys. Rev. C11 (1975) 1448.

[5] E. M. Henley and T. E. Keliher, Nucl. Phys. A189 (1972) 632; M. K. Banerjee, University of Maryland Technical Report No. 75-050 (1975).

BINDING ENERGY CALCULATION OF TRITON WITH REALISTIC POTENTIALS

Ismat Zakia and S. A. Afzal, Atomic Energy Centre, Dacca, Bangladesh.

In the past several binding energy calculations[1-3] were done in the group theoretical formalism with the states of triton and alpha particle obtained according to Englefield Selection rules[4] and reasonably good binding was obtained in each case. In all these calculations generally attractive forces of central and tensor components were applied. But we know that nucleon-nucleon scattering data demands the inclusion of repulsive core at small distances of order 0.4 - 0.5 fm. Among the several methods through which one can incorporate this effect in this type of calculation one of them is to use soft core nucleon-nucleon interaction. So in the present study we have calculated the binding energy of triton with the Englefield states with a series of potentials mostly soft core potentials. Further in the group theoretical formalism there appear a large number of states with different symmetries that may contribute to the binding energy of the triton ground state. Again for each state of lowest configuration in a particular symmetry several states of higher configuration mixing will come up. As a consequence a large number of states become involved and in general the Hamiltonian matrix elements become unmanagably large. In order to tackle this problem Englefield evolved a selection rule[5] by which states that contribute significantly to the binding energy may be found out. According to this selection rule all those states and their higher configuration mixing will contribute most to the binding energy which will give non-zero matrix element with the lowest configuration namely $|(0S)a\ (0S)b\rangle$ of the ground state of triton. In the earlier calculation with attractive nucleon-nucleon forces the reasonable binding energy has been obtained. So the present calculation will give an idea how well the Englefield selection rule works when one uses a realistic force.

Let $\vec{r}_1$, $\vec{r}_2$ and $\vec{r}_3$ be the position vectors of the three particles. We write $\vec{\rho}_i = \frac{\vec{r}_i}{a_o}$ with a_o as the oscillator well parameter so that $\vec{\rho}_i$ becomes a non-dimensional vector. In order to eliminate the motion of the centre of mass of the system we define a set of vectors

$$\vec{a} = \frac{1}{\sqrt{6}}(2\vec{\rho}_1 - \vec{\rho}_2 - \vec{\rho}_3)$$

$$\vec{b} = \frac{1}{\sqrt{2}}(-\vec{\rho}_2 + \vec{\rho}_3)$$

$$\vec{S} = \frac{1}{\sqrt{3}}(\vec{\rho}_1 + \vec{\rho}_2 + \vec{\rho}_3)$$

which are normalised and orthogonal to each other and $\vec{S}$ represents the centre of mass motion.

The space and spin-isospin wave function of triton have been constructed on the group theoretical method[6] in which the symmetric group S_3 has been employed. The irreducible representation of a symmetric group Sn are characterised by different possible partition $[\lambda] = [\lambda_1, \lambda_2, .., \lambda k]$ of n where $\lambda_1\ \lambda_2, \ldots, \lambda k$ are all positive integers and satisfy the condition $\lambda 1 + \lambda 2 + \ldots + \lambda k = n$ and $\lambda_1 \geq \lambda_2 \geq \ldots \geq \lambda k$. In the case of S_3 there are three possible partitions namely [3], [21] and [111]. The element in the sth row and rth column of a matrix representing the permutation P in the representation $[\lambda]$ is denoted by the symbol, $\theta^{[\lambda]}_{rs}(P)$ which satisfies the orthogonality relation

$$\sum_P \theta^{\lambda}_{rs}(P)\, \theta^{\lambda'}_{s'r'}(P^{-1}) = \frac{n!}{f^{\lambda}}\, \delta(\lambda,\lambda')\, \delta(r,r')\, \delta(s,s')$$

where P^{-1} is the inverse of P and $f^{[\lambda]}$ is the dimension of the matrix in the representation $[\lambda]$. The Young operators are defined by the relation $O^{[\lambda]}_{s,r} = \frac{f^{[\lambda]}}{n!} \sum_P O^{[\lambda]}_{rs}(P^{-1})\, P$ which satisfies the relation

$$O^{[\lambda]}_{s,r}\, O^{[\lambda']}_{s',r'} = \delta(\lambda,\lambda')\, \delta(s,s')\, O^{[\lambda]}_{r',r}$$

The Young operators belonging to the irreducible representation [3], [21] and [111] of the symmetric group S_3 have been expressed in a convenient factorial form by Jahn[6]. Now applying Young operator $O^{[\lambda]}_{r,s}$ to some suitable generating function, space function for a partition of $[\lambda]$ and dimension of r can be constructed as

$$\Theta_r^{[\lambda]} = \sum_s O_{r,s}^{[\lambda]} \phi$$

similarly the spin-isospin function $\Gamma_{\tilde{r}}^{[\tilde{\lambda}]}$ for the corresponding adjoint partition of $[\tilde{\lambda}]$ with the dimension of $\tilde{r}$ can be constructed.

In the case of triton Englefield S and D states are constructed following the partition [3] and [21] respectively. The basis function has been chosen as $\phi_{nlm}(a_0\vec{\rho})$ which is a stationary solution of the Schrodinger equation belonging to the eigenvalue $\frac{\hbar^2}{m a_0^2}(n+3/2)$ with potential $V(\vec{\rho})\, \frac{1}{2}\frac{\hbar^2 \rho^2}{m a_0^2}$. Then the orbital states of internal motion of a three particle nucleus are constructed from the products $\phi_{n_a l_a m_a}(\vec{a})\, \phi_{n_b \ell_b m_b}(\vec{b})$ vector coupled to a resultant angular momentum in L. The N quantum oscillator - well wave function of the internal motion for a three particle nucleus belonging to a particular representation $[\lambda]$ of S_n and characterised by the Yamanouchi symbol r can be written as $|(N)[\lambda]\, rLML\rangle$ $= \sum_{\substack{n_a l_a \\ n_b l_b}} |n_a \ell_a, n_b \ell_b; LML\rangle\, (n_a \ell_a, n_b \ell_b; L|[\lambda]\ rL\rangle$

with $n_a + n_b = N$ and $(n_a \ell_a, n_b, \ell_b; L\,|\,[\lambda]\ rL\rangle$ are the orbital (n/n-2,2) fractional parentage coefficients. Similarly spin-isospin function can be constructed. Then the totally antisymmetric N quantum states belonging to the representation $[\lambda]$ can be written as

$$\Psi((N)[\lambda]\ \ TM_T\ \ SM_S\ \ LM_L) = \frac{1}{\sqrt{f^{[\lambda]}}} \sum_r |(N)[\lambda]\ r_L\ \ M_L\rangle\, |[\tilde{\lambda}]\ \ \tilde{r}\ \ TM_T\ \ SM_S\rangle$$

In detailed notation a complete set of antisymmetric states for a system of three nucleons can be written as

$$|[\lambda]\,NSL\,JT\rangle = \frac{1}{\sqrt{f^{[\lambda]}}} \sum_{\substack{S_{23}T_{23}\\ n_a l_a n_b l_b}} |S_1 T_1, S_{23} T_{23}|\}[\tilde{\lambda}]\ \ \tilde{r}ST]$$
$$\times [n_a \ell_a, n_b \ell_b\ \ |\}[\lambda]\ \ rNL]\ \{|S_1 T_1, S_{23} T_{23}; ST\rangle\, |n_a \ell_a, n_b \ell_b; ML\rangle\}^J$$

In this way we have constructed S and D states with [3] and [21] symmetries respectively, which according to Englefield selection rule, contribute most. Regarding states of higher configuration mixing, states of upto eight quanta of excitation in both S and D have been taken into account.

In all, six different nucleon-nucleon potentials have been taken, of which the first one is purely central of Baker, Gammel et al[7], second one by Volkov[8], third and fourth given by Afnan and Tang[9] and fifth one by Eikemeier and Hackenbroich[10] are soft core central potentials and finally the sixth one is the Goldhammer[11] soft core potential with both central and tensor components present. In all cases radial part of the interaction has been taken as of Gaussian shape. Then the matrix elements of kinetic energy, central potential energy and tensor potential (where required) have been calculated for each potential with all the S and D states. In treating the soft core it is necessary to introduce short range correlation effect in the wave function. But we have not taken it because we think that the inclusion of states of higher configuration mixing are sufficient to take care of that effect. Results are given in Table I.

In the case

Potential	Binding Energy (MeV)
I	7.30
II	6.79
III	1.13
IV	0.71
V	2.21
VI	2.90

Table I: Calculated binding energy of triton with different nucleon-nucleon interactions.

of first potential we have obtained the binding almost 90% of the experimental one. For the Volkov potential the percent is 83. In the cases of potentials, third to sixth the results are not good. For the third and fourth potentials the results are rather poor. For the fifth potential the percentage is 21 while for the last one the calculated binding energy is about 35% of the experimental one. The first two results demonstrate nicely that Englefield states are quite sufficient to reproduce the binding energy, but for other cases this is not true. There are various reasons one may attribute for not obtaining good binding with these soft core interactions. Among them the following points are worth considering:

(i) We have taken both for S and D states upto eight quanta of excitation. This may not be sufficient. States of higher than eight quanta of excitation may be important.

(ii) We have dealt the soft core nucleon-nucleon interaction but short range correlation effect which is needed to treat this interaction, is not taken on the understanding that higher configuration mixing and fully antisymmetrization will to a great extent take care of that. This may not be sufficient.

and (iii) We have taken Englefield states alone but Non-Englefield states may have appreciable contribution in this case.

Now let us look at these points critically. Regarding the inclusion of states of much higher configuration mixing it may be mentioned that Jackson and Elliott[1] have calculated the triton binding with potential deduced directly from the experimental nucleon-nucleon phase shift and they have taken upto eight quanta. In the case of alpha particle different authors[1,2] have used upto eight quanta also. In all these calculations it has been found that binding energy curve versus quantum number shows the near converging tendency after eight quanta. Similiar tendency has been observed here also. That only points to the fact that mere addition of states of higher quanta does not improve the situation. Then the question of introducing short range correlation comes up. Although short range correlation may be successfully utilised in the wave function of lower quanta, for the states of higher configuration mixing the introduction is not that easy, due to sheer volume of works. Finally large number of Non-Englefield states are excluded. It may be that some are important. But to trace them one needs a selection rule.

Work is in progress in some of the above suggested lines.

REFERENCES

(1) A.D. Jackson and J.P. Elliott, Nucl. Phys. A125 (1969)276
(2) S.A. Afzal, Phys. Letts. 39B(1972)478
(3) M. Rahman, Ph.D. Thesis, Southampton, (1965)
(4) M.J. Englefield, Ph.D. Thesis, Southampton(1956)
(5) G. Derrick and J.M. Blatt, Nucl. Phys. 8(1958)310
(6) H.A. Jahn, Group Theory and Structure of light nuclei, paper 1 (unpublished)
(7) G.A. Baker jr., J.L. Gammel, B.L. Hill and J.G. Willes, Phys. Rev. 125(1962)1754
(8) A.B. Volkov, Nucl. Phys. 74(1965)33
(9) I.R. Afnan and Y.C. Tang, Phys. Rev. 175(1968)A1337
(10) H. Eikemeier and H.H. Hackenbroich, Zeits Phys. 195(1966)412
(11) P. Goldhammer, Phys. Rev. 116(1959)676.

ANALYSIS OF THE SHAPE OF THE TRINUCLEON WAVE FUNCTION+

J.L. BALLOT and M. FABRE DE LA RIPELLE
Institut de Physique Nucléaire, Division de Physique Théorique*
91406 - ORSAY - France

For calculation of reactions like electron scattering photonuclear effects etc... various trinuclear wave functions are commonly used. We intend to study to what extent a model way describe properly the trinucleon.

The criterion we will adopt for defining a good trinucleon wave function is the following : This function must be a solution of the Schrödinger equation for a N-N potential giving the experimental binding energy and reproducing as far as possible the charge form factor, at least for the low momentum transfer insuring a good r.m.s. radius.

The earlier trinuclon wave functions have been chosen for the sake of simplicity of the general form $\Psi(r)$ where

$$r^2 = 2\sum_{i=1}^{3} (\vec{x}_i-\vec{X})^2 = \frac{2}{3} \sum_{i,j>i} (\vec{x}_i-\vec{x}_j)^2 \tag{1}$$

$\vec{X}$ is the centre of mass and $\vec{x}_i$ the coordinates of the nucleons.

It describes the completely symmetrical state S only. The familiar Gaussian, Irving and Irving-Gunn functions are of this type.

Gibson [1] described an S state with a similar wave but included mixte symmetry S' and D states in his earlier work

We solve the Schrödinger equation by the hyperspherical harmonic expansion method. The wave function is expanded into hyperspherical harmonics (h.h.)

$$\Psi(r,\Omega) = \sum_{K=0}^{\infty} P_{2K}(\Omega)\ u_K(r)\ r^{-5/2} \tag{2}$$

Ω is a set of 5 angular coordinates and r, the hyperradius, is related to the Jacobi coordinates

$$\vec{\xi}_1 = \vec{x}_2-\vec{x}_1 \qquad \vec{\xi}_2 = \sqrt{3}(\vec{x}_3-\vec{X}) \tag{3}$$

by

$$r^2 = \xi_1^2 + \xi_2^2 \tag{4}$$

$P_{2K}(\Omega)$ is an element of the potential basis of grand orbital 2K and $u_K(r)$ is the related partial wave.

In order to test the sensitivity of the wave function, and for the sake of simplicity, we have neglected the non central forces. We used two model potentials : the potential V^x which has a too weak repulsive core (40 MeV at the origin) and the potential G^2 which has much too strong repulsive core (2.5 GeV at the origin) with respect to a phase shift analysis of the N-N scattering (Rome 1972).[2]

+Presented by M. Fabre de la Ripelle

*Laboratoire associé au C.N.R.S

Both potentials give the experimental binding energy and r.m.s. radii for ^{3}H and ^{3}He. The charge form factor stands over the experimental data for the V^x potential but is very well reproduced till the minimum (12 fm^{-2}) for the G2 potential for both ^{3}H and ^{3}He. We expect in doing this way to determine a domain near which the realistic predominant partial wave $u_0(r)$ may remain.

The Schrödinger equation has been treated numerically and the convergence reached after solving 12 coupled equations (Rome 1972).

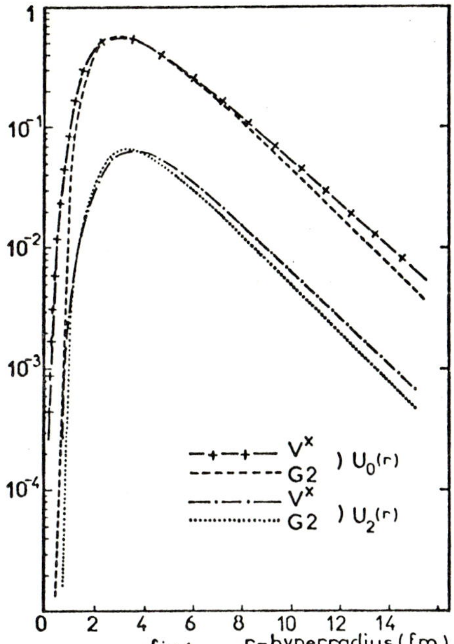

fig 1.

In figure 1 the partial wave $u_0(r)$, and $u_2(r)$ relative respectively to the completely symmetrical and mixed symmetry states, have been plotted for both potentials in semi logarithmic scale.

Their shape are very similar because the potentials have been chosen in order to give the same binding energy E and r.m.s. radius and that

i) the theory predicts the same asymptotic expression

$$u_K \sim \exp - \sqrt{\frac{mE}{\hbar^2}}\ r$$

for all the partial waves,

ii) the position of the maximum of the wave $u_0(r)$ is closely related to the r.m.s. radius

iii) the magnitude of $u_2(r)$ determines the difference between the charge form factor of the isospin doublet ^{3}H - ^{3}He.

The differences appear near the origin only where the wave proceeding from the strongly repulsive G2 potential drop down actually much more quickly than the one of the V^x potential. The asymptotic trend is well illustrated by the straight behaviour of $u_0(r)$ for r > 6 fm.

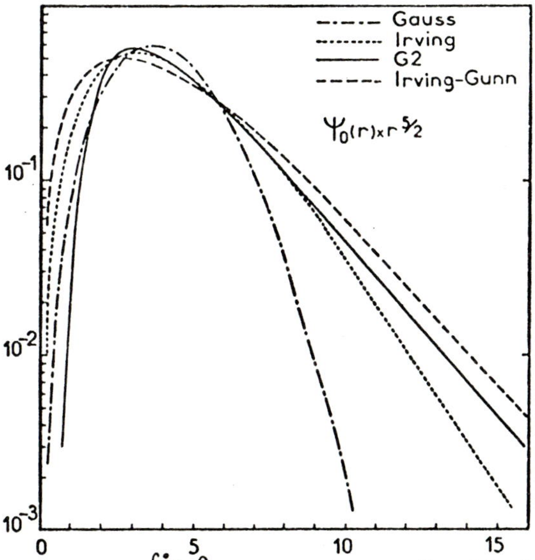

Fig 2.

The figure 2 shows a comparison between $u_0(r)$ of G2 and the Gaussiann Irving and Irving-Gunn functions fitted to the same r.m.s. radius.

None of these can raisonably well describe a trinucleon. The is still the Irving one. On the contrary the Irving-Gunn function does not reproduce properly a trinucleon. The S wave of Gibson [1] has been constructed in order to fit the charge form factor and the two body photodisintegration. Its shape is surprisingly near the solution of V^x as is shown on fig.3. We therefore expect that a solution of the Schrödinger equation reproducing the charge form factor will fit

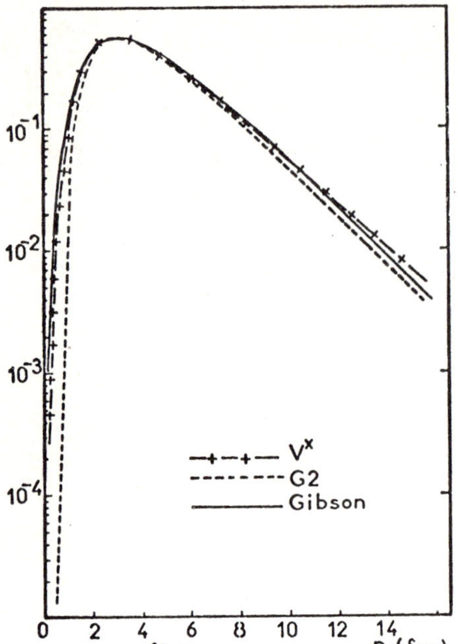

Fig 3.

equally well the photodisintegration. Starting from the wave function $\Psi(\vec{\xi}_1,\vec{\xi}_2)$ ones defines two densities :

$$\left.\begin{aligned} R_1(\xi_1) &= \int |\Psi|^2\, d^3\xi_2 \\ R_2(\xi_2) &= \int |\Psi|^2\, d^3\xi_1 \end{aligned}\right\} \qquad (5)$$

R_1 is the two body correlation function and R_2 is, except a scalling factor $\sqrt{3}$, the one body density. r is invariant when exchanging ξ_1 and ξ_2 therefore using the first partial wave $u_0(r)$ only gives

$$R_1(\xi)=R_2(\xi)=\frac{4}{\pi}\int_0^{+\infty} |u_0(r)r^{-5/2}|^2 x^2 dx$$

$$r^2 = x^2+\xi^2$$

The two densities are the same.

The derivation with respect to ξ

$$\frac{dR}{d\xi} = -\frac{2}{\pi^2}\,\xi\int_0^{+\infty} |u_0(r)r^{-5/2}|^2\, dx$$

is always negative. The density which start horizontally for $\xi = 0$ is a monotonously decreasing function, therefore $u_0(r)$ cannot describe any correlation arising from a strongly repulsive core potential like G2 which cancel the two body density near the origin. These correlations are taken into account by the other h.h. in the expansion (2). When exchanging ξ_1 and ξ_2 , $P_{2K}(\Omega)$ has the parity $(-1)^K$. The odd harmonics alone are responsible for the difference between the correlation function and the one body density. They have been plotted for the V^x and G2 potentials on Fig.4.

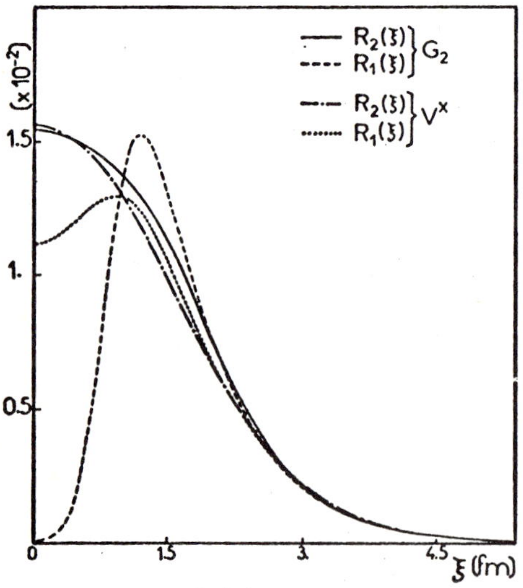

Fig 4.

The strong repulsion cancels practically the two body correlation function for the G2 potential near the origin but the one body density has no hole. The main difference between the one body densities for the V^x and G2 potentials lay in smaller curvature near the origin in the last case.

It is customary to agree that when a nucon goes away from the center of mass its wave decreases asymptotically like

$$\xi_2^{-1} \exp - \sqrt{\frac{m}{\hbar^2} E_s}\; \xi_2$$

where E_s is the separation energy. This statement which proceed from an independent particle model point of view imply that the one body density behaves asymptotically like

$$\xi_2^{-2} \exp\left(-\,2\sqrt{\frac{m}{\hbar^2} E_s}\;\xi_2\right)$$

instead of the hyperspherical model prediction $\xi_2^{-7/2} \exp(-2\sqrt{\frac{mE}{\hbar^2}}\, \xi_2)$ where E is the total binding energy.

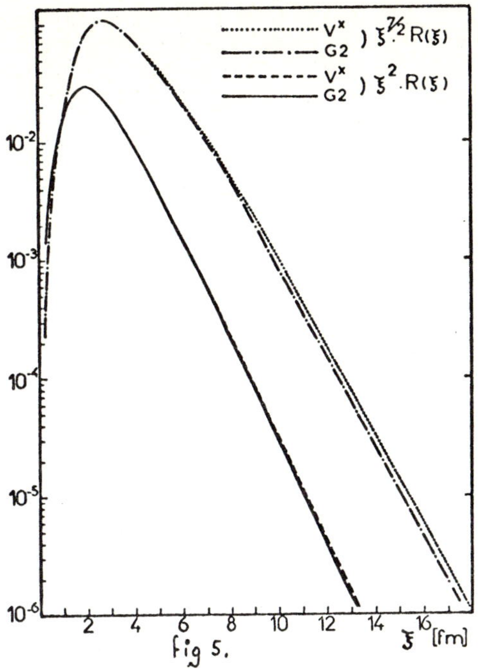

fig 5.

The two curves $\xi^{7/2} R_2(\xi)$ and $\xi^2 R_2(\xi)$ are shown on Fig.5 in semi-logarithmic scale.

Both curves lead to a straight line in the asymptotic region. The slope of the first curve corresponds to the total binding energy expected but the slope of the other corresponds to 17 MeV which is actually very much over the value of about 6 MeV predicted by the independent particle model. The overlapping function $R_{23}(\xi_2)$ between deuteron and trinucleon is given in terms of an h.h. expansion by

$$R_{23}(\xi_2) = \int \phi_d(\vec{\xi}_1) \sum_{K=0}^{+\infty} P_{2K}(\Omega) \times$$

$$\times\ r^{-5/2}\, u_K(r)\, d^3\xi_1 \qquad (6)$$

where $\phi_d(\vec{\xi}_1)$ is the deuteron wave function.

Asymptotically u_{2K} behaves like $\exp(-\sqrt{\frac{mE}{\hbar^2}}\, r)$ leading to

$$\xi^{5/2} R_{23}(\xi) \sim C^{te} \exp(-\sqrt{\frac{mE}{\hbar^2}}\, \xi_2) \qquad (7)$$

instead of the expression $\xi R_{23}(\xi) \sim C \exp(-\sqrt{\frac{mE_s}{\hbar^2}}\, r)$ provided by the independent particle model (e.g. Kim-Tubis)[3].

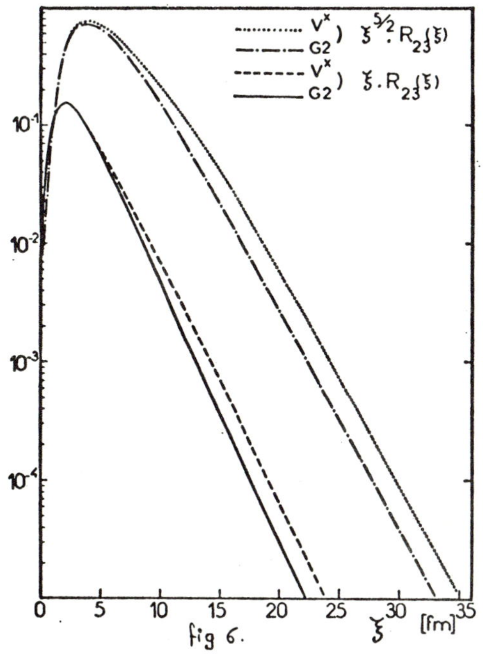

fig 6.

The two curves $\xi^n R_{23}(\xi)$, n = 1, 5/2 are plotted on Fig.6 in semi-logarithms scale for the potentials V^x and G2. The deuteron wave function is extracted from the triplet even part of the G2 potential giving a binding energy of - 2.7 MeV. The slope of the curve for n = 1 corresponds to an energy of about 11 MeV between 10 and 20 fm which does not agree with the separate energy E_s = 5.6 MeV. For n = 5/2 one finds, as expected a curve decreasing exponentially like (7) in which E is the total binding energy of the trinucleon.

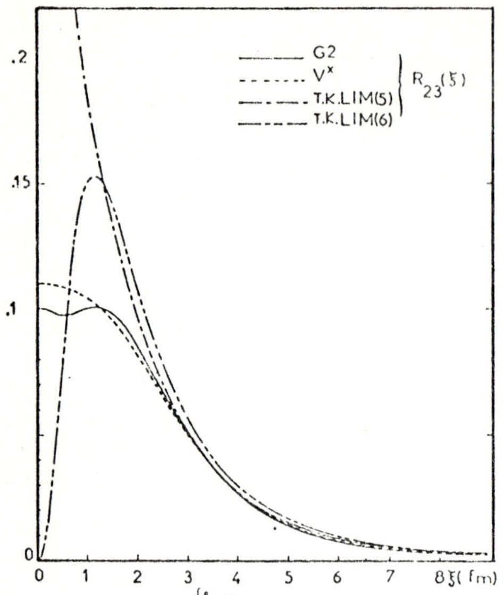

fig 7

The overlapping functions R_{23} between the deuteron wave extracted from the Vinh Mau[4] potential and He calculated for both potentials V^x and G2 are shown on fig. 7 together with the Hulten type function [5] and the Eckhart form [6] which according to T.K. Lim may reproduce their shape (with $C^2(^3He) = 3$).

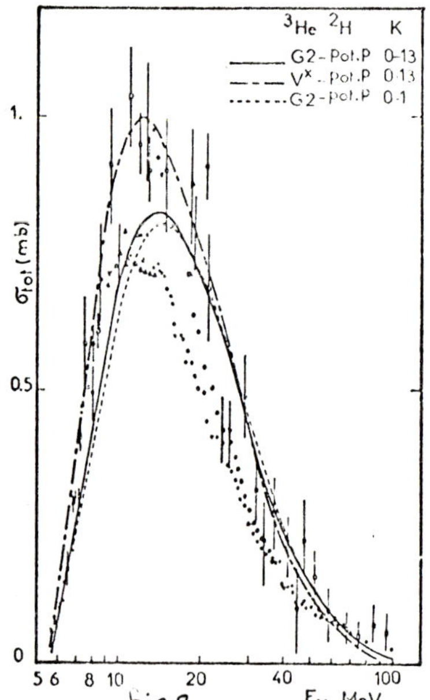

Fig 8

The R_{23} function has been used to calculate the two body dipole photodisintegration of ^3He in the Born approximation. The figure 8 reproduces the total cross section for both potentials V^x and G2 in terms of the photon energy E_γ. The experimental data are taken from G. Ticcioni et al. [7].

References

[1] B.F. Gibson, Nucl. Phys. B2 (1967) 501

[2] J.L. Ballot, M. Beiner and M. Fabre de la Ripelle, Proceedings of the Int. Conf. on the Nuclear Many-Body Problem, Roma Sept. 1972

[3] Y.E. Kim and A. Tubis, Phys. Rev. Lett. 29 (1972) 1017

[4] M. Lacombe, B. Loiseau, J.M. Richard, R. Vinh Mau, P. Pirès and R. de Tourreil, Phys. Rev. 12D (1975) 1495

[5,6] T.K. Lim, Phys. Rev. Lett. 30 (1973) 709 ; Phys. Lett. 43B (1973) 349

[7] G. Ticcioni, S.N. Gardiner, J.L. Matthews and R.O. Owens, Phys. Lett. 46B (1973) 369.

COULOMB ENERGY OF ^{3}HE

H. Zankel, H. Zingl
Institut für Theoretische Physik, Universität Graz
A-8010 Graz, Austria

The AGS (1) quasiparticle method provides an elegant way to include the Coulomb interaction in the three-nucleon system, especially, as demonstrated by Alessandrini (2) and Zankel (3), if only two charged nucleons are present in the three-body system. The method takes into account the Coulomb force exactly and so we are faced with the usual Coulomb singularity trouble. Nevertheless, since there is still no managable exact treatment (4) of the Coulomb many-body system available, we circumvent the Coulomb singularity by using a Yukawa type cut off. But performing several numerical tests it turned out that for the bound state problem we would not need a cut off-procedure. Considering the small inter-nucleon distances in the three-nucleon bound state it seems to be plausible that the long range Coulomb singularity rather weakly influences the three-nucleon binding energy.

Switching off all Coulomb contributions in the ^{3}He we obtain assuming charge symmetry of the nuclear forces a model for the Triton. The binding energy difference of ^{3}H and ^{3}He is then the Coulomb energy (E_C) of ^{3}He. Figure I represents graphically the

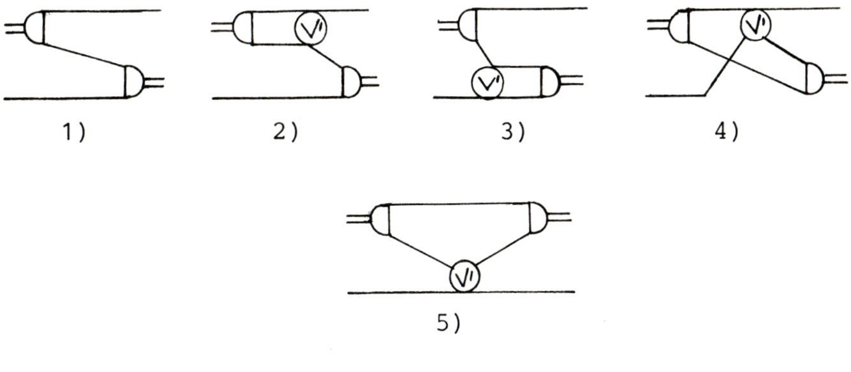

Figure 1: Effective potential of ^{3}He

effective potential of ^3He. The contribution of each term to E_C was examined with the result that graph 2 and 3 cover half of the Coulomb energy (0.3 - 0.34 MeV), whereas graph 4 contributes 0.2 MeV and graph 5 0.1 MeV. These model calculations were performed using separable potentials of rank two as given in ref. 5.) and shown in Table 1.

Wave	Charge	Potential	a(fm)	r(fm)	E_C
$^1S_0$	n-p	Hamman	-23,78	2.61	
	p-p	Graz a	- 7.82	2.78	
	p-p	Graz b	- 7.83	2,72	
	n-n	Graz a	-16.40	2.93	
	variation	-16.4 to -17.2 in a			0.62-0.65
	variation	2.80 to 2.93 in r			
	n-n	Graz b	-16.80	2.85	0.64
$^3S_1$	n-p	Mongan	5.39	1.72	

Table 1

Because the p-p and, after removal of the Coulomb contributions, also the charge symmetric n-n interactions are the relevant ones for our Coulomb energy we have studied their influence on E_C in more detail. Using a GRAZ (6) type potential we constructed two sets of p-p parameters adjusted to reproduce a) the low energy parameters exactly and the phase shifts only approximately or b) the phase shifts and low energy parameters approximately. After extracting the Coulomb contribution, which seems to be a not too strong model dependent procedure (7) we obtain sets of n-n data. The binding energy difference (E_C) remains for our different sets of parameters within the interval 0.62 MeV < E_C < 0.67 MeV. The restriction that E_C < 0.76 MeV and $|a_{nn}| < |a^s_{pp}|$ is only fullfilled if $r_{nn} < r^s_{pp}$ as found by Stephenson (8) with a different method did not ensue from our model. Recent calculations with a rank one p-p potential exactly fitted to the low energy data and the phase shifts up to 100 MeV indicated the possibility to narrow the interval for the a_{nn} and r_{nn} values. But the results as shown in Table 1 do not allow a tight restriction on the neutron-neutron low energy parameters.

References

1) E.O.Alt, P.Grassberger and W.Sandhas, Nucl.Phys.2B (1967) 167.

2) V.A.Alessandrini, C.A.Garcia and H.Fanchiotti, Phys.Rev.170 (1968) 935.

3) H.Zankel and H.Zingl, Acta Phys.Austr.44, to be published.

4) A.G.Gibson and C.Chandler, J.Math.Phys.15 (1974) 1366. They show a way how to treat the Coulomb problem in an exact time independent theory regardless of the practical calculations.

5) H.Zankel, C.Fayard and G.H.Lamot, Lett.Nuovo Cim.12 (1975) 221.

6) L.Crepinsek et al., Acta Phys.Austr.42 (1974) 139.

7) H.de Groot, private communication.

8) B.F.Gibson and G.J.Stephenson Jr., Phys.Rev.C11 (1975) 1448.

THREE-ALPHA MODEL CALCULATIONS OF ^{12}C IN THE FADDEEV FORMALISM

Y. Fujiwara and R. Tamagaki
Department of Physics, Kyoto University
Kyoto, Japan

1. Introduction

Study with the viewpoint of "α-like four body correlations and molecular structures in light nuclei" has revealed the possible existence of multi-cluster states near or above the dissociation energy of the α-particle.(1) Microscopic treatments usually adopted for studying these states, such as the generator co-ordinate method, take the Pauli principle into account exactly but relative motion is solved in more or less restricted way. Effective interaction between α-clusters is weak compared with internal binding one, because of the exchange repulsion due to the Pauli principle.(2) Therefore it is important to solve relative motion exactly in the sense of few-body problems.

We study the structure of ^{12}C by adopting the 3α boson model in the Faddeev formalism, where we employ the α-α effective interaction reproducing the characteristics obtained microscopically and deal with the isospin T=0 states which can be treated as bound states when the Coulomb effect is switched off. We have obtained the several important low-lying states owing to the roles of the partial waves up to G contributing in a characteristic way for each state. In particular, this treatment can predict two 0^+ states with quite different characters; the ground state with the triangle form where the S, D and G waves contribute equally in the binding and the well-developed cluster state with the weakly coupled $\alpha+{}^8Be(0^+)$ feature near the α-threshold.(3)

2. α-α effective potential

Microscopic study of the α-α interaction has revealed the existence of the energy-independent damped inner oscillation of the relative wave functions.(2) The α-α potential is required to reproduce this inner oscillation as the reflection of the Pauli principle in addition to the 2α data. These requirements can be satisfied with the one-term nonlocal separable potential by adopting the form factors $g_\lambda(k)$ of the harmonic oscillator type, as shown by Kukulin and Neudatchin;(4) that is, in momentum representation,

$$v_\lambda^{\alpha-\alpha}(k,k') = -(\hbar^2/M_\alpha)\,\xi_\lambda\, g_\lambda(k)\, g_\lambda(k')$$

with $g_\lambda(k) = (k^2+K_0^2)(k^2-K_1^2)(k^2-K_2^2)e^{-1/2\cdot\beta^2k^2}$ for $\lambda\leq 4$ and $g_6(k) = (k^2+K_0^2)k^6e^{-1/2\cdot\beta^2k^2}$ where M_α is the α-particle mass, and the α-α relative angular momentum λ is even only. The parameters ξ_λ, K_0^2, K_1^2, and K_2^2 and β are determined for each partial wave (in Table I). These values are slightly modified from the original ones(4) to get more reasonable fitting; to reproduce the 2α binding energy -1.4 MeV (without the Coulomb energy), the Coulomb switched-off "experimental" α-α phase shifts and the inner nodal point of the α-α relative wave functions in almost energy-independent manner.

λ	β (fm)	ξ_λ^{-1} (fm^{-13})	K_0^2 (fm^{-2})	K_1^2 (fm^{-2})	K_2^2 (fm^{-2})
0	1.02	0.134×10^3	8.14	0.54	3.3
2	1.02	0.230×10^3	7.68	3.2	0
4	1.02	0.487×10^4	22.3	0	0
6	1.02	0.974×10^5 (fm^{-17})	10.		

Table I. Parameters of the α-α effective potential.

In the Faddeev formalism the total wave function is obtained by superposing functions $\psi^{J^\pi}(\mathbf{k},\mathbf{q})$ with three possible Jacobi coordinates: $\Psi = \psi_\alpha + \psi_\beta + \psi_\gamma$. $\psi_\alpha \equiv \psi^{J^\pi}(\mathbf{k},\mathbf{q})$ can be expanded in the partial wave, according to Harrington:(5)

$$\psi^{J^\pi}(\mathbf{k},\mathbf{q}) = \sum_{\lambda,\ell} [\gamma^2+k^2+\tfrac{3}{4}q^2]^{-1}\, g_\lambda(k)\,\tau_\lambda(\gamma^2+\tfrac{3}{4}q^2)\, f^{J^\pi}_{\lambda\ell}(q)\,[Y_\lambda(\hat{\mathbf{k}})Y_\ell(\hat{\mathbf{q}})]_J ,$$

where $\mathbf{q}$ is the relative coordinate between α and 2α's and ℓ the corresponding angular momentum, γ^2 is related to the binding energy of the 3α system, $E=-\hbar^2\gamma^2/M_\alpha$, and $\tau_\lambda(\gamma^2+3q^2/4)$ denotes the α-α reduced t-matrix. We call $f^{J^\pi}_{\lambda\ell}(q)$ the reduced wave function, which obeys the angular momentum coupled integral equation.(5)

3. Energy spectra

The calculated energy spectra are shown in Fig.1, where we can see the change of the spectra according to the increase of the partial waves (λ,ℓ) taken into account; $(\lambda,\ell) \leqq (\lambda_M, \lambda_M+1)$ with λ_M= 0, 2, 4 and 6.

The large energy gain of 0_1^+ and 2_1^+ states takes place when the partial wave $(\lambda,\ell)=(2,2)$ and (4,4) are taken into account. This feature is due to the characteristics of this potential to produce the inner oscillation, and different from the results by using the α-α forces with repulsive core.(6) This result is consistent with those of the orthogonality condition model recently performed by Horiuchi,(7) and by Smirnov et al.(8)

The 2_1^+ state has the reduced wave functions $f^{2_1^+}_{22}(q)$ and $f^{2_1^+}_{44}(q)$ similar to $f^{0_1^+}_{22}(q)$ and $f^{0_1^+}_{44}(q)$ respectively, and it seems to have the same intrinsic structure as the 0_1^+ state. The level distance between 2_1^+ and 0_1^+ is in good agreement with the experimental value.

We have obtained the 0^+ states in the region near the α-threshold energy, where in shell model calculations such low 0^+ states have not been obtained. For negative parity states, the 3^- state and 1^- state are predicted at 12.5 MeV and 11.1 MeV excitation energy respectively.

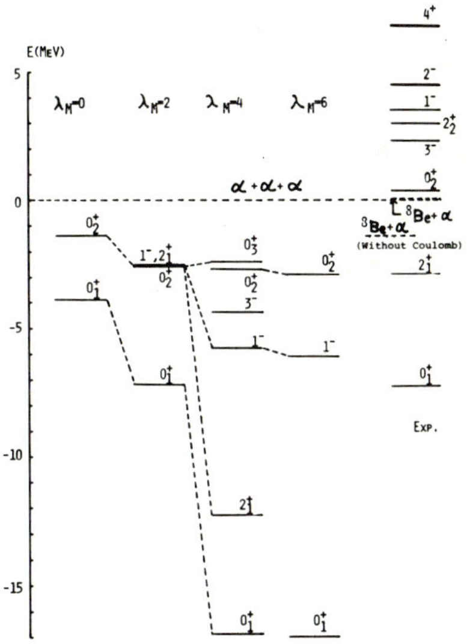

Fig.1. Energy spectra when partial waves (λ,ℓ) are contained up to (λ_M, λ_M+1).

4. Properties of the 0^+ states

Calculated reduced width amplitudes of $\alpha+{}^8Be(0^+)$ are shown in Fig.2 (see the next page), where the S, D and G partial waves are successively contained. From this figure we can see the 0_2^+ state (λ_M=2) connects with the 0_3^+ state (λ_M=4). The 0_1^+ state is the compact 3α-system with shell-model-like structure. On the contrary, the 0_2^+(λ_M =0 or 2) or the 0_3^+(λ_M=4) state has a very large reduced α-width, which is close to the Wigner limit $\sqrt{3/\rho}$ for the 0_3^+ state (λ_M=4).

Further, we examine the implication of the wave function and the potential energy, through $\langle\Psi|\Psi\rangle = 3\langle\psi_\alpha|\psi_\alpha\rangle + 3\langle\psi_\alpha|\psi_\beta+\psi_\gamma\rangle$ and $\langle\Psi|v_\alpha+$

$v_\beta + v_\gamma |\Psi\rangle = 3\langle\psi_\alpha|v_\alpha|\Psi\rangle + 3\langle\psi_\beta+\psi_\gamma|v_\alpha|\Psi\rangle$, where we call the first terms the direct type and the second the exchange one.

Mechanism of energy gain of the 0_1^+ state is explained by the fact that each overlap component of the exchange form between the S-D, S-G and D-G waves ($3\langle\psi_\alpha(S)|\psi_\beta(D)+\psi_\gamma(D)\rangle$, $3\langle\psi_\alpha(S)|\psi_\beta(G)+\psi_\gamma(G)\rangle$ and $3\langle\psi_\alpha(D)|\psi_\beta(G)+\psi_\gamma(G)\rangle$) is large (5~10%). This internal correlation of partial waves brings about the large exchange-type potential energy ($3\langle\psi_\beta+\psi_\gamma|v_\alpha(S)|\Psi\rangle$, $3\langle\psi_\beta+\psi_\gamma|v_\alpha(D)|\Psi\rangle$ and $3\langle\psi_\beta+\psi_\gamma|v_\alpha(G)|\Psi\rangle$ $\cong$ -15∿ -18 MeV for each partial wave).

On the other hand, the 0_3^+ state (λ_M=4) with the weakly coupled $\alpha+{}^8$Be(0^+) feature is mainly composed of the S wave; that is, the S wave direct overlap ($3\langle\psi_\alpha(S)|\psi_\alpha(S)\rangle$) is 56%, and the S-S exchange overlap ($3\langle\psi_\alpha(S)|\psi_\beta(S)+\psi_\gamma(S)\rangle$) is 36%. Orthgonality to the 0_1^+ and 0_2^+(λ_M=4) with compact structure is essential.

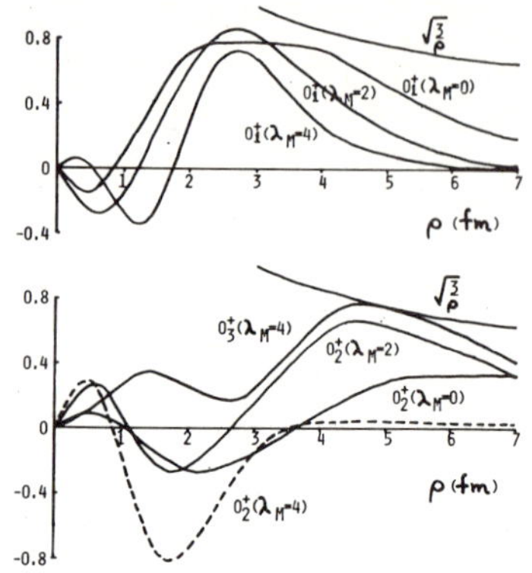

Fig.2. Reduced $\alpha+{}^8$Be (0^+) width amplitudes of the 0^+ states, when partial waves $\lambda=\ell$ are contained up to λ_M; λ_M=0, 2 and 4.

As for another 0_2^+ state (λ_M=4), its reduced wave functions $f_{\lambda\lambda}^{0_2^+}(q)$ have one node for λ=0 and no node for λ=2, and correspond to the total quanta N=2 if they are described approximately in the harmonic oscillator basis. Therefore we can say that this 0_2^+ state (λ_M=4) corresponds to a forbidden state when the Pauli principle is treated exactly.[7] This indicates that for some states it is necessary to supplement proper consideration concerning the forbidden states in the boson-model approach to three-cluster systems when we adopt the effective inter-cluster interaction to reproduce the inner oscillating feature.

The authors thank Dr. H. Horiuchi for enlightening discussions about the 3α problem.

References

1. H. Horiuchi, K. Ikeda and Y. Suzuki, Prog.Theor.Phys.Suppl. No.52 (1972), 89.
2. J. Hiura and R. Tamagaki, Prog.Theor.Phys.Suppl. No.52(1972), 25.
3. Y. Suzuki, H. Horiuchi and K. Ikeda, Prog.Theor.Phys. 47(1972), 1517.
4. V.I. Kukulin and V.G. Neudatchin, Nucl.Phys. A157(1970), 609.
5. D. Harrington, Phys.Rev. 147(1966), 685.
6a. H. Hebach and P. Henneberg, Z.Phys. 216(1968), 204.
b. J. Fulco and D. Wong, Phys.Rev. 172(1968), 1062.
c. Y. Kawazoe, T. Tsukamoto and H. Matsuzaki, Prog.Theor.Phys. 51 (1974), 428.
d. T.K. Lim, Nucl.Phys. A158(1970), 385.
7. H. Horiuchi, Prog.Theor.Phys. 51 (1974), 1266 and 53(1975), 447.
8. Yu.F. Smirnov, I.T. Obukhovsky, Yu.M. Tchuvil'sky and V.G. Neudatchin, Nucl.Phys. A235(1974), 289.

A STUDY OF $^{6}_{\Lambda\Lambda}He$ WITH TWO-BODY LOCAL POTENTIALS IN FADDEEV'S FORMALISM

H. Roy-Choudhury and V.P. Gautam
Department of Theoretical Physics,
Indian Association for the Cultivation of Science
Calcutta-32, India.
and
D.P. Sural
Department of Physics, Jadavpur University,
Calcutta-32, India.

In this paper we propose to study the properties of the $^{6}_{\Lambda\Lambda}He$ hyperfragment using the three-body $\alpha\Lambda\Lambda$ model and the Faddeev method [1]. We have used realistic local two-body potentials. The $\Lambda\Lambda$ interaction is represented by an exponential potential with a hard core of radius a,

$$V(r) = +\infty \qquad r \leq a$$
$$= -V_0\, e^{-\beta(r-a)} \qquad r > a \tag{1}$$

The $\Lambda\Lambda$ potential parameters a and β, [Eq.(1)] used by us are the same as those of Tang and Herndon [2]. Different values for the strength parameter V_0 have been taken as shown in Table 1.

The $\alpha\Lambda$ potential has been taken to be of exponential form without any core,

$$V'(r) = -V_0'\, e^{-\beta' r} \tag{2}$$

The parameters of this potential have been determined with the constraint that V'(r) gives rise to the same $\alpha\Lambda$ low energy parameters as could be obtained from the Gaussian potential B of Tang and Herndon [2].

The three-body Faddeev equations as such give rise to a set of two dimensional integral equations, but, suitable separable expansions for the off-shell two-body t-matrices can reduce them to a system of one dimensional integral equations. In the two-body channels we are dealing with, the $\Lambda\Lambda$ interaction has a repulsive core and the $\alpha\Lambda$ possesses the bound state. Hence the unitary pole approximation, which corresponds to one term approximation of the unitary pole expansion for the off-shell two-body t-matrix, will be a very good approximation [3].

The unitary pole expansion for the s-wave two-body t-matrix can be found for an exponential potential with hard core by solving the homogeneous Lippman-Schwinger equation which yields the form factors [4]. For the general case, let us assume the existence of a two-body bound state with binding energy ϵ. Then the nth form factor can be written as,

$$v_{n,l=0}(k,-\epsilon) = \frac{1}{k}\sqrt{\frac{\beta}{\mu\pi}}\left[\sin ka - \frac{y_n I}{J_{\frac{2k}{\beta}+1}(y_n)}\right] \tag{3}$$

where

$$I = \int_0^1 t \sin\left(ka - \frac{2k}{\beta}\ln t\right) J_{\frac{2k}{\beta}}(t y_n)\, dt \tag{4}$$

Here k and μ are respectively the relative momentum and the reduced mass of the two-body system, y_n is the nth zero of the Bessel function $J_{2k/\beta}$ (y) and $k^2 = 2\mu\epsilon$. In terms of these form factors the two-body s-wave t-matrix can be expressed as,

$$t(k,k';s) = 2\pi^2 \sum_{n,m}^{N} v_{n,l=0}(k,-\epsilon)\, \Delta_{n,m}(s)\, v_{m,l=0}(k',-\epsilon) \quad (5)$$

where

$$[\Delta^{-1}(s)]_{n,m} = -\lambda_n \delta_{n,m} - \int_0^\infty dk \frac{k^2 v_{n,l=0}(k,-\epsilon)\, v_{m,l=0}(k,-\epsilon)}{s - \frac{k^2}{2\mu}} \quad (6)$$

with $\lambda_n = \beta^2 y_n^2 / 8\mu V_0$.

In Eq.(5) N is a finite integer. This is the well known unitary pole expansion. These equations can be easily converted to the forms suitable for application to our $\Lambda\Lambda$ and $\alpha\Lambda$ channels.

The integral I in Eq.(4) has been evaluated numerically by Mathur et al [4] by using the Gauss-Legendre quadrature method. However, in course of application to the three-body problem, large values of k also appear. In those cases, because of the presence of oscilatory sine function in the integrand, convergent results for the integral can not be obtained even by using a very large number of Gaussian points. We have, therefore, developed an alternative method to evaluate the integral (4). In this method the Bessel function $J_{2k/\beta}(ty_n)$ within the range of the integral has been approximated by a six degree Chebyshev polynomial using a method due to Lanczos [5]. The integral (4) can then be worked out analytically. With our approach, the subsequent integration required for the evaluation of $[\Delta^{-1}(s)]_{n,m}$ [Eq.(6)] can also be done analytically. In this way we have been able to increase the accuracy of the numerical results while reducing the computer time involved.

Starting from the Lippman-Schwinger equation for the hadron-deuteron scattering, Hetherington and Schick [6] have arrived at the three-body Faddeev equations for the auxiliary operators in terms of which the total T-matrix for the above process could be expressed. These Faddeev equations are three sets of three coupled equations and become a set of matrix-integral equations after taking matrix elements between the relevant three-particle states. For binding energy calculations the inhomogeneous parts of the integral equations are to be dropped. If we number the particles Λ, α, and Λ as 1, 2, and 3 respectively, make a partial wave analysis of the homogeneous set of integral equations and retain only the s-wave parts, we arrive at the following pair of integral equations,

$$X_0(q,q') = \int_0^\infty dq'' K_0^{21}(q,q'')\, \tau_1(q'')\, Y_0(q'',q') \quad (7)$$

$$Y_0(q,q') = \int_0^\infty dq'' K_0^{12}(q,q'')\, \tau_2(q'')\, X_0(q'',q') + \int_0^\infty dq'' K_0^{13}(q,q'')\, \tau_1(q'')\, Y_0(q'',q') \quad (8)$$

with

$$K_0^{ij}(q,q') = c^{ij} \frac{q'^2}{4\pi^2} \int_{-1}^{1} d\omega \frac{v_1^i(|\vec{q}' + \frac{m_j}{M_i}\vec{q}|)\, v_1^j(|\vec{q} + \frac{m_i}{M_j}\vec{q}'|)}{E - \frac{q^2}{2m_i} - \frac{q'^2}{2m_j} - \frac{(\vec{q}+\vec{q}')^2}{2m_{ij}}} \quad (9)$$

and $C^{12} = C^{21} = \sqrt{2}\, C^{13} = \sqrt{2}$.

Here v_1^i 's are the form factors (arising from the interaction between two particles other than the ith one) introduced earlier in Eq.(3), E is the total three-body energy in the centre of mass system and m_i is the mass of the ith particle. Further, $M_i = M - m_i$ and $m_{ij} = M - m_i - m_j$, where M is the sum of the masses of all three particles. Gauss Legendre quadrature method has been used to convert the Eqs. (7) and (8) to a set of linear algebraic equations, the coefficients of which depend upon K_0^{ij}'s [Eq.(9)]. The K_0^{ij} 's have been evaluated numerically by similar quadrature method. To achieve convergence in the determination of the K_0^{ij}'s, quadrature points ranging from 16 to 60 are required. For finding a solution of Eqs.(7) and (8), 16 point quadrature has been found to be sufficient for producing convergent results. Thus from Eqs. (7)and (8), we arrive at a Fredholm determinant $|I - K_0\tau|$ of the order 32 x 32. This determinant is evaluated for different values of the three-body energy, E. The particular value $-E_0$, for which $|I - K_0\tau|$ vanishes, is found out. Then E_0 is nothing but $B_{\Lambda\Lambda}$, the $\Lambda\Lambda$ binding energy in the system $^6_{\Lambda\Lambda}He$.

Table 1. The $\Lambda\Lambda$ interaction strength parameters V_0, and the corresponding binding energies $B_{\Lambda\Lambda}$ due to Tang and Herndon and present calculation. The other two-body potential parameters are,
$a = 0.4$ fm, $\beta = 5.0590\,fm^{-1}$, $V_0' = 0.7157\,fm^{-1}$ and $\beta' = 1.6250\,fm^{-1}$.

Sets	$V_0(fm^{-1})$	$B_{\Lambda\Lambda}$ (MeV)	
		Tang and Herndon	Present calculation
1	5.3233	11.75 ± 0.12	12.19
2	5.0698	10.74 ± 0.11	11.53
3	4.7859	9.72 ± 0.01	10.85
4	4.7656	9.65 ± 0.10	10.80

In Table 1 we have given the two-body potential parameters and the corresponding $B_{\Lambda\Lambda}$ values. We have also shown the $B_{\Lambda\Lambda}$ values found by Tang and Herndon [2] using the variational method. The experimental value for $B_{\Lambda\Lambda}$ is 10.8 ± 0.6 MeV [7]. The value 10.80 MeV is obtained from our calculation for $V_0 = 4.7656\ fm^{-1}$.

REFERENCES

1. L.D. Faddeev, Soviet Phys. - JETP 12 (1961) 1014;
Doklady 6 (1961) 384; Doklady 7 (1963) 600.
2. Y.C. Tang and R.C. Herndon, Nuovo Cimento 46B (1966) 117.
3. E. Harms, Phys. Rev. C1 (1970) 1667;
E. Harms and V. Newton, Phys. Rev. C2 (1970) 1214.
4. V.S. Mathur, A.V. Lagu and C. Maheshwari,
Nucl. Phys. A178 (1972) 365.
5. C. Lanczos, Journ. Math. Phys. 17 (1938) 123.
6. J.H. Hetherington and L.H. Schick, Phys. Rev. 137 (1965) 935.
7. D.J. Prowse, Phys. Rev. Letters 17 (1966) 782.

STRUCTURE OF THE ALPHA PARTICLE
BASED ON REALISTIC TWO- AND THREE-BODY INTERACTIONS

Y. Akaishi
Department of Physics, Hokkaido University, Sapporo, Japan

To solve few-nucleon problems on the basis of realistic interactions is a fruitful subject. In particular, realistic wave function of the alpha particle is of great use for investigations of properties of strongly correlated nucleon systems.

The ATMS method

A new variational method has been presented by Akaishi et al.[1], which makes it possible to treat the alpha particle with realistic potentials as well as the triton. The variational wave function is constructed by amalgamating two-nucleon correlation functions into the multiple scattering process. We call this method of the construction ATMS. ATMS consists of three steps; 1) to make an initial wave function Φ which represents the symmetry character of the nucleon system, 2) to obtain the on-shell correlation function η and the off-shell one ζ by solving the Bethe-Goldstone equation, and 3) to construct the multiple scattering correlation function F from the two-body functions $u^{(on)}=1-\eta$ and $u=1-\zeta$. The essential point of ATMS exists in the fact that the multiple scattering process in the nucleus can be solved with two kinds of the two-nucleon correlation functions.

The ATMS wave function has the form

$$\Psi = F\Phi ,$$

$$F = \frac{1}{D}\sum_{(ij)}\Big[\Big\{\prod_{(kl)}^{\neq(ij)} u(kl)\Big\}\Big\{u^{(on)}(ij) - \frac{(A+1)(A-2)}{A(A-1)}u(ij)\Big\}\Big] ,$$

$$D = \sum_{(ij)}\Big[\Big\{\prod_{(kl)}^{\neq(ij)} u(kl)\Big\}\Big\{1 - \frac{(A+1)(A-2)}{A(A-1)}u(ij)\Big\}\Big] ,$$

where (ij) denotes a nucleon pair and A is the mass number. The on-shell function is taken to be state-dependent and in the case of the alpha particle

$$u^{(on)}(ij) = {}^1u_S\,{}^1P_E(ij) + {}^3u_S\,{}^3P_E(ij) + {}^3u_D S_{ij} ,$$

where P's are the projection operators to the spin-isospin states of two nucleons and S_{ij} is the tensor operator. The last term induced by the strong tensor force in the realistic potential produces the D state of the total system.

Before going to results for the alpha particle, we demonstrate the power of ATMS in the case of the system of three α-particles interacting with the Ali-Bodmer (d) s-state α-α potential[2]. This system is simple and convenient for the comparison of ATMS with other methods. Two kinds of two-body correlation functions are determined by the Euler equation (ATMS-Euler) derived from ATMS under Ritz's variational principle. In Table I, the upper-bound and the Temple's lower-bound energies are shown together with the results by Lim [3] and Visschers et al.[4]. Two bounds in ATMS are surprisingly close compared with the case of Lim. This fact clearly shows that the ATMS wave function is very accurate. The energy difference 1.06

Table I. Energies of 3-α system (in MeV).

	Upper	Lower	Pot.energy
Lim [3]	-3.3	-46.7*	-8.05*
K-harmonics[4]	-4.06		
ATMS	-5.12	-5.8	-12.78

* calculated by the present author.

MeV between K-harmonics and ATMS is fairly large, because it amounts to about 10 % of the total potential energy. The ATMS method is very powerful in solving few-body bound-state problems.

The ground state of the alpha particle

Variational calculation by ATMS was made for the alpha particle by Sakai et al.[5]. The Hamada-Johnston potential (H-J)[6] and the Tamagaki potentials (OPEH and OPEG)[7] are employed as realistic potentials. H-J has the hard core of the large radius 0.485 fm and the strong tensor force in the triplet-even state. OPEH has the hard core of 0.42 fm and the weaker tensor force than H-J. OPEG is characterized by a Gaussian type soft core of 2 GeV height. Coulomb potential is not included in the calculation. The search for energy minimum is done with respect to two parameters introduced into the ATMS wave function.

In the calculation the multiple integration must be carried out numerically. In the case of the alpha particle the multiplicity is nine. The quasi-random number method (QRN) for numerical integration was developed to the practical use in the nuclear system by Tanaka et al.[8]. While the integration error of the Monte Carlo method is proportional to the inversed square root of the sampling number N, that of QRN is proportional to the squared inverse of N. The computation time is considerably saved by QRN.

The results for the alpha particle is shown in Table II. The upper-bound energies are -20.6 MeV for H-J, -23.3 MeV for OPEH and -22.5 MeV for OPEG. It should be emphasized that the large negative values are obtained by ATMS even for the potentials with the hard core and the strong tensor force.

The kinetic energies in Table II are about two or three times as large as that of a harmonic oscillator model which is $(9/4)\hbar\omega = (81/32)(\text{rms radius})^{-2} \simeq 50$ MeV. This fact indicates that the alpha particle is the strongly correlated system. It is also noted that the kinetic energy sensitively depends on the realistic potential employed. Regarding the potential energies, the largest contribution comes from the tensor force in the triplet-even state through the coupling between the S and the D states. The strong tensor force is one of the characteristics of the realistic potential and therefore the D-state wave function must properly be constructed. ATMS gives the reasonable D-state function and is powerful in this respect. The D-state probability in the alpha particle amounts to 13 % for H-J.

		H-J	OPEH	OPEG
Energy		-20.6	-23.3	-22.5
Kinetic energy		131.1	101.6	70.6
Potential energy		-151.7	-124.9	-93.1
Central	1E	-51.3	-42.8	-29.2
	3E	-26.2	-40.0	-20.8
	1O	0.003	0.001	0.002
	3O	-0.4	-0.2	0.3
Tensor	SD	-79.3	-46.4	-46.1
	DD	9.1	3.4	3.5
LS		-0.7	0.4	0.6
Quadratic LS		-2.9	0.5	-1.5
rms radius (fm)		1.65	1.65	1.66
State probability (%)	P(S)	87.1	89.9	90.0
	P(S')	0.1	0.1	0.2
	P(D)	12.8	9.8	9.8

Table II. Energies of the alpha particle (in MeV).

The charge form factor of the alpha particle for elastic electron scattering is well reproduced with the ATMS wave function in the low momentum transfer region as shown in Fig.1. But in the high momentum region the calculated peak is about one fourth of the experimental one. The situation is almost the same as that of 3He. The three-body force discussed below hardly improves the situation (Katayama et al.[9]). It may be probable that the discrepancy in the form factor can not be removed as far as the alpha particle is treated as an aggregate of nucleons.

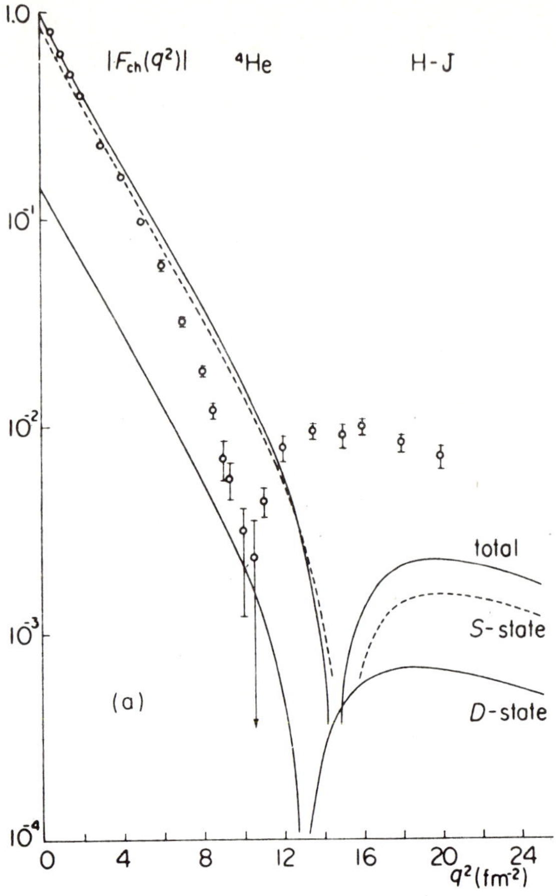

Fig.1.The charge form factor of 4He

Effect of the three-body force

How should the experimental energy -28 MeV of the alpha particle be understood? Realistic two-body potentials yield -20~ -23 MeV and deficiency of several MeV probably remains. The contribution of the three-body force to the alpha particle energy was investigated by Sato et al.[10].

The three-body force considered is the two-pion-exchange one (Fig.2) which was derived by Fujita et al.[11] in the static limit. The pionic form factor used by Brown et al.[12] and Nogami et al.[13] are not adopted, because they greatly reduces the one-pion-exchange potential (OPEP) even around 2 fm and changes the sign of OPEP at 1.4 fm when they are introduced into the two-body potential. The strength C_p of the three body potential was reexamined with the current data on the π+-N scattering and determined to be

$$C_p = 0.45 \text{ MeV}$$

by Sato et al.[10]. This value is three fourths of the currently used 0.61 MeV which was roughly estimated.

The expectation values of the three-body force with respect to the ATMS wave functions are given in Table III. The case of the triton is also included in the table. The value -1.16 MeV in the triton is consistent with -1.38 MeV by Pask [14] and -1.98×(0.45/0.61) MeV by

Table III. Contributions of the three-body force.

Nucleus	4He		3H
Wave function	H-J	OPEH	H-J
Total contribution	-7.53	-8.81	-1.16
SS	-2.12	-2.91	-0.31
SD (MeV)	-5.44	-5.98	-0.84
DD	0.03	0.07	-0.01
P(D) (%)	16.1	12.1	8.59
rms radius (fm)	1.61	1.49	1.95

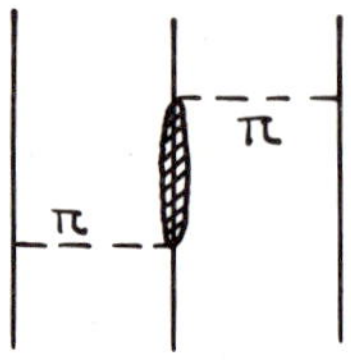

Fig.2. The two-pion-exchange three-body force.

Yang [15] (rms radius = 1.60 fm). The contribution of the three-body force to the alpha particle amounts to about 30 % of the binding energy. It is very large but is not unreasonable in comparison with the potential energy from the two-body force. It is known that the SD coupling term is large, because the three-body force has strong tensor character. The ratio of the contribution in the alpha particle to that in the triton is rather large 6∼7 and is almost unchanged by introducing the pionic form factor. The number 6∼7 is larger than the number of the trios of nucleons in the alpha particle. The contribution has a sensitive dependence on the nuclear density.

Full variational calculation with the two- and the three-body potentials was performed (Sato et al.[10]). The energy curves versus rms radius are shown in Fig.3. In the case of H-J, the rms radius moves from 1.61 fm to 1.47 fm and the contribution of the three-body force to the binding energy increases from 7.5 MeV to 10.2 MeV. It may be concluded that the three-body force has the large effects in the alpha particle, that is, about 10 MeV contribution to the binding energy and the favorable reduction of the rms radius.

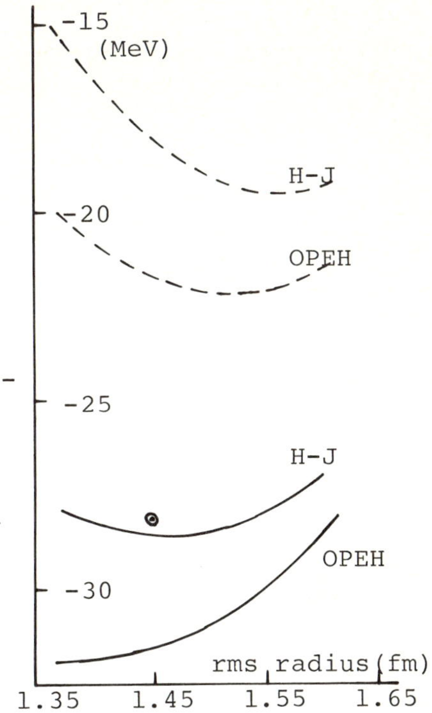

Fig.3. The energy curves of the alpha particle.

Excited states of the alpha particle

As the multiple scattering correlation function F is scalar and totally symmetric, the spin, isospin and parity assignment in the initial function Φ is retained. Therefore, we can apply the ATMS method to some excited states by taking initial functions with different quantum numbers.

The T=0, odd-parity states 0^-, 1^-, 2^- of the alpha particle were calculated by Sakai et al.[16]. They were treated as bound-state problem and H-J was used. The energy spectrum obtained is shown in Fig.4. The correct level order is reproduced, though the mechanism of the level splitting is rather different from that in the 1p-1h shell model. Energy shifts of about 1 MeV in the excited states and about 8 MeV in the ground state are, however, necessary for the fitting to the observed spectrum. As seen from Table IV, each term of the excited-state energies has a close resemblance to that of the triton. This fact seems to suggest a p-3H and/or n-3He structure in

Table IV. Energies of the T=0, 0^-, 1^-, 2^- states (in MeV).

Nucleus	4He				3H
J^π	0^-	1^-	2^-	ground 0^+	$1/2^+$
$E(J^\pi)$	-6.4	2.1	-2.9	-20.6	-6.0
Kinetic energy	74.8	76.0	75.4	131.1	65.9
Potential energy	-81.2	-73.9	-78.3	-151.7	-71.9
rms radius (fm)	2.26	2.29	2.28	1.65	1.95

the excited states. Then, the contribution of the three-body force in the excited states may be expected to be the nearly same value as that in the triton. These energy shifts may be regarded as a few data with which we can estimate phenomenologically the strength of the three-body force. calculation is now in progress. There is a possibility to attain a comprehensive understanding of the T=0 states including the ground state.

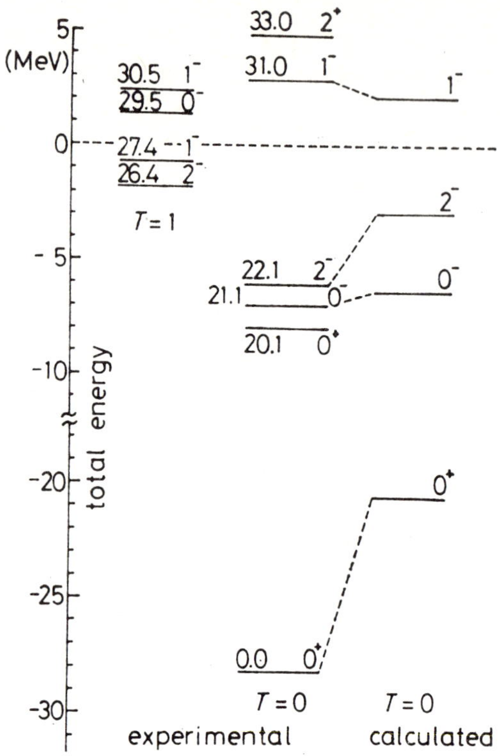

Fig.4. Energy spectrum of the alpha particle.

Concluding remarks

ATMS makes it possible to understand the structure of the alpha particle comprehensively on the basis of the realistic interaction. The three-body force gives the considerable effects on the binding energy and the nuclear density of the alpha particle.

It may be expected that the understanding the structure of the alpha particle is one step to the theory based on the realistic interaction.

Acknowledgements

I would like to express thanks to Professor H. Tanaka for his encouragement and valuable discussions, and Dr. Sakai, Professor M. Sato and authors of Prog. Theor. Phys. Suppl. No.56 for collaborations.

References

[1] Y. Akaishi, M. Sakai, J. Hiura and H. Tanaka, Prog. Theor. Phys. Suppl. No.56 (1974),6.
[2] S. Ali and A. R. Bodmer, Nucl. Phys. 80 (1966),99.
[3] T. K. Lim, Nucl. Phys. A158 (1970),385.
[4] J. L. Visschers and R. Van Wageningen, Phys. Letters 34B (1971), 455.
[5] M. Sakai, I. Shimodaya, Y. Akaishi, J. Hiura and H. Tanaka, Prog. Theor. Phys. Suppl. No.56 (1974), 32.
[6] T. Hamada and I. D. Johnston, Nucl. Phys. 34 (1962), 382.
[7] R. Tamagaki, Prog. Theor. Phys. 39 (1968),91.
[8] H. Tanaka and H. Nagata, Prog. Theor. Phys. Suppl. No.56 (1974), 121.
[9] T. Katayama, M. Fuyuki, Y. Akaishi, S. Nagata and H. Tanaka, ibid., 54.
[10] M. Sato, Y. Akaishi and H. Tanaka, ibid., 76.
[11] J. Fujita and H. Miyazawa, Prog. Theor. Phys. 17 (1957),360.
[12] G. E. Brown and A. M. Green, Nucl. Phys. A137 (1969), 1.
[13] R. K. Bhaduri, Y. Nogami and C. K. Ross, Phys. Rev. C2 (1970), 2082.
[14] C. Pask, Phys. letters 25B (1967),78.
[15] S. N. Yang, Phys. Rev. C10 (1974),2067.
[16] M. Sakai, Y. Akaishi and H. Tanaka, Prog. Theor. Phys. Suppl. No.56 (1974),108.

MULTI-CLUSTER PROBLEM BASED ON THE MICROSCOPIC THEORY

Hisashi Horiuchi

Department of Physics, Kyoto University,
Kyoto 606 JAPAN

Introduction

Recent development of the cluster model study of the light nuclei [1] is revealing the existence and the importance of the multi-cluster structure. One of the most typical examples is the 3α structure [1∿4] of the excited levels in ^{12}C, namely the 0_2^+ state at 7.6 MeV and the broad 2_2^+ state [4] at 10.3 MeV.

For the investigation of these multi-cluster structure, it is strongly urged to solve the inter-cluster motion exactly as in the case of the few-nucleon problem. The reason is as follows: As is seen in the small binding energy (unbound by 95 keV) of the α-α system, the inter-cluster interaction is generally weak and so we can expect the existence of the structure of multi-clusters whose relative motions are widely spread. The microscopic studies [5] of the inter-cluster interaction based on the resonating group method (RGM) show as the general feature of the composite particle force, the strong inner exchange repulsion and the weak outer attraction.

The classical α particle model [6] predictions of the excited state properties contradict to the present experimental data in many important facets such as the impossiblility of reproducing the excited rotational band upon 6.05 MeV 0_2^+ in ^{16}O. The failure of this model is considered to be due to its wrong fundamental assumption; it postulates for all the low-lying excited states the small vibration of the α particles around (or the rotation of) the ground-state-arrangement of the α's. This is hardly justified from the above-mentioned character of the elementary inter-cluster interaction.

The purpose of this paper is to show that, if we use the proper inter-cluster interaction which are based on the microscopic nucleon-cluster model study, the multi-cluster model revives again as a promising approach in getting the unified understanding of the light nuclear structure. This time we report the results of the study of the 3α system which is a typical case of multi-cluster systems since the α-α interaction is well understood from the microscopic studies and as stated before the levels with the well-developed 3α structure are observed in ^{12}C. The study of 3α system is a starting point into more complex systems.

The Orthogonality Condition Model

In our study of the many-cluster problem, we adopt the orthogonality condition model (OCM) [7] to describe the inter-cluster interaction. The OCM has been introduced by Saito for the α-α system in order to extract the essential roles of the exchange kernels of the RGM [5,8∿10] which has succeeded not only in reproducing the α-α experimental phase shifts but also in giving the microscopic interpretations of the characteristic features of the phenomenological local potentials proposed to fit the α-α data.

The OCM equation of α+α system is,

$$(T+V^D-E)\chi(\vec{r})=0 \ , \quad \langle\chi(\vec{r})|\text{Forbidden States}\rangle=0 \ ,$$

where V^D is the effective α-α potential which is here chosen to be the folding potential with use of two-nucleon force. The forbidden states $\chi_F(\vec{r})$ are defined as the solution of

$$\mathcal{A}\{\chi_F(\vec{r})\phi(\alpha_1)\phi(\alpha_2)\}=0 \ ,$$

and when we adopt the harmonic oscillator wave function with oscillator parameter $\nu(=m\omega/2\hbar)$ for $\phi(\alpha)$, χ_F are the harmonic oscillator function $R_{N\ell}(r)Y_{\ell m}(\hat{r})$ with number of oscillator quanta $N=2n+\ell$ less than 4. (Needless to say, all odd partial wave states are forbidden.) Then for S wave 0S, 1S states are forbidden, for D wave 0D is forbidden but for other partial waves all states are allowed. Due to the orthogonality condition (OC) to the forbidden states, the α-α relative wave functions for

S and D waves have the almost energy-independent inner oscillation for a wide energy range. Since inner oscillation causes a large amount of kinetic energy, it works like as the repulsive force in this inner region. The position of the outermost nodal point is just near the radius of the repulsive core of the phenomenological potentials [11]. This is the explanation by Tamagaki-Tanaka [9], Okai-Park [10] and Saito [7] why we need the repulsive core in the phenomenological potentials. Fig.1 shows the good fit of the OCM calculation by Saito to the experimental α-α scattering phase-shifts. The OCM has been also applied with fruitfull successes to the $\alpha+{}^{16}O$ and $\alpha+{}^{12}C$ systems [1].

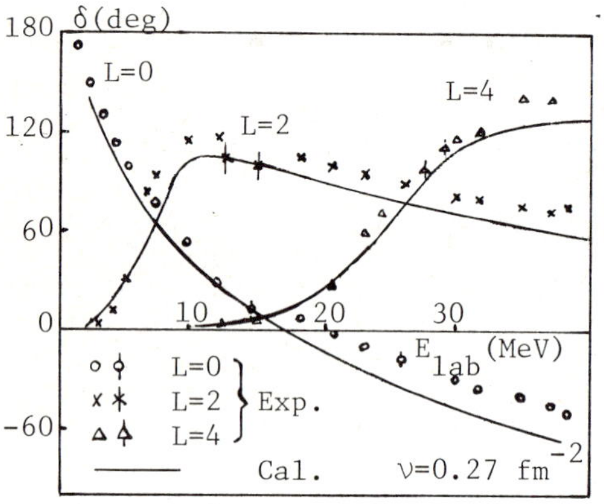

Fig.1 α-α phase-shifts by OCM
Two-nucleon force is Schmidt-Wildermuth one with Serber mixture.

It is easily understood to be plausible in quantum system that the Pauli exclusion principle causes the forbidden states to occupy instead of the forbidden area (core area) in configuration space. Compared to the static nature of the potential core, the repulsive function of OCM is more dynamical; for example, at high energy where the relative motion wave length is fairely shorter than the forbidden state spread, the orthogonality to the forbidden states is automatically satisfied, which means the repulsive core effect of OCM has melted. We will see later in some situation of the multi-cluster system the repulsive core effect of OCM also melts away.

Formulation of 3α OCM

By employing the OCM to describe the α-α interaction, the 3α equation of motion is written,

$$(\Sigma T_i - T_G + \sum_{i>j} V^D_{ij} - E)\chi(\vec{s},\vec{t})=0 , \quad \vec{s}=\vec{X}_3-(\vec{X}_1+\vec{X}_2)/2, \ \vec{t}=\vec{X}_2-\vec{X}_1,$$

$$\langle\chi(\vec{s},\vec{t})\,|\,\text{any } \alpha\text{-}\alpha \text{ pair forbidden states}\rangle=0 .$$

The functional space of 3α can be devided into two mutually orthogonal subspaces; one is composed of the 3α wave functions which satisfy the above orthogonality condition (OC) to the 2α forbidden states between any α-α pair of 3α system, and is called as the 3α allowed space. The other is the subspace orthogonal to the 3α allowed space and is called as the 3α forbidden space. It is easy to prove that any function $\chi_F(\vec{s},\vec{t})$ of the thus-defined 3α forbidden space satisfies the equation $\mathcal{A}\{\chi_F(\vec{s},\vec{t})\phi(\alpha_1)\phi(\alpha_2)\phi(\alpha_3)\}=0$. (Inversely the solution of this equation can be shown to belong to the above-defined 3α forbidden space [3].) The 3α allowed states (which are of course totally symmetric) can be classified by the number of the harmonic oscillator quanta N and by the Elliott SU(3) group [3], as shown in Table I.

N	$(\lambda,\mu)^n$
8	(0,4)
9	(3,3)
10	(6,2),(2,4)
11	(9,1),(5,3),(3,4)
12	(12,0),(8,2),(6,3),(4,4),(0,6)
13	(11,1),(9,2),(7,3),(5,4),(3,5)
14	(14,0),(12,1),(10,2),(8,3),(6,4)^2,(2,6)
15	(15,0),(13,1),(11,2),(9,3)^2,(7,4),(5,5) (3,6)

Table I. SU(3) classification of 3α allowed states

This table shows a good correspondence of the 3α allowed states with the shell model classification; for N<8 there is no state allowed by the Pauli principle and for N=8 only $\Phi\{(0s)^4(0p)^8;(0,4)J\}$ is allowed as [444] symmetry shell model states. The explicit relation between our 3α allowed states $\chi_A((\lambda,\mu)\kappa J)$ and the ^{12}C shell model states can be obtained by constructing the 12-nucleon wave function $\mathcal{A}\{\chi_A((\lambda,\mu))\phi(\alpha_1)\phi(\alpha_2)\phi(\alpha_3)\}$; for example $n_A\mathcal{A}\{\chi_A((0,4)J)\phi(\alpha_1)\phi(\alpha_2)\phi(\alpha_3)\}=\Phi\{(0s)^4(0p)^8;(0,4)J\}$.

Results and Comparison with Experiments

The 3α OCM equation has been solved by diagonalizing the Hamiltonian matrix with the 3α allowed states [3]. We have used following Ref.[7] Schmidt-Wildermuth two-nucleon force with Serber mixture to get V^D. Oscillator parameter $\nu=0.275$ fm^{-2}. Fig.2 shows the calculated energy spectra which are compared with the observed ^{12}C energy levels below 15.11 MeV T=1 $J^\pi=1^+$ state, except the 14.71 MeV level whose T

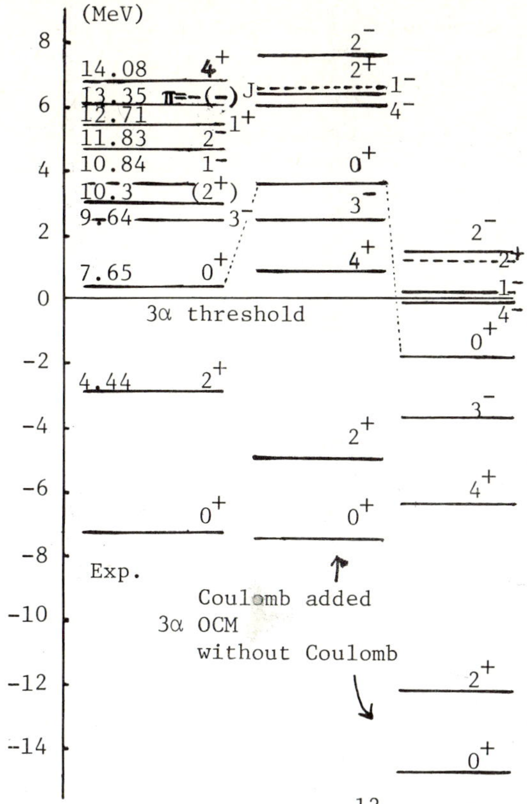

Fig.2 Energy spectra of ^{12}C by OCM 0_2^+ energy is obtained by adopting all the allowed 3α translationally invariant harmonic oscillator states up to N =30. It is almost convergent.

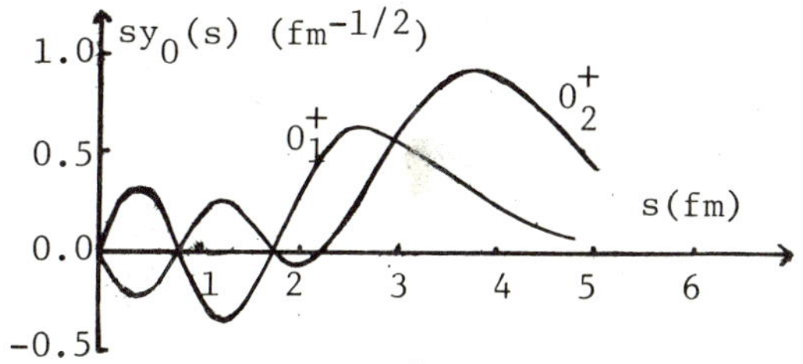

Fig.3 Reduced α+^{8}Be(0^+) width amplitudes of 0_1^+ and 0_2^+ states. S^2 values are 0.62 and 1.40 for 0_1^+ and 0_2^+, respectively.

N	8	10	12	14	16	18	20	22
0_1^+	0.80	0.08	0.08	0.02	0.01	0.01	<0.01	<0.01
0_2^+	0.13	0.13	0.12	0.17	0.14	0.12	0.08	0.05

Table II. Squared norms of the expansion components by the number of oscillator quanta of 0_1^+ and 0_2^+ by OCM.

and J^π are unknown. All these observed levels have T=0 and we consider that our calculation has reproduced whole these levels except the 1^+ level at 12.71 MeV. This level consists mainly of the spin angular momentum S=1 configuration and can not be represented by α cluster model.

First we consider the positive parity levels. The calculated binding energy of the ground state 0_1^+ is seen to be very good. We will discuss this point later. The moment of inertia of the calculated ground band (0_1^+, 2_1^+, 4_1^+) is however too large compared with experiments. We see in Fig. 2 the second 0^+ is reproduced near 3α breakup threshold. Fig. 3 shows the reduced width amplitudes (RWA) of α+^{8}Be(0^+) breakup for 0_1^+ and 0_2^+. (Here ^{8}Be(0^+) wave function for RWA evaluation is calculated by solving 2α OCM.) We clearly see that 0_2^+ is the state with large α+^{8}Be(0^+) clustering at surface [2]. This is consistent with large experimental α+^{8}Be(0^+) decay reduced width of the observed 0_2^+ near Wigner limit value. (θ^2_{exp}=1.0±0.3.) This result differs from the usual postulation of the geometrical linear chain configuration of 3α's to this level. To see the characters of 0_1^+ and 0_2^+ further, the squared norms of the expansion components by the number of oscillator quanta are given in Table II. The ground state (0_1^+) is seen to have a large (0,4) (N=8) component, showing a good correspondence with the p-shell shell model description of the ground state. On the contrary, the expansion coefficients of 0_2^+ spread over many higher N, which also indicates the well-developed clustering character of this level. In spite of this large difference between two structures, the monopole transition matrix element between them is obtained to be 6.1 fm^2 which is close to the observed value 5.8 fm^2. As for 2_2^+ state the convergence of our calculation is not full and in Fig. 2 the extrapolated energy position is shown by dotted line. This 2_2^+ has a similar character as 0_2^+ and is considered to correspond to the observed 10.3 MeV level with large α width. The 2^+ assignment [4] to this 10.3 MeV level is also supported by the fact that our calculation gives 0_3^+ and 0_4^+ levels which lie several MeV higher than 2_2^+ and have small α widths (S^2<0.4).

Next we investigate the negative parity states. We can see in Fig. 2 the level position of the calculated 3_1^-, 1_1^-, 2_1^- correspond fairly well with the observed ones. Our calculation gives a low-lying 4_1^- level near 1_1^-. We suggest that this state corresponds to the observed level at 13.35 MeV which has unnatural parity but J assignment is uncertain experimentally. As for the other candidates of 1^+, 2^-, 3^+, our calculation gives 2_2^- and 3_1^+ several MeV higher than 4_1^- and very high excitation energy >30 MeV to 1_1^+ which reflects that 1^+ is an unfavorable coupling in 3α model (allowed 1^+ 3α state does not exist for N<12 and for N≥12 there are few allowed 1^+ states for each N.) The microscopic α model calculation with Brink-type wave function [12] also gives low lying 4_1^- near 1_1^-. These negative parity levels seem to constitute $K^\pi=3^-$ and 1^- rotational bands reflecting the trance of the SU_3 (3,3) representation of N=9 states.

Comparison with the 3α Calculation with the Phenomenological α-α Force with Repulsive Core

One of the characteristic features of 3α OCM results is the large binding energy of the ground state 0_1^+, which is remarkably different from the previous 3α calculations with phenomenological α-α forces. Similar results are also obtained recently with use of α-α forces to reproduce inner oscillating behavior [13]. In Table III we compare our OCM results with the 3α results [15] with use of the Ali-Bodmer α-α force [11]. In order to investigate the reason for this large difference, we adopt the following identity relation,

$$\langle\Phi_1(3\alpha)|(\Sigma T_i - T_G + \Sigma_{i>j} V_{i,j})|\Phi_2(3\alpha)\rangle = 2\langle\Phi_1(3\alpha)|(T_{1,2}+1.5V_{1,2})|\Phi_2(3\alpha)\rangle$$

which is valid for any totally symmetric three-body wave functions Φ_1 and Φ_2. This equation simply reflects the fact that in n-body system there are n (n-1)/2 pairs of two-body interaction terms while the number of kinetic energy terms increases linearly as (n-1). When 3α's come near each other, the above increase of the effective number of interaction bonds over that of kinetic energy terms causes the strengthening of the repulsive effect in the inner region for the case of the α-α force with inner repulsive core. On the contrary, in the OCM case, the (deep) attractive potential is deepend by 1.5 times while the 2α orthogonality condition remains unchanged, thus resulting in an increase of the attractive effect in the whole potential range for the 3α allowed compact state such as (0,4) (N=8). The above mentioned situation can be seen quantitatively in Table IV, where the eigen-energies of the fictitious 2α Hamiltonian $T_{1,2}+1.5V_{1,2}$ are given.

	OCM			Ali-Bodmer(e)	
$\Sigma_\ell V_\ell(\alpha-\alpha)$	S+D	S+D+G	full	S+D	S+D+G
0_1^+	-2.42	-14.43	-14.68	-6.64	-7.40
0_2^+	5.98	-1.11	-1.26	-1.57	-2.01

Table III. Comparison of 0_1^+ and 0_2^+ energies by OCM with those calculated with the Ali-Bodmer force(e). (Coulomb force is neglected in both cases.)

	OCM	Ali-Bodmer(e)
0^+	-13.18	-4.45
2^+	-9.57	-2.52
4^+	-1.36	-12.73

Table IV. Lowest three eigenenergies of the fictitious 2α Hamiltonian $T_{\alpha\alpha}+1.5V_{\alpha\alpha}$.

It should be noted that the so-called "core" region of the inter-cluster interaction is very large; in 2α case almost half the radial spread of ^{8}Be density is in the "core" region ($\gamma_{core}\lesssim 2$fm). Thus the opposite actions originating from the different treatments of the inner "core" region part of the α-α interaction can cause such large difference of the 0_1^+ energies in Table III. In 0_1^+ state by OCM, the increase of the attractive effect in the inner region overwhelms the large kinetic energy due to the inner oscillation of α-α wave function. This leads to the large amplitude of the inner oscillation of α-α waves which means the invalidity of the substitution of the repulsive core in place of the orthogonality condition.

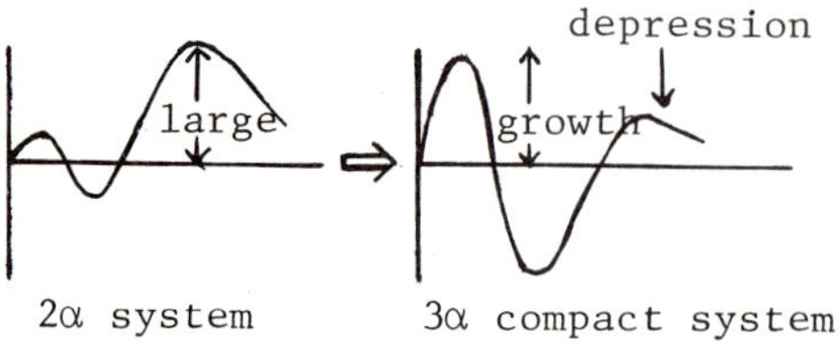

Coexistence of Shell- and Molecule-like Structures

Since, as seen above, the inner repulsive function of OC can melt away in 3α system, it is no more self-evident that we have obtained the well-developed 3α molecule-like structure of 0_2^+. The mechanism of the coexistence of the different structures 0_1^+ and 0_2^+ will be given by the following conjecture. The second 0^+ state must be orthogonal to the ground 0_1^+ state. Since the 0_1^+ state has a compact 3α structure, the orthogonality requirement of 0_2^+ to this 0_1^+ makes at least one of the 3α's in 0_2^+ go apart from others. Under this situation, the 2α OC recovers its function as the repulsive core. These two kinds of repulsive effect generate the molecular structure of 0_2^+ bound by the α-α tail attraction. According to the above conjecture, the inter-relation between 0_1^+ and 0_2^+ is close in spite of their apparent large difference. This close relation between 0_1^+ and 0_2^+ explains why the observed monopole transition rate is so large between two 0^+ state with very different structures. The fast β-decay rates to 0_2^+ or 2_2^+ from ^{12}B (or ^{12}N) are also considered to reflect the interplay character predicted by our OCM. Since our above argument originates from the fundamental character of OCM for the general inter-cluster interaction, we can expect similar explanation as in ^{12}C will apply to the coexistence phenomena in other nuclei. The non-small monopole transition rate between the 0_2^+ with cluster structure and the ground 0_1^+ with shell structure in ^{16}O may be another example of the dynamical interplay effect predicted by our OCM approach.

Concluding Remarks

From the success of the 3α OCM we can expect the many-cluster problem with use of the inter-cluster interaction described by OCM gives fruitful results for the unified understanding of the light nuclear structures. It can treat not only the developed cluster states but also the compact shell model states in a same framework. This makes us possible to investigate the coexistence mechanism of different structures. The 3α OCM results show clearly the indispensablility of the proper treatment of the interaction between composite particles based on the microscopic model and present a good starting point for the studies of more complex systems such as $^{16}O+2\alpha$ [14] and $^{12}C+2\alpha$ which are now in progress giving interesting results.

Acknowledgement

The author wishes to thank Profs. R. Tamagaki and K. Ikeda, Drs. S. Saito and Y. Suzuki and Messrs. K. Kato and Y. Fujiwara for many valuable discussions on many-cluster problem.

References

[1] Proceedings of the INS-IPCR International Symposium on Cluster Structure of Nuclei and Transfer Reactions Induced by Heavy Ions (1975)
[2] H. Horiuchi, K. Ikeda and Y. Suzuki, Prog. Theor. Phys. Suppl. No.52 (1972) 89.
[3] H. Horiuchi, Prog. Theor. Phys. 51 (1974) 1266 and 53 (1975) 447.
[4] H. Morinaga, Phys. Letters 21 (1966) 78.
[5] J. Hiura and R. Tamagaki, Prog. Theor. Phys. Suppl. No.52 (1972) 25.
[6] D.M. Dennison, Phys. Rev. 96 (1954) 378 and A.E. Glassgold and A. Galonsky, ibid 103 (1956) 701.
[7] S. Saito, Prog. Theor. Phys. 41 (1969) 705.
[8] K. Wildermuth and W. McClure, Springer Tracts in Modern Physics, 41 (Springer Verlag Berlin, 1966).
[9] R. Tamagaki and H. Tanaka, Prog. Theor. Phys. 34 (1965) 191.
[10] S. Okai and S.C. Park, Phys. Rev. 145 (1966) 787.
[11] S.A. Afzal, A.A.Z. Ahmad and S. Ali, Rev. Mod. Phys. 41 (1969) 247.
S. Ali and A.R. Bodmer, Nucl. Phys. 80 (1966) 99.
[12] N. de Takacsy, Nucl. Phys. A178 (1972) 469.
[13] Y. Fujiwara and R. Tamagaki, private communication.
Yu.F. Smirnov, I.T. Obukhovsky, Yu.M. Tchuvil'sky and V.G. Neudatchin, Nucl. Phys. A235 (1974) 289.
[14] K. Kato and H. Bando, private communication.
[15] J.L.Visschers and R.van Wageningen, Phys.Letters 34B (1971),455.

THE DEUTERON WAVE FUNCTION FROM ND SCATTERING EXPERIMENTS AND NN POTENTIAL

V. V. Komarov

Institute of Nuclear Physics, Moscow State University;
Moscow, U.S.S.R.

As is known in quantum mechanics, the behaviour of the amplitude of two-body potential scattering off the energy shell and the dependence of the bound state wave function for two particles on the relative momentum of these particles is largerly determined by the form of the interaction potential function.

In the paper /1/ it was discussed that from some three-particle scattering experiments in nuclear physics one can extract experimental information on a drift of the two-particle scattering amplitude off the energy shell and, hence, obtain on important evidence for the structure of the two-particle interaction potential.

In the present report, in an effort to look into nucleon-nucleon forces, we have analyzed the dependence of the vertex function of the deuteron decay into two nucleons $G_d(P)$ on the relative momentum P of nucleons . The vertex function $G_d(P)$ was extracted from the differential cross sections of ND elastic scattering at large angles.

The function $G_d(P)$ being investigated is related to the deuteron wave function in the momentum representation $\varphi_d(P)$ by the relation

$$\varphi_d(P) = G_d(P)(P^2+\alpha^2)^{-1}, \tag{1}$$

where α^2/m is the deuteron binding energy.

As is known, in the spin space of two nucleons the operator $\hat{\varphi}_d(P)$ can be represented as

$$\hat{\varphi}_d(P) = \frac{1}{4\pi}\left\{u(P) + \frac{1}{\sqrt{8}}\hat{S}_{12}(P)\,w(P)\right\}, \tag{2}$$

where $\hat{S}_{12}(P)$ is the tensor force operator and $u(P)$ and $w(P)$ are the scalar functions corresponding to the S and D deuteron states.

Similarly, the operator $\hat{G}_d(P)$ in the spin space is represented as

$$\hat{G}_d(P) = \frac{1}{4\pi}\left\{g_0(P) + \frac{1}{\sqrt{8}}\hat{S}_{12}(P)\,g_2(P)\right\}. \tag{3}$$

Here the functions $g_0(P)$ and $g_2(P)$ are the contributions from the S and D states to the deuteron vertex function. For determining the functions $g_0(P)$ and $g_2(P)$, there exist equations of the form:

$$g_0(P) = -4\pi\int_0^\infty \frac{V_0(P,K)\,g_0(K)}{K^2+\alpha^2}K^2dK, \quad g_2(P) = -4\pi\int_0^\infty \frac{V_2(P,K)\,g_2(K)}{K^2+\alpha^2}K^2dK, \tag{4}$$

where

$$V_0(P,K) = \int_0^\infty j_0(P,r)\,V(r)\,j_0(K,r)\,r\,dr$$

$$V_2(P,K)=\int_0^\infty j_2(P,r)V(r)j_2(K,r)\,r\,dr. \tag{5}$$

As follows from eqs. (4), the properties of the functions $g_0(P)$ and $g_2(P)$ are mainly determined by the kernel of these equations or by the properties of the functions $V_0(P,K)$ and $V_2(P,K)$ with respect to the variable P.

Since equations (4) can be solved analytically only for the square well potential it is impossible to obtain the functional dependences of $g_0(P)$ and $g_2(P)$ on the function $V(r)$. Therefore, for the analysis of the phenomenological vertex function of the deuteron with a view to reproducing the structure of the nucleon-nucleon potential $V(r)$, equations (4) have been solved by numerical methods. In the present report it was done for a number of different functions $V(r)$: (a) the Hülthen potential /2/; (b) the potential with a finite repulsive core /3/; (c) the Hamada-Johnston potential /4/; (d) the Gaussian potential; (e) the potential of the Yale university group /5/; (f) the Saxon-Woods potential with a repulsive core /6/; (g) the potential suggested in the work /7/ with a finite repulsive core; and a square-well potential with a repulsive core. All the functions used in this calculations made it possible to reproduce the phase-shift curves in nucleon-nucleon scattering and the low-energy parameters with the same accuracy. Below, the results of one calculations will be given in $g_0^2(P)$ and $g_2^2(P)$ terms for the direct comparison with a fenomenalogical deuteron wave function, which we extract from the experiments.

Our mathematical experiment has shown that as expected in the region of small momentum P values, $P \leqslant 1$ fm^{-1}, the function $g_0^2(P)$ is far greater than the function $g_2^2(P)$. Next, in this region the behaviour of $g_0^2(P)$ is dependent on $V(r)$ behaviour when $r \to \infty$, and on the mean radius of the potential but is independent of the structure of $V(r)$ at small r comparable with the nucleon radius ($r_N \approx 0.5$ fm). Thus the function $g_0^2(P)$ is independent of the form of the core, when $P \leqslant 1$ fm^{-1}.

Fig.1 presents, some of the calculated curves which demonstrated the dependence of the function $g_0^2(P)$ on the form $V(r)$ for $r \to \infty$.

In the range of variation in the momentum P from 1 fm^{-1} to 2 fm^{-1} the behaviour of $g_0^2(P)$, is determined by the definition of the function $V(r)$ at small distances. If the potential $V(r)$ is assumed to be of an attractive character, then the function $g_0^2(P)$ rapidly decreases with increasing P. If the repulsive core in the nuclear potential is assumed to be located at a distance approximately equal to the nucleon radius, then the function $g_0^2(P)$ in the indicated region has a minimum whose position depends on the core radius r_c and on the finite or infinite value of the core V_c. The function $g_0^2(P)$ in the covered range of variation in the momentum P exhibits a maximum for the potentials with a repulsive core (Fig. 1) whose width and height are connected with the definition of a specific form of the core. The decrease in the function $g_2^2(P)$ when $P > 3$ fm^{-1} is also largerly associated with the form of the core. These characteristic features of the behaviour of the functions $g_0^2(P)$ and $g_2^2(P)$ can be illustrated with some examples given in Fig. 1. Now, it is interesting to compare the above mentioned behaviour of $g_0^2(P)$ and $g_2^2(P)$ in dependence of $V(r)$ to the values of the deuteron vertex function obtained from the experimental data.

As noted previously, the dependence of the vertex of the deuteron $G_d(P)$ on the relative momentum of two nucleons can be obtained from the analysis of the differential cross sections for N + d reactions. We have suggested that the angular distributions of protons elastically scattered from deuterons at angles of 140°-180° be used for this purpose.

In a theoretical analysis of the angular distributions it was proposed to use the representaion of the elastic scattering amplitude as the infinite series of nonrelativistic graphs (Fig. 2a), which was obtained by the graph Summation method. The contributions of the graph A in Fig. 2a represent the process of nucleon pick-up from the deuteron by a proton, B represent the process of direct scattering of a proton from a nucleon of the deuteron. It is known that the sum of the graphs $A_1 + B_1$ represents the amplitude in the impulse approximation. The other graphs of the series represent single ore more rescattering amplitudes. These graphs are correction to the impulse approximation for the case of high energy initial protons and can be estimated for this case. The momenta K_0 and K on the graphs of Fig. 2a denote the initial and final proton momenta in the c.m. system.

When $K \approx -K_0$, i.e. for large angles of elastically scattered protons and at a proton energy far in excess of the binding energy of nucleons in the deuteron the amplitude of the p + d → p + d reaction is mainly determined by the contribution from the graph A_1.

$$A_1(\vartheta_{\bar{K}_0\bar{K}}) \approx G_d^2\left(K_0\sqrt{\tfrac{5}{4} + \cos\vartheta_{\bar{K}_0\bar{K}}}\right)\left[\alpha^2 + K_0^2\left(\tfrac{5}{4} - \cos\vartheta_{\bar{K}_0\bar{K}}\right)\right]^{-1} \tag{6}$$

The basic dependence of the amplitude of the (p + d → p + d) reaction on angle $\vartheta_{\bar{K}_0\bar{K}}$ between the directions $\bar{K}_0$ and $\bar{K}$ in the region $\vartheta_{\bar{K}_0\bar{K}} \approx 180^\circ$ is determined by the vertex function $G_d^2(P)$ and propagator. It thus appears that the behaviour of the proton angular distribution at large angles can provide information on the behaviour of the function $G_d^2(P)$ for the momentum values $P = K_0(5/4 + \cos\vartheta_{\bar{K}_0\bar{K}})^{1/2}$. In what follows we shall be interested in variations of $\cos\vartheta_{\bar{K}_0\bar{K}}$ from - 0.8 to 1.0 Hence, the function $G_d^2(P)$ will be reproduced from the experimental data in the range of variation in P from $K_0 0.5\ \mathrm{fm}^{-1}$ to $K_0 0.67\ \mathrm{fm}^{-1}$.

The angular distributions of protons elastically scattered by deuterons for the primary proton energies in the lab. system 10 MeV /8/ and 30 MeV /9/ have been analysed for investigating the vertex function in the region of small momentum values $P \leq 1\ \mathrm{fm}^{-1}$. For this energies of initial protons the differential cross section calculations should be based on the exact integral equations for the proper amplitudes. We used for this purpose the method of solution given in ref. /10/, where was used a separable potential of exponential form. This potential provided the same behaviour of the off-shell two-nucleon amplitudes and the deuteron vertex in the region under consideration as OPEP. The good agreement of the theoretical and experimental differential cross sections for 10 and 30 MeV demonstrated that the OPEP tail is more preferable than other potentials' tails (Fig. 1a).

Next, we have investigated the proton angular distributions from the (p + d → p + d) reaction at the primary proton energies 150 MeV /11/, 198 MeV /12/ and 185 MeV /13/. Our theoretical analysis of the differential cross sections for protons elastically scattered from deuterons at large angles was based on the representation of the reaction amplitude by the contributions from the graphs A_1 A_2 and B_1, B_2 Fig. 2b. This description of the amplitude of the elastic (p + d) scattering was obtained correct

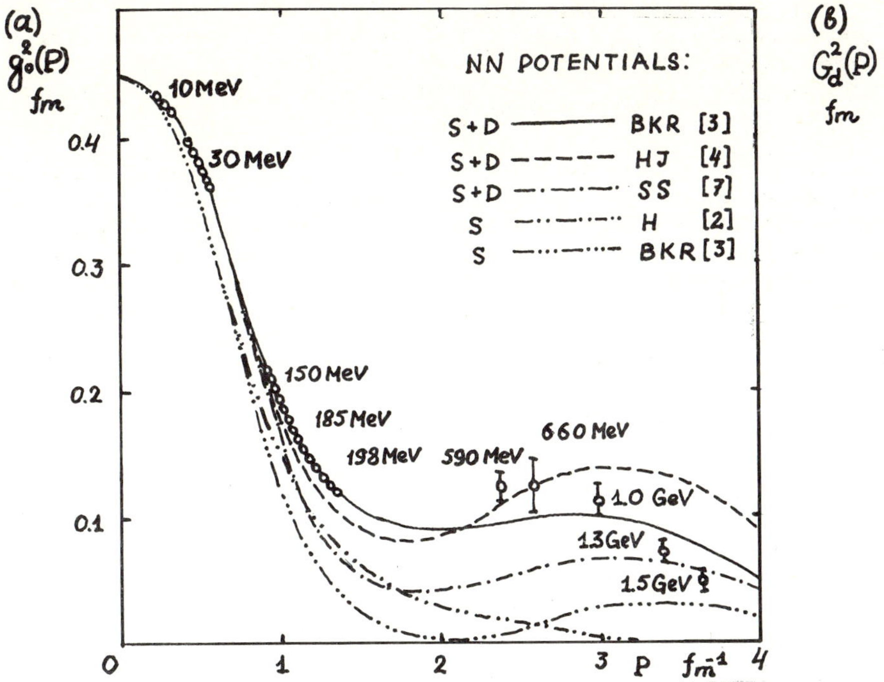

Fig.1. The square of the deuteron vertex function $G_d^2(P)$, calculated for different NN potentials. a(S-state , $g_0^2(P)$, b) S + D state, $G_d^2(P)=g_0^2(P)+g_2^2(P)$ Points respresent $G_d^2(P)$ extracted from ND experiments.

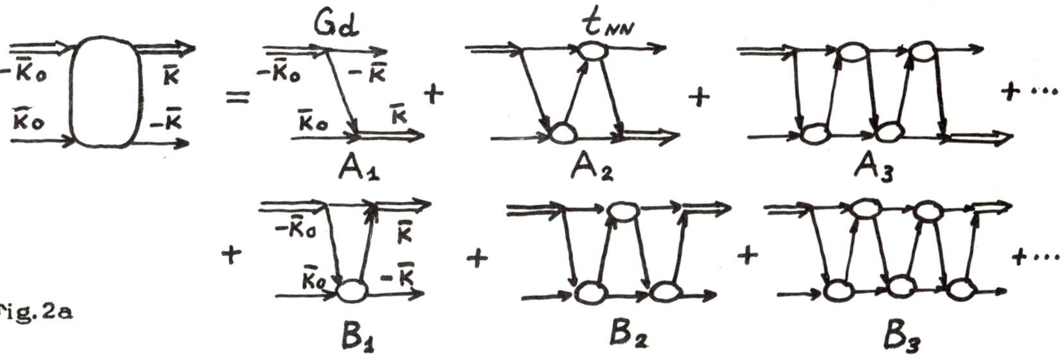

Fig. 2a

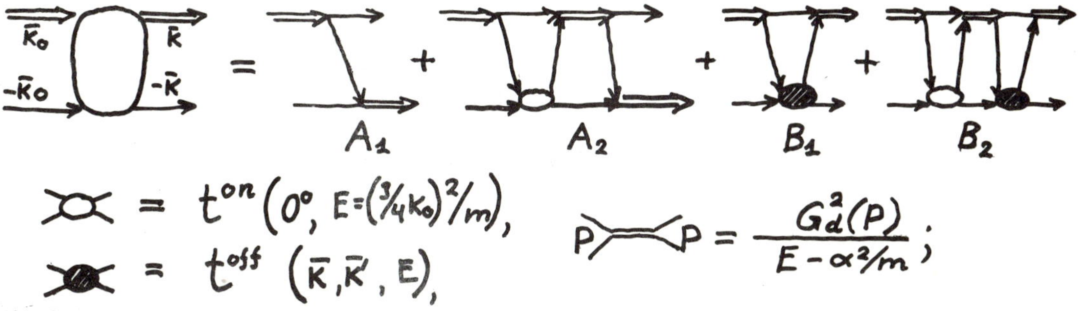

Fig. 2b

to the terms $(\gamma/K_0)^4$. The momentum γ in the small parameter $(\gamma/K_0)^4$ is a mean momentum of nucleon in the deuteron. It can be determined from the expression

$$\int_0^{\gamma} \varphi_d^2(P)\, dP \approx 1$$

If the function $\varphi_d(P)$ is Hülthen function, then $\gamma \approx 5\alpha$. The above analysis was based on convergence properties the series corresponding to the contributions from the graphs A and B, Fig. 2a /14/. The second and the follows intermediate rescatterings by this comsideration include quasi free two-nucleon scatterings at high relative energies and scatterings at low relative energies with a deuteron formation. In proposed description for the (p + d → p + d) reaction differential cross section at large proton emission angles the following expression was obtained

$$\frac{d\sigma}{d\Omega} = \frac{1}{6}\left\{\sum_{SPIN} F(\vartheta)F^*(\vartheta)\right\}; \quad \vartheta = \vartheta_{\bar{K}_0 \bar{K}},$$

$$F(\vartheta) = \left\{ G_d^2\left(K_0\sqrt{\tfrac{5}{4} - \cos\vartheta}\right)\left[\alpha^2 + K_0^2\left(\tfrac{5}{4} - \cos\vartheta\right)\right]^{-1} + \right.$$

$$\left. + 4\Phi_d\left(\frac{K_0+K}{2}\right)\left[t_{np}^{off}\left(\tfrac{3}{4}K_0; \bar{K} + \tfrac{\bar{K}_0}{2}; E = \left(\tfrac{3}{4}K_0\right)^2/m\right) + t_{pp}^{off}\left(\tfrac{3}{4}K_0; \bar{K} + \tfrac{\bar{K}_0}{2}; E = \left(\tfrac{3}{4}K_0\right)^2/m\right)\right]\right\} F_2,$$

$$F_2 = \frac{1}{3\pi}\left\{1 + \frac{\gamma^2}{2\pi K_0}\left[t_{np}^{on}\left(\vartheta = 0^\circ; E = \left(\tfrac{3}{4}K_0\right)^2/m\right) + t_{pp}^{on}\left(\vartheta = 0^\circ; E = \left(\tfrac{3}{4}K_0\right)^2/m\right)\right]\right\}.$$

The functions t_{np}^{off}, t_{pp}^{off}, G_d, and Φ_d (deuteron form-f.) have been calculated for BKR (S+D), HJ (S+D), and SS (S+D)

(7)

Here t_{np} and t_{pp} are the (n,p) and (p ,p) scattering amplitudes at zero angle and energy $(3/4k)^2/m$. We note that our representation (7) of the differential cross section for the reaction in question differs from those known in the literature /15/ in that the contribution to the amplitude of the graph A_2 and B_2 (Fig. 2a) was taken into account. This contribution is a correction to the amplitude of the impulse approximation in the description of nucleon elastic scattering at large angles. An analysis of the experimental angular distribution of protons emitted at large angles from the (p + d → p + d) reaction for the proton energies 150 MeV, 198 MeV and 185 MeV using eq. (7) has made it possible to choose the values of the deuteron vertex function $G_d^2(P)$.

Comparing the phenomenological behaviour of the function in the range of variation in the momentum P from 0 to 1.5 fm^{-1} with the curves calculated for the different functions $V(r)$ one may draw a conclusion that the best representation for the two-nucleon interaction potential is the function $V(r)$ whose behaviour at large radii is similar to that of the OPEP whereas at small radii in the region $r \approx 0.7 - 0.5$ fm it has a repulsive core.

Information on the structure of a repulsive core can be extracted from the analysis of the behaviour of the function $G_d^2(P)$ in the range of variation in p from 2 fm^{-1} to 4 fm^{-1}. Therefore we have investigated the angular distributions of protons from the (p + d → p + d) reaction at large angles for the following primary proton energies 590 MeV /12/, 660 MeV /12/, 1 GeV /16,17/, 1.3 GeV /16/ and 1.5 GeV /16/.

Using the graph summation method for the theoretical description of these angular distributions we have obtained a representation for the

(p+ d → p+ d) scattering amplitude taking into account two- and three-body forces whose effect in the given kinematic region of interaction of three nucleons is significant. The contributions from the graphs which, except for the terms of the order $(\gamma(\kappa_0))^4$, should be included in the calculation of the reaction amplitude are shown in Fig. 3. Here the graphs A_1 and A_2 represent the process of nucleon pick-up by a proton from the deuteron determined by two-body forces whereas the graphs C_1 and C_2 represent this process due to three-body forces. This way of implementation of the two and three body forces based on the assumption that nucleon-nucleon interaction can be described in the model of one and two-pion exchange /19/. The exchange of barions by that is included in the intermediate states of the two-pion exchange.

Using the procedure for calculating the contribution from the graph C_1 proposed in the work /18/, we have obtained the following representation for the (p + d → p + d) reaction differential cross section at large angles at the primary proton energies 0.5 - 1.5 GeV

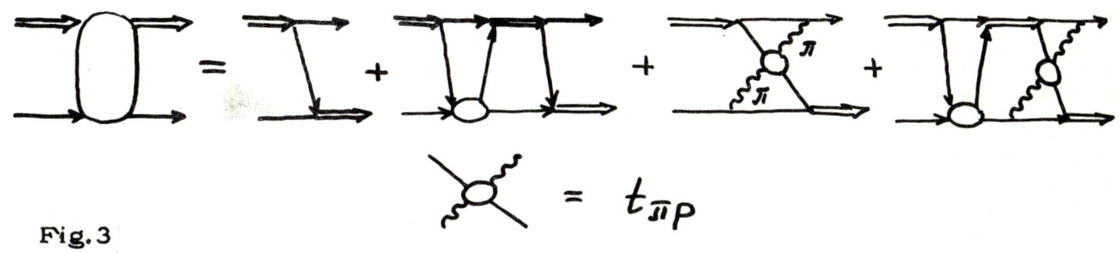

Fig. 3

$$\frac{d\sigma}{d\Omega} = \frac{1}{6}\Big\{\sum_{spin} T(\vartheta)T^*(\vartheta)\Big\};\ T(\vartheta) = \frac{1}{3\pi}\Big\{G_d^2(m^2-u^2)^{-1} + F_1(\vartheta)\Big\}F_2, \tag{8}$$

u - four momentum transfer.

Here F_1 is the amplitude of the process (graph C_1 in Fig.3) taken from /18/.

The values of the function $G_d^2(P)$ in the range of P variation from 2.4 to 4 fm^{-1} obtained from the analysis of the angular distributions using eq. (8) are given in Fig. 1b. One can see from this figure that in this region the function $G_d^2(P)$ decreases as $1/P^4$. This decrease cannot be explained in terms of the representation of the two-nucleon scattering potential for small distances by an infinite repulsive core. A reasonable agreement can be obtained if one chooses a finite core. Thus, the among the phase-equivalent potentials the potential with a finite repulsive core and OPEP tail turns out to be most efficient for describing the behaviour of the deuteron wave function. The internal structure of a finite core can apparently be determined from the analysis of the effects associated with drift off the energy shell of the two-particle scattering amplitude.

REFERENCES

1. V.V. Komarov, Invited talk, Proc. Int. Conf. Few Body Problems in Nuclear and Particle Physics, Quebec, Canada, 1974.
2. L. Hülthen, M. Sugawara, Handb. der Phys. 39(1957)74.
3. C.N. Bressel, A.K. Kerman, B. Rouben, Nucl.Phys. A124(1969)1624.
4. T. Hamada, I.D. Johnston, Nucl.Phys. 34(1962)382.
5. K.E. Lassila, M.H. Hull, Jr., H.M. Ruppel, F.A. McDonald, and G. Breit, Phys.Rev. 126(1962)881.
6. M.I. Zurina, A.M. Popova, I.T. Obukhovski , K.A. Ter-Martirosyan, Izvestia AN SSSR, ser.fiz. 35(1970)1789.
7. D.W.L. Sprung and M.K. Srivastava, Nucl.Phys. A139(1969)605.
8. D.C. Kocher and T.B. Clegg. Nucl.Phys. A132(1969)455.
9. A.R. Johnston, W.R. Gibson, E.A. McClatchie, J.H.P.C. Megaw, G.T.A. Squier, R.J. Griffiths and F.G. Kingston, PLA Progress Report 1966, PHEL/R 136, p.37.
10. J. Bruinsma, W. Ebenhöh, J.H. Stuivenberg and R. Van Wageningen. Nucl.Phys. A228(1974)52.
11. H.Postma, R. Wilson, Phys. Rev. 121 (1961)1229.
12. J.D. Seagrave, Proc. Int. Conf. Three Body Problem in Nuclear and Particle Phys., North-Holland, 1970.
13 P.C. Gugelot, J. Källne and P. Renberg, Physica Scripta 10 (1974) 252.
14. T.V. Gaivoronskaya, V.V. Komarov, A.M. Popova. Izvestia AN SSSR, phys. 25(1971)146.
15. H. Kottler and K.L. Kowalski. Phys.Rev.138(1965)B619.
16. E. Coleman et al. Phys. Rev.Lett. 16(1966)761.
17. G. Bennett et al. Phys. Rev.Lett. 19(1967)387.
18. N.S. Craigie and C. Wilkin. Nucl.Phys. B14(1969)477.
19. D.O. Riska and G.E. Brown. Nucl. Phys. A153(1970)8, G.E. Brown and A.M. Green. Nucl.Phys. A137(1969)1.

THREE-BODY VERTICES WITH TWO-BODY TECHNIQUES*

A.N. Mitra and V.K. Sharma+
Department of Physics, University of Delhi, Delhi-110007, India
+Department of Physics, Dharma Samaj College, Aligarh 202001, India

It has long been recognized [1-4] that vertex functions for few particle systems provide a convenient medium for the analysis of reactions in the language of Feynman diagrams, analogously to elementary particle processes. The development of three-particle theory during the last decade [5-7] has provided considerably more impetus for the use of the language of three-body vertex functions through the possibility of their "exact" evaluation with only two-body input. While three-body vertices are probably superfluous for the description of only three-body processes (for which exact amplitudes are already available) their practical usefulness often extends to reactions involving more than three-particle systems (for which "exact" amplitudes are still a distant goal), as long as such systems can be meaningfully described in terms of not more than three particles playing the active role. Indeed recent efforts [7-8] at extraction of few-body vertices from transfer reactions already indicate enough consistency in the estimation of coupling constants from the asymptotic normalizations of the involved wave functions. A fuller use of the available three and four body [9,10] techniques should provide a wider logical basis for evaluation not only of these "coupling constants" but also of the corresponding "form factors", without recourse to empirical parametrization of the wave functions.

The object of this paper is to investigate a simplified construction of three-body vertices. This must check against their standard definition as overlap integral [4]. Unfortunately this definition involves a non-trivial normalization of three-body wave functions with realistic N-N potentials, and has little practical scope for extension beyond A = 3.

A more promising candidate is the spectator (Nd) function, arising in the separable approximation to a three-body problem. Since the "spectator" is an effective two-body system which, after construction, does not "remember" the manner in which it was formed out of the (original) 3N system, it is ideally suited for the construction of the vertex function with an effective two-body normalization. While there is no apriori reason to justify such an optimism, we find from the study of an idealized 3-boson system that a two-body normalization of the N-d system, together with a simple combinatorial factor (arising from the identity of the particle involved) provides an excellent simulation of the actual vertex function defined as on overlap integral [4]. Indeed the method seems to work so well for the 3-boson problem that we feel quite optimistic about its practical usefulness for more complicated 3N systems (with realistic N-N potentials), and possibly also for the simulation of 4N vertices [9] without having to face their formidable normalization problems.

The essential steps of the calculation are as follows. The wave function for the relative motion between the clusters S_1 and S_2 in the break-up $S_3 \rightleftarrows S_1 + S_2$, can be expressed in momentum (p) space as

---

* This paper was declared as an Invited Talk by the Discussion Leader (L.P. Kok)

\# A preliminary result on these lines had been submitted to the Munich Conference (1973) but a numerical error detected in those calculations had necessitated a full re-investigation, resulting in the present study.

$$I_{12}^{(3)}(p) = \left(\frac{A_3!}{A_1!\,A_2!}\right)^{1/2} \int d\vec{\xi}_1 d\vec{\xi}_2 \Psi_3^*(\vec{\xi}_1,\vec{\xi}_2;\vec{p})\,\Psi_1(\vec{\xi}_1;\vec{p})\,\Psi_2(\vec{\xi}_2;\vec{p}) \tag{1}$$

where $(\vec{\xi}_1,\vec{\xi}_2)$ are the internal coordinates and Ψ_i $(i = 1,2,3)$ are the normalized wave functions of the three clusters. The corresponding vertex function $\gamma_{12}(p)$ is defined as

$$\gamma_{12}^{(3)}(p) = (2\mu_3)^{-1}(2\pi)^{3/2}(p^2 + \alpha_3^2)\, I_{12}^{(3)}(p) \tag{2}$$

where μ_3 is the reduced mass of S_1 and S_2 with c.m. momenta $\pm p$,

$$\alpha_3^2 = 2\mu_3 (B_3 - B_1 - B_2) \tag{3}$$

and B_i is the binding energy of S_i. The normalization is such that the contribution of the S_3-pole to the $S_1 - S_2$ scattering cross-section at c.m. momentum $\vec{k}$ is:

$$\left(\frac{d\sigma}{d\Omega}\right)_3 = \frac{1}{\pi^2}\left[\mu_3 \gamma_{12}^{(3)}(p)\right]^4 \times (k^2 + \alpha_3^2)^{-2} . \tag{4}$$

For the dNN vertex with two identical bosons (N,N), Eq.(2_) yields:

$$\gamma_{NN}^{d}(p) = (2\mu_{NN})^{-1}(2\pi)^{3/2}(2)^{1/2}\Psi_d(p) \times (p^2 + \alpha_d^2) \tag{5}$$

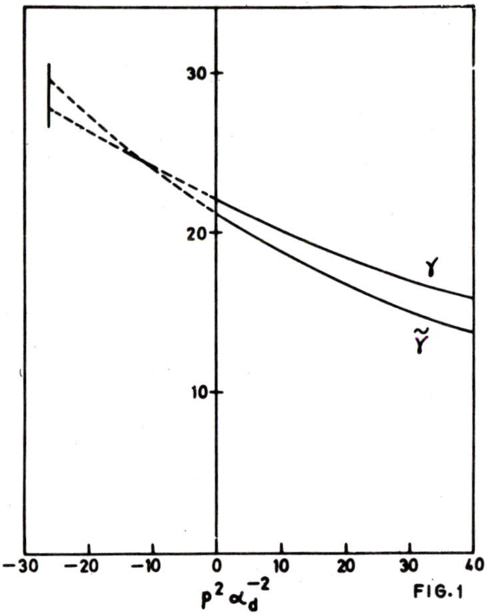

Fig.I: Exact (γ) versus simulated ($\tilde{\gamma}$) Ndt vertex functions.

where the normalized deuteron function $\Psi_d(p)$ is related to $I_{NN}^{d}(p)$ by $I_{NN}^{d}(p) = (2)^{1/2}\Psi_d(p)$. Further, the form factor $g(p)$ of an S-wave Yamaguchi potential [11] is related to Ψ_d by:

$$(p^2 + \alpha_d^2)\,\Psi_d(p) = N_d g(p) , \tag{6}$$

$$N_d^{-2} = \int d\vec{q}\, g^2(q)(q^2 + \alpha_d^2)^{-2} . \tag{7}$$

For the ideal 3-boson problem via pair-wise separable potentials with form factor g(p), the overlap integral, Eq.(1), is expressible in terms of the unnormalized 3N wave function Ψ_t as

$$I^t_{Nd}(P_3) = (3)^{1/2} N_t \int d\vec{q}\, \Psi^*_t(\vec{q},\vec{P}_3)\, \Psi_d(\vec{q}) \tag{8}$$

where [6]

$$\Psi_t(\vec{p}_{12},\vec{P}_3) = \sum_{123} (p^2_{12} + \tfrac{3}{4} P^2_3 + \alpha^2_t)^{-1} g(\vec{p}_{12}) G(\vec{P}_3) \tag{9}$$

$$\vec{p}_{12} = \tfrac{1}{2}(\vec{P}_1 - \vec{P}_2) \quad , \quad -\vec{P}_3 = \vec{P}_1 + \vec{P}_2 \tag{10}$$

and $$N_t^{-2} = \int d\vec{p}\,d\vec{q}\, \left|\Psi_t(p,q)\right|^2 . \tag{11}$$

The normalized Ndt vertex function $\gamma^t_{Nd}(p)$ defined by the substitution of Eq.(8) in Eq.(2) is plotted in Fig. I for the binding and Yamaguchi parameters [11]:

$$M^{-1} \alpha^2_d = 1.1 \text{ MeV} , \quad \beta = 11.1\, \alpha_d \quad . \tag{12}$$

As to the more interesting possibility of simulating the vertex function directly from the spectator N-d function $G(\vec{P})$, we first note that it satisfies an integral equation of the form [6]

$$(\tfrac{3}{4} P^2 + \alpha^2_t - \alpha^2_d) W^2(P) G(P) = \int d\vec{q}\, K(\vec{P},\vec{q}) G(\vec{q}) \tag{13}$$

where the kernel $\underline{K}$ is symmetric, and

$$W^2(P) = \int d\vec{q}\, g^2(q)(q^2 + \tfrac{3}{4}P^2 + \alpha^2_t)^{-1}(q^2 + \alpha^2_d)^{-1} . \tag{14}$$

A more symmetrical form of the spectator function is

$$\tilde{G}(\vec{P}) = W(P) G(\vec{P}) \tag{15}$$

which obeys a two-body normalization given by

$$N^{-2}_{Nd} = \int d\vec{q}\, G^2(\vec{q}) W^2(q) \tag{16}$$

The normalized N-d wave function $\tilde{G}(\vec{P}) N_{Nd}$ now serves the role of the overlap integral (8), except for a combinatorial factor analogous to the N-N-d case. The latter works out as $(3/2)^{1/2}$, where the numerator $(3)^{1/2} = (3!/1!2!)^{1/2}$ would arise from a full 3-body treatment, while a division by $(2)^{1/2} = (2!/1!1!)^{1/2}$ is necessitated by the fact that the 'deuteron' in the Nd system which has been counted as 'elementary' in the spectator description, is actually a composite whose combinatorial effect has already been taken into account in the calculation of the numerator. Thus the simulated vertex function $\tilde{\gamma}(P)$ finally works out as,

$$\tilde{\gamma}(P) = (2\mu_3)^{-1}(2\pi)^{3/2}\ (P^2 + 4/3\,\alpha_t^2 - 4/3\,\alpha_d^2\ )(3/2)^{1/2}\ N_{Nd}\tilde{G}(P) \qquad (17)$$

which is also plotted in Fig.1 alongside the exact one.

The quality of the fit may be judged by its accuracy within 10% over a considerable range of P^2 varying from the pole i.e. $P^2 = -4/3(\alpha_t^2 - \alpha_d^2) \approx -26\alpha_d^2$ in this model, to about $+15\alpha_d^2$ covering the region of primary physical interest for most transfer reactions. At the pole the ratio of these two quantities is exactly $2(2/3)^{1/2}\ N_t N_{Nd}^{-1} \approx 0.9$. We therefore feel that this result is interesting enough for the possibilities of practical applications to more realistic vertex functions with not more than the input N-N potential.

For a four system the problem of normalization of the corresponding wave functions is so much more involved that at the present state of computer technology there appears little chance of the exact evaluation of αdd and αtN vertex functions in terms of overlap integrals if one is to use wave functions calculated entirely in terms of N-N input potentials. On the other hand, it is tempting to speculate that these vertices can be simulated by the spectator functions of d-d and t-N scattering which satisfy two coupled one dimensional integral equations[12] which appear in the process of successive reduction of the four body problem to an effective two-body problem [9]. In that event, the combinatorial factors which are now needed for αdd and αtN vertices are $(6/4)^{1/2}$ and $(6/3)^{1/2}$ respectively. Details of these calculations will be published separately.

REFERENCES

1. R. Blankenbecler, M.L. Goldberger and F.R. Halpern, Nucl. Phys. 12 (1959) 1100.
2. R.D. Amado, Phys. Rev. Letter 2 (1959) 399.
3. I. Shapiro, in Selected Topics in Nuclear Theory (IAEA Vienna,1963).
4. A.S. Rinat et al, Nucl. Phys. A190 (1972) 328.
5. L.D. Faddeev, Soviet Phys. JETP 12 (1962) 529.
6. A.N. Mitra, in Advances in Nucl. Phys. Vol. 3 (Plenum Press, 1969).
7. Y.E. Kim et al , Annual Rev. Nucl. Sci. Vol. 24 (1974) 69.
8. T.K. Lim, Phys. Letts. 55 (1975) 252;
   A.G. Baryshnikov et al Phys. Letts. 51B (1974) 432.
9. E.O. Alt, P. Grassberger and W. Sandhas, Phys. Rev. C. 1 (1970) 85.
10. T. Sasakawa, Contribution to this Conference.
11. Y. Yamaguchi, Phys. Rev. 95 (1954) 1728.
12. A preliminary account of these calculations is given in A.N. Mitra, Flinders University Report (1973) - unpublished.

PROTON-INDUCED ^{2}H AND ^{3}He BREAK-UP AT 156 MeV.

T. Yuasa
Institut de Physique Nucléaire, Université de Paris-Sud, 91406, ORSAY, France.

For several years the break-up of ^{2}H and of ^{3}He induced by protons of 156 MeV has been investigated at Orsay (France) by detecting simultaneously the (p, 2p) and (p, pn) reactions. The purpose of this work was to study nucleon-nucleon (or nucleon-particle) multiple scattering beyond first order, especially in kinematic regions favouring N-N or 3N final-state interactions. [1], [3].

1. 1. The ^{2}H break-up by ^{2}H(p, 2p)n and ^{2}H(p, pn)p. - The p-p and p-n quasi-free scattering (QFS) and the p-p and p-n final-state interactions (FSI).

These reactions have been investigated in kinematically complete experiments by using a liquid target and by choosing the kinematical conditions so as to obtain the $d^3\sigma/d\Omega_1 d\Omega_2 dE_1$ spectra which pass through either the QFS peak ($E_3 \sim 0$) or the p-p or p-n FSI peaks ($E_{13}^{CM} \sim 0$) where the labels 1, 2 and 3 mean the nucleons detected in the first, second telescope and the recoil nucleon. The experimental results and their comparison with theoretical calculations have been reported. [1], [3], [4].

Recently, L'Huillier, Benoist and Ballot have studied theoretically the sensitivity of proton-induced deuteron-break-up reactions in the range of a few hundred MeV to two-nucleon off-shell t-matrix elements when the multiple scattering of orders equal to two or higher are added. They have attempted to estimate the error due to the truncation of the Faddeev series at second order. In order to do it, they have introduced a model called "the fixed-scatterer approximation (FSA)" used in elastic nucleon-nucleus scattering. [5]. In this model the relative motion of the two target nucleons during the collision time is neglected. The comparison of our experimental cross-sections $d^3\sigma$ divided by the kinematic factor with the FSA cross-sections (divided by the kinematic factor) up to second order J_1 and $J_{12}^{\pm}$ and the integral solutions $J^{\pm}$ are presented in Fig. 1. The script $\pm$ refers to the two different phase shift solutions. The multiple-scattering consideration corrects in the expected way the second order evaluations $J_{12}^{\pm}$. In particular J^{+} reproduces the experimental shape but overvalues the FSI peak, while J^{-} gives better account of the mean magnitude although it flattens the structure. It may be that the order-higher-than-two contribution is not negligible with the employed realistic-local potentials.

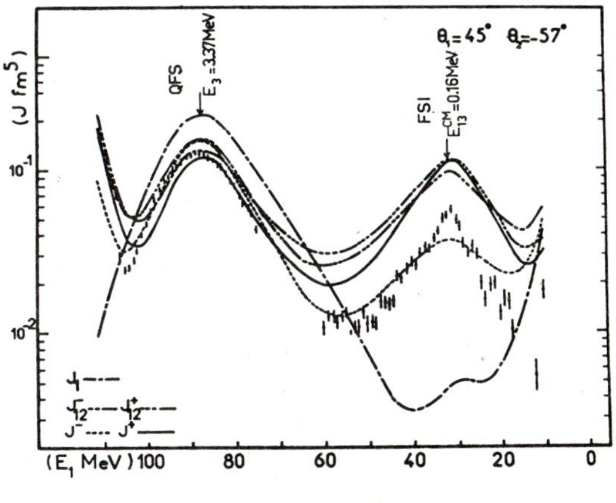

Fig. 1. The ^{2}H(p, 2p)n reaction at 156 MeV, $\theta_1 = 45°$, $\theta_2 = -57°$. The meaning of J_1, J_2, J_{12} and J is explaned in the text. The potential used is that of Hamada-Johnston.

1.2. The ^2H break-up study for special kinematic conditions which favour the detection of the deviation from the theoretical approximate calculation of the Faddeev theory.

At the International Conference in Québec, Lambert et al. [6] have presented their experimental results obtained for the ^2H (p, 2p)n reaction in the special kinematic conditions corresponding to the recoil nucleon at rest in the total center-of-mass system. We hoped to test this experiment at 156 MeV [7]. The kinematic conditions satisfying the above conditions are for E_0=156MeV, $\theta_1 = -\theta_2 = 58.3°$: E_1 = 68MeV, E_2 = 68.571 MeV, E_3 = 17.049 MeV, θ_3=0.388°, E_{12}^{CM} = 99.760 MeV, E_{13}^{CM} = 24.506 MeV, E_{23}^{CM} = 25.088 MeV, E_3^{CM} = 0.

The experimental set-up was the same as that used in the ref. [8] except for the target which was liquid ^2H. The $d^3\sigma_{exp}$ spectrum has been compared with the calculation made by L'Huillier with the FSA approximation (Fig. 2). The absolute experimental cross-sections at E_1 = 30 MeV and at E_1 = 68 MeV are 19 $\pm$ 2μb sr^{-2} MeV^{-1} and 21 $\pm$ 3μb sr^{-2} MeV^{-1} respectively [7]. Our results do not contradict the results of ref. [6]. Detailed discussions are given in Ref. [7].

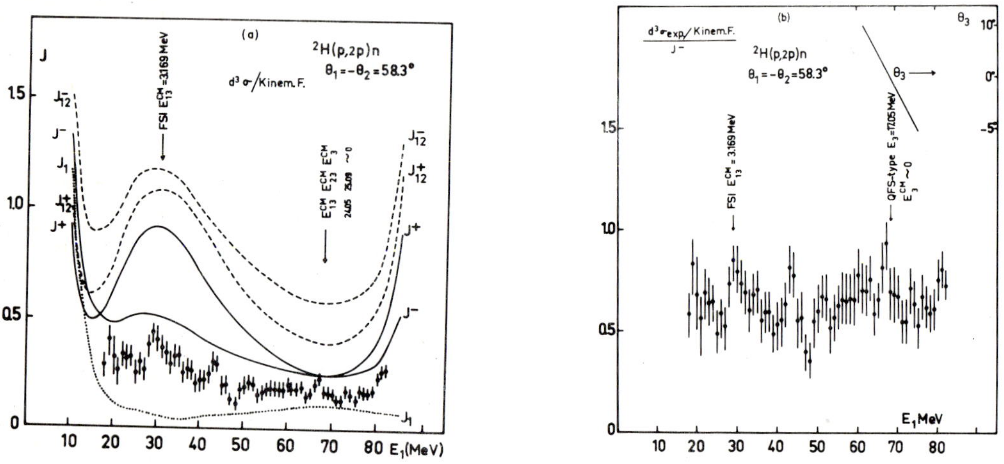

Fig. 2. The ^2H(p, 2p)n reaction spectra for $\theta_1 = -\theta_2 = 58.3°$ with E_0=156 MeV.
(a) : J as a function of E_1. (b) $d^3\sigma_{exp}$/kinem. F. /J^-.
The meaning of the terms is given in reference [5].

2. The break-up of ^3He induced by 156 MeV protons through the ^3He (p, 2p)d, d, the (p, pn)2p and the (p, dp) p reactions. - The quasi-free scatterings p-p, p-n, and p-d (QFS) and the three-nucleon final-state interactions (FSI).

After the QFS and FSI studies of ^2H break-up which gave proof of the existence of p-n and p-p FSI enhancements for their minimum relative energy and of the Coulomb effect in the latter case when E_{133}^{CM} is almost zero, we have extended our study to the four-nucleon system. The ^3He break-up induced by 156 MeV protons was studied by using a liquid target and by adjusting the kinematic conditions for p-p, p-n and p-d quasi-free scatterings (QFS) or for three-nucleon final-state interactions (FSI) [2], [8].
The study of the (p, 2p) and (p, pn) differential cross-section spectra ($d^3\sigma$) in the QFS regions and those calculated with PWIA have been published [2]. In the second series we studied FSI more precisely. We communicate here a preliminary analysis of the present status of this work, in particular, including our results on the (p, dp) FSI [8].

2.1. The ^3He (p, dp)p reaction.

The experimental lay-out was similar to that of ref. [2] except for a small plastic scintillation counter which was added to the second telescope to define $\Delta\theta_2$ better.

This reaction, giving real three particles in the final-state, can be investigated in a kinematically complete experiment by fixing two scattering angles θ_1, θ_2 and by measuring a proton energy E_1. But we have recorded 4 parameters ΔE_1, E_1, ΔE_2 and E_2 for each correlated event to eliminate any competitive reaction events. The differential cross-section $d^3\sigma/d\Omega_1 d\Omega_2 dE_1$ spectrum obtained for θ_1=42.6° and θ_2=-71° which passes through the point corresponding to E^{CM}_{p-d}=0.013MeV for E_1 = 40 MeV is presented in ref. [8] (Fig. 1).

The QFS-type peak is at E_1 = 79 MeV (E_3 = 0.966 MeV) having $d^3\sigma$ = 0.107 ± 0.002 mb sr^{-2}MeV^{-1}. The spectrum shows very clearly the proton-deuteron FSI contribution around E_1 = 40.0 MeV (E^{CM}_{13} = 0.013 MeV). The $d^3\sigma$ has a minimum at this energy due to the Coulomb repulsion and two maxima, one at E_1 = 35.5 MeV (E^{CM}_{13} = 0.40MeV) and another at 44.5 MeV (E^{CM}_{13} = 0.40 MeV). A PWIA calculation made by using a Hulthèn-type wave function for the deuteron and an Irving-Gunn-type one for ^3He and by summing the matrix elements for p-p and p-d scattering is also shown in Fig. 1 of Ref. [8] in the dashed curve D. The Watson-Migdal enhancement factor with a Coulomb field multiplied by the $d^3\sigma$ calculated with PWIA (curve D) is also shown in the same figure by using $^4r_{p-d}$=1.89 fm and $^4a_{p-d}$= 11.9 fm (fine dashed curve B [9]); $^4r_{p-d}$= 1.50 fm and $^4a_{p-d}$=11.9fm (dot-dashed curve A) and $^4r_{p-d}$= 2.8 fm, $^4a_{p-d}$= 12.2 fm (dotted curve C [10]).

The $^4r_{p-d}$ and $^4a_{p-d}$ values used in ref. [9] give better agreement with the experimental results than the other two sets. But since the PWIA -calculation normalization may not be precise in this region, this agreement should not be used to decide the best $^4r_{p-d}$ and $^4a_{p-d}$ values.

2.2. The ^3He (p, 2p)d and $\bar{d}$ reactions.

These reactions were detected simultaneously with the (p, dp) reaction at θ_1=42.6° and θ_2= -71° [8]. The (p, dp) reaction was discriminated very clearly by the difference of the energy loss in the ΔE_1 detector in the first telescope. The discrimination of the (p, 2p)d reaction was not possible due to the poor E_2 resolution (ΔE_2 = 3.5 MeV). Because of the importance of this reaction relative to the (p, 2p)$\bar{d}$ reaction, it was necessary to study the $d^4\sigma/d\Omega_1 d\Omega_2 dE_1 dE_2$ spectrum by using the E_1-E_2 bi-parametric spectrum. The ratio of the (p, 2p)d reaction to the (p, 2p)$\bar{d}$ reaction is about 2.9 ± 0.5 at E_2 = 68.2 MeV. In analysing the $d^4\sigma$ spectrum, we have also found the contribution of the pick-up reaction (p, $\widehat{2p}$, p)n or (p, $\bar{d}$, p)p. We present in Fig. 3 the $d^3\sigma/d\Omega_1 d\Omega_2 dE_1$ spectrum obtained by the projection of the E_1-E_2 spectrum onto the E_1-axis. In order to indicate the complexity of the $d^3\sigma$ spectrum, we present in Fig. 4 the locus of all the reactions that may have contributed to it. (The (p, dp) reaction has been excluded but shown as reference).

The $d^3\sigma$ spectrum presents a peak near E_1 = 21 MeV. It corresponds probably to the FSI between p-$\bar{d}$ or $\widehat{2p}$-n, because E^{CM}_{13} ~ 0.03 MeV at this E_1 value. [In reality $d^3\sigma$ presents a minimum at this energy and a maximum at E_1 ~ 24.5 MeV (E^{CM}_{13} ~ 0.4 MeV) for (p, 2p) $\bar{d}$ reaction]. The pick-up reactions also present FSI in this region. There is a hump at E_1 ~ 58 MeV which corresponds to E^{CM}_{13} ~ 15 to 16 MeV in the (p, 2p)d and $\bar{d}$ reactions (Fig. 4 curves a, a'). The $d^4\sigma$ spectrum obtained with the E_1-E_2 spectrum at E_2 = 77 MeV manifests effectively this peak. This $d^4\sigma$ spectrum shows also another peak near E_1 ~ 35 to 40 MeV due to the pick-up reaction (Fig. 4, curve c) which also corresponds to $E_{13}^{CM} \sim 15$-16 MeV. This peak might be due to the excitation state of the 2p-n system with $E_{exc} \sim 20$ to 22 MeV. The spectrum $d^4\sigma/d\Omega_1 d\Omega_2 dE_1 dE_2$ obtained by the projection of the events on the E_1-E_2 locus of the pick-up reaction onto the E_2-axis presents this peak at $E_2 \sim 77$ MeV ($\Gamma \sim 5$MeV), (and another one at $E_2 \sim 81$ MeV corresponding to $E_{13}^{CM} \sim 9$ MeV and $\Gamma \sim 4$ MeV) (Fig. 5).

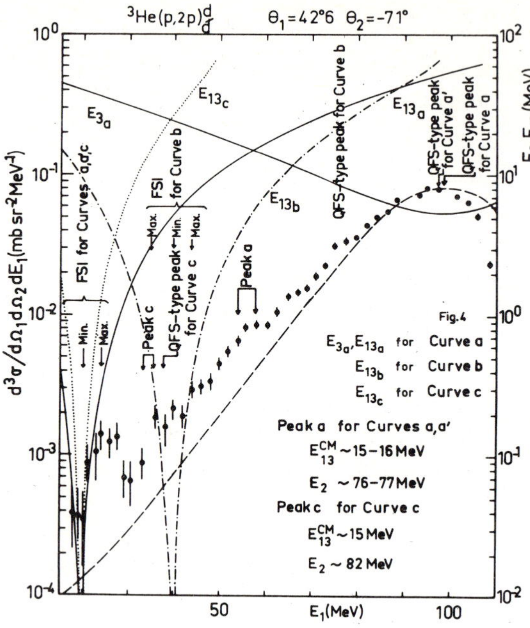

Fig. 3. The $d^3\sigma/d\Omega_1 d\Omega_2 dE_1$ spectrum for the (p, $\widehat{2p}$)d and d^* reactions (mixed with other reactions). The dashed curve represents the PWIA calculation with the normalization factor N = 0. 94

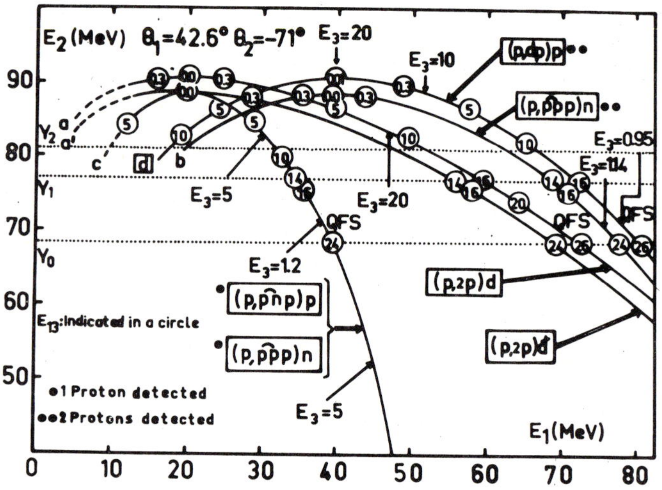

Fig4. E_1-E_2 bi-parametric spectra for the all reactions that may have contributed to the $d^3\sigma$ spectrum shown in Fig. 3. For the (p, pn)$\widehat{2p}$ reaction the second proton must be replaced by a neutron and d(or d^*)by $\widehat{2p}$.

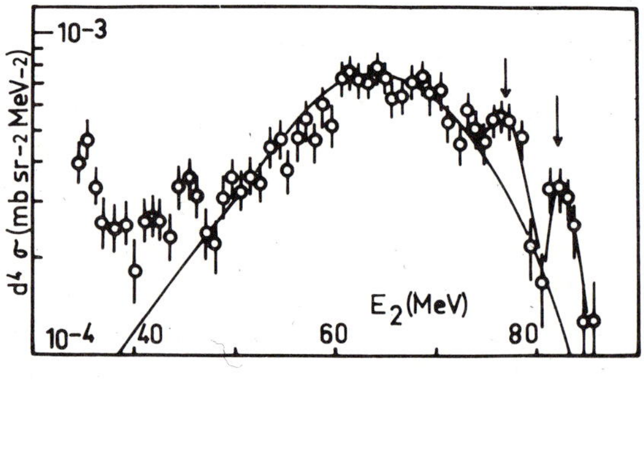

Fig. 5. $d^4\sigma/d\Omega_1 d\Omega_2 dE_1 dE_2$ spectrum for the (p, $\widehat{2p}$, p)n or (p, d^*, p)p reactions projected onto E_2-axis (ΔE_1 = 10 MeV). The continuous curve represents the PWIA calculation with N = 0. 42

2.3. The ^{3}He (p, pn)$\widehat{2p}$ and (p, $\widehat{2p}$, n)p reactions.

In the same experiment we detected the (p, pn)2p reaction by measuring neutrons with the second telescope in which the S2 and NE213 counters were coupled in anticoincidence [8]. For the analysis of this reaction we were obliged to utilise the E_1-T_{12} bi-parametric spectrum because the energy resolution of the NE213 counter for neutrons was not good.

The statistics are not sufficient in spite of the long run (200 hours). The spectrum $d^3\sigma/d\Omega_1 d\Omega_2 dE_1$ is shown in Fig. 6. The p-$\widehat{2p}$ FSI near E_1 = 21 MeV may be sure. The peak at $E_1 \sim 58$ MeV is real. So if the peak provides from excited state of the system p-$\widehat{2p}$, the l-value of this state could not be zero.

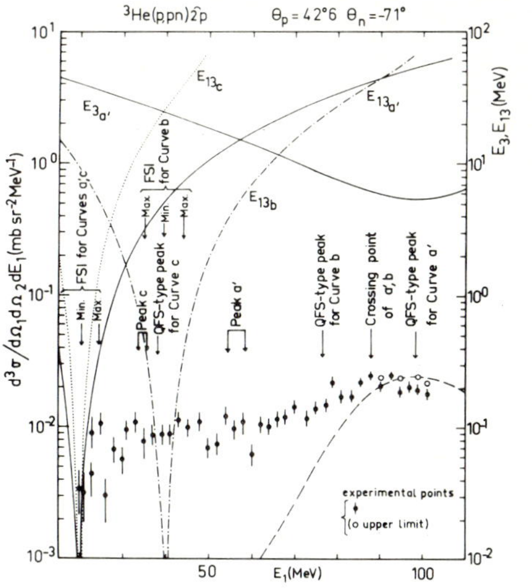

Fig 6. The $d^3\sigma/d\Omega_1 d\Omega_2 dE_1$ spectrum for the (p, pn)$\widehat{2p}$ reaction mixed probably with the (p, $\widehat{2p}$, n)p reaction. The dashed curve represents the PWIA calculation.

References

[1] J. P. Didelez, I. D. Goldman, E. Hourany, H. Nakamura-Yokota, F. Reide and T. Yuasa, Phys. Rev. C 10 (1974) 529 (The preceeding references are included)

[2] J. P. Didelez, R. Frascaria, N. Fujiwara, I. D. Goldman, E. Hourany, H. Nakamura-Yokota, F. Reide and T. Yuasa, Proceedings of Int. Conf. at Québec (Laval Univ. Press 1975,) 667 and Phys. Rev. C (1975) (to be published).

[3] T. Yuasa, Proc. of Int. Conf. at Québec (loc. cit.) 430.

[4] M. L'Huillier, Thesis of Doctorat d'Etat (Paris VII, 1974) not published. J. L. Ballot, M. L'Huillier and P. Benoist-Gueutal, Phys. Rev. C 12 (1975) 725 (The preceeding references are included).

[5] P. Benoist-Gueutal, Journ. de Phys. 34 (1973) 943 (The preceeding references are included). M. L'Huillier, P. Benoist-Gueutal and J. L. Ballot, Phys. Rev. C 12 (1975)948.

[6] J. M. Lambert, P. A. Traedo, R. G. Allas, L. A. Beach, R. O. Bondelid and E. M. Diener, Int. Conf. at Québec (Laval Univ. Press 1975, p. 531). W. M. Kloet, P. H. D. Univ. of Utrecht (1973)

[7] N. Fujiwara, E. Hourany, H. Nakamura-Yokota, F. Reide and T. Yuasa, Present Conference.

[8] N. Fujiwara, E. Hourany, H. Nakamura-Yokota, F. Reide , V. Valković and T. Yuasa, Present Conference.

[9] W. T. Van Oers and J. D. Seagrave, Phys. Lett. 24 B (1967) 562.

[10] J. Arvieux, A. Fiore, J. Jungerman and N. Van Sen, Proc. of the Int. Conf. Los Angeles (North Holland /American Eslevier 1972) 519.

PROTON INDUCED DEUTERON BREAK-UP AT E_p=12.5 MeV

Đ. Miljanić
Institute Ruđer Bošković, Zagreb, Yugoslavia
E. Andrade* and G.C. Phillips
T.W. Bonner Nuclear Laboratories, Rice University,
Houston, Texas 77001, U.S.A.

This paper represents a new analysis of the data[1] on proton induced deuteron break-up at E_p=12.5 MeV. The aims of this analysis are: 1. to compare the data with the exact three-body calculations, 2. to obtain information on the relation between the exact three body calculation and the simple theoretical approximations: the Watson-Migdal model (WMM) for the final state interaction (FSI) and the plane wave impulse approximation (PWIA) for the quasi-free scattering (QFS). The analysis is done using Ebenhöh code (EC)[2], which solves the three particle Faddeev equations exactly for separable, S-wave, spin dependent N-N interactions. The N-N potential is of the Yamaguchi form.

The figures 1,2,3 show some typical examples of the comparison of the EC calculations and the data on np FSI, pp QFS and pn QFS, respectively. No normalization factor has been applied to the calculated values. In the most other cases, not shown here, the EC calculations also agree with the measured cross sections, both in shape and absolute magnitude. Fig. 4. shows the comparison of the calculated and measured cross sections for the kinematical conditions, where the energies of the both detected particles are equal. While the PWIA[1] gives an order of magnitude bigger cross section and fails to give the correct ratio of the pn and pp QFS peak heights, EC calculations give good quantitative agreement. Differences between pp and pn QFS cross sections are mainly a consequence of the effect of the Pauli exclusion principle as it was pointed out by Cahill[3].

Fig. 5. shows pn FSI data measured at $\Theta=32.5^\circ$ compared with the prediction of both WMM and EC calculations. In the WMM case the relative values of the singlet and triplet contributions have

---

* Permanent address: Instituto de Fisica, Universidad Nacional Autonoma de Mexico, Mexico 20, D.F.

been found by least square fit calculations. The predictions of both calculations do not differ much in this case. Recently, Kluge et al.[4)] have analyzed the data for deuteron break-up on protons for E_d=52 MeV. They have found, that the ratios of the singlet to triplet contributions, obtained for E_{np}=0 using EC calculations, do not differ from those extracted from data using WMM. Fig. 6. shows the results of the analysis of the data for E_p=12.5 MeV. Vertical lines represent the ratios of the singlet to triplet contribution for E_{np}=0 (with estimated uncertainties), extracted using WMM, while the curve represents the same ratios given by EC calculations. At this energy the agreement is only qualitative. The reason for this one could find in the low incident energy. At this low energy the relative energies in other two pairs of nucleons are not high in some regions of the FSI peak, especially at forward angles. At the

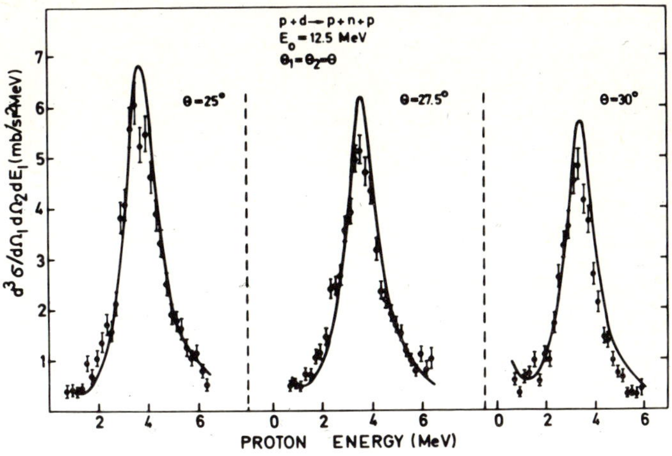

Fig. 1.

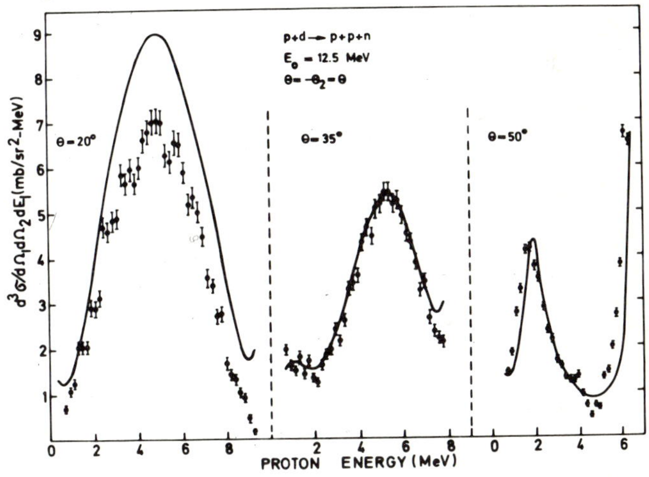

Fig. 2.

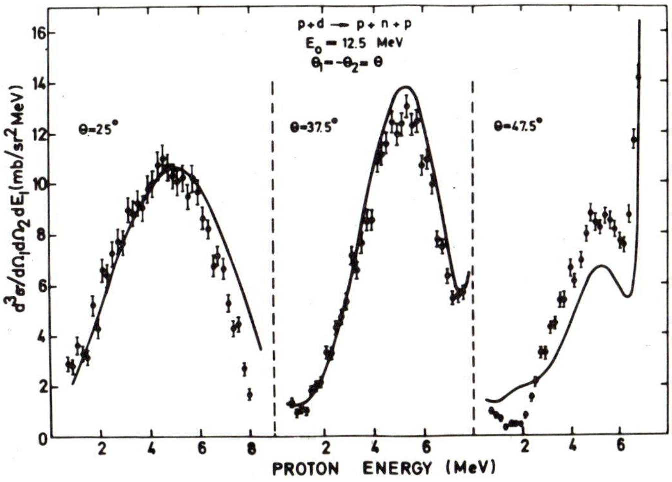

Fig. 3.

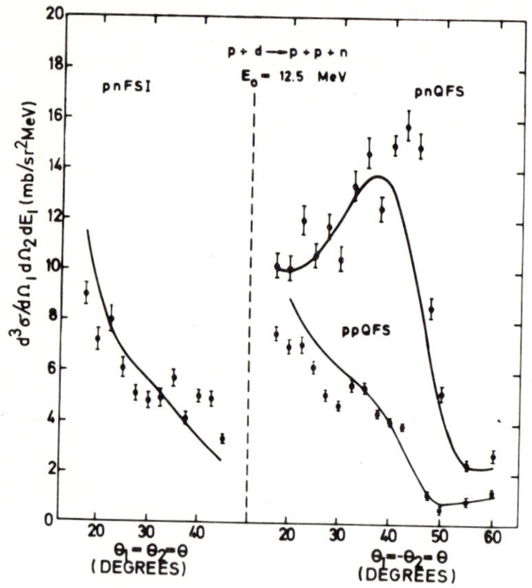

Fig. 4.

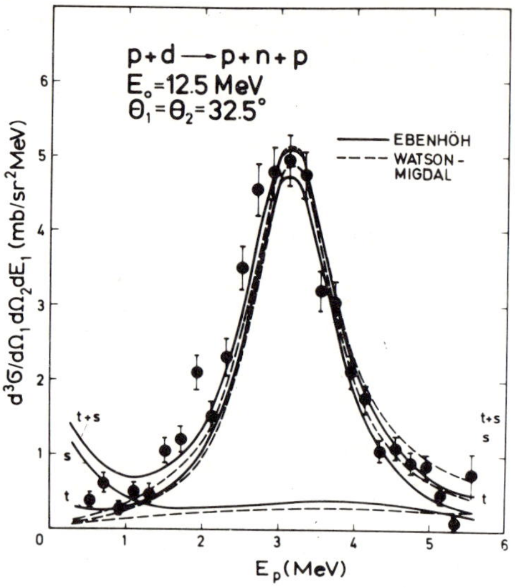

Fig. 5.

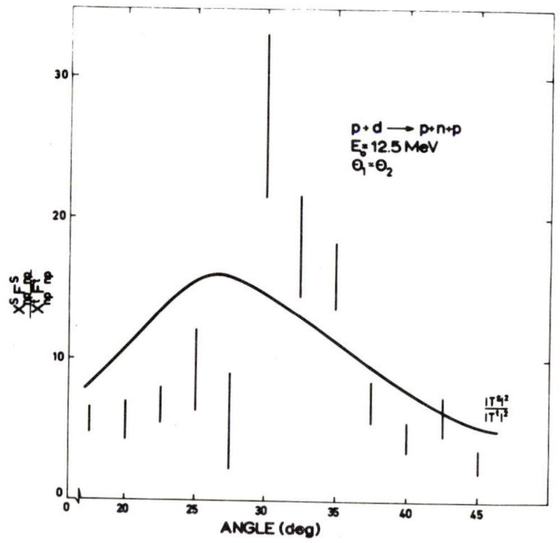

Fig. 6.

same time the undetected particle energy is not high enough, that QFS contributions could be neglected. Because of that the main assumption of WMM cannot hold in this case.

This analysis shows, that only the dinamically complete break-up model could give satisfactory results at low energies.

References

1. E. Andrade, V. Valković, D. Rendić, and G.C. Phillips, Nucl. Phys. A183 (1972) 145.
2. W. Ebenhöh, Nucl. Phys. A191 (1972) 97.
3. R.T. Cahill, Nucl. Phys. A185 (1972) 236.
4. W. Kluge, R. Schlüfter and W. Ebenhöh, Nucl. Phys. A228 (1974) 29.

COMPARISION OF DEUTERON BREAKUP SPECTRA AT 8.5 MEV WITH FADDEEV CALCULATIONS USING DIFFERENT SEPARABLE POTENTIALS

B. Kühn, H. Kumpf, J. Mösner, W. Neubert and G. Schmidt
Zentralinstitut für Kernforschung der Akademie der Wissenschaften der DDR, Rossendorf, DDR 8051 Dresden, PSF 19

Our kinematically complete spectra of the deuteron breakup by protons measured at E_p = 8.5 MeV 1), have been compared with the results of Ebenhöh calculations using the following six formfactors of the separable potential: Yamaguchi (Y), exponential (E), quadratic (Q), linear (L), modified exponential (M) and S-shape (S) 2). Figures 1 and 2 demonstrate that the absolute cross section is given with moderate accuracy by theory, while the shape of the spectra is reproduced with different success. In the case of QFS the Y-formfactor rather perfectly fits the experiment (fig. 1).

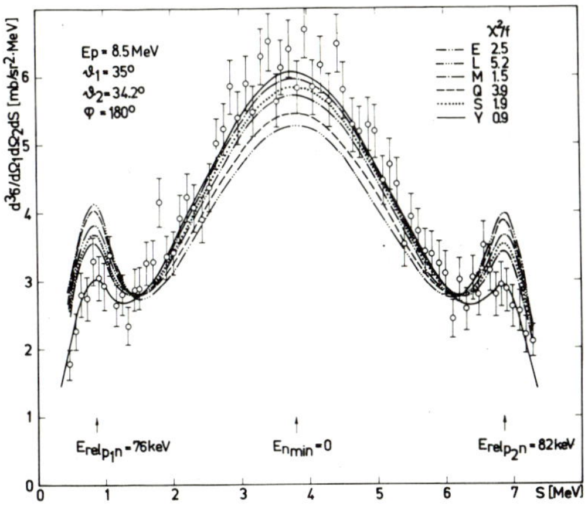

Fig. 1 Deuteron breakup spectrum, projected on the kinematic locus S, for QFS kinematics

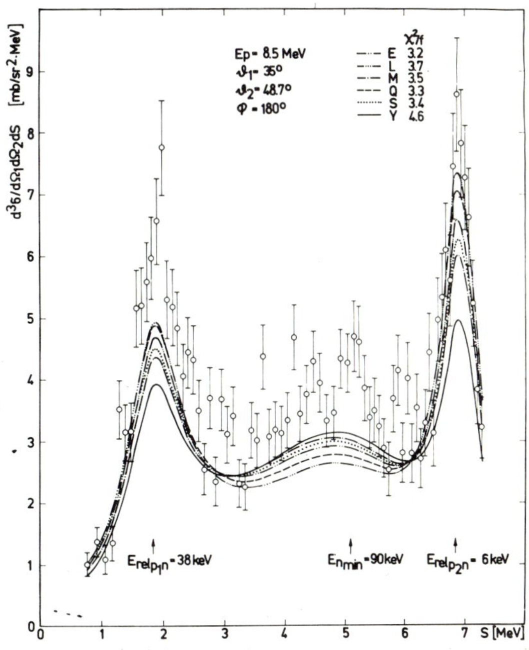

Fig. 2 Deuteron breakup spectrum, projected on the kinematic locus S, for FSI kinematics

Note: The theoretical curves are shown without any normalisation!

For the kinematic conditions of np-FSI great discrepancies as well between theory and experiment as between calculations with different formfactors are observed. Looking at the whole body of experimental spectra and the corresponding calculations one cannot select any of these formfactors which provides a better overall agreement than another, but it can be seen, that obviously kinematical regions of FSI are especially sensitive to the choice of the potential.

Petersen et al. 3) and Kluge et al. 4) have shown, that the angular distribution of the FSI cross section is a sensitive test for the NN-potential. We have calculated the angular distributions for an energy of bombarding protons of 8.5 MeV using Y- and E-formfactors. The resulting curves exhibit a remarkable variation with the used potential (fig. 3). Furthermore a rather great discrepancy to the two experimental points must be emphasized.

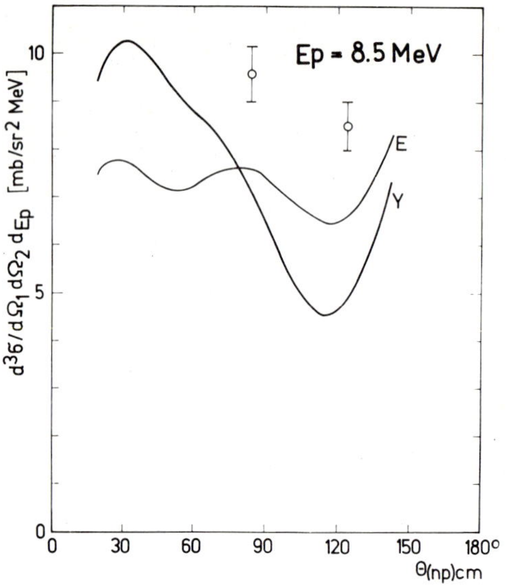

Fig. 3 Angular distributions of the FSI cross sections calculated for Y and E formfactors for E_p = 8.5 MeV. On the abscissa is plotted the C.M. angle of the (np)-pair in FSI.

We are aware of the approximations involved in the Ebenhöh code and of the insufficient phase equivalence of the used potentials. Nevertheless the rather great variations seem to be a hint of a high sensitivity of the angular distributions to the actual NN-interaction also in this region of small energies. Therefore a measurement of the FSI angular distribution at 8.5 MeV is in progress in our laboratory.

1) B. Kühn, H. Kumpf, J. Mösner, W. Neubert, G. Schmidt, Nucl. Phys. A247 (1975) 21
2) J.Bruinsma, W. Ebenhöh, J. H. Stuivenberg, R. van Wageningen, Preprint Natuurkundig Lab.d.Vrije Universiteit Amsterdam, Febr. 1974
3) E.L.Petersen et al., Proc. of the int. conf. on Few Body Problems in Nuclear and Particle Physics, Quebec 1974, p. 395
4) W. Kluge, R. Schlüfter, W. Ebenhöh, loc. cit. p. 432

THE BREAKUP REACTION $p_0+D \rightarrow p_1+p_2+n$ AT $E_0 = 20$ MeV

J. Doornbos, W. Krijgsman, C.C. Jonker

Natuurkundig Laboratorium der Vrije Universiteit,
Amsterdam, The Netherlands

We report results of complete experiments on the breakup reaction $p_0+D \rightarrow p_1+p_2+n$ with detection of the two protons. Former experiments of our group studied this reaction at an energy of 12.9 MeV with the deuteron as a projectile [1].
From the analysis of these measurements it became clear that two regions of phase space were sensitive to the nucleon-nucleon (N-N) potentials used : those of proton-neutron Final State Interaction (p-n FSI) and those where the quartet contribution is zero. This was confirmed by a reanalysis of the data on p-n FSI and pp Quasifree scattering (QFS) of Klein [2] at 16 MeV by Bruinsma and Stuivenberg [3].
All the above data were compared with calculations based on the Faddeev equations, using one term separable N-N potentials with either a Yamaguchi (CYY) or an exponential (CEE) formfactor, correctly taking into account the charge dependence of the N-N force [4].

The comparison of data on FSI and QFS with theory is usually hampered by the inaccuracy and the lack of systematics of existing data, and by the different ways of analyzing the data. Therefore we decided to measure the angular distribution of the p-n FSI as a function of the production angle in the center of mass (θ_{d^*cm}) and also the pp QFS for symmetrical angles where the quartet contribution is zero. This study is made as a function of the bombarding energy, while special attention is given to obtaining good statistics and minimizing the error ($\sim$ 5%).

The experimental methods comprised the following points :
The collected data were analyzed off-line. The particle identification was by time of flight below and by the ΔE-E method above 4 MeV. The random pp coincidences were subtracted by comparing results of two consecutive bursts of the cyclotron. The experimental spreads (angular resolution, target thickness etc.) were taken into account with a Monte Carlo method. The protons of the elastic (p,d)-scattering were counted in a monitor detector to obtain the absolute normalization. Moreover as the single spectra of the two detectors were accumulated continuously, they could be used as a check on the normalization.

The experimental and theoretical FSI crossections were, for a comparison, integrated over a region of relative p-n energy of 50 keV (typical number of counts so obtained is 3000) and the ratio determined. Then the theoretical peak-value was multiplied with this ratio to yield the experimental peak-crossections. For the QFS data a similar procedure was followed.
In all experiments we measured the crossection along the whole kinematical locus. A typical spectrum is given in fig. 1, with S the arclength along the kinematical curve. The error-bars represent the statistical errors ($\sim$ 5% in the top of the peak) while the error in the normalization is 2%.

From these measurements we obtained the following preliminary results (fig. 2 to 4) :
Klein [2]) has measured the p-n FSI at 16 MeV from $\theta_{d^* cm}$ = 140° to 60°. His experimental crossections, also analyzed as stated above, were smaller than the theoretical (CEE) ones, the difference increasing to 20% at 60°. Therefore we extended his measurements to more forward angles ($\theta_{d^* cm}$ = 56.4°, 51.1°, 45.0°) and repeated his experiment at 80° and 70°. We found a systematic difference with his results, i.e. our crossections are 10% larger (fig. 2 and 3).
At 20 MeV two runs were made. In the first run p-n FSI was measured at 55.4°, 74.5°, 89.9°, 124.7° and pp QFS at the angle combinations (θ_1 = 42.35°, θ_2 = 41.35°) and (36.35°, 36.50°). In the second run we obtained FSI data at 48.2°, 63.9°, 108.0°, 134.0° and for a check again at 89.9°. In the last case the FSI crossection differed only one percent from that measured in the first run. pp QFS was measured for (32.2°, 32.0°) and (28.5°, 27.6°).

The theoretical and experimental crossections for FSI are plotted in fig. 2 against $\theta_{d^* cm}$. In fig. 3 their ratio is given for CEE. The experimental QFS peak-crossections are compared with theory in fig. 4. Along the horizontal axis the value of the c.m. angle of one of the protons is given. In the figure the point of zero momentum transfer to the neutron is indicated by the arrow.

This investigation forms a part of the research program of the Foundation for Fundamental Research of Matter (F.O.M.) which is financially supported by the Netherlands Organization for Pure Scientific Research (Z.W.O.).

References

1) P.H. Schram, Thesis, Vrije Universiteit Amsterdam, 1975.
2) H. Klein, Nucl. Phys. A199 (1973) 169.
3) J.H. Stuivenberg, J. Bruinsma, R. van Wageningen, Proc. Int. Conf. on few body problems in nuclear and particle physics, ed. R.J. Slobodrian et al. (les presses de l'université Laval, Québec, 1975) p. 505.
4) These calculations were performed in close cooperation with the theoretical group of Professor R. van Wageningen.

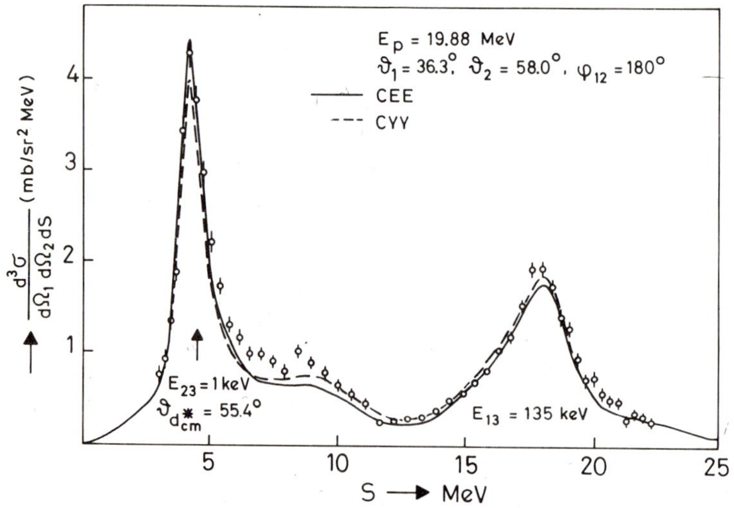

fig. 1.

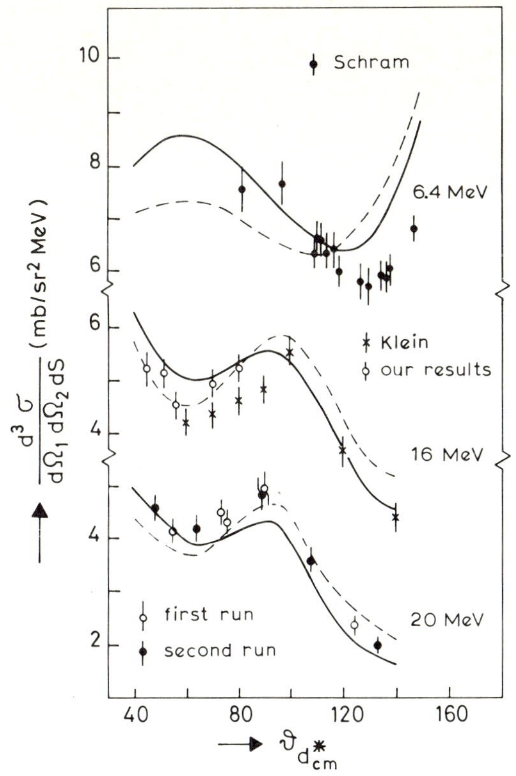

fig. 2.
—— CEE
- - - CYY

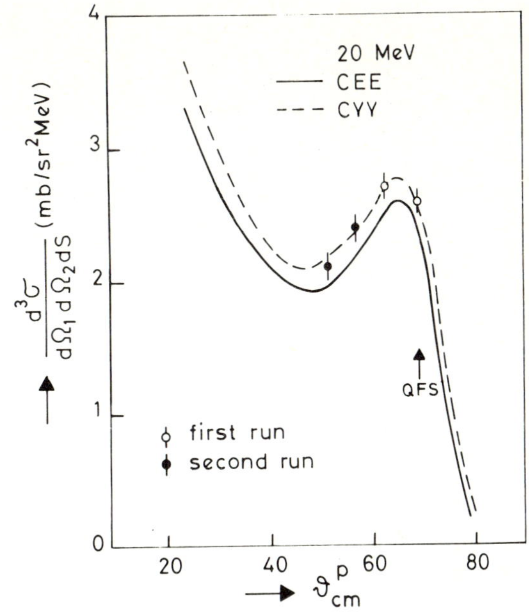

fig. 4.

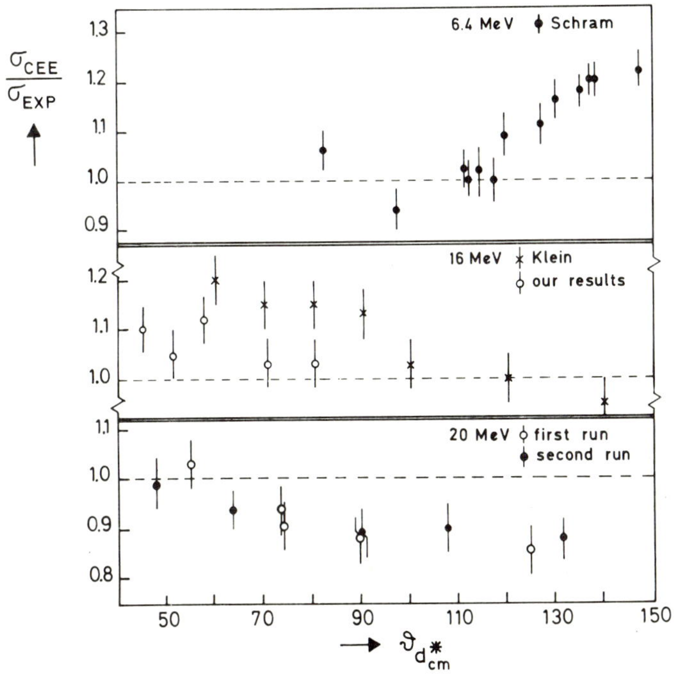

fig. 3.

THREE-BODY BREAK-UP REACTIONS

R. BOUCHEZ
Institut des Sciences Nucléaires & Centre d'Etudes Nucléaires, Grenoble

Nucleon-deuteron break-up reactions are extensively studied [1] both theoretically [2-7] and experimentally [8-18,26]. From deviations between the calculated and experimental data, off energy-shell effects in the two-nucleon interaction and dynamics effects in the three-nucleon system are measured. The data are mainly studied close to the NN fsi and qfs singularities to measure the NN interaction parameters.

At low energy, below $\approx$ 20 MeV, exact calculations of nd break-up amplitudes have been performed by Cahill [2] and Ebenhöh [3] from the Faddeev formalism including S-wave NN separable potential models, and by Kloet [4] applying Padé technic to the multiple scattering series with S-wave NN local potential models. The amplitudes close to the nn fsi and qfs singularities are very sentive to the ^1S_o nn interaction parameters. Therefore the nn fsi data are used to extract the a_{nn} scattering length, while the r_{nn} effective range parameter is fixed at its charge-symmetric value of 2.8 fm, because at a so small ($E_{nn} < 1$ MeV) relative energy the fsi amplitudes are less sensitive to r_{nn}.

The more reliable value of a_{nn} has been extracted from the nn fsi data of the complete experiment [8] by the Hambourg group ; when the exact Ebenhöh calculations [3] are applied a value of $a_{nn} = -17.1 \pm .8$ fm was obtained. Taking into account two other measurements of a_{nn} from the $\pi^- d \rightarrow nn\gamma$ and $3(3,\alpha)nn$ experiments, Henley and Wilkinson [19] suggested the averaging value of $-16.4 \pm .9$ fm.

However some other measurements have given a smaller scattering length value. First, the nn and np fsi data of the complete experiment [9] by the Grenoble group have been analysed with both the exact Cahill' [2] and Kloet' calculations [4]. Both analysis showed there is an intermediate (≈ -20 fm) value of a_{nn} that may reproduce the right shape of the nn peak, but not the right magnitude (calculated nn and np peaks are a factor 1.8 higher). Therefore some instrumental uncertainty could exist, and this measurement of a_{nn} cannot challenge the -16.4 fm value.

Smaller values of a_{nn} have also been extracted from the proton spectra in incomplete experiments. In the experiment [12] by a Tokyo group, a value of -18.3 ± .2 fm has been obtained when the Cahill' calculated spectra are folded in the instrumental resolution. Then again in the experiment [13] by a Russian group, a still smaller value of -22 ± 2 fm has been extracted [20] from proton data applying the Popova parametrisation of multiple scattering series. This value has been confirmed [23] when exact Cahill'calculations are compared with the unfolded spectrum [13] close ($E_{nn} \lesssim$ 730 keV) to the proton peak. Thus YT charge-independent model ($a_{nn} = a_{np} \approx -24$ fm), with a Tabakin potential for the ^1S_o state instead a Yamaguchi potential, and the YY charge-independent model give the same shape as the experimental spectrum. Whereas the YYY charge-dependent model ($a_{nn}= -16$ fm, $a_{np}= -24$ fm) gives a quite different shape from the experimental shape [13]. Also the magnitude of the experimental peak (28.7 mb/MeV. sr) is fairly well reproduced by the YY (25.17) or the YT (26.9) model, but not ($\approx$20) by the YYY model.

As such an extraction of a_{nn} from proton spectra can be hampered by instrumental [22] and theoretical uncertainties [20], the extraction process was reexamined. For that purpose a new experiment [21] was performed with multi-wire improved detectors [25] giving each proton trajectory with a good ($\approx$1°) precision. The proton data are displayed in the fig.1 and compared with the data from [12,14], obtained with a good resolution ($\Delta E \approx$ 400 keV) and statistics ($\approx$1%). First the unfolding procedure was applied to the data. But the fluctuations of the data induced too large oscillations to the unfolded spectrum (fig.2); therefore the folding procedure should give more reliable values. Also this study showed that the fluctuations of the experimental data are still too large, but resolution and statistics are nearly satisfactory. Then the comparison of data with folded and calculated spectra showed that a large energy-range ($E_{nn} \lesssim$ 2 MeV) is to be used, in order to take into account the background effect below the proton peak. Effectively the proton peak close to the nn fsi singularity involves more background amplitudes from far away singularities (qfs, np fsi) than the nn peak incomplete experiments. Unfortunately, the relative magnitude of the background is not independent of the model [12,20]. Thus, to reduce this model dependence the relative magnitude of the background can been taken [12] as a fitting parameter (just as a_{nn}); then a_{nn} $\approx$-18 fm was obtained. Keeping the background fixed from the calculated

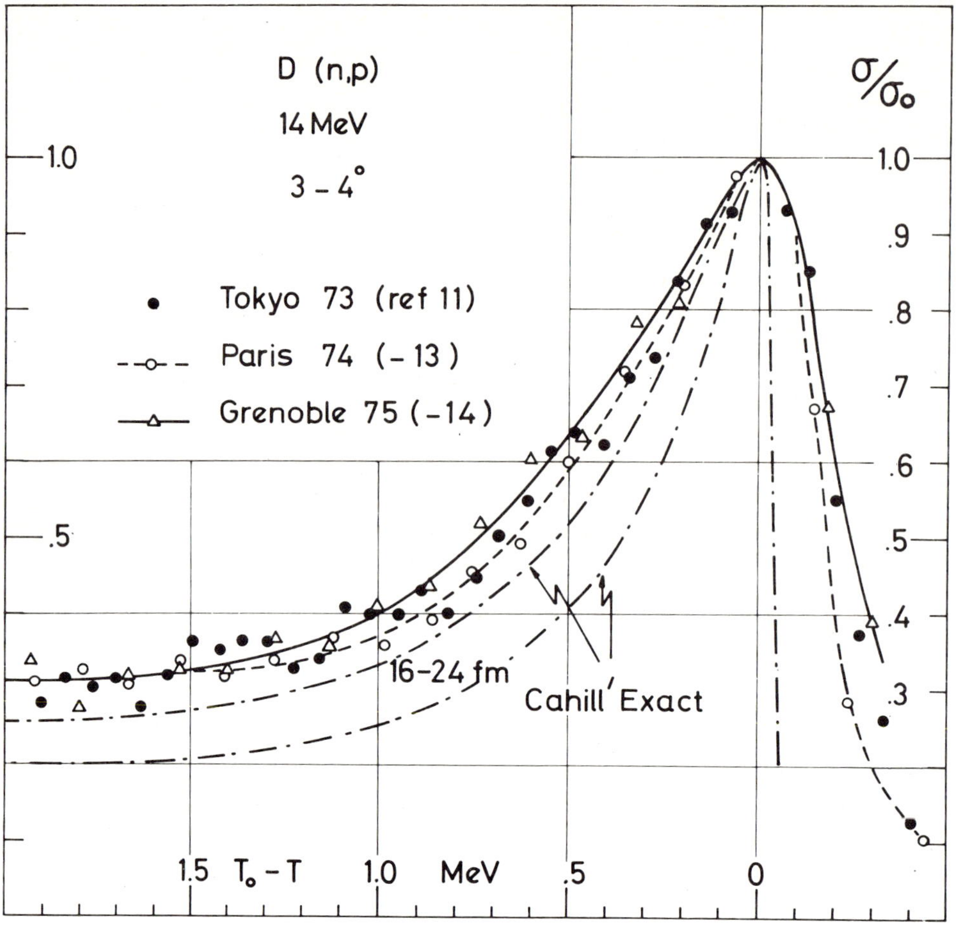

Fig.1 - Normalised proton data from ref.[12,14,21], and spectra from exact Cahill' calculations [2] for a_{nn} = - 16 and - 24 fm. Normalisation (σ/σ_o vs T_o - T) is made at the proton peak.

spectra (fig.1,2) for a_{nn} = - 16 and - 24 fm, the folding procedure, applied on a E_{nn} energy range of 2 MeV, gives an intermediate value of a_{nn}, in agreement with the Tokyo measurement.

Proton spectra measured [18] at higher energy (E $\approx$ 50 MeV) would allow [22] to determine a precise value of a_{nn}, taking into account instrumental and theoretical developments. At such an energy the multiple scattering series converge, but higher partial waves (mainly P and D-wave) have to be introduced [5] in the NN potential model.

The r_{nn} effective range parameter for the ^1S_o state could be measured [24] from the nn qfs data for a larger E_{nn} relative energy of $\approx$10 MeV, when a_{nn} is fixed at his value measured from nn fsi data. The nn qfs

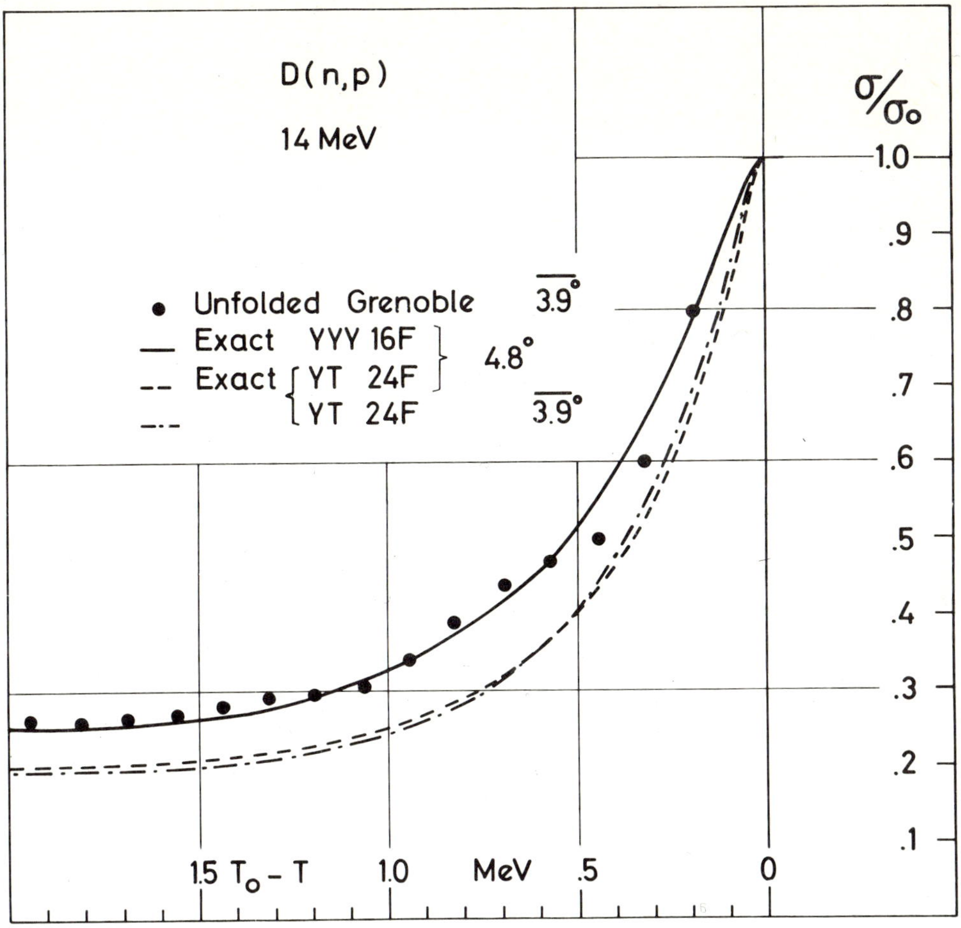

Fig.2 - The unfolded proton data from ref.[21] get rid of instrumental effects. The calculated spectra are as in fig.1.

---

amplitude is more sensitive to r_{nn}, but is also model dependent. Therefore, so long as NN models are not consistent with the magnitude of the NN fsi and qfs data, a measurement of r_{nn} should be extracted from the relative magnitude of qfs peaks measured for different E_{nn} relative energy.

Lastly, the three-nucleon break-up has to be studied in the whole three-nucleon phase space, not only close to the NN singularities, but far from the singularities. Thus, the symmetric case where all nucleon pairs have the same relative energy ($E_{ij} \approx E/2$) the three-nucleon character of the system should be the most pronounced [4], and the break-up amplitudes not very sensitive to the NN interaction. Therefore the three-nucleon interaction effects should be studied in this symmetric case, where the three-nucleon system radius is smallest and the exchange currents should produce three-nucleon forces. Practically this study should be made at relatively high relative energy (E_{ij} > 100 MeV) when should happen [27] off-shell effects due to new singularities as the Δ resonance.

References

1. S. Fiarman and S.S. Hanna, Nucl.Phys. A251 (1975) 1
2. R.T. Cahill, Phys.Rev. C9 (1974) 473
3. W. Ebenhöh, Nucl.Phys. A191 (1972) 97
4. W.M. Kloet and J.A. Tjon, Proc.Int.Conf. on few-particle problems, Los Angeles, North-Holland, 1972, 380; Nucl.Phys. A210 (1973) 380
5. M. Durand, Nucl.Phys. A201 (1973) 313
6. M. l'Huillier, Thesis, Orsay, 1974
7. J.M. Wallace, Phys.Rev. C8 (1973) 1275
8. B. Zeitnitz, R. Maschuw, P. Suhr and W. Ebenhöh, Phys.Rev.Lett.28 (1972) 1656 ; Los Angeles Conference (op.cit.) 1972, 117
9. R. Bouchez, S. Desreumaux, J.C. Gondrand, C. Perrin, P. Perrin and R.T. Cahill, Nucl.Phys. A185 (1972) 166
10. W.H. Breunlich, S. Tagesen, W. Bertl and A. Chaloupka, Los Angelès Conference (op.cit.), 1972, 100
11. A.I. Sankov, Yu.I. Chernukin, E.M. Oparin and G.A. Gornitsyn, Sov. J.Nucl.Phys. 14 (1972) 157
12. S. Shirato, K. Saitoh, N. Koori, R.T. Cahill, Nucl.Phys.A215 (1973) 277
13. B. Skorodumov, G. Raduk, A. Mikulskii and Yu. Talanin, XXe Meeting on Nuclear Spectroscopy, Leningrad 47 (1970)
14. F. Coçu, G. Ambrosino, D. Guerreau, Proc.Int.Conf. on Few Body Problems, Québec, 1974 (to be published)
15. E. Graf, C. Lunke, J. Rossel, R. Wenusch and C. Zangger, Helv. Phys. Acta, 39 (1966) 578
16. E. Bovet, F. Foroughi and J. Rossel Helv.Phys.Acta 48 (1975) 137
17. W. Ebenhöh, B. Sundgvist, A. Johansson, L. Amten, L. Glantz, L. Gönczi and I. Koersner, Phys.Lett. 49B (1974) 137
18. A. Stricker, Y. Saji, Y. Ishizaki, J. Kokama, H. Ogata, T. Suehiro, I. Nonaka, Y. Sugiyama, S. Shirato and N. Koori, Nucl.Phys.A190 (1972) 284
19. E.M. Henley and D.H. Wilkinson, Los Angeles Conf.(op.cit.) 1972, 242
20. R. Bouchez, Symp. on three-body problem, Budapest 1971, Acta Physica Academiae Scientiarum Hungaricae, 33 (1973) 151
21. S. Desreumaux Cen-Grenoble 1975 (to be published)
22. J.C. Davis, J.D. Anderson, S.M. Grimes and C. Wong, Los Angeles Conf. (op.cit.) 1972, 104
23. R.T. Cahill and R. Bouchez, Note Cen-Grenoble, PhN 71/8S
24. D. Vranić, I. Šlaus, G. Paić and P. Thomaš, Québec Conf. 1974 (to be published)
25. S. Desreumaux, A. Chisholm, P. Perrin and R. Bouchez, Communication at this Conference
26. M.W. Mc Naughton, R.J. Griffiths, I.M. Blair, B.E. Bonner, J.A. Edgington, M.P. Mayand, N.M. Stewart, Nucl.Phys.A239 (1975) 29
27. R. Bouchez and J. Jungerman, Ann. Phys. 7 (1972) 443

A TEST OF THE PROTON-INDUCED ^2H BREAK-UP STUDY FOR SPECIAL KINEMATIC CONDITIONS.

N. Fujiwara*, E. Hourany**, H. Nakamura-Yokota***, F. Reide and T. Yuasa
Institut de Physique Nucléaire, Université de Paris-Sud, BP N°1, 91406 ORSAY.

In the last 15 years the break-up of the deuteron by nucleons has been measured in various kinematic conditions. The recent development of theoretical treatments of the three-body problem permits to fit the available data up to about 50 MeV except for the polarization data, using separable S-wave potentials. From this fact, it seems that the great part of the three-nucleon data are not so sensitive to the details of the nucleon-nucleon interaction. At incident energies greater than about 100 MeV, it is rather difficult to describe the break-up of the deuteron in large momentum transfer regions even by the modified separable two-body potentials [1]. The contribution of the scattering matrix elements for two-nucleon scatterings up to the second order were calculated using the off-shell t-matrix elements with various local potentials and including the D wave component of deuteron wave function in the proton-induced ^2H break-up at 156 MeV [2] and compared with the experimental results [2], [3]. Though the calculated cross-sections were a little larger than the experimental ones at QFS and FSI regions and several times larger at the intermediate region, they explained fairly well the experimental results. Recently, the contribution of higher order scatterings were estimated in the so-called FSA model and a fairly good fit was obtained at 156 MeV incident energy [3]. The effects of the higher order scatterings may be important even at high incident nucleon energies.

We have made an experiment on the proton-induced ^2H break-up at 156 MeV with the detector angles θ_1 and θ_2 proposed by Lambert et al. [4] to satisfy the kinematic conditions in which the spectator nucleon is at rest in the total center of mass system. These conditions which correspond to destructive interference minima have been found very sensitive to the N-N potential through 3 body Faddeev analysis. We hoped to test these conditions at 156 MeV. This experiment was carried out by using 156 MeV stochastically extracted proton beam of the Orsay synchrocyclotron. The angles of the detectors relative to the beam direction were $\theta_1 = -\theta_2 = 58.3°$. The used liquid deuterium target and the detection system were identical to those described in ref. [2] and [5]. This time the (p, pn) cross-sections were not measured with acceptable statistical errors.

The obtained triple differential cross-sections $d^3\sigma/d\Omega_1 d\Omega_2 dE_1$ as a function of the detected proton energy E_1 are shown in Fig. 1 for the (p, 2p) reaction. The $d^3\sigma$ denoted as J_{exp} does not vary very much in all the energy range whilst the phase space factor changes rather rapidly. The error bars indicate only the statistical errors. Small peaks due to the FSI and QFS-type (minimum E_3 value) are observed at E_1 = 30 MeV and 68 MeV respectively. For the latter peak kinematic energy of the spectator neutron E_3 is 17 MeV and the relative energies E_{13}^{CM} & E_{23}^{CM} are 24.5 MeV and 25.1 MeV respectively (the suffixes 1, 2 and 3 indicates the nucleons detected by the detectors 1 and 2 and the spectator one). Therefore this peak is far from FSI or real QFS.

The theoretical calculation with the FSA was made by L'Huillier for the present geometry. It is also shown in Fig. 1. The calculated cross-sections with the first order scattering J_1 are very small in comparison with those calculated onto the second order scattering, J^+_{12} and J^-_{12}. The estimated values J^+_{12} or J^-_{12} are too much larger than the experimental values. This fact might indicate the importance of higher order scatterings even at 156MeV. The calculation of the higher order terms is carried out by using a modified S wave separable potentials [3]. The result depends on the modification factor F, and we can see that the experimental values are reproduced very well by putting F = 0. 523 + i 0. 1517 fm.

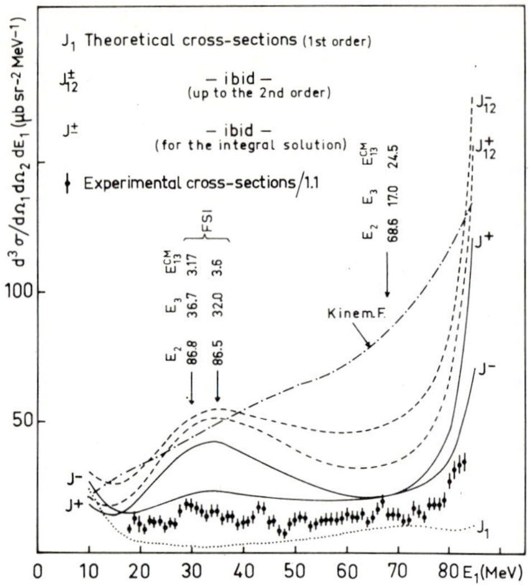

Fig. 1. $d^3\sigma$ spectrum and the calculated J curves. ^2H (p, 2p)n, $\theta_1 = -\theta_2 = 58.3°$.

In conclusion, the cross-section of the deuteron break-up at 156MeV might be explained by the FSA-type three-body scattering theory even for the data in the regions far from QFS and FSI regions, by using the off-shell t-matrix element of the realistic potential for the first (and second) order scatterings, and the separable potentials for higher order scatterings. The second order scattering calculation results obtained by using the realistic potentials and off-shell t-matrix elements determine the unknown coefficients introduced for the used separable potentials. When the model is refined, it might be possible, to examine the effects of the three-body force and the off-shell t-matrix elements. At present time, we can say that our experimental results are not incompatible with those of Lambert et al.

References

[1] J. M. Wallace, Phys. Rev. C7 (1973)10.

[2] J. P. Didelez, R. Frascaria, N. Fujiwara, I. D. Goldman, E. Hourany, H. Nakamura-Yokota, F. Reide and T. Yuasa, Phys. Rev. C10 (1974) 529. Thesis of M. L'Huillier (Orsay 1974).

[3] M. L'Huillier, P. Benoist-Gueutal and J. L. Ballot, Phys. Rev. C12(1975)948. J. L. Ballot, M. L'Huillier and P. Benoist-Gueutal, loc. cit. p. 725.

[4] J. M. Lambert, P. A. Treado and R. G. Allas, L. A. Beach, R. O. Bondelid and E. M. Diener, Proc. Int. Conf. (Québec, August 1974) (LavalUniv. Press (1975) 531).

[5] H. Nakamura-Yokota, F. Reide and T. Yuasa, Nucl. Instr. and Meth. 108 (1973) 509.

* Permanent address : Kyoto University, Kyoto, Japan.

** " " : Lebanese University, Faculty of Sciences, Haddat-Beyrouth, Lebanon.

*** " " : Tokyo Institute of Technology, Meguro, Tokyo, Japan.

THE N-P FSI ANGULAR DISTRIBUTION IN
DEUTERON BREAKUP AT 45 MeV

E. L. Petersen, M. I. Haftel, R. G. Allas, L. A. Beach
and R. O. Bondelid
Naval Research Laboratory, Washington, D. C. 20375 U. S. A.
and
P. A. Treado and J. M. Lambert
Georgetown University, Washinton, D. C. 20057 U. S. A.
and Naval Research Laboratory

The Final State Interaction (FSI) region of phase space is interesting because of the sensitivity of the cross section to both on-shell and off-shell effects. We have pointed out previously that the FSI region is sensitive to off-shell effects and that the sensitivity increases with energy (Haftel [1-3]). The present experiment was performed to provide an FSI angular distribution at an energy higher than previously available. The experiment was performed with the NRL Sector Focusing Cyclotron using standard coincidence electronics for kinematically complete experiments.

The data is compared with the results of calculations using several S-wave separable potentials tailored to test for sensitivity to off-shell and on-shell effects. These potentials have been described previously (Haftel [1-3]). The potentials have two different form factors A and B: A has off-shell behavior similar to that of the Reid soft core potential while B has off-shell behavior resembling the softer Yamaguchi. The potentials fit two different sets of phase shifts: set 1 corresponding nearly to experimental p-p 90° cross sections and set 2 corresponding to cross sections less than the p-p experimental values at energies above 10 MeV. The potentials also predict two different triton binding energies: 8.3 and 11.0 MeV. The calculations are performed using the Ebenhöh code as modified by Haftel. Rather than the original "hybrid" calculation a full three coupled channel calculation was performed.

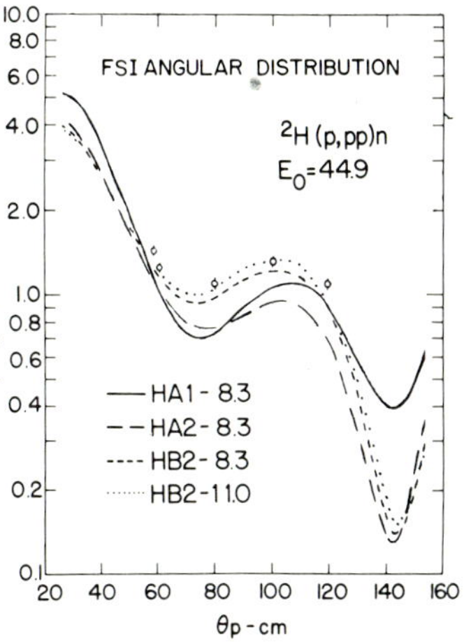

Figure 1. Angular dependence of the FSI peak. A comparison of the predictions of the various potentials shows sensitivity due to:

a) on-shell changes HA1-8.3 - HA2-8.3
b) off-shell changes HA2-8.3 - HB2-8.3
c) triton binding energy HB2-8.3 - HB2-11.0

The data appears to favor the HB set of potentials. This is surprising as the HA1-8.3 potential has characteristics closer to "realistic" potentials. On the other hand, the result that the measured cross section is larger than expected is consistent with previous comparisons: Our previous results at 23 MeV (Petersen [4]), even when compared with a 3 channel calculation; several experiments discussed by

Kluge, Schlufter and Ebenhöh [5]; and work at 8.5 MeV by Kühn, Kumf, Mosner, Neubert, and Schmidt [6]. The experimental results at 8.5 MeV were in fact higher than any of the potentials used. The general trend of data being higher than theory is somewhat disconcerting as we have nearly convinced ourselves that deuteron breakup calculations at low energies are insensitive to the form factor as long as the potentials are phase shift equivalent and predict the same triton properties. This experiment appears to favor soft core potentials (similar to Yamaguchi); but the general trend at lower energies says we must be very cautious about such conclusions.

We have to repeat the cliche that the present results call for more theoretical work. The present experiment by itself would suggest investigation of effects that should become more important at higher energies, such as P waves and tensor forces. The remaining obvious effect, which will enter at all energies, is the Coulomb force in the intermediate state. Approximations of the type made in reference [5] only include the Coulomb effect in the final state.

Our results, when put in the context of previous experiments and calculations, emphasize that there are still some basic factors not being considered in deuteron breakup calculations.

References

[1] M. I. Haftel and E. L. Petersen, in "Few Body Problems in Nuclear and Particle Physics," edited by R. J. Slobodrian, B. Cujec and K. Ramavataram (University of Laval Press, Quebec 1975), pp 748, 746, 752.

[2] M. I. Haftel and E. L. Petersen, Phys. Rev. Lett. 33 (1974) 1229.

[3] M. I. Haftel, E. L. Petersen and J. M. Wallace, to be published.

[4] E. L. Petersen, et al. in "Few Body Problems in Nuclear and Particle Physics," (see Ref 1) p. 395.

[5] W. Kluge, R. Schlufter, and W. Ebenhöh, Nuc. Phys. A228 (1974) 29.

[6] B. Kuhn, H. Kumpf, J. Mösher, W. Neubert and G. Schmidt, Nuc. Phys. A247 (1975) 21.

OFF-SHELL EFFECTS IN NUCLEON-DEUTERON SCATTERING

J.H. Stuivenberg and R. van Wageningen

Natuurkundig Laboratorium der Vrije Universiteit
Amsterdam, The Netherlands

During the last few years considerable progress has been made in the calculation of three-nucleon observables with realistic nucleon-nucleon interactions. In the case of nucleon-deuteron scattering above the breakup threshold however until now rather simple interactions have been used (1) due to the complexity of the Faddeev equations. The calculations of elastic phase shifts and breakup cross sections show some sensitivity to the interactions used, but as these interactions are in general not phase equivalent, it is hard to draw conclusions about off-shell effects. Only Haftel (2) has reported calculations with phase equivalent rank one separable interactions, which are however energy dependent.

Therefore we have performed calculations with phase equivalent separable potentials of rank two. With these interactions of the form

$$V(p,q) = -g(p)g(q) + h(p)h(q)$$

one has the freedom to choose the attractive formfactor g(p) rather arbitrarily. By means of inverse scattering theory a repulsive formfactor is then (numerically) constructed in such a way that V(p,q) reproduces exactly a given set of (experimentally determined) phase shifts. Moreover it is always possible to bring these potentials in the socalled "limiting form" where the attractive formfactor itself reproduces the (anti) bound state pole (3). This formfactor is then directly related to the (anti) bound state wave function. Thus it is possible to construct families of phase equivalent potentials with prescribed bound state wave functions.

As these potentials are not given in analytical form the normal way of solving the Faddeev equations by means of contour deformation is not suitable here. Therefore we developed a new computer code following an idea by Fuda (4). Essentially the integral equations are iterated once and can then be solved along the real axis.

We report here some calculations of nucleon-deuteron elastic phase shifts and breakup cross sections at E_n = 16 MeV. Only the $^1S_0$ and $^3S_1$ interactions have been taken into account and charge independence is assumed. For both spin states we have constructed two phase equivalent potentials which are fitted to the $^1S_0$ and $^3S_1$ phase shifts of Arndt and Macgregor (5). In figure 1 and 2 the attractive formfactors are given for these interactions.

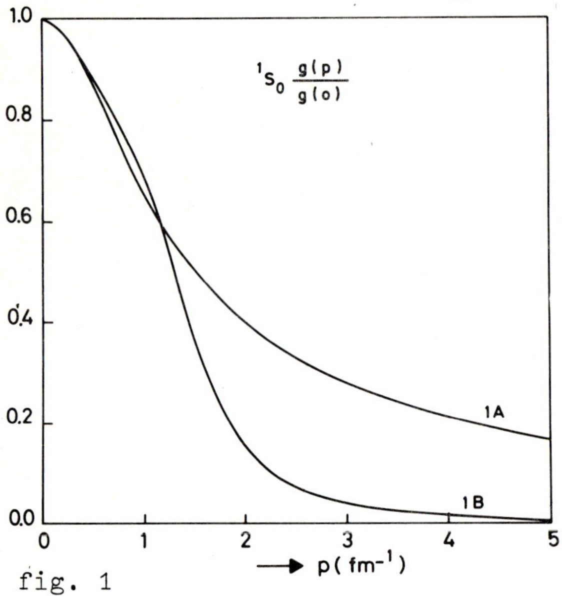

fig. 1

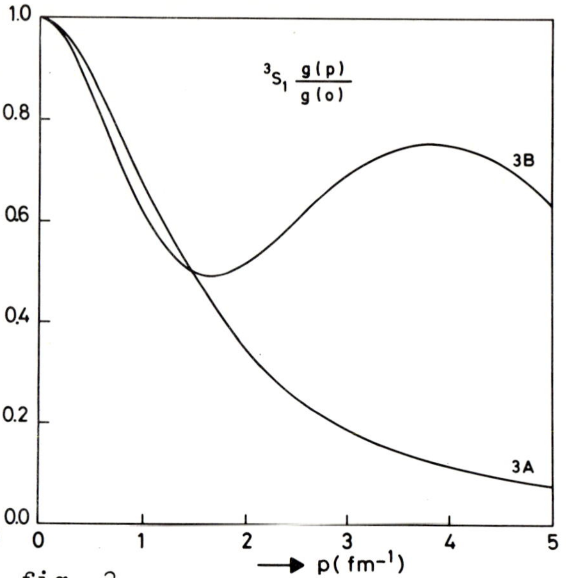

fig. 2

The formfactors 1A and 2A are Yamaguchi-like. All potentials are in the limiting form. In figure 3 a typical breakup cross section is shown for the potential combinations 1A + 3A and 1B + 3B.

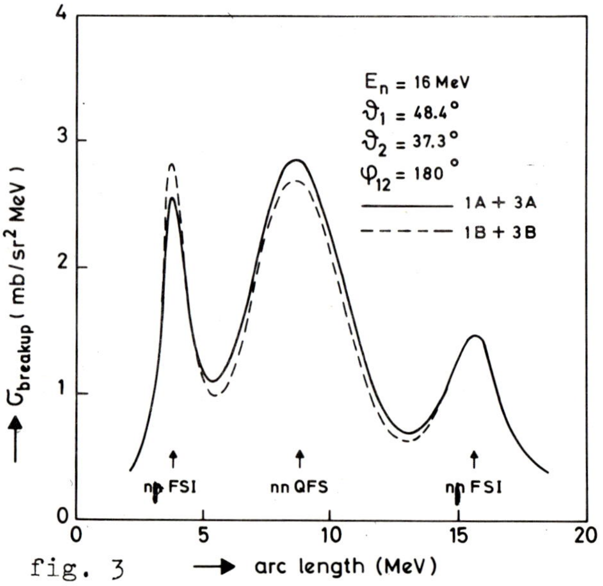

fig. 3

The off-shell differences in the final state and quasifree scattering regions are of the order of 10% at this energy. In table 1 the $^2\delta_0$ phase shift is given for various potential combinations. Again it is found that the differences between 1A and 1B depend rather strongly on the choice of the triplet interaction and vice-versa. A systematic investigation of these off-shell effects is in progress. In particular it will be interesting to see how much off-shell freedom there exists with a reasonable choice of the two-body bound state wave functions.

	1A	1B
3A	123.1°	115.4°
3B	118.9°	105.5°

Table 1

1) J. Bruinsma, W. Ebenhöh, J.H. Stuivenberg and R. van Wageningen, Nucl.Phys. A228 (1974) 52.

2) M.I. Haftel and E.L. Petersen, Phys.Rev. Lett.33 (1974) 1229.

3) H. Fiedeldey and N.J. McGurk, Nucl.Phys. A189 (1972) 83.
H. de Groot, thesis, Vrije Universiteit, Amsterdam (1975).

4) M.G. Fuda, Phys.Rev. Lett.32 (1974) 620.

5) M.H. Macgregor, R.A. Arndt and R.M. Wright, Phys.Rev. 82 (1969) 1714.

NUCLEON-DEUTERON BREAKUP QUANTITIES CALCULATED WITH SEPARABLE INTERACTIONS INCLUDING TENSOR FORCES AND HIGHER PARTIAL WAVES

J. Bruinsma and R. van Wageningen

Natuurkundig Laboratorium der Vrije Universiteit
Amsterdam, The Netherlands

Here we report nucleon-deuteron breakup calculations at 22.7 MeV nucleon bombarding energy with separable interactions including tensor forces. We solved the Faddeev equations with the Fuda-Stuivenberg method (1,2). The angular momentum analysis has been done in the same way as Doleschall (3). We performed the calculations with charge independent interactions: Y-Y7, Y-Y0, E-E0 and Y-E0 (the first letter indicates the singlet formfactor, the second the triplet one and the number gives the percentage D-state).
Y denotes the Yamaguchi formfactor, E the exponential one (4). All potentials have been fitted to the following low energy scattering parameters: a_s = -20.34 fm, r_s = 2.7 fm, a_t = 5.397 fm, r_t = 1.722 fm. The Y-Y7 potential gives Q_d = 0.283 fm^2. The equations with total angular momentum and parity J^π up to $^{19}/_2{}^+$ were solved.

In figure 1 we compare the n-d breakup cross sections for the Y-Y7 and Y-Y0 potential. The p-d experimental points are from Petersen et al. (5). One should be careful to compare the theoretical curves with the experimental points since it has been shown that charge dependent calculations give slightly different results (6).

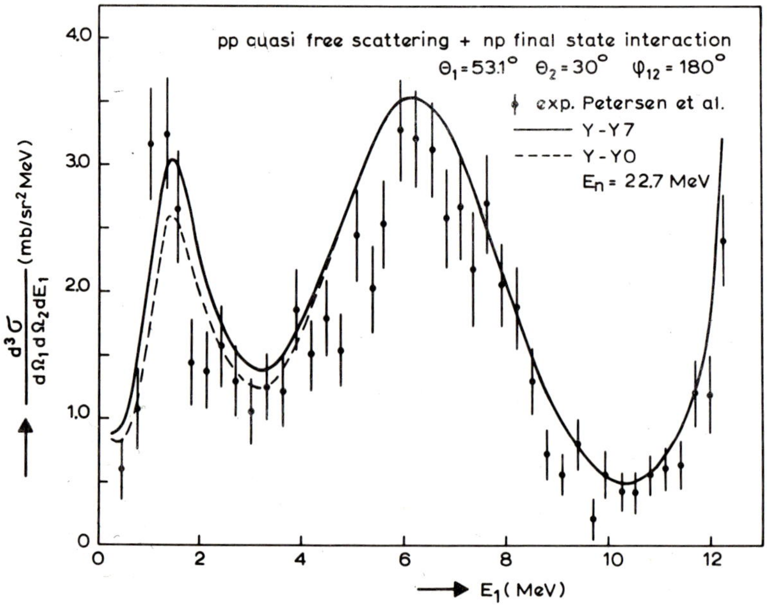

fig. 1

Table I

The differences in the peaks for the final state interaction breakup cross sections for several production angles Θ_{d*} (see ref. 4 figure 12)

Θ_{n_1}	Θ_{n_2}	Θ_{d*}	$\frac{YY7-YY0}{YY0}(\%)$	$\frac{EE0-YY0}{YY0}(\%)$	$\frac{YE0-YY0}{YY0}(\%)$
53.1	-30.0	45.7	24.6	10.2	5.4
50.0	-49.8	74.5	.2	.9	-.5
65.0	-40.7	94.4	-5.2	-8.3	-4.0
90.0	-27.4	122.8	-4.4	-17.3	-6.2
120.0	-15.4	148.0	.3	-16.8	-5.0

Comparison made at all angle combinations from Petersen shows that the quasi free scattering peaks are little influenced by the tensor force. The final state peak is more sensitive. In table I we compare the fsi peaks for the chosen interactions. We notice that the differences between the tensor potential and Y-Y0 behave qualitatively in the same way as the differences between other S-wave potentials.
Because the choice of a more realistic formfactor has been shown to be important (4), it is desirable to reanalyse the experimental results with realistic potentials including tensor forces.

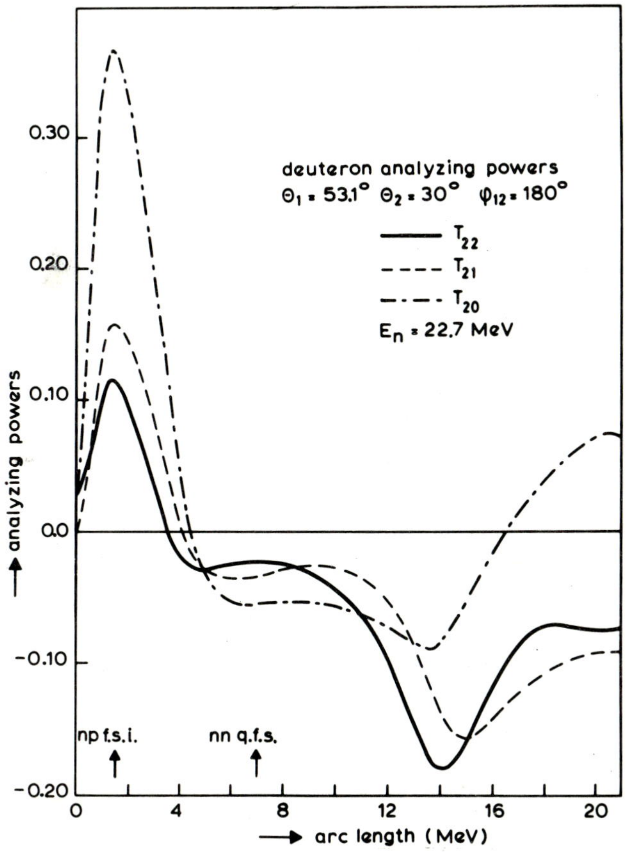

fig.2

Figures 2 and 3 show the deuteron tensor analyzing powers along the kinematical curve for two angle combinations. The T_{20} is large in the final state region (first peak). So it seems feasible to determine it experimentally.

The deuteron vector analyzing powers (A_y) and neutron polarizations (P_n) are less than 5 percent for the angle combinations of figures 2 and 3. As in the elastic scattering we expect that the P-wave contributions will be important here. To describe the experimentally measured A_y and P_n (see eg. Conzett (7)) we now add P-wave interactions in our breakup calculations.

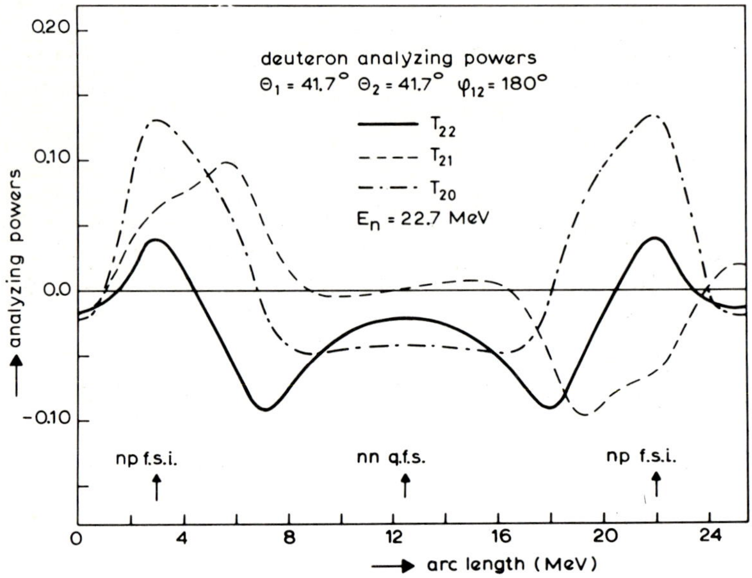

fig. 3

This investigation forms a part of the research program of the Foundation for Fundamental Research of Matter (F.O.M.) which is financially supported by the Netherlands Organization for Pure Scientific Research (Z.W.O.). We would like to thank Dr. B.L.G. Bakker and J.H. Stuivenberg for their help and cooperation.

References

1) M.G. Fuda, Phys.Rev.Lett. 32 (1974) 620
2) J.H. Stuivenberg, Thesis, Amsterdam, to be published
3) P. Doleschall, Nucl.Phys. A201 (1973) 264
4) J. Bruinsma, W. Ebenhöh, J.H. Stuivenberg, R. van Wageningen, Nucl.Phys. A228 (1974) 52
5) E.L. Petersen, M.I. Haftel, R.G. Allas, L.A. Beach, R.O. Bondelid, P.A. Treado, J.M. Lambert, M. Jain, J.M. Wallace, Phys.Rev. C9 (1974) 508
6) J.H. Stuivenberg, J. Bruinsma, R. van Wageningen, Proc.Int.Conf. on few body problems in nuclear and particle physics, ed. R.J. Slobodrian et al. (les presses de l'université Laval, Quebec, 1975) p. 505
7) H.E. Conzett, Proc.Int.Conf. on few body problems in nuclear and particle physics, ed. R.J. Slobodrian et al. (les presses de l'université Laval, Quebec, 1975) p. 566

DETERMINATION OF THE TOTAL NUCLEON-DEUTERON BREAKUP CROSS SECTION AT E_d = 26.5 MeV

R. van Dantzig, B.J. Wielinga and G.J.F. Blommestijn
Institute for Nuclear Physics Research (IKO), Amsterdam
and
I. Slaus
Institute "Ruder Boskovic", Zagreb, Yugoslavia

In this paper we present data on the total nucleon-deuteron breakup cross section at E_d = 26.5 MeV or corresponding incoming proton energy (in the anti-lab system) of E_p = 13.25 MeV. The results have been obtained in an 80 hour run using the synchro-cyclotron and the BOL-multi-detector system (ref. 1) at IKO.

Coincident protons from the p+d → ppn reaction were detected and the events were written on magnetic tape. Details and other results of the same set of data are given elsewhere (ref. 2). The present measurement is a sample covering roughly 70% of the total phase space with a resolution in angle of typically 1.5^o - 2° and in energy of ∿ 150 keV. From this experimental material, after proper calibration, particle identification, and appropriate kinematical transformations, we obtained the neutron and proton angular distributions multiplied by the geometrical efficiency function of the BOL scattering chamber. The dependence on geometrical variables is then reduced to a minimum by accumulating similar angular distributions based on simulated events, generated in a Monte Carlo procedure in which the phase space is homogeneously sampled. In the simulated data all relevant instrumental effects are included. The data are given in fig. 1 and 2. To get the final angular distributions, the BOL-data are divided by the simulated data and then again multiplied by the full phase space factor.

The BOL-data are compared with a Faddeev calculation using a local S-wave potential (ref. 2, 3) which also includes all instrumental effects. Thus the data and the histograms in the figures can be directly compared. The accuracy of the absolute normalization is 20%. In fig. 1 the proton data of Van Oers (ref. 4) at 24.35 MeV are given together with our present results, for comparison. It should be noted that the cross section is defined here as the breakup-cross section (not the reaction cross section which is twice the breakup cross section because of the two protons in the final state). From the figure we conclude that the data of Van Oers markedly deviate both from our data and from the theoretical prediction. It would be interesting to know whether this deviation is caused by a strong dependence of the incoming energy. The difference between our data and the calculation is discussed in ref. 2.

In fig. 2 the neutron angular distribution is given, which is obtained by calculating the neutron CM-angle from the proton coincidences. The dashed line in this figure gives the "full" theoretical angular distribution deduced from the calculations by Kloet and Tjon (ref 2, 3), while the histogram and the data are both (equally) effected by the geometrical efficiency of the detection system. Using the data in fig. 2 we have calculated the integrated breakup cross section for the backward and forward (relative to the incoming proton direction) hemisphere. The contribution to the total breakup cross section at each angle is obtained by calculating

$$(\sigma_{4\pi}^{th}(\vartheta)/\bar{\sigma}_{BOL}^{th}(\vartheta)) \times \bar{\sigma}_{BOL}^{exp}(\vartheta)$$

and integrating this quantity over solid angle. ($\sigma_{4\pi}^{th}(\vartheta)$ represents the unconstrained theoretical differential cross section, $\bar{\sigma}_{BOL}(\vartheta)$ contains all instrumental effects including geometrical detection efficiency). The results are given in table I.

Our total breakup cross section result nicely fits in the series of measurements of Catron et al. (ref. 5) (interpolation of these data gives ∿ 167 mb). The backward hemisphere disintegration cross section strongly differs from the result given by Van Oers at E_p = 12.2 MeV (230 ± 18 mb) which is to be expected in view of the different angular dependence shown in fig. 1.

References

1. L.A.Ch. Koerts, K. Mulder, J.E.J. Oberski, R. van Dantzig, Nucl. Instr. and Meth. 92 (1971) 157 and subsequent articles.
2. B.J. Wielinga et al., Lett. Nuovo Cim. 11 (1974) 655
   B.J. Wielinga et al., preprint Dec. 1975, submitted to Nucl. Phys.
3. W.M. Kloet, J.A. Tjon, Nucl. Phys. A210 (1973) 380
   W.M. Kloet, J.A. Tjon, Ann. Phys. 79 (1973) 407
4. W.Th.H. van Oers, Ph.D.thesis 1963, University of Amsterdam.
5. H.C. Catron, M.D. Goldberg, R.W. Hill, J.M. Le Blanc, J.P. Stoering, M.A. Williamson, Phys. Rev. 123 (1961) 213

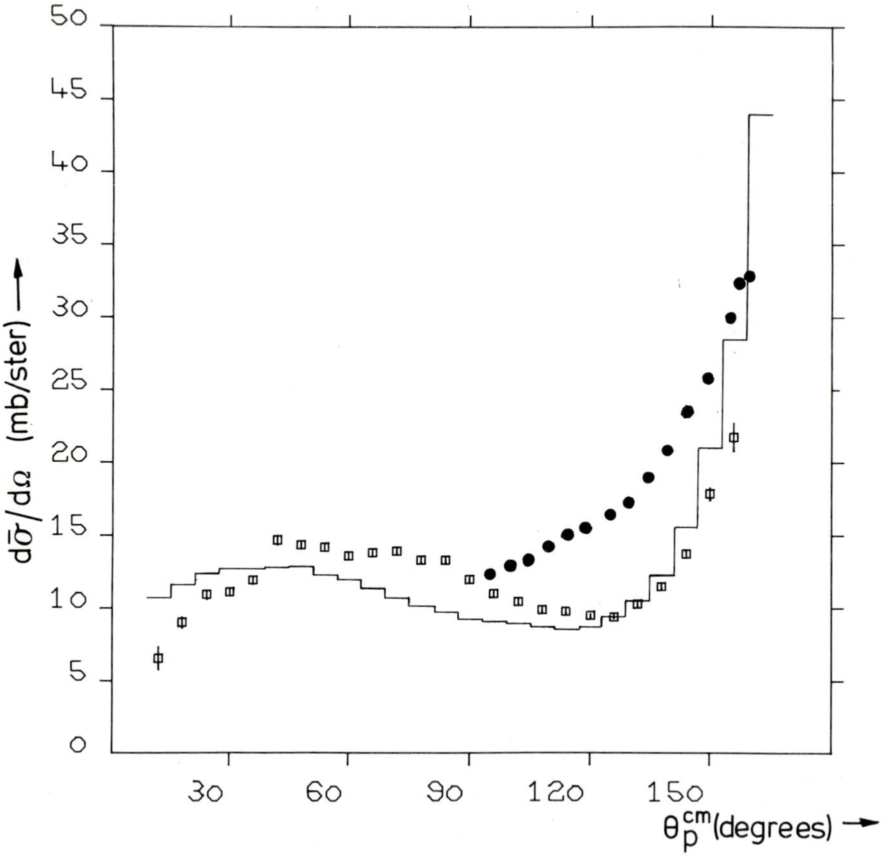

fig. 1 Theoretical and experimental proton-deuteron breakup cross section ($\bar{\sigma}$), as a function of proton CM angle both containing the geometrical detection constraints of the BOL system (resp. histogram, squares). The dots represent data from Van Oers (ref. 4).

Table I

	definition	cross section (mb) at E_p=13.25 MeV	
		experimental	theoretical
FORWARD PROTON BREAKUP	$\int_0^{\pi/2}\int_0^{E_{max}} \frac{d^2\sigma^{breakup}}{d\Omega dE} d\Omega dE$	91 ± 18	94*
BACKWARD PROTON BREAKUP	$\int_{\pi/2}^{\pi}\int_0^{E_{max}} \frac{d^2\sigma^{breakup}}{d\Omega dE} d\Omega dE$	78 ± 18	69
TOTAL PROTON BREAKUP	FORWARD + BACKWARD BREAKUP	169 ± 36	163

*Graphically obtained from the theoretical curve with an accuracy of 3%.

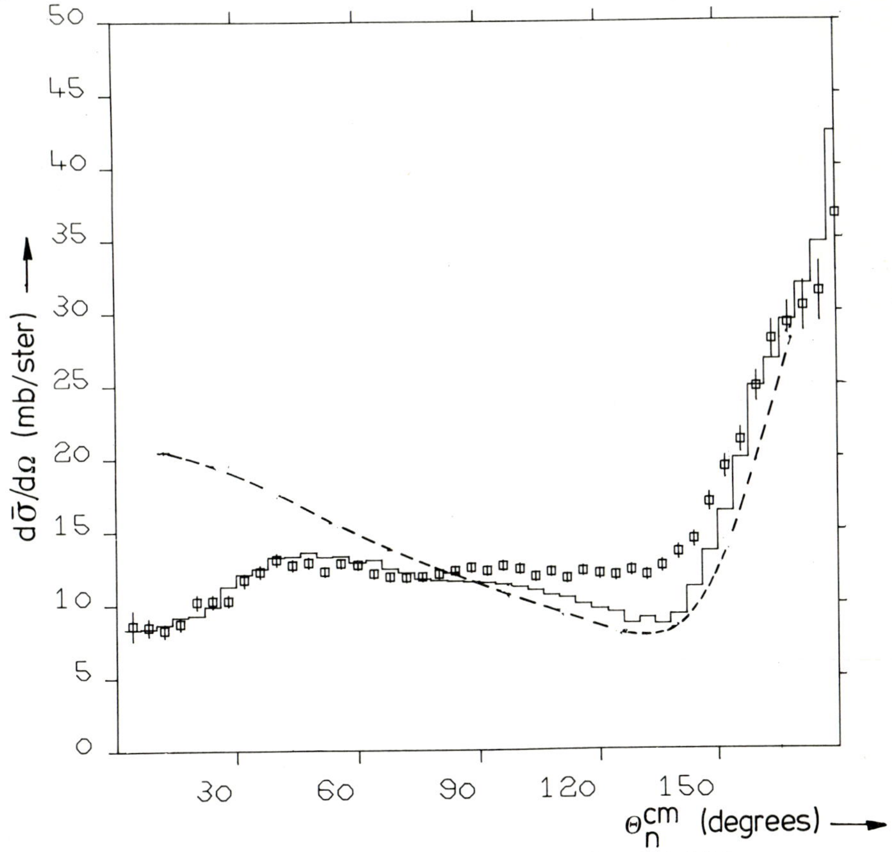

fig. 2 Same as fig. 1. The dashed curve gives the "full" cross section (4π detection of the coincident proton). Differences between data and calculations are discussed in ref. 2.

STUDY OF DOUBLET PROTON-DEUTERON BREAKUP AT E_p = 50 MeV

G.J.F. Blommestijn, Y. Haitsma, R. van Dantzig
Institute for Nuclear Physics Research (IKO), Amsterdam
and
I. Slaus
Institute "Ruder Boskovic", Zagreb, Yugoslavia

In this note we present some first results on the p+d → ppn reaction at E_p = 50 MeV, obtained with the BOL multidetector system (ref. 1) and the synchro-cyclotron at IKO. The data selected here concern a very special region of the phase space in the final state, namely the one near the point where the p-p quartet amplitude vanishes. In some recent theoretical investigations, using various potentials for the nucleon-nucleon interaction, Kloet and Tjon found (ref. 2) that in this region the theoretical n-d breakup cross section exhibits a deep minimum, which is sensitively dependent on the core region of the potential, used in their Faddeev calculation. This can be understood because the doublet amplitude tends to be more sensitive to the inner part of the N-N potential than the quartet amplitude because the Pauli priciple is less operative in the doublet configuration.

Proton-proton coincidences, recorded on magnetic tape during 60 hours of measuring, were afterwards identified and calibrated in energy and angle. We selected the data near the theoretical minimum given by Kloet (ref. 2) as close as our BOL geometry allows ($\theta_{p1}^{LAB} = \theta_{p2}^{LAB} = 33^o45'$, $\Delta\phi = 129^o35'$ and by E_{p1} = 21 MeV (ref. 2)). The data have been investigated near the indicated point of phase space varying the proton angles in two degree steps. Typical examples are given in fig. 1. The angular acceptance chosen here was $\pm 3^o$ for θ and $\pm 10^o$ for $\Delta\phi$. The accuracy of the absolute normalization of the data is about 25%. The depth of the experimental minimum in the cross section does not vary much by changing $\theta_{p1} = \theta_{p2}$ from 34° to 40°, while we have found that the theoretical minimum is very sensitive to kinematics close to the symmetry line $\theta_3 = \theta_4$, $T_3 = T_4$. The theoretical curves have been averaged over the experimental angular and energy acceptance. The difference between point and finite geometry calculations is shown in fig. 1.

Fig. 2 shows the data at $\theta_{p1} = 34^o \pm 1^o$, $\theta_{p2} = 34^o \pm 3^o$ and $\Delta\phi = 140^o \pm 10^o$ in steps of 4 MeV along the E_{p1}-axis. This channel width results from a compromise between angular resolution and statistics. The theoretical curves, which are Faddeev calculations with local S-wave potentials of Kloet (ref. 2) at the above mentioned kinematical region, have been smoothed over 4 MeV in energy. Experimental and theoretial cross sections at these kinematical conditions differ by factors 2 - 4, while the shapes of energy spectra are essentially similar. The predictions for potentials I - III and I - IV calculated in finite geometry do not differ significantly (see fig. 2. In fig. 1 predictions for these potentials differ by less than 15%, and only I - III prediction is plotted). The ratio of the cross sections predicted by potentials I - III and I - IV is very sensitive to θ_3, θ_4 and $\Delta\phi$ and varies from a factor of 5 (ref. 2) to 1 over $\Delta\theta \sim 3^o$ and for $\Delta\theta \sim 10^o$.

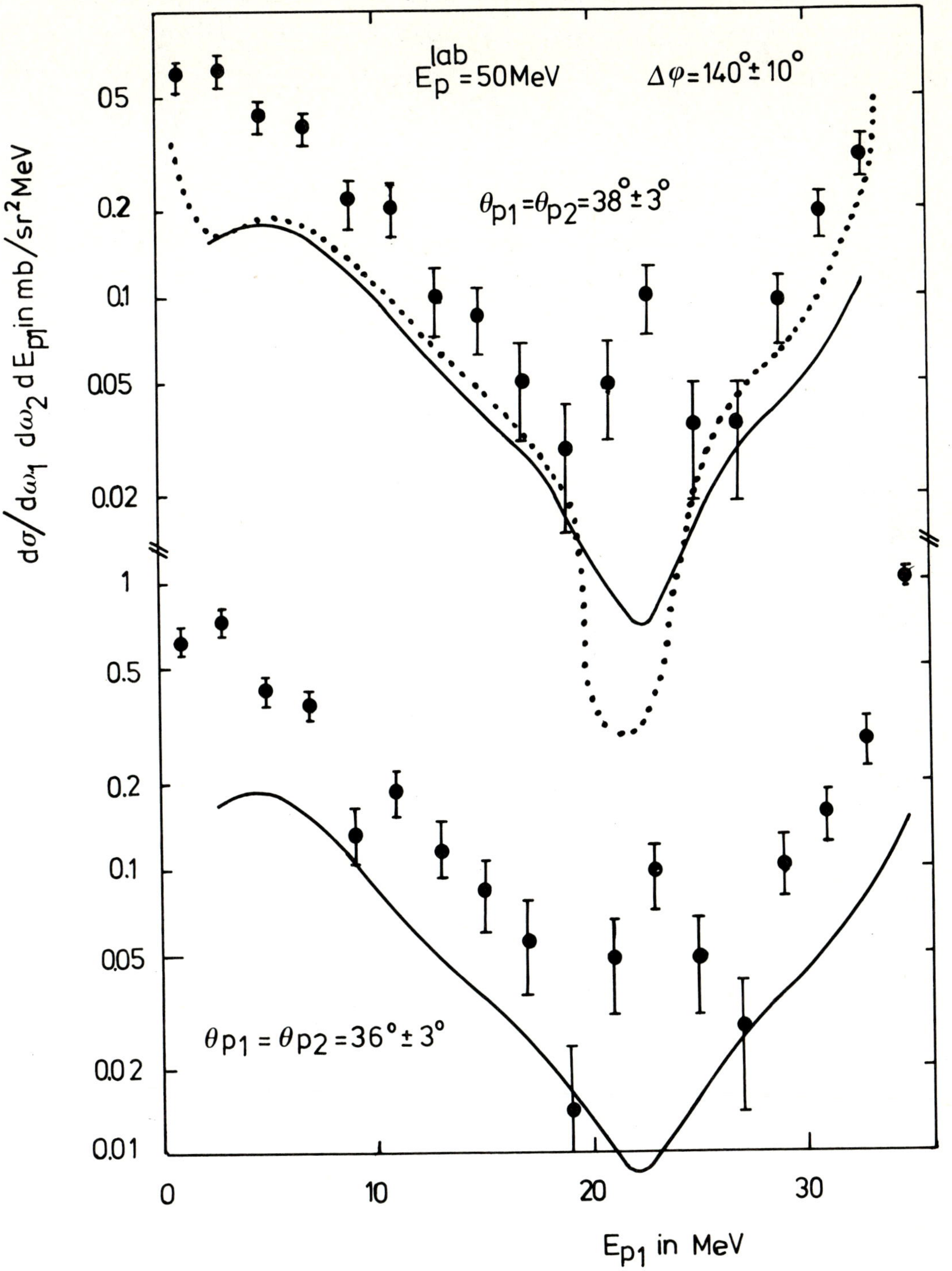

fig. 1 Deuteron breakup cross section at E_p^{LAB} = 50 MeV and $\Delta\phi = 140^o \pm 10^o$; for $\theta_{p1} = \theta_{p2} = 36^o \pm 3^o$ and $\theta_{p1} = \theta_{p2} = 38^o \pm 3^o$.
Solid curves: the results for potential I - III - finite geometry.
Dotted curve: the result for the point geometry for potential I - III.

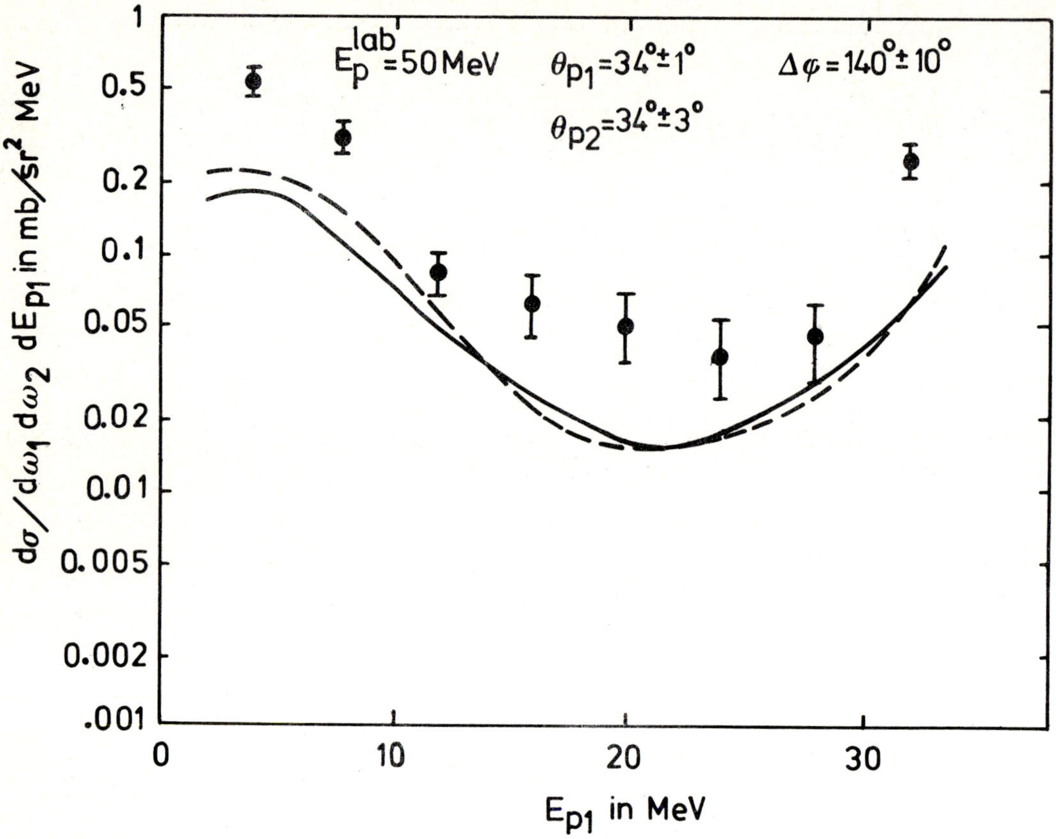

fig. 2 As fig. 1, for $\theta_{p_1} = 34^{\circ} \pm 1^{\circ}$, $\theta_{p_2} = 34^{\circ} \pm 3^{\circ}$.
The curves are:
Solid curve : the result for potential (I - III)
Dashed curve : " " " " (I - IV)
both for finite geometry.

References

1. L.A.Ch. Koerts, K. Mulder, J.E.J. Oberski, R. van Dantzig Nucl. Instr. and Meth. 92 (1971) 157 and subsequent articles
2. W.M. Kloet, J.A. Tjon, Nucl. Phys. A210 (1973) 380
W.M. Kloet, J.A. Tjon, Ann. Phys. 79 (1973) 407

NEUTRON EXPERIMENTS AT LAMPF[†]

Mahavir Jain[*]

Texas A&M University
College Station, TX, 77843, U.S.A.

The problem of the nucleon-nucleon force is the most fundamental in nuclear physics and is basic to particle physics. However, in the energy range from pion production threshold to 1 GeV, the N-N interaction is rather poorly determined. In general, at these energies, there is no unique set of phase shifts and coupling parameters; the I=1 parameters are known at least qualitatively, but the I=0 parameters are not even known qualitatively. This is illustrated by the variation of $^3S_1$ phase shift from -17° to 35° in the three solutions of the energy-independent nucleon-nucleon phase shift analysis of Glonti [1]. In addition, these results are in considerable disagreement with the analyses of MacGregor [2]. This is due to the paucity of the n-p scattering data including polarization and triple scattering parameters. Furthermore, as will be shown later, there is considerable disagreement between the results, from different groups in the intermediate energy region of even so basic an observable as the n-p differential cross section. Therefore, we have started a long range program for the definitive determination of the n-p interaction at LAMPF energies (300 to 800 MeV). This is an ambitious project which will ultimately require the performance of many experiments. Each successive set of measurements will clarify our understanding of the n-p system to some degree and simplify the problems of the subsequent measurements.

In this communication we will give a general description of the experimental set-up and the various neutron experiments performed at LAMPF. In most of the work to be described here, the collaborators and technical persons have been: M. L. Evans, G. Glass, J. C. Hiebert, R. A. Kenefick, L. C. Northcliffe (Texas A&M University); J. G. Boissevain, B. E. Bonner, D. Brown, J. H. Fretwell, A. C. Niethammer, J. E. Simmons, K. D. Williamson (Los Alamos Scientific Laboratory); H. C. Bryant, C. G. Cassapakis, S. Cohen, B. Dieterle, C. P. Leavitt, D. M. Wolfe (University of New Mexico), C. W. Bjork, P. J. Riley (University of Texas); and D. W. Werren (University of Geneva).

EXPERIMENTAL SETUP

Fig. 1 shows a schematic drawing of the apparatus. Most of the charged particles in the neutron beam were swept by magnet M_1; the rest were vetoed by scintillator S_o. The neutron flux ($\gtrsim$ 250 MeV) was monitored by scintillator telescopes near the collimator exit, viewing a CH_2 target at $\pm$ 25°. The primary proton flux was obtained from a toroidal current monitor. The neutron beam was incident on a liquid-hydrogen (LH_2) or an appropriate scattering target. The resulting charged particles were detected by a multi-wire proportional chamber (MWPC) spectrometer. The spectrometer consisted of a large bending magnet and four MWPC (W_1-W_4), each of which measured x and y coordinates of a hit. Typical bend angles were ~ 22°, which gave momentum determination accuracy of ~ 0.5%. Another counter telescope, set an angle of 45° with respect to the LH_2 target, monitored the neutron flux x target thickness. Multiple scattering was minimized by placing He bags between W_1 and W_4. An acceptable event satisfied overlap coincidence between $S_o.S_1.S_{2L}.S_{2R}$ and a majority coincidence of both the horizontal and vertical MWPC's, and data for it was written on tape.

[†]Work performed under the auspices of the U.S. Energy Research and Development Administration.

[*]Currently in residence at the Los Alamos Scientific Laboratory, Los Alamos, NM 87545, U.S.A. Travel funds received under N.S.F. Grant No. OIP76-06014.

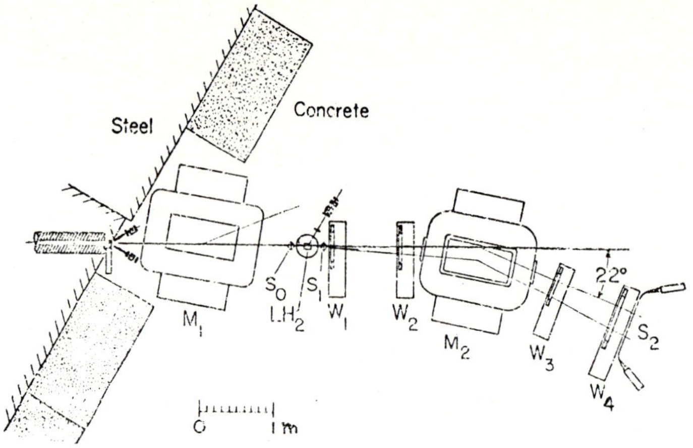

Fig. 1. Layout of Apparatus.

The off line analysis of data was performed by numerical integration of the particle trajectories using known magnetic field values The optimum momentum, P, was obtained by a χ^2 minimization. This momentum, scattering angles θ and ϕ, and other data for each event were stored for further analysis. The protons, deuterons and pions were separated by placing cuts in the time of flight between S_1 and S_2 vs momentum scatter plot. The geometric acceptance of the spectrometer was empirically determined. The discrete coordinate information provided by MWPC caused artificial discontinuity in the θ and ϕ distributions. This "granularity" was removed by randomly redistributing the hit wire coordinates to about $\pm$ 1/2 wire spacing. The neutron data were derived from the recoil proton spectra using the LH_2 radiator. Corrections due to the energy variation of n-p CEX cross section were applied from interpolation of the available data. Background subtraction was made from the target empty runs. The data are also corrected for the inelastic processes in the LH_2 radiator.

RESULTS

0°Neutron Spectra for ^2H, ^9Be, ^{12}C and ^{27}Al at 647 MeV

A reasonably monoenergetic high energy neutron beam of known momentum spectrum is a prerequisite for the detailed investigation of the n-p interaction. Experimenttal neutron spectra in this energy range were poorly known. Therefore, we made precise measurements of the neutron spectra from proton bombardment of liquid-deuterium, ^9Be, ^{12}C and ^{27}Al targets. The LAMPF proton beam of 647 MeV was used and the neutrons were collimated to $\pm$0.1° at 0°. The neutron spectra from deuterium at 0° is shown in Fig. 2 (Bjork [3]). It is dominated by a nearly monoenergetic intense high energy neutron peak

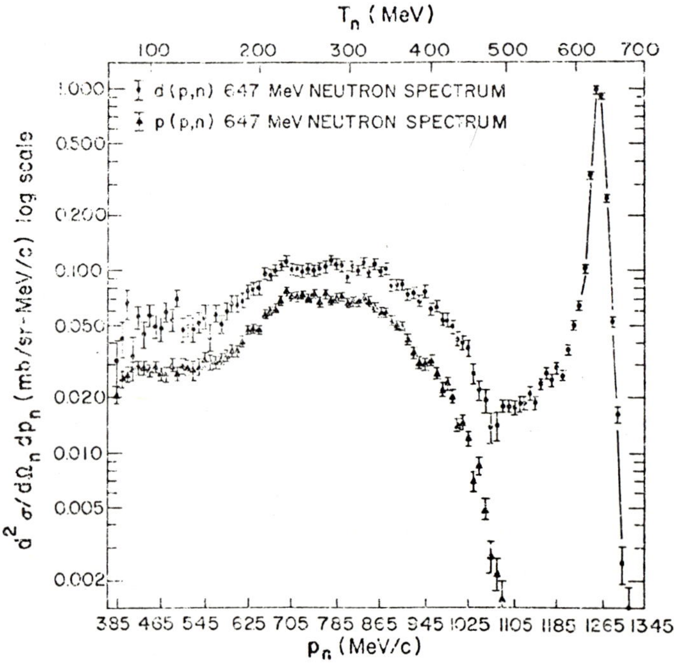

Fig. 2. Neutron Spectra at 0° from ^1H & ^2H

($T_n \sim 643$ MeV, $\delta T_n \sim 13$ MeV FWHM), from the quasifree p-n scattering coupled with the final state interaction of the two protons (Cromer [4]). The peak is relatively isolated from the small background of lower energy neutrons. The ratios of the maximum and integrated counts in the peak and valley regions are 70 and 30, respectively, which makes this neutron beam an ideal n-p probe. The integrated peak cross section is 28.5 mb/sr, which is 0.58 of the free n-p CEX cross section. This reduction in the yield is due to the non-availability of some states because of strong FSI of 2 protons in ^1S_o state (Larson [5]).

The neutron momentum spectra from ^9Be, ^{12}C and ^{27}Al exhibit a sharp high energy peak due to (p,n) CEX reaction (Cassapakis [6]), but the energy resolution of ~ 15 MeV is not sufficient to resolve nuclear levels. The integrated peak cross sections for ^2H, ^9Be, ^{12}C and ^{27}Al are in the ratio of 1:1.26:0.66 and 1.84 respectively. These can be qualitatively understood by the number of unpaired neutrons outside a closed shell and reduced contribution of core neutrons (Palevsky [7]). On the lower side of the peak a shoulder occurs very close to the position of giant resonances observed at lower energies. A broad peak at ~ 900 MeV/c is observed due to pion production. Integrated cross sections in this region display $\sim A^{2/3}$ dependence. Also the internuclear cascade calculations (Bertini [8]) averaged over 0° to 3° give fits for ^{27}Al in this region, but underestimate the CEX peak. These spectra at 800 MeV show similar features.

n-p CEX Cross Section at 647 MeV

The 3 previous measurements of the elastic n-p differential cross sections near 650 MeV are shown by open symbols in Fig. 3. The absolute values of the cross sections differ beyond normalization errors and the relative shape of the angular distributions are different. To resolve the discrepancy, we have measured $d\sigma/d\Omega$ for n-p CEX with high precision at 647 MeV (Evans [1]). Data were taken for $51° < \Theta_{CM} < 180$. A background subtraction for each run was made by using corresponding LH_2 target empty run. The relative normalizations of runs from the monitors were consistent to 1.6%. The deuterons from the $^1H(n,d)\pi^o$ reaction were simultaneously detected ($\Theta_{Lab} < 13.2°$). From isotopic invariance, this cross section is half of the cross section for the $^1H(p,d)\pi^+$ reaction, which is known to 5% (Richard-Serre [13]). The neutron flux was calculated from the deuteron yield and provided absolute normalization of the n-p cross sections to an accuracy of $\sim 7\%$. The cross section values shown in Fig. 3 are in reasonable agreement with the Saclay values but predict a different backward peak. In this energy region the pole-extrapolation method of Chew [14] to extract the π-n coupling constant, f^2, is reasonably accurate (Cziffra [15]) and therefore provides a test of the angular distribution near 180°. The present data gave $f^2 = 0.073 \pm 0.003$, but the same method did not give a reasonable value for the Saclay data, unless the data for $\Theta_{CM} > 176°$ was excluded. The above fact coupled with the high accuracy and internal consistency of our data shows the reliability of our results, which should be of great value in the phase shift analyses in this energy region.

$p+p \to n+p+\pi^+$ Reaction at 645, 764 and 798 MeV

The pion production in N-N collisions is the primary inelastic process in the medium energy range and its understanding is vital to the strong interactions. It is the simplest system which displays the interaction of π-N and N-N forces intimately, yet is not well understood. The Mandelstam model [16] includes the N-N FSI explicitly and fits the data below 600 MeV well. At higher energies (0.8 to 3 GeV) the effects of the baryon resonances, especially the (3,3) resonance, require a phenomenological approach of OPE model (Ferrari [17]). Thus data in the energy range of 600 to 800 MeV are very desirable to test the theoretical models. Previous experiments have only detected charged particles and are few in number. We have made precise measurements of the neutron momentum spectra in this reaction for the first time (Glass [18]). Since a given neutron momentum corresponds to a unique invariant mass of the π^+p system, the data can be simply related to the main process of Δ^{++} formation.

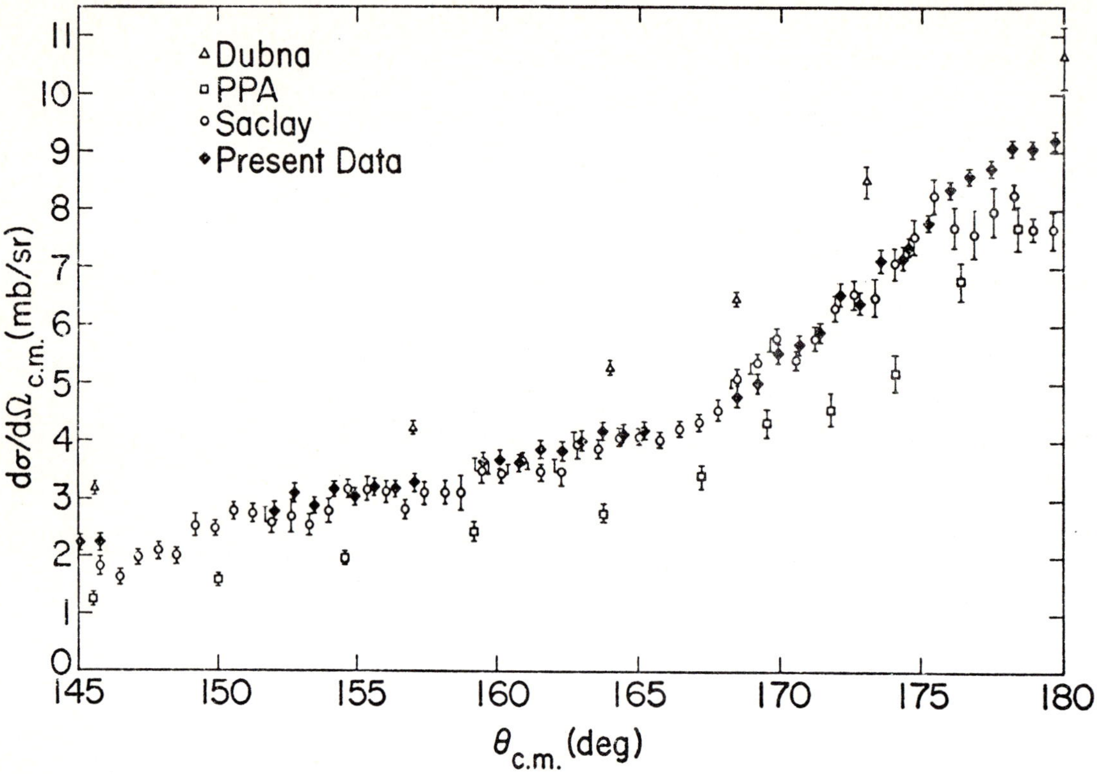

Fig. 3. Differential cross section data for n-p charge exchange scattering near 640 MeV; Dubna (630 MeV, Ref. 9); PPA (640 MeV, Ref. 10); Saclay (645 MeV, Ref. 11); Present data (647 MeV. Ref. 12).

The 0° neutron momentum spectra at 764 MeV (Fig. 4), 645 MeV (Fig. 1) and 800 MeV show a broad high energy peak. The peaks correspond to an invariant mass of π^+p system ~ 1200 MeV. This shift from the 1232 MeV mass of Δ^{++} can be understood in OPEM [17]. The square symbols (Fig. 4) show the cross sections expected from the reflection symmetry in the c.m. and are in agreement with the measured values for $P_n > 400$ MeV/c. The solid curve is a theoretical prediction of Gibbs [19]. The calculations include second order diagrams and use 3 free parameters; 2 for π-N off-shell effects and 1 for pion reabsorption term. The fit is good in the region of Δ^{++}, but underestimates the cross section ~ 780 MeV/c, where relative n-p energy is close to zero. The enhancement ~ 700 MeV/c due to N-N FSI at 647 MeV is more striking (Fig. 1), indicating the need for more detailed calculations. In addition the importance of the nonresonant part of the pion production transition amplitude can be deduced by the deviation of $\sigma(pp\to n)/\sigma(np\to p)$ from 3.

n-d reactions at 800 MeV

The N-d elastic scattering and in particular the backward peaking, which is a characteristic of baryon exchange, has generated intense theoretical interest (reviewed by Simmons [20]). The salient points of different calculations are; the existence of the N^* (1688, J=5/2+) components in the deuteron wave function at 1% level (Kerman [21]); the significant role of $p+p \to d+\pi^+$ process or the so called triangle diagrams (Barry [22]); and single scattering + n-exchange with a postulated large q^2 deuteron form factor (Gurvitz [23]). The available data on p-d elastic scattering is usually smoothly extrapolated to 180°. With a neutron beam, the cross section can be directly measured at 180°. The angular distributions of nd → dx and nd → px reactions were measured at 0°, 4°, 8°, 12° and 16°. The analysis of the data is in very preliminary form. The deuteron spectra at 0°

show the expected elastic peak ~1815 MeV/c and a large broad peak ~ 1600 MeV/c. The latter is due to quasifree deuteron production from nucleons in deuterium (n'N' →dπ). The analysis also suggests flattening of n-d elastic cross sections for 160° < Θ_{CM} < 180°. The proton spectra at 0° are very similar to the neutron spectra measured from the charge conjugate reaction D(p,n)2p, despite differences in the FSI of 2 nucleons (Phillips [24]). The space limitations preclude further discussion of the experiments. The (n,charged particle) reactions on Li, Be and C were studied at 800 MeV. Data on np→π will be discussed by David Wolfe at this conference. We have made good progress on the construction of a polarized target for n-p polarization measurements. Lastly, the Universities of New Mexico and Temple collaboration is analyzing the angular distribution for pp → n reaction and also working on small angle n-p scattering.

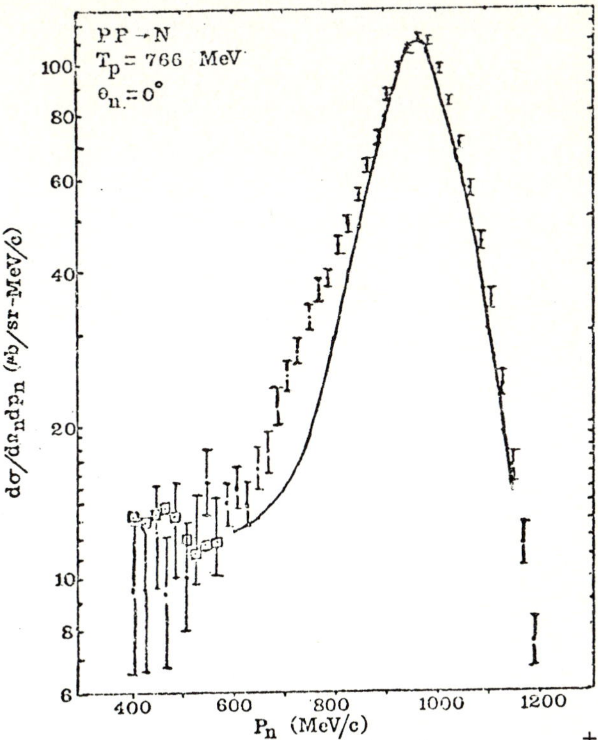

Fig. 4 Cross sections for the $p+p \rightarrow n+p+\pi^+$ reaction vs neutron momenta (lab). Theoretical fit is from Ref. 20.

We thank M.L. Evans, G. Glass, L.C. Northcliffe and J.E. Simmons for many helpful discussions.

REFERENCES

[1] L.N. Glonti et al., Report P1-6387, J.I.N.R. Dubna (1972).
[2] M.H. MacGregor, R.A. Arndt and R.M. Wright, Phys. Rev.173 (1968) 1272.
[3] C. Bjork et al., Proc. Intl, Conf. on Few Body Problems in Nucl & Part. Phys., Quebec (1974) 443.
[4] A.H. Cromer, Phys. Rev. 129 (1963) 1680.
[5] R.R. Larson, Nuovo Cimento 18 (1960) 1039.
[6] C. Cassapakis et al., Bull. Am. Phys. Soc. 20 (1975) 83.
[7] H. Palevsky et al., Phys. Rev. Lett. 9 (1962) 503.
[8] H.W. Bertini, Phys. Rev. C6 (1971) 631.
[9] N.S. Amaglobeli et al., Report P-340, J.I.N.R. Dubna (1959), See also Ref. 1.
[10] P.F. Shepard et al., Phys. Rev. D10 (1975) 2735.
[11] G. Bizard et al., Nucl. Phys. B85 (1975) 14.
[12] M.E. Evans et al., Submitted for Publication to Phys. Rev. Lett.
[13] C. Richard-Serre et al., Nucl. Phys. B20 (1970) 413.
[14] G. F. Chew, Phys. Rev. 112, (1958) 1380.
[15] P. Cziffra and M.J. Moravcsik, Phys. Rev. 116 (1959) 226.
[16] S. Mandelstam, Proc. Roy. Soc. (London) A244 (1958) 491.
[17] E. Ferrari and F. Selleri, Nuovo Cimento 27 (1963) 1450.
[18] G. Glass et al., to be published: see also High Energy Phys. & Nucl. Structure Conf. (Contributed Papers), Santa Fe (1975) 249.
[19] W.R. Gibbs, B.F. Gibson and G.J. Stephenson, Jr., ibid, p. 254.
[20] J.E. Simmons, Proc. High En. Phys. & Nucl. Struc. Conf., Santa Fe (1975) 103.
[21] A.K. Kerman and L.S. Kisslinger, Phys. Rev. 180 (1969) 1483.
[22] G.W. Barry, Ann. of Phys. 73 (1972) 482.
[23] S.A. Gurvitz, Y. Alexander and A.S. Rinat, Preprint (Nov. 1975).
[24] R.J.N. Phillips., Nucl. Phys. 53 (1964) 650.

A NEW EXPERIMENT ON THE INTERACTION OF 14 MeV NEUTRONS WITH TRITONS

S. Desreumaux[+], A. Chisholm[×], P. Perrin[+] and R. Bouchez[+†]

The $^3H(n,n)$ elastic scattering and the $^3H(n,2n),(n,3n)$ triton break-up reactions at 14 MeV have been extensively studied [1 to 6]. The results have stimulated controversy as to the possible binding of the three-neutron system. The experimental evidence for or against the existence of the tri-neutron ^3n consists of the presence or absence of a proton peak in the $^3H(n,3n)p$ break-up reaction. Such a proton peak would be expected above 4.7 MeV for a ^3n bound state. Only one experiment [4] reported a $\sim$ 6 MeV unidentified proton peak which could be associated with a bound ^3n system. The other data [5,6] have indicated the non-existence of ^3n, but do not exclude the small ($\lesssim$ 0.1 mb/sr. MeV) theoretical estimates for ^3n.

The present experiment was performed with improved techniques using large ($\sim$ 20°) acceptance-angle multi-wire and ΔE-E silicon-junction detectors. A long run was made giving 9.10^4 elastic tritons ($\sim$ 320 mb/sr) at the $\sim$ 6°5 mean angle, and 10^4 break-up deutons ($\sim$ 35 mb/sr) from the $^3H(n,d)^2n$ reaction. Also was observed the deuton peak ($\sim$ 2 mb/sr) from $^3He(n,d)d$. The proton spectrum shows (fig.2) the 14 MeV elastic peak, the 14.7 MeV peak from $^3He(n,p)T$ due to $\sim$ 10% ^3He in the tritium target. The 9.5 MeV peak was produced by fortuitous coïncidences associated with the huge triton peak. Finally the proton spectrum shows a small 6 MeV proton peak (0.7 ± . 3 mb/sr) which contains about (200 ± 100) events and seems to correspond to the 6 MeV proton group ($\sim$ 3.8 mb/sr) observed in the Zagreb experiment [4]. The present proton group appears as a 15% modulation of the proton background, and is partially produced by fortuitous events associated with the big deuteron peak.
Besides the proton background is relatively high ($\sim$ 3 mb/sr) due to a too high pressure (Argon + 20% CO_2, 600 mm Hg) in the 4 multiwire detectors. Therefore a 6 MeV proton group associated to a 1 MeV ^3n bound state, if it does exist, can be estimated $\lesssim$ 0.3 mb/sr MeV.

+ Centre d'Etudes Nucléaires/DRF, Grenoble, France.
† Institut des Sciences Nucléaires, Université de Grenoble.
× Department of Physics, University Auckland, New Zealand.

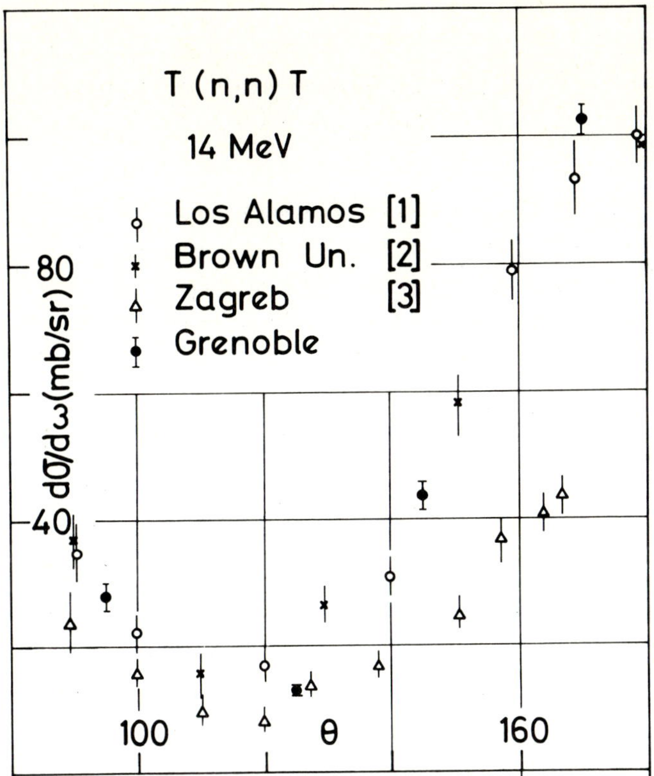

Fig.1a - Experimental cross-sections (cm) from ref [1,2,3] and from Table 1 below

Table 1

θ_{cm} degree	σ_{cm} mb/sr	Scaling factor %
95±1	27.6±1.9	8.5
125±1	13.0± .7	7.3
145±1	43.6±2.5	7.5
170±1	102.7±2.3	5.4

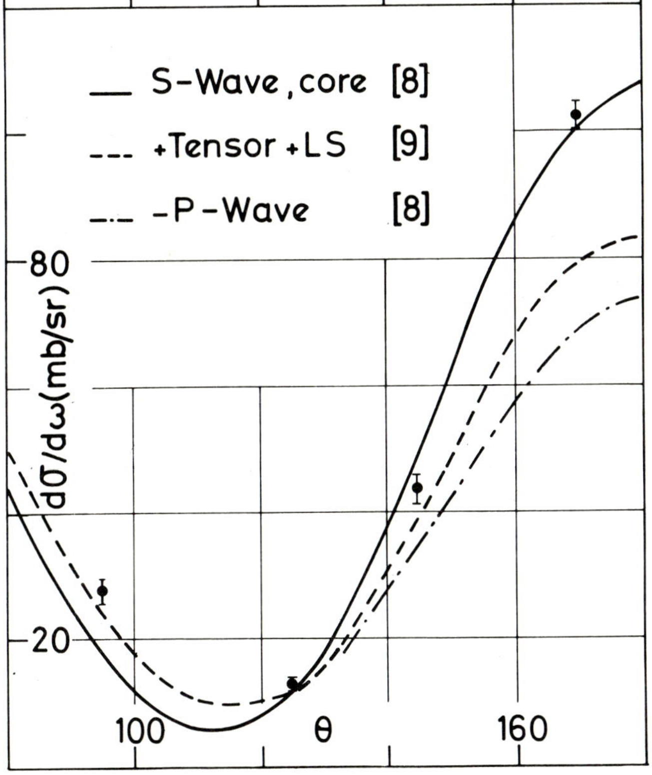

Fig.1b - Calculated cross-section (cm) from ref [8] and [9] for various NN potential models

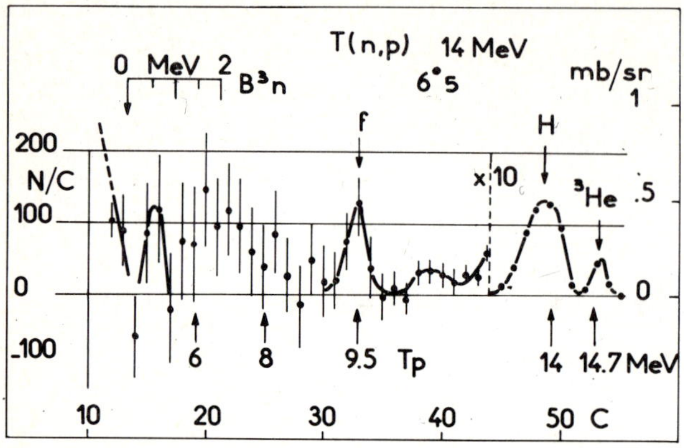

Fig.2 - Proton spectrum observed in the interaction of 14 MeV neutrons on a tritium target. Proton associated with 1 MeV bound ^{3}n should appear at $\approx$ 6 MeV.

The ^{3}H(n,n) backward elastic scattering cross-section was also independently measured. A special care has been taken to determine the scaling factor in the absolute cross-section measurements with neutrons. Then reliable values have been obtained when measurements are made relative to the H(n,n) elastic scattering cross-section for measuring the neutron flux, and relative to the ^{3}H(p,p) cross-section [7] for measuring the amount of tritium present in the reaction target. The results are displayed in the fig.1a and the Table 1. The calculations of the ^{3}H(n,n) cross-section have been made by Eekhof [8] with the resonating group structure method, using realistic S-wave spin-dependent Yukawa-type NN interaction with a core. These calculated backward cross-sections fit very well (fig.1b) the present data, but this agreement is accidental because when tensor and LS forces are included [9] the values become $\approx$ 20% smaller. However a good fit should be restored when P-partial wave is included in the NN interaction model, according to Eekhof'calculations [8] made with an extreme odd NN potential.

References

1. J.H. Coon, C.K. Bockelman, H.H. Barschal, Phys. Rev. 81 (1951) 33
2. J.M. Kootsey, Nucl. Phys. A113 (1968) 65
3. I. Basar, M. Cerineo, P. Tomaš and D. Miljanić, Fizika 1 (1968)105
4. V. Ajdačić, M. Cerineo, B. Lalović, G. Paić, I. Šlaus, P. Tomaš, Phys. Rev. Lett. 14 (1965) 444
5. G. Paić, Int. Conf. on few-particle problems, Los Angeles (1972) 539
6. S. Fiarman and S.S. Hanna, Nucl. Phys. A251 (1975) 1
7. J.L. Detch, R.L. Hutson, N. Jarmie and J.H. Jett, Phys. Rev. C4 (1971) 52
8. H.R. Eekhof, Thèse Vrije Universiteit te Amsterdam 1970
9. H.H. Hackenbroick and P. Heiss, Z. Phys. 242 (1971) 352

PROTON-DEUTERON FINAL-STATE INTERACTIONS IN PROTON INDUCED ^{3}He BREAK-UP

N. Fujiwara*, E. Hourany**, H. Nakamura-Yokota***, F. Reide, V. Valkovic+ and T. Yuasa.

Institut de Physique Nucléaire, Université de Paris-Sud, 91406, ORSAY, France.

The ^{3}He break-up reaction induced by 156 MeV protons has been studied at Orsay in order to obtain detailed information on the final-state interaction(FSI) between a proton and a deuteron. Up to the present, no data have existed on the p-d FSI at this energy. The scattering length and the effective-range of the p-d scattering were determined from elastic scattering in the proton-energy range above 500kev [1], [2]. These parameters ^4a and ^4r_o given in ref. [1] and [2] are widely different from each other. Information about the p-d interaction at very low relative energies may be obtained also from a study of the FSI.

The experimental lay-out was similar to the one described in ref. [3] and in ref. [4]. A liquid ^{3}He target ($\emptyset$ = 30 mm and the thickness = 3 mm) and the 156 $\pm$ 0.5 MeV proton beam were used. The two telescopes were positioned at $\theta_1 = 42.6°$ ($\Delta\Omega_1 = 7.60 \times 10^{-4}$ sr) and $\theta_2 = -71°$ ($\Delta\Omega_2 = 5.48 \times 10^{-4}$ sr). The first ΔE_1-E_1 telescope was placed at 60 cm from the target and was coupled in anticoincidence with a thin ring plastic scintillation counter which was destinated to eliminate background counts due to a slit defining $\Delta\Omega_1$. The ΔE_1 detector was a surface barrier Si detector of 300 μm thick and the E_1 detector was a NaI (Tl) crystal of $\emptyset$ = 5.08 cm and L = 7.62 cm with a 0.2 mm Be window. The second telescope was placed at 6 m from the target and was composed of a plastic scintillation counter, S_2 ($\emptyset$ = 18.5 cm, thickness = 2mm) and a NE213 liquid scintillation counter [3] ($\emptyset$ = 18.5 cm, L =25.4 cm) (designated NE213). The telescope detected charged particles when S2 and NE213 were coupled in coincidence and neutrons (or photons) when S2 and NE213 were coupled in anticoincidence. In order to eliminate background counts for the charged particles, a thin plastic scintillation counter, $S2_T$, (5.0 cm $\times$ 7.5 cm $\times$ 0.1 cm) was placed just outside the reaction chamber and was coupled in coincidence with S2 and NE213. This counter was ignored for neutron detection.
Five parameters E_1, ΔE_1, E_2, T_{12} and charge identification signal for each event have been recorded. (The notation of the terms can refer to ref. [4]).
The $d^3\sigma$ spectrum of the ^{3}He (p, dp) p reaction at $\theta_1 = 42.6°$ and $\theta_2 = -71°$ are shown in Fig. 1. The spectrum has been obtained by projecting the E_1 - E_2 two-dimensional spectrum onto the E_1-axis.
The recoil proton energy E_3 (in the laboratory system), the relative energy between the first particle (deuteron) and the recoil proton E_{13}^{CM} are presented also in the same figure. The peak around E_1 = 79 MeV (E_3 = 0.966 MeV) corresponds to the p-d quasi-free-type scattering (QFS). The cross-section at E_1 = 79 MeV is 0.107 $\pm$ 0.002 mb sr^{-2} MeV^{-1}. The spectrum shows very clearly

* Permanent address : Kyoto University, Kyoto, Japan.
** " " : Lebanese University, Faculty of Sciences, Haddat-Beyrouth, Lebanon.
*** " " : Tokyo Institute of Technology, Meguro, Tokyo, Japan.
+ " " : The Institute Ruder Boskovic, Zagreb, Yugoslavia.

the deuteron and proton FSI contribution around E_1 = 40.0 MeV (E_{13}^{CM}= 0.013 MeV). The $d^3\sigma$ has a minimum at this energy probably due to Coulomb repulsion and two maxima at E_1=35.5 MeV (E_{13}^{CM}= 0.4 MeV) and at 44.5 MeV (E_{13}^{CM} = 0.40 MeV). In an attempt to determine the scattering parameters, we have compared our experimental $d^3\sigma$ s, with the calculated ones based on the plane-wave impulse approximation (PWIA) for the initial-state interaction [5] by using a Hulthèn-type wave function for the deuteron and an Irving-Gunn-type one for ^3He wave function. Final-state interaction effects are incorporated by the multiplication of the cross-sections by a Watson-Migdal enhancement factor for charged particles [6] . The effective-range parameters ^4a = 11.9fm and ^4r_o=1.89 fm give the best fit to the experimental spectrum. This agreement could be accidental because of the use of the simple impulse approximation for the description of the basic spectrum, but it is obvious that the shape of the spectrum is very sensitive to the values of the effective-range parameters in the FSI region. We have also studied simultaneously the (p, 2p) and (p, pn) reactions. [8]

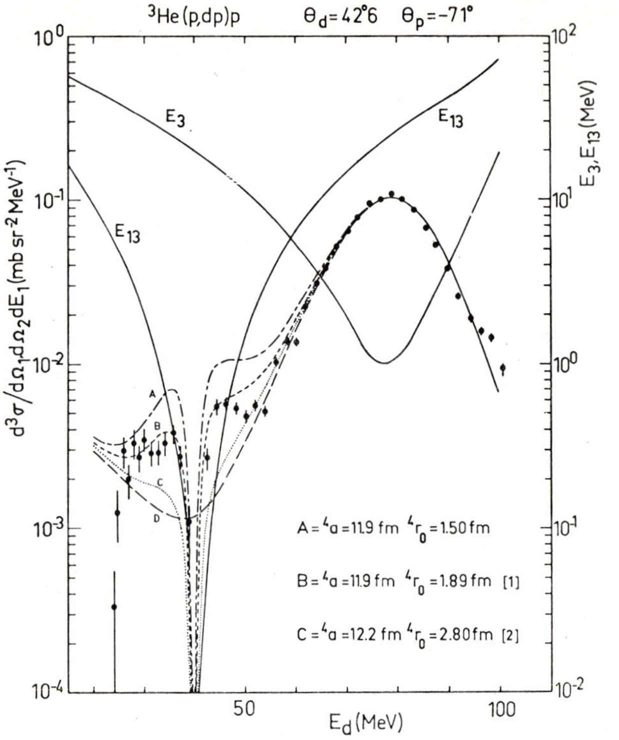

Fig. 1 : The $d^3\sigma$ experimental spectrum of the ^3He (p, dp)p reaction and the calculated one. N (Normalization factor at QFS peak) = 0.617.

References

[1] W. T. H. Van Oers and J. D. Seagrave, Phys. Lett. 24B (1967) 562.

[2] J. Arvieux, A. Fiore, J. Jungerman and N. Van Sen, Proceedings of the Int. Conf. of Los Angeles (North-Holland, Publ. Comp. 1972) p. 519.

[3] H. Nakamura-Yokota, F. Reide and T. Yuasa, Nucl. Instr. Meth. 108 (1973)531. J. P. Didelez, R. Frascaria, N. Fujiwara, E. Hourany, H. Nakamura-Yokota, F. Reide and T. Yuasa, Proceedings of the Int. Conf. of Québec (Laval Univ. Press 1975) p. 667.

[4] J. P. Didelez, R. Frascaria, N. Fujiwara, E. Hourany, H. Nakamura-Yokota, F. Reide and T. Yuasa, to be published in Phys. Rev C(under press 1975).

[5] A. F. Kuckes, R. Wilson,and P. F. Cooper, Ann. of Phys. 15 (1961) 193.

[6] H. Bruckmann, W. Gehrke, W. Kluge, H. Matthäy, L. Schänzler and K. Wick, Proceedings of the Int. Conf. of Birmingham (North-Holland, Publ. Comp. 1970) p. 230.

[7] Y. Avishal, W. Ebenhöd and A. S. Rinat-Reiner, Ann. of Phys. 55 (1969) 341

[8] T. Yuasa, Present Conference.

A PROPOSED EXPERIMENT FOR DETERMINING THE NEUTRON-NEUTRON EFFECTIVE RANGE BY MEANS OF DOUBLY QUASI-FREE SCATTERING IN THE D+D REACTION

H. Kumpf

Zentralinstitut für Kernforschung der Akademie der Wissenschaften der DDR, Rossendorf, DDR 8051 Dresden, PSF 19

The present contribution deals with a proposal for determining the effective range r_{nn} and possibly phases at higher energy for the n-n interaction. The reason for the need of measured n-n parameters is the dependence of the Coulomb corrections on the off-shell-behaviour of the N-N-interaction as demonstrated by the author 1) and by P. U. Sauer 2). If one now takes charge symmetry for granted, certain restrictions for the off-shell-behaviour can be derived from a comparison of p-p with n-n parameters 3).

In principle D(n,2n)p is the simplest source of two interacting neutrons. As r_{nn} governs the scattering process in the region of several MeV relative energy one would chose kinematical conditions corresponding to quasi-free scattering of the two neutrons 4). But the form of resulting spectra is rather insensitive to r_{nn}, whereas the absolute value of the cross section is influenced more strongly. However precision absolute measurements with neutrons as incident particles are very difficult to achieve experimentally. The task would be very much facilitated if charged particles directly from an accelerator could be used. Now one can find kinematic conditions for quasifree scattering of two neutrons in the D(d,2n)2p reaction with zero momentum transfer to each of the remaining protons (doubly quasifree scattering DQFS). If the detectors are placed on opposite sides of the beam on equal angles this condition is:

$$\cos \vartheta = 1\sqrt{2 - 8E_B/E_{lab}} \quad ; \; E_B = -2.2245 \text{ MeV}$$

This shows that the DQFS mechanism is possible at $E_{lab} > -8 \cdot E_B$ = 17.8 MeV. In order to get a certain impression about the expected spectra some calculations in the impulse approximation were done. The cross-section is then represented by

$$d\sigma/d\Omega_1 d\Omega_2 dE_1 dE_2 = |f_n|^2 \phi_p^2 \phi_t^2 \rho$$

Here f_n is the n-n scattering amplitude, ϕ_p and ϕ_t are wave functions of incident and target deuterons, ρ is the appropriate phase-space factor. Two variables remain undetermined in a two-dimensional measurement of four outgoing particles. They are integrated over. Actual calculations took antisymmetrization into account.

The figure shows the result for 35 MeV incident energy. The yield is enhanced around the spot where the momenta transferred to the protons are zero. Again the form of the spectrum is insensitive to the effective range, information on r_{nn} beeing contained in its absolute magnitude. Unfortunately the impulse approximation does not reproduce the absolute yield. Thus one can only invite theorists to develop procedures for solving the four particle equations in this special case. Mean-while experimentalists should already do precision measurements. Finally experiments on a_{nn} had started also before computer codes for the three particle equations were developed.

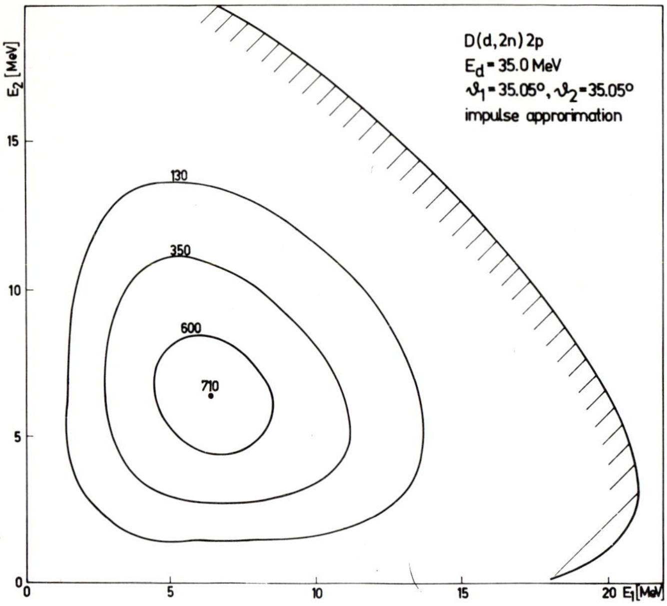

Figure. Impulse approximation for D(d,2n)2p reaction. On the axis are the energies of the detected neutrons. Events are restrained to the interior of the hatched line. Values of the cross-section $d\sigma/d\Omega_1 d\Omega_2 dE_1 dE_2$ indicated on the isohypses are in arbitrary units.

References

1) H. Kumpf, Yad. Fiz. 17 (1973) 1156
2) P. U. Sauer, Phys. Rev. C7 (1973) 943
3) D. W. L. Sprung and M. K. Srivastava, Nucl. Phys. A244 (1975)315
4) I. Slaus, Czechosl. J. Phys. B24 (1974) 1255

QUASIFREE PROCESSES IN THE REACTION $^3He + ^3He$ FOR $E_{^3He}$ = 50, 65 and 78 MeV

R. G. Allas, L. A. Beach, R. O. Bondelid and E. L. Petersen
Naval Research Laboratory, Washington, D. C. 20375 U.S.A.

and

P. A. Treado, J. M. Lambert, R. A. Moyle and L. T. Myers
Georgetown University and Naval Research Laboratory
Washington, D. C. 20057 U.S.A.

and

I. Slaus
Ruder Bošković Institute
Zagreb, Yugoslavia

There are several three-particle breakup combinations that can occur when ^3He bombards ^3He. Measurements for a number of quasifree scattering (QFS) and quasifree reactions (QFR) interactions for $^3He + ^3He$ have been obtained in kinematically complete experiments. Charged particle pairs are detected and conditions are chosen favoring the observation of quasifree processes. The data are analyzed in terms of the plane-wave-impulse-approximation with the Hulthén wave function. The magnitude and shape of the predicted cross section is compared to the experimental cross section for a given interaction. The momentum wave function is extracted for small values of the **spectator** momentum. These measurements of $^3He + ^3He$ are compared to measurements for $^3H + ^3He$ obtained at identical ^3He beam energies that we have previously published [1]. The free cross sections change significantly with energy and angle and play an important role in the predicted $d^3\sigma$. It may be possible to gain insight into the structure of the ^3He and triton ground state wave functions by comparison of the reactions involving $^3He \Rightarrow pd$, $^3He \Rightarrow n(2p)$, $^3He \Rightarrow p(pn)$, $t \Rightarrow nd$, $t \Rightarrow p(2n)$, and $t \Rightarrow n\ (np)$ vertices as a function of bombarding energy.

A gas cell target and fixed detector-aperture geometry was used. The effective target thickness and detector solid angles were calculated [2]. The calculations were compared to measurements of the product of these parameters from $^2H(p,p)^2H$ and $^2H(p,pp)n$ two- and three-body reactions for E_p = 22 MeV where the $d^2\sigma$ and $d^3\sigma$ are well known [3,4]. Beam alignment was monitored and both on-line hardwired and playback software data analysis was utilized. Figure 1 shows the 50 MeV d + h QFS data.

Table I presents the results of the $^3He + ^3He$ measurements and Table II presents the $^3H + ^3He$ values [1]. The results in Table I must be considered as preliminary.

Table I

The Ratios of the Experimental to Theoretical Cross Sections for the Quasifree Processes Observed in $^3He + ^3He$ and for Small Values of **Spectator** Momentum ($< 0.3\ fm^{-1}$)

Reaction	Type	50 MeV	65 MeV	78 MeV
$^3He(^3He,dh)p$	QFS	0.2	0.3	0.3
$^3He(^3He,ph)d$	QFS	0.3	0.2	0.4*
$^3He(^3He,dd)2p$	QFR	0.5	0.2	
$^3He(^3He,pt)2p$	QFR		0.2	

$^*Q_{trans.} = 0.4 - 0.5\ fm^{-1}$

Table II

The Ratios of the Experimental to Theoretical Cross Sections for the Quasifree Processes Observed in ^3H + ^3He and for Values of **Spectator** Momentum < 0.8 fm^{-1}

Reaction	Type	50 MeV	65 MeV	78 MeV
$^3H(^3He,dh)n$	QFS	0.24	0.23	0.2
$^3H(^3He,ph)nn$	QFS	0.07	0.07	0.075
$^3H(^3He,pt)d$	QFR	0.80	0.30	0.20
$^3H(^3He,dd)d$	QFR	0.52	0.45	0.21
$^3H(^3He,dt)p$	QFR	1.7	1.0	0.76

There appears to be more fluctuation in the ^3H + ^3He ratios than those for ^3He + ^3He. However, it must be pointed out that a larger range of **spectator** momentum (Q_t) is used for the ^3H + ^3He values. A recent analysis of the 50 MeV $^3H(^3He,dt)p$ QFR data using only the < 0.3 fm^{-1} Q_t region suggests the value of 1.7 for $^3H(^3He,dt)p$ QFR is only 0.7. Using the smaller Q_t range for the ^3He + ^3He data gives preliminary results that do not change as a function of reaction and, for QFS, as a function of bombarding energy. There is a suggestion of QFR variation with bombarding energy, just as is indicated in Table II for ^3H + ^3He QFR processes. Clearly, further data reduction is needed to compare the mass-3 reaction vertices listed above.

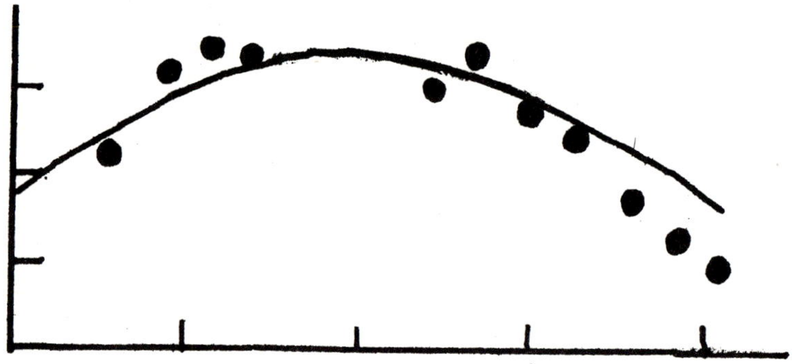

Figure 1. The extracted Fourier transform as a function of $Q_{\mathbf{spectator}}$ for the h + d QFS with 50 MeV and an angle pair of 30.1-30.1.

References:

1. I. Slaus, R.G. Allas, L.A. Beach, R.O. Bondelid, E.L. Petersen, J.M. Lambert and D.L. Shannon, Phys. Rev C8 (1973) 444
2. E. Bar-Avraham and L.C. Lee, Nucl. Instr. and Meths. 64 (1968) 262
3. S.N. Bunker, J.M. Cameron, R.F. Carlson, J.R. Richardson, P. Tomas, W.T.H. Van Oers and J.W. Verba, Nucl. Phys. A113 (1968) 461
4. M.I. Haftel and E.L. Petersen, private communication

DOUBLE SPECTATOR PROCESS IN THE ^{3}He(^{3}He,dd)pp REACTION

J. M. Lambert, P. A. Treado and L. T. Myers
Georgetown University and Naval Research Laboratory
Washington, D. C. 20057 U.S.A.
and
R. O. Bondelid, R. G. Allas, L. A. Beach and E. L. Petersen
Naval Research Laboratory, Washington, D. C. 20375 U.S.A.
and
I. Slaus
Ruder Bošković Institute
Zagreb, Yugoslavia

Measurements of nuclear systems with three or four particles in the final state can be influenced by a number of processes: quasifree scattering (QFS), quasifree reactions (QFR), final state interactions (FSI), simultaneous breakup, which results in "phase space" distribution of events, and second-order effects. In this paper, the four-body continuum for the reaction ^{3}He(^{3}He,dd)pp is investigated with two detector-telescopes in coincidence to determine whether the double quasifree scattering (DQFS) process (in which both undetected protons are spectators) is observable. The ^{3}He(^{2}H,dd)p and ^{2}H(^{3}He,dd)p reactions are also measured for DQFS-equivalent d-d center of momentum conditions with proton spectators in the target and projectile, respectively. The DQFS reaction was investigated at two bombarding energies, 50 and 78 MeV. The 50 MeV equivalent reactions were measured at $E_d = 22.3$ MeV and $E_{^3He} = 33.4$ MeV. Data analysis provides information that is compared to two plane-wave-impulse-approximation (PWIA) predictions. The first prediction is based on calculating the cross section for a kinematically complete experiment, then averaging over the projectile spectator momentum using the equivalent (three-body) reaction data. The second is based on an integration over both undetected (spectator) particles. The ratio of the experimental cross section to the PWIA prediction is the same, 0.10, for both 50 MeV equivalent QFS reactions and there is no contribution in the DQFS experiment from the p-d FSI; thus, the PWIA analyses provide insight into the observability and the possible spectral structure of the DQFS process.

One question that can be addressed with the DQFS measurements is whether the ^{3}He has a fairly strong p + d cluster configuration when interacting in a four-body breakup. Assuming this to be true, the DQFS interaction is that of the deuteron in the projectile with the deuteron in the target. The T-matrix can be written and for the first prediction can be separated to give a delta-function representing conservation of momentum of the entire system, an effective d-d scattering matrix element, and two Fourier transform terms, for the projectile and target spectator, respectively. The similarity of the DQFS T-matrix and the T-matrix for the three-body equivalent reaction with the spectator in the target is apparent. This allows the DQFS prediction to be based on the three-body data. The spectator Fourier transforms are peaked and the allowed region of phase space contributing to the DQFS process is severely limited. To a first approximation and because of the conservation of momentum, the wave functions contribute to the peak cross section by convolution of the two Fourier transforms. In this approximation, the cross section at the peak is given by $d^4\sigma = d^3\sigma\ \Phi_p^2(0)\ \Delta\bar{k}_p N_c$, where N_c is a convolution factor, $\Delta\bar{k}_p$ is an average of the phase space for the projectile spectator, Φ_p^2 is the square of the projectile spectator Fourier transform, and $d^3\sigma$ is the PWIA equivalent target spectator predicted QFS cross section. Prior to predicting a DQFS cross section, the choice of the Hulthén wave function is made. Ignoring distortion leads to the over estimation of $d^3\sigma$ by the PWIA and must be accounted for by using the ratios of the experimental to predicted cross sections for the three-body equivalent QFS reactions. This gives $d^4\sigma = d^3\sigma\ \Phi_p^2(0)\ \Delta k_p \Delta\Omega_p k_{po}^2\ N_c N_3^2$, where the subscript "po" denotes the initial projectile spectator (energy and momentum), and

N_3 refers to the equivalent reaction ratios mentioned above, which are equal. The second and more rigorous approach to obtaining a PWIA prediction for the DQFS process again uses conservation of momentum of the entire system and separates out the effective d-d scattering matrix element, but integrates over all possible proton momenta combinations for each point in the four-body continuum.

Table I presents the relevant experimental information.

Table I

Target	Beam	Energy	Angles	Reaction	$d^3\sigma$ or $d^4\sigma$	N_3 or N_4
^2H	^3He	33.4	35.4° - 35.4°	QFS	10 ± 1	.11 ± .03
^3He	^2H	22.3	35.3° - 35.3°	QFS	3 ± .4	.09 ± .03
^3He	^3He	50	30.1° - 30.1°	DQFS	60 ± 15	.9 ± .4
^3He	^3He	78	37.0° - 37.0°	DQFS	100 ± 30	1.4 ± .6

In Table I, the values of N_4 are the ratios of the experimental to the theoretical prediction (first method) of the DQFS cross section and for the position in the four-body continuum of maximum enhancement; the $d^3\sigma$ and $d^4\sigma$ are in mb/MeV str^2 and μb/MeV^2 str^2, respectively, and are experimental values. Figure 1 shows the extracted Fourier transform of the 22.3 MeV equivalent reaction as a function of the **spectator** momentum. Figures 2 and 3 depict the 50 and 78 MeV DQFS data and the PWIA predictions. The predictions are normalized to the experimental peak values by the N_4's given in Table I.

The data suggest that the DQFS process is observable, and the comparison of the data to PWIA predictions indicate that ^3He has a reasonably strong p + d cluster configuration and that a PWIA analysis is as appropriate for DQFS processes as it is for QFS processes.

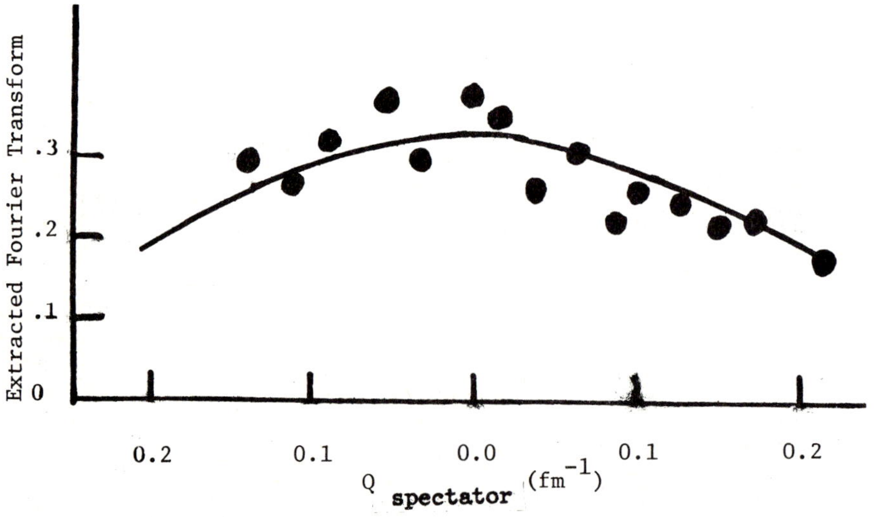

Figure 1. The extracted Fourier transform as a function of $Q_{\mathbf{spectator}}$ for the d + d QFS equivalent reaction with E_d = 22.3 MeV.

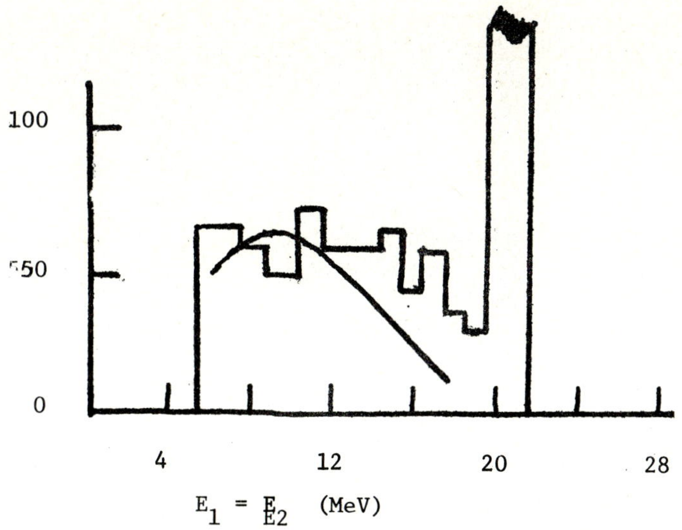

Figure 2. The DQFS 50 MeV data with a possible PWIA fit (from the first method, see text). Data are projection along the four-body continuum diagonal, where $E_d = E_d$.

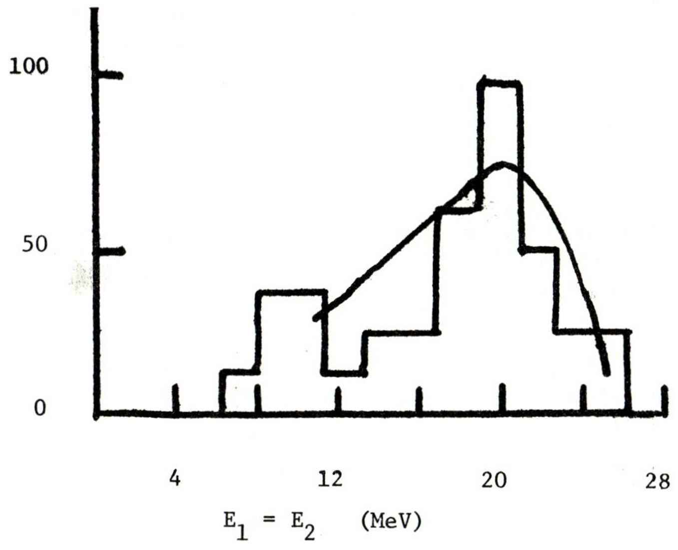

Figure 3. The DQFS 78 MeV data with a possible PWIA fit (from the first method, see text). Data are projection along the four-body continuum diagonal, where $E_d = E_d$.

QUASIFREE PROCESSES IN THE $^2H + ^3He$ REACTIONS

P. A. Treado, J. M. Lambert, R. A. Moyle and L. T. Myers
Georgetown University and Naval Research Laboratory
Washington, D. C. 20057 U.S.A.
and
R. G. Allas, L. A. Beach, R. O. Bondelid and E. L. Petersen
Naval Research Laboratory, Washington, D. C. 20375 U.S.A.
and
I. Šlaus
Ruder Bošković Institute
Zagreb, Yugoslavia

In the interaction between ^3He and ^2H there are several three-particle breakup combinations possible. Kinematically complete measurements were obtained with ^2H beam energies of 22.3 and 35 MeV and ^3He beam energies of 30.0, 33.4, and 52.5 MeV. Charged particle pairs were detected with conditions favoring the observation of quasifree processes, both quasifree scattering (QFS) and quasifree reactions (QFR). The d + d, p + d, and ^3He + p QFS interactions and the $^3He + n \Rightarrow p + t$, $d + d \Rightarrow p + t$, and $^3He + n \Rightarrow d + d$ QFR interactions have been observed. The data for a given interaction are analyzed and compared to a plane-wave-impulse-approximation (with a Hulthén wave function) prediction. The momentum wave function is extracted for small values of **spectator** momentum.

Data were acquired primarily with a gas cell target having either ^2H or ^3He gas and fixed detector-aperture geometry. The effective target thickness and detector solid angles were calculated [1] and the calculations compared to measurements of these parameters from $^2H(p,p)^2H$, and $^2H(p,pp)n$, two- and three-body reactions for E_p = 22 MeV where the $d^2\sigma$ and $d^3\sigma$ are well known [2,3]. A final check was obtained by using a CD_2 target and measuring the ^3He + p QFS and $^3He + n \Rightarrow p + t$ QFR for identical conditions as had been used with the gas cell; $E_{^3He}$ = 33.4 MeV with 16.2° and 45° detector angles. Beam alignment was monitored throughout the data accumulation. On-line analysis was available with hardwired-particle-identification which was stored on magnetic tape. The two-particle coincident ΔE and E signals from each telescope also were stored on magnetic tape. Playback software analysis is presented. For those data compared, the software analysis was within 10% of the hardwired information and generally higher.

Table I presents the results. Those entries in parenthesis should be considered to be preliminary results. The (CM) values listed in the table refer to the outgoing QF interaction products, and the $Q_{\mathbf{spectator}}$ is the momentum of the **spectator** particle in fm^{-1}. The data tend to indicate that: a) there is little difference in $d^3\sigma$ (exper)/$d^3\sigma$ (PWIA) as a function of beam energy (incoming-channel CM energy), and this ratio does not depend on whether the spectator is in the projectile or in the target, b) the cross section ratio usually decreases as the θ(CM) decreases, and c) the cross section ratio for regions where $Q_{\mathbf{spectator}}$ is > 0.2 fm^{-1} can be significantly different for the region of $Q_{\mathbf{spectator}} \leq 0.2\ fm^{-1}$.

Figures 1 and 2 depict typical extracted Fourier transform data as a function of $Q_{\mathbf{spectator}}$ Experimental data for a QFS and a QFR are shown with the PWIA prediction for each.

TABLE I

Beam	Energy	Angles	Reaction	Spectator	T(CM)	θ(CM)	$\frac{d^3\sigma(\text{exper.})}{d^3\sigma(\text{PWIA})}$	Q spectator
Quasifree Scattering:								
^3He	30	34 -17.4	^3He + p	Targ. (T)	5.6	107	0.20 ± 0.05	≤0.2
^3He	33.4	45 -16.2	^3He + p	T	6.8	83	0.078 ± 0.010	<0.2
^3He	52.5	34 -18.3	^3He + p	T	11.9	109	0.21 ± 0.08	<0.2
^3He	52.5	40 -17.9	^3He + p	T	11.6	96	0.16 ± 0.08	<0.16
^3He	52.5	52 -15	^3He + p	T	11.6	71	0.12 ± 0.06	<0.15
^2H	22.3	35.3-35.3	d + d	T	5.6	90	0.114 ± 0.034	<0.2
^2H	35	27 -18	d + d	T	6.3	130	(0.11)	≥0.25
^2H	35	32 -32	d + d	T	8	90	(0.13)	<0.15
^2H	35	43 -34	d + d	T	11.5	78	(0.14)	>0.2
^3He	33.4	35.4-35.4	d + d	Proj. (P)	5.6	90	0.090 ± 0.028	<0.15
^3He	33.4	45 -16.2	d + d	P	4.2	120	0.06 ± 0.03	.17-.30
^3He	52.5	34 -18.3	d + d	P	3	~110	0.11 ± 0.06	.3 -.6
^3He	52.5	40 -17.9	d + d	P	5	~110	0.08 ± 0.04	.3 -.6
^3He	52.5	52 -15	d + d	P	8.5	~110	0.06 ± 0.04	.25-.45
^2H	22.3	35.3-35.3	p + d	T	5.8	113	0.24 ± 0.07	.28-.48
^2H	35	43. -34	p + d	T	11.5	78	(0.14)	~0.27
^2H	35	32 -32	p + d	T	5	73	(0.23)	<0.2
^2H	35	27 -18	p + d	T	7.5	40	(0.16)	<0.3
^3He	52.5	52 -15	p + d	P	3.4	73	0.14 ± 0.06	.37-.5
^3He	52.5	52 -15	p + d	P	13.7	44	0.19 ± 0.06	.5 -.65

TABLE I (Cont'd.)

Beam	Energy	Angles	Reaction	Spectator	T(CM)	θ(CM)	$\frac{d^3\sigma(\text{exper.})}{d^3\sigma(\text{PWIA})}$	Q spectator
Quasifree Reactions:								
^2H	35	65 -46.1	^3He + n to p + t	P	11.3	84	(0.18)	<0.13
^3He	30	34 -17.4	^3He + n to p + t	T	5.9	107	0.25 ± 0.08	<0.2
^3He	33.4	45 -16.2	^3He + n to p + t	T	6.8	83.4	0.252 ± 0.042	<0.17
^3He	52.5	34 -18.3	^3He + n to p + t	T	11.4	109	0.26 ± 0.06	<0.3
^3He	52.5	40 -17.9	^3He + n to p + t	T	11.5	96	0.23 ± 0.07	<0.2
^3He	52.5	52 -15	^3He + n to p + t	T	11.6	70	0.21 ± 0.04	<0.3
^2H	22.3	35.3-35.3	d + d to p + t	T			0.72 ± 0.27	<0.2
^2H	35	34 -43	d + d to p + t	T	16	113	(0.46)	<0.2
^2H	35	32 -32	d + d to p + t	T	15	128	(0.42)	~0.2
^2H	35	18 -27	d + d to p + t	T	14	150	(0.27)	<0.1
^3He	52.5	34 -18.3	^3He + n to d + d	T		126	0.28 ± 0.10	.1-.2
^3He	52.5	40 -17.9	^3He + n to d + d	T		132	0.30 ± 0.10	.2-.3

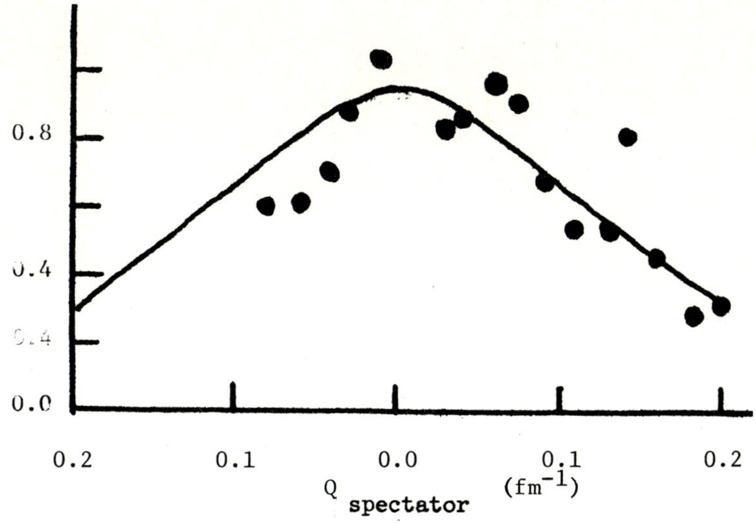

Figure 1. The extracted Fourier transform as a function of $Q_{spectator}$ for the h + p QFS with 33.5 MeV and an angle pair of 16.2-45.

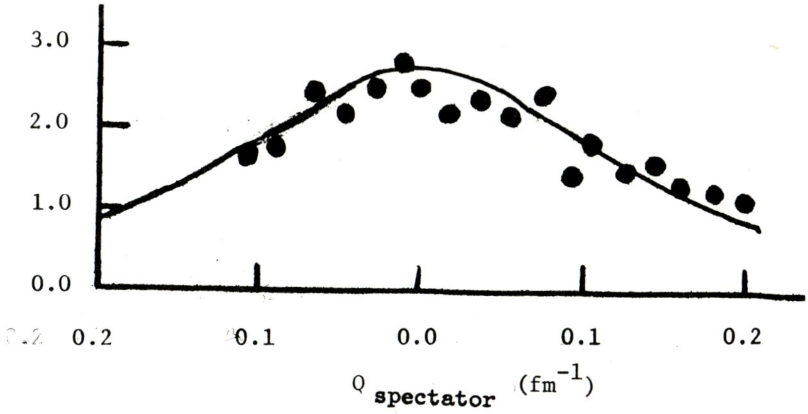

Figure 2. The extracted Fourier transform as a function of $Q_{spectator}$ for the h + n to p + t QFR with 33.5 MeV and an angle pair of 16.2-45.

References:
1. E. Bar-Avraham and L.C. Lee, Nucl. Instr. and Meth. 64 (1968) 262
2. S.N. Bunker, J.M. Cameron, R.F. Carlson, J.R. Richardson, P. Tomas, W.T.H. Van Oers and J.W. Verba, Nucl. Phys. A113 (1968) 461
3. M.I. Haftel and E.L. Petersen, private communication

FURTHER EVIDENCE OF HIGH ENERGY DEUTERONS

IN THE REACTION $^3He + ^3He$ *

R.J. Slobodrian, R. Pigeon, M. Irshad, S. Sen and J. Asai
Université Laval, Département de Physique, Laboratoire de Physique Nucléaire
Québec P.Q., G1K 7P4, Canada

The reaction

$$^3He + ^3He \rightarrow ^2H + ^4He + e^+ + \nu \qquad (1)$$

is in principle of the same type as

$$^1H + ^1H \rightarrow ^2H + e^+ + \nu \qquad (2)$$

The latter being of foremost interest for astrophysics. Reaction (1) has been studied recently [1] and positive results were reported [2] in the literature. Subsequently experiments have been performed with the purpose of improving statistics and also to study possible processes which might yield spurious deuterons, i.e. not originated from the ^3He target. The experimental procedure was similar to that reported in Ref. 2, and fig. 1 shows a two-dimensional spectrum with a significant statistical improvement over previous data : there are 48 events satisfying the two necessary conditions for acceptance, that is, of having given a deuteron identification pulse (crosses), and of following the law of energy loss characteristic of deuterons (events in the band). Both criteria are applied to on-line data obtained with a triple telescope in coincidence followed by a fourth detector in anticoincidence. The average cross-section at 20° is $\frac{d^2\sigma}{dEd\Omega} = 2.17$ nanobarn MeV^{-1} sr^{-1}, in agreement with the published value [2], particularly taking into consideration the statistical errors.

The study of processes which might produce spurious deuterons was accomplished by several means. Firstly, elastically scattered ^3He might induce (^3He, d) reactions in impurities, foils, etc. Thus the target cell was filled with air and also with ^4He, with a negative result. The possibility of a (p,d) reaction induced by the protons from ^3He (τ,p) ^5Li was also ruled out by using ^2H as target and producing a sharp proton group from the reaction ^2H (τ,p) ^4He, no deuterons were observed. Tritons which may be produced in the Havar foil would yield deuterons below the sensitivity of the system.

In summary the deuterons originate in the $^3He + ^3He$ interaction. The implications of this result have been reported already in reference 2. Figure 1 shows also a curve obtained assuming that a diproton (^1S_o p-p) spectrum of shape as given by the Migdal-Watson theory, decays into deuterons. There is within statistics a reasonable resemblance of the theoretical curve and the experimental spectrum. This seems to favour a reaction mechanism in two steps with the intermediate formation of a diproton.

Future experiments on this interesting reaction could consist of measurements employing different techniques than the one reported in the present series of experiments, for example, using the magnetic analysis of reaction products which could eliminate easily the elastic ^3He particles. A correlation with the α-particles would be interesting, but counting rates would become extremely low.

* Work supported in part by the Atomic Energy Control Board of Canada and the Ministery of Education of Québec.

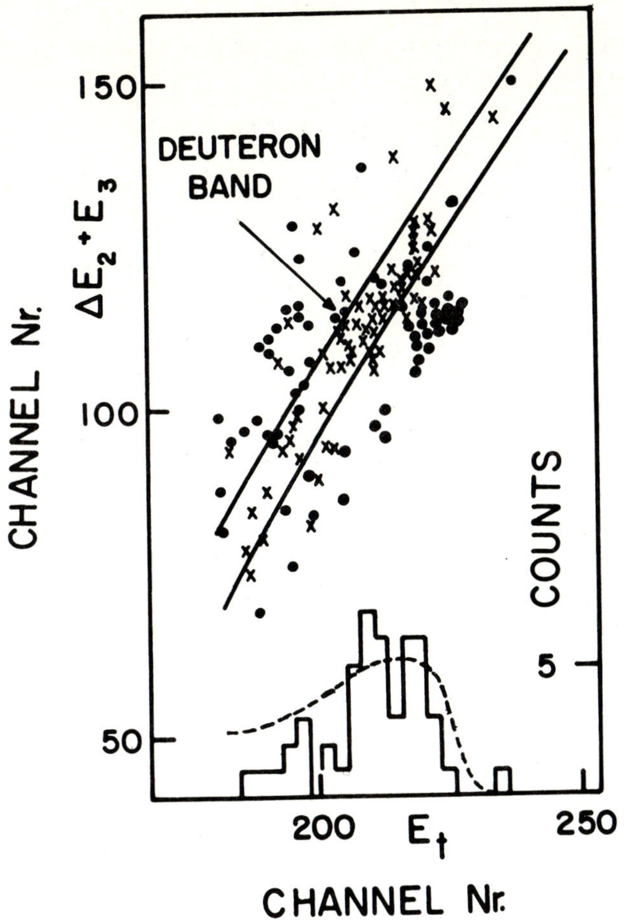

FIG. 1

Spectrum at 20° Lab. Abscissas and ordinates are given as channel numbers of total energy (E_t) and residual energy ($\Delta E_2 + E_3$) after passing the first detector. Scales are approximately 100 KeV Ch^{-1}. The histogram is a projection of deuterons. The dashed line shows a deuteron spectrum assuming that the reaction produces diprotons ($^1S_0$ p-p) which is in turn transform into deuterons.

REFERENCES

1) R. Pigeon, U. von Moellendorff, C.R. Lamontagne, M. Irshad and R.J. Slobodrian, Few Body Problems in Nuclear and Particle Physics, Eds. R.J. Slobodrian, B. Cujec, K. Ramavataram, p. 434, Les Presses de l'Université Laval, Québec (1975).

2) R.J. Slobodrian, R. Pigeon and M. Irshad, Phys. Rev. Lett. 35 (1975) 19.

PRODUCTION OF HIGH ENERGY DEUTERONS IN THE $^3He+^3He$ REACTION AND THE SOLAR NEUTRINO PROBLEM - A COMMENT

Đ. Miljanić
Institute Ruđer Bošković, Zagreb, Yugoslavia

In a recent study of $^3He+^3He$ reactions at E_o=13.6 MeV Slobodrian et al.[1] observed deuterons with energies above 20 MeV, which they attributed to the $^3He+^3He \rightarrow d+^4He+e^++\nu$ reaction (hereafter referred to as reaction (1)). From the data measured over the angular interval between 17° and 30° they extracted an average value for the cross section $d^2\sigma_1/dEd\Omega$ of 3.4 nb $MeV^{-1}sr^{-1}$. This result, if true, might have very significant consequences, and because of that it is important to prove or disaprove the validity of the data. In this note we want to point out one of the possible less exotic sources of deuterons, contribution of which has to be taken into account in any experiment of this kind. A correlation measurement is proposed as the alternative approach to the experiment.

Even if the separation of deuteron pulses from other pulses was unambiguous, even if deuterons originated in the ^3He gas, and even if their energy at Θ_{LAB} = 20° was higher than 19.5 MeV, that does not necessarily mean that deuterons were a consequence of reaction (1). A very low secondary triton "beam" of a comparable energy might cause the same effect through the $t+^3He \rightarrow d+\alpha$ reaction (reaction (2)). Sources of continuous energy tritons are (^3He,t) reactions (reaction (3)) on the slits (S_o) and on the gas cell (GC) wall (see Fig. 1). Whether the contribution from the slits was significant, it depended mainly on the material from which the slits were made, on their quality, on their distance from the gas cell, and on the amount of the beam hitting them. Since details of the slits are not mentioned in Ref. 1, the other source of tritons,

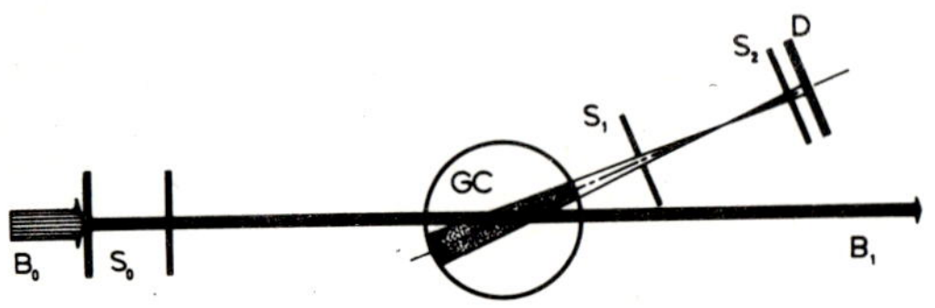

Fig. 1. Schematic of the experimental apparatus. S_o, S_1, S_2 - slits, B_o, B_1 - beam before and after the S_o slits, GC-gas cell, D-detectors.

namely the gas cell wall from Havar foil (2.1 mg cm^{-2}) will be discussed in this note.

The most abundant element in Havar is cobalt (42.5%). Other elements present in Havar are Cr, Fe, Ni, W, Mo, Mn. Q-value for reaction (3) on cobalt (^{59}Co) is -1.09 MeV, while Q-values for reactions (1) and (2) are 13.79 and 14.32 MeV, respectively. From these values it is obvious that the end point of the deuteron energy spectrum from reaction (1) and the end point of the spectrum from reaction (2), induced by secondary tritons, coincide (within the experimental resolution) and follow the same angular dependence. There are no data on reaction (3) for most elements and for the energy and angular region of interest here. However, from the existing data for reaction (3) on other $1f_{7/2}$ nuclei[2] at somewhat higher energies and having in mind high density of states of residual nuclei (e.g., there are 80 levels up to 5 MeV excitation in ^{59}Ni), one can put limits to the average value for the cross section in the very forward direction to be $0.3 < \overline{d^2\sigma_3/d\Omega\, dE} < 3$ mb $sr^{-1}MeV^{-1}$.

Data on the inverse reaction of reaction (2) exist for the center-of-mass energies E_{CM} = 7.11 and 2.31 MeV in the ^3He - t system[3]. These values are close to the upper (6.7 MeV) and lower (3.7 MeV) center-of-mass-energy limits of interest here. Since there are no strong resonances at the corresponding excitation in ^6Li, one can put limits to the average laboratory cross section in the very forward direction to be $2 < \overline{d\sigma_2/d\Omega} < 12$ mb sr^{-1}.

The target thickness for reaction (2) could be larger by more than one order of magnitude than the thickness for reaction (1). The target for reaction (2) is the whole shaded cone in Fig. 1, while for reaction (1) it is only the region of the same cone cut out by the ^3He beam. The ratio of the target thicknesses is then $3 < x_{21} < 20$. The average solid angle for the triton "beam" can be estimated to be $0.04 < \overline{\Delta\Omega_3} < 0.4$ sr.

Putting all these numbers together, one can see that the following equation could be satisfied ($\Delta\Omega_1 \approx \overline{\Delta\Omega_2}$):

$$\frac{d^2\sigma_1}{d\Omega\, dE} \approx N_{Hav} \cdot \overline{\frac{d^2\sigma_3}{d\Omega\, dE}} \cdot \overline{\Delta\Omega_3} \cdot x_{21} \cdot \overline{\frac{d\sigma_2}{d\Omega}} \approx 1 \text{ nb sr}^{-1} \text{ MeV}^{-1}$$

N_{Hav} is the number of "active" atoms per unit area in Havar. This consideration shows that deuterons in the experiment[1] might originate from reactions (2) and (3). It also indicates the ways how the difficulties could be overcome in a repeated experiment. The best way would be the use of a differentially pumped ^{3}He target. However, even the replacement of Havar by a nickel foil will lower the yield of secondary tritons in the energy region of interest by at least one order of magnitude. These problems might be further reduced by minimizing x_{21} and $\Delta\Omega_3$ by proper collimation inside the gas cell.

As an alternative approach to this experiment one can consider the correlation measurement, in which both deuterons and α particles will be detected in coincidence. The feasibility of the correlation measurement could be shown by comparison with the experiment on ^{3}He+^{4}He bremsstrahlung[4]. The momenta of leptons in reaction (1) are low compared with those of deuterons and α particles, and thus the kinematics of process (1) restricts the available phase space similarly as in bremsstrahlung. Due to this fact, it is easy to show, that in order to check the results quoted in Ref. 1, in the correlation experiment one has to be able to measure the cross section of the same order of magnitude as the measured bremsstrahlung cross section (10μb/sr^2). Taking proper care of secondary tritons, the results of a correlation measurement will be much more reliable owing to the more severe restrictions on data, and will firmly prove or disprove the results of Ref. 1.

The author is pleased to acknowledge valuable discussions with Professor I. Šlaus.

References

1. R.J. Slobodrian, R. Pigeon, and M. Irshad, Phys. Rev. Lett. 35 (1975) 19.
2. H. Rudolph and R.L. Mc Grath, Phys. Rev. C8 (1973) 247 and references therein.
3. E.E. Gross, E. Newman, M.B. Greenfield, R.W. Rutkowski, W.J. Roberts, and A. Zucker, Phys. Rev. C5 (1972) 602.
4. J. Birchall, B. Frois, R. Roy, and R.J. Slobodrian, in Proc. Int. Conf. on Few Particle Problems in the Nuclear Interaction, Los Angeles 1972, ed. I. Šlaus, S.A. Moszkowski, R.P. Haddock, W.T.H. van Oers (North Holland 1972) p.61.

RESONANT SUBSYSTEMS AS A CHALLENGE TO NUCLEAR REACTION THEORY

H.H. Hackenbroich, W. Schütte and H. Stöwe
Institut für Theoretische Physik der Universität zu Köln
Köln, West Germany

One of the main goals of nuclear few body theory has been the formulation of exact integral equations governing the motion of three and more nucleons (cf. e.g., the review by Sandhas [1]). It seems worthwhile to confront these theories with the results of other approaches in which approximations were made in the very first step - i.e. in the basic equations - but not in the subsequent ones (except from well controlled mathematical ones). Thus one may estimate the influence of specific parts of the internucleon interaction and of reaction mechanisms. In this talk we want to discuss one aspect of this problem - the importance of unbound but resonant substructures.

We start this discussion by mentioning two well-known facts: Firstly, there is no difficulty in describing scattering states which show two fragments in each channel only. In this case we may express wave functions as (cf. in this context [2])

$$\Psi_{\varkappa}^{J\pi} = \mathcal{A}\left\{\bar{\Psi}_{\varkappa}^{J\pi}\right\}$$

$$\bar{\Psi}_{\varkappa}^{J\pi} = \sum_{\lambda=1}^{N} \sqrt[4]{\frac{M_\lambda}{2\hbar^2\varepsilon_\lambda}}\left[\left[\phi_\lambda^{j_1}\bar{\phi}_\lambda^{j_2}\right]^{S(\lambda)}\frac{1}{R_\lambda}Y_{L(\lambda)}(\Omega_{\vec{R}_\lambda})\right]^{J}$$

$$\times\left(\delta_{\varkappa,\lambda}F_{L(\lambda)} + a_{\varkappa,\lambda}G_{L(\lambda)}^{reg} + \chi_\lambda^{\varkappa}\right) + \Lambda_{\varkappa}^{J\pi}$$

Here ϕ_λ and $\bar{\phi}_\lambda$ are the wave functions of the two fragments in channel λ ; $S(\lambda)$ is the channel spin, M_λ is the reduced mass, ε_λ the asymptotic kinetic energy and the distance vector between the fragments in channel $\varkappa$. F_L is the regular and G_L^{reg} the "regularized" irregular Coulomb function; the function χ_λ is square integrable and expresses the difference between the "true" channel function and the sum of the Coulomb functions - which is the true function in the asymptotic region only. $\Lambda_\varkappa$ is a square integrable A particle function accounting for closed channels and distortion processes. It is possible in principle to write down similarly an exact expression for a wave function which describes three fragment break-up processes too using the correct asymptotic behavior [3]; but such a formula must contain integrals over subsystem energies (or at least infinite sums if a complete set of 5-dimensional angular functions is used). Thus any - rigorous or approximate - solution of the corresponding equation of motion is much more complicated than in the two fragment case.

Secondly, we know that light nuclear systems show low energy resonances which are rather narrow. Examples are the α^* (the first excited state of the α particle) and the $3/2^-$ and $3/2^+$ states of the 5 nucleon systems. Consequently we have to demand that a few nucleon theory should take into account explicitely not only bound states of subsystems but these resonant states also. On the other hand, the main effect of these resonances must be retained if a Watson-Migdal type approximation is used because the corresponding resonance energies are approximately real.

This simple idea can be checked immediately by testing its simple consequences. The most simple systems which may be used for this purpose are the unstable systems ^{5}He and ^{5}Li. For low energies only the fragmentations α + nucleon are possible. At higher energies we see deuteron + ^{3}H (or ^{3}He) channels; 2.22 MeV above their thresholds three fragment break-up can occur. If the assumption expressed above is correct we should see not only one sequence of $J^\pi = 3/2^-, 1/2^-$ resonances in those systems (connected with α + nucleon channels) but also a second one which may be characterized as "potential" resonances in α^* + nucleon channels. This is due to the fact that the α^* is a breezing mode of the α and has the same quantum numbers; therefore the interaction between nucleons and α or α^* should be rather similar. The observation of these states is facilitated because both α and α^* have zero spin; nevertheless it is still difficult since we have four other resonances in this energy region. The second $3/2^-$ resonance has been established experimentally [4]; the $1/2^-$ one could not be seen yet due to its larger width. These states have been shown to exist theoretically by Heiss and Hackenbroich [5] some years ago by a variational calculation (cf. [2],[6]) in which the α^* wave function has been approximated by a bound state function; this calculation started from a realistic model for the nucleon-nucleon interactions. The calculation also showed that the S-matrix elements for the two fragment processes are strongly influenced by this second doublet; the omission of these resonances will lead to qualitative errors.

The second check for the importance of resonant subsystems is provided by the ^{6}Li scattering system. Above 15.795 MeV excitation energy we have open channels of α + d and ^{3}He + ^{3}H structure plus break-up into α + p + n. Several MeV above this energy also the break-up into ^{3}H + d + p and ^{3}He + d + n becomes possible. ^{3}He + d (and ^{3}H + d) have a $3/2^+$ resonance close to threshold; these resonances - which are split in energy due to Coulomb effects - will influence the ^{6}Li system. Obviously this should result in scattering states which are not eigenstates to isospin in this energy region. Thus, if we have a α + d fragmentation in the incoming channel (T = 0) we do not expect T = 0 outgoing channels only. If we observe ^{3}He + ^{3}H, we cannot conclude that we have either S = 1 and L even or S = 0 and L odd demanded by T = 0 and by the assumption that ^{3}H and ^{3}He differ in isospin orientation only. So the reaction cross section should not be symmetric around 90° (Barshey-Temmer theorem) in this energy region. This is seen from fig. 1 which shows absolute values for reaction S-matrix elements taken from a calculation performed by us [7]; this figure shows that also isospin violating processes occur with rather large probability in this energy region, as known from experiment for several years [8]. It should be mentioned again that a consistent calculation for the ^{6}Li system has to include these sequential decay modes in order to obtain reasonable results even for the two fragment reactions [9], [10]; calculations of this kind reproduce not only two fragment reaction data, but three body break-up processes too.

For other systems resonant substructures are important too even if their effects can be checked by elaborate calculations only. However, all the (variational) calculations performed so far use Watson-Migdal-type approximations; i.e., for the first step of the reactions the wave functions of the subsystem resonances are approximated by square integrable ones. In better calculations one has to get rid of this approximation; but one immediately realizes that an expansion of the wave function in terms of the complete set of five dimensional angular functions is badly convergent due to the existence of these resonances. It might be much better to use suitably chosen wave packets constructed with the help of channel Hamiltonians (cf. Schmid and Schwager [11]). Equivalent statements should be made about future approximate solutions of rigorous A body integral equations.

It should be noted that the existence of substructure resonances does not spoil one important simplification of few nucleon calculations, namely, that approximately only one irreducible representation of the permutation group is contained in spatial (and, consequently, spin-isospin) parts of the wave functions (cf. [12], [2]). As an example we consider the ^{6}Li system: The d, ^{3}H, ^{3}He and α wave functions can be considered to have spatial parts invariant under particle permutations (which is true up to $\approx$90%). Thus, if we have α + d or ^{3}H + ^{3}He fragments in incoming channels, the A particle wave functions should contain large spatial parts only which are in the Littlewood products of  and  namely

 (I)

and

 (II)

the 3/2$^+$ five nucleon substructure has a spatial part with Young diagram

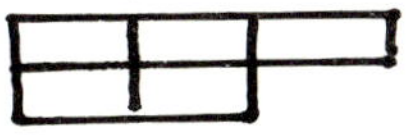

In the 6 nucleon system channel functions describing one nucleon plus this 5 nucleon subsystem have spatial parts of Young diagrams

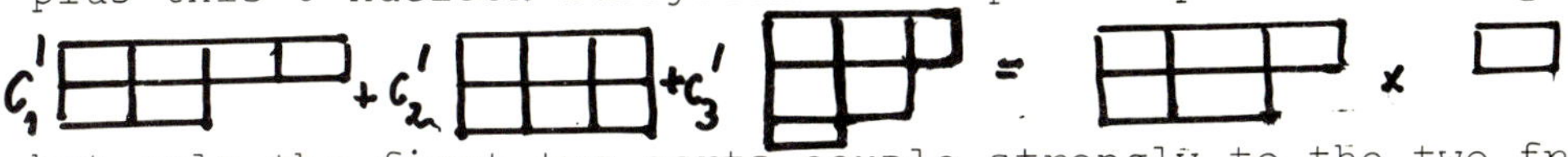 (III)

but only the first two parts couple strongly to the two fragment channels. C_1 and C_1' are not zero for S = 1, T = 0 or S = 0, T = 1 only, C_2 and C_2' vanish for these T,S combinations; here S denotes not the channel spin but the ordinary spin which is approximatively equivalent.

As a consequence of the supermultiplet assumptions described above together with the fact that S = 0 transitions are most favoured, we conclude that isospin conserving α + d $\leftrightarrow$ ^{3}He + ^{3}He reactions occur predominantly in even parity states. On the other hand, the isospin non conserving part of this reaction - which is mainly a two-step process via the intermediate resonant substructures - should result mainly from processes in odd states. This shows that in first crude approximation only parts of the possible amplitudes need consideration.

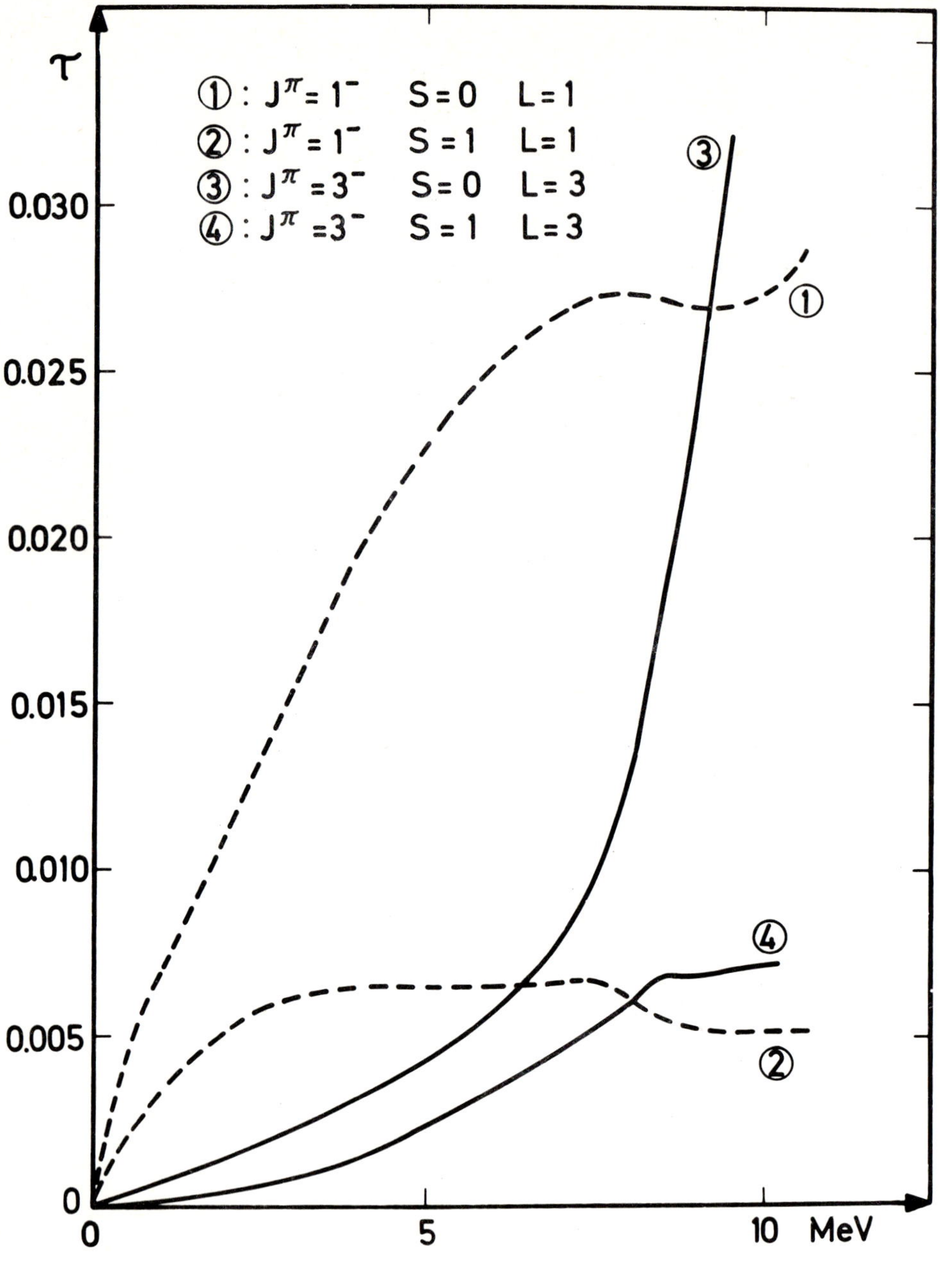

Fig. 1. Absolute values for ^2H + $^4He \rightarrow$ ^3He + ^3H reaction S matrix elements. S and L denote channel spin and relative orbital angular momentum for ^3He + ^3H channels. 1 and 3 denote isospin allowed, 2 and 4 isospin forbidden transitions.

References

1 W. Sandhas, Proceedings of this conference
2 H.H. Hackenbroich, in The Nuclear Many Body Problem (Editrice Compositori, Bologna) 1972
3 W. Glöckle, Z. Physik 271 (1974) 31
4 H. Schröder et al., Phys. Lett. 48 B (1974) 206
5 P. Heiss and H.H. Hackenbroich, Nucl. Phys. A162 (1071) 530
6 H.H. Hackenbroich, in Clustering Phenomena in Nuclei (Univ. of Maryland Press, College Park, Md.) 1975
7 W. Schütte and H.H. Hackenbroich, to be published
8 V. Nocken et al., Nucl. Phys. A213 (1973) 97
9 H.H. Hackenbroich et al., Nucl. Phys. A221 (1974) 461
10 P. Heiss, Z. Physik A272 (1975) 267
11 J. Schwager and E.W. Schmid, Nucl. Phys. A205 (1973) 168
12 P. Kramer, in Clustering Phenomena in Nuclei (Univ. of Maryland Press, College Park, Md.) 1975

REACTIONS $^{10}B + d$, $^{9}Be + {}^{3}He \rightarrow 3\alpha$ AND EXCITED STATES IN ^{12}C. †

R. Roy, University of California, Lawrence Berkeley Laboratory, California, U.S.A.
R.J. Slobodrian, Laboratoire de Physique Nucléaire, Département de Physique, Université Laval, Québec G1K 7P4 Canada.
G. Goulard, C.M.R., Département de Physique, St-Jean, J0J 1R0, Québec, Canada.

The main features of the $^{9}Be + {}^{3}He$ and $^{10}B+d$ reactions going to $^{8}Be+\alpha$ and subsequently to $\alpha+\alpha+\alpha$, are not inconsistent with a direct mechanism over a wide energy range (1). However, it was found at low energies that some features of the experimental data were not explained in this context. Some excited states of ^{12}C have been reported in the region between 29 and 31 MeV (2), attainable also through the reactions mentioned above. Clearly in such case a compound nucleus mechanism and sequential decay should be adequate to describe both reactions, as follows:

$$A + b \rightarrow {}^{12}C^* \rightarrow \alpha_1 + {}^{8}Be \rightarrow \alpha_1 + \alpha_2 + \alpha_3$$

Measurements have been carried out using the van de Graaff accelerator at Université Laval. The experimental technique, described in detail elsewhere, (3) consisted in the measurement of α-α correlation spectra. The salient feature of the detection system consisted in the particle identification carried out at one of the correlation angles, and also in the determination of the coincidence time spectrum, which permitted an unusually clean operation through a precise selection of the true coincidence region. The accumulation of correlation spectra was effected on-line using a PDP-9 computer with a triple parameter system, the time spectrum being recorded as third parameter.

For the reaction ^{9}Be ^{3}He, the ^{3}He beam energies were fixed at 4.4 MeV, 4.96 MeV and 5.6 MeV with several detection angles for the final state alphas, the corresponding excitation energies of the ^{12}C compound nucleus being at 29.6 MeV, 30 MeV and 30.5 MeV. In order to get the same excitation energies in $^{12}C^*$ with the other entrance channel $^{10}B+d$, the incident energies of the deuteron beam were fixed at 5.3 MeV, 5.77 MeV and 6.37 MeV. The energy levels found by Jacquot et al. (2) are around these values for the ^{12}C compound nucleus, i.e. et 29.8 MeV (spin-parity 1^- or 3^+), 30.5 MeV (2^+) and 32.5 MeV (1^+).

The theoretical expressions of the cross-section are the same as those we already used (4) for other levels in ^{12}C with some success. The main characteristics are as follows : both nuclei ^{12}C and ^{8}Be are considered as compound nuclei, the R-matrix formalism is used to obtain the scattering amplitude and the identity of the three particles in the final state leads to a symmetrical amplitude.

Computations have been done for different values of the spin and parity of ^{12}C ^{12}C (0^+, 1^+, 2^+, 3^+, 4^+). The ^{8}Be levels which can be included are 0^+ (ground state), 2^+ and 4^+ and the first doublet 2^+, some higher levels could also contribute to the yield but not very significantly.

---

† Supported in part by the Atomic Energy Control Board and the National Research Council of Canada.

Corresponding parameters are given elsewhere (4). We have found that in the case of the reaction $^{9}Be+{}^{3}He$, it seems impossible to reproduce the experimental data at both incident energies 4.4 MeV and 4.96 MeV. For any ^{12}C spin-parity values we always get the peaks at the left (figs. 1 and 2) smaller than those at the right. It looks like there is no resonant ^{12}C state for this ingoing channel, but a positive result is that the interaction between two alphas in the final state arises where it is predicted by kinematics (figs. 1 and 2). At the ^{3}He third incident energy, some good fits are possible if we assume the values 3^+ or 2^+ for the compound state (figs. 3 and 4). For the reaction $^{10}B+d$, the number of ingoing channels becomes rather large higher (several quantum numbers for the channel spin and the orbital momentum) so that no very clear results are obtained. However, assuming that the reaction is going through a ^{12}C compound state, computed cross-section curves for $J^{\Pi} = 3^+$ and 2^+ are very similar to experimental data (figs. 5 and 6). Therefore, it seems possible to conclude that some excited states of ^{12}C exists in this excitation energy region.

In conclusion it seems that the properties of unbound three body states can be deduced from correlation spectra followed by an appropriate analysis.

Fig. 1 :

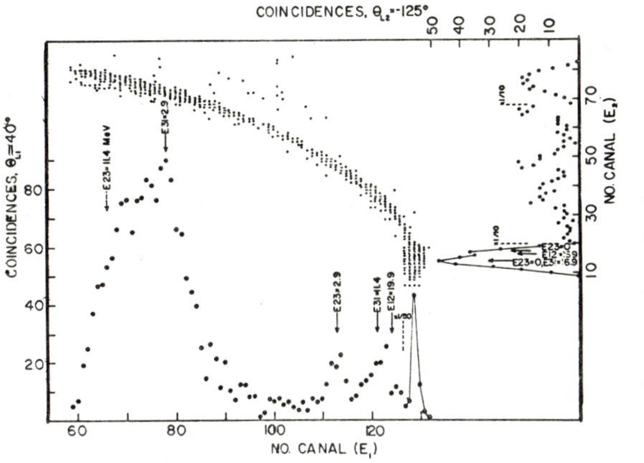

α-α correlation spectrum for $E_{\tau} = 4.424$ MeV

Fig. 2 :

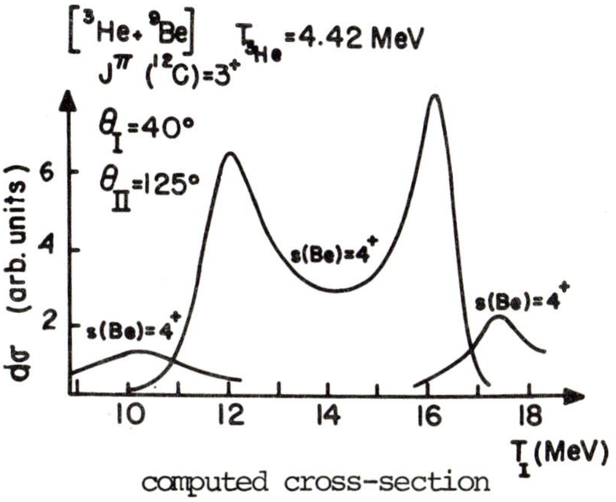

computed cross-section

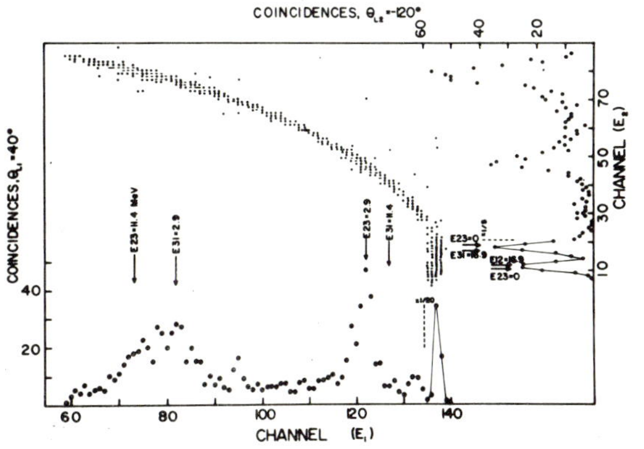

Fig. 3 :

α-α correlation spectrum for

$E_{\tau} = 5.624$ MeV.

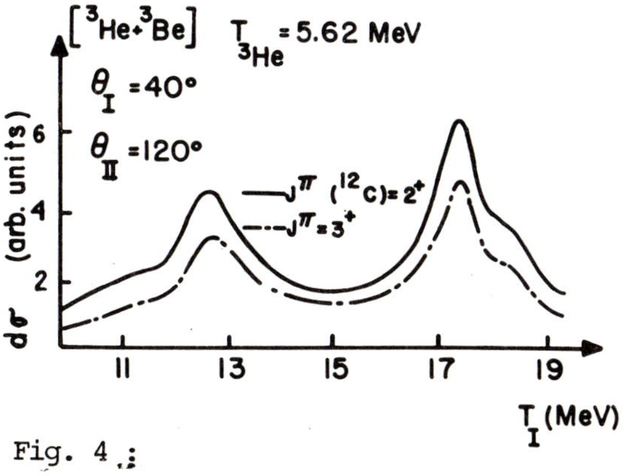

Fig. 4 :

computed cross-section with the partial width $^{12}C \to \alpha + {}^{8}Be(4^+)$ equal to the one of the channel $^{12}C \to \alpha + {}^{8}Be(2^+)$

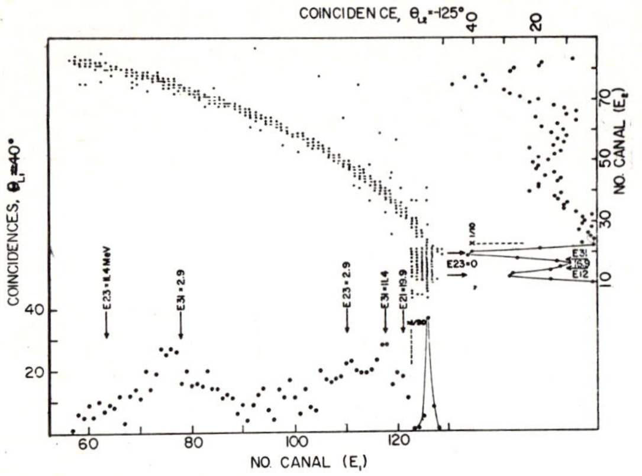

Fig. 5 :

α–α correlation spectrum for

E_d=5.294 MeV.

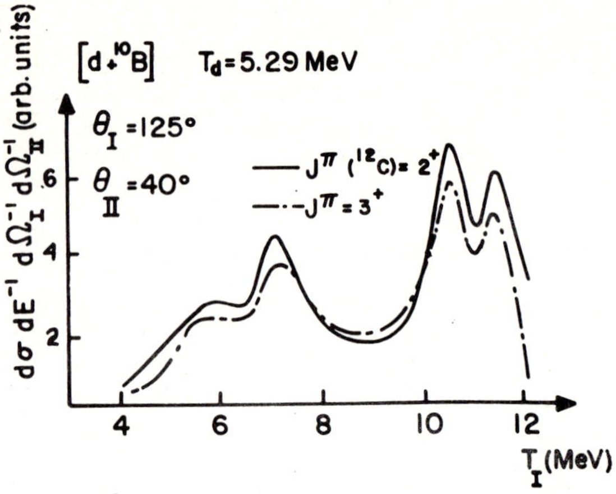

Fig. 6 :

computed cross-section for E_d=5.294 MeV with $\Gamma[{}^{12}C \rightarrow \alpha + {}^{8}Be(4+)] = 3\Gamma[{}^{12}C \rightarrow \alpha + {}^{8}Be\ (2^{+})]$.

REFERENCES

(1) R. Roy : Ph. D. thesis, Laval University, Québec, 1974.

(2) C. Jacquot, Y. Sakamoto, H. Jung and L. Girardin : Nucl. Phys. A201, 247 (1973).

H.D. Shay, R.E. Peschel, J.M. Long and D.A. Bromley, Phys. Rev. C9, 76 (1974).

(3) R. Roy, J. Birchall, B. Frois, U. von Moellendorff, C.R. Lamontagne and R.J. Slobodrian, Nucl. Phys. A245, 87 (1975).

(4) R. Roy, G. Turcotte, R.J. Slobodrian and G. Goulard, Proceedings of the International Conference on Few Body Problems in Nuclear and Particle Physics, les presses de l'Université Laval, 1975, ed. by R.J. Slobodrian, B. Cujec and K. Ramavataram; G. Goulard, Can. J. of Phys. 51, 2233 (1973).

FOUR-BODY DESCRIPTION
OF $d + d \longrightarrow p + t$ AND $p + {}^3He \longrightarrow p + {}^3He$ REACTIONS

M. Sawicki and J.M. Namysłowski
Institute of Theoretical Physics, Warsaw University

Abstract

Four body integral equations based on the Alt-Grassberger Sandhas approach are applied to the four-nucleon system. Our principal aim is the evaluation of cross sections for processes like $d + d \rightarrow p + t$, and $p + {}^3He \longrightarrow p + {}^3He$. A nonlocal, separable, spin-dependent nucleon-nucleon interaction, acting in the S-wave is assumed. The three-body, and two-deuteron internal scattering amplitudes are also made separable due to the application of Bateman method. The partial wave projection of the three-body potentials is made in an approximate way, following AGS. Due to good co-operation of several approximations we find three-body and four-body bound state energies very close to the experimental data. A simple relation between the three-body and the four-body bound state energies, noticed by Tjon is confirmed. Integral equations in the positive energy region are solved in the second K-matrix Born approximation. We get a very good agreement with the experimental data for the scattering processes mentioned in the title at energies 9.75, 13.8, 25.3 and 51.5 MeV.

The scattering of deuteron on deuteron is the first example of scattering of one composite object on another composite object. In particular, the rearrangement process $d+d \rightarrow p+t$ is a master example of the one-nucleon transfer reaction in the collision of two composed nuclei. This process, and the companion elastic scattering process $p + {}^3He \rightarrow p + {}^3He$ are most often the lacking points in the literature concerned with the four-body problem. Usually, the four-body bound state problem has been solved in different approximations while the four-body scattering problem has only been treated in 1969 by Alt, Grassberger and Sandhas [1] in the K-matrix Born approximation. The aim of our calculations is to put in one doable framework both the rearrangement, elastic scattering and bound state problems. We emphasize more the positive-energy four-body problem than the bound state problem. The last one gave us results quite close to the experimentally observed values only by an accidental cooperation of several approximations, made in order to have a numerically doable scheme. We accept these approximations as redefining us, in accordance with the experimental data, the kernel of the four-body integral equation in the negative energy range. This part of the kernel is also felt most essentially in the positive, low energy range. For that reason our rearrangement and elastic scattering results, given in Fig's 1 and 2, are so close to the experimental data.

Our scheme of calculations is a particular version of the Alt, Grassberger and Sandhas [1] generalized separable method. The most important differences between our calculations and that given in reference [1] are: 1° the application of the Bateman method instead of the Sturm one, and 2° the 2-nd order K-matrix Born approximation which respects the principal value part of integration instead of the 1-st order K-matrix Born approximation which ignores this part. The applicability of the Bateman method for finding the bound state energies and scattering lengths has been demonstrated by Kharchenko and collaborators [2]. For us, the most important advantage of the Bateman method over the Sturmian functions is the possibility of an explicit analytic continuation, needed in our calculations of the principal part of integration. We have two Bateman points, one at zero, and other at a value corresponding to the extremal binding energy.

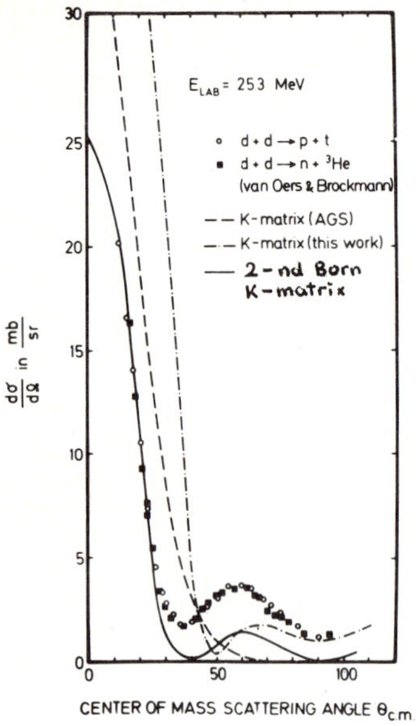

Fig.1

W.T.H.van Oers and K.W. Brockman, Nucl.Phys. 48, 625 /1963/

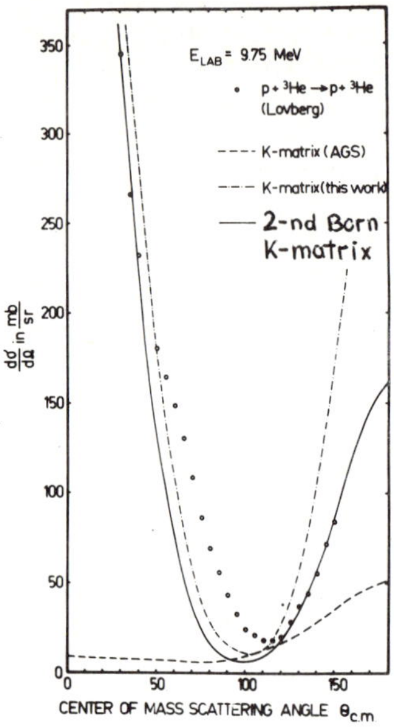

Fig.2

R.H. Lovberg, Phys.Rev. 103, 1393 /1957/

The two body input is taken as the S-wave, Yamaguchi separable interaction, fitting the following scattering lengths and effective ranges in the I=0, S=1 and I=1, S=0 states, respectively, a_d = 5.416 fm, a_ϕ =-23.68 fm, r_d= 1.75 fm, and r_ϕ =2.67 fm. The deuteron binding energy is 2.225MeV. On the three-body level, we make an approximate S-wave projection of the Lovelace potential Z, and folowing reference [1] we ignore in the denominator terms proportional to $\cos\theta$. This approximation has been tested in ref. [1] , and for small momenta and Yukawa type interaction it represents a good approximation. Another approximation made for calculational simplicity is the use of the "diagonal" Bateman formfactors also for the "off-diagonal" potentials. It means, that the three-body potentials, now denoted as $V_{dd}^{\frac{1}{2}\frac{1}{2}}$, $V_{dd}^{\frac{3}{2}\frac{3}{2}}$, $V_{\phi\phi}^{\frac{1}{2}\frac{1}{2}}$, and $V_{\phi\phi}^{\frac{3}{2}\frac{1}{2}}$ are represented as they should be, for example,

$$V_{dd}^{\frac{1}{2}\frac{1}{2}}(q',q;E) = \Lambda_{dd}^{\frac{1}{2}\frac{1}{2}} \cdot \sum_{\nu=1}^{2} F_\nu^d(q';E) \cdot F_\nu^d(q;E) \quad , \qquad /1/$$

where Λ_{nm}^{IS} are the spin-isospin recoupling coefficients, and F_ν^m are the Bateman formfactors. However, for the "off-diagonal" potentials $V_{d\phi}^{\frac{1}{2}\frac{1}{2}}$, and $V_{\phi d}^{\frac{1}{2}\frac{1}{2}}$ we put, for example

$$V_{d\phi}^{\frac{1}{2}\frac{1}{2}}(q',q;E) = \Lambda_{d\phi}^{\frac{1}{2}\frac{1}{2}} \cdot \sum_{\nu=1}^{2} F_\nu^d(q';E) \cdot F_\nu^\phi(q;E) \quad , \qquad /2/$$

with the same formfactors F_ν^m as in Eq./1/. This cuts down the number of formfactors, and consecutively the number of auxiliary amplitudes. Similar approximation is also made for the "potential" corresponding to the nucleon-nucleon interaction in the presence of two other nucleons forming d or ϕ states.

The above approximations cooperate, giving us the triton binding energy 7.36 MeV, the pole of the deuteron + deuteron propagator at 4.30 MeV, and the binding energy of the α -particle at 24.8 MeV. Neglecting spin and isospin, we get the bound state of three bosons at - 17.1 MeV, and for four bosons we have a ground state at - 44.75 MeV and an excited state at-23.1 MeV. Our results, denoted by SN, the results of Kharchenko and collaborators [3] /K/, Narodetskii [4] /N/, and Tjon [5] /T/ are plotted in Fig.3. They confirm an observation made earlier by Tjon [5], that all results lye on a straight line, passing through the experimental point.

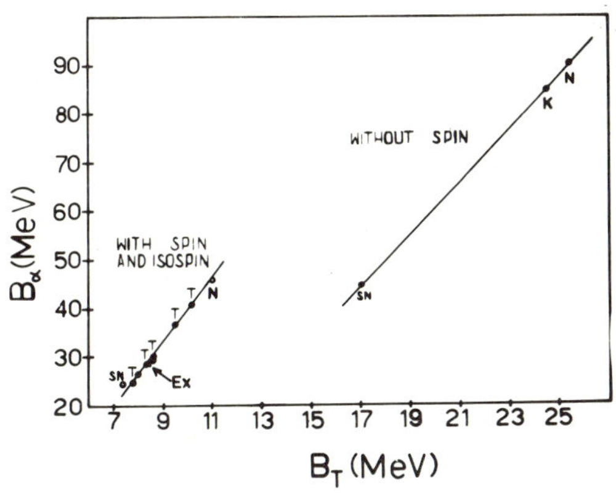

Fig.3

References

1. E.O.Alt, P.Grassberger and W. Sandhas, Phys.Rev.C1/1970/85
2. V.F. Kharchenko, V.E. Kuzmichev, Nucl.Phys. A183/1972/ 606; ibid. A196 /1972/636; Phys.Lett., 42B /1972/ 328; Yadern.Fiz. 17 /1973/ 975; V.F. Kharchenko, V.E. Kuzmichev and S.A.Storozhenko, Nucl.Phys., A188 /1972/ 609.
3. V.F. Kharchenko, V.E. Kuzmichev and S.A. Shadchin, Nucl.Phys., A226/1974/ 71.
4. I.M. Narodetskii, Nucl.Phys., A221 /1974/ 191.
5. J.A. Tjon, Phys.Lett., 56B /1975/ 217.

INVESTIGATION OF FEW NUCLEON SYSTEMS USING POLARIZED DEUTERONS

W. Grüebler

Laboratorium für Kernphysik, Eidg. Technische Hochschule
8049 Zürich, Switzerland

The use of polarized deuterons as a probe to investigate systems involving few nucleons has revealed rich information about the structure of such systems. In the last decade, an enormous amount of experimental data has been accumulated by this technique mainly due to the development and improvement of polarized ion sources. A review of the more recent work was given last summer at the Fourth International Symposium on Polarization Phenomena in Nuclear Reactions [1]. The present paper is concerned with some considerations concerning the advantages and problems using polarized deuterons.

The unique determination of the scattering matrix elements requires a so-called complete experiment, which means the measurement of a complete set of independent observables for each angle and energy. In principle, Simonius [2] has shown the different type of experiments which are necessary in order to obtain a complete set of observables. Apart from the analysing power measurements, the determination of polarization- transfer- and correlation coefficients are also required. The measurement of these coefficients involves double scattering of the polarized beam and scattering of the polarized beam by a polarized target.

Firstly, let us consider the simplest case of the elastic scattering of deuterons from a spin-0 target. This can be described by four independent complex amplitudes (M-matrix elements) so that there are seven independent real numbers to be measured at each angle and energy whereby an over-all phase is not taken into consideration. Parity conservation and time reversal invariance have already been taken into account. The left hand side of table 1 summarizes the possible independent measurements. The type of experiments are ordered according to increasing complexity.
N denotes the number of new linearly independent observables not contained in any measurement above it, T is the total number of linearly independent measurements listed so far and P the total number of measurements which are actually independent. The latter determine P independent parameters of the amplitudes. The difference T-P gives the number of additional relations between observables of a more complicated character. From this table it is obvious that the largest amount of information is gained from the analysing powers measurement. Apart from the cross section measurement it is also the simplest experiment and can therefore be carried out over large energy and angular ranges. This advantage also holds for more complicated spin structures like the elastic scattering of deuterons on spin 1/2 particles, which involves the scattering from protons and ^{3}He in the few nucleon system. A

Table 1. Measurements in Deuteron Elastic Scattering

Target Spin Examples	0 d-α N	 T	 P	1/2 d-p, d-^{3}He N	 T	 P
Diff. cross section	1	1	1	1	1	1
Pol. of recoil particle	0	1	1	1	2	2
Analysing powers	4	5	5	4	6	6
Correlation coefficients	0	5	5	12	18	18
Pol. transfer coefficients	11	16	7	20	38	23

similar summary for this case is listed on the right hand side of table 1. It is interesting to note again that from the easier experiments the measurement of the deuteron analysing powers contribute the most to new information.

At the present time, one is able to carry out complete experiments only at a few angles and energies, since the more complicated experiments involving the determination of polarization-transfer- and correlation coefficients are extremely time-consuming. Also, the accuracy of their results is limited. This situation is mainly due to technical difficulties which most probably will be the limiting factor for yet a while. Hence, the question arises, how much input data are needed for a phenomenological analysis in order to understand few nucleon systems? What are, for example, the ambiguities and the accuracy of the results of a phase shift analysis, if an incomplete set of observables is used? In the past, many such attempts have been made with varying orders of success.

In a recent phase shift analysis of d-α scattering [3], the differential cross section and the vector- as well as the three tensor analysing powers are used as input data. With an incomplete data set, the analysis was carried out up to 17 MeV deuteron energy. No ambiguity was observed during the course of the analysis. The final phases have been used to calculate the polarization-transfer coefficients via the M-matrix elements, and are compared with the measurement of 14 different coefficients [4]. In general, the agreement between prediction and experiment is excellent [3]. This suggests that the phase-shifts found are generally correct and only a larger amount of polarization-transfer data over a more extensive angular and energy region could nail down the phase shifts more precisely. The reason for this somewhat surprising result is that in the analysed energy range, the restriction to lower partial waves ($\ell \leqslant 4$) is a good approximation.

This, in turn, is encouraging, and has led us to try the case of systems with more complicated spin structure. A first attempt with d-^{3}He scattering and the same observables as in d-α scattering failed. It clearly showed the need for more accurate data and measurement in extended angular ranges. A new set of high precision data was measured in the energy range between 2 and 12 MeV in steps of one MeV. It was possible to extend the measurement in general to c.m. angles between 12° and 165°, each angular distribution of an analysing power component containing up to 50 data points.

Fig. 1 shows some of the results. The dots are the actual measurements. A phase shift analysis with an automatic search program gave the first preliminary results. The solid lines in fig. 1 are fits obtained so far. The parametrization of the analysis allows for the splitting of the phase shifts and the possible mixing of two states. At energies above the threshold for break-up of the deuteron, the phase-shifts are considered to be complex. The orbital angular momenta with all degrees of freedom are restricted to $\ell \leqslant 2$. The F- and G-waves, assumed to be small, are taken into account as unsplit phase shifts without mixing. At the present stage of the analysis, the preliminary results are very promising. No ambiguities have been observed, the fits are still improving and the resulting phase parameters show a smooth behaviour as a function of energy, although the data set at one energy has been analysed independently of the others.

We will now turn to the most interesting problem that of the three nucleon system, for which a phase shift analysis had been completed in our laboratory [5]. Six observables, corresponding to the differential cross section, the proton analysing power and the four deuteron analysing powers, have been used as input data. With the same parametrization as in the d-^{3}He scattering, the final parameters behave satisfactorily for incident proton energies up to 5.75 MeV. The strongest mixing is observed between the $^2S_{1/2}$ and $^4D_{1/2}$ states in which case both orbital angular momentum and channel spin are not conserved. A study of the sensitivity of the polarization transfer-coefficients for the proton-to-proton polarization transfer was found to be small to changes in the mixing parameters and the sequences of the

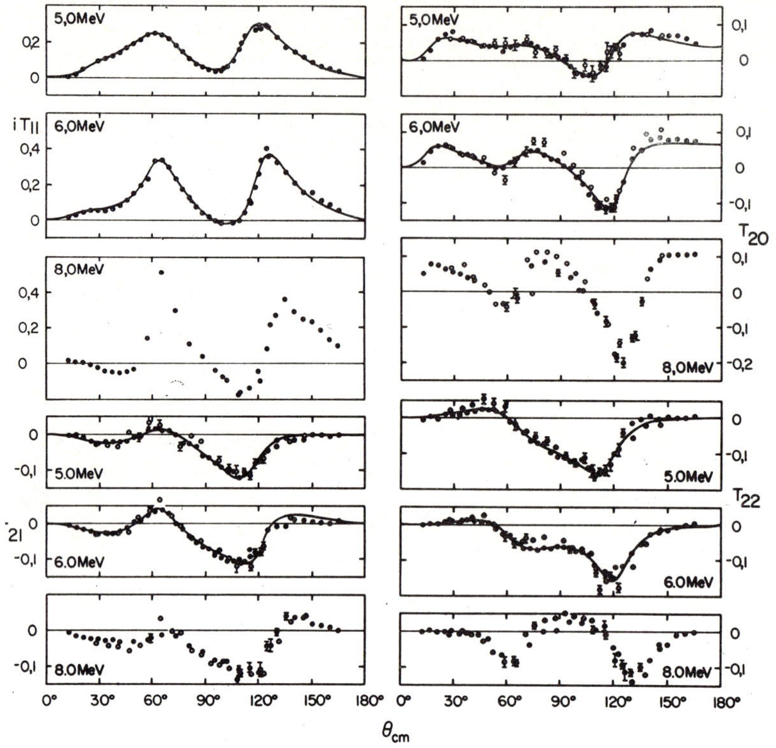

Fig. 1. Analysing powers for the d-^{3}He (solid dots) and the d-^{3}H (open circles) elastic scattering. The solid lines are fits of a phase shift analysis.

multiplets. The proton-to-deuteron vector polarization transfer coefficients have fairly large values but their sensitivity to the changes considered in the phase shifts and mixing parameters is relatively small. The predicted vector-to tensor polarization transfer coefficients are generally smaller than 0.1 in the energy range of the analysis. They are comparable with the measurements of Mitchell and Ohlsen [6], but in most cases, since the errors of the experimental data are as large as the predicted effects, little information on the quality of the predictions can be gained.

After completion of the phase-shift analysis, measurements of angular distributions of the deuteron analysing powers of the d-p elastic scattering have been carried out in Los Alamos up to 16 MeV, corresponding to an incident proton energy of 8 MeV. An attempt has been made to extend the analysis to this higher energy. Preliminary fits are in good agreement with the new measurements [7]. The phase shifts obtained are compatible with the results at lower energy but because of the limited number of data points and limited angular range covered by the measurement they are still very inaccurate. It is therefore seen how important it is to measure complete angular distributions of observables which can be measured easily. Only after having obtained data at extreme forward and backward angles, experimentally more complicated observables should be attempted to be measured. The information in the accurate data of a nearly complete angular distribution seems to be larger than in that of a few inaccurate data points of observables which are technically more difficult to obtain. The most difficult polarization measurements should then carried out at angles and energies where the theory or an analysis predicts the highest sensitivity of the parameters.

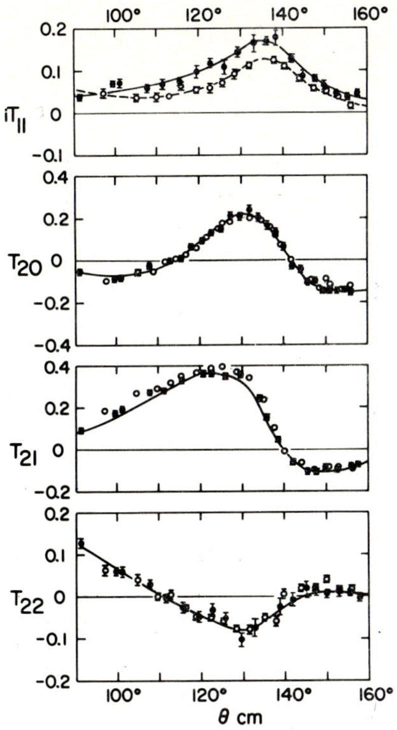

Fig. 2. Comparison of the analysing powers for the $^2H(d,p)^3H$ reaction at 10 MeV (solid dots) and the $^2H(d,n)^3He$ reaction at 11.5 MeV (open circles). The solid and dashed lines are Legendre polynomial fits.

We do not want to discuss here all possible symmetry violation tests; a review of the present state is given in ref.[8]. We want however to discuss a certain observation made in mirror reactions. A difference in polarization between the emitted proton and neutron in the mirror reactions $^2H(d,p)^3H$ and $^2H(d,n)^3He$ has been reported by Hardekopf et al. [9]. It was found that the discrepancy disappears if they compared the polarization at the same exit channel energies. This is a fairly crude correction for the difference in the Coulomb forces. A similar significant and energy dependent difference was found in the vector and two tensor analysing powers of the same mirror reactions [10]. These measurements are more reliable since the same reaction is always used for both cases to determine the polarization of the incoming polarized deuteron beam. The trend of the dissimilarity in the exit channel, which increases with decreasing energy, appears in the entrance channel too. When the analysing power results are corrected by shifting the $^2H(d,n)^3He$ energy scale to the same outgoing proton energy of the mirror reaction allowing in this way for the difference in the reaction Q-value, the descrepancy in the tensor components disappears, but the agreement for the vector analysing power also deteriorates. A comparison is shown in fig. 2, where, in a limited angular range in the backward hemisphere, the result of the $^2H(d,p)^3H$ reaction at 10 MeV (solid dots) is compared with the data of the mirror reaction $^2H(d,n)^3He$ at 11.5 MeV (open circles). This energy difference corresponds fairly well to the Q-value difference.

A further interesting comparison can be made between the elastic scattering of polarized deuterons on ^3H and ^3He. Here again we see the advantage of the use of polarized deuterons since four independent observables, which depend on different scattering amplitude combinations, can be compared. Fig. 1 shows the results of these two elastic scattering at the same incident deuteron energy. While the measurements agree for some components at certain energies, significant deviations are observed at others. Only part of the occurring discrepancies can be attributed to the 0.24 MeV higher excitation of the ^5He nucleus than the mirror nucleus ^5Li, resulting from the different threshold energies. Phase shift analysis which are in progress in our laboratory for both scatterings will reveal more information. A final decision whether these observed deviations are due to a charge symmetry violation can only be made after a careful investigation of the effects from Coulomb forces. Recently, Simonius [11] has investigated whether the differences between the $^2H(d,p)^3H$ and $^2H(d,n)^3He$ reactions could be explained by the long range part of the Coulomb interaction. Assuming that the wave functions are independent of the Coulomb interaction within the range of the nuclear forces, and that non-diagonal matrix elements (produced by mixing parameters) contribute only to the first order of the correction, he found that the complicated matrix elements

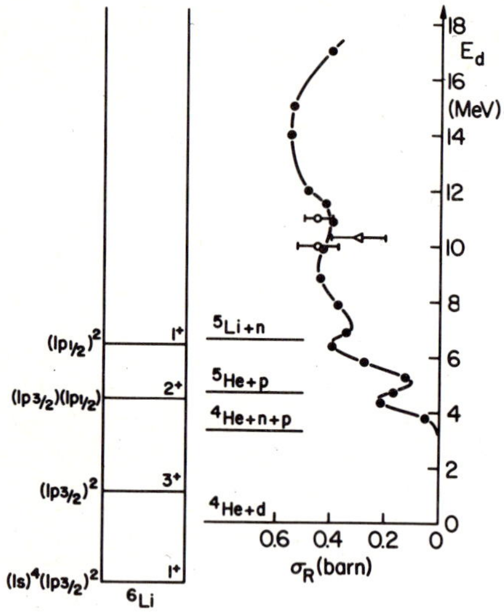

Fig. 3. Total reaction cross section of the d-α entrance channel

simplify greatly and that the partial wave amplitudes differ only by a correction factor. Numerical calculations for $\ell = 2$ give no final answer, since the elastic phase shifts in the exit channels are not known well enough. More experimental information is needed in this case.

Finally we want to discuss an example which shows that besides the well known resonant behaviour of elastic partial waves further information about the structure of light nuclei can be gained by taking the inelastic channels into consideration. On the right hand side of fig. 3 the total reaction cross section has been calculated from the d-α phase shifts: the dots represent results of the calculations, open circles and triangles are the experimental data [12,13]; the solid line is a curve just connecting the calculated results. The thresholds of the inelastic channels and the ^{6}Li level scheme are shown to the left. It is interesting to note that the 2^+ and 1^+ resonances in ^{6}Li affect the reaction cross section quite strongly. In fact, these resonances show up more clearly in the inelastic rather than in the elastic cross section. The structures found for the ^{6}Li level assuming jj coupling are indicated in fig. 3. They are consistent with the calculations of Hackenbroich et al. [14] who found it necessary to include, besides the α-n-p channel, not only the ^{5}He+p and ^{5}Li+n fragmentations, but also the ^{5}He*+p and ^{5}Li*+n channels in order to obtain agreement with the experimental results.

At higher deuteron energies two broad peaks appear in the reaction cross section suggesting two further resonances in ^{6}Li, but no resonating real part of the phase shifts has been observed in this energy region. This fact can be explained by the weakness of these resonances and by the large absorption in this energy region indicating that the decay occurs mostly through the open inelastic channels, including not only the decay through the ground states of ^{5}He and ^{5}Li but also that through the first excited states of the same nuclei. An indication of such a resonance is given bei the typical resonant behaviour of the Argand plot in the energy region between 12 and 15 MeV [3] suggesting a further 1^+ resonance.

In conclusion, we have shown some examples which demonstrate the importance of the use of polarized deuteron beams in the investigation of few nucleon systems. It is also seen that in spite of the overwhelming amount of data there is still need for further experiments, which, considering the improvement of technical devices, will no doubt be carried out in the near future.

References

[1] Proc. Fourth Int. Symp. on Polarization Phenomena in Nuclear Reactions, Zürich (Birkhäuser Verlag Basel, in press)

[2] M. Simonius, Third Int. Symp. on Polarization Phenomena in Nuclear Reactions (University of Wisconsin Press 1971) p. 401

[3] W. Grüebler, P.A. Schmelzbach, V. König, R. Risler and D. Boerma, Nucl. Phys. A242 (1975) 265

[4] G.G. Ohlsen, G.C. Salzmann, C.K. Mitchell, W.G. Simon and W. Grüebler, Phys. Rev. C8 (1973) 1639

[5] P.A. Schmelzbach, W. Grüebler, R.E. White, V. König, R. Risler and P. Marmier, Nucl. Phys. A197 (1972) 273

[6] C.K. Mitchell and G.G. Ohlsen, Proc. Int. Conf. on Few Particles Problems in Nuclear Reactions (North-Holland 1972) p.460

[7] P.A. Schmelzbach, W. Grüebler, V. König and R. Risler, Fourth Int. Symp. on Polarization Phenomena in Nuclear Reactions, Zürich (Birkhäuser Verlag, Basel, in press) p. 485

[8] M. Simonius, Proc. Fourth Int. Symp. on Polarization Phenomena in Nuclear Reactions, Zürich (Birkhäuser Verlag, Basel, in press) p. 75

[9] R.A. Hardekopf, R.L. Walter and T.B. Clegg, Phys. Rev. Lett. 28 (1972) 760

[10] W. Grüebler, V. König, P.A. Schmelzbach, R. Risler, R.E. White and P. Marmier, Nucl. Phys. A193 (1972) 129

[11] M. Simonius, Jahresbericht Lab. für Kernphysik ETHZ, 1972, p. 138

[12] G.G. Ohlsen and P.G. Young, Nucl. Phys. 52 (1964) 134

[13] J.C. Allred, D.K. Froman, A.M. Hudson and L. Rosen, Phys. Rev. 82 (1957) 786

[14] H.H. Hackenbroich, P. Heiss and Le-Chi-Niem, Nucl. Phys. A221 (1974) 461

NEUTRON POLARIZATION STUDIES ON REACTIONS BETWEEN LIGHT NUCLEI*

L. C. Northcliffe
Texas A&M University
College Station, Texas, USA

The ability to make good fast-neutron polarization measurements has been linked to the development of apparatus and techniques to supplement the standard time-of-flight measurements of fast-neutron physics. Examples are the development of good polarization analyzers and of spin-precession techniques for the elimination of false asymmetries. The recent development of polarized ion sources for use with particle accelerators to provide intense beams of high-energy polarized particles has made feasible the performance of neutron polarization measurements which are more accurate, more comprehensive and more complex. At Texas A&M these developments have been combined in the construction of a facility for fast-neutron polarization measurements. The purpose of this paper is to present a brief description of the facility, the techniques for analysis of polarization data and the experimental program which has evolved there. The work, of course, is that of many people. Among them are J. D. Bronson, E. P. Chamberlin, B. Craft, M. L. Evans, R. G. Graves, M. Hamm, J. C. Hiebert, M. Jain, H. D. Knox, C. W. Lewis, J. M. Moss, W. H. Peeler, J. G. Rogers, R. C. Rogers, D. P. Saylor, J. W. Watson, H. Wolverton and R. L. York.

EXPERIMENTAL SETUP AND DATA-TAKING METHODS

The polarized ion source is a commercial (ANAC) positive ion source of the atomic-beam type. It is mounted on the shielding roof over the cyclotron and the ion beam is injected down into the center of the cyclotron through an axial hole in the pole of the cyclotron magnet. An electrostatic buncher in the axial-injection line is used to peak the injected beam at the optimum portion of the cyclotron rf period for acceleration. The beam is inflected into the median plane of the cyclotron with an electrostatic mirror and accelerated conventionally thereafter. Typical ion beam currents from the source are 10 μA and typical values for extracted cyclotron beams are 0.5 μA, although currents as high as 1.6 μA have been extracted on occasion. The beam energies are variable, up to 50 MeV for protons and up to 60 MeV for deuterons. The vector polarization of the beams is ∿75% for protons and ∿55% for deuterons. The polarization of the beam is along the injection axis (vertical) and can be reversed at the source, giving either up (+) or down (-) polarization for the cyclotron beam. As also is done with unpolarized beams, the cyclotron is tuned slightly off-resonance so as to narrow the time-width of the beam micropulse, thus facilitating time-of-flight measurements.

After momentum analysis in a 157° analyzing magnet the beam passes through a monitor polarimeter consisting of a He-gas target and a pair of detectors placed to determine the left-right asymmetry in the $\vec{p}$-He or $\vec{d}$-He elastic scattering. The beam is then refocussed and deflected into the neutron-beam line where it passes through the neutron-production target and after deflection through 90° is collected in a heavily-shielded Faraday cup. The neutron-production target usually used is a liquid-nitrogen-cooled high-pressure (15 atm) gas target in which the energy loss is ∿1.5 MeV. The neutrons emerge through the shielding wall surrounding the target by way of collimator channels at 0° and 18°. Built into the 18° channel are two identical transverse-field magnets in tandem which can be used to precess the neutron spin when run in parallel (p) but give no net precession when run in antiparallel (a) mode.

For measurement of the neutron polarization a liquid-helium (ℓHe) neutron polarimeter is placed, on either collimator axis, 4.5 m downstream from the target. The polarimeter consists of a ℓHe scattering sample on the collimator axis and four NE102-scintillator neutron detectors, each subtending a solid angle of 0.061 sr, placed at symmetrical left- and right-scattering angles of ±78° and ±125°. The ℓHe is contained in an alumina ceramic cup of volume ∿160 cm^3 and is viewed by a photomultiplier tube so as to observe the scintillations produced in the ℓHe by the He-recoils from n-He elastic scattering. The ℓHe cell is kept filled by gravity feed from a ℓHe reservoir of 10.5 liter volume. One filling of the reservoir suffices for ~160 hr of operation. The ℓHe cell and ℓHe reservoir are isolated from the surroundings by a heat shield at 77°K.

A near coincidence of signals from the ℓHe cell and one of the "side detectors" (NE102 scintillators) signifies an event, and for each event four parameters of data and a tag identifying the side detector involved are read by an on-line computer and recorded on magnetic tape. The recorded data are the pulse height (H) from the ℓHe cell, the pulse height (P) from the side detector, the time-of-flight (t) from production target to ℓHe cell and the time of flight (δt) between the ℓHe cell and the side detector.

Since the cyclotron-beam polarization can be reversed (+ or -) or alternatively there is the option of precessing the neutron spin (a or p) only one of the four side detectors is essential. The use of four not only quadruples the counting efficiency but also provides well-known means for the elimination of geometric and other systematic errors as well as a variety of checks on internal consistency. In order to minimize errors due to long-term drifts in any part of the apparatus, runs are made in the sequence $(^+_p)(^-_p)(^-_a)(^+_a)(^+_a)(^-_a)(^-_p)(^+_p)$, where (^+_p) for example signifies a run with cyclotron-beam polarization "up" and spin-precession magnet fields "parallel". Since four running modes are possible $[(^+_p),(^-_p),(^-_a),(^+_a)]$ there are four yield expressions (Graves et al.[1]) which are functions of four unknowns, each of which is a function of the neutron energy T_n: the transverse polarization transfer coefficient $K_y^{y'}(T_n)$; the polarization parameter $P_{y'}(T_n)$ for the $(p,\vec{n})$ or $(d,\vec{n})$ reaction; the analyzing power $A_y(T_n)$ of the $(\vec{p},n)$ or $(\vec{d},n)$ reaction; a constant N_o proportional to the cross section for the reaction without polarization. The values of $K_y^y(T_n)$, $P_{y'}(T_n)$, $A_y(T_n)$ and N_o are obtained by simultaneous solution of the four linear yield equations.

The foregoing discussion is applicable for the case of a "polarization transfer" experiment. It is also necessary, of course, to perform less complex experiments, such as measurements of the neutron polarization $P_{y'}(T_n)$ in $(p,\vec{n})$ or $(d,\vec{n})$ reactions in which the ℓHe polarimeter is used but the cyclotron beams are unpolarized. Further details can be found in recent publications by Knox et al.[2] and Rad et al.[3][4]. Analyzing power measurements are made by using polarized cyclotron beams with a simple NE102-scintillator neutron detector at 18° and determining the "up-down" asymmetry from sets of runs made in the sequence (+)(-)(-)(+). Measurements of the neutron production cross section $d^2\sigma/d\Omega dT_n$ are made by a two-parameter time-of-flight technique (Northcliffe et al.[5]).

OFF-LINE ANALYSIS OF THE MULTIPARAMETER DATA

In the off-line analysis, the ability to study and manipulate the four parameters of information (H,P,t and δt) provides powerful methods for discriminating against background and improving experimental resolution. An interesting example of this is demonstrated in Fig. 1, which shows steps in the analysis of the data tagged by one of the side detectors in one set of runs. The correlation spectrum of t vs δt for the raw data is shown in Fig. 1(a), while Fig. 1(b) shows this spectrum after corrections have been made for the "time-walk" in the discriminator signals for both the ℓHe and side detectors. (The former correction was found by examining the t-H dependence for the prompt γ-ray peak, which was made prominent by replacing the cooled-gas target with a solid target; after correction for the ℓHe signal time-walk, a display of P vs δt for a narrow band of t-values permitted a determination of the side-detector time-walk.) The sharpening of the spectrum produced by the elimination of time-walk is obvious. In Fig 1(c) the spectrum is shown after another correction has been made, for the variation of the time of flight δt with neutron energy T_n. In all three spectra a faint band of events can be seen to the left of the main group. These have been identified as false background events caused by γ-ray-producing reactions of the incident neutrons with the ceramic walls of the ℓHe cell. They are easily eliminated by imposing the δt-gate G shown in Figs. 1(c) and 1(d). Note that the separation of real and false events is not as clean in the raw δt spectrum (dotted line of Fig. 1(d)). An additional basis for

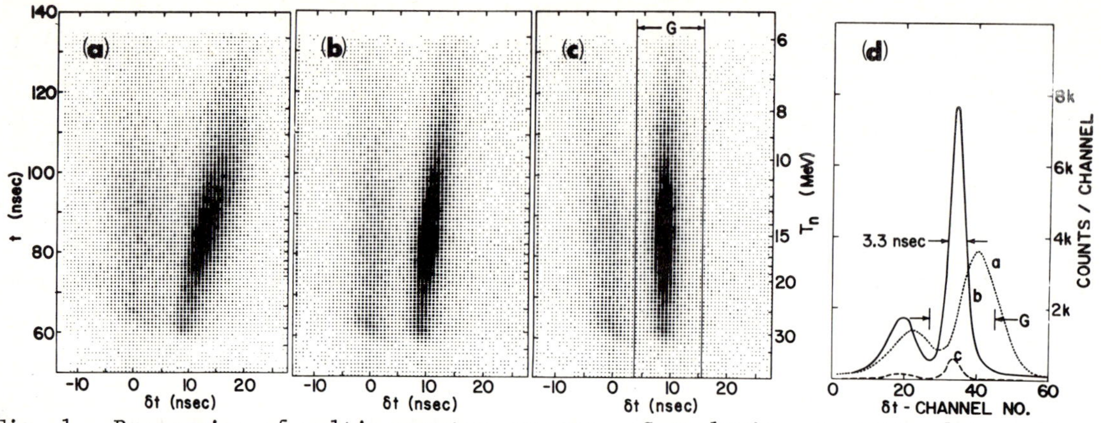

Fig. 1. Processing of multiparameter spectra. Correlation spectrum of t vs δt: (a) for raw data; (b) after correction for time-walk in scintillator signals; (c) after further correction for known variation of δt with T_n. (d) The δt-spectra for the raw data (curve a), the fully-corrected data (curve b) and the (fully-corrected) background obtained with an empty gas-target cell.

background elimination is provided by the observation of a correlation between the group of false events in Figs. 1(a)-1(c) and a band of abnormally small H-values in the spectrum of t vs H. Many of the false events are eliminated by setting the H-threshold above this band.

EXPERIMENTAL RESULTS

p-d Breakup at T* = 14 MeV.

A major effort has been devoted to the study of the neutrons produced at 0° and 18° in the p-d breakup reaction at center-of-mass energy $T^* \simeq 14$ MeV, using alternately 21.4 MeV protons and 42.8 MeV deuterons to initiate the reaction. In this way, four different kinematic regions in the center-of-mass system are explored. Our measurements of the cross sections $d^2\sigma(T_n)/d\Omega dT_n$ for all four regions already have been reported (Graves et al.[6]). They were found to be in good agreement with the predictions of a three-body-breakup code (Jain and Doolen[7]). The code is based on an exact solution of the three particle Faddeev equations for separable spin-dependent S-wave nucleon-nucleon potentials. The cross section measurements have been supplemented by measurements of $P_{y'}(T_n)$ at 18° for both reactions (Rad et al.[3][4]). The ℓHe polarimeter was used with unpolarized beams to get these results. Since the three-body-breakup code only uses S-waves it can only give zero for the polarization. The observed values are small but non-zero, and are similar in magnitude but opposite in sign to that which would be expected from quasifree p-n or n-p scattering. Recently we have used the polarized beam and a simple detector at 18° to measure $A_y(T_n)$ with high statistical accuracy. Analysis of the data is incomplete. Finally, to complete this set of p-d breakup measurements we have used polarized beams and the ℓHe polarimeter to obtain data on $K_y^{y'}(T_n)$ for all four regions.

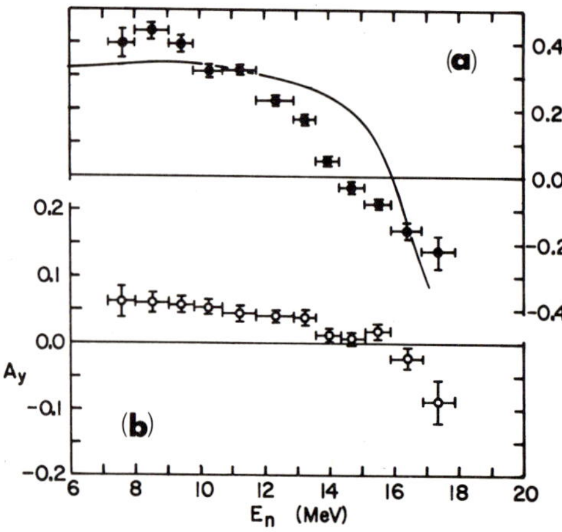

Fig. 2. Spin observables for the reaction $^2H(p,n)pp$ at $\theta_n=18^o$ for $T_p= 20.4$ MeV. (a) Measured values of the parameter $K_y^{y'}(T_n)$ (solid symbols) and the prediction (solid line) of a three-body separable-potential model; (b) measured $A_y(T_n)$ values (open symbols).

The results for $^2H(\vec{p},\vec{n})pp$ at 18°, the first of their kind, were recently published (Graves et al.[1]) and are reproduced in Fig. 2, which also shows the $A_y(T_n)$ values deduced from the same data. The prediction for $K_y^{y'}(T_n)$ given by the breakup code is shown by the curve in Fig. 2(a). Since only S-waves are used in the code the prediction for $A_y(T_n)$ necessarily is zero. The need for refinement of the code to include higher partial waves is evident. The data obtained in the other three kinematic regions are not yet completely analyzed.

Polarization Transfer in the $^{11}B(\vec{p},\vec{n})^{11}C^{gs}$ Analog Reaction.

It has been pointed out by Madsen et al.[8] that measurements of $K_y^{y'}$ at 0° in certain charge-exchange reactions may make it possible to distinguish between central spin-spin and tensor two-body forces using direct reaction theory. They have shown in calculations for the $^{11}B(\vec{p},\vec{n})^{11}C^{gs}$ analog reaction that $K_y^{y'}(0°)$ is very sensitive to the relative contributions of the individual components of the spin-dependent force. Their predictions are reproduced in Fig. 3, in which our measurements also are shown. The predictions correspond to various combinations of potentials used in the effective central-plus-tensor two-body interaction. The charge-exchange part of this interaction is given by

$$V_e = \vec{\tau}_o \cdot \vec{\tau}_i [f_c(\vec{r}_{oi})(V_\tau + V_{\sigma\tau}\vec{\sigma}_o \cdot \vec{\sigma}_i) + V_T S_{12} f_T(\alpha',\beta,\vec{r}_{oi})]$$

where, respectively, τ, σ and S_{12} are the isospin, spin and two-body tensor operators, f_c and f_T are the central and tensor radial form factors, and V_τ, $V_{\sigma\tau}$ and V_T are the isospin, spin-isospin and tensor-isospin potentials. Curve A is the prediction when only L = 0 orbital angular momentum transfer is considered. Curves B to F correspond to various values of the potentials V_τ, $V_{\sigma\tau}$ and V_T. For example, curves B, D and E are predictions for purely central forces ($V_T = 0$). Curves C and F correspond to central-plus-tensor interactions with slightly different values of the potentials. Comparison of curves A and B shows the effect of higher L transfers. Comparison of curves C and B shows the effect of including the tensor force. Our measurements are in best agreement with the simple L = 0 central force prediction, curve A. This is surprising in view of the existing evidence in favor of a substantial tensor force in the effective two-nucleon interaction (Crawley et al. [9]). It would be premature to conclude that the strength of the tensor force is substantially less than in current estimates. The effect of an energy dependence in the two-body force itself has not been included in the calculations shown in Fig. 3. The present results do show, however, that Knowledge of $K_y^{y'}$ places a substantial constraint on the allowed strength and form of the spin-dependent parts of the effective interaction. Finally, it should be noted that our measured values of $K_y^{y'}$ join smoothly to an extrapolation of the values measured at lower energy by Lisowski et al.[10].

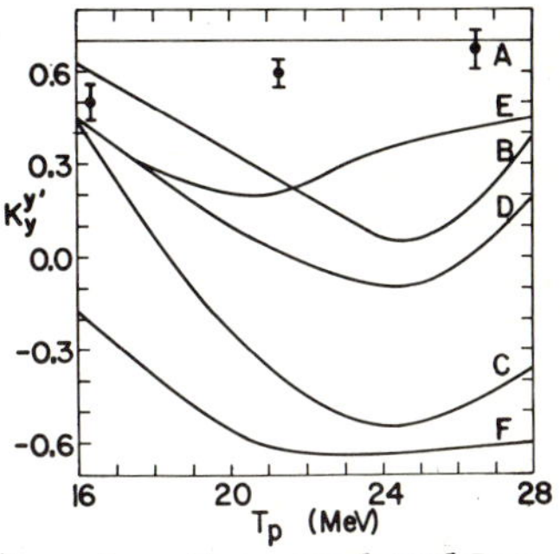

Fig. 3. Measured values of $K_y^{y'}(T_p)$ at $\theta_n=0°$ for the analog reaction $^{11}B(\vec{p},\vec{n})^{11}C^{gs}$ compared with predictions (solid lines) based on direct-reaction theory with a central-plus-tensor two-body potential.

Polarization in the $^2H(\alpha,\vec{n})\alpha p$ Reaction at 39.4 MeV.

Before acquiring our polarized ion source we had speculated on the possibility of obtaining polarized neutrons by quasifree n-He scattering using the $^2H(\alpha,\vec{n})\alpha p$ reaction. We have measured the neutron polarization (Knox et al.[2]) and the results are shown in Fig. 4. along with the predictions of this simple model (square symbols). While substantial polarizations are observed they are not as large as predicted on the basis of quasifree scattering. Also shown in the figure is a prediction given by the modified impulse approximation of Nakamura[11]. The agreement is reasonably good in that the sign and trend of the polarization is reproduced even though the magnitude is not precisely duplicated.

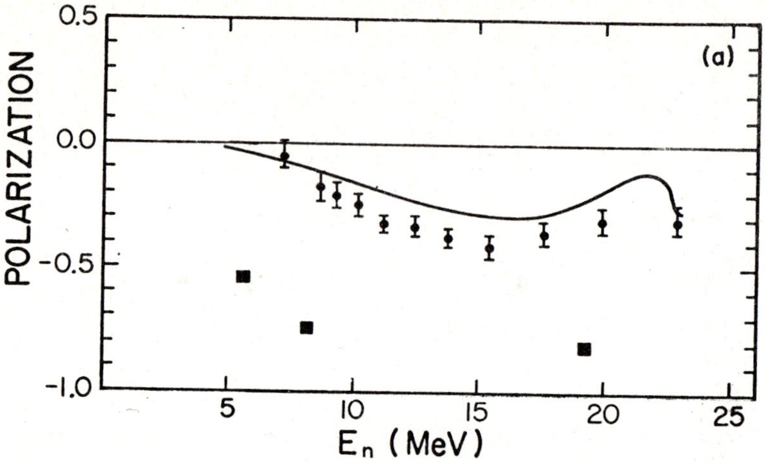

Fig. 4. Polarization in the $^2H(\alpha,\vec{n})\alpha p$ reaction at 39.4 MeV. The error bars are statistical only. The square symbols are predictions of a simple model of quasifree scattering. The curve is a prediction by the modified impulse approximation.

CONCLUSION

It is hoped that the existence of a comprehensive set of p-d breakup data at a common center-of-mass energy will stimulate theorists to perform three-body-breakup calculations using more realistic nucleon-nucleon forces, and that the ^{11}B polarization transfer data will lead to a better understanding of the effective nucleon-nucleon interaction.

REFERENCES

*This work was supported in part by the U. S. Atomic Energy Commission and the U. S. National Science Foundation.

[1] R. G. Graves, M. Jain, H. D. Knox, E. P. Chamberlin and L. C. Northcliffe, Phys. Rev. Lett. 35(1975)917.
[2] H. D. Knox, R. G. Graves, F. N. Rad, M. L. Evans, L. C. Northcliffe, H. Nakamura and H. Noya, Phys. Lett. 56B(1975)33.
[3] F. N. Rad, L. C. Northcliffe, D. P. Saylor, J. G. Rogers and R. G. Graves, Phys. Rev. C 8(1973)1248.
[4] F. N. Rad, L. C. Northcliffe, J. G. Rogers and D. P. Saylor, Phys. Rev. Lett. 31(1973)57.
[5] L. C. Northcliffe, C. W. Lewis and D. P. Saylor, Nucl. Instr. Meth. 83(1970)93.
[6] R. G. Graves, M. Jain, L. C. Northcliffe and F. N. Rad, Proc. Int'l. Conf. on Few Body Problems in Nucl. and Part. Phys., Quebec (1974)754.
[7] M. Jain and G. Doolen, Phys. Rev. C 8(1973)124.
[8] V. A. Madsen, J. D. Anderson and V. R. Brown, Phys. Rev. C 9(1974)1253.
[9] G. M. Crawley, S. M. Austin, W. Benenson, V. A. Madsen, F. A. Schmittroth and M. J. Stomp, Phys. Lett. 32B(1970)92.
[10] P. W. Lisowski, R. L. Walter, T. B. Clegg and C. E. Busch, Bull. Am. Phys. Soc. 19(1974)477.
[11] H. Nakamura, Nucl Phys. A208(1973)207; Nucl. Phys. A223(1974)599.

POLARIZATION IN FINAL STATE INTERACTIONS : $^3He\ (^3He, \vec{p})\ p\alpha$ and $^6Li\ (^3He, \vec{p})\alpha\alpha$ *

R.J. Slobodrian, M. Irshad, R. Pigeon, C. Rioux, J. Asai and S. Sen
Université Laval, Département de Physique, Laboratoire de Physique Nucléaire
Québec, P.Q. G1K 7P4 Canada

The first measurements of polarization in these two final state interaction reactions to date have become possible with the sucessful operation of the polarimetry facility at Université Laval [1]. Figure 1 shows a view of the scattering chamber, the polarimeters and charge collection cup (consisting of two halves for the monitoring of the beam direction). Figure 2 shows a diagram of the electronics and the computer associated with one polarimeter. On line operation allows a particle identification of the detected events on the basis of the distinctive loans on a ΔE-E two-dimensional display. The routine operation consists of two measurements with the polarimeters alternating their positions with respect to the beam. This procedure has proven to be particularly useful with solid targets.

The experimental results on the reaction $^3He\ (^3He, \vec{p})\ p\alpha$ correspond to the ground state of ^5Li, which is broad and unbound, lying near zero energy in the p-α system. It is a Breit-Wigner resonance of $J^\pi = 3/2^-$, coinciding with shell model expectations. In this model, its configuration is $(1s_{1/2})^4\ (1p_{3/2})^1$, and it is well separated from other levels. This precludes ambiguities due to mixing of configurations. Hopefully, the reaction observables are then directly related to the reaction mechanism.

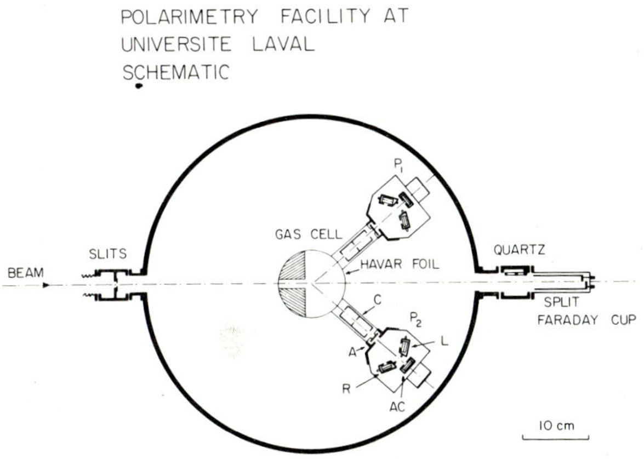

P_1, P_2 :POLARIMETERS ; C:GAS COLLIMATOR
A:ANALYZERS ; L:LEFT DETECTOR ; R:RIGHT DETECTOR
AC:ANTICOINCIDENCE DETECTOR

FIG. 1

A first approach to the interpretation of the experimental results has been carried out using the global analysis with the optical model [2] and the DWBA theory using a code for two particle transfer [3].

No symmetrisation of amplitudes has been implemented yet, and it is believed that its effects are small for angles smaller than 60°. Within the usual DWBA treatment the possible combinations of (JLST), transferred total, orbital, spin angular momenta and iso-spin respectively are : (2110), (1110) and (1101), L being odd because of parity considerations. The spectroscopic amplitude $S_{AB}^{1/2}$ reduces to a Racah coefficient, as one of the angular momenta in the 9-j symbol is zero. Figure 3 shows the cross section and polarization curves calculated compared to the experimental data in dash-dot lines. Another approach to the reaction is to consider the transfer of a deuteron cluster instead of a general two nucleon transfer. In this case the deuteron spin has to be aligned antiparallel to the ^3He spin to permit the s-shell closure and the orbital momentum of the proton in the deuteron has to be parallel to the deuteron spin to reach the $3/2^-$ state in ^5Li. Therefore it follows that $J = 2$. The value $J = 1$ is not excluded by general angular momentum coupling rules, but is incon-

* Work supported in part by the Atomic Energy Control Board of Canada and the Ministery of Education of Québec.

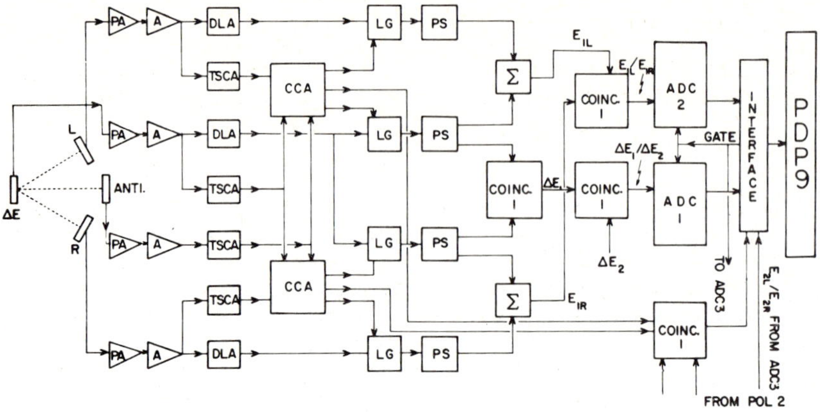

FIG. 2

Block diagram of the electronics for one polarimeter. The symbols used are : ΔE-analyser silicon detector, L - left silicon detector, R - right stopping silicon detector, ANTI-anticoincidence silicon detector, PA - Preamplifier, A - amplifier, DLA - delay amplifier, TSCA - timing single channel analyser, CCA - coincidence - anticoincidence circuit, LG - Linear gate, PS - Pulse stretcher, Σ - sum circuit, COINC 1 - coincidence circuit set to coincidence requirement 1.

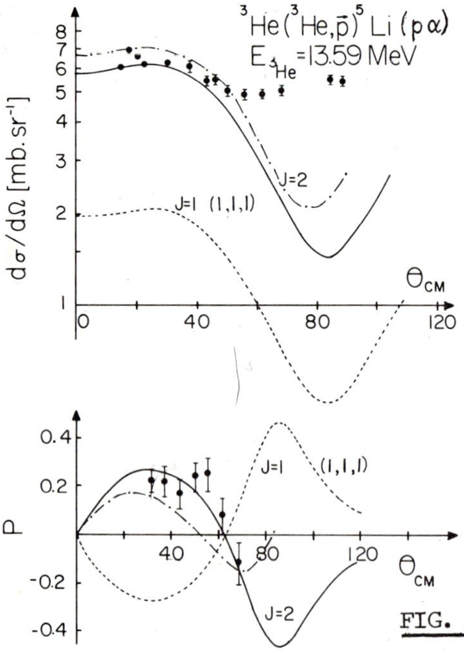

FIG. 3

sistent with the data and the above arguments concerning the closure of the s-shell. The theoretical curves for (2110) and (1110) transfer are shown by solid and dashed lines respectively. The best fit to the polarization data is produced unquestionably by the (2110) curve. In addition there is a clear selection of angular momentum transferred in this two nucleon transfer reaction, analogous to the single nucleon transfer [4]. The reaction ^{6}Li (^{3}He, $\vec{p}$)αα populates mainly the states of ^{8}Be. The ground state and first excited state polarizations have been measured. These are even-spin and parity states, which have large widths for decay into α+α, and should not be adequately represented in terms of the shell model, at variance width the ^{5}Li case. Figure 4 shows experimental results together with preliminary theoretical curves, calculated also in the context of the DWBA including finite range interactions and non-locality corrections in the incoming and outgoing channels. The proton - ^{8}Be and ^{3}He-^{6}Li optical model parameters were obtained from the literature [5,6]. For the transition to the ^{8}Be ground state the neutron-proton pair was assumed to be in a $(P_{3/2})^2$ configuration, sharing equally the binding energy. The JLST combinations are thus restricted to (1010) and (1210) and a coherent sum over L = 0 and L = 2 was taken. For the transition to the first excited state, the n-p pair was assumed to be in $p_{3/2}$ and $p_{1/2}$ orbits respectively, restricting the JLST combinations to (1010), (1210) and (2210), a coherent sum was taken over the

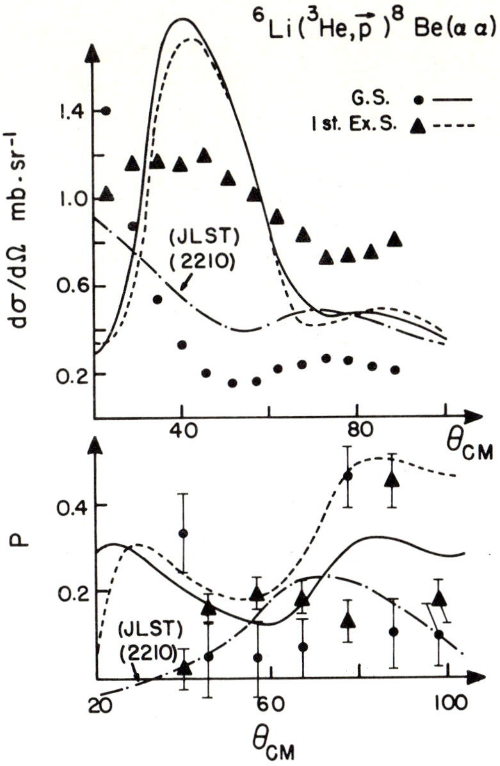

FIG. 4.

first two J transfer values. The data seem to indicate that the J = 2 transfer is favoured over the J = 1 transfer.

The final state interaction peaks seem to show a variation of the polarization across their width, although our statistics so far does not warrant a cathegorical statement. Such variation is also predicted by the DWBA code varying the Q-value in order to sweep through the resonance. Both reactions yield protons and alpha particles in the final state and to the extent that the latter can be considered "elementary" [7] a three body final state theory should be applicable to this rearrangement reaction which the authors propose to implement as second step in the analysis of the present results. It is clear that S-wave separable interactions in the final state, often used in the context of three nucleon systems, will not suffice for the interpretation of the polarization. Further experimental and theoretical work on both reactions is still in progress. In particular the elastic channel ^3He+target and the $(^3He,\vec{p})$ reaction channel will be studied as a function of energy.

REFERENCES

1) B. Frois, J. Birchall, R. Lamontagne, R. Roy and R.J. Slobodrian, Nucl. Instr., and Meth. 96 (1971) 431.

M. Irshad, S. Sen, R. Pigeon and R.J. Slobodrian, Fourth International Symposium on Polarization Phenomena in Nuclear Reactions, Zurich (1975) p.J 7.

2) J. Raynal, code Magali D. Ph T169, 42 (CEN, Saclay).

3) J.M. Nelson and B.E.F. Macefield, paper 74-9 University of Birmingham, unpublished.

4) T.J. Yule and W. Haeberli, Nucl. Phys. A117 (1968) 1.

5) M.F. Werby, S. Edwards and V.J. Thomson, Nucl. Phys. A169 (1971) 81.

6) H. Ludecke, T. Win-Tjin, H. Werner and J. Zimmerer, Nucl. Phys. A109 (1968) 676.

7) R. Roy, J. Birchall, B. Frois, U. von Moellendorff, C.R. Lamontagne and R.J. Slobodrian, Nucl. Phys. A245 (1975) 87.

EXCITED STATES OF ^{4}He STUDIED WITH THE POLARIZED PROTON CAPTURE REACTION

S. S. Hanna, H. F. Glavish, G. King, J. R. Calarco, D. G. Mavis, V. K. C. Cheng, and E. Kuhlmann

Department of Physics, Stanford University, Stanford, California 94305 USA*

The lowest excited states of ^{4}He obtained from resonance analyses of (p,p) and (p,n) reactions have been summarized by Fiarman [1] and Meyerhof. The negative parity levels are consistent in spin and approximate location with the $1p1s^{-1}$ states predicted by the shell model. Positive parity states are expected to be composed partly of 1p-1h and partly of 2p-2h excitations. For a self-conjugate nucleus, such as ^{4}He, E1 transitions to the ground state are allowed only from $T = 1, 1^-$ levels. Hence the two $T = 1, 1^-$ levels postulated at 27.4 and 30.5 MeV are expected to dominate the characteristics of the E1 radiation in nucleon capture reactions which form ^{4}He. These states comprise the giant E1 resonance of ^{4}He. The presence of third and fourth degree Legendre polynomials P_3, P_4 in the gamma-ray angular distribution [see eq. (1) below] is evidence for E2 radiation which is thought to reflect a $T = 0, 2^+$ state [1] in the region of 33 MeV in ^{4}He. Presumably, this state comprises the giant isoscaler E2 resonance of ^{4}He.

The unpolarized angular distribution measurements [2] show that for E1 radiation the $S = 1$ component is strongly suppressed. Crone [3] and Werntz have shown that this is strong evidence for dominant singlet strength in the lower $T = 1, 1^-$ level. However this is contradicted by the ^{3}He + p elastic scattering measurements [4] which imply the dominant singlet strength is in the upper $T = 1, 1^-$ level, at least in ^{4}Li. The key to resolving this problem is to determine the phases as well as the amplitudes of both the $S = 0$ and $S = 1$ components of the S-matrix elements for proton capture and to relate [5] these through the unitarity of the S-matrix to the $S = 0$ and $S = 1$ phase shifts and mixing parameter obtained from ^{3}H + p elastic scattering measurements [6] .

In order to determine the phases as well as the amplitudes of the capture S-matrix elements, the reaction $^3H(\vec{p},\gamma)^4He$ has been studied at eight proton energies between 6.0 and 16.0 MeV. At each energy the unpolarized yield $d\sigma(\theta)/d\Omega$ and the analyzing power $A(\theta)$ were measured simultaneously at several angles using a polarized proton beam. The polarized beam was provided by a polarized ion source of the atomic beam, sextupole magnet type and was accelerated by the Stanford FN tandem van de Graaff. The beam current at the target was about 10 nA. The ^{3}H target was 3 $mg \cdot cm^{-2}$ (3 Curie) ^{3}H-erbium on a 2 $mg \cdot cm^{-2}$ platinum backing. The γ-rays from the reaction were detected with the Stanford 24 cm x 24 cm NaI spectrometer.

The spin direction for the protons could be set either up or down by switching the rf transitions in the polarized source. Reversing the spin did not significantly change the magnitude of the proton polarization. At a given energy and angle a measurement of the γ-ray yield in the capture reaction was made by alternating runs with proton spin up with runs with proton spin down. The analyzing power $A(\theta)$ was then determined from the expression $A(\theta) = P^{-1}(N_u - N_d)/(N_u + N_d)$, where P is the magnitude of the proton polarization, and N_u and N_d are the γ yields at an angle θ for spin up and down, respectively. The cross section $d\sigma/d\Omega$ is simply proportional to $N_u + N_d$. The unpolarized angular distribution $d\sigma/d\Omega$ and the analyzing power may be expanded as follows:

$$d\sigma(\theta)/d\Omega = A_o\left[1 + \sum_{k=1} a_k P_k(\cos\theta)\right] \qquad (1)$$

$$A(\theta)d\sigma(\theta)/d\Omega = \sum_{k=1} b_k P_k^1(\theta) \qquad (2)$$

At each energy, sufficient angles were chosen to enable both the E1 and the E2 matrix elements to be extracted. Typical results for $d\sigma/d\Omega$ and $Ad\sigma/d\Omega$ are shown in fig. 1. The coefficients obtained by fitting the data with eqs. (1) and (2) are plotted in Fig. 2. For orientation the yields σ(E1) and σ(E2), as obtained from the unpolarized measurements [2], are shown at the top of fig. 2. The amplitudes and phases obtained from these data from least squares fits, are shown in fig. 3. The quantities $^1P_1$ and $^3P_1$ are the singlet and triplet P wave (E1) matrix elements while $^1D_2$ and $^3D_1$ refer to the E2 radiation. The phases of $^3P_1$ and $^1D_2$ relative to that of $^1P_1$ are plotted at the bottom. A preliminary analysis suggests that the capture S-matrix elements are consistent with the ^{3}H + p elastic scattering phase shifts extracted by Hardekopf [6] et al. with a large mixing parameter.

The E2 strengths derived from the E2 amplitudes $^1D_2$ and $^3D_2$ are plotted at the top of fig. 3. and compared with the earlier result of Gemmell and Jones (see ref. 2) obtained from unpolarized measurements. In the region of overlap the agreement is quite good, but in the higher energy region the new results depart from the trend of the earlier measurements [2] shown by the dashed line in fig.2. The general picture that emerges from the new results is that of a very broad E2 resonance covering the range of excitations from 24 to 32 MeV (and presumably beyond). It is not possible to determine the isospin of this E2 strength from these measurements, nor to establish the degree of isospin mixing.

REFERENCES

* Supported in part by the National Science Foundation.

1 S. Fiarman and W. E. Meyerhof, Nucl. Phys. A206 (1973) 1.

2 W. E. Meyerhof, M. Suffert, and W. Feldman, Nucl. Phys. A148 (1970) 211.

3 L. Crone and C. Werntz, Nucl. Phys. A134 (1969) 161.

4 D. H. McSherry and S. D. Baker, Phys. Rev. C1 (1970) 888.

5 H. F. Glavish, C. C. Chang, J. Calarco, S. S. Hanna, R. Avida and W. E. Meyerhof, Few Particle Problems in the Nuclear Interaction, eds. I. Slaus, S. A. Moskowski, R. P. Haddock and W. T. H. Van Oers, (North Holland, Amsterdam, 1972) p. 632.

6 R. A. Hardekopf, P. W. Lisowski, T. C. Rhea, R. L. Walter and T. B. Clegg, Nucl. Phys. A191 (1972) 481.

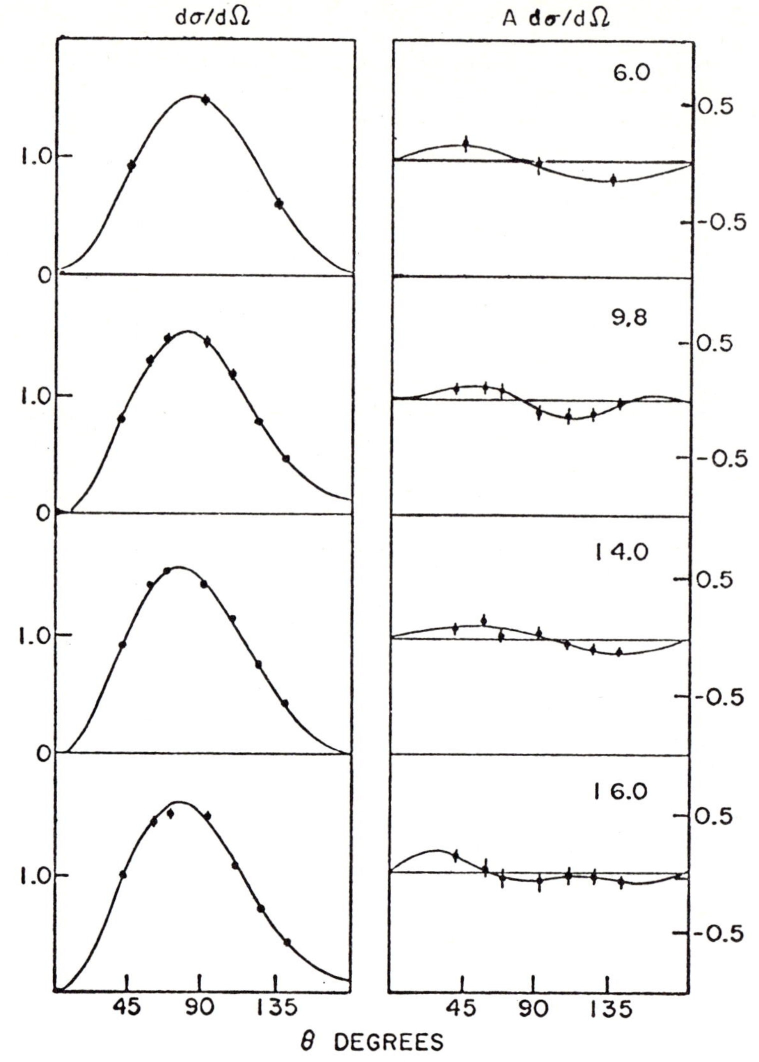

Fig. 1. Angular distributions of cross sections and analyzing powers measured in the reaction $^3H(p,\gamma)^4He$.

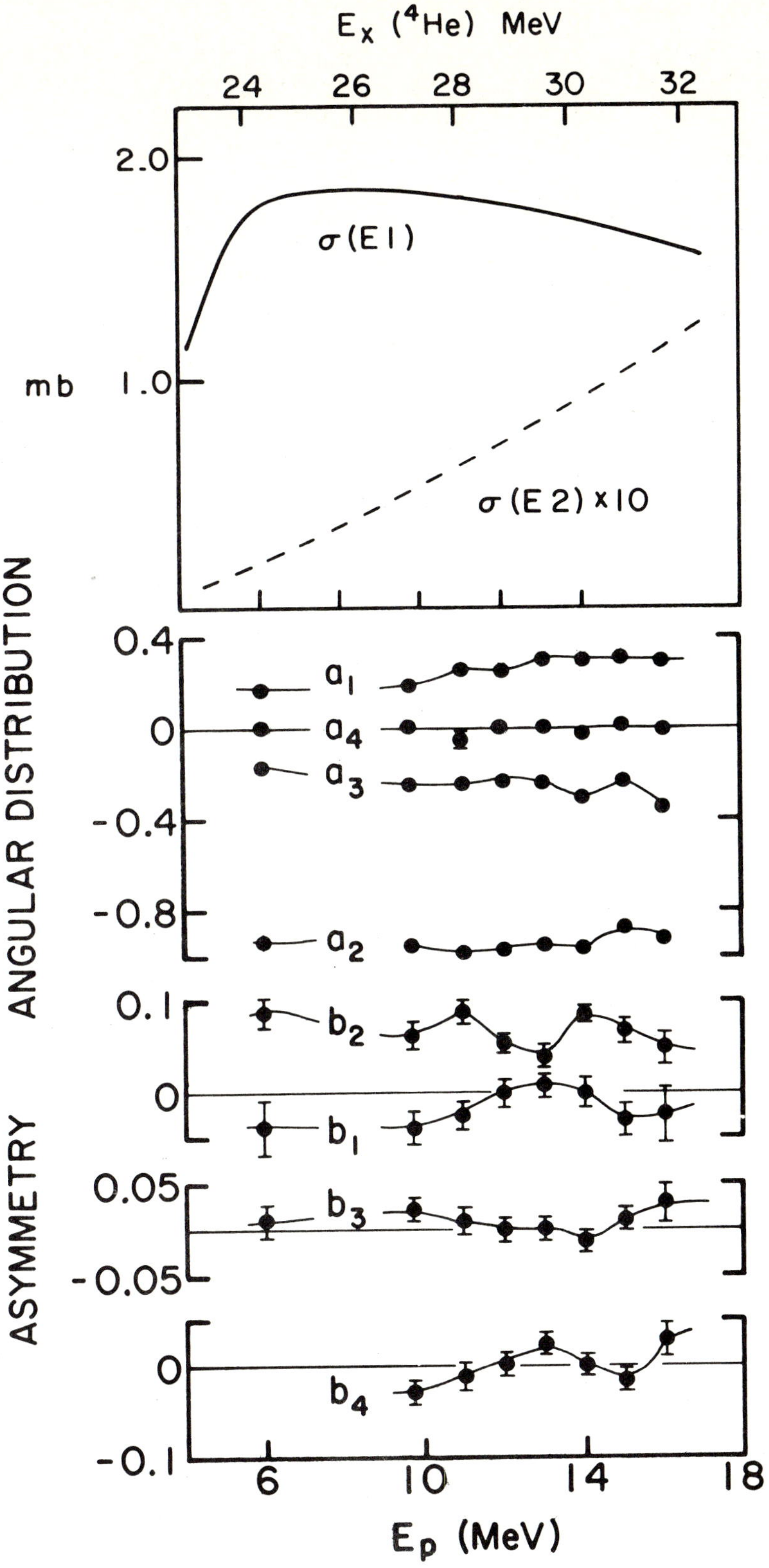

Fig. 2. Top: E1 and E2 cross sections from $^3H(p,\gamma)^4He$ (ref. 2). Below: coefficients extracted from the angular distribution data.

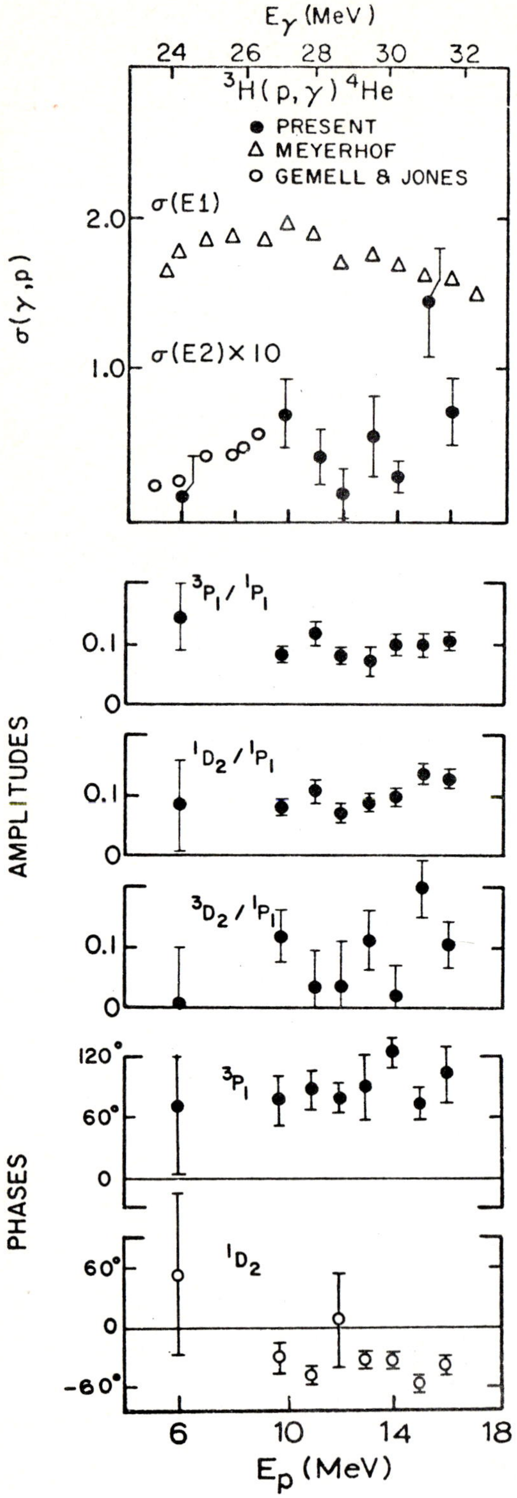

Fig. 3. Top: E1 and E2 cross sections from $^3H(p,\gamma)^4He$ as measured by several authors. Below: amplitudes and phases extracted from the coefficients in fig. 2.

A MICROSCOPIC DESCRIPTION OF POLARISATION PHENOMENA IN THE FOUR NUCLEON SYSTEM

S. Ramavataram, K. Ramavataram and C.L. Rao
Département de Physique et Laboratoire de Physique Nucléaire
Université Laval, Québec, Canada G1K 7P4.

In earlier calculations Ramavataram [1,2] we have demonstrated that a continuum treatment of the $(1s_{1/2}^{-1}\ 1p)$ excitations in the mass-4 system using the coupled-channels method provides a satisfactory description of different aspects of the available experimental data. In Ramavataram [1], it was shown that the energies of the T=1, negative parity states of the A=4 system and more particularly the observed distribution of the E1 transition strength can be well reproduced in such calculations. Extension of the model Ramavataram [2] to the case of the T=0 components arising from the above configurations, provided not only a good description of the data on particle channels but also suggested a broad 1^-, T=0 resonance around 23 MeV of excitation. Bound state calculations Barrett [3] had predicted this non-spurious resonance around 31 MeV (ie) above the T=1 resonances.

We report here the results obtained from the coupled-channels model for the energy dependence of the neutron polarisation P_n observed Haight [4] in the reaction $t(p,n)^3He$. Below 3 MeV proton energy our calculations confirm that the value of P_n is determined essentially by the interference effects between the $^3P_0$ and $^3P_2$ amplitudes. In the 3 to 5 MeV region, the cross-over to negative values takes place at an energy below that observed experimentally (curve 1, fig. 1). However if the 1^-, T=0 resonance is shifted to about 24 MeV (curve 2, fig. 1), the agreement with experiment is considerably improved. Beyond 5 MeV, calculations involving only the negative parity resonances gave values of P_n which were generally well below experiment. A positive parity T=0 resonance is reported experimentally Fairman [5] around 25 MeV in ^4He and a 2^+, T=0 resonance has been proposed Werntz [6] around 29 MeV. Including the appropriate 1p-1h configurations in our calculations we obtained a 1^+, T=0 resonance at 25 MeV and a 2^+, T=0 resonance around 40 MeV of excitation. Shifting the 2^+ resonance to 29 MeV and including these two resonances we obtained the values of P_n shown in curve 2 of fig. 1. In fig. 1 we have given the results (curve 3) of an R-matrix analysis Haight [4] which includes all resonances considered here except the 1^-, T=0 and 1^+, T=0 states. Note that in curve 3 the cross-over to negative values takes place at a higher energy compared to experiment. Beyond 6 MeV meaningful comparison between the prediction of the coupled-channels method and the R-matrix analysis is not possible till the position of the 2^+, T=0 resonance is determined more realistically in the former case. Improvements to the model are under consideration.

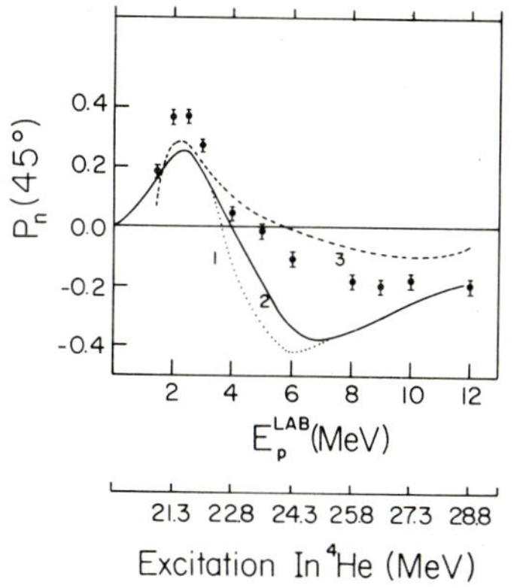

Figure 1.

1. S. Ramavataram, C.L. Rao and K. Ramavataram, Nucl. Phys. A226 (1974), 173.

2. S. Ramavataram, C.L. Rao and K. Ramavataram, Nucl. Phys. (to be published).

3. B.R. Barrett, Phys. Rev. 154 (1967) 955.

4. R.C. Haight et al., Phys. Rev. Lett. 28 (1972) 1587.

5. S. Fairman and W.E. Meyerhoff, Nucl. Phys. A206 (1973) 1.

6. C. Werntz and W.E. Meyerhoff, Nucl. Phys. A121 (1968) 38.

A STUDY OF THE ^{6}Li (^{3}He, pα)α REACTION

K. Ramavataram, R. Larue and R. Gagnon
Département de Physique, Université Laval, Québec, Canada

G.C. Ball, W.G. Davies and A.J. Ferguson
Chalk River Nuclear Laboratories, Chalk River, Ontario, Canada
and
R.E. Warner
Department of Physics, Oberlin College, Oberlin, Ohio, U.S.A.

The study of the resonance structure of the very light nuclear systems from the point of view of determining experimentally the energies, the spins and the parities of the resonances is in the nature of a challenge principally because of the large widths associated with them. In the Mass-5 system several positive parity resonances Ramavataram (1) have been predicted in the region of 20 MeV of excitation energy, but direct experimental evidence for the existence of these resonances has been sparse Ramavataram (2), King (3).

In the ^{6}Li (^{3}He, pα)α reaction there are two principal ways in which a resonant state in ^{5}Li could be formed. The incident ^{3}He could pick up a neutron from the target having a residual ^{5}Li system. Alternately, in view of the strong probability for α+d clustering in the ground state of ^{6}Li, the projectile could form a ^{5}Li system by the pick-up of a deuteron from the target nucleus.

We have performed a kinematically complete experiment to study the three-body pick-up in the ^{6}Li (^{3}He, pα)α reaction by bombarding an isotopically enriched target with a 36 MeV ^{3}He beam of the Chalk River MP tandem accelerator. The p+α channel was chosen to study the breakup of ^{5}Li. Protons and alphas were detected in coincidence for different coplanar geometries of the two counters which enabled us to examine different regions of excitation in the mass-5 system. By kinematically tracking the resonance peaks over the different geometries used here, we

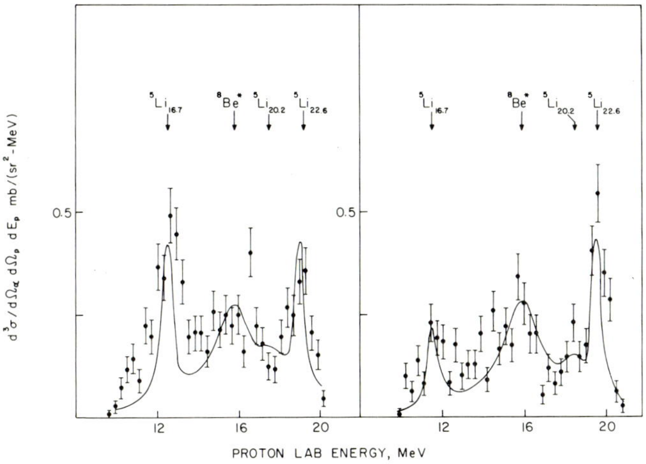

FIGURE 1 - Coincidence spectra between protons and alpha particles obtained in the ^{6}Li (^{3}He, pα)α reaction for two different geometries.

have been able to identify the peaks in the spectra which correspond to the 16.7, 20.2 and 22.6 MeV states of ^{5}Li and unresolved states around 20 MeV of excitation in ^{8}Be. Spectra obtained at two different geometries are shown in fig. 1. The solid lines are the results of a theoretical calculation assuming a sequential decay mechanism in which the ^{5}Li compound system formed by a direct process breaks up into a proton and an alpha particle. The energies and the widths obtained in the present experiment are in good agreement with earlier calculations Ramavataram (1, 4).

1. K. Ramavataram and S. Ramavataram, Nucl. Phys. A147 (1970) 293.

2. K. Ramavataram, D.J. Plummer, T.A. Hodges and D.G. Montague, Nucl. Phys. A174 (1971) 204.

3. H.T. King, W.E. Meyerhof and R.G. Hirko, Nucl. Phys. A178 (1972) 337.

4. S. Ramavataram and K. Ramavataram, Can. J. Phys. 49 (1971) 1798.

UNITARITY, ANALYTICITY AND THE RELATIVISTIC THREE BODY PROBLEM

I. J. R. Aitchison
Department of Theoretical Physics
12 Parks Road
Oxford OX1 3PQ
England.

1. INTRODUCTION

Reactions of the type a+b → 1+2+3 are commonly analysed by using the isobar model, in which one writes the amplitude schematically as

$$F(s_1,s_2,s_3;s) = C_1(s)/D_1(s_1) + C_2(s)/D_2(s_2) + C_3(s)/D_3(s_3) \tag{1}$$

where s_i is the invariant subenergy of the j-k pair and s is the square of the c.m.s. energy of the three final particles (spin and angular momentum labels are here ignored). The D_i's are the denominator functions of the j-k two body amplitudes, assumed known. The C_i's are found from fitting the data.

By assuming the form (1) for different J^P states, it has been possible to extract [1] the variation with s of the relative phases of the C_i's for different J^P states decaying to 1+2+3. In this analysis, it is crucial that the phase variation in the variables s_i is entirely given by the known functions D_i^{-1}, since the extraction of the phases of the C_i's depends on interference effects involving the D_i's.

Amplitudes of the form (1), however, are in principle incorrect, since they fail to satisfy unitarity, either in the two body channels [2] (s_i variables), or in the three body channel (s) [3]. Unitarity corrections spoil the simple "factorisation" property of (1); the C_i's have to be replaced by functions ϕ_i which depend on at least one of the s_i, as well as on s. Nor are the ϕ_i completely harmless: as Amado has emphasised [4], they are forced by elastic unitarity in the subenergy channels to have the normal threshold branch point present in D_i^{-1}, and so they are certainly complex, and possibly rapidly varying, above threshold.

I want to discuss briefly some approaches to the problem of unitarising the isobar model (i.e. finding satisfactory ϕ_i's). But it turns out that, in trying to solve the unitarisation problem in the cheapest possible way consistent with analyticity, we are led to a simple model for the relativistic three body problem - one, in fact, that generalises the old Khuri-Treiman [5] model. Thus, I want to place the discussion in the context of a search for good relativistic three particle equations. I shall emphasise the necessity of having to attend to both analyticity and unitarity, and the difficulty of doing so in a completely satisfactory way.

Since the early work on K-T models [6][7], more extensive investigations have been done by R. and J.-Y. Pasquier [8]. I shall frequently refer to their substantial contributions, some of which, unfortunately, are still only available in thesis form, and merit wider appreciation.

2. THE ISOBAR MODEL

(a) Unitarity. The necessity for unitarity corrections to the isobar model can be easily understood. Just as, in the two body problem, single particle exchange Born terms are unitarised by "rescattering" them to form loops, so in this case, to the isobar amplitude (a) must be added rescattering corrections, of which the simplest is (b). The final factor here, D_i^{-1}, takes care of the two body intermediate

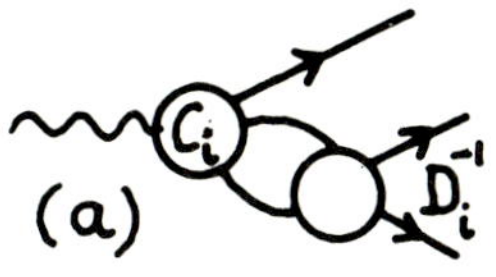

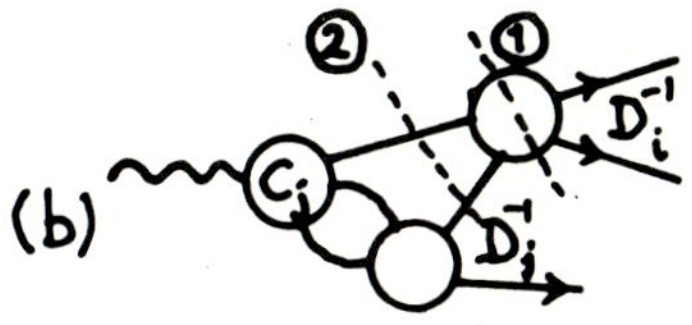

states found by cutting the final blob (cut ①); but the graph has a further two body intermediate state (cut ②) which is not taken care of by D_i^{-1}. The discontinuity across this second cut in s_i is, in this case, clearly proportional to the amplitude $C_j D_j^{-1}$ deriving from the other channel j. Thus, the presence of a third strongly interacting particle causes a two body cut additional to that dealt with by the functions D^{-1}.

Indeed, one can show [9] that, if F is written as (1) with $C_i(s)$ replaced by $\phi_i(s_i,s)$, (as would be appropriate for the decay of a $J^P=0^+$ state into three spinless particles interacting in s-waves only), the two body discontinuity formula for F implies that ϕ_i has a non-zero discontinuity across the elastic two body (j-k) cut given by

$$\mathrm{disc}_i\ \phi_i = 2i\rho_i(s_i)\ N_i(s_i) \sum_{j\neq i} \frac{1}{2}\int_{-1}^{1} dx_i\ \phi_j(s_j,s)/D_j(s_j) \tag{2}$$

Here N_i is the numerator function of the j-k two body amplitude, ρ_i is the phase space factor, and x_i is the cosine of the angle between the momenta of particles i and j in the j-k c.m.s. In (2) the role of the parallel channels j in the discontinuity of ϕ_i is clearly seen.

(b) Analyticity. How can (2) be implemented? One obvious way to satisfy it is to notice that ρ_i itself has a square root branch point at the threshold of the elastic cut, so that disc $\rho_i = 2\rho_i$; hence the ansatz

$$\phi_i = C_i(s) + i\rho_i(s_i)N_i(s_i) \sum_{j\neq i} \frac{1}{2}\int_{-1}^{1} dx_i\ \phi_j/D_j \tag{3}$$

will certainly satisfy (2) if $C_i(s)$ is, as indicated, independent of s_i (or, more generally, has no elastic cut in s_i). Such an ansatz is analogous to the K-matrix parametrisation in the two body problem: there, elastic unitarity for the 2→2 amplitude M says that disc $M^{-1}=2i\rho$, which can be satisfied by taking $M^{-1}=K^{-1}+i\rho$ if K is regular at threshold. This leads to $M=K(1-i\rho K)^{-1}$, the usual K matrix parametrisation. In fact, (3) is exactly what is obtained by a K-matrix approach to the decay problem [10]; in this case an integration is involved, because of the extra (continuous) degree of freedom present in the three body problem (the other subenergy variable).

Eq.(3) is a set of integral equations for the ϕ's, the solutions to which will satisfy (2). This does not mean that they are the correct functions to use in phenomenological analyses! In fact, they are deficient for another reason - analyticity. The analytic properties of the functions ϕ_i generated by (3) do not agree with those given by Feynman graphs (field theory): they contain additional singularities that should not be present in the physical amplitude. A student at Oxford, Richard Golding, and I have recently argued [11] that in the A1→3π case such spurious singularities are the cause of some rapid variation in the corrections to the isobar model proposed by Goradia et al. and by Ascoli and Wyld [12], based on equations of the form (3). This probably explains why Ascoli and Wyld found that when using such ϕ's the χ^2 fit to their data was worsened. Similar spurious fluctuations are probably present in the analogous calculations reported by Aaron et al. [13] for the ππN case.

Thus, analyticity is very important, even phenomenologically. If we again adopt the cheapest possible way of including it, we may restrict $\phi_i(s_i,s)$ to have only the elastic two body cut in s_i, and no other singularities. Assuming adequate convergence, we can then write a dispersion relation for ϕ_i of the form

$$\phi_i = C_i(s) + \frac{1}{\pi}\int_{\substack{\text{elastic}\\ \text{cut}}} \frac{ds_i'}{s_i'-s_i-i\varepsilon}\ \rho_i(s_i')\ N_i(s_i') \sum_{j\neq i} \frac{1}{2}\int dx_i\ \phi_j/D_j \tag{4}$$

This equation for the ϕ's is the simplest one for generating corrections to the isobar model which are consistent with subenergy unitary and analyticity. It is exactly the old Khuri-Treiman model [5], first introduced for $K\to 3\pi$ decay.

3. THE SINGLE VARIABLE INTEGRAL EQUATION GENERATED BY MINIMAL UNITARITY AND ANALYTICITY IN THE TWO BODY CHANNELS

As it stands, (4) is inconvenient because the right hand side involves the ϕ's under two integrals. But it can be transformed into an equation involving only a single integration [6]. Taking now the case of spinless identical particles of mass μ, the result is

$$\Phi(s_1,s) = C(s)/D(s_1) + \frac{1}{D(s_1)}\int_{-\infty}^{(\sqrt{s}-\mu)^2} \Delta^{(1)}(\lambda^2,s,s_1)\Phi(\lambda^2,s)d\lambda^2$$
$$+ \frac{1}{D(s_1)}\int_{-\infty}^{0}\left[\Delta^{(2)}(\lambda^2,s,s_1) + \Delta^{(3)}(\lambda^2,s,s_1)\right]\Phi(\lambda^2,s)d\lambda^2 \qquad (5)$$

where the $\Delta^{(i)}$ are certain kernels we shall return to later, and $\Phi = \phi/D$. This equation has some remarkable properties. Consider first the case N = constant (δ-function forces). Then, the non-relativistic limit of (5) reduces [6] to exactly the Faddeev equations appropriate to zero range interactions [14]. It is worth emphasising that this result is remarkable, because (5) was derived solely from considerations in the two body (subenergy) channels, while the Faddeev equations are of course explicitly constructed in the three body sector. This result encourages one to hope that (5) actually represents an interesting model for the relativistic three body problem.

The first property one would like to verify, if (5) is to be a candidate for predicting properties in the three body channel (s), is that of three body unitarity. It turns out, indeed, that Φ defined by (5) does satisfy appropriate three body discontinuity equations [7]. Thus [7] the ingredients of two body unitarity and analyticity have generated amplitudes which automatically satisfy the three body discontinuity relations.

This, however, is not quite enough. The Pasquiers pointed out [8] that the $3\to 3$ amplitude contained in the model at this stage, though it satisfied the three body discontinuity relations consistently, could nevertheless not be regarded as a fully satisfactory $3\to 3$ amplitude, because it is not symmetric (time reversal invariant). One simple reason for this can be seen [8] by considering the $3\to 3$ amplitude contained in the graph (c). If we are to obtain a true $3\to 3$ amplitude by cutting this graph at ①, we should get the Feynman graph (d).

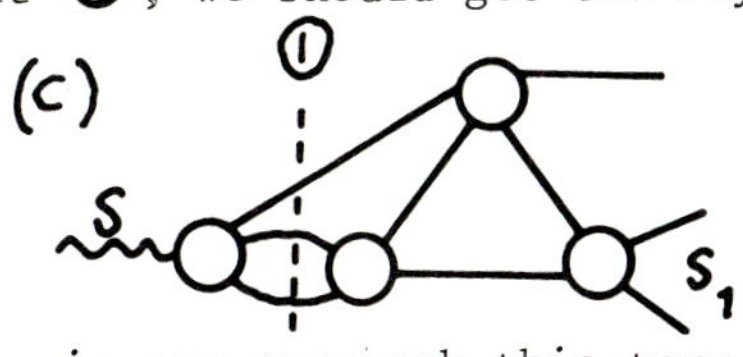

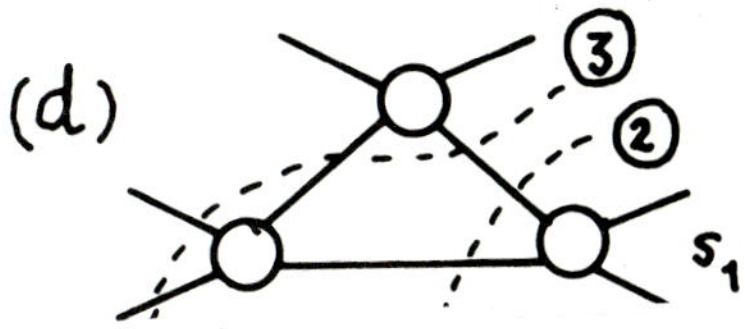

However, in our approach this term has effectively been constructed by dispersing in the two body variable s_1, including only the elastic cut ②. It is well known that if the full Feynman graph is to be reconstructed by dispersing in one channel, all cuts must be included. This would mean including the cut ③ in this case. This, however, corresponds to a singularity at $s_1=(4\mu)^2$ i.e. an inelastic threshold whereas we have neglected everything except the elastic threshold $s_1=(2\mu)^2$ (cut ②). Thus the $3\to 3$ amplitude contained in the model is not, in fact, quite equal to the expected sum of Feynman graphs, and turns out to be not symmetric. Thus one is led to seek modifications of (5) which embody a symmetric $3\to 3$ amplitude.

4. MODIFICATIONS OF THE K-T MODEL

For $N\neq$ constant, none of the kernels $\Delta^{(i)}$ is symmetric. Even for N = constant, $\Delta^{(2)}$ and $\Delta^{(3)}$ are not symmetric. However, by multiplying $\Delta^{(1)}$ by simple kinematical factors it is possible to make it symmetric if N = constant; in fact, it is then simply proportional to the J=0 projection of the OPE graph (with pointlike

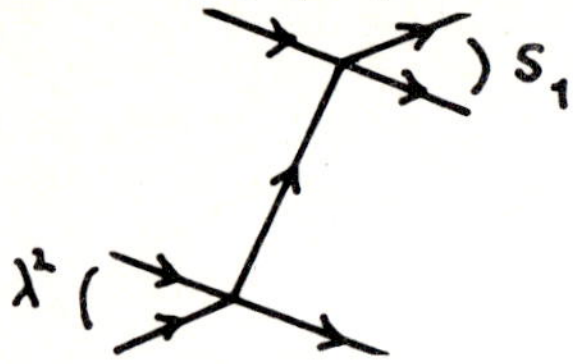

vertices). Now it is actually only $\Delta^{(1)}$ which is relevant to the three body discontinuity calculation, since the three body cut $\sqrt{s} \geq 3\mu$ corresponds to a part of the λ^2 integration in (5) that involves only $\Delta^{(1)}$. Thus it is tempting to suggest that (5) might be replaced by (for N = constant)

$$\Phi(s_1,s) = C(s)/D(s_1) + \frac{1}{D(s_1)} \int_{-\infty}^{(\sqrt{s}-\mu)^2} \Delta^{(1)}(\lambda^2,s,s_1)\Phi(\lambda^2,s_1)d\lambda^2 \qquad (6)$$

This equation indeed looks like an obvious relativistic generalisation of the Faddeev equations for zero range forces (and could be generalised to other J states by replacing $\Delta^{(1)}$ by the full unprojected OPE term). However, the fact that (5) incorporated the statement that Φ had only the normal threshold singularity $s_1=4\mu^2$ should lead us to suspect that the singularity structure of (6) is unsatisfactory, and this is so.

The trouble is that the kernel $\Delta^{(1)}$, as well as being singular on the boundary of the physical decay region, is also singular on another curve inside the domain of the variables s_1 and λ^2 covered by (6). The reason is that the OPE graph, as well as representing a physical process in the decay region, can also be physical for negative values of s_1 and λ^2. In that case these variables are momentum-transfer like, and the process is (below left):

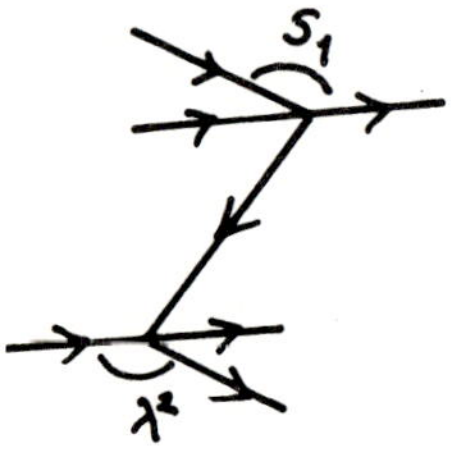

In this process the internal particle is "backward going" i.e. in non-covariant language it has negative energy. Because of the additional imaginary part (which is absent in the non-relativistic limit) $\Phi(s_1,s)$ as given by (6) will be complex even for s_1 below threshold [8]; furthermore, the trace of the kernel will be complex for $\sqrt{s}$ below 3μ. In the full equation (5), however, $\Delta^{(2)}$ acquires a compensating imaginary part for negative s_1 and λ^2 which just cancels that of $\Delta^{(1)}$, so that the reality properties of (5) are satisfactory.

One might consider eliminating this difficulty with (6) by simply restricting the λ^2 integral to run from α to $(\sqrt{s}-\mu)^2$ only, where $\alpha \geq 0$. Such "truncated" equations were considered by the Pasquiers [8], who concluded that $\alpha=0$ was the best choice. In that case the additional singularities in s_1 that one introduces lie in the region $s_1 \geq (\sqrt{s}+\mu)^2$. As the Pasquiers remark, this is in the inelastic region, so that we see again that symmetrisation of the kernel seems to be connected with the inclusion of inelastic states in the subenergy channel.

Another way out of these difficulties with (6) would be to simply eliminate the negative energy part of the OPE propagator. This produces an equation analogous to the Alessandrini-Omnes [15] relativistic version of the Faddeev equations, as modified by Basdevant and Kreps [16], and recently proposed again by Amado [17]. We will call this the "relativistic Faddeev" approach. The difficulty with it, as was pointed out by Basdevant and Omnes [18], is that now the analytic properties in E are unsatisfactory - specifically, there is a branch point at E=0 ($E = \sqrt{s}$).

There are indications that this singularity at E=0 can be important. Pasquier has used (5) (suitably generalised to include angular momentum effects) to look for the ω as a 3π resonance, including only the ρ state in the $\pi\pi$ channels. With one adjustable parameter, he succeeded in generating a narrow (~15 MeV width) ω state. It is noteworthy that Basdevant and Kreps [16] failed to find any sign of an ω using the "positive energy" propagator; Basdevant and Omnes [18] attribute this at least partly to the spurious singularity at E=0, and the numerical work of Pasquier supports this interpretation [19].

One hopes that in the one case the lack of symmetry, and in the other the singularity at E=0, will be unimportant in cases where the contribution from the decay region is enhanced by two body resonances.

I now want to report briefly on the non-zero angular momentum case.

5. ANGULAR MOMENTUM COMPLICATIONS

In their extension of the K-T equations to include angular momentum effects [20], the Pasquiers followed a method that had been used in the spinless case: the $1\to 3$ amplitude was considered as the analytic continuation in $\sqrt{s}$ of a $2\to 2$ process. Standard techniques can be applied to the latter, and the continuation performed to the decay region; in this way, kinematical constraints and singularities can be identified.

This method does not, however, lead immediately to an isobar model structure in the decay region, though the amplitudes can eventually be cast into that form. The point I wish to make here is that it is perhaps simpler to assume an isobar model structure for the barrier and angular factors in the first place, and study the two body unitarity and analyticity requirements on the resulting amplitudes, just as was done above for the spinless case. In this way we shall end up with equations for amplitudes $\Phi^{J\cdots}(s_1,s)$, where the ... indicates other angular momentum and spin labels. It is important to realize straightaway that this point of view is different from the conventional Faddeev one: the isobar angular momentum decomposition is done for each J^P separately, and there is no guarantee that the resulting equations for the $\Phi^{J\cdots}$'s are exactly the J^P projection of a single equation. In simple cases this does happen: the quantity $\Delta^{(1)}$ introduced above is the J=0 projection of the simple OPE graph, and if the three mesons only interacted in s-waves the kernels for the other J states would in that case be the appropriate J-projection of this graph. In general, however, even the "$\Delta^{(1)}$"-like pieces of the kernel are not precisely the appropriate projections of the indicated OPE graph.

We shall consider here only spinless particles, but include all other relative and total angular momentum effects. Using the Pasquier's notation, we write the amplitude for a+b→3 , with momenta $\underline{p}_1+\underline{p}_2 \to \underline{q}_1+\underline{q}_2+\underline{q}_3$, as

$$F = \sum_J (2J+1)\Big\{\sum_{\Lambda_1\ell_1} \mathcal{D}^{J*}_{\Lambda_1 0}(\beta_1,\alpha_1,0)(2\ell_1+1)d^{\ell_1}_{\Lambda_1 0}(\theta_{12})F_1^{J\ell_1\Lambda_1}(s_1,s)$$

$$+ \sum_{\Lambda_2\ell_2} \mathcal{D}^{J*}_{\Lambda_2 0}(\beta_2,\alpha_2,0)(2\ell_2+1)d^{\ell_2}_{\Lambda_2 0}(\theta_{23})F_2^{J\ell_2\Lambda_2}(s_2,s)\Big\}$$

where for simplicity we have assumed only two of the final state pairs interact strongly. Here ℓ_i is the j-k pair partial wave, the Λ_i are helicity labels, and $\alpha_{1(2)}$, $\beta_{1(2)}$ are the polar and azimuthal angles of $\underline{p}_1$ in the 3π c.m.s., with the z axis along $\underline{q}_{1(2)}$, $-\underline{q}_{2(3)}$ being in the x-z plane with positive x-component. θ_{12} is the angle between the momenta of particles 1 and 2 in the 2-3 c.m.s.

The discontinuity relation across the s_1 normal threshold cut is

$$\text{disc}_1\, F_1^{J\ell_1\Lambda_1}(s_1,s) = 2i\rho(s_1)\, M_1^{\ell_1 *}(s_1)F^{J\ell_1\Lambda_1}(s_1,s)$$

where $M_1^{\ell_1}$ is the two body elastic amplitude in the ℓ_1-wave, and

$$F^{J\ell_1\Lambda_1}(s_1,s) = \int d^{\ell_1}_{\Lambda_1 0}(\theta_{12})\mathcal{D}^J_{\Lambda_1 0}(\beta_1,\alpha_1,0)F\cdot\frac{\sin\alpha_1 d\alpha_1 d\beta_1}{4\pi}\,\frac{d\cos\theta_{12}}{2} \tag{7}$$

(note the vital distinction between $F^{J\ell_1\Lambda_1}$ and $F_1^{J\ell_1\Lambda_1}$!) Clearly, the right hand side of (7) involves a simple term $F_1^{J\ell_1\Lambda_1}$ (by orthogonality), but there is another term, the contribution, of course, of the other channel. To evaluate it, we have to find the relation between the angles α_2,β_2 and α_1,β_1. Actually, this is quite simple: the systems of axes in which the two sets of angles are defined are related to each other by a rotation. It turns out that we can write

$$\mathcal{D}^{J*}_{\Lambda_2 0}(\beta_2,\alpha_2,0) = \sum_M \mathcal{D}^{J*}_{\Lambda_2 M}(\chi_{12})\mathcal{D}^{J*}_{M0}(\beta_1,\alpha_1,0)$$

where χ_{12} is the angle between $-\underline{q}_1$ and $-\underline{q}_2$ in the three body c.m.s. Using this relation in (7) we find

$$\text{disc}_1\, F_1^{J\ell_1\Lambda_1}(s_1,s) = 2i\rho(s_1)M_1^{\ell_1 *}(s_1)F_1^{J\ell_1\Lambda_1}(s_1,s)$$

$$+\ 2i\rho(s_1)\, M_1^{\ell_1 *}(s_1) \sum_{\Lambda_2\ell_2} \int C^J_{\ell_1\Lambda_1\ell_2\Lambda_2} F_2^{J\ell_2\Lambda_2}(s_2,s)\frac{d\cos\theta_{12}}{2} \qquad (8)$$

where the recoupling coefficient $C^J_{\ell_1\Lambda_1\ell_2\Lambda_2}$ is given by

$$C^J_{\ell_1\Lambda_1\ell_2\Lambda_2} = (2\ell_2+1)\, d^{\ell_1}_{\Lambda_1 0}(\theta_{12})\, d^J_{\Lambda_2\Lambda_1}(\chi_{12})\, d^{\ell_2}_{\Lambda_2 0}(\theta_{23}) \qquad (9)$$

This is proportional to the Wick recoupling coefficient [21] for this case.

The first term of the right hand side of (8) can as usual be dealt with by factoring out $\left(D_1^{\ell_1}\right)^{-1}(s_1)$: writing

$$F_1^{J\ell_1\Lambda_1}(s_1,s) = \phi_1^{J\ell_1\Lambda_1}(s_1,s)/D_1^{\ell_1}(s_1) \qquad \text{we find}$$

$$\text{disc}_1\, \phi_1^{J\ell_1\Lambda_1}(s_1,s) = 2i\rho N_1 \sum_{\Lambda_2\ell_2} \int C^J_{\ell_1\Lambda_1\ell_2\Lambda_2}\phi_2^{J\ell_2\Lambda_2}\Big/D_2^{\ell_2}\cdot\frac{d\cos\theta_{12}}{2} \qquad (10)$$

A similar analysis can be done using the L-S scheme instead of helicity [12]. Here S is the relative orbital angular momentum of one pair (the previous ℓ) and L is the orbital angular momentum of the third particle relative to the c.m.s. of that pair. The relation between these two schemes is easily found, for the spinless case, by using the formulae given in Jacob and Wick [22]. As is always the case, the L-S formalism involves more Clebsch-Gordon sums - for instance, in the recoupling coefficient, which, as is clear from (9), is quite simple in the helicity formalism.

Following the analogy of the spinless case, the next step would be the insertion of (10) into a dispersion relation, so as to build in analyticity. But here there is a well-known snag - in reactions involving non-zero spins and angular momenta there are kinematical singularities and constraints to contend with. These difficulties were handled by the Pasquiers in the helicity formalism, using the standard analysis invented for the $2\to2$ reactions, and continuing it to the $1\to3$ situation. It turns out, however, that the threshold factors (particularly those involving L) that one would naturally use in the L-S scheme give exactly the same results, rather more physically. For example, the constraint found by the Pasquiers on a certain combination of the amplitudes F^{111} and F^{110} (that it had to vanish like $(s_1-(\sqrt{s}\pm\mu)^2)^1$ as $s_1 \to (\sqrt{s}\pm\mu)^2$) is simply the L=2 barrier factor.

Such barrier factors (in both S and L), when combined with the d functions, generally correspond exactly to factors extracted from the analysis of kinematic singularities and constraints. Discontinuity relations for the residual "kinematic-singularity-free" amplitudes can easily be derived from (10), and inserted into dispersion relations. The transformation to single variable form is most easily done by the method introduced by the Pasquiers [8]. One finds that the "$\Delta^{(1)}$" part of the kernel is not in general the J^P projection of the relevant OPE graph. At all events, the K-T kernel provides an unambiguous prescription in each case, and one that respects analyticity. There is, after all, no particular reason why the "$\Delta^{(1)}$" part of the kernel <u>should</u> be the J^P projection of an OPE process, and there are the "$\Delta^{(2)}$" and "$\Delta^{(3)}$" parts as well, in any case.

References

[1] G. Ascoli et al., Phys.Rev.D7(1973)669; Yu. M. Antipov et al., Nucl.Phys.B63 (1974)1.

[2] R. Aaron and R. D. Amado, Phys.Rev.Lett.31(1973)1157.

[3] G. N. Fleming, Phys.Rev.135(1964)B551; I. J. R. Aitchison and R. Pasquier, Phys.Rev.152(1966)1274; I. J. R. Aitchison, Il Nuov.Cim.51(1967)A249, 272.

[4] R. D. Amado, Phys.Rev.C11(1975)719.

[5] N. N. Khuri and S. B. Treiman, Phys.Rev.119(1960)1115.

[6] I. J. R. Aitchison, Phys.Rev.137(1965)B1070.

[7] I. J. R. Aitchison and R. Pasquier, Phys.Rev.152(1966)1274.

[8] R. Pasquier and J. Y. Pasquier, Phys.Rev.170(1968)1294; <u>ibid</u>.177(1969)2482; R. Pasquier, Thèse, Université Paris-Sud, Orsay, 1973.

[9] I. J. R. Aitchison, in <u>Three Particle Phase Shift Analysis and Meson Resonance Production</u>, Daresbury Study Weekend, 1-2 February 1975, Ed. J. B. Dainton and A. J. G. Hey, pp.81-95.

[10] M. O. Taha, Nuov.Cim.42A(1966)201; P. R. Graves-Morris, Nuov.Cim.50A(1967)681; Y. Goradia et al., LBL preprint; G. Smadja, LBL-382 (unpublished); G. Ascoli and H. W. Wyld, Illinois preprint, and talk presented by G. Ascoli at the 4th Int. Conf. on Exp. Meson Spectroscopy, Boston, 1974, APS No. 21.

[11] I. J. R. Aitchison and R. J. A. Golding, Phys.Lett.59B(1975)288.

[12] Cited in reference [10].

[13] R. Aaron et al., Northeastern University preprint, NUB 2251, April 1975.

[14] G. V. Skornyakov and K. A. Ter-Martyrosyan, J.E.T.P.4(1957)658.

[15] V. A. Alessandrini and R. L. Omnès, Phys.Rev.139(1965)B167.

[16] J. L. Basdevant and R. E. Kreps, Phys.Rev.141(1966)1398, 1404 and 1409.

[17] R. D. Amado, Phys.Rev.12C(1975)1354.

[18] J. L. Basdevant and R. L. Omnès, Phys.Rev.Lett.17(1966)775.

[19] R. Pasquier, Thèse, cited in [8].

[20] See the second of the references cited in [8].

[21] G. C. Wick, Ann.Phys.18(1962)65.

[22] M. Jacob and G. C. Wick, Ann.Phys.7(1959)404.

RESONANCE AMPLITUDES IN LIGHT NUCLEI*

G. Igo
Univ. of Calif., Los Angeles, CA 90024 USA

ABSTRACT

Experimental evidence for the existence of resonances in nuclei is tenuous with one exception. In this paper, recent attempts are described to obtain direct evidence in light nuclei from experiments on large angle elastic p-d scattering, and p-^{4}He elastic scattering in the interference region at relatively small momentum transfers.

Large Angle p-d Scattering: One of the first speculations that there might be on N* component in nuclei was made by Kerman and Kisslinger[1] in 1969 (K-K model). They noticed that the S and D amplitudes for nucleon pickup based on the Chew-Goldberger[2] model were inadequate to explain the magnitude of the cross section observed at 1 GeV[3] at backward angles in p-d elastic scattering. With the inclusion of a small N* component, they were able to fit the data qualitatively.

In order to test the validity of this description, we have recently made an analysis[4] of lower energy p-d large angle elastic scattering data. Based on the Kerman and Kisslinger (KK) model[1] and the KK model with second order corrections[5], we carried out the analysis of the results of a deuteron-proton (d-p) elastic scattering experiment[6] at large angles. The data shown in fig. 1 were obtained at deuteron energies of 291, 362, and 433 MeV, as well as additional proton-deuteron data at 140 MeV and 316 MeV. The possibility of determining the D state and D*-state probabilities from the fit to these data has been investigated. Most of the information on the probability of the D state P_D comes from phenomenological potentials which fit nucleon-nucleon data. Table I lists phenomenological potentials[7-11] considered in this work and the corresponding values of P_D, which vary from 4.5% to 7.5%. In addition the influence of possible N* components in the deuteron was assessed.

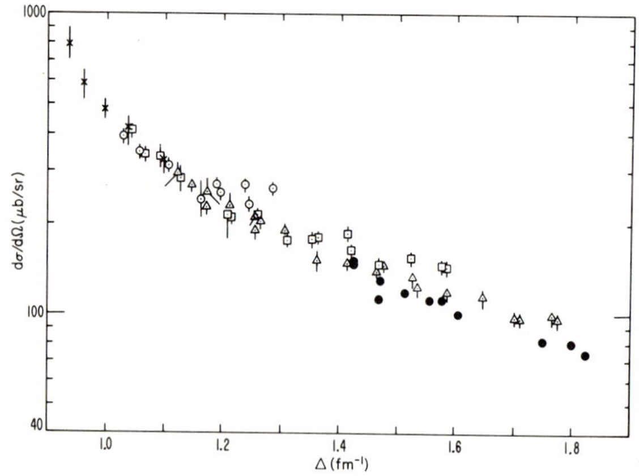

Fig. 1. A plot of $(d\sigma/d\Omega)_{c.m.}$ versus Δ for 140 MeV protons (crosses), 316 MeV protons (closed circles), 291 MeV deuterons (open circles), 362 MeV deuterons (squares), and 433 MeV deuterons (triangles) at large scattering angles in the c.m. system. The quantity Δ is the momentum transfer variable characteristic to the pick-up process.

Phenomenological potential	D-state probability P_D(%)
Bressel-Kerman-Rouben (BKR)	6.5
Hamada-Johnson (HJ)	7.0
Reid (soft core) (RS)	6.5
Lomon-Feshbach	
LF1	4.6
LF2	4.9
LF5	5.2
LF12	5.1
LF15	7.55

TABLE 1

Comparison of the prediction of modern phenomenological potentials

*Supported in part by ERDA.

Before describing the results of the calculations, it is appropriate to summarize the general features of large angle elastic scattering in the proton-deuteron system at energies above 150 MeV. Near 1 GeV, the backward rise beyond 130° in the d-p elastic scattering differential cross section is larger than would be predicted by single nucleon exchange (fig.2)[3,6] if the deuteron wavefunction consists of the triplet S- and D-states only. Kerman and Kisslinger proposed an explanation

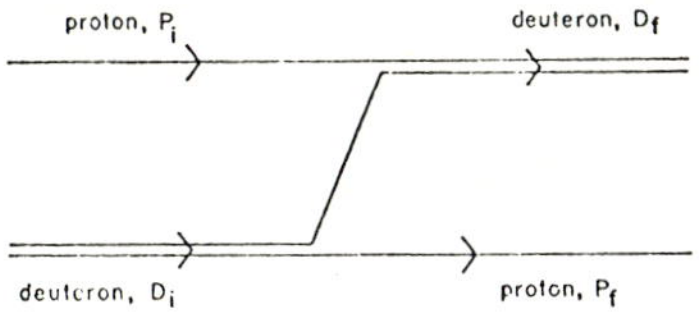

Fig. 2. The diagram represents single nucleon exchange.

Fig. 3. The diagram of the exchange of a neutron or an N*(1688).

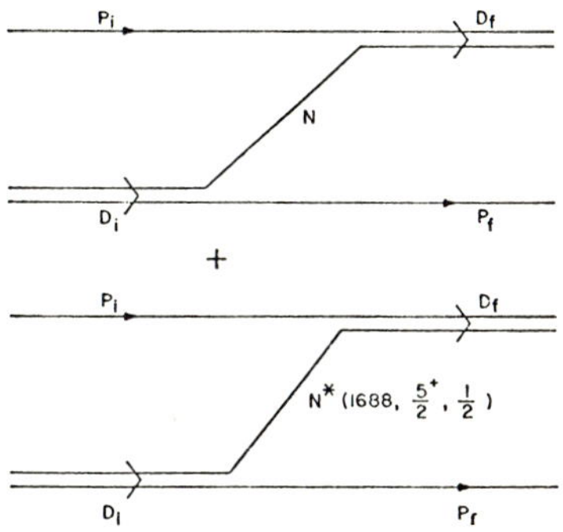

of this rise in the differential cross section. They assume that the ground state of the deuteron contains an admixture of a $|D^*\rangle$ configuration containing an N* (1688 MeV, spin = $\frac{5}{2}^+$, T = $\frac{1}{2}$). Particular importance is attached to the value of $\frac{5}{2}^+$ for the spin of N* (1688) since it leads to the L = 2 configuration. This makes the N*(1688) resonance important for the transfer process near 1 GeV. Based on a static field theory, Kerman and Kisslinger estimated the probability of the $|D^*\rangle$ in the ground state of the deuteron to be 1.0% - 2.0%. An expression for the differential cross section in the c.m. system was obtained by the generalization of the single nucleon exchange process to an exchange of either a neutron or N*(1688) (fig.3).

It has been pointed out[5)] that distorted waves (see fig. 4) in the initial and final channels would lead to a second order correction to the KK model which reduce the theoretical value of the differential cross section at 1 GeV. Since the KK model without distorted waves is markedly unsuccessful in fitting the data, we have investigated the effect of these corrections.

The differential cross sections for large angle d-p elastic scattering at 291, 362, and 433 MeV deuteron energies show the general behavior expected on the basis of the KK model[6)]. Data obtained at proton energies of 140 MeV and 316 MeV and at deuteron energies of 291, 362, and 433 MeV illustrated in fig. 1, are sensitive to Δ = 0.9 to 1.8 fm^{-1} (Δ is the momentum transfer in the pickup reaction). The general behavior of the Fourier-Bessel transformation of the deuteron wavefunction [$I_S(\Delta)$, $I_D(\Delta)$ and $I_{D^*}(\Delta)$] is that the S-state dominates the region of $\Delta < 1.3$ fm^{-1},

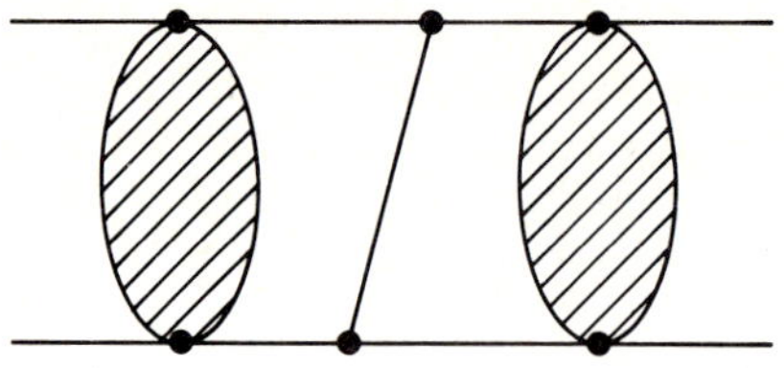

Fig. 4. The distorted wave diagram.

the D-state dominates the region between 1.3 and 1.8 fm^{-1}, and the D* state becomes increasingly significant at large values of Δ.

The calculated cross section, based on KK model employing (i) a deuteron wave function derived from one of various phenomenological potentials (with varying D-state probability) and (ii) a |D*> component of 2.0% show a characteristic deviation from the data in the region of Δ where the S-state is dominant (see for example fig.5, solid line). This is a general feature found for all the phenomenological potentials that we tried: RS, BKR, LF1, LF5, and LF15 (identified in table 1) and MFIII. MFIII is the Moravcsik analytic fit to the Gartenhaus wavefunction, the third approximation[10].

The deuteron wavefunctions for S-and D-states derived from LF1 and LF15 with a D-state probability of 4.6% and 7.55% respectively and MFIII for S- and D-state (probability of 6.7%) are used in all calculations. The calculted differential cross sections and the three deuteron wavefunctions, mentioned before, are shown in figs. 5-7 by a solid line, with a D* component of 2.0%. These figures (5-7, solid line) show without doubt that the discrepancy at small Δ(<1.3 fm^{-1}) exists for a variety of deuteron wave functions.

It is pointed out above that second order corrections tend to reduce the theoretical cross section calculated on the basis of first order considerations. It is shown, in ref. 5 that the cross section at 1.0 GeV calculated with second order corrections is reduced by about 20%. Therefore the second order correction might be expected to be large at lower energies.

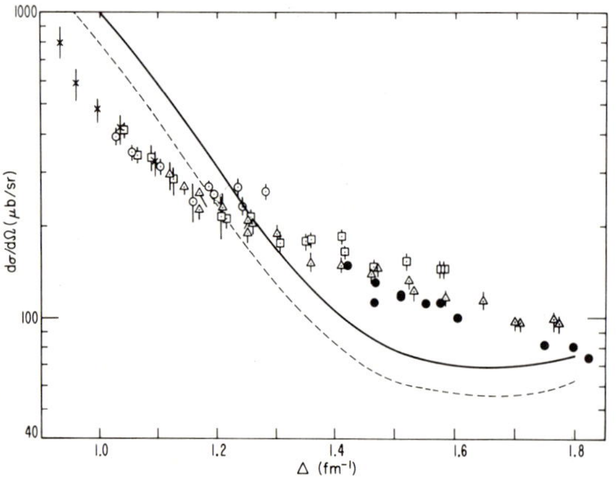

Fig. 5. A plot of (dσ/dΩ)c.m., which had been calculated with use of the deuteron wavefunction derived from LF1 with a D-state component of 4.6% and D* state component of 2.0%, versus Δ. The solid line represents the first order calculations only, while the dashed line includes the second order corrections.

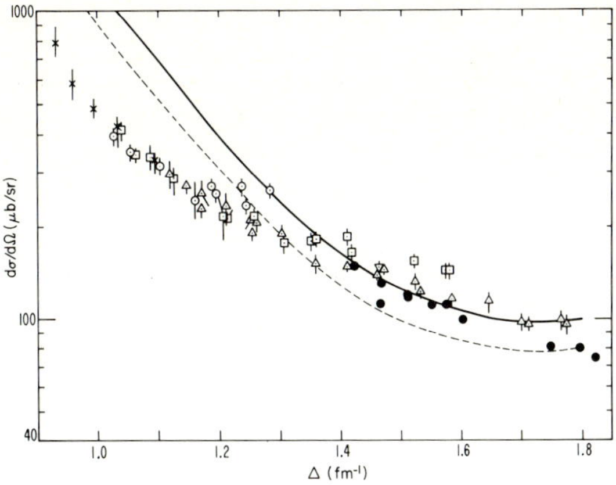

Fig. 6. A plot of (dσ/dΩ)c.m., which had been calculated from Moravcsik fit III wavefunction with a D-state component of 6.7% and D* state component of 2.0%, versus Δ. The solid line represents the first order calculation only, while the dashed line includes the second order corrections.

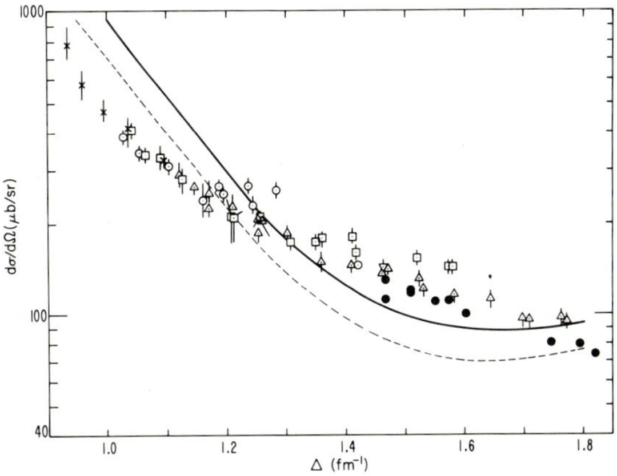

Fig. 7. A plot of (dσ/dΩ)c.m., which had been calculated with use of the deuteron wavefunction derived from LF15 with a D-state component of 7.55% and D* state component of 2.0%, versus Δ. The solid line represents the first order calculations only, while the dashed line includes the second order corrections.

Figs. 5-7 (dashed line) show the calculated cross section with second order corrections taken into account. The deuteron wavefunctions derived from LF1 and LF15 are employed in the calculation presented in figs. 5 and 7 respectively. The MFIII deuteron wavefunction is used in the calculation presented in fig. 6. The introduction of second order corrections into the calculation of the differential cross section tends to destroy the fit in the region of large Δ (1.3-1.8 fm^{-1}) and, though it reduces the cross section in the region of small Δ (1.3-1.3 fm^{-1}), the discrepancy there still exists.

The effect of changing the probability of the D* state is seen to influence the cross section at the large values of Δ (1.4-1.8 fm^{-1}) as depicted in figs. 8-10 for the representative wavefunctions [LF1 (P_D = 4.6%), MFIII (P_D = 6.7%), LF15 (P_D = 7.55%)]. We conclude that the Kerman-Kisslinger model employing a deuteron wavefunction derived from phenomenological potential is unsuccessful in fitting the large angle d-p elastic scattering data at deuteron energies of 291, 362, and 443 MeV and at proton energies of 140 and 316 MeV. The inclusion of second order corrections is not sufficient to remove the significant discrepancy at small values of Δ, and thus we are unable to assign unambiguous magnitudes to the D and D* components in the deuteron.

<u>P-^{4}He Elastic Scattering at Small Momentum Transfer</u>: Palevsky et al.[12] and Baker et al.[13] have measured p-^{4}He scattering near 1 GeV using large magnetic spectrometers. More recently absolute values for the differential cross sections, which were not reported in reference 2, have been obtained by the Saclay group[14]. The two measurements[12,14] agree rather well except in the region of the interference minimum at $-t \approx 0.25 (GeV/c)^2$. In view of the theoretical interest in p-^{4}He scattering in the 1 GeV region[15-19], we have studied the t dependence of p-^{4}He elastic

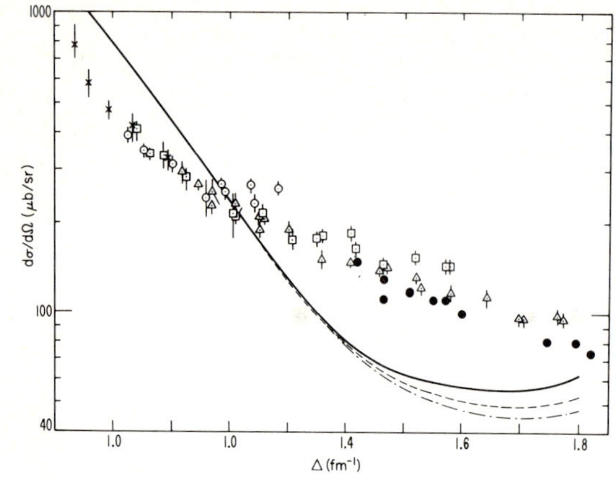

Fig. 8. A plot of (dσ/Ωd)c.m., which had been calculated with use of the deuteron wavefunction derived from LF1 with a D-state component of 4.6% and the second order corrections included, versus Δ. The solid line represents a D* state component of 2.0%. The dashed line represents a D* state component of 1.8%. The dashed-dotted line represents a D* state component of 1.7%.

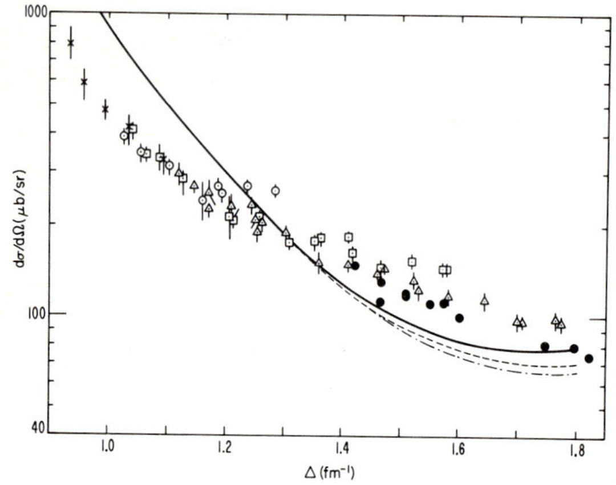

Fig. 9. A plot of (dσ/dΩ)c.m., which had been calculated with use of Moravcsik (fit III) wavefunction with a D-state component of 6.7% and the second corrections included, versus Δ. The solid line represents a D* state component of 2.0%. The dashed line represents a D* state component of 1.8%. The dashed-dotted line represents a D* state component of 1.7%.

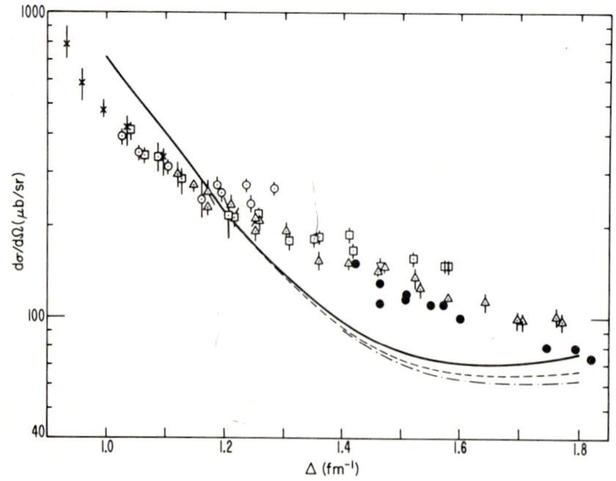

Fig. 10. A plot of (dσ/dΩ)c.m. versus Δ. This was calculated with use of the deuteron wavefunction derived from LF15 with a D-state component of 7.55% and the second order corrections included. The solid line represents a D* state component of 2.0%. The dashed line represents a D* state component of 1.8%. The dashed-dotted line represents a D* state component of 1.7%.

scattering in the interference region at 0.58 and 0.72 GeV using a recoil alpha particle technique and proton beams from the 184" synchrocyclotron. In addition we have completed the data taking (but not the analysis) of p-He^4 elastic scattering at 400 MeV, 1 GeV, 2.68 GeV, and 4.89 GeV at the Bevatron. In the region of the first minimum [$t \sim 0.25$ $(GeV/c)^2$], a number of other measurements have been made recently. From SACLAY, there is new data at 0.35, 0.65, and 1.54 GeV (absolute cross sections)[14].

The UCLA-LBL recoil measurements[20] near 1 GeV referred to earlier were directed to the question of the depth of the minimum of the 1 GeV region. In Fig. 11. The measurement at 0.59 GeV is compared with the older data from SREL[21)] at that energy. There is fair agreement between the data sets, particularly when the uncertainty in the absolute normalization is considered. The absolute uncertainty in the cross sections for the UCLA-LBL data is 5% (gas target). Since a liquid target was used in the SREL experiment, the uncertainty in the absolute scale may be larger.

Pursuing the question of the depth of the minimum at 1 GeV, it is convenient to remove some of the dependence on bombarding energy or on the invariant quantity s, the total energy. If the invariant $d\sigma/dt$ versus t is plotted, explicit dependence on beam momentum disappears in multiple scattering theory formulations. Only the rather slowly varying effects associated with the s-dependence of the nucleon-nucleon amplitude cause difference. This is therefore a convenient representation when it is desired to compare p-He^4 elastic data at different energies. The data available on the s-dependence of p-He^4 elastic scattering near the first minimum is summarized in Fig. 12. The SREL data has not been included because the error uncertainties are larger than in the 590 MeV UCLA-LBL data set. It is seen that the UCLA-LBL data at 0.72 GeV also displays a relatively shallow minimum in good agreement with the trend seen in the Saclay data and considerably shallower than the BNL data.

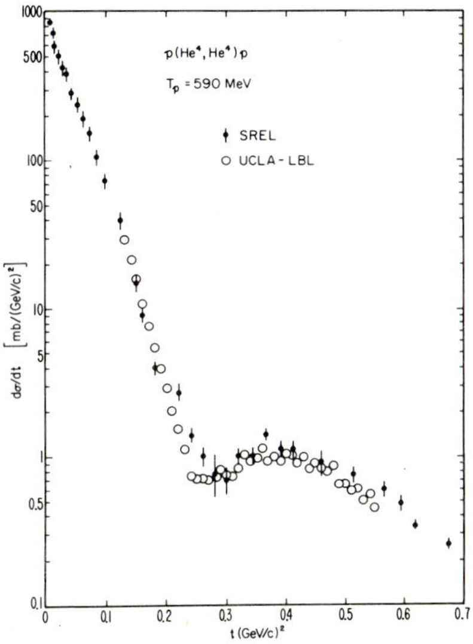

Fig. 11. p(^{4}He,^{4}He)p at 0.59 GeV.

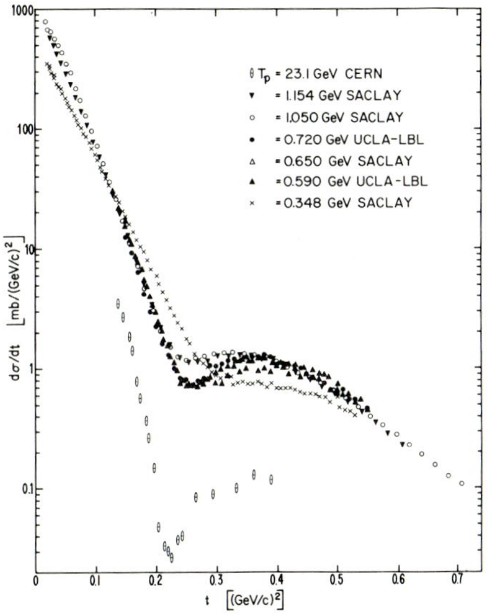

Fig. 12. The dependence of elastic scattering of protons from helium on bombarding energy. The Saclay data has been normalized by a factor of 0.9.

The well documented shallow minimum observed in p-He^4 scattering at 1 GeV is a very interesting phenomena. In looking closely at the data (Fig. 12), the ratio R of the height of the second peak to the depth of the dip reaches a maximum near .65 GeV. Someplace between 1.15 GeV and 23.1 GeV, the ratio increases again. Thus measurements in progress above 1 GeV at the Bevatron are of particular interest. In Fig. 13, the difference between the smoothly varying 0.35 GeV cross section and the cross sections at the same t at 1.15, 1.05, 0.72, 0.65 and 0.59 GeV normalized by the cross section at 0.35 GeV are plotted versus t. The minimum near 0.23 $(GeV/c)^2$ is associated with the depth of the first minimum and the maximum near 0.33 $(GeV/c)^2$ is associated with the second maximum (at t_{max}) in the cross section. The trend of R and t_{max} are evident in this plot.

The cross section has been calculated[22] using the Glauber Approximateion[18] at five energies. The object was to see if the behavior of R and t_{max} could be reproduced qualitatively. In the calculations a spin averaged nucleon-nucleon amplitude of the form

$$f_{NN}(t) = i\frac{\sigma_{NN}}{4\pi}(1-i\,\alpha_{NN})\,\exp\,[b_{NN}t] \qquad (1)$$

has been adopted, using the empirical energy dependence of the slope parameter b_{NN}, σ_{NN}, and α_{NN}. The helium form factor of Frosch et al.[23] has been used to represent He^4. The results are shown in Fig. 14. The behavior of R is qualitatively reproduced. The first minimum is shallow at 1.15 and 1.05 GeV and also at 0.35 GeV. It is much deeper at 0.65 and 0.59 GeV. It is also evident that the fits are far from adequate.

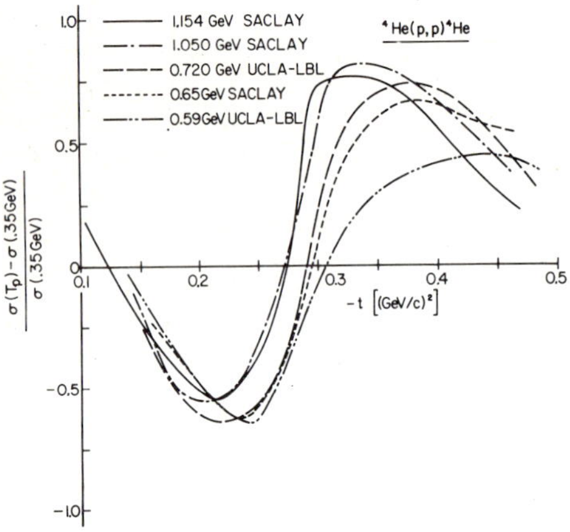

Fig. 13. The differences between the cross section at 1.15, 1.05, 0.72, 0.65, and 0.59 GeV and the cross section at 0.35 GeV normalized by the latter.

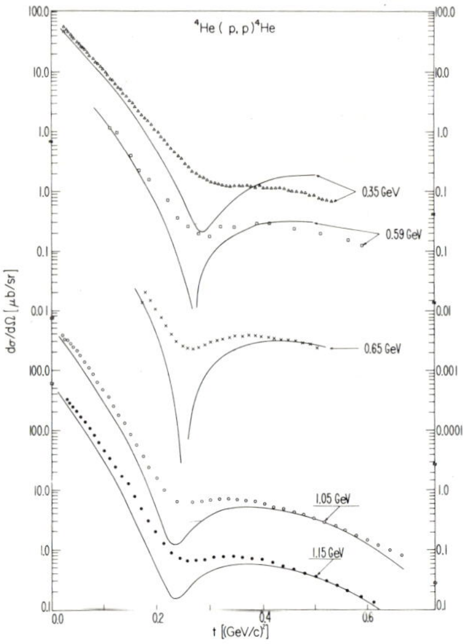

Fig. 14. Comparison of calculations using the Glauber Approximation and the p-He^4 elastic scattering data.

The theoretical picture on p-He^4 is not clear. Most of the early theoretical effort was made to fit the 1 GeV data from the Cosmotron or the Saclay data without absolute normalization; and therefore do not satisfactorily explain the new data. The importance of polarization in p-He^4 scattering in this energy region is indicated by the large polarizations which have been measured to date, up to about 600 MeV.[24] Boridy and Feshbach[25] have recently included the spin-spin term in the nucleon-nucleon scattering amplitude in the first order Raleigh-Lax term in the optical potential. It has been shown earlier that the spin-orbit potential is important (Franco,[26] Kujawski,[27] Lambert and Feshbach,[28] Lykasov and Tarasov[29]). This is illustrated in Fig. 15 from the work of Lambert and Feshbach. The dashed curve is calculated with a spin-averaged nucleon-nucleon amplitude. Including spin dependence substantially increases the cross section in the region where double scattering predominates. The effect of the density-dependent term is clearly significant in the same region of t. The dot-dashed line in Fig. 15 is calculated without it. Although spin dependent effects, two-body and three-body correlations are important, they generally fail to reduce R sufficiently to match the data. Recent calculations with potential models by Coker and Hoffman[30] and Clark et al.[31] reinforce this conclusion. It is necessary to make a disclaimer here, because the spin dependence of the nucleon-nucleon amplitudes are quite ambiguous near 1 GeV.

The effect of N* amplitudes have not been considered until recently.[32,33] Wallace has discussed their influence on the depths of the minimum in a recent contribution.[32]

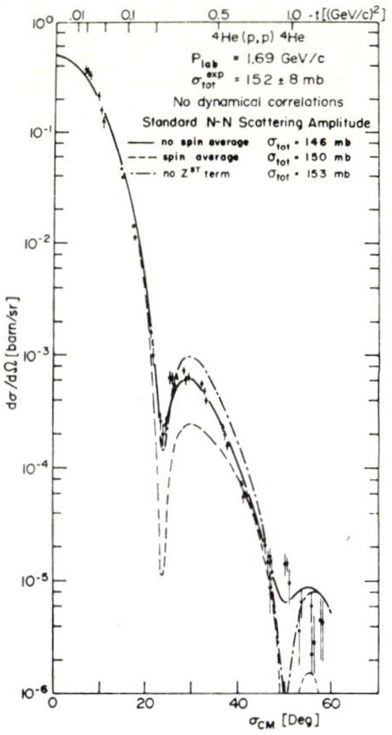

Fig. 15. The differential cross section in p-He^4 scattering at 1 GeV.

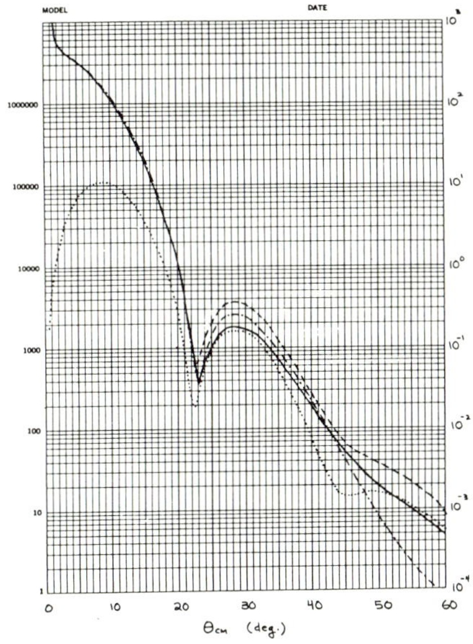

Fig. 16. The four curves of $d\sigma/d\Omega$(right vertical scale, units mb/sr) versus θ_{cm} are: -•-•- Glauber spin non-flip amplitude $|F_o|^2$; ———Glauber plus leading corrections spin non-flip amplitude $|F_o+F_1|^2$; •••••••••Glauber spin-flip amplitude $|G_o|^2$; --------- resultant total elastic scattering $|F_o+F_1|^2 + |G_o|^2$.

The High Energy Expansion[34]) is used which has been shown to be valid over an angular range easily encompassing the region of the interference minimum at 1 GeV. The charge form factor of helium is represented by the sum of two Gaussian distributions adjusted to reproduce the minimum at -t -0.4 $(GeV/c)^2$. Fig. 16 illustrates the importance of spin-flip and corrections to Glauber Theory. None of these curves has either the N* amplitude or short range correlations. Fig. 17 shows the N* effect. The N* amplitude is based on σ_R = 23 mb which means the N* production is strong enough to account for 23 mb inelastic N-N cross section. If only part of inelastic scattering is coherent N*, this should be reduced. Fig. 18 shows the effect of reducing σ_R to 10 mb. Another way of filling in the minimum is by exploiting the phase uncertainty in the nucleon-nucleon amplitudes. The effect of exploiting phase uncertainty of elementary nucleon-nucleon amplitudes to fit Saclay data is now shown. The dot-dash arrow is total scattering with no N* amplitude but with a phase factor, $e^{-i1.61t}$(-t in $(GeV/c)^2$) in the pp and pn amplitudes. This curve shows a fit can be obtained to the shoulder at 25° by choice of phase but that near 40° the previous good fit is lost.

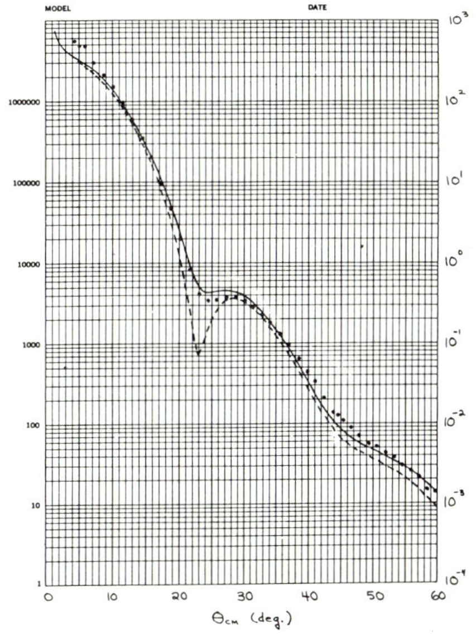

Fig. 17. The right vertical scale is $d\sigma/d\Omega$ in mb./sr. and the curves are: ------- same as resultant scattering of Fig. 2; ——— resultant scattering including N* amplitude of Ikeda: $|F_o+F_1+F^*|^2 + |G_o|^2$; ••••••••• Saclay data using original normalization.

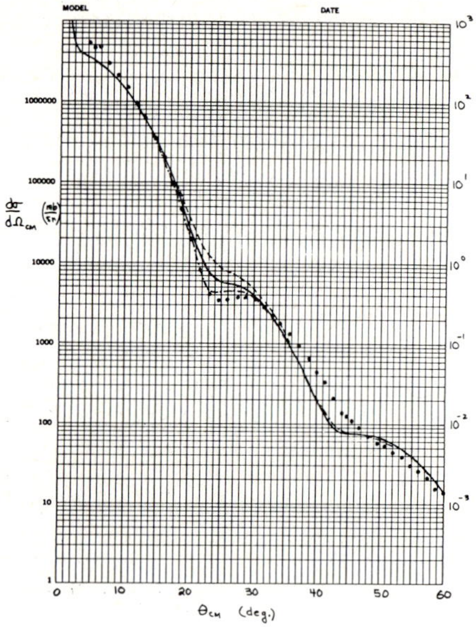

Fig. 18. Effect of a phase uncertainty —·—·—·, N* giving σ_R = 10 mb ————, and N* giving σ_R = 23 mb -------.

References

[1] A. K. Kerman and L. S. Kisslinger, Phys. Rev. 180 (1969) 1483.

[2] G. Chew and M. Goldberger, Phys. Rev. 77 (1950) 470.

[3] G. W. Bennett, J. L. Friedes, H. Palesky, R. J. Sutter, G. J. Igo, W. D. Simpson, G. C. Phillips, R.L.Stearns and D. M. Corley, Phys. Rev. Lett. 19 (1967) 387.

[4] M. Nasser, G. Igo, V. Perez-Mendez, Nucl. Phys. A229, 113 (1974).

[5] T. Fukushima, Ph.D. dissertation, Carnegie-Mellon University, 1972, unpublished.

[6] G. J. Igo, J. C. Fong, S. L. Verbeck, M. Goitein, D. L. Hendrie, J. B. Carroll, B. Macdonald, A. Stetz and M. C. Makino, Nucl. Phys. A195 (1972) 33.

[7] C.N. Bressel, A.K. Kerman and B. Rouben, Nucl. Phys. A124 (1969) 624.

[8] R.V. Reid, Ann. of Phys. 50 (1968) 511.

[9] T. Hamada and I.D. Johnston, Nucl. Phys. 34 (1962) 382.

[10] M.J. Moravcsik, Nucl. Phys. 7 (1958) 113.

[11] E.L. Lomon and H. Feshbach, Ann. of Phys. 48 (1968) 94.

[12] H. Palevsky, J.L. Friedes, R.J. Sutter, G.W. Bennett, G.J. Igo, W.D. Simpson, G.C. Phillips, D.M. Corley, N.S. Wall, R.L. Stearns, and B. Gottschalk, Phys. Rev. Letters 18 (1967) 1200.

[13] S.D. Baker, R. Beurtey, G. Bruge, A. Chaumeaux, J.M. Durand, J.C. Faivre, J.M. Fontaine, D. Garreta, D. Legrand, J. Sandinos, J. Thirion, R. Bertini, F. Brochard, and F. Hibou, Phys. Rev. Letters 32 (1974) 839.

[14] E. Aslanides, T. Bauer, R. Bertini, R. Beurtey, A. Boudard, F. Brochard, G. Bruge, A. Chaumeaux, H. Catz, J.M. Fontaine, R. Frascaria, P. Gorodetzky, J. Guyot, F. Hibou, M. Matoba, Y. Terrien, and J. Thirion, Contribution IV. A14, Sixth International Conference on High Energy Physics and Nuclear Structure, Santa Fe, New Mexico, June 1975.

[15] K.M. Watson, Phys. Rev. 89 (1953) 575; and Rev. Mod. Phys. 30 (1958) 565.

[16] A.K. Kerman, H. McManus, and R.M. Thaler, Ann. Phys. (N.Y.) 8 (1959) 551.

[17] H. Feshbach and J. Hufner, Ann. Phys. (N.Y.) 56 (1970) 268; H. Feshbach, A. Gal, and J. Hufner, ibid. 66 (1971) 20; E. Lambert and H. Feshbach, ibid. 76 (1973) 80; H. Feshbach and J.J. Ullo, ibid. 82 (1974) 156.

[18] R.J. Glauber, in High Energy Physics and Nuclear Structure, edited by G. Alexander (North-Holland, Amsterdam, 1967), p. 311; and in Lectures in Theoretical Physics, Vol. 1, edited by W.E. Brittin and L.G. Dunham (Wiley-Interscience, New York, 1959), p. 315.

[19] R.H. Bassel and C. Wilkin, Phys. Rev. 174 (1968) 1179.

[20] S.L. Verbeck, J.C. Fong, G. Igo, C.A. Whitten, Jr., D.L. Hendrie, Y. Terrien,

V. Perez-Mendez and G.W. Hoffmann, Nucl. Phys. accepted for publication.
[21] E.T. Boschitz, W.K. Roberts, J.S. Vincent, M. Blecher, K. Gotow, P.C. Gugelot, C.F. Perdrisat, L.W. Swenson, and J.R. Priest, Phys. Rev. C6 457 (1972).
[22] G.D. Alkhasov, private communication (1975).
[23] R.F. Frosch et al., Phys. Rev. 160, 874 (1967).
[24] W.T.H. van Oers, Proc. of the International Conference on Few Body Problems, Quebec, 1974.
[25] E. Boridy and H. Feshbach, Contribution to the VI International Conference on High Energy Physics and Nuclear Structure, Santa Fe, June 9-14, 1975.
[26] V. Franco, Phys. Rev. Letters 21, 1360 (1968).
[27] E. Kujawski, Phys. Rev. C1, 1651 (1970).
[28] E. Lambert and H. Feshbach, Ann. Phys. 76, 80 (1973).
[29] G.I. Lykasov and A.V. Tarasov, Sov. J. Nucl. Phys. 17, (153 (1973).
[30] W.R. Coker and G.W. Hoffman, Bull. APS II 20, 1166 (1975).
[31] B.C. Clark, Bull. APS II 20, 1192 (1975).
[32] S.J. Wallace, Bull. APS II 20, 1194 (1975).
[33] M. Ikeda, Phys. Rev. C6, 1608 (1972).
[34] S.J. Wallace, Phys. Rev. C12 (1975) 179.

N*'s IN THE DEUTERON GROUND STATE.

H.J. Weber
University of Virginia, Charlottesville, Va. 22901, USA

Abstract:
The development over the past year of spectator experiments in search for isobars in the deuteron ground state is reviewed as well as nonrelativistic coupled-channel calculations of isobar configurations and a relativistic three-body model.

In the past few years models involving nucleon resonances, i.e. isobars or N^*'s, as possible nuclear constituents have been studied vigorously [1]. These models may be regarded as descriptions of the off-mass-shell behavior of bound nucleons and are attempts to shed light on the structure of nuclei in the regime of small inter-nucleon distances which may be studied with present medium and high energy facilities. In such kinematic configurations one expects indeed to measure the mutual polarization of nucleons in close collisions which, via the processes $NN \rightarrow (NN^*$ or $N^*N^*) \rightarrow NN$, reveal their internal structure and complex excitation spectrum.

Experiments.

There is some, albeit indirect, evidence for the presence of $\Delta_{33}(1236)$ in the deuteron from electromagnetic processes, e.g. radiative neutron capture near threshold [2]. Yet, one would like to prove directly the presence of N^*'s in the ground state by shaking loose an N^* that pre-existed in the deuteron. Several such spectator experiments that were designed to probe the ($\Delta\Delta$) configuration have recently been reported. M. Goldhaber et al. [3] analyzed quasi-elastic scattering of $\pi^\pm$ and K^+ mesons off the Δ^- constituent in $d=\Delta^{++}\Delta^-$ at 15(12) GeV/c beam momentum. Similar studies were then done for $\pi^+ + (d=\Delta\Delta) \rightarrow \pi^+\Delta\Delta$ at 4 GeV/c [4], $\bar{p} + (d=\Delta\Delta) \rightarrow \bar{p}\Delta\Delta$ at 5.5 GeV/c [5] and $p + (d=\Delta\Delta) \rightarrow \Delta NN$ at 3.33 GeV/c [6]. Each experiment recorded a significant number of events involving a (πN) system in the $\Delta(1236)$ mass region recoiling backward in the lab. A backward Δ in the lab was thought very unlikely to be produced in the reaction so that such Δ's might have pre-existed in the deuteron. Thus it came as a stunning surprise when the angular distribution of the πN decay products in the Δ-spectator rest frame consistently revealed considerable forward-backward asymmetry, thereby demonstrating strong background contamination to any genuine, possibly observed Δ-spectator signal. Clearly then, backward production of a Δ in the lab by an incident high energy projectile necessarily occurs, yet is not sufficient as proof that the Δ pre-existed in the deuteron. Thus only an upper limit of 0.7 to 0.9% total ($\Delta\Delta$) probability could be extracted from the data.

Two methods have since been suggested to suppress the background of nonresonant (πN) systems. First, nucleon spectators in this background have the momentum distribution of the deuteron (np) wave function which peaks at zero, while the ($\Delta\Delta$) distribution is broader and centered at $\sim$ 0.4 GeV/c. Hence eliminating all nucleons with momentum $\lesssim$ 0.2 GeV/c will not cause a significant loss of Δ-spectator events but reduces the (πN) noise drastically [7]. This technique was successfully used by Benz and Söding [8] who studied the inclusive reaction $\gamma d \rightarrow \Delta^{++}(1236)$ + anything with the DESY bremsstrahlung beam at 5.5 GeV maximum energy, the signature being a Δ^{++} recoiling backward in the lab. In fact, a whole ridge of concentrated background events in a mass versus momentum plot of ($p\pi^+$) spectator events disappears when spectator protons with momentum $\lesssim$ 0.2 GeV/c are eliminated. The remaining events are taken to represent the mass distribution of Δ^{++} spectators; it has a maximum at 1.2 GeV/c^2 and its shape agrees with the $\Delta\Delta$ model. The Δ-spectator azimuthal and polar decay angular distributions now are nearly isotropic rather than asymmetric. It is also consistent with a spectator interpretation that there is no correlation between the Δ^{++} momentum, its mass or decay angles, or the incident photon energy. The measured momentum distribution is in reasonable agreement with the ($\Delta\Delta$) wave function. However, an isobar probability of $\sim$ 3% in the deuteron is extracted from the data: The inclusive reaction measures not only the ($\Delta\Delta$) probability but also a weighted sum of all other (NN^*) configurations (with qualitatively similar wave functions) whose N^* decays via

Δ^{++} which then is counted as a genuine Δ-spectator.

The second method consists in choosing kinematic regions where the nonresonant background is minimal and can be separately observed. This has been done [9] for forward proton production in π^--deuteron collisions, $\pi^- + (d=\Delta^{++}\Delta^-) \rightarrow p\Delta^-$ whose Δ^--spectator signal hinges on Δ^{++} exchange [10], while the background is mostly caused by quasi-elastic backscattering of the incident π^- from the bound proton in (d=np). Thus a backscattered π^- measured in coincidence with the incident-π^--forward-proton event is counted as background, whereas the absence of such a π^- is taken as the signature of a Δ^- spectator event. The Saclay π^- beam was used at $\sim$ 1.1 GeV/c where the π^-p backward elastic cross section has a sharp minimum which remains when the Fermi motion of the bound neutron is included. The ($\Delta\Delta$) configuration would fill the dip. The observed differential cross section peaks at 0° proton lab angle in agreement with the ($\Delta\Delta$) model [10]. In contrast, the free π^-p distribution whose shape closely resembles the quasi-elastic cross section from a deuterium target has a minimum for 0° protons. It appears that the background and Δ^-spectator signal have indeed been separated. This leads to a ($\Delta\Delta$) probability of 0.4%, as yet uncorrected for π^- multiple scattering. At 2.1 GeV/c where the π^-p backward elastic cross section has its deepest minimum the situation would look even better considering that the background could be further suppressed by eliminating all events with a slow spectator neutron. An analysis of $\pi^-d \rightarrow p\Delta^-$ at 1.68 GeV/c by a Dubna group [11] remains inconclusive because neither the crucial slope of the proton angular distribution nor the energy dependence were measured.

Nonrelativistic Theory.

The observed ($\Delta\Delta$) and (NN^*) probabilities between 0.4 and 3% rely on theoretical predictions for their wave functions which vary appreciably and thus increase the uncertainty of measured N^* probabilities. Estimates for the dominant ($\Delta\Delta$) configuration are based on the impulse approximation and nonrelativistic kinematics in conjunction with a OPE ($NN \rightarrow \Delta\Delta$) transition potential [12]. Angular momentum, isospin and parity selection rules determine the ($\Delta\Delta$) and (NN^*) (L,S) components. E.g. for ($\Delta\Delta$) there is an S-wave (0,1), two D-waves (2,1), (2,3) being dominant and a G-wave (4,3). Compared with the (np) deuteron S and D states, the isobar wave functions peak at higher momentum ($p_F \sim 0.35$ to 0.6 GeV/c) and fall off much more slowly at high momentum. Inclusion of ρ-meson exchange [13] weakens the tensor transition potential, decreases the ($\Delta\Delta$) probability and removes much of the strong dependence on short-range cut-off form factors. The medium range attraction of the NN potential involved must now be modified due to the additional attraction provided by the dispersion contribution with one or two intermediate Δ's. Recently nonrelativistic coupled-channel calculations have been carried out for (NN^*), ($\Delta\Delta$) and (np) configurations in the OPE approximation, coupling a single LS-channel of one isobar configuration to the ^3S and ^3D states of the (np) configuration at a time [14]. The conclusion is that the impulse approximation (IA) overestimates by a factor of 10 the single-$\Delta\Delta$-channel calculation (SCC). However, this depends crucially on the too strong diagonal $\Delta\Delta$ repulsion used involving a ($\Delta\Delta\pi$) coupling larger by a factor of 25 than the quark model prediction. Arenhövel [15] has done a full coupled-channel calculation and finds that the SCC approximation is grossly misleading in this case: The tensor force is now so strong in the full coupled channel case that the deuteron would consist of Δ's to 99% with a binding energy of 10^4 GeV! With quark model estimates for the relevant coupling constants, IA and CC actually give very similar results. From Table 1 we see that both ρ-exchange and the diagonal $\Delta\Delta$ interaction which is repulsive in all $\Delta\Delta$ channels decrease the $\Delta\Delta$ probability. The medium range attraction of the central Reid soft-core NN potential has been varied to fit the deuteron binding energy. The (np)^3D state probability is $\sim$ 6.4%.

NN→ΔΔ	*ΔΔ→ΔΔ*		*method*	*ΔΔ-prob. [%]*				
ρ	π	$\rho+\omega$		^3S	^3D	^7D	^7G	*total*
—	—	—	*IA*	.038	.012	.878	.073	1.00
—	—	—	*CC*	.036	.013	.856	.062	.97
—	√	—	*CC*	.023	.014	.649	.060	.74
√	—	—	*IA*	.075	.014	.476	.040	.60
√	—	—	*CC*	.089	.014	.458	.036	.60

✓		✓	✓	CC	.059	.017	.344	.034	.45

Table 1. (ΔΔ) probabilities from different NN→ΔΔ and ΔΔ→ΔΔ potentials including (marked by ✓) π, ρ, ω exchange.

A Relativistic Model for (NN^*) Configurations.

Estimating the (ΔΔd) vertex by a nonrelativistic wave function with four LS components involves taking the Δ's to have equal energy and the physical mass m_Δ = 1.236 GeV/c^2, whereas in a spectator reaction the recoiling Δ(k) is on its mass shell and the exchanged Δ has a mass $(m_d^2 + m_\Delta^2 - 2m_d k_0)^{1/2} < m_d - m_\Delta$, clearly a large extrapolation. Then, by Lorentz and reflection invariance there are five independent (ΔΔd) helicity amplitudes. Finally, retardation effects may be sizeable because a covariant pion propagator in a NN→ΔΔ transition has a pole near the physical region just where the isobar wave functions have their maximum. Thus nonrelativistic potential theory may be quite inadequate. These problems can be practically studied in terms of a covariant, linear, off-shell equation [16]

$$\langle p|T(s)|q\rangle = \langle p|V(s)|q\rangle + \int d^4k_1 \langle p|V(s)|k_1,Q_1\rangle G \langle k_1,Q_1|T(s)|q\rangle \tag{1}$$

for a relativistic system of three basic particles: two nucleons $N(k_1)$, $N(k_3)$ and a pion $\pi(k_2)$. Using a quasi-particle approximation, an off-shell N and N^* are admitted as (π,N) composites in a relativistic three-body scattering problem of an N from a quasi-particle N or N^* system proceeding (see Fig.1) via intermediate three-body, i.e. quasi-two-body ($N(k_1)$, $N^{(*)}(K_1)$ = (π,N)) systems to the final quasi-two-body states ($N(k_3)$, $N^{(*)}(K_3)$ = (π,N)). This system and Eq.(1) are extrapolated from the on-shell (such quantities are generally denoted by a bar) total energy $\bar{s} = \bar{W}^2 = \bar{K}^2 = (\bar{k}_1 + \bar{k}_2 + \bar{k}_3)^2$ to the bound-state deuteron pole at $s = W^2 = m_d^2$

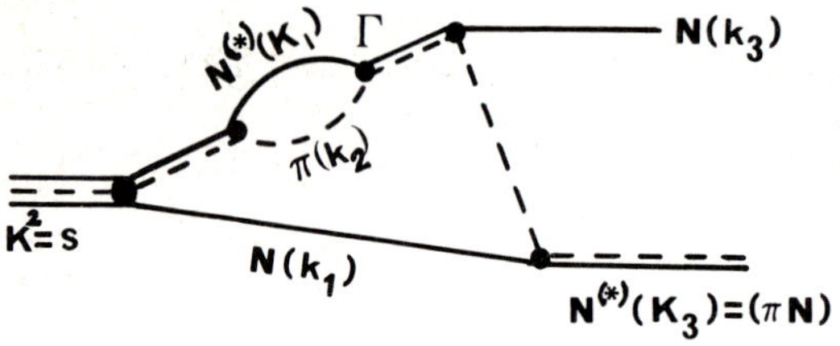

Fig.1. $NN^{(*)}$ configuration as a relativistic π,N,N system.

so that the energy components of momenta become $k_\alpha^\circ = \bar{k}_\alpha^\circ\, W/\bar{W}$, etc. This prescription is unambiguous [17] and derives from time-reversal invariance (more precisely CT, or W→−W) when the cluster properties are correctly enforced on the basis of the relative (Wightman-Gårding) momenta [18]

$$q_1 = (k_2-k_3)/2 - (m_2^2-m_3^2)K_1/2\bar{K}_1^2, \quad K_1 = k_2+k_3, \tag{2}$$
$$Q_1 = (k_1-K_1)/2 - (m_1^2-\bar{s}_1)K/2\bar{W}^2, \quad \bar{s}_1 = \bar{K}_1^2, \quad K = k_1+k_2+k_3 \quad ,$$

which reduce to three-vectors in the two- and three-body c.m. systems, resp., i.e. $q_1 \cdot K_1 = 0 = Q_1 \cdot K$. The two-body interaction V in Eq.(1) consists of pion exchange between (π,N) = $N^{(*)}$ quasi-particles, and the separable $NN^{(*)} \to NN\pi$ amplitude

$$\langle p|T(s)|NN\pi\rangle = \langle p|T(s)|N(k_1),\big(\pi(k_2)N(k_3)\big) = N^{(*)}\rangle\, g(K_1^2)\, \Gamma \tag{3}$$

involves the quasi-particle decay vertex Γ and propagator $g(s_1)$. The cluster variables $s_\alpha = K_\alpha^2 = \bar{Q}_\alpha^2 + s(\bar{s}_\alpha - \bar{Q}_\alpha^2)/\bar{s}$ do not introduce any spurious singularity at s = 0 which has plagued recent work on πN[16] and πd[19] scattering. If $\mathbf{Q}_1$, $\mathbf{Q}_3$ are chosen as independent variables,

$$\bar{W} = (m_1^2+\mathbf{Q}_1^2)^{1/2} + (m_2^2+(\mathbf{Q}_1+\mathbf{Q}_3)^2)^{1/2} + (m_3^2+\mathbf{Q}_3^2)^{1/2} \quad , \tag{4}$$

where $\mathbf{Q}_\alpha^2 = -\bar{Q}_\alpha^2$ in the three-body c.m.s.

The three-particle propagator G in Eq.(1) and the transition potential V are determined from dispersion relations à la Blankenbecler and Sugar [20] using their discontinuities which follow from imposing elastic and three-particle unitarity. The discontinuity of T from Eq.(11) may be written here as

$$T^+ - T^- = T^+(G^+-G^-)T^- + T^+G^+(V^+-V^-)G^-T^- \quad . \tag{5}$$

The unitarity relation is restricted to two- and three-body $NN^{(*)}$ and $NN\pi$ states,

$$T^+ - T^- = i \sum_{n=2,3} T^+ \rho_n T^- \quad , \tag{6}$$

where ρ_n is the n-body phase space volume element. Inserting Eq.(3) in Eq.(6) and comparing the latter with Eq.(5) yields

$$G^+ - G^- = 4im_1 m^{(*)}(2\pi)^2 \delta^+(s_1 - m^{(*)2})\delta^+(k_1^2 - m_1^2) - \pi^{-1} 2im_1 m_3 g^+ g^- \int d^4 q_1 <\Gamma^2> \prod_{j=1}^{3} \delta^+(k_j^2 - m_j^2), \tag{7}$$

where the first term is due to elastic unitarity and the second to (πN) quasi-particle break-up such that $\pi(k_2)$ forms an $N^{(*)}$ with the same $N(k_3)$ in T^- and T^+. When the pion combines with a different nucleon in T^- and T^+ of Eq.(6), the exchange term of Eq.(5) is generated,

$$G^+(k_3)(V^+ - V^-)G^-(k_1) = 4m_1 m_3 i(2\pi)^3 \prod_{j=1}^{3} \delta^+ (k_j^2 - m_j^2) \, \Gamma g^+ g^- \Gamma \quad . \tag{8}$$

Eq.(7), (8) admit a solution of the spectator form, prescribed by $W \to -W$ reflection,

$$G(k_i) = 4\pi m_i g(s_i) \cdot \delta^+(\bar{k}_i^2 - m_i^2)\frac{\bar{W}}{W} = \frac{2\pi m_i}{k_i^{\circ}} g(s_i)\, \delta\,(k_i^{\circ} - \frac{W}{\bar{W}}\bar{k}_i^{\circ}) \propto \delta(Q_i^{\circ}) \tag{9}$$

where g is the dressed $N^{(*)} = (\pi N)$ propagator of Eq.(3) reducing Eq.(1) to three dimensions. If we now use this in Eq.(8) together with

$$\prod_{j=1}^{3} \delta^+(k_j^2 - m_j^2) = \bar{s}^{\frac{1}{2}} \delta(s - \bar{s})\delta(Q_1 \cdot K/\sqrt{s})\delta(Q_3 \cdot K/\sqrt{s})/4\bar{k}_1^{\circ} \bar{k}_2^{\circ} \bar{k}_3^{\circ} \tag{10}$$

in the three-body c.m.s. where $\bar{k}_2^{\circ} = (m_2^2 + (\underset{\sim}{Q}_1 + \underset{\sim}{Q}_3)^2)^{\frac{1}{2}}$, etc. and

$$\delta^+(k_1^2 - m_1^2)\delta^+(k_3^2 - m_3^2) = (\bar{s}_1 + \bar{s}_3 - m_1^2 - m_3^2)\delta(Q_1 \cdot K/\sqrt{s})\delta(Q_3 \cdot K/\sqrt{s})/8\bar{k}_1^{\circ}\bar{k}_3^{\circ}\bar{s}, \tag{11}$$

this yields

$$V^+ - V^- = 2\pi i \delta(s - \bar{s})\Gamma\rho\Gamma, \quad \rho = \frac{2\bar{W}}{\bar{k}_2^{\circ}} \; \frac{\bar{s}}{\bar{s}_1 + \bar{s}_3 - m_1^2 - m_3^2} \tag{12}$$

A dispersion relation in the total energy variable s, to which the three constituent particles N,N,π contribute equally, then yields $V(s) = \Gamma_\rho \Gamma/(s - \bar{s})$ with $\bar{s}$ given in Eq.(4). Thus it is unitarity for the intermediate N,N,π of positive energy in conjunction with the cluster properties that eliminate retardation effects in relativistic three-body theory and the foregoing pole of the OPE transition potential. Similarly in the two-body c.m.s., i.e. the quasi-particle rest frame,

$$\delta^+(k_2^2 - m_2^2)\delta^+(k_3^2 - m_3^2) = \sqrt{\bar{s}}_1 \; \delta(s_1 - \bar{s}_1)\delta(q_1^{\circ})/2\omega_2\omega_3 \; , \tag{13}$$

where $\omega_i = (m_i^2 - \bar{q}_1^2)^{\frac{1}{2}}$, $\bar{q}_1^2 = -\underset{\sim}{q}_1^2$ so that Eq.(7) may be written as

$$g(s_1^+) - g(s_1^-) = 4\pi i m^{(*)} \delta(s_1 - m^{(*)2}) - g(s_1^+)g(s_1^-) \int \frac{d^3 q_1}{(2\pi)^3} \; \frac{m_3\sqrt{\bar{s}}_1}{\omega_2\omega_3} <\Gamma^2> 2\pi i \delta(s_1 - \bar{s}_1). \tag{14}$$

The resulting unsubtracted dressed $N^{(*)}$ propagator takes the form

$$g(s_1) = 2m^{(*)}/\left[m^{(*)2} - s_1 - \int \frac{d^3 q_1}{(2\pi)^3} \; \frac{2m_3 m^{(*)}}{\omega_2\omega_3} \; \frac{(\omega_2 + \omega_3)<\Gamma^2>}{(\omega_2 + \omega_3)^2 - s_1}\right] \quad . \tag{15}$$

If we now introduce wave functions $\tilde{\psi}_{NN^{(*)}} \propto g<NN^{(*)}|T|d>$ in the homogeneous version of Eq.(1), extrapolated to $s = m_d^2$, we obtain a set of coupled channel equations which reduce to coupled one-dimensional integral equations for the radial components upon performing the usual partial-wave helicity analysis. If a phenomenological (np) deuteron wave function is used as input, they can be solved by matrix inversion; when the (np) deuteron wave function is determined via a OBE potential V, they may be solved by diagonalization. The $NN^*(1520)$ configuration in the deuteron ground state has the three (LS) components (1,1), (1,2) and (3,2): the negative N^* parity requires odd L and the N^* spin 3/2 gives S = 1,2. Proper antisymmetrization generates symmetric kernels. The numerical results for the three (five upon including the (np) S and D states) coupled channels are qualitatively similar to the nonrelativistic impulse approximation.

References.

[1] For reviews, see H. Arenhövel and H.J. Weber, Springer Tracts 65 (1972) 58, H. Arenhövel, in Proc. Sympos. Interaction Studies in Nuclei, Mainz (1975), H. Jochim and B. Ziegler, eds. (North-Holland); H.J. Weber, ibid.; R. Beurtey, in Proc. Sixth Int. Conf. on High Energy Phys. and Nucl. Struct., Santa Fe (1975). D. Nagle et al., eds. (AIP, NY).
[2] D.O. Riska and G.E. Brown, Phys. Lett. 38B (1972) 193.
[3] C.P. Horne, et al., Phys. Rev. Lett. 33 (1974) 380.
[4] M.J. Emms, et al., Phys. Rev. Lett. 52B (1974) 372.
[5] H. Braun, et al., Phys. Rev. Lett. 33 (1974) 312.
[6] B.S. Aladashvili, et al., Nucl. Phys. B89 (1975) 405.
[7] N.R. Nath, P.K. Kabir, and H.J. Weber, Phys. Rev. D10 (1974) 811.
[8] P. Benz and P. Söding, Phys. Lett. 52B (1974) 367
[9] R. Beurtey, J. Duchazeaubeneix, J. Faivre, J. Lugol, J. Saudinos, L. Goldzahl, C. Cvijanovich, L. Dubal, H.H. Duhm, C. Perdrisat, Coll. William and Mary preprint (1975).
[10] N.R. Nath, H.J. Weber, and P.K. Kabir, Phys. Rev. Lett. 26 (1971) 1404.
[11] B. Abramov et al., Moscow preprint ITEP-38 (1974).
[12] H. Arenhövel, M. Danos, and H.T. Williams, Nucl. Phys. A162 (1971) 12, N.R. Nath and H.J. Weber, Phys. Rev. D6 (1972) 1975.
[13] H. Arenhövel, Phys. Lett. 53B (1974) 224, P. Haapakoski and M. Saarela, ibid. p. 333.
[14] S. Jena and L.S. Kisslinger, Ann. Phys. (N.Y.) 85 (1974) 251, E. Rost, Nucl. Phys. A249 (1975) 510.
[15] H. Arenhövel, Zeitschr. f. Phys., in press.
[16] R. Aaron, R. Amado, and J. Young, Phys. Rev. 174 (1968) 2022.
[17] This lack of arbitrariness is important since corrections to nonrelativistic nuclear theory from the off-shell behavior contribute already to the lowest order relativistic corrections. See, e.g., D. Drechsel and H.J. Weber, Nucl. Phys. (1976), in press.
[18] J.M. Namyslowski, Nuovo Cim. 57A (1968) 355, U. Weiss, Nucl. Phys. B44 (1972) 573.
[19] R. Woloshyn, E. Moniz, and R. Aaron, MIT preprint #497 (1975).
[20] R. Blankenbecler and R. Sugar, Phys. Rev. 142 (1966) 1051.

Acknowledgement.

It is a pleasure to thank I.R. Afnan, H. Arenhövel, J.M. Eisenberg, M. Goldhaber, L.S. Kisslinger, T.I. Kopaleishvili and C.F. Perdrisat for useful discussions.

STRUCTURE IN THE BACKWARD PROTON-DEUTERON SCATTERING IN THE 40-1000 MEV REGION

Shin-ichi MORIOKA and Tamotsu UEDA*
Department of Nuclear Engineering, Osaka University, Suita, Osaka
*Department of Applied Mathematics, Faculty of Engineering, Osaka University, Toyonaka, Osaka

We would like to talk about p-D backward scatterings at 40-1000 MeV. The p-D backward differential cross sections show characteristic behavior in energy dependence[1,2]; a dip at 300 MeV and a bump at 500 MeV, and also a sharp drop with increasing incident proton energy above 1 GeV. In order to interpret these features, two kinds of models have been proposed. One relates the p-D scatterings to $pp \rightarrow \pi D$ reactions[2], and the other involves contributions from observed $I=1/2$ nucleon resonances in addition to the nucleon[3]. All of them reject an old non-relativistic theory due to Chew and Goldberger[4], because the non-relativistic theory predicts too small p-D backward differential cross sections for incident energies above 1 GeV (actually it begins to fail at about 400 MeV).

However Chew-Goldberger pickup (CGP) amplitude is very instructive. CGP amplitude is written in c.m.s. at 180° as

$$M_{CGP} \propto [\ \tfrac{1}{4}\ \vec{p}_E^{\,2}/M + |E_B|\,]\ \phi_F^*(\tfrac{1}{2}\ \vec{p}_E)\ \ \phi_E(\tfrac{1}{2}\ \vec{p}_E)\,, \tag{1}$$

where $\vec{p}$, M, E_B and ϕ are the nucleon momentum, the nucleon mass, the binding energy of the deuteron and the wavefunction of the deuteron respectively and the subscripts E and F denote the entering and the final states. As is seen in the eq.(1), the incident momentum is an argument of the deuteron wavefunctions, and hence the high-energy p-D backward scatterings reveal the high momentum part, namely the inner part in the configuration-space, of the deuteron wavefunctions. Therefore one may consider at first that the failure of CGP is caused by ambiguities of the inner part of the deuteron, and hence it is natural to apply the CGP picture to this problem with relativistic corrections.

Recently, Gross has derived P-state wavefunctions for deuteron, which is considerably large in the deuteron inner part; being comparable to the S- and D-state wavefunctions. Also in this Gross's model[5], a relativistic N-exchange amplitude is decomposed into terms of the CGP type and its relativistic corrections. Contributions due to the P-state wavefunctions are larger in the relativistic terms than in the CGP type term. In this paper we show a possible interpretation of the dip-bump structure in terms of the N-exchange model with full relativistic corrections.

First we write down the relativistic N-exchange amplitude with Blankenbecler-Cook vertex function Γ[6] as

$$M_{EX} = \bar{u}^c(p_E)\ \ \varepsilon_F\Gamma_F\ \frac{1}{\gamma k-M}\ \ (\gamma k-M)\ \frac{1}{\gamma k-M}\ \ \varepsilon_E\Gamma_E\ \ u^c(p_F)\,, \tag{2}$$

$$\Gamma = F\gamma + Gp/M + (M-\gamma k)/(2M)(H\gamma + Ip/M)\,, \tag{3}$$

where u^c, ε, p, k and γ are the charge-conjugated spinor, the deuteron polarization vector, the nucleon four momentum, and the four momentum transfer between the entering nucleon and the deuteron respectively.

The four Lorentz-invariant functions F, G, H and I in Γ involves an argument $q=|\vec{q}|$ and related to the deuteron wavefunctions as

$$F = (2E-M_D)/(2N_0)[\sqrt{2}\phi_0 + \phi_2 + \sqrt{3}M/q\phi_{11}]$$

$$G = (2E-M_D)M/\{2N_0(E+M)\}[\sqrt{2}\phi_0 - (2E+M)/(E-M)\phi_2 + \sqrt{3}(E+M)/q\phi_{11}],$$

$$H = -\sqrt{3}EM/(N_0 q)\phi_{11},$$

$$I = M^2/(N_0 M_D)[\sqrt{6}M_D/q\phi_{10} - (2E-M_D)/(E+M)\{\sqrt{2}\phi_0 + (E+2M)/(E-M)\phi_2\}],$$

where E, M_D, q and N_0 are the nucleon energy, the deuteron mass the relative momentum of the bound nucleon (in c.m.s. $q=1/2p_E$) and a normalization constant, and ϕ_0 and ϕ_2 are the S- and D-state wavefunctions in the momentum-space, and ϕ_{10} and ϕ_{11} denote the singlet and triplet P-state wavefunctions in the momentum-space, respectively. An example of the P-state wavefunctions is given by Hornstein and Gross [5](hereafter refered to as HG). When ϕ_0, ϕ_2, ϕ_{11} and ϕ_{10} are given, the p-D differential cross sections can be calculated.

Our numerical results are shown in Fig.1a, b and 2. Here we neglected ϕ_{10} because it is small, compared with other wavefunctions [5]. We employed the Hulthén wavefunction[7] for the S and D states and a function almost equal to HG for the P state. They give 6.7 and 0.53% for the D- and P-state probabilities respectively. Fig.1a shows comparisons of relativistic results with non-relativistic one(CGP calculation) in total contribution and in partial contribution from each state. We note that the relativistic contribution from the D state is by far larger than the corresponding non-relativistic one. The P-state contribution, which is just relativistic effect, is appreciable. As the P-state wavefunction has a peak at 200 MeV in the

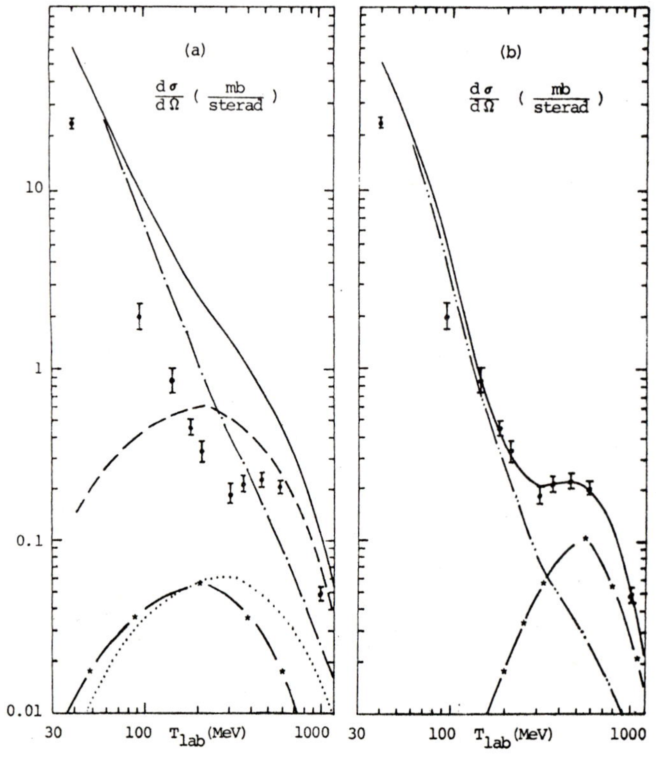

Fig.1 The relativistic and non-relativistic(CGP) theoretical p-D elastic cross sections at the c.m.s. angle 180° are compared with the data[1,8]. The abscissa shows the proton laboratory energy T_{lab}. Hulthén and HG wavefunctions are employed in Fig.1a and the modified ones are used in Fig.1b. —— represents the total relativistic results; — * — the contribution from the P-state component; — · — the total non-relativistic(CGP) result; .... and ---- the contributions due to the D-state component in the CGP and the relativistic theory respectively; — ·· — represents the relativistic total contributions from the S- and D-state components.

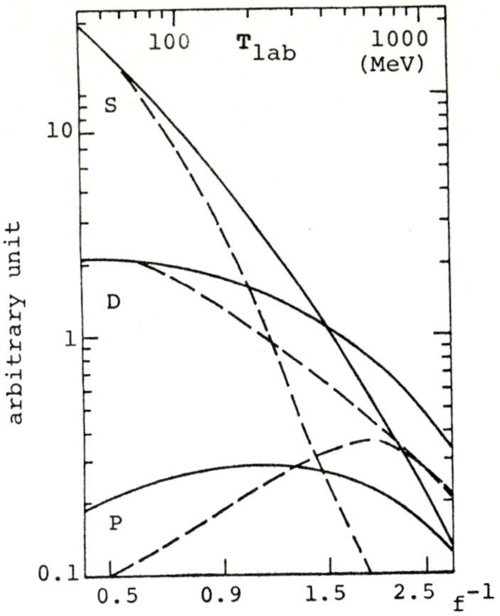

Fig.2
Deuteron wavefunctions in momentum-space. The solid and broken lines show Hulthén and HG wavefunctions and its modification respectively.

momentum-space, its contribution to the cross sections also shows the peak at 200 MeV. The peak is due to j_1 involved in ϕ_{11} and characteristic feature of the P-state contribution. If the inner part of the P-state wavefunction in the configuration-space is changed to be larger than that of HG, the peak of its contribution to the cross sections moves to higher energies.

For a fit to the experimental data, we have to modify the deuteron wavefunctions as is shown in Fig.2. Fig.1b shows theoretical predictions with employing these modified wavefunctions. Though we have some uncertainty in the modification, in any case the D-state wavefunction must be modified for the too large effects.

Without the P-state wavefunction ϕ_{11}, we find difficulty to describe the dip-bump structure with reasonable S- and D- state wavefunctions used. Then, as is seen in Fig.1a, the P-state component is necessary for describing the dip-bump structure. In the regions of higher energies, the contributions from the S- and D-state components become negligible, and hence the sharp drop in this energy region is reproduced by the form of the P-state wavefunction.

Considerable changes from Hulthén and HG wavefunctions are necessary in the large momentum or inner part of the deuteron wavefunctions for a reasonable fit to the data. So the changes of the wavefunctions in this region may be allowable.

To summarize, the relativistic effects are considerable for p-D backward differential cross sections. Especially the effect from the D-state component is remarkable, and P-state component plays an important role in reproducing the dip-bump structure and the sharp drop in energy dependence. One of the authors(S.M.) thanks Prof. T.Sekiya for his encouragement.

References

[1] J. C. Alder et al., Phys. Rev. C6(1972)2010.
[2] B. S. Bhakar, Nuovo Cimento 18A(1973)737.
[3] J. S. Sharma et al., Nucl. Phys. B35(1971)466.
[4] G. F. Chew and M. L. Goldberger, Phys. Rev. 77(1950)470.
[5] F. Gross, Phys. Rev. D10(1974)223.
J. Hornstein and F. Gross, Phys. Lett. 47B(1973)205.
E. A. Remler, Nucl. Phys. B42(1972)56 and B42(1972)69.
[6] R. Blankenbecler and L. F. Cook, Phys. Rev. 119(1960)1745.
[7] M. J. Moravcsik, Nucl. Phys. 7(1958)113.
[8] J. H. Williams and M. K. Brussel, Phys. Rev. 110(1958)136.
O. Chamberlain and M. O. Stern, Phys. Rev. 94(1954)666.
H. Postoma and R. Wilson, Phys. Rev. 121(1961)1229.
G. Igo et al., Nucl. Phys. A195(1972)33.
E. Coleman et al., Phys. Rev. Lett. 16(1966)761.

THE PROCESS $pp \rightarrow d\rho^+$ IN A MODEL OF N AND N^* EXCHANGES

J.S. Sharma
Department of Physics and Astrophysics, University of Delhi, Delhi-110007, India.

During recent years, there has been a revival of interest in the study of the deuteron-stripping reaction [1]. Several mechanisms have been proposed to explain their main experimental features [2-6]. The mechanism of one nucleon exchange which offers a simple understanding of the sharp drop in the u-variable in terms of the deuteron's form factor does not unfortunately generate an adequate magnitude for the cross sections. A second mechanism first proposed by Yao [2], and investigated in considerable detail by Cragie and Wilkin [3] is the so-called triangle (or pion-exchange) diagram which reproduces the variations of the cross sections with energy and angle fairly well, but the absolute magnitudes of the cross sections remain rather low. A novel suggestion for enhancing the cross sections through the exchange of resonances (N^*) in the u-channel (apart from the N-exchange) was made by Kerman and Kisslinger [4], who considered a Reggeized N-exchange mechanism so as to include the $F_{15}(1688)$ resonance in the exchange. This model was criticized by Cragie and Wilkin on the ground that Reggeization at an energy of around 1.0 GeV where the rest masses themselves involved are of a similar magnitude could be two optimistic.

Sharma et al [5,6] proposed a mechanism of N and N^* exchanges in a model of higher baryon couplings with pions developed by Mitra and Coworkers [7] in the recent years and applied to several meson-baryon process [8]. They took a fairly long list of N^* exchanges in the GeV region, and not merely the F_{15}, since in the GeV region several N^* resonances could make comparable contributions. Subsequently some improvements were incorporated in this model and the refined model was applied to two related processes, viz. $pd \rightarrow dp$ and $pp \rightarrow d\pi^+$ [6]. The model was found to be very successful in reproducing the experimental data both in absolute magnitude as well as in the shape. Being encouraged by the earlier results, we report in this paper the results of this model for the process $pp \rightarrow d\rho^+$.

To evaluate the amplitude for this process via N and N^* exchanges, we need (a) dNN vertex (we use the same extended structure as in [6] (b) $\bar{N}N\rho$ form factor for the N exchange and (c) dNN^* coupling and $\bar{N}N^*\rho$ coupling [8] for the N^* exchange contribution. The $\bar{N}B_LV$ couplings needed for this purpose are governed by the Clebsch Gordan coefficients of SU(6) x O(3) and are of the following relativistic forms (for notations see [8])

$$A_L = \beta\, \bar{u}^{L+\frac{1}{2}}_{\mu_1 \cdots \mu_L}(P)\, i\sigma_{\mu\nu} q_\nu\, q_{\mu_1} \cdots q_{\mu_L} V_\mu\, u(p)\, f_L$$

$$B_L = \beta\, \bar{u}^{L+\frac{1}{2}}_{\mu_1 \cdots \mu_L}(P)\, i m_V \gamma_\mu\, q_{\mu_1} \cdots q_{\mu_L} V_\mu\, u(p)\, f_L$$

$$E_L = \beta\, \bar{u}^{L+\frac{1}{2}}_{\mu_1 \cdots \mu_L}(P)\, V_{\mu_1} q_{\mu_2} \cdots q_{\mu_L}\, u(p)\, f_L$$

$$q_\mu = \frac{1}{2}(k_\mu - p_\mu)$$

$$f_L(P^2, p^2, k^2) = \text{Form Factor [8]} \qquad (1)$$

Here P_μ, p_μ, k_μ are the four momenta of N^*, the nucleon and the vector meson respectively; m_V and V_μ represent the mass and the polarization vector of the vector meson. $\beta = 5\sqrt{2}/3$ and 4/3 for even-L-(56) and odd-L-(70)N^* states respectively. To calculate the contribution of N^* exchanges to the amplitude for this process, we use dNN^* coupling which we calculate by using (i) dNN extended vertex (ii) $\bar{N}N^*V$ couplings and (iii) $\bar{N}NV$ coupling. The method of calculation has been given explicitly in ref. [6] and we obtain the dNN^* vertex as

$$\bar{\Gamma}_L^{(\mu)}(d,P) = \frac{3}{16}\cdot\frac{G_2}{\pi m}\left(m+\frac{1}{2}i\gamma\cdot d\right)\frac{4\pi}{m}\bar{N}_\gamma\,\gamma\cdot\varepsilon\,\frac{m_d - i\gamma\cdot d}{2\sqrt{2}\,m_d}$$
$$\times\left[\frac{1}{\beta+\alpha}+\frac{t}{\sqrt{2}(\gamma^2-\alpha^2)^2}\left(\alpha^3+\frac{1}{2}\gamma\left(\gamma^2-3\alpha^2\right)\right)\right]\left(m-\frac{1}{2}i\gamma\cdot d\right)$$
$$\times\,\beta\left(m_\rho^2-2mT\right)^{-1} f_L\, V_\mu\, i\sigma_{\mu\nu}q_\nu \langle q_{\mu_1}\cdots q_{\mu_L}\rangle_{AV} \tag{2}$$

The 4-momenta of the particles involved in the reaction $pp\rightarrow d\rho^+$ are shown in Fig. 1. The invariant amplitude may be written as follows:

$$A(p,p') = T(p,p') - T(p',p) \tag{3}$$

where the first term on the right corresponds to Fig. 1 and the second term is obtained from the first by interchanging the protons. In our model

$$T(p,p') = T_N(p,p') + \sum_{N^*} T_{N^*}(p,p')\ , \tag{4}$$

where T_N and T_{N^*} represent the contributions to the amplitude from N and N^* exchanges respectively [6]. The cross section is

$$\left(\frac{d\sigma}{d\Omega}\right)_d = \frac{m^2}{16\pi^2}\,\frac{2m_d}{s}\left(p_c^f/p_c^i\right)\frac{1}{4}\sum_{\text{spins}}|A|^2 \tag{5}$$

where $s = -(p+p\ )^2$ and $p_c^{i,f}$ are the initial (final) c.m. momenta. The experimental data to test our model is taken from H.L. Anderson et al [9] who depict the variation of cross section with energy at essentially forward-backward directions. The corresponding theoretical cross sections, Eq.(5), are shown in Fig. 2, along with the data [9].

As in the cases of $pd\rightarrow dp$ and $pp\rightarrow d\pi^+$, the main contributors (apart from the N-exchange), are N^* ($J = L+\frac{1}{2}$) exchange $P_{11}(1470)$, $P'_{11}(1750)$, $D_{13}(1520)$ and $F_{15}(1688)$ while $J = L - \frac{1}{2}$ exchanges play a negligible role.

In producing agreeement with the experimental data, N^*-exchanges ($J = L+\frac{1}{2}$) have played a significant role by exchanging the cross sections by 15 to 20 percent. The antisymmetrization in the initial pp state which produces enhancement of the cross sections of about 25% as a result of the interference between 'direct' and'exchange' terms, Eq.(3), helps produce the correct magnitude for the cross section.

The calculated curve according to Eq.(5) seems to reproduce the data in sufficient details, in terms of both the absolute magnitude as well as the shape up to centre-of-mass energy square (s) = 20 $(GeV)^2$. Above this energy range the calculated cross-section falls short of the experimental data by ∼10%. We believe that in this energy region, some other higher resonances which we have not taken into account, may play a role.

REFERENCES

1. E. Coleman et al, Phys. Rev. Letters 16 (1966) 761.
   R.M. Heinz et al, Phys. Rev. 167 (1968) 1232.
2. T. Yao, Phys. Rev. 134B (1964) 454.
3. N.S. Cragie and C. Wilkin, Nucl. Phys. 14B (1969) 477.
4. A. Kerman and L. Kisslinger, Phys. Rev. 180 (1969) 1483.
5. J.S. Sharma et al, Nucl. Phys. 35B (1971) 466;
   Phys. Letters 46B (1973) 44.
6. J.S. Sharma and A,N. Mitra, Phys. Rev. 9D (1974) 2547.
7. A.N. Mitra, Ann. Phys. 67 (1971) 518;
   A.N. Mitra, Nuovo Cimento 64A (1969) 603.
8. Sudhir Sood and A.N. Mitra, Phys. Rev. D7 (1973) 2111.
   S.A.S. Ahmed and A.N. Mitra, Phys. Rev. D7 (1973) 2125.
9. H.L. Anderson et al, Phys. Rev. Letters 26 (1971) 108.

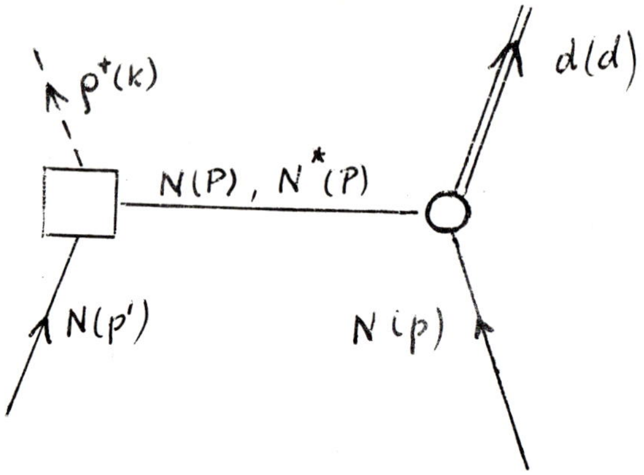

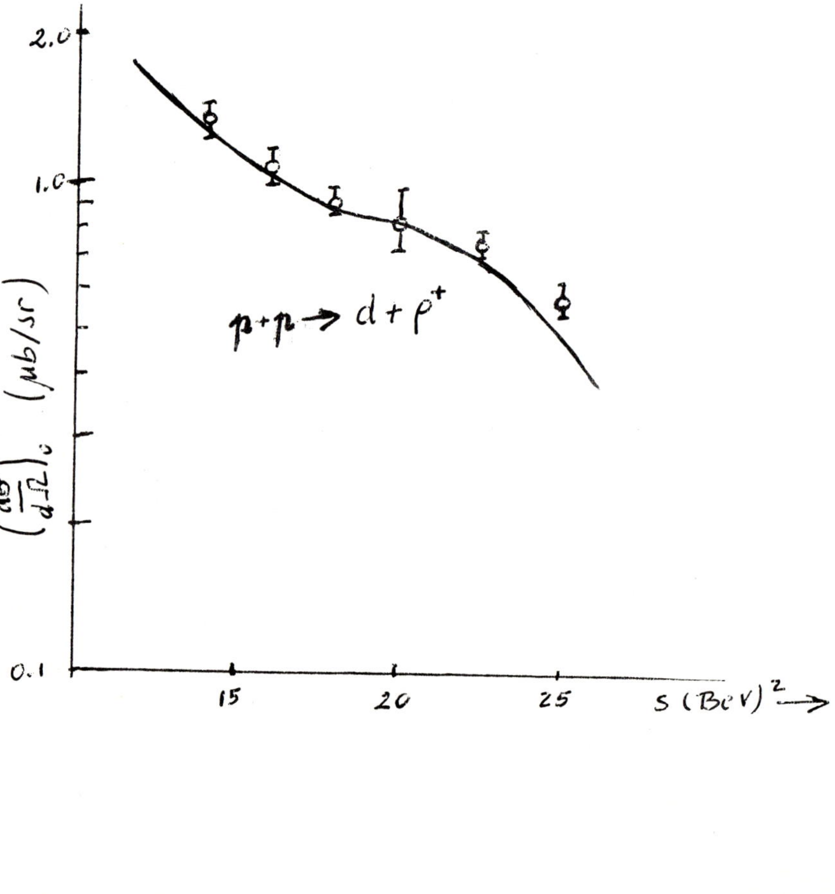

PION PRODUCTION IN NEUTRON-PROTON COLLISIONS

D.M. Wolfe, C. Cassapakis, B.D. Dieterle, C.P. Leavitt, W.R. Thomas, Univ. of New Mexico*; G. Glass, M. Jain, L.C. Northcliffe, Texas A&M Univ.*; B.E. Bonner, J.E. Simmons, LASL*.

Introduction

The production of pions in nucleon-nucleon collisions is an integral part of the strong interaction picture. Despite its importance, however, the experimental situation is poor. Some data exist [1,2] for p-p collisions but there is very little for the n-p case. There are two experiments [3,4], using approximately 600 MeV neutrons in the reactions

$$n + p \rightarrow nn\pi^{+} \qquad (1)$$

$$n + p \rightarrow pp\pi^{-} \qquad (2)$$

but the incident neutron spectrum is known poorly, impairing the usefulness of the data. A bubble chamber experiment has studied reaction (2) but using neutrons from 400 to 970 MeV and with poor statistical accuracy.

The 800 MeV neutron beam at LAMPF allows us to investigate a particularly interesting theoretical situation. The Mandelstam model very probably fails to describe pion production in this region [7]. Two of the basic assumptions of the model, that only a few partial waves are present and that the production matrix elements are independent of energy, are expected to fail at 700-800 MeV. Another basic tenet of the theory is that pion production takes place via the Δ (1236) resonance. Since the pion and one of the outgoing nucleons must then be in a (3/2,3/2) state, only incoming np states of T=1 should be allowed. In the notation of Rosenfeld [8], the cross-section we are looking at is $\sigma\ (np\rightarrow\pi^{\pm}) = 1/2\ (\sigma_{11} + \sigma_{01})$ where the first index is the isospin of the initial state and the second index the isospin of the final state nucleons. σ_{11} is completely determined by the reaction $pp \rightarrow pp\pi^{0}$. A search for non-resonant $\sigma=0$ contributions would thus provide a stringent test of the model.

The one pion exchange model (OPEM) of Ferrari and Selleri [9], valid from about 800 MeV to 1500 MeV, fails in the same region due to the neglect of nucleon-nucleon final state interactions [7]. The present np data may be used to test these models in the region of overlap and, along with new elastic scattering data [10,11] should provide the input for calculations of np phase shifts in the inelastic region.

Experimental Method

The neutron beam is produced at 0° from the reaction $p + D \rightarrow n + X$ using the LAMPF 800 MeV proton beam. The neutron spectrum obtained is shown in fig. 1. Approximately 50% of the neutron flux is in the peak at 800 MeV (13MeV FWHM). We are able to utilize the microstructure of the LAMPF proton beam (one .25 ns pulse every 5 ns) to eliminate most of the lower energy neutrons. We accept only those events with a neutron flight time between the deuterium production target and the LH2 target, shown in fig. 2, corresponding to $\beta_n = .85$.

Due to the 5ns ambiguity inherent in such timing and to our .6ns timing resolution, we are also sensitive to a neutron band $470 \leq T_n \leq 510$ MeV. This band corresponds to an intensity of approximately 10% of the 800 MeV neutrons. As most of the energy spectrum available to pions from 800 MeV neutron production is not

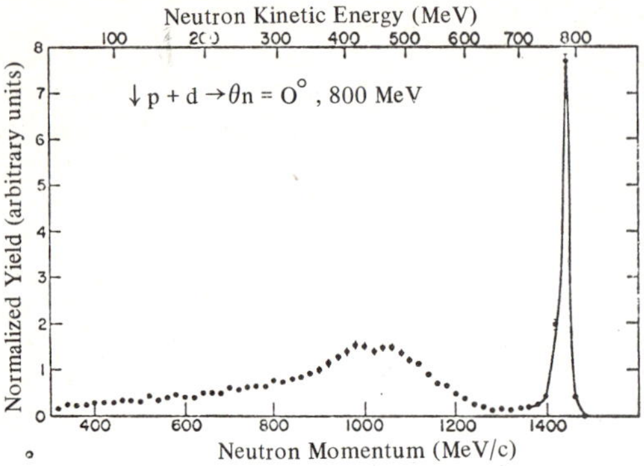

figure 1

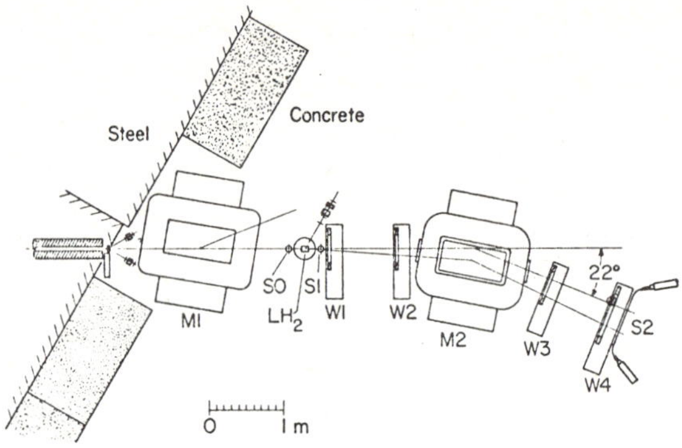

figure 2

available to those produced by a 500 MeV beam this background is reduced to <5%. The ratio of production cross sections at 500 and 800 MeV reduces it to a negligible level. Similarly, double pion production is estimated [5] to be <2%. Thus the pion events are produced entirely by 800 MeV neutrons.

The pions are detected by means of a single arm multiwire proportional chamber (MWPC) spectrometer [12] shown in figure 2. In order to collect both $\pi^{\pm}$ spectra, the current in magnet M2 was reversed approximately every two hours. Helium bags were placed between W2-W3 and W3-W4 to reduce multiple scattering. A one atmosphere isobutane Cerenkov counter, between W1 and W2 was used to count electrons. Except at 0° pion production angle the electron background was <1%. At 0° there was a significant number of electrons due to photons in the beam. These were eliminated by Pb in the neutron collimator hole, timing with the beam RF signals (the electrons arriving at $\beta = 1$ time, the neutrons at $\beta = .85$ time), and the Cerenkov counter. Pions were identified by their times of flight between S1 and S2 and by their momenta as determined from orbit information provided by the MWPC's. At positive magnet currents protons were easily separable. There was very little background for the negative currents. At each angle, elastic proton scattering data were taken which, by comparison with a recent measurement of the elastic np cross-section using the same spectrometer, [11] will allow us to get absolute normalization values for the $\pi^{\pm}$ production cross-sections.

Analysis

Data were taken over the entire pion momentum spectrum at 0, 8, 16, 24, 30, 36, 42, 48, 56 and 72 degrees lab. We report here on 10-20% of the data at 0, 8, 16, 30, 56 and 72 degrees for negative pions only. After the subtraction of electron and target empty backgrounds, the only significant corrections remaining are due to pion decay in flight through the spectrometer. Since the path length is nearly independent of angle, this calculation is straightforward. A calculation of the probability that a decay muon will be accepted in the spectrometer solid angle, where we are unable to distinguish it from a pion by timing, has been made by means of a Monte Carlo program which includes the effects of multiple scattering. This program indicates that, as expected, most π-μ decays may be eliminated on the basis of a poor χ^2 fit to a single particle trajectory through the spectrometer, and that the probability of confusing a muon with a pion is small and nearly independent of particle momentum. Some lab momentum spectra are shown in fig. 3.

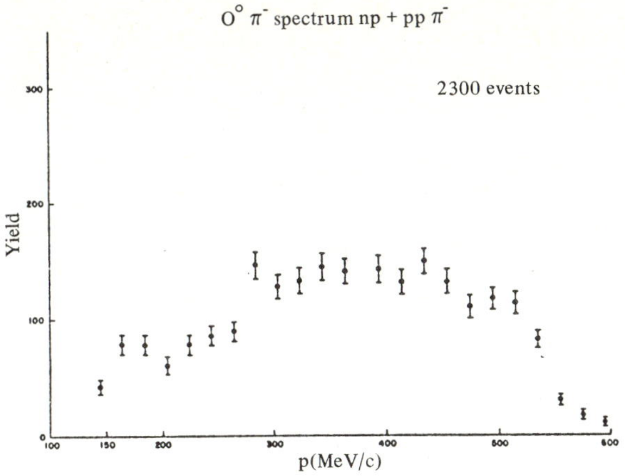

fig. 3a

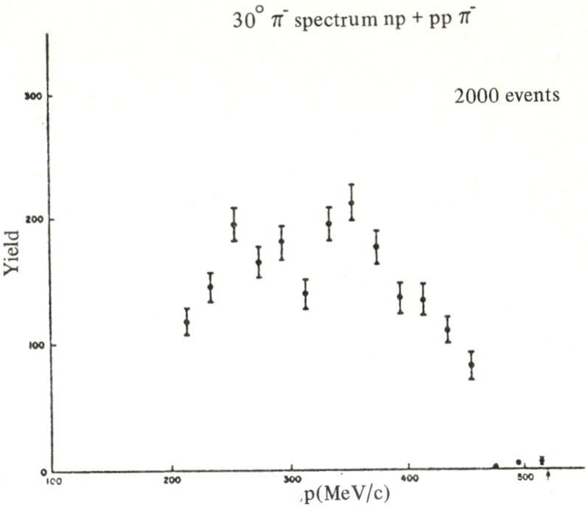

fig. 3b

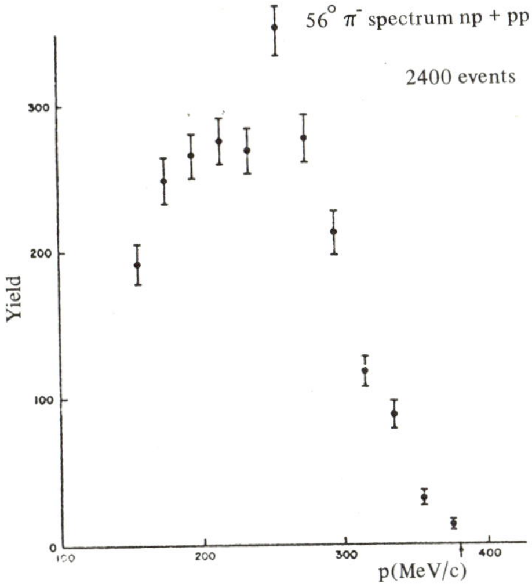

fig. 3c

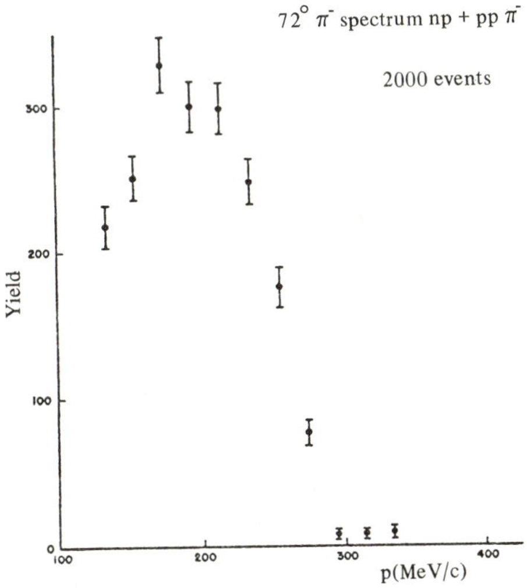

fig. 3d

Our spectrometer acceptance does not extend much below 50 MeV pion kinetic energy. These spectra may then be transformed into the center-of-mass and are shown in figure 4. These spectra are all seen to peak at 2/3 of the maximum momentum. There is also a large content of low energy mesons. Both of these effects have been seen in previous experiments [3,5], although direct comparisons are difficult. Rushbrooke et al. [5] report on spectra for a neutron energy interval similar to ours, but their statistics in the interval are poor. Dzhelepov et al [3] have a lower neutron energy with a broad energy spread. At the larger c-m angles we see less of the pion spectrum due to the fixed lab momentum cutoff. Since the peak positions and shapes are approximately independent of angle, we are able to extrapolate to 0 momentum. We then use these complete spectra to calculate the angular distribution shown in figure 5.

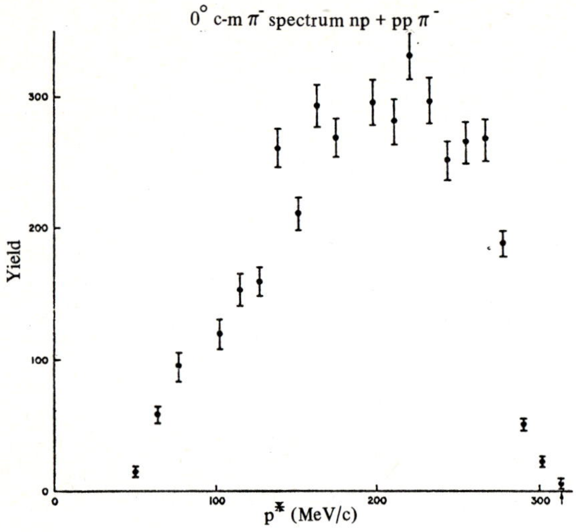

fig. 4a

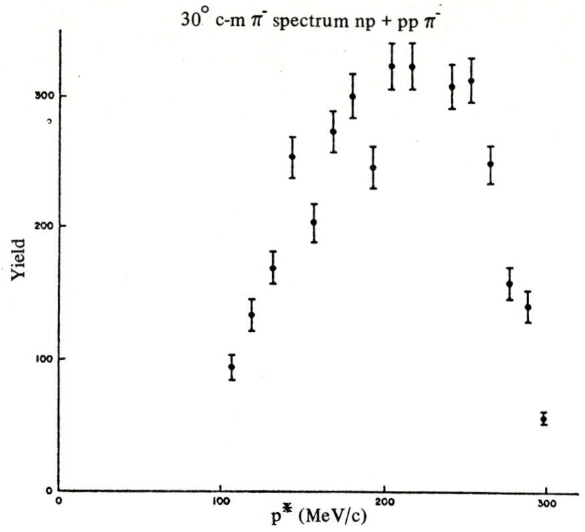

fig. 4b

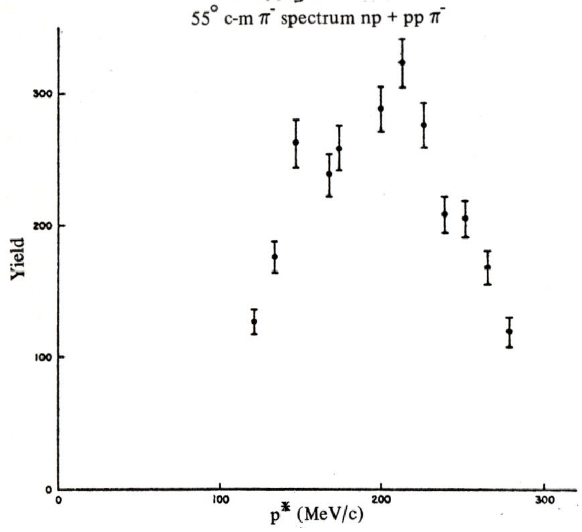

Fig 4c

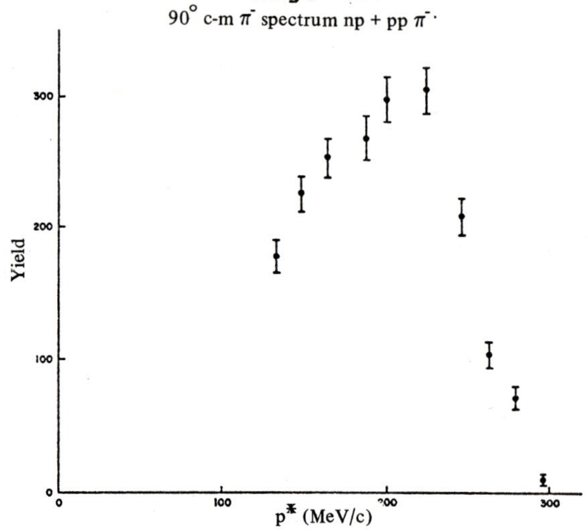

fig. 4d

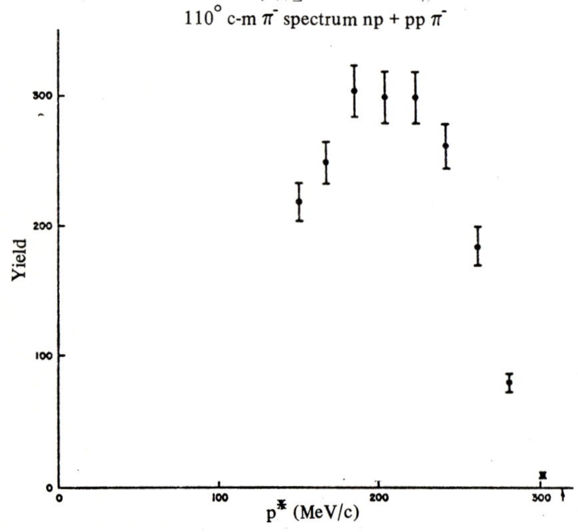

fig. 4e

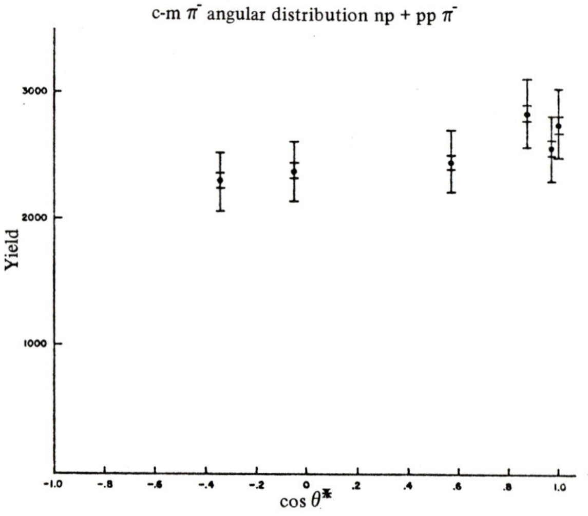

fig. 5

The small error bars in figure 5 represent the statistical errors, the larger ones our present uncertainty in calculating corrections and in making extrapolations. We fit this curve to $A(1 + (.22 \pm .18)\cos^2\theta)$, using the large error bars. The results of Dzhelepov et al [3] are $B(1 + (.39 \pm .1)\cos^2\theta)$. It is to be noted that at the time of this Conference, the data are only six weeks old and represent but a small fraction of the total. We are confident that both statistical and systematic uncertainties will be reduced to the 5-10% level. Finally we note that under charge symmetry $np \rightarrow pp\pi^-$ becomes $pn \rightarrow nn\pi^+$. Thus our data, when analyzed, should allow us, under the charge symmetry assumption to describe a complete c-m angular distribuiton.

References

[1] J. Pratt et al. Contributed paper to the VI International Conference on High Energy Physics and Nuclear Structure, 1975.

[2] G. Glass et al. Contributed paper to the VI International Conference on High Energy Physics and Nuclear Structure, 1975.

[3] V.P. Dzhelepov et al., Soviet Physics JETP 23, 993 (1966).

[4] Yu. Kazarinov and Yu. Simonov, Soviet Journal of Nuclear Physics 4,100 (1967).

[5] J.G. Rushbrooke et al, Il Nuovo Cimento XXXIII, 1509 (1964).

[6] S. Mandelstam, Proc. Roy. Soc. A244,491 (1958).

[7] U. Amaldi, Jr., Rev. Mod. Phys. 39, 649 (1967).

[8] A.H. Rosenfeld, Phys. Rev 96, 139 (1954).

[9] E. Ferrari and F. Selleri, Il Nuovo Cimento XXVII, 1450 (1963).

[10] G. Bizard et al., Nucl. Phys. B85, 14 (1975).

[11] LASL, U. New Mexico, Texas A & M, U. Texas collaboration, to be published.

[12] D.W. Werren et al, LASL Report LA-5396-MS (1973).

*Supported by U.S. ERDA

(π , NN) REACTIONS FOR NUCLEAR STRUCTURE STUDIES

B. K. Jain
Nuclear Physics Division, Bhabha Atomic Research Centre
Bombay-400 085, India

Pions differ from conventional nuclear probes in that they can be absorbed by the nucleus; thus they can transfer their rest mass of 140 MeV to the nucleus without transferring the corresponding momentum. Kinematically this means that if the absorption takes place on a single nucleon in the nucleus, that nucleon must have momentum of the order of 500 MeV/c. Consequently the study of the reaction such as (π , N) provides a handle to the investigation of the least known high momentum components of the nuclear wave function. This information is complementary to that obtained in the (p, 2p) reaction where low momentum components contribute most to the cross-section. However, since in the shell model picture of the nucleus the maximum momentum a nucleon in the nucleus can have is only of the order of 300 MeV/c, the probability for the (π , N) reactions to occur is very small. More probable absorption processes are these where more than one nucleon participate. One of these modes is the (π , NN) reaction in which the pion is absorbed on a correlated pair of nucleons. As the most probable momentum of the nucleon pair in the nucleus is low, to satisfy kinematic requirements the outgoing nucleons preferentially move in opposite directions. Experimental measurements on these reactions contain two types of information:
(i) as the energy of the each of the two outgoing nucleons is about 70 MeV, their relative momentum is around 700 MeV/c. This corresponds to a separation of about 0.5 fm. of two nucleons in the nucleus. The study of (π , NN) reactions may, therefore, be utilized to study two body short range correlations (SRC) in the nucleus.
(ii) The emission of two nucleons creates a two hole state in the nucleus. Since it is known from the studies of Moszkowski and Scott (1) that the short range ($\approx$ 0.5 fm.) part of the nucleon-nucleon interaction does not influence the average shell model potential, the absorption of pions on a correlated pair of nucleons should produce two hole states very similar to the states produced in the (p, 2p) and the (p, pd) reactions. The missing energy spectrum and the recoil momentum distribution of (π , NN) reaction then provide a further test of the truth of the shell model. This potential of pion absorption studies was pointed out earlier by Ericson (2), and Wilkinson (3) and recently reemphasized by Jain and Banerjee (4).

Quantitative extraction of the above information is however, a little difficult due to the uncertainty in the description of the absorption vertex and the existance of the three body final state. Nevertheless, various attempts (5, 6) have been made which use either the one nucleon or the two nucleon model for the absorption vertex. In my talk I shall make a comparative study of these models and discuss the utility of the two nucleon model based on the phenomenological interaction.

ONE-NUCLEON MODEL

In the one-nucleon model the pion is assumed to be absorbed by a single nucleon. The relevant interaction, $H_{\pi N}$ is assumed to be the non-relativistic pseudoscalar-pseudovector pion-nucleon interaction, i.e.

$$H_{\pi N} = \frac{f}{m}\left[\underset{\sim}{\sigma}\cdot\underset{\sim}{\nabla}\; \underset{\sim}{\tau}\cdot\underset{\sim}{\varphi} - \frac{m}{M}\, \underset{\sim}{\tau}\cdot\underset{\sim}{\varphi}\; \underset{\sim}{\sigma}\cdot\underset{\sim}{\nabla}_N \right] , \qquad (1)$$

where m and M are the masses of pion and nucleon respectively; σ and τ are spin and isospin Pauli matrices; φ is the pionic field; f is the coupling constant, which is determined by fitting p-wave pion nucleon scattering data. In practical calculations with this model the relative wave function of the pair of nucleons in the nucleus is modified at short distances by multiplying the wave function by a correlation function of the Jastrow type (7). The interaction between the outgoing nucleons is described by a realistic nucleon-nucleon interaction. Due to the compexity of the formalism in this model the scattering of the out going nucleons with the residual nucleus is generally neglected. Studies with this model have shown that the pion absorption probability is sensitive to the correlations present in both the initial and final states. However, there does not seem any reliable way of separating one from the other. Therefore, by comparing the experimental and computed results effectively only one parameter is determined which represents the combined effect of the correlations in the initial and final states. Furthermore, the correlation function deducd in this manner suffers from the uncertainties due to the incomplete treatment of the final state interaction and the uncertainties in the absorption vertex. In fact, using the full three body solution for final state, Garcilazo and Eisenberg (8) have shown that the interaction of the outgoing nucleons with the residual nucleus is very important. Also, the calculations of Koltun and Reitan (9) on the ^{6}Li(π^- , nn) reaction, where they consider the scattering of s-wave pion on one nucleon before it is absorbed by another, show that the one nucleon model itself may be an inadequate description of the absorption vertex. In conclusion, therefore, it seems that though the (π , NN) reaction contains information about the SRC and the two hole states, their extraction using the one nucleon model is highly unreliable.

TWO-NUCLEON MODEL

In the two-nucleon model, the pion is absorbed by a pair of nucleons. Based on symmetry principles, the pion absorption vertex can be described by a phenomenological interaction (6),

$$H_{\pi NN} = \left[g_0 \left\{ \tfrac{1}{2}(\underset{\sim}{\tau}_1 - \underset{\sim}{\tau}_2)\cdot\underset{\sim}{\varphi}\, \tfrac{1}{2}(\underset{\sim}{\sigma}_1 + \underset{\sim}{\sigma}_2)\cdot\underset{\sim}{k} \right\} + g_1 \left\{ \tfrac{1}{2}(\underset{\sim}{\tau}_1 + \underset{\sim}{\tau}_2)\cdot\underset{\sim}{\varphi}\, \tfrac{1}{2}(\underset{\sim}{\sigma}_1 - \underset{\sim}{\sigma}_2)\cdot\underset{\sim}{k} \right\} \right] \delta(\underset{\sim}{r}_1 - \underset{\sim}{r}_2) \qquad (2)$$

where $\underset{\sim}{k}$ is the relative momentum of the emitted nucleons. This interaction assumes that the pions are absorbed in an s-state; hence it is good for low energy pions. In the formalism based on this model the scattering of the outgoing nucleons with the residual nucleus is included through the optical potential. Contrary to the one-nucleon model, the SRC in the initial and final states are not introduced explicitly. The information about them is contained in the parameters g_0 and g_1. If one assumes that the SRC in the nucleus are the same as those in the free nucleon-nucleon system, $|g_0|$ and $|g_1|$ can be determined from the analysis of the elementary reactions

$$\pi^+ d \rightarrow p + p \qquad ; \qquad p + p \rightarrow \pi^0 + p + p$$

Any deviation from this hypothesis may be measured by renormalizing the parameters $|g_0|$ and $|g_1|$ for a nucleus by fitting the pion absorption data on it. Since the scattering the continuum nucleons with the residual nucleus is taken care of in this model, the renormalized interaction parameters should also provide a relative measure of the SRC in different nuclei.

The two-nucleon model based on the phenomenological interaction should also be useful for studying unambiguously the two hole states. The reasons for this expectation arises from the fact that, independent of the absorption vertex which is mainly determined by the short range (≈ 0.5 fm.) part of the nucleon-nucleon interaction, the two hole states physics is mainly determined by the long range part of the nucleon-nucleon interaction.

The disadvantage of the Hamiltonian (equ. 2) is that the detailed form of the correlation function is not included in it. This, however, may not be bad if the pion absorption data do not depend sensitively on the detailed form of the correlation function. This point is investigated later on. Also to have confidence in the deductions based on the Hamiltonian (eqn. 2) it is necessary to compare its predictions with the experiments on nuclei of known nuclear structure. With this motivation we have computed the recoil momentum distribution and the one neutron energy spectrum for the stopped pions in the (π^-, nn) reaction on ^{6}Li and ^{12}C nuclei. For ^{6}Li we have used the cluster model wave function and for ^{12}C the intermediate coupling wave function. The parameters for these wave functions are taken such that they fit the (p, 2p), (p, pd) and the elastic electron scattering data on these nuclei.

The distorted waves for the continuum neutrons are described in terms of partial waves. The parameters of the optical potential, which are required in the neutron energy region of 10-70 MeV, are assumed to vary linearly with energy. The potentials and their variation with energy are computed from the sparsely available parameters reported in literature (10).

For pions in the 1s atomic orbit we have used the wave function of Seki and Cromer (11). These wave function are obtained, in an approximate way, by solving a set of coupled equations in which the effect of the strong pion-nucleon interaction is simulated by a complex square-well potential and that of the finite charge distribution of nucleus by a uniformly charged sphere. For the 2P orbit we used the hydrogenic wave function.

^{6}Li

The α-d cluster model wave function for ^{6}Li is written as

$$|^6\mathrm{Li}| = N\,\Psi_\alpha \Psi_d \Psi_{\alpha d}\, \xi(1234;56) , \qquad (3)$$

where Ψ_x is the internal wave function of the cluster x, $\Psi_{\alpha d}$ is the inter-cluster wave function, ξ is the spin wave function. Following Schmidt et al. (12) and Jain et al. (13), $\Psi_{\alpha d}$ is given by

$$\begin{aligned}\Psi_{\alpha d}(R) &= R^2\left[\exp(-C_1R^2) + C_2\exp(-C_3R^2)\right] , \text{ for } R \leq R_m,\\ &= B\exp(-CR)/R , \text{ for } R > R_m.\end{aligned} \qquad (4)$$

The constant C is determined by the α-d binding energy. Other parameters appearing in equ. 4 are

$$C_1 = 0.18\ \mathrm{fm}^{-2},\ C_2 = 0.25,\ C_3 = 0.065\ \mathrm{fm}^{-2},\ R_m = 6.13\ \mathrm{fm}.$$

The other cluster wave functions used in our calculations are those of Tang et al. (14) and Alberi et al. (15). The Tang et al. wave function is given as

$$\Psi_{\alpha d}(R) = R^2 \exp(-2\beta R^2/3) \quad , \quad \text{for } R \leq 3.0 \text{ fm}$$
$$\propto \exp(-CR)/R \quad , \quad \text{for } R > 3.0 \text{ fm} \qquad (5)$$
$$\beta = 0.329 \text{ fm}^{-2}$$

The wave function of Alberi et al. is

$$\Psi_{\alpha d}(R) = R^2 \exp(-2\beta R^2/3) \; ; \; \beta = 0.18 \text{ fm}^{-2} \qquad (6)$$

The wave function of Schmidt et al. and Tang et al. fit the B. E., electron scattering and the (p, pd) data, while that of Alberi et al. fits only the electron scattering data.

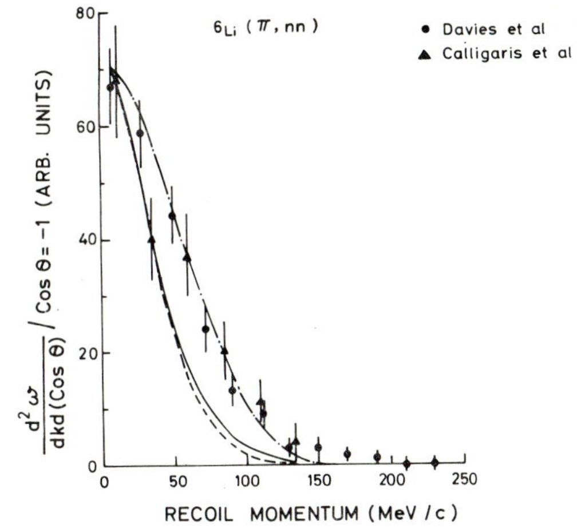

Fig. 1. Recoil momentum distribution. Curves (---) (—), (-•-) correspond respectively to the Schmidt et al., Tang et al. and Alberi et al. wave functions.

The recoil momentum distribution is plotted in fig. 1 along with the experimental data (16). As we seen from this figure the calculated distributions are in more or less accord with the experiments.

^{12}C

For the absorption of pions on $(1p)^2$ nucleons in ^{12}C, a peak is observed(17) in the summed energy spectrum centred around 5.0 MeV excitation energy of ^{10}B. In the theoretical calculations for this peak we have included the contribution from the states of ^{10}B upto 6.0 MeV excitation. The strength of the transitions to various states is calculated using the intermediate coupling wave functions of Boyarkina (18). The radial part of the bound state wave functions for the neutron and the proton are taken as the solution of the Woods-Saxon potential. The parameters of this potential are taken from the work of Elton and Swift (19) on the analysis of the electron scattering. These wave functions also give a good description of the (p, 2p) and (p, d) reactions on ^{12}C (ref. 20). In fig. 2 we have plotted the calculated one neutron energy spectrum along with the experimental data.

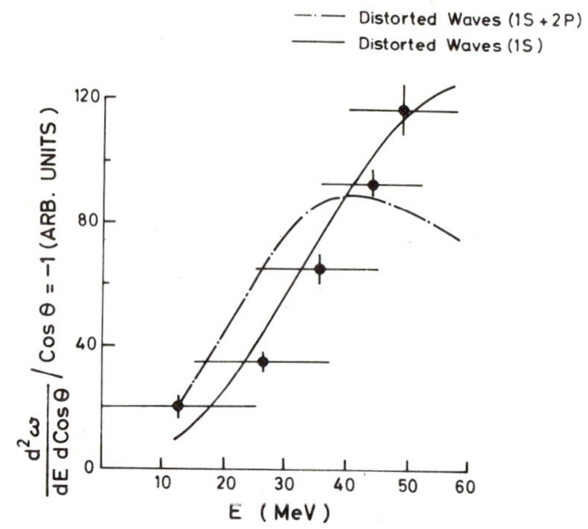

Fig. 2. One neutron energy spectrum for ^{12}C for back to back emission of neutrons.

The solid line corresponds to the contribution of 1S pion while the dot-dashed curve represents the combined contribution of (1S+2P) pion. In calculating the combined strength the relative population of pions in 1S and 2P states is taken as 10;90. This is suggested by the X-ray data (21). As we see from this figure the 1S contribution agrees very well with the experiments. The combined strength, however, disagrees with the experiments.

MODEL DEPENDENCE

In order to investigate the sensitivity of the (π, NN) reaction to the correlation function, in fig. 3 we have compared the recoil momentum distribution for ^6Li in one nucleon model using different parameters for the correlation function and with that in the two-nucleon model. In the one-nucleon model the correlations in the initial and final states are introduced through the Jastrow factors F and G (for simplicity it is assumed that F=G). For the actual form of F, we take the one parameter dependence suggested by Danos (22), i.e.

$$F(r) = 1 - j_0(qr),$$

where j_0 (qr) is the spherical Bessel function and q has the interpretation of the momentum exchanged between the interacting nucleons. The inter-cluster wave function is given by equ. 4.

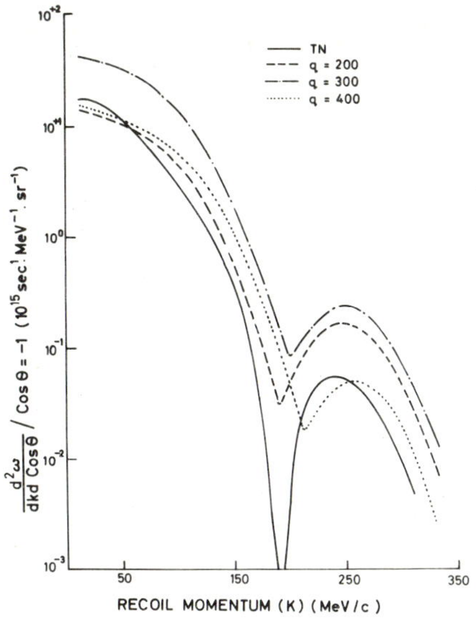

Fig. 3. Recoil momentum distribution for ^{6}Li. The continous curve corresponds to the two-nucleon model.

Different curves in fig. 3 correspond to q=200, 300 and 400 MeV/c. It is observed that the shape of the distribution is about the same for various values of q and in two-nucleon model. The absolute magnitude, however, differ.

Incidentally, the q dependence of the absorption strength in fig. 3 also shows a kinematical effect that it goes through a maximum around $q \approx 300$ MeV/c. This value of q corresponds to the relative momentum of the emitted nucleons equal to about 600 MeV/c. As indicated earlier, this is the matching condition required for the pion absorption.

The author thanks Dr. N. Sarma for his help in preparing the manuscript of this paper and useful discussions.

REFERENCES

(1) S. A. Moszkowski and B. L. Scott, Ann. Phys. 11 (1960) 65.
(2) T. E. O. Ericson, Phys. Lett. 2 (1962) 278.
(3) D. H. Wilkinson, Comm. in Nucl. and Particle Phys. 2 (1968) 48.

(4) B. K. Jain and B. Banerjee, Nuovo Cim. 69 (1970) 419;
B. K. Jain, Bhabha Atomic Research Centre, Bombay, 1970, report-512, unpublished.

(5) J. M. Eisenberg and J. Letourneurx, Nucl. Phys. B3 (1967) 47;
H. J. Weber, Ann. Phys. 53 (1969) 93;
D. S. Kotun, Adv. Nucl. Phys. 3 (1969) 71.

(6) S. G. Eckstein, Phys. Rev. 129 (1963) 413.

(7) R. Jastrow, Phys. Rev. 98 (1955) 14379.

(8) H. Garcilazo and J. M. Eisenberg, Nucl. Phys. A220 (1974) 13.

(9) D. S. Koltun and A. Reitan, Phys. Rev. 155 (1967) 1139.

(10) P. H. Bowen et al. Nucl. Phys. 22 (1961) 640; P. E. Hodgson, Proc. Conf. on Direct Reaction Mech., Padua, 1962, p. 103;
D. T. Winner and R. M. Drisko, Tech. Rep. Dept. Phys., Sarah Meller Scaife Radiation Lab., June 1965, unpublished.

(11) R. Seki and A. H. Cromer, Phys. Rev. 156 (1967) 93,

(12) E. W. Schmidt, Y. C. Tang and K. Wildermuth, Phys. Lett. 7 (1963) 263.

(13) A. K. Jain, N. Sarma and B. Banerjee, Nucl. Phys. A142 (1970) 330;
A. K. Jain and N. Sarma, Phys. Lett. 33B (1970) 271.

(14) Y. C. Tang, K. Wildermuth and L. D. Pearlstein, Phys. Rev. 123 (1961) 548.

(15) G. Alberi and C. Cioffi degli Atti, Proc. Int. Conf. Clustering Pheno. in Nuclei, Bochum, 1969, p. 239.

(16) H. Davies et al., Nucl. Phys. 78 (1966) 663; F. Calligaris et al. Nucl. Phys. A126 (1969) 209.

(17) D. L. Cheshire and S. B. Sabbotka, Phys. Lett. 30B (1969) 244.

(18) A. Boyarkina, Bull. Acad. Sci., U. S. S. R. (phys. ser.) 28 (1964) 255.

(19) L. R. B. Elton and A. Swift, Nucl. Phys. A94 (1967) 52.

(20) R. Shanta and B. K. Jain, Nucl. Phys. 175 (1971) 417;
I. S. Towner, Nucl. Phys. A93 (1967) 145.

(21) D. K. Anderson et al., Phys. Rev. Lett. 24 (1970) 71.

(22) M. Danos, Lecture Notes, Les Houches Summer School, 1968, p. 543.

ON THE ENERGY DEPENDENCE OF ($\pi^{\pm}$, ^{4}He) AND (π^-, ^{12}C) BACKWARD ELASTIC SCATTERING CROSS SECTIONS

F. Balestra, E. Bollini, L. Busso, R. Garfagnini, G. Piragino, A. Zanini
Istituto di Fisica dell'Università, INFN - Sezione di Torino, Italy

C. Guaraldo, R. Scrimaglio
Laboratori Nazionali di Frascati, Frascati, Italy

I. V. Falomkin, M. M. Kulyukin, R. Mach, F. Nichitiu, G. B. Pontecorvo, Yu. A. Shcherbakov
Joint Institute for Nuclear Research, Dubna, USSR

We have measured the large angle elastic scattering cross section for the reaction (π^-, ^{12}C), in the energy range 60 - 90 MeV (1), at 30 MeV for the reaction (π^+, ^{4}He) and at 70 MeV for the reaction (π^- , ^{4}He), using a 180 streamer chamber magnetic spectrometer (2), exposed to the beams of Frascati Laboratory. In Fig. 1 the experimental data for (π^-, ^{12}C) backward elastic scattering cross section are compared with the optical model predictions, using Mach potential (3). This potential differs from the originally proposed gradient potential of Kisslinger (4) cointaining an additional term proportional to the Laplacian of nuclear density. This term, obtained taking into account the Fermi motion of target nucleons, because of its surface-peaked nature, mainly affects the large angle scattering. The experimental data show a maximum value at about 75 MeV, while the theoretical prediction is at about 85 MeV. In both cases it results shifted respect to the maximum value (at about 140 MeV) of the total elastic scattering cross section (5).

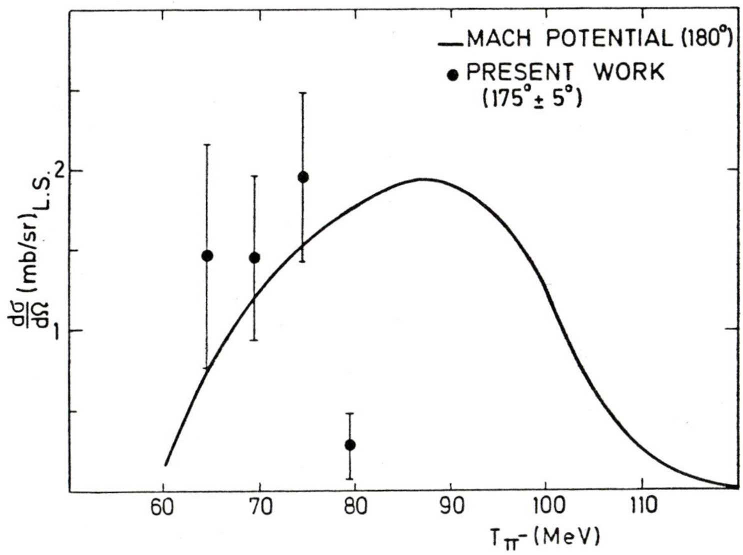

Fig. 1

Fig. 2 shows the comparison between the existing experimental data (6) of ($\pi^{\pm}$, ^{4}He) large angle (> 160°) differential elastic scattering cross section and the optical model predictions, using Mach and Laplacian potentials, as made by Dubna-Torino collaboration (7) for the ($\pi^{\pm}$, ^{4}He) elastic scattering experiment . As for (π^-, ^{12}C) scattering the maximum value of ($\pi^{\pm}$, ^{4}He) backward elastic scattering cross section is shifted respect to the maximum value of the total elastic scattering cross section (at about 150 MeV). These preliminary results show that the study of backward scattering at low pion energy, as also pointed out by Hüfner (8), can certainly give more informations about pion-nucleus scattering.

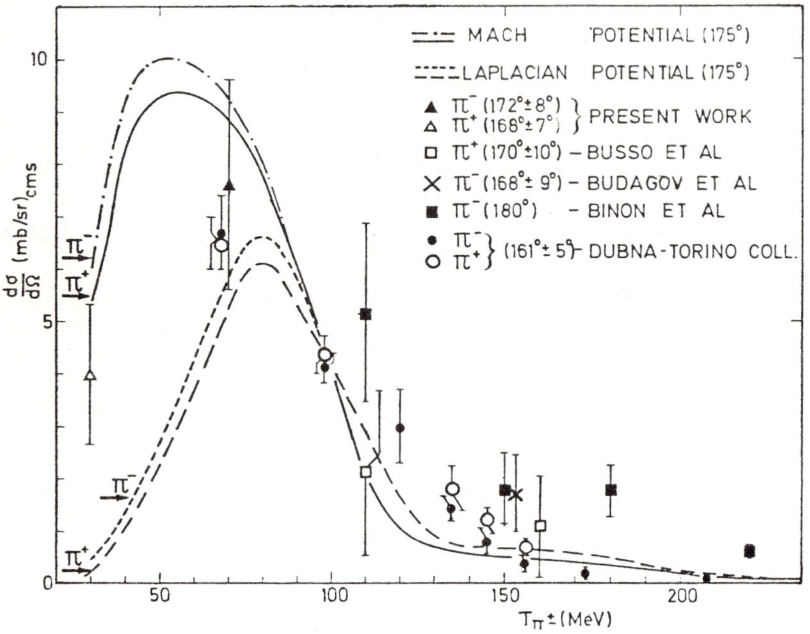

Fig.2

References

1. R. Barbini, C. Guaraldo, R. Scrimaglio, F. Balestra, L. Busso, R. Garfagnini and G. Piragino, Lett. Nuovo Cim., 12 (1975) 359
2. F. Balestra, L. Busso, R. Garfagnini, G. Perno, G. Piragino, R. Barbini, C. Guaraldo, R. Scrimaglio, I. V. Falomkin, M. M. Kulyukin, G. B. Pontecorvo and Yu. A. Shcherbakov, Nucl. Instr. Meth., 119 (1974) 374; 125 (1975) 157
3. R. Mach, Nucl. Phys., A205 (1973) 56
4. L. S. Kisslinger, Phys. Rev., 98 (1955) 761
5. F. Binon, P. Duteil, J. P. Garron, J. Gorres, L. Hugon, J. P. Peigneux, C. Schmit, M. Spighel and J. P. Stroot, B17 (1970) 168
6. L. Busso, S. Costa, R. Garfagnini, G. Piragino, R. Barbini, C. Guaraldo and R. Scrimaglio, "Few particles problems", North-Holland P. C. (1972) 866; Yu. A. Budagov, P. F. Ermolov, E. A. Kushnirenko and V. I. Moskalev, Sov. Phys. JETP, 15(1962) 824; F. Binon, P. Duteil, M. Gouarère, L. Hugon, J. Jansen, J. P. Lagnaux, H. Palevsky, J. P. Paigneux, M. Spighel and J. P. Stroot, Phys. Rev. Lett., 35 (1975) 145
7. Yu. A. Shcherbakov, T. Angelescu, I. V. Falomkin, M. M. Kulyukin, V. I. Lyashenko, R. Mach, A. Mihul, Nguyen Minh Kao, Nichitiu, G. B. Pontecorvo, V. K. Sarychieva, M. G. Sapozhnikov, M. Semerdjieva, T. M. Troshev, N. I. Trosheva, F. Balestra, L. Busso, R. Garfagnini and G. Piragino, Nuovo Cim. (in press)
8. J. Hüfner, Proc. of VIth Int. Conf. on High Energy Physics and Nuclear Structure, Santa Fe and Los Alamos (1975)

ON COHERENT NEUTRAL PION PHOTOPRODUCTION FROM THE DEUTERON

K.Srinivasa Rao and K.Venkatesh
MATSCIENCE, The Institute of Mathematical Sciences, MADRAS
INDIA
and
S.Srinivasa Raghavan
Physics Department, College of Engineering, MADRAS, INDIA

****

In our earlier studies[1,2], we have shown that the cross sections for

$$\gamma d \longrightarrow d\pi^0 \tag{1}$$

at fixed momentum transfers, computed in the impulse approximation, using the amplitudes of Chew et. al.[3], taking into account deuteron wave functions obtained from "realistic" nucleon-nucleon potentials[4], are in reasonable agreement with the Orsay experimental Data[5]. In this contribution, we report of the differential cross sections for $\gamma d \rightarrow d\pi^0$ with different "realistic" potentials and compare the same with the experimental data[6] in the first pion-nucleon resonance region.

Our impulse approximation theory clearly exhibits the resonant structure of the cross section observed for the first time by Bouquet et. al.[5]. In the fixed momentum transfer region, 240 MeV/c $< k <$ 443 MeV/c, we find[2] that the inclusion of the terms involving the deuteron structure functions:

$$F_{SD} = \int_{x_0}^{\infty} u(x) j_2(\tfrac{1}{2} kx) \; w(x) dx = F_{DS}, \tag{2}$$

and

$$F_{DD}^{(\ell)} = \int_{x_0}^{\infty} w^2(x) j_\ell (\tfrac{1}{2} kx) \; dx, \; (\ell = 0,2), \tag{3}$$

where u(x) and w(x) are the S- and D- state radial wave functions of the deuteron, respectively, enhances the cross section (obtained with the F_{SS} term only) by about 10 to 30%, over the incident photon energy range of 170 MeV $< E_\gamma <$ 500 MeV. The $F_{DD}^{(2)}$ terms, in fact, reduce the cross section only by 0.3 to 2%.

In table 1, we show the theoretical differential cross sections, for E_γ = 280 MeV, obtained with several "realistic" nucleon-nucleon potentials. A comparison of these values with the corresponding ones obtained with the Hulthen wave function, clearly reveals the importance of the effect of the explicit inclusion of the D-state admixture in the ground state wave function of the deuteron, on (1). However, we note that the small differences in the "realistic" nucleon-nucleon potentials - due to the nature of the "core", etc. - do not play a significant role in the study of coherent neutral pion, Photoproduction from the deuteron. We observe that the inclusion of the F_{SD} and $F_{DD}^{(\ell)}$ terms tend to diminish the cross sections below $\theta_{c.m} = 25^\circ$ but change sign thereafter and enhance the cross section by about 15%.

Table 1

Angular distribution for $\gamma d \rightarrow d\pi^0$ at E_γ = 280 MeV

$\theta_{c.m.}$	Differential cross section in μb/sr.					
	Hulthen	Yale	MHJ	RSC	dTRS	Expt.[6]
0°	16.97	12.92	12.93	13.15	13.44	-
30°	19.99	16.23	16.28	16.52	16.59	-
60°	18.35	14.92	15.00	15.28	14.99	-
90°	11.74	8.98	9.03	9.25	8.91	-
100°	9.54	7.11	7.15	7.33	7.03	7.18 ± 0.28
110°	7.56	5.47	5.51	5.65	5.41	4.83 ± 0.18
120°	5.85	4.11	4.14	4.26	4.07	3.84 ± 0.14
130°	4.43	3.01	3.03	3.13	2.99	2.90 ± 0.11
140°	3.29	2.15	2.17	2.25	2.16	2.63 ± 0.09
150°	2.42	1.51	1.54	1.61	1.55	2.08 ± 0.08
160°	1.82	1.08	1.10	1.16	1.13	1.79 ± 0.07
180°	1.34	0.74	0.77	0.82	0.81	-

The agreement obtained with our no parameter impulse approximation theory at E_γ = 280 MeV is good but the fit to the experimental data deteriorates as the incident photon energy approaches the first pion-nucleon resonance region ($E_\gamma \sim$ 320 MeV). This is to be expected since multiple scattering corrections have not been taken into account. In recent times, Pazdzerskii[7], Senyushkin[8] and Lazard et. al.[9] have studied this aspect of the problem.

Pazdzerskii[7] states that the multiple scattering correction decreases the differential cross section at all angles, by 25 - 30% for E_γ = 285 MeV and by 30 - 35% for E_γ = 345 MeV, making use of a Hulthen wave function for the deuteron. On the other hand, Senyushkin[8] finds that the multiple scattering contribution enhances the total cross section below (at E_γ = 285 MeV) and above the resonance region and reduces the total cross section in the resonance region by about 30%. Senyushkin employs a phenomenological deuteron wave function having a hard-core radius as well as a D-wave contribution (P_D = 7 %). The angular distributions obtained by Senyushkin have a different shape when compared to those in refs. 1,5 and 9 and further, he makes no comparison with the available experimental data. Senyushkin's results show a slightly enhanced differential cross section at all angles for E_γ = 235 MeV, which contradicts the results of Pazdzerskii. But, the most recent study of Lazard et. al.[9], using the multiple scattering correction in the lowest order i.e. taking to account only the pion rescattering contribution - and the MHJ deuteron wave functions, reveals that, at E_γ = 320 MeV, though the differential cross section is reduced upto about $\theta_{c.m.} \sim 110^o$, as the angle increases, the rescattering contributions become predominant and enhance the cross section. This is not supported by the studies of Pazdzerskii and Senyushkin who has plotted the differential cross section for E_γ = 315 MeV. We therefore wish to emphasize that it is necessary to make consistant studies of multiple scattering corrections to the $\gamma d \to d\pi^o$ cross sections.

The authors are thankful to Professor Alladi Ramakrishnan and Professor R.Vasudevan for their interest in this work. One of us (K.S.R.) wishes to thank Professor D.W.L.Sprung, McMaster University, Hamilton, Canada, for readily making his computer code available to get the deuteron structure factors for various nucleon nucleon potentials.

References

1. K.Srinivasa Rao and S.Srinivasa Raghavan, Prog. Theor. Phys. 52(1974) 578. Also, K.Srinivasa Rao, R.Parthasarathy and V.Devanathan, J. de Physique 34 (1973) 683.

2. K.Srinivasa Rao, K.Venkatesh and S.Srinivasa Raghavan, to be published.

3. G.F.Chew, M.L.Goldberger, F.Z.Low and Y.Nambu, Phys. Rev. 106 (1957) 1345.

4. K.E.Lassila, M.H.Hull, Jr., H.M.Ruppel, F.A.McDonald and G.Breit, Phys. Rev. 126 (1962) 881, R.V.Reid, Jr., Ann. Phys. 50 (1968) 411, J.W.Humberston and J.B.G.Wallace, Nucl. Phys. A141 (1970) 362, R.de Tourreil, B.Rouben and D.W.L.Sprung, Orsay preprint IPNO/TH. 74-14 (1974). These are referred to as Yale, Reid, Modified Hamada-Johnston (MHJ) and dTRS potentials, respectively, in the text.

5. B.Bouquet, J.Buon, B.Grelaud, H.Ngynen Ngoc, P.Petroff, R. Riskalla and R.Tchapoutian, Phys. Letts. 41B (1972) 536, Nucl. Phys. B79 (1974) 45.

6. G.von Holtey, G.Knop, H.Stein, J.Stumpfig and H.Wahlen, Z.Phys. 259 (1973) 51.

7. V.A.Pazdzerskii, Sov. J. Nucl. Phys. 6 (1962) 278.

8. V.B.Senyushkin, Sov. J. Nucl. Phys. 18 (1974) 270.

9. C.Lazard, P.Petroff, R.J.Lombard and Z.Maric, Orsay preprint IPNO/TH.74-18 (1974).

MODELS FOR PION PRODUCTION BY PROTONS IN LIGHT NUCLEI; pd → tπ AT MEDIUM ENERGIES

V.S. Bhasin
Department of Physics and Astrophysics, University of Delhi, Delhi-110007, India

Increasing efforts are being made to understand the role of pions in nuclear reactions. In particular, pion production in the proton bombardment of light nuclei ranks as a most fundamental problem in intermediate energy research. Recent data [1] in pd → tπ, taken at two different energies, together with some earlier data [2] provide a survey of pion angular distributions over the range of incident proton energies 300-600 MeV. The availability of this data for this reaction thus puts a new and hopefully important constraint on the theoretical models which can be used to describe such reactions as pd → πt, pt → πα etc. in the medium energy range.

The simplest model is that of "pionic proton stripping" or the so-called single nucleon mechanism illustrated in Fig. 1(a) in which the pion is assumed to be emitted directly by the incoming nucleon which is then absorbed by the nucleus. For low energy pions (~50 MeV) adequate results are obtained [3] but they are very sensitive to the details of the nuclear wave function and to the distortion of the pion wave. For medium energy pions (~200 MeV) agreement is much less satisfactory since high momentum transfers and short range (~0.33 F) components of the nuclear wave function are involved. The main advantages of this approach are its simplicity and compatibility with existing DWBA codes. Moreover it does not lead to a π^- production which provides a natural explanation for the suppression of π^-s observed experimentally[4]. The serious disadvantage of this mechanism, however, is that it requires that the entire momentum transfer, which is quite large, be absorbed by a single nucleon. Thus the results become very sensitive to the large momentum components of the wave function, which are in general, known very poorly.

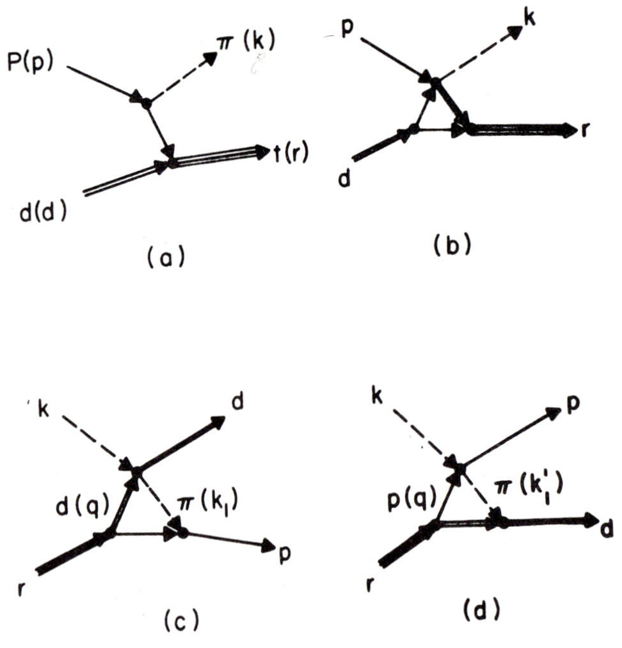

Fig.1(a). Feynman diagram for "pionic proton stripping" process in pd → tπ. (b) Feynman diagram for Ruderman model. (c) Feynman diagram for time reversed Process πt → pd in Yao model. (d) Feynman diagram in Yao model expressing pd → tπ in terms of πp → πp and π+ (singlet deuteron) → deuteron.

A second approach originally suggested by Ruderman [5] assumes that the basic mechanism for pion production in nuclei is the same as that in p-p collisions and, in particular, because of kinematic constraints, the same as in p p → π d. Thus, as shown in Fig. 1(b), the incident nucleon interacts with a nucleon within the target nucleus causing a pion to be emitted with the incident nucleon being captured by the target nucleus. Obviously since two nucleons are involved the momentum transfer can be shared among the nucleons and thus the result is not quite so sensitive to the unknown large momentum components of the wave functions. Based on the model, and using the independent particle model of the nucleus, Ingram et al [6] have benn able to relate in general, the reaction $A(p,\pi)A+1$ to the "elementary reaction" $p(p,\pi^+)d$. They thus study the pion production from a number of light nuclei and find a good fit (within a normalization factor $\sim$2) to the energy dependence of the $pd \rightarrow t\pi$ forward cross section but achieve what they call a limited success, at the level of orders of magnitude, in predicting the angular distribution.

Recently Fearing [7] obtained significant improvement over the earlier calculations by including effects of distortion or multiple scattering and by treating the overlap of the initial and final wave functions more accurately. Even with these modifications the model is not found to be adequate especially at large angles and at relatively high energies, in so much so that the backward peak at 470 MeV is not at all reproduced. The reasons for the absence of this backward peak could be twofold: i) the nuclear wave function still greatly inhibits large momentum transfers; and ii) the multiple scattering effects within deuteron by the incoming proton which could play an important role at large angles are not adequately taken into account in the model.

Keeping essentially these limitations of the Ruderman model in view, Duck and myself [8] proposed a possible extension of the Yao model [9] for the reaction $pp \rightarrow d\pi$ to a process $pA \rightarrow \pi(A+1)$ by supposing that the momentum is essentially transferred by the exchange of a single pion far off the mass-shell and the nucleon scattering takes place at low relative momentum so that the long range part of the nuclear wave function also makes a significant contribution. Yao first proposed the model for $pp \rightarrow \pi d$ in the GeV range but Barry [10] has shown that it can be extended to lower energies provided oneuses a normalization factor (related to the spread of relative momentum of the virtual nucleons) in evaluating the sum over intermediate states.

The calculations in this model as applied to the reaction $pd \rightarrow \pi t$ can be greatly felicitated by working in the rest frame of triton and visualizing it from the time reversed process $\pi t \rightarrow pd$ as illustrated in Figs. 1c and 1d. In Fig. 1c the amplitude, which we find to be dominant, arises from the scattering of the pion of momentum k from a deuteron of momentum q in the triton (of momentum r) to the final deuteron of momentum d and a virtual pion of momentum k_1 which is absorbed by the spectator nucleon leading the final proton of momentum p. Thus the matrix element M, for the process shown in Fig. 1(c) is

$$M = \int d\vec{q}\; G_{tdN}\, G_o\, M_{\pi d}\, G_{N\pi N}\, F(k_1^2)/(k_1^2-\mu^2) \quad , \tag{1}$$

where the nucleons constituting deuteron and triton are taken to be non-relativistic while pions are treated relativistically. In eq.(1) k_1 is the four momentum of the off-shell pion and $F(k1)$ is the Ferrari-Sellari form factor introduced to account for the off-mass shell pion; G_o is the Green's function for the free n-d system, $G_o = 2\mu/q^2 + \alpha^2$ (μ being the reduced mass (=2m/3)) and α the triton binding energy parameter). $M_{\pi d}$ is the matrix elemnt for π-d scattering.

The evaluation of the integral in Eq.(1) is simplified by observing that with a rapid fall off provided by the Greens function Go and the momentum dependence implied in the vertex G_{tdN} it is reasonable to assume that the virtual pion momentum k_1 is slowly varying for momenta q so that the pion propagator can be taken out of the integral and q-dependence of the factors in the integrand can be neglected. The integration over virtual momenta q then simply involves integral ove r the Green's function Go multiplied by the vertex G_{tdN} i.e. the integral over the Fourier transform of the triton wave function. This corresponds to the triton wave function evaluated at the origin $\Psi(0)$ entering the amplitude as a constant parameter. With these simplifying assumptions the expression for the $pd \to t\pi$ c.m. differential cross-section is obtained as

$$\frac{d\sigma}{d\Omega} = \frac{k}{p}\left[\Psi(0)\, g\, \frac{F(k_1^2)}{k_1^2-\mu^2}\right]^2 \frac{k_1^2}{6}\, \frac{s_{\pi d}}{s}\left(\frac{d\sigma}{d\Omega}\right)_{\pi d} . \tag{2}$$

In fact, Eq.(2) involves the πd cross-section for which one of the pions is off-the-mass-shell. The various symbols such as s, $s_{\pi d}$ etc. are defined in terms of of the Mandelstam variables, viz.

$$s = (p+d)^2 = (t+r)^2; \quad t = (k-p)^2 = (d-r)^2; \quad u = (k-d)^2 = (r-p)^2. \tag{3}$$

Also we take $s_{\pi d} = \frac{2}{3}s + \frac{1}{3}\mu^2 - 2$ ($\hbar$ = c = nucleon mass = 1) to be the square of the πd energy in the πd cm system. This value of the scattering angle has the virtue that it correlates very closely the pion production angle in the over all c.m. system.

Another apparently competing process within the frame work of triangle diagram is the one shown in Fig. 1(d) in which the pion scatters from a nucleon of the triton and then the virtual pion is absorbed by a spin singlet and isotriplet deuteron (d^*) going to the final deuteron. We thus encounter in this mechanism the πd^*-d vertex and the π-N amplitude. In evaluating the former we suppose that the virtual pion with momentum k_1 is absorbed by d^* in two steps: each static nucleon absorbing momentum $k_1/2$. This leads to involve P_{11} π-N amplitude (with pions off the mass-shell and at energy below threshold) which is calculated using the Chew-Low model. With this factor Fig. 1(d) contributes a cross section $\lesssim 1\%$ of Fig. 1(c). We shall therefore drop this term and concentrate only on the mechanism given in Fig. 1(c).

To present the results as obtained from Eq.(2) for $pd \to t\pi$ differential cross-section at a few incident proton energies, we evidently need information on πd scattering. In Ref. [8] we employed a rather simplified model for the elastic π-d scattering by observing that the experimental data at incident pion energies from 90 MeV to $\sim$250 MeV can be explained, at least in qualitative terms, under impulse approximation. The results thus obtained for the total cross-section as well as the angular distribution were depicted in Figs. 2 and 3 of Ref.[8]. While the results were found to be in overall agreement with experiment, it was, however, noticed that the differential cross-section at large angles was rather large. The apparent discrepancy was primarily attributed to the model employed for π-d scattering.

In view of the rather simplified model used for π-d scattering on which the results were found to depend critically we have here, as an alternative, fed the experimental data, where available, for π-d scattering directly in Eq.(2). In this way we are able to test, in direct terms, the validity of Yao-Barry model. Fig. 2 depicts the differential cross-section when the experimental data [13] for elastic π-d scattering is employed for incident energy of pions around 180 MeV which, in fact, effect, corresponds to the emitted pion energy when the protons are bombarding deuterons at 470 MeV. The experimental value of the forward differential cross-section in π-d which is about 80 mb/sr determined $\Psi(0)$ = 0.012 which, indeed, is closer to the one given by Lim [12]. The agreeement with experiment is rather excellent especially at large angles. This is very encouraging in view of the fact that other models of this reaction are particularly plagued by extremely small cross-sections at large angles. It may be stressed that the rescattering effects within the deuteron via π-d scattering by feeding directly the experimental differential cross-section from 0° to 180° have been adequately incorporated - a feature which is directly responsible for reproducing the backward peak.

It would have been very desirable to obtain the π-d data at somewhat higher pion energies ($\sim$250 NeV) to further check the consistency of this model especially at 590 MeV.proton energy in this reaction.

The results in this model are thus encouraging enough to apply it to the process pt$\rightarrow\pi\alpha$ and study especially the large angle cross-section. By analogy with pd calculations one might expect the pt$\rightarrow\pi\alpha$ angular distribution to be determined principally by elastic scattering. It may be a added that we are using here a rather simplified version of the model in so far as we ignore the effect of inelastic and charge-exchange processes involved in the scattering of pions on the targets. Experimental evidence both in the case of π-d and π-t scattering shows that these contribute too little to the differential cross-section especially at small angles and at medium energies. Therefore, their effect may only be marginal on pion production in pd and pt collisions. Another gross simplification that seems to be unavoidable so far is to ignore any dependence of the pion-target cross-section on the mass of the virtual pion involved.

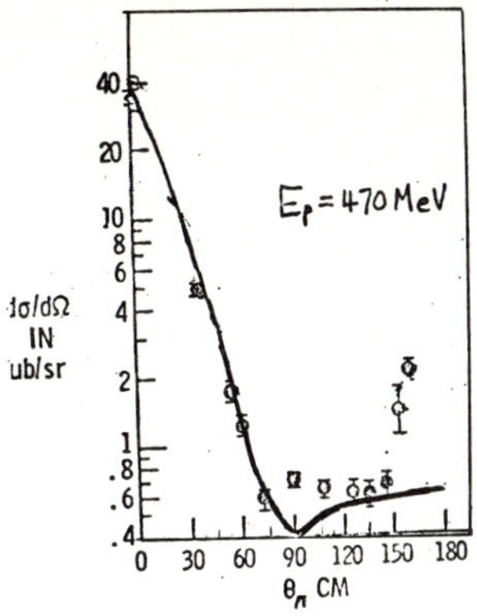

Fig.2. Differential cross-section for pd$\rightarrow$tπ at proton incident energy 470 MeV. The input π -d cross-section is fed from experimental data[13].

Another area where this model may be directly applied is in studying a structure in the missing mass spectrum of processes like pd$\rightarrow$pX, pd$\rightarrow He^3X^0$ etc. in the GeV region. The reaction pd$\rightarrow$pX(where X denotes the combination of particles with total baryon number 2) with incident proton beam 10 to 26 GeV/c momentum range was studied by Belletini et al [14] who had established the existence of a well pronounced maximum of width $\sim$250 MeV in the missing mass spectrum for m_X. The experiment was carried out with small momentum transfers in the range, $10^{-3} < t_{pp}$ $(GeV/c)^2 < 10^{-1}$. A theoretical investigation of this process based on a triangle diagram was carried out by Dalkarov [15] who attributed the observed structure to a moving complex singularity. However, in order to rule out the possibility of alternative mechanisms, further experimental data covering wide range of momentum transfers would have to be obtained. The process pd$\rightarrow X^0He^3$, on the other hand, where detailed experimental information on the missing mass spectrum is fortunately available, should find an important place for investigating the role of triangular mechanisms.

REFERENCES

1. W. Dollhopf, et al, Nucl. Phys. A217 (1973) 381.
2. K.R. Chapman, et al, Nucl. Phys. 57 (1964) 499.
3. E. Rost and P.D. Kanz, Phys. Lett. 43B (1973) 17.
4. J.J. Domingo et al, Phys. Letters 32B (1970) 309.
   J. Rohlin et al, Phys. Letters 40B (1972) 539.
5. M. Ruderman, Phys. Rev. 87 (1952) 383.
6. C.H.Q. Ingram et al, Nucl. Phys. B31 (1971) 331.
7. H.W. Fearing, Phys. Rev. C110 (1975) 1211.
8. V.S. Bhasin, I.M. Duck, Phys. Lett. 46B (1973) 309.
9. T.Yao, Phys. Rev. 134 (1964) B454.
10. G. Barry, Ann. Phys. 73 (1972) 482.
    G. Barry, Phys. Rev. D7 (1973) 1441.
11. A. Ramakrishnan, et al, Nucl. Phys. 29 (1962) 680.
12. T.K. Lim, Phys. Rev. Lett. 30 (1973) 709.
    L.P. Kok and A.S. Rinat (Reiner), Nucl. Phys. A156 (1970) 593.
13. J. Norem, Nucl. Phys. B33 (1971) 512.
14. G. Belletini et al, Phys. Lett. 18 (1965) 167.
15. Dal'Karov, JETP Lett . 3 (1966) 150.
    See also I.S. Shapiro, Soviet Uspekhi Vol. 10 (1968) 515.
16. J. Banaigs et al, Nucl. Phys. B67 (1973) 1.
    See also, J.C. Arjos et al, Nucl. Phys. B67 (1973) 37.

CURRENT ALGEBRA AND LOW ENERGY πN SCATTERING

Michael Scadron, Physics Department, University of Arizona, Tucson, Arizona USA

I. INTRODUCTION

The recent accurate low energy πN data [1] makes it possible to test in detail the various predictions of current algebra. While the data is slightly above the threshold, the tests are at threshold and in the sub-threshold region $\nu < \mu$, $|t| < 4\mu^2$, where ν is the pion lab energy, μ is the pion mass and t is the invariant momentum transfer. The result is that current algebra does indeed map out the physical amplitudes [2,3], but one must be careful to appreciate the delicate and sometimes accidental cancellations between the various amplitudes, e.g. at threshold. One can also use the current algebra representation to extrapolate the πN amplitudes off the pion mass shell $q^2 = \mu^2$. This has important applications in nuclear physics, including descriptions of the three body binding energies in nuclear matter [4-6], three body forces in low Z nuclei [7] and pion condensates in neutron stars [8].

II. THE CURRENT ALGEBRA WARD IDENTITIES

One begins by considering the axial-vector nucleon amplitude

$$M^{ij}_{\mu\nu} = i \int d^4x\, e^{iq'\cdot x} \langle N|[A^i_\mu(x), A^j_\nu(0)]|N\rangle\, \theta(x_0) \tag{1}$$

where i,j (μ,ν) are isospin (space-time) indices, and then computes its double divergence keeping the equal time commutator terms (ETC) while discarding surface terms:

$$q'^\mu M^{ij}_{\mu\nu} q^\nu = -q'^\mu \langle N|[A^i_\mu, Q^j_5]_{ETC}|N\rangle + \langle N|[Q^i_5, i\partial\cdot A^j]_{ETC}|N\rangle$$
$$+ i \int d^4x\, e^{iq'\cdot x} \langle N|[\partial\cdot A^i(x), \partial\cdot A^j(0)]|N\rangle\, \theta(x_0) . \tag{2}$$

The algebra of currents implies

$$\langle N|[A^i_\mu, Q^j_5]_{ETC}|N\rangle = i\varepsilon^{ijk} \langle N|V^k_\mu|N\rangle$$
$$= i\varepsilon^{ijk}\, \tfrac{1}{2}\tau^k\, \bar{u}[F^V_1(t)\gamma_\mu + F^V_2(t)\, i\sigma_{\mu\nu}(q-q')^\nu]u, \tag{3}$$

$F^V_{1,2}(t)$ being the isovector nucleon form factors (F^p-F^n) and the πN σ term is defined by

$$\langle N|[Q^i_5, i\partial\cdot A^j]_{ETC}|N\rangle = \delta^{ij}\, \sigma_N(t)\, \bar{u}u . \tag{4}$$

One then extracts the four πN amplitudes $[(S-1)_{fi} = i\bar{u} Mu\, \delta^4(P_{fi})]$, defined by $M^{ij} = \delta^{ij}M^{(+)} + i\varepsilon^{ijk} M^{(-)}$ and ($m=m_N$)

$$M^{(\pm)} = A^{(\pm)} + B^{(\pm)}\gamma\cdot(q'+q)/2 = F^{(\pm)} - B^{(\pm)} [\gamma\cdot q', \gamma\cdot q]/4m \tag{5}$$

where $F = A + \nu B$, $\nu = (s-u)/4m$, from the Ward identity (2) by using the PCAC relation $\partial\cdot A = f_\pi\mu^2\phi_\pi$ in the third term on the RHS of (2) and combining it with pion pole part of the LHS of (2), denoted by $M^\pi_{\mu\nu}$. Further removing the nucleon (s and u channel) poles from $M_{\mu\nu}$ and M according to

$$M_{\mu\nu} = M^\pi_{\mu\nu} + M^N_{\mu\nu} + \bar{M}_{\mu\nu};\ M = M^N + \bar{M} , \tag{6}$$

where $\overline{M}_{\mu\nu}$ and $\overline{M}$ are the remaining background amplitudes, and expanding (2) in powers of q' and q, one is led to the current algebra representation of the on-shell πN background amplitudes

$$\overline{F}^{(+)}(\nu,t) = \sigma_N(t)/f_\pi^2 + C^{(+)}(\nu,t) \tag{7}$$

$$\nu^{-1}\overline{F}^{(-)}(\nu,t) = F_1^V(t)/2f_\pi^2 - g^2/2m^2 + \nu^{-1}C^{(-)}(\nu,t) \tag{8}$$

$$\nu^{-1}\overline{B}^{(+)}(\nu,t) = \nu^{-1}D^{(+)}(\nu,t) \tag{9}$$

$$\overline{B}^{(-)}(\nu,t) = (F_1^V(t) + F_2^V(t))/2f_\pi^2 - g^2/2m^2 + D^{(-)}(\nu,t) \tag{10}$$

(the sign change of σ_N in (7) depends upon the soft limit to be discussed later). Here f_π is 92 MeV, $g^2/2m^2$ ($g^2/4\pi \approx 14.3$) is the non-pole background difference between $q'^\mu M^N_{\mu\nu} q^\nu$ and M^N in (6) for the πNN coupling $g\gamma_5$, and $C^{(\pm)}$ and $D^{(\pm)}$ respectively represent the $F^{(\pm)}$ and $B^{(\pm)}$ type background axial-vector amplitudes $q'^\mu \overline{M}_{\mu\nu} q^\nu f_\pi^{-2}$. To form the complete current algebra amplitudes, one adds (7) - (10) together with the (pseudoscalar coupling) nucleon pole terms

$$F_P^{(+)}(\nu,t) = \frac{g^2}{m}\,\frac{\nu_B^2}{\nu_B^2-\nu^2}\ ,\quad \nu^{-1}F_P^{(-)}(\nu,t) = \frac{g^2}{m}\,\frac{\nu_B}{\nu_B^2-\nu^2}$$

$$\nu^{-1}B_P^{(+)}(\nu,t) = \frac{g^2}{m}\,\frac{1}{\nu_B^2-\nu^2}\ ,\quad B_P^{(-)}(\nu,t) = \frac{g^2}{m}\,\frac{\nu_B}{\nu_B^2-\nu^2}\ . \tag{11}$$

It should be stressed that $F_P^{(+)}$ here is evaluated according to dispersion theory; the numerator of $A_P^{(+)}+\nu B_P^{(+)}$ being ν^2 is evaluated at the residue of the pole, $\nu = \pm\nu_B$, where $\nu_B = -q'\cdot q/2m = (t-2\mu^2)/4m$. In effect, the background amplitude $\overline{F}_P^{(+)} = \overline{A}^{(+)} + \nu\overline{B}^{(+)} - g^2/m$ corresponds to $F_P^{(+)}$ being computed via pseudovector coupling $g(\mu/2m)i\partial\cdot\gamma\gamma_5$.

To obtain a complete representation of the πN amplitudes one must find $C^{(\pm)}$ and $D^{(\pm)}$. Since these latter amplitudes stem from $q'^\mu \overline{M}_{\mu\nu} q^\nu$, it is possible to describe them at low energy corresponding to "small" values of q'_μ and q_ν. In this case one expects $\overline{M}_{\mu\nu}$ to be dominated by the Δ isobar [9], including a non-resonant Δ background part analogous to the $g^2/2m^2$ nucleon background parts in (8) and (10). Because of the ambiguous nature of the spin 3/2 Δ propagator away from resonance, it is non-trivial to obtain this latter Δ background part. A straightforward but tedious determination of the dispersion-theoretic amplitudes in $M_{\mu\nu}$ free of kinematic singularities allows one to use the (unique) on-shell spin 3/2 projection operator (at resonance), leading to [2] ($h_{1,2} \to h_{0,1}$ in (33) of ref. 2)

$$C_\Delta^{(+)} = \frac{2g^{*2}}{9m^2}\,\frac{\nu_\Delta\alpha}{\nu_\Delta^2-\nu^2} - \frac{g^{*2}(M+m)}{9M^2m}\,[(M-m)(M+m)^2 + 2\mu^2(2M+m) - q'\cdot q\,(2M-m+2\mu^2(M+m)^{-1})] \tag{12}$$

$$\nu^{-1}C_\Delta^{(-)} = -\frac{g^{*2}}{9m^2}\,\frac{\alpha}{\nu_\Delta^2-\nu^2} + \frac{g^{*2}}{9M^2}\,((M+m)^2 - q'\cdot q) \tag{13}$$

$$\nu^{-1}D_\Delta^{(+)} = \frac{2g^{*2}}{3m}\,\frac{\beta}{\nu_\Delta^2-\nu^2} \tag{14}$$

$$D_\Delta^{(-)} = -\frac{g^{*2}}{3m}\frac{\nu_\Delta \beta}{\nu_\Delta^2-\nu^2} + \frac{2g^{*2}m}{9M}\left(1+\frac{\nu_\Delta}{M}\right) \tag{15}$$

where [10] $g^{*2}/4\pi \approx 0.264\mu^{-2}$, $M \approx 1231$ MeV and (12)-(15) are functions of ν and t (and q^2) with $\alpha = (E_\pi^2-q'\cdot q)((M+m)^2-q'\cdot q) + (\mu^2-q'\cdot q)((M+m)\ E_\pi-\frac{1}{2}(\mu^2+q'\cdot q))$, $\beta = E_\pi^2 - \frac{1}{3}(M+m-E_\pi)^2 - q'\cdot q$, $2m\nu_\Delta = M^2-m^2-q'\cdot q$ and $2ME_\pi = M^2 - m^2 + \mu^2$. The first terms in (12)-(15) are the unique pole contributions and the second terms in (12), (13) and (15) are the dispersion-theoretic background parts, with g* assumed to constant, independent of q^2. Including q^2 variations of g and g* correspond to Goldberger-Treiman corrections of O(5%), the same size as most other N* contributions [2,3,13]. Thus, to first approximation, (12)-(15) represent the bulk of the background terms of (7)-(10). It should be noted that there is an almost total cancellation in C_Δ^+ between the pole and non-pole terms at $\nu = q'\cdot q = 0$. This occurs because the structure of $q'^\mu M_{\mu\nu}^{(+)} q^\nu$ dictates that in general $C^{(+)}$ must have the form [11] $a\nu^2 + bq'\cdot q$ plus correction terms $O(\mu^4)$.

It also is possible to compute the non-pole parts of (12)-(15) for various field theory models of the spin 3/2 Δ propagator, [3,12,13] characterized by an arbitrary parameter Z. While Z = 1/2 seems to be favored on theoretical grounds [13], Z ~ 0 to -1/2 appears more in line with the data [3].

Finally, the t dependence of the first terms in (7)-(10) is needed to fully test the current algebra structure. Experiment gives $F_1'^V(0) = .046\ \mu^{-2}$ and $F_2'^V(0) = .220\ \mu^{-2}$, and we parameterize the t dependence of $\sigma_N(t)$ as

$$\sigma_N(t) = \sigma_N\,[1 + \beta(t/\mu^2 - 2)] \tag{16}$$

where the $t/\mu^2 - 2$ structure follows from the soft limit to be discussed later.

III. COMPARISON WITH ON-SHELL DATA

First one investigates the sub-threshold region $\nu < \mu$, $|t| < 4\mu^2$, $q^2 = q'^2 = \mu^2$ in terms of an expansion about the symmetry point [10] $\nu = t = 0$,

$$\bar{F}^{(+)}(\nu,t) = f_1^+ + f_2^+ t + f_3^+\nu^2 + f_4^+\nu^2 t + f_5^+\nu^4 + f_6^+\nu^4 t + f_7^+ t^2 + \ldots \tag{17}$$

with similar expansions for $\nu^{-1}\bar{F}^{(-)}$, $\nu^{-1}\bar{B}^{(+)}$ and $\bar{B}^{(-)}$ with coefficients f_i^-, b_i^+ and b_i^-, respectively. Owing to the recent accurate low energy data [1], these 28 coefficients have been obtained from subtracted, fixed t dispersion relations [14]. The results are listed in Table I (in units of $\mu = 139.6$ MeV, $\bar{u}u \to 2m$) and have roughly 5% errors in the first column:

TABLE I - EXPERIMENT	1	2	3	4	5	6	7
f_i^+	-1.45	1.18	1.14	.17	.20	.04	.035
f_i^-	-.52	-.10	-.18	-.04	-.04	-.01	.01
b_i^+	-3.49	.17	-.98	.09	-.30	.04	-.01
b_i^-	8.34	.20	1.04	-.06	.27	-.03	.07

Before presenting the current algebra predictions, one first determines the σ term at the point [11] $\nu = 0$, $t = 2\mu^2$ where the background $C^{(+)}(0,2\mu^2)$ must be small, $O(\mu^4)$. Applying (16) and Table I to (7) we see that

$$\bar{F}^{(+)}(0,2\mu^2) = \sigma_N/f_\pi^2 + O(\mu^4) \approx 1.05\ \mu^{-1}, \tag{18}$$

corresponding to $\sigma_N \approx 65$ MeV. Other recent determinations [2,15] of σ_N range from roughly 55 to 75 MeV. (The value of σ_N has significant implications for the theory of chiral symmetry breaking, such as a non-strange quark mass of 140 MeV). The slope parameter β can be found by computing the f_1^+ and f_2^+ coefficients from (7) and comparing with Table I, giving an approximate value [3,5] of $\beta \approx .4$. Thus, using $\sigma_N \approx 65$ MeV and $\beta \approx .4$, the current algebra predictions (7)-(10) with the Δ dominated background amplitudes of (12)-(15) give the results listed in Table II:

TABLE II -

THEORY	1	2	3	4	5
f_i^+	-1.35	1.20	.73	.11	.12
f_i^-	-.59	-.09	-.17	-.02	-.03
b_i^+	-3.54	.15	-.64	.05	-.12
b_i^-	8.31	.20	.75	-.05	.14

While the first two columns in Table II are in excellent agreement with Table I, the last three columns of Table II are typically 30-40% smaller in magnitude than in Table I. Since these latter columns arise from higher order terms in power series expansions of the pole denominators $(\nu_\Delta^2 - \nu^2)^{-1}$, it is reasonable to assume that these discrepancies are due to finite Δ width corrections to such isobar denominators. Thus we conclude that the current algebra expansions (7)-(10) correctly map out the on-shell sub-threshold πN amplitudes.

At threshold, $\nu = \mu$, $t = 0$, one can predict scattering lengths from (7)-(10). For the s-waves one has $(3a_0^{(+)} = a_1 + 2a_3$, etc.)

$$4\pi(1 + \mu/m)a_0^{(+)} \approx -g^2\mu^2/4m^3 + (1-2\beta)\sigma_N/f_\pi^2 + C^{(+)}(\mu,0) \quad (19)$$

$$4\pi(1 + \mu/m)a_0^{(-)} \approx g^2\mu^3/8m^4 + \mu/2f_\pi^2 + C^{(-)}(\mu,0), \quad (20)$$

and with $C_\Delta^{(-)}(\mu,0) = -.04\ \mu^{-1}$, (20) gives $a_0^{(-)} \approx .078\ \mu^{-1}$, compared to the measured value $.087\ \mu^{-1}$. The discrepancy for $a_0^{(+)}$ is far more significant, because (19) gives, with $\beta \approx .4$ and $C_\Delta^{(+)}(\mu,0) \approx -.68\ \mu^{-1}$, a value for $a_0^{(+)}$ eight times experiment, $-.005 \pm .002\ \mu^{-1}$. In effect, the first two terms in (19) almost cancel for $\beta \approx .4$, leaving a large $C_\Delta^{(+)}$ to enhance $a_0^{(+)}$. The problem in this case is that $C_\Delta^{(+)}(\mu,0)$ ought to be much smaller, reflecting the suppression of f_3^+ and f_5^+ in Table II. This problem is further enhanced by the accidental suppression of the σ term and its cancellation against the Born term at threshold. In the field theory approach of Ref. 3, this problem is resolved by adjusting Z and by including a diffractive high energy contribution to $\overline{F}^{(+)}$ and $\overline{B}^{(+)}$. The p-wave scattering lengths, however, encounter no such problem because the first two columns of Table II dominate and there is no accidental cancellation among the various terms. In particular, since the Chew-Low approach teaches us that the p-wave amplitudes are primarily controlled by the nucleon poles, it suffices to consider combinations where the Born terms are effectively small,

$$a_{1-}^{(+)} + (2 + 3\mu/m)a_{1+}^{(+)} \approx (2\pi)^{-1}\,[g^2/4m^3 + (\partial/\partial t + \mu/2m\ \partial/\partial\nu^2)\overline{F}^{(+)}(\mu,0)] \quad (21)$$

$$a_{1-}^{(-)} - a_{1+}^{(-)} \approx (8\pi m)^{-1}[-g^2\mu/4m^3 + (1 + \kappa^V)/2f_\pi^2 + D^{(-)}(\mu,0)]. \quad (22)$$

Applying (7), (12) and $D_\Delta^{(-)}(\mu,0) \approx 5.62\ \mu^{-2}$, (21) and (22) become $.24\ \mu^{-3}$ and $.064\ \mu^{-3}$, comparing favorably with the measured values, $.27\ \mu^{-3}$ and $.068\ \mu^{-3}$, respectively. Finally, s-wave effective ranges can be investigated in the current algebra scheme [9]. Once again the situation is complicated by the large cancellations between the Born contributions and the background [8-10]; the data is somewhat uncertain as well [16]. Consequently, although current algebra qualitatively explains the s-wave scattering lengths and effective ranges [17,9], in a quantitative sense the delicate cancellations at threshold place a severe demand on the theory. This is less a statement on the drawbacks to the theory and more a conclusion that threshold is indeed a "funny" point.

IV. PCAC AND OFF-SHELL EXTRAPOLATIONS

Defining $F(\nu,t) = F(\nu,t;\ q^2 = \mu^2,\ q'^2 = \mu^2)$, it is possible to extrapolate the 4 πN amplitudes to $q^2 = q_0^2 - \vec{q}^2 \neq \mu^2$. In particular, the soft pion theorems of Adler [18] and Adler-Weisberger [19] imply

$$\bar{F}^{(+)}(0,\mu^2;\ 0,\mu^2) = 0 \tag{23}$$

$$\bar{F}^{(+)}(0,0;\ 0,0) = -\sigma_N/f_\pi^2 \approx -1.05\ \mu^{-1} \tag{24}$$

$$\nu^{-1}\bar{F}^{(-)}(0,0;\ 0,0) = (1-g_A^2)/2f_\pi^2 \approx -.62\ \mu^{-2}, \tag{25}$$

where (23) corresponds to one soft pion, (24) and (25) to two soft pions. That q^2 corrections from mass shell to (23)-(25) are small (PCAC) can be seen by comparing (23) to (Table I)

$$\bar{F}^{(+)}(0,\mu^2;\ \mu^2,\mu^2) = f_1^+ + f_2^+ + f_7^+ \approx -.23\ \mu^{-1} \tag{26}$$

and by respectively comparing (24) and (25) to

$$\bar{F}^{(+)}(0,0;\ \mu^2,\mu^2) \approx -1.45\ \mu^{-1};\ \nu^{-1}\bar{F}^{(-)}(0,0;\ \mu^2,\mu^2) \approx -.52\ \mu^{-2}. \tag{27}$$

Such corrections are reasonably small on the scale of the maximum sub-threshold values $2\mu^{-1}$ or $2\mu^{-2}$, consistent with the 6% PCAC corrections to the Goldberger-Treiman relation, $f_\pi g = m g_A$, involving one pion.

There is, however, a rapid (sign) variation in $\bar{F}^{(+)}$ between (18) to (23) to (24). This can be understood from the off-shell version of (7) provided

$$\sigma(t;\ q^2,q'^2) = \sigma_N\ [(1-\beta)(\frac{q^2+q'^2}{\mu^2} - 1) + \beta(\frac{t}{\mu^2} - 1)]\ , \tag{28}$$

consistent with (16), (23) and (24) and corresponding to a Weinberg-type [17] power series expansion in q and q'. One can even assume (28) for extrapolations $O(|q^2| \sim 10\mu^2)$ because the corrections are presumably $O(q^2q'^2/m^4)$ -- smaller in fact than those occuring in Weinberg's successful explanation of low energy ππ scattering. This sign variation has been verified directly by the estimate [20] of σ_N(66 ± 18 MeV) from the off-shell expression (24).

Given the off-shell version of (7) (i.e., (7) with $q^2 \neq \mu^2$) along with (28), the $a\nu^2 + bq'\cdot q$ structure of $C^{(+)}$ allows us to simply represent [5,6] the off-shell behavior of $\bar{F}^{(+)}$ at the point $\nu = 0$ as

$$\bar{F}^{(+)}(0,t;\ q^2,q'^2) = (t/\mu^2 - 1)\ \sigma_N/f_\pi^2 + 2\ \bar{F}^{(+)}(0,\mu^2)q'\cdot q/\mu^2, \tag{29}$$

valid at least for extrapolations $|q^2| \leq 10\mu^2$, where $\bar{F}^{(+)}(0,\mu^2)$ is the on-shell analogue of the Adler amplitude, (26), and $2q'\cdot q = q^2 + q'^2 - t$. At points other than $\nu = 0$, more than 2 parameters enter the off-shell versions of (7)-(10) and so the extrapolations in q^2 are less reliable than (29).

Consider, for example the 3 body contribution to the binding energy of nuclear matter. The dominant term contains an internal isospin even πN amplitude $\bar{F}^{(+)}$ with off-shell, spacelike pions, $q^2 \leq 0$, $q_0 = 0$ and $\nu = 0$ [4-6]. Then (29) corresponds to extrapolating from the nearby symmetry point $s = m^2 \approx 45\mu^2$ down to the region of interest, $s = m^2 - |\vec{q}|^2 \sim 40\mu^2$. The extrapolation errors are presumably much less than those of Δ isobar models (assuming one knows the correct form of the off-shell isobar propagator) where one begins at $s \sim M^2 \sim 78\mu^2$, or those of threshold extrapolation models where $s = (m+\mu)^2 \sim 60\mu^2$. Furthermore, in the latter case, there are accidental cancellations in the s-wave scattering length and effective range, making such an extrapolation extremely unreliable. It is useful to appreciate that $\bar{F}^{(+)}(0,\mu^2)$ and σ_N in (29) in effect accurately represent delicate cancellations between large terms. In the former case, $\bar{F}^{(+)}(0,\mu^2) = \bar{A}^{(+)} - g^2/m \sim 27\mu^{-1} - 27\mu^{-1}$, consistent with the Adler zero; in the latter case $\bar{F}^{(+)}(0,2\bar{\mu}^2) = 2g_{\sigma\pi\pi}g_{\sigma NN}(m_\sigma^2 - 2\mu^2)^{-1} - g^2/m \sim 28\mu^{-1} - 27\mu^{-1}$ in the σ model, implying $\sigma_N \approx g_A^2(\mu/m_\sigma)^2 m_N \sim 65$ MeV for $m_\sigma \sim 700$ MeV.

REFERENCES

[1] J. Carter, D. Bugg and J. Carter, Nucl. Phys. B58 (1973) 378.
[2] M. Scadron and L. Thebaud, Phys. Rev. D9 (1974) 1544.
[3] M. Olsson and E. Osypowski, U. of Wisc. preprint, Aug. 1975; M. Olsson, L. Turner and E. Osypowski, Phys. Rev. D7 (1973) 3444.
[4] G. Brown and A. Green, Nucl. Phys. A137 (1969) 1.
[5] S. Coon, M. Scadron and B. Barrett, Nucl. Phys. A242 (1975) 467.
[6] S. Coon, B. Barrett, M. Scadron, D. Blatt and B. McKellar, contributed paper to this conference.
[7] E. Harper, Y. Kim, P. McNamee and M. Scadron, to be published.
[8] D. Campbell, R. Dashen and J. Manassah, Phys. Rev. D12 (1975) 1010.
[9] H. Schnitzer, Phys. Rev. 158 (1967) 1471; K. Raman, ibid. 164 (1967) 1736.
[10] G. Höhler, H. Jakob and R. Strauss, Nucl. Phys. B39 (1972) 237.
[11] T. Cheng and R. Dashen, Phys. Rev. Lett. 26 (1971) 2153.
[12] R. Peccei, Phys. Rev. 176 (1968) 1812; L. Brown, W. Pardee and R. Peccei, ibid, D4 (1971) 2801.
[13] L. Nath, B. Etemadi and J. Kimel, Phys. Rev. D3 (1971) 2153; L. Nath and A. Sarker, Trieste preprint, 1974.
[14] H. Nielsen and G. Oades, Nucl. Phys. B72 (1974) 310.
[15] Y. C. Liu and J. Termaseren, Phys. Rev. D8 (1973) 1602; G. Hite and R. Jacob, Phys. Lett. 53B (1974) 310; Y. Chao et al., ibid. 57B (1955) 150.
[16] P. Achuthan, G. Hite and G. Höhler, Z. Physik 242 (1971) 167.
[17] S. Weinberg, Phys. Rev. Lett. 17 (1966) 616.
[18] S. Adler, Phys. Rev. 137 (1965) B1022.
[19] S. Adler, Phys. Rev. Lett. 14 (1965) 1051; W. Weisberger, ibid. 14 (1965) 1047.
[20] M. Scadron and H. Jones, Phys. Rev. D10 (1974) 967.

PARITY-NONCONSERVATION EFFECTS IN THE TWO-NUCLEON SYSTEM-SENSITIVITY TO THE SHORT-RANGE STRONG NUCLEAR FORCE

B.A. Craver,* E. Fischbach†, Y.E. Kim* and A. Tubis†
Purdue University, West Lafayette, Indiana 47907 U.S.A.

The effects of parity nonconserving (PNC) nuclear interactions have been studied in recent years in the hope of discriminating among various models of the weak interaction.[1-4] Since the weak N-N interaction is of short range, its calculated effects are expected to be sensitive to the short range correlations of the nuclear wave function. Consequently, the connection between a given weak-interaction Hamiltonian and experiment has very little significance in the absence of a careful study of the sensitivity of PNC effects to the assumed short range behavior of the strong parity conserving (PC) nuclear forces.

In this paper P_γ, the photon circular polarization in np radiative capture at thermal energies, is calculated for a class of weak-interaction Hamiltonians and several short range non-local phase-equivalent transformations (PET's)[5] of the Reid soft-core potential (RSCP)[6]. The calculations are done in coordinate space and are exact except for the usual first-order approximation in the weak and electromagnetic interactions. The calculations are greatly simplified by the use of Green's functions for the full strong Hamiltonian. For scattering states, this Green's function is most conveniently calculated from the physical radial scattering solution and a solution which has a pure outgoing-wave asymptotic form. Once this Green's function is calculated, P_γ for various weak Hamiltonians and (phase-equivalent) strong Hamiltonians may be calculated by a direct quadrature. We have not taken into account the modification of the electromagnetic operators due to the velocity dependence of the phase-equivalent strong Hamiltonians. Since these modifications are of very short-range ($r \leq 1$fm), they should not contribute much to P_γ (which depends mainly on the long range parts of the strong wave functions).

For the Cabibbo weak Hamiltonian and the RSCP, we find $P_\gamma = 2.28 \times 10^{-8}$ in good agreement with the value of 2.1×10^{-8} of Desplanques [7] and 2.28×10^{-8} of Lassey [8] and McKellar. The sensitivity of P_γ to PET of the strong potential is indicated in Table 1.

Table 1

P_γ (in units of 10^{-8}) caleulated for different PET of the RSCP and the Cabibbo weak Hamiltonian. UT101, UT102, and UT103 refer to the form factors of the PET given by Vary [5]. The letters A,B and C indicate that the PET has been applied to the $^3S_1$- $^3D_1$, $^1S_0$ strong potential (A), the $^3P_0$, $^3P_1$ strong potential (B) and the entire RSCP (C).

	UT101	UT102	UT103
A	0.13	0.41	1.17
B	0.60	1.11	1.56
C	9.5	8.4	25.8

Calculations with other weak Hamiltonians give similar results.

In view of the limited number of PET's considered, our results certainly do not preclude the possibility of finding a PET which could reconcile the theoretical value of P_γ with the experimental value $-(1.30 \pm 0.45) \times (10)^{-6}$[9]. The significance of such a result would, in itself, be questionable unless the implication of the modified strong interaction for the properties of nuclear systems with $A \geq 3$, and for parity-violating effects in heavier nuclei, were determined at the same time. For the present our results simply demonstrate the dangers entailed in trying to discriminate among various weak Hamiltonians solely on the basis of calculations of P_γ for a given strong nuclear potential.

References

1. E.M. Henley, Ann. Rev. Nucl. Sci. 19 (1969) 367.
2. B.H.J. McKellar, in High Energy Physics and Nuclear Structure, edited by S. Devons (Plenum, New York, 1970), p. 682.
3. E. Fischbach and D. Tadić, Phys. Rep. 6C (1973) 123.
4. M. Gari, Phys. Rep. 6C (1973) 317.
5. J.P. Vary, Phys. Rev. C7 (1973) 521.
6. R.V. Reid, Jr. Ann. Phys. 50 (1968) 411.
7. B. Desplanques, Nucl. Phys. A242 (1975) 423.
8. K.R. Lassey and B.H.J. McKellar, Phys. Rev. C11 (1974) 349.
9. V.M. Lobashov, D.M. Kaminker, G.I. Kharkevich, V.A. Kniazkov, N.A. Lozovoy, V.A. Nazarenko, L.F. Sayenko, L.M. Smotritsky, and A.I. Yegorov, Nucl. Phys. A197 (1972) 241.

---

*Supported by the U.S. National Science Foundation.

†Supported by the U.S. Energy Research and Development Administration.

THE PARITY VIOLATING ASYMMETRY IN THE RADIATIVE CAPTURE OF POLARIZED NEUTRONS BY PROTONS

B.A. Craver,† Y.E. Kim† and A. Tubis*
Purdue University
W. Lafayette, In. 47907
P. Herczeg*
Los Alamos Scientific Laboratory
P. Singer
Technion-Israel Institute of Technology, Haifa, Israel

In addition to the recently observed photon circular polarization for unpolarized thermal neutrons, another parity violating effect of interest in the $n+p \rightarrow d+\gamma$ reaction is the asymmetry α in the photon angular distribution with respect to an initial neutron polarization.

We report here a calculation of α, using a two-nucleon strong interaction given by the Reid soft-core potential and several of its phase equivalents [1]. The calculations, which are exact with respect to the strong interaction, are facilitated by the use of Green's function techniques [2].

The asymmetry can be decomposed as $\alpha = \alpha_0 + \alpha_1$, where α_0 and α_1 are, respectively, due to the isoscalar and isovector components of the parity violating Hamiltonian. (There is no contribution from an isotensor term in the Hamiltonian if the electromagnetic current has only isoscalar and isovector components and if the small parity conserving capture from the ^3S-state is neglected.) In the impulse approximation, the mixing of opposite parity states in the initial n-p states and the deuteron states contributes only to α_1 [3]. Nonvanishing contributions to α_0 arise from exchange-current terms [4] which could be significant in models of weak interactions, such as the Cabibbo theory, where the isovector weak coupling is much smaller than the isoscalar one. The knowledge of the magnitude of α_0 is important even if α would be dominated by α_1, since it determines the level beyond which α can be used as a test for the existence of a $\Delta S = 0$, $\Delta I = 1$ nonleptonic weak interaction.

The various possible types of contributions to α_1 and α_0 can in general be summarized as follows:

$$\alpha_1 = \alpha_1^{\text{mixing}} + \alpha_1^{\text{p.v. exch}} + \alpha_1^{\text{p.c. exch}} , \qquad (1)$$

$$\alpha_0 = \alpha_0^{\text{p.v. exch}} + \alpha_0^{\text{p.c. exch}} . \qquad (2)$$

In (1) and (2), α_1^{mixing}, $\alpha_0^{\text{p.v. exch}}$ and $\alpha_{0,1}^{\text{p.c. exch}}$ refer, respectively, to the contribution of the parity violating force in the impulse approximation, of the parity violating two-body electromagnetic operators evaluated between states of definite parity, and of the parity conserving two-body electromagnetic operators evaluated between mixed parity states.

α_1^{mixing} has been calculated using the parity violating potential described in ref. [5]. We find $|\alpha_1^{\text{mixing}}| \approx 5\text{-}8\times10^{-9}$ for the Cabibbo theory, and $|\alpha_1^{\text{mixing}}| \approx 1\text{-}2\times10^{-7}$ for theories in which the isovector coupling is larger by an order of magnitude [6].

The most important contributions to $\alpha_0^{\text{p.v. exch}}$ arise from meson-exchange diagrams corresponding to the parity conserving electromagnetic transitions $\rho^{+,-,0} \rightarrow \pi^{+,-,0}\gamma$ and the parity violating weak-electromagnetic transitions $A^{+,-,0} \rightarrow \pi^{+,-,0}\gamma$ [4]. The matrix elements of the resulting parity violating two-body operators vanish between ^3S-states, and consequently

$$\alpha_0^{\text{p.v. exch}} \approx \alpha_0^{\text{p.v. exch}}(^3\tilde{S} \rightarrow {}^3D) + \alpha_0^{\text{p.v. exch}}(^3\tilde{D} \rightarrow {}^3S) , \qquad (3)$$

where $^3\tilde{S}$, $^3\tilde{D}$ and ^3S, ^3D are the n-p scattering and the deuteron states respectively. We find that for each strong potential used, the two contributions in (3) are of opposite sign and approximately equal in magnitude: $|\alpha_0^{\text{p.v. exch}}(^3\tilde{S} \rightarrow {}^3D)| \approx |\alpha_0^{\text{p.v.exch}}(^3\tilde{D} \rightarrow {}^3S)| \approx 4 \times 10^{-10} - 1 \times 10^{-9}$ or $2 \times 10^{-9} - 6 \times 10^{-9}$, depending on whether the weak matrix elements are calculated in the factorization approximation or taken to be about six times larger, as suggested by another estimate [7]. As a result of the cancellation, one obtains $|\alpha_0^{p.v.exch}| \approx 5 \times 10^{-11}$ or 5×10^{-12}. The same cancellation occurs also for $\alpha_1^{p.v.\ exch}$. We are presently evaluating the contribution of the $^3\tilde{D} \rightarrow {}^3D$ transition which was ignored in equation (3), and also investigating the contributions to $\alpha_0^{p.c.exch}$. The latter could be of the same order of magnitude as the individual terms in (3) and thus comparable to $\alpha_1^{mixing,\ Cabibbo}$. On the other hand, $\alpha_1^{p.c.exch}$(and also $\alpha_1^{p.v.exch}$) is expected to be in general at least an order of magnitude smaller than α_1^{mixing} in a given theory.

References

[1] J.P. Vary, Phys. Rev. C7 (1973) 521.

[2] B.A. Craver, E. Fischbach, Y.E. Kim, and A. Tubis, submitted to Phys. Rev. D.

[3] G.S. Danilov, Phys. Letters 18 (1965) 40.

[4] E.M. Henley, Phys. Rev. C7 (1973) 1344; P. Herczeg and L. Wolfenstein, Phys. Rev. D11 (1975) 611 and references therein.

[5] E. Fischbach, D. Tadić and K. Trabert, Phys. Rev. 186 (1969) 1688.

[6] It should be noted that the possibility of isovector parity violating couplings 2-3 orders of magnitude larger than in the Cabibbo model, and thus values of $|\alpha|$ as large as 5×10^{-6} - 1×10^{-5} are not completely excluded. See P. Herczeg in the <u>Proceedings of the VIth International Conference on High Energy Physics and Nuclear Structure</u>, Santa Fe and Los Alamos, June 9-14,1975.

[7] B.H.J. McKellar and P. Pick, Phys. Rev. D6 (1972) 2184.

---

* Supported by the U.S. Energy Research and Development Administration.

† Supported by the U.S. National Science Foundation.

REMARKS ON THE LEVEL SCHEME OF F^{19}

by K. Bleuler

Institut für Theoretische Kernphysik der Universität Bonn
Nussallee 14-16, D-53 B O N N, W.-Germany

The level scheme of F^{19} plays by now an important role in the experimental determination of the so-called weak parity violation. This is related to the fact that the low lying part of this spectrum has a most characteristic structure: It consists of a relatively dense system of co-called parity dubletts, i.e. on 3 pairs of levels with the same angular momentum but with opposite parity ($1/2^{\pm}$, $5/2^{\pm}$, $3/2^{\pm}$). These assignments are in perfect contradiction to the conventional sequence expected from shell structure which, in turn, would yield only the levels with positive parities in this energy region.

In order to understand this particular system of assignments, we propose here a special coupling scheme, namely the so-called Parity Mixing[1]. Although designed originally for heavier structures (where it is in fact not applicable) this method might be suitable for this special case of a lighter nuclear structure because we have to deal here with a particular situation: The levels of the valence nucleons in the fields of the O^{16} core, i.e. the low lying states of F^{17}, from which the F^{19} nucleous has to be constructed, have the special property (which, in turn, does not arise in heavier nuclei) of exhibiting from the outset a pair of levels ($1/2^{\pm}$) with opposite parity and relatively small energy separation. (For an interpretation of this level scheme compare an earlier paper of the author[2]).

The main idea of this scheme is the following: Within the framework of a spherical Hartree-Fock treatment, we introduce generalized single particle states allowing for an admixture between states with the same angular momentum but with opposite parities. Although the 2-body forces to be introduced are, of course, parity conserving, such a selfconsistent generalized Hartree-Fock solution might yield a lower total energy for this new kind of parity admixed single particle states. This fact is mainly due to the action of a 2-body tensor force which contains an operator O_{12} between particle 1 and 2 of the type

$$O_{12} = (\vec{r}^{\,1}\vec{\sigma}^{1})(\vec{r}^{\,2}\vec{\sigma}^{2})$$

which corresponds to a parity exchange, i.e. which yields - being the product of two pseudo-scalars - a simultaneous transition of both particles to states with opposite parity and the same angular momentum.

In this first step we thus obtain (if the relevant single particle levels are close enough) a selfconsistent solution built on parity mixed states, i.e. without any sharp parity at all, apparently in contradiction to the parity conservation of the Hamiltonian. (This fact has a certain analogy to the deformed Hartree-Fock solutions which, in turn, give lower energies for the breaking of spherical symmetry of the single particle states). On the other hand (again in analogy to the deformed case), this new solution is degenerate - degeneracy 2 - whereby the second state is obtained by operating with the reflection operator (which commutes with the Hamiltonian) on the original state. With the help of this degeneracy the physical states with sharp parities are then obtained by the well-known projection procedure (in analogy to the case of deformation in space where the projection yields the rotational states). This last step finally yields two (probably narrow lying) states with opposite parity and the same angular momentum. The transitions between them might be visualized as a so-called parity vibration. (Things could be even better understood if the action of the tensor force is replaced by the corresponding pseudo-scalar pion field: The parity violating solutions are those with the non-zero pseudo-scalar field of a given sign, introducing in this way a definite sense of rotation). It is this a scheme which seems to produce in a most natural way the characteristic parity dubletts of the level scheme of F^{19}.

---

1) K. Bleuler, Parity Mixing in Spherical Nuclei, Report of the Varenna School XXXVI (Academic Press, New York, 1966, p.464)

2) Correlation in the Structure of Light Nuclei, Conference Report Correlations in Nuclear Physics, Balatonfüred, Hungary 1973

ELECTROMAGNETIC PROPERTIES OF TRI-NUCLEON SYSTEMS WITH SEPARABLE POTENTIALS

V.K. Gupta and S.S. Mehdi

Department of Physics and Astrophysics, University of Delhi, Delhi-110007, India.

I would like to present in this talk a brief report of the study of the electromagnetic properties of nuclear three body systems with separable potentials. Despite the advent of large computers and the availability of Faddeev formalism for exact calculations with local potentials, separable potentials have retained their importance as is evident from a large number of papers dealing with this topic that have been submitted to this conference. They have a strong appeal because of their comparative simplicity and the ease with which 3-nucleon systems can be solved for fairly realistic potentials.

A rather thorough investigation of the 3-nucleon systems has been made over the years with fairly comprehensive separable potentials (including a tensor force term in the triplet and a repulsive term in the singlet state)[1]. The results of this study on the triton binding energy and the doublet n-d scattering length are in excellent agreement with experiments for several sets of potentials, all of which reproduce the N-N phase shifts fairly well upto about 300 Mev lab energies [2]. The binding energy calculation also provides the triton wave function, which can then be gainfully employed for a study of the finer effects, like the coulomb energy of ^{3}He and the electromagnetic form factors of ^{3}H and ^{3}He, etc.

The interest in the coulomb energy of ^{3}He (E_c) is due to its ramifications on the charge symmetry of nuclear forces. A large number of variational calculations have yielded $E_c \sim 0.6$-0.65 MeV. Calculations via the solution of Faddeev equations have yielded $E_c \sim 0.62$-0.69 MeV. However, the experimental difference in the binding energies of ^{3}H and ^{3}He is $E_B = 0.764$ MeV. The gap $\Delta = E_B - E_c$ which is ~ 0.1-0.16 MeV for the above mentioned calculations is usually attributed to the charge asymmetry of nuclear forces.

On the other hand the interest in the trinucleon electromagnetic form factors is mainly due to their relatively recent measurements upto rather high momentum transfers ($k^2 = 20$ fm^{-2})[3]. The main feature of these measurements is the appearance of a minimum in the charge form factor of ^{3}He at $k^2 = 11.6$ fm^{-2} and a rather strong secondary maximum at $k^2 \gtrsim 14$ fm^{-2}. The occurence of the minimum in the charge form factor suggests the existence of strong correlations in the 3-body wave function at short distances. No such minimum appears in the magnetic form factor of ^{3}He for k^2 upto ~ 16 fm^{-2}. The ^{3}H form factors are known only upto 8 fm^{-2}, and do not show any interesting structure. The same form factor data give for the rms radii the following figures: $r_{ch}(^3\mathrm{He}) = 1.88 \pm 0.05$ fm, $r_{mag}(^3\mathrm{He}) = 1.95 \pm 0.11$ fm; $r_{ch}(^3\mathrm{H}) = r_{mag}(^3\mathrm{H}) = 1.70 \pm 0.05$ fm.

The earlier calculation on E_c [4] had been performed with separable potentials which did not include a repulsive core in the singlet state, and had yielded $E_c = 0.84$ MeV. A similar calculation of the electromagnetic radii [5] had given results to within 15% of the then experimental values. These gaps in E_c and the radii had been sought to be filled by the inclusion of the repulsive term in the singlet state, which is expected to make a significant contribution to the electromagnetic properties.

We have recently calculated these quantities with separable potentials which do include a repulsive term in the singlet-state [6]. We have also taken this opportunity to perform the calculations with two different potential shapes- one is the traditional shape for which 3-body calculations have been performed (shape-1), and the other we have now proposed (shape-2). The general form of the rank-2 (central) separable potential in the singlet state is

$$-M\langle p|V_s|p'\rangle = \lambda_s[s(p)s(p') - h(p)h(p')]$$

where h(p) provides the necessary repulsion. The traditional forms (i.e., shape-1) for s(p), h(p) and c(p) (the similar form factor for the (central) triplet interaction) are

$$s(p),c(p) = (p^2+\beta_{s,c}^2)^{-1} \quad ; \quad h(p) = np^2(p^2+\beta_h^2)^{-2}$$

We have now proposed the alternative forms (shape-2)

$$s(p),c(p) = (p^2+\beta_{s,c}^2)^{-2}; \; h(p) = np^2(p^2+\beta_h^2)^{-2}$$

One motivation for introducing a new form for these functions is to study the off-shell effects in the potential. Since for a given shape not much variation in the parameters is permitted due to the constraint of providing a fit to the two-body data, the variation in the shapes should allow a study of the off-diagonal elements of the potential matrix to some extent. Further, we have obtained sets fitted to two different values of the singlet effective range, r_{os}, so as to study the effect of variation in this parameters on the binding energy as well as the electromagnetic properties of ^{3}H and ^{3}He; this variation, as we shall presently see, is quite important for the coulomb energy problem.

For the coulomb energy of ^{3}He, calculations have so far been performed only for shape-1 potentials. We include in the tablethe result for set G_1 in the notation of Ref.1, which yields E_C=0.77 MeV. This is ambarrassingly close to the experimental value of 0.764 MeV for E_B. This striking agreement, however, is not to be taken too literally. For completeness one must include (at least) a tensor term in the triplet potential which tends to reduce the coulomb energy. Once the repulsive term has been included, the tensor term is expected to further reduce E_C by 3-5%, which should bring it down to at best 0.7-0.73 MeV. This is much higher than $E_C \sim 0.6$ MeV obtained from variational calculations by Okamoto et al [7].

Table: The electromagnetic radii (in fermi), the ^{3}He coulomb energy (in MeV) and the triton binding energy (also in MeV) for various potential sets: Set I corresponds to shape-1 while sets II-IV correspond to shape-2. Set II corresponds to a rank-1 singlet potential in which no repulsion is included, whereas all the other sets include a repulsive term (denoted by H). Sets III and IV are fitted to r_{os}=2.7 fm and 2.8 fm, respectively.

Set	Potential	$r_{ch}(^3He)$	$r_{mag}(^3He)$	$r_{ch}(^3H)$	$r_{mag}(^3H)$	E_C	E_t
I	$C_Y^{eff}+(S+H)_{G_1}$	1.63	1.58	1.56	1.62	0.77	9.01
II	$C_{MG}+S_{MG}$	1.93	1.67	1.81	1.76	--	9.90
III	$C_{MG}+(S+H)_{MG_1}$	1.96	1.68	1.83	1.79	--	9.49
IV	$C_{MG}+(S+H)_{MG_2}$	1.97	1.68	1.84	1.79	--	9.34
	Experimental values	1.87±0.05	1.95±0.11	1.70±0.05	1.70±0.05	Δ=0.76	8.49

Though there is some evidence for charge symmetry breaking in nuclear forces, and on theoretical grounds such symmetry breaking must exist, the question is only about its magnitude in the 2-nucleon potential and its effect on the 3-nucleon system. If the variational calculations are to be believed, this symmetry breaking effect in ^{3}He could be as high as 15-20%, whereas the separable potential model calculations would have it to be 5-10%. The truth may lie some where around 10%. Now from the sign of Δ, whatever its magnitude, it follows that the n-n potential is on the "average" some what stronger than the p-p potential. This has, however, to be reconciled with the fact that according to the recent estimates, the n-n scattering length seems to be slightly less negative than the (coulomb)corrected) p-p scattering length:

$$a_{nn} = -16.4 \pm 0.9 \text{ fm}; \quad a_{pp} = -17.1 \pm 0.2 \text{ fm}.$$

Explanations of $\Delta \sim 0.15$ MeV from charge asymmetric potentials had so far always assumed $|a_{nn}| > |a_{pp}|$ [7]. It may perhaps be difficult to reconcile "large" positive Δ with $|a_{nn}| < |a_{pp}|$. Of course, the N-N potential is not characterized by the scattering length alone, and one possible way to explain this anomaly is to have the n-n effective range r_{nn} to be greater than the p-p effective range r_{pp} [8]. Experimentally r_{nn} is too poorly determined to warrant any such assertion. By how much r_{nn} will have to be greater than r_{pp} to provide the right sign and magnitude (?) for Δ will again depend upon the relative sensitivity of the binding energy to the singlet effective range (as compared to its sensitivity to the singlet scattering length), a question which has yet not been settled. Thus we feel that the problem of the coulomb energy of ^{3}He and the charge asymmetry of nuclear forces is still not properly understood, and requires for its satisfactory explanation further analysis of the off-shell effects of the N-N potential, the effects of the change in singlet effective range on the binding energies and of course a much better determination of the low energy n-n parameters.

As for the electromagnetic form factors and radii, calculations have been performed for both the potential shapes mentioned earlier. The main result of the calculation is negative and rather disappointing, viz., we do <u>not</u> find any minimum in the charge form factor of ^{3}He even upto $k^2 = 20\ \text{fm}^{-2}$. This is true for both the potential forms that have been considered. Of course, the tensor force contribution has yet to be added, but it will be too much to expect the tensor force to provide the necessary minimum when the much more important repulsive term seems to have failed. In fact, one of the potential sets considered by us does include the tensor force, though in an indirect manner. The results with this set are qualitatively no different from those of the others.

On a closer examination, this failure to reproduce the minimum in the charge form factor seems to be due to the nature of the repulsive term in the separable potential model. The repulsive core of this potential model is too soft to keep the nucleons from coming close enough to each other and thereby provide a strong short range correlation. This can be seen from an examination of, say, the two-body bound state wave function. Increasing the "rank" of the potential does not change the qualitative nature of the wave function. For a potential of rank-2 which includes a repulsive term, the two-body bound state wave function has the form

$$u(r) = N \int_0^\infty p \sin(pr)\left[s(p) - Kh(p)\right] (p^2 + \alpha^2)^{-1}\, dp\ ,$$

where K is a constant depending upon integrals involving s(p) and h(p), while for a rank-1 potential the corresponding expression is

$$u(r) = N \int_0^\infty p \sin(pr)\, s(p)\, (p^2 + \alpha^2)^{-1}\, dp.$$

As functions of <u>r</u> the two expressions have very nearly the same behaviour unless of course one chooses certain very special forms for s(p) and h(p). It seems, therefore, that the simple minded separable potentials which have served so well in our understanding of the 3-nucleon system are not adequate if we want to probe too deep inside the nucleus. The situation could of course be radically different (hopefully for the better) with "hard core" potentials.

After having said this, let us study some other aspects of the calculation. A comparative study of the results for the two shapes (see table) indicates that even for a rank-1 potential, shape-2, the one we have now proposed, gives much better results than the earlier shape-1, the improvement in the radii alone being $\sim$15-20%. This is very much to be expected, since $c(p) \to 0$ as $p \to \infty$ much faster for shape-2 than for shape-1. Correspondingly, the (two-body) wave function is hollower near the origin for this shape. Thus, in a way shape-2 simulates the effect of a repulsive term. In fact the addition of a repulsive term to this rank-1 potential hardly brings about any further improvement in the radii. Thus shape-2 seems to be better suited for simple minded test calculations with model rank-1 potentials.

Another interesting aspect of the calculation with this shape is that it provides $r_{ch}(^3H) > r_{mag}(^3H)$, in contrast to the results obtained by Kim et al and the results of other earlier calculations. Of course the magnetic radii are well known to be sensitive to meson-exchange effects (much more than the charge radii), but all the same it would be interesting to see if more accurate experiments could detect a difference in the two radii.

As we have discussed a little while ago, the sensitivity of the triton binding energy to the singlet effective range r_{os} is of crucial importance to the ^{3}He coulomb energy question. Recently Gibson and Stephenson [8] found that the triton binding energy, E_T is very sensitive to this parameter, in contrast to the conclusion Mitra [1] reached a few years ago. We find that though E_T is not as sensitive to r_{os} as reported by the authors of Ref. 8, it is not completely insensitive either. The effective range does not, therefore, seem to be a hot candidate for resolving the anomaly in the ^{3}He coulomb energy problem, unless new experiments give an r_{nn} which is significantly greater than r_{pp}.

In the light of this discussion it will not be unreasonable to assume that the unexpectedly large difference in the electromagnetic properties as obtained from the two shapes is at least partly to be attributed to their off-energy shell differences. From a large number of earlier variational and "Faddeev" calculations of the binding energy and the electromagnetic form factors of ^{3}H and ^{3}He, Haftel [9] and Tjon et al [10] have concluded that a consistent explanation of both binding energy and form factors requires a substantial contribution from the three-body forces. Once, again, such forces are certainly predicted from the fundamental theory, as we shall learn from Prof. McKellars talk tomorrow. But estimates of the effects of three-body forces on the binding energies vary considerably [11]. Perhaps a lot more needs to be known about the off-shell behaviour of the two-body potential before one can decide as to how much three-body force contribution is required for a consistent explanation of all the 3-nucleon data.

REFERENCES

1. A.N. Mitra, Adv. Nucl. Phys. 3 (1969) 1.
2. G.L. Shrenk, V.K. Gupta, and A.N. Mitra, Phys. Rev. C1 (1970) 895.
3. J.S. McCarthy et al, Phys. Rev. Letters 25 (1970) 884.
M. Bernstein et al, Lett. Nuovo Cimento 5 (1972) 431.
4. V.K. Gupta and A.N. Mitra, Phys. Letters 24B (1967) 27.
5. V.K. Gupta, B.S. Bhakar, and A.N. Mitra, Phys. Rev. 153 (1967) 1114.
6. S.S. Mehdi and V.K. Gupta, Phys. Rev. Letters 33 (1974) 709.
7. K. Okamoto and C. Pask, Ann. Phys. (NY) 68 (1971) 18.
8. B.F. Gibson and G.J. Stephenson, Jr., Phys. Rev. C11 (1975) 1448.
9. M.I. Haftel, Phys. Rev. C7 (1973) 80.
10. J.A. Tjon, B.F. Gibson and J.S. O'Connell, Phys. Rev. Letters 25 (1970) 540.
11. S. Barshay and G.E. Brown, Phys. Rev. Letters 34 (1975) 1106.

INTEGRAL VERSUS FRACTIONAL CHARGES OF QUARKS - A VIEW THROUGH GAUGE THEORY

G. Rajasekaran
Tata Institute of Fundamental Research, Bombay 400005, India

Do quarks have fractional charges (Gell-Mann - Zweig scheme) or integral charges (Han-Nambu scheme)? This is one of the important questions in quark physics at the present time. The purpose of this talk is to focus attention on some remarkable consequences of gauge theory which have a bearing on this question.

The quarks in either scheme can be written in matrix forms:

Gell-Mann - Zweig (G-Z) Han-Nambu (H-N)

$$\begin{pmatrix} p_1^{2/3} & p_2^{2/3} & p_3^{2/3} \\ n_1^{-1/3} & n_2^{-1/3} & n_3^{-1/3} \\ \lambda_1^{-1/3} & \lambda_2^{-1/3} & \lambda_3^{-1/3} \end{pmatrix} \qquad \begin{pmatrix} p_1^{0} & p_2^{1} & p_3^{1} \\ n_1^{-1} & n_2^{0} & n_3^{0} \\ \lambda_1^{-1} & \lambda_2^{0} & \lambda_3^{0} \end{pmatrix}$$

Here the superscripts denote the electrical charges. The subscripts 1, 2, 3 spanning the horizontal direction denotes the new degree of freedom 'colour' which actually had its origin in old quark physics - namely the conflict of the apparent total symmetry of the three-quark wavefunction in the baryonic ground state with Fermi-Dirac statistics (Greenberg [1], Dalitz [2], Mitra and Majumdar [3]).

The colour degree of freedom plays an increasingly important role in the dynamical theory of hadrons. This extra degree of freedom has been exploited for the construction of unified gauge models of strong, electromagnetic and weak interactions (Weinberg [4]). Whereas the gauge group of the weak and electromagnetic interactions is supposed to act in the space of the old quantum numbers such as isospin and strangeness, the gauge group of the strong interactions is taken to be SU'(3) acting on the three colour indices. The motivation for going for a gauge theory of strong interactions is the desire for asymptotic freedom - namely the hope that asymptotically free nonabelian gauge theory of strong interactions would explain why the constituents (partons) of the hadrons behave like free objects in the deep inelastic lepton-hadron scattering processes.

For definiteness let us consider a gauge theory of strong, electromagnetic and weak interactions based on the group

$$G = SU'(3) \times SU_L(2) \times U(1) \tag{1}$$

where SU'(3) is the colour gauge group associated with strong interactions and $SU_L(2) \times U(1)$ is the Weinberg-Salam group used to generate weak and electromagnetic interactions. The colour gauge group SU'(3) introduces an octet of gauge vector bosons G^i_μ ($i = 1 \ldots 8$) called gluons which mediate the strong interactions. The weak* and electromagnetic interactions are mediated by the three massive weak bosons W_μ^+, W_μ^- and Z_μ, together with the photon.

Either the fractionally charged G-Z quarks or the integrally charged H-N quarks can be used in the construction of the above gauge theory. However, certain special problems arise in the H-N case because of the structure of the charge matrix Q which can be written as

$$Q = Q_0 + Q_8 \tag{2}$$

where Q_0 is the colour singlet part which involves the generators of $SU_L(2) \times U(1)$ and Q_8 is the colour octet part consisting of the generators of $SU'(3)$. The numerical values are $Q_0 = (2/3, -1/3, -1/3)$ for p_i, n_i, λ_i where $i = 1, 2, 3$ and $Q_8 = (-2/3, 1/3, 1/3)$ for q_1, q_2, q_3 where $q = p, n, \lambda$. Note that for G-Z quarks $Q = Q_0$ and so it is the colour octet piece Q_8 which distinguishes H-N quarks. Because of the mixing between the generators of $SU_L(2) \times U(1)$ and $SU'(3)$ introduced by Q in the H-N case, the exclusive association of $SU_L(2) \times U(1)$ with weak and electromagnetic interactions and $SU'(3)$ with strong interactions breaks down. Thus, all the three interactions - strong, electromagnetic and weak mix with each other.

The complex of mixings in the Han-Nambu case have generally discouraged a serious study of the problem and most people have followed the Gell-Mann - Zweig scheme. An important exception is the work of Pati and Salam [5] who proposed this gauge model with H-N quarks three years ago. However, until recently nobody tried to see through the maze of the mixing equations and carry the analysis to its logical conclusion. When this is done, the results turn out to be unexpected and interesting (Rajasekaran and Roy [6], Pati and Salam [7]). Here I shall try to show how the main consequences of the gauge theory with Han-Nambu quarks can infact be derived by simple physical considerations.

We shall first concentrate on the electromagnetic interactions. Since gluons induce horizontal transitions in the quark matrix and since the charges of the H-N quarks do vary along the horizontal direction, it is clear that some of the gluons are charged. From this simple observation that in the H-N scheme, the gluon octet contains charged gluons, a string of consequences follow. Consider the electromagnetic vertex of the charged gluon G^+ (Fig. 1). Where can such a vertex come from? In the gauge theory, this has to be generated out of the Yang-Mills vertex shown in Fig. 2 (which is a part of the full $SU'(3)$ vertex), where f is the strong coupling constant. The neutral gluon G^0 and the photon A have to mix in such a way as to produce the physical gluon $\tilde{G}^0$:

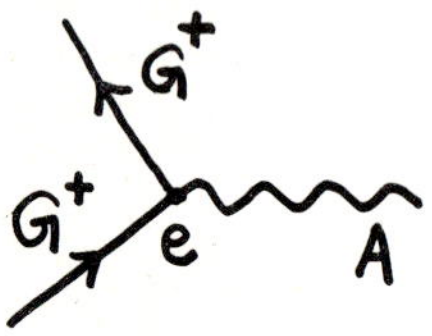

Fig. 1

$$\tilde{G}_\mu^0 = G_\mu^0 - \frac{e}{f} A_\mu . \tag{3}$$

For then, on reexpressing the G^0 line of the vertex in Fig. 2 in terms of $\tilde{G}^0$ and A lines with the help of eq. (3) we get the two vertices of Fig. 3 where the second diagram is the desired electromagnetic vertex. But then, we have to buy an additional consequence. The kinetic term $-\frac{1}{4}\left(\partial_\mu G_\nu^0 - \partial_\nu G_\mu^0\right)^2$

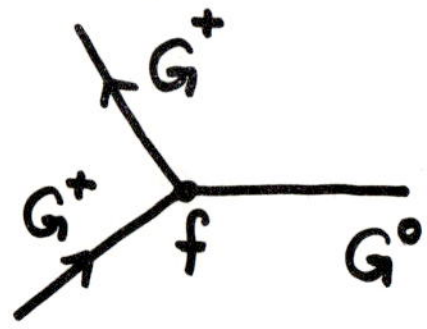

Fig. 2

---

* For consistency with the observed selection rules of weak interactions, the vertical dimension of the quark matrix has to be enlarged to include at least one more row (called 'charm').

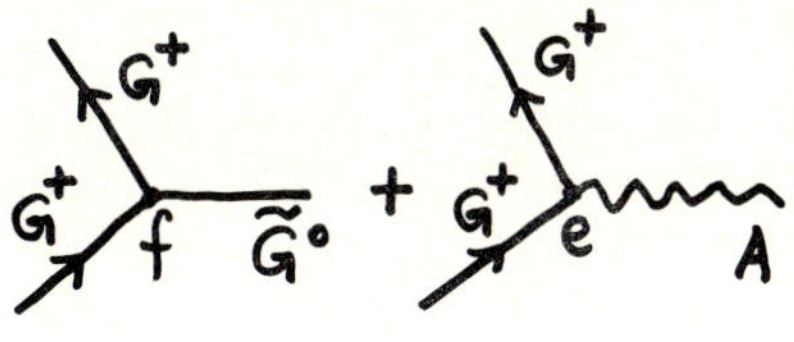

Fig. 3

in the Lagrangian gives rise to the cross term

$$-\frac{1}{2}\frac{e}{f}\left(\partial_\mu \tilde{G}^0_\nu - \partial_\nu \tilde{G}^0_\mu\right)\left(\partial_\mu A_\nu - \partial_\nu A_\mu\right)$$

when eq. (3) is used. We thus get a quadratic vertex (Fig. 4) which plays a crucial role in what follows. *

Actually, among the eight gluons G^i_μ ($i = 1\ldots 8$), there are two pairs of charged gluons analogous to $\rho^\pm$, $K^{*\pm}$ and to generate the correct electromagnetic interactions of both these pairs, we need both the following mixing equations:

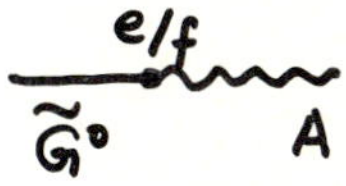

Fig. 4

$$\tilde{G}^3_\mu = G^3_\mu - \frac{e}{f} A_\mu\,; \qquad \tilde{G}^8_\mu = G^8_\mu - \frac{e}{\sqrt{3}f} A_\mu \tag{4}$$

and hence, Fig. 4 should in fact be replaced by two quadratic vertices ($\tilde{G}^3$ A) and ($\tilde{G}^8$ A).

Next, consider the electromagnetic interaction of the quarks (to the lowest order in e), shown in Fig. 5. To the direct diagram (a) we have to add the extra diagram (b) with the gluon intermediate state since that is also of order e. The matrix element of diagram (b) involves the following factor:

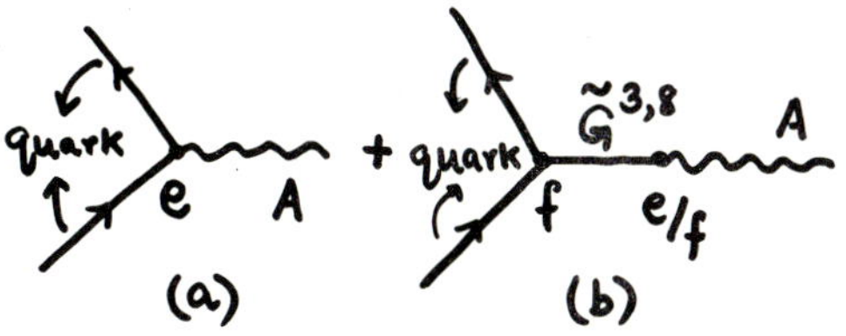

Fig. 5

$$f\gamma^\lambda \left\{\frac{-(g_{\lambda\nu} - q_\lambda q_\nu/m_g^2)}{q^2 - m_g^2}\right\} \frac{e}{f}\left(q^2 g^\nu{}_\mu - q^\nu q_\mu\right)$$

$$\longrightarrow e\gamma_\mu \left(\frac{-q^2}{q^2 - m_g^2}\right)$$

where q and m_g are the momentum and mass of the gluon and the second line is obtained by using the fact that the $q_\mu q_\nu$ terms do not contribute when sandwiched between the quark spinors $\bar{u}$ and u. Hence, the complete matrix element of the current is (noting that the charge matrix $Q_0 + Q_8$ occurs for the direct diagram (a) but only the colour octet part Q_8 is coupled at the strong vertex of diagram (b))

$$\begin{aligned} \langle \text{quark}|\, e j^{em}_\mu \,|\text{quark}\rangle &= e\bar{u}(Q_0 + Q_8)\gamma_\mu u + e\bar{u} Q_8 \gamma_\mu u \left(\frac{-q^2}{q^2 - m_g^2}\right) \\ &= e\bar{u}\left\{Q_0 + Q_8\left(\frac{-m_g^2}{q^2 - m_g^2}\right)\right\}\gamma_\mu u. \end{aligned} \tag{5}$$

We thus see that the effective charge of the quark can be written as

$$Q_{eff} = Q_0 + Q_8\left(\frac{-m_g^2}{q^2 - m_g^2}\right). \tag{6}$$

At this point, let us note that all the low-lying hadrons which are so far identified are colour singlets so that the matrix elements of Q_8 between these states are zero. The colour octet part Q_8 is expected to manifest itself only above a

---

* We do not define a physical photon field orthogonal to $\tilde{G}^0$. If one does that, then this quadratic vertex will be absent, but instead there will be a direct gluon coupling to the leptons. The final results are the same in either approach. We should also note that in our approach, eq. (3) leads to a multiplicative factor for the kinetic term of the photon. This can be absorbed by suitable definition of 'renormalized' fields and coupling constants.

certain colour threshold. But the new result given in eq. (6) is that even above colour threshold, for asymptotic q^2, i.e. $q^2 \gg m_g^2$, the colour octet contribution is damped out. The colour singlet part Q_0 does not distinguish between G-Z and H-N quarks. So, in the asymptotic region the Han-Nambu quarks appear to have the same fractional charges as the Gell-Mann - Zweig quarks. And, in any case, only in the asymptotic region - i.e. q^2 very large compared to the masses in the problem - that the quark charges can be measured through the deep inelastic electron-hadron scattering experiments. For only in this asymptotic region the simple parton picture is applicable. Thus, Nature appears to have engineered a conspiracy by which it seeks to prevent us from knowing the true nature of the quarks. The quarks may really be integrally charged but since the coloured part of the charge is quenched, we will see only the uncoloured part which is fractional.

However, colour is not blacked out completely. There is still the gluon contribution - since, as we already noted, the gluons are charged and hence can be seen by the photon. So, let us look at the electromagnetic interaction of the gluon (Fig. 6). Again we have to add the extra diagram (b) with gluon-intermediate state which leads to the same over-all factor $m_g^2(m_g^2 - q^2)^{-1}$ for the complete matrix element. However, the important point is that, in this case, this factor does not kill the contribution completely. The reason is that the direct Yang-Mills coupling (Fig. 6a) alone leads to a contribution in deep inelastic lepton-hadron scattering that increases with $|q^2|$ and hence violates Bjorken scaling strongly. So, all that the factor $m_g^2(m_g^2 - q^2)^{-1}$ accomplishes is to restore Bjorken scaling to the gluon-contribution. Explicitly, looking at the gluon contribution to the deep inelastic electromagnetic structure function F_2, we get

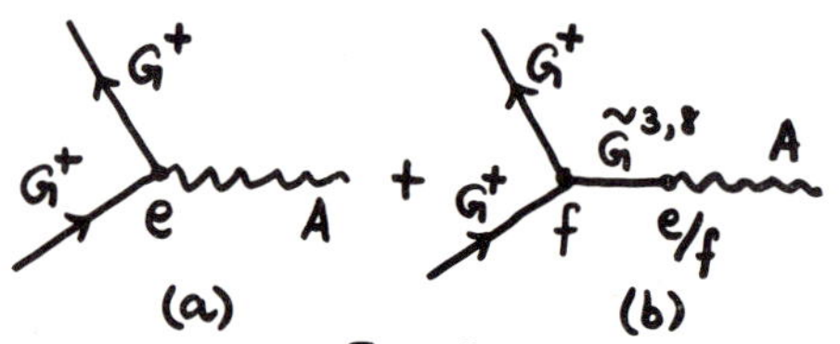

Fig. 6

$$F_{2\,EM}^{gluon} = \left\{ x g(x) \left(4 + \frac{4|q^2|}{3m_g^2} + \frac{|q^4|}{3m_g^4} \right) \right\} \left(\frac{-m_g^2}{q^2 - m_g^2} \right)^2 \xrightarrow[|q^2| \to \infty]{} \frac{1}{3} x g(x) \qquad (7)$$

where x is the fractional momentum carried by the gluon-parton and g(x) is the probability function for gluon inside the nucleon. The expression within the curly bracket is the contribution from Fig. 6a and it blows up in the limit $|q^2| \to \infty$, but the full contribution scales. On the other hand the direct contribution (Fig. 6a) to F_1 contains only the first power of q^2 and hence gets damped out to zero in the scaling limit. Hence, in particular, the Callan-Gross relation $F_2 = 2xF_1$ will be violated.

The magnitude of this effect can be estimated (Rajasekaran and Roy [8]) from experimental data at lower q^2 (i.e. SLAC and CERN regions) which may be assumed to be below colour threshold. The well-known agreement of the SLAC electron-scattering experiments and the CERN neutrino-scattering experiments with the expectations of the G-Z quark-parton model is the motivation for this assumption. With this assumption, the CERN data on neutrino scattering implies that only about half of the momentum of the nucleon resides in the quark-partons so that the other half should be carried by the unexcited gluons. Using this fact, one finds that gluons contribute about 15% to the integral of $F_{2\,EM}$. It is conceivable that the effect is more significant for small x. Experimental detection (or otherwise) of this gluon contribution presumably at the higher energy regions being explored at Fermilab can serve to distinguish between Han-Nambu and Gell-Mann - Zweig schemes. (This numerical estimate is not affected much even if we assume that gluons contribute at lower energies.)

In the case of $e^+ e^-$ annihilation into hadrons also, only the G-Z part of the quark-contribution survives asymptotically and the asymptotic contribution from the two pairs of charged gluons turns out to be 1/8.

We may now include the weak interactions via the $SU_L(2) \times U(1)$ gauge theory. Then, the mixing equations which define the physical gluon fields $\tilde{G}^i$ are

$$\begin{aligned} \tilde{G}^i_\mu &= G^i_\mu - \frac{g}{f} W^i_\mu \qquad (i = 1,2,3) \\ \tilde{G}^8_\mu &= G^8_\mu - C\frac{g'}{f} U_\mu \\ \tilde{G}^i_\mu &= G^i_\mu \qquad (i = 4,5,6,7). \end{aligned} \tag{8}$$

Here, W^i_μ (i = 1, 2, 3) and U_μ are the gauge bosons of the $SU_L(2)$ and U(1) groups respectively and g and g' are the corresponding semiweak coupling constants; C is an arbitrary constant. These are the most general equations one can write at the level of exact $SU_L(2) \times U(1)$ symmetry.

After the $SU_L(2) \times U(1)$ symmetry is broken, we may rewrite W^3_μ and U_μ in eq. (8) in terms of Z_μ and A_μ as in the Weinberg-Salam model and thus get

$$\begin{aligned} \tilde{G}^\pm_\mu &= G^\pm_\mu - \frac{g}{f} W^\pm_\mu \qquad (\pm \equiv 1 \pm i2) \\ \tilde{G}^3_\mu &= G^3_\mu - \frac{g}{f}\cos\theta\, Z_\mu - \frac{e}{f} A_\mu \\ \tilde{G}^8_\mu &= G^8_\mu + C\frac{g}{f}\sin\theta\,\tan\theta\, Z_\mu - C\frac{e}{f} A_\mu \end{aligned} \tag{9}$$

where

$$e = gg'(g^2+g'^2)^{-1/2} \quad ; \quad \tan\theta = g'/g. \tag{10}$$

To generate the correct electromagnetic interactions for all the charged members of the octet, we have to set $C = 1/\sqrt{3}$. These are the mixing equations which replace eqs. (3) or (4). Thus, the charged gluons $G^\pm$ mix with the charged weak bosons $W^\pm$ and the neutral gluons $G^{3,8}$ mix not only with the photon, but also with Z.

We can follow an argument exactly parallel to that already used for electromagnetism. We then find that the weak currents of the quarks also have a colour singlet part plus a colour octet part and the colour octet part is damped by the same factor $m_g^2 (m_g^2 - q^2)^{-1}$ and hence vanishes asymptotically, but the asymptotic gluon contributions to the weak structure functions are nonvanishing and are given by

$$\begin{aligned} F_{2\,CC}^{gluons} &\longrightarrow \frac{1}{2}\, x\, g(x) \qquad \text{(charged current)} \\ F_{2\,NC}^{gluons} &\longrightarrow \frac{1}{4}\left(1+\frac{1}{3}\tan^4\theta\right) x\, g(x) \qquad \text{(neutral current)}. \end{aligned} \tag{11}$$

The gluon contributions given in (11) amount to only a few percent in the integral.

I want to close with two questions. 1) Note that the gluon mass m_g disappears in the final results. What happens if $m_g = 0$? 2) Can one really rule out the possibility of colour excitation even at CERN/SLAC energies?

References

1 O. W. Greenberg, Phys. Rev. Lett. 13 (1964) 598.
2 R. H. Dalitz, High Energy Physics, (Gordon and Breach, 1965) 253.
3 A. N. Mitra and R. Majumdar, Phys. Rev. 150 (1966) 1194.
4 S. Weinberg, Rev. Mod. Phys. 46 (1974) 255.
5 J. Pati and A. Salam, Phys. Rev. D8 (1973) 1240.
6 G. Rajasekaran and P. Roy, TIFR preprint TH/75-38 (Pramana, in press).
7 J. Pati and A. Salam, Trieste preprint IC/75/95, Maryland preprint 76-061.
8 G. Rajasekaran and P. Roy, TIFR preprint TH/75-42 (Phys. Rev. Lett. in press).

MESON-EXCHANGE CURRENTS IN ELECTRON-DEUTERON SCATTERING +

M. Gari* and H. Hyuga
Institut für Theoretische Physik, Ruhr-Universität Bochum, Germany

Very recently the elastic electron-deuteron scattering cross section has been measured at high momentum transfer with a very surprising result [1]. The comparison with the few available calculations in this momentum range seems to indicate that a conventional meson-exchange treatment of the deuteron formfactor (FF) at high momentum q is by orders of magnitudes off. Obviously the "flattening" out of the deuteron-FF as has been predicted [2, 3, 4] does not occur. At least not in the region of present experiments. The manifestation of these results would be very surprising as we know that the discrepancies between experiment and impulse approximation in thermal n-p capture [5, 6] as well as in electrodisintegration of the deuteron at threshold [7] can be removed completely by inclusion of meson-exchange currents.

In previous calculations also the exchange currents have been calculated only partially. While Jackson et al. [9] have taken into account only pion-exchange processes for $q^2 < 35\ fm^{-2}$, Chemtob et al. [3] consider $\rho\pi\gamma$ and $\sigma\pi\gamma$ contributions and Blankenbecler and Gunion [2] ρ-exchange. A complete analysis including all established processes has not been done. In the present note we give our results of a conventional exchange current calculation for the deuteron-FF where we include all established exchange currents (figures 1 - 4). The treatment is conventional in so far as we do not consider relativistic effects [10] or baryon resonance [11] admixtures to the deuteron. The main points of our calculation can be summarized as follow:

(i) exchange currents: π, ρ, ω, $\rho\pi\gamma$ (finite decay width of the ρ-meson is also taken into account),

(ii) two types of photon-nucleon FF are used (a) ampirical dipole fit, (b) Iachello et al. FF [12]. The $\rho\pi\gamma$ vertex FF is taken from ref. [3] but corrected for the now measured decay width [8],

(iii) the pion-nucleon FF is taken from Schmit [13] in the monopole form. The ρ and ω meson-nucleon FF's are taken from ref. [12] in the dipole form. For zero decay width of the ρ-meson a simple monopole form is assumed.

In our notation the elastic electron-scattering cross-section is given by:

$$\frac{d\sigma}{d\Omega} = \left(\frac{d\sigma}{d\Omega}\right)_{Mott} \left[F_E^2(q^2) + (1 + 2 \tan\frac{\theta}{2}) \frac{q^2}{6M^2} F_M^2(q^2) \right]$$

with

$$F_E^2(q^2) = F_C^2(q^2) + \frac{q^4}{18} F_Q^2(q^2)$$

F_C, F_Q and F_M denote the charge, quadrupole and magnetic formfactors, respectively.

Our results for the formfactor F_E^2 are given for two ranges of momentum transfer.

Figure 3 : $0 \leq q^2 \leq 60$ fm^{-2} and

figure 4 : 50 fm$^{-2} \leq q^2 \leq 200$ fm^{-2}.

The results presented in the diagrams (figures 1 - 4) are obtained by the use of deuteron wavefunctions derived from the Reid-soft-core potential. We calculated the FF's also for Hamada-Johnston and super-soft-core wavefunctions. The results are very similar. For high momentum transfer (figure 4) we show also the effect of a non-vanishing ρ-meson decay width. We realize that this is a non-negligible effect. A more detailed and complete discussion of our results is given in ref. [15].

Already at low momentum transfer we realize the limits of the calculations due to the freedom in the choice of the photon-nucleon-FF. Although the meson exchange contributions are almost the same for both photon-nucleon-FF's we note that in the momentum range $q^2 \simeq 40$-50 fm^{-2} it is already difficult to disentangle both effects. The only region where meson-exchange contributions are not masked by the freedom in the choice of the FF's is for 20 fm$^{-2} \leq q^2 \leq 40$ fm^{-2}.

For high momentum transfer the situation changes considerably (figure 4). Here already the impulse approximations show considerable differences. This is mostly due to the minimum in the FF of Iachello et al. (IJL). The IJL impulse approximation is in strong disagreement with the present experiment. The empirical dipole fit is very close to the experiment as noted also in ref. [1]. Looking at the formfactor $F_E^2(q^2)$ with exchange current contributions we realize that F_E^2 calculated with the IJL-formfactor is very close to the impulse approximation calculated with the empirical dipole-FF. F_E^2 with exchange currents for the empirical dipole-FF gives results a bit larger than the one obtained with IJL up to momentum transfer $q^2 \sim 100$ fm^{-2}. It seems to us that this range of momentum transfer is not very promising for obtaining information on the neutron-FF as different effects mask each other. The comparison of our results with the experimental data shows interesting features. We emphazise that our treatment of the exchange currents is well able to agree with the experiment. From the present comparison one would rather believe to obtain some information on the neutron-FF at higher momentum transfer. For this purpose we calculated the deuteron-FF up to $q^2 = 200\ fm^{-2}$. Here we realize a dramatic change in the deuteron-FF. A very interesting point, however, is that the total deuteron formfactor $F_E^2(q^2)$ is very different for different neutron-FF's. In the present case the large difference arises from the fact that the IJL-formfactor has a minimum at about $q^2 \sim 80\ fm^{-2}$ while the dipole-FF doesn't. So the IJL-formfactor changes sign and so leads to very different results for the total formfactor $F_E^2(q^2)$, as compared to the dipole case. This actually results from the fact that not all exchange contributions depend on the isoscalar nucleon-FF so while some contributions add in the IJL case they substract in the dipole case. For this reason it seems to us that the range of interest for the deuteron-FF is above $q^2 \sim 100\ fm^{-2}$. In this region one can hope to obtain valuable information on the neutron-FF.

As far as the magnetic formfactor is concerned only the low momentum transfer region shows interesting features (figure 2). The impulse FF has a minimum at $q^2 \gtrsim 40\ fm^{-2}$ for both types of photon-nucleon FF's. In the total magnetic FF calculated with the inclusion of meson-exchange currents this minimum disappears completely. This region of momentum transfer, however, seems worthwhile to investigate.

In conclusion we note an overall good agreement of our deuteron-FF (including exchange currents) with the present experiments.

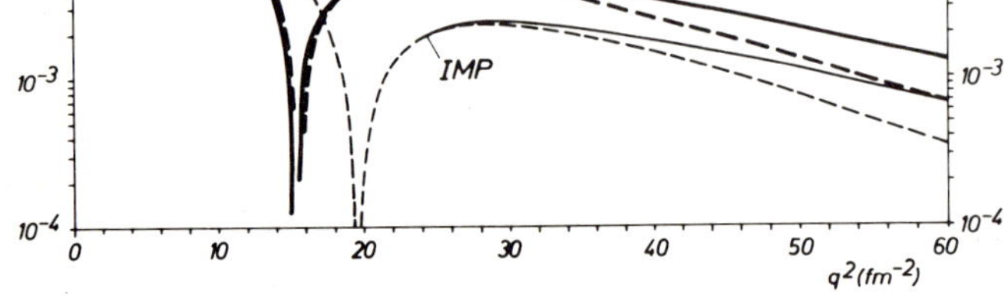

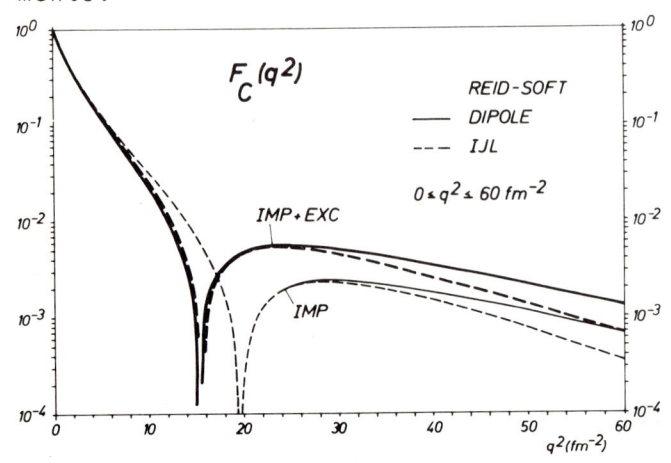

Figure 1
Charge formfactor (absolute value) of the deuteron

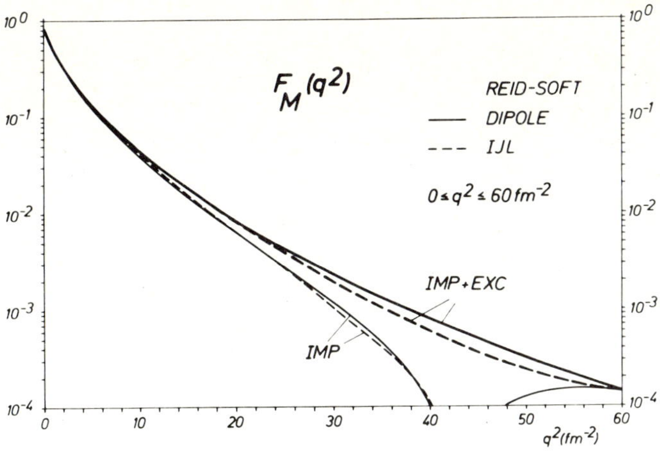

Figure 2
Magnetic formfactor of the deuteron

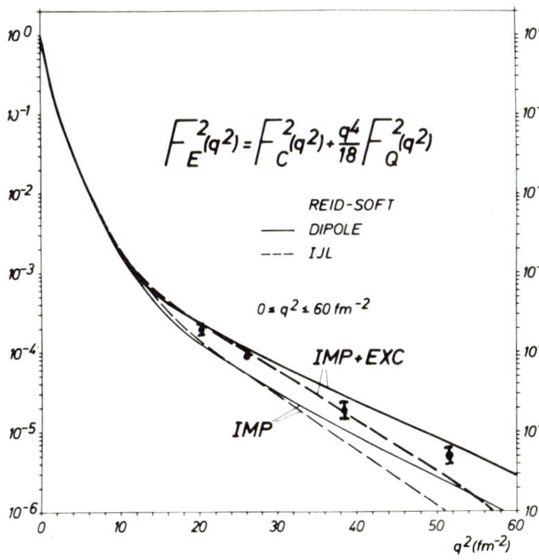

Figure 3
Electric deuteron formfactor
at low momentum transfer

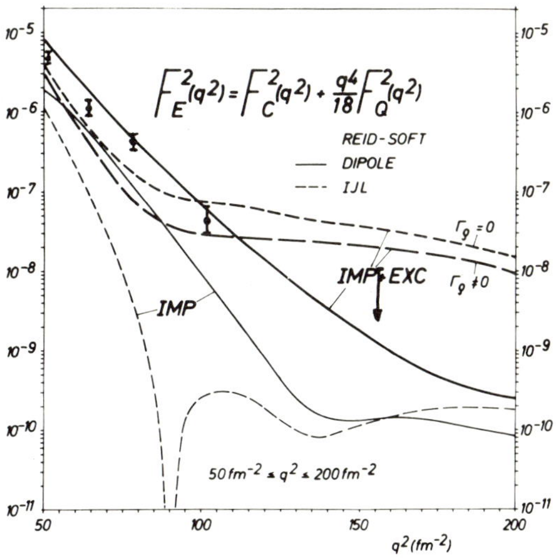

Figure 4
Electric deuteron formfactor
at high momentum transfer

REFERENCES

[1] R.G. Arnold et al., Phys. Rev. Letters 35 (1975) 776
[2] R. Blankenbecler and J. Gunion, Phys. Rev. D4 (1971) 718
[3] M. Chemtob, E. Moniz and M. Rho, Phys. Rev. C10 (1974) 335
[4] R. Adler, Phys. Rev. 141 (1966) 1499
[5] D.O. Riska and G.E. Brown, Phys. Letters 38B (1972) 193
[6] M. Gari and A. Huffmann, Phys. Rev. C7 (1973) 994
[7] J. Hockert, D.O. Riska, M. Gari and A. Huffman, Nucl. Phys. A217 (1973) 14
[8] B. Gobbi et al., Phys. Rev. Letters 33 (1974) 1450
[9] A. Jackson, A. Landé and D.O. Riska, Phys. Letters 55B (1975) 23
[10] F. Coester and A. Ostebee, Phys. Rev. C11 (1975) 1836
[11] H. Arenhövel and H. Miller, Z. Physik 266 (1974) 13
[12] F. Iachello, A. Jackson and A. Landé, Phys. Letters 43B (1973) 191
[13] C. Schmit, Phys. Letters 52B (1974) 381
[14] J. Elias et al., Phys. Rev. 177 (1969) 2075
[15] M. Gari and H. Hyuga, to be published

+ Supported by the Minister für Wissenschaft und Forschung des Landes Nordrhein-Westfalen
* Invited talk

THE THREE NUCLEON BOUND STATE PROBLEM AS A PROBE OF THE INTERNUCLEON FORCE

Michael I. Haftel
Naval Research Laboratory
Washington, D. C. 20375 USA

For the past decade, the three-nucleon (3N) problem has been regarded as a potentially ideal probe of those features of the nuclear force unobtainable from two-nucleon (N-N) experiments. The 3N system is the simplest nuclear system where the effects of both the off-shell behavior of the N-N force and possible three-body forces would show up. Thanks to the Faddeev equations, one can obtain an essentially exact solution of the 3N dynamics, at least in the bound states, and hence study unambiguously the above effects. In view of the computational accuracy in solving the 3N bound-state problem, it is disturbing, yet challenging to us, that not one model of the N-N force has yet been found that satisfactorily predicts the static and electromagnetic properties of the 3N bound states. Realistic (local) N-N potentials underbind the triton by 1-2 MeV [1],[2]. These potentials predict the position of the diffraction minimum of the ^{3}He charge form factor ($F_{ch}(q^2)$) at $q_o^2 \approx$ 13-17 fm^{-2} (experimentally $q_o^2 \approx 11.5$ fm^{-2}) and predict the height of the second maximum about an order of magnitude too low. If one considers off-shell variations of local potentials [3], E_T may increase by as much as ~ 2 MeV, but only at the cost of ruining the fit to $F_{ch}(q^2)$ (^{3}He) even more.

One of the main conclusions of the Laval conference [4] was that the effects of the "meson degrees of freedom" must be taken into account if one desires to thoroughly understand few-nucleon systems and the internucleon force. The failure of the non-relativistic 3N Schroedinger equation with any N-N force to predict the 3N bound state properties could well be due to the neglect of such effects.

The presence of meson degrees of freedom present at least two relatively unexplored features into the 3N problem: the appearance of three-body forces and corrections to the electromagnetic form factors from meson exchange currents. This paper reassesses the role of the 3N bound state problem in obtaining information about the short-range internucleon dynamics in view of the possible importance of these features of the meson degrees of freedom. To accomplish this, we calculate the triton binding energy (E_T) and the ^{3}He, ^{3}H charge form factors, both without and with exchange current corrections, for various phenomenological models of the internucleon force, one of which includes a 3N force.

Meson field theory predicts that 3N forces should be present. Yang [5] estimates that they contribute about 2.3 MeV to E_T. I [3] have phenomenologically hypothesized their presence from the E_T - $F_{ch}(q^2)$ anomaly with two-body potentials, while Brayshaw [6] has estimated their contribution to E_T, on a similar basis, at about 2.5 MeV. The evidences of the presence of 3N forces on the basis of these form factor calculations [3],[6] are suspect since meson-exchange-current contributions were not included. Kloet [7] and Tjon have shown that in the momentum transfer region critically involved in the 3N force hypothesis ($q^2 > 6$ fm^{-2}) exchange current corrections are large. In fact, for the Reid soft-core potential, they largely resolve the discrepancy between theory and experiment in ^{3}He. Obviously, conclusions concerning 3N forces as well as the usefulness of the 3N bound state problem to probe the internucleon dynamics must be re-examined in light of Kloet and Tjon's calculation.

Calculations of 3N bound state properties with 3N forces or meson exchange currents are rather sparse. Delves [8] variationally has employed an explicit 3N force in solving the 3N Schroedinger equation. Kloet and Tjon's calculation remains the best treatment of exchange currents in the 3N system. This paper, to my knowledge, is the first that includes both the effects of 3N forces and meson exchange currents, as well as possible off-shell effects. The main questions it attempts to answer are:

1. To what extent can 3N bound state calculations (of E_T, $F_{ch}(q^2)$) be used to elucidate off-shell and three-body features of the nuclear force?
2. Can effects due to the strong interaction (off-shell, 3N forces) be distinguished from the electromagnetic effects (exchange currents)?
3. Is a satisfactory, presently calculable theory available to explain the 3N bound state properties? If so, what are the features of such a theory, especially with regards to 3N forces and exchange currents?

To answer these questions we consider three different N-N potentials: the Reid [9] soft-core potential (R), the de Tourreil-Sprung A [10] super soft-core potential (S) and the Graz (preliminary version) [11] separable potential (GRP). We also consider a potential (GRPA), generated from the GRP potential by a method to be described shortly, that contains a 3N force. All of these potentials are approximately equivalent up to 350 MeV in N-N scattering, except for the $^3S_1$ + $^3D_1$ mixing coefficient (ϵ_1). The two-body effective range parameters and static deuteron properties, as well as the triton binding energy, appear in Table I.

Table I

Two-body effective range parameters, deuteron (E_d) and triton (E_T) binding energies. Model GRPA parameters are the same as GRP.

Potential	a_s(fm) / r_s(fm)	a_t(fm) / r_t(fm)	E_d(MeV) / E_T [a)] (MeV)	P_D(%) / Q(fm^2)
Reid (R)	-17.1	5.39	2.2246	6.47
	2.80	1.72	6.34 (7.0)	.280
de Tourreil-Sprung A (S)	-17.3	5.50	2.2237	4.43
	2.84	1.85	7.02 (7.64)	.261
Graz Prelim (GRP)	-17.7	5.45	2.2248	3.69
	2.71	1.79	7.94 (8.5)	.313

a) Values not in parentheses are obtained from 3 channel calculations using ^1S_o and $^3S_1$ + $^3D_1$ interactions. Values in parentheses are for more complete calculations in refs. 2 except for GRP where the result of a full Faddeev calculation is estimated.

One interesting fact, not completely obtainable from Table I, is that despite large variations in P_D, all potentials give very close to the same deuteron electric form factor [12]. Potential S gives almost perfect agreement with experiment for $q^2 < 35$ fm^{-2} with all other potentials giving results fairly close to the experimental data. The main factor that accounts for S giving .7 MeV more binding than R for E_T is the lower P_D value (4.4% vs. 6.4%). The explicit presence of the super soft core plays essentially no role since the off-shell behaviors of R and S are nearly identical at least for momentum components of $k \leq 2$ fm. For GRP both the lower P_D and softer off-shell behavior play roles in the increase of E_T over the S value [12]. The properties of GRP in Table I (and the deuteron form factor results) indicate that it is possible to fit the experimental E_T with a two-body force without grossly violating any direct two-body experimental constraints.

To generate three-body forces (e.g., model GRPA) we first consider a three-body Hamiltonian with only two-body forces, i.e., $H = \sum_i T_i + \sum_{i<j} V_{ij}$. In the three-body system, this Hamiltonian yields a triton binding energy (E_T), wave function (ψ_T) and electromagnetic form factors (collectively referred to as $F(q^2)$). We then introduce a two or three-body unitary operator U and define a transformed

Hamiltonian $\tilde{H} = UHU^+$. By the unitarity, the Hamiltonian $\tilde{H}$ preserves the spectrum of H but changes the wave functions and form factors (i.e., if $H\psi = E\psi$, $\tilde{H}\tilde{\psi} = E\tilde{\psi}$ where $\tilde{\psi} = U\psi$). Hence if $H \to E_T, \psi_T, F(q^2)$, then $\tilde{H} \to E_T, \tilde{\psi}_T, \tilde{F}(q^2)$ where $\tilde{\psi}_T = U\psi_T$, and $\tilde{F}(q^2) \neq F(q^2)$.

In general, the transformed Hamiltonian may be written

$$H = UHU^+ = \Sigma T_i + \sum_{i\,j} \hat{V}_{ij} + V^{(3)} \tag{1}$$

where $\hat{V}_{ij}$ is a (perhaps altered) two-body potential, and $V^{(3)}$ represents all terms that depend on the coordinates of all three particles, i.e., three-body force terms. Sáenz [13] and Zachary have derived general conditions on U such that $\tilde{H}$ is equivalent to H for three (or many) body scattering and $\hat{V}_{ij}$ is equivalent to V_{ij} for two-body scattering. The type of transformation we consider satisfies their conditions for two and three-body scattering equivalence.

In this work, we consider a transformation of the hyperspherical type, i.e.,

$$\langle Ry|U|Ry'\rangle = U(R,R')\delta(y-y') \tag{2}$$

where $R^2 = 2/3\,\Sigma|\vec{x}_i - \vec{x}_j|^2$ and y represents all additional coordinates needed to specify the three-body system. Hyperspherical transformations preserve the two-body force ($\hat{V}_{ij} = V_{ij}$) and, if $U(R,R') \xrightarrow[R \text{ or } R' \to \infty]{} \delta(R-R')/R^5$ they lead to 3 body forces in the usual sense - i.e., $V^{(3)}$ vanishes when <u>any</u> interparticle distance becomes large. In our three-body model GRPA, we take

$$\begin{aligned} U(R,R') = {} & \delta(R-R')/R^5 - (1-\cos\theta)[g_1(R)g_1(R') + g_2(R)g_2(R')] \\ & + \sin\theta\,[g_1(R)g_2(R') - g_2(R)g_1(R')] \end{aligned} \tag{3}$$

where the unitarity condition is

$$\frac{\pi}{16}\int_0^\infty R^5 g_i(R)g_j(R) = \delta_{ij}.$$

For GRPA we take $g_i(R) \sim e^{-\alpha_i R}(\beta_{0i} + \beta_{1i}R + \beta_{2i}R^2)$ with $\alpha_1 = \alpha_2 = 2.1\ \text{fm}^{-1}$, $\beta_{01} = \beta_{02} = 0$, $\beta_{11} = -1.0\ \text{fm}^{-1}$, $\beta_{12} = 1.5873\ \text{fm}^{-1}$, $\beta_{21} = 0.3\ \text{fm}^{-1}$, $\beta_{22} = -1.0\ \text{fm}^{-1}$ and $\theta = 11^\circ$. The starting two-body potential is the GRP potential since GRP predicts close to the experimental E_T. This binding energy is preserved in GRPA and this potential may then provide a test whether one can fit both E_T and $F_{ch}(q^2)$ if a three-body force is included.

We now consider the following exchange current diagrams.

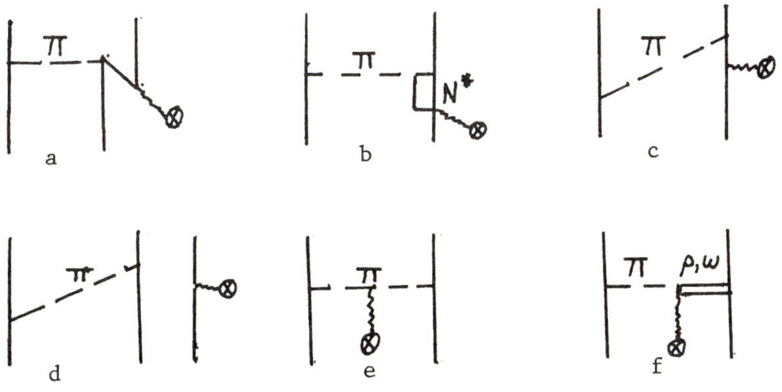

Fig. 1 - Meson exchange currents

Diagrams (a) and (b) have previously been considered for the Reid potential by Kloet and Tjon and are calculated for all the potentials we consider. Diagrams (c) and (d) are the "recoil" diagrams discussed by Jackson [14], Landé and Riska in the deuteron case and are calculated here for ^{3}He and ^{3}H. Diagram (e) does not contribute to the charge form factor while Kloet and Tjon have shown (f) to be negligible compared to (a) for the Reid potential. We ignore (f) in the present calculation. The calculations of the exchange currents were done by W. M. Kloet following the details [7] of his previous calculations. An error occurred in the calculation of diagram (b) previously, which has been corrected (it turns out that (b) is negligible).

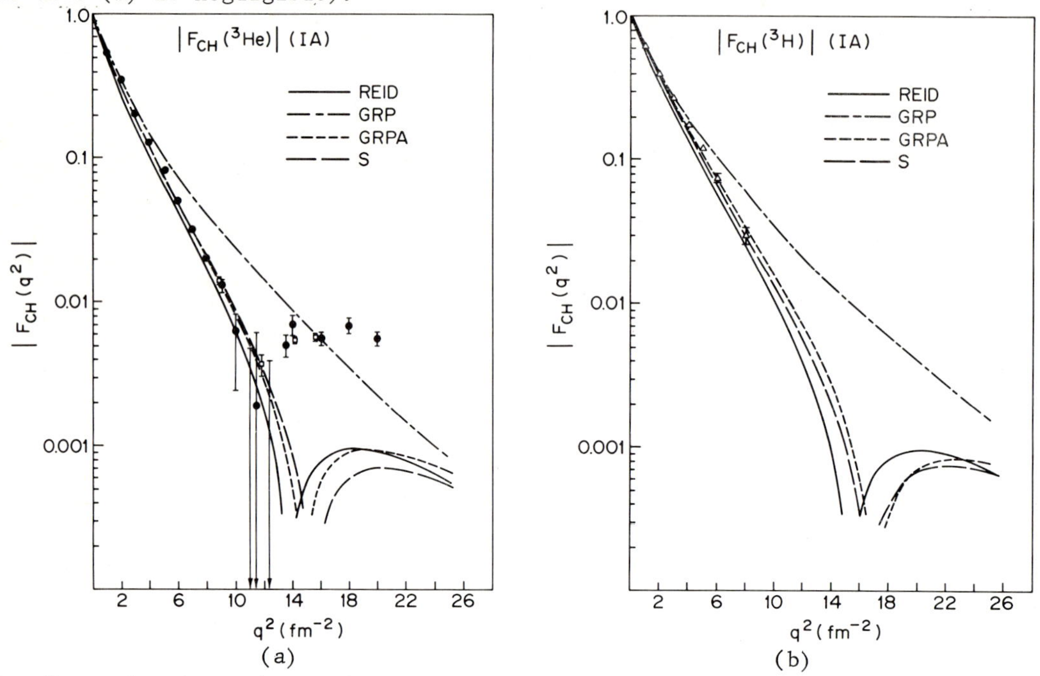

Fig. 2. Trinucleon charge form factors in the impulse approximation (IA)

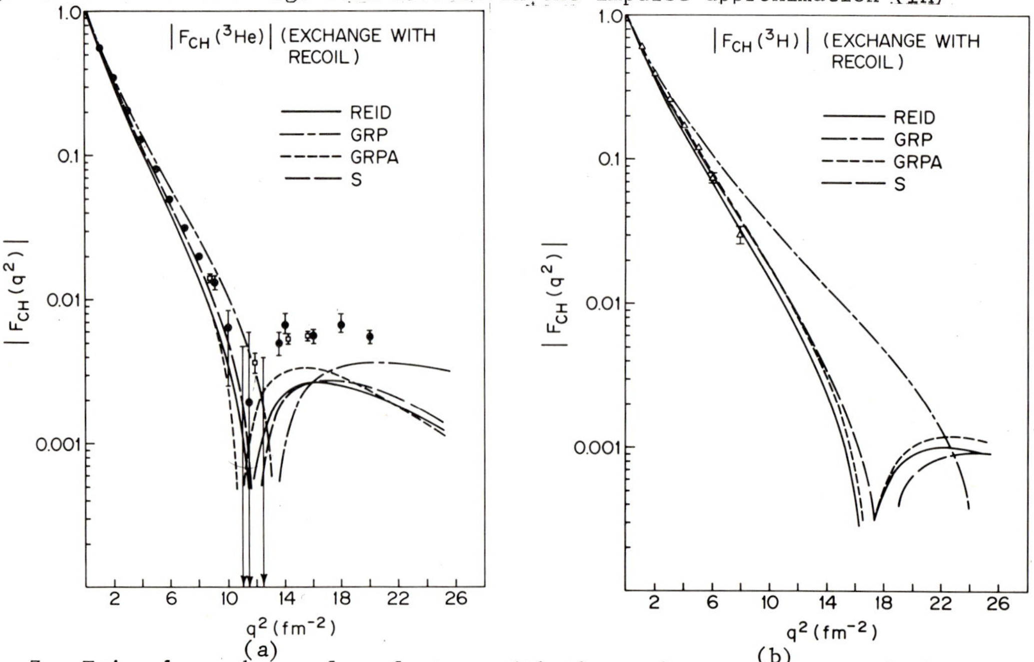

Fig. 3. Trinucleon charge form factors with the exchange currents of Fig. 1(a)-(d).

Figure 2(a) illustrates $F_{ch}(q^2)$ (^3He) in the impulse approximation (IA). For the two-body interactions there is significant potential sensitivity, especially for $q^2 > 10\ fm^{-2}$. Of the two-body models, S fits the data well for $q^2 < 10\ fm^{-2}$, but it underbinds the triton by about 1 MeV. The two-body force that fits E_T badly overestimates $F_{ch}(q^2)$ for all but very low q^2. Model GRPA, which includes a three-body force, fits $F_{ch}(q^2)$ well for $q^2 < 10\ fm^{-2}$ while retaining the correct E_T. In the IA none of the models give a good fit for $q^2 > 10\ fm^{-2}$; all models have the minimum too far out and the second maximum far too small. Figure 2(b), which shows $F_{ch}(q^2)$ (^3H), exhibits similar trends between potentials. Here S and GRPA obtain good agreement with the available data.

Figure 3(a) illustrates $F_{ch}(q^2)$ (^3He) with the exchange current corrections. The exchange corrections, while quite potential dependent, are such as to markedly narrow differences between potentials for the total $F_{ch}(q^2)$. For any model the exchange currents allow the prediction of the diffraction minimum to be nearly correct. In all cases the height of the second maximum is within a factor of two of experiment. Moreover, for ^3He, with exchange there is very little model dependence in $F_{ch}(q^2)$ and all models essentially predict agreement with experiment.

Figure 3(b) illustrates $F_{ch}(q^2)$ (^3H) with the exchange current corrections. The exchange corrections are small and $F_{ch}(q^2)$ is still fairly model dependent. Again the data favors either a two-body force that underbinds the triton (like S) or a combination of two and three-body forces that fit E_T. The inadequacy of the two-body force that fits E_T, however, is clearly manifested only at the $q^2 = 6\ fm^{-2}$ and $q^2 = 8\ fm^{-2}$ points. More measurements, especially at larger q^2, would be desirable to confirm the presence of 3N forces.

In conclusion, both exchange currents and 3N forces play roles here in satisfactorily explaining the 3N bound state data. Exchange corrections alone essentially account for $F_{ch}(q^2)$ (^3He) while $F_{ch}(q^2)$ (^3H), due to its exchange current insensitivity, is a good testing ground for nuclear forces, especially 3N forces.

References

[1] E. P. Harper, Y. E. Kim, A. Tubis, Phys. Rev. Lett. 28 (1972) 1533.

[2] C. Gignoux and A. Laverne, Phys. Rev. Lett. 29 (1972) 436; A. Laverne and C. Gignoux, Nuc. Phys. A203 (1973) 597.

[3] E. P. Harper, Y. E. Kim and A. Tubis, Phys. Rev. C6 (1972) 1601; M. I. Haftel, Phys. Rev. C7 (1973) 80.

[4] H. P. Noyes, Few Body Problems in Nuclear and Particle Physics, (R. J. Slobodrian, B. Cujec and K. Ramavataram, Les Presses de L'Université Laval 1975) 823.

[5] S. N. Yang, Phys. Rev. C10 (1974) 2067.

[6] D. D. Brayshaw, Phys. Rev. C7 (1973) 1731.

[7] W. M. Kloet and J. A. Tjon, Phys. Lett. 49B (1974) 419; W. M. Kloet and J. A. Tjon, Few Body Problems in Nuclear and Particle Physics, (R. J. Slobodrian, B. Cujec and K. Ramavataram, Les Presses de L'Université Laval 1975) 523.

[8] L. M. Delves and M. A. Hennell, Nucl. Phys. A168 (1971) 1347.

[9] R. V. Reid, Ann. Phys. (N.Y.) 50 (1968) 411.

[10] R. de Tourreil and D. W. L. Sprung, Nucl. Phys. A201 (1973) 193.

[11] L. Crepinsek, H. Oberhummer, W. Plessas and H. Zingl, Acta Phys. Austr. 39 (1974) 345; L. Crepinsek, C. B. Lang, H. Oberhummer, W. Plessas and H. Zingl, Universität Graz preprint.

[12] M. I. Haftel and W. M. Kloet (to be published).

[13] A. W. Sáenz and W. W. Zachary, Phys. Lett. 58B (1975) 13.

[14] A. D. Jackson, A. Lande and D. O. Riska, Phys. Lett. 55B (1975) 23.

NEUTRON DEUTERON COLLISION AND THE CHARGE FORM FACTOR OF TRITON

S. N. Banerjee
Physics Department, Jadavpur University
Calcutta - 32, India.

Several works have been performed to investigate the nucleon deuteron collision problem with the two body non-local separable potential of Yamaguchi[1)] type. With the advent of the Faddeev equations, calculations have also been carried out on the three nucleon problem with local potentials. To investigate the static properties of the three nucleon systems, a model has been suggested in which the deuteron has been described as an elementary isoscalar vector particle, regardless of its structure, in the neighbourhood of the elastic threshold of the neutron deuteron scattering. The present author[2)] has also investigated the neutron deuteron collision in a N/D model with the idea that the deuteron behaves as an elementary isoscalar vector particle. Morioka and Ueda[3)] have also discussed the neutron deuteron scattering in a model in which the inner part of the deuteron is regarded as an elementary particle and a field theoretical model is applied to study it, obtaining good agreement with experiment at forward and backward scattering angles. Our present model of the description of the deuteron as an elementary vector particle essentially reduces the three-body collision problem to a two-body problem as the collision between a neutron and the deuteron. With two body non-local one term separable potential of Yamaguchi type to parametrise the interaction due to nucleon exchange between the neutron and the deuteron, we write the two-body Schrodirger equation as

$$(\alpha^2 + p^2)\,\psi(p) = \lambda\, g(p) \int g(p')\,\psi(p')\,dp'$$

Here $\frac{\alpha^2}{\mu}$ represents the binding energy of triton (^3H), all other symbols have their usualmeanings. With $g(p) = N\cdot(p^2+\beta^2)^{-1}$ and assuming the same β for the doublet and the quartet channels in our model, we have adjusted λ and β so as to represent the binding energy of ^3H as 8.5 Mev and the doublet scattering length $^2a \sim 2.2$ fm. Now we can adjust λ for the quartet channel so as to give a fit to the quartet scattering length $^4a \sim 6.4$ fm. Our adjusted value of $\beta \simeq 5.0$ fm^{-1} comes approximately the Compton wave length $\frac{\hbar}{mc}$ of the exchanged particle, the proton. Hence the range of the nucleon exchange force is correctly reproduced in our one term separable potential model.

The charge form factor of ^3H becomes in ourmodel as

$$F_{ch}(q^2) = \frac{4\pi\gamma}{q} F_p(q^2) \left[\tan^{-1}\frac{q}{2\alpha} + \tan^{-1}\frac{q}{2\beta} - 2\tan^{-1}\frac{q}{(\alpha+\beta)} \right]$$

with $\gamma = \frac{1}{2\pi} \cdot \frac{\alpha\beta(\alpha+\beta)}{(\beta-\alpha)^2}$.

Expanding this for small values of q^2 and comparing it with $F_{ch}(q^2) = 1 - q^2\langle r_{ch}^2\rangle\frac{1}{6}$, we get the root mean square charge radius $\langle r_{ch}^2\rangle^{1/2}$ of triton as 1.82 fm compared to the experimen -tal value (1.70 ± .05) fm of Collard et al. The charge form factor of ^3H has also been evaluated which is found to compare favourably with the experimental values for small values of the square of momentum transfer ($q^2 \leq 4\ fm^{-2}$).

In view of the approximation made in representing the deuteron by an elementary isoscalar vector particle and using only one separable protential term, reasonably accurate values are obtained for the charge form factor of ^3H ($q^2 \leq 4\ fm^{-2}$), particularly for the root mean square bharge radius of ^3H.

References

1. Y. Yamuguchi, Phys. Rev 95 (1954) 1628
2. S. N. Banerjee, Prog. Theor. Phys. (communicated)
3. S. Morioka and T. Ueda, Prog. Theor. Phys. 51 (1974) 1103
4. H. Collard, R. Hofstadter, E.B. Hughes, A. Johansson, M.K. Yearian, R. B. Day and R.T. Wagner, Phys. Rev. 138B (1964) 802

TRINUCLEON PHOTOEFFECT USING COUPLED HYPERSPHERICAL HARMONICS

K.K. FANG and J.S. LEVINGER
Rensselaer Polytechnic Institute Troy, N.Y. 12181, U.S.A.

and

M. Fabre de la Ripelle
Institut de Physique Nucléaire, Division de Physique Théorique*
91406 - ORSAY - France

We extend the calculation of Fabre and Levinger[1] for triton E-1 transitions to final isospin 3/2, by using an additional grand orbital L = 3, besides L = 1, in the hyperspherical harmonic (h.h.) expansion. We calculate the total photodisintegration cross section $\sigma(E_\gamma)$ of H, using the model of Volkov's two-body potential, assumed to have Wigner character : $V(r) = 144.86 \exp(-(r/0.82)^2) - 83.34 \exp(-(r/1.60)^2)$. Equation 2.22 of FL is replaced by the two coupled differential equations :

$$(T_1 - E)u_1(r) = -3V_0(r)u_1(r) + 3V_4(r)u_3(r)$$

$$(T_3 - E)u_3(r) = 3V_4(r)u_1(r) - 3V_0(r)u_3(r)$$

with the kinetic energy operators

$$T_1 = -(\hbar^2/m)(d^2/dr^2) + (\hbar^2/m)(35/4r^2)$$

$$T_3 = -(\hbar^2/m)(d^2/dr^2) + (\hbar^2/m)(99/4r^2)$$

Here $V_0(r)$ and $V_4(r)$ are multipoles of the potential. (See FL eq. 2.24). We find eigenphase solutions $u_1^\alpha(r)$, $u_3^\alpha(r)$ and $u_1^\beta(r)$, $u_3^\beta(r)$. We then use these solutions to find overlap integrals and the photo-effect cross section.

The table gives the uncoupled result[2,3] (L = 1 only, using the formalism of FL) and the coupled result (L = 1 and 3) using our formalism for ten different photon energies. We also show the corresponding integrated cross section σ_{int}, for comparison with the Thomas-Reiche-Kuhn value of 19.9 MeV-mb for a Wigner potential.

We also show in Table 1 the phase shift δ for the uncoupled case ; the Blatt-Biedenharn[4,5] eigenphases $\delta_\alpha, \delta_\beta$ and the mixing parameter ε for the coupled case. We have calculated the energy spectrum for the emitted odd nucleon, using our final state wave function. At 12.33 MeV, where the mixing parameter is small, we find an energy distribution close to that found by Delves[6] for the single grand orbital, L = 1.

The key result is the demonstration of the rapid speed of convergence of the h.h. expansion for this example. Examine the comparison of our integrated cross sections with the value 19.9 MeV-mb from the TRK rule. The discrepancy of 6 % for the uncoupled case is reduced to only 3 % by including one higher grand orbital! Of course, our assumption of a pure Wigner force gives poor agreement with experiment.

---

*Laboratoire associé au C.N.R.S.

^3H Photoeffect, Volkov Potential

	Uncoupled		Coupled			
E_γ (MeV)	σ(mb)	tan δ	σ(mb)	tan δ_α	tan δ_β	tan ε
8.83	0.009	0.26	0.011	0.28	0.04	-0.39
9.51	0.18	0.54	0.26	0.65	0.10	-0.13
10.61	1.23	1.23	1.73	2.11	0.17	-0.03
12.33	3.14	20.5	3.01	-3.82	0.24	0.06
15.02	2.04	-1.63	1.69	-1.18	0.33	0.15
19.38	0.66	-0.94	0.61	-0.81	0.47	0.22
26.98	0.14	-0.91	0.16	-0.85	0.67	0.27
41.79	0.018	-1.23	0.024	-1.03	1.06	0.31
76.92	0.003	-2.93	0.002	-2.1	1.86	0.39
201.7	0.0	2.60	0.0	5.43	1.72	0.85
σ_{int}	21.1 MeV-mb		20.5 MeV-mb			

Table 1

We extend our model calculation to the spin-dependent V^x potential acting in (1+) states, namely

$$V(r_{12}) = 130 \exp(-(r_{12}/0.80)^2) - 65.3 \exp(-(r_{12})1.5)^2)$$

We assume zero potential for (3-) states. We obtain the following two coupled differential equations :

$$(T_1 - E)u_1(r) = -(3/2)(V_0(r) - V_2(r))u_1(r) + (3/2)(V_4(r) - V_2(r))u_3(r)$$

$$(T_3 - E)u_3(r) = -(3/2)(V_0(r) + V_2(r)/3)u_3(r) + (3/2)(V_4(r) - V_2(r))u_1(r)$$

Using the same Runge-Kutta numerical method, we solve these c.d.e. Table 2 gives the phase shifts and total cross sections for both uncoupled and coupled channels. We note that the coupled calculation with 2 h.h. is close to the uncoupled result, using the formalism of FL. Results for coupled h.h. agree with Fetisov's experiment for photon energies above 20 MeV, but disagree below 20 MeV.

^{3}He Photoeffect, V^x Potential

	Uncoupled		Coupled			
E_γ (MeV)	σ (mb)	tan δ	σ (mb)	tan δ_α	tan δ_β	tan ε
8.13	0.0008	0.0004	0.0008	0.002	0.03	-0.005
8.81	0.01	0.002	0.01	0.003	0.06	-0.01
9.91	0.08	0.006	0.08	0.01	0.08	-0.03
11.63	0.24	0.016	0.25	0.02	0.11	-0.06
14.32	0.52	0.035	0.52	0.04	0.14	-0.11
18.78	0.81	0.07	0.78	0.08	0.18	-0.24
26.29	0.88	0.15	0.82	0.15	0.25	-0.36
41.09	0.55	0.28	0.67	0.25	0.37	-1.16
76.22	0.10	0.45	0.10	0.51	0.37	+0.40
201.02	0.0002	0.37	0.0009	0.38	0.46	0.84

Table 2

References

[1] M. Fabre de la Ripelle and J.S. Levinger, Nuov. Cim. 25A, 555 (1975).

[2] K. Myers, private communication.

[3] K.K. Fang, J.S. Levinger and M. Fabre de la Ripelle, Bull. Amer. Phys. Soc. 20, 577 (1975).

[4] J. Blatt and L. Biedenharn, Phys. Rev. 86, 399 (1952).

[5] D. Drechsel and L. Maximon, Ann. of Phys. 49, 403 (1968).

[6] L.M. Delves, Nucl. Phys. 20, 268 (1962).

TOTAL CROSS SECTION OF THE (γ,d) DISINTEGRATION OF THE α-PARTICLE

H.L. Yadav* and B.K. Srivastava
Department of Physics, Indian Institute of Technology, Kharagpur, India

It is by now well established [1-3] that the cross section for the reaction $^4He(\gamma,d)d$ is consistent with the $^1S_0 \longrightarrow {}^1D_2$ electric quadrupole transition so that the wave function R_{dd} for the relative motion of the two deuterons must describe a wave in the D-state. Mainly because of the presence of a resonant state in ^4He at about 30 MeV excitation, the plane wave approximation for the relative motion of the two deuterons in the reaction is very poor [4,5] and final state interactions must be taken into account. For this purpose we use the nonlocal separable potential approach - a method successfully applied by Rahman et al [6] in the $^3H(\gamma,n)d$ reaction - to calculate the final state d + d relative wave function.

The cross section for the $^4He(\gamma,d)d$ reaction is given by [7]

$$\sigma = \frac{8\pi^3 \nu}{c} \sum_{m=-1}^{1} \left| M_{if}^m \right|^2 \tag{1}$$

in which the matrix element

$$M_{if}^m = e \int \phi_\alpha \left(\sum_p Q_p\right) \phi_d \, \phi_d \, R_{dd} \, d\tau \tag{2}$$

where the symbols have the meaning found in the literature [7] .

The general approach for the calculation of the relative motion wave function assuming a nonlocal separable potential of the form

$$K_l(r,r') = \lambda q_l(r) \, q_l(r') \tag{3}$$

through fitting the l phase shifts is given in the literature [8]. We apply it to the calculation of R_{dd} for

$$q_2(r) = r^2 \, e^{-\beta r} \tag{4}$$

by fitting the (d,d) phase shifts for the l = 2 state **given** by Thompson [4] .

In our calculation of $^4He(\gamma,d)d$ cross section we choose the wave function of the α-particle to have the familiar Gaussian and Irving forms while the deuteron is described by the Gaussian and Hulthen wave functions. We determine the single unknown parameter in these wave functions by fitting the charge radius of the respective systems. For the sake of calculational convenience we use the Gaussian deuteron with the Gaussian alpha and the Hulthen deuteron with the Irving alpha. Finally for these wave functions eqs.(1) and (2) give the matrix elements M_{if}^m and the total cross section $\sigma(\gamma,d)$.

* Present address: Institut für Kernphysik der Kernforschungsanlage Jülich, West Germany

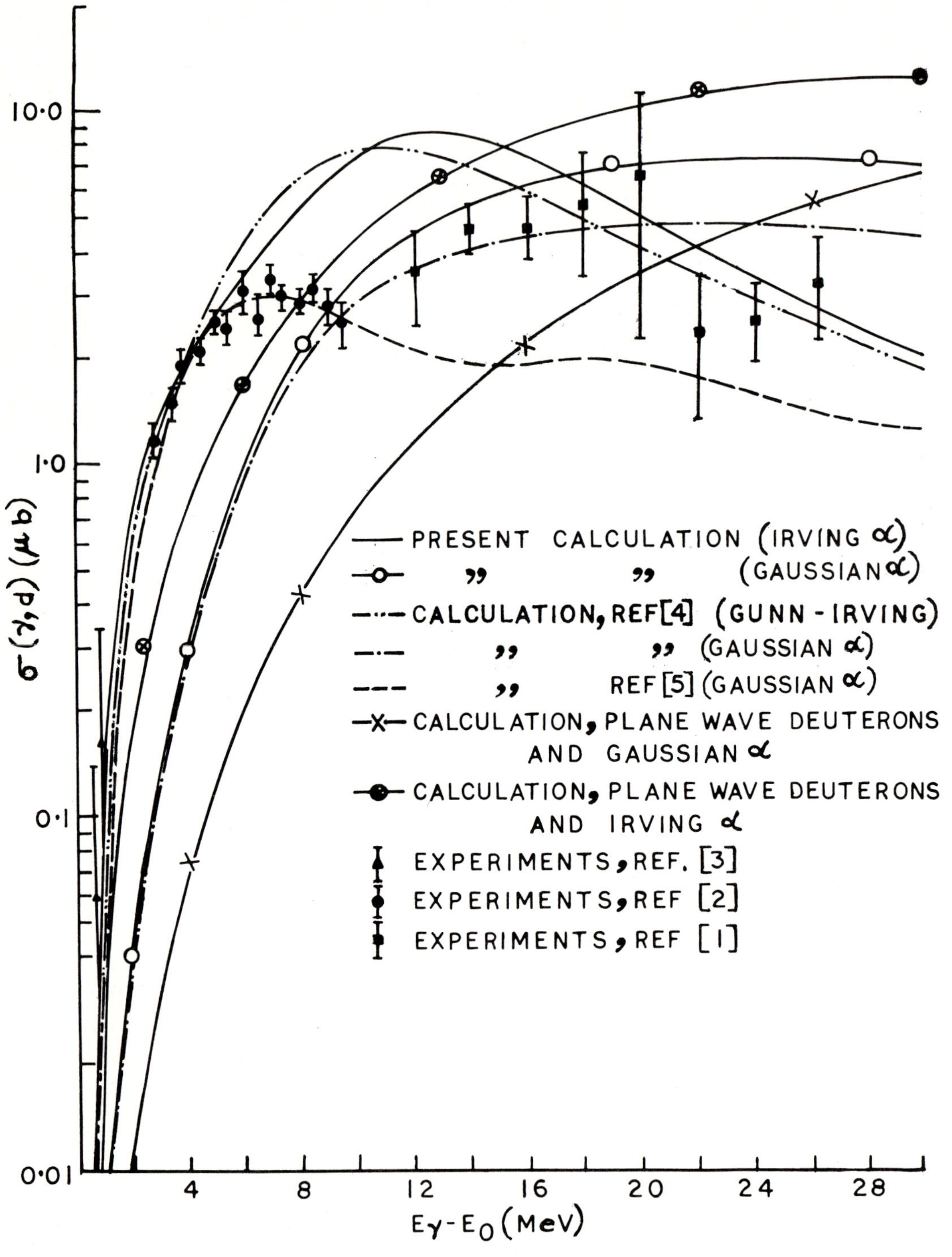

FIG. 1 $^4He(\gamma, d)d$ REACTION CROSS SECTION WITH GAUSSIAN AND IRVING TYPE ^4He FUNCTIONS.

In fig.1 we show our results for the Gaussian alpha and the Irving alpha along with the theoretical results of Thompson [4] and Erdas et al [5] , and the experimental values of references [1-3] . For purposes of comparison the figure also gives curves for the plane wave deuterons for (i) Gaussian alpha and (ii) Irving alpha.

From fig.1 we find that the inclusion of final state interactions between the deuterons through nonlocal separable potentials results in considerable improvement over the plane wave approximation. Since in our calculations we have used the same phase shifts as obtained by Thompson [4] , close resemblance of our curve for Gaussian alpha to that of Thompson [4] for the Gaussian wave function shows that the method used in our calculations to incorporate the final state interactions is quite satisfactory.

A comparison of our results and also of Thompson [4] with those of Erdas et al [5] , who have employed dispersive approach in their calculations, shows that the results of Erdas et al [5] are in better agreement with experiments in the low energy region but fail to do so at higher energies.

We conclude that the use of nonlocal separable potential to describe the final state interactions of the deuterons in the calculation of $^4He(\gamma,d)d$ cross sections is quite satisfactory and provides the same result as the resonating group method [4] . Also we find that the experiments show a definite preference for the Irving over the Gaussian alpha.

References

1. D.M.Skopik and W.R.Dodge, Bull.Am.Phys.Soc. 15 (1970) 23; D.M.Skopik, W.R.Dodge and B.F.Gibson, Bull.Am. Phys.Soc.15 (1970) 481
2. W.E.Myerhof, W.Feldman, S.Gilbert and W.O'Connell, Nucl. Phys. A131 (1969) 489
3. R.W.Zurmuhle, E.W.Stephens and H.H.Staub, Phys.Rev. 132 (1963) 751
4. D.R.Thompson, Nucl. Phys. A154 (1970) 442
5. F.Erdas, A.Pompei, P.Quarati and B.Mosconi, Nucl. Phys. A174 (1971) 657
6. M.Rahman, H.M.Sengupta and D.Hussain, Nucl.Phys. A168 (1971) 314
7. B.H.Flowers and F.Mandl, Proc. Roy.Soc. A206 (1951) 131
8. D.Husain and S.Ali, Am. J. Phys. 38 (1970) 597.

ELECTRON SCATTERING AND CORRELATION STRUCTURE OF LIGHT NUCLEI

M.A.K. Lodhi
Department of Physics, Texas Tech University
Lubbock, Texas 79409 USA

It has been known for some time that the short-range correlations due to the repulsive part of the nuclear interaction is exhibited in the nuclear form factors as obtained from high energy electron scattering. Accordingly these effects have been investigated by several workers in this field. In most of the calculations the Gaussian-type correlation function has been invoked in the single-particle wave functions. This type of correlation function causes the exchange of both small and large momenta, thus simultaneously mixing up the effects due to long- and short-range correlations with the independent particle motion. Hence for the independent particle shell model purpose one still has a job of separating the effects exclusively due to the short-range correlations. It has, however, been pointed out if the correlation parameter in the Gaussian type function is sufficiently small, it can ensure the modification of the two-body wave function only at large relative momenta. [1]

In this work the harmonic oscillator basis functions are used. The nuclear form factors as obtained from elastic electron scattering are calculated, with Jastrow's technique by means of the cluster expansion of Iwamoto Yamada, [2] in the Born approximation. The correlated wave function is given in the form:

$$\Psi(\underline{r}_1, \underline{r}_2, \ldots, \underline{r}_A) = \psi(\underline{r}_1, \underline{r}_2, \ldots, \underline{r}_A) \prod_{\substack{i = 1 \\ i < j}}^{A} f_\nu(r_{ij}) \tag{1}$$

where ν, in general, denotes the dependence of f on ℓ-, spin, and isospin quantum numbers, and

$$f_\nu(r_{ij}) = f_\nu(|\underline{r}_1 - \underline{r}_j|) = f_\nu(r) = 1 - \int j_0(k'_\nu r)\, w(k'_\nu)\, dk'_\nu . \tag{2}$$

The forms of $w(k')$ are assumed to be Gaussian and delta types. For the delta type function $w(k')$ leads to:

$$f(r) = 1 - j_0(kr) \tag{3}$$

This type of correlation function is simulated to exchange a definite momentum $\hbar k$ between a pair of nucleons (ij) which are otherwise assumed to be independent. Actually the assumption that only one momentum $\hbar k$ is being exchanged between a nucleon pair is a gross approximation of the fact that "the momentum to be exchanged between the pair is spread." Calculations performed with a Gaussian spread of $\hbar k$ produce no qualitative change on the result. [3]

Following the approach taken by Danos and Maximon [4] the correlation factor f(r) is transformed to the center of mars system of the nucleus. The results for nuclear form factors calculated with the wave function (1) are presented for some light nuclei.

For 1s nuclei they are as follows:

$$F_{^2H} = \frac{1 - 2CG^2}{1 - 2G^2} F_0 \tag{4}$$

$$F_{^3H} = F_{^3He} = \frac{1 - 2(1+2C)G^2}{1 - 6G^2} F_0 \tag{5}$$

$$F_{^4He} = \frac{1 - 6(1+C)G^2}{1 - 12G^2} F_0 \tag{6}$$

where

$$F_0 = \int |\psi|^2 \exp[i\, \underline{q} \cdot \underline{r}_i]\, d\underline{r}_1 \ldots d\underline{r}_A = \exp[-q^2/4\alpha^2] \tag{7}$$

$$F_{ij} = \int |\psi|^2 \exp(i\, q \cdot \underline{r}_i)\, j_0(kr_{ij})\, d\underline{r}_1 \ldots d\underline{r}_A = CGF_0 \tag{8}$$

$$G^2 = \int |\psi|^2 j_0(kr_{ij})\, d\underline{r}_1 \ldots d\underline{r}_A = \exp[-k^2/2\alpha^2], \tag{9}$$

and

$$C = \frac{\sinh\left(\frac{qk}{2\alpha^2}\right)}{\left(\frac{qk}{2\alpha^2}\right)} \tag{10}$$

Among the 1p nuclei the results for the form factor of ^6Li are presented here:

$$F_{el} = \frac{1}{3N^2}[2F_1 + F_5 - 2\{6G_s(1 + C_s) F_1 + F_5) + 4(F_{sp} + F_{ps})$$
$$+ 4G_{sp}(3F_1 + F_5) + 2G_pF_1 + F_{pp}\}], \tag{11}$$

where

$$N^2 = 1 - 2(6G_s + 8G_{sp} + G_p), \tag{12}$$

$$G_s = \exp[-k^2/2\alpha_s^2] \tag{13}$$

$$G_p = \frac{1}{9}[(3 - \frac{k^2}{2\alpha_p^2})^2 + \frac{k^4}{10\alpha_p^4}] \exp[-k^2/2\alpha_p^2] \tag{14}$$

$$G_{sp} = (3 - \frac{k^2}{2\alpha_p^2}) \exp[-\frac{k^2}{4}(\frac{1}{\alpha_s^2} + \frac{1}{\alpha_p^2})] \tag{15}$$

$$F_1 = \exp[-q^2/4\alpha_s^2] \tag{16}$$

$$F_5 = (1 - \frac{1}{6}\frac{q^2}{\alpha_p^2}) \exp[-q^2/4\alpha_p^2] \tag{17}$$

$$F_{sp} = C_s G_{sp} F_1 \tag{18}$$

$$F_{ps} = [C_p \left(1 - \frac{q^2 + k^2}{2\,\alpha_p^2}\right) + 2\cosh\left(\frac{qk}{2\alpha_p^2}\right)] \exp\left[-\left(\frac{k^2}{4\alpha_s^2} + \frac{q^2 + k^2}{4\,\alpha_p^2}\right)\right], \tag{19}$$

and

$$F_{pp} = \frac{1}{9}\left[\left(3 - \frac{k^2}{2\,\alpha_p^2}\right)\left(C_p - \frac{q^2 + k^2}{2\alpha_p^2}\,C_p + 2\cosh\frac{qk}{2\alpha_p^2}\right)\right.$$

$$\exp\left(-\frac{q^2 + 2k^2}{4\,\alpha_p^2}\right) + \sum_{\ell} \frac{i^{\ell}\theta_{\ell}\sqrt{\pi}\,q^{\ell}k^4}{2^{\ell+2}\,\Gamma\left(\frac{2\ell+3}{2}\right)\alpha_p^{\ell+4}} \exp(-k^2/4\alpha_p^2) \sum_{m=0}$$

$$\left.\left(-\frac{k^2}{4\alpha_p^2}\right)^m \frac{\Gamma\left(m + \frac{\ell+7}{2}\right)}{m!\;\Gamma\left(m+\frac{7}{2}\right)} \times {}_2F_1\left(-m,\ -\frac{2\ell+1}{2} - m;\ \frac{7}{2};\ \frac{q^2}{k^2}\right)\right] \tag{20}$$

with

$\ell = 0$, 2 and 4 only, and

$\theta_0 = \frac{1}{5}$, $\theta_2 = \frac{2}{7}$, and $\theta_4 = \frac{18}{35}$

As an example, the elastic charge form factor of ^{4}He is presented in the accompanying figure. The experimental data are taken from Frosch et al. [5] It is evident that the effects due to the short range correlations are dominant in the high momentum region.

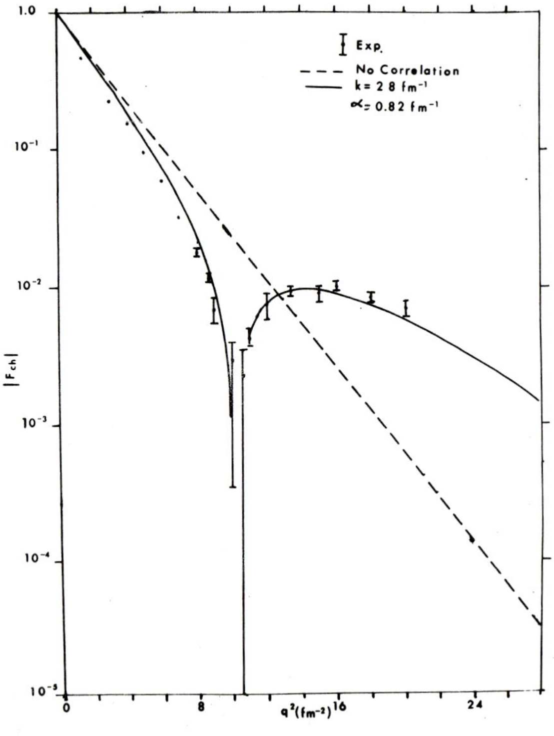

Fig.: ^{4}He charge form factor as obtained from elastic electron scattering.

1. C. Ciofi Degli Atti and N. M. Kabachink, Phys. Rev. C1 (1970) 809.
2. F. Iwamoto and Yamada, Prog. Theo. Phys. 17 (1967) 543.
3. S. T. Tuan, L. E. Wright, and M. G. Huber, Phys. Rev. Lett. 23 (1969) 174; K. Chung, M. Danos, and M. G. Huber, Phys. Lett. 29B (1969) 265.
4. M. Danos and L. C. Maximon, J. Math. Phys. 6 (1965) 766.
5. R. F. Frosch et al., Phys. Rev. 160 (1967) 874.

POSSIBLE EVIDENCE FOR ENHANCED $\Delta S = 0, \Delta T = 2$
WEAK INTERACTION IN PARITY MIXING IN NUCLEI

B.H.J. McKellar[*+], K.R. Lassey[+±], M.A. Box[+]
and
A.J. Gabric[+]

Many authors have despaired of finding a parity non-conserving (PNC) nucleon-nucleon potential which fits all the available data on parity mixing in nuclei. In this paper we present a solution of the problem. The key to the solution is the introduction of a strong $T = 2$ component to the potential. If one accepts the present experiments and most recent calculations in the literature, then one is forced to introduce such a potential.

The reasoning is as follows:

1) The experiment of Waffler and his collaborators [1] on the α decay of the 2^- state at 8.88 MeV in ^{16}O is in essential agreement with the calculations [2] using a "standard potential" (i.e. a ρ exchange term calculated by factorization, and π exchange calculated from current algebra. See ref. 3 for a discussion of this and other PNC potentials calculated from an assumed weak interaction Hamiltonian). This experiment is sensitive only to the $\Delta T = 0$ term in the potential, and therefore fixes it to be in the vicinity of that of the standard potential.

2) The experiment of Adelberger et al. [4] on the parity mixing of the 110 keV γ of ^{19}F is also in essential agreement, in magnitude but not sign, with the calculations [5] using a standard potential. This transition is sensitive to both the $\Delta T = 0$ and $\Delta T = 1$ parts of the PNC potential. One concludes that either the $\Delta T = 0$ potential is of the standard strength but the opposite sign as the standard potential, and that the $\Delta T = 1$ potential is not greatly enhanced, or that the $\Delta T = 0$ potential is of standard strength and sign and the $\Delta T = 1$ potential is enhanced, giving a contribution to A_γ of the opposite sign to, and dominating

---

* Service de Physique Theorique, CEN-Saclay, BPn°2 Gif-sur-Yvette, France.
+ School of Physics, University of Melbourne, Parkville, Vic., Australia.
± School of Physical Sciences, Flinders University of South Australia, Bedford Park, S.A., Australia.

the $\Delta T = 0$ contribution.

3) The experiment of Lobashov et al. [6] is in violent disagreement with the present calculations [7] with the standard potential. The observed circular polarisation of the γ from $n+p \rightarrow d+\gamma$ is $-(1.3\pm0.45)\times10^{-6}$, but the standard calculated value is $+(3\pm0.6)\times10^{-8}$! A detailed analysis of the calculation [7] shows that, while the experiment is sensitive to the $\Delta T = 0$ and $\Delta T = 2$ parts of the potential, the contribution of the $\Delta T = 0$ part is very small.

To obtain agreement with the datum by enhancement of the $\Delta T=0$ term requires an enhancement of 10^3, in clear contradiction to cases (1) and (2). However we can enhance the $\Delta T = 2$ part by a factor of about 30 and obtain agreement for $n+p \rightarrow d+\gamma$ without disturbing the agreement found in ^{16}O and ^{19}F. One can also adjust the $\Delta T=2$ potential to obtain the correct sign for the circular polarisation.

4) There is an experimental limit on the parity mixing in p-p scattering at 15 MeV [8]. This is just consistent with our enhanced $\Delta T=2$ potential, but if it were lowered significantly would exclude the potential in its present form.

5) In heavier nuclei calculations are much more difficult. However there is a persistent trend that the PNC effects are calculated to be too small in magnitude (by a factor of 4 or 5) and to have the wrong sign [9]. This can also be corrected by the enhancement of the $\Delta T=2$ potential. It could also be accounted for by an enhanced $\Delta T=1$ potential.

Thus we are forced to one of two conclusions:

Either A: Some of the existing experiments and calculations are in error e.g. all calculations except $n+p \rightarrow d+\gamma$ could be wrong, and the $\Delta T=0$ potential could really be enhanced by a factor of 1000, or the $n+p \rightarrow d+\gamma$ experiment could be wrong and the $\Delta T=0$ potential and $\Delta T=2$ potential have about the standard strength, with the $\Delta T=1$ potential enhanced.

or B : The experiments and calculations are correct and the $\Delta T=2$ potential is strongly enhanced.

Since either conclusion has very important implications for the theory of weak interactions we strongly urge the following experiments and calculations be undertaken:

i) The Lobashov experiment should be repeated

ii) The asymmetry experiment in p-p scattering should be repeated to lower the limit or find an effect.

iii) The asymmetry in n-p scattering should be measured. An asymmetry of order 10^{-6} at 15 MeV is required for consistency with the Lobashov experiment [10].

iv) The calculation of the ^{16}O α decay should be repeated.

v) The asymmetry of the photon in n-p photocapture should be measured. The T=1 enhancement explanation of the ^{19}F and heavy nuclei results predicts an asymmetry of order 10^{-7}.

REFERENCES

[1] N. Nenbeck, H. Schober and H. Waffler, Phys. Rev. C10, 320 (1974).
[2] M. Gari and H. Kummel, Phys. Rev. Lett. 23, 26 (1969) and E.M. Henley, T.E. Keliher and D.U.L. Yu, Phys. Rev. Lett. 23, 941 (1969).
[3] M.A. Box, B.H.J. McKellar, P. Pick and K.R. Lassey, J. Phys. G1, 493 (1975).
[4] E.G. Adelberger, H.E. Swanson, M.D. Cooper, J.W. Tape and T.A. Trainor, Phys. Rev. Lett. 34, 402 (1975).
[5] M.A. Box and B.H.J. McKellar, Phys. Rev. C11, 1859 (1975); M. Gari, A.H. Huffman, J.B. McGrory and R. Offerman, Phys. Rev. C11, 1485 (1975); M.A. Box, A.J. Gabric and B.H.J. McKellar, to be published.
[6] V.M. Lobashov et al., Nucl. Phys. A197, 241 (1972).
[7] K.R. Lassey and B.H.J. McKellar, Phys. Rev. C11, 349 (1975) and Saclay preprint DPh-T/75/74 (1975), to be published; M.L. Rustgi and H.J. Pirner, Phys. Rev. C10, 2099 (1974), and Nucl. Phys. A239, 427 (1975); B. Deshplanques, Nucl. Phys. A242, 423 (1975).
[8] J.M. Potter et al., Phys. Rev. lett. 33. 1307 (1974).
[9] See e.g. B. Desplanques, Thesis Orsay 1975.
[10] B.H.J. McKellar, Saclay preprint DPh-T75/7 (1975), Nucl. Phys. to be published.

APPLICATION OF THE CHANNEL COUPLING ARRAY THEORY TO SOME ATOMIC THREE-BODY SYSTEMS

F. S. Levin*
Physics Department
Brown University
Providence, Rhode Island 02912

The channel coupling array method is a procedure for deriving sets of coupled integral equations with connected kernels [1] for the T operators T_{jk} whose on-shell matrix elements give the amplitudes for transitions from an initial state of n distinguishable particles in (a 2-body) channel k to states in the various final (m-body, $m \geq 2$) channels j. The initial calculations [2,3] were performed for low energy $e^- + H$ scattering; unitary results were obtained using the integral equation

$$\mathcal{T}^{\pm} = (V_1 W_{11} \pm P_{21} V_2 W_{21})[1 + G_1 \mathcal{T}^{\pm}] \quad , \tag{1}$$

where $\mathcal{T}^{\pm}$ is a symmetrized T operator; +(-) refers to singlet (triplet) spin states; $V_i = -e^2/r_i + e^2/r_{12}$ is the interaction of electron i with the proton and electron j (i and j = 1,2); W_{11} and $W_{21} = 1 - W_{11}$ are elements of the channel coupling array W; P_{21} is a 2-particle transposition operator; and $G_1 = (E - H + V_1 + i0)^{-1}$ is the channel-1 outgoing wave Green's function, with H the $e^- + H$ Hamiltonian. The scattering lengths $a^{\pm}$ and S-wave phase shifts $\eta^{\pm}$ obtained by solving (1) in the approximation that only the 1s ground state occurs in the spectral representation of G_1 are in good agreement with those calculated by other, more complicated, methods. Eq. (1) is most accurate when the preferred [1] channel permuting array (CPA) choice of W is made, that is, $W_{11} = 0$, $W_{21} = 1$. Subsequent to these initial computations, the channel coupling array theory has been used as the basis for other atomic and molecular calculations, including variational calculations for $e^- + H$ and bound state calculations for H_2^+ and H^-; some of these are discussed in the following.

Although variational principles have been derived for the T operator integral equations [4], it was decided to use a differential equation form of (1) as a means for performing variational calculations for $e^- + H$ phase shifts $\eta^{\pm}$. This was done in order to compare the calculated values of $\eta^{\pm}$ with those of Massey and Moiseiwitsch [5] as well as with the $\eta^{\pm}$ obtained from previous work. To this end, the variational wave function ψ_{MM} employed by Massey and Moiseiwitsch was used as the trial function:

$$r_1 \psi_{MM} = \varphi_{1s}(2)\, k^{-1/2}[\sin kr_1 + (a + be^{-r_1})(1 - e^{-r_1}) \cos kr_1] \quad , \tag{2}$$

where $\varphi_{1s}(i)$ is the hydrogenic ground state for particle i, k is the wave number, and a and b are the variational parameters ($a = \tan \eta_{trial}$).

Eq. (1) may be written in differential form as

$$[E - H_1 - (V_1 W_{11} \pm P_{21} V_2 W_{21})]\, \Psi(1,2) \equiv L^{\pm}\Psi(1,2) = 0 \quad , \tag{3}$$

where $H_1 = H - V_1$ is the channel-1 Hamiltonian operator and $E \propto k^2$. A Kohn variational principle [6] is obtained from (3) by considering $I^{\pm} = (\Psi_t, L^{\pm}\Psi_t)$, where for each partial wave, the trial function Ψ_t asymptotically contains the trial K matrix element, in this case $\tan \eta_t$. Using ψ_{MM} for Ψ_t and solving the coupled equations for a and b obtained from $\partial I/\partial a = -1$ and $\partial I/\partial b = 0$, $\tan \eta^{\pm}$ was determined from $\tan \eta^{\pm} = a + I^{\pm}$. Results for the choice $W_{11} = 0$ are compared in Table I with those obtained (a) variationally by Massey and Moiseiwitsch using the symmetrized state $\Psi^{\pm}_{MM} = (1 \pm P_{12})\ \psi_{MM}/\sqrt{2}$ and the operator L given by L = E - H; (b) by solving the integral equation (1) using the 1s approximation as noted above [3]; (c) by Burke and collaborators [7] from an approximate solution of the Schrödinger equation $(E - H)\Psi = 0$ in which Ψ is approximated by $\Psi \simeq (1 \pm P_{12}) \sum_n \varphi_n(1)F_n(2)$, where φ_n is a hydrogenic state ($n\ell$ = 1s, 2s, 2p) and F_n is a scattering coefficient which yields, asymptotically, the relevant matrix elements involving $\eta^{\pm}$ (i.e., $\exp(i\eta^{\pm}) \sin \eta^{\pm}$ or $\tan \eta^{\pm}$); and finally (d) by Schwartz [8] using the multi-parameter wave function whose results are generally accepted as the most accurate available.

The object of the variational calculation was three-fold: first to set up a Kohn-type variational principle for multi-channel scattering using the W-array formalism; second to test the accuracy of a simple wave function such as that of Eq. (2); and third, to determine if Lippmann's relation [9], which plays a vital role in both the derivation and the solution of the integral equations of the channel coupling array theory, also is maintained in the variational procedure.

This latter point is an important one, and can be easily understood, within the context of Eq. (1). It was noted above that the CPA choice $W_{11} = 0$ is preferred; the reason is that Eq. (1) then has a connected kernel. But when $W_{11} = 0$, it appears at first glance that the solution to (1) has no direct Born term, i.e., no matrix element of V_1. Using Lippmann's relation [9], viz., $G_j V_j = G_j V_k + 1$, $j \neq k$ = the 2-body initial channel, valid when acting on states of energy E in channel k, it can be shown that even with $W_{11} = 0$, there is always a full V_1 occurring in the solution. Consider the two lowest order terms in the iterated solution to (1) ($W_{11} = 0$): $\pm P_{21}V_2 + P_{21}V_2G_1P_{21}V_2 = \pm P_{21}V_2 + P_{21}V_2P_{21}G_2V_2 = \pm P_{21}V_2 + V_1G_2V_2 = \pm P_{21}V_2 + [V_1 + V_1G_2V_1] = [V_1 \pm P_{21}V_2] + V_1G_2V_1$, the first bracketed term being the result of Lippmann's identity for the present case, the second showing that precisely the direct and exchange Born terms occur with the correct phase relation. Similarly, the higher order terms also occur correctly. Since the solution to (1), whether approximate or numerical, will always involve G_1, it is clear that the effects of V_1 will be present. This is the key to understanding the accuracy of the results of the approximate solution to (1) with $W_{11} = 0$. On the other hand, when $W_{11} = 0$ is used in a variational calculation based on (3), G_1 does not manifestly enter the solution, and hence it is of interest to determine how accurately a simple trial function used in conjunction with $L^{\pm}$ of (3) can reproduce more exact results.

We see from Table I that the variational result based on Eq. (3) does reasonably well for the singlet case and $k \leq .6a_0^{-1}$ as compared with both the Massey-Moiseiwitsch and Schwartz calculations, although it remains consistently larger in magnitude than any of the other sets of results. The triplet results are not as good as the singlet. Neither the variational nor the exact $W_{11} = 0$ results have been shown to be bounds, however, so the fact that they can be larger than Schwartz's values is not necessarily meaningful. The fact that the variational results are also larger than the exact ones would seem to be an artifact of the method, as we also note below in the discussion of the bound state calculations. We conjecture that approximate solutions, such as those obtained variationally, will always be greater in magnitude than those obtained from solving the W-equations exactly. At present, this is a conjecture based on numerical calculations. An analytic proof would be a useful and interesting result.

As discussed elsewhere [10], the differential form of the channel coupling array equations used so far corresponds to expanding the Schrödinger state Ψ via

$$\Psi = \sum_n W_{\ell n}\Phi_n \quad , \tag{4}$$

and then deducing the following equations for the channel scattering states Φ_n (here the particles are assumed distinguishable):

$$(E - H_n)\Phi_n = V_n \sum_m W_{\ell m} \Phi_m \quad . \tag{5}$$

In this equation the partionings $H = H_j + V_j$ are used, where V_j is the set of inter-fragment interactions for the (bound) fragments or clusters forming channel j and $H_j = H - V_j$, while the sums on n and m in (4) and (5) run over channel labels up to N, and ℓ is a free index to be chosen as is convenient within the requirement of yielding a W which leads to a connected kernel integral equation for the transition operators T_{jk}. One example is $\ell = n$, W_{nm} then being given by a CPA. In this case it has been shown [11] that $\Phi_m = \Psi$ for all m, i.e., that the label is simply a means for selecting a complete set in which to expand Ψ that is appropriate to the channel Hamiltonian H_m.

An alternate procedure has also been given [12]. Here Ψ is expanded first via

$$\Psi = \sum_n \psi_n \quad , \tag{6}$$

and then a new set of equations for the channel components ψ_n is derived:

$$(E - H_n)\psi_n = \sum_m W_{\ell m} V_m \psi_m \quad . \tag{7}$$

Notice that the structures of (5) and of (7) are quite different, even though both contain a sum on channel labels and the factor $W_{\ell m}$. Solution of (7) leads to transition operators $\tilde{T}_{jk}$ which are the transpose of the T_{jk} discussed above. While exact solutions of the equations for the T_{jk} and the $\tilde{T}_{jk}$ would yield identical transition amplitudes (assuming the phases of time-reversal invariance), approximate solutions are not expected to be the same.

The initial application to bound state problems of the channel coupling array theory was made by Krüger [13] to the H_2^+ system using Eq. (7). We first indicate one form of the general CPA case and then specialize to this system. Choosing $\ell = n$ in (7) and then using $W_{nm} = \delta_{m,n+1}$[mod N], we have a typical set of equations for the ψ_n:

$$(E - H)\psi_n = V_{n+1} \psi_{n+1} \quad , \; 1 \le n \le N \text{ [mod N]} \quad . \tag{8}$$

When n = 2, (8) reduces to

$$(E - H_1)\psi_1 = V_2\psi_2 \tag{9a}$$

and

$$(E - H_2)\psi_2 = V_1\psi_1 \quad , \tag{9b}$$

and (6) becomes

$$\Psi = \psi_1 + \psi_2 \quad . \tag{10}$$

Let us now consider the H_2^+ system, consisting of protons p_1 and p_2 and an electron. This is a two-center problem, with $\vec{r}_i$ labelling the position of the electron w.r.t. proton i, and $\vec{R}$ being the coordinate of proton 2 relative to proton 1. This system has two 2-body channels in a scattering theory sense: $p_i + e^-$ in a bound state and p_j free, $i \neq j$, $i = 1,2$. Correspondingly, the Hamiltonian

$H = K - e^2/r_1 - e^2/r_2 + e^2/R$ (in the Born-Oppenheimer approximation) has 2 partitions of the form $H = H_j + V_j$, where

$$H_j = K - e^2/r_j \quad , \quad j = 1,2 \quad , \tag{11a}$$

and

$$V_j = \frac{e^2}{R} - \frac{e^2}{r_i} \quad , \quad i \neq j \quad , \tag{11b}$$

with

$$K = \frac{\hbar^2}{2m} \nabla^2 \quad . \tag{11c}$$

To determine the ground state energy, each ψ_i in (10) was approximated by the hydrogenic ground state for an H-atom consisting of $p_i + e^-$: $\psi_i = a_i \varphi_{1s}(\vec{r}_i)$. Hence, $\Psi = a_1\varphi_{1s}(\vec{r}_1) + a_2\varphi_1(\vec{r}_2)$. Substitution into Eqs.(9) gives a secular equation for the energy $E = E(R)$. Minimizing w.r.t. to R gives the equilibrium separation R_e and the ground state energy $E_b = E(R_e)$. The results of the calculation are given in Table II, where they are compared with the results of a standard Rayleigh-Ritz (RR) calculation using a trial function identical to the Ψ of the channel coupling array calculation [Eq. (10)] and (b) the exact numerical calculations of Wind based on a separation of the Schrödinger equation in confocal elliptic coordinates [14]. While the R-R calculation yields E_b accurate to about 6% (all comparisons are with Wind's numerical results), R_e is determined to only 25%. In contrast to this, the channel coupling calculation gives an accuracy of about 2% for both E_b and R_e! The method thus proves to be extremely good, reminiscent of the similar results obtained for $e^- + H$ scattering in the 1s approximation [3]. Not only are the values of E_b and R_e close to those found from a much more sophisticated calculation, but the exact and approximate $E = E(R)$ curves are found to be in close agreement for all R, as indicated in Figure I (courtesy of H. Kruger [13]). Also, as noted above for

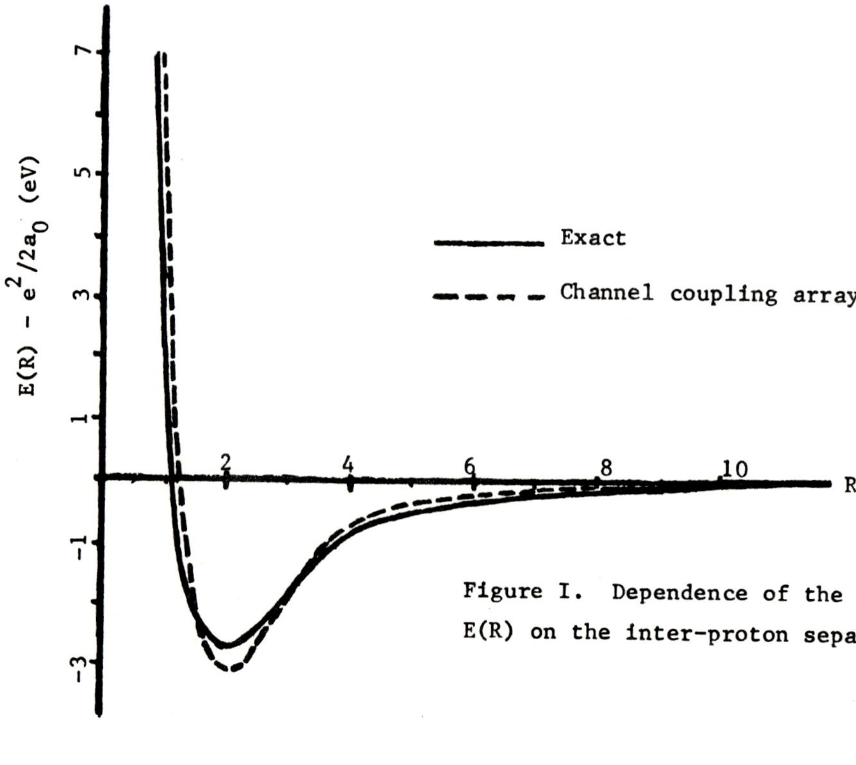

Figure I. Dependence of the H_2^+ energy E(R) on the inter-proton separation R.

the e^- + H variational calculations, E_b as computed using Eq. (9) is larger in magnitude than in the calculation of Wind. It is an interesting aspect of the present procedure that, while the probability density is the same as obtained from the R-R calculation, the equilibrium separation is much closer to the exact numerical value than in the R-R method.

All the results obtained so far have been based on what appears to be a very crude approximation, viz., retaining only the hydrogenic ground state as an expansion basis. The obvious implication is that the channel coupling array method allows the dynamics of the problem to be taken account of in more detail than does the straightforward solution of the Schrödinger equation, at least when the same crude approximations are used. As a further investigation of this implication, the method has been applied to calculating the affinity of the H^- atom. The value of 0.7 eV for the affinity [4] is a clear indication that the proton is very well shielded by one of the electrons, leaving relatively little coulomb attraction to bind the second. Hence, this case should be a sensitive test of the channel coupling array method. This system has been well studied using the Rayleigh-Ritz method, as noted in Ref. [6], and we present here only the results of calculations based on the channel coupling array theory.

The initial calculation was performed in analogy to that for H_2^+, i.e., using Eqs. (9) with $H_i = K_1 + K_2 - e^2/r_j$, $V_i = -e^2/r_i + e^2/r_{12}$, $K_i = -\hbar^2\nabla_i^2/2m$, and $\psi_i = a_i\varphi_{1s}(\vec{r}_1)\varphi_{1s}(\vec{r}_2)$. The result was a failure to bind by 3.4 eV. In view of the preceding comments on screening, this is not a surprising result. Although it is possible to set up more elaborate approximations using Eqs. (9), it was decided at this point to explore an alternate approach. This is to solve Eq. (3) extended to negative energies. Since for $E > 0$ solution of (3) describes e^- + H scattering, then for $E < 0$ it must describe H^- bound states. Indeed, since (3) gives the matrix elements of the symmetrized operator $\mathcal{T}^{\pm}$ for $E > 0$, it is clear that (3) will give the H^- bound states because these are just the poles of $\mathcal{T}^{\pm}$.

From the e^- + H scattering length calculations [1,3], it is known that the preferred CPA choice $W_{11} = 0$ gives the best results. Hence, we seek to solve Eq.(3) for this choice of W_{11}. With or without this choice, (3) is an integro-differential equation. In accord with the positive energy case [1,3] and the above calculations, we look for bound state solutions of the form $\Psi = \varphi_{1s}(2)\psi_b(1)$, where $\psi_b(r) \to 0$, $r \to \infty$. Since the equation for ψ_b is integro-differential, it is useful to estimate the binding energy of 1 that would be obtained by solving for ψ_b. This is done very easily using the scattering length approximation [15], which should be valid because the binding (0.7 eV) is so small [the analogy with n + p triplet scattering [15] is obvious]. The binding energy (the affinity) may be expressed as $E_b = \hbar^2k_b^2/2m$. We expect only the space symmetric (spin singlet) state to be bound. Previous calculations yielded a value of $a^+ = 6.320$ (in a.u.) for the singlet scattering length. Using the approximation $k_B \simeq 1/a^+$, we find for the affinity $E_b \simeq 0.34$ eV, in reasonable agreement with the value of 0.7 eV. Exact solution of the relevant equation for ψ_b may be expected to give an even better result for E_b. Since a direct calculation of E_b in this manner is not yet available, it is of interest to consider an approximation to E_b obtained by using a Rayleigh-Ritz variational procedure applied to Eq. (3). A trial function of the simple form $\Psi_t = \varphi_{1s}(2)\, e^{-\alpha r_1}$ was used, and both signs (±) occurring in (3) were retained. Only the positive sign (singlet) gave binding for $\alpha \geq 0$. The value determined by minimizing was $E_b \simeq 2.04$ eV, obtained for the value $\alpha \simeq 0.5$ (a.u.). The results of all these calculations of E_b are summarized in Table III.

There are two interesting points to this third calculation of E_b. First, we obtain a bound H^- system, in contrast to the calculation based on Eq. (3). Considering the extremely simple form of the trial wave function and the very low actual binding energy, the result of 2.04 eV from such a crude approximation may be considered to be less objectionable than implied by its being a factor of 3 too large. Indeed, such a result, taken in conjunction with the scattering length calculation of E_b, is encouraging, and suggests that once again the 1s approximation is rather accurate. The second point is that the approximate value of 2.04 eV seems to be larger than the actual E_b that will result from an exact solution of (3) when $\psi = \varphi_{1s}(2)\psi_b(1)$ is used, assuming (as we do) that the scattering length approximation is a reasonably accurate estimate of E_b. This again is suggestive of the idea that with the channel coupling array method, approximate results may in general be larger in magnitude than exact results.

We may summarize the results of these sets of calculations as follows. The channel coupling array method, at least in its application to simple three-body atomic and molecular systems, yields reasonably accurate output for rather crude initial input. The results so far obtained are generally more accurate than would be obtained by solving the Schrödinger equation using such crude wave functions. Although calculations using the method with more sophisticated wave functions are required, the method appears to be a very promising one. More sophisticated calculations are in progress; it is hoped that they will fulfill the initial promise of the method.

References

[1] D. J. Kouri and F. S. Levin, Bull. Am. Phys. Soc. 19 (1974) 489; Physics Letters 50B (1974) 421; Phys. Rev. A10 (1974) 1616; Nuclear Phys. A250 (1975) 127 and in press; Proceedings of the VIth International Conference on Few-Body Problems (Laval University Press, Quebec, 1975) p. 47. W. Tobocman, Phys. Rev. C9 (1974) 2466.

[2] M. Baer and D. J. Kouri, J. Math. Phys. 14 (1973) 1637. This was the first application of equations based on a channel coupling array procedure. However the equations used led to non-unitary amplitudes and have not been considered further. An analysis of this non-unitary result is given in the second of references [3] below.

[3] D. J. Kouri, M. Craigie and D. Secrest, J. Chem. Phys. 60 (1974) 1851; D. J. Kouri, F. S. Levin, M. Craigie and D. Secrest, ibid 61 (1974) 17, and in press.

[4] W. Tobocman, Phys. Rev. C 10 (1974) 60; D. J. Kouri and F. S. Levin, Phys. Rev. C 11 (1975) 352.

[5] H. S. W. Massey and B. L. Moiseiwitsch, Proc. Roy. Soc. A205 (1951) 483.

[6] A comprehensive discussion is given by B. L. Moiseiwitsch, Variational Principles (Interscience, John Wiley and Sons, New York, 1966).

[7] P. G. Burke and K. Smith, Rev. Mod. Phys. 34 (1962) 458.

[8] C. Schwartz, Phys. Rev. 124 (1961) 1468.

[9] B. A. Lippmann, Phys. Rev. 102 (1956) 256.

[10] D. J. Kouri and F. S. Levin, Nuclear Phys. (in press); Y. Hahn, D. J. Kouri and F. S. Levin, Phys. Rev. C10 (1974) 1615.

[11] See the first reference cited in [10], and also D. J. Kouri and F. S. Levin, to be submitted for publication.

[12] Y. Hahn, D. J. Kouri, and F. S. Levin, Phys. Rev. C10 (1974) 1620; D. J. Kouri, H. Krüger and F. S. Levin, submitted for publication.

[13] H. Krüger, private communication.

[14] The results for the Rayleigh-Ritz and Wind calculations are taken from F. L. Pilar, Elementary Quantum Chemistry (McGraw-Hill, New York, 1968).

[15] J. M. Blatt and V. F. Weisskopf, Theoretical Nuclear Physics (John Wiley and Sons, New York, 1952); Y. Ohmura and H. Ohmura, Phys. Rev. 118 (1960) 154.

* Work supported in part by the U. S. E.R.D.A.

Table I. Comparison of e^- + H Phase Shifts
(upper portion - η^+; lower portion - η^-)

k	$W_{11} = 0$ Kohn Variat.	Eq. (1), 1s approx.	Massey-Moiseiwitsch Variat.	Burke and Smith 1s,2s,2p	Schwartz Variat.
0.1	2.623	2.541	2.299	2.491	2.553
0.2	2.156	2.039	1.792	1.974	2.067
0.3	1.772	1.646	1.498		1.696
0.4	1.476		1.315		
0.5	1.263	1.080	1.218	1.082	1.202
0.6	1.120	0.871	1.199		1.041
0.1	2.995	2.935	2.908	2.936	2.939
0.2	2.892	2.737	2.679	2.715	2.717
0.3	2.773	2.554	2.463		2.500
0.4	2.664		2.257		
0.5	2.573	2.264	2.070	2.096	2.105
0.6	2.517	2.170	1.901		1.933

Table II. Comparison of H_2^+ Ground State Energies and Equilibrium Separations.

	Exact Calculation	Rayleigh-Ritz	Channel Coupling Array Theory
R_e (a.u.)	2.02	2.5	2.07
$E_b = E(R_e)$ (eV)	-16.398	-15.39	-16.69

Table III. Comparison of H^- Affinities from Different Channel Coupling Array Calculations.

Eqs. (9)	Scattering Length Approximation to Eq. (3)	Variational Estimate for Eq. (3)
no binding	0.34 eV	2.07 eV

THE OPTICAL POTENTIAL IN THE CHANNEL COUPLING ARRAY METHOD: A MODEL STUDY

by

Donald J. Kouri*
Departments of Chemistry and Physics
University of Houston
Houston, Texas 77004

and

Rudolf Goldflam†
Department of Physics
University of Houston
Houston, Texas 77004

---

*Supported in part by R.A. Welch Foundation Grant E-608 and by a Limited Grant in Aid from the University of Houston

†Supported in part by NSF Grant MPS 74-13719.

The problem of rigorously treating collisions of three(or more) particle systems was pioneered some years ago by Faddeev.[1] Since his groundbreaking work on three body scattering, there has been considerable effort expended in trying to develop other alternative formulations of few body collisions which might be somewhat easier to apply and generalize to more complex situations.[2-13] Recently, we have reported a new approach to such collision problems which leads to new sets of coupled equations for the transition operators associated with a three (or more) body system.[14] A key feature of the method is the introduction of an array whose purpose is to permit one to couple different arrangements in a general fashion.[14-17] Once the general coupled equations for the $T_{\beta\alpha}$ are obtained(where $T_{\beta\alpha}$ is the transition operator whose matrix elements give the scattering amplitude for going from arrangements α and β), the channel coupling array may then be used to ensure that the kernel of the equations, when iterated a sufficient finite number of times, contains no disconnected diagrams.[14,16-17] For sufficiently well behaved potentials, this leads to a continuous kernel. This then ensures that finite rank approximations to the kernel are elements in a convergent sequence. In the present discussion, we wish to focus on a specific choice of the channel coupling array, called the channel permutting array, which leads to equations whose iterated kernal is connected. For purposes of simplicity, we shall consider a model situation in which only two arrangements are present. Then the general coupled equations may be written as[14,16-17]

$$T_{\beta\alpha} = V_\beta W_{\beta\alpha} + V_\beta \sum_\lambda W_{\delta\lambda} G_\lambda^+ T_{\lambda\alpha} \tag{1}$$

where V_β is the full perturbation in arrangement β, defined in terms of the full Hamiltonian H and unperturbed β arrangement Hamiltonian H_β by

$$V_\beta = H - H_\beta, \tag{2}$$

G_λ^+ is the unperturbed Green's operator defined by

$$G_\beta^+ = 1/(E - H_\beta + i\varepsilon), \tag{3}$$

and $W_{\delta\lambda}$ is the δ,λ element of the channel coupling array. The δ index is free and may be chosen in several ways.[16] For our purpose, we shall set δ equal to β, and then Eq.(1) reads

$$T_{\beta\alpha} = V_\beta W_{\beta\alpha} + V_\beta \sum_\lambda W_{\beta\lambda} G_\lambda^+ T_{\lambda\alpha}. \tag{4}$$

This corresponds to a specific off-shell extension of the transition operator and holds only for two body initial states. As has been shown before, the channel permuting array choice of $W_{\beta\lambda}$ leads to a kernel which, upon sufficient iteration, becomes connected. For the specific two arrangement model, this choice of the W array is taken as:

$$W = \begin{pmatrix} 0 & 1 \\ 1 & 0 \end{pmatrix}, \tag{5}$$

where upon Eq.(4) is explicitly

$$T_{\alpha\alpha} = V_\alpha G_\beta^+ T_{\beta\alpha}, \tag{6}$$

$$T_{\beta\alpha} = V_\beta + V_\beta G_\alpha^+ T_{\alpha\alpha}. \tag{7}$$

These equations are also equivalent to simultaneous equations for the full scattering wavefunction, since the transition operators satisfy the relation

$$T_{\lambda\alpha}|\phi_\alpha(i)\rangle = V_\lambda|\psi_\alpha^+(i)\rangle. \tag{8}$$

Combining Eqs.(6)-(8) leads to[16-18]

$$|\psi_\alpha^+(i)\rangle = |\phi_\alpha(i)\rangle + G_\alpha^+ V_\alpha \psi_\alpha^+(i)\rangle, \tag{9}$$

$$|\psi_\alpha(i)\rangle = G_\beta^+ V_\beta |\psi_\alpha^+(i)\rangle. \tag{10}$$

Now it has been noted that substitution of Eq.(10) into Eq.(9) leads to[16-19]

$$|\psi_\alpha^+(i)\rangle = |\phi_\alpha(i)\rangle + G_\alpha^+ V_\alpha G_\beta^+ V_\beta |\psi_\alpha^+(i)\rangle, \tag{11}$$

which is recognized as a type of Lippmann-Schwinger equation where a new complex potential U_α, given by[21]

$$U_\alpha = V_\alpha G_\beta^+ V_\beta, \tag{12}$$

has been introduced. It is interesting to note that the three arrangement analogue of Eq.(11) for three particles involves a kernel which is completely connected.[14-19] It is also clear that Eq.(11) would result if first we had substituted Eq.(7) into Eq.(6) and then used Eq.(8) specialized to λ equal to α. Thus, Eq.(11) is a direct consequence of having iterated the coupled channel T operator equations for a specific choice of the channel coupling array. In this work, we wish to examine the consequences of using Eq.(11) (or equivalently Eqs.(6) and (7)) in formulating the optical potential. For notational simplification, we rewrite Eq.(11) (using Eq.(12)) as

$$|\psi_\alpha^+(i)\rangle = |\phi_\alpha(i)\rangle + G_\alpha^+ U_\alpha |\psi_\alpha^+(i)\rangle. \tag{13}$$

We then introduce a projector P onto the elastic scattering state, and its complement Q, so that[20]

$$P + Q = 1, \tag{14}$$

$$PQ = QP = 0. \tag{15}$$

Then the components of $|\psi_\alpha^+(i)\rangle$ in P space and Q space satisfy[20]

$$|\psi_{\alpha P}^+(i)\rangle = |\phi_\alpha(i)\rangle + PG_\alpha^+ U_\alpha |\psi_\alpha^+(i)\rangle, \tag{16}$$

and

$$|\psi_{\alpha Q}^+(i)\rangle = QG_\alpha^+ U_\alpha |\psi_\alpha^+(i)\rangle, \tag{17}$$

and we note that

$$|\psi^{+}_{\alpha}(i)\rangle = |\psi^{+}_{\alpha P}(i)\rangle + |\psi^{+}_{\alpha Q}(i)\rangle \tag{18}$$

If we solve Eq.(17) for $|\psi^{+}_{\alpha Q}(i)\rangle$ and substitute the result into Eq.(16), we obtain (upon simplification)

$$\tilde{U}_{\alpha} = PU_{\alpha}P + PU_{\alpha}Q(E-H_{\alpha} - U_{\alpha} + i\varepsilon)^{-1}QU_{\alpha}P, \tag{19}$$

as a resulting expression for the optical potential $\tilde{U}_{\alpha}$. We can contrast this result with the more standard one given by[20]

$$\tilde{U}'_{\alpha} = PV_{\alpha}P + PV_{\alpha}Q(E-H_{\alpha} - V_{\alpha} + i\varepsilon)^{-1}QV_{\alpha}P. \tag{20}$$

We compare the first terms in each expression by noting that in $PU_{\alpha}P$, for an energy E above the threshold for reaction, $\mathrm{Im}PU_{\alpha}P$ is nonzero. This implies the opening of channels besides the elastic one. By contrast, $PV_{\alpha}P$ has no energy dependence and one must depend on the structure of the $Q(E-H_{\alpha}-V_{\alpha}+i\varepsilon)^{-1}Q$ factor in the second term of Eq.(20) to provide information that the rearrangement channel is open. Furthermore, this information is not present in any readily transparent form since it involves the full Hamiltonian rather than the rearrangement unperturbed Hamiltonian. In $PU_{\alpha}P$, we note that it is $(E-H_{\beta} + i\varepsilon)^{-1}$ that enters so that the appearance of a complex part is directly indicated whenever E is greater than that necessary to produce a β arrangement state.

It is possible to use the first term approximation to $\tilde{U}_{\alpha}$ to gain some direct insight into factors governing precisely where rearrangement occurs. Because of the fact that $PU_{\alpha}P$ is a nonlocal interaction (due to the presence of G^{+}_{β} in U_{α}), we shall consider the effect of $PU_{\alpha}P$ on the initial state for the following model. (Note, this approximation is similar to a Born type approximation but can also be developed in terms of the distorted wave approximation. To do the latter, one should employ a distorted wave Green's operator G^{+}_{β}, the V_{γ} are replaced by the difference of the channel perturbation V_{γ} and the distortion potential v_{γ}, and the initial state is taken to be a distorted wave state for the initial arrangement). We take our system to be

$$A + BC \rightarrow \begin{cases} A + BC\ (\alpha) \\ AB + C\ (\beta), \end{cases} \tag{21}$$

where we constrain all three particles to a line, and we assume particle B to be infinitely massive. Thus, we have the nonreactive channel α and the reactive one β. Then we examine

$$\langle\chi_{\alpha}(n_o)|V_{\alpha}G^{+}_{\beta}V_{\beta}|\phi_{\alpha}(i)\rangle = \langle\chi_{\alpha}(n)|V_{\alpha}G^{+}_{\beta}V_{\beta}|\chi_{\alpha}(n_o)\xi_{k_{no}}\rangle, \tag{22}$$

where $|\chi_{\alpha}(n_o)\rangle$ is the initial internal state of BC and $|\xi_{k_{no}}\rangle$ describes the motion of A relative to B. Here, we have used the fact that

$$P|\chi_{\alpha}(n_o)\rangle = |\chi_{\alpha}(n_o)\rangle. \tag{23}$$

Finally, we take the potentials V_{γ} to be of a separable form in configuration space, so that

$$V_{\gamma}(x,y) = \sum_{\ell} V^{(1)}_{\gamma\ell}(x)V^{(2)}_{\gamma\ell}(y), \tag{24}$$

with x being the A-B distance and y being the BC distance.

Then it is easy to verify that

$$\begin{aligned}\mathrm{Im}\langle\chi_{\alpha}(n_o)|V_{\alpha}G^{+}_{\beta}V_{\beta}|\chi_{\alpha}(n_o)\xi_{k_{no}}\rangle \\ = -\sum_{n}\frac{1}{k_n}\chi_{\beta}(n|x)\sum_{\ell}V^{(1)}_{\alpha\ell}(x)\Big[\int_0^{\infty}dy\,V^{(2)}_{\alpha\ell}(y)\chi_{\alpha}(n_o|y)\sin k_n y\Big] \\ \sum_{\ell'}\Big[\int_0^{\infty}dy'\sin k_n y'\,V^{(2)}_{\beta\ell'}(y')\chi_{\alpha}(n_o|y')\Big]\Big[\int_0^{\infty}dx'\chi_{\beta}(n|x')V^{(1)}_{\beta\ell'}(x')\sin k_{n_o}x'\Big],\end{aligned} \tag{25}$$

which is obviously of the form

$$\mathrm{Im}\langle\chi_\alpha(n_o)|V_\alpha G_\beta^+ V_\beta|\chi_\alpha(n_o)\xi_{k_{no}}\rangle$$

$$= -\sum_{\substack{\text{open } n,\\ \ell}} \frac{1}{k_n}\chi_\beta(n|x)V^{(1)}_{\alpha\ell}(x)B_{\ell n}. \tag{26}$$

The $B_{\ell n}$ are constants which depend on the energy under consideration. We can make some general remarks without specifying the model any further. For simplicity, however, let us consider a collision system A + BC where the energy is such that only the ground internal states of the AB and BC clusters are accessible. Then Eq.(26) involves only a single $\chi_\beta(n|x)$. We recall that it is simply the bound internal state of the final cluster. Thus, we expect the right hand side to be larger as $\chi_\beta(n|x)$ is larger. Since the imaginary part of the optical potential tells where one has sinks (or sources), it follows that we expect reaction to be enhanced where $\chi_\beta(n|x)$ is larger. This has a simple physical interpretation. As particle A comes toward particle B in the BC cluster, the reaction is favored when A is roughly in the region where the AB cluster wavefunction is large. Of course, this view has to be qualified because one also has the x-dependent factor $\Sigma V^{(1)}(x)B_{\ell n}$ to consider. The effect of this is to simply say that one requires forces to be present to cause the rupture of the BC bond and to remove the energy from the A-B relative motion. This effect is present through the $\Sigma V^{(1)}_{\alpha\ell}(x)B_{\ell n}$ factor and it can cause a shift of where the AB cluster wavefunction is largest. Finally, the factors $1/k_n B_{\ell n}$ contain an energy dependence so that the imaginary part of $PU_\alpha P$ can increase as the energy changes. This can result both in scale changes and in shifts of the region of x where reaction occurs. Thus, the resulting picture of the reaction is one where the formation of AB and destruction of BC is governed by the AB cluster wavefunction, the region where the interaction is strong, and a complicated energy dependence which can result in drastic changes in how much reaction occurs and where it occurs.

We remark that if one is at an energy where more than one state of the AB cluster is open, then one will have several regions where reaction can occur. It is clear that these regions will be associated with the different internal state AB cluster wavefunctions becoming large. In this context, it is of interest to note that it may be for some potentials that a particular excited AB internal state may precisely match the region where the $\Sigma V_{\alpha\ell}(x)B_{\ell n}$ is also large. In that case, we will expect reaction to go preferably into that internal state of AB. This can produce population inversion and thus may possibly be a source of lasing action. The present method would provide a relatively inexpensive way to determine whether this is indeed the case.

Our final comment regards the case where more than one internal state of the BC molecule is energetically accessible. In that case, there will be a contribution to Im $\tilde{U}_\alpha$ which comes from the loss of flux into the α-arrangement excited states. We expect these will not be well represented by the first term approximation $PU_\alpha P$. This is because these excited α arrangement states enter U_α only through the dissociative continuum. From a computational standpoint, such states would be neglected. Thus, we expect the approximation of $\tilde{U}_\alpha$ by $PU_\alpha P$ to be useful in describing the loss of flux to reactive channels only. For this reason, it has been called the arrangement channel optical potential.[21] Computational studies of this formalism for some simple models are currently under study.

References

[1] L.D. Faddeev, JETP (Sov. Phys.) 12, 1014 (1961); Sov. Phys. Dokl. 6, 384 (1961) and 7, 600 (1963); Mathematical Aspects of the Three Body Problem (Davey, N.Y., 1965). See also the earlier work by K.M. Watson, Phys. Rev. 89, 575 (1953) and 103, 489 (1956).

[2] C. Lovelace, Phys. Rev. 135, B1225 (1964); and in Strong Interactions and High Energy Physics, ed. R.G. Moorhouse (Oliver and Boyd, London, 1964) p.437.

[3] R.L. Omnes, Phys. Rev. 134, B1358 (1964).

[4] S. Weinberg, Phys. Rev. 133, B232 (1964).

[5] P. Federbush, Phys. Rev. 148, 1551 (1966).

[6] R.G. Newton, Phys. Rev. 153, 1502 (1967); J. Math. Phys. 8, 851 (1967).

[7] Y. Hahn, Phys. Rev. 169, 794 (1968); Y. Hahn and K.M. Watson, Phys. Rev. C5, 1718 (1972).

[8] E.O. Alt, P. Grassberger and W. Sandhas, Nucl. Phys. B2, 167 (1967).

[9] O.A. Yakubovskii, Sov. J. Nucl. Phys. 5, 937 (1967).

[10] I.H. Sloan, Phys. Rev. C6, 1945 (1972).

[11] B.R. Karlson and E.M. Zeiger, Phys. Rev. D9, 1761 (1974).

[12] Gy. Bencze, Nucl. Phys. A210, 568 (1973).

[13] E.F. Redish, Nucl. Phys. A225, 16 (1974); Phys. Rev. C10, 67 (1974); Nucl. Phys. A235, 82 (1974).

[14] D.J. Kouri and F.S. Levin, Bull. Am. Phys. Soc. 19, 489 (1974); Phys. Rev. A10, 1616 (1974); Phys. Letts. 50B, 421 (1974).

[15] M. Baer and D.J. Kouri, J. Math. Phys. 14, 1637 (1973). See also M. Baer and D.J. Kouri, Phys. Rev. A4, 1924 (1971).

[16] D.J. Kouri and F. S. Levin, Nucl. Phys. A250, 127 (1975) and A253, 395 (1975).

[17] W. Tobocman, Phys. Rev. C9, 2466 (1974).

[18] Y. Hahn, D.J. Kouri and F.S. Levin, Phys. Rev. C10, 1615 and 1620 (1974).

[19] For the case of 3 particles, this equation was independently obtained by several other investigators. See, e.g., W. Glöckle, Nucl. Phys. A141, 620 (1970); V. Vanzani, Letts. Nuovo Cimento 10, 610 (1974).

[20] H. Feshbach, Ann. Phys. 5, 357 (1958); Ann. Rev. Nucl. Sci. 8, 49 (1958).

[21] So far as we are aware, this was first derived in J.-L Lin and D.J. Kouri, J. Chem. Phys. 60, 303 (1974).

HIGH ENERGY ELECTRON SCATTERING FROM A TWO CENTER POTENTIAL

S. S. Chandel and S. Mukherjee
Department of Physics
Himachal Pradesh University
Simla, India

I. INTRODUCTION

The scattering of high energy electrons from a two center potential is of considerable interest in nuclear and molecular physics. The Schroedinger equation for the problem is separable in spheroidal coordinates and the scattering cross-section can be given in terms of spheroidal phase shifts. The scattering problem with a relativistic electron has not yet been studied. The relevant Dirac equation is not separable and an approximation seems to be the only way out. The purpose of this paper is to present an approximate solution of the problem. Following Mukherjee (1), we shall apply a generalized form of Sommerfeld - Maue(2) approximation so that the solution of the Dirac equation can be constructed from that of a Schroedinger-like equation. The latter permits a phase shift analysis and the relativistic scattering cross-section is then obtained in terms of these spheroidal phase shifts. In application to physical problems, one may have to consider a large number of randomly oriented two center scatterers (nuclei) . The cross-section, therefore, has to be averaged over all nuclear orientations.

II. SOMMERFELD-MAUE APPROXIMATION

The solution of the Dirac equation with a screened two center potential V can be obtained, in the Sommerfeld-Maue approximation, from the solution ψ_0 of the equation

$$(\nabla^2 + p^2 + 2\epsilon V) \psi_0 = 0 \quad . \tag{1}$$

Assuming the form

$$\psi_0 = U(\vec{p})\, e^{i \vec{p}.\vec{r}} f(\vec{r}, \vec{p}) \, , \tag{2}$$

we get for the solution of the Dirac equation

$$\psi(\vec{r}) = e^{i\vec{p}.\vec{r}} \left[f(\vec{r}, \vec{p}) - \frac{i}{2\epsilon} \vec{\alpha}.\nabla f(\vec{r}, \vec{p}) \right] U(\vec{p}) . \tag{3}$$

We now introduce a prolate spheroidal coordinate system (ξ , η , φ) fixed with the nucleus. In the asymptotic region,

$$(\psi_0)_{scatt} \xrightarrow[\xi \to \infty]{} U(\vec{p}) \frac{p}{c\xi} e^{ic\xi} F(\eta, \varphi ; \vec{p}) , \tag{4}$$

where $C = \frac{1}{2} PR$, R being the separation between the charge centers. It then follows from (2) and (4) that

$$f(\vec{r},\vec{P}) \xrightarrow[\xi\to\infty]{} \frac{P}{c\xi} F(\eta,\varphi;\vec{P}) \exp\left[\frac{iR\xi}{2}\left\{P - \sqrt{1-\eta^2}\,(P_1\cos\varphi + P_2\sin\varphi) - P_3\eta\right\}\right], \quad (5)$$

where (P_1, P_2, P_3) are the cartesian components of $\vec{P}$. The amplitude of the scattered current density is now obtained from (3) and (5) :

$$|J_{scatt}| \xrightarrow[\xi\to\infty]{} \frac{P^2}{2c^2\xi^2}\left|F(\eta,\varphi;|P|,\omega,\alpha)\right|^2\left(1 + \eta\cos\omega + \sqrt{1-\eta^2}\sin\omega\cos(\varphi-\alpha)\right), \quad (6)$$

where (ω,α) give the polar and the azimuthal angles of the vector $\vec{P}$. In the asymptotic region, $r^2|J_{scatt}|$ gives the differential scattering cross-section in the direction given by angles $\cos^{-1}\eta$ and φ. The problem, therefore, reduces to the determination of $F(\eta,\varphi;|P|,\omega,\alpha)$ for a given two center potential.

III. SPHEROIDAL PHASE SHIFT ANALYSIS

The equation (1) is separable in the spheroidal coordinates provided the potential is of the form

$$V(\xi,\eta) = \frac{U(\xi)+W(\eta)}{\xi^2-\eta^2}. \quad (7)$$

For simplicity, we shall take $W(\eta) = 0$. The angle function $S_{m\ell}(c,\eta)$ is then independent of the potential and can be normalized as

$$\int d\eta\, S_{m\ell}(c,\eta)\, S_{m\ell'}(c,\eta) = \delta_{\ell\ell'} N_{m\ell}(c). \quad (8)$$

The function $S_{m\ell}$ can be expanded as

$$S_{m\ell}(c,\eta) = \sum' d_n(c|m,\ell)\, P^m_{m+n}(\eta), \quad (9)$$

where the summation is over even or odd integers n depending on whether $(\ell-m)$ is even or odd.

It is now possible to make a spheroidal phase shift analysis to determine the scattering amplitude. The relativistic total scattering cross-section for a given direction of the nuclear axis is given by

$$\Sigma_t(\omega,\alpha) = \frac{1}{2p^2}\int d\eta\, d\varphi \left[1+\eta\cos\omega+\sqrt{1-\eta^2}\sin\omega\cos(\varphi-\alpha)\right]$$

$$\sum_{m,m',l,l'} \frac{1}{N_{ml}N_{m'l'}}(2-\delta_{m0})(2-\delta_{m'0})\cos m(\varphi-\alpha)$$

$$\cos m'(\varphi-\alpha)\left(e^{2i\sigma_{ml}}-1\right)\left(e^{-2i\sigma_{m'l'}}-1\right)S_{ml}(c,\eta)$$

$$S_{ml}(c,\cos\omega)\, S_{m'l'}(c,\eta)\, S_{m'l'}(c,\cos\omega)\,. \qquad (10)$$

The integral has been evaluated by making use of the expansion (9). On averaging over a random distribution of $\vec{R}$, one obtains finally $\langle \Sigma_t \rangle$ in terms of the constants $N_{ml}(c)$, $d_n(c|m,l)$, and the spheroidal phase shifts σ_{ml}. The calculation follows the method of Chandel (3).

To apply these results to a non-spherical nucleus, giving a screened spheroidal Coulomb potential V, it is useful to note that the constant $C \lesssim 1$ for an electron of energy $\epsilon \lesssim 50$ MeV. The corresponding $d_n(c|m,l)$ decrease rapidly as n increases and it is sufficient to consider in that case the first few terms in the expression for $\langle \Sigma_t \rangle$. Also, the spheroidal phase shifts σ_{ml} become independent of m when l is large and almost equal to the phase shifts σ_l for the corresponding spherical problem ($R \to 0$).

REFERENCES

1. S. Mukherjee and S. D. Majumdar, Ann. Physik (Leipzig) 16 (1965) 360.
2. A. Sommerfeld and A. W. Maue, Ann. Physik (Leipzig) 22 (1935) 629.
3. S. S. Chandel and S. Mukherjee, to be published in Prog. Theo. Physics (Kyoto).

MULTIMAGNON BOUND STATES

Chanchal K. Majumdar
Physics Department, Calcutta University
Calcutta-700 009 I N D I A

Nearly forty years after the theoretical prediction by Bethe[1], the magnon bound states were detected in $CoCl_2.2H_2O$ (CC2) by Torrance and Tinkham[2] . At first 2, 3, 4 and 5 - magnon bound states were found. Recently with excitation by HCN laser Nicoli and Tinkham[3] have detected up to fourteen magnon bound states. The CC2 system is almost Ising-like and a simple interpretation is possible, but it has also opened up an interesting branch of few-body systems.

The magnon bound states consist of several overturned spins travelling together in an otherwise aligned spin assembly. In the linear chain of spin-$\frac{1}{2}$ particles with the Hamiltonian

$$H = -J \sum_{i=1}^{N} \vec{S}_i \cdot \vec{S}_{i+1} \qquad (J > 0,\ N+1 \equiv 1) \tag{1}$$

their eigenvalues ε_B^r are [1]

$$\varepsilon_B^r = \frac{1}{r} J (1 - \cos K), \tag{2}$$

K being the centre-of-mass momentum of the r-spin unit. Orbach solved the 2-magnon problem in the linear chain with longitudinal anisotropy [4]

$$H = -J \sum_{i=1}^{N} \left[S_i^z S_{i+1}^z + \sigma \left(S_i^x S_{i+1}^x + S_i^y S_{i+1}^y \right) \right]. \tag{3}$$

The 2-magnon problem in two and three dimensions for the Heisenberg Hamiltonian

$$H = -\frac{1}{2} J \sum_{i,\delta} \vec{S}_{\vec{i}} \cdot \vec{S}_{\vec{i}+\vec{\delta}} \tag{4}$$

(i labels the lattice sites; δ joins a site to its nearest neighbours) was discussed by Dyson[5] and solved later completely [6]. Majumdar[7] showed that the three magnon bound state problem could be exactly formulated for one, two and three dimensions in terms of the Faddeev equations[8]. For(1), he solved the integral equation numerically to get Bethe's result. Later Majumdar and Mukhopadhyay[9] obtained numerically the eigenvalues for the corresponding integral equation with longitudinal anisotropy. An analytical expression for these eigenvalues was found by Gochev[10] using a more specialized technique for (3). Van Himbergen and Tjon [11] have discussed the Faddeev equations (see also Millet and Kaplan [12]) and found some unphysical solutions. They have numerically tackled the two dimensional problem [13] . Here we give a simple derivation of all these unphysical solutions in one dimension. The method might be useful in obtaining an analytic derivation of the physical solution as well. Finally, we comment on the fully anisotropic case.

The discussion of the dynamics of (4) is simplified by using Dyson's transformation to ideal spin waves. Eq.(4) is transformed into a pseudo-Hamiltonian in terms of ideal spin wave creation and annihilation operators.

$$H = E_0 + \sum_\lambda \epsilon_\lambda \alpha_\lambda^* \alpha_\lambda - \frac{1}{4N} J \sum_{\tau,\rho,\lambda} \Gamma_{\rho\tau}^\lambda \, \alpha_{\tau+\lambda}^* \alpha_{\rho-\lambda}^* \alpha_\rho \alpha_\tau \tag{5}$$

with

$$\epsilon_\lambda = \tfrac{1}{2} J (\gamma_0 - \sigma \gamma_\lambda) , \quad \gamma_\lambda = \sum_\delta \exp(i \vec{\lambda} \cdot \vec{\delta}) , \tag{6}$$

$$\Gamma^\lambda_{\rho\tau} = \gamma_\lambda - \sigma\gamma_{\lambda-\rho} - \sigma\gamma_{\lambda+\tau} + \gamma_{\lambda+\tau-\rho} . \tag{7}$$

E_0 is the ground state energy, ϵ_λ is the spin wave energy for wave vector λ. The last term in (5) is the spin wave interaction and happens to be non-hermitian. This is why the unphysical solutions appear.

The two particle t-matrix can be explicitly calculated. For Eq. (4) in one dimension, it is

$$(\lambda\mu|t(E)|\lambda\rho) = -\frac{4J}{N} \frac{\cos\mu(\sigma\cos\lambda - \cos\rho)\sigma^2\cos^2\lambda\left[(1-z)^2 - \sigma^2\cos^2\lambda\right]^{\frac{1}{2}}}{(1-z)\left[\{(1-z)^2 - \sigma^2\cos^2\lambda\}^{\frac{1}{2}} - (1 - z - \sigma^2\cos^2\lambda)\right]} \tag{8}$$

$z = E/2J$, 2λ is the centre-of-mass momentum. This t-matrix has a pole at the physical bound state $E = J(1-\sigma^2\cos^2\lambda)$ but also an unphysical pole at $E = 2J$.

Note that (8) is separable . With this the three magnon Faddeev equation in one dimension can be reduced to a linear integral equation

$$\Psi(p_1) = \frac{\sigma^2\cos^2\frac{1}{2}p_1}{\pi D} \int_{-\pi}^{\pi} dp_2 \frac{\left[\sigma\cos\frac{1}{2}p_1 - \cos(K - \frac{1}{2}p_1 - p_2)\right]\cos(K - p_1 - \frac{1}{2}p_2)\ \Psi(p_2)}{\omega - \frac{3}{2} + \frac{1}{2}\sigma\{\cos(K-p_1) + \cos(K-p_2) + \cos(K-p_1-p_2)\}} \tag{9}$$

with $\omega = E/2J$, (10)

$$D = \left[1 - \omega + \tfrac{1}{2}(1 - \sigma\cos(K-p_1))\right]\left(1 - \frac{1 - \omega + \frac{1}{2}(1 - \sigma\cos(K-p_1)) - \sigma^2\cos^2\frac{1}{2}p_1}{\left[\{1 - \omega + \frac{1}{2}(1 - \sigma\cos(K-p_1))\}^2 - \sigma^2\cos^2\frac{1}{2}p_1\right]^{\frac{1}{2}}}\right) . \tag{11}$$

In higher dimensions, a set of coupled linear integral equations appear [7].

The physical eigenvalue for the three magnon bound states is [10]

$$E \equiv \varepsilon^3_B = J\left[1 - \sigma^2 \frac{2 + \sigma\cos K}{4 - \sigma^2}\right] . \tag{12}$$

For $\sigma = 1$, the form of the eigenfunction is known [11] but a complete analytic derivation of both the eigenfunction and the eigenvalue for (9) has not been given.

As for unphysical solutions [14] , there are two sets. With the transformation

$$x = \tan\tfrac{1}{2}p_1 , \qquad y = \tan\tfrac{1}{2}p_2 , \tag{13}$$

Eq.(10) becomes

$$\Psi(x) = \frac{2}{\pi D}\int_{-\infty}^{\infty} dy \frac{(yf+g)}{(y^2+1)^{\frac{3}{2}}(x^2+1)^{\frac{3}{2}}}\left[1 - \frac{(y^2+1)\,v}{(x^2+1)(y-z)(y-\bar{z})}\right]\Psi(y), \quad (14)$$

with

$$f = (x^2-1)\sin K + 2x\cos K, \quad g = (x^2-1)\cos K - 2x\sin K,$$

$$a = 3-(E/J), \quad a' = a/\sigma, \quad v = x^2 - 2x\frac{\sin K}{a'+\cos K} + \frac{a'-2\sigma-\cos K}{a'+\cos K},$$

$$D = \frac{a'+\cos K}{2(1+x^2)}\left(x^2 - 2x\frac{\sin K}{a'+\cos K} + \frac{a'-\cos K}{a'+\cos K}\right)\left(1-\frac{v}{d}\right),$$

$$z, \bar{z} = -\frac{2(x\cos K - \sin K)}{(x^2+1)(a'+\cos K)} \pm i\frac{d}{(x^2+1)},$$

$$d^2 = x^4 - 4x^3\frac{\sin K}{a'+\cos K} + 2x^2\frac{a'^2-3\cos^2 K}{(a'+\cos K)^2} - 4x\frac{\sin K(a'-\cos K)}{(a'+\cos K)^2} + \frac{(a'-\cos K)^2-4}{(a'+\cos K)^2}. \quad (15)$$

The first set can be shown to have the solution

$$\Psi_1(x) = \frac{1}{(1+x^2)^{\frac{1}{2}}}\cdot\frac{x^2 - 6x\frac{\sigma\sin K}{a+3\sigma\cos K} + \frac{a-3\sigma\cos K}{a+3\sigma\cos K}}{x^2 - 2x\frac{\sigma\sin K}{a+\sigma\cos K} + \frac{a-\sigma\cos K}{a+\sigma\cos K}}; \quad (16)$$

the eigenvalues are given by the roots of the cubic

$$a^3 - a^2\left(2+\frac{7}{8}\sigma^2\right) + 3a\sigma^2 - \frac{9}{8}\sigma^4 = 0. \quad (17)$$

In general there is a real and a pair of complex conjugate eigenvalues. The second set gives a single eigenvalue:

$$\Psi_2(x) = \frac{1}{(1+x^2)^{\frac{1}{2}}}\cdot\frac{x^3 - x\frac{a+3\sigma\cos K}{a+\sigma\cos K} + \frac{2\sigma\sin K}{a+\sigma\cos K}}{x^2 - 2x\frac{\sigma\sin K}{a+\sigma\cos K} + \frac{a-\sigma\cos K}{a+\sigma\cos K}}, \quad (18)$$

$$E/J = 1 - \frac{1}{8}\sigma^2. \tag{19}$$

Eqs.(13) and (14) are likely to give an analytic derivation of the physical solution as well, but the algebra appears still formidable.

For Eq.(3), Gochev has found recursion relations [10] to calculate higher magnon bound states, but the experiments[2, 3] also require a knowledge of the spectrum of the completely anisotropic Hamiltonian

$$H = -\sum_{i=1}^{N} \left[J_x S_i^x S_{i+1}^x + J_y S_i^y S_{i+1}^y + J_z S_i^z S_{i+1}^z \right] - g m H_e \sum_{i=1}^{N} S_i^z. \tag{20}$$

Approximate analytic calculations have been attempted [15] .Exact numerical computations for finite chains ($N \leqslant 10$) could be done to estimate the critical field beyond which the bound magnon spectrum can be clearly seen. As for higher dimensions there is numerical work on two dimensions[13],but little hope of progress in the analytical attack on the three dimensional equations.

References.

1. H. Bethe, Z Physik 71 (1931) 205.
2. J.B.Torrance and M.Tinkham, Phys.Rev. 187(1969) 587, and 187(1969) 595.
3. D.F.Nicoli and M.Tinkham Phys. Rev. B9(1974) 3126.
4. R.Orbach, Phys.Rev. 112 (1958) 309.
5. F.J.Dyson Phys.Rev. 102(1956) 1217.
6. J.Hanus, Phys.Rev.Lett. 11(1963) 336. M.Wortis Phys.Rev. 132(1963) 85, N.Fukuda and M.Wortis,J.Phys.Chem.Solids 24(1963) 1675.
7. C.K.Majumdar, Phys. Rev. B1(1970) 287.
8. L.D.Faddeev,Zh.Eksp.Teor.Fiz. 39(1961) 1459-English translation, Sov.Phys. JETP 12 (1961) 1014.
9. C.K.Majumdar and G.Mukhopadhyay, Phys. Lett. 31A(1970) 321; C.K.Majumdar, G.Mukhopadhyay and A.K.Rajagopal Pramāṇa 1(1973) 135.
10. I.G.Gochev, Zh. Eksp. Teor.Fiz. 61(1971) 1674-English translation Sov.Phys.JETP 34 (1972) 892.
11. J.Van Himbergen and J.Tjon, Physica 76 (1974) 503.
12. H.Millet and T.Kaplan Phys.Rev. B 10(1974) 3923.
13. J.Van Himbergen and J.Tjon Phys. Lett. 50A(1974) 189 and to be published.
14. C.K.Majumdar, J. Math. Phys. (to appear in 1976); S.K.Mukhopadhyay and C.K.Majumdar, J.Math. Phys. (to appear in 1976).
15. H.C.Fogedby Phys. Rev. B10(1974) 4000.

SOLVABLE FEW BODY MODELS AS LABORATORIES FOR NUCLEAR REACTION THEORIES

Bruce H.J. McKellar

Service de Physique Théorique, CEN-Saclay, BPn°2, 91190 Gif-sur-Yvette, France and School of Physics, University of Melbourne, Parkville, Vic. 3052, Australia*

1. INTRODUCTION

We have found that the study of simple solvable models of a nuclear reaction process is a very useful technique for obtaining insight into the process, and for testing the validity of the necessarily approximate theories which can be applied to real reaction processes. All of the models we describe are of the separable potential type, applied in progressively more complex situations, namely single channel scattering [1], many channel scattering [2], and three body many channel scattering [3,4,5].

2. SINGLE CHANNEL SCATTERING [1]

We begin by considering an energy dependent separable potential of the form

$$\langle \underset{\sim}{p}'|V(E)|\underset{\sim}{p}\rangle = \mu^{-1}(E+i\varepsilon)\; g(p)\; g(p') \tag{2.1}$$

where $\mu(z)$ is a real analytic function of z, meromorphic in the z plane cut from E_1 to ∞. For example, the optical potential of the many channel problem of section 3 is of this type. The solution of the Schrodinger equation proceeds in the usual way. However because of the energy dependence of the solutions,

$$\psi_{\underset{\sim}{k}}^{(\pm)}(\underset{\sim}{p}) = \delta(\underset{\sim}{k}-\underset{\sim}{p}) + \frac{g(p)\; g(k)}{k^2-p^2\pm i\eta}\, X^{-1}(k^2+i\eta,\, k^2\pm i\eta) \tag{2.2}$$

where

$$X(z_1,z_2) = \mu(z_1) - \int dp\, \frac{g^2(p)}{z_2-p^2} \; , \tag{2.3}$$

the wavefunctions for different k are not orthonormal, and the scattering matrix elements $S_{k,k'}$ are not $\langle\psi_{\underset{\sim}{k}}^{(-)}|\psi_{\underset{\sim}{k}'}^{(+)}\rangle$. These facts are of course general results for any energy dependent potential. The advantage of the present model is that it demonstrates these difficulties analytically, and so permits explicit construction of the biorthogonal set of states $\tilde{\psi}_{k}^{(\pm)}$, with the properties that, in the absence of bound states,

$$\langle\tilde{\psi}_{\underset{\sim}{k}}^{(\pm)}|\psi_{\underset{\sim}{k}'}^{(\pm)}\rangle = \delta(\underset{\sim}{k}-\underset{\sim}{k}') \tag{2.4}$$

$$\int d\underset{\sim}{k}\; \psi_{\underset{\sim}{k}}^{(\pm)}(\underset{\sim}{p})\; \tilde{\psi}_{\underset{\sim}{k}}^{(\pm)*}(\underset{\sim}{p}') = \delta(\underset{\sim}{p}-\underset{\sim}{p}') \tag{2.5}$$

and

$$S_{k,k'} = \langle\tilde{\psi}_{\underset{\sim}{k}}^{(-)}|\psi_{\underset{\sim}{k}}^{(+)}\rangle \tag{2.6}$$

The biorthogonal wavefunctions are most easily constructed from the equation

$$\tilde{\psi}_{\underset{\sim}{k}}^{(\pm)} = \psi_{\underset{\sim}{k}}^{(\pm)} + \int dk\; \tilde{\psi}_{\underset{\sim}{k}'}^{(\pm)}\; \Delta^{(\pm)}(\underset{\sim}{k}',\underset{\sim}{k}) \tag{2.7}$$

where

$$\Delta^{(\pm)}(\underset{\sim}{k}',\underset{\sim}{k}) = \delta(\underset{\sim}{k}'-\underset{\sim}{k}) - \langle\psi_{\underset{\sim}{k}'}^{(\pm)}|\psi_{\underset{\sim}{k}}^{(\pm)}\rangle \tag{2.8}$$

Explicitly

$$\Delta^{(+)}(\underset{\sim}{k}',\underset{\sim}{k}) = g(k)\; g(k')\, \frac{[\mu(k'^2-i\eta)-\mu(k^2+i\eta)]}{k^2-k'^2+i\eta} \tag{2.9}$$

$$X^{-1}(k'^2-i\eta,k'^2-i\eta)\; X^{-1}(k^2+i\eta,k^2+i\eta)$$

If the energy dependence of the potential is weak, we can readily solve the equation (2.7) for $\tilde{\psi}_{\underset{\sim}{k}}^{(\pm)}$ iteratively, expecting Δ to be small and the iterative solution to converge rapidly.

The alteration of the fundamental relation for $S_{\underset{\sim}{k},\underset{\sim}{k}'}$ has a number of practical consequences. Among the most important are

(i) the modification of the two potential formula, which now becomes

$$T_{fi} = T_{fi}^{(1)} + \langle\tilde{\phi}_f^{(-)}|H-H_1|\psi_i^{(+)}\rangle$$

where $V=V_1+V_2$, ϕ are eigenstates of $H_1=H_o+V_1$, and $H = \int d\underset{\sim}{k}\ E_k\psi_k\rangle\langle\tilde{\psi}_k$; $H_1 = \int d\underset{\sim}{k}\ E_k\phi_k\rangle\langle\tilde{\phi}_k$. This modification has consequences of a fundamental rather than practical nature for the theory of (d,p) reactions - the Goldberger and Watson [6] derivation of the usual DWBA expression is preferred and the more usual derivation, quoted e.g. by Hodgson [7], breaks down because of the energy dependence of the optical potential. But one can still use the standard codes.

(ii) One can extend these methods to construct the two-body T matrix off energy shell for an energy dependent potential. This can then be used as input for three-body equations [5].

3. MANY CHANNEL PROBLEMS [2]

Now we take a somewhat more complicated separable potential appropriate to a many channel problem

$$\langle \underset{\sim}{p}'|V|\underset{\sim}{p}\rangle = M^{-1}\ g(p)\ g(p') \tag{3.1}$$

where M is an N×N matrix connecting the N channels of the problem, the target with Hamiltonian H_T having N states $\Phi_1,\ldots,\Phi_N$ with energies $\varepsilon_1,\ldots,\varepsilon_N$.
The solution of the Schrodinger equation is immediate, it is

$$\Psi_{i\underset{\sim}{k}}^{(\pm)}(\underset{\sim}{p}) = \Phi_i\ \delta(\underset{\sim}{p}-\underset{\sim}{k}) + g(\underset{\sim}{p})\ G_{oi}^{(\pm)}(\underset{\sim}{p},k^2)\ g(\underset{\sim}{k})\ F^{-1}(k^2+\varepsilon_i\pm i\eta)\ \Phi_i \tag{3.2}$$

for the scattering states generated by a particle of momentum $\underset{\sim}{k}$ incident or the target in the state Φ_i. The matrix Green's function is

$$G_{oi}^{(\pm)}(\underset{\sim}{p},k^2) = \frac{1}{k^2+\varepsilon_i-p^2-H_T\pm i\varepsilon} \tag{3.3}$$

and

$$F(z) = M - \int d\underset{\sim}{p}\ g^2(p)\ \frac{1}{z-p^2-H_T} \tag{3.4}$$

The bound states occur at the energies E_n at which F(z) is singular. The wave-function is given by

$$\Psi_n(\underset{\sim}{p}) = \frac{1}{E-p^2-H_T}\ \frac{g(p)}{\left[\frac{d}{dE}\det F(E)\Big|_{E=E_n}\right]^{1/2}}\ B_n \tag{3.5}$$

where the vector B_n is defined by

$$B_n\ B_n^+ = \operatorname{adj} F(E_n) \tag{3.6}$$

and we use the notation adjM for the adjoint matrix to M [8].

A surprise was in store for us when we evaluated the spectroscopic factor sum rule. For a particular single particle state $u_\alpha(p)$, this states that

$$\sum_n |\langle\Phi_i u_\alpha|\Psi_n\rangle|^2 + \int d\underset{\sim}{k} \sum_i |\langle\Phi_i u_\alpha|\Psi_{i\underset{\sim}{k}}\rangle|^2 = 1 \ . \tag{3.7}$$

Suppose we consider the case where there is just one bound state, Ψ_1, whose major component is the particle bound to the ground state Φ_1 of the target, and we take

$u_\alpha = N \langle\Phi_1|\Psi_1\rangle$, where N is a normalization constant, and i = 1. Then $|\langle\Phi_1 u_\alpha|\Psi_1\rangle|^2$ is almost unity. In fact if γ is a measure of the size of the off diagonal elements of M,

$$|\langle\Phi_1 u_\alpha|\Psi_1\rangle|^2 = 1 - O(\gamma^2) \tag{3.8}$$

The rest of the spectroscopic factor strength is of course in the continuum. However the surprise in the evaluation is that

$$|\langle\Phi_1 u_\alpha|\Psi_1\rangle|^2 + \int d\underset{\sim}{k} \sum_{i\neq 1} |\langle\Phi_1 u_\alpha|\Psi_{i\underset{\sim}{k}}\rangle|^2 = 1 - O(\gamma^4) \;, \tag{3.9}$$

i.e. that most of the strength is in those continuum states in which the particle is incident on the excited states of the target. Once again, we expect this result to be general [2].

4. THE THREE BODY PROBLEM

(i) Spectroscopic factors [3]

The three body problem with separable potentials has long been used as a test piece for theories of rearrangement collisions [9]. However, one of the interesting questions in the theory of (d,p) reactions is the reliability with which the theory can extract spectroscopic factors. This cannot be tested in the usual calculations, because the spectroscopic factors are always unity. However if the proton-core and neutron-core interactions are of the many channel separable type described in the previous section, spectroscopic factors less than unity, and hence a meaningful test, are available. A typical result is shown in figure 1.

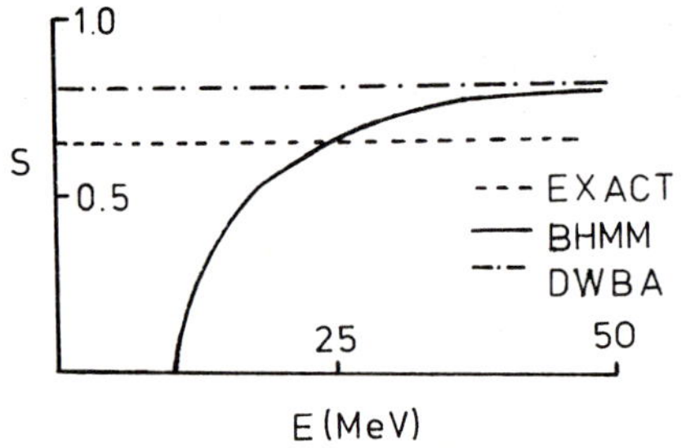

Fig.1 - Spectroscopic factor (left) and angular distributions (below) for exact and approximate theories. (From ref. [3])

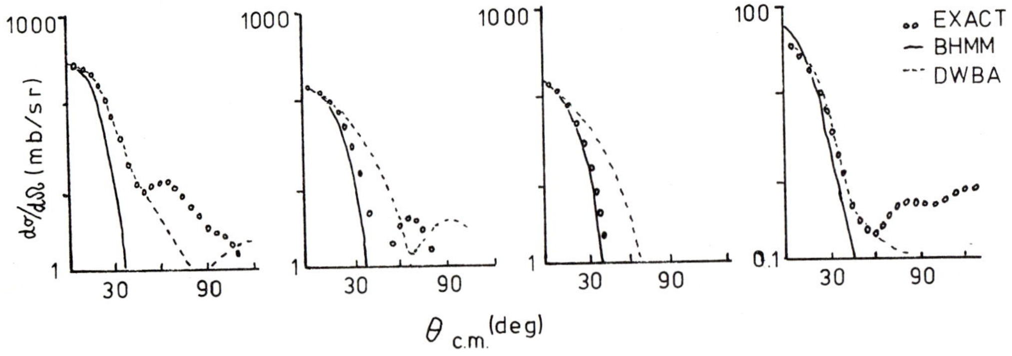

The conclusions are:

a) BHMM and DWBA theories both give fair results for the angular distribution.
b) The BHMM spectroscopic factor is energy dependent, as has also been found empirically.
c) Both theories approach the same overestimate of the spectroscopic factor at high energies.

Searching for an explanation of the latter point, we decided that the most likely one is that two step processes are important.

(ii) Two step processes [4]

One of course can include two step processes in the theory by doing a multichannel calculation but it is convenient also to have an approximate theory. In looking for such a theory we noticed, by comparing results with our models, some deficiencies in Robson's approach [10]. Basically he criticises the standard method because of the lack of completeness of optical model wavefunctions discussed in section 2. He goes on to construct a model based on the exact many channel Green's function, but omits the contributions from states of the type of Ψ_{ik} , $i \neq 1$, of section 3. Since it is just these states which are responsible for the non-completeness, Robson is guilty of the same fault that he criticises. A consistent theory of the two step process along the lines of Robson's, but taking now completeness and the modifications to the two potential formula into account, has been constructed [4], but no numerical results are available yet.

(iii) Energy dependent potentials [5]

The construction of the three body equations for energy dependent potentials is the subject of a contribution to this conference [5]. A possible future application of the model described in this section is the comparison of results obtained using exact multichannel n-target and p-target potentials, and those obtained using the optical potentials in the formalism of ref. [5].

ACKNOWLEDGEMENTS

It is a pleasure to thank my collaborators, Y.E. Kim, K. King and Colette McKay for their contribution to the work reported here; S.T. Butler, N. Austern and B.M. Spicer for helpful conversations, and R.C. Johnson for a useful correspondence. My thanks are also due to Dr C. De Dominicis and my colleagues of the Service de Physique Théorique for their kind hospitality.

REFERENCES and FOOTNOTES

* Permanent address

[1] C.M. McKay and B.H.J. McKellar, to be published.
[2] B.H.J. McKellar, J. Phys. G1 (1975) 180; Phys. Lett. B32 (1970) 246.
[3] K. King and B.H.J. McKellar, Phys. Rev. Lett. 30 (1973) 562; Phys. Rev. C9 (1974) 1309.
[4] C.M. McKay and B.H.J. McKellar, to be published.
[5] Y.E. Kim, C.M. McKay and B.H.J. McKellar, Contribution to this conference and to be published.
[6] M.L. Goldberger and K.M. Watson, "Collision Theory", J. Wiley and Sons, N.Y. (1964).
[7] P.E. Hodgson, "Nuclear Reactions and Nuclear Structure", Clarendon Press, Oxford (1971).
[8] A.C. Aitken, "Determinants and Matrices", Oliver and Boyd, Edinburgh (1967).
[9] See for example A.N. Mitra, Phys. Rev. 139 (1965) B1472; P.E. Shanley, Ann. Phys. (N.Y.) 44 (1966) 363.
[10] D. Robson, Phys. Rev. C7 (1973) 1.

FEW-BODY APPROACHES TO DIRECT NUCLEAR REACTIONS

V. VANZANI

Istituto di Fisica dell'Università - 35100 Padova - Italy

1.- GENERAL DYNAMICAL EQUATIONS

In principle, nuclear reactions and scattering processes involving composite particles should be treated in a N-body multichannel context. Recently there has been an increased interest in studying the N-body scattering problem and several formalisms have been developed for treating in an exact way the many-body dynamical aspects of the problem. Among these, some integral equation approaches are particularly interesting, namely the ones in which the unknown quantities are the transition operators (that are closely related to physically observable quantities).

In order to facilitate the study of the general structure of the above integral equations it is convenient to use the AGS symmetric off-shell extension $U^{b_j a_i}(z)$ of the on-shell transition operators [1]

$$U^{b_j a_i}(z)=G^{-1}_{b_j a_i}(z)\bar{\delta}_{b_j a_i}+V^{b_j a_i}+V^{b_j}G(z)V^{a_i}=G^{-1}_{a_i}(z)\bar{\delta}_{b_j a_i}+V^{b_j}G(z)G^{-1}_{a_i}(z) \tag{1.1}$$

from the configuration characterized by the partition a_i of the N particles into i clusters to the configuration characterized by the partition b_j. In (1.1) $\bar{\delta}_{b_j a_i} = 1-\delta_{b_j a_i}$, $V^{a_i} = V-V_{a_i}$. V is the sum of all the interactions; V_{a_i} and V^{a_i} are, respectively, the internal and external interactions in channel a_i; $V_{b_j a_i}$ $(V^{b_j a_i})$ denotes the interactions that are internal (external) in both the channels a_i and b_j, while $V^{b_j}_{a_i}$ denotes the interactions which are internal in channel a_i but external in channel b_j. G and G_{a_i} are the total and the a_i- channel resolvent operator, respectively, and $G_{b_j a_i}$ is the resolvent of the Hamiltonian constructed with $V_{b_j a_i}$.

By means of suitable procedures, based on appropriate decompositions of the channel external interactions one arrives at the following two different sets of equations for transitions from an initial two-cluster channel to another two-cluster channel (elastic and rearrangement transitions)[2]

$$U^{b_2 a_2}= (G^{-1}_{a_2}+V^{b_3}_{a_2}\delta_{a_2 \supset b_3})\bar{\delta}_{b_2 a_2} + \sum_{\substack{c_2 \neq b_2\\ c_2 \supset b_3}} V^{b_3}_{c_2} G_{c_2} U^{c_2 a_2} \quad , \tag{1.2}$$

$$U^{b_2 a_2}=G^{-1}_{b_2 a_2}\bar{\delta}_{b_2 a_2} + \sum_{i=3}^{N-1} \sum_{\gamma_i (c_i \not\supset a_2)} T^{\gamma_{i+1}}_{c_i} G_o G^{-1}_{a_2} + \sum_{\gamma_2} T^{\gamma_3}_{c_2} G_o U^{c_2 a_2} \tag{1.3}$$

with $c_{N-1} \not\supset b_2$ and

$$T^{\alpha_{i+1}}_{a_i} = V_{a_{N-1}} G_{a_{N-1}} V^{a_{N-1}}_{a_{N-2}} G_{a_{N-2}} \ldots V^{a_{i+1}}_{a_i} G_{a_i} G^{-1}_{o} \quad . \tag{1.4}$$

In (1.3) and (1.4) α_i denotes a sequence of partitions $\alpha_i=(a_i,a_{i+1},\ldots,a_{N-1})$ with $a_i \supset a_{i+1} \supset \ldots \supset a_{N-1}$. Transition operators describing breakup processes are given in terms of $U^{b_2 a_2}$ by the interconnecting relations (j>2)

$$G_{b_j} U^{b_j a_2} = G_{b_2} U^{b_2 a_2} + \delta_{b_2 a_2} \tag{1.5}$$

Note that different equations of the kind (1.2) can be written in correspondence to different choices of b_3. It is convenient to denote the transition operator written in the form (2) as $U^{b_2 a_2}_{b_3}$. By further decomposing $V^{b_3}_{c_2}$ into $V^{b_4}_{c_3}$ and so on we arrive at a set of integral equations for $U^{b_2 a_2}_{\beta_3}$ whose kernel becomes connected after N-2 iterations. These equations are closely related to the ones obtained by Alt, Grassberger and Sandhas in their matrix formulation of the N-body theory [3], by Karlsson and Zeiger [4] and in ref.[5] in the language of two-sided operators. Eqs.(1.3) correspond to the ones derived by Sloan, Bencze and Redish[6].

The kernel of eqs.(1.3) becomes connected after a single iteration. However, it appears more complicated than the previous one. Furthermore the inhomogeneous term of (1.3) is not so simple as in (1.2) : a suitable off-shell transformation is required in order to simplify it; but the new off-shell continuation leads to predictions different than before, if approximate treatments are made.

The symmetric transition operators (1.1) satisfy the generalized LS equations (with the cluster indices omitted)

$$U^{ba} = G_a^{-1}\,\overline{\delta}_{ba}+V^b\delta_{ca}+V^bG_cU^{ca} \qquad (1.6)$$

that have been introduced in refs.[7]. A set of values for b and c can be chosen so that the result is a set of coupled integral equations [8]. By introducing a channel coupling array whose elements W_{ba} obey $\sum_a W_{ba}=1$ one obtains the symmetric transition operator version of the Kouri-Levin equations, namely

$$U^{ba}=G_a^{-1}\,\overline{\delta}_{ba}+V^bW_{ba}+V^b\sum_c W_{bc}G_cU^{ca} \qquad (1.7)$$

If the coupling array elements are chosen as in ref.[9] the kernel of (1.7) be - comes connected after N-1 iterations.

The above N-body integral equations provide a rigorous basis for the study of the nuclear reactions. However, they involve such a large number of continuos variables that their application is feasible only if their dimension can be considerably reduced. This can be made by introducing generalized separable approximations or taking account only of the dominant few-cluster parts of the spectral expansion of the kernel [1,10,11]. Some preliminary promising calculations based on a generalized separable method have been carried out for the four-nucleon system[12,13].

It is worthwhile outlining an alternative procedure which makes the problem approximately tractable in some specific cases. It is based on an appropriate employment of the many-body Gell-Mann-Goldberger transformation for the symmetric transition operators [14]

$$U^{ba}=(W^a+W^a\overline{G}_aW^a)\delta_{ba}+(1+W^b\overline{G}_b)\overline{U}^{ba}(1+\overline{G}_aW^a) \qquad (1.8)$$

where

$$\overline{U}^{ba}=\overline{G}_a^{-1}\,\overline{\delta}_{ba}+\overline{V}^b+\overline{V}^bG\overline{V}^a \qquad (1.9)$$

To obtain (1.8) the channel external interactions have been splitted into $W^a+\overline{V}^a$ and the resolvent equations for G and $\overline{G}_a$ have been used ($\overline{G}_a$ is constructed similarly to G, with W^a instead of V^a). W^a can be regarded as a distorting potential and $\overline{V}^a$ as a residual interaction. Note, however, that the eq.(1.8) is an exact expression, which holds for arbitrary splitting of V^a and V^b. The first term of (1.8) represents the pure distorting potential contribution. The second term is constructed by sandwiching between distorting wave operators the operators $\overline{U}^{ba}$, which are generated by the residual interactions in the field of the distorting potentials. The latter operators have the same structure as U^{ba}. Therefore, by introducing suitable decompositions of the residual interactions, one can formally derive integral equations with connected iterated kernel for the residual transition operators $\overline{U}^{ba}$. If one considers reactions proceeding via direct mechanisms one can suitably combine the integral equation formalism with the DW method and assume that, in the spirit of the DW approximation, it is possible to choose the distorting potentials in such a way that the residual interactions become small enough to allow iterative solution of the integral equations for the residual operators. A first attempt to apply this procedure has been made by Lovas [15] in a three-body context.

The exact N-body equations take account of all the degrees of freedom of the colliding systems. However, there exists a great variety of nuclear processes where only a small number of degrees of freedom participates actively to the reaction. It seems physically reasonable to treat these reactions as effective few-body problems. In particular single particle (or single cluster) transfer or exchange reactions can be described as effective three-body problems. If these reactions involve an excitable nucleus, two-body multichannel interactions should be introduced in the three-body context. In a certain sense these multichannel interactions simulate some many-body aspects of the problem.

In the following we limit ourselves to describe a multichannel separable potential approach to nucleon-nucleus scattering and to discuss some three-body distorted-wave methods.

2. - A MULTICHANNEL SEPARABLE POTENTIAL APPROACH TO NUCLEON-NUCLEUS SCATTERING

Let us now present a two-body multichannel scattering formalism with separable short range potentials and Coulomb interactions (16). This formalism can be used for the study of elastic and inelastic scattering processes between a structureless and a composite excitable particle when rearrangement effects are expected to be small. It could be suitably extended to the off-shell case and employed in few-body problems with one excitable particle.

Let V_c and V_s be, respectively, the Coulomb and short range interactions. Then the total transition operator T takes the form

$$T = T_c+T_{sc}=T_c+(1+T_cG_o)T_{s(c)}(1+G_oT_c) \tag{2.1}$$

with $T_c=V_c+V_cG_cV_c$ and $T_{s(c)}=V_s+V_sGV_s$ [see (1.8)]. Sandwiching (2.1) between eigenstates $|i\rangle$ of the composite particle Hamiltonian and, assuming that V_c acts only in the channel relative motion space, one obtains

$$T_{ji}\equiv \langle j|T|i\rangle = T_{c,i}\delta_{ji} + (1+T_{c,i}G_o)T_{s(c),ji}(1+G_oT_{c,i}) . \tag{2.2}$$

The matrix elements $T_{s(c),ji}$ satisfy the set of coupled integral equations

$$T_{s(c),ji} = V_{s,ji} + \sum_r V_{s,jr}G_{c,r}T_{s(c),ri} , \tag{2.3}$$

where $G_{c,r}$ is the effective Coulomb resolvent in the r-channel.

If one assumes that the potential matrix elements $V_{s,ji}$ are separable in each partial-wave state, namely

$$V_{s,ji}= - \sum_{\ell m}|\chi_{\ell m,j}\rangle \lambda_{\ell,ji} \langle\chi_{\ell m,i}| , \tag{2.4}$$

the two-body multichannel problem becomes exactly soluble in presence of Coulomb interactions too. The solution can be written in form

$$T_{s(c),ji}= - \sum_{\ell m}|\chi_{\ell m,j}\rangle t_{c\ell,ji}\langle \chi_{\ell m,i}| \tag{2.5}$$

with

$$t_{c\ell,ji}=\left[(\mathbf{1}+\boldsymbol{\lambda}_\ell\mathbf{g}_{c\ell})^{-1}\boldsymbol{\lambda}_\ell\right]_{ji} ; \mathbf{g}_{c\ell,ji}=g_{c\ell,i}\delta_{ji}=\langle\chi_{\ell m,i}|G_{c,i}|\chi_{\ell m,i}\rangle \delta_{ji} . \tag{2.6}$$

Note that the set of equations satisfied by $t_{c\ell,ji}$

$$t_{c\ell,ji}=\lambda_{\ell,ji}-\sum_r \lambda_{\ell,jr}g_{c\ell,r}t_{c\ell,ri} \tag{2.7}$$

can be alternatively solved by means of a suitable channel-decoupling procedure, precisely by removing one after the other all the n-1 channels different from j. One obtains (n=j)

$$t_{c\ell,ni}=\lambda^{(n-1)}_{c\ell,ni}-\lambda^{(n-1)}_{c\ell,nn}\, g_{c\ell,n}t_{c\ell,ni}\equiv\lambda^{(n)}_{c\ell,ni} \tag{2.8}$$

if one introduces the recurrence relations ($m\leq n$)

$$\lambda^{(m)}_{c\ell,ni} =\lambda^{(m-1)}_{c\ell,ni} -\lambda^{(m-1)}_{c\ell,nm}\, g_{c\ell,m}\left(1+\lambda^{(m-1)}_{c\ell,mm}\, g_{c\ell,m}\right)^{-1}\lambda^{(m-1)}_{c\ell,mi} ;\lambda^{(0)}_{c\ell,ni}=\lambda_{\ell,ni}. \tag{2.9}$$

Eq.(2.8) has the same formal structure as the single-channel equation for t (but in the multichannel case the renormalized coupling parameters (2.9) are energy - -dependent). On the energy-shell one has [16]

$$\langle \vec{k}'_j | T_{sc,ji} | \vec{k}_i \rangle = -\sum_{\ell m} e^{(\sigma_{\ell,j}+\sigma_{\ell,i})} u_{c\ell,j}(k'_j) t_{c\ell,ji}(k_i) u^*_{c\ell,i}(k_i) Y^m_\ell(\hat{k}'_j) Y^{m*}_\ell(\hat{k}_i) \quad (2.10)$$

with some self-explanatory notation ($u_{c\ell,i}$ are Coulomb-modified form factors). In our numerical calculations we have chosen the functional form

$$v_\ell(r)=\exp(-\beta_\ell r)r^{\ell-1}, \quad u_\ell(k)=\left(\frac{2}{\pi}\right)^{\frac{1}{2}}(2\ell)!!k^\ell(k^2+\beta_\ell^2)^{-\ell-1} \quad (2.11)$$

for the form factors. This choice leads to simple analytic expressions,e.g.

$$u_{c\ell}(k)=\left(\frac{2}{\pi}\right)^{\frac{1}{2}}(2\ell+1)!C_\ell(\eta)k^\ell(k^2+\beta_\ell^2)^{-\ell-1}\exp\left[2\eta\operatorname{arctg}\frac{k}{\beta_\ell}\right]. \quad (2.12)$$

We have studied the elastic scattering of neutrons and protons by ^{4}He, ^{12}C and ^{16}O targets. We have carried out some preliminary single-channel calculations. With a single separable term in each partial wave we have obtained a good fit of the experimental phase shifts. The neutron and proton scattering data relative to the same state have been reproduced by almost the same values of the parameters. A detailed account of these first results is given in ref.[17]. We are now analysing some typical effects of several overlapping or compound resonances in the general multichannel context.

3.- ON THE THREE-BODY DISTORTED-WAVE METHODS

Let us now discuss some three-body distorted-wave formalisms for rearrangement processes a+(b+c)→b+(a+c). We begin by considering single integral equations with connected kernel. Dodd and Greider started from the DW representation [18]

$$U^{-ba}=(1+W^b\bar{G}_b)(\bar{V}^a+\bar{V}^b G\bar{V}^a)(1+\bar{G}_a W^a)+G_b^{-1}\bar{G}_a W^a \quad (3.1)$$

for the conventional asymmetric transition operator U^{-ba}. Since they choose W^a so as not to lead to rearrangement, they neglect the last term in (3.1). Thus they implicitly introduce a new operator $\tilde{U}^{-ba}$ (which coincides with U^{-ba} on the energy-shell only). They derive the integral equation

$$\tilde{U}^{-ba}=(1+W^b\bar{G}_b)(\bar{V}^a+\bar{V}^b\bar{G}^b\bar{V}^a)(1+\bar{G}_a W^a)+(1+W^b\bar{G}_b)\bar{V}^b\bar{G}^b(V_b+W^b)G_b\tilde{U}^{-ba}, \quad (3.2)$$

where $\bar{G}^b$ is the resolvent constructed with the interactions which are complementary to V_b+W^b (namely with V^b-W^b). It is easily seen that the original operator U^{-ba} satisfies the equation [with the same kernel K^b as in (3.2)]

$$U^{-ba}=(1+W^b\bar{G}_b)(\bar{V}^a+\bar{V}^b\bar{G}^b\bar{V}^a)(1+\bar{G}_a W^a)+G_b^{-1}\bar{G}_a W_a-(1+W^b\bar{G}_b)\bar{V}^b\bar{G}^b(V_b+W^b)\bar{G}_a W^a+K^b U^{-ba} \quad (3.3)$$

which, with respect to (3.2), contains an extra inhomogeneous term that does not generally vanish on the energy-shell. The reason for such a difference is the following: in deriving (3.3) one must use (3.1) off-shell, $G_b^{-1}\bar{G}_a W^a$ is multiplied by the kernel of (3.2) and the result is not generally zero. A third different result can be obtained by starting from the symmetric operator U^{ba}. One has

$$U^{ba}=(1+W^b\bar{G}_b)(\bar{G}_b^{-1}+\bar{V}^a+\bar{V}^b\bar{G}^b\bar{V}^a)(1+\bar{G}_a W^a)-(1+W^b\bar{G}_b)\bar{V}^b\bar{G}^b(V_b+W^b)(1+\bar{G}_a W^a)+K^b U^{ba} \quad (3.4)$$

The inhomogeneous term in the well-behaved eqs. (3.2)-(3.4) consists of various parts, only one of which corresponds to the usual DWBA $[(1+W^b\bar{G}_b)\bar{V}^a(1+\bar{G}_a W^a)]$. The complete inhomogeneous term could be regarded as an approximation better than the DWBA. It is referred to [in [18,19]] as a mathematically meaningful first-order approximation (MMFA). Note that we have different possible MMFAs- corresponding to the above different off-shell continuations of the on-shell transition operators - and consequently different predictions. (Eqs.(3.2) - (3.4) lead to the same result only in the case of an exact treatment). Furthermore it should be reminded that the connectedness of the kernel is not itself sufficient to guarantee the convergence of the series generated by the above equations. If the iterative solution exists, then the rate of convergence of such a solution will be a direct measure of the validity of the MMFA. Unfortunately, on one hand the study of the convergence properties is hampered by the complicate nature of the kernel and on the other hand the numerical evaluation of higher order terms requires a hard work.

A different generalized DW formalism can be introduced by starting from the FAGS equations, formally eliminating the components from implicit channels- channels different from the initial and final ones - and introducing in a suitable way distortion operators in the explicit channels. If U^{ca} is eliminated in the FAGS system for U^{ba}, U^{aa}, U^{ca} and U^{bc} is eliminated in the FAGS system for U^{ba}, U^{bb}, U^{bc}, one obtains the following exact expression

$$U^{ba}=(1+U^{bb}G_oT_bG_o)(G_o^{-1}+T_c)(1+G_oT_aG_oU^{aa})-U^{ba}G_oT_aG_cT_bG_oU^{ba} . \tag{3.5}$$

Inserting in (3.5) separable expansions $\sum_{rs}|\alpha r\rangle t_{\alpha,rs}\langle\alpha s|$ for the explicit channel scattering operators T_α, one has

$$T^{bn,am}=\langle\Phi_{bn}|U^{ba}|\Phi_{an}\rangle=\sum_{rs}\{\langle F^-_{bn,bs}|\langle bs|G_c|ar\rangle|F^+_{ar,am}\rangle-\langle F^-_{bn,as}|\langle as|G_c|br\rangle|F^+_{br,am}\rangle\} \tag{3.6}$$

where

$$|F^+_{\beta j,\alpha i}\rangle = \left(\delta_{\beta\alpha}\delta_{ji}+\sum_r t_{\beta,jr}u_{\beta r,\alpha i}\right)|\vec{p}_\alpha\rangle = \lambda_{\beta j}\langle\beta j|\Psi^+_{\alpha i}\rangle \tag{3.7}$$

and $u_{\beta r,\alpha i}=\langle\beta r|G_oU^{\beta\alpha}G_o|\alpha i\rangle$. Practically three-body states are expanded in products of two-body states: the form factor state vectors $|\beta j\rangle$ belonging to the internal β-two-body subspace and the generalized spectator functions or DW states $|F^+_{\beta j,\alpha i}\rangle$ belonging to the β-channel relative motion space (see the last expression in (3.7) (where $|\Psi^+_{\alpha i}\rangle$ are total scattering states))(10,20). These generalized DW states reduce to the usual optical wave states for $\beta=\alpha$ and $\lambda_{\alpha i}\langle\alpha i|=\langle\psi_{\alpha i}|$ (the i-th bound state of the α-subsystem). This latter condition is equivalent to the replacement of the free three-body resolvent G_o by the free two-body one $\hat{G}_o$ ($\hat{G}_o|\alpha i\rangle=|\psi_{\alpha i}\rangle$).

Eq. (3.6) can be easily interpreted in a graphic language. Each term

$$\langle F^-_{bn,bs}|\langle bs|G_c|ar\rangle|F^+_{ar,am}\rangle=\langle F^-_{bn,bs}|\langle bs|(G_o+G_oT_cG_o)|ar\rangle|F^+_{ar,am}\rangle \tag{3.8}$$

represents the amplitudes of the following two diagrams between generalized DW states: a polar diagram, describing the transfer or exchange of c and a triangular diagram, describing in addition to the c-transfer or exchange the a-b off-shell scattering (7). Some insight into the relative role played by these diagram mechanisms can be obtained from the study of the corresponding singularities (see Sect. 5.2.3 in ref.(7)). The last part in (3.6) corresponds to rather complex diagrams (involving three successive rearrangements). Their singularities are expected to be rather far from the physical region boundary and their contributions to be generally isotropic (21). Thus, by disregarding these contributions of not direct nature, we are left with a generalized DW amplitude, which includes also inelastic scattering in the initial and final channels. The rearrangement is effected by the "effective transition potential" $\langle bs|G_c|ar\rangle$, which has a structure more general than in the DWBA (7). Our generalized DWA has a structure similar to the amplitude of the so-called Feynman-diagram summation method (FDSM), based on the above polar and triangular diagrams (22). Note, however, that generalized DW states, instead of the optical ones, appear in our context. In other words the FDSM as well as the DWBA neglect the contributions from three-body intermediate states involved in the spectator states.

Finally we consider some relevant simplifications which occur in heavy-target approximation. If one deals with a stripping process it is convenient to start from the exact formula ($T_{c(a)}=V_c+V_cG^aV_c$, G^a constructed with V^a)

$$U^{ba}=(1+T_{c(a)}G_oT_bG_o)(G_o^{-1}+T_c)(1+G_oT_aG_oU^{aa}) \quad , \tag{3.9}$$

In heavy-target approximation $W^b \simeq V_c(\bar{G}_b \simeq G^a)$.
Approximating $\langle\Phi_{bn}|(1+T_{c(a)}G_oT_bG_o) = \langle\Phi_{bn}|(1+V_cG^a)V_bG_o$ by $\langle F_{bn}^{-opt}|\langle\psi_{bn}|V_bG_o$ one has

$$T^{bn,am} \simeq \langle F_{bn}^{-opt}|\langle\psi_{bn}|V_bG_cV_a\sum_r|\psi_{ar}\rangle|F^+_{ar,am}\rangle \qquad (3.10)$$

Since $|F_{\alpha i}^{+opt}\rangle|\psi_{\alpha i}\rangle = G_o(V_a+W^a)|F_{\alpha i}^{+opt}\rangle|\psi_{\alpha i}\rangle$ it follows that

$$T^{bn,am} \simeq \langle F_{bn}^{-opt}|\langle\psi_{bn}|(1-V_cG_o)G_o^{-1}G_cV_a\sum_r|\psi_{ar}\rangle|F^+_{ar,am}\rangle = \langle F_{bn}^{-opt}|\langle\psi_{bn}|V_a\sum_r|\psi_{ar}\rangle|F^+_{ar,am}\rangle. \quad (3.11)$$

Thus, in this case, the polar and triangular diagram contributions involved in $\langle bn|G_c|ar\rangle$ sum up to give the usual effective transition potential $\langle\psi_{bn}|V_a|\psi_{ar}\rangle$ for stripping processes[23]. For deuteron stripping the generalized spectator states $|F^+_{ar,am}\rangle$ can be constructed according to the prescriptions of the unitary pole approximation. Then the above formulation becomes closely equivalent to the Johnson-Tandy treatment of deuteron stripping [24].

REFERENCES

(1) E.O.Alt, P.Grassberger and W.Sandhas, Nucl.Phys.B2(1967) 167; P.Grassberger and W.Sandhas, Nucl.Phys.B2(1967) 181
(2) V.Vanzani, to be published
(3) E.O.Alt, P.Grassberger and W.Sandhas, JINR, E4-6688 (Dubna,1972); W.Sandhas, Acta Phys.Austr.Suppl.13(1974) 679
(4) B.Karlsson and E.M.Zeiger, Phys.Rev.D9 (1974) 1761 and D10(1974) 1291
(5) V.Vanzani, Nuovo Cim. 2A (1971) 525
(6) I.Sloan, Phys.Rev. C6(1972) 1945; Gy.Bencze, Nucl.Phys.A210 (1973) 568; E.F.Redish, Nucl.Phys. A225(1974) 16
(7) V.Vanzani, Proc.Ext.Sem.Nucl.Phys.,IAEA (Trieste,1973) Vol.II,287; Lett.Nuovo Cim. 10(1974) 610
(8) W.Tobocman, Phys.Rev. C11(1975) 43; C12 (1975) 741 and 1146
(9) D.J.Kouri and F.S.Levin, Phys.Lett. 50B(1974)421; Phys.Rev.A10 (1974) 1616; Nucl.Phys. A250 (1975) 127
(10) P.Grassberger and W.Sandhas, Z.Physik 217(1968) 9 and 220(1969) 29
(11) E.F.Redish, Nucl.Phys. A235(1974) 82
(12) E.O.Alt, P.Grassberger and W.Sandhas, Phys.Rev. C1(1970) 85
(13) M.Sawicki and J.M.Namyslowski, Preprint IFT/14/75 (Warsaw,1975)
(14) G.Cattapan and V.Vanzani, Lett.Nuovo Cim., in press
(15) I.Lovas, Ann.of Phys. 89(1975) 96
(16) G.Cattapan, G.Pisent and V.Vanzani, Z.Physik A274(1975) 139
(17) G.Cattapan, G.Pisent and V.Vanzani, Nucl.Phys. A241 (1975) 204
(18) L.R.Dodd and K.R.Greider, Phys. Rev. 146(1966) 675
(19) B.L.Gambhir and J.J.Griffin, Phys.Rev. C7(1973) 1006
(20) A.N.Mitra, Adv.Nucl.Phys. 3(1969) 1; D.P.Bouldin and F.S.Levin, Nucl.Phys.A 189 (1972) 449; V.Vanzani, Nuovo Cim. 16A(1973) 449
(21) V.M.Kolybasov, G.A.Leskin and I.S.Shapiro, Sov.Phys.- Usp.17(1974) 381
(22) E.I.Dolinskiĭ, L.D.Blokhintsev and A.M.Mukhamedzhanov, Nucl.Phys. 76(1966) 289
(23) R.Anni, L.Taffara and V.Vanzani, Nuovo Cim. 23A (1974) 431
(24) R.C.Johnson and P.C.Tandy, Nucl.Phys. A235(1974) 56

NUCLEAR REACTIONS AND SCATTERING
IN THE 3BODY AND EIKONAL FORMALISM

J. M. Namysłowski
Institute of Theoretical Physics, Warsaw University, Poland

Abstract
A unified scheme of evaluating transfer reactions and elastic scattering of a bound system on a target is proposed, particularly in the view of its applicability to the heavy-ion processes. We assume, that three bodies are involved in these processes, and use the exact definition of transition operators in the 3-body framework, an adiabatic approximation, and a modification of the eikonal method to evaluate a correction to the exactly solvable local potentials.

Transfer nuclear reactions and elastic scattering of a compound system on a target involve many bodies, but we restrict ourselves to three bodies and write

$$\begin{aligned} (c_1,t) &+ c_2 \longrightarrow c_1 + (c_2,t) \,, \\ (c_1,t) &+ c_2 \longrightarrow (c_1,t) + c_2 \,, \\ c_1 &+ (c_2,t) \longrightarrow c_1 + (c_2,t) \,. \qquad (1) \end{aligned}$$

All these processes we want to consider simultaneously in one scheme, assuming local 2-body interactions V_{12}, V_{t2}, and V_{1t}. Our main interest is in processes in which C_1, C_2, (C_1,t) and (C_2,t) are heavy ions. Therefore, we pay most attention to the interaction V_{12} which must have a local character.

To evaluate cross sections for processes (1) we can choose in practice among four doable schemes: i/ DWBA and optical potential, ii/ 3-body integral equations, with 2-body separable potentials, iii/ eikonal approaches, and iv/ semiclassical methods. We shall not discuss the fourth scheme, and among the first three we emphasize some advantages of a particular version of the eikonal method.

The DWBA and the 3-body integral equation methods are complementary in several respects. The first lacks its basic support, and even it is known to be a first term of a divergent series [1]. The second follows from the basic Faddeev equations, but in practice it is usually restricted to separable interactions, and very low partial waves. Numerically, the second method is quite involved, because of delicate procedures [2] required by singularities in the kernel of these equations. Several attempts were made to understand good results of DWBA in terms of the more basic 3-body approach[1,3,4,5,6]. Assuming separable interactions, acting like a projection operator, it is possible to reformulate DWBA for stripping reactions[3,5] in such a way that it becomes an exact scheme in the separable model. However, in details[5], it can not be concluded that DWBA is a good approximation, in spite of producing reasonable angular distributions in this simple model. For break up proceses, again assuming separable interactions, it was found[6], that below 100 MeV the DWIA yields a reasonable shape of differential cross section, but misses the magnitude by a factor 2, and the convergence of the multiple scattering series is very slow[1,6]. An unpleasant feature of these studies is, that in practice in the DWBA scheme all interactions are local, while the practical 3-body integral equation method heavily explores the mathematical simplification granted by the separable nature of interactions.

The eikonal approach[7] stays with the local interactions, and offers the possibility of summing an infinite series in interactions. Similarly to the DWBA scheme, the eikonal method gives directly the scattering amplitude, without any necessity of solving either an integral or differential equation. A very simplified application of the eikonal method was found[8] to work surprisingly well for scattering of heavy ions. However, the eikonal method is an approximation, known to work well for small scattering angles and large energies. Therefore our aim is to reformulate[9] the eikonal scheme in such a way that it gives a result as close as possible to the exact answer, and enables us to sum up in a simple way an infinite series in interactions.

We present our modification of the eikonal method first on the level of the 2-body subsystem. The scattering amplitude we evaluate by an explicit, numerical integration in the matrix element $\langle \phi | V | \psi \rangle$, with the full wave function ψ determined from a special 1-st order differential equation. Close to the actual potential V we find a potential V_{exact}, for which we know an explicit solution. For example, if V is a screened Coulomb potential, then V_{exact} is a point-like Coulomb interaction. If V is a superposition of a Woods-Saxon and a screened Coulomb potential, then V_{exact} is a superposition of an appropriate square well and Coulomb potential, joint at a given point. In such a case ψ_{exact} is obtained by a numerical adjustment of constants at the joining point. Having ψ_{exact} we define a 1-st order differential equation, corresponding to the proper Schrödinger equation, with such an additional term that Ψ_{exact} remains the exact solution also of the 1-st order differential equation. We write $\Psi_{ex} \equiv f_{ex} \exp(i\vec{k}\cdot\vec{r})$, and have the following Schrödinger and the 1-st order differential equations

$$\left(\tfrac{1}{2} m^{-1} \nabla^2 + i m^{-1} \vec{k}\cdot\nabla - V_{ex}\right) f_{ex} = 0 , \qquad (2)$$

$$\left(A + i m^{-1} \vec{k}\cdot\nabla - V_{ex}\right) f_{ex} = 0 , \qquad (3)$$

where A must insure the fulfilment of Eq.(3), with f_{ex} found from Eq.(2). For V close to V_{exact} we assume that an appropriate f is found from an equation like Eq.(3), with V_{exact} and f_{ex} replaced by V and f, respectively, but with the same A. Then, it is straightforward to get the full wave function for V as

$$\Psi = \Psi_{ex} \exp\left[-i m k^{-1} \int_{-\infty}^{z} (V - V_{ex})\, dz\right]. \qquad (4)$$

So, in our approach, the eikonal approximation is used only for the evaluation of a correction to the known, exact result. The main variation with angle is given by Ψ_{exact}, while a correction is generated by the small difference between V and V_{exact}. Some details of this method can be found in reference [9].

For three bodies C_1, C_2, and t we introduce the standard relative position and momentum variables, and let our bodies to interact via local potentials $V_{12}(\vec{R} - m_t(m_1+m_t)^{-1}\vec{r})$, $V_{t2}(\vec{R} + m_t(m_1+m_t)^{-1}\vec{r})$, and $V_{t1}(\vec{r})$. The total 3-body wave function, taken for example in the channel (1,t) 2, should obey the Schrödinger equation. However, we approximate this equation in the spirit of the adiabatic approximation[4], replacing $-\frac{1}{2}(m_1+m_t)(m_1 m_t)^{-1}\nabla_r^2 + V_{t1}(\vec{r})$ by -B, the binding energy of t in the (t,1) subsystem. Thus we have

$$\left[-\tfrac{1}{2}\mu^{-1}\nabla_R^2 + V_{12}(\vec{R} - m_t(m_1+m_t)^{-1}\vec{r}) + V_{t2}(\vec{R} + m_t(m_1+m_t)^{-1}\vec{r}) - B - E\right]\Psi_{(1,t)2} = 0, \qquad (5)$$

where $\mu = m_2(m_1+m_t)(m_1+m_2+m_t)^{-1}$, and the variable $\vec{r}$ is treated as a parameter. For the infinite separation of the bodies (1,t) and 2, i.e. $R \to \infty$, the dependence of $\Psi_{(1,t)2}$ on $\vec{r}$ is put as the bound state wave function of the (1,t) subsystem, while for finite values of R it is modified, because of the dependence of V_{12} and V_{t2} on $\vec{r}$.

The ratio $m_t(m_1+m_t)^{-1}$ is small in most cases of transfer nuclear reactions, particularly in the heavy-ion processes. Thus V_{12}, and V_{t2} depend mainly on $\vec{R}$, and we can consider an exact potential $V_{exact}(\vec{R}) \equiv V_{12}^{ex}(\vec{R}) + V_{t2}^{ex}(\vec{R})$, found from a sum of square wells and Coulomb potentials. We choose it close to the sum of the actual potentials appearing in Eq.(5). Similarly as in the 2-body case, we get Ψ_{ex}, which we write as $f_{ex} \exp(i\vec{Q}\cdot\vec{R})$, and define a 1-st order differential equation. By definition f_{ex} has to be its solution. We have

$$\left[i\mu^{-1}\vec{Q}\cdot\nabla_R - V_{ex}(\vec{R}) + D\right] f_{ex} = 0 . \qquad (6)$$

Then, for the sum of the actual potentials $V_{12}(\vec{R} - m_t(m_1+m_t)^{-1}\vec{r}) + V_{t2}(\vec{R} + m_t(m_1+m_t)^{-1}\vec{r})$ we assume a similar equation as Eq.(6), and solving it, we eliminate the above introduced function D. We get

$$\Psi_{(1,t)2}(\vec{R},\vec{r}) = \varphi_{(1,t)}(\vec{r})\, f_{ex}(\vec{R}) \exp\left[i\vec{p}_{lab}\cdot\vec{R}\left(1+\frac{m_2}{m_1+m_t}\right)^{-\frac{1}{2}}\left(1+\frac{m_2}{m_{(1,t)}}\right)^{\frac{1}{2}}\right] * \qquad (7)$$
$$* \exp\left\{-ip_{lab}^{-1} m_2 \left(1+\frac{m_2}{m_1+m_t}\right)^{-\frac{1}{2}}\left(1+\frac{m_2}{m_{(1,t)}}\right)^{-\frac{1}{2}} \int_{-\infty}^{z}\left[V_{12}(\vec{R}-m_t(m_1+m_t)^{-1}\vec{r}) + V_{t2}(\vec{R}+m_t(m_1+m_t)^{-1}\vec{r}) - V_{ex}(\vec{R})\right]dz\right\},$$

where p_{lab} is the laboratory momentum of the projectile (1,t) having mass $m_{(1,t)}$.

The scattering amplitude for the 1st process listed in Eq.(1) we get from the matrix element

$$T_{(1,t)+2\to(2,t)+1} = \left\langle \phi_{(2,t)1} \middle| V_{1t}(\vec{r}) + V_{12}(\vec{R} - m_t(m_1+m_t)^{-1}\vec{r}) \middle| \Psi_{(1,t)2} \right\rangle, \qquad (8)$$

where $\phi_{(2,t)1}$ is the bound state wave function for the (2,t) subsystem times a plane wave in the relative variables between (2,t) and 1. To evaluate the r.h.s. of Eq.(8) we must perform numerically a 6-th dimensional integration over $\vec{R}$ and $\vec{r}$, and one integration in Eq.(7). In comparison with the full DWBA matrix element, which includes recoil[10], we have one more integration. However, there are several advantages of our scheme: i/ there is no need of an optical potential, required by the DWBA scheme, ii/ all processes can be calculated in terms of the basic interactions V_{12}, V_{t2}, and V_{1t}, correlating transfer processes and elastic scattering in a unique scheme, iii/ the most important interaction V_{12} is included exactly, at least as far as the dependence on $\vec{R}$ is considered, iv/ the effect of the continuum states in the (1,t) subsystem is included in the adiabatic approximation, and v/ there is no need to solve any differential equation to evaluate $\Psi_{(1,t)2}$, while in the DWBA scheme one has to evaluate separately the initial and final distorted waves.

In comparison with the separable 3-body integral equation method we have the following advantages: i/ our potentials are local, as commonly used in the description of the nuclear reactions, ii/ there is no restriction on the number of partial waves, and iii/ there is no necessity of solving an integral equation with a singular kernel. However, there are also some drawbacks: i/ the potential V_{1t} and the motion in the (1,t) subsystem are only included approximately, and ii/ there is the necessity of making evaluation in the position space, and then evaluating 6-dimensional integrals to get to the momentum space.

An important advantage of our scheme is the lack of the partial waves, so if energy increases we are not troubled by a large number of partial waves, but to the contrary we get even better chances for our scheme to work well. To improve the eikonal method toward the low energies we apply Tikochinsky method [11] and work with an integral representation of a potential, for which we make the eikonal approximation. For some class of potentials this method enables us to move down the energy for which the eikonal approximation works well.

References

1 K.R. Greider and L.R. Dodd, Phys.Rev. 146 /1967/ 671, I.A. Sloan, Phys.Rev. 185 /1969/ 1361.

2 J.H. Hetherigton and L.H. Schick, Phys.Rev. 137 /1965/ 935B.

3 R.D. Amado, Phys.Rev. 132 /1963/ 485; A.N. Mitra, Phys.Rev. 139 /1965/ B1472; A.S. Reiner and A.I. Jaffe, Phys.Rev. 161 /1967/ 935; A. Aaron and P.E. Shauley, Phys.Rev. 142 /1966/ 608 and Ann. of Phys. 44 /1967/ 363; J.V. Noble, Phys.Rev. 157 /1967/ 939; A.I. Baz, Nucl. Phys. 51 /1964/ 145; T.G. Efimenko, B.N. Zakhariev and V.P. Zhuzunov, Ann. of Phys. 47/1968/ 275; E.O. Alt, P. Grassberger and W. Sandhas, Nucl.Phys. A139 /1969/ 209; D.P. Bouldin and F.S. Levin, Phys.Lett. 37B /1971/ 145 and Nucl. Phys. A189 /1972/ 449. V. Vanzani, Nuovo Cimento 16A /1973/ 449.

4 R.C. Johnson and P.R. Soper, Phys.Rev. CI /1970/; J.D. Harvey and R.C. Johnson, Phys.Rev. C3 /1971/ 636.

5 D.P. Bouldin and F.S. Levin, Phys.Lett. 42B /1972/ 167.

6. S.K. Young and E.F. Redish, Phys.Rev. C10 /1974/ 498.

7 G. Moliere, Z. für Naturforshung 2A /1947/ 133; J.R. Glauber in Lectures in Theoretical Physics, Boulder /1958/, Interscience Publishers, Inc., New York, London 1959, page 315.

8 Z. Kirzon and A. Dar, Phys.Lett. 37B /1971/ 166 and Nucl.Phys. A237 /1975/ 319.

9 E.A. Bartnik, Z.R. Iwiński and J.M. Namysłowski, Phys.Lett. 53A /1975/ 5 and Phys. Rev. A 12 /1975/ 1785.

10 K.S. Low and T. Tamura, Phys.Rev. C11 /1975/ 789.

11 Y. Tikochinsky, Phys.Lett. 29B /1969/ 270.

THREE-BODY APPROACH TO THE NUCLEON-NUCLEUS OPTICAL POTENTIAL*

P. C. Tandy[†], E. F. Redish, and D. Bollé[††]
University of Maryland, College Park, MD 20742 USA

In the Watson single scattering theory of the optical potential [1] it is customary to approximate the propagators by two-body Green functions in order to simplify calculations. The reaction mechanism being described, however, is decidedly three-body in character. The central difficulty in building three-body models for nucleon-nucleus elastic scattering is to find the proper way of imbedding the superposed three-body reaction mechanisms in the many-body problem without introducing serious overcounting effects. One would also like an explicit description of the intermediate state processes responsible for absorption.

We present here a three-body approximation to the optical potential theory which overcomes the overcounting problem and is capable of including the following effects: (1) the proper kinematics of the struck nucleon, (2) its binding potential, (3) the identity of target nucleons, and (4) realistic nuclear wave functions and spectroscopic factors. The three-body model for the optical potential can be extended using unitarity methods to yield a unified three-body-like model of elastic scattering, pickup, and single nucleon knockout.

We consider the scattering of a projectile from a target of A identical nucleons. The operator describing elastic scattering from the optical potential U(E) is given by the two-body Lippmann-Schwinger (L-S) equation

$$T(E) = U(E) + U(E)\, P_e\, G(E)\, T(E) \tag{1}$$

where $G(E) = (E + i\varepsilon - K_o - H_T)^{-1}$ is the propagator for a free projectile with kinetic energy K_o and H_T is the target Hamiltonian. The propagator P_e projects onto the target ground state and the subscript e labels the elastic channel. The non-elastic channels described by $Q_e = 1 - P_e$ are included formally in the optical potential, U(E), which has the multiple scattering expansion [1]

$$U(E) = \sum_{i=1}^{A} \tau_e^i + \sum_{i\neq j}^{A} \tau_e^i\, Q_e\, G\, \tau_e^j + \sum_{i\neq j\neq \varepsilon}^{A} \cdots \tag{2}$$

The interaction potential v_i between the projectile and the i-th target nucleon has been eliminated in favor of the scattering operator τ_e^i which is given by

$$\tau_e^i = v_i + v_i \frac{Q_e}{E + i\varepsilon - K_o - H_T}\, \tau_e^i\ . \tag{3}$$

The single scattering approximation truncates this series after the first term. With this Eq. (1) becomes

$$T = \sum_{i=1}^{A} \tau_e^i (1 + P_e\, G\, T) = \sum_{i=1}^{A} T^i\ . \tag{4}$$

Our three-body model is specified by ignoring all many-body intermediate states in (3) except the continuum single-particle excitations of the i-th target nucleon. Accordingly, following the methods used for the bound state, we approximate the target Hamiltonian in Eq. (3) by $H_T = K_i + K_{C_i} + H_{C_i} + u_i$. The parts of H_T are, respectively, the kinetic energy of the i-th nucleon, the center of mass kinetic energy of the residual nucleus, its internal Hamiltonian, and the single particle potential felt by the i-th nucleon. The residual interaction of the i-th nucleon gives rise to intermediate states of higher order and has been omitted. The free three-body Hamiltonian is defined as $H_o = K_o + K_i + K_{C_i} + H_{C_i}$. For convenience of expression we here suppress the internal states of H_{C_i}.

The sum of the three-body mechanisms is imbedded in the full (A+1)-body problem by summing over the struck nucleons and defining a new elastic transition operator [3]

$$T_e = \sum_{i=1}^{A} f_i \, P_i \, T^i \tag{5}$$

where T^i is defined in (4). The operator P_i is a permutation of the coordinates of the target nucleons which maps the i-th configuration into a reference configuration, and f_i is the sign of the permutation. The operator T_e has the same physical matrix elements as T, and maintains the proper symmetry in the three-body equations for the optical potential operators. The L-S equation for T_e is

$$T_e = A \tau_e (1 + P_e \, G \, T_e) \tag{6}$$

where

$$\tau_e = v + v \frac{Q_e}{E + i\varepsilon - H_o - u} \tau_e \, . \tag{7}$$

No label is necessary to identify the struck nucleon in Eq. (7).

The solution of (7) requires three-body methods. Expressing this equation in three-body form using standard operator manipulations yields [3]

$$\tau_e = t_p \, G_o \, \tau_p \tag{8}$$

$$\tau_p = G_o^{-1} + \{t_e \, G_o - G_o^{-1} \, P_e \, G\} \, \tau_e \tag{9}$$

where $G_o = (E + i\varepsilon - H_o)^{-1}$ and the two-body T matrices are given by

$$t_e = u + u \, G_o \, t_e \tag{10}$$

$$t_p = v + v \, G_o \, t_p \, . \tag{11}$$

The subscript p labels the pickup channel. Equations (8) and (9) represent a practical realization of the Watson single scattering optical potential operator τ_e in terms of three-body integral equations of the AGS form [4]. The kernel contains an overcounting correction, $G_o^{-1} P_e G$, which arises from the presence of the operator Q_e in Eq. (7) and results in a proper counting of the intermediate state P_e. Were this correction not present, the elastic intermediate state would be present in each of the A three-body problems and would therefore be severely overcounted.

Unitarity relations for τ_e can be readily obtained using standard three-body methods. Combining this with Eq. (6) allows us to extract the full unitarity relation satisfied by T_e, namely,

$$T_e - T_e^{\dagger} = T_e^{\dagger} \Lambda_e T_e + T_p^{\dagger} \Lambda_p T_p + T_o^{\dagger} \Lambda_o T_o \, . \tag{12}$$

The operators Λ_e, Λ_p, and Λ_o are projectors onto the on-shell states of the elastic, pickup, and knockout channels corresponding to states of the Hamiltonians $H + u$, $H_o + v$, and H_o, respectively. The projectile-nucleus operators T_p and T_o are the transition operators for the pickup and knockout reactions and are given by

$$T_p = \sqrt{A} \, \tau_p (1 + P_e \, G \, T_e) \tag{13}$$

$$T_o = (1 + t_p \, G_o) \, T_p \, . \tag{14}$$

Thus, in this three-body model, the use of the unitarity relation permits us to identify the reactive processes leading to the optical potential absorption. The relations (13) and (14) allow calculation of the amplitudes for these processes directly.

If the three-body equations for the optical potential operator (Eqs. (8) and (9)) are multiplied on the right by $\sqrt{A}$ $(1 + P_e \, G \, T_e)$ then using Eqs. (6) and (13) three-body-like equations are obtained for the transition operators directly, viz.

$$T_e = \sqrt{A}\, t_p\, G_o\, T_p \tag{15}$$

$$T_p = \sqrt{A}\, G_o^{-1} + \frac{1}{\sqrt{A}} \left(t_e\, G_o - (A-1)\, G_o^{-1}\, P_e\, G\right) T_e \, . \tag{16}$$

Equations (15) and (16) represent integral equations of a modified AGS form for nucleon-nucleus elastic scattering, pickup, and (via Eq. (14)) knockout amplitudes. The modifications include normalization factors ($\sqrt{A}$) and an overcounting correction (of the form $G_o^{-1}\, P_e\, G$) arising from the many-body nature of the true problem.

This model treats the identity of target nucleons correctly and when matrix elements of (15) and (16) are taken, realistic nuclear form factors may be introduced. Spectroscopic factors therefore enter in a natural way.

Beginning with the Watson single scattering term for the optical potential and construcing a three-body approximation for the reaction mechanism, we arrive at a unified three-body model of the elastic, pickup, and knockout processes. This model should be valuable whenever three-body reaction mechanisms are important and need to be summed to all orders. Some examples are: (1) the contribution of deuteron breakup to (d,p) and (p,t) reactions [5], (2) multistep breakup contributions to knockout reactions at energies below 100 MeV [6], and (3) the off-shell contribution of the effective interaction for optical potential calculations below 100 MeV [7].

References

[1] K. M. Watson, Phys. Rev. 89 (1953) 575; M. Goldberger and K. M. Watson, Collision Theory (Wiley, N.Y., 1964).

[2] A. K. Kerman, H. McManus, and R. M. Thaler, Ann. Phys. (N.Y.) 8 (1959) 551.

[3] P. C. Tandy, E. F. Redish, and D. Bollé, Phys. Rev. Lett. 35 (1975) 921 and in preparation.

[4] E. O. Alt, P. Grassberger, and W. Sandhas, Nucl. Phys. B2 (1967) 167.

[5] R. C. Johnson and P. J. R. Soper, Phys. Rev. C1 (1970) 976; R. C. Johnson and P. C. Tandy, Nucl. Phys. A235 (1974) 56; P. D. Kunz and L. D. Rickertsen, Bull. A.P.S. Ser. II, 20 (1975) 666.

[6] S. K. Young and E. F. Redish, Phys. Rev. C10 (1974) 498.

[7] G. M. Lerner and E. F. Redish, Nucl. Phys. A193 (1972) 565.

*Work supported in part by U. S. E. R. D. A.

†Present address: Department of Theoretical Physics, Research School of Physical Sciences, Australian National University, Canberra, Australia.

††Onderzoeker IIKW, Belgium. Work supported in part by NATO. On leave from Institute for Theoretical Physics, University of Leuven, Heverlee, Belgium. Present address: Department of Physics, University of Western Ontario, London, Ontario, Canada.

THREE-BODY K OPERATORS AND UNITARY APPROXIMATIONS: APPLICATION TO MODEL (d,p) AND (d,d) PROCESSES

F. S. Levin*
Physics Department
Brown University
Providence, Rhode Island 02912

Three-body systems have been widely used in studies of nuclear reactions, both as model problems from which insight might be gained concerning nuclear reaction mechanisms and as a tool for analyzing specific nuclear reaction data. The need for such studies is evident, since scattering theory can only be applied in approximate form to nuclear reactions, and neither the meaning nor the validity of the approximations so used are generally well understood. In the present work, we discuss a simple means for introducing unitary-type approximations for the three-body problem, compare results calculated using a form of this approximation with exact numerical solutions for (d,p) and (d,d) reactions in a 3-body model, and then indicate how the method can be generalized to introduce similar approximations for use in analyzing actual nuclear reaction data.

The unitary approximation method discussed herein was recently developed for the 3-body problem by Kouri, Levin and Sandhas [1]. It is based on the channel coupling array theory of many-body scattering [2], which we briefly summarize as a preliminary to introducing the approximation scheme.

Consider a system of n distinguishable particles which can be observed in various arrangement channels [3,4] labelled j, k, ℓ, etc. Corresponding to these channels are different partitions of the full n-body Hamiltonian H into a channel Hamiltonian H_ℓ and a channel perturbation V_ℓ: $H = H_\ell + V_\ell = H_m + V_m = \ldots$. H_ℓ describes internal (bound) and relative motion states of the fragments or clusters forming channel ℓ, and V_ℓ is the set of inter-fragment interactions. The eigenstates of the H_ℓ are the asymptotic states of the system. Associated with each H_ℓ are ingoing (-) and outgoing (+) wave Green's functions $G_\ell(\pm) \equiv (E \pm i0 - H_\ell)^{-1}$, where E is the total energy.

The main quantities of interest are the transition operators $T_{jk}(\pm)$. When k is a 2-body channel, on-shell matrix elements of $T_{jk}(+)$ are the usual amplitudes for transitions from a state in channel k to a state in an arbitrary m-body ($m \geq 2$) final channel j. When k is an m-body channel, m > 2, then on-shell matrix elements of $T_{jk}(+)$ must be replaced by on-shell matrix elements of the operator $U^{+}_{jk}(+) = \sum_n T_{jn}(+)G_n(+)G_k^{-1}(+)$ in order to obtain the transition amplitudes [5]. We return to this point below. The $T_{jk}(\pm)$ obey

$$T_{jk}(\pm) = V_j W_{\ell k} + V_j \sum_{n=1}^{N} W_{\ell n} G_n(\pm) T_{nk}(\pm) \quad , \qquad (1)$$

where the sum is over the channels of interest [5] and $W_{\ell k}$ is an element of the NxN channel coupling array W. The channel index ℓ in (1) may be chosen as is convenient, while W itself is to be selected so as to guarantee that (1) can be transformed by iteration to a connected kernel set of equations. Methods for choosing ℓ and then W have been discussed elsewhere [2]; an interesting recent development [6] is a choice of W for the case of arbitrary n that leads to the Bencze-Redish equations [7].

We now specifically consider the case n = 3, for which there are 3 classes of W that lead to connected kernel equations [1,2]. None of the general results derived below is dependent on a particular choice of W, and we therefore work with the general equation (1), which can be expressed in matrix form as

$$T(\pm) = \mathcal{V} + \mathcal{V}\, G(\pm)T(\pm) \quad , \tag{2}$$

where $\mathcal{V}_{jk} = V_j W_{\ell k}$ and $G_{jk}(\pm) = G_k(\pm)\delta_{jk}$.

To introduce unitary approximations, one must first show that (2) leads to the correct discontinuity relation for $DT \equiv T(+) - T(-)$, first derived by Lovelace [9]. Defining the discontinuity DG by $DG = G(+) - G(-)$, Eq. (2) leads to

$$DT = T(+)[DG]T(-) \quad . \tag{3}$$

For simplicity, we now assume (a) that the right hand subscript in DT_{jk} will always denote a 2-body channel and (b) that both sides of Eq. (3) will always be half-on-shell (to the right). Relaxation of (a), as discussed in [1], leads to no changes in our conclusions, but relaxation of (b) gives contributions which vanish on-shell.

There are contributions to $DG_\ell = -2\pi i\delta(E - H_\ell)$ from both 2-body bound states and breakup states. Denoting the breakup channel by the index 0, collecting all terms contributing to (3) from DG, using the relations $U^{+}_{j0}(+) = \sum_m T_{jm}(+)[1 + G_m(+)V^{(m)}]$ and $T_{0j}(\pm) = [1 + V^{(j)}G_j(\pm)]T_{jk}(\pm)$, where $V^{(m)}$ is the usual pair interaction in channel m ($V^{(0)} = 0$) and the latter relation is valid on shell, DT_{jk} of (3) becomes

$$DT_{jk} = -2\pi i \sum_{\ell=1}^{3} T_{j\ell}(+)\ \Delta_\ell(E)\ T_{\ell k}(-) - 2\pi i\ U^{+}_{j0}(+)\ \Delta_0(E)\ T_{0k}(-). \tag{4}$$

Eq. (4) is the proper discontinuity relation [9] with $\Delta_\ell(E) = P(\ell)\delta(E - H_\ell)$ and $\Delta_0(E) = \delta(E - H_0)$, where $P(\ell)$ is the projection operator onto the 2-body bound states in channel ℓ.

Having established that (2) obeys (4), the next step is to introduce K operators in analogy with the reaction operators of 1-body scattering theory. We wish to do this in such a way that the operators K_{jk} will have zero discontinuity [i.e., $DK = 0$], while their relation to the $T_{jk}(+)$ via a damping equation will contain the singularity structure. This is easily accomplished if it is noted that $G_\ell(\pm)$ is related to DG_ℓ by

$$G_\ell(\pm) = G^P_\ell \mp \frac{1}{2} DG_\ell \tag{5}$$

where $G^P_\ell = \mathcal{P}/(E - H_\ell)$, $\mathcal{P}$ denoting principal value. The matrix of K operators is now defined by

$$K = \mathcal{V} + \mathcal{V}\, G^P K \quad , \tag{6}$$

where $G^P_{jk} = G^P_k\ \delta_{jk}$. That is, K is defined in a manner analogous to the definition (2) of the $T(\pm)$, except that $G_\ell(\pm)$ of (2) is replaced by its "real part" G^P_ℓ. Because $DG^P = 0$ and $D\mathcal{V} = 0$ by construction, (6) implies $DK=0$, as required.

Equations (2) and (6) may be combined to yield the desired damping equation:

$$T(\pm) = K \pm \frac{1}{2} K\ DG\ T(\pm) \quad . \tag{7}$$

The final, key point in the analysis is now to recognize that because DK = 0, K can be eliminated from the (±) pair of equations (7): on doing so we regain Eq. (3). From (3) we obtain (4), the desired discontinuity relation. We may draw two conclusions from these results: (1) Eq. (7) guarantees the proper singularity structure for the operators $T_{jk}(\pm)$; (2) this singularity structure (discontinuity relation) is maintained if the exact K of (6) is replaced in (7) by an arbitrary K' as long as DK' = 0. This second conclusion immediately provides a basis for introducing unitary approximations, since (7) will always lead to (4) as long as the K appearing in (7) satisfies DK = 0, even though it may not satisfy (6).

The advantages of using unitary approximations are clear, although the accuracy of the resulting T_{jk} remains (in each case) to be established. In the present instance, the advantage of only having to solve the simple, non-singular equation (7) as compared to solving or attempting to solve (1) or (2) is, in our opinion, reflected in the non-necessity of having to solve a hierarchy of equations in order to introduce unitary approximations, as in the work of Kowalski [10] or Cahill [11]. This results from the fact that the driving term $\mathcal{V}$ in (2) and (6) has zero discontinuity. It is amusing to compare these results with their 1-body analogs. In the 1-body case (6) and (7) are replaced by linear (rather than matrix) equations. One can then "unitarize" the Born approximation by replacing the 1-body K in the analog of (7) by the Born term V, leading to the matrix element relation $T_\ell = V_\ell/(1 - iV_\ell)$, where ℓ here means orbital angular momentum. No such simple relation occurs for all the T_{jk} in the 3-body case, as we shall see.

Eqs. (6) and (7) have been used to approximate the (d,p) and (d,d) partial wave amplitudes for the Mitra 3-body model of stripping [12], and the results have been compared with the exact numerical results of Bouldin and Levin [13]. In this model, an infinitely heavy, structureless core interacts via separable S-wave potentials with a model neutron-proton pair initially bound via another S-wave separable potential to form a model deuteron with reduced mass M_P and binding energy 2.225 MeV. In these calculations, Yamaguchi form factors were used for the separable interactions, while the nucleon-core binding energy was taken to be 3 MeV. This model produces [13] exact (d,p) and (d,d) angular distributions having shapes and magnitudes characteristic of stripping and elastic scattering on light nuclei, eg., $d + O^{16}$. The model also exhibits some angular momentum localization, a feature characteristic of many-body DWBA calculations.

The structure of the equations to be solved depends on the choice of W. In the example considered herein, the 3x3 Faddeev-Lovelace choice [2] was used, as follows. The channel perturbation V_j may be expressed in terms of the pair interactions by $V_j = \sum_n \bar{\delta}_{jn} V^{(n)}$, where $\bar{\delta}_{jn} = 1 - \delta_{jn}$. Then $V_j W_{\ell m}$ is given by $\sum_n \bar{\delta}_{jn} V^{(n)} W_{\ell m}$, and for each n we choose $\ell = n$ and then set $W_{nm} = \delta_{mn}$, giving $\mathcal{V}_{jm} = V^{(m)}\bar{\delta}_{jm}$. Hence, (2) and (6) become

$$T_{jk}(\pm) = \bar{\delta}_{jk} V^{(k)} + \sum_m \bar{\delta}_{jm} V^{(m)} G_m(\pm)\, T_{mk}(\pm) \tag{8a}$$

and

$$K_{jk} = \bar{\delta}_{jk} V^{(k)} + \sum_m \bar{\delta}_{jm} V^{(m)} G_m^P K_{mk} \quad , \tag{8b}$$

where j, k and m run from 1 to 3.

The approximation employed consists first of retaining only the first term on the right-hand side of (8b):

$$K_{jk} \simeq \bar{\delta}_{jk} V^{(k)} \tag{8c}$$

Substitution of (8c) into (7) and comparison with (2) shows that the use of (8c) is the same as solving (2) with $G(\pm)$ replaced by its on-shell part $\pm$ 1/2 DG. The second aspect of the approximation is to neglect the contribution of the breakup cut in (7), when it is energetically allowed. Hence, (7) becomes a purely algebraic equation for the partial wave amplitudes, which are obtained by taking the relevant on-shell matrix elements of the $T_{jk}(+)$.

In the form of the Mitra model solved by Bouldin and Levin, it was assumed for simplicity that the neutron-core and proton-core interactions are the same. Hence, the (d,p) and (d,n) amplitudes are identical, and only 2 amplitudes characterize the reaction (breakup is ignored): $A_{dp}^{(\ell)}$ and $A_{dd}^{(\ell)}$, where the superscript refers to orbital angular momentum. Correspondingly, only 2 interaction matrix elements are important: $V_{np}^{(\ell)}$ and $V_{pc}^{(\ell)}$, where V_{np} and V_{pc} are the neutron-proton and the proton-core interactions. The $A_{\alpha\beta}^{(\ell)}$ and $V_{\alpha\beta}^{(\ell)}$ are obtained from Legendre polynomial expansions of on-shell, plane wave matrix elements of $T_{\alpha\beta}$ and $V_{\alpha\beta}$. Solution of the appropriate form of Eq. (7) yields

$$A_{dp}^{(\ell)} = V_{np}^{(\ell)}/D^{(\ell)} \tag{9a}$$

and

$$A_{dd}^{(\ell)} = -2\pi i \frac{M_p k_p}{\hbar^2} V_{pc}^{(\ell)} A_{dp}^{(\ell)} \quad , \tag{9b}$$

with

$$D^{\ell} = 1 + \left(\frac{2\pi M_p}{\hbar^2}\right)^2 k_p k_d V_{pc}^{(\ell)} V_{np}^{(\ell)} + \frac{\pi i M_p k_p}{\hbar^2} \bar{V}_{pc}^{(\ell)} \delta_{\ell 0} \quad , \tag{9c}$$

where k_α is the wave number of particle α and $\bar{V}_{pc}^{(\ell)}$ is a diagonal, deuteron to deuteron elastic scattering matrix element of V_{pc}.

From (9a) we see that $A_{dp}^{(\ell)}$ is proportional to matrix elements of V_{np} modified by the denominator D^{ℓ}, which at the higher energies acts like a damping or optical potential absorption factor. On the other hand, $A_{dd}^{(\ell)}$ is essentially proportional to on-shell matrix elements of $V_{np}G_p(V_{nc}+V_{pc})/D^{(\ell)}$, except that G_p is replaced by its on-shell part, $\frac{1}{2}$ DG_p. Here G_p is the Green's function for the channel in which p is free and n is bound to the core. Because G_p is replaced by $\frac{1}{2}$ DG_p, $V_{np}(\frac{1}{2}\, DG_p)(V_{nc} + V_{pc})$ cannot be transformed into a Born term plus a second term via Lippmann's identity [14], as in exact treatments of the channel coupling array theory [2]. Even if it could, however, it should be clear that neither such a transformed term nor the original term can provide a very accurate description of elastic scattering. We would expect many more terms or perhaps the inclusion of breakup effects to modify the results. Notice also that while $A_{dp}^{(\ell)}$ has the appearance of a unitarized Born approximation, $A_{dd}^{(\ell)}$ does not.

In Figure I, a diagrammatic representation of Eqs. (9a) and (9b) is given. In this figure the ovals represent schematically the effect of the denominator $D^{(\ell)}$.

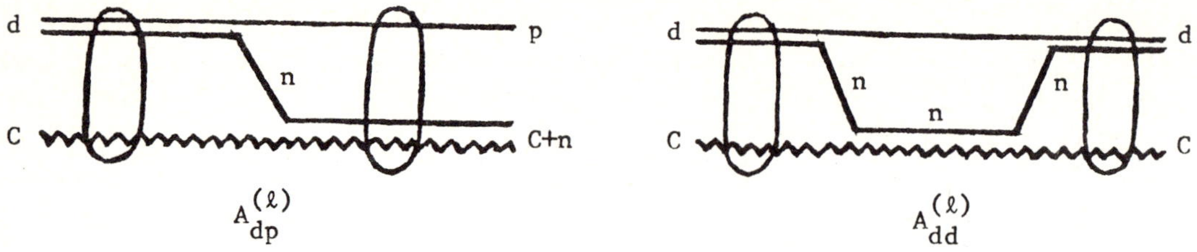

Figure I. Schematic representation of partial wave approximate (d,p) and (d,d) amplitudes.

It should be clear that $A_{dd}^{(\ell)}$ cannot be well represented by the simple on-shell process of two successive nucleon exchanges.

In Figures II and III are comparisons of the (d,p) and (d,d) angular distributions for incident energies of 6.7 and 15.12 MeV.

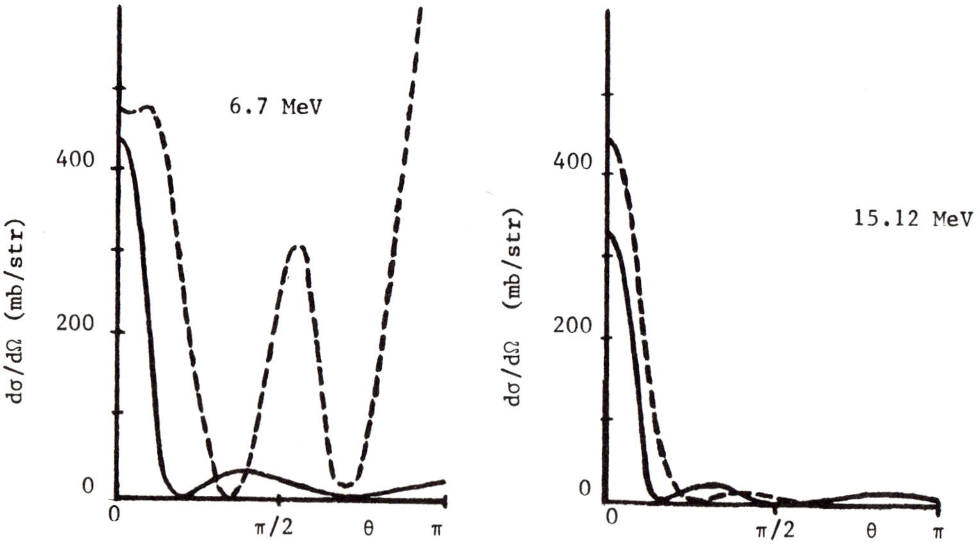

Figure II. Exact (-) and approximate (--) stripping cross sections.

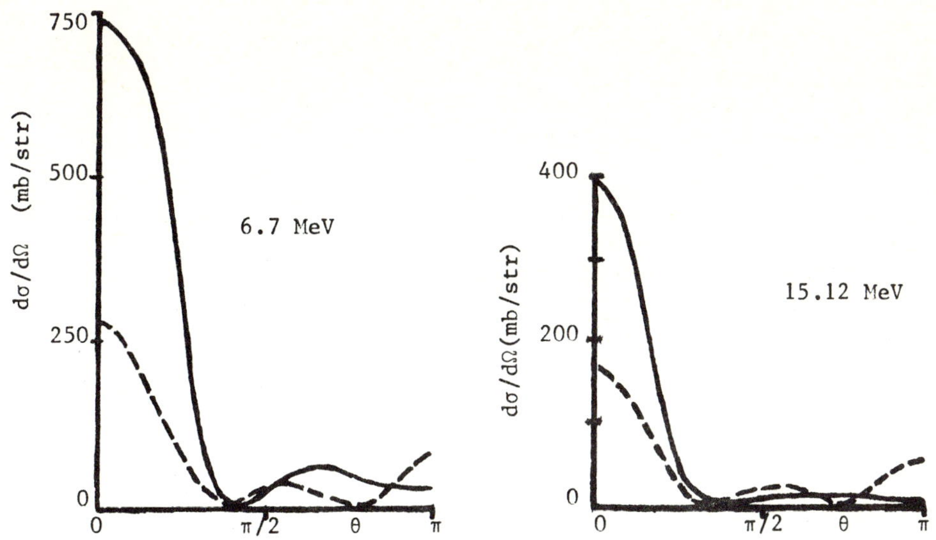

Figure III. Exact (-) and approximate (--) deuteron elastic scattering cross sections.

From these results, it is evident that at the higher energy of 15.12 MeV, A_{dp} provides a reasonable approximation to the exact stripping amplitude. A similar result holds for the 11.2 MeV case. The important role of the denominator $D^{(\ell)}$ as a damping factor cannot be stressed too strongly here, since it is known from the earlier calculations [13] that V_{np} by itself produces amplitudes that are much larger than the exact ones, just as in (d,p) calculations for actual nuclear stripping reactions. However, neither the low energy (d,p) results (1.78 MeV case as well as the 6.7 MeV case) nor any of the (d,d) results are in good agreement with the exact calculations of Bouldin and Levin. In view of the above remarks on A_{dd}, the (d,d) results are not surprising. The failure of the low energy (d,p) calculations is not understood. It is hoped that further calculations will provide an explanation for as well as improvements to the present angular distributions. There are two obvious means of proceeding. One is to use higher order terms from Eq. (6) in the approximation to K_{jk}; the other is to include the effects of the breakup cut in Eq. (7). Calculations involving these corrections are in progress.

As a final point, we examine the possible extension of this method to the case of actual nuclear rearrangement reactions such as one-nucleon stripping and pickup, multiparticle transfer, or heavy-ion reactions. In such cases, there seems to be no need to use an approximate amplitude which satisfies the many-body discontinuity relation, since nuclear reaction cross sections are generally much smaller than either elastic or total cross sections. We therefore can consider the generalization of Eqs. (2), (6) and (7) to the many-body case, but with truncations, so that only channels of (apparent) immediate physical interest are retained. Optical potentials and their concomitant wave functions can as usual be used to include effects of non-strongly coupled channels. As an example suppose channels j, ℓ and m are assumed to be coupled strongly to the initial channel k. Then Eqs. (2), (6) and (7) would be approximated by a set of equations for four operators each, and only channels j, k, ℓ and m would enter into the intermediate channel sums in the approximations to Eqs. (2), (6) and (7). Matrix elements would now be taken between distorted wave rather than plane wave states. Furthermore, if in addition to k, channels j, ℓ and m are 2-body channels, then (7) reduces to a set of algebraic equations for the partial wave matrix elements of the T operators expressed in terms of the corresponding partial wave matrix elements of the operator K. These latter, depending on the approximations used in Eq. (6), could also be determined algebraically. Notice that such a procedure would allow for (2-body) multistep corrections to DWBA to be included in an extremely simple fashion. Under such approximations, and because the matrix elements of T and K would be related by algebraic equations, the major computational problems would be in evaluating distorted wave matrix elements of the K operators. This should be no more difficult than currently experienced for DWBA or coupled channel calculations.

The many-body extension thus appears promising for three-reasons: First, it is an extension of a method which is moderately successful in calculating the higher energy (d,p) approximate cross sections. Second, optical model wave functions will occur in the many-body calculations and should play the same role (damping factors, etc.) as seen physically in the present model results and seen in reality in calculations based on other treatments of nuclear reactions. Finally, the method allows for multistep contributions to be approximated in an extremely simple way, i.e., through the solution of purely algebraic equations when the selected channels are of the 2-body type. It may be worth noting here that the method is in no way restricted to separable potentials, as in the model calculations.

References

[1] D. J. Kouri, F. S. Levin and W. Sandhas, submitted to Physical Review.
[2] D. J. Kouri and F. S. Levin, Bull. Am. Phys. Soc. 19 (1974) 489; Physics Letters 50B (1974) 421; Phys. Rev. A10 (1974) 1616; Nuclear Phys. A250 (1975) 127, and in press; Proc. of the VIth International Conference on Few-Body Problems (Laval University Press, Quebec, 1975), p. 47.
W. Tobocman, Phys. Rev. C9 (1974) 2466.
[3] H. Ekstein, Phys. Rev. 101 (1956) 880.
[4] R. G. Newton, Scattering Theory of Waves and Particles (McGraw-Hill, New York, 1966).
[5] The sum must include at least all of the 2-body channels if the correct discontinuity relation is to be satisfied, as shown by P. Benoist-Gueutal, Physics Letters 56B (1975) 413.
[6] F. S. Levin, unpublished; M. L'Huillier, unpublished (private communication from E. F. Redish).
[7] Gy. Bencze, Nuclear Phys. A210 (1973) 568; E. F. Redish, Nuclear Phys. A225 (1974) 16.
[8] E. O. Alt, P. Grassberger and W. Sandhas, Nuclear Phys. B2 (1967) 167.
[9] C. Lovelace, in Strong Interactions and High Energy Physics, ed. by R. G. Moorhouse (Oliver and Boyd, London, 1964), and Phys. Rev. 135 (1964) B1225.
[10] K. L. Kowalski, Phys. Rev. D5 (1972) 395; D6 (1972) 3705; D7 (1973) 1806.
[11] R. T. Cahill, Nuclear Phys. A194 (1972) 599. See also P. C. Tandy, Ph.D. thesis, The Flinders University of South Australia (1972, unpublished).
[12] A. N. Mitra, Phys. Rev. 139 (1965) B1472.
[13] D. P. Bouldin and F. S. Levin, Physics Letters 37B (1971) 145; 42B (1972) 167; Nuclear Physics A189 (1972) 449; Proc. of the Vth International Conference on Few-Body Problems (North-Holland -- American Elsevier, New York, 1973), p. 949.
[14] B. A. Lippmann, Phys. Rev. 102 (1956) 256.

* Work supported in part by the U. S. E.R.D.A.

A MODEL FOR DEUTERON STRIPPING AND BREAK-UP REACTION

SUPROKASH MUKHERJEE and SUBRATA RAY
Saha Institute of Nuclear Physics
Calcutta-700009, India

and

SANTOSH K. SAMADDAR
Chandernagore College
Hooghly, West Bengal
India

In the adiabatic theory of Johnson 1) et al one replaces $f(r_n, r_p)$ in deuteron stripping matrix element $\langle \chi_p^{(-)}(r_p) F_n(r_n) | v_{np}(r) | f(r_n, r_p) \rangle$ by the averaged function $f(R) = \xi(R) + B(R)$ where $\xi(R)$ and $B(R)$ are, respectively, the elastic scattering and averaged break up wave-functions of deuteron. With a minimal set of approximations we obtain 2)

$$B(R) = \int G_c(E_B; R, R') \left[\bar{V}(R') - V_{op}^{(d)}(R') \right] \xi(R') \, d^3R'$$

$$\bar{V}(R) = \frac{1}{g} \langle k=0 | v_{np} \tau | \chi_0 \rangle$$

$$V_{op}^{(d)}(R) = \langle \chi_0 | \tau | \chi_0 \rangle$$

$$g = \langle k=0 | v_{np} | \chi_0 \rangle$$

$G_c(E_B, R, R')$ is the Coulomb Green function with the energy parameter E_B; the integrations in equations above are over relative co-ordinates 1), and $|\chi_0\rangle$ is deuteron ground state. The operator τ can be expanded 3) in terms of the optical potentials of neutron, $V_{op}(r_n)$ and proton, $V_{op}(r_p)$, as $\tau = V_{op}(r_n) + V_{op}(r_p) + C$. The role of C in deuteron optical potential, $V_{op}^{(d)}(R)$, is discussed in ref.3. Now, C = 0 gives $\bar{V}(R) = V_J(R)$, the adiabatic potential of Johnson et al 1). Again, the equation for $B(R)$, is different from that of Johnson et al 1) and does not lead to a homogeneous equation for $f(R)$, as obtained by Johnson et al 1), who also approximated $E_B = E_d$, the energy of deuteron. The effect of the extra inhomogeneous terms was tested by applying our theory to (p,d) reaction on ^{208}Pb for 22 MeV proton, and it was found to be very large. It was also noted that the other two approximations of Johnson et al i.e. $\bar{V}(R) = V_J(R)$ or C = 0 and $E_B = E_d$ were poor. The form of τ of ref.3 was also found inadequate for this purpose. In absence of any direct method of computing C, we obtained $\bar{V}(R)$ in a parametric form with form factors like those of $V_{op}^{(d)}(R)$, by an automatic search code so as to obtain a minimum $\chi^2 \simeq 3$ for the angular distribution of (p,d) in ^{208}Pb from p1/2 state. With a $\bar{V}(R)$ thus obtained the angular distributions of stripping from other states of ^{208}Pb were calculated and the results are shown by the solid curves in the Fig.1 (lower part). The parameters of $\bar{V}(R)$ and other potentials are given in Table I. It is clear from the Table I that $\bar{V}$ is very different from $V_J(R)$. The solid curve for p1/2 distribution is for $\bar{V}(R)$ and the dotted curve gives what it is for $\bar{V}(R) = V_J(R)$. Our best fit gives

$E_B = \frac{1}{2} E_d$ approximately, and the fit worsens considerably if Johnson's value of $E_B = E_d$ is taken. Although $\overline{V}(R)$ depends on $V_{op}^{(d)}(R)$ taken, the goodness of fit* determines $\overline{V}(R)$ more or less uniquely. It is noted that $B(R)$ is out of phase with $\xi(R)$ in lower partial waves, a typical example is shown in the upper part of the figure, for $\ell = 2$. This results into a cancellation between DWBA matrix elements and those arising from break up terms and a natural suppression of the radial integrals of stripping matrix elements for lower partial waves and this suppression takes place although the radius of $\overline{V}(R)$ is less than that of deuteron optical potential. This is opposite of what adiabatic theory of ref.1 would predict.

Table I

Potential parameters, Depths in MeV, Lengths in Fermis

	V	V_J	$V_{op}^{(d)}$	$V_{op}^{(p)}$
V_o	79.2	102.0	97.5	51.8
r_o	1.093	1.25	1.16	1.25
a_o	0.86	0.68	0.80	0.65
W_o	14.0	13.5	15.8	10.0
r'_o	1.19	1.25	1.27	1.25
a'_o	1.22	0.78	0.77	0.76

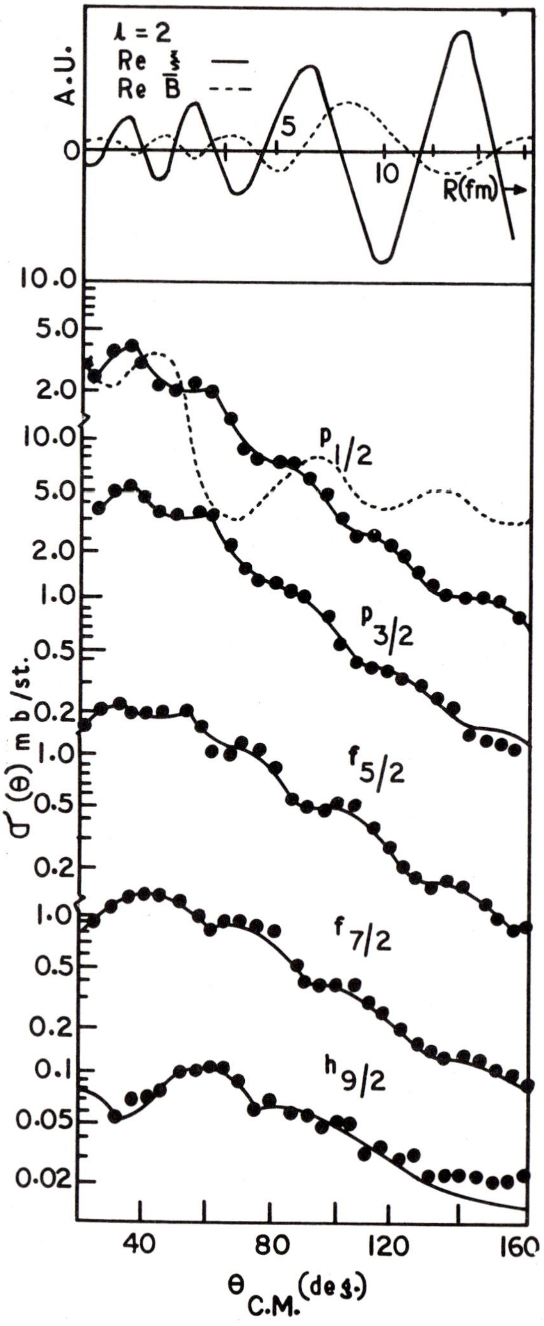

The amplitude for (d,pn) reaction is given by 4)

$$T(d,pn) = \langle \eta^{(-)}(\underset{\sim}{K}) | M(\underset{\sim}{k}, R) | \xi \rangle$$

where

$$M(\underset{\sim}{k}, R) = \langle \chi(\underset{\sim}{k}, r) | \tau | \chi_0(r) \rangle$$

is the matrix element of τ between ground state χ_0 and continuum state $\chi(\underset{\sim}{k}, r)$ of deuteron, and $\eta^{(-)}(\underset{\sim}{K}, R)$ is the Coulomb distorted wave of centre-of-mass motion, $\underset{\sim}{k}$ and $\underset{\sim}{K}$ are the wave vectors of the relative and centre-of-mass motion of outgoing neutron and proton. The approximate form of τ used in ref.3 does not produce the observed angular distribution of (d,pn) on ^{208}Pb at 12 MeV. The results for other forms of τ will be reported elsewhere.

1) R. C. Johnson and P. J. R. Soper, Phys.Rev. C1(1970)976
J. D. Harvey and R. C. Johnson, Phys.Rev. C3(1971)636
2) S. Ray, S. K. Samaddar and S. Mukherjee, Nucl. and Solid State Physics (India) 15B(1972)77, ibid 16B(1973)42
3) S. Mukherjee, Nucl.Phys.A118(1968)423
4) S. Mukherjee, S. Ray and S. K. Samaddar. To be published in Rept.Progr.Phys.

* The following Table shows the various sets of the phenomenological $V_{op}^{(d)}$ and the corresponding $\bar{V}$ obtained by minimising angular distribution of $p_{1/2}$ state of ^{208}Pb, the goodness of fit being given by χ^2_{min}. The depths are in Mev and ranges in fermis.

Set	V_o	r_o	a_o	W_o	r_o	a_o	χ^2_{min}
$V_{op}^{(d)}(R)$	97.5	1.16	0.80	15.8	1.27	0.77	4.19
$\bar{V}(R)$	79.2	1.093	0.86	14.0	1.19	1.22	
$V_{op}^{(d)}(R)$	79.8	1.34	0.796	16.6	1.47	0.598	56.0
$\bar{V}(R)$	45.5	1.307	0.333	23.2	1.379	0.756	
$V_{op}^{(d)}(R)$	101.5	1.11	0.92	11.8	1.39	0.76	28.6
$\bar{V}(R)$	94.4	1.03	1.469	18.2	1.406	0.67	
$V_{op}^{(d)}(R)$	81.5	1.32	0.764	15.2	1.409	0.562	42.1
$\bar{V}(R)$	82.9	1.442	0.798	12.0	1.440	0.37	

N-BODY INTEGRAL EQUATIONS AND ORTHOGONALITY SCATTERING

E.W. Schmid and H. Ziegelmann
Institut für Theoretische Physik der Universität Tübingen,
D-74 Tübingen, Germany

N-body integral equations furnish a rigorous basis in non-relativistic scattering. This feature may be welcome in formal applications. In practical applications, rigour is paid for by numerical complication and there is no hope to find a full solution when N is larger than three or four. Chances are better when we use some other method to determine gross features of the scattering process, insert this knowledge into the integral equation and use the latter only to determine some fine details. Or, when we study details of the N-body reaction while using only crude approximations for the interaction of the subsystems.

What are such crude features of scattering ? We may ask as well, what is the strongest part of the interaction between composite particles. The answer is funny: it is no interaction at all, it is the Pauli exclusion principle ! The next strongest interaction is the one which causes bound states, the third would be the one which causes resonances.

How can we insert our knowledge of the crude effects of the Pauli principle and of bound states into an N-body integral equation ? There are two ways.

1) Let

$$f = h + Kf \tag{1}$$

be an N-body integral equation in shorthand notation. And suppose that a second integral equation

$$f_c = h_c + K_c f_c \tag{2}$$

has nothing in it but crude features of the scattering process. Eq. (2) is meant to be simple enough to have a known resolvent $R_c = (1-K_c)^{-1}$. Following the old method of Schmidt [1] we subtract $K_c f$ from each side of Eq. (1) and multiply by R_c from the left. We get

$$f = R_c h + R_c (K-K_c) f, \tag{3}$$

which is an equation for the "detail" because the crude features are subtracted out.

2) When equations like Eq. (2) have been written down for the N-1, N-2, .. -body system, we can use their solutions as input into the N-body equation.

What can we take as Eq. (2) ? A resonating-group equation, written in the form (2) would already be much simpler than a rigorous integral equation and still be a good approximation. But resonating-group theory is too complicated, already, for our purpose. The crude effects mentioned above are strongly connected to the generalized Levinson theorem. The Levinson theorem follows from orthogonality constraints in the continuous spectrum, and therefore we suggest to take as Eq. (2) the equations for orthogonality scattering.

Orthogonality scattering has been studied by Saito [2] and many other people. Let's forget for the moment the direct interaction between composite particles and let us look at orthogonality scattering with respect to only one state $|u\rangle$. Saito writes down the equation

$$(E-PH_0P)\psi = 0 \ , \tag{4}$$

with the projection operator

$$P = 1 - |u\rangle\langle u| \ .$$

Demanding that $\langle u|\psi\rangle = 0$ he arrives at

$$(E-H_0)|\psi\rangle = -|u\rangle\langle u|H_0|\psi\rangle \ , \tag{5}$$

with $-|u\rangle\langle u|H_0$ as a separable rank-one potential which produces scattering functions which are orthogonal to $|u\rangle$. We want to approximate a kernel of resonating-group theory and are looking for a separable and symmetric orthogonality potential. We first notice that the Saito potential is not uniquely defined by the condition $\langle u|\psi\rangle = 0$. Take an arbitrary state $|w\rangle$, with $\langle u|w\rangle = N \neq 0$. Then

$$(E-H_0)|\psi\rangle = -|w\rangle \frac{1}{N}\langle u|H_0|\psi\rangle \tag{6}$$

also describes orthogonality scattering (multiply with $\langle u|$ from the left and get $E\langle u|\psi\rangle = 0$, as in case of Eq. (5)!). Choosing $|w\rangle = H_0|u\rangle$ we get a symmetric potential. Neither Eq. (6) nor (5) has a unique solution at $E = 0$. We can change that also by writing

$$(E-H_0)|\psi\rangle = |u\rangle(E+C)\langle u|\psi\rangle - H_0|u\rangle \frac{1}{N}\langle u|H_0|\psi\rangle \ . \tag{7}$$

Depending on what we choose for C we can have a negative-energy, zero-energy or positive-energy bound state, a redundant state or a unique scattering solution with no bound state at finite energies. This suggests to write down the most general separable potential with form factors $|u\rangle$ and $H_0|u\rangle$ and look for constraints on the coefficient matrix which make it a rank-one, symmetric, orthogonality-scattering potential. The resulting Schrödinger equation is

$$(E-H_0)|\psi\rangle = [C(E)-H_0]|u\rangle\,\lambda(E)\langle u|[C(E)-H_0]|\psi\rangle \ , \tag{8}$$

with

$$\lambda(E) = \langle u|[C(E)-H_0]|u\rangle^{-1} \tag{8a}$$

as the condition for orthogonality scattering. In order to get orthogonality scattering without the presence of a bound state at finite energies we choose $C(E)$ to be a hyperbola which either approaches E as $E \to -\infty$ and $\langle u|H_0|u\rangle$ as $E \to +\infty$, or $\langle u|H_0|u\rangle$ as $E \to -\infty$ and E as $E \to +\infty$. The value of $C(E)$ at $E = 0$ is a degree of freedom (in addition to the choice of $|u\rangle$) which allows us to approximate a given phase shift. With $C(E_0) = E_0$ we get a bound state $|u\rangle$ at the energy E_0. This energy may be negative, zero, or positive.

The Lippmann-Schwinger equation which corresponds to Eq. (8), with a suitable choice of $|u\rangle$ and $C(E)$ can now serve as part of our "crude" equation (2). There will be, of course, several physical two-body channels, and Eq. (8) will thus yield only a submatrix on the main diagonal of the full kernel K_c; channel-channel coupling is neglected in Eq. (2).

Orthogonality scattering is not only a good tool approximately to describe Pauli exclusion. It can also approximate the influence of a bound state on the scattering amplitude. After all, a bound state implies an orthogonality constraint in the scattering region. This sheds new light on the unitary pole approximation. Because, if we only demand orthogonality of the scattering solution to some bound state $|u\rangle$ with energy E_0, the potential of Eq. (8) with $C(E) \equiv E_0$ will be just the usual separable potential which produces the bound state $|u\rangle$. And the T-matrix resulting from Eq. (8) will be the usual bound-state-pole approximation. $C(E_0) = E_0$ instead of $C(E) \equiv E_0$ leaves us an extra degree of freedom (namely an energy dependent form factor) to approximate a given phase shift.

Generalizations are straightforward. Orthogonality constraints with respect to N states $|u_1\rangle$, $|u_2\rangle$ ... $|u_N\rangle$ lead to a separable rank-N orthogonality potential $\Sigma_i\ |v_i\rangle\langle v_i|$ with form factors $|v_i\rangle =$ $= \Sigma_j\ a_{ij}[C_j(E)-H_0]|u_i\rangle$. The C_j are functions of energy which are arbitrary, up to the condition that they should not lead to singular coefficients a_{ij}. The coefficients a_{ij} are determined by the orthogonality conditions.

We could include a direct part in the crude interaction of composite particles. Formally, this would only replace H_0 by H (= H_0 + direct interaction) in the discussion above. Numerically, it would be rather unpleasant. Also, we would no longer have the possibility of putting $h_c = h$ in Eq. (2) with the consequence of $R_{ch} = f_c$ in Eq. (3). As we will see in the example of (n-d) quartet scattering, our extra degree of freedom in the rank-one orthogonality-scattering potential allows us to approximate a given (non-resonant) phase shift rather well without a direct interaction.

Resonances are not represented by our crude Eq. (2). Only a few compound resonances will be present with zero width, as bound states of closed channels. In order adequately to include resonances we would have to increase the rank of our separable potential. This would lead us to something similar to the quasiparticle method of Weinberg [3]. Our kernel K_c is a chess board of kernels and we would be playing the Weinberg game in a slightly modified way and only in the subkernels along the diagonal. We do not want, however, to go that far. We only want to subtract out the crudest influences on scattering: Pauli repulsion and bound states.

As an example will show, the choice of our potential does not seem to be very "ideal" in the sense of Weinberg. Nevertheless, we are confident that our subtraction makes sense, because we are following physical arguments: an orthogonality-scattering potential yields a reasonable on-shell amplitude and it cannot be too wrong off-shell, since orthogonality scattering is known to produce reasonable wave functions in the reaction region. Therefore, the second term on the right-hand side of Eq. (3) cannot be too large in the neighborhood of the elastic region.

Let us consider n-d quartet scattering as an example. In this case we are able to solve Eq. (1) when a separable interaction of Yamaguchi type is assumed for the nucleon-nucleon interaction. We also know the integro-differential equation of resonating-group theory. In the so-called oscillator limit, this equation has a redundant state of gaussian shape and the scattering wave functions can be made orthogonal to this state. We pick this state, with a suitable width, as state $|u\rangle$ in the orthogonality-scattering potential of Eq. (8). With $|u\rangle$ fixed, we can alter the phase shift only by changing $C(E)$.

Fig. 1 shows several l = O phase shift curves. All of them result from the orthogonality constraint with respect to our given state $|u\rangle$. In all cases the function C(E) is a hyperbola, as mentioned above, but with different values for C(O). For C(O) $\to \infty$ we get the same phase shift curve as the one for the Saito potential. We notice that the orthogonality constraint leaves us pretty much freedom in the phase shift.

The oscillator limit of a resonating-group equation without distortions is a rather poor approximation when a loosely bound deuteron is involved. Therefore, our state $|u\rangle$ is not well determined. If we want to express a given phase shift by an orthogonality-scattering potential for the purpose of using the latter as interaction within a subsystem, we should feel free to adjust both $|u\rangle$ and C(E). In Fig. 2 we have tried to approximate the (real part of the) l = O quartet phase shift which has been calculated from the (n-d) integral equation. With $\langle\vec{p}|u\rangle = (1 + (p/\beta)^2 + \alpha(p/\beta)^4)^{-1}$, $\alpha = 0.0021$, $\beta = 0.36\ \mathrm{fm}^{-1}$ and C(O) = 544 MeV we get a reasonable fit, which could still be improved by a more careful search.

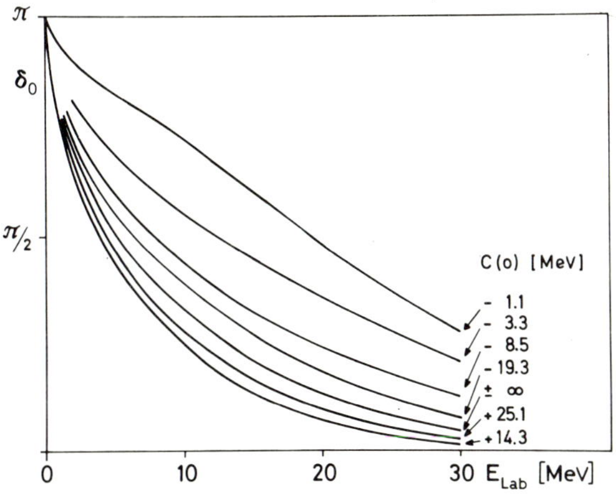

Fig. 1. Phase shifts of orthogonality scattering from a given state $|u\rangle$ with different values of C(O).

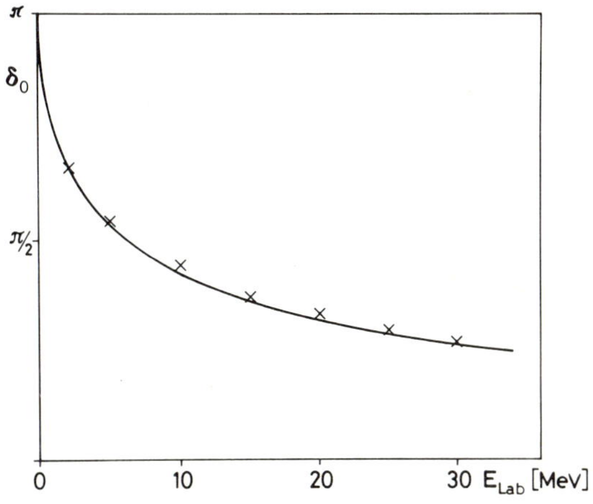

Fig. 2. Phase shifts of l = O (n-d) quartet scattering (crosses) fitted by an orthogonality-scattering potential.

It seems interesting to see how "ideal" our choice of an orthgonality-scattering potential is in the sense of Weinberg [3]. Alt, Sandhas and Strauss [4] have calculated the eigenvalues of the kernel K of (n-d) quartet scattering at the elastic threshold. They obtained a value of - 0.91515 for the eigenvalue largest in magnitude. Fig. 3 shows a trajectory of the eigenvalue which corresponds to the curve with C(O) = 14.3 MeV of Fig. 1. We see that our choice does not seem to be very "ideal". But, at least the sign of the eigenvalue at E = O is right. For comparison, we depict in Fig. 4 the trajectory of the Saito potential, with the same state $|u\rangle$ as in Fig. 3. It behaves in a qualitatively different manner (as a result of the presence of a zero-energy bound state). Following the usual definition, one might even say that the Saito potential is an attractive potential while the potential which produces the trajectory of Fig. 3 is a repulsive one. Thinking in terms of phase shifts, the classification

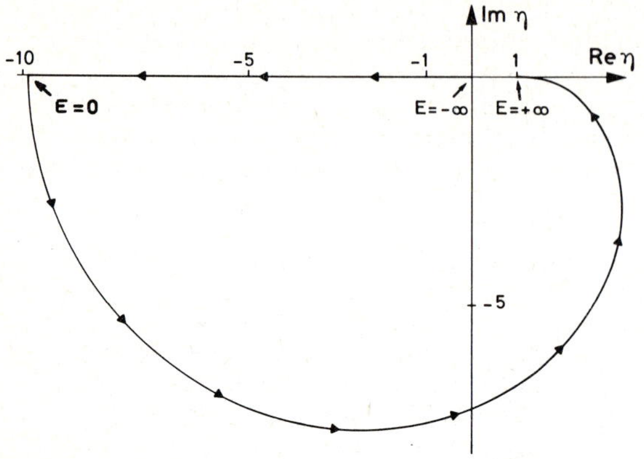

Fig. 3. Trajectory of the eigenvalue η(E) of the kernel K_C for the lowest phase shift curve in Fig. 1.

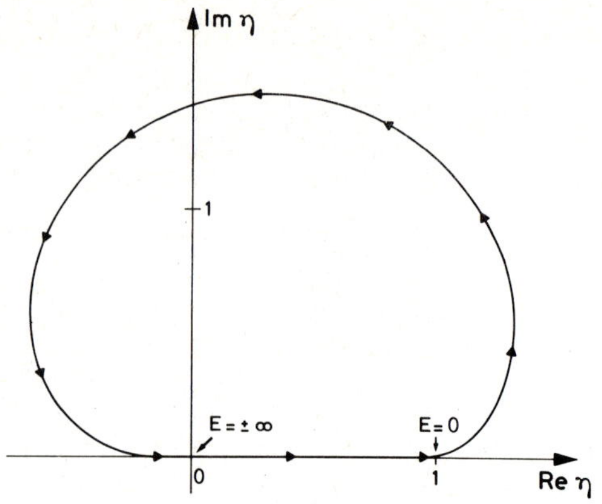

Fig. 4. Trajectory of the eigenvalue η(E) of the kernel K_C for the Saito potential; the state |u> is the same as in Fig.1 and 3.

of orthogonality-scattering potentials into attractive and repulsive ones is not very meaningful. We have seen that the bound state can be shifted with little or no influence on the phase shift. However, when such a potential is used in an N-body integral equation to describe Pauli repulsion within a subsystem, it may become quite important whether the potential produces a bound state or not.

Further investigations will be necessary to get more insight into the relationship between the special choice of an orthogonality-scattering potential and the properties of the "subtracted" N-body integral equation (3).

References

[1] E. Schmidt, Math. Ann. 63 (1907) 433.

[2] S. Saito, Progr. Theor. Phys. 40 (1968) 893; 41 (1969) 705.

M.J. Englefield and H.S.M. Shoukry, Progr. Theor. Phys. 52 (1974) 1554.

C.M. Shakin and M.S. Weiss, Phys. Rev. C11 (1975) 756.

[3] S. Weinberg, Phys. Rev. 131 (1963) 440.

[4] E.O. Alt, W. Sandhas, and W. Strauss, unpublished.

A MICROSCOPIC APPROXIMATION FOR SEQUENTIAL DECAY PROCESSES IN LIGHT NUCLEAR SYSTEMS

P. Heiss
Institut für Theoretische Physik, Universität Köln, West-Germany

It is well known that the calculation of three-body break-up reactions poses considerable difficulties. Only recently solutions of the Faddeev integral equations have been obtained for the three-nucleon system using realistic nucleon-nucleon potentials [1]. A corresponding integral-equation formalism can be written down also for systems of four and more particles [2]. However, it seems to be unrealistic to expect presently the actual solution of these equations for more than the simplest physical problems as e.g. the four-nucleon ground state problem.

The treatment of the few-nucleon problem by variational methods, as the cluster model formalism, has been very successful for bound states and two-fragment scattering cases. Nevertheless, these methods have not been applied to three-body break-up reactions, and probably this cannot be done in the near future.

However, it is well known from the three-nucleon problem that quasi two-particle methods as the quasi-free-scattering and the final state interaction (FSI) approximation give very satisfactory results for appropriate kinematical conditions. In the following I will show, how the FSI approximation can be incorporated in the cluster model reaction formalism.

For this purpose we consider a three-fragment break-up reaction which proceeds via sequential decay

$$A - B \rightarrow C - D \rightarrow C - E - F .$$

As an example we take

$$\alpha - d \rightarrow p - {}^{5}He \rightarrow p - \alpha - n .$$

Here we assume that the excited nucleus D decays to E and F only when its interaction with the nucleus C can be neglected. In this case the first step of the reaction namely A - B → C - D can be approximately described by using a quasibound wave function for the excited nucleus D. The quasibound function which vanishes asymptotically should be chosen to be a good approximation of the true scattering function of D inside the interaction region. Using this approximation one can apply the usual multichannel cluster model reaction theory [3] to describe the first step of the reaction.

For the description of the second step, namely the decay of D, we have to consider two points. Firstly we have to introduce the finite energy width of the excited nucleus D, which is neglected in the quasi-bound approximation. Secondly we have to treat the influence of a possible spatial orientation of the spin J_D of the particle D.

In the case of the ^{5}He nucleus we have a spin of 3/2; hence the spatial orientation of ^{5}He can be described by its tensor polarizations of ranks less or equal 3. Assuming a sequential decay mechanism one obtains with the help of some angular momentum algebra the following expression for the cross section of the break-up process $A - B \to C - D \to C - E - F$

$$\frac{d\sigma}{d\Omega_1 d\Omega_2 dE} = \rho(E) \sum G(\ell,\ell';J_D,s,q,\gamma) a_{\ell s}(E) a^*_{\ell' s}(E) t(q,\gamma,\Omega_1) Y^*_{q\gamma}(\Omega_2) \qquad (1)$$

The sum is over ℓ, ℓ', s, q, γ and $\rho(E)$ is a phase space factor. The $t(q,\gamma,\Omega_1)$ are the tensor polarizations of the intermediate nucleus D and depend only on the dynamics and geometry of the first step. The quantum numbers ℓ, ℓ', s, J_D correspond to the relative motion of the particles E and F and the spin and the total angular momentum in the subsystem D. The symbol G denotes a geometrical coefficient which does not depend on the dynamics of the system and the $a_{\ell s}$ are amplitudes corresponding to the decay of D in channels with E and F having different angular momentum quantum numbers ℓ, s, where of course always $\ell + s = J_D$. These amplitudes depend only on the properties of the second step of the process.

In order to take into account the finite energy width of the subsystem E - F we want to use a Watson-Migdal type approximation for the break-up matrix element. Such an approximation has already been indicated in the expression for the cross section (1). For this purpose we compare the exact form of the break-up T-matrix element in partial wave decomposition and the T-matrix element for the first step of the process in the quasi-bound approximation.

$$T \sim \langle \Phi(C)\phi^-(D)\chi_0(C-D)|V^{CD}|\Psi^+\rangle$$
$$T_q \sim \langle \Phi(C)\ \Phi(D)\chi_0(C-D)|V^{CD}|\Psi^+_q\rangle \qquad (2)$$

Here Ψ^+ is the total wave function of the system, V^{CD} is the total interaction minus the internal interaction of the subsystems C and D. χ_0 is the free relative motion function between C and D. $\Phi(C)$ is the internal wave function of particle C. The essential difference between these expressions consists in the replacement of the scattering function $\phi^-(D)$ for the subsystem D by the quasi-bound function $\Phi(D)$ in the lower expression. It can be shown that T can be approximated by restricting the integration to the interaction region. Now we exploit that in the vicinity of the resonance energy of the subsystem D and inside the interaction region the energy-dependence and the spatial dependence of $\phi^-(D)$ can be factorized in good approximation. By making the obvious assumption that $\phi^-(D)$ contains the most important energy dependence of T, we arrive at the approximation

$$T = a(\varepsilon)\ T_D \qquad (3)$$

Here $a(\varepsilon)$ is the energy dependent amplitude of $\phi^-(D)$ and T_D is the T-matrix element taken at the resonance energy of the subsystem D. The factor $a(\varepsilon)$ can be easily obtained by a scattering calculation for the subsystem D.

In the final step of our approximation we replace T_D by T_q. This should be reasonable if the properties of the quasi-bound function $\Phi(D)$ are taken into account. It is obvious that the main effort of our method lies in the calculation of the T-matrix elements T_q for the first step of the process.

Next I want to present some results for the ^{6}Be and ^{6}Li scattering systems. For the ^{6}Be system a multichannel calculation has been performed which includes the two-body fragmentation ^{3}He-^{3}He, and fragmentations with quasi-bound nuclei: α-(pp), ^{5}Li($3/2^-$)-p, ^{5}Li($1/2^-$)-p, ^{5}Li($3/2^+$)-p. We used nucleon-nucleon forces which have been described in Ref. [3]. The internal wave functions for the various fragments were obtained from a Ritz variational calculation as a superposition of Gaussians. The parameters of these functions are given in Refs. [4,5] . For these fragmentations we have calculated the S-matrix and observables for the elastic ^{3}He-^{3}He scattering and the reaction ^{3}He-^{3}He $\rightarrow$ ^{5}Li($3/2^-$)-p. The results for elastic ^{3}He-^{3}He scattering are in reasonable agreement with experimental data [5]. Cross sections and polarization values for the reaction ^{3}He-^{3}He $\rightarrow$ ^{5}Li-p are shown in Fig. 1 together with experimental points taken from Refs. [6,7]. There are some discrepancies which will be investigated in a further analysis including also the calculations of the second step of the process, namely the decay of the ^{5}Li system.

A similar calculation has been performed for the more complex ^{6}Li scattering system. Here I will show some results for the reaction α-d $\rightarrow$ ^{5}He-p $\rightarrow$ α-n-p. The S-matrix elements for the first step have been obtained from a multichannel calculation which included α-d, ^{3}H-^{3}He, ^{5}Li-n and ^{5}He-p fragmentations. This calculation has been performed for the energy region below the ^{3}H-^{3}He threshold. It has been shown in Ref. [8] that the results of this calculation for the elastic α-d scattering are in excellent agreement with experimental data for the differential cross section and for polarization quantities.

The experiments to which we compare our break-up results have been performed in Cologne with a 22.8 MeV α-particle beam [9]. The results are shown in Fig. 2. For these curves one common normalization factor has been used, as the experimental data were not given in absolute units. For the kinematical situations shown in Fig. 2 the two-step process under discussion is well isolated from other reaction mechanisms. As can be expected, the agreement is less satisfactory when these other reaction mechanisms give significant contributions.

In conclusion I want to point our that because of the comparatively large spins involved in the intermediate states this type of reactions seems to be suited to gain additional information about the noncentral parts of the nucleon-nucleon interaction.

The author thanks H.H. Hackenbroich for his helpful cooperation.

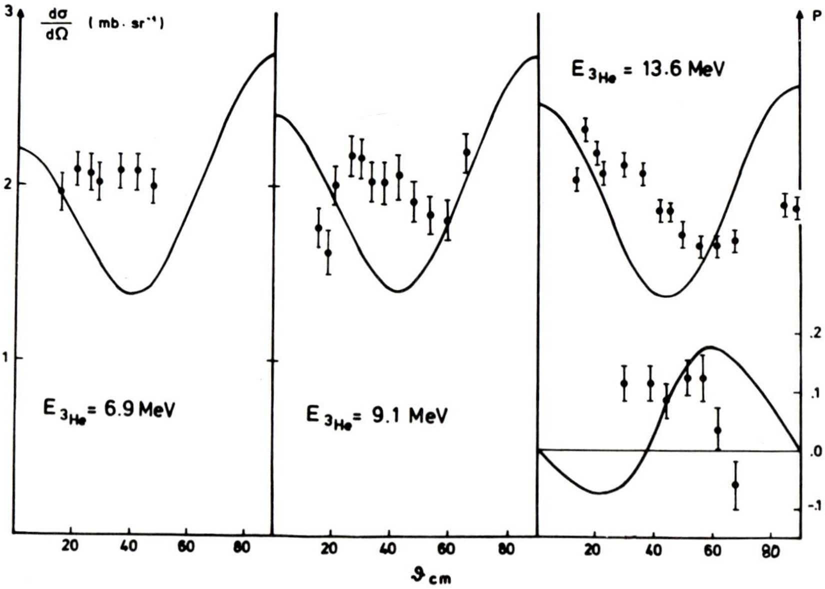

Fig. 1. Results for the ^{3}He-^{3}He → p-^{5}Li transition. Experimental cross-section values for 6.9 MeV and 9.1 MeV are taken from Ref. [6]. For 13.6 MeV the experimental cross sections (taken from Ref. [7]) have to be multiplied by a factor 3.

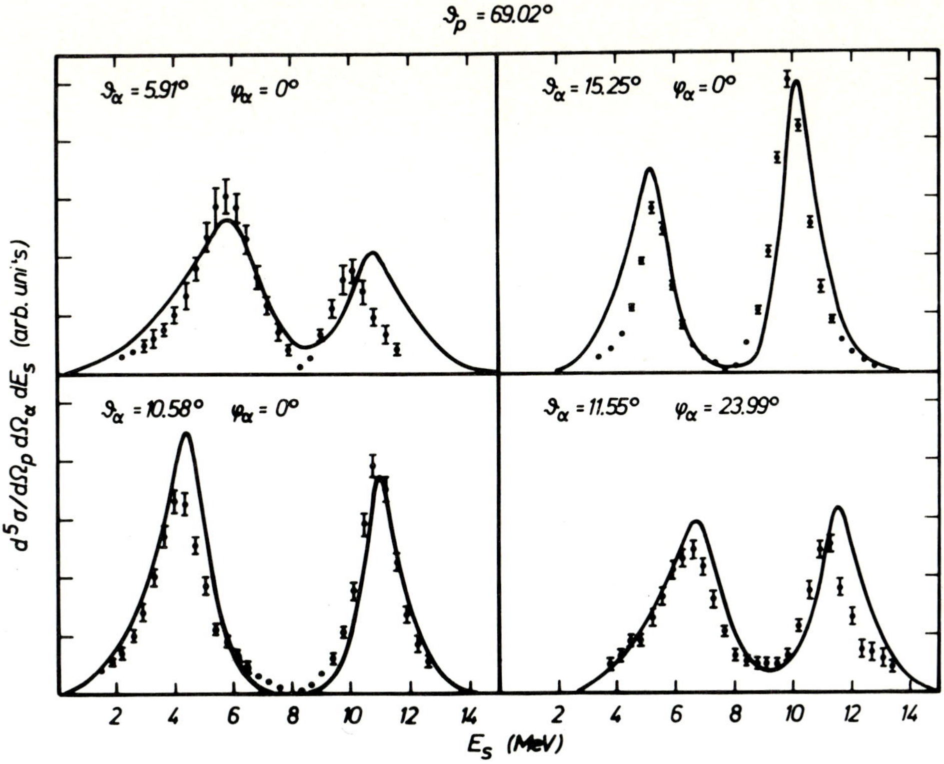

Fig. 2. Break-up cross sections for the α-d → ^{3}He-p → α-n-p reaction, projected onto the kinematical curve. E_α= 22.5 MeV. Angles are given in the lab. system. Experimental points are taken from Ref. [9].

References

[1] W.M. Kloet, J.A. Tjon: Nucl. Phys. A210, 380 (1973)
[2] W. Sandhas: "Exact N-Body Integral Equation", Symposium on Few-Particle Problems in Nuclear Physics, Univ. Tübingen 1975
[3] H.H. Hackenbroich, in: The Nuclear Many-Body Problem, Ed. F. Calogero, C. Ciofi Degli Atti, p. 706, Bologna
[4] P. Heiss; Z. Physik A272, 267 (1975)
[5] P. Heiss and H.H. Hackenbroich; submitted to Phys. Letters B
[6] M.L. Slobodrian et al.: Nucl. Phys. A194, 577 (1972)
[7] J. Asai and R.J. Slobodrian: IV. Int. Symposium on Polarization Phenomena, Zürich 1975
[8] H.H. Hackenbroich, P. Heiss and Le-Chi-Niem: Nucl. Phys. A221, 461 (1974)
[9] K. Prescher; to be published

ON THEORY OF DEUTERON STRIPPING REACTION

S.N. Mukherjee
Physics Department, Banaras Hindu University
Varanasi-221005, India

I. INTRODUCTION

Distorted wave Born approximation (DWBA) theory with a skilful choice of optical potential parameters has provided good fits to deuteron scattering and deuteron stripping data. However, efforts are still made to develop an improved theory which would not only provide more reliable analysis of experiment, but also afford a more significant test of the mathematical approximations on which DWBA is based.

Several attempts have recently been made to include the effect of deuteron break up on elastic scattering of deuterons from nuclei. Most of them do not attempt to include full three-body treatment, but rather try to stay closer to two body optical model formalism. Particularly the work of Johnson and Soper (1) on deuteron break up effects has raised a good deal of interest.

The effect of stripping channel on deuteron elastic scattering from nuclei has been investigated by Rawitscher and Mukherjee (2) by introducing a set of coupled equations which describes the dissociation (and recombination) of incident deuteron into (and from) the stripping channel. Coupling of the reaction channels mentioned above to the incident deuteron channel provides accurate deuteron scattering wavefunction and hence improves DWBA predictions of stripping and pick up reactions.

II. COUPLED CHANNEL ANALYSIS OF ^{52}Cr (d,p) REACTION

Here we will analyse deuteron elastic scattering and deuteron stripping data for ^{52}Cr target employing a coupled channel (C.Ch.) method in which $\Delta\ell=1$ stripping channel has been coupled to the incident deuteron channel. The incident deuteron energy ranges from 7 to 11 MeV. The results of C.Ch. calculations are compared with DWBA and experimental data of Kocher and Haeberli (3) for 10 MeV incident deuteron energy in Fig.1. It is evident that the predictions of both C.Ch. theory (CC1) and DWBA for $p_{3/2}$ and $p_{1/2}$ stripping transitions are larger than the corresponding experimental values. One minimum around $\theta=105°$ present in the(d,p) transition to $p_{3/2}$ state, which does not appear in DWBA calculation is reproduced by C.Ch. theory.

The C.Ch. deuteron potential parameters which reproduce d-^{52}Cr elastic scattering data for coupling constant N=2 and incident deuteron energy in the range 7-11 MeV are given in Table I.

Table I. Energy independent deuteron potential parameters used in C.Ch. calculation

Nucleus	Real			Imaginary			Spin Orbit		
	V	r_V	a_V	W_{Vol}	r_W	a_W	V_{SO}	r_{SO}	a_{SO}
^{52}Cr	91.0	0.915	0.830	22	1.4	0.90	7.5	0.85	0.60
^{48}Ca	120.7	0.906	0.846	60	1.4	0.90	7.5	0.85	0.60

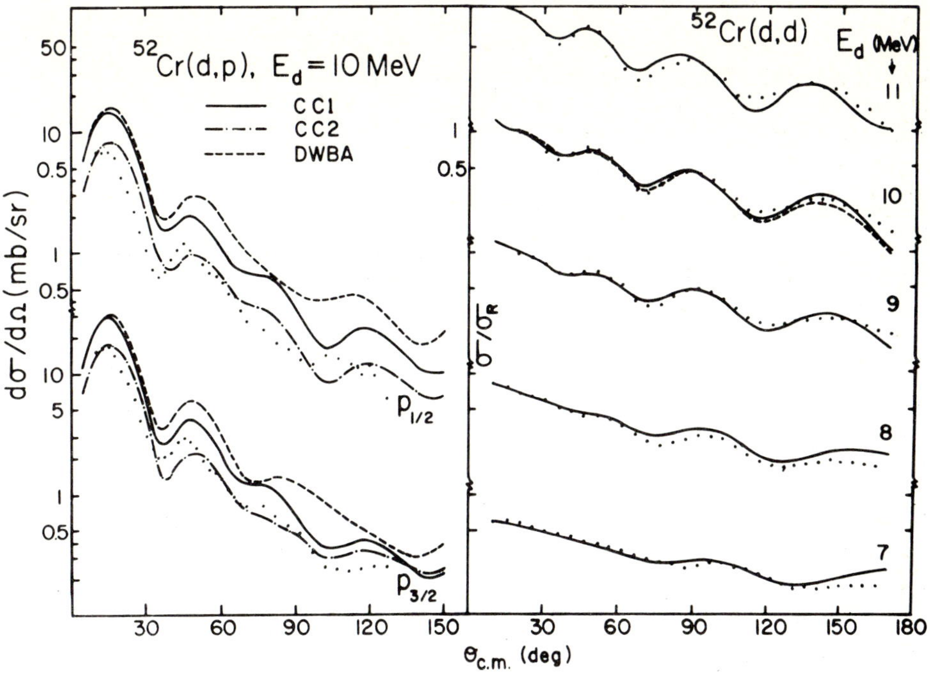

Fig.1. Angular distributions of (d,d) and (d,p) cross-sections for ^{52}Cr target.

The proton optical potential parameters used here are taken from the detailed analysis of p-^{52}Cr elastic scattering data by Becchetti and Greenlees(4) and the neutron potential parameters are obtained by fitting the last neutron finding energy. It may be remarked that the C.Ch. calculation of two ℓ=1 stripping angular distribution can be made comparable with experiment by decreasing the neutron radius from 1.25 fm to 1.04 fm (CC2). The depth of the imaginary (volume type) part of the deuteron optical potential used in the analysis of (d,d) and (d,p) data for ^{52}Cr target is found to be nearly one third of that used recently by Mukherjee and Shyam(5) for similar studies in ^{48}Ca target. One of the reasons may be that in chromium the stripping transitions are not as strong as they are in the case of calcium target.

III. ENERGY DEPENDENCE OF DEUTERON OPTICAL POTENTIAL

In this section we estimate the stripping channel contribution to the nonlocality of deuteron optical potential. The C.Ch.treatment described previously is used for this purpose. The case we are considering is one for which $\Delta\ell$=1 stripping is the most probable. ^{48}Ca is chosen as the target. The incident deuteron energy varies from 4 to 16 MeV. The procedure of calculation is as follows:

The C.Ch.predicted deuteron elastic scattering angular distribution at each incident deuteron energy is fitted by a conventional d-^{48}Ca optical model potential. The energy independent d-^{48}Ca potential-parameters used in C.Ch. calculation are shown in Table I. Starting with the same parameter set for each incident deuteron energy preliminary searches show that r_V and a_W are fairly energy independent.

Next four parameter searches on V, a_V, W and r_W were carried out and a_V is found to be weakly energy independent. Employing r_V=1.0fm, a_W=0.5fm, a_V=0.97fm, three parameter searches on V, W and r_W are performed and the final values are plotted in Fig.2 against incident deuteron energy. It is seen from Fig.2 that the effect of coupling ℓ=1 stripping channel is to produce a nonlocality in the d-^{48}Ca optical potential which gives the right type of energy dependence for the equivalent local optical potential. The energy dependence of the real well depth from present analysis is approximated by a straight line of gradient -0.16. Rawitscher's(6) analysis of d-^{40}Ca optical potential shows much weaker energy dependence.

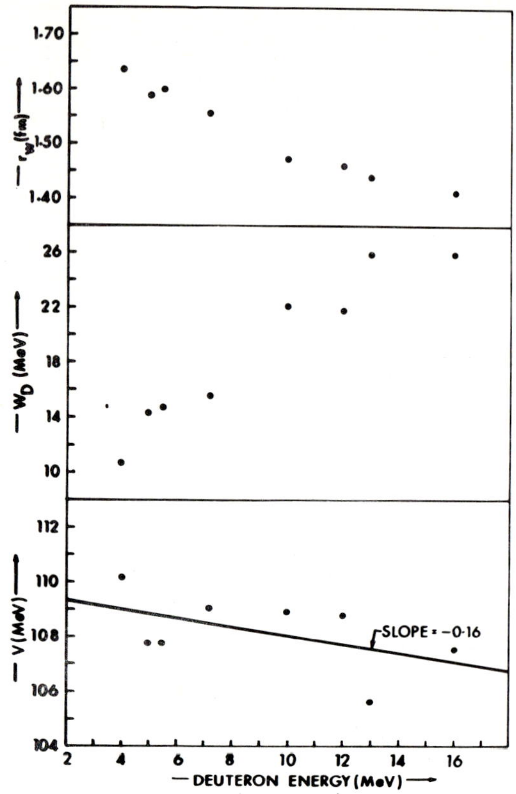

Fig.2. Energy dependence of d-^{48}Ca optical potential parameters.

The phenomenological deuteron optical potential has a linear energy dependence of slope -0.5 for its real well depth. This indicates that the coupling of the p-stripping channel to the elastic channel is responsible for approximately 30% of the observed energy dependence of the real well depth in the case of ^{48}Ca. The remaining energy dependence can be attributed to the deuteron break up channel, other non-elastic channels and intrinsic energy dependence of the two-body effective interaction. The radial variation of the absorbing potential at different energies in the tail portion suggests that the stripping probability is nearly the same for all energies in this region but increases with increasing penetration at higher energies. [Fig 3]

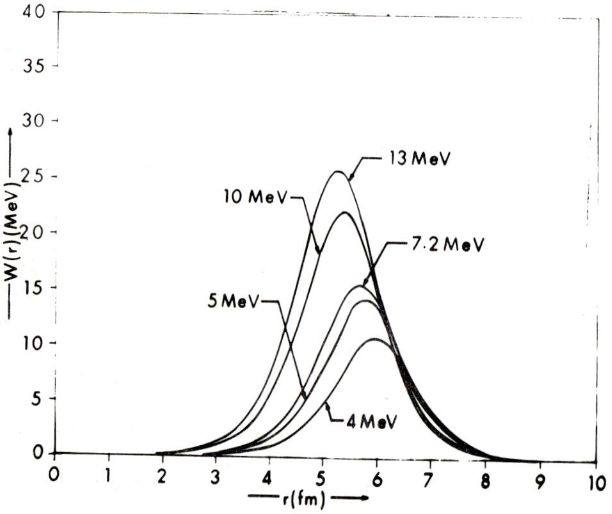

Fig.3. Radial variation of absorbing potential.

IV. DEUTERON STRIPPING TO CONTINUUM

Recent experiment by Fuchs et al. (7) on deuteron stripping(d,p) to continuum shows the following two features:

1. In general, there is a surprising parallelism in the shape between (d,p) and (n,n) cross section for the same target.

2. High 'ℓ' values are enhanced in stripping as compared to neutron elastic scattering.

Most of the theories proposed so far for (d,p) reaction to continuum are the extension of DWBA theory (8). In the present text we will use DWBA theory in its usual form under certain simple assumptions, to explain 'parallelism' and 'ℓ-effect'. If one assumes that the contribution from nuclear interior is negligible and unscattered part of the neutron wavefunction is also relatively weak near a isolated resonance, then one can write cross section for the A(d,p)B* reaction for particular (ℓj) as

$$\frac{d^2\sigma}{d\Omega_p dE_p} = \sin^2\delta_{\ell j} \left(\frac{d^2\sigma}{d\Omega_p dE_p}\right)_{s.p.} \quad .. (1),$$ when the neutron is not observed.

$\left(\frac{d^2\sigma}{d\Omega_p dE_p}\right)_{s.p.}$ is the energy differential cross section that would be calculated by the usual DWBA formula replacing bound neutron wavefunction by a scattering wavefunction as follows: A real Woods-Saxon potential well is adjusted in depth so that it produces an (ℓj) orbit resonating (i.e., having phase shift $\pi/2$) at the proper neutron energy.

Again, we write for spin zero target the total neutron scattering cross section near a resonance as

$$\sigma_{tot}^{(n,n)} \simeq \frac{2\pi}{q_n^2}(2j+1)\sin^2\delta_{\ell j} \quad (2)$$

where j and q_n denote respectively the spin of the resonance and neutron momentum. From the above two expressions one can obtain the ratio

$\frac{d^2\sigma}{d\Omega_p dE_p} / \sigma_{tot}$ (n,n) which

is generally referred to as 'stripping enhancement factor' in the literature. Fig.4 shows the calculated values for stripping enhancement factor for ^{32}S(d,p) reaction to continuum at an incident deuteron energy of 12 MeV for different resonance angular momentum ' ' The experimental values Bommer et al.(9) are shown with

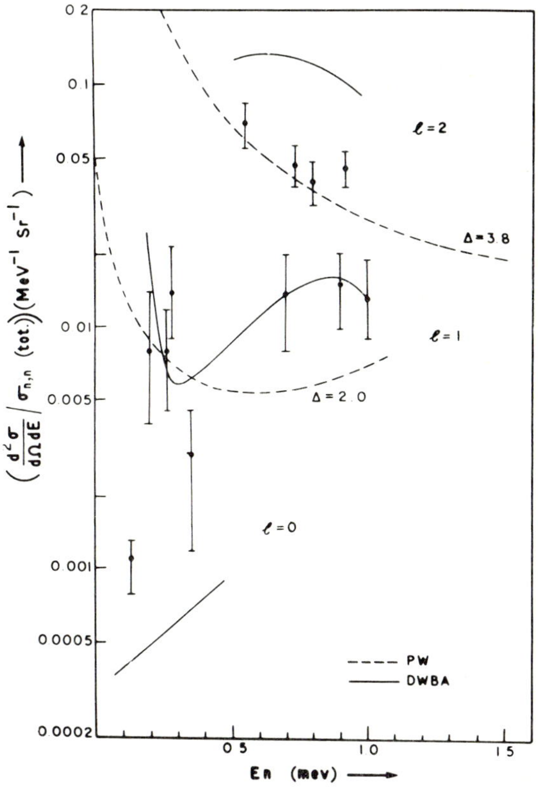

Fig.4. Stripping enhancement factor.

error bars. The curve with solid line represents DWBA calculation using resonating wavefunction. The dotted curve represents plane wave calculation using Butler cut-off radius $R_0=(1.4\ A^{1/3}+\Delta)$fm as suggested by Baur and Trautmann (10).

The problem of slow convergence of the radial integral is dealt in DWBA calculation by the method suggested by Berggren (11) in which the integral can be defined as

$$\lim_{\alpha \to 0} \int_0^\infty [\text{Integrand (r) dr}]\, e^{-\alpha r^2}$$

In the numerical calculation of energy differential cross section one computes with several values of α to extrapolate the integrated value to $\alpha=0$. The stability of the computed values against change of upper limit of radial integration has been ensured. In fact this consideration limit one in taking α arbitrarily small as ideally should be the case. Typical values of α used in the calculation are within 0.001 to 0.005. The small variation of computed energy differential cross sections against α (in the range used) encourages one to claim that the extrapolated value must be very close to mathematically defined limiting value. Thus the DWBA method using resonating wavefunction is quite successful in reproducing the observed ℓ-enhancement in stripping to unbound states (12).

In addition, extension of DWBA theory for stripping to unbound states, can supplement the existing methods of spin-parity assignment to resonances.

To conclude we believe that the theory of deuteron stripping reaction is in an interesting state. It is also believed that a three body model of deuteron stripping and break-up will describe the deuteron nucleus system more accurately(13). The work presented here is mostly done in collaboration with R.Shyam, D.C.Agrawal, S.Pal, D.K.Srivastava and N.K.Ganguly. The author is grateful to Mr. Tara Pada Bhattacharya for inspiration.

REFERENCES

1. R.C. Johnson and P.J.R. Soper, Phys. Rev. C1 (1970) 976.
2. G.H.Rawitscher and S.N.Mukherjee, Annals of Phys. 68 (1971) 57.
3. D.C.Kocher and W.Haeberli, Nucl.Phys. A196 (1972) 225.
4. F.D.Becchetti and G.W.Greenlees, Phys.Rev. 182 (1969) 1190.
5. S.Mukherjee and R.Shyam, Phys.Rev. C11 (1975) 476.
6. G.H.Rawitscher, Phys.Rev.Lett. 20 (1968) 673.
7. H.Fuchs, H.Homeyer, Th.Lorenz and H.Oeschler, Phys.Lett. 37B (1971) 285, Nucl.Phys. A196 (1972) 286, Phys.Lett. 52B (1974) 421.
8. F.S.Levin, Annals of Phys. 46 (1968) 41, J.Bang and J.Zimanyi, Nucl.Phys. A139 (1969) 534, C.M.Vincent and H.T.Fortune, Phys.Rev. C2 (1970) 782, Phys. Rev. C8 (1973) 1084, R.Huby and D.Kelvin, J.Phys. G1 (1975) 203.
9. J.Bommer, Inaugural Ph.D. Dissertation, Free University of Berlin (1974) unpublished.
10. G.Baur and D.Trautmann, Z.Phys. 267 (1974) 103.
11. T.Berggren, Nucl.Phys. A109 (1960) 265.
12. S.N.Mukherjee, R.Shyam, S.Pal and N.K.Ganguly (To be published).
13. J.P.Farrell, Jr., C.M.Vincent and N.Austern (Preprint).

SOME ASPECTS OF STUDIES OF INTERACTIONS BETWEEN TWO LIGHT NUCLEAR SYSTEMS

S. Ali, Atomic Energy Centre, Dacca, Bangladesh.

Few-nucleon scattering data have been useful in deriving interactions between light nuclear systems. In such derivations, one is guided by a faith in the non-relativistic approach and in the concept of potentials between systems. Although models have been proposed which are devoid of Hamiltonians and can describe reasonably well the scattering of light nuclear systems, we shall nevertheless assume the validity of a two-body potential description of particles and systems of particles. The purpose of this talk is to review some of the approaches made so far in understanding the features of interactions V_{AB} between two light nuclear systems A and B. (e.g., N-α, α-α, α-nucleus, ^{16}O-^{16}O etc.).

First comes the phenomenological approach in which one collects the available experimental data on angular distributions, cross-sections, polarization etc. and constructs a potential through the intermediacy of the phase shifts. While such phenomenological potentials are useful in many calculations, they have been subject to criticism. Firstly there is the question of arbitrariness - every author has his own potential. Secondly these phase-equivalent potentials sometime give rise to the well-known problem of spurious states and one needs special care in projecting out these states. Nevertheless, the phenomenological potentials still remain the most practical to deal with and when carefully constructed, they could be made to incorporate certain physical facts in order to fix more or less uniquely certain parts of the interaction potential. For example, in the case of N-N interaction, the coupling constant of OPEP had been taken long ago from phase shifts for very high ℓ-values using essentially a 'Centrifugal Barrier' philosophy which states that as ℓ increases, the phase shifts approach the values for OPEP. Such an idea was also used in the case of the α-α interaction[1)] and it was gratifying to note that the phenomenological α-α potential obtained this way was quite in accord with the one obtained from basic studies. Thus phenomenological potentials can indeed be quite suitably constructed both in the case of two light bodies (e.g. the Reid soft core potential for the N-N interaction) as well as in the case of two 'heavies'. Of the various types of phenomenological interactions, the optical model potentials including the microscopically substantiated local ones[2)] have enjoyed considerable success in the scattering of light nuclei.

Formulations have been made to construct unique interaction potentials from the inverse scattering theory, given the phase shifts at all energies. The inverse problem has been thoroughly studied for local potentials[3)] and investigated also for separable ones[4)]. The usual difficulty with potential construction from the inverse theory, apart from the computational one is that the experimental phase shifts are mostly given in a rather limited energy interval and that too with some inaccuracies and sometimes inconsistencies. This may lead to unphysical behaviour in the constructed potentials[5)] and as a result, analytical expressions which are desirable in many cases cannot be fitted to these potentials. It is because of these practical problems that the inverse theory has not been applied to very many real collision processes.

An important approach in understanding the interactions between two light nuclear systems has been the calculation from first principles. Here one builds an interaction between two few-nucleon systems in terms of interactions between the building blocks namely the nucleons. Such a microscopic study is actually provided by the Resonating Group Method (RGM)[6)] according to which one writes the wave function of the entire system as an antisymmetrized product of wave functions of the individual systems times a wave function of relative motion. More explicitly one writes $\Psi = \mathcal{A}[\Phi_A \cdot \Phi_B \cdot F_{AB}(r)]$ where $\mathcal{A}$ is the usual antisymmetrization operator and Φ's are the fragment wave functions. $F_{AB}(r)$ describes relative motion and is determined from the variational principle $\delta \int \Psi^*(H - E')\Psi \, d\tau = 0$ where H is the Hamiltonian of the system and E' the total energy. One is finally led to an integrodifferential eqn

$$\left[\frac{\hbar^2}{2\mu}\nabla^2 + E - V_D(r)\right] F(\underline{r}) = \int K(\underline{r}, \underline{r}') F(\underline{r}') \, d\underline{r}'$$

where E is the relative energy, V_D (r) is the direct part of the 2-system interaction involving no nucleon exchange between them. The effects of nucleon exchanges are all contained in the kernel K (r, r') which incorporates the character of a repulsion and is a complicated function of E and the parameters of the nucleon-nucleon potential and fragment wave functions. The equation for F is converted into a sum difference equation (Robertson method) and is solved for either phase shifts or bound states of the A-B system with appropriate boundary conditions. The kernel K can be evaluated analytically provided one assumes that the fragment wave functions and all interaction potentials are given as Gaussian or sums of Gaussians (allowing for the description of soft core). In almost all calculations involving RGM, these assumptions have, in fact, been made. An inconsistency that prevailed in most of the RGM calculations[7,8] is that one fixed the parameters of the fragment wave functions not from the binding energy calculation of the fragments in a variational way as should have been the case but from their r.m.s. radius. However, calculations in a refined cluster model framework[9-11] using saturation forces have used fragment wave functions which have been determined by a variational procedure which starts from the same assumption about the internucleon forces as made in the scattering calculation. More recently, in a RGM calculation on α-α scattering[12] it has been shown that even if the N-N forces are non-saturating, one can still fix the α-particle wave function parameters variationally without there being appreciable collapse. In fact the r.m.s. radius of the α-particle determined by the variational parameters is found to be only 10-15% smaller than the experimentally observed value. At the same time, the value of the α-particle parameter thus determined together with the two-body force parameters gives an adequate description of the α-α scattering data.

The one channel RGM interaction has been known to reproduce only the gross features of A-B scattering but not finer details which need the consideration of reaction channels and refined N-N interaction etc. One can take account of the reaction channels either by adding a phenomenological imaginary potential[13] to the one obtained from first principles or preferably by doing a rigorous multichannel calculation. Although the latter is a quite involved one, efficient computational techniques have now been evolved (e.g. the Cologne group has apparently developed programmes for taking account of as many as 14 coupled channels) and many microscopic calculations are currently being performed[8),11)]. It still remains to be seen what effect the momentum dependent N-N potentials would have on the over-all features of the A-B interaction. The difficulty of examining these and other realistic N-N potentials is mainly computational. One of the major problems of the microscopic calculations is that they become increasingly unmanageable when the number of nucleons in A and/or B becomes large. On the other hand, one would be interested to see the relevance of first-principle-calculations to heavy ion scattering. Although the usual RGM calculations may be prohibitive in such cases,* one may note that the A-B elastic scattering could be caused by a potential having an outer part proportional to the direct component of the A-B potential which can be easily generated from the matter density distributions of A and B and the N-N potential. The potential in the inner region which suffers from many ambaguities (the fairly unambiguous part of the A-B interaction corresponds to those separation distances for which $r_{AB} \gtrsim (r_A + r_B)$, the sum of the fragment radii below which one just has an overlapping (A+B) system) can be replaced by a hard core. Such semi-phenomenological ideas for obtaining effective interactions between systems have, in fact, been pursued by Tang and

* In fact, attempts have been made[14] to examine whether, in order to ease computation, a reduction in the number of exchange integrals occurring in the case of RGM calculations for heavy ions is possible. It has been shown, for example, in the case of ^{6}Li -α scattering that at subcoulomb energies, drastic reductions are possible by a systematic selection of the contribuding integrals. Very recently the RGM has, however, been extended to the case of α- ^{40}Ca scattering[15] thus invoking further work on heavy ion scattering.

his collaborators at Minnesota but a systematic application to heavy ion scattering has yet to be made.

A method related to the RGM is the generator co-ordinate method (GCM) in which one solves an integral Hill - Wheeler equation which involves the microscopic A-nucleon Hamiltonian. By imposing proper scattering boundary conditions on this equation, derivations have been made[16)] of an effective nucleus - nucleus potential directly from the basic two nucleon Hamiltonian. Non-locality and energy dependence appear in the effective potential which takes care of Pauli exchange effects and occurrence of nuclear shape, size etc during the collision. The GCM and RGM are found to be essentially equivalent and are unified by an integral transformation. A number of papers have appeared in the literature on derivation of effective nucleus - nucleus potentials with applications to $\alpha - \alpha$ and $^{16}O - ^{16}O$ systems, using of course a simplified two-nucleon potential. These papers differ in calculational procedures and approximations. We refer to the recent paper of Beck et al[17)] on the calculation of scattering phase shifts by GCM, which contains references to earlier work.

An approach which seems somewhat favoured by the features of the RGM potentials is the use of non-local separable ones (NLS) in describing the A-B scattering data. The microscopic theory already suggests that the A-B interaction is non-local and energy dependent. To what extent, these interactions are separable remains a matter of examination. Although no direct link exists between NLS potentials and the microscopic ones, it has been shown in the case of $\alpha - \alpha$ interaction by Leung and Park[18)] that for Gaussian shapes of N-N and α-particle wave functions, the resultant kernel of the $\alpha - \alpha$ interaction can be recast, using a formula due to Hill, into a sum of separable terms. Such an observation lends some support to the assumption of separable form for $\alpha - \alpha$ potential. A similar situation may obtain for other 2-system interactions+ . In fact, separable potentials have recently been used for a number of systems like N-N, α-N, $\alpha - \alpha$ [19),20)] etc and in each case the separable representation has been found to be quite satisfactory. Incidentally it is worth mentioning that such 2-body analyses with separable potentials become impressively tidy if one uses a co-ordinate representation in which it becomes possible to take into account the Coulomb effect in a simple but exact way.[20)] Also in co-ordinate representation, one can construct equivalent local interactions from separable ones using some suitable prescriptions.[21)] The disquieting feature of equivalent interactions is that they depend not only on ℓ but also on k.

Encouraged by the success of separable potential description of light nuclear systems, one may try such analysis for heavy ions. When reaction channels open, the separable potentials can be made complex and the analysis just goes through. Thus one may forget the local optical model analyses of 2-body systems and use, instead, a separable optical model in co-ordinate space with the added advantage that one now obtains closed expressions summed over ℓ for various amplitudes in terms of the form factors of separable interactions. It may be advisable to use the separable optical model analysis in conjunction with the semi-classical description of heavy ion scattering which is often helpful in knowing the 'ℓ' values that dominate the scattering amplitudes.

Coming to local two-system potentials, a discussion of folding model seems to be in order. In 1968, Greenlees, Pyle and Tang[22)] developed a reformulation of the optical model in which real parts of the nucleon-nucleus potential were derived from nuclear matter distribution and the nucleon-nucleon force. In computing the interaction potential between

+ If one could apply similar arguments in the case of the N-N interaction, then the separability of the N-N interaction would imply some sort of internal structure of the nucleons, e.g., quarks. However, it may not be easy to apply such reasoning to the case of N-N interaction being developed from the interaction between quarks for which a potential description is not very clear. Nevertheless, the idea seems rather amusing that there might be some connection between separability and compositeness.

the incident nucleon and the target nucleus, they neglected the effect of polarization of the incident nucleon and the effect due to exchange of the incident nucleon with the nucleons in the target. To allow for corrections for these effects, the depth and range of the folded potential were left adjustable. Calculations with this model agreed well with those for phenomenological optical model analysis. It was suggested that the model may be extended to the analysis of elastic scattering of complex particles by nuclei. In fact α-nucleus potentials have been derived using this model by folding an effective α-N interaction into the density distribution of the nucleus.[23-24] The fits to α-particle scattering from a number of nuclei using the folded model are improved appreciably over those produced by the standard optical model and are sensitive to the choice of matter distribution parameters. The r.m.s. matter radii deduced have also been in good agreement with other estimates. In all these calculations, it has been assumed that the interaction between the α-particle projectile and the nucleus is actually replaced by a sum of interactions of the α with each individual nucleon, thus avoiding many body features. For a struck nucleon which is bound, one needs to consider the off-energy shell matrix element of the α-nucleon interaction. However, the α-particle has been found to be a strongly absorbed particle with the effect that the interaction takes place in the surface region where the matter density is low. Thus in the absence of substantial clustering, the effects of multiple scattering and the Pauli principle constraints should be reduced compared to nucleon scattering. In that case, the effective interaction between an α-particle and a bound nucleon would tend to approach the free alpha-nucleon interaction. It is indeed amazing to find that such a simple model giving a first order treatment of the problem works so beautifully in describing α-nucleus scattering. It would be interesting to explore the degree of success of the folding model in describing the scattering of α's or other light projectiles from light target nuclei. This, in turn might shed light on the cluster structure of the targets. In particular, an amusing and extreme test case of the folding model would seem to be to generate an α-α potential from α-nucleon one and reproduce the α-α experimental data. Since the nucleons in the α-particle are held together extremely tightly even a qualitative fit to the α-α data would support the ability of the free nucleon-α interaction to describe α-nucleus scattering. Such a calculation has just been completed at Dacca[25] and the results indicate that the effective Sack-Biedenhara - Breit type α-N interaction determined from a reasonable fit to the α-α data are only somewhat different from the one that reproduces the non-resonant features of the scattering of nucleon from α-particles. The difference may be attributed to the fact that the target nucleus itself is an α-cluster. This thus gives evidence to earlier observations[23] namely that a consistent picture of α-nucleus scattering can be obtained and that the effective α-N interaction is quite well determined. Very recently, the folding model for α-nucleus scattering has been used in a slightly different approach. A real potential obtained by folding the nucleon-nucleus phenomenological potential with α-particle form factor is found to describe elastic α-nucleus scattering data over a wide range bombarding energy (20 - 160 MeV) and target mass number (A=24-90)[26]. The observed dependence of the α-nucleus potential strength on alpha-energy has also been accounted for by the model.[26),23]

The folding model has been extended to the case of interactions between two heavy ions.[27] Also as a departure from the folding model, density dependent two-body effective interactions have been used by Sinha[28] to construct the real part of the nucleus-nucleus optical potential. Density dependent forces, are known to take account of the saturation property of the two-body interaction which prevents the nuclear density from increasing beyond a certain magnitude. In the usual folding model, this property is ignored. However, Sinha's results indicate that the repulsion produced by the density dependent part of the interaction is important when the densities of the nuclei overlap considerably but is not so for distances greater than the touching radius. Also the exchange effects are found to be not so important.

Besides the approaches discussed above for studying interactions between two light nuclear systems, attention is however, drawn to the recent methods of studies of the interaction energy between two heavy ions e.g. the method of determination of interaction potential by minimisation with respect to the density parameters and the method of determination of interaction between two overlapping nuclei from moleculer or shell model orbitals and an effective two-body interaction. Recently using an approach similar to the above mentioned two methods, Brink and Stancu[29] have derived the interaction potential between two nuclei from the Skyrme interaction. Employing a two-centre harmonic oscillator potential to construct the density and kinetic energy density of the ground state of the combined system and of the separated nuclei, they have used a Skyrme density-dependent force to calculate the interaction potential between two ^{16}O nuclei. The latter was shown to be an energy dependent one and a large part of the exchange contribution to the potential was shown to come from modifications produced by antisymmetrization on the kinetic energy density. For properties of heavy ion interaction potentials calculated in the energy density formalism see the recent article of Ngo et al[30].

In conclusion, we remark that the ways of handling interactions between two light nuclear systems still remain far from being unique. Each has its own price and convenience. Nevertheless it is expected that by studying these interactions from different points of view, one would come closer and closer to the actual features which characterize such interactions.

REFERENCES

(1) S.Ali and A.R.Bodmer, Nucl.Phys.80(1966)99
(2) V.I.Kukulin, V.G.Neudatchin and Yu.F.Smirnov, Nucl.Phys.A245(1975)429
(3) R.G.Newton, J.Math Phys.1(1960)319
(4) See e.g. R.L.Mills and J.F.Reading, J.Math Phys.10(1969)32
(5) J.Benn and G.Scharf, Helv.Phys.Acta 40(1967)271
(6) Y.C.Tang, Proc.Int.Conf.on Clustering phenomena in Nuclei, Bochum, Germany, 1969.
(7) F.S.Chwieroth, R.E.Brown, Y.C.Tang and D.R.Thompson, Phys.Rev.C8(1973)938 and references therein.
(8) F.S.Chwieroth, Y.C.Tang and D.R.Thompson, Phys.Rev.C9(1974)56
(9) Le-Chi-Niem, P.Heiss and H.H.Hackenbroich, Z.Physik 244(1971)346
(10) P.Heiss and H.H.Hackenbroich, Nucl.Phys.A202(1973)335
(11) H.H.Hackenbroich, P.Heiss and Le-Chi-Niem, NP.221(1974)461
(12) S.A.Afzal and S.Ali, Contributed paper presented at this Conference.
(13) D.R.Thompson, Y.C.Tang, J.A.Koepke and R.E.Brown, Nucl.Phys.A201(1973)301
(14) D.Clement, E.J.Kanellopoulos and K.Wildermuth, Phys.Letters 55B(1975)19
(15) H.Freidrich and K.Langanke NP.A252(1975)47
(16) F.Tabakin, Nucl.Phys.A182(1972)497
(17) R.Beck, J.Borysowicz, D.M.Brink and M.V.Mihailovic, Nucl.Phys.A244(1975)58
(18) C.C.H.Leung and S.C.Park, Phys.Rev.187(1969)1239
(19) G.Cattapan, G.Pisent and V.Vanzani, Nucl.Phys.A241(1975)204
(20) A.A.Z.Ahmad, S.Ali, Nasima Ferdous and Masuma Ahmad, Nuovo.Cimento (in press)
(21) S.Ali and D.Husain, unpublished.
(22) G.W.Greenlees, G.J.Pyle and Y.C.Tang, Phys.Rev.171(1968)1115
(23) D.F.Jackson and R.C.Johnson, Phys.Lett.49B(1974)249 and references therein.
(24) A.Budzanowski et al, Particles and Nuclei, 6(1974)97
(25) Nasreen Gul, Ferdous, M.Sc.Thesis, Dacca University 1975(unpublished)
(26) P.P.Singh, P.Schwandt and G.C.Yang, Phys.Lett.59B(1975)113 and references therein.
(27) D.M.Brink and N.Rowley, Nucl.Phys.A219(1974)79 and references therein.
(28) B.Sinha, Phys.Rev. C11(1975)1546
(29) D.M.Brink and Fl.Stancu Nucl.Phys.A243(1975)175
(30) C.Ngo, B.Tamain, M.Beiner, R.J.Lombard, D.Mass and H.H.Deubler, Nucl.Phys.A252 (1975)237.

SCATTERING OF MASS-3 PROJECTILES FROM HEAVY NUCLEI

Sheela Mukhopadhyay, D. K. Srivastava and N. K. Ganguly
VEC Project, Bhabha Atomic Research Centre
Trombay, Bombay-400 085 (India)

1. INTRODUCTION

The interaction between heavy ions is a subject of great interest. It is well known that α-particle scattering shows most of the features which are observed in heavy ion scattering. In as much as mass-3 system is intermediate between heavy and light particles it will be interesting to investigate the scattering of mass-3 projectiles to see if it is possible to extend it to study the heavy ion scattering. Indeed; we have seen that the 'molecular type' potentials[1], with a soft repulsive core and a shallow attractive well used for heavy ion collisions can be used to fit the elastic scattering data of mass-3 projectiles also. In the first part of this paper, we have given a description of how this potential is generated with a special emphasis on saturation and second order effect through a density dependent interaction between nucleon and mass-3 projectiles. In the second part, we show that the asymmetry dependence observed in the potential describing the scattering of mass-3 particles from heavier nuclei actually originates from the isospin interaction, when triton and helion are treated as two members of an isospin doublet.

2. INTERACTION POTENTIAL-EFFECT OF DENSITY DEPENDENCE

A first approximation to composite particle optical potential is given by

$$V_{op}^{c}(R) = \int V_{op}^{n}(|\bar{R}+\bar{r}'|)\, \rho_c(r')\, d\bar{r}' \qquad (2.1)$$

where V_{op}^{c} and ρ_c are respectively the optical potential and density distribution for composite particle and V_{op}^{n} is the nucleon optical potential. This kind of folding model largely overestimates the interaction potential not only at the centre but also near the strong absorption radius, where the densities of the target and the projectile hardly overlap[2]. However, if V_{op}^{n} in eq. (2. 1) is determined by fitting the experimental nucleon-nucleus cross-section data, a part of the second order and exchange effects are already simulated. The region explored by elastic scattering is generally limited to the surface of the nucleus and a definite conclusion about the central value of the potential should be drawn with reservations. However, for $\alpha-\alpha$ scattering, Ali and Bodmer[3] got a repulsive core. Moreover, liquid drop model calculations to fit the fusion cross-sections for heavy ion collisions give potentials which indicate a tendency to become repulsive as the two nuclei approach closer and closer[4]. Also Michaud et al[5] found it necessary to have a soft repulsive core in order to fit the giant resonances simultaneously for the three systems, ^{12}C-^{12}C, ^{12}C-^{16}O and ^{16}O-^{16}O. Thus it seems reasonable that a properly determined potential should not only have a correct slope near the strong absorption radius but also must have a repulsive core. One of the most important correction to eq. (2. 1) is the exchange effect. Pauli's principle limits the number of states available for any interaction and thus produces saturation as the density increases. This considerably reduces the potential at lower energies. Besides, the second order processes describing the energy non-conserving virtual transitions to excited states also reduce the potential. We find that these effects together for mass-3 particles amount to nearly 40% of the first order effect[6]. To introduce saturation, it is convenient to use a density dependent two body interaction which becomes effectively weaker as the nuclear density increases[7]. We used a nucleon-helion (or triton) interaction proposed by Park and Satchler[8] and introduced saturation through a density dependent multiplicative term. The modified interaction has the form

$$V_{n-h(\text{or } t)}(r) = [22.5 - 2.5(\tau_N \cdot \tau_3 + \sigma_N \cdot \sigma_3 + \tau_N \cdot \tau_3\, \sigma_N \cdot \sigma_3)]\, e^{-0.2(\bar{r}-\bar{r}')^2} \times \left(1 - c\rho^{2/3}\left(\frac{\bar{r}+\bar{r}'}{2}\right)\right) \quad (2.2)$$

Here ρ is the local density distribution calculated by employing sudden approximation[9]. The helion-nucleus potential was obtained by folding the interaction given in eq. (2.2) with the density distribution of the targets. Employing nuclear matter theory, Bruecker et al[10] computed heavy-ion potentials which were molecular type. On the contrary, Fliessbach[11] showed that non-equilibrium matter densities which give rise to repulsive core, are no longer obtained if properly antisymmetrized wave functions are used to compute local densities. However, recently Brink and Stancu[12] obtained a molecular type potential with a soft repulsive core for ^{16}O-^{16}O system even after including antisymmetrization. We find that by varying the value of c in eq. (2.2) from 0 to 4.0, we can introduce this effect in the potential (see fig. 1). The curve for c = 1.0 in fig. 1 resembles closely the potentials obtained by us in phenomenological analysis[6]. Since c is a measure of saturation, beyond a certain value of c, the potential shows a soft repulsive core. The tails of the families of potentials remain nearly the same for all values of c, and consequently they fit only the forward scattering data equally well. It may be possible to have further selection only if extensive backward data are available. It is interesting to note the close similarity between these potentials for mass-3 projectile and that obtained by Brink and Stancu[12] for two ^{16}O nuclei.

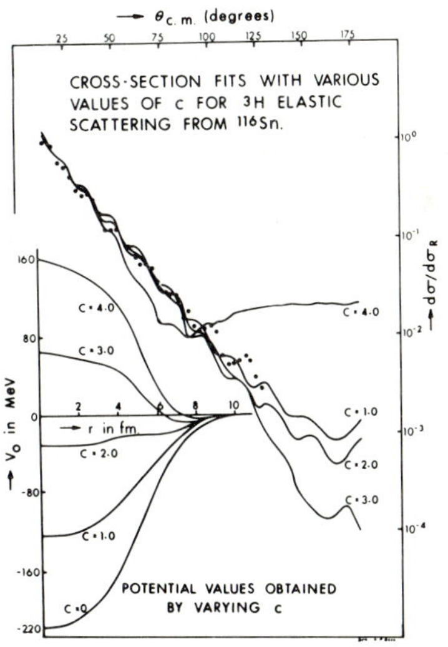

Fig. 1 Potential curves for different values of c in eq. (2.2) and the cross-section fits to data obtained from E. R. Flynn.

3. ISOSPIN FOR MASS-3 PARTICLES

Our analysis proves that similar to nucleons[14], the asymmetry dependence observed for mass-3 potentials can indeed be explained as a consequence of a complex isospin interaction, treating triton and helion to form an isospin doublet. This contradicts the recent findings of some authors regarding isospin of mass-3 particles[13]. The phenomenological potential depths for triton was found to have the following asymmetry dependence[6]

$$V_0^{3H} = 133.45 - 29.07 \frac{N-Z}{A} ; \quad W_0^{3H} = 22.81 - 19.64 \frac{N-Z}{A} \qquad (3.1)$$

In case the second terms in Vo and Wo are really isospin dependent, then their signs are expected to be negative for ^{3}H and positive for ^{3}He. By phenomenological analysis of helion scattering data, it was observed that if the same form factors are chosen for ^{3}H and ^{3}He, eq. (3.1) with sign reversed for the asymmetry term fits the ^{3}He elastic scattering data even better consistently for all the targets, indicating that the isospin interaction of mass-3 particles is complex and is responsible for the observed asymmetry dependence. Also a comparison of ^{3}He potentials calculated in section 2 coupled with those for ^{3}H taken from ref. (6) give values for isospin terms which agree with the phenomenological values strikingly well. Finally, ^{3}H and ^{3}He potentials were calculated using formalism of Samaddar et al[15] but treating n and p as two members of an isospin doublet and taking care of the isospin dependence of both the first and second order nucleon potentials. It was observed that to the first order, isospin potential for mass-3 projectiles equals that for nucleons but the second order terms decrease it.

We are thankful to Mr. S. Pal for helpful discussions and to Mrs. K. Varadarajan for numerical assistance. One of the authors (SM) is grateful to Dr. M. K. Mehta for his keen interest and also wishes to acknowledge the National Science Talent Search Scholarship awarded by NCERT.

REFERENCE

1. B. Block and F. B. Malik, Phys. Rev. Lett. 19 (1967) 239
2. G. R. Satchler, Phys. Lett. 59B (1975) 121
3. S. Ali and A. R. Bodmer, Nucl. Phys. 80 (1966) 99
4. R. Bass, Nucl. Phys. A231 (1974) 45
5. G. J. Michaud, Phys. Rev. C8 (1973) 525;
   G. J. Michaud and E. W. Vogt, Phys. Rev. C5 (1972) 350
6. Sheela Mukhopadhyay, D. K. Srivastava and N. K. Ganguly, Nucl. Phys. to be published.
7. G. E. Brown, Comm. on nucl. and Part. Phys. 4 (1970) 75
8. J. Y. Park and G. R. Satchler, Particles and Nuclei 1 (1971) 233
9. K. A. Brueckner, J. R. Buchler, S. Jorna and R. J. Lombard, Phys. Rev. 171 (1968) 188
10. K. A. Brueckner, J. R. Buchler and M. M. Kelly, Phys. Rev. 173(1968)944
11. T. Fliessbach, Z. Phys. 238 (1970) 329
12. D. M. Brink and Fl. Stancu, Nucl. Phys. A243 (1975) 175
13. P. P. Urone, L. W. Put, H. H. Chang and B. W. Ridley, Nucl. Phys. A163 (1971) 225;
    J. Nurzynski, Nucl. Phys. A246 (1975) 333
14. G. R. Satchler, Nucl. Phys. A91 (1967) 75
15. S. K. Samaddar, R. K. Satpathy and S. Mukherjee, Nucl. Phys. A150 (1970) 655

WORK SHOP ON EXPERIMENTAL TECHNIQUES

SESSION III

Organized by
R. O. Bondelid
Naval Research Laboratory, Washington, D. C.

Discussion Leaders

Part 1. R. O. Bondelid
Part 2. J. R. Richardson
Part 3. M. K. Mehta

Scientific Secretaries

M. K. Mehta (BARC, Bombay)
S. C. Pancholi (U. of Delhi)

Rapid technological developments, the advent of the new meson factories and the decreasing cost of mass storage devices provide new tools for the nuclear physics experimenter. The workshop on experimental techniques was a unique effort to provide a forum of discussion among experimentalists regarding the techniques presently used as well as future developments for accumulating few particle interaction data. Having a workshop exclusively devoted to experimental techniques was a new experience and its success indicates that future conferences should profit from including discussion sessions of this nature.

In keeping with the "bridge" theme of the conference a variety of topics was covered even including the relevance of few particle measurements to nuclear power generation. After each talk there was a lively discussion some of which is included in these proceedings at the end of each paper. During the session experimental techniques ranging from novel applications of photographic emulsions, particle identification, CAMAC technology to applications of microprocessors in nuclear physics experiments were discussed. The higher energies available from the meson factories present a variety of particle detection problems, some of which are economic. The various methods used for detection of several hundred MeV particles were discussed in detail.

The talk on the performance of the transplanted cyclotron at Chandigarh aroused spirited discussion about the desirability of transplanting accelerators which are nearing the end of their scientific usefulness at the parent institution. Some arguments in support of such transplants held that training gained by making the accelerator operational was invaluable to the participants and would highly raise their morale as well as provide a modest research and teaching tool. On the other hand the argument was advanced that a somewhat obsolescent machine needs an unusually imaginative and experienced group of people to produce quality research.

A survey of the A>3 data revealed a lack of consistancy in presenting experimental results. The evidence is that possible trends in energy dependence and angular distributions are lost by experimenters using a variety of analyses in presenting their results. The suggestion was made that the few particle community agree upon a method of presenting and preserving data so that trends in data can be related to any future theoretical developments.

SURVEY OF FEW NUCLEON REACTIONS FOR A > 3

J. M. Lambert, P. A. Treado, L. T. Myers and E. I. Karaoglan
Georgetown University, Washington, D. C. 20057 U.S.A.

A survey of recent publications on experiments in few-nucleon reactions will be discussed. This presentation is that of an experimentalist, and assesses the development and problems in this field. The scope of the survey is limited in the following ways: the total mass of the interacting system is from A = 4 to A = 12, the publications considered are from 1970 to the present and only coincidence experiments are included, and from only four sources (Physical Review, Nuclear Physics, Physical Review Letters, and Physics Letters). This survey does not include papers from previous conference proceedings because these are readily available, and also because many are in print. Undoubtedly, there are a few papers that have been missed in this survey. Also, the judgment of what constitutes an experimental paper is not a singular one.

This presentation focuses on two aspects of the problem. First, discussion of experiments concerned with a given total number of nucleons, and secondly some discussion of specific areas of interest appropriate to the field in general is given. Of the more than 70 papers used, about half are in Nuclear Physics, one-quarter are in Physical Review, and the rest are in the Letters. The A = 3 system has been excluded because this is one area in which a reasonably exact theory exists. It is sincerely hoped that extensions of this theory will soon be available for systems of A ≥ 4.

In terms of A, the total number of nucleons involved, Fig. 1 shows the energy range of most of the experiments to be discussed, and shows the systems as a function of the center-of-mass (CM) energy of the initial system on a logarithmic scale. (The different symbols indicate the mass of the projectile.) The first thing that is obvious is that above about 50 MeV, essentially all of the experiments are performed with protons. A second comment is that almost all of the experiments involve two major features; quasifree (QF) interactions or final state interactions (FSI) which includes sequentially decaying systems. These two processes virtually dominate experimental data.

A = 4

These experiments include N-trion and d-d interactions that are dominated by N-N FSI or N-N and N-d QF mechanisms with a 4-nucleon breakup background. One recurring question is whether the isospin forbidden p-n singlet FSI is observed in the d-d reaction. To within a percent or so it appears that the tail of the triplet FSI can account for structure associated with this portion of phase space. Experimental comparisons of (p,2p) and (p,pd) on ^{3}He have led to discrepancies that invoked the possibility of direct pick up to help explain spectral structure, while in contrast, the ratio of d(d,pd) to d(d,nd) have been constant contrary to the situation for p + d. This has also been a prolific area of investigation of trion wave functions, with the question of the best choice still left open. Finally, 4-body investigations have shown that phase space will not explain the continuum, as expected. N-N FSI contributions have been isolated, and in the 4-body case the singlet N-N FSI is no longer forbidden. Also, we will be reporting at this conference on contributions from double QFS.

A = 5

The majority of these experiments are being done to investigate QF processes with all possible pairs being observed, including both quasifree scattering (QFS) and quasifree reactions (QFR). An important question being addressed is to what extent distorted waves improve the impulse approximation predictions.

In some cases substantial improvements are obtained using distorted waves, but in the tightly bound alpha particle, evidence that plain waves are as appropriate as distorted waves exists. An experimental situation that can cause problems is the spectral structure due to ^{4}He excited states. This can cause severe problems with n-n scattering-length experiments, and the success or failure of these experiments can depend upon preselecting the most advantageous regions of phase space. Also, Coulomb FSI occurs in some of these experiments and this FSI can produce dramatic reductions in the observed cross sections. This effect occurs when two charged particles have almost zero relative energy and no threshold resonance structure, e.g. p-d or p-t. In this situation the Coulomb repulsion can dominate, causing a striking dip in the cross section, and can interfere with the investigation of other processes. Fortunately, this Coulomb FSI is limited to a rather small region of phase space and can usually be avoided. This region has been utilized as a testing ground for models of energy dependent attenuation of QF processes, as well as for the necessity of using fully antisymmetrized wave functions.

A = 6

These experiments include trion-trion and d-α experiments, split about equally between investigations of QF and FSI reactions. Again, for the d-α reactions, the singlet p-n FSI is isospin forbidden and has been investigated to check isospin conservation assumptions. An experiment that contributed to the understanding of the QF process involved the use of the spinless alpha particle on a deuteron target to investigate the ratio of α-p to α-n QFS. This ratio was found to be essentially constant in strong contrast to the situation in which a proton was used as the projectile. Thus, the importance of spin and statistics in the QF process was indicated. For experiments in this A = 6 region, care must be taken to understand all of the contributions to the data from sequential ^{4}He and ^{5}He breakup.

A = 7

These experiments are dominated by p + ^{6}Li measurements of QF processes to investigate the structure of ^{6}Li. This nucleus is expected to have a strong d-α cluster configuration because of its low binding energy as well as a t - ^{3}He cluster structure which has a much larger binding energy. This nucleon system has allowed extensive testing of the impulse approximation predictions and has led to a number of very interesting results. The method of defining the alpha clustering probability has led experimentalists to get numbers ranging from a few percent to 100%. Test of the Treiman-Yang conditions have indicated the dominance of the α-d pole contribution. In addition, shell model considerations indicated experimental momentum distributions should show a deep minimum at the (p,2p) "QF condition" as is found in the case of (p,2p) experiments on ^{7}Li. Very careful experiments have been performed, and it has now been confirmed in a reasonable manner that the knocked out proton has predominantly s-wave characteristics rather than the p-wave structure predicted by the shell model. The ability to obtain unexpected structure information about the light nuclei is an excellent example of one of the more important successes of multiparticle breakup experiments. More will be said of the cluster probability later. Another interesting aspect which appears at higher energies of the impulse approximation has been observed. There, the width of the momentum distribution is apparently increasing which is rather interesting because it has been quite characteristic of low energy QF experiments (below 100 MeV) to have momentum widths noticeably narrower than predicted by theory. In terms of sequential information, this region, as well as a number of others, has been investigated in order to try to establish the predicted $1/2^+$ first excited state of ^{5}He which still remains an enigma.

A = 8, A = 9

These experiments are almost all performed with Li targets and d or ^{3}He projectiles. The A = 8 experiments have been dominated by QF reactions, while for A = 9, low energy sequential reactions have been the focus of interest. Here and in other regions, angular correlation symmetry axes have been investigated to interpret reaction mechanisms. This is a region that was earlier investigated for "proximity scattering," but this does not appear to be a reaction which produces an observable enhancement, and certainly is not a dominant type reaction.

A = 10

The data on this nucleon system comes from two main reactions: p - ^{9}Be and α - ^{6}Li. An unusual (non-coincidence) experiment uses the ^{9}Be(p,2n)^{8}B reaction in an attempt to produce a calibration source for the solar neutrino experiments. Another is the α - ^{6}Li reaction that confirmed the final state energy prescription for this QFS reaction. This is an example of using the highly oscillatory α-α cross sections to advantage. QF processes predominate, with attempts at estimating clustering probabilities. One of these predictions is based on simple shell model and cluster wave functions for ^{6}Li and appears to obtain much better results with cluster functions and greater than 90% α-α cluster probability. Another very interesting method of analysis (which is also applied to A = 7 and A = 8) matches internal oscillator wave functions to external spherical Hankel functions. This model appears to be capable of fitting quite a number of experimental data, and of giving good physical insight into the problem without the use of distorted waves.

A = 11, A = 12

The nucleon systems of A > 10 are complex enough that sequential processes tend to dominate. Attempts have been made to perform angular correlation analysis in the CM of the sequentially decaying system, but a number of problems still exist with this type of analysis. If it were possible to do these calculations, especially with absolute cross sections, it would not only be possible to extract more information about particle unstable systems, but it would also be possible to extract these peaks as background in QF analysis. Some attempts at studying QF process vertices with A = 5 particles involved have been productive.

COMMENTS FOR EXPERIMENTALISTS

As experiments become more difficult and finer spectral detail is required, more care should be used in the planning of experiments. If the final state contains alpha particles and nucleons, deuterons, tritons, or another alpha particle, there is a good chance of getting strong sequential FSI contributions from any combinations of these pairs. Confidence can be expected in the final results only if efforts to avoid regions where there is going to be interference between different processes are successful. We have good experience now from the exact 3-body solutions to know that interference terms can be quite large and very difficult to predict without a rigorous theory. At this point it is probably much more reliable to extract a QF background from sequential data than visa-versa, especially because of the lack of knowledge of the angular correlation functions for sequential breakup. When investigating QF processes, judicious choice of angle pairs and redundancy of particle identification can pay big dividends in obtaining useful data. In addition, trying to obtain QF information away from the momentum distribution peak can sometimes prove misleading.

A final comment concerning cluster probabilities. The lack of agreement on how this information should be extracted is distressing. The fact that different experiments will quote probabilities from a few percent to almost 100% for the alpha clustering in ^6Li surely indicates the need for some method of comparing various experiments. We have been trying to correlate these data for a large number of experiments, and in many cases, there is no way to use the data without doing a complete re-analysis, assuming enough data are available. Not only are there the problems of comparing and/or using plane and distorted wave analyses, but the variety of wave functions used in each seems to be unending, even without the complication of using cutoff procedures. Insight into the overall usefulness and limitations of the QF process may be available if, in addition to whatever method was used for analysis, each paper would also include the parameters from an agreed upon simple plane wave formalism. Even if it were not overly meaningful in a theoretical sense, it would allow data from the whole range of experiments in Fig. 1 to be compared or utilized for checking advances in the theory.

For example, it would be very interesting to have the ratio of experimental cross section to plane wave prediction, N, for a large energy range, and all done with the same analysis. Similarly, following the peak width as a function of energy and CM angles might prove to be extremely valuable. In earlier searches, we have found that there appears to be a fairly strong indication that N depends on the relative velocities of the scattering particles in the final state and not on the energies, momenta, or scattering cross sections, but it is difficult to extract equivalent data from the literature to confirm this conjecture.

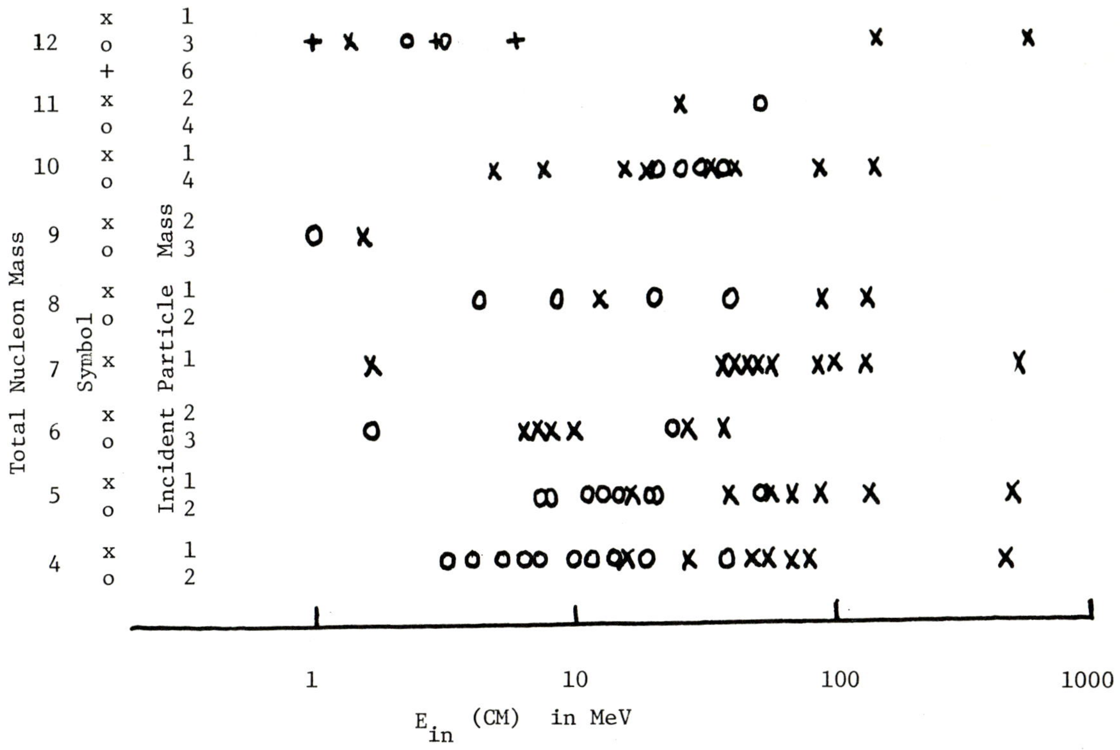

Figure 1. Energy range of experiments discussed. The horizontal axis is the center of mass energy of the initial system on a logarithmic scale. The symbols in each line indicate the mass of the projectile.

REFERENCES

A = 4

1. E. Andrade, V. Valkovic, D. Rendic and G.C. Phillips, Nucl. Phys. A183 (1972) 145
2. D.I. Bonbright, R.G. Allas, R.O. Bondelid, E.L. Petersen, A.G. Pieper, R.B. Theus and I. Slaus, Phys. Rev. C8 (1973) 114
3. J.P. Burg, J.C. Cabriooat, M. Chemarin, B. Ille and G. Nicholai, Nucl. Phys. A179 (1972) 385
4. A.A. Cowley, P.G. Roos, H.G. Pugh, V.K.C. Cheng, and R. Woody III, Nucl. Phys. A220 (1974) 429
5. I.M. Duck, V. Valkovic, and G.C. Phillips, Phys. Rev. Letts. 29 (1972) 875
6. M.B. Epstein, I. Slaus, D.L. Shannon, J.R. Richardson, J.W. Verba, H.H. Forster, C.C. Kim and D.Y. Park, Phys. Letts. 36B (1971) 305
7. M.B. Epstein, I. Slaus, D.L. Shannon, H.H. Forster, M. Furic, C.C. Kim and D.Y. Park, Nucl. Phys. A199 (1973) 225
8. R. Frascaria, V. Comparat, N. Marty, M. Morlet, A. Willis and N. Willis, Nucl. Phys. A178 (1971) 307
9. M.J. Fritts and P.D. Parker, Nucl. Phys. A198 (1972) 109
10. P. Kitching, G.A. Moss, W.C. Olsen, D.R. Lehman, J.R. Priest, W.J. Roberts, J.C. Alder, W. Dollhopf, W.J. Kossler and C.F. Perdrisat, Phys. Rev. C6 (1972) 769
11. H.G. Pugh, P.G. Roos, A.A. Cowley, V.K.C. Cheng and R. Woody III, Phys. Letts. 46B (1973) 192
12. I. Slaus, M.B. Epstein, G. Paic, J.R. Richardson, D.L. Shannon, J.W. Verba. H.H. Forster, C.C. Kim, D.Y. Park and L.C. Welch, Phys. Rev. Letts. 27 (1971) 751
13. V. Valkovic, I. Duck, W.E. Sweeney and G.C. Phillips, Nucl. Phys. A183 (1972) 126
14. V. Valkovic, N. Gabitzsch, D. Rendic, I. Duck and G.C. Phillips, Nucl. Phys. A182 (1972) 225
15. B.J. Wielinga, A.D, IJpenberg, K. Mulder, and R. Van Dantzig, Phys. Rev. Letts 27 (1971) 1229
16. W. Von Witsch, G.S, Mutchler, G.C. Phillips and D. Miljanic, Phys. Rev. C8 (1973) 403
17. R.E. Warner, S.P., DiCenzo, G.C. Ball, A.J. Ferguson, J.S. Forster, Nucl. Phys. A243 (1975) 189
18. W. Von Witsch, W. Viefers, H. Mommsen, P. David and F. Hinterberger, Nucl. Phys. A195 (1972) 617

A = 5

19. P.A. Assimakopoulos, E. Bearsworth, D.P. Boyd and E.F. Donovan, Nucl. Phys. A144 (1970) 272
20. L.A. Charlton, S.M. Kelso and R.E. Warner, Nucl. Phys. A178 (1971) 39
21. R. Frascaria, P.G. Roos, N. Marty, M. Morlet, A. Willis, V. Comparat and N. Frijiwara, Phys. Rev. C12 (1975) 243
22. A. Niiler, R.J. Spiger, W. Von Witsch and G.C. Phillips, Nucl. Phys. A197 (1972) 263
23. J.G. Rogers, G. Paic, J.R. Richardson and J.W. Verba, Phys. Rev. C2 (1970) 828
24. P.G. Roos, Phys. Rev. c9 (1974) 2437
25. I. Slaus, R.G. Allas, L.A. Beach, R.O. Bondelid, J.M. Lambert, E.L. Petersen and D.L. Shannon, Phys. Rev. C8 (1973) 444
26. R.E. Warner, G.C. Ball, W.G. Davies, A.J. Ferguson and J.S. Forster, Phys. Rev. Letts 27 (1971) 961
27. R.E. Warner, B.E. Corey, E.L. Petersen, R.W. Bercaw and J.E. Poth, Nucl. Phys. A148 (1970) 503
28. R.E. Warner, G.R. Flierl, W.G. Davies, G.C. Ball, A.J. Ferguson, J.S. Forster, and S.A. Gottlieb, Nucl. Phys. A192 (1972) 341
29. R.E. Warner, S.A. Gottlieb, G.C. Ball, W.G. Davies, A.J. Ferguson and J.S. Forster, Nucl. Phys. A221 (1974) 593
30. R.E. Warner and F.W. Vogt, Nucl. Phys. A204 (1973) 433

A = 6

31. R.G. Allas, D.I. Bonbright, R.O. Bondelid, E.L. Petersen, A.G. Pieper and R.B. Theus, Phys. Rev. Letts. 28 (1972) 569

32. E. Hourany, H. Nakamura, F. Takeutchi and T. Yuasa, Nucl. Phys. A222 (1974) 537
33. B. Kuhn, H. Kumpf, S. Parzhitsky and S. Tesch, Nucl. Phys. A183 (1972) 640
34. R. Larose-Poutissou and H. Jeremie, Nucl. Phys. A218 (1974) 559
35. E. K. Lin, R. Hagelberg and E.L. Haase, Nucl. Phys. A179 (1972) 65
36. T. Rausch, H. Zell, D. Wallenwein and W. Von Witsch, Nucl. Phys. A222 (1974) 429

A = 7

37. J.C. Alder, W. Dollhopf, W. Kossle, G. F. Perdrisat, W.K. Roberts, P. Kitching, G.A. Moss, W.C. Olsen and J.R. Priest, Phys. Rev. C6 (1972) 18
38. R.K. Bhowmik, C.C. Chang, P.G. Roos and H.D. Holmgren, Nucl. Phys. A226 (1974) 365
39. W. Dolhopf, C. Lunke, C.F. Perdrisat, P. Kitching, W.C. Olsen, J.R. Priest and W.R. Roberts, Phys. Rev. C8 (1973) 877
40. M. Jain, P.G. Roos, H.G. Pugh and H.D. Holmgren, Nucl. Phys. A153 (1970) 49
41. R.B. Liebert, K.H. Pulser and R.L. Burman, Nucl. Phys. (1973) 335
42. C.A. Miller, J.W. Watson, D.I. Bonbright, F.I.S. Wilson and D.O. Wells, Phys. Rev. Letts. 32 (1974) 684

A = 8

43. B. Antolkovic, Nucl. Phys. A219 (1974) 332
44. M. Furic, R.K. Cole, H.H. Forster, C.C. Kim, D.Y. Park, J. Rucker, H. Spitzer and C.N. Waddell, Phys. Letts. 39B (1972) 629
45. J.Y. Grossiord, C. Coste, A. Guichard, M. Gusakow, A.K. Jain, J.R. Pizzi, C. Bagieu and R. deSwiniarski, Phys. Rev. Letts. 32 (1974) 173
46. R. Hagelberg, E.L. Haase and Y. Sakamota, Nucl. Phys. A207 (1973) 366
47. S. Kato, H. Orihara, S. Kubono, J. Kasagi, H. Ueno, T. Nakagawa and T. Tohei, Nucl. Phys. A195 (1972) 534
48. S. Kohmoto and Y. Sakamoto, Phys. Rev. Letts. 34 (1975) 550
49. D. Miljanic, T. Zabel, R.B. Lievert, G.C. Phillips and V. Valkovic, Nucl. Phys. A215 (1973) 221

A = 9

50. K.H. Bray, J.M. Cameron, G.C. Neilson and T.C. Sharma, Nucl. Phys. A181 (1972) 319
51. J.C. Heggi and P.W. Martin, Phys. Letts. 43B (1973) 289
52. J.C. Heggie and P.W. Martin, Nucl. Phys. A212 (1973) 78
53. D.T. Thompson and G.E. Tripard, Phys. Rev. C6 (1972) 452

A = 10

54. N.F. Davison, M.J. Canty, D.A. Dohan and A. McDonald, Phys. Rev. C10 (1974) 50
55. G. Deconninck, A. Giorni, J.P. Longequeue, J.P. Maillard and T.U. Chan, Phys. Rev. C3 (1971) 2085
56. S.T. Emerson, V. Valkovic, W.R. Jackson, C. Joseph, A. Niiler, W.D. Simpson, and G.C. Phillips, Nucl. Phys. A169 (1971) 317
57. P. Gaillard, M. Chevalier, J.Y. Grossiord, A. Guichard, M. Gusakow and J.R. Pizzi, Phys. Rev. Letts. 25 (1970) 593
58. J.M. Lambert, R.J. Kane, P.A. Treado, L.A. Beach, E.L. Petersen and R.B. Theus, Phys. Rev. C4 (1971) 2010
59. B. Mithra and R. Laverriere, Phys. Letts. 58B (1975) 17
60. J.R. Quinn, M.B. Epstein, S.N. Bunker, J.W. Verba and J.R. Richardson, Nucl. Phys. A181 (1972) 440
61. Y. Sakamoto, P. Cüer and F. Takeutchi, Phys. Rev. C11 (1975) 668

A = 11

62. J.M. Lambert, P.A. Treado, L.A. Beach, R.B. Theus and E.L. Petersen, Nucl. Phys. A152 (1970) 516
63. M.A.A. Sonnemans, J.C. Wall and R. Van Dantzig, Phys. Rev. Letts. 31 (1973) 1359

A = 12

64. L.L. Gadeken and E. Norbeck, Phys. Rev. C6 (1972) 1172
65. J. Kasagi, T. Nakagawa, N. Sekine and T. Tohei, Nucl. Phys. A239 (1975) 233
66. J.M. Lambert, P.A. Treado, D. Haddad, R.A. Moyle and J.C. Sessler, Phys. Rev. Letts. 27 (1971) 820

67. M. C. Taylor, V. Valkovic and G. C. Phillips, Nucl. Phys. A182 (1972) 558
68. P. A. Treado, J. M. Lambert, V. E. Alessi and R. J. Kane, Nucl. Phys. A198 (1972) 21
69. W. Von Witsch, M. Ivanovich, D. Rendic, V. Valkovic and G. C. Phillips, Nucl. Phys. A180 (1972) 402

DISCUSSION

N.Sarma: Invariably the data is tied up with the analysis. It would be nice if raw data would be presented in terms of cross section in addition to cross section divided by phase space, etc.

J. Lambert: I certainly agree. However, in some cases the amount of data is too large to publish. In this case, an agreed upon method of presentation would be very helpful.

M. Jain: I do not see any great difference in presenting QFS data either as a cross section or cross section divided by the phase space factor, since the latter is well known. Also on not seeing a dip in the 6Li(p,2p)5He reaction, the Maryland data at 100 MeV clearly shows the dip due to the p-state knockout.

J. Lambert: I meant to indicate that the 6Li(p,2p) shows nothing like the dip expected for p-state protons as seen clearly in 7Li(p,2p) experiments. It appears that substantial amount of S wave is needed to explain the results.

H. Pugh: I agree it would be very desirable to publish raw data with no theoretical interpretation. This does have some problems, however, in that it is usually necessary to correct for finite angular resolution before comparison with theory. I would also like to point out that there is a recent paper from Maryland submitted to Physical Review by Goldberg et al. which compares 6Li(p,pd)4He and 6Li(p,pt)3He at 100 MeV and finds using a DWIA analysis that the 3He-t cluster probablility is somewhat larger than the d-4He cluster probability, as predicted by cluster model calculations.

R. Allas: Publication of raw data leads to practical problems., e.g., frequently the amount of data is too much for an article.

G. Igo: At the Berkeley 184-inch we have measured the $(\alpha,2\alpha)$ reaction at 850 MeV and 650 MeV. This data has been analyzed using DWIA. The effect of distortion is of the order of a factor of ten. On the other hand, distortion effect at 80 MeV [$(\alpha,2\alpha)$ experiment measured at LBL] is a factor of 1000 greater. It is difficult to trust results where the distortion reaches such a large value.

E. Petersen: An additional complication in using the Simple Impulse Approximation has shown up in experiments comparing p-pp and p-pn processes in 2H and Li. These processes show very different quasi-free angular distributions, and thus do not easily fit in to the SIA viewpoint.

J. Lambert: Even though the limitations of SIA are well known and in each case the best theory should be used, some consistent method of searching for systematic properties is needed.

NEUTRON-PHYSICS TECHNIQUES

Đ. Miljanić
Institute Ruder Bošković, Zagreb, Yugoslavia

Experiments performed with neutrons still give a challenge to many experimental groups, although they represent a difficult task. The reasons for this interest in many-body neutron-induced reactions stem from the fact that these reactions are one of the suitable ways to learn something on the nucleon-nucleon interaction, to shed light on the structure of atomic nuclei, as well as on some specific features of three-body reactions. The aim of this paper is to review existing experimental approaches, as well as to point out some possible new ways of measuring reactions induced by neutrons of energies below 50 MeV.

The cross section for the reaction of the type $n+T \rightarrow 1+2+3$ depends on five independent kinematic variables. Measurements in which all five independent kinematic variables are determined are often referred to as complete experiments, in contrast to incomplete ones, in which less than five variables are measured.

Fig. 1 shows two different types of incomplete experiments. S and T denote the source of neutrons and the target, respectively, while D´s are complete detection and identification systems. The first system is the most commonly used experimental set-up, where one of three outgoing particles is detected and its energy and scattering angle are determined. The first attempt to determine the neutron-neutron scattering length from neutron-induced deuteron break-up was performed in this way, namely by measuring the proton energy spectrum /1/. After this, many similar experiments were performed and the quality of data was considerably improved. The Livermore group /2/ reports the use of a special new spectrometer, which gives a large signal-to-background ratio and has good energy and angular resolution (ΔE=200 keV, $\Delta\theta<3^o$). It should be pointed out, however, that even with such quality of data, there is a problem of quantitative interpretation and extraction of a_{nn}. This is mainly due to kinematic incompleteness. Because of the integration over various kinematic conditions, the sensitivity of the spectrum to the scattering parameters is low. The data from this kind of measurement should be used as a testing ground for the overall completeness of any theory rather than for a particular purpose. Similar measurements were also used to study other three-particle reactions on light nuclei (see e.g. /3/).

The second type of incomplete experiment, shown in Fig. 1b, has not been used, as far as we know (except in fission studies). The main characteristic of this experiment is that the neutron source is close to the target, and so the large fraction of the neutron beam is used. However, in this way the information on incident direction is lost and although one detects two outgoing particles, the only kinematic variables that are determined are energies (E_1,E_2) and the angle between the detectors (θ_{12}). From these variables one can uniquely determine the energy and the scattering angle of the third, undetected particle. This experimental arrangement could be particularly useful for obtaining information on different decay modes of a given nuclear state relatively high in excitation. In a single measurement, this arrangement could also give information on the angular dependence of the cross section $d^2\sigma/d\Omega_3 dE_3$ over the relatively wide angular region. Fig. 2 shows two schematics for kinematically complete experiments. Set-up a) was used to study different aspects of deuteron break-up reactions, namely nucleon-nucleon quasi-free scattering (QFS) and final-state interactions (FSI). Three different versions have been used so far:

1. D_1 and D_2 are neutron counters, while the target is at the same time a proton detector /4,5/.
2. D_1 and D_2 are neutron counters - proton is not detected /6/.
3. D_1 is a neutron counter, D_2 is a proton counter, the target is thin /7/.

In the first two versions, the measurement of redundant kinematic variables was used for elimination of the background. The accuracy of several measurements in the FSI region was such that from the extracted values of a_{nn} it was possible to conclude on the breaking of charge independence. However, due to experimental as well as theoretical errors no conclusion could be drawn on charge-symmetry breaking. The cross section in the QFS region is more sensitive to the other effective range parameter r_{nn}. However, high statistical errors of the performed QFS measurement inhibit any conclusion on the accurate value of r_{nn}. There is an interesting proposal /8/ for the determination of r_{nn} using the ratio of the cross sections measured under two different kinematic conditions in the QFS region. However, to achieve the desired accuracy in a reasonable measuring time, it seems imperative to use a multidetector system.

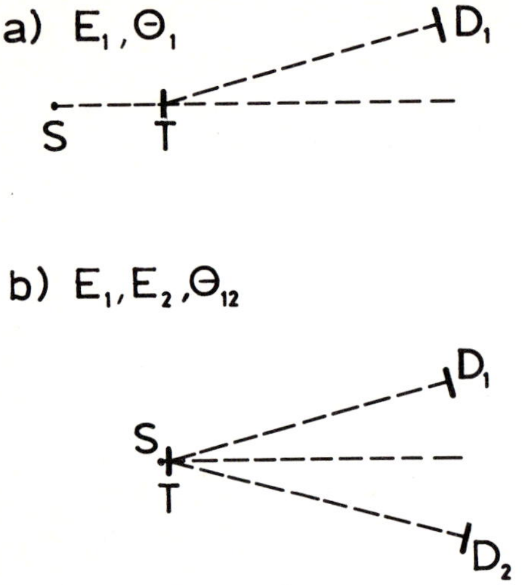

Fig.1. Incomplete experiments.

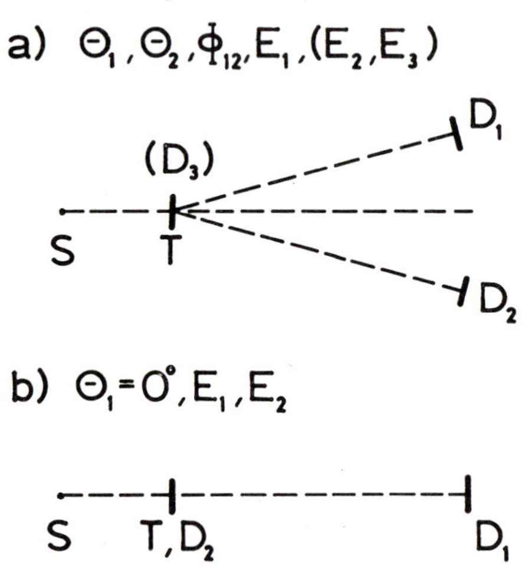

Fig.2. Complete experiments.

Fig. 2b shows another type of complete experiment, where two of three outgoing particles have to be neutral and one of them has to be detected at 0° or 180°. In the only case measured until now /9/ (the deuteron break-up) the target was at the same time a proton detector. Due to the geometrical restriction ($\Theta_1 = 0^o$) the kinematics does not depend on any azimuthal angle and the events are coplanar. Since the target is also the detector, a large part of available phase space is explored, both the regions where quasi two-body processes are present (QFS and both n-n and n-p FSI) and regions far from their influence. Another interesting feature of this kind of experiment is the possibility of measurement along the constant relative energy locus, where it is hoped that the cross section is sensitive to the off-shell effects /10/. In many respects this type of experiment seems to be superior to all other complete experiments mentioned above. If the neutron detector (in the case 2b) is moved from 0° or 180°, complete experiment degenerates into a special type of incomplete experiment, which was also used in the measure-

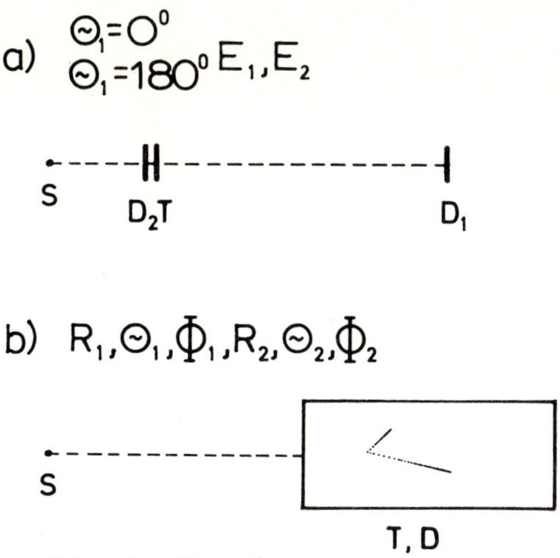

Fig.3. Complete experiments.

ment of the neutron energy spectrum from a deuteron break-up reaction (see e.g. /11/).

Fig. 3 shows two other set-ups,where only charged particles are detected. Fig. 3a is similar to Fig. 2b except that the target does not act as a detector, but is very close to one of the detectors. Again, because of the detection at 0° or 180°, the same kinematic conditions are fulfilled. Also, a large part of available phase space is explored. Our preliminary investigations with a counter telescope as D_1 and a silicon detector as D_2 gave promising results.

Nuclear emulsions (Fig. 3b) loaded with light elements are also used in the study of three-body reactions (see e.g. /12/). Such emulsions are at the same time targets for neutrons and detectors of charged particles. One can deduce complete kinematics of the event from the "star" produced by charged particles in the emulsion. Similarly,they could be used for the study of other many-body reactions (n>3), the measurement of which is almost impossible by counter techniques. However, complications in the analysis of events from these reactions arise when some of the particles are neutral and when incident neutrons are not monoenergetic. Advantages of nuclear emulsions are experimental simplicity and 4π geometry. However, serious shortcomings are poor energy resolution and tedious analysis. Even so, these measurements can give valuable information on mechanisms and properties of nuclear states.

At the end one has to mention possible ways to improve angular and energy resolution and statistics, which were very poor in many neutron experiments. The use of position sensitive detectors and multidetector systems could answer to these needs.

References

1. K.Ilakovac, L.G.Kuo, M.Petravić, I.Šlaus and P.Tomaš, Phys.Rev. Lett. 6 (1961) 356
2. R.C.Haight, S.M.Grimes and J.D. Anderson, Bull.Am.Phys.Soc. 20 (1975) 1194
3. V.Valković, I.Šlaus, P.Tomaš and M.Cerineo A98 (1967) 305
4. I.Šlaus, J.W.Sunier, G.Thompson, J.C.Young, J.W.Verba, D.J.Margaziotis, P.Doherty, R.T.Cahill, Phys.Rev.Lett. 26 (1971) 789
5. B.Zeitnitz, R.Maschuw, P.Suhr, W. Ebenhöh, J.Bruinsma and J.H. Stuivenberg, Nucl.Phys. A231 (1974) 13 and refs. therein
6. E. Bovet, F. Foroughi and J.Rossel, Helv.Phys.Acta 48 (1975) 137
7. V.Valković, M.Furić, Đ.Miljanić and P. Tomaš, Phys.Rev. C1 (1970) 208
8. D.Vranić, I.Šlaus, G.Paić and P.Tomaš, in Proc.Int.Conf. Few Body Problems in Nuclear and Particle Physics, Quebec 1974, ed.R.J.Slobodrian et al. (Les Presses de l'Université Laval, Quebec, 1975) p. 660.
9. J. Kecskeméti, T. Czibók and B. Zeitnitz, Nucl.Phys.A254 (1975)110
10. M.Jain, J.G.Rogers and D.P.Saylor, Phys.Rev.Lett. 31 (1973) 838
11. E.Arai, Nucl.Phys.A160 (1971) 161; H.Grässler and R.Honecker, Nucl.Phys. A136 (1969) 446
12. B. Antolković, Nucl.Phys. A231 (1974) 29

THE STUDY OF FEW PARTICLE SYSTEMS WITH PHOTOGRAPHIC EMULSIONS

B. Bhowmik

Department of Physics and Astrophysics
University of Delhi, Delhi, INDIA

Λ-N INTERACTION

Although Photographic Emulsions have been in use for more than 25 years their usefulness in some special areas still remains supreme. For instance, the bulk of our knowledge regarding Λ-N and ΛΛ forces has been derived essentially from hypernuclear studies in nuclear emulsions produced by strange particle beams. It was however realized in the early 60's that for an accurate estimation of the binding energy B_Λ of Λ-hyperon, systematic errors such as uncertainty of Range Energy relation, the stopping power and shrinkage factor, may play a dominant role. One way out was to estimate the mass of Λ-hyperon in the same stack of emulsion which could provide a yard stick for B_Λ estimation, thus eliminating the errors mentioned above. The work of the Delhi group [1] was unsatisfactory because of poor beam intensity. Only 123 events were used and by combining the then available world data (317 events) the mass value obtained was $M_\Lambda = (1115.40 \pm 0.13)$ MeV. The statistical error was (0.05) MeV and the rest stems from the above mentioned systematic errors. Using this value of Λ-mass, the binding energy for the lightest hyper-nucleus "hypertriton" (Λnp) was obtained to be B_Λ (3H) = (0.14 ± 0.14) MeV on the basis of 29 events. Obviously the accuracy was poor.

Another point arising out of the possible difference in binding energy of 4H and 4He could not be ascertained with confidence with the then available accuracy.

With the improvement of beam intensity and purity, investigations were launched in a big way by the European collaborators in the early 70's. Jurie et al [2] have collected 37,000 π^--mesic hypernuclei of which 4,000 were uniquely identified from decay kinematics. The B_Λ for the lightest hypernuclei system was obtained as B_Λ (3H) = (0.13 ± 0.05) MeV (204 events). The mass of the Λ-hyperon was determined in the same stack, and the value was $M_\Lambda = (1115.57 \pm 0.03)$ MeV (490 events). The stopping power was calibrated from the measured range of (972) protons in $\Sigma^+ \to p + \pi^0$ decay. The suspicion that the Delhi compilation of Λ-mass could be in error due to an uncertainty in range-energy relations arose from Schmidt's [3] measurement of Λ-mass in a bubble chamber using momentum measurements. His M_Λ value was $M_\Lambda = (1115.61 \pm 0.07)$ MeV.

Let us now consider,

B_Λ (4H) = (2.04 ± 0.04) MeV (155 events).
B_Λ (4He) = (2.39 ± 0.04) MeV (279 events).

The difference of binding energies, $\Delta B_\Lambda = B_\Lambda$ (4He) $- B_\Lambda$ (4H) $= (0.35 \pm 0.05)$ MeV, is now clearly established and indicates a significant breakdown of charge symmetry in Λ-hyperon interaction. Such a difference cannot be attributed to known Coulomb effects [4] which were estimated to be ~ -0.25 MeV. The most important source of charge symmetry breaking in the Λ-nucleon interaction is likely to be iso-spin mixing of baryons and mesons (ΛΣ mixing and $\pi^0\eta$ mixing).

Λ-Λ INTERACTION

The Knowledge of the Λ-Λ force is very meagre as there are only two [5] examples of double hyperfragments observed so far, $_{\Lambda\Lambda}$10Be and $_{\Lambda\Lambda}$6He. The second event allows

$B_{\Lambda\Lambda}$ estimation from both the production and decay kinematics and one gets a value $B_{\Lambda\Lambda} = (10.8 \pm 0.5)$ MeV, $\Delta B_{\Lambda\Lambda} = B_{\Lambda}({}_{\Lambda\Lambda}6He) - 2B_{\Lambda}({}_{\Lambda}5He) = (4.7 \pm 0.5)$ MeV. Clearly, more examples are needed to augment our library of double hyperfragments.

The two observed events of double hyperfragments were produced by the capture of Ξ^- - hyperon at rest. However, slow Ξ^--hyperons are hard to get as they generally decay in flight due to their short lifetime, about 10^{-10} s. Only hyperons produced backward in the c.m. system are likely to be slow enough to survive. Besides, they may be slowed down due to passage through some heavy material like the walls of shielding material or the pole piece of a magnet. Incidentally, the ${}_{\Lambda\Lambda}6He$ was produced by a Ξ^- -hyperon coming from outside the photo-emulsion stack which was receiving exposure to a 5 GeV/c K^- beam parasiting behind the Brookhaven bubble chamber commissioned for the second run for the Ω^- experiment.

LOADING TECHNIQUE

For future experiments, therefore, it is worthwhile to employ a method of quickly slowing down the hyperons by using some heavy material as a moderating medium. The loading technique seems to be an answer to this problem. For instance, if the photo-emulsion is loaded with tungsten powder of 5-20 μm size, the Ξ^--hyperon produced in such a target may be slow enough to give rise to a Ξ^- capture producing two Λ-hyperons, $\Xi^- + p \rightarrow \Lambda + \Lambda + 28.5$ MeV. Both the hyperons may remain trapped in the fragments of the capture star. The actual possibility of tungsten loading has been demonstrated by Lord [6].

REFERENCES

1. B. Bhowmik and D. P. Goyal, Nuovo Cim., 28 (1963) 1494.
2. Jurie et al. Nucl. Phys. B52 (1973) 1.
3. P. Schmidt, Nevis Report (1965) 140.
4. R. H.Dalitz and Von Hippel, Phys. Lett. 10 (1964) 153.
5. M. Danysz et al., Phys. Rev. Lett. 11 (1963) 29.
6. J. J. Lord, NALREP October 1974.

DISCUSSION

Y. Prakash: I have two comments to make. First is that for high energy physics work emulsions have been succesfully used in association with spark chambers. The spark chambers are used to locate the origin of desired events. This reduces the scanning efforts. Recently a proposal for the search for charmed quarks has been submitted to the FNAL where this technique has been exploited through neutrino interactions. The second comment is that one has to be rather cautious in using emulsions for low energy physics where one has to find energies of ions whose ranges are < 20 μm. In this region accurate range energy relations are not available. Further, calibrations can not possibly be relied upon when taken from one batch of emulsion to another or from one charge ion to another. This is of particular importance for experiments of the type reported by Miljanic.

A SHORT REPORT ON THE VARIABLE ENERGY CYCLOTRON
AT CHANDIGARH

H. S. Hans
Department of Physics, Panjab University, Chandigarh, India

The variable energy cyclotron, which was transferred from the department of Physics, Rochester University, Rochester (N.Y.), USA, is now functioning satisfactorily at Chandigarh. The machine is housed in an underground building of about 80′ x 80′ (Figure). Apart from repairing, renovating and suitably modifying the various components, a few major modifications have been made, which have improved the functioning of the machine. (1) The Dee and the central stem of the cavity have been directly connected together electrically and not through a press contact as before. This change has reduced the radiofrequency heating due to contact resistance to a very large extent. (2) Another 12" diffusion pump with 2000 litres/sec capacity has been added to increase the pumping speed. This additional pump has resulted in a very stable operation of the oscillator. (3) The vacuum chamber containing the Dee has been supported from inside with brass supports, and further spacers have been provided between the chamber and the pole pieces. These spacers helped us to obtain undisturbed vacuum with or without the main magnetic field. (4) A closed circuit chilled water system has been provided to supply chilled water at 30 lbs/sq. in. pressure to the cooling coils of the magnet, the oscillator tube and the vacuum pumps.

We have obtained a maximum internal current of >10 microamperes of protons at about 4 MeV energy at the outermost orbit inside the chamber. The energy will be further increased to near the full value of 8 MeV with the availability of a spare oscillator tube (RCA 5771). Normally, with an internal current of 4-6 microamperes at the outermost orbit, it is possible to obtain about 0.5 microampere at the target about 4 meters away after the analysing magnet. The oscillator frequency, the amplitude of the Dee-voltage, the main magnetic field and the analysing magnetic field are stable enough to provide a reasonably stable beam on the target. Some of the characteristics of the machine are: Magnetic field 14 kiloGauss, power consumption in magnet-400 amperes at 100 volts, frequency range of the oscillator-10 to 20 MHz, maximum oscillator power-20 kVA, Dee voltage-30 kV, ion source-hooded arc type, particles and energies-protons 2 to 8 MeV, deuterons 2 to 4 MeV, alphas 2 to 8 MeV, 3-He 22 to 11 MeV, total power consumption-60 to 70 kVA. The details of the functioning of the machine will be published shortly.

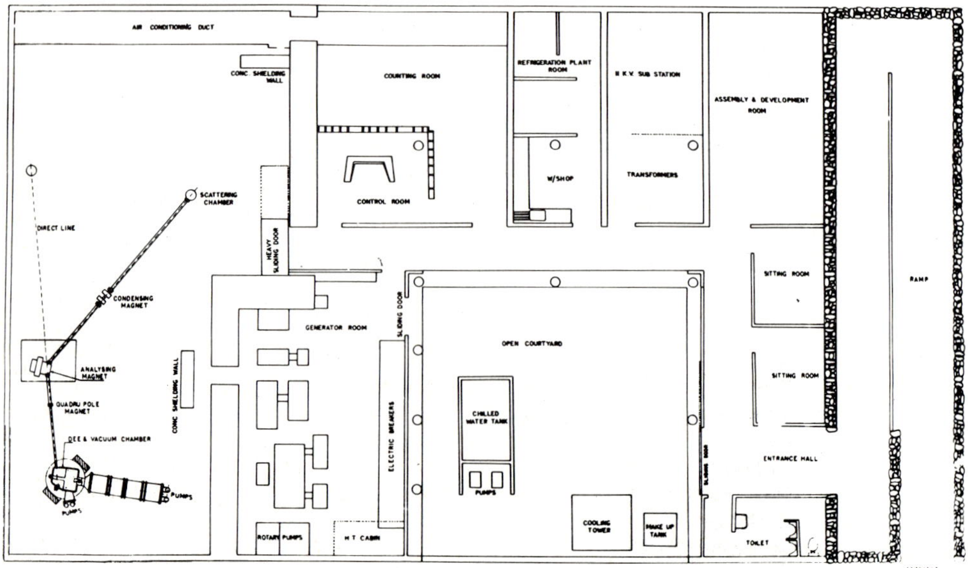

SOME TECHNIQUES OF CORRELATION EXPERIMENTS

Richard G. Allas
Naval Research Laboratory, Washington, D. C. 20375 U.S.A.

Since this is a workshop on experimental techniques, I shall not discuss any results or information to be gleaned from correlation experiments. In the brief time available, I shall limit myself to general remarks concerned with correlation techniques. In particular, I shall not discuss manufacturers and their equipment by name, except where needed to make a general point. I also would like to leave time for discussion of systems and procedures that people may have developed and might want to present. Most of my remarks will be limited to correlated pairs of charged particles. If a neutron or γ-ray are detected, some of the remarks will not apply, at least not in that leg.

Let us first look at a relatively simple system $^3He + ^3H = A + B + C$ where we detect all possible charged particle pairs A,B and leave particle C undetected, either charged or neutral. Table I gives the results. There are 18 possible charged particle pairs. A few (3) are two-body reactions and thus do contribute only at particular angle pairs; some may not be kinematically allowed at all angle pairs; some may be created in only small amounts. Thus, by judicious choice of angle pairs, bombarding energy, detectors (thickness, veto) one may be able to eliminate some of the possible charged particle combinations. But, in most instances, the elastic scattering of the beam itself from the target will create an undesirable chance coincidence background. Thus, we are lead to the almost universal requirement of particle identification. In some cases identification of one particle may be sufficient, but in general both pairs will have to be identified. Figure 1 is an attempt to illustrate all the salient

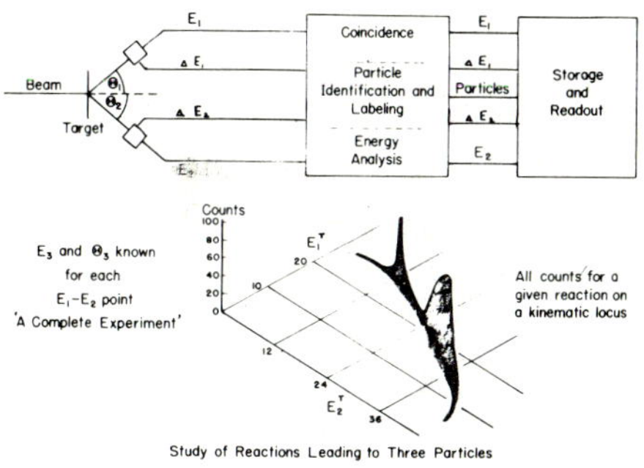

Fig. 1. Schematically illustrates a two-body coincidence measurement set up. If the detected particles are from a three-body breakup; the expected results are sketched in the lower part of the figure.

features of a three particle system where two particles are detected. If only groups of properly identified particles are selected and one plots the energy E_1 of each particle in one group vs. the energy E_2 of the particle in coincidence with it in the second group, one gets a locus as sketched in the lower part of Fig. 1. The pronounced peaks are shown to illustrate some dynamical situations like final state interactions or quasifree scattering which, however, are not part of this talk. Such individual loci are generated for all allowed particle pairs in a three-body reaction and may cross each other, or be so close together as to be experimentally very difficult to distinguish. In addition, accidentals and genuine four and more body breakup reactions may give an undesired background.

I shall briefly comment on particle identification techniques (PID). Recently, an excellent review article by F. Goulding and B. Harvey [1] has been published with a good biography for further details. It is a "must" reading for all not already familiar with the latest techniques in particle identification, in particular heavy ion identification. Except for photographic techniques, time-of-flight (TOF), and separation in magnetic fields, practically all other methods rely on measuring the ΔE and E of the particle. The photographic emulsion technique is discussed in a separate paper and I shall say no more about it. Using the $\Delta E, E$ measurement one employs some algorithm to identify the particle, e.g. $C_p = f(\Delta E, E)$. Here C_p stands for a particular identified particle species. All algorithms are usually based on the Bethe-Block type energy loss equation. One of the more common algorithms is based on

$$C_p = (E + \Delta E)^k - E^k. \qquad (1)$$

Here k is a constant of the order $k \sim 1.7$. I should point out that more accurately the above equation does not represent a particle species directly, but rather is proportional to Mq^2_{eff}. For light ions the distinction is not important, but must be recognized for heavy ions. Many hardwired particle identifiers (some commercially available) solve Eq. (1) in an analog way. Computer based systems can either solve the above equation or relay on sets of range/energy tables and calculate $C_p = [R_p(E + \Delta E) - R_p(E)]$. Here $R_p(E)$ is the range of particle "p" with energy E in the detector. Using suitable look-up techniques one can identify a large variety of particles without turning to an identification algorithm. Especially taxing problems are encountered if heavy ions have to be separated, e.g. ^{12}C, ^{13}C, ^{14}N,... In some cases one has to split the energy detector into several detectors, e.g. $E_T = \Delta E_1 + \Delta E_2 + E$. Obviously, if a mixture of particles of varying ranges is encountered E_2 could be the ΔE_2 detector for some low range particle. By arranging a set of detectors of various (increasing) thicknesses and summing signals from several thin ΔE detectors in the stack one can achieve a "thick" ΔE detector needed for high range particle identification in the presence of less penetrating heavy ions. If one is not interested in high range particles one uses a "veto" detector as the final detector in the stack. Analog circuitry has been developed for triple detector identifier systems. In general, this method has practical limitations because of the number of ADC's available (or inputs to multiplexed ADC's) since each detector signal has to be digitized separately, but in coincidence with all other signals. After the particles have been identified, the identification and energy must be organized and stored for data retrieval.

Also measuring the TOF of a particle through a known distance, combined with an energy measurement, will determine the mass. Nonrelativistically we have

$$v^2 = 2E/M \text{ or } M = 2E(t/d)^2, \qquad (2)$$

where t is the time and d is the distance traveled. Unfortunately, long flight path necessarily involve serious efficiency problems due to the poor collection geometry. However, short path length require very fast timing circuitry. Since energy and distance travel usually are quite well known, the main uncertainty is in the time measurementand and we get

$$\Delta M/M = 2.8(E/M)^{\frac{1}{2}} \Delta t/d . \qquad (3)$$

Some of the best timing achieved to date is of the order of 100 ps.

Bending in a magnetic field also can be used for partical identification since (nonrelativistically) $B\rho \sim Mv/q$. However, for heavy ions in particular, the effective charge frequently is not Z but $q_{eff} \leq Z$ since charge changing collisions can occur in the residual gas. Heavy ions also can be emitted from the target in a variety of charge states.

At NRL we use both hardwired particle identifiers and computer generated particle identification. We have six ADC's interfaced to our computer. All coincidence determination and timing is done by external circuitry. In the usual set-up two of the ADC's are fed the summed ($\Delta E + E$) signals from the two detectors. These are combined into a 24-bit word for transmission to the computer. The lower 12 bits are fed by one detector and the upper 12 bits are fed by the other detector. We normally operate in 10 bit (1024 channels) resolution. This leaves 4 bits in the 24-bit word. One bit is reserved for special computer use. Thus, 3 bits are left for particle identification tagging. This gives eight possible combinations, however, only seven are used because all 0 is not an acceptable condition due to presence of non-tagged events. The computer monitors two sets of particle pairs selected by their tag bits, assembles the information into E_1 vs. E_2 matrixes and displays the matrixes in two 64 * 64 arrays for on-line monitoring purposes. In addition, all coincidence $\Delta E_1, E_1$, $\Delta E_2, E_2$ signals are fed to the remaining four ADC's. Each $\Delta E, E$ signal is digitized and assembled into a 24-bit computer word. All six digital outputs from the six ADC's are transmitted as a group to the computer. Thus, both hardwired particle identification and $\Delta E, E$ signals for each event are stored. Off-line analysis generate particle identification and total energy of all possible charged particle combinations, especially those that are not available in the seven hard-wire tagged set. A scaler bank also is read into the computer at periodic intervals.

Sometimes one can combine pre-selection of interesting events for later computer analyses. An interesting technique is to combine TOF and $\Delta E, E$ measurements. The ΔE and E detector are separated by a flight path and the ΔE detector is used

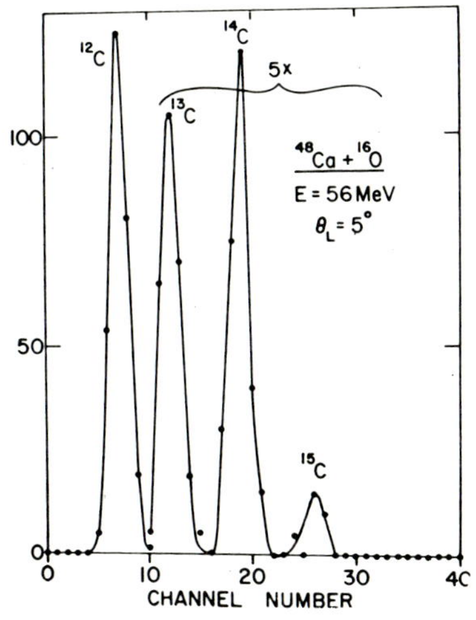

Fig. 2. Shows separation of C isotopes by combining TOF and $\Delta E, E$ techniques. This figure is taken from reference [2].

to generate the start pulse for the TOF. Combining these techniques the excellent particle identification spectrum for heavy ions shown in Fig. 2 was obtained [2].

After particle identification one must store the information for analysis and retrieval. Use of multichannel analyzers limits one to only a few particle pairs. Sometimes a magnetic tape is attached to the multichannel analyzer for event by event recording for later analysis. However, most accelerator laboratories have on-line computers and the information is transferred to them. Here the question is how to provide the interface from the nuclear instruments to the computer. One route is to go directly to an individual computer interface implemented in-house. Or, one may decide to go CAMAC. My remarks on CAMAC are by necessity very limited and I certainly cannot do justice to this subject. An indication of the scope of this topic is that recently (14-16 October 1975) the Second International Symposium on CAMAC in Computer Applications was held in Brussels. There are many excellent articles published on the subject and some of the latest results will be available in the proceedings. Meanwhile, I would like to call your attention to the CAMAC Bulletin (Commission des Communautés Européennes, DG XIII, 29 rue Aldringen, Luxembourg). Recently [3] it has published two supplements. Supplement A called "For Newcomers Only" by Hans-Joachim Stuckenberg of DESY, Hamburg, Germany, gives an introduction to CAMAC, both hardware and software. Supplement B contains a bibliography assembled by the same author and organized by subject matter such as, e.g. CAMAC Applications Notes in Environmental Control.

Basically, CAMAC is an all purpose interface (hardware and software) for exchanging data and control information between a computer and its on-line data inputs and outputs. It is designed to be independent of the host computer, or in special cases to operate on a stand alone bases. Recently, some computer manufacturers have made available special crate controllers which interface directly from the I/O bus of their computer to the CAMAC highway. Many different manufacturers are supplying various devices to fit into the CAMAC crates. The CAMAC bulletin carries a CAMAC product guide. A total system based on a PDP 11/05 computer is shown in Fig. 3. This figure is taken from

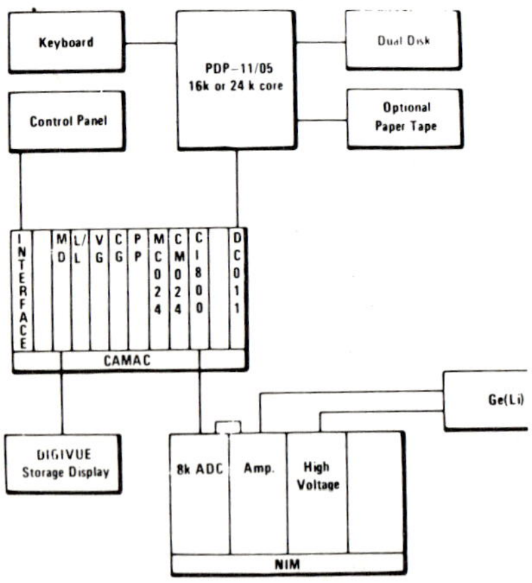

Module Diagram

Fig. 3. Illustrates one possible computer-CAMAC-NIM self-contained system.

applications note #1 of the bulletin [3]. The system is commercially available, but used here only as an illustration of what is available. Finally: CAMAC stands for Computer Aided Measurement And Control or, in a lighter vein, Confuse All Measurements in Any Case [4].

One of the newest developments in instrumentation that is undoubtedly going to find many applications in the laboratory is the microprocessor. Some microprocessors can be combined with CAMAC as shown in Fig. 4. Some probably will be developed to take over many of the functions presently handled by hardwired instrumentation, e.g. in coincidence circuitry, ADC controllers to identify and format coincidence events in many detector environments and possibly even "firmware" particle identifiers. However, for the most part they still are a "do it yourself" project.

Since this is a workshop in experimental techniques, I should like to point out a possible source of trouble and error that may be easily overlooked [5]. This

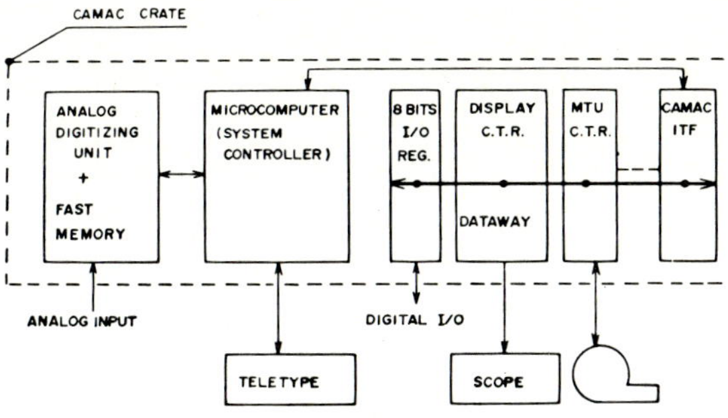

System Structure

Fig. 4. A self-contained microprocessor controlled CAMAC crate with I/O devices. This figure was taken from [3] Application Note #2.

source of trouble, namely the detector itself, came as a mild surprise to us. We recently had reason to question some α-α coincidence scattering measurements. An investigation showed that one of the detectors had a different pulse shape (rise time) depending on the position were the incident particle struck. But, let me describe what we did and what we found.

A 13.7 MeV-proton beam from the NRL Cyclotron was focused and collimated down to a 1-mm spot size on a gold target. The detector was mounted on the upper turntable in a 30-inch ORTEC scattering chamber. An aperture of 1-mm diameter

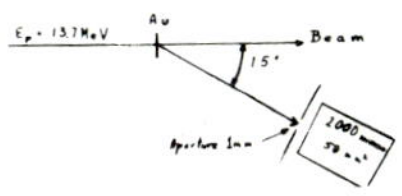

Fig. 5. Simple set up for checking detectors.

was mounted on the lower turntable so that it could be rotated in front of the detector. The small aperture was moved in front of the detector and held fixed while the detector was rotated so as to expose different areas of the detector to elastically scattered protons at 15°. (Fig. 5). The results are shown in Fig. 6. Here we plot counts/integrator vs. displacement (rotation through an angle has been converted to displacement) measured from the geometrical center of the detector. The upper curve is for scattering from gold, the lower curve is for scattering from carbon. The two curves have different and arbitrary normalizations. The important points to note are that the center of the active area and the geometrical center do not coincide and that the active area of the detector is only about one-half of the nominal 50 mm^2 (should be $\pm$ 4-mm displacement). A similar situation was observed in five out of six detectors tested.

In Fig. 7 we show some other features found in a "bad" detector. The upper curve is counts integrator measured as described above. The second curve is the peak channel vs. position. The shift in peak channel is $\sim$ 150 keV or $\sim$ 1%. The third curve is the FWHM as a function of position and it changes from $\sim$ 40 keV to $\sim$ 100 keV (3 channels to 7 channels). Thus, we see that a single counting test, such as frequently done with an α-source, would not reveal defects of this nature.

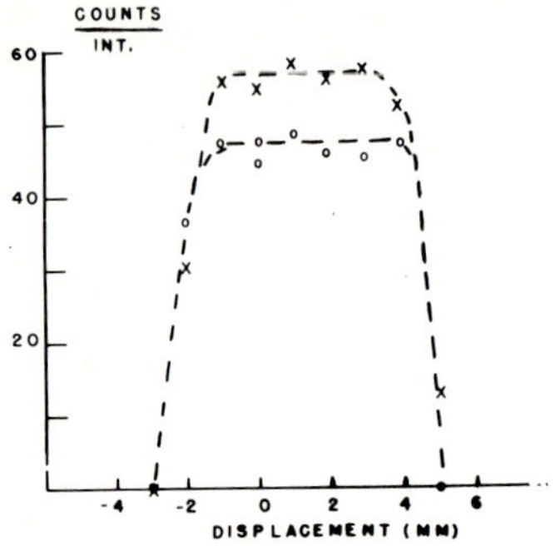

Fig. 6. Counts/Integrator vs. position in the detector obtained with the set up of Fig. 5.

The effect on a coincidence experiment using such a detector can be devastating and is illustrated in Fig. 8. Here we used a mylar target to detect p-p

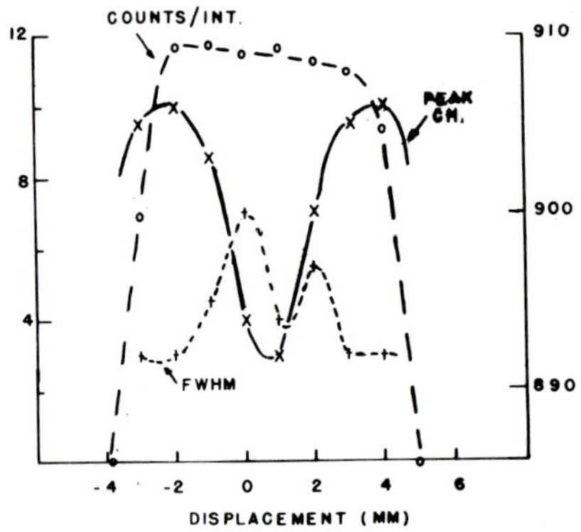

Fig. 7. Here we show the various parameters measured and their variation with displacement. The right-hand scale is for the peak channel. The left-hand scale is relative counts/Int. or absolute in channels for the FWHM.

coincidences in the $^1H(p,p)^1H$ reaction at 45° - 45°. One detector was kept fixed and the other rotated in a similar way as for the single counter tests.

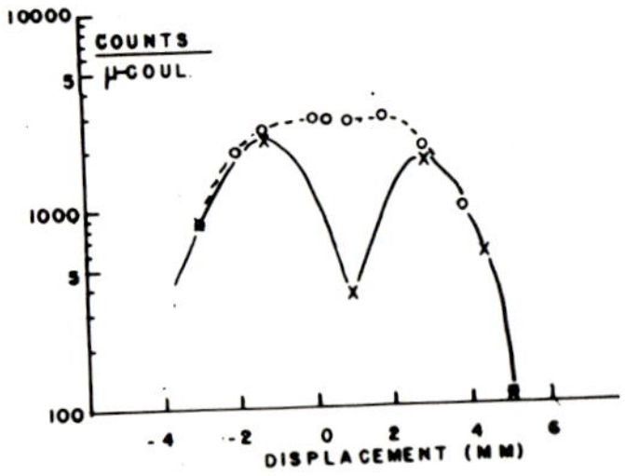

Fig. 8. The results of a coincidence experiment. The upper curve is for two good detectors. The lower curve is for one bad detector.

The upper curve is for a good detector, the lower curve is for the previously described "bad" detector. Note the logarithmic scale. This count loss does not show in a simple single counter test. If such a "bad" detector were used in a coincidence experiment it could cause trouble or even highly unreliable results. The problem, as previously mentioned, was traced to changes in rise time and pulse shape of the detector depending on the position in the active area were the detected particle struck. This particular detector was a used detector. We did not have any new detectors on hand to see if such problems can be encountered with brand new detectors. In the future, we plan to test all our new detectors in a way similar to the one described.

Table I

$^3He + ^3H \rightarrow A + B + C$ (undetected)

	$A(\theta_A)$	$B(\theta_B)$	C
a)	p	p	^4H
b)	p	d	t
c)	p	t	d
d)	p	h	2n
e)	p	α	n
f)	d	p	t
g)	d	d	d
h)	d	t	p
i)	d	h	n
j)	d	α	-
k)	t	p	d
ℓ)	t	d	p
m)	t	h	-
n)	h	p	2n
o)	h	d	n
p)	h	t	-
q)	α	p	n
r)	α	d	-

References

1. Fred S. Goulding and Bernard G. Harvey, Ann. Rev. Nucl. Sci. 25 (1975) 167.
2. B. Zeidman, W. Henning and D. G. Kovar, Nucl. Instr. Methods 118 (1974) 361.
3. CAMAC Bulletin 13 (1975) - distributed by: Commission des Communautés Européennes, 29, rue Aldringen, Luxembourg.
4. CAMAC Bulletin 13 (1975) Supplement A, 20.
5. A. G. Pieper and L. A. Beach, private communication.

DETECTION TECHNIQUES AT INTERMEDIATE ENERGIES

J.M. Cameron
Department of Physics, University of Alberta, Edmonton, Alberta, Canada

Today many more people are entering the field of intermediate energy nuclear physics--some because of their own preference and some perhaps of necessity! Many of us undoubtedly repeat the same development projects. I plan to review briefly only some examples of systems planned and hope that you can contact the appropriate source to get details. Because of time constraints I choose not to discuss the use of high resolution spectrometers; a very comprehensive account of their use has recently been presented.[1])

The cheapest detection system one can use is probably a range-counter telescope. As an example, I will describe one system being used in early experiments involving few nucleon systems at TRIUMF.[2]) Each spectrometer consists of a 5" x 5" multiwire proportional chamber, two thin 5" x 5" plastic scintillators separated by a copper degrader (8" x 8") and, finally, a NaI (Tℓ) detector which is 3" thick and 5" in diameter (see figure 1). These NaI detectors should be capable of a resolution better than 1%; however, to date our measurements have been limited by beam energy spread and kinematic broadening, resulting in 3.5% FWHM as shown in figure 1. When used with 0.25" plastic scintillators separated by 1 meter flight path, the system is capable of very good particle identification as seen in figure 2. The resolution of such a system and the energy bite attainable depend on the choice of the thicknesses of the absorber and the NaI. It turns out that the resolution of the system is, within wide limits, independent of the energy deposited in the NaI counter. As a general rule, for high energy protons in copper, the energy loss should be kept larger than 10% and smaller than 75%.[3]) However, a reasonable energy bite may dictate a thicker NaI than this. The resolution of such a system is then close to 1% for quite a range of parameters as shown in figure 3.

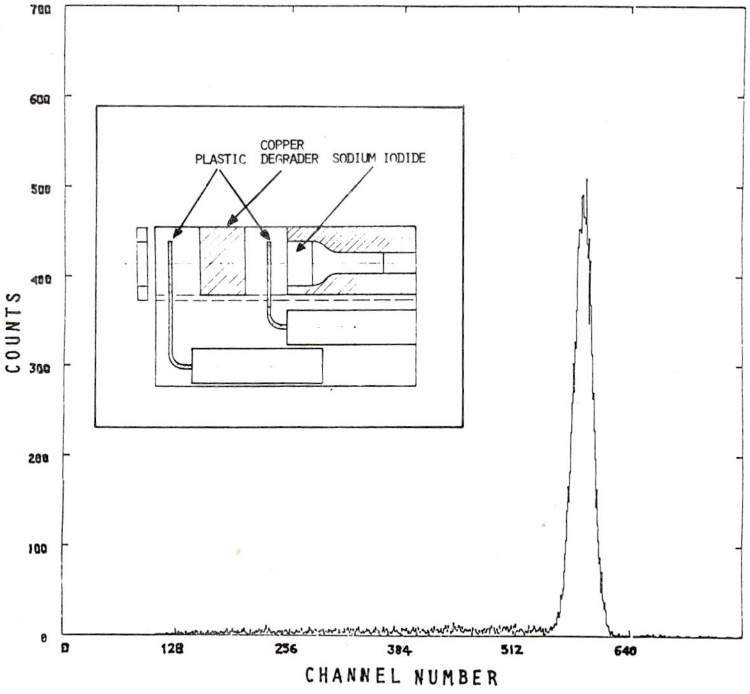

Figure 1. Energy spectrum of 150 MeV protons in 5" ϕ x 3" NaI detector.
Inset: Geometry of detector telescope.

Counter telescopes, such as the ones described above, do however suffer from large inefficiencies due to nuclear reactions in the absorber or NaI and due to scattering out effects. The former correction can be estimated from the available total reaction cross section data but the latter must be determined for each new geometry. It may be worthwhile to standardize on a number of basic geometries so that everyone need not repeat such tedious measurements. The results for our geometry are given in figure 4. It may be noticed that above 470 MeV there is no appreciable portion of the detector over which the efficiency is flat, as is the case at 252 MeV.

The need for good angular resolution dictates the use of multiwire proportional

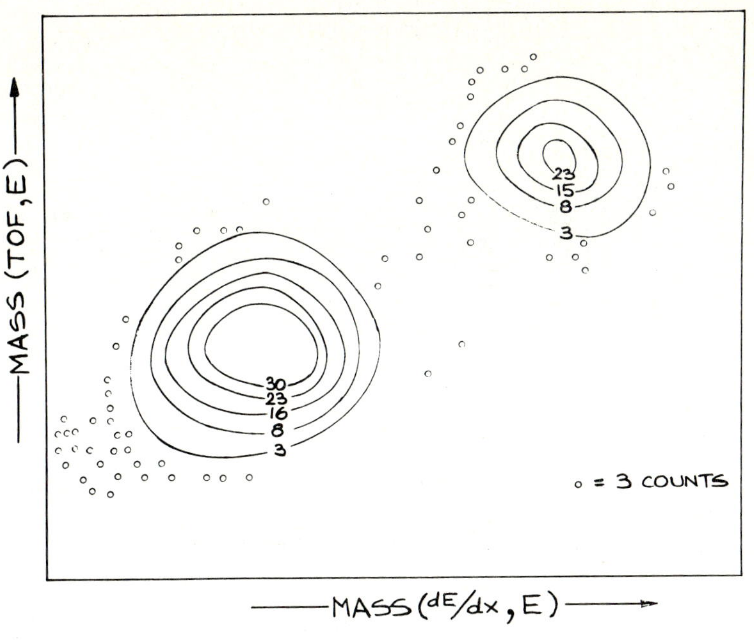

Figure 2. Particle identification spectrum obtained from 400 MeV protons on CH_2 target at a scattering angle of 46°.

chambers. There are many good, and some not so good, systems operating today and a limited selection is available commercially. Such chambers consist of planes of fine parallel wires mounted orthogonally. The wire planes are inserted between grids held at about 4500 volts. A wire spacing of 2 mm seems typical with some problems still being encountered when 1 mm spacing is attempted. Our 5" x 5" chambers work with the standard "magic" gas mixture bubbled through methylal.[4]) The latter step is necessary to prevent excessive current through the chamber in high intensity fluxes. The higher data rates available from newer machines force one to use a readout system with amplifiers and one shots on individual wires; however, in some circumstances delay-line readouts may still be more practical. Recent developments at Stanford of a dead timeless readout system [5]) with a pulse pair resolution of 125 nsec should be very useful to users of pulsed machines.

To obtain better resolution a Bifilar Helical system has been developed at LAMPF and 0.33 mm spatial resolution was obtained.[6]) The ultimate in resolution today seems to be the domain of the drift chamber, spectacular results of 100 µm being obtained with 10 cm detectors.[7]) The Saclay group use 50 cm drift chambers, shown in figure 5, with a resolution of 1.5 mm being attained.[8])

A number of investigations have recently been made of small angle elastic scattering using solid state detectors to detect the

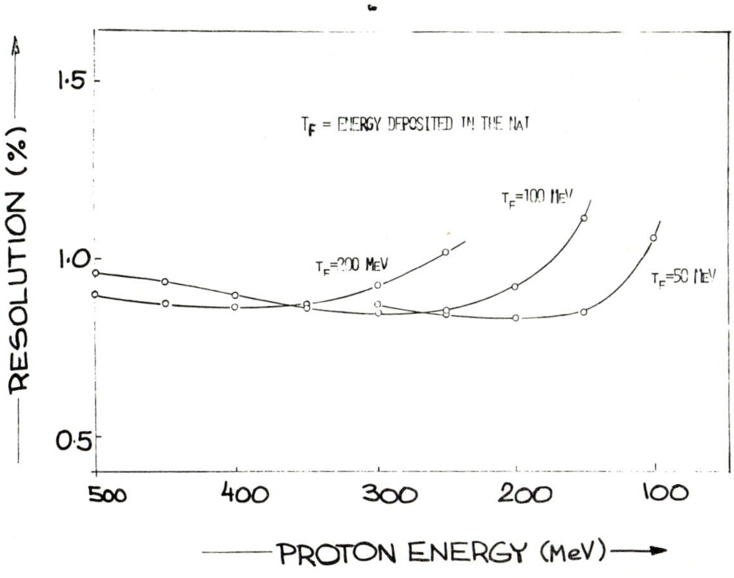

Figure 3. Energy resolution of counter telescope as a function of incident beam energy.

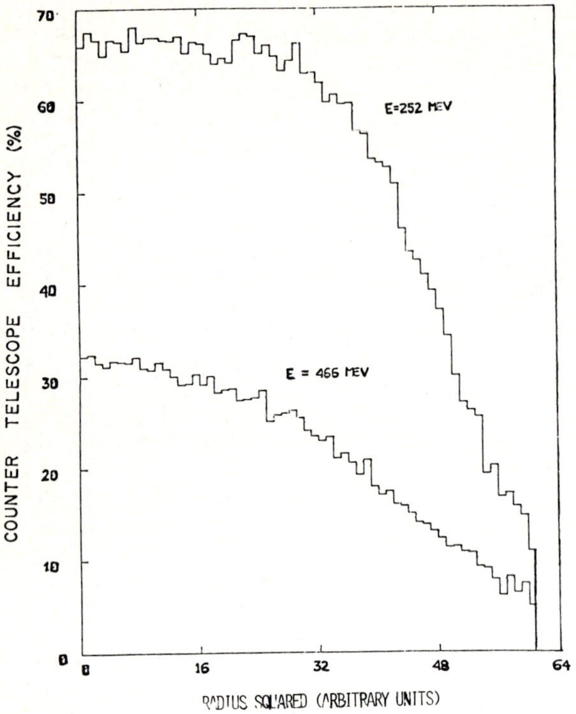

Figure 4. (a) The efficiency of the counter telescope as a function of radius squared.

(b) The peak efficiency as a function of incident proton energy.

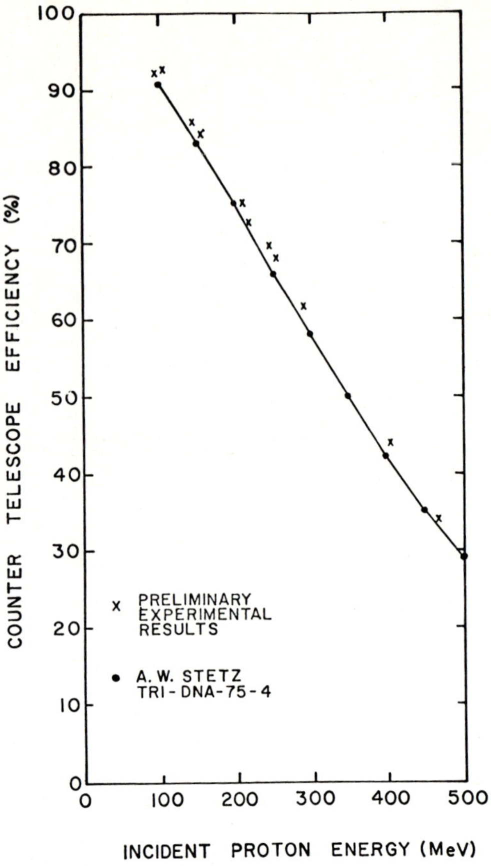

Figure 4 (b)

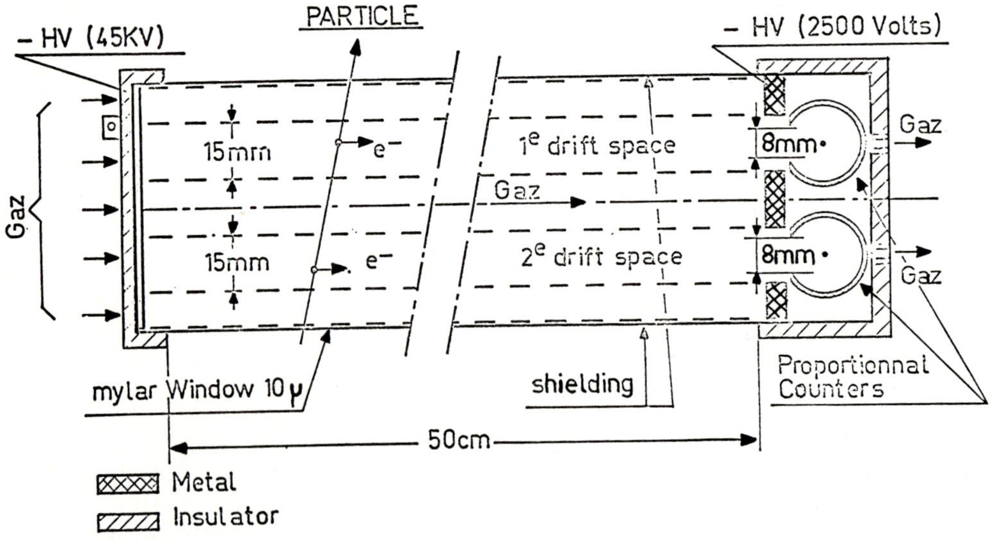

Figure 5. The 50 cm drift chamber used by the Saclay group.

recoiling nucleus.[9]) We have extended this technique by detecting the protons' position using a MWPC in coincidence with the low energy recoil detector and have obtained measurements down to 3°(Lab) at 200 MeV. Some of these results are presented elsewhere in these proceedings.[10])

A number of experiments involving an outgoing neutron have been performed as part of the initial program in the nucleon-nucleon area at LAMPF. The system used to analyze the neutrons consists of a proton recoil telescope in which a 10 cm liquid hydrogen target is used as a radiator. The protons produced by the n-p charge exchange are analyzed using a large solid angle (~ 5 msr) proton magnetic spectrometer. The system used is shown in figure 6 together with two spectra measured using it.[11]) For energies below 100 MeV it is still feasible to use reasonable (10 m) flight paths with time-of-flight techniques. The (p,pn) experiments being mounted at TRIUMF will do this and a resolution of 1.5% at 100 MeV is expected with present timing performance.

The use of high-purity germanium detectors for charged particle detection at intermediate energies is also a possibility.[12]) Again the same geometry considerations apply as in the case of range telescopes and one is restricted to a useful depth which is about twice the diameter. Currently germanium crystals are available up to about 4 cm in diameter, thus stopping 200 MeV protons may be considered as an upper limit. For heavier ions one is of course much better off. A major limit on the application of these detectors is radiation damage. A flux of about 10^9 n/cm^2 seriously degrades the gamma ray resolution but if the resolution is determined by other experimental parameters (e.g. target thickness, beam energy spread), as is often the case for charged particle work, levels above

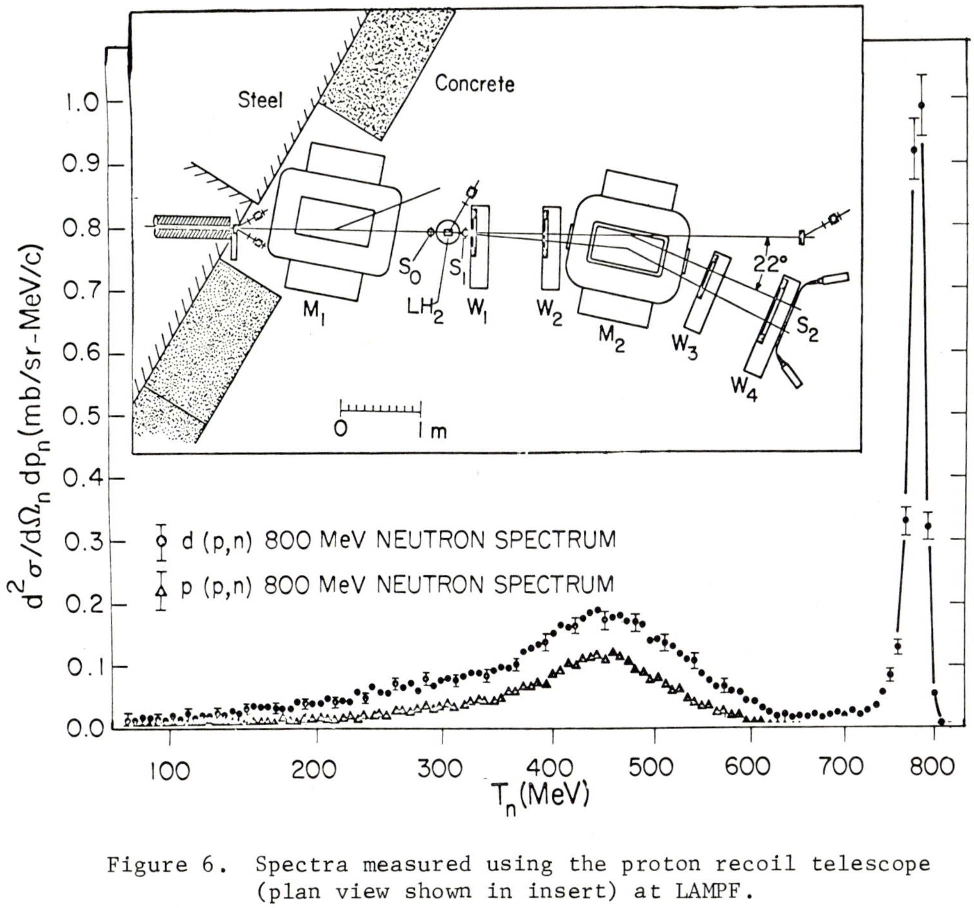

Figure 6. Spectra measured using the proton recoil telescope (plan view shown in insert) at LAMPF.

this may be tolerated. The detectors can be repaired by annealing them; this may even be carried out with the detector in the cryostat. The major problem at present is that the thickness of the n-type contact, made by drifting Lithium, also increases considerably during the annealing stage. This will be a serious drawback if a number of detectors is to be used in a telescope. These detectors should give an energy resolution better than 0.1%. The spectrum in figure 7 was obtained with 102 MeV proton using the p^3 channel at LAMPF. The resolution of 510 keV is due

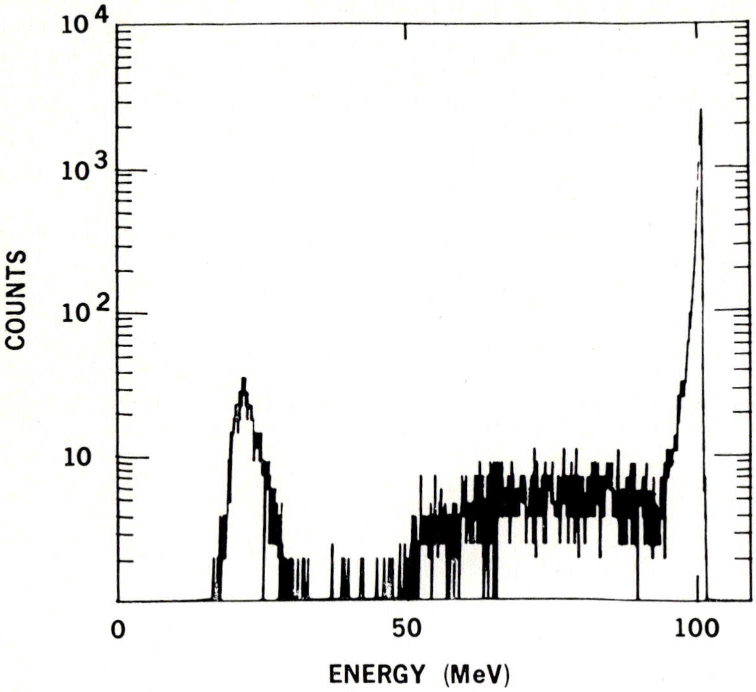

Figure 7. Energy spectrum taken with intrinsic germanium detectors of 102 MeV protons from the p^3 channel at LAMPF.

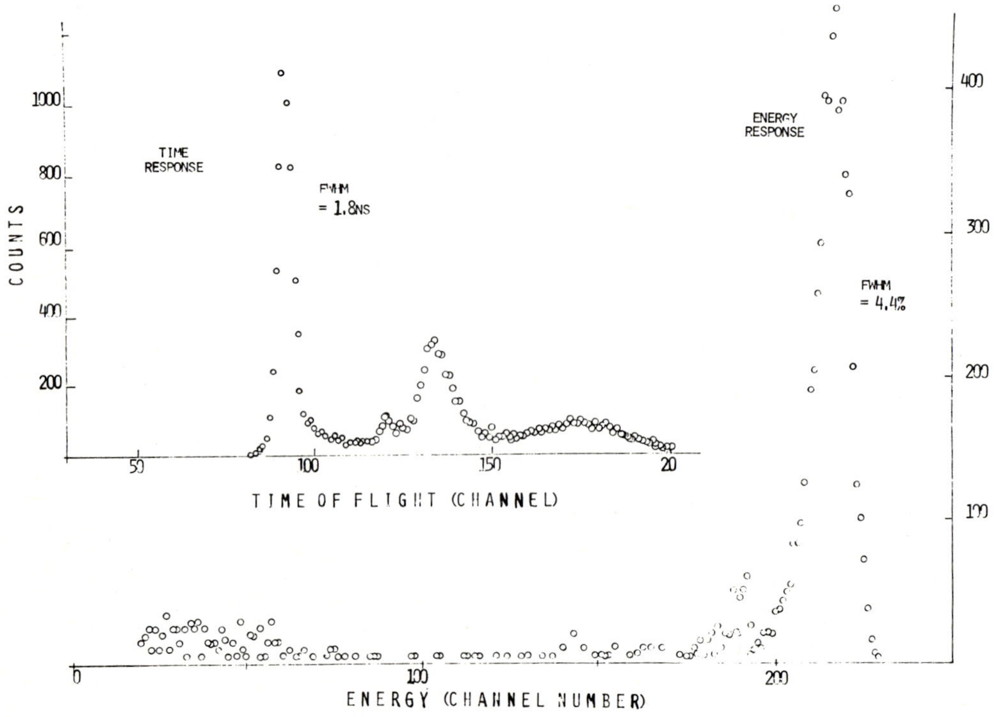

Figure 8. Spectra taken with an 18" ϕ x 20" NaI crystal at TRIUMF. The channel was tuned for 144 MeV electrons.

mainly to the energy resolution of the channel. A number of possibilities are being explored to replace the Li junctions, including ion implantation. It appears that these detectors will be very useful for the detection of charged particles up to a few hundred MeV.

Finally, a few words about high energy photon detectors. If resolution is not of importance, it is probably easiest to use lead-glass Cherenkov detectors of the type used at Orsay [13]) or Berkeley.[14]) Such detectors have a very high efficiency, approaching 100% for photons above 50 MeV, but the resolution is very poor, typically 30 to 50%. The use of large sodium iodide detectors is also becoming more common with the advent of intense secondary beams. The 18" ϕ x 20" "TINA" crystal has recently been tested at TRIUMF with 144 MeV electrons and an energy resolution of 4.4% obtained [15]) as shown in figure 8. Perhaps more impressive is the time resolution of 1.8 nsec which allows these detectors to be used in time-of-flight studies. It may also be possible to perform pulse shape discrimination to separate events due to high energy neutrons and gamma rays in these detectors as has been demonstrated to be the case at lower energies.[16])

I hope this review will give you some idea of the capabilities of detection systems now in use and perhaps serve as a starting point for those who want to produce new ones.

1. R. Beurtey, Lectures presented at the Summer School on Nuclear and Particle Physics at Intermediate Energies, Brentwood, British Columbia (1975). To be published.
2. W.T.H. van Oers, Proc. Inter. Conf. on Few Body Problems in Nuclear and Particle Physics, Les Presses de l'Université Laval, Québec (1975) 307
   J.M. Cameron, *Ibid.* p.807.
3. A.W. Stetz, TRIUMF Design Note, TRI-DNA-75-4 (1975).
4. G. Charpak, H.G. Fisher, C.R. Gruhn, A. Minten, F. Sauli, G. Plch, and G. Flügge, Nucl. Instr. and Meth. 99 (1972) 279.
5. S.L. Shapiro, M.G.D. Gilchriese, and R.G. Friday. To be published in Proc. of IEEE Nuclear Science Synposium (1975).
6. E.R. Flynn, S. Ottesen, N. Stein, H.A. Thiessen, D.M. Lee, and S.E. Sobottka, Nucl. Instr. and Meth. 111 (1973) 67.
7. A. Breskin, G. Charpak, F. Sauli, M. Atkinson, and G. Schulta, Nucl. Instr. and Meth. 124 (1975) 189.
8. J. Sandinos, J.C. Duchazeaubeneix, C. Laspalles, and R. Chaminade, Nucl. Instr. and Meth. 111 (1973) 77.
9. J.C. Fong, *et al.* To be published in Proc. of VI Inter. Conf. on High Energy Physics and Nuclear Structure, Sante Fe (1975).
10. J.R. Richardson, The TRIUMF Experimental Program. To appear elsewhere in these proceedings.
11. J.E. Simmons, private communication (1975).
12. J.F. Amann, P.D. Barnes, S.A. Dytman, J.A. Penkrot, A.C. Thompson, and R.H. Pehl, Nucl. Instr. and Meth. 126 (1975) 193.
13. A. Willis, V. Comparat, R. Frascaria, N. Marty, M. Morlet, and N. Willis, Phys. Rev. Letters 28 (1972) 1063.
14. D.I. Sober, M. Arman, D.J. Blasberg, R.P. Haddock, K.C. Leung, and B.M.K. Nefkeus, Nucl. Instr. and Meth. 108 (1973) 573.
15. M.D. Hasinoff, private communication (1975).
16. C.M. Bartle, Nucl. Instr. and Meth. 124 (1975) 547.
    J.D. Kurfess, G.H. Share, and R.B. Theus, Bull. Am. Phys. Soc. 20 (1975) 562.

DISCUSSION

R. Slobodrian: The lower spectrum of your slide on the Los Alamos neutron detection system corresponds to p+p going to what products?

M. Jain: The two spectra shown in the slide are zero degree neutron momentum spectra from bombardment of deuterium and hydrogen by 800 MeV protons. The spectrum from D(p,n)2p reaction exhibits a sharp almost monoenergetic high energy peak, separated by a valley and a continuum of low energy neutrons. The spectrum from hydrogen exhibits a similar broad peak, which is essentially due to $p+p \rightarrow n+p+\pi^+$ reaction as the cross sections due to two pion production and bremsstrahlung are very small. These will be discussed later in the conference.

N. Ganguly: Does the use of intrinsic germanium detectors offer any difficulty up to a point when annealing is required? Could you please give an order of magnitude of radiation could be used before annealing would be required.

J. Cameron: It appears that about 10^9 neutrons per cm^3 is sufficient to cause discernable deterioration of the resolution.

G. Igo: How much data do you have on the resolution of the NaI detectors? Isn't the limit on the detector thickness really associated with nuclear absorption rather than the diameter of the detector since you can make a correction with spacial detectors? When complex ions are incident on the Cu-NaI detectors do you have difficulty due to stripping?

J. Cameron: Our measurements of the resolution have been limited by kinematic broadening, when this is folded out the error is large but it must be better than 1%. Of course, with the positional information we can correct for geometrical dependence, this, however, takes quite a bit of computer time so it is easier if one has a plateau. So far we have no information on efficiency for heavier ions.

P. Singh: What range of energy deposited in the NaI crystal was used in the efficiency vs. energy curve that you presented? What is a typical dynamic range of the NaI cum copper degrader system ?

J. Cameron: The range of energy deposited in the NaI was from 80 to 150 MeV. The dynamic range decreases rapidly with increasing energy going from 120 MeV at E(p)=200 MeV to about 30 MeV at E(p)=500 MeV.

RELEVANCE OF FEW-NUCLEON PROBLEMS TO NUCLEAR POWER

A. S. Divatia
VEC Project
Bhabha Atomic Research Center, Bidhan Nagar, Calcutta, India 700064

INTRODUCTION

It is well known that the study of few-nucleon problems did not specifically start because they were relevant to nuclear power. However, as the need for power has become more urgent and the systems which may generate nuclear power in the future are likely to be highly complex, it has become necessary to examine the question of relevance of few-nucleon problems to nuclear power.

In order to study this question, we must examine the nuclear data needs for nuclear power, and see what fraction of the needs pertain to few-nucleon problems. It is also necessary to restrict the scope of study to nuclei up to carbon, since we want to consider only few nucleon systems. The nuclear data needs for nuclear power have been studied exhaustively by many groups all over the world and The International Atomic Energy Agency, operating through the International Nuclear Data Committee and their Nuclear Data section, have compiled and evaluated these nuclear data needs. It is therefore possible to draw upon the various studies and compilations of the IAEA for examining the question of relevance. We shall examine the relevant nuclear data needs for fission reactors, fusion reactors and nuclear safeguards programmes.

FISSION REACTORS

The world request list for Nuclear Data Measurements WRENDA-75, published in June 1975 [1] contains over 1200 requests for improved nuclear data needed in support of the fission reactor development programmes of 21 member states of the IAEA and one other International Organization. These requests deal with nuclei ranging from hydrogen to carbon. A few examples of the requests are given here:

Target	Projectile	Reaction	Energy Range	Accuracy	Priority
1-H	n	diff. elastic	0.7 - 20 MeV	20%	1
	n	elastic	1.0 eV - 1.0 keV	0.5 - 1%	1
6-Li	n	(n,x)	1.0 keV- 3.0 MeV	1.0%	1
10-B	n	(n,x)	1.0 keV- 1.0 Mev	2.0%	1

Thus, only about 3% of the total requests concern few-nucleon systems and therefore the role to be played by the experimenter is not quantitatively large. However, considering the accuracies of the data required, the role to be played by the experimenter is quite significant and important. The experimenter should therefore concentrate on improving the measurement techniques or on devising more ingenious ways of measuring the required data. A useful purpose can also be served if it becomes possible to calculate the cross-section to the required accuracy, and in particular to determine the energy dependence of the cross-section. Thus, a challenge is posed to both the experimenters and the theoreticians.

FUSION REACTORS

WRENDA-75 also contains nuclear data requests for fusion reactor development. There are 329 data requests from five member states. However, it is noted in WRENDA-75 without a model which permits determination of the consequences of uncertainties in nuclear data on the performance of a fusion reactor, data needs cannot be established by rigorous means. Of the 329 requests, 62 pertain to few-nucleon systems. Some of these concern cases where there is a 3-body breakup. A few examples of the requests are given:

Target	Projectile	Reaction	Energy Range	Accuracy	Priority
2-H	n	(n,2n)	<15 MeV	20%	2
3-He	n	(n,p)	1.0 keV - 15 MeV	10%	1
3-He	3-He	(3-He,2p)	0.1 MeV - 5 MeV	15%	3
6-Li	n	(n,nd)	<15 MeV	10%	1
7-Li	n	(n,nt)	3 MeV - 14 MeV	5%	1

Thus, about 20% of the total requests concern few-nucleon systems. However, these requests pertain to the basic energy producing reactions which are of vital importance. The accuracies required are not high.

When fusion reactions are considered, charged particle nuclear data become important. IAEA carried out a survey of the charged particle nuclear data for controlled thermonuclear research [2] and based on this, J. R. McNally, Jr. of Oak Ridge National Laboratory has prepared a list of important reactions. Some of these reactions are listed:

Reaction	Energy	Priority
6-Li(d,alpha)alpha + 22.4 Mev	0.1 - 5 MeV	1
6-Li(d,d′)6-Li* - d + alpha - 15 MeV	3.0 - 6 Mev	1
11B(p,alpha)2alpha + 8.7 MeV	0.01 - 5 MeV	1

NUCLEAR SAFEGUARDS

There are 130 requests from four countries pertaining to nuclear safeguards in WRENDA-75. Out of these only 1 request concerns few-nucleon systems. The importance of few-nucleon systems in nuclear safeguards is thus limited.

CONCLUSION

This limited survey indicates that the study of few-nucleon systems has the strongest relevance to fusion reactor programs, as compared to other applications. There is considerable scope for experimental and theoretical work in this direction.

REFERENCES

1. WRENDA-75, IAEA Report No. INDC (SEC)-46/U+R+F+S. June 1975.
2. Survey of current and future needs for charged particle and photonuclear reaction data, A. Calamand, IAEA Report No. INDC(NDS)-62/W+ special July 1, 1974.

DISCUSSION

R. Bondelid: In your study for preparing for this talk did you find other areas of study in which few nucleon studies are relevant?

A. Divatia: In looking through the INDC publication there was no obvious relevance, but it is not ruled out.

H. S. Hans: Does the information required correspond to the total cross section or angular distribution or both?

A. Divatia: This depends on the problem. Sometimes it is the total cross section and sometimes the differential cross section, as shown in the tables.

B. Kuhn: Does anyone know of measurements of the total breakup cross section, for instance of d(n,nn)p?

A. Divatia: CINDA-75 lists quite a few entries concerning the d(n,nn)p reaction; many of them are for E(n)=14 MeV. I have not evaluated the entries to determine how many pertain to total break up cross section measurements. In any case, as mentioned in my talk, better accuracies are required.

PART 3

INVITED PAPERS AND RAPPORTEURS' TALKS AT PLENARY SESSIONS

SOME TOPICS IN THE NUCLEON-NUCLEON INTERACTION

R. VINH MAU

Division de Physique Théorique*, Institut de Physique Nucléaire, Paris**
& Laboratoire de Physique Théorique des Particules Elémentaires, Paris**

In this talk, I do not intend to cover the latest developments which have occured in the field of nucleon-nucleon interaction since the Quebec Conference. I would rather try to explain some of my prejudices on this subject hoping that you will share some of them.

Forty years have now elapsed since Yukawa published his paper on the meson theory of nuclear forces, yet it is quite fair to recognize that the only truly unquestioned result is the one given by Yukawa himself, namely the one pion exchange (OPE) potential which describes the long range part of the interaction very well. However, it is also unquestionable that the OPE is only a first approximation and must be supplemented by further contributions. Now what should be added to the OPE contribution is still subject to controversy.

I do not wish to hark back again to the pure phenomenological analyses. Various phase shift analyses(1) and phenomenological potentials(2) have been discussed at length at many meetings on nuclear forces in the last few years. The common feature of most phenomenological potentials is that they try to fit the same data but have different shapes. It is hoped that off shell constraints will reduce the ambiguities. Unfortunately, at present, the prescriptions for the off shell extrapolation seem far from being unique.

The OPE contribution can be considered, in a perturbation theory, as a second order process of a fundamental ΠNN interaction vertex and of course calculations of the next order processes, in particular the 4th order ones (fig. 1) have been attempted(3) long

* Laboratoire associé au C.N.R.S.

** Postal address : Université Pierre et Marie Curie, Tour 16 - E1
4, Place Jussieu, 75230 Paris Cedex 05 - France

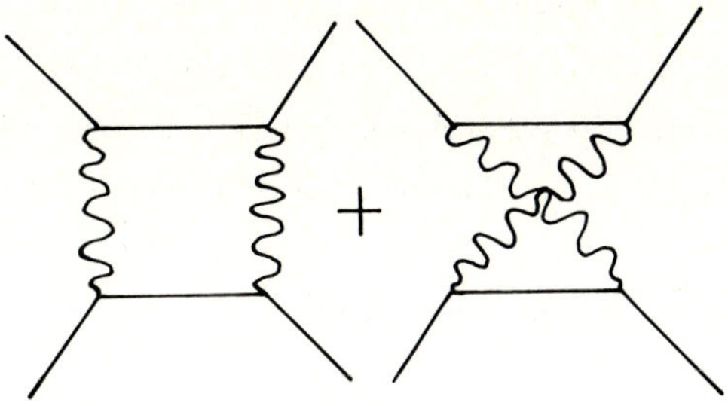

fig. 1

ago. Although the early calculations performed in a perturbative framework contain ambiguities and are not sufficient to fit the data, it should be stressed that these contributions are important and should be accounted for by any realistic theoretical model of nuclear forces.

Before going into the details of specific interaction models, let us recall that the two nucleon interaction can be considered as mediated by the exchange of particles or systems of particles in the crossed (nucleon-antinucleon) channels. This is represented in figure 2.

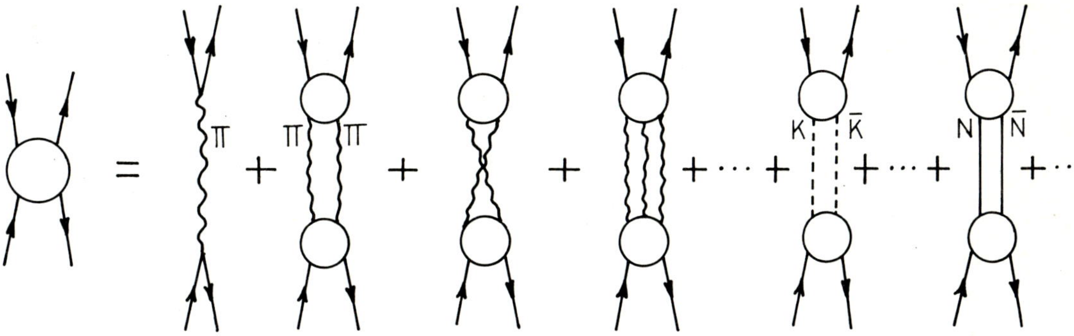

n_1, p_1 and n_2, p_2 are the 4 momenta of the initial and final nucleons

fig. 2

In this figure, the blobs of each term can contain anything allowed by conservation laws. For example, in the 2 pion exchange term, each blob could contain one nucleon, one nucleon + any number of pions, a nucleonic resonance, etc...

In the configuration space, the interaction* can be written as a series of terms :

$$V \simeq \frac{g^2 e^{-\mu r}}{\mu r} + \int_{4\mu^2}^{\infty} \rho_{2\Pi}(t)\, \frac{e^{-\sqrt{t}\,r}}{r}\, dt + \int_{9\mu^2}^{\infty} \rho_{3\Pi}(t)\, \frac{e^{-\sqrt{t}\,r}}{r} dt$$

$$+ \text{ etc...}$$

* I disregard spin and isospin complications throughout this talk, for the sake of simplicity of presentation.

which correspond to the one pion exchange (OPE) contribution, the two pion exchange (TPE) contribution etc... The actual expressions for the spectral functions $\rho_{2\Pi}$, $\rho_{3\Pi}$ etc... depend, of course, on the contents of the blobs in fig. 2. For example, if one keeps only the one nucleon state in each blob of the TPE contribution one gets the 4th order perturbation theory we mentionned previously.

Although at present no definite a priori statement can be made about the relative sizes of the various spectral functions, one can say, using phase space factor arguments, that the slope of the functions ρ_i's, at various thresholds, decreases with the mass of the exchanges systems. Consequently, if no unexpected strong enhancement occurs in the many particle exchanges, the spectral functions ρ_i's may have the shapes represented in figure 3.

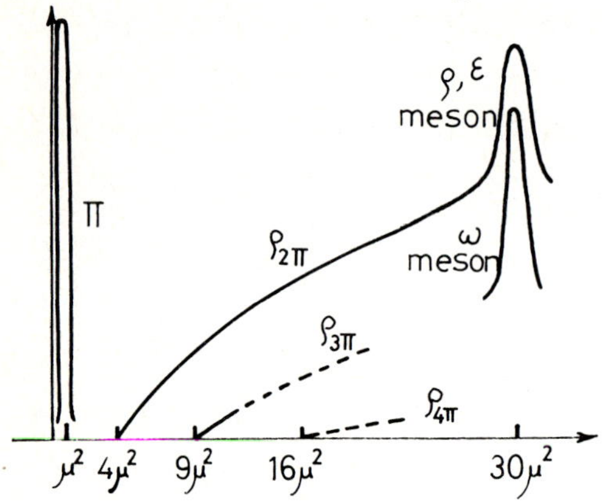

The spectral functions ρ_i's

fig. 3

It is worth noting that the series represented in figure 2 is not a perturbation expansion but a series of contributions that have shorter and shorter ranges.

The One Boson Exchange (OBE) Models(4)

These models pick out from the series of figure 2 only the resonant intermediate states : ρ and "σ" resonances for the 2 pion state, ω resonance for the 3 pion state, Φ resonance for the $K\bar{K}$ state etc... (figure 4) and treat them in the zero width approximation, i.e. as particles which are coupled directly to the nucleon.

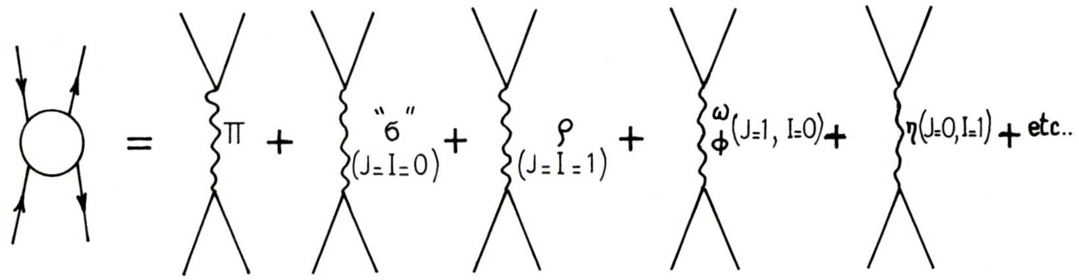

fig. 4

This leads to a scattering amplitude :

$$M(s,t) = \sum_{i=\Pi,\rho,\text{ etc..}} \frac{g_i^2}{t - m_i^2}, \quad (s = (p_1 + n_1)^2,\ t = (p_1 + p_2)^2)$$

or a potential :

$$V = \sum_{i=\Pi,\rho,\text{ etc..}} \frac{g_i^2}{m_i} \frac{e^{-m_i r}}{r}$$

The coupling constants g_i are, in general, left free and subsequently adjusted to fit the NN data, the masses m_i are fixed at the resonance positions with, however, an exception for the scalar and isoscalar "σ meson" : all the fits need $m_\sigma \simeq 450 - 500$ MeV while experiments show a strong interaction at $\sqrt{t} \simeq 800 - 1000$ MeV with a large width in the pion-pion S wave. Such a low mass "σ particle" is needed in the OBE models to provide a necessary medium range attraction since the contribution from the other bosons add up to an interaction which is too repulsive.

In Table I, are shown the results as well as the coupling constant values obtained by the Bonn group[(4)].

COUPLINGS CONSTANTS

	g_Π^2	g_η^2	g_σ^2	g_δ^2	g_ρ^2	f_ρ/g_ρ	g_ω^2
HM1	14.2	2.2	5.9	8.7	1.4	4.5	25
HM2	14.2	2.	5.7	0.8	0.5	6.2	10

NN-SCATTERING

	a_s	r_s	a_t	r_t	χ^2/data
HM1	-23.7	2.68	5.50	1.86	3.0
HM2	-23.7	2.68	5.45	1.79	2.6

DEUTERON

	E	Q	P_D
HM1	2.2241	0.2840	5.75
HM2	2.2246	0.2864	4.32
	(MeV)	(fm^2)	(%)

Table I

The fit is satisfactory although the necessity for a fictitious "σ particle" remains a major difficulty and spoils the beauty of this simple and elegant extension of the OPE.

Besides the mesonic resonance contributions it was recognized, shortly after the discovery of the Δ(1236) resonance, that pion-nucleon resonances play also an important role in the nucleon-nucleon interaction(5). In the OBE models, there is no room for such effects since the mesonic resonances, considered as particles, are coupled to nucleons through point like interaction.

The most complete way to include the nucleonic resonance effects as well as realistic mesonic resonance effects in the nucleon-nucleon interaction is to use information provided to us by accurate studies on pion-nucleon and pion-pion interactions. The natural framework of this kind of approach is based on dispersion relations, and for convenience I will refer to this as the dispersion relation approach.

The Dispersion Relation Approach

As already mentioned, the long range forces are correctly described by the OPE and the next longest range forces are due to the two pion exchange (TPE). Here the underlying belief is that most of the medium range forces should be correctly given by the TPE when the latter is properly determined. Also the hope is that the long and medium range forces if accurately known would provide strong enough constraints to leave little freedom for an eventual phenomenological determination of the short range part of the interaction.

The main task is therefore to relate the TPE contribution with known properties of pions, their interactions with themselves and with nucleons.

Barring heavy technical algebraic details(5,6), and taking liberties with mathematical rigor, a quick way of doing this is to proceed with the following steps :

i) Write a fixed energy dispersion relation

$$M(s,t) = \frac{g^2}{t-\mu^2} + \frac{1}{\Pi}\int_{4\mu^2}^{\infty}\frac{\rho(s,t')}{t'-t} + (s \leftrightarrow u) \qquad (1)$$

with $s = (p_1 + n_1)^2, \quad t = (p_1 - p_2)^2, \quad u = (p_1 - n_2)^2$

ii) Split the absorptive part $\rho(s,t')$ into different contributions according to those of figure 2, namely

$$\rho(s,t') = \rho_{2\Pi}(s,t') + \Theta(t-9\mu^2)\rho_{3\Pi}(s,t') + \Theta(t-16\mu^2)\rho_{4\Pi}(s,t) + \ldots$$

iii) In the t channel ($N\bar{N} \to N\bar{N}$), perform a partial wave expansion of

$$\rho_{2\Pi} = \sum (2\ell + 1) \text{ Im } F_\ell^{N\bar{N}\to 2\Pi \to N\bar{N}}(t') \quad P_\ell(\cos \Theta_s) \qquad (2)$$

where $F_\ell^{N\bar{N}\to 2\Pi\to N\bar{N}}(t')$ are partial wave amplitudes for $N\bar{N}$ scattering with two pions as intermediate state.

iv) Express Im $F_\ell^{N\bar{N}\to 2\Pi\to N\bar{N}}(t')$ in terms of $N\bar{N}\to 2\Pi$ helicity amplitudes $f_\ell^{N\bar{N}\to 2\Pi}$ through the unitarity condition

$$\text{Im } F_\ell^{N\bar{N}\to 2\Pi\to N\bar{N}}(t'') = \begin{pmatrix}\text{kinematic}\\ \text{factor}\end{pmatrix} \left| f_\ell^{N\bar{N}\to 2\Pi}(t'') \right|^2 \qquad (3)$$

which can be assumed to hold for $4\mu^2 \leqslant t'' \leqslant 50\mu^2$ since inelasticity is very weak up to the $K\bar{K}$ threshold.

v) Use the analyticity properties of $f_\ell^{N\bar{N}\to 2\Pi}$ to write the dispersion relation

$$\ell n \left| f_\ell^{N\bar{N}\to 2\Pi}(t') \right| = \frac{1}{2\Pi} \int_{4\mu^2}^{50\mu^2} \frac{\delta_\ell^{\Pi\Pi\to\Pi\Pi}(t'')}{t'' - t'} \, dt'' + \frac{1}{\Pi} \int_{-\infty} \frac{\Phi_\ell^{N\bar{N}\to 2\Pi}(t'')}{t'' - t'} \, dt'' \qquad (4)$$

where $\Phi_\ell^{N\bar{N}\to 2\Pi}$ is defined by $f_\ell^{N\bar{N}\to 2\Pi} = \left| f_\ell^{N\bar{N}\to 2\Pi} \right| e^{i\Phi_\ell^{N\bar{N}\to 2\Pi}}$ and where the unitarity condition has been used again to identify $\Phi_\ell^{N\bar{N}\to 2\Pi}$ with the $\Pi\Pi$ phase shifts $\delta_\ell^{\Pi\Pi\to\Pi\Pi}$, for $4\mu^2 \leqslant t'' \leqslant 50\mu^2$.

In the first integral of eqn.(4), the low angular momentum (S and P waves) $\delta_\ell^{\Pi\Pi}$ can be taken from theoretical models or from experimental studies ($\Pi N\to\Pi\Pi N$, $K_{\ell 4}$ decay, $e^+e^-\to\Pi\Pi$, nucleon form factors etc...) and the high angular momentum (D or higher waves) are known to be negligible.

In the second integral of eqn.(4), $f_\ell^{N\bar{N}\to 2\Pi}$ can be obtained from the Froissart-Gribov formula since $t''<0$, namely

$$f_\ell^{N\bar{N}\to 2\Pi}(t'') = \frac{1}{\Pi}\int_{m^2}^{\infty} Q_\ell\left(\frac{2w+t''-2m^2-2\mu^2}{2(t''-4m^2)^{1/2}(t''-4\mu^2)^{1/2}}\right)\mathrm{Im}\, M^{\Pi N\to\Pi N}(w,t'')\, dw$$

where $M^{\Pi N\to\Pi N}(w,t'')$ is the physical ΠN scattering amplitude since $w>m^2$ and $t''<0$.

Schematically, one can summarize iii) to v) by :

$$\rho_{2\Pi}(s,t') \sim \left|f_0^{N\bar{N}\to 2\Pi}(t')\right|^2 + \left|f_1^{N\bar{N}\to 2\Pi}(t')\right|^2 \cos\Theta_s + \sum_{\ell\geq 2}(2\ell+1)\left|f_\ell^{N\bar{N}\to 2\Pi}\right|^2 P_\ell \cos\Theta_s$$

(first term ↑) $\Pi\Pi$ S wave phase shift (J=I=0) + ΠN phase shifts

(second term ↑) $\Pi\Pi$ P wave phase shift (J=I=1) + ΠN phase shifts

(third term ↑) ΠN phase shifts

Note, that in the OBE models, the last sum is set to zero and $\left|f_0^{N\bar{N}\to 2\Pi}\right|^2$ and $\left|f_1^{N\bar{N}\to 2\Pi}\right|^2$ are approximated respectively by $g_\sigma^2\delta(t'-m_\sigma^2)$ and $g_\rho^2\delta(t'-m_\rho^2)$.

The same procedure could, in principle, be repeated for $\rho_{3\Pi}$ and $\rho_{4\Pi}$ with, however, formidable complications. In the case of $\rho_{3\Pi}$, the ω meson exchange can be easily taken into account, other specific processes can also be included[7].

The set of equations (1) to (4) forms the bulk of the relationshipswhich via analyticity properties, unitarity and crossing connect the TPE contribution with the pion-nucleon and pion-pion interactions. The pion-nucleon scattering is very accurately known by various phase shift analyses[8]. The S and P wave pion-pion interaction has also been extensively studied these last few years[9].

By using these analyses as inputs, one automatically includes all the nucleonic resonances (through ΠN phase shifts) as well as the realistic S and P wave pion-pion interaction, and besides these the smooth background of both pion-nucleon and pion-pion scattering is also taken into account.

The ability of the OPE, of the above calculated TPE and of the ω exchange contributions (as part of the three pion exchange) to describe the long and intermediate range forces can be checked by calculating the low energy (up to 330 MeV) peripheral ($J>2$) NN phase shifts(10) and by comparing them with the empirical ones. The agreement is satisfactory.

At this stage, there are two possibilities :

i) for the nucleon-nucleon scattering itself, one can perform a phase shift analysis similar to those done by the Livermore group(1), now with extra constraints on lower angular momentum phase shifts from the TPE and ω exchange. This program needs a long and exacting labour but is worth while in view of the coming experimental results from LAMPF, TRIUMPF, SIN etc... It is under investigation by the Paris group. It also is worth noting that the TPE contribution is imaginary and can provide inelasticity parameters.

ii) for further applications to nuclear many body systems, it is useful to define an equivalent potential which, once supplemented by a short range core, can be used in various nuclear calculations. Such a program has been carried out at Stony Brook(11) and in Paris(12):

The TPE of the Stony Brook potential contains the 4th order contribution, the ρ exchange and a $f_0^{N\bar{N}\to 2\Pi}$ found by educated guess. The ω exchange is included with a coupling constant $g_\omega^2/4\Pi=6.24$ This potential is regularised for high momentum transfer t (or small r) by a function obtained from certain multiple neutral vector meson exchange processes treated in the eikonal approximation. The calculated phase shifts are in good qualitative agreement with the experimental results. Some results are displayed for illustration in figure 5. The deuteron wave function is also computed and indicates a weaker short range repulsion than most phenomenological potentials.

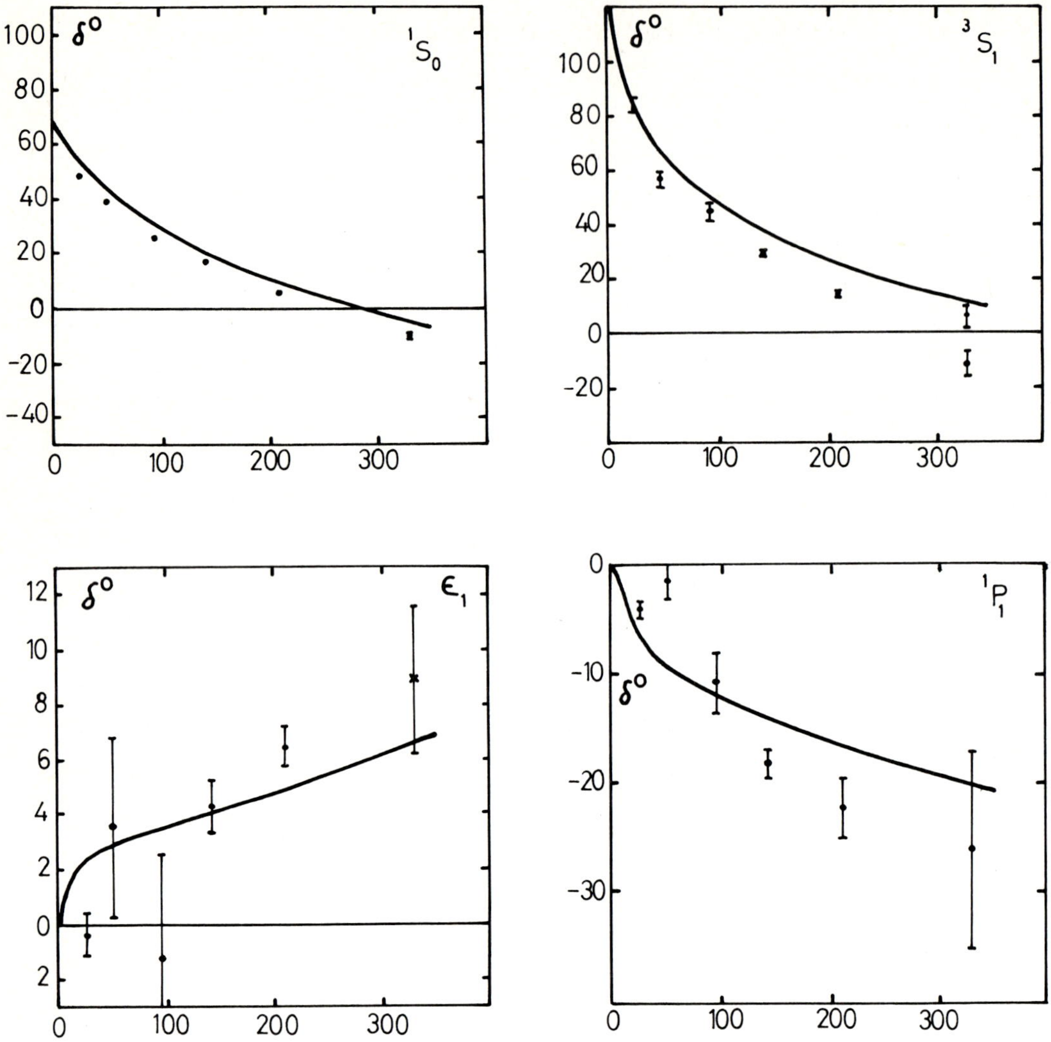

δ°
$^1S_0$
100
80
60
40
20
0
-20
-40
0
100
200
300
δ°
$^3S_1$
100
80
60
40
20
0
-20
0
100
200
300
δ°
ϵ_1
12
10
8
6
4
2
0
2
0
100
200
300
0
$^1P_1$
-10
δ°
-20
-30
0
100
200
300

fig. 5

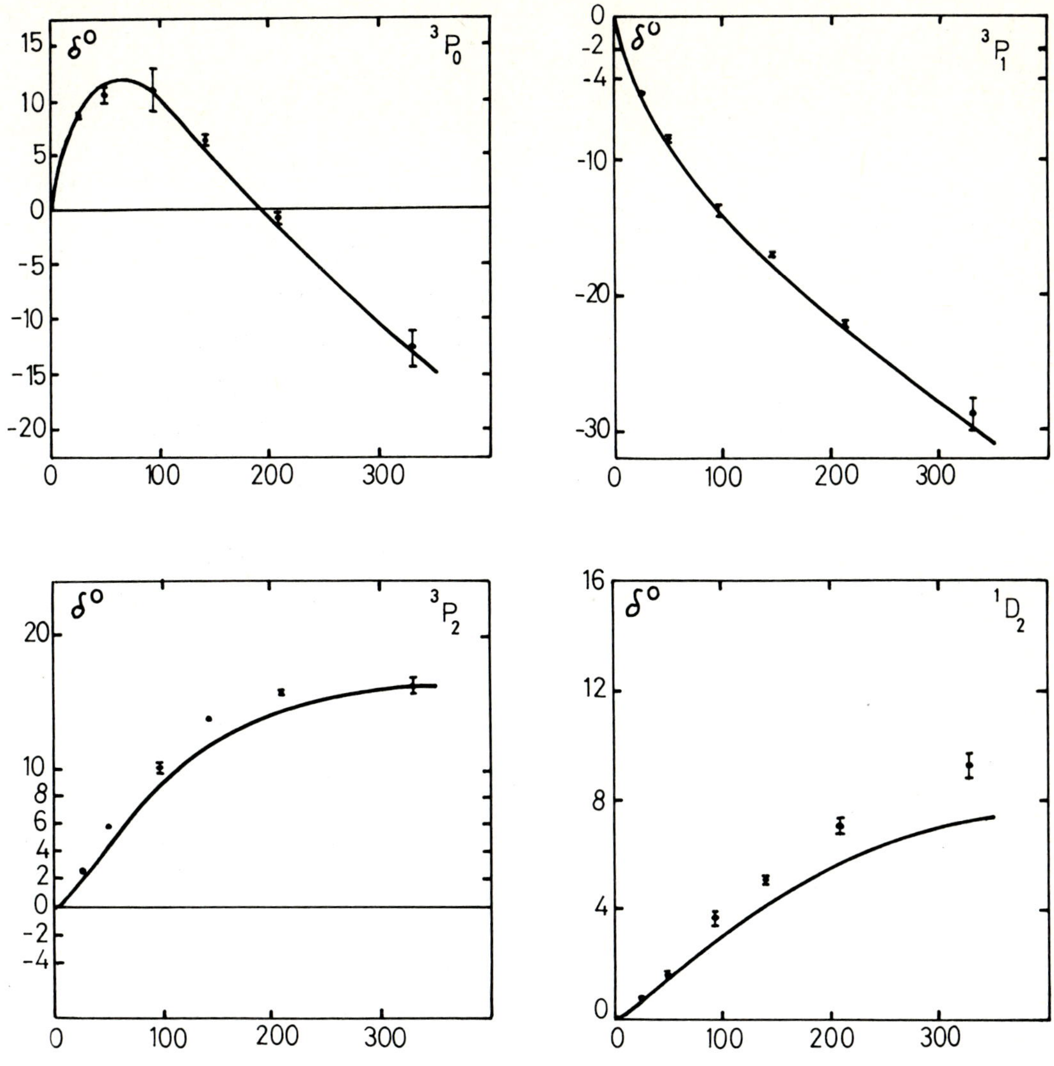
δ°
³P₀
15
10
5
0
-5
-10
-15
-20
0
100
200
300
δ°
³P₁
0
-2
-4
-10
-20
-30
0
100
200
300
δ°
³P₂
20
10
8
6
4
2
0
-2
-4
0
100
200
300
δ°
¹D₂
16
12
8
4
0
0
100
200
300

fig. 5

The Paris group(12) calculated the TPE potential from the complete ΠN amplitude as known in terms of phase shifts determined by the Glasgow group(8). The helicity amplitudes $f_0^{N\bar{N}\to 2\Pi}$ and $f_1^{N\bar{N}\to 2\Pi}$ used here are consistent with observed ΠΠ S and P waves. The ω coupling constant is taken to be $g_\omega^2/4\Pi=9.52$. This somewhat larger value is due to the fact that the nucleonic resonance contributions are attractive and also that the $f_0^{N\bar{N}\to 2\Pi}$ used here induces more attraction than in the Stony Brook model. This theoretical long and medium range potential is cut off at $r \sim 0.8$ fm and replaced for $r \lesssim 0.8$ fm by a simple phenomenological soft core which contains six parameters (in each isospin state) adjusted to fit the NN phases for all $J \leqslant 6$ as well as the deuteron parameters. In spite of the small number of adjustable parameters, the χ^2/data obtained are 2.5 for proton-proton scattering and 3.7 for neutron-proton scattering as good as are given by the well known purely phenomenological fits. The results of the fit are shown in figure 6 and in Table II.

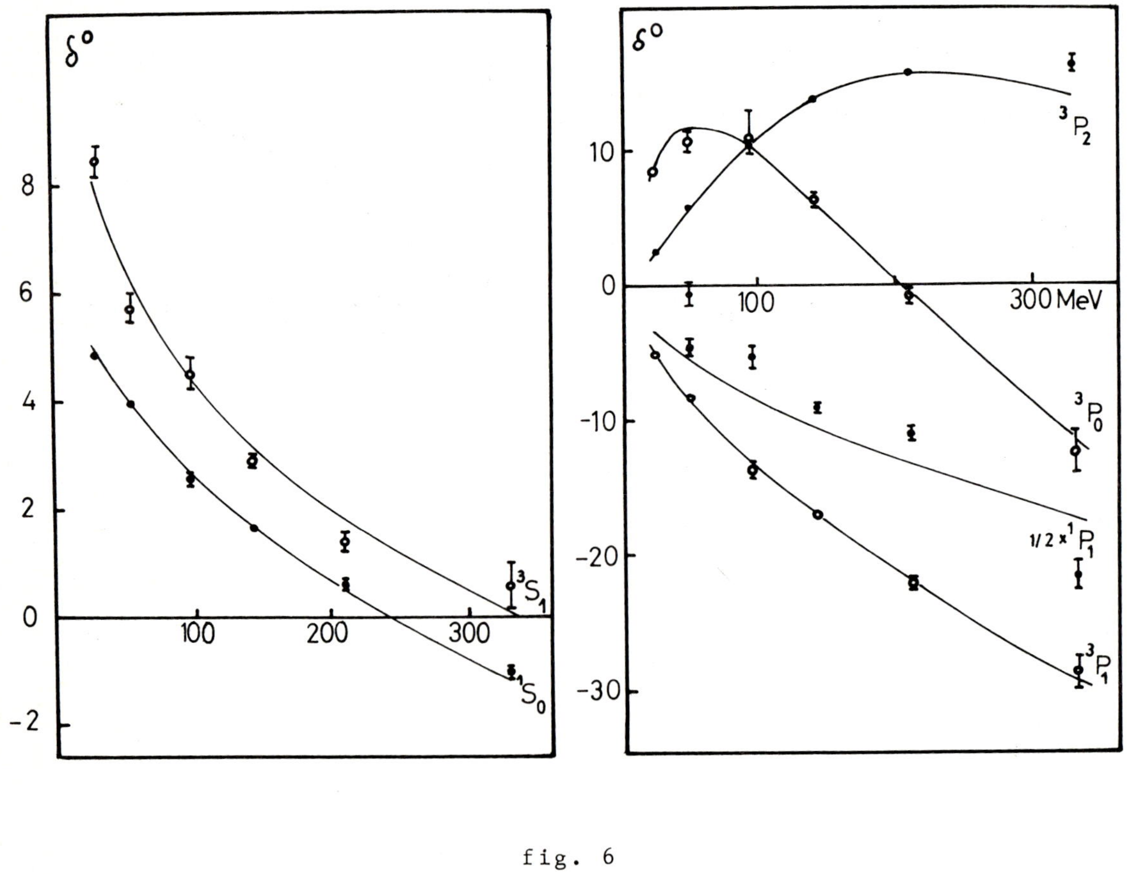

fig. 6

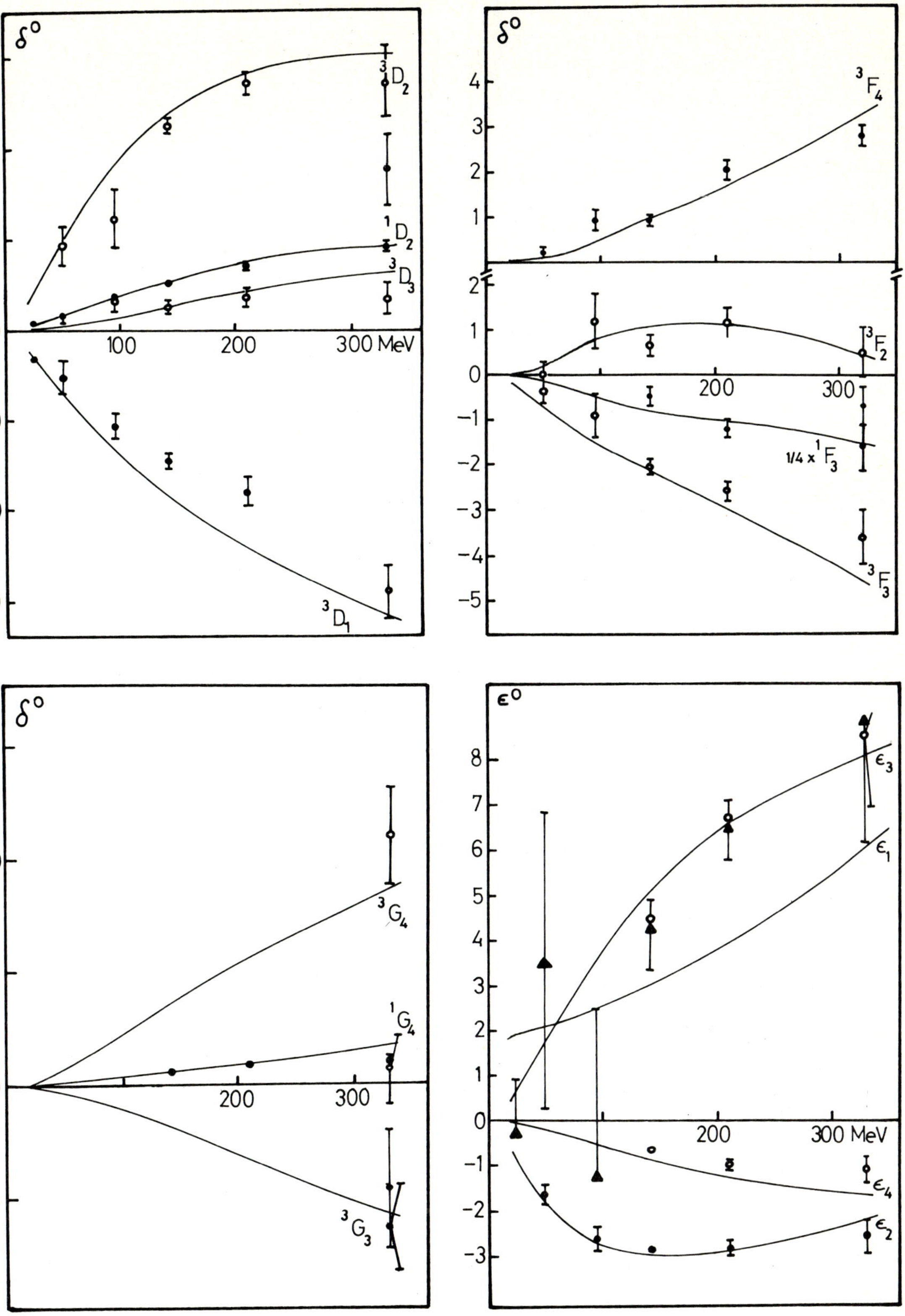
δ°
30
20
10
0
-10
-20
-30
100
200
300 MeV
$^3D_2$
$^1D_2$
$^3D_3$
$^3D_1$
δ°
4
3
2
1
0
-1
-2
-3
-4
-5
200
300
$^3F_4$
$^3F_2$
$1/4 \times {}^1F_3$
$^3F_3$
δ°
15
10
5
0
-5
200
300
$^3G_4$
$^1G_4$
$^3G_3$
ε°
8
7
6
5
4
3
2
1
0
-1
-2
-3
200
300 MeV
ϵ_3
ϵ_1
ϵ_4
ϵ_2

fig. 6

E_D	Q_D	$P_D\%$	μ_D
-2.2246	.290	6.75	.8392
(-2.2246 ± .001)	(.2875 ± .002)		(.8574 ± .000006)

a_{np}	r_{np}	a_{pp}	r_{pp}
5.4179	1.753	-7.817	2.747
(5.413 ± .005)	(1.748 ± .005)	(-7.823 ± .01)	(2.794 ± .015)

The deuteron and effective range parameters. Experimental results are given in brackets.

Table II

The potential is found to be significantly energy dependent in its central component which can be interpreted as some kind of non locality. However, this energy dependence is only linear and can be easily converted into simple p^2 or velocity dependence.

The next step is to use these potentials in nuclear calculations (finite nuclei, infinite nuclear matter, neutron stars, etc..). If the outcome turns out to be sensible, one of the goals of this Conference will have been achieved, namely a bridge would have been laid between particle physics and nuclear physics.

<u>The multichannel approach</u>.

Another possible way of accounting for the nucleonic resonances effects in the NN interaction is to treat the elastic NN→NN scattering as well as inelastic processes like $NN \to NN^*$, $NN \to N^*N^*$ etc... in a multichannel formalism(13) (fig.7). In this procedure, iterations in the Schroedinger equation will give rise to ladder diagram contributions of the type illustrated in fig. 8.

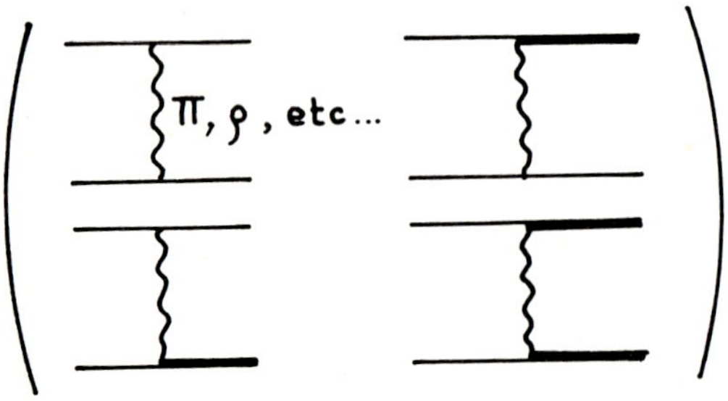

—— : nucleon

∿∿ : pion or mesonic resonance

▬ : nucleonic resonance

fig. 7

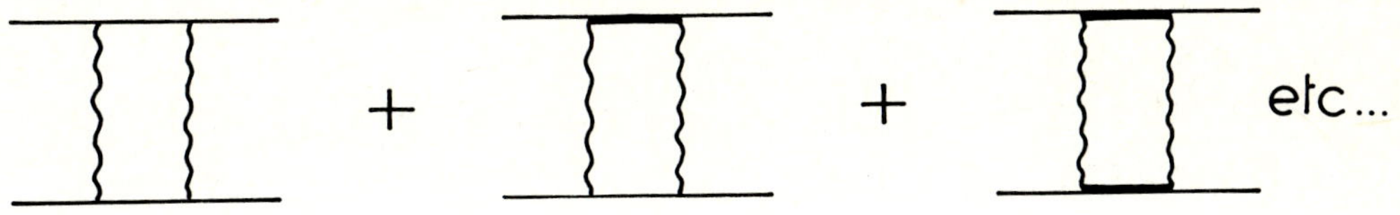

fig. 8

Such contributions are similar to those described in the previous section with however a difference : the intermediate nucleonic states are treated here non relativistically. On the other hand, the present approach leaves out crossed diagrams like the ones represented in fig. 9 which, from the proceeding section analyses, are very important.

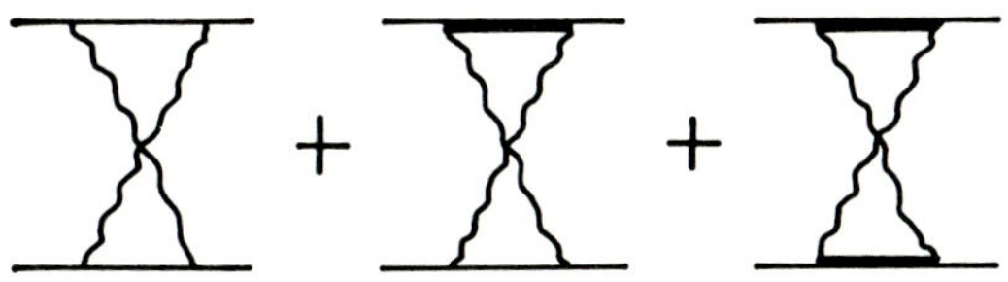

fig. 9

Up to now, only the Δ(1236) resonance has been includes in the coupled channel analysis[14]. The higher nucleonic resonances should also be included since, again from the previous section,we know that they contribute significantly.

In the alloted time, I was unable to report on the interesting papers submitted to the session SVM of this Conference. I would like to apologize to the authors of the following papers :

- Completeness Relations in Scattering Theory for Non-Hermitian Potentials by Y.E. Kim and A. Tubis.
- Analyticity Constraints of the Nucleon-Nucleon Transition Matrix by S.K. Mukhopadhyay.
- Two Body Bound State and Parametrization of Half-Off-shell t-Matrix by A.V. Lagu and C. Maheshwari.
- Parametrisation of the Half-shell t-Matrix when the Two-body Bound-state is delineated by V.S. Mathur, A.V. Lagu and C. Maheshwari
- A Possible Phenomenological Form for the Two Body Local Potential by A.V. Lagu.

- The Physical Basis of the UPE-Some Remarks by A.N. Mantri and A.V. Lagu.

- A simple Method of Calculating Scattering Parameters for separable Potentials with Coulomb Interaction by H. Kumpf.

- An OBE-Model Including the NΔ-Interaction ; Two-nucleon and Nuclear Matter Results by K. Bleuler, R. Machleidt and K. Holinde.

- A Momentum-Space OBEP including the N-Δ-Interaction by K. Holinde and R. Machleidt.

- Alpha Alpha Scattering Phaseshift Analysis by a New Method by B. Deo and C.C. Hazra.

- More Consistent Resonating Group Calculation of α-α Interaction by S.A. Affzal and S. Ali.

REFERENCES

(1) M.H. Mac Gregor, R.A. Arndt and R.M. Wright, Phys. Rev. 182 (1969) 1714 and references cited herein
P. Signell and J. Holdeman, Phys. Rev. Lett. 27 (1971) 1393
R.A. Arndt, R.H. Hackman and L.D. Roper, Phys. Rev. C9 (1974) 555.

(2) See, for example, P. Signell in the Proceedings of the International Conference on Few Particle Problems in Nuclear Interaction - Los Angeles 1972, North Holl. & Amer. Elsevier Publ.

(3) M. Lévy, Phys. Rev. 88 (1952) 725
A. Klein, Phys. Rev. 90 (1953) 1101
See also, F. Partovi and E.L. Lomon for a complete list of references.

(4) K. Erkelenz, Phys. Reports 13C (1974) 191
K. Holinde and R. Machleidt, Nucl. Phys. A247 (1975) 495 referred to as HM1, K. Holinde and R. Machleidt referred to as HM2
M.M. Nagels, T.A. Rijken and J.J. de Swart, Phys. Rev. D12 (1975) 744

(5) D. Amati, E. Leader and B. Vitale, Phys. Rev. 130 (1963) 750
W.N. Cottingham and R. Vinh Mau, Phys. Rev. 130 (1963) 735

(6) M. Chemtob, J.W. Durso and D.O. Riska, Nucl. Phys. B38 (1972)141
W.N. Cottingham, M. Lacombe, B. Loiseau, J.M. Richard and R. Vinh Mau, Phys. Rev. D8 (1973) 800
G. Bohannon and P. Signell, Phys. Rev. D10 (1974) 815
G.B. Epstein and B.H.J. Mc Kellar, Phys. Rev. D10 (1974) 1005

(7) B. Loiseau and W. Nutt to be published in Nucl. Phys.

(8) See Particle Data Group, UCRL Report N° 20030

(9) See for example J.L. Basdevant, B. Bonnier, C.D. Froggatt, J.L. Petersen and C. Schomblond in the Proceedings of the 2nd Aix en Provence International Conference on Elementary Particles 1973

(10) R. Vinh Mau, J.M. Richard, B. Loiseau, M. Lacombe and W.N. Cottingham, Phys. Lett. 44B (1972) 1

(11) A.D. Jackson, D.O. Riska and B. Verwest, Nucl. Phys. A249 (1975) 397

(12) M. Lacombe, B. Loiseau, J.M. Richard, R. Vinh Mau, P. Pires and R. de Tourreil, Phys. Rev. D12 (1975) 1495

(13) H. Sugawara and F. von Hippel, Phys. Rev. 172 (1968) 1764

(14) S. Jena and L.S. Kisslinger, Ann. of Phys. 85 (1974) 251
A.M. Green and P. Haapakoski, Nucl. Phys. A221 (1974) 429

THE COULOMB PROBLEM AND THE SEPARATION OF ELECTROMAGNETIC EFFECTS IN FEW-NUCLEON SYSTEMS

Peter U. Sauer
Theoretical Physics, Technical University, Hannover, West Germany

1. INTRODUCTION

The fact, that nucleons also have electromagnetic (e.m.) properties, can usually be turned to our advantage. Think only of all the information on nuclear structure gathered from electron scattering. However, the e.m. interaction of nucleons exhibits unwanted characteristics, too. It is often a stumbling-block for theorists and furthermore a nuisance. The e.m. interaction is a stumbling-block for theorists, since the infinite range of the Coulomb force creates particular formal difficulties in scattering problems. The e.m. interaction is also a nuisance, since it interferes with the strong interaction. Its contributions to nuclear phenomena are always present and can cover up fine details of the unknown strong interaction, which we would like to study unambiguously. I was asked to discuss these two unpleasant and entirely different aspects of the e.m. interaction in two- and three-nucleon systems.

The talk is organized as follows. I shall briefly recall the known e.m. properties of nucleons and of the two-nucleon interaction. I shall then review the conceptual problems in describing proton-proton (PP) and proton-deuteron (PD) scattering. Finally, I shall talk on the separation of e.m. contributions from some sensitive few-nucleon properties, specifically, from the effective-range parameters of low-energy nucleon-nucleon scattering and the mass difference between ^{3}He and ^{3}H.

2. ELECTROMAGNETIC NUCLEON FORM FACTORS AND THE TWO-NUCLEON ELECTROMAGNETIC INTERACTION

Nonrelativistic quantum mechanics is used. The two-nucleon hamiltonian should therefore retain terms of low order only in the ratio of nucleon velocity $\hbar k/M$ over c. The e.m. properties of nucleons are contained in the electric and magnetic Sachs form factors,

$$G_E(q^2) = F_D(q^2) - \left(\frac{\hbar q}{2Mc}\right)^2 \kappa F_P(q^2) \qquad (2.1a)$$

$$G_M(q^2) = F_D(q^2) + \kappa F_P(q^2) , \qquad (2.1b)$$

which are determined by electron scattering. I choose them to depend on the three-momentum transfer $q^2 = \vec{q}^2$. κ is the anomalous magnetic moment, $\mu_P = 1 + \kappa_P$ and $\mu_N = \kappa_N$ are the total proton and neutron magnetic moments in units of the nuclear magneton $\mu_o = e_P \hbar/(2M_Pc)$. e_P is the proton charge, M the nucleon mass. F_D and F_P are the Dirac and Pauli form factors respectively with the normalization $F_D(0) = F_P(0) = 1$. Knowledge of the form factors up to momentum transfers $(Mc/\hbar)^2 \sim 25\ fm^{-2}$ should suffice for a nonrelativistic theory. In this region both proton form factors and the magnetic neutron form factor are known with satisfying accuracy, whereas the neutron charge form factor $G_{EN}(q^2)$ is only poorly determined allowing a wide range of parametrizations [1]. Its experimental indeterminacy has sizable effects on the three-nucleon charge form factors [2] . In Fig. 2 the

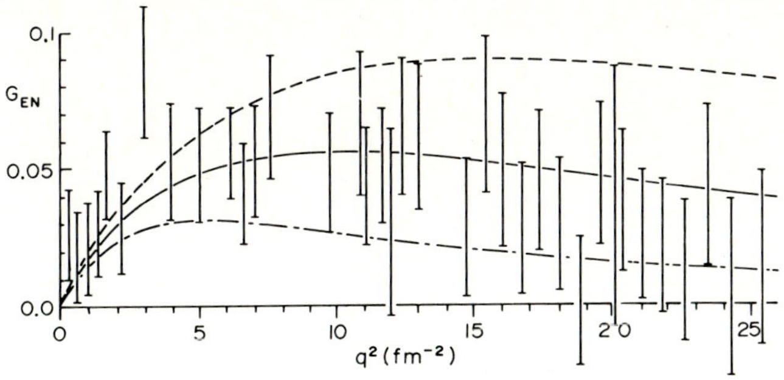

Fig. 1. Various parametrizations of the neutron charge form factor according to Ref. [1]

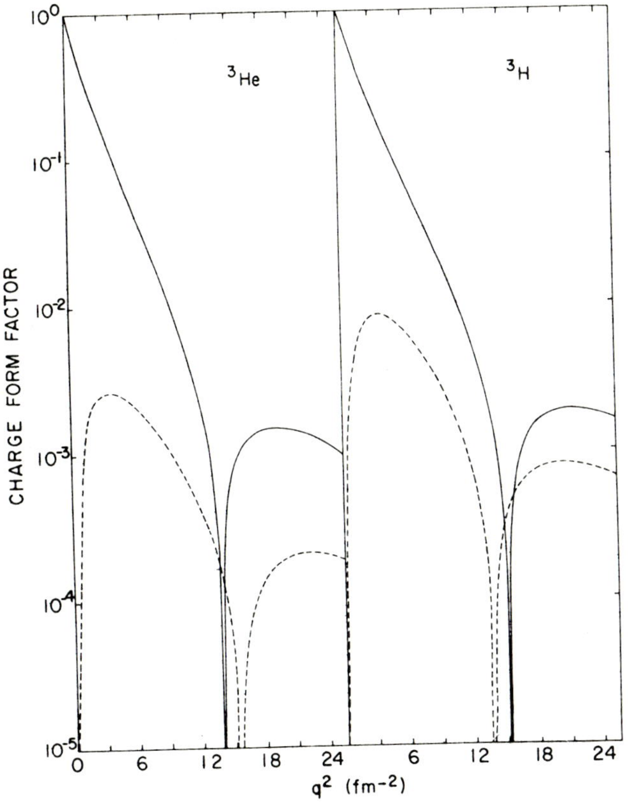

Fig. 2. ^{3}He and ^{3}H charge form factors. The dashed curve is the difference between the three-nucleon form factors when the maximal and minimal parametrizations of the neutron charge form factor according to Fig. 1 are used.

form factors calculated with a standard $G_{EN}(q^2)$ are plotted along with the difference between these form factors when the maximal and minimal parametrizations of $G_{EN}(q^2)$ are used. Already for intermediate momentum transfers between 6 and 12 fm^{-2}, the theoretical ^{3}H charge from factor is uncertain by approximately 10-15%, the ^{3}He charge form factor by 4-9%.

An accurate determination of the neutron charge form factor remains a challenge to experimentalists. Its present experimental errors are, however, rather inconsequential for the direct e.m. interaction w, where the form factors are needed to take into account the finite extent of the nucleonic charge e and magnetization μ,

$$e_i = G_{Ei}(q^2) \ , \ \mu_i = G_{Mi}(q^2) . \tag{2.2}$$

The e.m. interaction consists of the one-photon exchange w_γ and vacuum polarization w_{vac}. The one-photon exchange has the shown momentum representation

$$2\pi^2 \langle \vec{k}' | w_\gamma | \vec{k} \rangle = e_P^2 \frac{e_1 e_2}{q^2} +$$

$$\mu_o^2 \left\{ 4 \frac{e_1 e_2}{q^2} \left[\vec{k}' \cdot \vec{k} - \frac{(\vec{k}' \cdot \vec{q})(\vec{k} \cdot \vec{q})}{q^2} \right] - e_1 e_2 - \frac{2}{3} \mu_1 \mu_2 \vec{\sigma}_1 \cdot \vec{\sigma}_2 \right\} + \tag{2.3a}$$

$$\mu_1 \mu_2 \left[\frac{(\vec{\sigma}_1 \cdot \vec{q})(\vec{\sigma}_2 \cdot \vec{q})}{q^2} - \frac{1}{3} \vec{\sigma}_1 \cdot \vec{\sigma}_2 \right] +$$

$$\left[4 (\mu_1 \vec{\sigma}_1 e_2 + e_1 \mu_1 \vec{\sigma}_2) - e_1 e_2 (\vec{\sigma}_1 + \vec{\sigma}_2) \right] i \frac{\vec{k}' \times \vec{k}}{q^2} + \mathcal{O}\left[\left(\frac{\hbar k}{M c} \right)^4 \right],$$

$$\vec{q} = \vec{k}' - \vec{k}. \tag{2.3b}$$

It is the hermitian reduction of the Breit interaction up to the order $(\hbar k/(Mc))^2$ by means of a Foldy-Wouthuysen transformation [3]. It can alternatively be derived from the relativistic one-photon exchange by eliminating the small spinor components [4]. The interaction is not Galilean invariant [5]. It is given and will be used for total pair momentum zero. The tensor and spin-orbit contributions are not operative in spin-singlet states (expect for the PN system, where the spin-orbit term contains a $\vec{\sigma}_1 - \vec{\sigma}_2$ piece). The first four terms survive in all partial waves, therefore also in ^1S_o with respect to e.m. contributions most important in two- and three-nucleon systems. The various terms are the static Coulomb, the orbit-orbit interaction of infinite range and manifestly nonlocal, the Darwin-Foldy and spin-spin contact terms. There is no additional contact term involving the anomalous magnetic moments, $-\mu_o^2 (\kappa_1 e_2 + e_1 \kappa_2)$, since Sachs form factors are employed [6]. This fact has been a source for double counting. When the one-photon exchange is used in the Blankenbecler-Sugar equation, the momentum-dependent corrections of the static Coulomb are modified [7]. The local e.m. potentials in ^1S_o are shown in Figs. 3 and 4. Our limited knowledge on the nucleonic form factors at large momentum transfer introduces only insignificant uncertainties at very small relative distances, consistent with the nonrelativistic expansion up to $(\hbar k/(Mc))^2$. As seen from the total local contribution, the magnetic terms add to the PP repulsion. The magnetic neutron-neutron (NN) interaction is repulsive, the magnetic PN interaction attractive. The neutron charge form factor has here been put to zero, yielding an error of 1% in the NN and an error of 10% in the PN e.m. local potentials.

The total two-nucleon interaction $v + w$ consists of two additional pieces, i.e., $v = v_h + v_{em}^{ind}$. The purely hadronic part v_h would remain, even if all e.m. properties of nucleons could be turned off. It is assumed to be charge-independent. Second, there is an isospin-

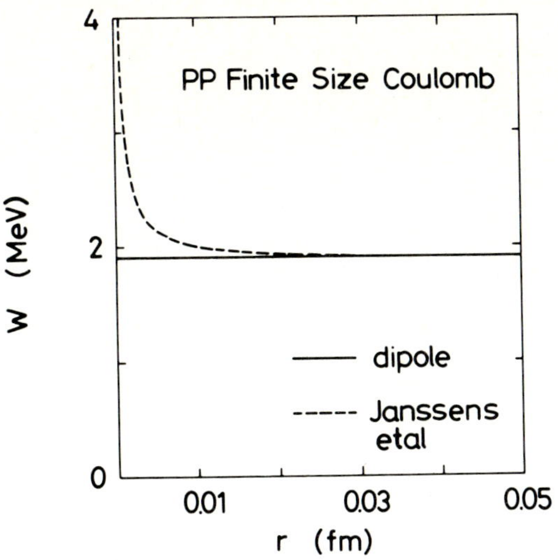

Fig. 3. PP Coulomb potential. Results with the proton charge form factors of Ref. [8] and [9] (almost indistinguishable, solid curve) and of Ref. [10] (dashed curve) are compared.

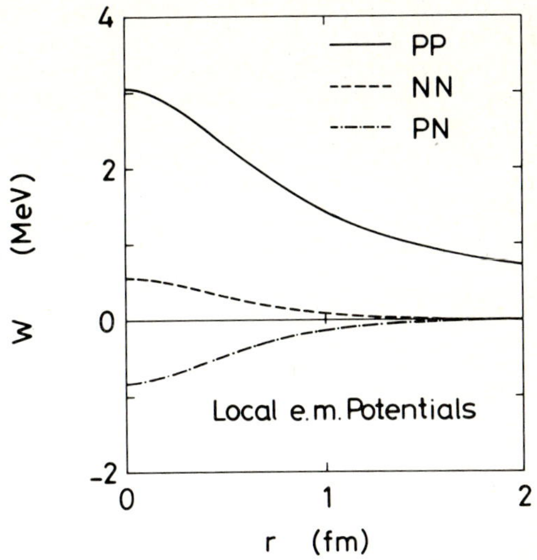

Fig.4. Local $^1S_0$ e.m. potentials. The nucleon form factors of Ref. [8] are used.

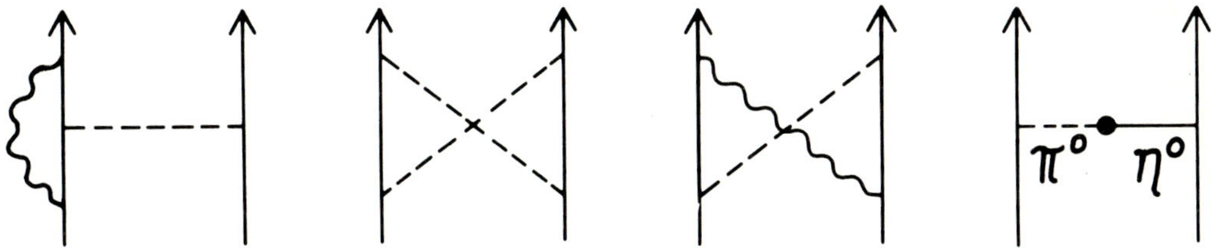

Fig. 5. Contributions to isospin-dependent part of nuclear force.

dependent modification v_{em}^{ind} of the hadronic interaction through e.m. effects on coupling constants (Fig. 5a), meson masses and nucleon masses in intermediate states (Fig. 5b), through simultaneous photon and meson exchanges (Fig. 5c), π^o-η^o (Fig. 5d) and ρ^o-ω^o-ϕ^o mixings. In contrast to the direct e.m. interaction w, the indirect one v_{em}^{ind} cannot yet be calculated with precision. Not as a substitute for these important first-principle calculations [11], but rather in order to set the goal for them, we would like to get experimental information on these indirect electromagnetic contributions to the nuclear potential. I shall take up this subject in the final part of the talk.

3. HOW TO INCLUDE COULOMB IN THE DESCRIPTION OF FEW-NUCLEON SYSTEMS

I now turn to the problem of setting up and solving singularity-free two- and three-body equations in the presence of Coulomb. I shall leave out vacuum polarization and the momentum-dependent corrections of infinite range to the Coulomb force. All other parts of the e.m. interaction are of short range and can in this context be treated together with the strong interaction.

The conceptual Coulomb problems are not simply an academic challenge, but a matter of practical importance. Why ? In the two-nucleon system there are sophisticated momentum-space one-boson exchange potentials which have never stood the test of low-energy PP scattering data [12,13].

In the three-nucleon system all PD experiments are more accurate than the corresponding ND experiments. If off-shell effects showed up at all in three-nucleon scattering [14], one would prefer to study them in PD rather than ND scattering. Furthermore, a succesful description of PD break-up would serve as a welcome reassurance that our method of extracting the NN scattering length from ND break-up data [15] is reliable. And finally, there might be valuable information on the charge symmetry of the nuclear force hidden in the difference between PD and ND effective-range parameters. We simply do not know yet.

Surely, Coulomb effects are expected to be small, but the details of the nuclear force which we want to study might even produce smaller effects. Therefore only a potentially exact treatment of Coulomb is of interest. What methods are available, where are the unresolved difficulties ? I shall distinguish between methods based on the Schrödinger equation and methods based on integral equations.

3.1 Schrödinger Equation

When describing the two- and three-nucleon systems by means of the Schrödinger equation, the coordinate-space form of the wanted solution has to be specified in the asymptotic region. If we knew it, all conceptual problems, not the practical problems, were over.

We surely know it for PP scattering, the ^{3}He bound state and PD scattering below break-up. However, in the break-up case, it is not known for all regions of configuration space. As in Fig. 6 I use

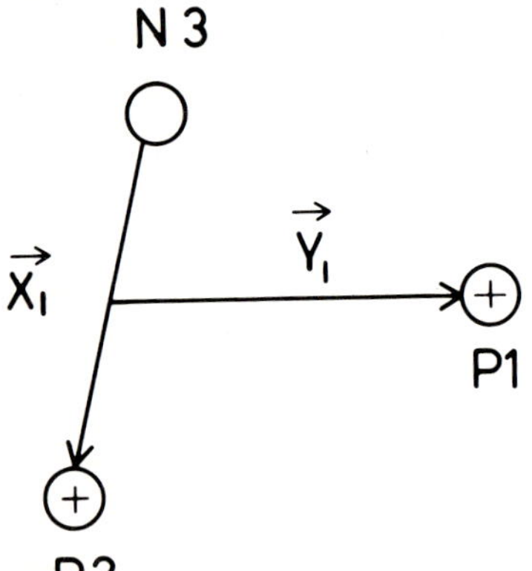

Fig. 6. Jacobi coordinates used in the description of the three-nucleon system.

Jacobi coordinates $(\vec{x}_i, \vec{y}_i)$ with corresponding momenta $(\vec{k}_i, \vec{q}_i)$ and kinetic energies $((\hbar k_i)^2/M,\ 3(\hbar q_i)^2/(4M))$ and define a six-dimensional hyperradius $\rho = x_i^2 + (4/3) y_i^2$. For a positive total energy E a total momentum κ, i.e. $E = \hbar^2\kappa^2/M$, is used. I assume that the nucleons are distinguishable, antisymmetrization can be carried out later. The Coulomb interaction acts between pair 3, the Coulomb parameter is denoted by $\eta = Me_p^2/\hbar^2$. The asymptotic break-up wave function

$$\langle \vec{x}_1 \vec{y}_1 | \Phi^{(+)}(\vec{q}) \rangle \underset{y_\beta \to \infty}{\sim} \langle \vec{x}_1 | D \rangle \langle \vec{y}_1 | \vec{q}^{\,c} \rangle \; + $$

$$\sum_{\beta=1}^{2} \langle \vec{x}_\beta | D \rangle \frac{e^{iqy_\beta}}{y_\beta} \, y_\beta^{-i\left[\eta/(\frac{3}{2}q)\right]} f^{(\beta)} + \langle \vec{x}_1 \vec{y}_1 | U \rangle \tag{3.1}$$

should have the elastic components, i.e., the full Coulomb wave function $\langle \vec{y}_1 | \vec{q}^c \rangle$ of the incident proton asymptotically taken with respect to the cm of the deuteron $\langle \vec{x}_1 | D \rangle$, outgoing spherical Coulomb waves, and it should have a proper break-up piece $\langle \vec{x}_1 \vec{y}_1 | U \rangle$ describing all three particles unbound, two of them still possibly experiencing the strong interaction. This term is known for ND scattering. It is a zero-energy two-nucleon scattering wave function for the pair β times an outgoing wave for the spectator nucleon [16]. To my knowledge, the corresponding piece for PD scattering is not written down in the literature. However, when all three particles are well separated, it takes the form [17]

$$\langle \vec{x}_\beta \vec{y}_\beta | U \rangle \underset{\substack{x_\beta \to \infty \\ y_\beta \to \infty}}{\sim} \rho^{-\frac{5}{2}} e^{i\kappa\rho} \rho^{-i\left[\eta/(2 \frac{x_3}{\rho} \kappa)\right]} A \quad . \tag{3.2}$$

Even if this last missing part in the asymptotic behavior were found, the troubles for the practical solution of the Schrödinger equation might not be over yet. The Kohn variational principle is usually employed, and usually requires the knowledge of the time-reversed scattering states which involve three incoming free particles [18]. The known results for these wave functions without Coulomb [18, 19] would have to be generalized to the Coulomb case as well. (Note, however, that Ref. [20] establishes the Kohn principle without the time-reversed scattering states.)

3.2 Integral Equations

In contrast to the Schrödinger equation, integral equations incorporate the asymptotic behavior of wave functions and are more convenient for momentum-space potentials. The Lippmann-Schwinger equation for PP scattering under the influence of an interaction $v + w$ is

$$|\phi^{(+)}(\vec{q})\rangle = |\vec{q}\rangle + g_o(e + io)(v + w)|\phi^{(+)}(\vec{q})\rangle, \tag{3.3}$$

where the free Green's function g_o refers to the kinetic energy operator k only, i.e., $g_o(\omega) = (\omega - k)^{-1}$. In case there were a bound state in the PP system, the corresponding homogeneous equation would not pose any problems, even if the interaction contained a piece of infinite range as the Coulomb interaction w. However, in the scattering situation the kernel with Coulomb is not completely continuous. As a consequence the corresponding momentum-space transition matrix has singularities [21]. They occur (1) at all available energies in the half-off-shell and on-shell matrix elements. There has been intensive work on understanding the latter singularities and on possible cures [22]. These singularities clearly demonstrate, that the momentum-space transition matrix is not the appropriate operator for scattering problems with Coulomb. A natural remedy is to isolate the part $g_o(e + io)\, w|\phi^{(+)}(\vec{q})\rangle$ of the integral equation which creates the singularities

$$\left[1 - g_o(e + io)w\right]|\phi^{(+)}(\vec{q})\rangle = |\vec{q}\rangle + g_o(e + io)\, v\, |\phi^{(+)}(\vec{q})\rangle \tag{3.4}$$

and to invert $\left[1 - g_o(e + io)\, w\right]$ explicitly.

$$|\phi^{(+)}(\vec{q})\rangle = |\vec{q}^c\rangle + g_o^c(e + io)\, v\, |\phi^{(+)}(\vec{q})\rangle \tag{3.5}$$

The Coulomb interaction is added to the unperturbed hamiltonian k, Coulomb is included in the free Green's function $g_o^c(\omega) = (\omega - k - w)^{-1}$. The total scattering from the pure Coulomb potential is described by the inhomogeneous term $|\vec{q}^c>$ and is well known. The resulting integral equation only deals with the modification of pure Coulomb scattering by the nuclear interaction v. The nuclear phase shifts are contained in the on-shell transition matrix in Coulomb-wave representation $<\vec{q}^c|t(e+io)|\vec{q}^c>$. It is defined in terms of the two-nucleon resolvent

$$(\omega - k - w - v)^{-1} = g_o^c(\omega) + g_o^c(\omega) t(\omega) g_o^c(\omega) , \tag{3.6}$$

it is related to the nuclear interaction v by a Lippmann-Schwinger equation with the Coulomb-modified Green's function

$$t(\omega) = v + v\, g_o^c(\omega) t(\omega) \tag{3.7}$$

and can practically be obtained from it in terms of the two-nucleon potential in Coulomb-wave representation. For momentum-space nuclear potentials, the transformation to Coulomb waves has to be carried out, the problem being numerical accuracy only [23]. The Coulomb-modified integral equation for t is also the basis for the whole literature on separable potentials with Coulomb [24]. Clearly, separable forms of the potential can be chosen phenomenologically to have analytic representations in Coulomb-wave space. With such an analytically known input the integral equation can be solved very accurately.

The Lippmann-Schwinger equations which define the PD scattering wave function $|\Phi^{(+)}(\vec{q})>$ uniquely are [25].

$$|\Phi^{(+)}(\vec{q})> = |D>|\vec{q}>_1 + (E-K-v_1+io)^{-1}(v_2+v_3+w_3)|\Phi^{(+)}(\vec{q})> \tag{3.8a}$$

$$|\Phi^{(+)}(\vec{q})> = (E-K-v_2+io)^{-1}(v_3+w_3+v_1)|\Phi^{(+)}(\vec{q})> \tag{3.8b}$$

$$|\Phi^{(+)}(\vec{q})> = (E-K-w_3-v_3+io)^{-1}(v_1+v_2)|\Phi^{(+)}(\vec{q})> \tag{3.8c}$$

There are only outgoing waves in channels two and three. K is the total three-body kinetic energy. The Equs. (3.8a) and (3.8b) do not have a completely continuous kernel in the presence of Coulomb, the corresponding homogeneous equation for the bound state does not pose problems. The singular pieces $(E-K-v_{1(2)}+io)^{-1}w_3|\Phi^{(+)}(\vec{q})>$ of the scattering equation have to be isolated and inverted explicitly as in the two-nucleon case. One arrives at singularity-free equations,

$$|\Phi^{(+)}(\vec{q})> = \delta_{\beta 1}\,|D\vec{q}^c>_\beta + G_\beta^c(E+io) \sum_{\gamma \neq \beta} v_\gamma\, |\Phi^{(+)}(\vec{q})> , \tag{3.9a}$$

$$G_\beta^c(\omega) = (\omega - K - w_3 - v_\beta)^{-1} , \tag{3.9b}$$

for which the uniqueness proof of Ref. [25] also holds. Introducing channel amplitudes

$$|\Phi^{(\beta)}(\vec{q})> = (E - K - w_3 + io)^{-1} v_\beta |\Phi^{(+)}(\vec{q})>, \tag{3.10a}$$

$$|\Phi^{(+)}(\vec{q})> = \sum_\beta |\Phi^{(\beta)}(\vec{q})> \tag{3.10b}$$

yields coupled equations for the wave function

$$|\Phi^{(\beta)}(\vec{q})\rangle = \delta_{\beta 1}\,|D\vec{q}^{\,c}\rangle_\beta + G^c_\beta(E+io)\,v_\beta \sum_{\gamma\neq\beta} |\Phi^{(\gamma)}(\vec{q})\rangle . \tag{3.11}$$

The Coulomb interaction is included in all channel Green's functions $G^c_\beta(\omega)$. They contain interactions between different pairs and are therefore general three-body operators. The inhomogeneous term $|D\vec{q}^{\,c}\rangle_\beta$ is the pure Coulomb scattering of the proton from the deuteron including Coulomb break-up, a tough three-body problem itself, presumably as tough as the full problem. Its solution is not at all known in contrast to the two-nucleon case. As for two nucleons, the Equs. (3.11) only deal with the modification of pure Coulomb scattering by the nuclear interaction v. A transition matrix $T(\omega)$ is defined in terms of the resolvent

$$(\omega - K - w_3 - \sum_\beta v_\beta)^{-1} = G^c_o(\omega) + G^c_o(\omega)\,T(\omega)\,G^c_o(\omega), \tag{3.12a}$$

$$T(\omega) = \sum_\beta T^{(\beta)}(\omega). \tag{3.12b}$$

It satisfies Faddeev equations [26] with a continous kernel

$$T^{(\beta)}(\omega) = t_\beta(\omega) + t_\beta(\omega)\,G^c_o(\omega) \sum_{\gamma\neq\beta} T^{(\gamma)}(\omega). \tag{3.13}$$

The transition matrices $t_\beta(\omega)$ are given in terms of the two-nucleon potential v_β,

$$t_\beta(\omega) = v_\beta + v_\beta\,G^c_o(\omega)\,t_\beta(\omega). \tag{3.14}$$

The Coulomb-modified free Green's function $G^c_o(\omega)$,

$$G^c_o(\omega) = (\omega - K - w_3)^{-1}, \tag{3.15a}$$

$$G^c_o(\omega) = \int d^3k_3 d^3q_3 \left[\omega - \frac{\hbar^2}{M}(k_3^2 + \frac{3}{4}q_3^2)\right]^{-1} |k^c_3\rangle\langle k^c_3| \times |\vec{q}_3\rangle\langle\vec{q}_3|, \tag{3.15b}$$

has a simple representation in the pair and spectator momenta $(\vec{k}_3,\vec{q}_3)$. It becomes, however, an unpleasant three-body operator in the two other sets of coordinates. Only $t_3(\omega)$ is a two-body transition operator in Coulomb-wave representation as encountered in two-nucleon scattering, $t_1(\omega)$ and $t_2(\omega)$ are also general three-body operators.

Thus, we do have well-defined singularity-free integral equations for the interesting nuclear physics case of two charged particles. However, these equations appear still impracticable. In order to construct their kernel, one has to deal with auxiliary three-body problems. The Equs. (3.13), suggested a long time ago, have therefore been solved only under uncontrollable and for us therefore uninteresting approximations [27]. But since our technology on solving integral equations with two continuous variables has greatly improved, the experts should have another close look at these almost forgotten exact equations.

What alternatives do we have ?
(1) Instead of inverting the whole Coulomb contribution to the integral equation explicitly, Veselova [28] suggests splitting up the momentum-space Coulomb transition matrix and inverting only that part which generates the Coulomb singularities. Her equations have a kernel, which can be constructed from two-body transition matrices alone. They have not been generalized to the break-up case. Alt's alternative [29] is within the quasi-particle approach [30] using ideas similar to Veselova's. All long-range Coulomb effects are collected in a pure Coulomb scattering amplitude. The Coulomb-modified strong amplitude is obtained from a one-dimensional integral equation with an effective interaction which contains all short-ranged Coulomb effects. The effective interaction can be given exactly, if the Coulomb potential between pair 3 is the only local interaction. Whenever the UPE expansion of a realistic nuclear potential leaves local remainders, it has to be demonstrated that the effective interaction can still be computed with sufficient accuracy. This method might work for the break-up case as well, this has not been studied yet.
(2) The Faddeev equations for the wave function suggest a second alternative. By multiplying with the inverse of the appropriate channel Greens function, they are turned into configuration-space differential equations as used in Ref. [31],

$$(K + w_3 + v_\beta - E)\,|\phi^{(\beta)}(\vec{q})> \;=\; - v_\beta \sum_{\gamma \neq \beta} |\phi^{(\gamma)}(\vec{q})> \tag{3.16}$$

The Equs.(3.16) look friendlier, since they have lost the complicated information on boundary conditions. The boundary conditions have to be imposed as in the Schrödinger equation. For elastic PD scattering below break-up and for break-up except in the final-state interaction region they are known. This differential form of the Faddeev equations can therefore be applied. It should only have the same technical difficulties as for ND scattering [31] .
(3) For elastic PD scattering below break-up, even the ordinary momentum-space integral equations can be used in a simple-minded way. Cut off Coulomb at a large radius at which the strong interaction is neglegible and include it as a short-ranged interaction. A crime is then committed against the Coulomb tail, but this mistake can exactly be corrected. Since the asymptotic behavior of the scattering wave function for two-body final states is well known, the calculated and required asymptotic forms are matched at the cut-off radius and the elastic phase shifts for Coulomb of infinite range are extracted from the computed elastic phase shifts for a cut-off Coulomb. (This method [32] is also applicable for PP scattering.) The Coulomb potential acts in all two-body partial waves. In any computation according to methods (2) and (3) one therefore has to make sure that Coulomb in partial waves much higher than in those anyhow required for a good description of the strong two-body interaction is not important for the nuclear PD phase shifts. This possible practical problem does not arise for Alt's method.

3.3 Summary

Where do we stand theoretically ?
(1) There are no problems in PP scattering any longer even for realistic momentum-space potentials. The most accurate method available for them does not use the T matrix in Coulomb-wave representation, but employes the cut-off Coulomb with subsequent matching to the correct Coulomb tail. The one-boson-exchange potentials of Ref. [13] are

being refitted to the low-energy PP data this way.
(2) There are no problems for the ^{3}He bound state. In fact, Hennell and Delves [32] have included Coulomb in their variational solution of the Schrödinger equation, Gignoux and Laverne did so, when solving the differential form of Faddeev equations [31] .
(3) There are also no conceptual Coulomb problems for PD scattering below break-up. Since the asymptotic form of the wave function is known everywhere, the most promising methods for realistic potentials are: (i) the Schrödinger equation with the Kohn variational principle, (ii) the coordinate-space Faddeev equations [31] , and (iii) the momentum-space Faddeev equations with a cut-off Coulomb and subsequent repair of the Coulomb tail. Furthermore,the quasi-particle method [29] now works for elastic PD scattering below and even above break-up. The only realistic PD calculation available to date [34] is based on the Schrödinger equation. The result is indicated for the doublet S phase in Fig. 7, which is too low, since the ^{3}He binding energy of this calculation as everybody's result for realistic potentials is too small.

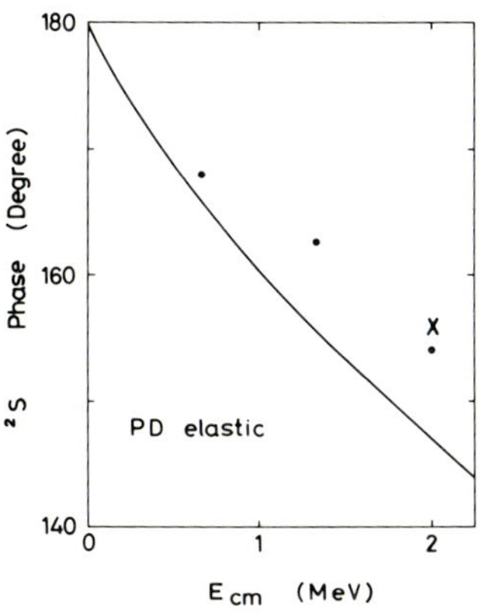

Fig. 7. Doublet S phase of PD elastic scattering according to Ref. [34]. The experimental points are from Refs. [35] (dot) and [36] (cross).

(4) Problems remain for PD break-up. The integral equations (3.11) and (3.13), which are even in the presence of Coulomb well defined and also hold in the break-up case, still appear impracticable. There is a formally correct method based on the ordinary momentum-space Faddeev equations [37]. Even with Coulomb, they yield convergent results off the energy shell. Since the on-shell singularities are known, they can be taken out from the off-shell transition operators, before they are analytically continued to the well-behaved physical amplitudes. To my knowledge, analytic continuation has not produced yet accurate results in atomic physics and it therefore should not be recommended for practical PD break-up calculations. The solution of the configuration-space Faddeev equations appears to me the most promising method for PD break-up, since it requires the same technical apparatus as already employed in ND break-up [31]. It is most urgent to establish the asymptotic behavior of the PD break-up wave function in the final-state interaction region. I also expect, that Alt's method and the momentum-space Faddeev equations with cut-off Coulomb and subsequent matching of the break-up wave function to the proper Coulomb tail have a good chance of success. On these grounds the Coulomb problem in three-nucleon scattering with two equally-charged particles does not look discouraging any longer.

4. SEPARATION OF ELECTROMAGNETIC EFFECTS

I now turn to a practical problem in which the e.m. interaction can be treated formally well, but becomes a nuisance in the analysis of data. Using experimental data with minimal additional theoretical input, we would like to establish unambiguous goals for the first-principle calculations of the indirect e.m. part v_{em}^{ind} in the nuclear force. The existing experimental data, which contain information best, are (1) the $^1S_0$ effective-range parameters of two-nucleon scattering and (2) the ^{3}He - ^{3}H binding-energy difference. What do they teach us quantitatively about the violation of isospin symmetries in the nuclear force ? The comparison of PN with PP and NN scattering is

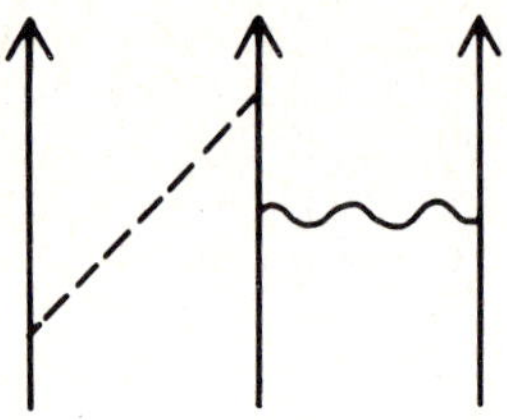

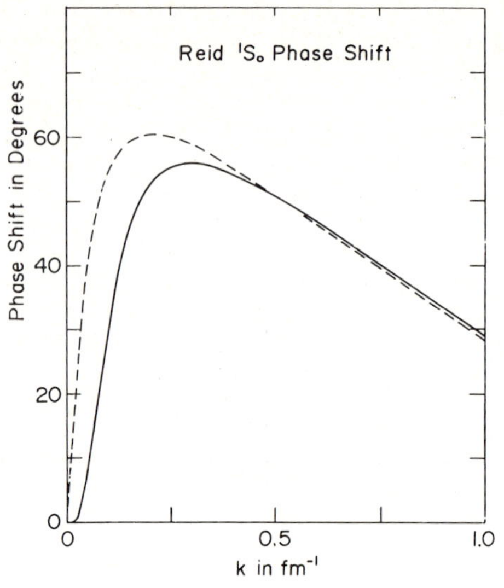

Fig. 8. Contribution to the three-body charge-asymmetric interaction.

Fig. 9. ^1S_o phase shift. The phase shifts calculated with Coulomb (solid curve) and without Coulomb (dashed curve) are compared.

affected by charge dependence. Otherwise both phenomena can only yield information on charge-symmetry breaking. The ^3He-^3H binding-energy difference may also receive a three-body charge-asymmetric contribution, e.g., from processes in which a photon and a pion are simultaneously exchanged between the three nucleons as in Fig. 8. But in both phenomena the charge-dependent direct e.m. interaction is also always present. Its contribution is unwanted and has to be separated out from the experimental data, before the nuclear charge-dependent and charge-asymmetric effects, we are really interest in, come to light. This subtraction of e.m. contributions can only be done theoretically. Can it be achieved in an unambignous manner, i.e., with results independent from models for the strong interaction ? To begin with the conclusion: The direct e.m. contributions to the ^3He-^3H binding-energy difference can be removed fairly cleanly. However, the same separation from the PP scattering data is model-dependent. An unambignous comparison of purely nuclear PP with NN scattering results is spoiled. I now present the evidence for these claims.

4.1. Two-Nucleon Scattering

In two-nucleon scattering e.m. contributions are most important for the PP system in ^1S_o. As an example, the ^1S_o phase of the Reid potential with and without point Coulomb is plotted in Fig. 9. At low energies their difference is sizable. There is a correspondingly strong shift from the experimental scattering length a_{PP}^{em} to its purely nuclear value a_{PP}. What is this value and its model-dependent uncertainty ? For static Coulomb this is a text-book problem with the text-book answer of approximately - 17 fm given long ago [38],

$$\frac{1}{a_{PP}} = \frac{1}{a_{PP}^{em}} - \eta \left[\ln \eta \, r_{PP}^{em} + 2\,\gamma_E - 0.824 \right], \tag{4.1}$$

but 40% of the total shift in this relation, i.e., η 0.824, arises from a model-dependent assumption on the zero-energy wave function, an assumption made with taste for good physics. In Equ. (4.1) r_{pp}^{em} is the experimental effective range, γ_E Euler's number. When removing the e.m. contributions in a proper calculation, the employed technique is standard. A nuclear potential v is fitted such that together with the direct e.m. interaction w it reproduces the experimental phase shifts. Then, the e.m. interaction is dropped and purely nuclear phase shifts are computed. When realistic local or weakly momentum-dependent potentials were employed and all refined direct e.m. effects were taken into account, the value of approximately - 17 fm, specifically - 17.1 ± 0.2 fm for a_{PP} [39] remained firm, still consistent with charge symmetry and the corresponding NN value [39, 15].

I got interested in this closed subject, since two-nucleon potentials with strong nonlocalities at small relative distances were in the meantime used in nuclear structure [40]. They should be tried in this context, too. Furthermore, I believed the spread of 0.2 fm in values for a_{PP} an overestimate of model-dependence, since results of data-nonequivalent potentials were compared. Instead the ^1S_o potentials $\tilde{v}$,

$$\tilde{v} = u(k + w - v)u^+ - k - w \ , \tag{4.2}$$

have the virtue to account - together with the e.m. interaction w - exactly for the same experimental scattering data at all energies as a given reference potential v. The potentials are related by unitary transformations u of its wave functions. u is of such short range that the experimental phase shifts and the theoretically firmly known exterior part of the nuclear potentials v and $\tilde{v}$ remain identical. The potentials v and $\tilde{v}$ differ by nonlocalities at small distances, where we have to treat the nuclear interaction phenomenologically anyhow. Surely, there is a latitude of potentials $\tilde{v}$ depending on u. The special unitary transformation u,

$$u = 1 - 2\,|g\rangle\langle g|\ , \quad \langle r|g\rangle = Nr(1 - \beta r)e^{-\alpha r}, \quad \langle g|g\rangle = 1\ , \tag{4.3}$$

is only one possible realization. α^{-1} controls the range of nonlocalities, β is another parameter to be varied. Scattering from the purely hadronic hamiltonians $k_h + \tilde{v}$ is then compared as a function of the parameters in u. The kinetic energy operator k_h may phenomenologically be corrected for e.m. contributions to the nuclear mass, M_h being the purely hadronic mass.

Phase-equivalent potentials were first used for this purpose in Ref. [41] without disturbing results, then by Kumpf [42] and independently by myself [43], and the latter two calculations with the static Coulomb interaction provided a shock. The Coulomb corrected ^1S_o PP effective range parameters, given in Fig. 10, turned out strongly model-dependent. The scattering length is especially sensitive, its model dependence is larger than the experimental error bars on the NN scattering length [39]. The effective range r_{PP} can drop below the experimental r_{PP}^{em}, believed impossible before. The model dependence decreases when the range of nonlocalities decreases, i.e. with larger α. It is not affected by the experimental uncertainty of the Coulomb potential at small relative distances. The remaining parts of the direct e.m. interaction add to the model dependence [44]. The accurate approximate relation between the nuclear scattering length a_{PP} of the phase-equivalent nuclear potentials $\tilde{v}$ and a_{PPR} of the purely nuclear reference hamiltonian $k_h + v$ is used,

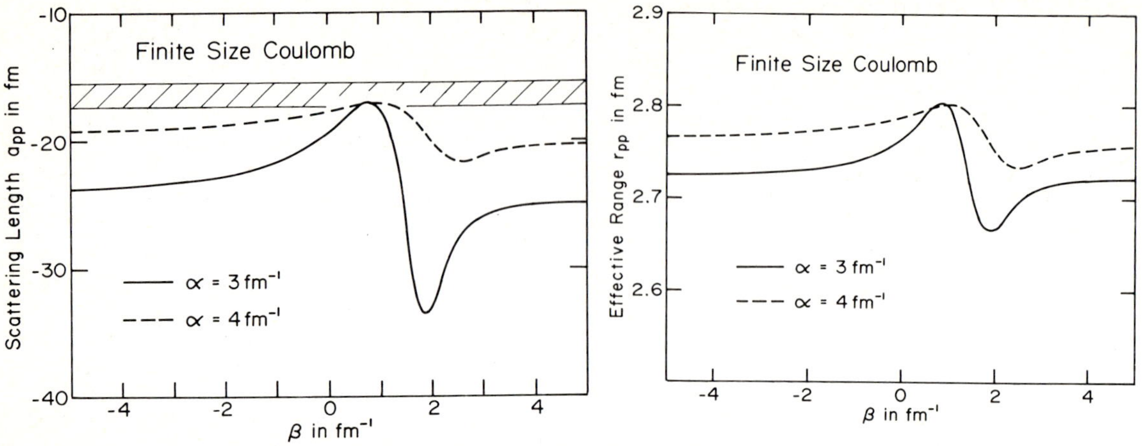

Fig. 10. PP purely nuclear ^1S_o effective range parameters. The potential models of (4.2) and (4.3) ared used. The Reid soft-core potential is the reference v. The cross-hatched area corresponds to the experimental NN scattering length.

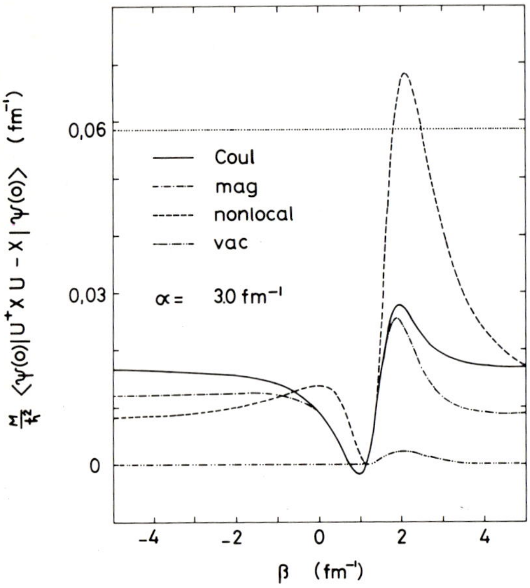

Fig. 11. Contributions to the matrix element (4.4). The dotted line is $(-1/a_{PPR})$. The results of this and the following figures refer to a Yamaguchi potential as reference v. The model dependence of the scattering length is thereby not affected [45] .

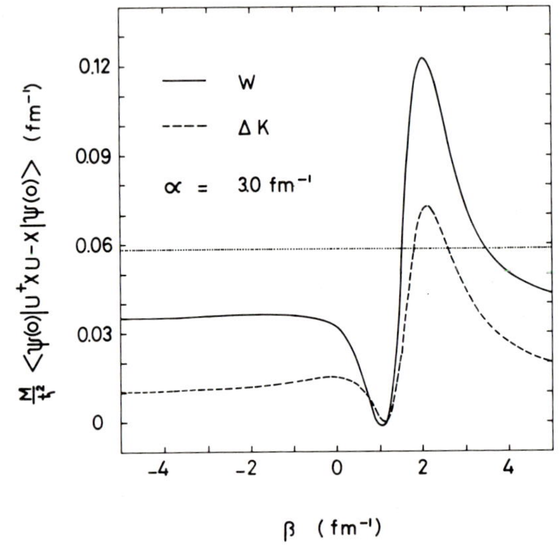

Fig. 12. Contributions to the matrix element (4.4). The contributions from the e.m. interaction w and the mass correction $\Delta k = k - k_h$ are compared. A Yamaguchi potential is the reference v.

$$\frac{1}{a_{pp}} = \frac{1}{a_{PPR}} + \frac{M_h}{\hbar^2} \langle\psi(o)\,|u^+(k - k_h + w)\,u - (k - k_h + w)\,|\psi(o)\rangle . \qquad (4.4)$$

$|\psi(o)\rangle$ is the zero-energy wave function of the reference hamiltonian. All effects of the e.m. separation enter linearly the matrix element of (4.4), relevant for model dependence. Individual contributions are shown in Fig. 11, they all have the same sign. There is obviously no concellation. A similar analysis in Ref. [46] claims cancellation, but it is based on erroneous plots of the local e.m. potentials [47]. In the parameter region, where the summed contributions to the matrix element increase beyond the value $(-1/a_{ppR})$, the scattering length a_{pp} even turns positive. There are nuclear $^1S_0$ potentials which even support a bound state, when the e.m. interaction is removed. This is a disaster, and we would desperately like to get rid of these obviously extreme parametrizations $\tilde{v}$ of the nuclear potential. Though nuclear matter is sensitive to off-shell effects, its calculation is still too inaccurate for the standards of few-nucleon theorists to serve as a constraint on $\tilde{v}$, the deuteron is an illegitimate $^1S_0$ constraint, since it is borrowed from another partial wave, and three-nucleon calculations have not been done for most of the potential models $\tilde{v}$ and, when done, have not produced so embarrassing off-shell effects [48, 43] that they could be held against the used potential models.

We note that the model dependence arising from the momentum-dependent part of the e.m. interaction is largest. Consistent with that fact, the e.m. mass correction, also a momentum-dependent correction, adds further significantly to the model dependence of a_{pp}. The N mass M_N is assumed to be the purely hadronic one, M_h. The e.m. mass contribution to the matrix element (4.4) is compared in Fig. 12 with the complete two-body e.m. contributions. Large momentum-dependent effects are plausible, since the zero-energy wave functions of the transformed potentials $\tilde{v}$ are exotically shaped as shown in Fig. 13. In con-

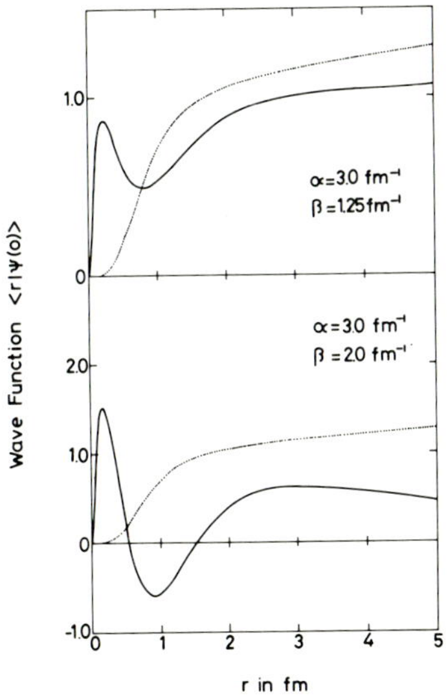

Fig. 13. Zero-energy wave functions of $k_h + \tilde{v}$. As dotted curve the corresponding wave function of the Reid soft-core potential is shown.

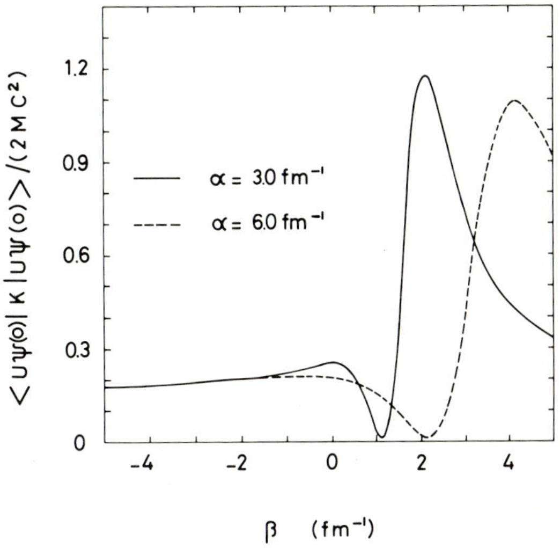

Fig. 14. Ratio of kinetic energy over rest mass contained in the zero-energy wave functions. Matrix element of dimension fm is divided by a suitable interaction length taken for simplicity to be 1 fm.

trast the traditional wave functions are smooth. The shape of the wave functions points to a disease of $\tilde{v}$ [44], $|\tilde{\psi}(o)\rangle = u|\psi(o)\rangle$ contains kinetic energy $\langle\tilde{\psi}(o)|k|\tilde{\psi}(o)\rangle$, which is often not small as compared to the rest mass $2Mc^2$ of the system. This ratio $\langle\tilde{\psi}(o)|k|\tilde{\psi}(o)\rangle/(2Mc^2)$, only about 2% even for strongly repulsive local potentials, can increase as in Fig. 14 to more than 100% due to nonlocalities. This is an unexpected warning that with general nonlocal potentials (as often used in nuclear structure to explore off-shell effects [41]) one might leave the realm of nonrelativistic quantum mechanics unknowingly. This legitimately disqualifies many of the potentials $\tilde{v}$ as test potentials in a nonrelativistic analysis. If one puts a limit of

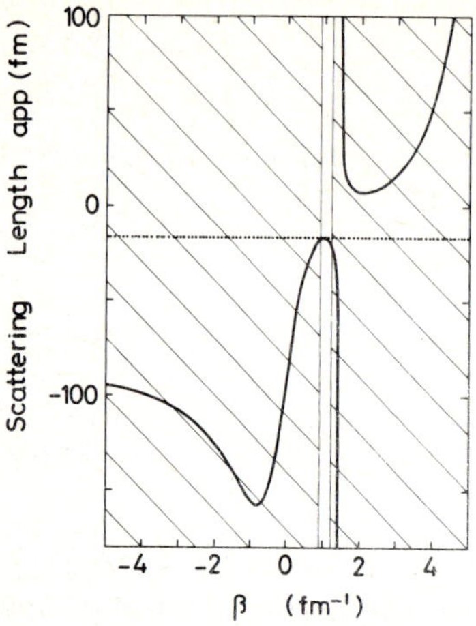

Fig. 15. Model dependence of a_{PP}. All contributions from w and $k - k_h$ are included. In the cross-hatched region the constraining ratio exceeds 10%. The dotted line denotes a_{NN}^{em}. v is a Yamaguchi potential.

acceptability on the ratio $<\tilde{\psi}(o)|k|\tilde{\psi}(o)>/(2Mc^2)$, the disastrous model dependence from all e.m. effects, shown for a_{PP} in Fig. 15, is severely reduced. A limit of 10% is imposed. But even with such a reasonable limit, the model dependence of a_{PP} is still too large to allow an unambiguous quantitative estimate on the amount of nuclear charge asymmetry. The model dependence of a_{PP} arising from the e.m. two-body contributions is plotted in Fig. 16 against the suggested constraint $<\tilde{\psi}(o)|k|\tilde{\psi}(o)>/(2Mc^2)$. Confining the nonlocalities in u to within 1.5 fm, i.e., $\alpha = 6$ fm^{-1}, and using a 10% constraint, the remaining model dependence of a_{PP} due to the e.m. two-body interaction w is 2.5 fm, it increases to 5 fm when the mass effect is included. There is also model dependence in PN and NN scattering, when e.m. contributions to their respective nuclear scattering lengths a_{PN} and a_{NN} are subtracted out. Using the same constraint as for PP scattering, the model dependence is so small, i.e., 0.75 fm for a_{PN} and 0.25 fm for a_{NN}, that a comparison of PN and NN scattering is not invalidated, still yielding a quantitative estimate for the nuclear charge dependence in two-nucleon scattering. The model dependence shown in Fig. 17 is, however,

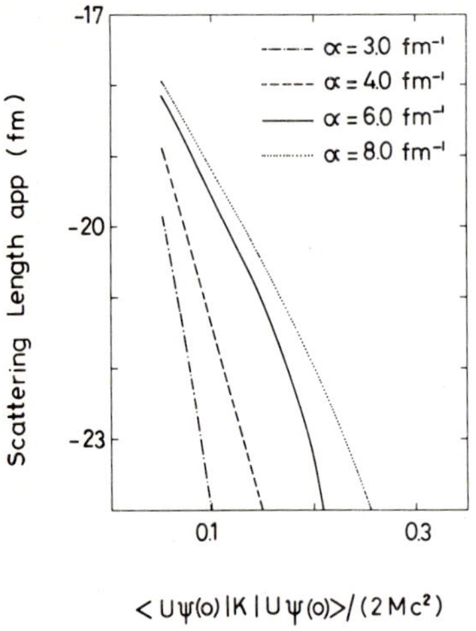

Fig. 16. Model dependence of a_{PP}. Only contributions from the e.m. interaction w are included. All values above curves are possible due to model dependence. A Yamaguchi potential is the reference v.

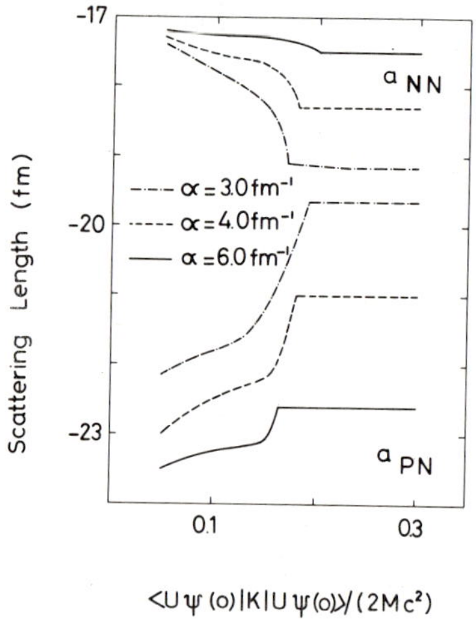

Fig. 17. Model dependence of a_{PN} and a_{NN}. Only contributions from w are included. The neutron charge form factor is put to zero. All values for a_{PN} (a_{NN}) below (above) curves are possible due to model dependence. A Yamaguchi potential is the reference v.

an underestimate, since the effect of the momentum-dependent e.m. potential, important for the size of the PP model dependence, has not yet been included and may contribute significantly despite the smallness of the neutron charge form factor. The model dependence arising from the e.m. mass contribution is also left out. Depending on the chosen nuclear standard M_h for the mass, it can show up in a_{PN} or a_{NN}.

I conclude. In two-nucleon scattering we cannot set unambiguous goals for the first-principle calculations of charge asymmetry:
(1) Due to our experience with local potentials, the value of -17 fm for the nuclear PP scattering length is still most probable, but its model dependence arising from different parametrization of the wave function at small relative distances is disappointingly large. Since the Coulomb correction is largest at small energies, the removal of e.m. effects becomes cleaner at higher energies [49]. The comparison of PN and NN scattering for estimating charge dependence is, however, still possible.
(2) The charge-dependent and charge-asymmetric components of the nuclear force will also contribute to the effective-range parameters in a correspondingly model-dependent way. First-principle calculations should therefore not focus on the theoretical nuclear effective range parameters a and r, but attempt a consistent description of the experimental values a^{em} and r^{em} in all three charge states, using reasonable models for the nuclear force and including the e.m. and nuclear isospin-dependent contributions simultaneously.
(3) The disappointing model dependence can be turned to our advantage. Instead of trying to establish the charge symmetry of the nuclear force experimentally, one may assume charge symmetry as a theoretical principle to hold with high accuracy. Charge symmetry then becomes an off-shell constraint [43] for nuclear potential models in ^1S_o, the only one we have. With the inclusion of all e.m. contributions, this constraint will be more effective than previously [50] when used with the static Coulomb potential alone.

4.2. ^3He - ^3H Binding-Energy Difference

With the separation of e.m. effects from the ^3He - ^3H binding-energy difference we enter friendlier water. This is surprising, since the three-nucleon system in general is more complicated. However, the allowed models of three-body wave functions are even in the nuclear interior restricted by the experimental knowledge of three-body electromagnetic form factors. In contrast, this is just the information we are lacking in two-nucleon scattering.

I choose to parametrize the experimental three-nucleon charge form factors $F_{CH,EXP}(q^2)$ [51] in terms of the proton and neutron body form factors of ^3H, $f_P(q^2)$ and $f_N(q^2)$,

$$F^{^3H}_{CH,EXP}(q^2) = G_{EP}(q^2)\, f_P(q^2) + 2\, G_{EN}(q^2)\, f_N(q^2), \tag{4.5a}$$

$$2\, F^{^3He}_{CH,\, EXP}(q^2) = 2\, G_{EP}(q^2)\, f_N(q^2) + G_{EN}(q^2)\, f_P(q^2). \tag{4.5b}$$

The same three-body wave function is assumed for ^3He. The body form factors are therefore determined by the experimental charge form factors. As observed by Fabre de la Ripelle [52] and by Friar [53], the ^3He - ^3H Coulomb energy difference, which arises from a two-body operator of the type

$$v_k(\vec{x}_k) = (2\pi)^{-3} \int d^3q\, A_i(q^2)\, A_j(q^2)\, V(q^2)\, e^{\,i\vec{q}\cdot\vec{x}_k}, \tag{4.6a}$$

$$A_i(q^2) = \frac{1}{2} \{ b_P(q^2) \left[1+\tau_Z(i) \right] + b_N(q^2) \left[1-\tau_Z(i) \right] \} , \tag{4.6b}$$

and the ^{3}He and ^{3}H charge form factors of Equ. (4.5), which arise from a one-body operator, are related in an almost model-independent way, an accident peculiar to the three-nucleon system. With the Jacobi variables $(\vec{x}_i, \vec{y}_i)$ of Fig. 6 the change of $\vec{x}_1$ into $-\vec{x}_1$, followed by the interchange of the two three-dimensional components $\vec{x}_1$ and $2\vec{y}_1/\sqrt{3}$ of the sixdimensional hyperradius ρ converts the coordinates around the mass center $\sqrt{3}(\vec{r}_i-\vec{r}_{cm})$, relevant for the form factors, into the internucleon distances $(\vec{r}_j-\vec{r}_k)$, relevant for the Coulomb energy. For general three-body wave functions using its symmetry properties alone, one arrives at the following perturbative estimate of the ^{3}He - ^{3}H Coulomb energy in terms of the experimental three-body charge form factors or in terms of the body form factors $f_P(q^2)$ and $f_N(q^2)$.

$$< {}^3He|v|{}^3He > - < {}^3H|v|{}^3H > =$$

$$\frac{1}{2\pi^2} \int_0^\infty q^2 dq \left[b_P^2(q^2) - b_N^2(q^2) \right] V(q^2) \left[\frac{4}{3} f_N(3q^2) - \frac{1}{3} f_P(3q^2) \right] \tag{4.7}$$

According to Ref. [52] the relation accounts for contributions from the totally symmetric S and mixed symmetry S' and D components of the three-nucleon wave function almost exactly up to about 1% and is proven there within the framework of hyperspherical harmonics. The relation is therefore highly accurate and model-independent and saves me from trying out a manifold of three-nucleon wave functions. As a perturbative result the relation (4.7) overestimates the Coulomb energy according to Erens [54] by about 1%. This relation, usually employed for the static Coulomb force only, is clearly more powerful than has been appreciated. It may be used [55] to estimate the contribution to the ^{3}He - ^{3}H binding-energy difference of any two-body interaction, which can be cast into the spin-independent form (4.6). And for the three-nucleon bound state all local terms in the direct e.m. interaction indeed have this form with high precision. Among the isospin-triplet partial waves ^1S_o is most important with a relative weight of 90% [31] . The odd spin-triplet partial waves therefore contribute rather little and are for the present purpose dropped introducing errors not larger than 10% in the magnetic contributions, which are anyhow small. In this spirit, the spin-spin contact interaction of Equ. (2.3) $-(2/3)\mu_o^2 \mu_1 \mu_2 \vec{\sigma}_1 \cdot \vec{\sigma}_2$ is replaced by and used in its spin-singlet form $2\mu_o^2 \mu_1\mu_2$. Furthermore, I note in passing that in the same approximation on odd spin-triplet partial waves the relation (4.7) also allows an instant model-independent test [55] of any nuclear charge asymmetric potential in the three-body system and such an estimate can be handled with ease even by non-threebody experts. There is only one draw-back this beautiful relation may carry: The experimental charge form factors are not solely one-body quantities. They contain exchange current admixtures $F_{CH,EXCH}(q^2)$, which have to be taken out,

$$F_{CH}(q^2) = F_{CH,EXP}(q^2) - F_{CH,EXCH}(q^2) , \tag{4.8}$$

since only the one-body part $F_{CH}(q^2)$ of the charge form factors is relevant for the ground-state expectation values (4.7). If we take the estimates of Ref. [56] for the exchange-current contibutions (Fig. 18) as representative, a correction of the binding-energy difference is found into the direction of decreasing the discrepancy with experiment.

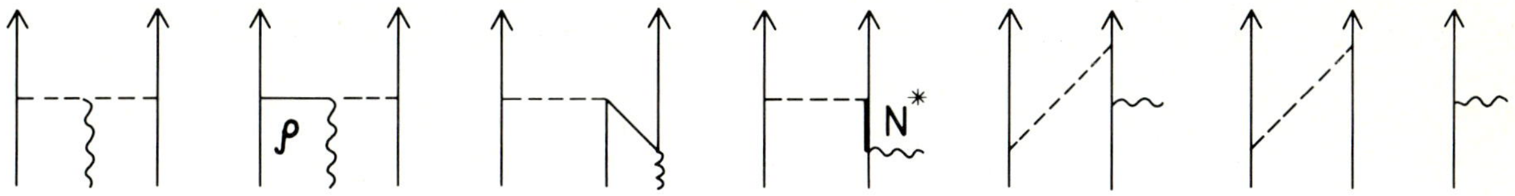

Fig. 18. Exchange-current contributions to the three-nucleon charge form factors.

Table. Direct e.m. contributions to ^3He-^3H binding-energy difference in keV.

	Model-independent estimates	
(I.1)	static Coulomb	638 ± 17
(I.2)	Foldy-Darwin	-2
(I.3)	magnetic	16 ± 2
(I.4)	vacuum polarization	4
	Model-dependent estimates	
(II.1)	momentum-dependent e.m.	15
(II.2)	P-N mass difference	23 ± 5
	Corrections	
(III.1)	perturbation	-6
(III.2)	exchange currents	10 ± 6
(III.3)	error in G_{EN}	± 1
	experiment	764
	total e.m. contribution	698 ± 31
	charge asymmetry	66 ± 31

The direct e.m. contributions to the ^3He - ^3H binding-energy difference [55] are summarized in the Table. The results (I), (III.2) and (III.3) are based on the relation (4.7). The grand total of nuclear charge asymmetry in the binding-energy difference is 66 ± 31 keV, which is substantially smaller than the previous estimate [57] of 100-150 keV. The reduction comes about by four small effects all working into the same direction:

(1) The trivial PN mass difference (II.2) whose size has always been known was purposely omitted in Ref. [57]. If we want to find the total charge asymmetry effect in the nuclear interaction, however, it has to be counted, as is done for the Coulomb anomaly of heavier nuclei. The specific value of 23 keV is only an estimate based on a total kinetic energy of 50 MeV obtained in Ref. [58] for ^3H and the

Reid soft-core potential.
(2) The contribution from the local magnetic interaction, (I.2) and (I.3), has increased as compared to [57], since double counting in the contact terms of Equ. (2.3) is removed.
(3) The effect of the momentum-dependent e.m. interaction (II.1) had never been included before.
(4) The removal of exchange-current effects from the experimental form factors (III.2) adds another 10 keV. If recoil corrections are left out, it even increases to 16 keV. This is the reason for the error of 6 keV assigned to (III.2).
The error in (I.1) arises from the uncertainties in the experimental charge from factors at small momentum transfer, only 4 keV of the error from the fact, that the experimental information on the triton form factor terminates at 8 fm^{-2}. The quoted error is hoped to be large enough to also account for the shortcomings of relation (4.7). The error due to the uncertainty in the neutron charge factor $G_{EN}(q^2)$, needed in (4.5) to extract the three-nucleon body form factors, is luckily very small (III.3).

I conclude: Obviously a quantitative estimate of nuclear charge asymmetry can be obtained better from the three-body system than from low-energy two-nucleon scattering.

The results of Chapters 2 and 4 were obtained in collaboration with R.A. Brandenburg, S.A. Coon and H. Walliser. Computations were performed at "Regionales Rechenzentrum für Niedersachsen in Hannover". In preparing the manuscript the author enjoyed discussions with W. Glöckle, H.H. Hackenbroich, W. Sandhas and E. Werner.

REFERENCES

[1] W.Bertozzi et al.,Phys.Lett.41B (1972)408.
[2] R.A.Brandenburg and P.U.Sauer,Phys.Rev.C12 (1975)1101.
[3] W.A.Barker and F.N.Glover, Phys.Rev.99 (1955)317.
[4] J.Schwinger,Phys.Rev.78 (1950)135.
[5] F.Close and H.Osborn,Phys.Rev.D2(1970)2127.
[6] D.Kiang,S.Machida and Y.Nogami,Can.J.Phys.51(1973)1120.
[7] M.Banerjee, preprint 1974.
[8] L.N.Hand,D.G.Miller,and R.Wilson,Rev.Mod.Phys.35 (1963)335.
[9] F.Iachello,A.Jackson, and A.Lande,Phys.Lett.43B (1973)191.
[10] T.Janssens et al.,Phys.Rev.142 (1966)922.
[11] D.O.Riska and Y.H.Chu,Nucl.Phys.A235 (1974)499; P.C.McNamee, M.D.Scadron, and S.A.Coon, Nucl.Phys.A249 (1975)483.
[12] A.Gersten,R.Thompson,and A.E.S.Green,Phys.Rev.D3 (1971)2076; T.Ueda,M.Nack, and A.E.S.Green, Phys.Rev.C8 (1973)2061.
[13] R.Machleidt and K.Holinde,Nucl.Phys.A247 (1975)495.
[14] W.M.Kloet and J.A.Tjon, Ann.Phys.79 (1973)407.
[15] B.Zeitnitz et al., Nucl.Phys.A231 (1974)13.
[16] J.Nuttall,Phys.Rev.Lett.19 (1967)473; J.Nuttall, and J.G.Webb, Phys.Rev.178 (1968)2226.
[17] R.W.Hart,E.P.Gray, and W.H.Guier, Phys.Rev.108 (1957)1512; R.K.Peterkop, JETP 43 (1962)616; M.R.H.Rudge and M.J.Seaton, Proc.Roy Soc.A283 (1964)262.
[18] S.P.Merkuriev,Nucl.Phys.A233 (1974)395.
[19] E.Gerjuoy,Phil.Trans.Roy.Soc.A270 (1971)197.
[20] M.Lieber,L.Rosenberg,and L.Spruch,Phys.Rev.D5 (1972)1330,1347; W.M.Bryce and F.Mandl,J.Phys.B5 (1972)912.
[21] J.C.Y.Chen and A.C.Chen, Adv.Atom.Mol.Phys.8 (1972)71.
[22] J.D.Dollard,J.Math.Phys.5 (1964)729;C.Chandler and A.G.Gibson, J.Math.Phys.15 (1974) 291 and 1366; H.van Haeringen, preprint 1975.

[23] P.U.Sauer and H.Walliser, in Few-Body Problems in Nuclear and Particle Physics, ed. by R.J.Slobodrian et al., Les Presses Université Laval 1975,p.108.
[24] G.Cattapan,G.Pisent and V.Vanzani,Z.Phys.A274 (1975)139; H. van Haeringen and R. van Wageningen,J.Math.Phys.16 (1974)1441.
[25] W.Glöckle,Nucl.Phys.A141 (1970)620,Z.Phys.271 (1974)31, and private communication.
[26] J.V.Noble,Phys.Rev.161 (1967)945; Gy.Bencze,Nucl.Phys.A196 (1972)135.
[27] Sh.Adya,Phys.Rev.166 (1968)991, 177 (1969)1406.
[28] A.M.Veselova,Theor.Math.Phys.3 (1970)542.
[29] E.O.Alt, contribution to this conference. See also W.Timm, M.Sc.thesis Münster 1975.
[30] E.O.Alt,P.Grassberger and W.Sandhas,Nucl.Phys.B2 (1967)167.
[31] C.Gignoux and A.Laverne, in Few Particle Problems in the Nuclear Interaction, ed. by I.Slaus et al., North-Holland 1972, p. 411, Nucl.Phys.A203 (1973)597; C.Gignoux,A.Laverne, and S.P.Merkuriev, Phys.Rev.Lett.33 (1974)1350.
[32] C.M.Vincent and S.C.Phatak, Phys.Rev.C10 (1974)391.
[33] M.A.Hennell and L.M.Delves, Nucl.Phys.A246 (1975)490.
[34] H.Stöwe,H.H.Hackenbroich, and P.Heiss, private communication.
[35] J.Arvieux,Nucl.Phys.A221 (1974)253.
[36] P.A.Schmelzbach et al., Nucl.Phys.A197 (1972)273.
[37] G.D.McCartor and J.Nuttall,Phys.Rev.A4 (1971)625.
[38] J.D.Jackson and J.M.Blatt,Rev.Mod.Phys.22 (1950)77.
[39] E.M.Henley and D.H.Wilkinson, in Few Particle Problems in the Nuclear Interaction, ed. by I.Slaus et al., North-Holland 1972, p.242.
[40] M.K.Srivastava and D.W.L.Sprung, in Advances in Nuclear Physics, Vol.8, ed. by M.Baranger and E.Vogt, Plenum 1975, p.121.
[41] M.D.Miller et al., Phys.Lett.30B (1969)157.
[42] H.Kumpf,Sov.J.Nucl.Phys.17 (1973)602.
[43] P.U.Sauer,Phys.Rev.Lett.32 (1974)626, Phys.Rev.C11 (1975)1786.
[44] P.U.Sauer and H.Walliser,to be published.
[45] H.de Groot and H.J.Boersma,Phys.Lett.57B (1975)21.
[46] C.W.Wong,S.K.Young, and K.F.Liu,Nucl.Phys.A253 (1975)96; C.W.Wong and S.K.Young, preprint 1975.
[47] M.S.Sher,P.Signell,and L.Heller, Ann.Phys.58 (1970)1,and erratum to be published.
[48] E.P.Harper,Y.E.Kim, and A.Tubis,Phys.Rev.C6 (1972)1601.
[49] W.Plessas,L.Streit, and H.Zingl,preprint 1974.
[50] A.W.Thomas and I.R.Afnan,Phys.Lett.55B (1975)425; N.J.McGurk and H.Fiedeldey,Can.J.Phys.53 (1975)1749.
[51] H.Collard et al.,Phys.Rev.138 (1965)B57; J.S.McCarthy et al., Phys.Rev.Lett.25 (1970)884;M.Bernheim et al., Nuovo Cim.Lett.5 (1972)431.
[52] M.Fabre de la Ripelle, Fizika 4 (1972)1.
[53] J.L.Friar,Nucl.Phys.A156 (1970)43.
[54] G.Erens,Ph.D. thesis Vrije Universiteit Amsterdam 1970.
[55] R.A.Brandenburg,S.A.Coon, and P.U.Sauer,to be published.
[56] W.M.Kloet and J.A.Tjon, preprint 1975.
[57] K.Okamoto and M.Horiike,Nuovo Cim.Lett.10 (1974)247, and references there.
[58] A.D.Jackson,A.Lande, and P.U.Sauer,Phys.Lett.35B (1971)365.

THREE BODY FORCES IN NUCLEI

Bruce H.J. McKellar

School of Physics, University of Melbourne, Parkville, Vict. 3052, Australia.

and

Service de Physique Théorique, CEN Saclay, B.P. No. 2, 91190 Gif-sur-Yvette, France.

1. INTRODUCTION

First we must understand what we are talking about - just what is a many body force? The quick answer is that it is a term in the potential energy of the system which depends on the coordinates of three or more particles. The concept of three body forces (to be specific) is unfortunately often confused with that of three body correlation contributions to the energy, which played an important role in the development of the theory of nuclear matter about 10 years ago [1], and it is worthwhile digressing to make the distinction clear.

Perhaps the simplest example is that of a real gas, whose equation of state is given by the well known virial expansion,

$$\frac{pV}{RT} = 1 + \frac{B_2}{V} + \frac{B_3}{V^2} + \ldots .$$

The first term arises when we neglect the interactions of the molecules entirely, the second if we consider the contribution of pairwise interactions, the third from interactions of triples, etc. Thus it is clear that 3 body forces contribute first to B_3. However, they are not the whole contribution; we can have interactions of triples through successive interactions of pairs, as the large number of you who have handled Faddeev equations know very well*. This contribution from pairwise interactions we will call the <u>three body correlation</u> contribution to the energy (or B_3). (In the shell model literature it is called an effective three body force.) The three body forces also contribute to B_3, and for the rare gases these contributions are approximately equal [3].

Three body forces are a complication, and as such have generally been ignored in nuclear physics, with such statements as "we will use two body forces only, until we are forced to use three body forces by some discrepancy with the data". As calculations in nuclear physics become more and more refined, we should ask the question "Have we yet found a discrepancy which forces us to consider three body forces?". The answer is, as it usually is in nuclear physics, maybe. There is for example a clear discrepancy of about 1 MeV between the calculated and observed binding energy of the triton [4]. And it seems impossible to move the binding energy curve of nuclear matter or finite nuclei [5] away from the Coester Line. However if one is skeptical of three body forces, he could simply say that this is a consequence of our lack of knowledge of the two body force. The skeptic would then go on to point out that in calculations of the energy levels of such nuclei as ^{19}O and ^{205}Pb three body forces were found to be unnecessary to reproduce the spectra [6], and that deviations from mass formulae deduced from pairwise interactions do not require three body forces [7]. However, a three body term in the energy (whether it arises from three body correlations or a three body force) did improve some shell model calculations [8]. Thus we conclude that there is a possible empirical need for many body forces of weaker strength

*In fact the work of Dashen, Ma and Bernstein [2] shows how to compute B_3 from the 3 particle → 3 particle amplitude obtained from the Faddeev equations, just as B_2 can be computed from the phase shift.

than the two body forces, and turn to the theoretical basis for such forces.

Almost as soon as the meson theory of nuclear forces was proposed, attempts were made to calculate three nucleon potentials arising from multiple pion exchange [9] Various types of three body forces considered in these pioneering works are shown in fig. 1. It is useful to classify such forces in terms of the number of

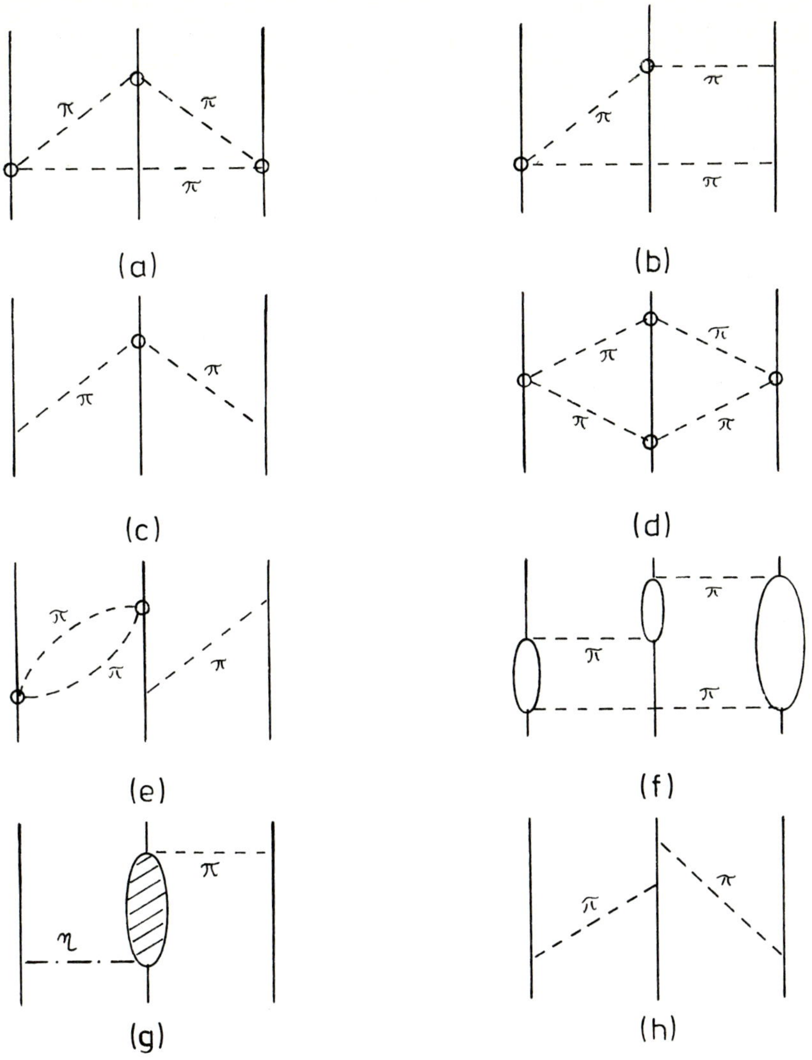

Figure 1

Various early 3 Body Forces

Figs. (a) to (e) arise in pair theory, the small circle representing the $NN\pi\pi$ pair interaction. The blobs in fig. (f) are reduced p wave πN amplitudes, and that of fig. (g) the reduced $\eta N \rightarrow \pi N$ amplitude. Fig. (h) is typical of many time ordered diagrams which give rise to three body forces.

exchanged pions, and to expect that forces with a smaller number of exchanged pions are more important because of their longer range. Because of this we concentrate most of our attention on the two pion exchange potential of fig. 2 [10]. More recently the three σ exchange potential of fig. 3 has also been considered [11] along with other three meson exchanges arising from triple meson couplings shown in fig. 4 [12]. At present there is no completely satisfactory evaluation

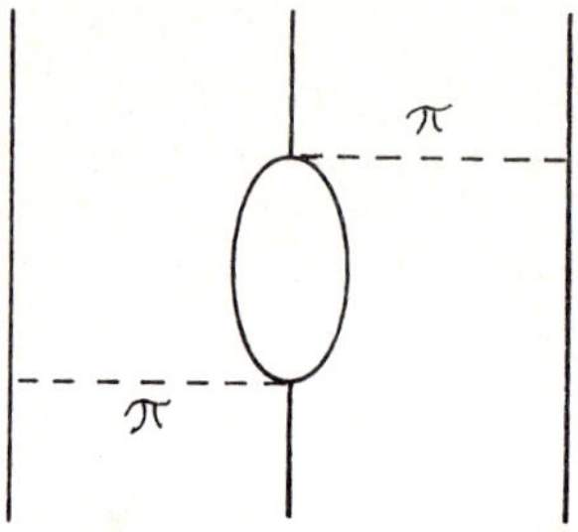

Figure 2

The Two Pion Exchange Three Body Force

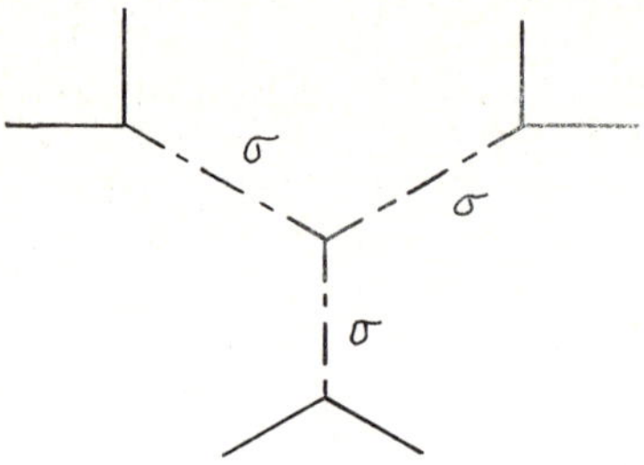

Figure 3

The Three Sigma Exchange Three Body Force

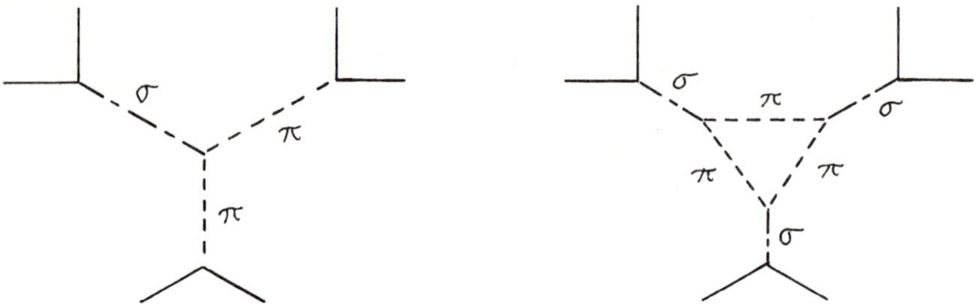

Figure 4

Two Three Body Forces resulting from $\sigma\pi\pi$ Coupling

of the energy contribution of such forces, even for the simplest systems, the triton and nuclear matter.

Theoretically, of course there is no reason to stop at three body forces. Especially at high densities many body forces become important [13] as do such exotica as pion condensates [14] and Lee-Wick states [15] with which they are intimately connected. However, it would take us too far from the theme of this conference to discuss such a system as high density nuclear matter.

In this talk I will first of all discuss the formal theory of calculations in nuclear matter and in the triton with three body forces. Then in section 3 I will discuss the two pion exchange 3 body force, which has received most attention. Section 4 is devoted to other three body forces.

2. FORMALISM WITH THREE BODY FORCES

2.1 The Three Body Problem

We consider first of all the problem of three particles, which we label as 1, 2, 3 which interact via two body potentials, V_{ij}, and via a three body potential, W. Soon after the Faddeev equations were developed for the three body problem, it was shown how they could be modified for the present problem, with the additional three body interaction [16]. We begin by writing the three body T matrix T as

$$T = T^{(0)} + T^{(1)} + T^{(2)} + T^{(3)} \tag{1}$$

where $T^{(i)}$ for $i \neq 0$ is the sum of all diagrams in which the last interaction is V_{jk} (i j k cyclic), and $T^{(0)}$ is the sum of all diagrams in which the last interaction is W, as illustrated in figure 5.

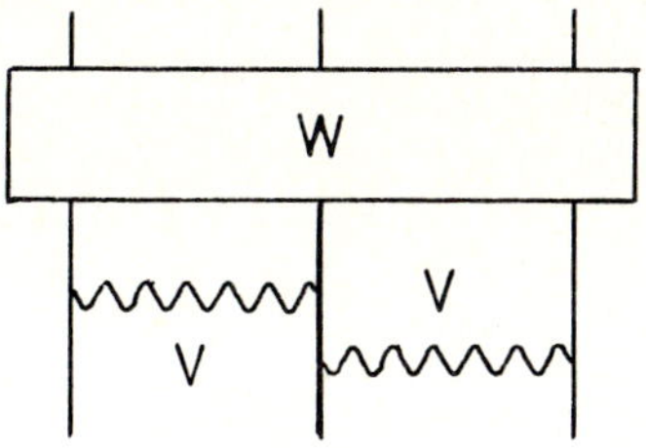

Figure 5

A typical contribution to $T^{(0)}$

Then one can derive three body equations for $W \neq 0$ of the symmetrical form

$$T^{(i)} = t^{(i)} + \sum_{j \neq i} t^{(i)} G_0 T^{(j)} \tag{2}$$

where $t^{(i)}$ for $i \neq 0$ are two body T matrices in 3 body space, and $t^{(0)}$ is the three body T matrix for interaction through W alone, derived from the Lippman-Schwinger equation

$$t^{(0)} = W + WG_0 t^{(0)} \tag{3}$$

(Since W connects all three particles, this equation does not suffer from the "disconnected diagram" disease). These equations may readily be cast into other forms, some of which were discussed in ref. [16].

As an example we generalise the Lovelace equations [17]. It is important to realise that they have the advantage that they are written for off shell extensions of the physical amplitudes. Since the three body force does not open any new channels, the introduction of additional Lovelace type amplitudes, as was done by Phillips [16] is an unnecessary complication. A much simpler generalisation is to introduce the notation $V_0 = W$ following Freedman et al. [16]. Then the Lovelace amplitudes

$$U^{+}_{\alpha\beta}(s) = \bar{\delta}_{\alpha\gamma} V_\gamma + \bar{\delta}_{\alpha\gamma} V_\gamma G_0(s) V_\delta \bar{\delta}_{\delta\beta}, \tag{4}$$

$$U^{-}_{\alpha\beta}(s) = V_\delta \bar{\delta}_{\delta\beta} + \bar{\delta}_{\alpha\gamma} V_\gamma G_0(s) V_\delta \bar{\delta}_{\delta\beta}, \tag{5}$$

(where $\bar{\delta}_{\alpha\beta} = 1 - \delta_{\alpha\beta}$ and we saw over repeated indices) have the same interpretation in terms of transition amplitudes, and they satisfy the equations

$$U^{+}_{\alpha\beta}(s) = \bar{\delta}_{\alpha\gamma} V_\gamma + U^{+}_{\alpha\delta}(s) G_0(s) t^{(\delta)}(s) \bar{\delta}_{\delta\beta} \tag{6}$$

$$U^{-}_{\alpha\beta}(s) = V_\delta \bar{\delta}_{\delta\beta} + \bar{\delta}_{\alpha\gamma} t^{(\gamma)}(s) G_0(s) U^{-}_{\gamma\beta}(s). \tag{7}$$

These have the same form as the original Lovelace equations, but now $V_0 = W$ and $t^{(0)} = W + WG_0 t^{(0)}$ as above.

However, to my knowledge these three body force equations have never been solved except for the case when W is separable. It is more usual to treat W as a perturbation, rather than to put it on the same footing as V. For this purpose it is more convenient to rewrite the equations (2) and (3) as

$$T = \tau + (1 + \tau G_0) W (1 + G_0 T) \tag{8}$$

where τ is the three body T matrix in the absence of the three body force W.

This equation is a well defined integral equation for T (because of W it has no "disconnected diagrams"), which can serve as the basis of a systematic perturbation expansion. To date, however, no calculation in the 3 body system to higher than first order in W has been done for realistic W. However, there is some hope that, in the future, calculations will be extended to higher orders, or all orders, in W. For such extensions the general theory of this section will be useful.

2.2 Nuclear Matter - The Bethe-Faddeev Equations

The properties of three body clusters in nuclear matter are described by the Bethe-Faddeev equations [1,19] and these were generalised to include the effects of three body forces by McKellar and Rajaraman [20]. The resulting equations have the same form as those of the preceding subjection, with the replacements $t^{(i)} \to g_{jk}$, where ijk are cyclic (and non zero) and g_{jk} is the Bruckner G matrix for the two particles j and k in the nuclear matter, and $G_0 \to -\frac{Q}{e}$ where e is the energy denominator. The most useful equation is the analogue of equation (8),

$$T = T_0 + (1 - T_0 \frac{Q}{e})W(1 - \frac{Q}{e} T) \tag{9}$$

where we use T for the sum of all three hole line diagrams in nuclear matter, and T_0 for the sum of all three hole line diagrams in the absence of three body forces. This relationship permits us to express the contribution of the three body forces to the binding energy of nuclear matter E_3 in the familiar way

$$E_3 = \sum_{\text{triples}} \sum_{P} \varepsilon_P \langle\Phi_{ijk}|W|\Psi_{P(ijk)}\rangle \tag{10}$$

where P(ijk) is a permutation of ijk, ε_P is the parity of the permutation, Φ_{ijk} is the wave function of the triples ijk in the absence of three body forces, Ψ_{ijk} is the wave function in the presence of three body forces, and the sums run all occupied triples (ijk) and all permutations P(ijk). The wave functions Φ_{ijk}, Ψ_{ijk} are determined from the generalised Bethe-Faddeev equations (9) by

$$\Phi = (1 - \frac{Q}{e} T_0)\chi; \quad \Psi = (1 - \frac{Q}{e} T)\chi \tag{11}$$

where χ is the plane wave wavefunction of the triple. The energy (9) can be expanded perturbatively in W if required.

Having now reviewed the formalism for calculating with a three body force, we turn to discussion of a particular three body force and its applications.

3. THE TWO PION EXCHANGE THREE BODY FORCE

3.1 The 3 Nuclear Potential

The type of force we have in mind here is shown in figure 2. Now of course we must define carefully just what the blob is. It is the πN scattering amplitude, continued off shell in the pion masses (because the pions are virtual), but after subtracting the so called "forward propagating Born terms", which represent the iteration of the two body force, as in figure 6.

This amplitude has been studied in various models.

Most of the calculations have been done retaining the P wave Δ(1236) contribution only following the work of Fujita and Miyazawa [10]. More recently much better models of the πN scattering amplitude have become available in the near threshold region [21] and these have been used by Coon, Scadron and Barrett [22], and by Coon, Scadron, Barrett, Blatt and McKellar [23] to construct three body forces.

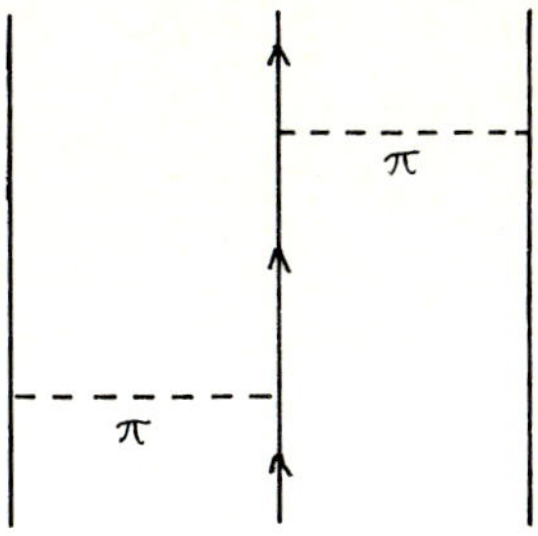

Figure 6

The Iterated Two Body Force, involving the Forward Propagating Born Terms for πN Scattering

These authors expand the amplitude in powers of ν and t, and q^2 and q'^2, and then adjust the coefficients in the expansion to fit

(a) the current algebra results when one or both q^2 vanish

(b) what is known about low energy scattering.

Restricting ourselves to the spin and isospin average amplitude, A, appropriate to nuclear matter, and noting that we need the amplitude only near $\nu = 0$ and $q_0 = q_0' = 0$, we approximate it by

$$A(\nu=0,t,q^2,q'^2)\Big|_{q_0 = q_0' = 0} = K(\vec{q}^2)K(\vec{q}'^2)[a + b\vec{q}.\vec{q}' + c(\vec{q}^2+\vec{q}'^2)+ \ldots] \quad (12)$$

and we find for the coefficients

$$a = \frac{\sigma}{f_\pi^2} = 1.10 \pm 0.05$$

$$b = -\frac{2\sigma}{f_\pi^2}\left(\frac{1}{\mu^2} + \frac{1}{4m^2}\right) + \frac{2}{\mu^2}\bar{F}^{(+)}(0,\mu^2,\mu^2,\mu^2) = -2.58 \pm 0.08$$

$$c = \frac{\sigma}{f_\pi^2}\left(\frac{1}{\mu^2} + \frac{1}{4m^2}\right) - \frac{g^2}{4m^3} + K'(0)\frac{\sigma}{f_\pi^2} = 1.03 \pm 0.05 \quad (13)$$

$K(q^2)$ is the πNN form factor discussed in more detail below, and $K'(q^2)$ is its derivative which can be estimated from the Goldberger-Treiman discrepancy. The numerical values differ slightly from those of ref. [22], because we have averaged a number of recent estimates of σ and $\bar{F}^{(+)}(0,\mu^2,\mu^2,\mu^2)$. For comparison we quote the widely used amplitude of Fujita and Miyazawa [10]. In the above notation it is

$$a = c = 0$$

$$b = -\frac{8}{9\mu^4}\int_0^\infty \frac{\sigma_{33}(p)}{p^2+\mu^2}\,dp = -1.40\ \mu^{-3},$$ with the numerical value used by Loiseau et al. [24] and Blatt and McKellar [25]. Loiseau et al. made an explicit model of the s wave terms a and c, corresponding to the diagrams of figure 7.

The parameters were chosen so that $a = 0$ and the effective range of s wave scattering is correct. This leads to the value $c = -0.021\ \mu^{-3}$. Such small values of a and c lead to a very small potential. The reason that these model values differ so strongly from ours is that they are obtained at the threshold $\nu = \mu$

Figure 7

The s wave πN Scattering Terms
(a) the pair term; (b) σ meson exchange

for real pions ($\vec{q}^2 = \vec{q}'^2 = \mu^2$). However the behaviour of the πN amplitude at this point is the result of a delicate cancellation [26]. If for example we take for the pair term the value given by the pseudoscalar theory (g^2/m), and then give m_σ a reasonable value of 5μ and demand $a = 1\mu^{-1}$, we obtain $c = 1.03\mu^{-3}$, in excellent agreement with the subthreshold expansion. We point out that it is however preferable to calculate a, b, and c directly, rather than calculating them as the difference of large, almost equal, contributions.

The discussion of the amplitude cannot close without a discussion of the πNN form factor $K(q^2)$ which appears in it. The favourite form factor for three body force calculations has been the Durr-Pilkuhn form factor [27] which has recently been shown by Pilkuhn [28] to give a good representation of high energy NN and $N\bar{N}$ scattering in the one pion exchange region. Some authors however have argued against this form factor because it changes the sign of the central OPEP force near 1.4 fm. However the OPEP with form factor is in fact nearer the phenomenological potential. Moreover careful analysis of theoretical Nucleon nucleon potentials discloses the need for pion form factor effects in this region of configuration space [29] and a calculation of the form factor on a theoretical basis [30] gives it a fall off approximated well in the region $q^2 < 0$ by the Durr-Pilkuhn form. Therefore we will use as the form factor for π exchange

$$H(q^2) = 1 + \zeta\frac{q^2-\mu^2}{q^2-\eta^2} \qquad (14)$$

(Strictly speaking $H = K^2(q^2)K_1(q^2)$ where K_1 is the off mass shell correction to the pion propagator $[(q^2-\mu^2)D(q^2) = K_1(q^2)]$ is the form factor appropriate to pion exchange between nucleons.) The parameters favoured by Pilkuhn and Loiseau and Nutt are $\zeta = 1$, $\eta = 3.2\ \mu$. This choice is called form factor III in the earlier literature [24, 25].

Having decided on a πN amplitude A appropriate for three body forces, one then has to obtain the three nucleon potential from it.

The procedure is adequately described in the literature [9,10] and we simply quote the results here. The potential may be conveniently considered in two parts - an (πN) s wave term coming from $a + c(\vec{q}^2+\vec{q}'^2)$, and a p wave term from $b\vec{q}.\vec{q}'$. The s wave potential is

$$V_s = \left(\frac{f}{4\pi}\right)^2(\underset{\sim}{\tau}_1.\underset{\sim}{\tau}_3)(\sigma_1.\nabla_1)(\sigma_3.\nabla_3)\{(a-2\mu^2c)Z_1(r_{12})Z_1(r_{23})$$
$$+ c[Z_0(r_{12})Z_1(r_{23})+Z_1(r_{12})Z_0(r_{23})] \qquad (15)$$

where $f^2/4\pi \approx 0.08$ is the pseudovector coupling constant, and

$$Z_n(r) = \frac{4\pi}{\mu} \int \frac{d^3q}{(2\pi)^3} \; e^{i\underset{\sim}{q}\cdot\underset{\sim}{r}} H(q^2)(q^2+\mu^2)^{-n}.$$

The p wave potential is

$$V_p = -\tfrac{1}{2}\left(\frac{f}{4\pi}\right)^2 b\mu^4(\underset{\sim}{\tau}_1\cdot\underset{\sim}{\tau}_3)(\sigma_1\cdot\nabla_1)(\sigma_3\cdot\nabla_3)(\nabla_1\cdot\nabla_3)Z_1(r_{12})Z_1(r_{23}) \qquad (16)$$

For the triton, which is not a spin-isospin saturated system, other amplitudes are also needed. They can be calculated from models [31] or derived from a sub-threshold expansion [32] of the type used here. However the major contribution to the energy comes from the potentials considered here.

3.2 Results in the Triton

These results are undoubtedly the most significant for the participants of the conference. However, I do not believe the present results, shown in table I, are reliable indications of the magnitude of the contribution of 3 body forces. A major difficulty is that none of these calculations included the effects of the

Table I

3 Body Force Contributions to the Binding Energy of the Triton	
Pask[33)]	- 2.44 (b=-1.06, a=c=0 other contributions)
Yang[31)]	- 2.32 MeV (b=-1.21, a=c=0 other contributions)
Tanaka et al.[31)]	- 1.16, -1.50 MeV (b=-1.03, a=c=0)

πNN form factor - they all set $H(q^2) \equiv 1$ or a constant. (Yang uses an average constant value of 0.78, but this does not introduce the main effect of H, the alteration of the short distance behaviour of the potential. This is accounted for in the value quoted for b.) This I believe to be a very poor approximation for the reasons outlined above. And we shall see that the effect of $H(q^2)$ is to substantially reduce the contribution of the three body forces in nuclear matter. Moreover, all these calculations omit the s wave potentials.

It is important to remark that all these calculations are by perturbation theory, to first order in W. Our nuclear matter experience suggests that this is probably a reasonable approximation, but it should be checked. Also one should notice that much of the contribution to the binding energy comes because W is a tensor like potential and connects S and D states in the triton. Early estimates, which omitted the triton D state wave function, gave much smaller results [10].

While these results are encouragingly of the order of the energy we seem to need, we nevertheless have to wait until a calculation of the three body energy is done using the potential as described above, with $H \neq 1$, in perturbation theory and hopefully in the exact equations also. I believe some calculations are in progress, so perhaps there will be some more definitive results by the time of the next Few Body Conference.

3.3 Nuclear Matter

The only other system for which serious calculations have been done is nuclear matter. Here the conventional wisdom was that there was a discrepancy of order 1 MeV between the observed binding energy and that calculated using two body potentials [5]. However this result has recently been called into question [34], and we should perhaps regard the present situation with regard to the two body force contribution to the energy of nuclear matter as uncertain. Unfortunately, precisely this argument over the binding energy of nuclear matter and the energy spectrum near the Fermi Surface produces a striking uncertainty in the three body force contribution, for reasons which are still not well understood.

Let me illustrate this by showing in Table 2 some results for the "old" 3 nucleon potential, of the above form with a = c = 0, b = $-1.40\ \mu^{-3}$. (In the notation of Loiseau et al., this means C_p = 0.61 MeV.) F.F.I means H = 1, F.F.III means $H(q^2)$ has the value described above (eqn 14).

Table 2

Calculations in Nuclear Matter with the Old Potential

Reference	Comments	Result
Loiseau, Nogami, Ross [24]	Cut off correlations, F.F.III m*=m, Δ = 0	- 1.77 MeV
	F.F.I	- 5.2 MeV
Blatt and McKellar I. [25]	Cut off correlations, H in OPEP, m*=m, Δ = 0 F.F.III	- 1.6 MeV
	F.F.I	- 5.2 MeV
Blatt and McKellar II	RSC correlations, m*=m, Δ=0, F.F.III	- 3.8 MeV
III	" m*=0.65m, Δ=0, F.F.III	- 2.5 MeV
Grangé et al. [35]	RSC correlations, Bruckner calculation (m*=0.65, Δ=0.9) F.F.III	- 0.7 MeV
Ueda et al. [12]	Hamada Johnson correlation, F.F.I Bruckner calculation	- 3.5 MeV
Kasahara et al. [36]	Smooth correlation, F.F.III, Bruckner calculation	- 0.4 MeV

Before discussing these results I should emphasise that an error in the Blatt-McKellar calculation for the Reid Soft Core potential has been corrected in this table. Contrary to the implication of reference [35] the large value of the 3 body contribution to the binding energy is not a consequence of this error, and there is still a substantial difference between cut-off correlation and more realistic correlation results.

Much more striking is the variation with respect to the spectrum of particles and holes, apparent from the last three entries in the table.

These calculations were made by deriving an effective two body potential and using this potential in a Bruckner calculation. It is by no means clear a priori that this is any better a procedure than simple perturbation theory. Indeed it can be justified as an approximation to the theory of section 2 only in terms of perturbation theory in W. However, it does provide, for those in possession of a conventional nuclear matter code, a convenient way of doing perturbation theory with an effective mass and energy gap for the hole state spectrum. Grangé et al. have verified that this hole state spectrum is the important ingredient in obtaining the spectacular reduction in binding energy by doing a perturbation calculation with such a spectrum.

The introduction of an energy gap sharply quenches the three body force contribution to the energy. But recent papers [34] suggest that the conventional large energy gap is not a good representation of the spectrum, and that it may be a better approximation to use the same effective mass for holes and particles, and no gap. This latter procedure ignores the enhancement of the effective mass near the Fermi surface. It is apparent that more work will be necessary before we understand this effect sufficiently to be able to give a reliable number for the contribution of three body forces to the energy of nuclear matter.

Moreover we have seen that the old potential is inadequate. For the potential which I now prefer, the results are given in table 3 for the case $m^* = 0.65m$, $\Delta=0$. As yet we unfortunately have no calculations with a non zero gap for this potential. I should warn you that in this case the results seem to be more sensitive to the details of the form factor than are the previous results. (This sensitivity occurs in the part of the potential proportional to c.)

Table 3

Comparison of the Three Body Force contribution to the Binding Energy of Nuclear Matter for the Old Potential and the New Potential (with $m^*/m = 0.65$, $\Delta = 0$)

Old potential	- 2.5 MeV
New potential	- 1.83 ± 0.25 MeV

The result is simply proportional to m^*/m (when $\Delta = 0$), so that answers for other values of this parameter are readily obtained. It will be noticed that the s wave potential leads to some reduction in the energy from three body forces, just as Coon, Scadron and Barrett found using the effective mass method.

Another way to illustrate the difference between these potentials is to look at the effective potential between two particles, when the intermediate particle off which the pion scatters is averaged over, including of course the correlations between this particle and the other two. This effective potential has an OPEP like form

$$V_{eff} = (\sigma_1 \cdot \sigma_2)(\tau_1 \cdot \tau_2)[V_C(r_{12}) + S_{12} V_T(r_{12})]$$

V_C and V_T are shown in figure 8, for both the old and the new potentials. It is immediately apparent that the old and new effective potentials are quite different and that it is impossible to guess the influence of the energy gap on the new potential - we will simply have to wait for the calculations now in progress to be completed.

4. OTHER THREE BODY FORCES

Many other three body forces have been proposed in the literature, as can be seen from a glance at figure 1. The first point to notice is that the result of the previous section is as small as it is as a consequence of cancellations - individual partial waves contribute up to a few MeV. In this situation forces of shorter range, which a priori would be expected to be small because of the short range correlations, may give significant contributions.

To begin with one can consider diagrams of the type of figure 2, with one or both pions replaced by heavier mesons. Choudry considered $\pi N \to \eta N$, and Green has emphasised, in another context, the importance of $\pi N \to \Delta \to \rho N$ and $\rho N \to \Delta \to \rho N$ in providing a more stable short distance behaviour. Such processes deserve more attention as contributors to many body forces.

Next we can consider 3 meson exchange forces. One of these, the 3σ force of figure 3 has featured predominately in discussions of the implications of the Lee-Wick abnormal state [15,37] for ordinary nuclei. Barshay and Brown [11] estimated

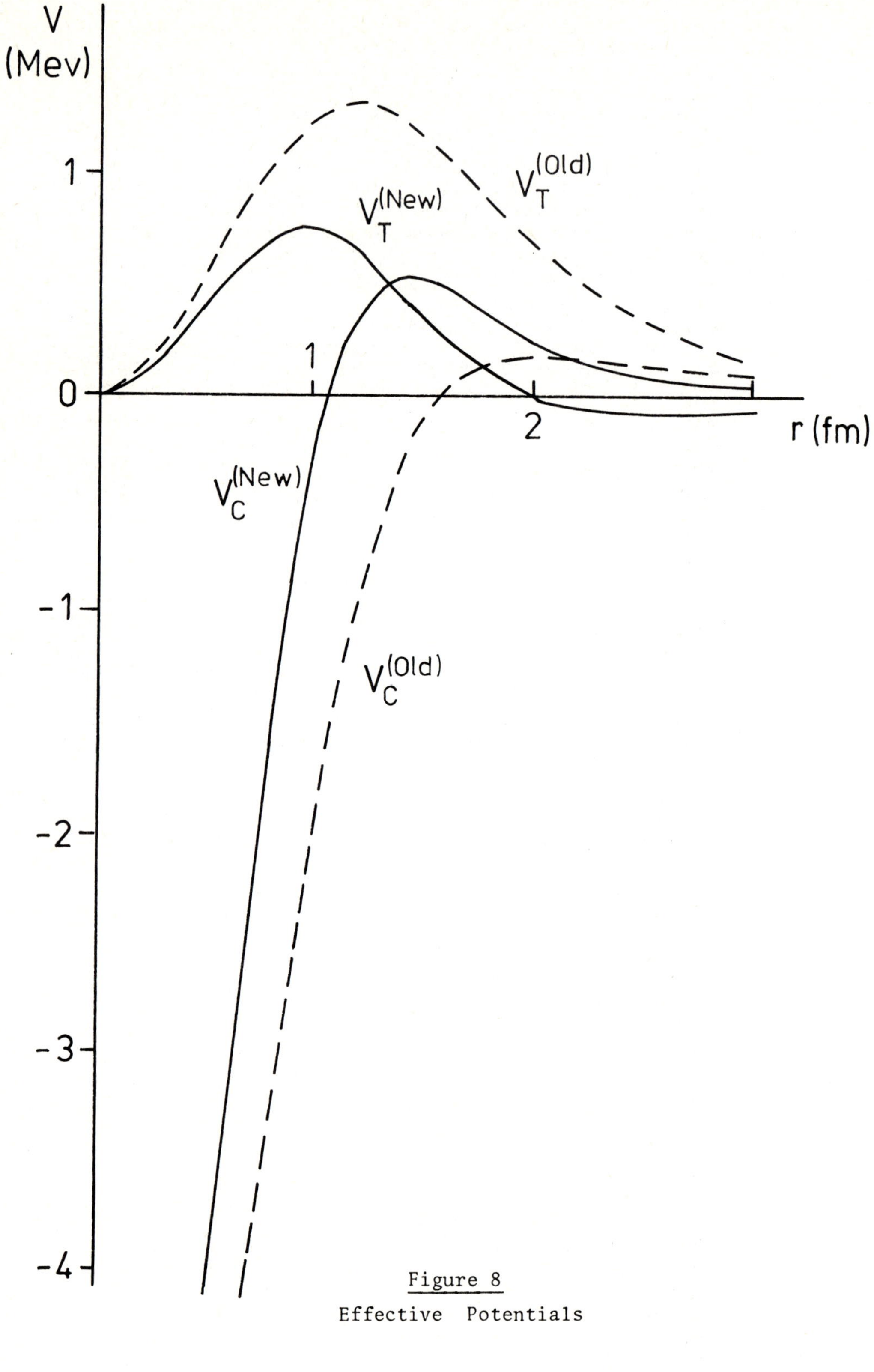

Figure 8

Effective Potentials

that for the parameters used in OBEP, the 3σ force would contribute about 15 MeV of attraction in nuclear matter. which they regarded as unacceptable. However their estimate needs to be improved on many grounds.

(i) Exchange effects were omitted. In a first crude approximation, simple combinatorics suggest that they will introduce a factor of $3/8$ [38], reducing the Barshay-Brown estimate to 6 MeV.

(ii) Barshay and Brown used a radial form which was a product of Yukawa forms in the interparticle distances. The radial dependence of the potential of figure 3 is much more complicated; it is a product of 3 Yukawas in the nucleon-3σ interaction point separations, integrated over the position of the 3σ interaction point, so it is not obvious that a product of Yukawas is even a useful approximation. Since they didn't really do any integrals anyway, this is probably nitpicking, but it should emphasise the fact that their estimate cannot be relied on to better than a factor of two or so.

(iii) Their numerical result cannot be taken as an argument against Lee and Wick, who do not use OBEP parameters for masses and coupling constants. Instead they use σ model type parameters, with a larger coupling constant and larger mass. With the large σ mass the effect is reduced because of the nucleon nucleon correlations as Barshay and Brown themselves point out. This further reduction is by a factor of two at least, even when the change in coupling constant is taken into account.

But now we have got the 3σ 3 body force contributing a couple of MeV to the binding energy of nuclear matter. Certainly there is room for such a force in nuclear physics, and it is now of the same order as other three body forces and cancellations (or enhancements) could occur. Certainly much more work is needed before three body forces in the Lee-Wick theory can be used to argue against it convincingly [37].

Barshay and Brown say that they prefer the light σ of OBEP because this reflects the broad width of the σ through its 2π coupling. However this then leads us to the diagram of figure 4b, which is calculated by Ueda et al. [12] to contribute 6.8 MeV attractive energy to the binding of nuclear matter. However the triple pair term of figure 1a which is repulsive is instrumental in cancelling the σ contribution*.

In fact, using the method of Drell and Huang (omitting exchange, and using a cut off at 0.8 fm), we find that the diagram of figure 1a contributes 17 MeV repulsion. This is for no form factor, whereas the Ueda et al. result included form factors, but we can nevertheless see that the 3σ-3π diagram and the pair theory 3π diagram will strongly cancel against each other, as we would have expected. Drell and Huang estimate that the exchange terms reduce the pair term to 64% of the value of the direct term only.

To return to the arguments of Barshay and Brown, we suggest that the two pion coupling to the σ will lead to the introduction of pair terms which will substantially reduce the 3σ contribution. A better estimate of the implications of

*Ueda et al. argue that the pair term will be suppressed because it requires initial and final nucleons to be present at the same time and they will vanish because of the exclusion principle. This is their reason for considering Fig. 4a above without Fig. 1c. This argument is suprious. In time ordered Feynman diagrams one should ignore Pauli principle restrictions in intermediate states [39]. The exchange terms are disconnected graphs, and if we take account of the Pauli principle in intermediate states we cannot restrict consideration to linked diagrams only.

the σ model-Lagrangian for ordinary nuclear matter is given by using the large σ mass of the model. Then the three body forces are no longer spectacularly large for the tree graphs, and the Nynman-Rho [37] argument of using the smallness of the 3 body force in normal nuclear matter as a criterion for determining the contribution of the loop diagrams in the abnormal nuclear state loses much of its force. They showed that it is possible, for a suitable choice of parameters, to have the abnormal state bound or unbound. But the three body forces in the normal state no longer prefer the unbound solution.

We can conclude this section on "other" three body forces with the simple remark that essentially no calculations have been done with them for the triton. We can scale Gelbard's [9] calculations for the 3π pair term to estimate that this plus the 3σ-3π term will give about 0.5 MeV in the triton, but this needs to be confirmed by further calculations.

5. CONCLUSION

Five months ago [40] I listed a series of areas of activity and controversy in the area of many body forces in nuclei. These were

(i) The influence of the πN σ term, and more generally s wave terms, on the 3 nucleon force.

(ii) The use of the effective two body potential in nuclear matter calculations.

(iii) The effect of short range correlations on the value of the three body force contribution to the energy.

(iv) The contribution of three nuclear forces in finite nuclei, the triton and shell model calculations.

(v) The "3σ" type of three body force.

(vi) N Body Forces ($N \geq 3$) and the binding energy of High Density Nuclear Matter.

In the meantime point (i) has been cleared up - calculations at least in nuclear matter have been done with a better potential, but only for the gapless spectrum. Point (ii) has been understood in the sense that an effective potential calculation in Bruckner Theory corresponds in first order in the three body potential to the first order solution of the equations of section 2. However the unexpected sensitivity of the result to a gap in the spectrum of hole states is as yet unexplained. No further work has been done on point (iii) - the discrepancy between cut off and more sophisticated calculations remains. The effects of three body forces in finite nuclei - point (iv) - remains an almost virgin field of study. Even the triton energy has not yet been calculated with a reasonable potential. And it is clear that 3 body forces could easily contribute a few tenths of an MeV to shell model matrix elements. Without them it is irresponsible to claim a better accuracy than that for shell model calculations from first principles. There has been some progress towards understanding the 3σ force of point (v) - at least there is no immediate justification of the alarm raised by Barshay and Brown. Point (vi), N Body Forces and high density matter has not been discussed here, but no further progress in this field has been made.

In all we still have more problems than solutions, so we can look forward to an interesting report at the next Few Body Conference.

ACKNOWLEDGEMENTS

I wish to express my gratitude to Professor A.N. Mitra and the organisers of this conference for their kind hospitality, and to my colleagues of the Service de Physique Théorique for useful discussions, and for their hospitality. Particular thanks are due to E.M. Nynman and M. Rho for educating me about Lee-Wick states.

Finally it is a pleasure to thank sincerely my various collaborators, R. Rajaraman, D.W.E. Blatt, S. Coon, M. Scadron and B.R. Barrett, who have contributed so much to my understanding of this subject.

REFERENCES

[1] R. Rajaraman and H.A. Bethe, Rev. Mod. Phys. 39 (1967) 745.
[2] R. Dashen, S.K. Ma and H.J. Bernstein, Phys. Rev. 187 (1969) 345.
[3] See e.g. B.M. Axilrod, J. Chem. Phys. 19 (1951) 719.
[4] Y.E. Kim and A. Tubis, Ann. Rev. Nucl. Sci. 24 (1974) 69.
[5] H.A. Bethe, Ann. Rev. Nucl. Sci. 21 (1971) 93.
D.W.L. Sprung, Adv. Nucl. Phys. 5 (1972) 225.
[6] J. Bloomquist, Nuclear Structure Symposium of 1000 Lakes, Joutsa, 1975.
[7] A. Faessler, S. Krewald and G.J. Wagner, Phys. Rev. C11 (1975) 2069.
[8] B.J. Cole, A. Watt and R. Whitehead, Phys. Letters 57B (1975) 24.
H. Dirim, J.P. Elliott and J.A. Evans, Nucl. Phys. A244 (1975) 301.
[9] H. Primakoff and T. Holstein, Phys. Rev. 55 (1938) 128.
L. Janossy, Proc. Cam. Phil. Soc. 35 (1939) 616.
G. Wenzel, Helv. Phys. Acta 15 (1942) 111; ibid 25 (1952) 569; Phys. Rev. 91 (1953) 1573.
S. Drell and K. Huang, Phys. Rev. 91 (1953) 1527.
A. Klein, Phys. Rev. 90 (1953) 1101.
K. Bruckner, C. Levinson and H.M. Mahmoud, Phys. Rev. 95 (1954) 217.
E.M. Gelbard, Phys. Rev. 100 (1955) 1530.
I. Fujita, M. Kawai and M. Tanifuji, Nucl. Phys. 29 (1962) 252.
K. Hasegawa, Prog. Theor. Phys. 30 (1963) 827.
S.R. Choudry, Phys. Rev. Lett. 22 (1969) 234.
[10] I. Fujita and H. Miyazawa, Prog. Theor. Phys. 17 (1957) 360.
F.M. Coury and W.M. Frank, Nucl. Phys. 46 (1963) 257.
M. Miyazawa, J. Phys. Soc. Japan 19 (1964) 1764.
B.A. Loiseau and Y. Nogami, Nucl. Phys. B2 (1967) 470.
G.E. Brown, A.M. Green and W.J. Gerace, Nucl. Phys. A115 (1968) 435.
B.H.J. McKellar and R. Rajaraman, Phys. Rev. Lett. 21 (1968) 450.
G.E. Brown and A.M. Green, Nucl. Phys. A137 (1969) 1.
R.K. Bhaduri, Y. Nogami and C.K. Ross, Phys. Rev. C2 (1970) 2082.
[11] S. Barshay and G.E. Brown, Phys. Rev. Lett. 34 (1975) 1106.
[12] T. Ueda, T. Swada and S. Takagi, paper SMCK2, this conference.
[13] B.H.J. McKellar and R. Rajaraman, Phys. Rev. Lett. 31 (1973) 1063, Phys. Rev. C10 (1974) 871, D.W.E. Blatt and B.H.J. McKellar, Phys. Rev. C12 (1975) 837.
[14] See for example, A.B. Migdal, Sov. Phys. (JETP) 36 (1972) 1052.
[15] T.D. Lee and G.C. Wick, Phys. Rev. D9 (1974) 2291;
T.D. Lee, Rev. Mod. Phys. 47 (1975) 267.
[16] A.C. Phillips, Phys. Rev. 142 (1966) 984.
D.Z. Freedman, C. Lovelace and J.M. Namyslowski, Nuovo Cim. 43A (1966) 1.
[17] C. Lovelace, Phys. Rev. 135 (1964) B1225.
[18] B.H.J. McKellar, to be published.
[19] H.A. Bethe, Phys. Rev. 138 (1965) B804.
[20] B.H.J. McKellar and R. Rajaraman, Phys. Rev. C3 (1971) 1877.
[21] M. Scadron and L. Thebaud, Phys. Rev. D9 (1974) 1544.
M. Olsson and E. Osypowski, to be published.
[22] S.A. Coon, M.A. Scadron and B.R. Barrett, Nucl. Phys. A242 (1975) 467.
[23] S.A. Coon, M.A. Scadron, B.R. Barrett, D.W.E. Blatt and B.H.J. McKellar, paper SMCK 1, this conference.
[24] B.A. Loiseau, Y. Nogami and C.K. Ross, Nucl. Phys. A165 (1971) 601.
[25] D.W.E. Blatt and B.H.J. McKellar, Phys. Lett. 52B (1974) 10, Phys. Rev. C11 (1975) 614.
[26] M.A. Scadron, invited paper, this conference.
[27] H.P. Durr and H. Pilkuhn, Nuovo Cim. 21 (1961) 1028.
[28] K. Bonyardt, H. Pilkuhn and H.G. Schlaile, Phys. Lett. 52B (1974) 271.
[29] R. Bryan, Phys. Rev. Lett. 35 (1975) 967.

[30] B.A. Loiseau and W.A. Nutt, to be published.
[31] S.-N. Yang, Phys. Rev. C10 (1974) 2067.
M. Sato, Y. Akaishi and H. Tanaka, Prog. Theor. Phys. Suppl. 56 (1974) 76.
[32] E. Harper, Y. Kim, P. McKamee and M.D. Scadron, to be published.
[33] C. Pask, Phys. Lett. 25B (1967) 78.
[34] V.R. Pandharipande, R.B. Wiringa and B.D. Day, Phys. Lett. 57B (1975) 205.
J.-P. Jenkenne, A. Lejeune and C. Mahaux, Nucl. Phys. A245 (1975) 411.
[35] P. Grangé, M. Martzolff, Y. Nogami, D.W.L. Sprung, and C.K. Ross, to be published.
[36] T. Kasahara, Y. Akaishi and H. Tanaka, Prog. Theor. Phys. Suppl. 56 (1974) 96.
[37] E.M. Nynman and M. Rho, to be published.
[38] R. Rajaraman and R.K. Trehan, Phys. Rev. A5 (1972) 392.
[39] R.P. Feynman, Phys. Rev. 76 (1949) 749.
[40] B.H.J. McKellar, Nuclear Structure Symposium of the 1000 Lakes, Joutsa, 1975.

RELATIVISTIC EFFECTS IN FEW BODY SYSTEMS

Franz Gross
College of William and Mary
Williamsburg, Virginia, 23185, U.S.A.

I. INTRODUCTION

While many of us have been discussing relativistic effects secretly in the corridors for several years, to my knowledge this is one of the first times a general talk has been devoted specifically to this subject. For this reason I will review work on this topic over the past 5-6 years. Also, in order to avoid being too abstract, I will introduce the subject by talking about specific physical processes.

First let me make a general comment concerning the question of defining what we mean by "relativistic effects". In the broadest sense, any process involving particle creation and annihilation can be regarded as a relativistic effect because these phenomena are a natural consequence of relativistic field theory. However, such a definition is too broad. We will take the point of view that any process which is more or less a separate physical mechanism is not a relativistic effect. This means that effects due to N*'s or Δ's will not be considered relativistic, but dynamical. Loosely speaking, a relativistic effect is one which must be added to a non-relativistic calculation in order to make it relativistic, and which does not correspond to a separate physical process. Hence relativistic effects are primarily kinematical; they arise only in a case when one has made a non-relativistic calculation of a particular physical process, and later wants to correct the calculation to include the effects of relativity.

A second general comment concerns the importance of relativistic effects. I believe that this field has reached a level of precision in both experiment and theory where relativistic effects should no longer be ignored. Whenever an anomaly arises, one should consider whether it could be a relativistic effect before one searches for new dynamical mechanisms. As the rest of this paper will show, there are a number of cases where relativistic effects have recently been found to be unexpectedly large, and where their inclusion eliminates most of a discrepancy or anomaly.

II. $n + p \rightleftarrows D + \gamma$

A. Radiative Neutron Capture

A striking example of the importance of relativistic effects, even at threshold, is provided by thermal neutron radiative capture. Prior to 1972, there had existed a 9.5% discrepancy between the theoretical and experimental values of the total cross section:

$$\begin{aligned} \sigma_{exp} &= 334.2 \pm 0.5 \text{ mb} \\ \sigma_{th} &= 302.5 \pm 4.0 \text{ mb (Noyes [1])} \end{aligned} \qquad (1)$$

This was resolved first by Riska and Brown[2] and then by Gari and Huffman[3]. Riska and Brown considered the time ordered diagrams shown in Fig. 1, which they considered to be meson exchange effects. Of the three diagrams, the largest contribution comes from the pair current, but in their paper they present only the combined contribution from the pair current and pion current. For the Δ contribution, they used the calculation of Stranahan[4]. Their results, together with other results discussed below, are summarized in Table I.

Table I - Corrections to the Total Cross Section from Various Processes (in %)

	Pair Current		Pion Current	Δ	Total
Riska-Brown	6.56			2.90	8.01
Gari-Huffman	HJ	7.30	-0.96	3.88	10.22
	RHC	7.22	-0.90	4.32	10.64
	RSC	7.62	-1.20	4.80	11.22
Dressler-Gross	3.72 + 4.50				

In the same year, Gari and Huffman did a similar calculation using a Chew Low model of pion photoproduction, which enabled them to separately calculate the Δ contribution. Instead of the γ^5 πNN interaction employed by Riska and Brown, they used a $\gamma^5\gamma^\mu$ interaction, which in the presence of electromagnetism generates a contact interaction equivalent in the non-relativistic limit to the pair term, see Fig. 2. (It also generates a pair term of its own, but this is vanishingly

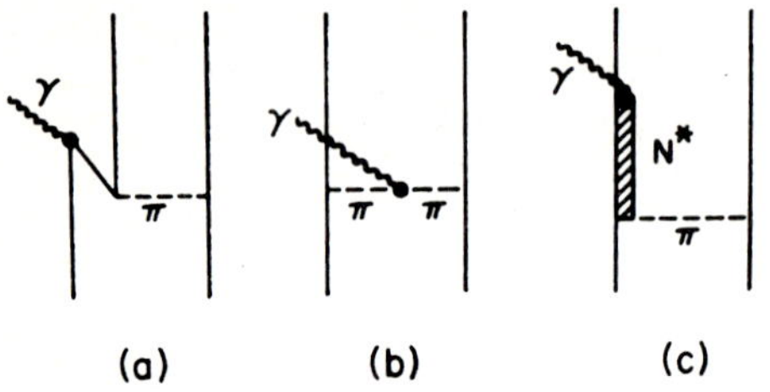

Fig. 1. The three time ordered meson exchange diagrams considered in Ref. [2] are (a) "pair" current, (b) "pion" current, and (c) Δ contribution (Taken from Ref. [9]).

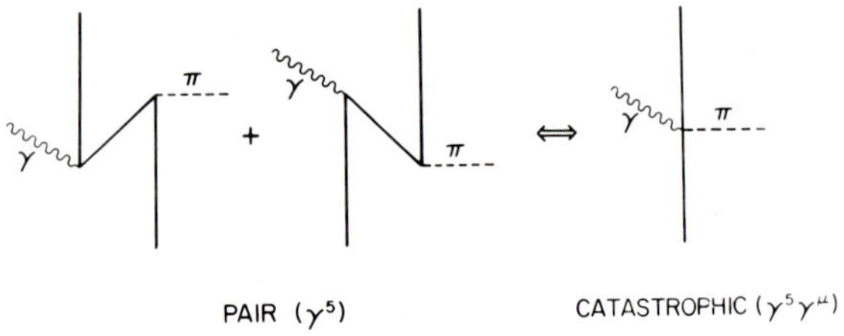

Fig. 2. Equivalence of the "pair" current and "catastrophic" current.

small in the non-relativistic limit.) They called the contact interaction a "catastrophic" current, but in view of its equivalence with the "pair" current we have labeled it as a pair current in Table I. They also investigated the sensitivity of their results to various nuclear models; the Hamada Johnson (HJ) and the Reid hard core (RHC) and soft core (RSC). Note that the results are not very sensitive to the model.

Table I shows that the bulk of the correction comes from the pair terms. Dressler and Gross[5] have examined these pair term contributions from another point of view which suggests that they might be better viewed as a relativistic effect instead of a meson exchange effect.

To see how this comes about, we must first recall that a single relativistic Feynman diagram is equivalent to a large number of time ordered (non-relativistic) diagrams. This is illustrated schematically in Fig. 3, where the virtual energies and momenta of the particles are labeled. If the initial and final state particles are physical, then their energies are

$$\begin{aligned} E &= \sqrt{M^2 + \vec{p}^{\,2}} \\ E_+ &\equiv E + \omega = \sqrt{M^2 + (\vec{k} - \vec{p})^2} \\ E_- &\equiv (E - q_0 - \omega) = \sqrt{M^2 + (\vec{p} - \vec{q} - \vec{k})^2} \end{aligned} \qquad (2)$$

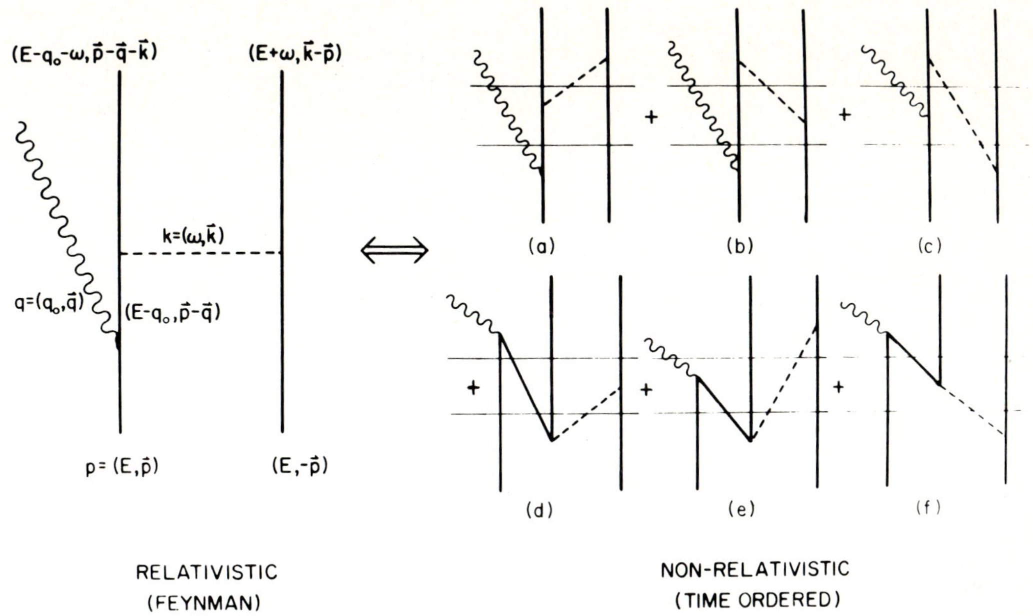

Fig. 3. Equivalence between a single relativistic Feynman diagram and a sum of non-relativistic time ordered diagrams.

The equivalence illustrated in Fig. 3 is displayed in detail in the following equation valid for particles with zero spin:

$$\frac{1}{M^2-(p-q)^2}\;\frac{1}{m_\pi^2-k^2} = \frac{1}{E_1^2-(E-q_0)^2}\;\frac{1}{\omega_k^2-\omega^2}$$

$$= \frac{1}{4E_1\omega_k}\left\{ \underset{(a)}{\frac{1}{(E_1-E+q_0)(\omega_k-\omega)}} + \left(\underset{(b)}{\frac{1}{(E_1-E+q_0)}} + \underset{(c)}{\frac{1}{\omega_k+\omega}}\right)\frac{1}{(E_1-E+q_0+\omega_k-\omega)} \right.$$

$$\left. + \left(\underset{(d)}{\frac{1}{E+E_1-q_0}} + \underset{(e)}{\frac{1}{\omega_k-\omega}}\right)\frac{1}{(E+E_1-q_0+\omega_k-\omega)} + \underset{(f)}{\frac{1}{(E+E_1-q_0)(\omega_k+\omega)}} \right\} \tag{3}$$

where the letter under each term tells which diagram that term comes from, and

$$E_1 = \sqrt{M^2 + (\vec{p} - \vec{q})^2}$$
$$\omega_k = \sqrt{m_\pi^2 + (\vec{k})^2} \tag{4}$$

are the mass-shell energies of the interacting particles. To obtain the various terms on the RHS of Eq. (3), we write the energy denominators corresponding to the cuts shown in Fig. 3, eliminate E_+ and E_- using Eq. (2), and use the fact that the total energy is 2E.

This decomposition shows that the "pair" terms considered by Riska and Brown are in fact part of the relativistic diagram shown in Fig. 4a. Furthermore, Fig. 4a is included in Fig. 4b (the relativistic impulse approximation, RIA) because the nucleon-nucleon bubbles satisfy a relativistic wave equation illustrated symbolically in Fig. 11 below. (An easy way to see this in a one boson exchange model is to realize that the bubbles in Fig. 4 are short-hand for an infinite sum of pion, rho, omega and other exchanges, so that the one extra pion exchange in

Fig. 4a is indistinguishable from part of the contributions already included in Fig. 4b.) We conclude that the pair terms, together with other relativistic corrections discussed below, will automatically be included when Fig. 4b is evaluated relativistically.

How does one evaluate such a relativistic diagram as 4b? I showed some time ago (Gross[6]) that this diagram is dominated by the approximation in which the spectator is on the mass shell. This means that if p is the 4 momentum of the spectator, then we make the approximation

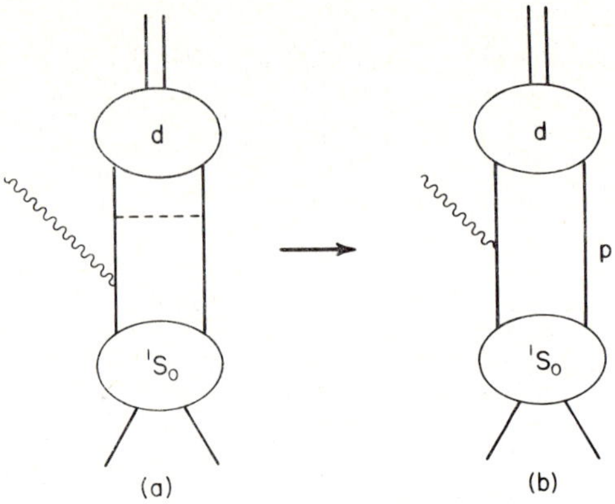

Fig. 4. One pion exchange (a) is absorbed into the relativistic impulse approximation (b). The bubbles are related to deuteron and $^1S_0$ wavefunctions as described below. The spectator 4 momentum is p.

$$\int_{-\infty}^{+\infty} dp_0 \int d^3p \frac{I(p_0,\vec{p})}{M^2-p^2-i\varepsilon} \rightarrow 2\pi i \int d^3p \frac{I(E,\vec{p})}{2E} \qquad (5)$$

where I is the rest of the integrand. Note that the approximate form is still relativistically invariant, and involves only an integration over 3 momentum just as in the non relativistic case. This approximation of placing the spectator on its mass shell has, of course, a very nice physical interpretation.

The next step is to write the approximate RIA in terms of time ordered diagrams as shown in Fig. 5. The small x on the spectator stands for the approximation Eq. (5), and u, w are the large, or positive energy deuteron wave functions, v_t, v_s are the small component or negative energy deuteron wave functions, y is the large, or positive energy $^1S_0$ wave function, and z is the small component $^1S_0$ wave function (see below).

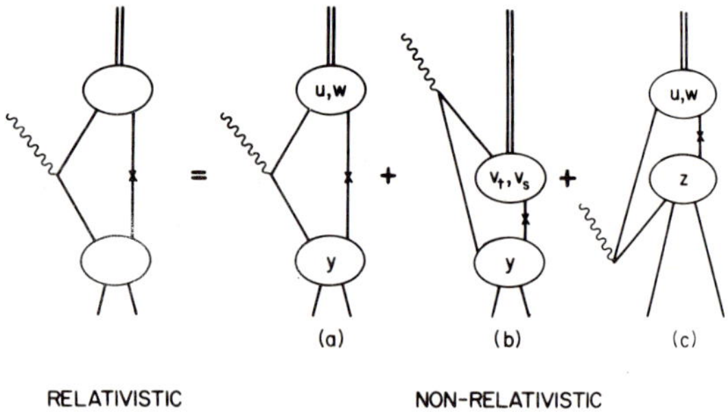

Fig. 5. Decomposition of RIA into 3 time ordered pieces involving large and small components of the nuclear wave functions.

Evaluating the diagrams at threshold, Dressler and Gross obtained,

$$M = (\mu_p-\mu_n)\underbrace{\int_0^\infty dr\, u(r)y(r)}_{(a)} + \underbrace{\frac{1}{\sqrt{6}}\int_0^\infty dr(Mr)y(r)v_t(r)}_{(b)} + \underbrace{\frac{1}{\sqrt{3}}\int_0^\infty dr(Mr)z(r)[u(r)+\frac{1}{\sqrt{2}}w(r)]}_{(c)} \qquad (6)$$

where the letter below each term corresponds to the diagram in Fig. 5 from which the term comes. When these were evaluated using the wave functions of Hornstein and Gross[7] and Delacroix and Gross[8], a 3.72% correction was obtained. This value is so small because the model used to compute the relativistic wave functions contains a mixture of 41% γ^5 and 59% $\gamma^5\gamma^\mu$ for the πNN coupling. When one adds 59% of the result of Gari and Huffman (which is the additional factor of 4.50% given in Table I) one obtains a result consistent with the other calculations. (See Table I.)

Our principle conclusion is that the large pair current terms can (and should) be regarded as a relativistic effect coming from the small components of the nuclear wave functions. I think it is a remarkable fact that one obtains such a large result - at threshold! When I first heard about Riska's result, I immediately understood that it was equivalent to the contributions from small component wave functions calculated in the OPE approximation, and for that reason I thought it was incorrect. Dressler and I did the calculation fully expecting to find that Riska had overestimated the result - but instead we confirmed Riska's result and learned that relativistic effects can be more important than one might expect a priori.

A second point should be emphasized. The Dressler-Gross calculation is equivalent to the Riska-Brown result only because, in this case, the OPE contribution to the small component wave functions is dominant. In other cases, contributions from other mesons may be important, particularly if the OPE contribution is suppressed or cancels. I will point out below that this is precisely what happens in the deuteron form factor. In general, it seems that it would be a more consistent procedure to express one's results in terms of small component wave functions, and then to determine the small component wave functions within the framework of a consistent dynamical scheme.

B. Backward Electrodisintegration of the Deuteron

Perhaps an even more striking example of the importance of relativistic effects occurs in electrodisintegration of the deuteron in the backward direction - i.e. where the electron scatters through 180°. In this case the virtual photon is magnetic, so that the process is almost identical to the time reversal of the radiative capture process, except that the photon is now virtual. In an important paper, Hockert, Riska, Gari and Huffman[9] calculated what they regarded as meson exchange current contributions to this process. They considered the same three contributions shown in Fig. 1, but again found that the major contribution came from the pair terms which, as we have seen, are part of the RIA.

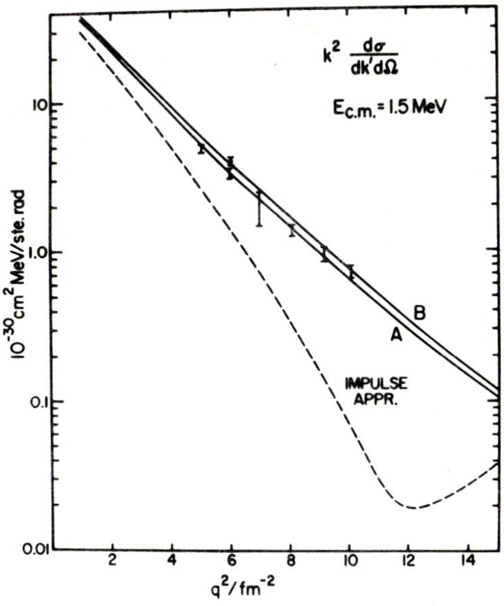

Fig. 6. Failure of the IA in 180° electrodisintegration (taken from Ref. [9]). Curve B includes the Δ contribution - curve A does not.

Their result for the non-relativistic impulse approximation (IA) for transitions to the $^1S_0$ final state with pair term corrections is

$$\left.\frac{d\sigma}{d\Omega dk'}\right|_{180°} = \frac{\alpha^2}{24\pi}\frac{pq^2}{Mk^2}[g+h]^2$$

$$g + h = G_M^V \int_0^\infty dr\, y(pr)[u(r)j_0(\tau) - \frac{1}{\sqrt{2}} w(r)j_2(\tau)]$$

$$+ G_E^V \int_0^\infty dr(Mr)y(pr)[j_0(\tau) + j_2(\tau)] \frac{I(z)}{3} (u(r) + \frac{1}{\sqrt{2}} w(r)) \qquad (7)$$

where G_M^V and G_E^V are the isovector magnetic and charge moments of the nucleon normalized to $G_E^V(0) = 1$, $q > 0$ is the magnitude of the virtual photon 4 momentum, p is the magnitude of the 3 momenta of the outgoing neutron and proton in their center of mass, k and k' are the incident and final electron energies

$$\tau = \frac{qr}{2}, \quad z = m_\pi r, \text{ and}$$

$$I(z) = \frac{g^2}{4\pi}\left(\frac{m_\pi}{M}\right)^2 \frac{e^{-z}}{z}\left(1 + \frac{1}{z}\right) \tag{8}$$

The term proportional to G_M^V is the IA, while the term proportional to G_E^V is the pair term. Figure 6, taken from their paper, shows how seriously the IA fails to fit the data, and how satisfactory the corrections make the final agreement.

Because the pair term is the bulk of the effect, the success of this calculation provides further evidence for the importance of relativistic effects. By examining Eqs. (6) and (7), and Refs. [7] and [8], it is easy to guess at the correct form for the relativistic correction term:

$$G_E^V \int_0^\infty dr\, Mr[j_0(\tau)+j_2(\tau)]\left[\frac{1}{\sqrt{6}}\, v_t(r)y(pr) + \frac{1}{\sqrt{3}}\left(u(r)+ \frac{1}{\sqrt{2}}\, w(r)\right)z(pr)\right] \tag{9}$$

This term reduces to the corrections given in Eq. (6) when $q \to 0$, and reproduces the pair terms in the OPE approximation. It would be interesting to calculate it with relativistic wave functions, and to confirm the correctness of (9) by explicit calculation of the RIA.

III. THE DEUTERON FORM FACTOR.

As we mentioned above, the pair term correction will agree with the RIA only when the OPE is the dominant mechanism. However, the OPE pair terms are purely isovector in the non-relativistic limit (c.f. Fig. 2 where we have seen that the pair terms are equivalent to a contact term which is purely isovector) so that they will only contribute to the deuteron form factor in higher order. To find the dominant corrections for this process, one would therefore have to consider the higher meson exchanges, and because of the ambiguities in how to handle these shorter range contributions to the nuclear force it is clearly superior to use the RIA, where the correction is expressed in terms of the small component wave functions, v_t and v_s, which can then separately be calculated carefully within the framework of a consistent dynamical model.

I will first discuss the magnetic moment, μ_d, and then discuss the results for relativistic corrections at low and high momentum transfer.

A. Corrections to μ_d

Corrections to the magnetic moment, μ_d, have recently been calculated (Gross[10]). If P_d is the D state probability, the results can be written

$$\mu_d = (\mu_p + \mu_n)\left(1 - \frac{3}{2} P_d\right) + \frac{3}{4} P_d + \Delta$$

$$\Delta = \int_0^\infty dr \frac{1}{\sqrt{3}} (Mr)\left\{v_t\left(\frac{u}{\sqrt{2}} - w\right) - v_s\left(u + \frac{w}{\sqrt{2}}\right)\right\} + \text{small terms} \tag{10}$$

If Hornstein and Gross (HG) wave functions are used, the value of Δ is .015 nuclear magnetons, which makes the experimental magnetic moment consistent with a 6 1/2% D state. (The HG D state is 6.4%.) In addition to these relativistic corrections, one would expect corrections due to electromagnetic modifications of the potential generated by momentum dependent couplings (Bergstrom[11], Scheerbaum[12]). The precise value of these corrections will depend on the detailed model employed, and estimates of such corrections for phenomenological non-relativistic potentials indicate that they could be as big as .0066 nuclear magnetons.

Because of these latter uncertainties, it is not possible to draw a sharp conclusion. The remarkable agreement obtained with the HG wave functions may be an accident, or it may be an indication that a reliable calculation of μ_d is now possible provided sufficient care is taken to obtain a reliable estimate of the intermediate range behavior of the small component wave functions through use of a self consistent relativistic model.

B. Low Momentum Transfer

We now turn to a discussion of non-static relativistic corrections to the deuteron form factor. This problem has been studied by a number of investigators, and is of particular current interest because of the new measurements by Arnold et al[13], which I will discuss below.

First I will review the calculations which obtain relativistic corrections to order $(v/c)^2$ or $(q/M)^2$. The differential cross section for electron deuteron scattering is (see Ref. [13])

$$\frac{d\sigma}{d\Omega} = \left.\frac{d\sigma}{d\Omega}\right|_{Mott} (A + B\tan^2\theta/2) \tag{11}$$

where θ is the electron scattering angle, $q > 0$ the magnitude of the 4-momentum transfered by the electron and

$$A = G_C^2 + \frac{q^2}{6M_d^2}G_M^2 + \frac{q^4}{18M_d^4}G_Q^2 . \tag{12}$$

B is proportional to the magnetic form factor, G_M, which will be ignored in the following discussion. The charge and quadrupole form factors can be written

$$G_{C,Q} = G_E^S(q^2)D_{C,Q}(q^2) + (2G_M^S(q^2) - G_E^S(q^2))D_{C,Q}^{SO}(q^2) \tag{13}$$

where G_E^S and G_M^S are the isoscalar charge and magnetic form factors of the nucleon normalized so that $G_E^S(0) = 1$. The structure functions D must be calculated theoretically. In 1973, Friar[14], using work of Krajcik and Foldy[15], obtained a very simple form for the relativistic corrections to the structure functions:

$$D_C = \left(1 - \frac{q^2}{8M^2} - \frac{q^4}{16M^2}\frac{d}{dq^2}\right)F_C$$

$$\frac{q^2}{6\sqrt{2}M_d^2}D_Q = \left(1 - \frac{q^2}{8M^2} - \frac{q^4}{16M^2}\frac{d}{dq^2}\right)F_Q$$

$$D_C^{SO} = \frac{q^2}{8M^2}\int_0^\infty dr[j_0 + j_2]\,w^2$$

$$D_Q^{SO} = \frac{3}{5}\int_0^\infty dr[j_0 + \frac{10}{7}j_2 + \frac{3}{7}j_4](\sqrt{2}\,rwu' - \sqrt{2}\,wu - rww') \tag{14}$$

where the arguments of the spherical Bessel functions are understood to be $\tau = q\,r/2$, the S and D state deuteron wave functions (u and w) are understood to be functions of r, the prime stands for differentiation with respect to r, and the non-relativistic form factors are

$$F_C = \int_0^\infty dr\; j_0[u^2 + w^2]$$

$$F_Q = \int_0^\infty dr\; j_2[uw - \frac{1}{\sqrt{8}}w^2] \tag{15}$$

so that all the other terms in (14) are relativistic effects. In obtaining these results, Friar observed that the much more complicated results I obtained earlier

(Gross[16]) reduced to (14) plus some additional terms which were added to D_C and D_Q

$$\Delta D_C = \frac{q^2}{M^2}\frac{d}{dq^2} I_C \; ; \qquad \Delta D_Q = \frac{6\sqrt{2}\, M_d^2}{M^2}\frac{d}{dq^2} I_Q$$

$$I_C = \int_0^\infty dr\, j_0 [u\hat{u} + w\hat{w}]$$

$$I_Q = \int_0^\infty dr\, j_2 [\tfrac{1}{2}(u\hat{w} + w\hat{u}) - \frac{1}{\sqrt{8}} w\hat{w}]$$

$$\hat{u} = (-\frac{d^2}{dr^2} + M\varepsilon)\, u$$

$$\hat{w} = (-\frac{d^2}{dr^2} + \frac{6}{r^2} + M\varepsilon)\, w \qquad (16)$$

where ε is the deuteron binding energy. If one used the Schrödinger equation, the terms involving $\hat{u}$ and $\hat{w}$ can be eliminated in favor of products of potentials times wave functions, so that the I terms are directly proportional to the nuclear potential. Hence, for purposes of this paper only, I will follow Ref. [17] and refer to these terms as "dynamic" corrections.

When Friar obtained these results, both he and I thought that the dynamic corrections were small, so that we were delighted with the approximate agreement obtained using two widely differing formalisms. However, Coester and Ostabee[17] have recently shown that the dynamic corrections are 1 to 3 times as big as Friar's results at small q^2, giving the bulk of the effect. (See Fig. 7.) Furthermore, they have again redone the calculation, using a formalism similar to that employed by Friar, and they obtain complete agreement with my results (Ref. [16]) for local potentials. If the potentials are not local, however, they obtain different results for the dynamic corrections. Finally, they assert that the dynamic corrections are required by Lorentz invariance.

The present status of relativistic corrections to the deuteron form factor is therefore somewhat confused. While there is considerable agreement between Coester and Ostabee and myself, there are still important differences between us for non local potentials. (A possible explanation for this difference may be a breakdown of the approximation described in Ref. [6].) And, while Coester and Ostabee seem to have found a mistake in the Friar (or Krajcik and Foldy) approach, I would like to be further convinced that it is really a mistake and not some intrinsic ambiguity lurking in the formalism.

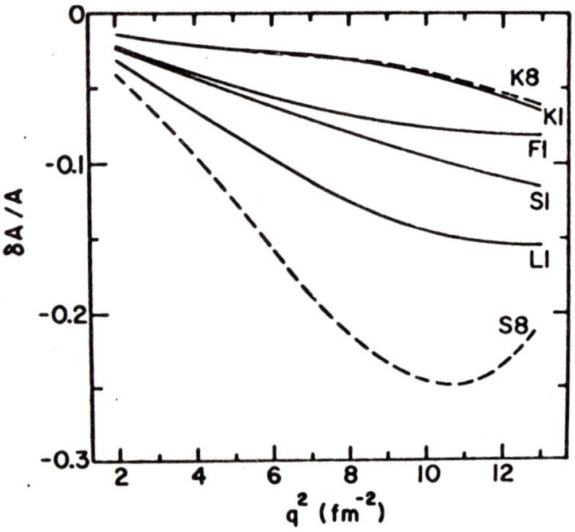

Fig. 7. Relativistic corrections at low q^2 (taken from Ref. [17]). Lines labeled K are Friar's corrections (for two different potentials), L is the Gross correction and F1 is the result for the Reid Soft core potential - taking its nonlocality into consideration. (S1 and S8 are results for two different potentials assuming Serber exchange.)

We now turn to the question of what other effects, if any, should be considered before one has a "complete" calculation of relativistic effects. Recently, Jackson, Lande, and Riska[18] have calculated corrections to the deuteron form factor from the "recoil" current (Fig.3c) and "pair" current (Fig. 3d-f).

However, in 1973 Thompson and Heller[19] already had emphasized that the "recoil" term should be regarded as included in the usual non-relativistic formalism (in the sense that when taking the non-relativistic limit of time ordered diagrams it is needed to reproduce the formalism with instantaneous potentials) so that to include it as a correction would be to double count. Similar conclusions have been reached recently by Woloshyn[20], Drechsel and Weber[21], and Friar[22], all of whom argue that the recoil term should not be added in as a separate relativistic effect. From the viewpoint developed in our discussion following Fig. 3, we see that all of these separate diagrams are included in the RIA, and that from the point of view of the RIA, the only correction terms not included in Eq. (14) and (16) are the contributions from the small component wave functions, which in the OPE approximation would be equivalent to the pair term corrections.

However, as we have mentioned already, the leading term from the pair term is isovector, and hence cancels in the deuteron form factor. This requires higher order terms be saved if the calculation is to be done correctly (see Ref. [21]), and even when finished one has no guarantee in this case that the heavier mesons will not contribute substantially to the result. I suggest that this is a case where the use of relativistic wave functions and the formalism described in Ref. [5] is really an advantage.

In this regard I would like to report on a calculation in progress by Arnold, Carlson and myself[23]. We have already obtained a formula for the corrections described above - due to the overlap between the large and small component wave functions. The formulae can be written as additional corrections to be added to D_C^{SO} and D_Q^{SO}:

$$\Delta D_C^{SO} = \frac{q^2}{12M^2}\int_0^\infty dr\; Mr(j_0+j_2)\left[\frac{v_s}{\sqrt{3}}(u-\sqrt{2}w) - \sqrt{\frac{2}{3}}\,v_t\left(u+\frac{w}{\sqrt{2}}\right)\right]$$

$$\begin{aligned}\Delta D_Q^{SO} = \int_0^\infty dr\; Mr\Big\{ & (j_0+j_2)\,u\left[\frac{v_t}{\sqrt{3}}+\frac{2}{\sqrt{3}}v_s\right] \\ & -\sqrt{3}\,wv_t\left(\frac{9}{35}(j_2+j_4)+\frac{4}{15}(j_0+j_2)\right) \\ & +\sqrt{\frac{3}{2}}\,wv_s\left(\frac{18}{35}(j_2+j_4)-\frac{2}{15}(j_0+j_2)\right)\Big\}\end{aligned} \qquad (17)$$

These terms have not yet been evaluated. In addition to these terms, there are other smaller corrections which are a measure of the theoretical error, so that when this work is finished we will have a built-in estimate of the uncertainties in the calculation.

Finally, there are two other types of effects which have not been included so far. The first of these are electromagnetic modifications of the potential, as illustrated in Fig. 8. These require a detailed knowledge of the relativistic dynamics of the two nucleon interaction. For example, if the crossed pion box is an important mechanism, as illustrated in Fig. 8a, then the diagram in Fig. 8b is also an important effect. However, since this corresponds to a more or less unique physical process, we would not regard this as a relativistic effect. Similarly, true meson exchange effects, where the photon interacts with one of the exchanged mesons, are of this type, and are not regarded as relativistic effects. (A contact term generated by a $\gamma^5\gamma^\mu$

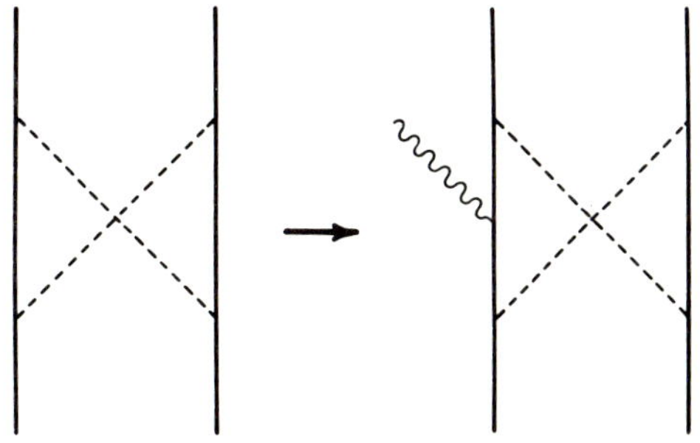

(a) (b)

Fig. 8. Example of a term in the potential which would generate an electromagnetic correction.

coupling as described in section II is also of this type.)

Finally, as pointed out in Refs. [20] and [21], changes in the nucleon current arising from the fact that the nucleons are off shell are relativistic effects (in our sense of the term) which have not yet been included in any of the calculations discussed so far.

C. High Momentum Transfer

We now turn to a discussion of relativistic effects at high momentum transfer. It is particularly important that these be calculated in view of the recent experiment of Arnold and Chertok (Ref. [13]). Preliminary results from this experiment show that the meson exchange calculations of Chemtob, Moniz and Rho[24] and Blankenbecler and Gunion[25] which predicted a larger cross section at high q^2 are not supported by the data. Instead, the non-relativistic model seems to do remarkably well (see Fig. 9), and relativistic corrections to the IA are therefore of considerable interest.

In Ref. [23] the relativistic effects due to the processes shown in Fig. 3 will be calculated for all momentum transfers. The formulae for this calculation are already available, but the numerical work is still in progress so that we cannot report on the results at this time. However, Arnold has evaluated the corrections in Eq. (14) and (16) at high q^2, and we show these in Fig. 10. Note that the corrections are moderate until about $q^2 = 120\ F^{-2}$.

Regarding the question of meson exchange effects, I would like to point out that the "pair" currents which gave such large results for neutron radiative capture (as discussed in Section II) and which have been called meson exchange currents by many people are not the same kind of meson exchange currents that were expected to show up in the deuteron form factor. The latter are of the pion current type shown in Fig. 1b, and for these there is very little concrete evidence. Hence, there is, at this point, no contradiction between the absence of exchange currents in the deuteron form factor and the importance of relativistic effects in radiative capture.

Finally, I mention (but have no space to discuss) the success of the quark model of Brodsky and Farrar[26] in anticipating the asymptotic behavior of the deuteron form factor. These results could suggest that the underlying quark structure of the deuteron is already appearing in the data of Ref. [13], and that examination of the deuteron in terms of nuclear wave functions and meson exchange currents is counterproductive at these momentum transfers.

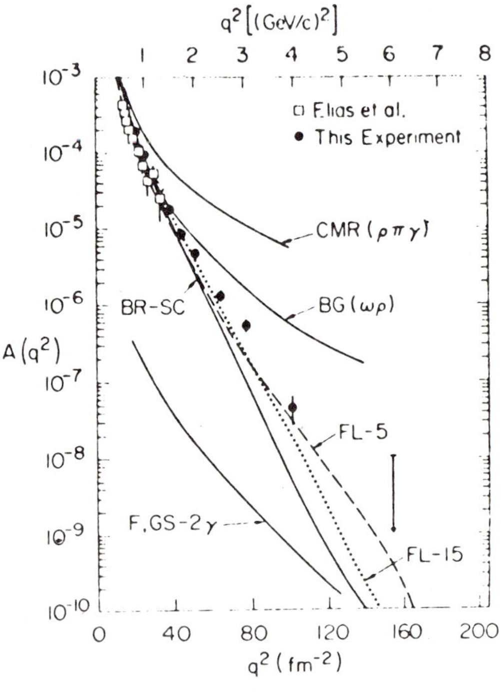

Fig. 9. New high q^2 deuteron form factor data from Ref. [13]. The curve labeled CMR refers to the meson exchange calculation of Ref. [24], while BG refers to Ref. [25]. The BR-SC, and FL curves are the non-relativistic form factor calculated with Reid Soft Core and Feshbach Lomon wave functions.

IV. RELATIVISTIC WAVE FUNCTIONS

By now it is appropriate to discuss the relativistic wave functions in somewhat more detail. I will make no attempt to review the several different relativistic models of the nuclear force now available (see Vinh Mau[27]), but will limit my discussion to the model I developed some years ago (Gross[28]) from which the current models of the small component wave functions referred to above were developed.

This model employs a wave equation in which one particle is restricted to its mass shell, as illustrated in Fig. 11. The model was originally introduced because the wave function one obtains has precisely the structure needed for the calculations described in sections II and III above.

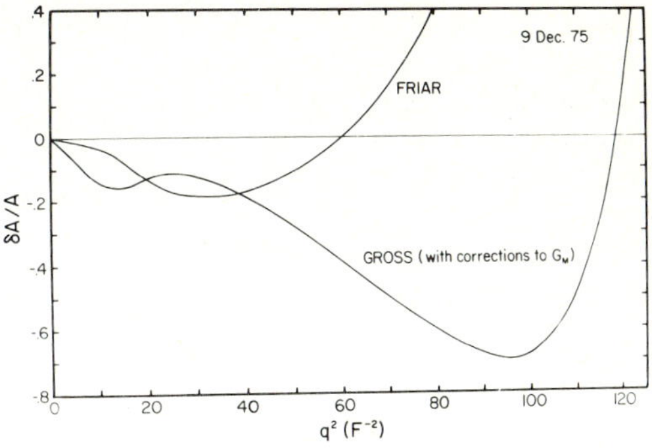

Fig. 10. Corrections given in Eq.(14) (labeled Friar) and Eq. (14) + (16) + corrections to G_M from Ref. [16] (labeled Gross). (Preliminary work by Arnold.)

The relativistic vertex function Γ is a form factor, and its connection with the positive (+) and negative (-) energy wave functions is given by the formulae below:

$$\psi^+_{s's}(\vec{p}) = \mathcal{N}\sqrt{\frac{M}{E_p}}\,\frac{\bar{u}(-\vec{p},s)\Gamma C\bar{u}^T(\vec{p},s')}{2E_p - W}$$

$$\psi^-_{s's}(\vec{p}) = -\mathcal{N}\sqrt{\frac{M}{E_p}}\,\frac{\bar{v}(-\vec{p},s)\Gamma C\bar{u}^T(\vec{p},s')}{W} \tag{18}$$

Where u and v are the dirac spinnors of the nucleon, $\mathcal{N}$ is a normalization constant, W the total energy and $E_p = \sqrt{M^2 + \vec{p}^2}$. When (18) is reduced to two component form, the result in position space is:

$$\psi^+ = \frac{1}{\sqrt{4\pi}}\left[\frac{u(r)}{r} + \frac{1}{\sqrt{8}}\frac{w(r)}{r}\left(\frac{3\sigma_1\cdot r\sigma_2\cdot r}{r^2} - \sigma_1\cdot\sigma_2\right)\right]\chi_{1m}$$

$$\psi^- = \frac{-i\sqrt{3}}{\sqrt{4\pi}}\left[\frac{1}{\sqrt{2}}\frac{v_t}{r}\frac{(\sigma_1+\sigma_2)\cdot r}{2r} + \frac{v_s}{r}\frac{(\sigma_1-\sigma_2)\cdot r}{2r}\right]\chi_{1m} \tag{19}$$

for the deuteron and

$$\psi^+ = \frac{1}{\sqrt{4\pi}}\frac{y(r)}{r}\chi_{00}$$

$$\psi^- = i\sqrt{\frac{3}{4\pi}}\,\frac{z(r)}{r}\,\frac{(\sigma_1-\sigma_2)\cdot r}{2r}\,\chi_{00} \tag{20}$$

for the $^1S_0$ state. The wave functions for the deuteron were estimated in Ref. [7], and have recently been calculated by solving the exact relativistic wave equation numerically for the simple case of exchange by a pion (pure γ^5 coupling) and a σ as illustrated in Fig. 11b (Buck and Gross[29]). In Fig. 12 and 13 these wave functions are shown.

The $^1S_0$ wave functions are estimated in Ref. [8], and are shown in Fig.14.

I wish to comment on some novel features of the BG wave functions of Ref. [29]. First, note that they exhibit a repulsive behavior at short distances. This repulsion must be due entirely to the structure of the wave equation and the treatment of the antinucleon degrees of freedom, because the usual non-relativistic limit of π and σ exchanges are known to be purely attractive in the deuteron channel. Secondly, the combined probabilities of the triplet and singlet P states (v_t and v_s) is about 2.7%, which is much larger than in Ref. [7] (where it was about 0.5%) and is due to the use of a pure γ^5 coupling which probably overestimates the contributions of the small components.

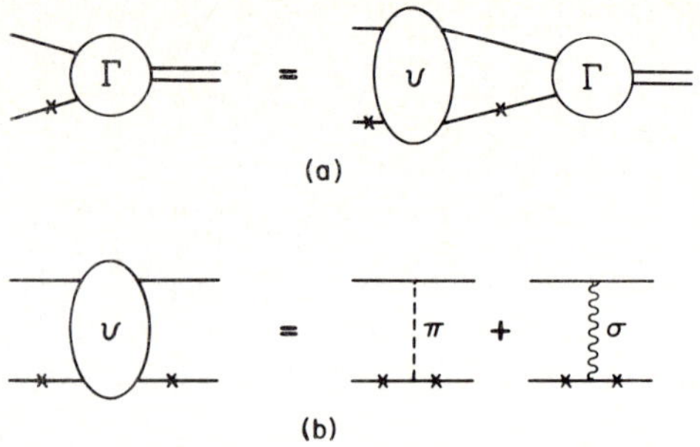

Fig. 11. (a) Representation of the wave equation of Ref. [28]. (b) The potential employed in Ref. [29].

How can these wave functions be tested? Are there physical processes where the small component wave functions give the major contribution, instead of the small corrections we saw in Sections II and III? The answer seems to be yes, and we now turn to a very exciting recent result.

V. BACKWARD n-d SCATTERING

The backward peak in n-d scattering has attracted interest for many years. The failure of the one nucleon exchange (ONE) approximation was the original motivation of Kerman and Kisslinger[30] for suggesting that N* exchange was important in this process, and since then many calculations including N*'s have been undertaken (see Sharma, Bhasin and Mitra[31]). While N*'s undoubtedly make

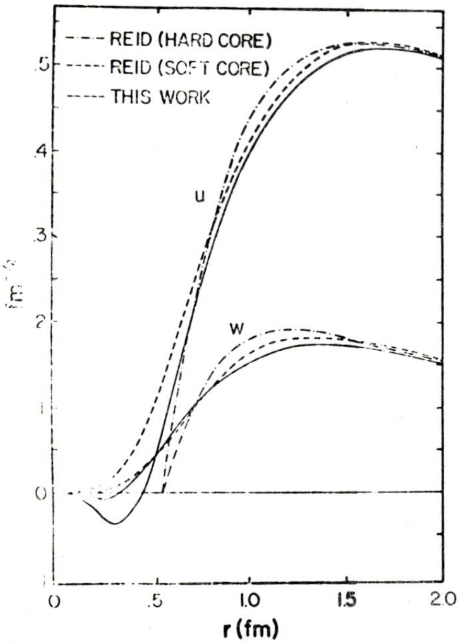

Fig. 12. The S and D state wave functions of Ref. [29] are compared with Reid wave functions at short range.

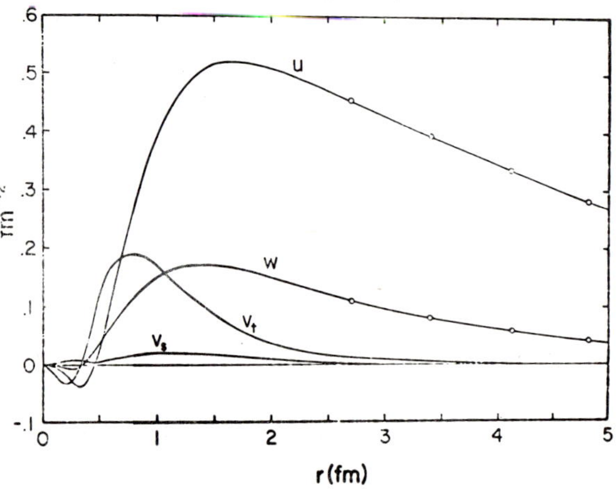

Fig. 13. The 4 deuteron wave functions of Ref. [29].

contributions to this process, it has recently been suggested that the evidence for their importance is no longer conclusive (Noble and Weber[32]).

It is not our intention here to review this process in its entirety. Rather, we wish to pursue the suggestion of Remler[33], that the small component wave function might be important in this process. That they are important was discovered by Morioka and Ueda[34] who reported on their work at this conference. These gentlemen were kind enough to forward an advance copy of their work to us, and when we received it Buck, Remler, and myself[35] redid the calculation. We obtained qualitatively similar results, but the quantitative details seem different. I will report on our calculation here.

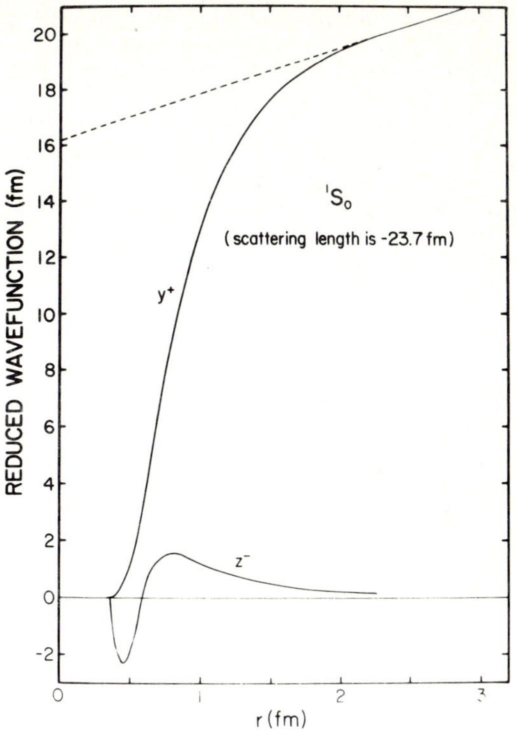

Fig. 14. The $^1S_0$ relativistic wave functions of Ref. [8].

For 180° scattering, we obtain a very simple formula:

$$\left.\frac{d\sigma}{d\Omega}\right|_{\substack{180^\circ \\ CM}} = 12\left(\frac{M^2+Q^2}{(E_p+E_d)M}\right)^2 \left\{ \left[\theta \mathcal{M}(u^2+w^2)\right]^2 + \left[\frac{M_d^3}{4\mathcal{M}}(v_t{}^2+v_s{}^2)\right]^2 + \frac{\theta M_d^3}{6}\left[(\sqrt{2}\,u-w)v_t - (u+\sqrt{2}\,w)v_s\right]^2 \right\} \tag{21}$$

where p is the magnitude of the 3 momentum of the proton in the center of mass system, $E_p = \sqrt{M^2+p^2}$, $E_d = \sqrt{M_d^2+p^2}$, and

$$Q = \frac{p}{M_d}(E_d - E_p)$$

$$\theta = 2E_Q - M_d = 2\sqrt{M^2+Q^2} - M_d$$

$$\mathcal{M} = \frac{1}{2}(E_d + M - E_p) \tag{22}$$

and all wave functions (which are in momentum space) have the Lorentz transformed Q as their arguments, $u = u(Q)$.

In Fig. 15, the results obtained from Eq. 21 with the BG wave functions are shown. Note that for $p > 400$ MeV/c, the small component wave functions give an enormous enhancement, which can be understood quite easily from Eq. (21) (see Ref. [35]). Because of the large P state probabilities, the BG wave functions probably overestimate the effect, and a more realistic calculation including a background from multiple scattering is underway. However, it is clear that it will no longer be possible to ignore the P state components of the deuteron wave function in any future calculation of backward nd scattering.

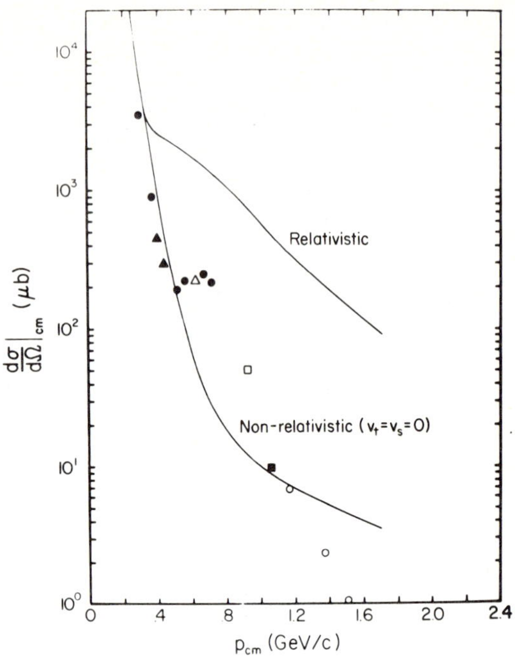

Fig. 15. The Eq. (21) evaluated with BG wave functions (Ref. [29]) compared with data from a number of experimental groups.

VI. EFFECTS IN ^{3}He AND ^{3}H

Finally, we turn to the question of ^{3}He and ^{3}H form factors. Space permits that we touch on this subject only very briefly.

Friar[14,36] has calculated relativistic effects similar to those calculated for the deuteron. He finds that they tend to shift the minimum further out, and hence do not help agreement with the data. However, his formulae do not include the "dynamic" corrections described in Section III, and hence it would be valuable to redo his calculation. Kloet and Tjon [37] have calculated the "meson" exchange contributions to the form factors. They found a large effect, which shifted the minimum in the ^{3}He form factor down by about 3 F^{-2} and raised the secondary maximum considerably, so that there is substantial improvement between theory and experiment. (See also Haftel[38].) As we discussed in Section II, the dominant contribution seems to come from the pair terms, and hence is also a relativistic effect, but of a different kind from that calculated by Friar. In view of the size of the Kloet-Tjon correction, it may be worth while to redo this calculation using small component ^{3}He wavefunctions, and include the dynamic corrections referred to above. Perhaps one of the major developments which can be expected of this subject in the next few years will be a more complete calculation of relativistic effects in three body systems.

REFERENCES

[1] H. P. Noyes, Nucl. Phys. 74 (1965) 508.
[2] D. O. Riska and G. E. Brown, Phys. Lett. 38B (1972) 193. This work is based to some extent on the important paper of M. Chemtob and M. Rho, Nucl. Phys. A163 (1971) 1.
[3] M. Gari and A. H. Huffman, Phys. Rev. C7 (1973) 994.
[4] G. Stranahan, Phys. Rev. 135 (1964) B953.
[5] E. T. Dressler and F. Gross, proceedings of the Quebec Conference (1974) 780, and submitted for publication.
[6] F. Gross, Phys. Rev. 140 (1965) B410.
[7] J. Hornstein and F. Gross, Phys. Lett. 47B (1973) 205.
[8] E. Delacroix and F. Gross, in preparation.
[9] J. Hockert, D. O. Riska, M. Gari, and A. Huffman, Nucl. Phys. A217 (1973)14.
[10] F. Gross, proceedings of the Quebec Conference (1974) 782.

[11] J. C. Bergstrom, Phys. Rev. C9 (1974) 2435.
[12] R. R. Scheerbaum, Texas A & M preprint.
[13] R. G. Arnold, B. T. Ckertok, E. B. Dally, A. Grigorian, C. L. Jordon, W. P. Schütz, R. Zdarko, F. Martin, B. A. Mecking, Proceedings of the Conference on High Energy Physics and Nuclear Structure held at Santa Fe (1975) 373; and Phys. Rev. Lett. 35 (1975) 776.
[14] J. L. Friar, Ann. of Phys. 81 (1973) 332.
[15] R. A. Krajcik and L. L. Foldy, Phys. Rev. D10 (1974) 1777; and Phys. Rev. Lett. 24 (1970) 545.
[16] F. Gross, Phys. Rev. 142 (1966) 1025, 152 (1966) 1517 E. See also B. M. Casper and F. Gross, Phys. Rev. 155 (1967) 1607.
[17] F. Coester and A. Ostebee, Phys. Rev. C11 (1975) 1836.
[18] A. D. Jackson, A. Lande, and D. O. Riska, Phys. Lett. 55B (1975) 23.
[19] R. H. Thompson and L. Heller, Phys. Rev. C7 (1973) 2355.
[20] R. M. Woloshyn, Phys. Rev. C12 (1975) 901.
[21] D. Drechsel and H. J. Weber, preprint.
[22] J. L. Friar, preprint.
[23] R. Arnold, C. Carlson, and F. Gross (in preparation).
[24] M. Chemtob, E. Moniz and M. Rho, Phys. Rev. C10 (1974) 344.
[25] R. Blackenbecler and J. Gunion, Phys. Rev. D4 (1971) 718.
[26] S. Brodsky and G. Farrar, Phys. Rev. D8 (1973) 826.
[27] R. Vinh Mau, invited talk presented to this conference.
[28] F. Gross, Phys. Rev. D10 (1974) 223; 186 (1969) 1448; proceedings of the Los Angeles Conference (1972) 38.
[29] W. Buck and F. Gross, submitted for publication.
[30] A. K. Kerman and L. S. Kisslinger, Phys. Rev. 180 (1969) 1483.
[31] J. S. Sharma, V. S. Bhasin, and A. N. Mitra, Nucl. Phys. B35 (1971) 466.
[32] J. V. Noble and H. J. Weber, Phys. Lett. 50B (1974) 233.
[33] E. A. Remler, Nucl. Phys. B42 (1972) 69; B42 (1972) 56; E. A. Remler and R. A. Miller, Ann. of Phys. 82 (1974) 189.
[34] S. Morioka and T. Ueda, paper submitted to this conference.
[35] W. Buck, F. Gross, and E. A. Remler (in preparation).
[36] J. L. Friar, Phys. Lett. 43B (1973) 108. Note errors in this paper as discussed by Friar[14].
[37] W. M. Kloet and J. A. Tjon, Phys. Lett. 49B (1974) 419.
[38] M. Haftel, contribution to this conference.

DISCUSSION

Slobodrian : How sensitive is the deep part of the core to the experimental data? Could you use a different shape than the flat one?

Vinh Mau : Yes, what I just showed you is the simplest possibility for the core. A fit with a superposition of Yukawa type short range terms, but constrained to give a finite value at r = 0, is in progress.

Bleuler : What about extension of one pion exchange to the N$\bar{N}$ system ? Repulsive core will be changed into attractive core whch in turn gives rise to new kinds of bound states of the N$\bar{N}$ system.

Vinh Mau : I have skipped this question because of lack of time. We know that the N$\bar{N}$ forces are the same as the NN forces if the exchanged system has even G parity, while they are of opposite sign for an odd G parity exchange system. Therefore, the TPE interaction remains the same whereas the omega exchange contribution (which is repulsive in the NN system) become attractive in the N$\bar{N}$ system. As we get the deuteron right with our potential, we expect to get (deuteron like) bound and/or resonant states near threshold in the N$\bar{N}$ system. Calculations in this respect are in progress in Paris. The results, when available, can be compared with the data obtained at Brookhaven in $\bar{p}$p and $\bar{p}$d reactions near threshold.

Sauer : Do you suggest use of your potentials in many-nucleon systems with the energy-dependence as it stands?

Vinh Mau : The energy dependence of the Paris potentials is linear, and linear energy dependence can be very easily transformed into velocity (or p^2) dependence. One can then use known methods for the many-nucleon problem. The energy dependence of the Stony Brook is more complicated.

Bleuler : Inclusion of the A-resonance in nuclear matter calculations yields a large decrease of binding; therefore additional binding from 3-body forces might still be welcome (all meson theoretical forces should be, as a matter of principle, included).

McKellar: I agree, if one is going to include the resonance explicitly it is necessary to include 3-body forces.

Fabre : Did you try to formulate your three body force in polar coordination in a 6 dimensional space which is the obvious generalization of the few body force in the 3 dimensional space and enables one to define unambiguously the tensor character of the various parts of the three-body potential?

McKellar : No, but it should certainly be look at.

Sauer : Shell model calculations use a finite model space which creates effective three-body forces. Can the contributions of the effective and fundamental three-body forces be disentangled?

McKellar : Empirical 3-body matrix elements of course contain both effective and genuine 3-body forces. Only first principle calculations can determine the contribution of each separately.

Mitra : From the diagrams you use for the definition of three-body forces, it appears that you can generate these diagrams by including $NN \to NN^*$ or $N\Delta$, etc. channels in association with the $NN \to NN$ channels (through pion exchange). Could one then say that "3-body forces" are essentially derivable from "two-body forces", with the definition of "nucleons" extended to incorporate "resonances" as well?

McKellar : That is part of the story. If one does a coupled channel calculation then the $N\Delta$ intermediate states in a three-body correlation give part of the three-body force. But one would still have a remainder after subtracting forward propagating Born terms for N and Δ poles from the N amplitude.

DYNAMICAL EQUATIONS AND APPROXIMATION METHODS

W. Sandhas

Physikalisches Institut der Universität Bonn, Bonn, Germany

1. INTRODUCTION

The integral equations approach to the three-body problem, decisively stimulated by Faddeev's formulation, provides the most powerful tool for studying the internal structure of this system. An essential step towards a detailed understanding of composite particle dynamics has been done in this way. The search for adequate extensions to the general N-body situation therefore represented, and still represents, a natural challenge. For various reasons this transition is non-trivial, and non-unique. Emphasizing different aspects of the three-body theory, different generalizations have been found.

In particular, it was the concept of connectedness of the (iterated) integral kernel which represented the guiding idea of almost all proposals. This rather formal concept allows for an arbitrary number of formulations, many of them being presumably only mathematically correct, but physically rather unsatisfactory. We, therefore, want to review the present status of the N-body theory in a less technical way. Starting from the basic, physically convincing definitions of scattering states, we replace the defining equations by more appropriate matrix relations. This is done in a reversible way, thus preserving in every step the original structure and information. In order to be as close as possible to the basic definitions, all relations are first derived for scattering states or half-on-shell transition amplitudes. The ambiguity in going over to corresponding operator identities (fully off-shell equations) is demonstrated.

Our procedure is most easily explained for the well known three-body theory, which will be recapitulated in an appropriate manner. The four-body problem shows already the typical difficulties of the general case. We, therefore, describe in detail the decisive transition from three to four particles. This is done in a way which exhibits the underlying structure of the generalization. The extension to arbitrary particle numbers should, therefore, become transparent, even without going into all the details of the general N-body notation which will be sketched only briefly.

Successive application of pole approximations reduces the general relations to effective two-body equations. Numerical results achieved in the four-nucleon case are finally summarized.

2. BASIC TWO- AND THREE-BODY EQUATIONS

Integral equations serve to combine the content of differential equations and boundary (asymptotic) conditions.

In the <u>two-body</u> case this is a well known concept. Let the Hamiltonian H be split into a kinetic energy part H_o and a (short range) potential V. Scattering states are then defined as those solutions of the stationary Schrödinger equation

$$(H_o + V)\,\psi^{(+)} = E\,\psi^{(+)} \tag{2.1}$$

which fulfil the asymptotic condition

$$\psi^{(+)}(\vec{r}) \underset{r\to\infty}{\sim} \phi(\vec{r}) + T(\theta)\,\frac{e^{ipr}}{r} \quad . \tag{2.2}$$

The full information of these equations is contained in the Lippmann-Schwinger equation

$$\psi^{(+)} = \phi + G_o V\,\psi^{(+)} \quad . \tag{2.3}$$

Here the plane wave solutions ϕ are eigenfunctions of H_o, characterizing the asymptotically free situation. The corresponding free Green's function is

$$G_o = (E + i\varepsilon - H_o)^{-1} \quad . \tag{2.4}$$

In the <u>three-body</u> case we proceed in an analogous way. Let the Hamiltonian be split into a kinetic energy part and a sum of two-body interactions $V_{ij} = V_\alpha$ with

$\alpha \neq i,j$:
$$H = H_o + \sum_{i<j} V_{ij} = H_o + V_\alpha + V_\beta + V_\gamma \; . \tag{2.5}$$
The definition of free states and, correspondingly, of "free" Hamiltonians depends on the asymptotically free configuration being considered.

We first study two-fragment channels, i.e., a situation where an elementary particle α is moving with relative momentum $\vec{q}_\alpha$, while the other two particles i,j form two-body bound states $\psi_{\alpha n}$ supported by the internal interaction $V_\alpha = V_{ij}$ (n collectively denotes the quantum numbers of $\psi_{\alpha n}$). The corresponding free states (channel states), replacing the plane waves of the genuine two-body case, are, therefore,
$$\Phi_{\alpha n} = e^{i\vec{q}_\alpha \cdot \vec{R}_\alpha} \psi_{\alpha n} \; . \tag{2.6}$$
$\vec{R}_\alpha$ is the position space variable conjugate to $\vec{q}_\alpha$. These free states are eigenfunctions of the "free" (channel) Hamiltonian
$$H_\alpha = H_o + V_\alpha \; . \tag{2.7}$$
Consequently, the total Hamiltonian (2.5) is split into a free part and the external interaction (often called residual or channel interaction),
$$H = H_\alpha + \overline{V}_\alpha \; , \tag{2.8}$$
the latter one acting between the colliding fragments,
$$\overline{V}_\alpha = \sum_{\gamma \neq \alpha} V_\gamma = \sum_\gamma (1-\delta_{\alpha\gamma}) V_\gamma = \sum_\gamma \overline{\delta}_{\alpha\gamma} V_\gamma \; . \tag{2.9}$$

Starting from an asymptotic configuration (2.6), with α = 1,2 or 3, fixed, the corresponding scattering state $\Psi^{(+)}_{\alpha n}$ allows for transitions into all channels. Hence, its asymptotic behaviour is
$$\Psi^{(+)}_{\alpha n} \underset{R_\beta \to \infty}{\sim} \delta_{\beta\alpha} \Phi_{\alpha n} + \underset{n'}{\sum\!\!\!\!\!\!\int} T_{\beta n', \alpha n} \, \psi_{\beta n'} \frac{e^{iq'_\beta R_\beta}}{R_\beta} \; . \tag{2.10}$$

It should be mentioned that for $\beta = \alpha$ an incoming free state appears, while for all re-arrangement channels, $\beta \neq \alpha$, only the outgoing spherical waves occur. The symbol $\sum\!\!\!\!\!\!\int$ indicates that we not only have to sum over the two-body bound states $\psi_{\beta n'}$ in channel β but also to integrate over two-body scattering states related to break-up processes. These considerations suggest that the generalization of (2.3) should be given by the three equations,
$$\Psi^{(+)}_{\alpha n} = \delta_{\beta\alpha} \Phi_{\alpha n} + G_\beta \overline{V}_\beta \Psi^{(+)}_{\alpha n} \; , \tag{2.11}$$
only one of them being inhomogeneous. As in (2.10), α is fixed while β takes the values 1,2 and 3. The "free" (channel) Green's functions associated with the "free" Hamiltonians H_β are now
$$G_\beta = (E + i\varepsilon - H_\beta)^{-1} \; . \tag{2.12}$$
Multiplying (2.11) by G_β^{-1} we immediately see that $\Psi^{(+)}_{\alpha n}$ fulfills the Schrödinger equation
$$H \, \Psi^{(+)}_{\alpha n} = E_{\alpha n} \Psi^{(+)}_{\alpha n} \; . \tag{2.13}$$
$\Psi^{(+)}_{\alpha n}$ moreover shows the asymptotic behaviour (2.10). This makes the uniqueness of (2.11) at least plausible.

For a rigorous proof we have to have in mind that the other two-fragment states $\Psi^{(+)}_{\gamma n}$, $\gamma \neq \alpha$, do not solve the set of equations (2.11), since for them the inhomogeneous terms occur for $\beta = \gamma \neq \alpha$. In order to show that the three-fragment (break-up) states $\Psi^{(+)}_o$ are also no solutions of (2.11), we mention that they fulfil three inhomogeneous equations with the kernels of (2.11),
$$\Psi^{(+)}_o = \Phi^{(+)}_\beta + G_\beta \overline{V}_\beta \Psi^{(+)}_o \; , \quad \beta = 1,2,3 \; , \tag{2.14}$$

but with inhomogeneous terms

$$\Phi_\beta^{(+)} = e^{i\vec{q}_\beta\cdot\vec{R}_\beta}\,\psi_\beta^{(+)} \quad , \tag{2.15}$$

where the two-body bound states occurring in (2.6) are replaced by scattering states $\psi_\beta^{(+)}$ orthogonal to them. This evidently means that (2.11) and (2.14) are two independent sets of equations which separately define the two-fragment and three-fragment scattering states. Their uniqueness is guaranteed if the states $\Psi_{\alpha n}^{(+)}$, $\alpha = 1,2,3$, together with $\Psi_o^{(+)}$ form a complete set in the continuous spectrum.

The equations (2.11) have been known for a long time[1]. They are also given, together with (2.14), in Faddeev's first paper on this subject, mainly in order to show that a single integral equation of the Lippmann-Schwinger type is non-unique[2]. Emphasizing the uniqueness of the whole set (2.11), Glöckle proposed a method for solving these equations directly[3].

3. MATRIX EQUATIONS

The usual procedures applied in the three-body theory are most easily understood as replacements of the three basic equations (2.11) by one equivalent matrix relation. In order to show this we introduce column vectors for the scattering states,

$$\Psi_{\beta\alpha} = \Psi_\alpha^{(+)} \quad , \tag{3.1}$$

and for the free states

$$\Phi_{\beta\alpha} = \delta_{\beta\alpha}\Phi_\alpha \tag{3.2}$$

(the index n labelling the various two-body bound states is suppressed for simplicity). Note that α is fixed, while $\beta = 1,2,3$ denotes the three elements in the respective columns. Defining furthermore the "free matrix Green's function"

$$\mathcal{G}_{o,\beta\alpha} = \delta_{\beta\alpha}\, G_\alpha \quad , \tag{3.3}$$

and a suitably chosen "potential matrix" $\mathcal{V}_{\beta\alpha}$, eq. (2.11) can be written in the form

$$\Psi_{\beta\alpha} = \Phi_{\beta\alpha} + \mathcal{G}_{o,\beta\beta}\sum_\gamma \mathcal{V}_{\beta\gamma}\,\Psi_{\gamma\alpha} \quad , \tag{3.4}$$

i.e., in form of a matrix Lippmann-Schwinger equation

$$\Psi = \Phi + \mathcal{G}_o\,\mathcal{V}\,\Psi \quad . \tag{3.5}$$

Eq. (3.4) is equivalent to (2.11) under the two conditions

$$\sum_{\gamma=1}^{3} \mathcal{V}_{\beta\gamma} = \overline{V}_\beta \tag{3.6}$$

and

$$\Psi_{1\alpha} = \Psi_{2\alpha} = \Psi_{3\alpha} \quad . \tag{3.7}$$

The latter requirement allows us to perform the sum in (3.4), reproducing in this way the original equations. Condition (3.7) is, of course, a consequence of the fact that according to (3.1) all elements of the column $\Psi_{\beta\alpha}$ are identical, the index β being artificially introduced in order to allow for the above matrix notation. Solving (3.4) under such a restriction contradicts, however, the general concept of integral equations. They should provide unique solutions without imposing any additional conditions. The decisive step in formulating adequate three-body matrix equations is, therefore, to choose the potential matrix in such a way that (3.7) is automatically implied.
Finally, we emphasize that the three-fragment (break-up) equations (2.14) can be replaced in the same way by matrix equations of the form (3.4) but with different inhomogeneous terms.

4. CHOICE OF POTENTIAL MATRICES

The simplest choice of $\mathcal{V}$, consistent with (3.6), is the diagonal matrix

$$\mathcal{V}_{\beta\alpha} = \delta_{\beta\alpha}\overline{V}_\alpha \quad , \tag{4.1}$$

i.e.,

$$\mathcal{V} = \begin{pmatrix} \overline{V}_1 & 0 & 0 \\ 0 & \overline{V}_2 & 0 \\ 0 & 0 & \overline{V}_3 \end{pmatrix}$$

This, however, causes a complete decoupling of the system (3.4),

$$\Psi_{\beta\alpha} = \Phi_{\beta\alpha} + G_\beta \overline{V}_\beta \Psi_{\beta\alpha} \; . \tag{4.2}$$

In view of the fact that condition (3.7) is no longer assumed, we have now only one equation for each component $\Psi_{1\alpha}$, $\Psi_{2\alpha}$ or $\Psi_{3\alpha}$ separately, while, according to (2.11), three relations are necessary to define a state uniquely. Thus, the original information contained in (2.11) is no longer preserved. In order to avoid such a decoupling, non-diagonal elements have to occur in each column of $\mathcal{V}$.

Kouri-Levin-Tobocman (KLT) choice: This is most easily fulfilled by a coupling scheme proposed by Kouri and Levin for three particles and extended to the general N-body case by Tobocman[4,5]. It corresponds to the definition of a potential matrix

$$\mathcal{V} = \begin{pmatrix} 0 & \overline{V}_1 & 0 \\ 0 & 0 & \overline{V}_2 \\ \overline{V}_3 & 0 & 0 \end{pmatrix} = \begin{pmatrix} 0 & V_2+V_3 & 0 \\ 0 & 0 & V_1+V_3 \\ V_1+V_2 & 0 & 0 \end{pmatrix} \tag{4.3}$$

which evidently couples all the occurring components of $\Psi_{\beta\alpha}$. For $\alpha = 1$ we have, e.g.,

$$\begin{aligned} \Psi_{11} &= \Phi_1 + G_1 \overline{V}_1 \Psi_{21} \\ \Psi_{21} &= \qquad\; G_2 \overline{V}_2 \Psi_{31} \\ \Psi_{31} &= \qquad\; G_3 \overline{V}_3 \Psi_{11} \; . \end{aligned} \tag{4.4}$$

It should be mentioned that this choice was originally introduced by first defining a channel coupling array, represented by the matrix

$$W = \begin{pmatrix} 0 & 1 & 0 \\ 0 & 0 & 1 \\ 1 & 0 & 0 \end{pmatrix} \; , \tag{4.5}$$

and subsequent multiplication by the channel interactions (the general ideas of this approach are summarized in an article by Kouri and Levin[6] presented at the Laval Conference 1974). To combine these two steps in the definition of a potential matrix represents only a small modification. The following choices of $\mathcal{V}$ can, therefore, be formulated as well in the language of the channel coupling array concept[7].

Faddeev-choice (exchange potentials): It is the characteristic feature of the KLT-choice (4.3) that the full external (channel) interactions $\overline{V}_\beta$ occur as elements of $\mathcal{V}$. A more symmetric form is obtained by spreading the two-body potentials contained in $\overline{V}_\beta$ over all non-diagonal elements of the corresponding row. This suggests the Faddeev choice

$$\mathcal{V} = \begin{pmatrix} 0 & V_2 & V_3 \\ V_1 & 0 & V_3 \\ V_1 & V_2 & 0 \end{pmatrix} . \tag{4.6}$$

The elements of $\mathcal{V}$ can be interpreted as exchange potentials,

$$\mathcal{V}_{\beta\alpha} = (1-\delta_{\beta\alpha}) V_\alpha = \overline{\delta}_{\beta\alpha} V_\alpha = \tag{4.7}$$

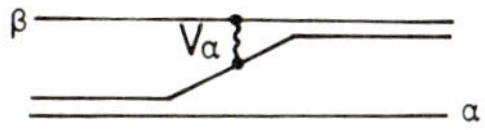

Comparison with (2.9) shows that condition (3.6) is satisfied.

Inserting (4.7) in (3.4) we arrive at the Faddeev-type equations

$$\Psi_{\beta\alpha} = \delta_{\beta\alpha} \Phi_\alpha + G_\beta \sum_\gamma \overline{\delta}_{\beta\gamma} V_\gamma \Psi_{\gamma\alpha} \; . \tag{4.8}$$

Being not identical with the original form of the Faddeev equations[2], they show their characteristic coupling scheme. For that reason (4.6) has been called Faddeev-choice. However, it should be mentioned, that this type of coupling scheme represents also the basis of various other independent or related methods, as, e.g., the ones by Watson, Skornyakov and Ter-Martirosyan, Mitra, Amado, Sitenko and Kharchenko, and Lovelace who was the first to point out the relationship between these approaches and the Faddeev theory[8,9].

Uniqueness: In order to prove the uniqueness of our matrix equations we have to show that their solutions fulfil condition (3.7). This is most easily done for the Faddeev-choice which also from this point of view shows its exceptional role. Multiplying (4.8) by $(1-G_oV_\beta)$ we find

$$\Psi_{\beta\alpha} = \sum_\gamma G_o V_\gamma \Psi_{\gamma\alpha} \tag{4.9}$$

which evidently means that $\Psi_{\beta\alpha}$ is independent of β and, therefore, according to the considerations of sec. 3, identical to the scattering states $\Psi_\alpha^{(+)}$. We note that this proof is easily extended to a mixture of the above choices, where two rows are chosen as in (4.6) and one as in (4.3).

Connectedness: The simplicity of our proof originates from the fact that the basic equations (2.11) are taken for granted which essentially means that completeness of scattering states and bound states is proposed. Their equivalence with the matrix equations is then mainly a simple algebraical consideration. In this way the underlying structures of the problem could be exhibited.

Usually the connectivity of the iterated kernels is emphasized. In fact this is achieved after two iterations for the KLT-choice and after one iteration for the Faddeev-choice, but in no order of iteration for the choice (4.1). In the Faddeev case it can be shown that after a sufficient number of iterations the kernel becomes even completely continuous. This allows for a direct proof of uniqueness. The advantage of such a procedure is that the completeness has not to be assumed but is directly proved[10].

Various different equations have been proposed which fulfil the connectivity property. Preserving less directly the original structures (2.11) they are not incorporated in the present review. Their properties and shortcomings are summarized, e.g., in an article by Amado[11].

Transition amplitudes: Up to now all equations have been given for scattering states. Multiplying (2.3) by V we get in the two-body case the half-on-shell equations

$$T\,\phi = V\,\phi + V\,G_o\,T\,\phi \tag{4.10}$$

for the scattering amplitudes T. Applying the same procedure onto (3.5) yields

$$\mathcal{T}\Phi = \mathcal{V}\Phi + \mathcal{V}\mathcal{G}_o\mathcal{T}\Phi , \tag{4.11}$$

where according to (3.6)

$$\sum_\gamma \mathcal{T}_{\beta\gamma}\Phi_{\gamma\alpha} = \sum_\gamma \mathcal{V}_{\beta\gamma}\Psi_{\gamma\alpha} = \overline{V}_\beta\Psi_\alpha^{(+)} . \tag{4.12}$$

The expression $\mathcal{T}\Phi$, therefore, really provides the transition amplitudes.

5. MODIFICATIONS

We should once more emphasize that the states $\Psi_{\beta\alpha}$ considered up to now are, despite of their matrix notation, the scattering states $\Psi_\alpha^{(+)}$. Multiplying eq. (4.8) from the left by G_oV_β, we arrive at the original form of the Faddeev equations

$$\psi_\alpha^{(\beta)} = \delta_{\beta\alpha}\Phi_\alpha + G_oT_\beta\sum_\gamma \overline{\delta}_{\beta\gamma}\,\psi_\alpha^{(\gamma)} . \tag{5.1}$$

Here the scattering states are replaced by the Faddeev states

$$\psi_\alpha^{(\beta)} = G_oV_\beta\Psi_{\beta\alpha} = G_oV_\beta\Psi_\alpha^{(+)} . \tag{5.2}$$

They evidently fulfil

$$\sum_\beta \psi_\alpha^{(\beta)} = G_o\sum_\beta V_\beta\Psi_\alpha^{(+)} = \Psi_\alpha^{(+)} , \tag{5.3}$$

i.e., provide a splitting of $\Psi_\alpha^{(+)}$ into three states with simpler asymptotic behaviour in position space[12].
In deriving (5.1) the original kernel is replaced according to

$$G_\beta V_\beta = G_o T_\beta . \tag{5.4}$$

The occurrence of the off-shell two-body T-matrix T_β in the Faddeev equations is often regarded as one of the most essential advantages of these relations. Indeed, unitary pole approximations are most easily performed in this formulation, due to the fact that separable potentials yield separable T-matrices[9].

For some questions the replacement (5.4) is, however, inconvenient. It has been mentioned, e.g., that inserting in $G_\beta V_\beta$ a complete set of two-body states, one directly sees that the kernel of (5.1) depends on two-body bound state poles and half-off-shell amplitudes only. This feature, while a priori clear, is made explicit in this way[13]. To leave the kernel in its original form $G_\beta V_\beta$ is of particular advantage for considerations of analyticity (unitarity relations, K-matrix equations[7]) since the whole energy dependence is contained in G_β.
It should be emphasized that (5.1) has still the structure of a matrix LS equation

$$\Psi = \Phi + \mathrm{G_o}\,\mathrm{V}\,\Psi \tag{5.5}$$

with modified free Green's functions and potential matrices[14]

$$\mathrm{G}_{o,\beta\alpha} = \delta_{\beta\alpha} G_o T_\alpha G_o \text{ , and } \mathrm{V}_{\beta\alpha} = \bar{\delta}_{\beta\alpha} G_o^{-1} . \tag{5.6}$$

The same modification replaces (4.11) by

$$\mathrm{T}\Phi = \mathrm{V}\Phi + \mathrm{V}\,\mathrm{G_o}\,\mathrm{T}\,\Phi \tag{5.7}$$

where, however, $\mathrm{T}\Phi = \mathcal{T}\Phi$.

6. OPERATOR IDENTITIES

In the two-body case the resolvent equation

$$G = G_o + G_o V G \tag{6.1}$$

and the operator LS equation

$$T = V + V G_o T \tag{6.2}$$

may be considered as off-shell extensions of (2.3) and (4.10). Operator identities for the above three-body matrix operators are introduced in an analogous way

$$\mathcal{G} = \mathcal{G}_o + \mathcal{G}_o \mathcal{V} \mathcal{G} \tag{6.3}$$

and

$$\mathcal{T} = \mathcal{V} + \mathcal{V} \mathcal{G}_o \mathcal{T} . \tag{6.4}$$

For the Faddeev choice (4.7) the full matrix Green's function is explicitly given by

$$\mathcal{G}_{\beta\alpha} = \delta_{\beta\alpha} G_o + G V_\alpha G_o . \tag{6.5}$$

Such off-shell continuations are, of course, non-unique. Starting from the modified forms (5.5) and (5.7), the relations

$$\mathrm{G} = \mathrm{G_o} + \mathrm{G_o}\,\mathrm{V}\,\mathrm{G} \tag{6.6}$$

and

$$\mathrm{T} = \mathrm{V} + \mathrm{V}\,\mathrm{G_o}\,\mathrm{T} \tag{6.7}$$

can be considered as the adequate analogues of (6.1) and (6.2).
According to (5.2) and (6.5) we have the relationship

$$\mathrm{G}_{\beta\alpha} = G_o V_\beta \mathcal{G}_{\beta\alpha} = G_o (V_\beta \delta_{\beta\alpha} + V_\beta G V_\alpha) G_o = G_o M_{\beta\alpha} G_o . \tag{6.8}$$

Inserting this expression and the definitions (5.6), we see that (6.6) is just the operator Faddeev equation

$$M_{\beta\alpha} = \delta_{\beta\alpha} T_\alpha + T_\beta G_o \sum_{\gamma \neq \beta} M_{\gamma\alpha} \tag{6.9}$$

which in this way turns out to be one of the possible three-body generalizations of the resolvent equation (6.1). Eq. (6.7) reads explicitly

$$U_{\beta\alpha} = \bar{\delta}_{\beta\alpha} G_o^{-1} + \sum_\gamma \bar{\delta}_{\beta\gamma} T_\gamma G_o U_{\gamma\alpha} . \tag{6.10}$$

These equations introduced by Alt, Grassberger and Sandhas (AGS)[15] represent, therefore, the corresponding analogue of the operator LS equation (6.2)[14].

Ambiguities: The above considerations show that the physical content of the basic equations (2.11) leads to various matrix equations. Different coupling schemes, represented by different choices of $\mathcal{V}$, may be used. Moreover, coupling of both, two-fragment and three-fragment channels, not considered here, has been proposed[4]. Different off-shell extensions of the resolvent or operator LS equations are possible. Multiplying (6.1) from the right by V the inhomogeneous term and the kernel become identical. The three-body analogue of this procedure is easily performed using the above matrix notation[16]. These ambiguities, already remarkable in the three-body case, have their counterparts in the N-body theory, where many formulations look confusingly different as long as the underlying structures are not exhibited.

7. BASIC N-BODY EQUATIONS

Our approach to the three-body theory can be generalized rather naturally. Again, we start from the consideration of two-fragment channels. In the four-body case seven partitions of the particles i,j,k,l into two fragments are possible, four of the form $\rho = (i,jkl)$, with i = 1,2,3 or 4, and three of the form $\rho = (ij,kl)$, with (ij) = (12), (13) or (14). In the following we also use σ and τ to label such partitions. Let us now proceed as in sec. 2. The full Hamiltonian

$$H = H_o + \sum_{i<j} V_{ij} = H_o + \sum_{\gamma=1}^{6} V_\gamma , \qquad (7.1)$$

with its six two-body interactions, is split in seven different ways into "free" (channel) Hamiltonians H_ρ and external (channel) interactions $\overline{V}_\rho$,

$$H = H_\rho + \overline{V}_\rho . \qquad (7.2)$$

Here

$$H_\rho = H_o + V_\rho \qquad (7.3)$$

is the Hamiltonian of two asymptotically free (composite) fragments. This means, V_ρ contains only the internal potentials V_{ij} acting between the particles in the clusters. In the case $\rho = (i,jkl)$ we have, e.g.,

$$V_\rho = V_{jk} + V_{jl} + V_{kl} \qquad (7.4)$$

and for $\rho' = (ij,kl)$

$$V_{\rho'} = V_{ij} + V_{kl} . \qquad (7.5)$$

The external (channel) interaction $\overline{V}_\rho$ is, on the other hand, just the interaction between the colliding fragments. In our above examples

$$\overline{V}_\rho = V_{ij} + V_{ik} + V_{il} \qquad (7.6)$$

$$\overline{V}_{\rho'} = V_{ik} + V_{il} + V_{jk} + V_{jl} . \qquad (7.7)$$

In analogy to (2.11) we have for each two-fragment scattering state $\Psi_\rho^{(+)}$ seven basic equations

$$\Psi_\rho^{(+)} = \delta_{\sigma\rho}\Phi_\rho + G_\sigma \overline{V}_\sigma \Psi_\rho^{(+)} . \qquad (7.8)$$

σ runs over the seven two-fragment partitions discussed above. G_σ is, of course, the "free" (channel) Green's function

$$G_\sigma = (E + i\varepsilon - H_\sigma)^{-1} . \qquad (7.9)$$

The channel states Φ_ρ are eigenfunctions of H_ρ characterizing the asymptotically free configuration corresponding to $\Psi_\rho^{(+)}$. In the case $\rho = (i,jkl)$ the channel state is a product of a three-body bound state and a plane wave describing the relative motion of particle i

$$\Phi_\rho = e^{i\vec{q}_i \cdot \vec{R}_i} \psi_{(jkl)} . \qquad (7.10)$$

For $\rho' = (ij,kl)$ we have a product of two-body bound states and their relative plane wave

$$\Phi_{\rho'} = \psi_{(ij)}\,\psi_{(kl)}\,e^{i\vec{q}_{\rho'}\cdot\vec{R}_{\rho'}} . \qquad (7.11)$$

Multiplying (7.8) by G_σ^{-1} we see that the solutions of this system obey the full Schrödinger equation. As in the three-body case the uniqueness of (7.8) can be made plausible by considering the asymptotic behaviour of $\Psi_\rho^{(+)}$. For a real proof we have to generalize the argumentation following eq. (2.13). It can be shown, indeed, that the three-fragment and the four-fragment states fulfil separately the seven equations (7.8) but with <u>different</u> inhomogeneous terms. In this way all occurring scattering states are uniquely defined (if their completeness in the continuous spectrum is assumed).

<u>Some details</u>: For a better understanding of our statements we sketch the situation in some more detail. In the three-fragment case $\overline{\rho} = (ij,k,l) = (\beta,k,l)$ we have six scattering states. Each of them satisfies seven equations with the kernels of (7.8)

$$\Psi_{\overline{\rho}}^{(+)} = \Phi_{\sigma\overline{\rho}}^{(+)} + G_\sigma \overline{V}_\sigma \Psi_{\overline{\rho}}^{(+)} \quad . \qquad (7.12)$$

It is essential that despite of the fact that we are now studying three-fragment states, the index σ labels the possible <u>two</u>-fragment configurations. The inhomogeneous terms $\Phi_{\sigma\overline{\rho}}^{(+)}$ vanish if the two-body cluster β in $\overline{\rho}$ is not contained in the fragments of σ. This evidently means that four of the equations (7.12) are homogeneous, three inhomogeneous. The non-vanishing terms are in the case $\sigma = (ijk,l) = (\beta k,l)$

$$\Phi_{\sigma\overline{\rho}}^{(+)} = \Psi_\beta^{(+)}\,e^{i\vec{q}_l\cdot\vec{R}_l} , \qquad (7.13)$$

i.e., products of the three-body scattering states given by (2.11) and the plane wave of particle l. In the case $\sigma = (ij,kl) = (\beta,kl)$ we have

$$\Phi_{\sigma\overline{\rho}}^{(+)} = \psi_\beta\,\psi_{(kl)}^{(+)}\,e^{i\vec{q}_\sigma\cdot\vec{R}_\sigma} \qquad (7.14)$$

where ψ_β is a two-body bound-state, $\psi_{(kl)}^{(+)}$ the scattering state of the two other particles. The plane wave describes their relative movement. Note that the inhomogeneous terms in (7.12) are orthogonal to the ones in (7.8).

An analogous situation holds in the four-fragment case $\hat{\rho} = (i,j,k,l)$. Again the corresponding scattering states obey eq. (7.8) with $\delta_{\sigma\rho}\,\Phi_\rho$ replaced by different inhomogeneous terms $\Phi_{\sigma\hat{\rho}}^{(+)}$. Here, as in (2.14), <u>all</u> the inhomogeneous terms are different from zero. Each of them is a product of a three-body break-up state $\Psi_0^{(+)}$ and the plane wave of the fourth particle if $\sigma = (ijk,l)$, or of two two-body scattering states and their relative plane wave if $\sigma = (ij,kl)$. The $\Phi_{\sigma\hat{\rho}}^{(+)}$ are therefore orthogonal to the inhomogeneous terms in (7.8) and (7.12).

In the general <u>N-body</u> problem the number of <u>two</u> fragment channels is[17]

$$n_2(N) = \frac{1}{2}\sum_{i=1}^{N-1}\binom{N}{i} = 2^{N-1} - 1 . \qquad (7.15)$$

Consequently, the 2-, 3-,... N-fragment channels are determined by independent sets of $n_2(N)$ equations with the kernels of the two-fragment case, and inhomogeneous terms orthogonal for different numbers of fragments. This can be shown directly by applying the resolvent equations (σ = two-cluster partition of the N bodies)

$$G = G_\sigma + G_\sigma \overline{V}_\sigma G \qquad (7.16)$$

onto the respective channel states. A more general consideration based on the principles of multichannel scattering theory has been given previously[18].

8. FOUR-BODY MATRIX EQUATIONS

As in the three-body case we may now replace the seven four-body equations (7.8) by matrix LS equations. Introducing a 7x7 potential matrix $\mathcal{V}^{\sigma\tau}$ with

$$\sum_\tau \mathcal{V}^{\sigma\tau} = \overline{V}_\sigma \qquad (8.1)$$

and considering the scattering states and channel states as column vectors

$$\Psi^{\sigma\rho} = \Psi_\rho^{(+)} , \tag{8.2}$$

$$\Phi^{\sigma\rho} = \delta_{\sigma\rho}\Phi_\rho , \tag{8.3}$$

our system (7.8) reads

$$\Psi^{\sigma\rho} = \Phi^{\sigma\rho} + G_\sigma \sum_\tau \mathcal{V}^{\sigma\tau}\,\Psi^{\tau\rho} . \tag{8.4}$$

Introducing furthermore as "free Green's function" the diagonal matrix

$$\mathcal{G}_o^{\sigma\rho} = \delta_{\sigma\rho}\, G_\rho , \tag{8.5}$$

we see that (8.4) has, indeed, the form of the two-body LS equation

$$\Psi = \Phi + \mathcal{G}_o \mathcal{V}\,\Psi . \tag{8.6}$$

As discussed in the analogous three-body situation this formulation is valid under the restriction

$$\Psi^{\tau\rho} = \Psi^{\sigma\rho} \quad \text{for arbitrary } \sigma,\tau \tag{8.7}$$

following from (8.2). This allows us, by summing over τ, to reproduce (7.8). If (8.7) is no longer required as an additional condition, the coupling of the system itself has to guarantee that the solutions automatically show this property. I.e., decoupling of the set (8.4) must be avoided.

KLT-choice: Denoting for the moment the seven two-fragment channels in an arbitrary way by $\rho,\sigma,\tau = 1,2,\ldots 7$, we have as the simplest possibility the analogue of (4.3)

$$\mathcal{V} = \begin{pmatrix} 0 & \bar V_1 & 0 & 0 & 0 & 0 & 0 \\ 0 & 0 & \bar V_2 & 0 & 0 & 0 & 0 \\ 0 & 0 & 0 & \bar V_3 & 0 & 0 & 0 \\ 0 & 0 & 0 & 0 & \bar V_4 & 0 & 0 \\ 0 & 0 & 0 & 0 & 0 & \bar V_5 & 0 \\ 0 & 0 & 0 & 0 & 0 & 0 & \bar V_6 \\ \bar V_7 & 0 & 0 & 0 & 0 & 0 & 0 \end{pmatrix} . \tag{8.8}$$

The characteristic feature of this choice is that it leads to a coupling of only two elements of Ψ in any step. We have, indeed, for $\rho = 1$

$$\begin{aligned} \Psi_{11} &= \Phi_1 + G_1\bar V_1\Psi_{21} \\ \Psi_{21} &= \quad\;\; G_2\bar V_2\Psi_{31} \\ &\;\vdots \\ \Psi_{71} &= \quad\;\; G_7\bar V_7\Psi_{11} . \end{aligned} \tag{8.9}$$

This means that only after inserting all these equations into each other, i.e., after $n_2(4)-1 = 6$ iterations of (8.6), a connected kernel is achieved. Written in the form

$$\mathcal{V}_{\sigma\rho} = \begin{cases} \bar V_\sigma \delta_{\sigma+1,\rho} & \text{for} \quad \sigma \neq n_2(N) \\ \bar V_{n_2(N)}\, \delta_{1\rho} & \text{for} \quad \sigma = n_2(N) \end{cases} \tag{8.10}$$

the KLT-choice is immediately applicable to the general N-body case[4,5].

Exchange potentials: The simplicity of the KLT-choice has to be contrasted with its considerable asymmetry. In the three-body case this shortcoming was easily cured by distributing the two-body potentials occurring in $\bar V_\beta$ over the non-diagonal elements of the 3x3 potential matrix $\mathcal{V}_{\beta\alpha}$ (see eqs. (4.3) and (4.6)). This is no longer an evident procedure for $N \geq 4$ where the number of two-body potentials $N(N-1)/2$ differs from the number $n_2(N)$ of two-fragment channels. In the four-body case $\bar V_\rho$ is, according to (7.6) and (7.7), a sum of only three or four potentials V_{ij} which have to be spread over a potential matrix which is at least 7x7. I.e., several zeros have to occur also in the non-diagonal elements. This requires an adequate prescription. A natural way is provided by generalizing systematically the concept of exchange potentials (Faddeev-choice) given in (4.7).

Starting from a <u>two-body</u> potential V_α, labelled by the two-body index α, the <u>three-body</u> exchange potential (4.7) was defined according to

$$\mathcal{V}_{\beta\alpha} = \bar{\delta}_{\beta\alpha} V_\alpha \quad . \tag{8.11}$$

Labelling now this potential matrix by an index ρ, denoting in which subsystem it acts, we introduce in an analogous way the <u>four-body</u> potential matrix

$$\mathcal{V}^{\sigma\rho} = \bar{\delta}_{\sigma\rho} \mathcal{V}^{\rho} \quad . \tag{8.12}$$

Repeating this procedure we evidently arrive, starting from V_α, after N-2 steps at an N-body exchange potential (see below).
Inserting in (8.12) the explicit form (8.11) of $\mathcal{V}^\rho$ we find

$$\mathcal{V}^{\sigma\rho}_{\beta\alpha} = \begin{cases} \bar{\delta}_{\sigma\rho} \mathcal{V}^{\rho}_{\beta\alpha} = \bar{\delta}_{\sigma\rho}\bar{\delta}_{\beta\alpha} V_\alpha & \text{if } \beta \subset \sigma, \text{ and } \beta,\alpha \subset \rho \\ 0 & \text{otherwise} \end{cases} \tag{8.13}$$

Due to the fact that $\mathcal{V}^\rho$ itself is a matrix in the three-body subsystems of $\rho = (i,jkl)$ or in the two-body subsystems of $\rho = (ij,kl)$, we get in the four-body case already an 18x18 potential matrix. The restrictions in the definition (8.13) ensure that the subsystems α,β have to be contained in the clusters of the partition ρ, otherwise the elements of $\mathcal{V}^{\sigma\rho}_{\beta\alpha}$ vanish. The expression (8.13) represents, indeed, what we would intuitively call an exchange potential. For $\sigma = (i,jkl)$ and $\beta = (kl)$ we have non-vanishing elements only if $\rho = (ij,kl)$ or $\rho = (ikl,j)$, i.e., only for those transitions where β occurs in one of the final clusters. The possible exchange diagrams are shown in Fig. 1.

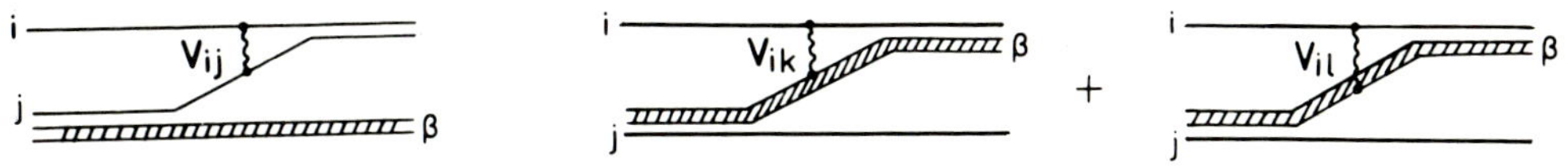

Fig. 1

From this graphical representation we easily read off the decisive property

$$\sum_{\alpha,\rho} \mathcal{V}^{\sigma\rho}_{\beta\alpha} = \bar{V}_\sigma \quad . \tag{8.14}$$

Hence, the definition of exchange potentials (8.13) represents, as in the three-body case, a natural way of spreading the potentials V_{ij} over the elements of an 18x18 matrix. We had even to blow up the dimension of the original 7x7 matrix in order to get a most symmetric form. The number 18 originates, of course, from the fact that there exist three possibilities to introduce two-body subsystems β in each of the four channels $\rho = (i,jkl)$ and two possibilities in each of the three channels $\rho = (ij,kl)$, i.e., 3x4 + 2x3 = 18. The concept of four-body exchange potentials has been first introduced by Grassberger and Sandhas (GS) with an argumentation similar to the one given above[19]. The algebraically highly sophisticated Yakubovsky theory[20], explained in detail by Faddeev[21] for N=4 at the Birmingham Conference, looks at first sight rather different. The underlying coupling schemes of both approaches are, however, identical. This has been shown by Alt, Grassberger and Sandhas[22] for N=4 and by Karlsson and Zeiger[16] for arbitrary N. The differences in detail are, of course, considerable (see the discussion following eq. (10.7)). Our present procedure differs also in some aspects from the previous ones. It provides a method to replace only the channel potential $\bar{V}_\sigma$ by the above potential matrix, introducing in this way the typical FY-GS channel coupling, preserving, however, the original structure of the basic equations.

According to these considerations we insert (8.14) into (7.8) regarding simultaneously the scattering states as elements of 18-dimensional column vectors $\Psi^{\sigma\rho}_\beta = \Psi^{(+)}_\beta$, and defining analogously $\Phi^{\sigma\rho}_\beta = \delta_{\sigma\rho}\Phi_\rho$ (if $\beta \subset \sigma$, zero otherwise). Eq. (7.8) then takes the matrix LS form

$$\Psi_\beta^{\sigma\rho} = \Phi_\beta^{\sigma\rho} + G_\sigma \sum_{\tau,\gamma} \mathcal{V}_{\beta\gamma}^{\sigma\tau} \Psi_\gamma^{\tau\rho} \quad . \tag{8.15}$$

A more appropriate representation is achieved if also G_σ is replaced by the matrix Green's function (6.5)

$$\mathcal{G}_{\beta\alpha}^{\sigma} = \delta_{\beta\alpha} G_o + G_\sigma V_\alpha G_o \quad . \tag{8.16}$$

Here an additional index σ denotes the subsystem where this operator acts ($\beta,\alpha \subset \sigma$). The important property

$$\sum_\alpha \mathcal{G}_{\beta\alpha}^{\sigma} = G_\sigma \tag{8.17}$$

of this matrix allows us to insert (8.16) in (8.15),

$$\Psi_\beta^{\sigma\rho} = \Phi_\beta^{\sigma\rho} + \sum_{\tau,\gamma,\delta} \mathcal{G}_{\beta\delta}^{\sigma} \mathcal{V}_{\delta\gamma}^{\sigma\tau} \Psi_\gamma^{\tau\rho} \quad . \tag{8.18}$$

Uniqueness: All these replacements are justified as long as, according to its definition, $\Psi_\beta^{\sigma\rho}$ is assumed to be independent of σ and β. When dropping this information, the independence has to be automatically guaranteed by the coupled equations, as discussed in the previous examples. In order to show that this is really the case, we proceed in complete analogy to the proof given in sec 4. Multiplying eq. (8.18) from the left by ($1 - \mathcal{G}_o \mathcal{V}^\sigma$), with $\mathcal{G}_o$ defined by (3.3), we immediately see that $\Psi_\beta^{\sigma\rho}$ is independent of σ. The independence of β then follows after multiplication by $(1 - G_o V_\beta)$. All summations in (8.18) can, therefore, be performed, so that via (8.14) and (8.17) the set of basic LS equations (7.8) is reproduced. The equivalence of the content of both sets is shown in this way.

The matrix equation (8.18) therefore represents only another way of reading (7.8). This, however, means that its mathematical structure is considered in a completely different manner, as demonstrated most easily by iterating (7.8) and (8.18). It is in particular the detailed subsystem structure made explicit in these equations which is of decisive advantage for many investigations of the internal dynamics of the problem, as well as for approximation methods. On the other hand it should not be overlooked that many conventional derivations of scattering theory are more easily based on the original form (7.8). In view of the equivalence established above no mathematical improvement is achieved in such cases by starting from (8.18) instead of (7.8).

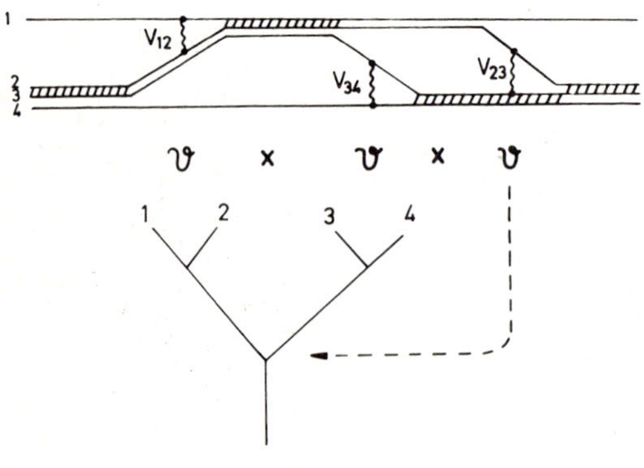

Fig. 2

Connectedness: Again it is a characteristic aspect of the FY coupling scheme that it yields connected kernels in a systematical way after (N-2) iterations. This is easily demonstrated in the GS-exchange potential formulation. Fig. 2 shows how $\mathcal{V}$ picks up a two-body interaction in every step. We see that after two iterations the kernel of (8.18) as well as the one of (8.15) becomes connected.

This situation can also be described as a systematical way of switching on two-body interactions. The graphical description by a connected tree[21,23] means that first particles 1 and 2 interact, afterwards particles 3 and 4. In a third step the interaction between the two clusters formed in this way is introduced.

9. N-BODY MATRIX EQUATIONS

The transition from the two-body to the three-body, and from the three-body to the four-body case, by defining appropriate exchange potentials, was performed in an algebraically equivalent manner (cf. (8.11) and (8.12)). Repeated application of this procedure yields potential matrices of the same structure for arbitrary particle numbers. The explicit form (8.13) exhibits clearly that here in contrast to the KLT-choice, the various elements of the potential matrix are labelled by two-fragment indices σ,ρ,τ as well as by the subsystem indices α,β,γ. Hence, in the general case an appropriate generalization of this notation is required. According to Yakubovsky[20] partitions of the N particles into k fragments are denoted by a_k, b_k, c_k . In the four-body case a_2,b_2,c_2 thus correspond to σ,ρ,τ and α,β,γ are replaced by a_3,b_3,c_3 .
As can be read off (8.13) certain partitions have to be contained in those with a smaller number of fragments. Following again Yakubovsky this is dealt with by defining chains. A chain

$$\alpha_i = (a_i,\ a_{i+1},\dots,\ a_{N-1}) \tag{9.1}$$

is a sequence of partitions a_k with the property $a_k \supset a_{k+1}$. I.e., a_{k+1} is the result of breaking up one fragment in a_k.
Generalizing (8.13) we now define potential matrices by the recurrence relation

$$\mathcal{V}^{a_i}_{\beta_{i+1},\alpha_{i+1}} = \bar{\delta}^{a_i}_{b_{i+1},a_{i+1}} \mathcal{V}^{a_{i+1}}_{\beta_{i+2},\alpha_{i+2}} \qquad i \leq N-3 \tag{9.2}$$

where the superscript of $\bar{\delta}^{a_i}_{b_{i+1},a_{i+1}}$ means that b_{i+1} and a_{i+1} must be contained in a_i, otherwise it is defined zero. For i = N-2 the potential is given by (8.11). Repeated application of (9.2) yields, with $\beta_3 = (b_3,\dots,b_{N-1})$

$$\mathcal{V}^{a_2}_{\beta_3\alpha_3} = \bar{\delta}^{a_2}_{b_3a_3} \cdots\cdots \bar{\delta}^{a_{N-2}}_{b_{N-1},a_{N-1}} V_{a_{N-1}} . \tag{9.3}$$

The essential point is now that the generalization of (8.14) holds (see also ref.16).

$$\sum_{a_2,\alpha_3} \mathcal{V}^{b_2a_2}_{\beta_3\alpha_3} = \sum_{a_2,\alpha_3} \bar{\delta}_{b_2a_2} \mathcal{V}^{a_2}_{\beta_3\alpha_3} = \bar{V}_{b_2} \tag{9.4}$$

where summation over a_2 and all partitions of α_3 is meant by this notation. As emphasized after eq. (7.15) we have for any two-fragment scattering state $\Psi^{(+)}_{a_2}$ a set of $n_2(N)$ LS equations of the form (7.8)

$$\Psi^{(+)}_{a_2} = \delta_{b_2a_2}\Phi_{a_2} + G_{b_2}\bar{V}_{b_2}\Psi^{(+)}_{a_2} . \tag{9.5}$$

The property (9.4) allows us to replace (9.5) by the matrix equation

$$\Psi^{b_2a_2}_{\beta_3} = \Phi^{b_2a_2}_{\beta_3} + G_{b_2} \sum_{c_2,\gamma_3} \mathcal{V}^{b_2c_2}_{\beta_3\gamma_3} \Psi^{c_2a_2}_{\gamma_3} \tag{9.6}$$

which evidently represents the N-body generalization of (8.15). Along the lines of the previous section we also may replace G_{b_2} by appropriately defined matrix Green's functions which obey ($a_3 \subset b_2$)

$$\sum_{\alpha_3} \mathcal{G}^{b_2}_{\beta_3,\alpha_3} = G_{b_2} . \tag{9.7}$$

This leads to the generalization of (8.18)

$$\Psi^{b_2a_2}_{\beta_3} = \Phi^{b_2a_2}_{\beta_3} + \sum_{\delta_3} \mathcal{G}^{b_2}_{\beta_3\delta_3} \sum_{c_2,\gamma_3} \mathcal{V}^{b_2c_2}_{\delta_3\gamma_3} \Psi^{c_2a_2}_{\gamma_3} . \tag{9.8}$$

Uniqueness: In analogy to the proofs given in secs 4 and 8 it can be shown that the solutions of (9.8) are independent of b_2 and β_3. Consequently (9.5) is immediately reproduced by performing the summations.

10. MODIFIED FOUR-BODY EQUATIONS

In the three-body case it was of advantage to multiply eq. (4.8) from the left by $G_o V_\beta$ (see sec.5). Applying the same procedure on (8.18) we get

$$\Psi^{\sigma\rho} = \Phi^{\sigma\rho} + G^\sigma \sum_\tau \bar{\delta}_{\sigma\tau} V^\tau \Psi^{\tau\rho} \quad . \tag{10.1}$$

Here, according to (6.8), the modified Green's function occurs which, apart from two factors G_o, is given by the Faddeev operators $M_{\beta\alpha}$. The "potential" (4.7) is furthermore replaced by the form given in (5.6).
In analogy to the procedure leading from (4.8) to (5.1), we now multiply (10.1) from the left by $G_o V^\sigma$, with G_o defined in (5.6). Introducing the four-body generalization of the Faddeev states (5.2)

$$\Psi^{(\sigma)\rho} = G_o V^\sigma \Psi^{\sigma\rho} \quad , \tag{10.2}$$

we get

$$\Psi^{(\sigma)\rho} = \Phi^{(\sigma)\rho} + G^\sigma V^\sigma \sum_\tau \bar{\delta}_{\sigma\tau} \Psi^{(\tau)\rho} \quad . \tag{10.3}$$

In (5.1) the subsystem transition operators T_γ were introduced by means of $G_\beta V_\beta = G_o T_\beta$. The corresponding three-body relation allows us to write (10.3) in the form

$$\Psi^{(\sigma)\rho} = \Phi^{(\sigma)\rho} + G_o T^\sigma \sum_\tau \bar{\delta}_{\sigma\tau} \Psi^{(\tau)\rho} \quad , \tag{10.4}$$

a notation exhibiting the striking analogy of the three- and four-body formulations discussed. With the explicit definition of the occurring quantities, eq. (10.2) reads

$$\psi_\beta^{(\sigma)\rho} = G_o T_\beta \sum_{\gamma \subset \sigma} \bar{\delta}_{\beta\gamma} G_o V_\gamma \psi_\rho^{(+)} \quad . \tag{10.5}$$

Summing over σ and β yields the scattering state $\psi_\rho^{(+)}$. This property, evidently, represents the generalization of (5.3).

Eq. (10.3) reads in detail

$$\psi_\beta^{(\sigma)\rho} = \Phi_\beta^{(\sigma)\rho} + G_o \sum_\gamma M_{\beta\gamma}^\sigma \sum_{\delta \subset \sigma} \bar{\delta}_{\gamma\delta} \sum_\tau \bar{\delta}_{\sigma\tau} \psi_\delta^{(\tau)\rho} \quad , \tag{10.6}$$

while the version (10.4) of the same equation yields

$$\psi_\beta^{(\sigma)\rho} = \Phi_\beta^{(\sigma)\rho} + G_o T_\beta G_o \sum_\gamma U_{\beta\gamma}^\sigma \sum_\tau \bar{\delta}_{\sigma\tau} \psi_\gamma^{(\tau)\rho} \quad . \tag{10.7}$$

The kernel of (10.6) is the one of the Faddeev-Yakubovsky four-body theory[20,21]. According to (10.7) this kernel is identical to the one proposed independently by Grassberger and Sandhas[19]. The latter form has the advantage that the two- and three-body transition operators explicitly occur. This makes (10.7) well suited for pole approximations (see sec. 12). In this context we recall that in the original GS approach pole approximations or their generalization, the quasiparticle method, have been applied already during the derivation. The GS-kernel was, therefore, originally not given in the form (10.7), but sandwiched between two- and three-body form factors. The effective two-body equations achieved in this way were directly applicable to practical calculations. Later on the whole derivation has been repeated by Alt, Grassberger and Sandhas performing first the operator algebra and applying afterwards the pole approximations. In this way the equivalence of the FY- and the GS-coupling schemes discussed above has been shown[22]. Two- and three-body pole approximations to the FY equations consequently reproduce the GS formalism.
Arbitrary particle numbers: It is one of the decisive advantages of the FY theory that the formalism was given from the very beginning for arbitrary N. Modifying the relations derived in sec. 9 along the lines of the present section, we get the

N-body generalization of (10.6) and (10.7). See in this context in particular ref.16, where the N-body extension of the AGS-four-body equations[22] is studied. We finally mention that all equations could have been given for scattering amplitudes instead of states (cf.sec. 4).

Operator identities: In analogy to the discussions in sec. 6, the above relations may be considered as half-on-shell restrictions of operator identities which are easily written down. The AGS four-body approach[22] as well as its N-body extension[16] were originally based on such operator identities. We give in sec. 12 only the LS-type equations following from (10.7) for the four-body transition operators.

11. KERNELS OF HIGHER CONNECTEDNESS

It is a characteristic feature of the above coupling scheme that the potential matrix contains the two-body interactions in a very simple and direct way. This, in particular, means that (8.13) is only two-body connected. Consequently two iterations of (8.15) are necessary to provide four-body connectedness. This situation is not changed if G_σ is replaced in (8.18) by the matrix $\mathcal{G}^\sigma$ or, as in (10.1), by the modified version G^σ. According to (6.6) and (6.9) we have

$$G^\sigma = G_o + G_o V^\sigma G^\sigma = G_o + \overline{G}^\sigma , \tag{11.1}$$

i.e., in explicit notation (after dropping two unessential factors G_o)

$$M^\sigma_{\beta\alpha} = \delta_{\beta\alpha} T_\alpha + T_\beta G_o \sum_{\gamma\neq\beta} M^\sigma_{\gamma\alpha} = \delta_{\beta\alpha} T_\alpha + \overline{M}^\sigma_{\beta\alpha} . \tag{11.2}$$

The latter form shows that only the second term $\overline{G}^\sigma$ on the right-hand side is three-body connected, a property most easily seen by iterating (11.2).

Four-body equations with three-body connected kernels are, however, immediately derived by multiplying (10.1) from the left by $G_o V^\sigma$,

$$G_o V^\sigma \Psi^{\sigma\rho} = G_o V^\sigma \Phi^{\sigma\rho} + \overline{G}^\sigma \sum_\tau \overline{\delta}_{\sigma\tau} V^\tau \Psi^{\tau\rho} . \tag{11.3}$$

After summation over σ, we find

$$\Psi^\rho = \Phi^\rho + \sum_\sigma \overline{G}^\sigma \sum_\tau \overline{\delta}_{\sigma\tau} V^\tau \Psi^\rho . \tag{11.4}$$

which explicitly reads

$$V_\beta \Psi^{(+)}_\rho = V_\beta \Phi_{\beta\rho} + \sum_\tau \left(\sum_\gamma \overline{M}^\tau_{\beta\gamma}\right) G_o \sum_{\delta\not\subset\tau} V_\delta \Psi^{(+)}_\rho \tag{11.5}$$

with $\Phi_{\beta\rho} = \Phi_\rho$ if $\beta \subset \rho$, zero otherwise. Here we had to have in mind that the algebraical structures of our three-body and four-body formulations are completely equivalent. In particular, the analogue of (5.3) was taken into account.

Thus, we arrived at six equations for the states $V_\beta \Psi^{(+)}_\rho$. The three-body connectivity of the kernel implies that only one iteration is necessary to achieve connectedness. Equations of this type have been derived by Rosenberg, by Mitra, Gillespie, Sugar and Panchapakesan, by Takahashi and Mishima, and by Alessandrini.[24] (see also ref. 23 and 25).

Sloan-Bencze-Redish (SBR) equations. The disadvantage of (11.5) is that it represents a set of equations for states $V_\beta \Psi^{(+)}_\rho$ not directly related to physical quantities. Summing, however, over $\beta \not\subset \sigma$ we get seven equations for the transition amplitudes (ρ fixed, σ arbitrary)

$$\mathcal{T}_{\sigma\rho} \Phi_\rho = \overline{\delta}_{\sigma\rho} (V_\rho - V^{(\sigma\rho)}) \Phi_\rho + \sum_{\tau\neq\sigma} \overline{M}^{\sigma\tau} G_o \mathcal{T}_{\tau\rho} \Phi_\rho . \tag{11.6}$$

Here

$$\overline{M}^{\sigma\tau} = \sum_{\beta\not\subset\sigma} \sum_\gamma \overline{M}^\tau_{\beta\gamma} . \tag{11.7}$$

$V^{(\sigma\rho)}$ is the potential contained in σ and in ρ. These equations derived by Sloan[26] have been generalized to arbitrary particle numbers by Bencze[27] and Redish[17,28].

Number of equations: In the four-body case we have 6 potentials, and 7 two-fragment channels. The latter number increases to 18 if a further labelling by two-body subsystem indices is introduced. Correspondingly we got the 6 equations (11.5), the seven KLT- and the seven SBR-equations, furthermore the 18 FY or GS equations. It should be mentioned that in the three-body case all these numbers reduce to 3. As we have seen it is not the number of potentials but the number of two-fragment channels which determines the number of basic equations. We regard, therefore, 7 or, in a more detailed description, 18 as the characteristic numbers of the four-body problem. Whether one considers 7 or 18 equations as the real generalization of the three Faddeev equations is a question which has no unique answer. As long as the subsystem information shall be taken into account most explicitly, the 18 equations are best suited (see next section). For more drastic approximations or for considerations of nuclear reaction theory, too detailed an information may be very inconvenient. Here the seven, or in general the $n_2(N)$ equations, are presumably preferable (see for example refs. 28 and 29).

Concluding remarks: To study 7 or 18 equations in the four-body case looked very natural in view of the fact that in this way the original information is most easily translated into the language of matrix equations. Inserting these equations into each other the number of coupled equations may be reduced, in the KLT case even to one. In this way, however, very complicated kernels are introduced. We should emphasize that it is one of the most characteristic features of recent few-body theory that the number of coupled equations is chosen sufficiently large in order to get a simple form of the potential matrix. In this sense 7 or 18 equations turn out to be optimal choices in the four-body case.

We have described those dynamical equations which can be understood as generalizations of the original set of defining LS equations. The resulting relations are physically easily interpretable. They moreover represent up to now the formulations most successfully applied. Similar results, derived however in a technically different way, e.g., via second quantization[30], or equations which almost exclusively emphasize the aspect of connectivity could not be included in our review without changing its general concept. Here we once more refer to other reports[11]. The list of references had to be restricted to articles of special relevance for our approach or to those publications where further references are found.

12. POLE APPROXIMATIONS

In the three-body case several efficient methods exist for solving the Faddeev equations. Most of them start from appropriate pole approximations. Improvement is, e.g., achieved by taking into account additional pole terms (pole expansion) or by handling the rest terms perturbatively (quasi particle method). These approaches are well known and have been summarized in various publications. The high number of variables makes the application of pole approximations even more essential for $N \geq 4$.

With respect to this general case we briefly recall the three-body procedure. Inserting in the AGS equations (6.10) pole terms of the (separable) form

$$T_\gamma \sim |g_\gamma > t_\gamma < g_\gamma| \quad , \tag{12.1}$$

and sandwiching (6.10) by the two-body form factors $<g_\beta|$ and $|g_\alpha>$ (multiplied by G_o) we arrive at effective two-body LS-equations[9]

$$\tilde{T} = \tilde{V} + \tilde{V}\tilde{G}_o\tilde{T} \quad . \tag{12.2}$$

Here the "transition operators", "potentials" and "free Green's functions" are

$$\tilde{T}_{\beta\alpha} = <g_\beta|\, G_o U_{\beta\alpha} G_o\, |g_\alpha> \; , \; \tilde{V}_{\beta\alpha} = \bar{\delta}_{\beta\alpha} <g_\beta|\, G_o |g_\alpha> \; , \; \tilde{G}_{o,\beta\alpha} = \delta_{\beta\alpha} t_\alpha \quad . \tag{12.3}$$

Our starting point (6.10) was one of the various off-shell continuations of the equations in secs 4 and 5. Four-body off-shell relations follow in the same way from the half-on-shell equations of secs 8 and 10. In particular we have the four-body AGS equations[22]

$$U^{\sigma\rho}_{\beta\alpha} = \bar{\delta}_{\sigma\rho}\, \delta_{\beta\alpha}\, G_o^{-1} T_\alpha^{-1} G_o^{-1} + \sum_\tau \sum_\gamma \bar{\delta}_{\sigma\tau}\, U^\tau_{\beta\gamma}\, G_o T_\gamma G_o U^{\tau\rho}_{\gamma\alpha} \quad . \tag{12.4}$$

The explicit occurrence of the two- and three-body transition operators T_γ and $U^\tau_{\beta\gamma}$ in the kernel makes them very appropriate for pole approximations (note that an index τ is necessary to label in which subsystem $U^\tau_{\beta\alpha}$ acts). Inserting (12.1) we reduce (12.4) in a first step to effective three-body equations

$$\tilde{U}^{\sigma\rho} = \bar{\delta}_{\sigma\rho} \tilde{G}_o^{-1} + \sum_\tau \bar{\delta}_{\sigma\tau}\, \tilde{T}^\tau\, \tilde{G}_o \tilde{U}^{\tau\rho} \quad . \tag{12.5}$$

Here the operators $\tilde{G}_o$ and $\tilde{T}^\tau$ are the ones defined in (12.2) and (12.3). The four-body transition operators, reduced in pole approximation to effective three-body operators, are evidently given by

$$\tilde{U}^{\sigma\rho}_{\beta\alpha} = \langle g_\beta |\, G_o U^{\sigma\rho}_{\beta\alpha}\, G_o\, | g_\alpha \rangle \quad . \tag{12.6}$$

It is the characteristic feature of our approach that the effective three-body equations (12.5) have the same structure as the genuine three-body AGS equations (6.10). This suggests to apply the usual three-body techniques also to (12.5). In analogy to (12.1), we approximate the amplitudes $\tilde{T}^\tau$ by pole terms

$$\tilde{T}^\tau_{\beta\alpha} = \langle g_\beta |\, G_o U^\tau_{\beta\alpha}\, G_o\, | g_\alpha \rangle \sim | \tilde{g}^\tau_\beta \rangle\, \tilde{t}_\tau\, \langle \tilde{g}^\tau_\alpha | \quad , \tag{12.7}$$

In the case $\tau = (i,jkl)$ such effective two-body poles evidently originate from three-body bound states and resonances. Now, repeating the procedure leading to (12.2) we reduce the effective three-body equations (12.5) to the effective two-body GS equations[19)]

$$\tilde{\tilde{T}} = \tilde{\tilde{V}} + \tilde{\tilde{V}}\, \tilde{\tilde{G}}_o\, \tilde{\tilde{T}} \quad . \tag{12.8}$$

The occurring operators are defined in strict analogy to (12.3) with $U_{\beta\alpha}$, G_o, t_α and $|g_\alpha\rangle$ replaced by the corresponding expressions in (12.5) and (12.7).

In other words, successive introduction of two- and three-body poles reduces the original four-body equations to effective two-body LS equations, a result first derived in ref. 19. The kernels of (12.4) and of the FY four-body equations are equivalent. It has been pointed out previously[22)] and was discussed also after eq. (10.7) that application of the above successive pole approximation scheme to the FY theory will only reproduce the GS result (12.8). Inspection of recent attempts[32,33)] demonstrates this also in detail.
Technically the pole approximations (12.1) are usually performed by expanding the two-body potentials V_γ into series of separable terms (Hilbert-Schmidt expansion, Bateman method). This suggests to expand also the effective potentials given in (12.3) in an analogous way, thus providing reliable pole terms (12.7).

First numerical solutions of (12.8) based on two-body Yamaguchi potentials have been performed by Alt, Grassberger and Sandhas[31)]. The discrepancy between the calculated ^{4}He binding energy of 50 MeV and the experimental value of 28,34 MeV is not unplausible in view of the overbinding of the triton obtained in the Yamaguchi case. This result is furthermore confirmed by the independent calculations of refs 32 and 33, listed in the following table. Sawicki and Namyslowski also started from the Yamaguchi potential. Lowering, however, the triton binding energy by their approximation technique, they reduced the ^{4}He value in a related way.[34)]

More reliable calculations have been performed by Tjon[35)], using the Malfliet-Tjon potential, and by Becker and Sandhas[36)], using separable potentials with Gauss form factors. The resulting agreement in the ^{4}He binding energy was not unexpected, since almost identical triton binding energies have been found by Bakker[37)] for these interactions.
In view of the fact that the Coulomb energy has not yet been taken into account, the agreement with the experimental value is also rather satisfactory.

Authors	^{4}He-energy(MeV)	^{3}H-energy(MeV)	2-body pot.
Alt, Grassberger, Sandhas[31]	50	10,3	
Kharchenko, Kuzmichev[32]	52,03	12,05	Yamaguchi
Narodetsky, Galpern, Lyakhovitsky[33]	45,73	11,03	
Sawicki, Namyslowski[34]	24,8	7,36	
Tjon[35]	29,6	8,56	Malfliet-Tjon
Becker, Sandhas[36]	31,5	8,7	Gauss (sep.)

First calculations of cross sections have also been given in ref. 31. There, instead of solving (12.8) exactly, a rather crude K-matrix approximation was used. Becker and Sandhas[36] recently found some improvement if in these calculations Gauss form factors are employed. Much better results are obtained by solving (12.8) in second order K-matrix approximation. Sawicki and Namyslowski[34] got in this way fairly good agreement with experiment for Yamaguchi form factors (Fig. 3). For higher energies the agreement is quite reasonable already in first order K-matrix approximation[36] (Fig. 4).

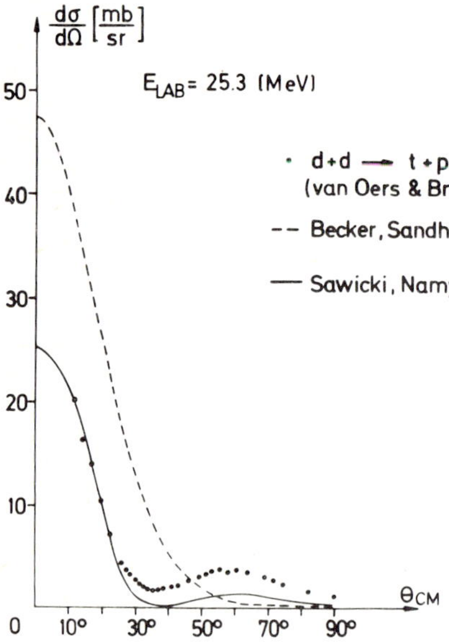

Fig. 3

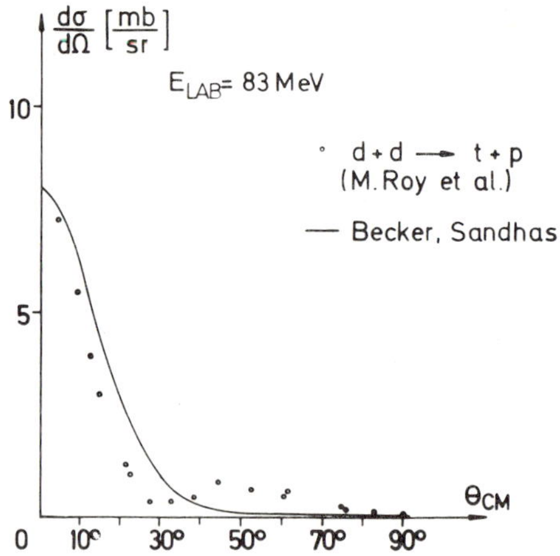

Fig. 4

These encouraging results show that the methods described are well suited for further refined investigations. Heavier nuclei may also be studied with such techniques by means of more drastic cluster approximations[38]. Moreover, it has been proposed by Mitra[39] to start from the very beginning from effective two-body equations of the form (12.8). Semiphenomenological vertex functions of the effective potentials have to be introduced in this approach. The detailed theory described supports this proposal, providing furthermore hints on reasonable choices of the vertex functions[40].

REFERENCES

1. B.A. Lippmann, Phys. Rev. 102 (1956) 264.
2. L.D. Faddeev, Zh. Eksperim. i Teor. Fiz. 39(1960)1459 (JETP 12 (1961) 1014).
3. W. Glöckle, Nucl. Phys. A141 (1970) 620.
4. D.J. Kouri and F.S. Levin, Phys. Lett. 50B (1974) 421.
5. W. Tobocman, Phys. Rev. C9 (1974) 2466.
6. D.J. Kouri and F.S.Levin, in Few Body Problems in Nuclear and Particle Physics, ed.: R.J. Slobodrian et al. (Quebec, 1975).
7. D.J. Kouri, F.S. Levin and W. Sandhas, to be publ. in Phys. Rev. C.
8. K.M. Watson, Phys. Rev. 89 (1953) 575; G.V. Skornyakov and K.A. Ter-Martirosyan, JETP 31 (1956) 775; A.N. Mitra, Nucl. Phys. 32 (1962) 529;R.D.Amado, Phys. Rev. 132 (1963) 141; A.G. Sitenko and V.F. Kharchenko,Nucl.Phys. 49 (1963) 15.
9. C. Lovelace, Phys. Rev. 135 (1964) B1225.
10. L.D. Faddeev, Mathematical Aspects of the Three-Body Problem in Quantum Scattering Theory (Leningrad, 1963)(Engl. transl.: Jerusalem, 1965).
11. R.D. Amado, in Elementary Particle Physics and Scattering Theory, Vol. 2, ed.: M. Chrétien and S.S. Schweber (New York, 1970).
12. H.P. Noyes, Phys. Rev. Lett. 23 (1969) 1201;D.D.Brayshaw,Phys.Rev.D8 (1973) 952; C.Gignoux and A.Laverne,Phys.Rev.Lett. 29 (1972) 436; C.Gignoux,A.Laverne and S.P.Merkuriev,Phys.Rev.Lett. 33 (1974) 1350;W.Glöckle,Z.Physik 271 (1974) 31.
13. B.R. Karlsson and E.M. Zeiger, Phys.Rev. D11 (1975) 939.
14. W.Sandhas, in Elementary Particle Physics, ed.: P.Urban (Acta Physica Austriaca, Suppl. IX (1972) 57).
15. E.O. Alt, P. Grassberger and W.Sandhas, Nucl. Phys. B2 (1967) 167.
16. B.R. Karlsson and E.M.Zeiger, Phys. Rev. D9 (1974) 1761;D10 (1974) 1291.
17. E.F. Redish, Nucl. Phys. A225 (1974) 16.
18. W.Sandhas,talk presented at the Symposium on Few Particle Problems in Nuclear Physics, Tübingen 1975; and to be publ.
19. P. Grassberger and W.Sandhas, Nucl. Phys. B2 (1967) 181.
20. O.A. Yakubovsky, Soviet J.Nucl. Phys. 5 (1967) 937;K.Hepp,Helv.Phys.Acta 42 (1969) 425.
21. L.D.Faddeev,in Three Body Problem in Nuclear and Particle Physics, ed.: J.S.C. McKee and P.M. Rolph (North-Holland, Amsterdam, 1970).
22. E.O. Alt, P.Grassberger and W.Sandhas, JINR Report E4-6688 (1972); and in Few Particle Problems in the Nuclear Interaction, ed.: I.Slaus et al.(North-Holland, Amsterdam, 1972); W.Sandhas in Progress in Particle Physics, ed.:P. Urban(Acta Physica Austriaca, Suppl.XIII (1974) 679);Czech.J.Phys.B25 (1975) 251.
23. R.Omnes, Phys. Rev. 165 (1968) 1265.
24. L. Rosenberg, Phys. Rev. 140 (1965) B217;A.N.Mitra,J.Gillespie,R.Sugar and N.Panchapakesan, Phys.Rev. 140 (1965) B1336;Y.Takahashi and N.Mishima, Progr. Theor. Phys. 34 (1965) 498;35 (1966) 440; V.A.Alessandrini,J.Math.Phys. 7 (1966) 215.
25. I.M. Narodetsky and O.A. Yakubovsky, Soviet J.Nucl.Phys. 14 (1972) 178.
26. I.H. Sloan, Phys. Rev. C6 (1972) 1945.
27. G.Bencze, Nucl. Phys. A210 (1973) 568;and contrib.to this conference.
28. E.F.Redish,Phys. Rev. C10 (1974) 67; Nucl. Phys. A235 (1974) 82.
29. W.Tobocman, Phys. Rev. C10 (1974) 61.
30. V.V. Komarov and A.M. Popova, Phys. Lett. 28B (1969) 476.
31. E.O. Alt, P.Grassberger and W.Sandhas, Phys. Rev. C1 (1970) 85.
32. V.F. Kharchenko and V.E. Kuzmichev, Phys. Lett. 42B (1972) 328.
33. I.M. Narodetsky, E.S. Galpern and V.N. Lyakhovitsky,Phys.Lett. 46B (1973) 51.
34. M. Sawicki and J.M. Namyslowski, preprint; and contrib. to this conference.
35. J.A. Tjon, Phys. Lett. 56B (1975) 217.
36. K. Becker and W. Sandhas, to be publ.
37. B.L.G. Bakker, thesis.
38. Y. Avishai, Phys. Rev. C6 (1972) 677.
39. A.N. Mitra, preprint; A.N. Mitra and V.K. Sharma, contr. to this conference.
40. H. Haberzettl and W. Sandhas,to be publ.

DYNAMICAL EQUATIONS AND APPROXIMATION METHODS†

Y.E. Kim
Department of Physics, Purdue University
West Lafayette, Indiana 47907, U.S.A.

Two major themes of Session I were, as implied by the title of the session and this talk, (1) non-relativistic dynamical equations for N(≥ 4) particle systems and (2) approximation methods for N(≤ 4) particle systems. Since the 1974 Quebec conference, a substantial number of new results on these topics have been reported in the literature and at this conference. Brief reviews of these topics and highlights of the reported results will be given below.

Dynamical Equations (N ≥ 4)

Since the pioneering work of Faddeev [1] on non-relativistic three-particle scattering theory, there have been many proposals to generalize the Faddeev-type equations to N-particle systems. Using a variety of different techniques for curing the problems of disconnected kernels, many authors have written down different sets of N-particle connected equations. Because of complexities of the N(≥ 4)-body problems due to the large number of degrees of freedom, there are, in principle, an infinite number of different ways to write the N-body equations, all of which have different structures and desired properties.

At this conference, interrelationships and even the equivalence of many different proposed N-particle equations were discussed and brought into focus. This is not surprising since the equivalence of these different formulations of the N-body problem is anticipated in the sense that the underlying Schroedinger equation is the same in all cases. One purpose for studying different approaches is to obtain practical approximation schemes for different physical processes. Approximate solutions are needed since the exact solutions of the integral equations for N(≥ 4)-particle systems are not feasible even with the presently available computing technologies.

In 1964, Weinberg [2] and von Winter [3] derived independently an N-particle dynamical equation as a sequence of linear integral equations for successively larger systems. Although the Weinberg-Winter (WW) equation is a single equation with connected kernels, the WW equation suffers (as pointed out by Federbush [4]) from the "Federbush disease" (existence of spurious solutions of the homogeneous integral equations which do not correspond to bound states of the underlying Schroedinger equation).

In 1965 Weyers (W) [5] obtained a set of integral equations for the four particle system containing six independent amplitudes (one for each particle pair). He subsequently reduced his equations to a set of coupled equations with only four independent amplitudes and later generalized his results to the case of N(≥ 5)-

†Rapporteur's talk given at the International Conference on Few Body Problems in Nuclear and Particle Physics, University of Delhi, Delhi, India, December 29, 1975 - January 4, 1976.

body systems [6]. In the same year of 1965, Rosenberg (R)[7], Mitra el al.(MGSP) [8], Takahashi and Mishima (TM)[9] and Alessandrini (A)[10] independently derived four-particle equations as sets of six coupled equations in which the kernels are the connected parts of the amplitudes for three-particle subsystems and also for the subsystems consisting of two non-interacting pairs. In contrast to the earlier equations of WW and W, the potentials do not appear explicitly in these equations.

The first integral equations for N(≥ 4)-particle systems, which were shown explicitly to be free of the Federbush disease, were derived in 1967 by Yakubovskii [11] who generalized the Faddeev integral equations to the case of an arbitrary number of particles. The kernels of the Faddeev-Yakubovskii (FY) equations are connected after (N - 2) iterations. The coupling (i.e. the number of coupled equations) is maximal for the FY equations. (For example, with N = 4, the FY equation is a set of 18 coupled integral equations [12].)

More recently, Alt, Grassberger and Sandhas (AGS) [13,14] have developed a formulation of the N-particle equations in terms of matrices of transition operators. The AGS N-particle equations were later shown to be equivalent to the FY equations by Karlsson and Zeiger [15]. Hence, the AGS equations are free of the Federbush disease and have the same maximal coupling as do the FY equations.

Another interesting approach is that of the minimal coupling scheme which was introduced by Sloan (SL) [16] for the four-particle system and was later generalized by Bencze [17] and Redish [18] (BR) to the case of arbitrary number of particles. For an N particle system, there are $2^{N-1} - 1$ coupled BR equations (one for each possible binary partition) which corresponds to minimal coupling. The BR equations become connected after a single iteration.

A more recent proposal is the channel coupling array method introduced by Kouri and Levin [19] and Tobocman [20] (KLT). Most recently, Vanzani has proposed yet another formulation [21] in which he uses appropriate splittings of the channel external interactions.

In Session I of this conference, Redish [22] summarized the properties of the BR equations, emphasizing the context in which the BR equations can provide starting points of approximation schemes for nuclear reaction theories. Bencze [22] expounded the BR formulation in a broader context by comparing it with other formulations. In particular, he indicated that the BR equations are equivalent to the KLT equation with a particular choice of the KLT coupling channel array. He also discussed a proof which showed that the BR equations are free of the Federbush disease. Levin reminded us, in his remark, that the KLT equations are free of the Federbush disease [24].

In one of the contributed papers, Sasakawa (SA) [25] described a new set of seven coupled equations (minimal coupling) for the wave function of a four particle system which are free of the Federbush disease. In addition, he shows the equivalence of the R, MGSP, TM, A, and SL equations to the FY equations, which suggests that all of the above four-particle equations are also free of the Federbush disease. Therefore, it now appears that most of the proposed N-particle equations discussed so far are free of the Federbush disease with the one definite exception being the WW equation. This is one example in which the efforts of studying different approaches have paid off in practical terms. We can now use the minimally coupled equations instead of the maximally coupled equations of FY or AGS in solving the bound-state problems for $N(\geq 4)$ particle systems. Up to now, the bound-state [26 - 31] and scattering [13, 32 - 35] problems for the four-particle system has been studied with either the F-Y or AGS equations. The minimally coupled equations should be preferably used in the future.

Karlsson [36] presented work which he did in collaboration with Zeiger. Their earlier three-particle equations [37] are generalized to the case of four particle systems using the FY equation. Their modified FY equations are constructed in such a way that only the half-off-shell two-and three-body inputs are required. In response to the remarks, made by Sandhas and Tandy, that the minimally coupled equations should be preferably used, Karlsson stated that their attempts to use the minimal equations had not been successful so far except for the case of the Sloan equation.

In a paper presented by Chandler, Chandler and Gibson [38] showed that their two-Hilbert-space formulation of time-independent multichannel scattering [39] can provide rigorous proofs of various N-particle integral equations such as those of AGS and others. In a response to a concern expressed by Levin, Chandler commented that the lack of rigorous interpretations for the unitarity of the scattering operator as described in their original paper [39] had not been resolved.

As is done in the three-particle case, it is sometimes useful to formulate the Faddeev-type equations for (real) K-matrix elements [40 - 43]. V.K.Sharma [44] discussed a set of Faddeev-type K-matrix equations for the four-particle system, which he derived from the N-body operators of Karlsson and Zeiger [15].

Coulomb Effects in the Three-Particle System

Only a few advances have been made in obtaining the exact solutions of the Coulomb effects in the three-particle system consisting of two charged particles plus a neutral particle or of three charged particles. In particular, the problem of three charged particles is not solved completely at present [45].

Except for the cases of three-body bound states and scattering from an electrically neutral bound state below the break-up threshold, the Faddeev equations must be modified in the presence of infinite-range Coulomb forces. In 1967, Noble [46] gave an approximate treatment in which the Coulomb force is cut off outside the range of nuclear interaction (screened Coulomb potential) and the free three-particle Green's function is replaced by the three-particle (screened) Coulomb Green's function. Although the use of a limiting procedure for screened Coulomb forces may be justified in practice (it is rigorously justified for the case of two-particle systems [47,48]), Noble's approach requires additional approximations for the three-particle Coulomb Green's function which is not known at the present time.

An exact procedure has been given by Veselova [49] in 1969 for correctly incorporating Coulomb forces in calculations of scattering from a charged bound state below the break-up threshold. At the 1969 Birmingham conference, Faddeev [12] suggested, with some optimism, that one promising approach may be the use of Dollard's [50] time-dependent formulation for the N charged-particle system. There have been some partially successful efforts in this direction [51,52].

At this conference, there is listed an abstract, submitted by Merkuriev [53], in which he claims to have solved the three charged-particle problem in configuration space. Unfortunately, the details of his work were not available at the conference. Since the numerical technologies for solving the three-particle scattering problem in the configuration space are now available [54,55], Merkuriev's result could provide a practical solution of the general three-body problem with Coulomb forces.

For the case of proton-deuteron scattering, Alt [56] reported a new formulation which is based on the quasi-particle approach of Alt, Grassberger and Sandhas [57] and on a screening technique similar to that of Veselova [49]. The strong nucleon-nucleon interaction is assumed to be separable and Coulomb interaction is treated exactly. Alt's integral equations are valid for all energies (i.e. below and above the break-up threshold). His calculated proton-deuteron quartet elastic scattering phase shifts at $E_{\ell ab}$ = 1,3, and 5 MeV are in excellent agreement with the experimental data of Arvieux [58]. This work is one of the first calculations in which the Coulomb effects are treated exactly. (There is another similar unpublished calculation by Hackenbroich et al. [59]). A formulation of the proton-deuteron scattering problem similar to that of Alt has been independently developed by Kharchenko and Storozhenko [60]. There is another formulation in the configuration space reported by Sasakawa and Sawada [55] at this conference.

Approximation Methods ($N \le 4$)

For $N(\ge 4)$ particle systems, some approximations are required to bring various dynamical equations discussed previously into numerically manageable forms for practical calculations. In Session I, two different approximation schemes were discussed, namely the separable expansion and the resonating group method.

One of the most widely used approximation methods for the three-particle problems is the separable expansion of the two-body t-matrix. There are a variety of different separable expansions in use, such as the Sturmian (or Weinberg) expansion, the unitary pole approximation (UPA), the unitary pole expansion, (UPE) and the separable expansion of Ernst, Shakin, Thaler and Weiss (ESTW) [61]. Recently, Adhikari and Sloan [62] have introduced a generalized separable expansion, of which the above various separable expansions are the special cases.

Oryu [63] proposed another generalized separable expansion. His method is based on the Noyes-Kowalski model [64] and the Bateman method [65]. It has the attractive feature that the unphysical poles, which appear in both the Noyes-Kowalski model and in the Bateman method, can be made to vanish simultaneously by an appropriate choice of a set of Bateman parameters. Furthermore, the expansion is applicable to all energies in contrast to the situation in previous applications of Bateman method [66,67]. Oryu [68] showed in his talk the rapid convergence of his separable expansion for a variety of different two-particle potentials (an important practical fact). He also discussed a generalization of his method to the case of the three-particle scattering amplitude [69]. This should be very useful in solving four-particle equations.

V.K. Sharma [70] reported his work on an approximation method for four-particle system. Using Wheeler's resonating group method, he was able to reduce the FY equation for a four-particle system to a set of coupled equations in one variable which is substantially simpler and more practical than the original FY equations. The kernels of his equations contain two- and three-nucleon bound and scattering states.

Another interesting study of the resonating group method was reported by J. Schwager [71]. He discussed what approximations and assumptions are needed to relate the integral equation approach to the resonating group equations, and he clarified his discussions by using the three-particle Faddeev integral equations.

Other Topics

Warke [72] discussed several physical properties of the two-body system, which are invariant under the phase-shift equivalent transformation. He showed that the Fredholm determinant and several other quantities are invariant in addition to the known invariants such as scattering phase shifts. The generalization of his work to the case of a three-particle system would be useful but very difficult. The only invariant quantity known for the three-particle case is, as Warke pointed out, the low energy elastic scattering amplitude for three free particles. This has been studied previously by Amado and Rubin [73], and by Adhikari [74].

McKellar presented work, which he did in collaboration with McKay and Kim [75]. The three-particle integral equations of Faddeev [1] and of Karlsson and Zeiger

[37] are generalized to include the case when the two-body potential is energy-dependent and is defined only for the positive energies (and for bound-state negative energies) as in the case of the field theoretic two-nucleon potentials developed by the Stonybrook group [76] and the Paris-Orsay group [77]. McKellar emphasized the neccessity of introducing the biorthogonal sets of two-body scattering states. The formal scattering theories for the case of effective energy-dependent (hence non-Hermitian) potentials have not been studied in detail until recently. McKellar and McKay [78] introduced the use of the biorthogonal bases for the two-particle continuum states. Biorthogonal bases have been used previously for the discrete bound-state spectrum [79]. Kim and Tubis [80] gave a perturbative proof of the completeness relation for the biorthogonal set. More work is required in formulating mathematically rigorous formal theories of scattering problem involving non-Hermitian potentials.

Unfortunately, the method of generalization reported by McKellar et al. is not unique. One way of avoiding this difficulty is to go back to the original formulation of three-body problem in which two-body interactions involved are Hermitian. As is well known, the effective energy-dependent interactions arise from the procedure in which some of the coupled channels involved are eliminated [81, 82]. For the three-nucleon system, it corresponds to eliminating the meson degrees of freedom to obtain the effective two- and three-body interactions which are dependent on the total energy of the three-nucleon system. It is reasonable to expect that a unique generalization can be found by investigating the three-particle dynamical equations in this context.

Before finishing this rapporteur's talk, I would like to bring to your attention the new recent experimental results of the Harvard group [83] for the doublet and quartet neutron-deuteron scattering lengths. These new results are in significant disagreement with the latest liquid mirror measurement of Dilg et al. [84]. Another accurate independent determination of these scattering lengths is needed to resolve these experimental inconsistencies.

REFERENCES

1. L.D. Faddeev, Zh. Eksp. Teor. Fiz. 39 (1960) 1459 [Engl. transl.: Sov. Phys. JETP 12 (1961) 1014].
2. S. Weinberg, Phys. Rev. 113 (1964) B232.
3. C. van Winter, Kgl. Danske Videnskab. Selskab, Mat.-Fys. Skrifter 2 (1964) Nr. 8.
4. P. Federbush, Phys. Rev. 148 (1966) 1551.
5. J. Weyers, Ann. Soc. Sci. Bruxelles Ser. I 79 (1965) 76.
6. J. Weyers, Phys. Rev. 145 (1966) 1236; ibid. 151 (1966) 1159.
7. L. Rosenberg, Phys. Rev. 140 (1965) B217.
8. A.N. Mitra, J. Gillespie, R. Sugar, and N. Panchapakesan, Phys. Rev. 140, (1965) B1336.
9. Y. Takahashi and N. Mishima, Progr. Theoret. Phys. (Kyoto) 34 (1965) 498; N. Mishima and Y. Takahashi, ibid. 35 (1966) 440.
10. V.A. Alessandrini, J. Math. Phys. 7 (1966) 215.
11. O.A. Yakubovskii, Yadern. Fiz. 5 (1967) 1312 [English transl.: Soviet J.

Nucl. Phys. 5 (1967) 937].

12. L.D. Faddeev, in Three Body Problem in Nuclear and Particle Physics, edited by J.S.C. McKee and P.M. Rolph (North-Holland, Amsterdam, 1970), p. 154.
13. E.O. Alt, P. Grassberger and W. Sandhas, JINR, E4-6688 (Dubna, 1972); in Few Particle Problems in the Nuclear Interaction, edited by I. Slaus et al. (North Holland, Amsterdam, 1972), p. 299.
14. W. Sandhas, Acta Phys. Austr. Suppl. 13 (1974) 679; Czech. J. Phys. B25 (1975) 251.
15. B. Karlsson and E.M. Zeiger, Phys. Rev. D 9 (1974) 1761; ibid. D10 (1974) 1291.
16. I. Sloan, Phys. Rev. C6 (1972) 1945.
17. Gy. Bencze, Nucl. Phys. A210 (1973) 568.
18. E.F. Redish, Nucl. Phys. A225 (1974) 16; ibid. A235 (1974) 82; Phys. Rev. C10 (1974) 67.
19. D.J. Kouri and F.S. Levin, Phys. Lett. 50B (1974) 421; Phys. Rev. A10 (1974) 1616.
20. W. Tobocman, Phys. Rev. C9 (1974) 2466; ibid. C12 (1975) 741 and 1146.
21. V. Vanzani, "A Direct Derivation of N-Body Scattering Equations", IFPD Preprint 17/75 (Padova, 1975).
22. E.F. Redish, "Generalized Faddeev Theory of Nuclear Reactions".*
23. Gy. Bencze, "Dynamical Equations for the N-Particle Transition Operators".*
24. Y. Hahn, D.J. Kouri and F.S. Levin, Phys. Rev. C10 (1974) 1620.
25. T. Sasakawa, "Four-Body Equation".*
26. V.F. Kharchenko and V.E. Kuznuchev, Nucl. Phys. A183 (1972) 606; ibid. A196 (1972) 636; Phys. Lett. 42B (1972) 328; Yad. Fiz. 17 (1973) 975; Czech J. Phys. B24 (1974) 1071.
27. V.F. Kharchenko, in Few Particle Problems in the Nuclear Interactions edited by I. Slaus et al. (North-Holland, Amsterdam, 1972), p. 663.
28. V.F. Kharchenko, V.E. Kuzmichev, and S.A. Shadchin, Nucl. Phys. A226 (1974) 71.
29. I.M. Narodetsky, E.S. Galpern and V.N. Lyakhovitsky, Phys. Lett. 46B (1973) 51.
30. I.M. Narodetsky, Nucl. Phys. A221 (1974) 191.
31. J.A. Tjon, Phys. Lett. 56B (1975) 217.
32. E.O. Alt, P. Grassberger and W. Sandhas, Phys. Rev. C1 (1970) 85.
33. W. Sanhas, an invited talk at this conference.*
34. V.F. Kharchenko and V.P. Levashev, Preprint ITP-75-107E (Kiev, 1975).
35. V.P. Levashev, Preprint ITP-75-22E (Kiev, 1975).
36. Bengt R. Karlsson, "Three- and Four-Body Equations with Half-off-Shell Input".*
37. B.R. Karlsson and E.M. Zeiger, Phys. Rev. D11 (1975) 939.
38. C. Chandler and A.G. Gibson, "On a Time-Independent Theory of Multichannel Quantum Mechanical Scattering".*
39. C. Chandler and A.G. Gibson, J. Math. Phys. 14 (1973) 1328.
40. I.H. Sloan, Phys. Rev. 165 (1968) 1587.
41. N. Mishima and M. Yamazaki, Progr. Theor. Phys. 34 (1965) 284.
42. K.L. Kowalski, Phys. Rev. D5 (1972) 395; ibid. D6 (1972) 3705; ibid. D7 (1973) 1806.
43. T. Sasakawa, Nucl. Phys. A186 (1972) 417; ibid. A203 (1973) 496.
44. V.K. Sharma, "K-Matrix Formalism for the Four-Body Problem and Bound State Scattering Theory".*
45. P.U. Sauer, an invited talk at this conference.*
46. J.V. Noble, Phys. Rev. 161 (1967) 945.
47. E. Prugovecki and J. Zorbas, Nucl. Phys. 213A (1973) 541.
48. J.R. Taylor, Nuov. Cim. B23 (1974) 313.
49. A.M. Veselova, Teor. Mat. Fiz. 3 (1970) 326; Preprint ITP-73-106P (Kiev,1973).
50. I.D. Dolllard, J. Math. Phys. 5 (1964) 729.
51. A.M. Veselova, Teor. Mat. Fiz. 13 (1972) 368.
52. A.G. Gibson and C. Chandler, J. Math. Phys. 15 (1974) 1366.
53. S.P. Merkuriev, "On the Three Charged Particle Scattering Theory", an abstract (S-V-3) submitted to this conference.
54. G. Gignoux, A. Laverne, and S.P. Merkuriev, Phys. Rev. Lett. 33 (1974) 1350.

55. T. Sasakawa and T. Sawada, contributed papers (S-II-7 and S-II-8) at this conference.*
56. E.O. Alt, "Local Strong and Coulomb Potentials in the Three-Nucleon System".*
57. E.O. Alt, P. Grassberger and W. Sandhas, Nucl. Phys. B2 (1967) 167; E.O. Alt and W. Sandhas, in Few-Body Problems in Nuclear and Particle Physics, edited by R.J. Slobodrian et al. (Les Presses de l'Universite Laval, Quebec, 1975).
58. J. Arvieux, Nucl. Phys. A221 (1974) 253.
59. H.H. Hackenbroich et al., quoted in reference [45].
60. V.F. Kharchenko and S.A. Storozhenko, "Integral Equations for Three-Nucleon Problem with Coulomb Interaction. Proton-Deuteron Scattering", Preprint ITP-75-53E (Kiev, 1975).
61. D.J. Ernst, C.M. Shakin, R.M. Thaler, and D.L. Weiss, Phys. Rev. C8 (1973) 2056; D.J. Ernst, C.M. Shakin, and R.M. Thaler, ibid. 8 (1973) 46.
62. S.K. Adhikari, Phys. Rev. C10 (1974) 1623; S.K. Adhikari and I.H. Sloan, Nucl. Phys. A241 (1975) 429.
63. S. Oryu, Progr. Theor. Phys. 52 (1974) 550.
64. H.P. Noyes, Phys. Rev. Lett. 15 (1965) 538; K. L. Kowalski, ibid. 15 (1965) 788.
65. H. Bateman, Proc. Roy. Soc. A100 (1921) 441; Messenger Math. 37 (1908) 179.
66. B. Akhmadkhodzaev, V.B. Belyaev, J. Wrzecionko and A.L. Zubarev, JINR Preprint E4-5763 (Dubna, 1971).
67. V.F. Kharchenko, S.A. Storozhenko, and V.E. Kuzmichev, Nucl. Phys. A188 (1972) 609.
68. S. Oryu and T. Ichihara, "Generalized Seperable Expansion Method of the Two-Body and the Three-Body Scattering Amplitudes".*
69. S. Oryu, "Theory of the Three-Body Separable Expansion Amplitude".*
70. V.K. Sharma, "K-Matrix Formalism for the Four-Body Problem and Bound State Scattering Theory".*
71. J. Schwager, "How Different is the Resonating-Group Method from the Integral-Equation Approach to Few-Particle Scattering?"*
72. C.S. Warke, "Physical Properties Invariant for Phase Equivalent Potentials".*
73. R.D. Amado and M.H. Rubin, Phys. Rev. Lett. 25 (1970) 194.
74. S.K. Adhikari, Phys. Rev. D8 (1973) 1195.
75. Y.E. Kim, C. McKay and B.H.J. McKellar, "The Three-Body Problem with Energy-Dependent Potentials".*
76. A.D. Jackson, D.O. Riska and W. Verwest, Nucl. Phys. A249 (1975) 397.
77. M. Lacome, B. Louiseau, J.M. Richard, R. Vinh Mau, and R. de Toureill, Orsay preprint (1975).
78. C. McKay and B.H.J. McKellar, "Scattering Theory for Energy Dependent Potentials", to be published.
79. See B.H. Brandow in the Proceedings of International Conference on Effective Interactions and Operators in Nuclei, University of Arizona, Tucson, Arizona, June 2-6, 1975. To be published in the Springer-Verlag "Lecture Notes in Physics" series.
80. Y.E. Kim and A. Tubis, "Completeness Relations in Scattering Theory for Non-Hermitian Potentials".*
81. P.-O Löwdin, J. Math. Phys. 3 (1962) 969, and references therein.
82. H. Feshbach, Ann. Phys. (N.Y.) 19 (1962) 287.
83. J. Callerame, D.J. Larson, S.J. Lipson, and R. Wilson, Phys. Rev. C12 (1975) 1428.
84. W. Dilg, L. Koester, and W. Nistler, Phys. Lett. 36B (1971) 208.

*See elsewhere in the proceedings of this conference.

DISCUSSION

Levin : Kouri-Levin equation is more general than that discussed. By choosing different channel coupling arrays one can get Faddeev equations for $n = 3$ and Bencze-Redish equations for arbitrary $n > 3$.

Sandhas : I agree that introducing potential matrices is closely related to choosing channel permuting arrays. Consequently almost all things described in my talk could also be formulated using this language. The reason for emphasizing the special Kouri-Levin-Tobocman choice was that just this choice was particularly characteristic for the channel permuting array.

COMPUTATIONAL METHODS IN FEW BODY SYSTEMS

J.A. Tjon
Institute for Theoretical Physics, University of Utrecht
The Netherlands.

In the past decade various successful computational methods have been proposed for solving the dynamical equations of few particle systems. As a result calculations with realistic two-nucleon interactions have become feasible on present-day computers for the tri-nucleon system. In this talk we would like to summarize and discuss some of these techniques which have successfully been applied to the bound state and scattering of three nucleons. In particular, we shall discuss the integral equation approach of Faddeev in detail. The last part of this talk will be devoted to the four-nucleon system where some encouraging results have been obtained recently within the framework of the Yakubovsky-Faddeev equations.

I. Bound State of Three Nucleons

In describing the various techniques for solving few particle equations we may distinguish between the variational methods [1-6] which mostly use the framework of the Schrödinger equation and the integral equation methods with the Faddeev equation as starting point [7-12]. Both approaches have yielded in the past few years successful calculations in the bound state problem of the tri-nucleon system, but with disturbingly different results for the dip and secundary maximum of the ^{3}He charge form factor (fig. 1).

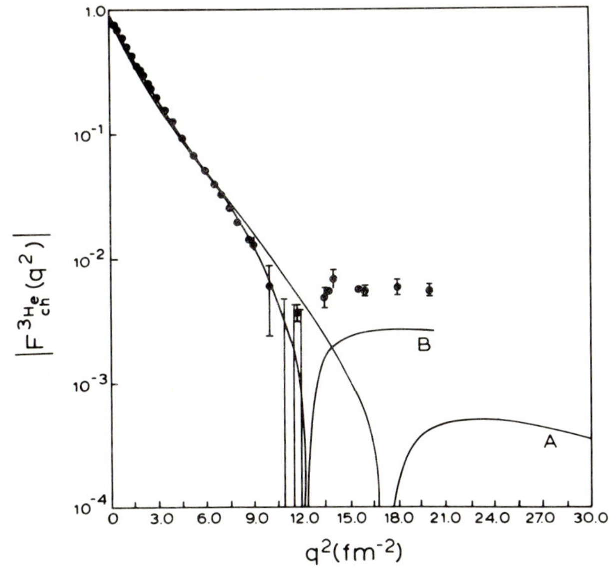

fig. 1 ^{3}He charge form factor for the complete Reid interaction. Curve A is the Faddeev result from ref. 10 and curve B the variational result from ref. 6.

In the variational calculations some basis set of states such as the harmonic oscillator or hyperspherical basis is used for the variational wave function. At this conference a study [13] was presented of the possible shape of the tri-nucleon wave functions constructed from rather different potentials with the hyperspherical method. The results suggest that as long as the binding energies of the tri-nucleon system and the charge form factors at low and intermediate momentum transfer are reproduced from a dynamical calculation, the resulting wave functions will not be very sensitive to the choice of the potential. Akaishi [14] discussed results for the triton and alpha particle for various realistic two-nucleon interactions using a new type of trial wave functions with two-nucleon correlations built in. For the Hamada-Johnston potential a binding energy was found of 6 MeV to be compared with the earlier results of 6.5 MeV from ref. 1. It would be interesting to apply this method to the case of the Reid soft core interaction.

Turning to the integral equation approach all of the calculations have been done using the Faddeev equations in the momentum representation with the exception of ref. 12. These authors used the configuration space in which the resulting equations are partial integro-differential equations [15] which are solved with the relaxation method. In the momentum representation the Faddeev equations are in general a coupled set of linear integral equations in two continuous variables. Due to the fact that the region of integration is dependent on the initial variables and the huge dimensionality of the resulting matrix after discretization of the integral equations the usual matrix inversion procedures work in general very poor or not at all. One of the methods which has been applied in the past few years with great success is related to the Padé approximant method [16], which is a effective way of summing the Neumann solution to the integral equations. Also the applicability has been explored of separable expansions of two-body T-matrices which I would like to discuss in more detail. The physical input to the Faddeev equations are usually the two-body T-matrices which satisfy a Lippmann-Schwinger equation (LS) in the case that the two-body interaction is describable with a potential. Assuming a s-wave interaction for simplicity, the LS-equation has the form

$$t(p,p';z) = V(p,p') + 4\pi \int_0^\infty \frac{p''^2 dp''}{z-p''^2} V(p,p'') \, t(p'',p';z) \tag{1}$$

If the potential V is of a separable form $V(p,p') = g(p)\, g(p')$, then it is easily seen from eq. (1) that t is also of a separable form. As a consequence, the Faddeev equations being in general integral equations in two continuous variables, reduce to integral equations in one continuous variable. It is precisely this reduction which makes it very attractive to study the possible convergence of separable expansions of two-particle interactions. The starting point is the eigenvalue problem

$$\lambda_n(z)\, \phi_n(p;z) = 4\pi \int_0^\infty \frac{p''^2 dp''}{z-p''^2} V(p,p'')\, \phi_n(p'';z) \tag{2}$$

with a given z. If the eigenvalues λ_n and eigenfunctions ϕ_n are known, then we may write for the two-body potential

$$V(p,p') = -\frac{1}{4\pi} \sum_n \lambda_n(z)\, \phi_n(p,z)\, \phi_n(p',z) \tag{3}$$

It should be noticed that for $z<o$, the eigenfunctions ϕ_n can be chosen real and that they satisfy the orthonormality condition

$$\int_0^\infty p''^2 dp''\, \phi_n(p'';z)\, \phi_m(p'';z)/(z-p''^2) = -\,\delta_{nm} \tag{4}$$

Eq. (3) can now be used as a separable expansion for the potential V. In the so-called unitary pole expansion method (UPE) [17] eq. (3) is used as the representation for V with the value of z kept fixed. The choice of z which has mostly been made is the position of the two-nucleon bound state. Using the representation (4) with a given number of terms, the two-body T-matrix can be determined for all off-shell values of the energy and it is of a separable form. The convergence rate of the UPE-method has been studied extensively by Harms and Newton [18] and more recently by Bakker and Ruig [19] for the local potentials I-III [20]. These potentials are of a Yukawa type and of the form

$$V(r) = -\lambda_A \frac{e^{-\mu_A r}}{r} + \lambda_R \frac{e^{-\mu_R r}}{r} \tag{5}$$

for both the $^1S_0$ and $^3S_1$ channels. The variations of the triton binding energy with the number of separable serms are shown in table 1.

Table 1

Convergence rate of the triton binding energy (MeV) as a function of number of terms N_t and N_s in the triplet and singlet channels (from ref. 19)

N_s \ N_t	1	2	3	5
1	8.63	8.61		8.52
2	8.51	8.50	8.48	
3			8.48	
6		8.56		

As a final result they find for these potentials 8.58 MeV binding for the triton.
For non-central interactions like the tensor force straightforward application of the UPE-method has encountered difficulties with convergence. Recently the UPE-method has been modified by Afnan and Read [21] for the case of a tensor force. It has been applied with success by them to the Reid interaction keeping only the $^1S_0$ and $^3S_1$-$^3D_1$ channels. In fig. 2 the results are shown.

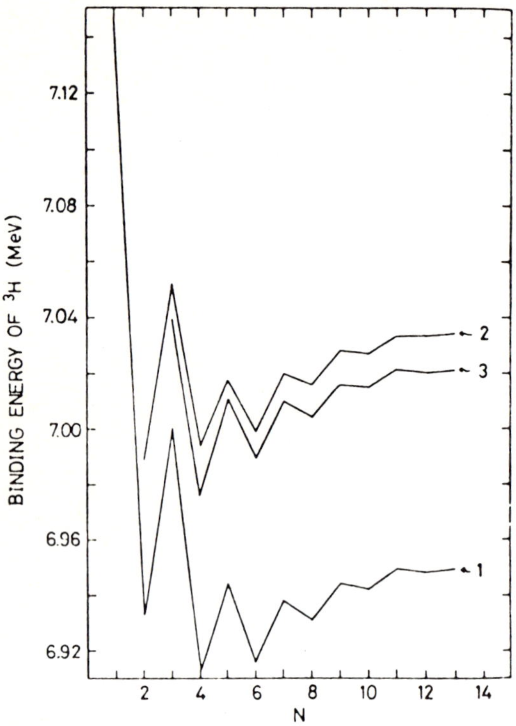

fig. 2
The triton binding energy for the Reid interaction as a function of the number of separable terms treated perturbatively. The different curves correspond to various choices of the part of the two-body T-matrix which is treated exactly in the Faddeev equation (from ref. 21).

In this calculation the Faddeev equations are solved exactly for a certain number of attractive and repulsive terms in the two-nucleon potential, while the other terms in the separable expansion are treated perturbatively. From these results the binding energy has been determined to be 7.02 MeV in reasonable agreement with other results. Since this kind of calculation can yield a more accurate result, it is interesting to extend these calculations to the higher components of the two-nucleon force and also to determine the ^{3}He form factor with this method. The complexity of the resulting integral equations need not be prohibitive since the Padé method could also be used here.

Another method which is also based on eq. (3) is the Hilbert-Schmidt method (HS). Here the value of z is not kept fixed in eq. (3) but is taken according to what is needed for the three-body equations. The solution of eq. (1) can readily be found to be

$$t(p,p';z) = \sum_n \phi_n(p;z)\ \phi_n(p';z)\ d_n(z) \tag{6}$$

where the propagator $d_n(z)$ is given by

$$d_n(z) = -\frac{1}{4\pi}\,\lambda_n(z)/(1-\lambda_n(z)) \tag{7}$$

I have recently studied the convergence rate of the expansion (6) for the potentials (5) and also for the Reid interaction with $^1S_0$ and $^3S_1$-$^3D_1$ components where the T-matrix elements with ℓ = 2 are dropped. The results are given in table 2 and 4.

Table 2

Convergence rate of the triton binding energy for the Reid interaction as a function of the number $N_\lambda^{(-)}$ of repulsive terms in the triplet channel. The number of attractive terms in the triplet channel is taken to be 2 while for the singlet channel we have $N_\lambda^{(+)} = N_\lambda^{(-)} = 2$.

$N_\lambda^{(-)}$	E_t (MeV)	$N_\lambda^{(-)}$	E_t (MeV)
2	8.1	6	7.2
3	7.7	7	7.0
4	7.2	8	6.9
5	7.3	9	6.9

From these calculations we may conclude that the convergence rate for the Reid interaction is slower than that for the simple Yukawa interactions. In particular, one needs a sizable number of repulsive terms for the Reid interaction. The final value found with this method for the Reid interaction is 6.8 MeV. We also see that in general the convergence rate is slower than in the UPE-method. This was already shown to be the case for the Hulthen potential [22]. The resulting ^{3}He charge form factor is shown in fig. 3 for the Reid interaction.

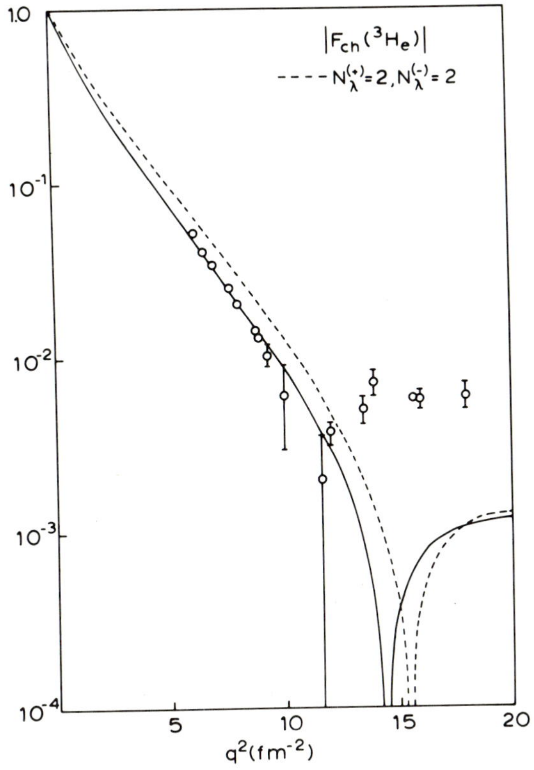

fig. 3

^{3}He charge form factor for the Reid interaction for the case that the total number of separable terms is 8. The final result (solid curve) is found with 18 terms. Experimental data are from ref. 23.

II. Scattering States of Three Nucleons

The main difficulty in the extension of bound state calculations to the continuum states using the momentum representation is due to occurrence of singularities in the path of integration. In particular, if we are above breakup threshold there is in addition to the two-body T-matrix singularities which arises from the presence of the two-body bound state also the Greens' function singularity. One way of getting rid of these singularities is a deformation of the integration path in the complex momentum plane. It has been applied with success in the case of separable interactions [24,25]. In the case that one uses the Padé method these singularities can readily be removed by introducing subtractions in the integrals [26]. Apart from the above complications the methods described in the bound state problem can be applied as well. In the configuration space approach the differential equations should be supplied with the appropriate asymptotic boundary conditions. Recently encouraging results have been reported for elastic n-d scattering [27]. At this conference a new procedure [28] has been suggested. It consists of separating out explicitly the asymptotic part of the wave functions. With respect to the solution of the Faddeev equations in the momentum representation the complete Reid interaction has been used to study the elastic n-d scattering [29]. Using the Padé technique the exact solutions of the three-body equations were constructed for the s-wave components of the two-nucleon T-matrix. Analogous to the separable model calculation of Pieper [30] the higher partial wave components of the Reid interaction were treated perturbatively. As an example the elastic n-d differential cross section

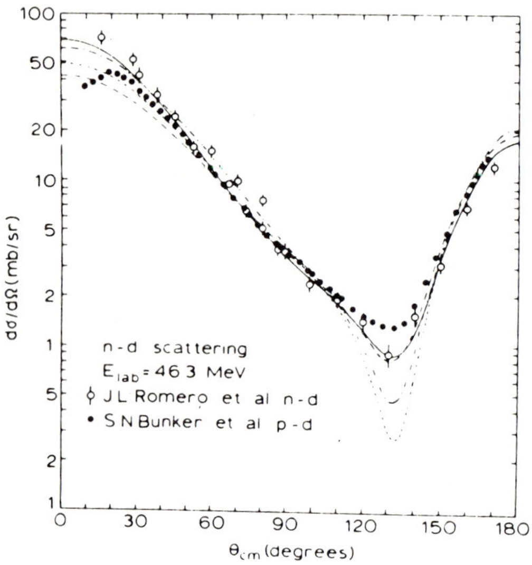

fig. 4
Elastic neutron-deuteron cross section for various choices of two-nucleon interactions (from ref. 29)
--- pure s-waves
-.- $^1S_0, {}^3S_1, {}^1P_1, {}^3P_0, {}^3P_1, {}^3P_2$
– – – $^1S_0, {}^3S_1$-$^3D_1$ and deuteron included d-wave component
—— full Reid potential

is shown in fig. 4 at 46.3 MeV. From this figure we see that keeping only the s-wave components in the calculation yield a very pronounced minimum. The same feature holds for other interactions [25,26]. Introducing the d-state components through the tensor force and the d-state of the deuteron give a significant filling up of the minimum, while the other components in the nuclear force donot affect this region sizably. A careful study of this region might give more information about the d-state in the deuteron. The overall agreement between the results of the complete Reid interaction and the experimental measurement is good up to 50 MeV.

From the computational point of view no essential new problems arise in the calculation of the breakup observables [31]. At this conference, phase equivalent separable potentials of rank 2 [32] have been used in the study of final state interactions. The potentials are constructed using inverse scattering theory. In principle these potentials may shed light on the question of off-shell sensitivity in certain regions of phase space [33,34]. It should however be pointed out that in the actual experimental situation the contributions from the higher partial wave components of the nuclear force might be dominant in these regions of phase space.

III. The four-nucleon system

Compared to the three-nucleon problem, almost no work has been done in studying the applicability of the integral equation approach to the alpha system. The starting point here is the Yakubovsky-Faddeev equation (YF) [35,36], which after partial wave decomposition is in general a coupled set of integral equations in three continuous variables. Two techniques have been explored to study these equations in the case that the two-body interaction is given by the separable Yamaguchi potential. One of them is the application of the Bateman method [37] while the other uses the HS-method [38] which has already been discussed for the three-nucleon case. In view of the promising results obtained recently with the HS-method we discuss this in more detail. From the kernel of the YF-equation two types of amplitudes are needed. One of them (the (3+1)-amplitude) is related to the contributions where three particles interact with each other while the fourth is free. The second amplitude (the (2+2)-amplitude) corresponds to the contributions where we have two pairs of mutually interacting particles while these pairs are not interacting with each other. The (3+1) and (2+2)-amplitudes satisfy linear integral equations. The method followed in ref. 38 consists now of using the HS representation analogous to eq. (6) for these amplitudes. Convergence with respect to the number of separable terms has been studied for the case of the Yamaguchi potentials.

Table 3

The 0^+ states of ^{3}He for the Yamaguchi potential for various numbers of separable terms (from ref. 38). $N_\eta^{(+)}$ and $N_\eta^{(-)}$ represent the number of terms in the (3+1)-amplitudes with pos. and neg. eigenvalues; N_ϕ is the number of terms in the (2+2)-amplitude

$N_\eta^{(+)}$	$N_\eta^{(-)}$	N_ϕ	E_1 (MeV)	E_2 (MeV)
1	0	1	-44.29	
1	0	3	-44.54	-11.06
1	0	4	-44.54	-11.06
2	1	4	-45.69	-11.39
3	2	4	-45.73	-11.63
4	3	4	-45.73	-11.69

The results are shown in table 3. From this we see that the convergence is very fast. In addition to the 0^+-ground state of ^{4}He at -45.73 MeV, an excited state was found at -11.73 MeV. In view of the triton binding energy of -11.03 MeV for the Yamaguchi potential, this excited state is very close to threshold. The above calculation has been extended to the case of the local interactions of eq. (5). In addition to the HS-expansion of the (3+1)- and (2+2)-amplitudes we also use the representation (6) for the two-body T-matrix. As an example the convergence rate with respect to the number of two-particle terms is shown in table 4 for the binding energy for the triton and the α-particle.

Table 4

The binding energies of the ground state of the tri-nucleon system (E_t) and the four-nucleon system (E_α) as a function of $N_\lambda^{(\pm)}$

$N_\lambda^{(+)}$	$N_\lambda^{(-)}$	E_t	E_α
1	0	9.47	36.6
1	1	7.96	26.4
1	2	7.76	24.7
2	0	10.13	40.8
2	1	8.60	30.2
2	2	8.38	28.5
3	2	8.53	29.5
4	2	8.56	29.6

Also in this case convergence in N_η and N_ϕ is very rapid. The final result is given by -29.6 MeV for the ground state of the alpha system. This should be compared with the experimental value of -28.34 MeV. Although there is also a stable excited state for the case we use one separable term for the two-body T-matrix, non was found on increasing the number of terms in the two-body T-matrix. The inclusion of more terms gives apparently rise to effective weakening of the interaction between the particles and as a result the excited state moves into the continuum. Finally the combined results for the binding energies for ^3H and ^4He have been plotted in fig. 5 for various approximations to the two-body T-matrix. From this we see that at least for this class of interactions there is a remarkable linear relation between these two observables. Assuming that this relation holds also for the Reid interaction considerable underbinding would be found for the ground state of ^4He. Taking a value of 7 MeV for the triton binding energy one finds 19 MeV for the alpha particle. This is in agreement with the results of Zabolitsky [40] who using a generalized Brueckner-Hartree-Fock-Faddeev method finds for the Reid interaction a value of 17.6 MeV on neglecting three-body terms. Inclusion of these gives a result of 20.1 MeV. Using the solutions of the YF-equations the ^4He-charge form factor can be calculated. The result for the potential (5) is given in fig. 6. Similar to the ^3He we find here a serious discrepancy with the experiments in the high momentum transfer region. Similar findings have been reported by Fink [41] and at this conference by Akaishi [14]. In fig. 7 the result of Akaishi for the Hamada-Johnston potential is given. It should however be noted that the results of refs. 14 and 41 are not in agreement with each other even at intermediate momentum trans-

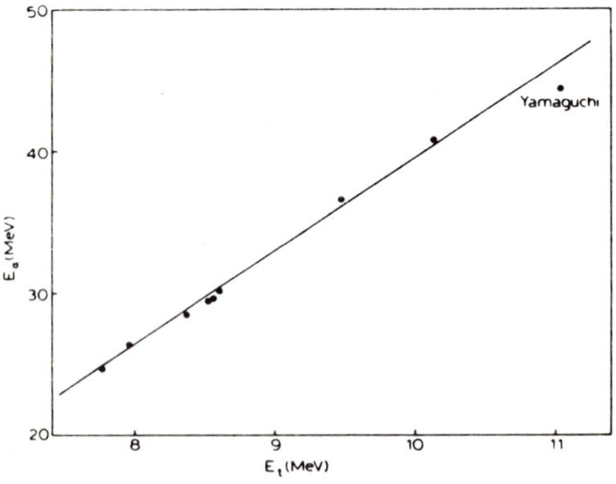

fig. 5 Compilation of the results for ^3H and ^4He-binding energies for various approximations to the local interactions.

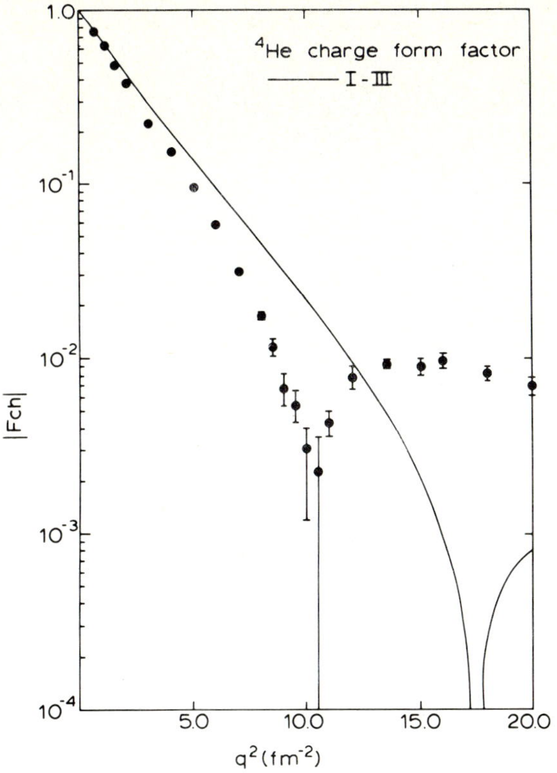

fig. 6

Charge form factor of ^{4}He for the potential set I-III The experimental data are from ref. 42.

fer. With respect to the scattering states some approximate results [43,44] have been reported recently, which look very encouraging. At this conference, p-^{3}He scattering calculations have been presented using the Bateman method [44].

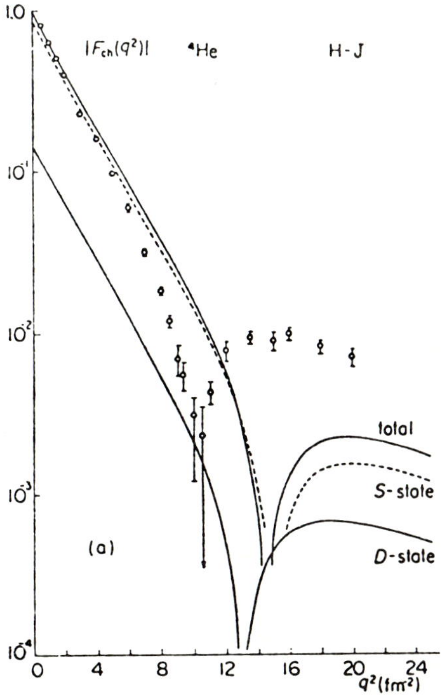

fig. 7

Charge form factor of ^{4}He for the Hamada-Johnston potential (from ref. 14).

To summarize, we have gradually learnt how to solve the three-body equations even for complicated realistic two-nucleon interactions. Much work is still needed with regard to the actual solution to the four-nucleon problem. In view of recent results, we may expect that the applicability of the YF-equations for realistic interactions will be demonstrated in the near future.

References

1. L.M. Delves and M.A. Hennell, Nucl.Phys. A168 (1971) 347; Phys.Lett. 40B (1972) 20.
2. A.D. Jackson, A. Lande and P.U. Sauer, Phys.Lett. 35B (1971) 365.
3. S.N. Yang and A.D. Jackson, Phys.Lett. 36B (1971) 1.
4. J. Bruinsma and R. Van Wageningen, Phys.Lett. 44B (1973) 221.
5. V.F. Demin, Y.E. Pokrovsky and V.D. Efros, Phys.Lett. 44B (1973)227
6. M.R. Strayer and P.U. Sauer, Nucl.Phys. A231 (1974) 1.
7. R.A. Malfliet and J.A. Tjon, Ann. of Phys. 61 (1970) 425.
8. J.A. Tjon, B.F. Gibson and J.S. O'Connell, Phys.Rev.Lett. 25 (1970) 540.
9. E.P. Harper, Y.E. Kim and A. Tubis, Phys.Rev.Lett. 28 (1972) 1533.
10. R.A. Brandenburg, Y.E. Kim and A. Tubis, Phys.Lett. 49B (1974) 205.
11. E. Harper, Phys.Rev.Lett. 34 (1975) 677.
12. G. Gignoux and A. Laverne, Phys.Rev.Lett. 29 (1972) 436.
13. J.L. Ballot and M. Fabre de La Ripelle, this conference.
14. Y. Akaishi, this conference.
15. H.P. Noyes in Proc. 1st Int. Conf. on the three-Body Problem in nuclear and particle physics, Amsterdam (1970) p. 2.
16. J.A. Tjon in Proc. of the Conf. on Few Body Problems in nuclear and particle Physics, Quebec (1974) p. 19.
17. E. Harms, Phys.Rev. C1 (1970) 1667.
18. E. Harms and V. Newton, Phys.Rev. C2 (1970) 1214.
19. B.L.G. Bakker and P. Ruig in Proc. on Few Particle Problems in the Nuclear Interaction. Los Angelos (1972) p. 357.
20. R.A. Malfliet and J.A. Tjon, Nucl.Phys. A127 (1969) 161.
21. I.R. Afnan and J.D. Read, Phys.Rev. C8 (1973) 1294.
22. L.P. Kok, thesis, University of Groningen (1969).
23. J.S. Mc Carthy et al, Phys.Rev.Lett. 25 (1970) 884.
24. R. Aaron and R.D. Amado, Phys.Rev. 150 (1966) 857.
25. R.T. Cahill and I.H. Sloan, Nucl.Phys. A165 (1971) 161.
26. W.M. Kloet and J.A. Tjon, Ann. of Phys. 79 (1973) 407.
27. G.Gignoux, A. Laverne and S.P, Merkuriev, Phys.Rev.Lett. 33 (1974) 1350.
28. T. Sasakawa and T. Sawada, this conference.
29. C. Stolk and J.A. Tjon, Phys.Rev.Lett. 35 (1975) 985.
30. S.C. Pieper, Nucl.Phys. A193 (1972) 529.
31. W.M. Kloet and J.A. Tjon, Nucl.Phys. A210 (1973) 380.
32. J.H. Stuivenberg and R. Van Wageningen, this conference.
33. D. Brayshaw, Phys.Rev.Lett. 32 (1974) 382; ibid 34 (1975) 1478.
34. M.I. Haftel and E.L. Peterson, Phys.Rev.Lett. 33 (1974) 1229; ibid 34 (1975) 1480.
35. O.A. Yakubovsky, J. Nucl.Phys. (USSR) 5 1967 1312.
36. L.D. Faddeev in Proc. 1st Int. Conf. on the Three-Body Problem in nuclear and particle Physics, Amsterdam (1970) p. 154.
37. V.F. Kharchenko and V.E. Kuzmichev, Phys.Lett. 42B (1972) 328.
38. I.M. Narodetsky, Nucl.Phys. A221 (1974) 191.
39. J.A. Tjon, Phys.Lett. 50B (1975) 217.
40. J.G. Zabolitsky, Nucl.Phys. A228 (1974) 272; ibid 285.
41. M. Fink, Nucl.Phys. A221 (1974) 163.
42. R.F. Frosch et al, Phys.Rev. 160 (1967) 874.
43. E.O. Alt, P. Grassberger and W. Sandhas, Phys.Rev. C1 (1970) 161.
44. M. Sawicki and J.M. Namyslowski, this conference.

BOUND STATES IN THE FEW BODY SYSTEMS

Tatuya Sasakawa

Department of Physics, Tohoku University
980 Sendai, Japan

1) Three Nucleon System

One of the problems in the three-nucleon bound system is the reconcilation of the binding energy (experimentally 8.4 Mev), the minimum (experimentally, $q^2 \sim 12$ fm^{-2}) and the height of the diffraction maximum of the electron scattering. Two contributions presented in this conference[1),2)] show that one can not get satisfsactory result in this respect with the one boson exchange potential.

Potential	RSC	OBE[1)] HM1	OBE[1)] EHM'	OBE[2)] A	OBE[2)] B_1	OBE[2)] B_2
E_T	7.0	7.2	6.5	6.75	6.77	7.48
q^2_{min}	13.9	15.3	15.5	16.8	14.0	13.7
R	3.3			8.8	7.2	7.6

Table I E_T, the triton binding energy ; q^2_{min} , the diffraction minimum; R , the ratio of the experimental value to the theoretical value of the diffraction maximum in ^{3}He.

Haftel[3)] has shown that the effect of the extended proton charge to the Coulomb energy difference (ΔE_c) of ^{3}H and ^{3}He is important. (Table I of Ref.3). He conjectures that the discrepancy between the calculated ΔE_c of $0.60 \sim 0.65$ Mev and the experimental one ($\Delta E_c =$ 0.76 Mev) may be due to the charge-symmetry breaking in the N-N interaction.

After the Los Angeles Conference[4)], we accept that a_{nn} and $a_{pp}^{nuclear}$ as about -17 fm. r_{nn} and r^s_{pp} (s;singlet) are not known so accurately as yet. Gibson-Stephenson[5)] has shown that only if $r_{nn} < r^s_{pp}$, $E_c < 0.76$ Mev and $|a_{nn}| < |a_{pp}^{nuclear}|$. Zankel and Zingl[6)] have shown for the Graz potential that $a_{pp}^{nuclear} = -17.0$, $a_{nn}=-16.5$, $r_{nn}=2.93$, $r^s_{pp} \sim 2.7$, and $\Delta E_c \sim 0.64$ Mev, thus contradicting the conjecture by Gibson and Stephenson.

Since the original idea of Mitra of this conference was the bridge concept, namely to construct bridges between neighboring regions, my talk will be presented on the bridge concept.

2) Bridge from nuclear matter to ^{4}He

In this conference Akaishi has presented a paper, in which the method of nuclear matter is applied to the binding energy and static properties of ^{4}He. He used the Watson expansion of the Bethe-Goldston equation, and assumed two kinds of correlation functions as parameters. So far ^{4}He has been solved involving tensor forces only by Demin et al.[7)] and Akaishi et al.[8)] Demin et al. used the Gaussian type potentials[9),10)], whereas Akaishi used the Hamada-Johnston and the Tamagaki potentials, which are realistic. Rather singular forces used by Akaishi caused larger values for kinetic energies and larger absolute values for potential energies than those of Demin et al. However, it is remarkable that all potentials yields almost the same binding energy ranging from about 21 Mev to 27 Mev. This fact may suggest the feature that we might not be able to select the "true" realistic potential from the binding energy of ^{4}He, as in the case of ^{3}H and ^{3}He.

	AKAISHI et al.[8]			DEMIN et al.[7]	
	HJ	OPEH	OPEG	EH	GPT
ENERGY	-20.6	-23.3	-22.5	-23.03	-26.8
K.E	131.1	101.6	70.6	77.95	57.64
P.E	-151.7	-124.9	-93.1	-100.98	-84.44

Table II The total energy, kinetic and potential energies of ^{4}He.

The situation about the charge form factor is almost the same as in ^{3}He ; the calculated diffraction minimum is shifted towards larger q^2 than the experimental value and the height of the second maximum is about one thirds of the experimental one.

Talking about the three-body force, Nogami et al. gives -0.8 Mev/A, Kasahara-Akaishi -0.8 Mev/A, Blatt-McKeller -4.0 Mev/A, for nuclear matter . It is now accepted that it is almost 1 Mev for ^{3}H, or slightly less than 1 Mev. For three-body force of the type  , Akaishi gives the following contributions from three-body forces : ^{4}He, -7.53 Mev for HJ wave function, -8.81 Mev for OPEH ; ^{3}H, -1.16 Mev for HJ. Due to the double tensor nature of the three-body force, it is natural that a large contribution comes from the coupling between S- and D- states. The big three-body force in ^{4}He might not be surprising, since ^{4}He contains four tritons which give about 4 Mev of three-body forces. For the two body potential energy of -151.7 Mev, the three body force of about -8 Mev is a small fraction. Any way, ^{4}He might be more interesting tool than ^{3}H for the investigation of nuclear force.

3) Two body clusters seen in nuclei from ^{3}H, ^{3}He, ^{4}He to ^{20}Ne

As one important bridge, I must remark that the two-body clusters are seen in three nucleon systems, in ^{4}He, and in heavier nuclei such as ^{20}Ne. As Merkur'ev[11] discussed in detail, triton has the asymptotic component of n-d channel and the break-up channel, although these are closed. Actually, we can see that the p-d channel is involved in ^{3}He from the phase shift analysis of p-^{3}He for large ℓ (peripheral phenomena). Let R be the radius of deuteron. We write the tail of the deuteron wave function as

$$\varphi = \frac{C}{(2\pi R)^{1/2}} \frac{e^{-r/R}}{r} \quad , \quad \text{with } R = 0.23 \text{ fm.}$$

If φ represents not only the asymptotic wave but the whole deuteron, then C=1. The calculation by the Reid potentialy yields $|C|^2 = 1.7$, whereas the analysis of the experimental result by Bolsterli and Hale[12] gives $|C|^2 = 2.2$. A similar value was given by Kim and Tubis[13], $|C|^2 = 2.8$, and also by Kisslinger.[14] Also, it is known that the sharp rise of the doublet phase shift of the n-d scattering is responsible to the virtual state of the n-d system involved in ^{3}H.[15),16]

Now, let us see the situation of ^{4}He. Table IV of the paper by Akaishi[17] shows that the excited states of ^{4}He with odd parity is very different from ^{4}He ground state with regards to the kinetic and potential energies and rather similar to ^{3}H. From this result, he conjectured that the excited states with odd parity consist of triton and the spectator particle with p-wave.

Now we go the bridge further and come to heavier nuclei. It is known that the levels with K=0$^-$ of ^{16}O and ^{20}Ne have large α-width, and these levels (9.85 (1$^-$), 11.62 (3$^-$), 16.90 (5$^-$), 20.9 (7$^-$) of ^{16}O and 5.80 (1$^-$), 7.17 (3$^-$), 10.30 (5$^-$), 13.43 (7$^-$) of ^{20}Ne) are conjectured as having the two-body cluster structure: $O^{16} = \alpha + C^{12}$, $Ne^{20} = \alpha + O^{16}$. The generator coodinate method[18] reproduces actually the alpha width of Ne^{20} for K = 0$^-$ band.

J	1^-	3^-	5^-	7^-
Cal	0.267	0.265	0.271	0.298
Exp	>0.14	0.26	0.31	0.25

Table III Reduces α-widths in unit of $3\hbar^2/2\mu a^2$ by $\alpha + {}^{16}O$ generator coordinate wave functions. Channel radius is taken as a=6.0 fm[18].

4) 3α-model of ^{12}C and a bridge to the shell model

The small binding energy (7.28 Mev) of the ground state of ^{12}C below the three-body break up threshold attracted many physicists to the 3α-particle model. Let us first summarize the results obtained so far.

J	0_1^+	2_1^+	0_2^+	reference
Exp. (Mev)	-7.28	-2.84	0.38	
No correlation	1.4 1.1			19) 20)
One term separable potential	-7.4 -6.8 -7.4	-1.16		21) 22) 23)
Separable approximation of soft core potential[24),25)]	-2.20 -2.79 -1.48		1.81 1.24	26) 27) 28) } involving S- and D-potentials

Table VI The energies in Mev of ^{12}C, the 3α-threshold being taken as 0 Mev. All calculated values take the Coulomb force into account.

The wave function without correlation yields unbound ^{12}C. The one term separable approximation[21),22),23)] gives good results. However, the potentials of this kind give too large binding energy for Be^8 : -2.9 Mev[21)] and -3.6 Mev[22)] in place of 0.1 Mev of the experimental value. The separable approximation of soft core potentials[24),25)] gives too small binding energy.[26),27),28)] The soft core potential taking account of the Pauli principle also yields small binding energy[29)]. The inclusion of the excited states makes the binding energy decrease (by 0.43 Mev for 0_1^+).[30)]

Fujiwara and Tamagaki[31)] has used the Kukulin-Neudatchin[32)] type $\alpha-\alpha$ potential, which has no core, but gives the second node around at 2 fm for the $\alpha-\alpha$ wave function. It is important to note that the position of this node does not change with energy (2 Mev ~ 26 Mev). The full wave with wave length of 2 fm inside of this node represents the 2 S-state, the 0S and 1S states being forbidden. Since this potential acts also at short distances of two alpha particles, the D- and G- wave potentials give very important contributions ; $W(0_1^+)$=-3.9 Mev for S wave potential only, and -7.2 Mev for S + D, -16.9 Mev for S + D + G and -17.0 Mev for S + D + G + I. Indeed, the binding energy obtained is too much, even when we take account of the Coulomb force of about 5 Mev. However, this model involves several interesting features. The α- width defined by $\gamma = \int j_\lambda(q\rho)\sqrt{3!} \langle \psi_{2\alpha}(k) Y_\lambda(\hat{q}) | \psi_{3\alpha}(\vec{k},\vec{q}) \rangle q^2 dq$ becomes maximum at ρ = 2.8 fm for 0_1^+, where γ is about 60% of the Wigner limit. It becomes maximum at ρ = 4.8 fm for 0_2^+ and reaches the Wigner limit. If the 0_2^+-state has a configuration like a linear chain as Hebach-Hennenberg[28)] demonstrated, the wave function involves approximately the δ-function $\delta(\theta-\pi)$. Since this δ-function consists of all partial waves with equal weights, the α-width for $\ell = 0$ must be small[33)]. Experimentally, the

0_2^+-state decays to 3α. Fujiwara-Tamagaki[31)] has shown that the 0_2^+-state is not a linear chain but consists mainly of relative S-waves of N=2 (N; the number of nodes of the relative motion.)

Now we go over to the SU(3) p-shell shell model in which the states forbidden for any pair of alpha particles are excluded. This model is called the orthogonality condition model (OCM). Horiuchi[34)] found that the 3α-configurations thus permitted are

N	(λ, μ)
8	(0 , 4)
9	(3 , 3)
10	(6 , 2) (2 , 4)
. .	. . . .

By employing these configurations with the Schmidt-Wildermuth two nucleon force, he got a very nice energies of ^{12}C. Also he got the reduced $\alpha + {}^8Be(0^+)$ width, which is almost the same as Fujiwara-Tamagaki.

N	8	10	12	14	16	18	20	22
0_1^+	0.80	0.08	0.08	0.02	0.01	0.01	<0.01	<0.01
0_2^+	0.13	0.13	0.12	0.17	0.14	0.12	0.08	0.05

Table V Configuration of 0_1^+ and 0_2^+ states of ^{12}C 34).

For the 0_1^+ state, N=8 (0,4) configuration is predominant, whereas for the 0_2^+ state the configurations spread over various N. Since N=8(0,4) represents the four hole p^{-4} configuration, namely, a 4-particle 4-hole configuration, the 0_1^+ state has the $\alpha + {}^8Be$ structure in the sense of the shell model. On the other hand, the shell model description of 0_2^+ may not be appropriate and the 0_2^+ state must be the "true" 3α-state. However, the situation is not so simple to say. There must be a strong overlap between these two states as indicated by the observed monopole transition matrix element of 5.8 fm^2, for which Horiuchi gives 6.1 fm^2. Concerning the cluster structure of 0_2^+, one open question may be the large $\ln$ ft of β-decay ($\ln$ft = 4) between the ground ^{12}B to 0_2^+ of ^{12}C[35)]. The ground state of ^{12}B is supposed to be a shell mode state. Since 0_2^+ is unbound the calculation of $\ln$ ft must be very hard, while the β-decay gives one of the most accurate informations in nuclear physics.

Here we note that Afzal and Ali[36)] is also calculating the $\alpha - \alpha$ scattering by the resonating group model.

5) OCM and three-nucleon system

The Tabakin potential has a spurious bound state at 240 Mev. This potential thus yields more than 300 Mev as the binding energy of a three-nucleon system. Englefield[37)] has applied the OCM to take off the spurious bound state and found no bound state for a three-nucleon system.

6) Three-particle vertex function

A very important proposal for the calculations of three-body vertex functions is presented by Mitra and Sharma[38)]. Their method must be useful in the future.

7) Bridge to other fields

Now we come across a bridge and go outside of nuclear physics. One is the ${}^6_{\Lambda\Lambda}He$ of $\alpha\Lambda\Lambda$ model by Choudhury et al.[39)] Using the unitary pole approximation for exponential potentials they got for some set of strengths and the range of the $\Lambda\Lambda$ potential a nice value for binding energy of ${}^6_{\Lambda\Lambda}He$, for which the experimental value is 10.8 ± 0.6 Mev[40)]. Their parameters are very close to those obtained by the variational calculation due to Tang and Hernodon.

The bridge to another field is to the molecular $^3(He)_4$ by Lim[41]. He found that $^3(He)_4$ exists and both the equivalent -two-body-method (ETBM) and the Faddeev UPA yields similar binding energies

	BE(^oK)	
ETBM	-0.073	for a hard core square well.
Faddeev UPA	-0.070	

8) In concluding my talk : So far we have seen many beautiful bridges between various regions. However, to my poor eyesight, the contemporary bridges appear very poor. The reason might be that I am looking things standing on the bank of the equation with connected kernels, the Faddeev-Yakubovski equation or its version.

For one thing, I can not understand why the variational wave functions, which have no correct asymptotic behavior, have shown so competitive results for all physical observables as the Faddeev wave function.

For another, it seems to me that the existing reaction theory and the resonating group theory have no relation with the Yakubovski equation. People usually use such an equation as $\psi = f_{\hat{i}}\varphi_{\hat{i}} + (E - T + i\varepsilon)^{-1}|\varphi_{\hat{i}}\rangle\langle\varphi'_{\hat{i}}|(V - V_i)|\psi\rangle$. The kernel of this equation is not compact as soon as $\varphi_{\hat{i}}$ is in continuum. The use of this equation might be serious for the resonating group (cluster) model, because the cluster model is expected to be valid at those energies where φ_i is near the breakup threshold, as 0_2^+ of ^{12}C.[42] In fact, the four body equation on the basis of the connected kernel is very different from the existing reaction theory or the resonating group theory.[43] Also, I always regard the shell model as a water-lily, which has no root on the ground, namely, on the Faddeev-Yakubovski type many-body theory. It is my pleasure to see that in this conference, a number of authors begin to work to construct a real bridge to these problems.

Finally, I would take a liberty of saying why I am working in this particular field of physics. This is because here we are handling linear algebra, for which the exact solution, namely the Fredholm solution, exists. On the other hand, the existing field theory is non-linear. For instance, the equation for the meson field interacting with nucleon involves the nucleon density, which is not linear with respect to the nucleon field. Indeed, the field theory was constructed on the basis of one-particle problem and in analogy with the equation of motion of classical physics. Physics we are dealing here is very different from existing physics in the sense that we can handle any kind of and any number of bound subsystems for which the old fashioned creation-annihilation operator description does not work. Also our physics involves nontrivial coordinate transformations. Again, this makes difficult the use of creation-annihilation operators. Because our physics is linear and need not introduce the second quantization, I think that what we are doing might be more suited for the description of Nature. In this sense, the approach by Noyes[44] seems to me very interesting. But it is still a rainbow bridge, far from getting a number. Therefore, we have a lot of things to do.

References

1) R.A.Brandenburg, P.U.Sauer and R.Machleidt, this conference.
2) E.Harper, this conference.
3) M.I.Haftel, this conference.
4) Few Particle Problems, (Ed. I.Slaus et al.) North-Holland (1972).
5) B.F.Gibson and B.G.J.Stephenson Jr., Phys.Rev. C11(1975),1448.
6) H.Zankel and H.Zingl, this conference.
7) V.F.Demin, Yu.E.Pokrovski, V.D.Efros, Phys.Lett.44B(1973),227.

8) Y.Akaishi et al. Prog.Theoret.Phys.Supple. 56(1974), and this conference.
9) H.Eikemeier and H.H.Hackenbroich, Nucl.Phys.A169(1971),407.
10) D.Gogny, P.Pires and R. de Tourrel, Phys.Lett.32B(1970),591.
11) S.P.Merkur'ev, Sov.J.Nucle.Phys. 19(1974),222.
12) M.Bolsterli and G.Hale, Phys.Rev.Lett. 28(1972),1285.
13) Y.E.Kim and A.Tubis, Phys.Rev.Lett.29(1972),1017.
14) L.S.Kisslinger, Phys.Lett.47B(1973),93.
15) G.Barton and A.C.Phillips, Nucle.Phys.132A(1969),97.
16) T.Sasakawa, Quebec Conference Proceedings (1974),p.561.
17) Y.Akaishi, this conference.
18) H.Horiuchi, Proceedings of INS-IPCR Symposium on Cluster Structure, Tokyo, 1975,p.41.
19) P.Darriulat, Nucle,Phys.76(1966),118.
20) T.K.Lim, Nucle.Phys. A158(1970),385.
21) D.R.Harrington, Phys.Rev.147(1966),685.
22) C.C.H.Leung and S.C.Park, Phys.Rev.187(1969),1239.
23) I.Duck, Nucle.Phys.84(1966),586.
24) S.Ali and A.R.Bodmer, Nucl.Phys.80(1966),99.
25) P.Darriulat, Phys.Rev.137(1965),B315.
26) J.L.Visschers and R.van Wageningen, Phys.Lett.34B(1971),455.
27) J.R.Fulco and D.Y.Wong, Phys.Rev.172(1968),1062.
28) H.Hebach and P.Hennenberg, ZS f.Physik,216(1968),204.
29) Y.Kawazoe, T.Tsukamoto and H.Matsuzaki, Prog. Theoret.Phys.51(1974),428.
30) R.J.Turner and A.D.Jackson, Nucle.Phys.A192(1972),200.
31) Y.Fujiwara and R.Tamagaki, this conference.
32) V.I.Kukulin and V.G.Neudatchin, Nucl.Phys.A157(1970),609.
33) M.Kawai, private communication.
34) H.Horiuchi, this conference.
35) H.Morinaga, private communication.
36) S.A.Afzal and S.Ali, this conference.
37) M.J.Englefield, this conference.
38) A.N.Mitra and V.K.Sharma, this conference.
39) H.Roy-Choudhury, V.P.Gautam and D.P.Sural, this conference.
40) D.J.Prowse, Phys.Rev.Lett.17(1966),782.
41) T.K.Lim, this conference.
42) T.Sasakawa, Proceedings of INS-IPCR Symposium on Cluster Structure, Tokyo, 1975, p.309.
43) T.Sasakawa, this conference and a preprint (Tohoku University).
44) P.Noyes, this conference.

DISCUSSION

Sauer (to Sasakawa) : You presented calculations for the coulomb energy difference between ^3He and ^3H. In order to estimate the nuclear charge asymmetry, you should also commit all the other direct electromagnetic contributors to the ^3He-^3H mass difference.

Sasakawa : Yes, of course.

Haftel (to Tjon) : Brayshaw and I don't disagree at 14 MeV. We found off shell sensitivity at higher energies where Brayshaw's method is untested.

Fabre (to Tjon) : When comparing the results given by the Faddeev equation with respect to those obtained by variational methods we have to keep in mind that these last methods fulfil the Ritz principle. Instead one does not know generally on what ide we are of the exact binding energy by solving the Faddeev equation. Therefore when this last method provides a binding energy lower than the one obtained by a variational calculation it does not mean at all that this result is better than the variational one. One must therefore be very careful when a comparison is done between the results obtained by both methods.

BREAKUP PROCESSES - A BRIDGE?

Ivo Šlaus
Institute "Ruđer Bošković", Zagreb, Yugoslavia

The purpose of natural philosophy is to understand nature. Even when one classifies merits of a specific scientific discipline in terms of internal (discoveries, new concepts, etc.) and external (interdisciplinarity, applications, socioeconomic benefits, etc.) values, one still just explores to what extent that discipline fits into our overall aim toward integral cognition. Having this in mind, let us attempt to review a small part of our own few-body field: breakup processes - its development, its present status, and its future - specifically, how it links with other research activities: nuclear structure and particle physics. We know that understanding in nuclear and particle physics does not come as one proceeds 1,2,3,4,...200,.. infinity, but we also know that few-body problems are a window into, maybe even the frontier of investigating nuclear interactions, reaction mechanisms, and models.

A distinct feature of a three-body problem is that it has been solved exactly, that there are theoretical predictions and by now even using fairly sophisticated interactions. What has the three-nucleon problem taught us about the nuclear interaction? i) Back in 1935 Thomas showed that the binding energy of ^{3}H would be infinite for an N-N force of zero range. ii) A long arduous work since 1961, involving n+d but also π^-+d and other breakup processes, has led to the value of a_{nn} /1/.

This might seem a rather modest output, especially when compared with our expectations: to learn about off-energy-shell interactions, three-body forces, mesonic degrees of freedom, quantitative determination of the breakdown of various symmetries, etc. (One should remember that a two-nucleon system was not much more generous in revealing the secrets of nature.) Yet, this endevour has taught us that:

1) Though even a rather naive nuclear potential (e.g., S-wave Yamaguchi) suffices to give an almost good fit to the breakup cross sections, it has been established that on-shell differences in N-N force affect predictions in all regions of phase space /2/ (Figs. 1 and 2). The potential that gives the best fit to the N-N data, Y1-11, generally gives the best fit to the breakup data. (The

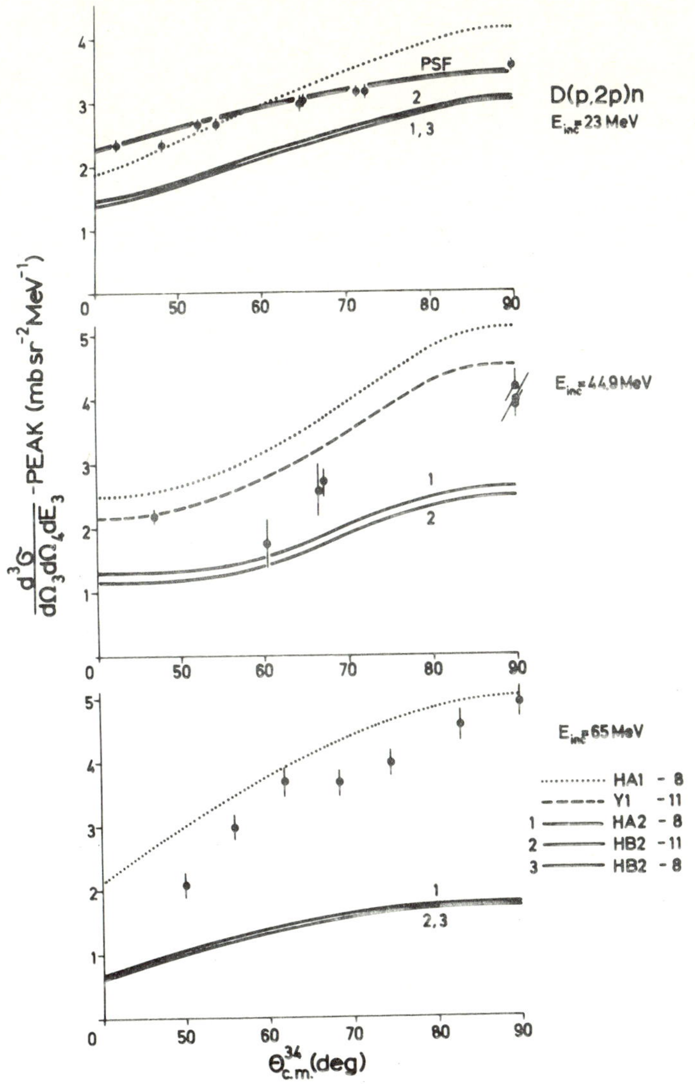

Fig. 1. The cross section at the QFS peak plotted as a function of angle for E_{inc}= 23 MeV /3/, 45 MeV /2/, and 65 MeV /4/. The curves are the Amado-model predictions using several different S-wave potentials: Y - Yamaguchi form factor, HA and HB Haftel form factors with k_c^2=4 fm^{-2} and -2.89 fm^{-2}, respectively, the fit to the N-N phase shifts is denoted by 1,2, and 3, the number after the dash indicates the triton binding energy. The curve labeled PSF is a phase space factor.

nuclear interaction used in these calculations is simple and therefore the degree of sensitivity can change when one includes a realistic N-N force.)

2) a) There is an amplification effect: QFS magnifies on-shell differences between potentials. This is caused by the interference between (n-p) FSI and (p-p) QFS mechanisms /2,3, 6/.

b) Unlike on-shell differences, short-range-interaction effects might show up only in restricted domains. Since QFS is a peripheral process, it is a result of a coherent contribution of many N-N^2 partial waves. The contribution of the L=0 part of the doublet amplitudes (which is sensitive to short range effects) to the QFS cross section, taking into account also the interference between the L=0 and L>0 amplitudes, is small (e.g., at 45 MeV, it is less than 10%).

Corollary: One can use QFS to determine the on-shell interaction: e.g., r_{nn} can be determined with an accuracy of 2-3% if the cross section is measured and calculated with an accuracy of 2-3% /7/.
3) It has been established that N-d polarization data are very sensitive to N-N higher partial wave interactions /8/. It is known that the region of low relative p-p energies (T_{pp}) is sensitive to the Coulomb force; however, at present the Coulomb force has been introduced into calculation in a very approximate way /9/. It is reasonable to expect that QFS is influenced by the Coulomb, tensor, and spin-orbit forces. As long as one uses the S-wave separable interaction, on-shell and off-shell variations do not change the shape of the predicted QFS angular distribution (Fig. 1). The fact that the shapes of the theoretical angular distributions disagree

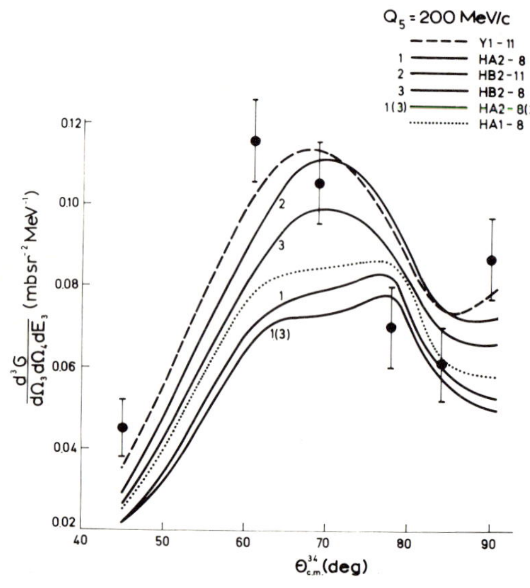

Fig. 2. Angular distribution for fixed momentum transfer at E_{inc}=45 MeV /5/. A symbol (3) means that the calculations are performed using three coupled equations and the correct charge-dependent interaction.

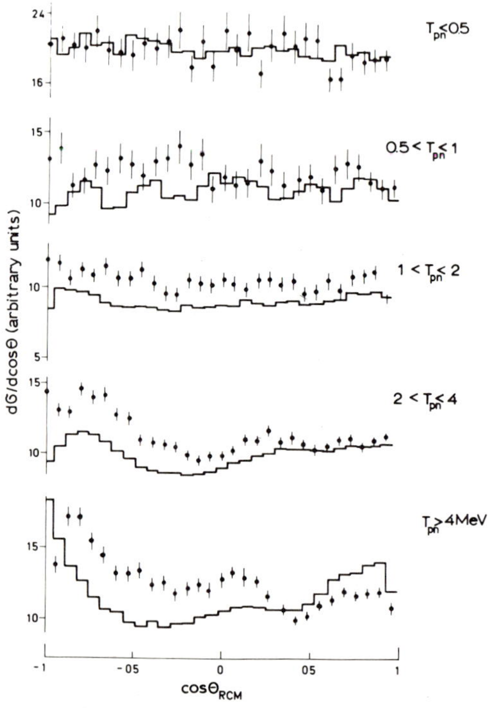

Fig. 3. T_{pp}> 1.5 MeV H(d,2p)n data /10/ at 26.5 MeV and model predictions as a function of cos Θ_{RCM} for five T_{pn} intervals.

with the data at 23 MeV, while the agreement improves with increasing energy, might indicate the importance of the Coulomb force.

An insight into the role of higher partial waves can be obtained by selecting those breakup data where Coulomb effects are small ($T_{pp} > 1.5$ MeV) and then analyzing those data in terms of T_{pn} and p-n recoil c.m. angle, Θ_{RCM}. The comparison of such data with the Faddeev calculation using the local Yukawa-type MT 13 potential is shown in Fig. 3. It is observed that the discrepancy increases with increasing T_{pn}. The RCM system is a useful description, when T_{pn} is small compared with the total c.m. energy. Thus, any argument in favor of possible effects of the $L > 0$ interaction can be given using the discrepancy only for $0.5 \lesssim T_{pn} \lesssim 2$ MeV.

4) Originally, it was hoped that N-N Bremsstrahlung is the ideal process to determine the off-shell behavior. However, our enthusiasm has diminished because of its weak sensitivity to off-shell variations and because of the difficulties in defining the electromagnetic current: velocity-dependent, relativistic, and exchange effects /11/. Since 1972 it has been hoped that three-body processes could provide information on the off-shell behavior, though it has never been clear whether one thinks about off-shell effects alone or one lumps together all short-range-interaction effects. If breakup processes are to provide an insight into the short-range force, it is felt that one should investigate areas away from

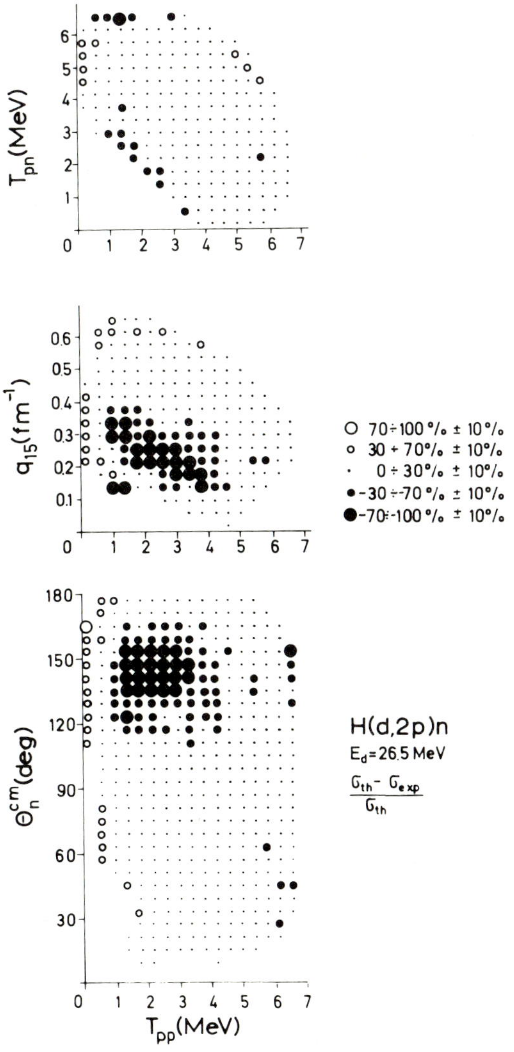

Fig. 4.

quasi-two-body processes, because there it would be easiest to separate on-shell and short-range effects.

Systems which simultaneously sample a major part of phase space (multi-detector systems, bubble chambers, nuclear emulsions, etc.) provide the possibility to search for interesting regions choosing physically most adequate variables. Experimental data and MT 13 potential predictions are compared /10/ (Fig. 4), projecting the data in the 5D space of independent kinematical variables on 2D spaces: a) T_{pp}-T_{pn}, b) T_{pp}-q_{15} (momentum transfer to the undetected neutron), and c) T_{pp}-Θ_n^{cm} (neutron c.m. angle with respect to the incoming proton). One concludes: a) the projection on the T_{pp}-T_{pn} space reveals a good fit of the calculation to the data. The discrepancy at T_{pp}< 0.5 MeV is due to the neglect of the repulsive Coulomb force. b) experiment and theory differ when projecting on either q_{15}-T_{pp} or Θ_n^{cm}-T_{pp} spaces. In the region $1 \lesssim T_{pp} \lesssim 4$ MeV, $0.15 \lesssim q_{15} \lesssim 0.3$ fm^{-1}, and $130^\circ \lesssim \Theta_n^{cm} \lesssim 155^\circ$, theory is lower than the data by a factor of nearly two. It is in this region of Θ_n^{cm} /12/ that the breakup cross sections at 14 and 50 MeV are very sensitive to the choice of the N-N forces. Note, however, that the

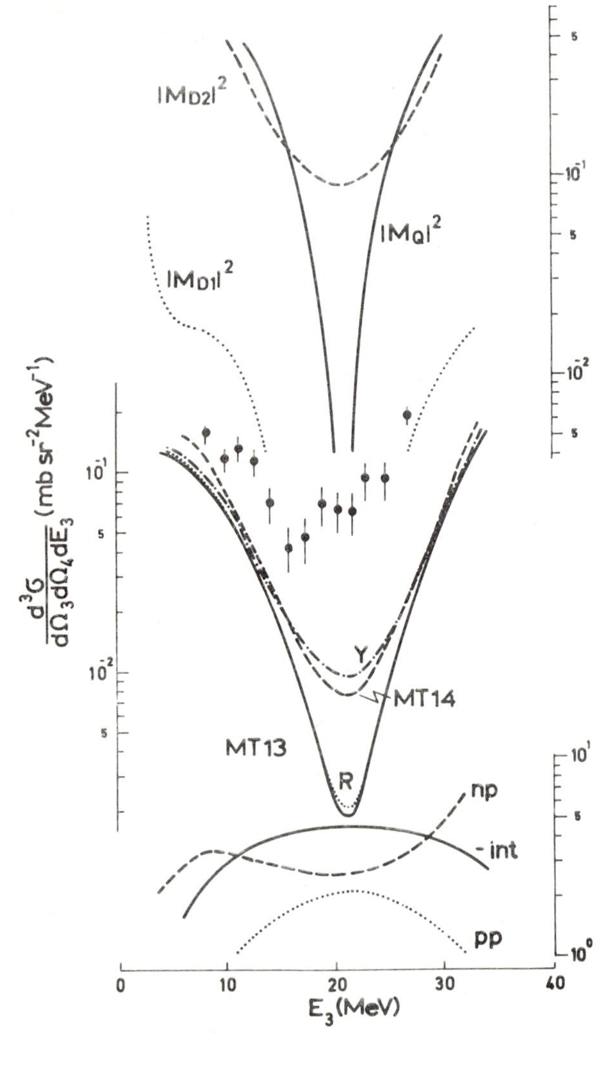

Fig. 5.

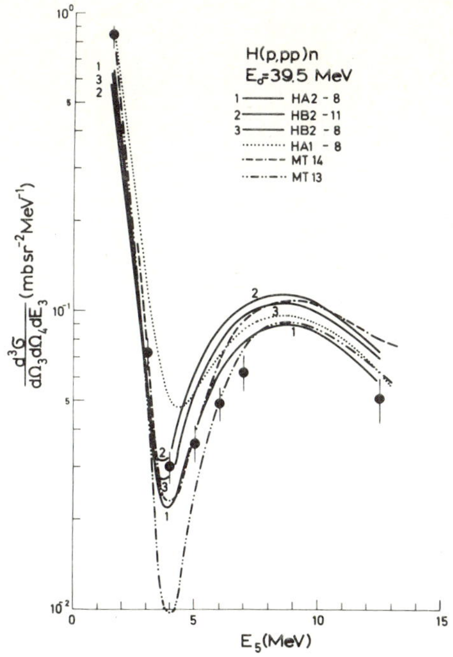

Fig. 6. Data /16/ and calculations /6,16/ along a constant relative energy locus. E_3=17.9 MeV, Θ_3=27.1°.

difference between data and theory (Fig. 4) is much larger /10/ than the difference between the two potentials: MT 13 and MT 14. Also, note that this discrepancy is partly correlated with the discrepancy in T_{pn} vs cos Θ_{RCM}. At E_p^{inc}=50 MeV, preliminary data /13/ are about ten times larger than the MT 13 prediction and three times larger than MT 14 (Fig. 5). The potentials MT 13 and MT 14 are on-shell and off-shell different. In order to understand the nature of this discrepancy, additional investigations were performed using separable potentials with different form factors /14/ (Yamaguchi and exponential form factors give differences comparable with MT 14 - MT 13 differences) and using phase-shift-equivalent (PSE) potentials /6/. Imposing the condition that PSE potentials should produce the same triton binding energy decreases the difference /6/. The cross section in the minimum is dominated by the M_{D2} amplitude, but the shape of the minimum is strongly influenced by M_Q (Fig. 5). The minimum corresponds to the pronounced destructive interference between the final n-p and p-p rescattering amplitudes /6/ (Fig. 5). This interference occurring along the "constant-relative-energy-loci" /15/, where the M_{D2} amplitude dominates again is also influenced by both on- and off-shell properties of the potential (Fig. 6). The angular variation of correlation spectra at symmetric coplanar angles samples a part of phase space where one gradually departs from on-shell. Fig. 7 shows data at 12.5 /17/, and 58.5 MeV /18/. The region between Θ_3=Θ_4=50° to 70° is quite sensitive to off- and on-shell variations, but due to strong interferences it might be difficult to isolate short-range effects. It is interesting to observe the inversion of the predicted cross section after one has crossed the interference minimum. It has been argued /6/ that the measurement of the FSI angular distribution is sensitive to short-range

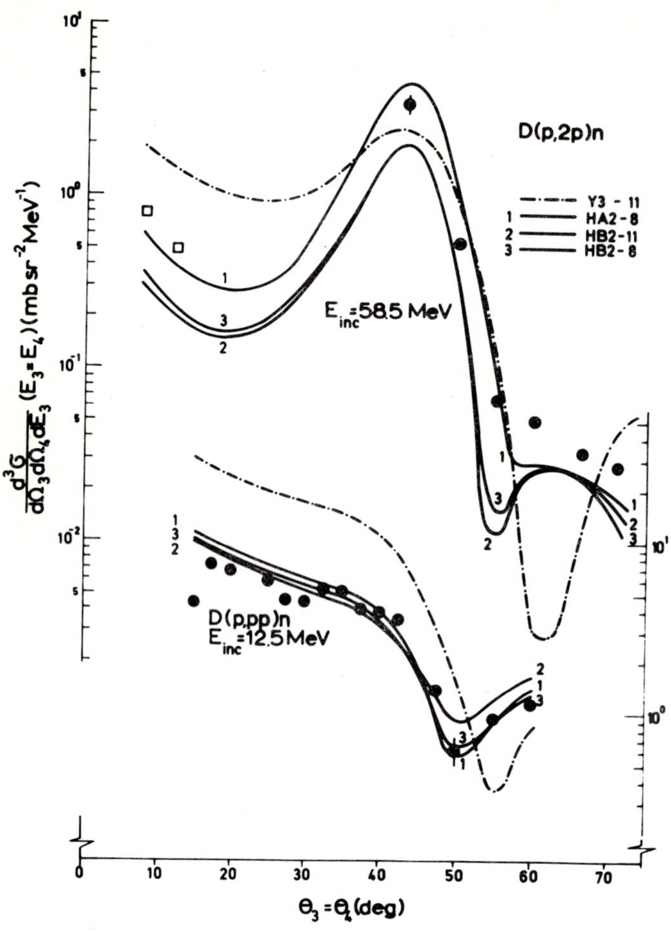

Fig. 7. The cross section at $E_3=E_4$ as a function of $\theta_3=\theta_4$. Open squares are data at 51.8 MeV /19/.

effects. The FSI is dominated by the M_{D1} amplitude, which turns out here to be the most sensitive one to short-range effects. A constraint on E_T considerably decreases the effect of short-range variations at E_{inc}=14 MeV, but at higher energies the effect seems to be appreciable.

During the last two years there has been considerable controversy concerning what can be learned about the off-shell interaction from breakup processes /20/. It has been recently shown /20/ that a complete knowledge of three-particle scattering observables is even in principle not sufficient to distinguish off-shell from three-body forces. This result is implicit in the interior-exterior separation of Noyes /21/ and it is related to the fact that a unitary transformation can be introduced that change both the off-shell behavior and the three-body force in an infinite number of ways and leaves three-body scattering observables unchanged /22/. Our efforts to understand short-range interactions rather than being terminated by this result are directed to solve: 1) whether the calculations allowing "maximum freedom" in the interaction can fit the breakup data (compare the 1.5 MeV missing in E_T); 2) in spite of the very weak sensitivity to short-range effects (off-shell and three-body force) - which, we all agree, increases with increasing incident energy - it is presumably necessary to under-

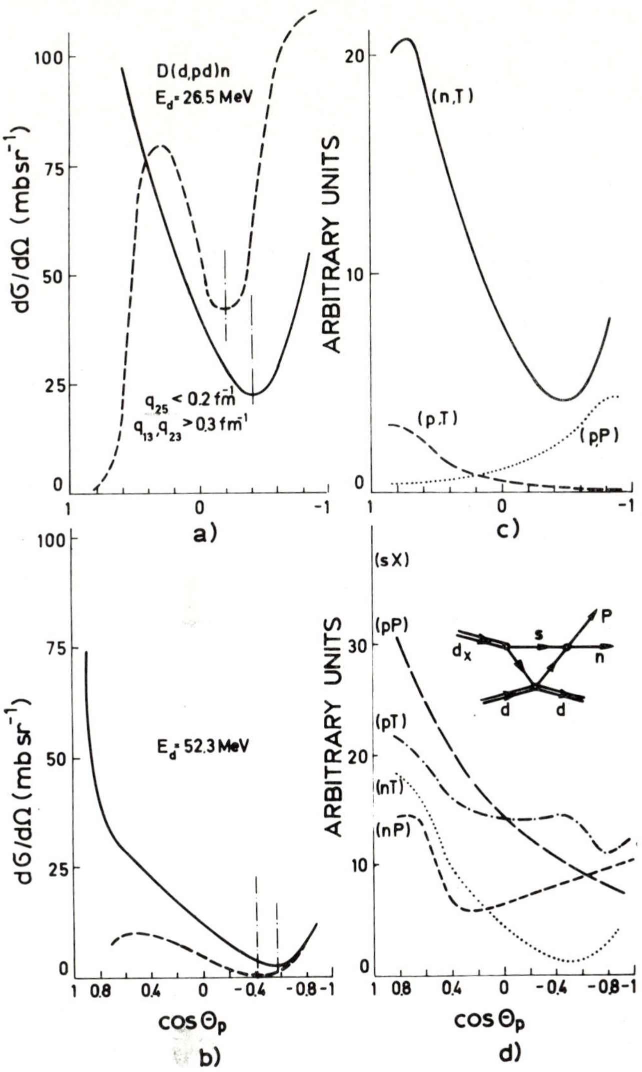

Fig. 8.a,b) Free (solid) and effective cross sections (dashed).
c,d) graphs for 26.5 MeV

take the effort to determine it experimentally; 3) comparison of these results with the off-shell interaction derived from two-hadron data (Bremsstrahlung, e^--d interaction /23/) and possibly on the other hand with N ≥ 4 data might help to separate the three-body force from off-shell effects.

The problem of more than three hadrons is considerably more complicated and richer than the one with three particles, but it has not yet reached the stage of development where the three-nucleon problem was a few years ago: in particular, there is disconnectedness between theory and experiment.

Most of our experimental research of breakup processes parallels the 1960-70 three-body work and it is directed to establish pronounced mechanisms. The following features have been established: 1) sequential decays, 2) quasifree scattering, 3) quasifree reactions, and 4) more complex quasi two-body processes.

Ad 2) D(d,pd)n QFS is one of the simplest breakup processes involving four nucleons because there is no strong FSI. However, there are four enhancements associated with four possible spectators. The energy and angular dependence of QFS have been studied /24-28/. The surprising minimum observed at 60 MeV has now

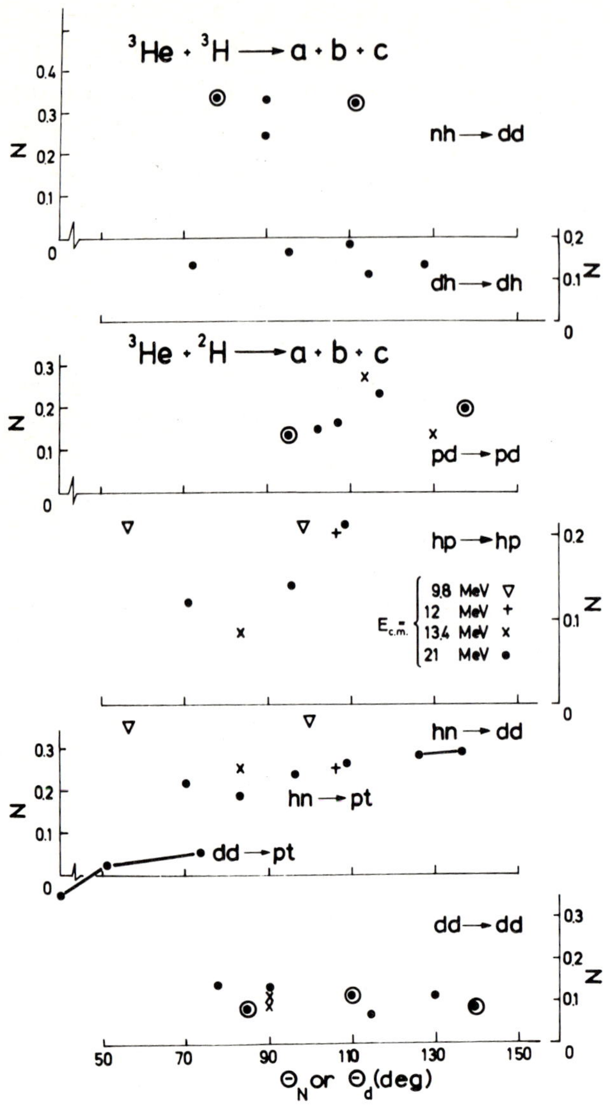

Fig. 9. Experimental/theoretical QF cross section as a function of angle. Double circles denote data for larger momentum transfer.

disappeared /28/ and QFS depends smoothly on incident energy. There are two angular-distribution measurements: at 26.5 MeV /24/ and 52.3 MeV /26/. Different from D(p,2p)n QFS angular distributions, the shape of which is well reproduced by the PWIA (Fig. 1), D(d,pd)n QFS angular distributions at both these energies depart from the PWIA (Fig. 8): the effective p-d cross section decreases sharply for small Θ_p^{cm} and its minimum is shifted. The second feature can be reproduced assuming that a deuteron interacting with a proton has shrunk. Fig. 8c and 8d display the contribution of pole and triangular graphs. The observed angular dependence can be produced by assuming that pole graphs corresponding to the neutron spectator in the target and the proton spectator in the projectile interfere constructively and all triangular graphs and the sum of higher-order graphs (which is probably isotropic) interfere destructively.

Ad 3) QFR are quite common in particle physics but the interest in these processes in nuclear physics started /29/ in 1972. Since then a systematic study of these processes - their energy, angular, and channel dependence - has been initiated /30-32/. It has been established that QFR enhancements do occur in many multiparticle reactions, e.g., $^3He(d,pt)p$, $^3H(^3He,dd)d$, $^3H(^3He,pt)d$, $^3H(\alpha,ht)n$, $^7Li(p,dd)\alpha$, $^6Li(d,pt)\alpha$, $^6Li(^6Li,\alpha\alpha)\alpha$, $^9Be(p,dh)^5He$, $^6Li(p,dh)^2H$ and that they can be described by the PWIA using some mechanism to reduce the absolute magnitude and to narrow the width. Part of this study is shown in Fig. 9, wherefrom the following features appear: a) The angular dependence for QFR and QFS in the $^2H+^3He$ and $^3H+^3He$ interaction seems to be well reproduced by the PWIA and the overall normalization factor N. However, the angular domain covered in these and all other QFR measurements is small, so one cannot reach any firm conclusions. b) Whereas the data on D(N,2N)N, $^3He(N,2N)D$, and $^3He(N,ND)N$ demonstrate that N increases with increasing incident energy, this is not true of QFR and QFS processes shown in Fig. 9. (see also ref. 31). 3) N depends on the reaction channel. To parallel this work with particle-physics research, Fig. 10 shows the spectator momentum distribution for various hadron-deuteron QF processes. It is known /33/ from D(p,2p)n that at high energies the impulse approximation using hard- or soft-core deuteron wave functions fails to explain the part of the distribution for the spectator momentum where one samples the region $r \lesssim 1$ fm and it is conceivable that the deuteron wave function consistent with N-N scattering data is inadequate. However, electron scattering and channel dependence /34-36/ for QFR shown in Fig. 10 demonstrate that it cannot be the entire explanation. The channel dependence can be summarized by defining the percentage f of events with $p \gtrsim 250$ MeV/c for each particular QF process. It is impossible to explain the momentum distribution by

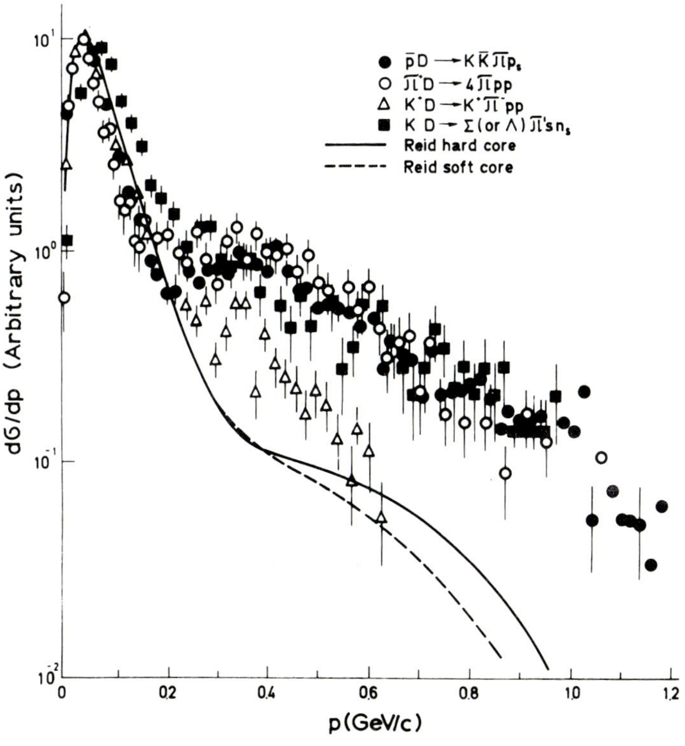

Fig. 10. Momentum distribution $|\Psi_d(p)|^2 p^2$

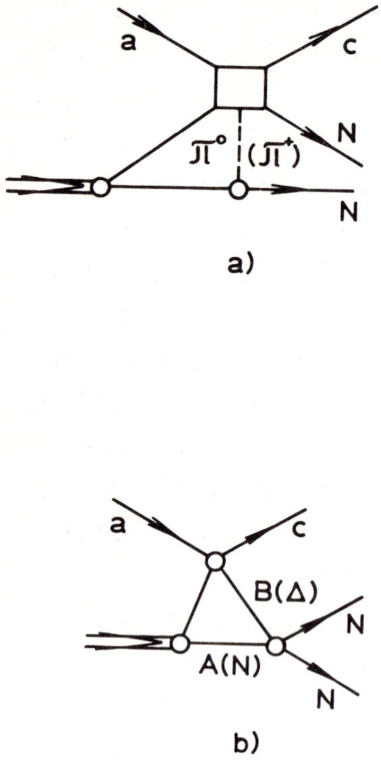

Fig. 11.

including double-scattering terms /36/ and the channel dependence corroborates it. It seems /36/ that the high-momentum excess, as well as its channel dependence can be explained by meson-exchange effects (Fig. 11a). Fig. 12. summarizes experimental f values with the prediction of the meson-exchange model (open circles connected with dashed lines) assuming the reabsorption probability, P_{π} of 20%. The model is capable of explaining the channel dependence. If one coherently adds the contributions of the exchange of charged and neutral pions (squares) one can use P_{π} = 15% and obtain even better agreement with the data. The only exceptions are $K^{-}D$ data which require P_{π} =100%. It seems that this can be explained by introducing triangular diagrams where the inside baryon B is different from the external one (Fig. 11b).

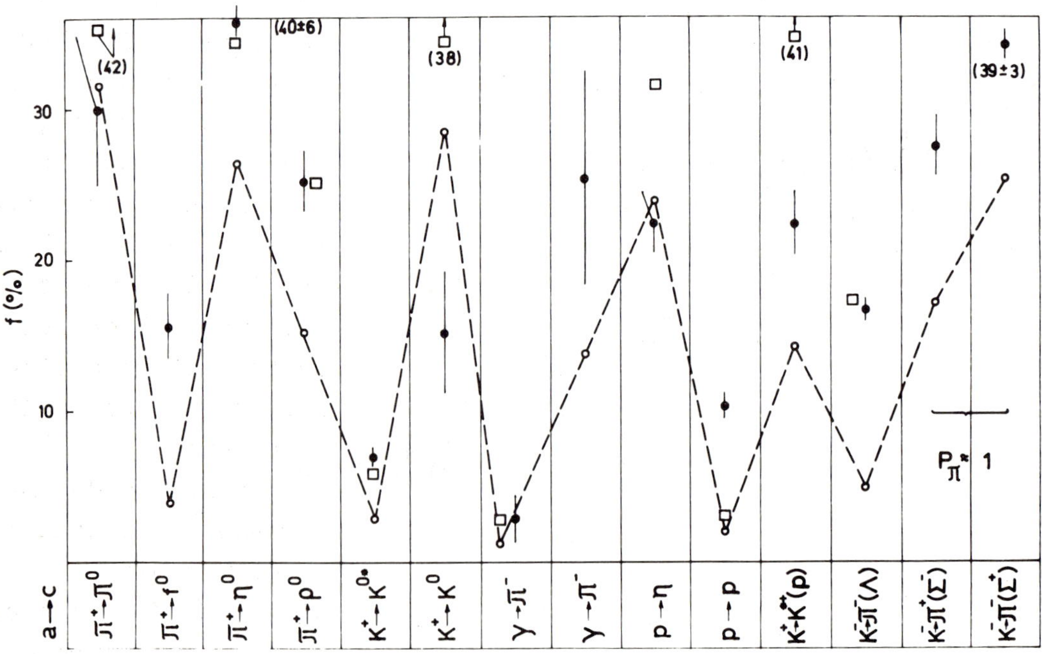

Fig. 12.

One of the reactions (K^-D, Fig. 10) shows a shift of the QF peak. Shifts were also observed in nuclear physics, e.g., $^3He(p,dp)p$ and explained /37/ in terms of FSI.

Ad 4) In discussing QF processes involving more than three hadrons, we have so far concentrated on reactions resulting in three particles. Processes resulting in four particles reveal the importance of quasi two-body processes. Examples of such quasi two-body processes are listed in Table 1 and discussed in contributed papers to this Conference.

A historical survey of the three-body problem would show three main avenues: development of the exact theory, theories where the emphasis is on the inclusion of the realistic NN force (e.g., variational calculations) and intuitive models. In 4,5,..-body problems one can now notice these same trends: 1) attempts to formulate the N-body extension /47/ of the Faddeev theory. 2) Microscopic model /48/ of Hackenbroich, where a sophisticated NN force is used and which can be applied to systems containing several nucleons, but has its own limitation among which, from the point of view of this survey, is that it is (at present) restricted to sequential decays. 3) Various intuitive models. There is no point to critically evaluate all these models, much of

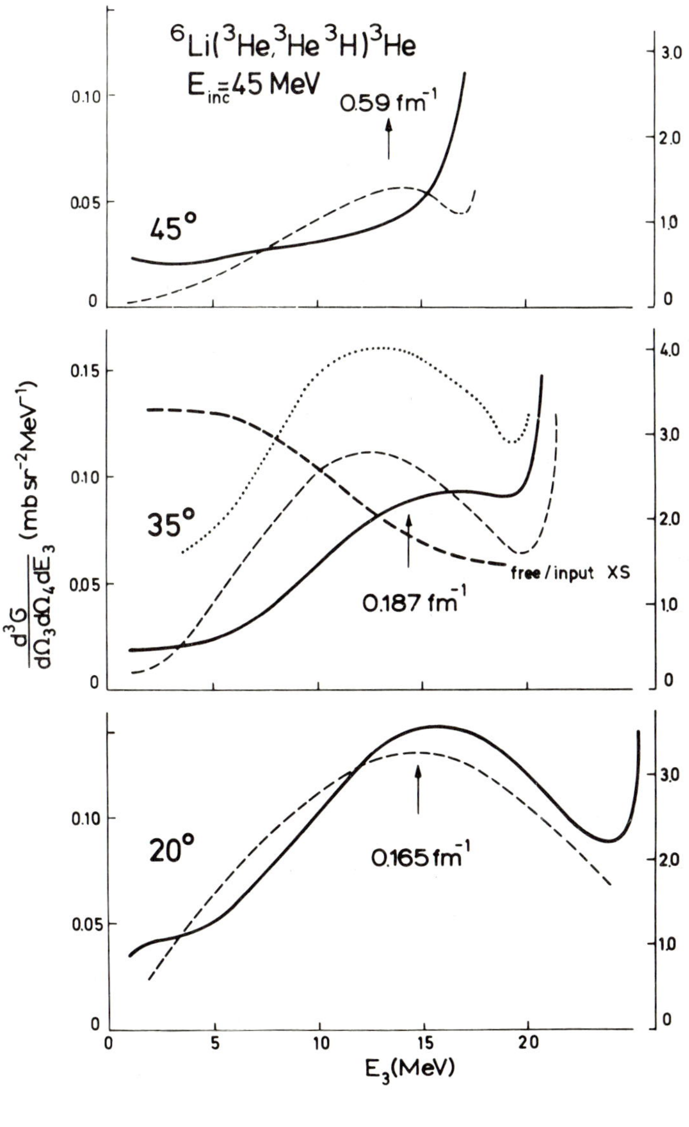

Fig. 13.

Table 1. TWO-BODY PROCESSES IN THE FOUR-BODY BREAKUP

Process	E_{inc} (MeV)	$\theta_3-\theta_4$ (deg)	Cross section ($\mu b/sr^2$ MeV)	Ref.
Double Spectator				
$D+D \rightarrow (p+p)_{QFS}+n_s^t+n_s^p$	34.7 80	34.8-34.8 43-43	70 50	38 39
$^3He+^3He \rightarrow (d+d)_{QFS}+p_s^t+p_s^p$	50 78	30.1-30.1 37-37	50 100	38 38
Double Final-State Interaction				
$D+D \rightarrow p^2+n^2$	23.15	18-36 25-43	840 630	40
$p+^3H \rightarrow d^*+d^*$	45.6	55-55	50	41
$n+^{10}B \rightarrow (\alpha,d)_{^6Li^*_{2.18}}+(n,\alpha)_{^5He_{gs}}$	14.4	4π	5.6%[&]	42
$n+^{12}C \rightarrow (\alpha,\alpha)_{^8Be_{gs}}+(\alpha,n)_{^5He_{gs}}$	14.4 18.2	4π 4π	no evidence no evidence	42 42
$n+^{14}N \rightarrow (\alpha,\alpha)_{^8Be_{gs}}+(\alpha,t)_{^7Li^*_{4.6}}$	18.2	4π	18%[&]	42
Crossing of QFS and FSI Bands				
$p+^3H \rightarrow (p+p)_{QFS}+(nn)_{FSI}$	20 45.6	21.3-21.3 38.7-38.7	 60	43 41
$p+^3He \rightarrow (p+p)_{QFS}+(pn)_{FSI}$	20	21.3-21.3 25.5-25.5		43
	35 45.6 100 156	37.5-37.5 38.7-38.7 42.5-42.5 peak	70 90 300 300	44 41 45 46
$^3H+^3He \rightarrow (d+d)_{QFR}+(pn)_{FSI}$	for several angles		80	
$\rightarrow (p+h)_{QFR}+(nn)_{FSI}$	at E_{inc} = 50,65		120	
$\rightarrow (p+t)_{QFS}+(np)_{FSI}$	and 78 MeV		100	30,31
Knock-out of $(np)_{^1S_0}$				
$p+^3H \rightarrow (p+d^*)_{QFS}+n$	45.6	47.5-47.5	30	41

[&] of the total four-body-breakup yield.

it has already been done /49/. Just two remarks concerning: 3.1. DWIA: The lesson learned from the three-body problem is that the PWIA is not correct. A straightforward correction /32/ is to include DW, but then one should worry about the meaning of the optical model for few-body systems. Surprisingly, but the optical model reasonably describes N-T /50/ and even N-D systems /51/. 3.2. Direct application of the Faddeev three-body theory (e.g., in the Amado version) to more complicated systems. Codes are being developed /52/ which will be able to treat three particles with arbitrary masses and spins and one could incorporate a reasonable interaction input. However, even now one can use the Ebenhöh code and calculate /53/, e.g., the ^{6}Li(^{3}He,ht)h breakup cross section. Fig. 13 shows the results of such calculations (solid curves - left-hand side ordinate) compared with the PWIA (dashed curves - right ordinate). The input interaction is the S-wave separable potential. This force is unrealistic, because even at 10 MeV the trion-trion cross section is anisotropic. The effect of a more realistic force could be studied by multiplying the "Amado model" curves with the ratio of the free t-h cross section as defined in the PWIA and the S-wave input force. At $\Theta_3=\Theta_4=35^{\circ}$, this ratio changes with E_3 and the resulting curve (dotted) has a larger cross section and its peak coincides with the PWIA peak. At $\Theta_3=\Theta_4=20^{\circ}$, the ratio is E_3-independent and it amounts to 1.5. Unfortunately, there are no data. The study of ^{6}Li(α,αh)t at 50 MeV

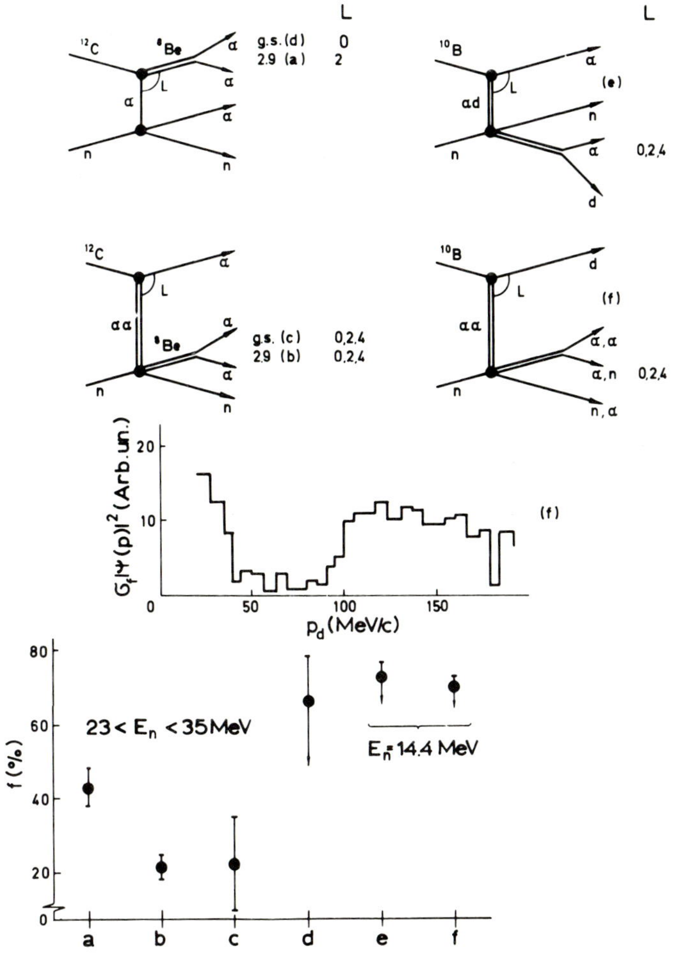

Fig. 14.

yields /54/ 0.7 mb/sr^2 MeV. The three-body approach to this problem is much too simplified. An adequate theory has to take into account internal degrees of freedom.

Why should we study these 4,5,..-hadron systems?

1) We have seen that two- and three-body systems cannot teach us all about nuclear forces. It is possible that 4,5,..-systems are more sensitive to some aspects of the interaction. A four-nucleon system is more tightly bound, it has known levels, and the mechanism of reactions is richer and potentially more sensitive to the force.

2) The interaction between unstable particles has always been the motive for few-body research. Insight into the interaction between more complex unstable particles, e.g., ^{8}Be+n, ^{6}Li*+n, etc., could be obtained from QF processes (Fig. 14). Preliminary results /42, 55/ seem to indicate the presence of some QF processes (Fig. 14f) and one can attempt to measure the presence of the ^{8}Be and ^{6}Li ground and excited states in ^{12}C and ^{10}B, respectively, investigating the contribution of higher relative angular-momentum components in the wave functions. Defining the percentage f of events beyond ~100 MeV/c, one obtains results which are not unreasonable in the spectator picture (except process d).

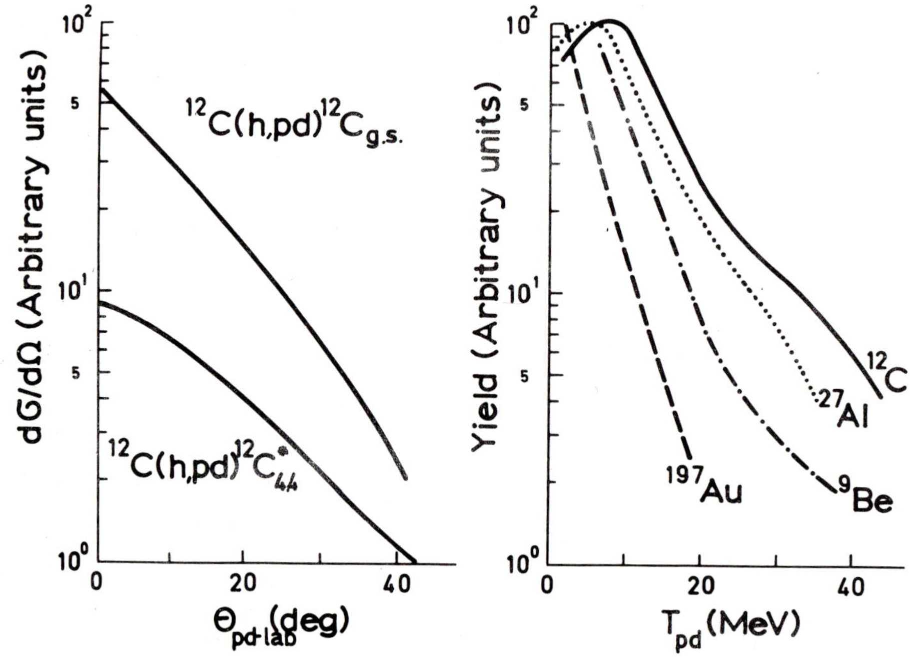

Fig. 15.

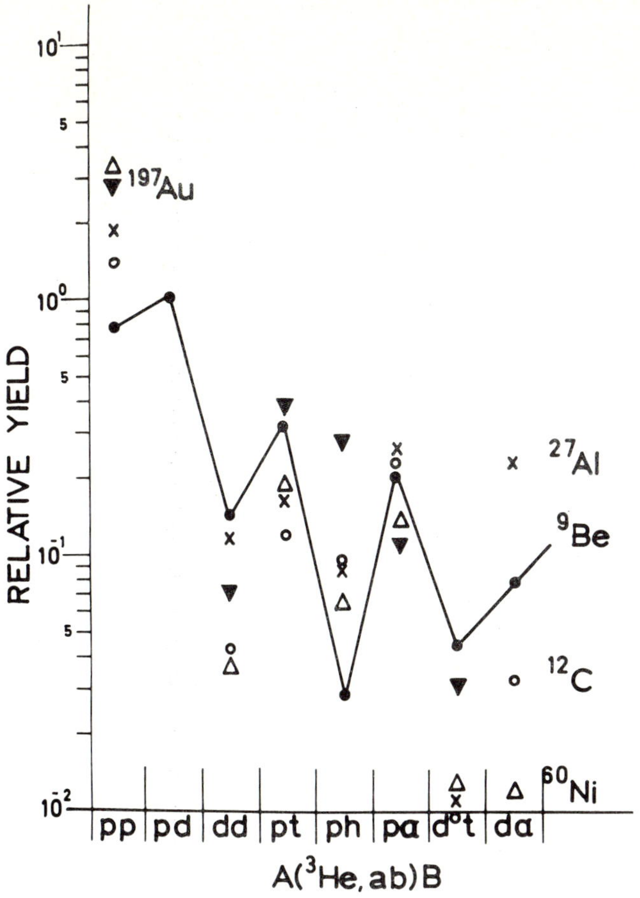

Fig. 16.

3) The presence of QFR implies that what has been assumed to be only a QFS involving complex particles is even in the picture of pole graphs a complex interplay of QFS and QFR. Instead of a transfer of D, T, or ^{4}He, one can have a correlated group of nucleons. If the factorization assumption is valid, then these processes could be distinguished through their different angular distributions /56/: ^{3}He(p,p)^{3}He vs ^{3}He(p,p)(pd), ^{3}He(p,d)(pp) as a function of T_{pp}, and ^{12}C(h,h)^{12}C vs ^{12}C(h,pd)^{12}C (Fig. 15). Thus, one could learn about the structure of nuclei. A careful choice of the final channel of the QF process can diminish the importance of FSI effects. All these features might allow a better insight into small components of wave functions, into the size of various clusters /57/, and into some specific mechanisms /58/.

4) Multinucleon removal induced by various projectiles is shedding /59/ new light on nuclear structure. The data are mostly derived from measurements of emitted gamma rays, consequently, one can lose information on the branch going to the ground state and one cannot distinguish the emission of, e.g., T from the 3N group. Techniques and models developed in few-body research are being applied /60/ to these processes. Preliminary results indicate that the sizeable fraction of the 69 MeV ^{3}He-induced two-fragment emission proceeds through the ground state of the residual nucleus (28% for ^{12}C, 18% for ^{27}Al, 20% for ^{60}Ni, and 30% for ^{197}Au). There is a definite trend to emit two fragments with the low relative energy and in the

forward direction (Fig. 15). The relative yield (Fig. 16) depends on the exit channel and on the target nucleus.

I have not even ventured to discuss the future of our field. Research has always been characterized by surprises. After all, what future would it be if it did not contain surprises.

References

/1/ K. Ilakovac et al., Phys. Rev. Lett. 6 (1961) 356; M. Cerineo et al., Phys. Rev. 133 (1964) B948; R.M. Salter et al., Few Particle Problems (ed. I. Šlaus et al., North-Holland Publ. Co., 1972) 112; B. Zeitnitz et al., Nucl. Phys. A231 (1974) 13

/2/ M.I. Haftel, I. Šlaus, D.L. Shannon, M.B. Epstein, W.T.H. van Oers, G. Anzelon, E.L. Petersen, W. Breunlich, to be published

/3/ E.L. Petersen et al., Phys. Rev. C9 (1974) 508

/4/ V.K.C. Cheng and P.A. Roos, Nucl. Phys. A225 (1974) 397

/5/ D.L. Shannon et al., Nucl. Phys. A218 (1974) 381

/6/ M.I. Haftel, E.L. Petersen and J.W. Wallace, preprint (1975)

/7/ D. Vranić et al., Few Body Problems (eds. R.J. Slobodrian et al., Les Presses de l'Univ. Laval, 1975) 660

/8/ P. Doleschall, Phys. Lett. 38B (1972) 298; C. Stolk and J.A. Tjon, Phys. Rev. Lett. 35 (1975) 985

/9/ W. Ebenhöh et al., Phys. Lett. 49B (1974) 137

/10/ B.J. Wielinga et al., Lett. Nuovo Cim. 11 (1974) 655; B.J. Wielinga et al., to be published

/11/ R.H. Thompson and L. Heller, Phys. Rev. C7 (1973) 2355

/12/ W.M. Kloet, Ph.D. thesis, Univ. Utrecht (1973)

/13/ R. van Dantzig et al., Priv. com. (1975)

/14/ J. Bruinsma, W. Ebenhöh, J.H. Stuivenberg and R. van Wageningen, Nucl. Phys. A228 (1974) 52

/15/ M. Jain, J.G. Rogers, D.P. Saylor, Phys. Rev. Lett. 31 (1973) 838

/16/ A.M. McDonald et al., Phys. Rev. Lett. 34 (1975) 488

/17/ E. Andrade et al., Nucl. Phys. A183 (1972) 145. These data should be renormalized up by ~10%.

/18/ J.L. Durand, O.M. Bilaniuk and C. Perrin, Nucl. Phys. A224 (1974) 77

/19/ J. Sanada et al., Few Body Problems (eds. I. Šlaus et al., North-Holland Publ. Co., 1972) 339

/20/ D.D. Brayshaw, Phys. Rev. Lett. 32 (1974) 382 and 34 (1975) 1478; M.I. Haftel and E.L. Petersen, Phys. Rev. Lett. 33 (1974) 1229 and 34 (1975) 1480; D.D. Brayshaw, Phys. Rev. C, to be published. Brayshaw also proves explicitly the connection between BC formalism and the Faddeev theory.

/21/ H.P. Noyes, Phys. Rev. Lett. 23 (1969) 1201

/22/ A.W. Saenz and W.W. Zachary, Phys. Lett. 58B (1975) 13

/23/ F. Coester and A. Ostbee, Phys. Rev. C11 (1975) 1836

/24/ A.D. IJpenberg et al., Few Body Problems (eds. I. Šlaus et al. North-Holland Publ. Co., 1972) 651

/25/ V. Valković et al., Nucl. Phys. A183 (1972) 126

/26/ J. Bialy et al., Few Body Problems (eds. I. Šlaus et al., North-Holland Publ. Co, 1972) 608

/27/ D.I. Bonbright et al., Phys. Rev. C8 (1973) 114

/28/ A. Hrehuss et al., Phys. Lett. 51B (1974) 454; G. Paić, priv. com. (1975)

/29/ M. Furić et al., Phys. Lett. 39B (1972) 629

/30/ I. Šlaus et al., Phys. Rev. C8 (1973) 444; R.G. Allas et al.,

Phys. Rev. C9 (1974) 787
/31/ R.G. Allas et al., Few Body Problems (eds. R.J. Slobodrian et al., Les Presses de l'Univ. Laval, 1975) 422 and 671
/32/ P.G. Roos and N.S. Chant, 2. Internat. Conf. Clustering Phenomena, Univ. Maryland (1975)
/33/ T.R. Witten et al., Nucl. Phys., to be published (1975)
/34/ e.g. D.W.L. Sprung, Few Body Problems (eds. R.J. Slobodrian et al., Les Presses de l'Univ. Laval, 1975) 475
/35/ B. Musgrave, Proc. Conf. on the Phenomenology of Particle Physics, Caltech, Pasadena, Calif. 1971
/36/ G. Alberi et al., Preprint 1975; G. Alberi et al., Phys. Rev. Lett. 34 (1975) 503
/37/ M.B. Epstein et al., Phys. Lett. 36B (1972) 305
/38/ J.M. Lambert et al., contribution to this Conference
/39/ Univ. Maryland Cyclotron Lab. Progress Report 1974, 44
/40/ R.E. Warner et al., Nucl. Phys. A243 (1975) 189
/41/ D.I. Bonbright et al., Phys. Rev. C, to be published
/42/ B. Antolković, M. Turk, priv. com. (1975)
/43/ M.J. Fritts and P.D. Parker, Nucl. Phys. A198 (1972) 109
/44/ C.Y. Hu et al., Few Body Problems (eds. I. Šlaus et al., North Holland Publ. Co., 1972) 648
/45/ Univ. Maryland Cyclotron Lab. Progress Report 1973, 73
/46/ R. Frascaria et al., Nucl. Phys. A178 (1971) 307
/47/ W. Sandhas, Czech. J. Physics B25 (1975) 251 and this Conference; O.A. Yakubovskii, Sov. J. Nucl. Phys. 5 (1967) 937
/48/ e.g., H.H. Hackenbroich, 4. International Symposium on Polarization Phenomena in Nuclear Reactions, Zürich 1975; also, 2. Internat. Conference on Clustering Phenomena, Univ. Maryland, 1975
/49/ I. Šlaus, Acta Physica Acad. Sc. Hung. 33 (1973) 191
/50/ H.S. Sherif and B.S. Podmore, Few Body Problems (eds. I. Šlaus et al., North-Holland Publ. Co., 1972) 691
/51/ R.M. DeVries et al., Nucl. Phys. A188 (1972) 449
/52/ P. Doleschall, priv. com. (1975)
/53/ D. Vranić and I. Šlaus, to be published
/54/ J.M. Lambert et al., Phys. Rev. C4 (1971) 2010
/55/ B. Antolković, I. Šlaus, M. Turk, P. Macq, to be published
/56/ B.J. Wielinga et al., Few Body Problems (eds. I. Šlaus et al., North-Holland Publ. Co., 1972) 687
/57/ J.Y. Grossiard et al., Phys. Rev. Lett. 32 (1974) 173. Though the interpretation of these particular data is presumably incorrect.
/58/ S. Kato et al., Nucl. Phys. A195 (1972) 534
/59/ O. Artun et al., Phys. Rev. Lett. 35 (1975) 773; V.G. Lind et al., Phys. Rev. Lett. 32 (1974) 479; D.R. Crisler et al., Phys. Rev. C5 (1972) 419
/60/ W. Hermens et al., Few Body Problems (eds. I. Šlaus et al., North-Holland Publ. Co. 1972) 968; R. van Dantzig et al., priv. com. (1975)

INCLUSIVE REACTIONS AND THE FEW-BODY PROBLEM

R. Rajaraman
Department of Physics and Astrophysics, University of Delhi, Delhi-110007, India

My motivation in giving a talk on this subject, derived in turn from the motivation of the Conference organisers in giving me this task is as follows. Although there is an intimate connection between the subject of inclusive reactions and the Few Body Problem, I am told that the former does not figure prominently in conventional Few-Body Problem literature, or conferences. Assuming that such is the case, I will venture to give a brief introduction to inclusive reactions, emphasing its connection to the Few Body Problem, but omitting topics of interest solely to high energy physicists. I will also try to assess the prospects for applying Few-Body-Problems expertise towards studying inclusive reactions.

To begin with, there is no such thing as an Inclusive reaction in the same sense that a reaction can be, say, inelastic or elastic, endothermic or exothermic. The distinction between "Inclusive" and "Exclusive" reactions lies not in the reaction itself, but in the eye of the beholder, so to speak. Consider for instance the high energy scattering of two particles a and b. They might scatter into several final channels denoted by

$$\begin{aligned} a + b &\longrightarrow a + b \\ &\longrightarrow c_1 + c_2 + \dots + c_n \\ &\longrightarrow d_1 + d_2 + \dots + d_m \quad \text{etc.} \end{aligned} \tag{1}$$

In this scattering processes, if the experimenters attempt to observe (or the theorists, to predict and explain) all the final state particles of a given channel, their momenta, spins, etc, they are said to be studying an "exclusive" reaction. By contrast, if in the same physical scattering phenomenon, they consider the momentum distribution of only one final state particle c (say, a π^+ meson) regardless of what else is produced with it, they are said to be studying an "inclusive" reaction. A mixture of all the channels containing at least one particle of type c will contribute to this inclusive reaction which is symbolically written as

$$a + b \longrightarrow c + X \tag{2}$$

where X stands for the unknown set of particles produced along with the observed particle c. Strictly speaking, this is a "single inclusive" reaction. Reactions such as

$$a + b \longrightarrow c_1 + c_2 + X \tag{3}$$

where two of the particles in the final state are observed are "double inclusive reactions", and so forth. For simplicity, let us start with single inclusive reactions in spinless (or spin averaged) processes. Then the relevant data is the momentum distribution of the observed particle c, i.e. $f(\vec{k},E) \equiv \omega_k \frac{d\sigma}{d\vec{k}}$, as a function of the total energy E of the pair (ab). It is evident that experimentally, this inclusive distribution is easier to measure than exclusive data. All one needs is a single detector capable of measuring the particle c, and the inclusive distribution can be obtained by moving the detector around. Consequently, inclusive distributions had been measured for quite some time, under the name of "single particle distributions". Interest in them was, however, limited because it was felt that from the point of view of theory, this distribution was a somewhat crude and average out quantity, given that a mixture of several final state channels could contribute to it, where every multiparticle channel is itself a difficult thing to analyse theoretically.

These single particle distributions enjoyed a revival around 1969 under the new name of inclusive distributions, thanks largely to theoretical notions that f(k,E) might have some simple behaviour, at least at very high energies. These ideas were championed in separate papers by Feynman [1] and by Yang and his collaborators [2]. Let me give just one example of their conjectures as illustration. They suggested that

$$\lim_{\substack{E \to \infty \\ \vec{k}\text{ fixed}}} f(\vec{k},E)_{lab.} = g(\vec{k}) \tag{4}$$

where $\vec{k}$, the momentum of the observed particle "c", and the total incoming energy E are measured in the lab. frame. Since the lab. energy of "c" is fixed as $E_{lab.}$ of the projectile is tending to infinity in equation (4), "c" is called a "fragment" of the target and the equation itself is called "limiting fragmentation". Feynman also suggested that

$$\lim_{\substack{E \to \infty \\ \vec{k}\text{ fixed}}} f(\vec{k},E)_{c.m.} = h(\vec{k}_\perp) \tag{5}$$

where $\vec{k}_\perp$ is the transverse momentum of "c". Note that equations (4) and (5) refer to distinct kinematic regions when the total energy tends to infinity. The latter, where $\vec{k}_{c.m}$ is kept finite is called the pionisation domain. Finally, by symmetry, a relation similar to equation (4)is also expected to hold in the frame where the projectile is at rest. This domain of fixed $\vec{k}$ in the projectile rest frame, called the projectile fragmentation region, is again distinct from the other two.

Much can be said about why such conjectures are by no means self evident and about the motivations of the authors in making them. Let me not go into that here as I want to establish the connection with the Few Body Problem as soon as possible. I will therefore go on to the work of Mueller [3] which not only pointed out this connection, but also put the Feynman-Yang conjectures on a firmer footing. It should however be mentioned that experimental data on a variety of high energy inclusive reactions supports the hypothesis of limiting fragmentation, aside from possible weak logE dependence.

The connection of the single inclusive reaction to the 3-body problem lies in the work of Mueller in 1970. Before resorting to precise equations, let me paraphrase the central idea. Consider the inclusive process

$$a(p_a) + b(p_b) \rightarrow c(k) + X \tag{6}$$

It's inclusive distribution $f(\vec{k},E)$ is clearly proportional to $\sum_{X_i} |\langle c(k),X_i|ab\rangle|^2$. Upon crossing the external line c, this is clearly related to $\sum_{X_i} |\langle X_i\,|\,\bar{c}(-k),a,b\rangle|^2$ where $\bar{c}(-k)$ is the antiparticle of c with the opposite four-momentum. At this stage, we exploit the fact that our ignorance of X_i is total, i.e. that $\sum_{X_i}$ goes over a complete set of states. Hence f(k,E) is related to the "total cross section" for the scattering of a,b <u>and</u> $\bar{c}$. Finally, by the obvious generalisation of the optical theorem to 3-body scattering, it is clear that this total cross section is in turn related to the appropriate generalization of the "imaginary part" of the forward scattering 3-body amplitude $a(p_a) + b(p_b) + \bar{c}(-k) \rightarrow a(p_a) + b(p_b) + \bar{c}(-k)$. This, in short, is the connection between the single inclusive reaction and the 3-body scattering problem. A more rigorous derivation of this idea is sketched below.

Note that $f(\vec{k},E)$ will in general be a function of the independent relativistic invariants $s \equiv (p_a+p_b)^2$, $t \equiv (p_b-k)^2$ and $M^2 \equiv p_X^2 = (p_a+p_b-k)^2$. M is the missing mass of the system X and s equals $(E_{c.m.})^2$ and is proportional to E_{lab} at high energies.

Now, $$\omega \frac{d\sigma}{d\vec{k}} \equiv f(\vec{k},E) = F \sum_{X_i} |\langle X_i c | ab\rangle|^2 \tag{7}$$
where F is the usual flux factor which behaves as $\frac{\text{const.}}{s}$ for large s.

By the usual LSZ reduction techniques

$$\langle X_i c | ab\rangle = \langle X_i | \hat{c}(k) | ab\rangle \tag{8}$$
$$= \int d^4y\, e^{iky} \langle X_i | j(y) | ab\rangle \tag{9}$$

where $\hat{c}(k)$ is the destruction operator of the particle c and j(y) is the current operator which forms the source term of the field equation for particle "c".

Thus $$f(k,E) \sim \frac{1}{s} \sum_{X_i} \int d^4y\, e^{iky} \langle ab | j(y) | X_i\rangle \langle X_i | j(o) | ab\rangle$$
$$= \frac{1}{s} \int d^4y\, e^{iky} \langle ab | j(y) j(o) | ab\rangle$$
$$\equiv \frac{1}{s} A(s,t,M^2) \tag{10}$$

But consider the $ab\bar{c} \to ab\bar{c}$ forward scattering process, where now M is the total c.m. energy. The same reduction techniques give

$$\langle ab\bar{c} | ab\bar{c}\rangle = i^2 \int d^4x \int d^4y\, e^{-ikx} \vec{K}_x\, \theta(x_0-y_0) \langle ab | [\phi(x),\phi(y)] | ab\rangle \overleftarrow{K}_y\, e^{+iky} \tag{11}$$

where
$$\vec{K}_x \phi(x) \equiv j(x)$$

Then usual maneouvers on the analyticity of this amplitude in the M^2 plane gives

$$A(s,t,M^2) \propto \text{disc}_{M^2} \langle ab\bar{c} | ab\bar{c}\rangle \tag{12}$$

which is the result anticipated by the earlier qualitative remarks. I have left out various well known constants of proportionality, volume factors etc., which are easily inserted into the above equations.

Similar relationship also exists between higher inclusive reactions and the corresponding Few-Body Problems. Consider for instance the double inclusive reaction
$$a + b \to c_1 + c_2 + X \tag{13}$$

where two particles in the final state are detected. The distribution measured will be

$$f(k_1,k_2,E) \equiv \omega_1 \omega_2 \frac{d^2\sigma}{d\vec{k}_1\, d\vec{k}_2} \tag{14}$$

Upon crossing both the c_1 and c_2 lines, this will be related to the forward scattering four-body process $a+b+\bar{c}_1+\bar{c}_2 \to a+b+\bar{c}_1+\bar{c}_2$ according to

$$f(k_1,k_2,E) \propto \frac{1}{s} \text{Disc}_{M^2} \langle ab\bar{c}_1(-k_1)c_2(-k_2) | ab\bar{c}_1(-k_1)\bar{c}_2(-k_2)\rangle \tag{15}$$

Again M here refers to the invariant "missing mass" of the unknown X.

This , then, is the major relationship of inclusive reactions to the Few-Body Problem. Some other facets of the inclusive reactions problem can also be related to the Few Body System, such as the triple-Regge coupling etc., but I will mention those later.

In order to predict inclusive distributions then, all we have to do is calculate the forward scattering amplitude of the corresponding 3 or 4 body system. Given the vast body of literature and arsenal of expertise already available in the area of Few Body Problems (as evidenced in the very conference), one might expect to directly apply all that to inclusive reactions. There is, however, a snag. We recall that the forward scattering of the 3 or 4 particles involved here is not a physical process. Some of the incoming particles (viz. those resulting from crossing) have negative energy, requiring some form of analytic continuation away from the physical region. Analytic continuation in itself will not deter 3-Body-Problem-theorists since even the physical non-relativistic 3-body problem, tackled a la Faddeev, requires knowledge of unphysical two-body amplitudes. As everyone in this gathering knows, the Faddeev technique splits the three-body scattering amplitude T(123) into

$$T(123) = T^{(1)} + T^{(2)} + T^{(3)} \qquad (16)$$

where $T^{(1)}$ stands for that part of the amplitude in which the pair (23) interacts last and so on. Then the $T^{(i)}$ obey the matrix integral equation (see for instance Gillespie [4]),

$$\begin{bmatrix} T^{(1)} \\ T^{(2)} \\ T^{(3)} \end{bmatrix} = \begin{bmatrix} t_{23} \\ t_{13} \\ t_{12} \end{bmatrix} + \begin{bmatrix} 0 & t_{23} & t_{23} \\ t_{13} & 0 & t_{13} \\ t_{12} & t_{12} & 0 \end{bmatrix} G_0 \begin{bmatrix} T^{(1)} \\ T^{(2)} \\ T^{(3)} \end{bmatrix} \qquad (17)$$

where t_{ij} is the two-body scattering amplitude between i and j, and G_0 is the Green's function $(E-H_0)^{-1}$. This formula assumes that there are only two-body forces. Its generalisation to include intrinsic 3-body forces has been worked out by Bruce McKellar and myself [5] but I wont go into that because I am sure Prof. McKellar will describe it in detail in his talk on 3-body forces. The important thing from our point of view is that even for calculating the on-energy-shell amplitude T(123) from the integral eqn.(16-17), one needs the two-body amplitudes t_{ij} off the energy shell, as input. One cannot directly plug them in from two body scattering data. Extrapolation from on-to off-energy-shell t_{ij} has to be done theoretically and depends on the underlying model. In non-relativistic problems this is most often achieved by using a suitable potential $v(r_{ij})$ to construct the t_{ij} by iteration. This leads to unique matrix elements $\langle p_i p_j | t_{ij} | q_i q_j \rangle$ even when the final and initial states have different energies. By this or related methods, people have for a long time now used reasonably well extrapolated t_{ij} off-the energy shell, plugged them into the integral equation and solved for the 3-body on shell scattering amplitude.

But the unphysical $\langle ab\bar{c} | ab\bar{c} \rangle$ amplitude needed for inclusive distributions calls for a more offensive form of analytic continuation. Whereas the intermediate states in the integral equation (17) did not have the same energy as the initial state, at least they comprised of particles of positive and physical energies, whereas in $\langle ab\bar{c} | ab\bar{c} \rangle$, the line $\bar{c}$ has negative energy $-\omega_k$, where ω_k is the energy of the observed particle c in the inclusive reaction. Even for a non-relativistic processes, this continuation in energy from ω_k to $-\omega_k$ is "relativistic" since ω_k includes the rest energy $(m_c)c^2$. Analytic continuation to this extent is, I personally feel, beyond the range of potential theory and similar non-relativistic models. Perhaps this is one reason why, understandably, not much work has been done on inclusive reactions based on conventional three-body problem techniques.

It would therefore be better to apply relativistic models to this 3-body problem.

Indeed, such methods have been used and have led to a limited amount of success in the inclusive problem. Let me divide these methods into two categories.

1. ANALYTICITY AND THE O(2,1) EXPANSION

This method, first employed by Mueller, has led to very useful results in ultra-relativistic limit, i.e. the total energy E tending to infinity. Consider, for instance, target fragmentation mentioned earlier which corresponds to fixed $(\vec{k})_{lab}$ as the total $E_{lab} \to \infty$. The inclusive distribution is a function of p_a, p_b and k which may be written in terms of angle-like O(2,1) variables η, ξ and ψ as

$$p_a = m_a\,(\cosh\eta,\ \sinh\eta, 0, 0)$$

$$p_b = m_b\,(1,\ 0,\ 0,\ 0)$$

$$k = m_c\,(\cosh\xi,\ -\sinh\xi\,\cos\psi,\ \sinh\xi\,\sin\psi, 0) \tag{18}$$

The fragmentation limit clearly corresponds to $\eta \to \infty$, as ξ and Ψ are kept fixed. The invariant function $A(s,t,M^2)$ to which $\omega \frac{d\sigma}{d\vec{k}}$ is related is also a function equivalently of η, ξ and ψ. It can be expanded in the complete set of O(2,1) harmonics $d^{\Lambda}_{om}(\eta)$ as

$$A(\eta,\xi,\psi) = \sum_m \int_{-\frac{1}{2}-i\infty}^{-\frac{1}{2}+i\infty} d\Lambda\ A^{\Lambda}_m(\xi,\psi)\ d^{\Lambda}_{om}(\eta)\ . \tag{19}$$

As $\eta \to \infty$, $d^{\Lambda}_{om}(\eta) \to (\cosh\eta)^{\Lambda}$

Thus, if the "partial wave" amplitude $A^{\Lambda}_m(\xi,\psi)$ has poles in the complex Λ plane, of which $\Lambda = \alpha$ has the largest real part, then

$$A(\eta,\xi,\psi) \xrightarrow[\eta\to\infty]{} \beta(\xi,\psi)(\cosh\eta)^{\alpha} \propto \beta(\vec{k})(E_{lab})^{\alpha} \tag{20}$$

One identifies the value of α by recognising that it is the leading Regge pole exchanged between the particle a and the pair $(b\bar{c})$ in the forward scattering process (see Fig.1). Note that as $\vec{k}_{lab}$ is fixed and $E_{lab} \to \infty$, the composite

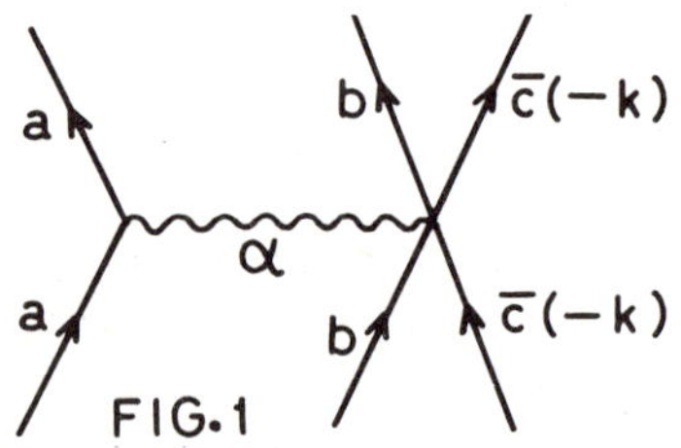

FIG.1

$(b\bar{c})$ is unaltered while the energyof "a" is increased to infinity. The situation is similar to two particle scattering, if $(b\bar{c})$ is viewed as a single system when $\vec{k}$ is fixed in the lab. frame. In conventional derivations of Regge theory, one expands the cross (t) channel two-body amplitude A(s,t) in terms of P $(\cos\theta_t)$:

$$A(s,t) = \sum_{\ell} a_{\ell}(t)\ P_{\ell}(\cos\theta_t) \tag{21}$$

The $P_\ell(\cos\theta_t)$ are harmonics of the rotation group O(3), which leaves the c.m. energy $\sqrt{t}$ invariant. One then analytically continues in ℓ, picks up the leading pole of $a_\ell(t)$. This same analysis could have been done in the s-channel. Then the little group of the Lorentz group which leaves a space-like momentum transfer invariant is the O(2,1) group, whose generator has continous eigen values along the imaginary line $-\frac{1}{2}-i\infty$ to $-\frac{1}{2}+i\infty$, instead of discrete integer values. This is what has happened in eq.(19). The O(2,1)expansion in (19) is thus essentially a Regge pole expansion, and the leading pole is expected to be the Pomeron with $\alpha = 1$. Thus, using (10) and (20),

$$\omega \frac{d\sigma}{d\vec{k}} \sim \frac{1}{s} A(s,t,M^2) \xrightarrow[s\to\infty]{} \frac{1}{s}\beta(\vec{k})(E_{lab.})^{\alpha} \tag{22}$$

or, since $\alpha = 1$ and $s \propto E_{lab.}$,

$$\omega\frac{d\sigma}{d\vec{k}} \xrightarrow[E\to\infty]{} \beta(\vec{k}) \tag{23}$$

We thus have the limiting fragmentation result (4) conjectured by Yang et al. and Feynman.

This then is a legitimate example of a three-body problem, although at ultra-high energies, where one can say something about the forward scattering amplitude, and relate it to single inclusive distributions. This technique can be extended to other kinematic domains and to double inclusive reactions also by making a multi-Regge expansion of the related four-body amplitude $\langle ab\bar{c}_1\bar{c}_2 | ab\bar{c}_1\bar{c}_2\rangle$.

2. PERTURBATIVE FIELD THEORY MODELS

Another relativistic model where one can expect to make the sort of analytic continuation required, is by summing Feynman diagrams of a relativistic field theory model underlying the interaction of particles a, b and $\bar{c}$. In taking this approach, there are two problems. Assuming that a,b and $\bar{c}$ are hadrons, one does not really know which particular field theory describes them, although there is a popular consensus that a model with fermion quarks and vector colour gluons in a non-Abelian gauge theory may turn out to describe hadrons (M. Gell-mann [6]). The second problem is that even given a field theory which can be perturbatively expanded in diagrams, it will be virtually impossible to even enumerate all the diagrams contributing to $\langle ab\bar{c} | ab\bar{c}\rangle$, let alone sum them. One must therefore resort to simplified models and pick some well defined subsequence of diagrams as an approximation. A simple example would be to treat all particles as being described by a single scalar field $\phi(x)$ with a $g\phi^3$ interaction, and sum the family of cross channel ladder diagrams shown in Fig. 2. This is a natural extension of a well known body of work in high energy two-body scattering, where the set of ladder diagrams with an arbitrary number of rungs is summed. That was first summed approximately, I believe, by Sawyer and latter using elegant Mellin transform techniques by Polkinghorne Bjorken and others [7]. The resulting two-body amplitude turned out to have the form $\beta(t)\, s^{\alpha(t)}$ as predicted by Regge theory. Corresponding analysis can be done for the $ab\bar{c} \to ab\bar{c}$ process, in both the pionisation and fragmentation regions by summing the ladder diagrams in Fig. 2. This is a somewhat more complicated situation, and arbitrary numbers of rungs must be included between the lines a, $\bar{c}$ and b. A student and I once did such summations [8]. We found that the amplitude for a given number of rungs ℓ, m and n as indicated in Fig. 2,

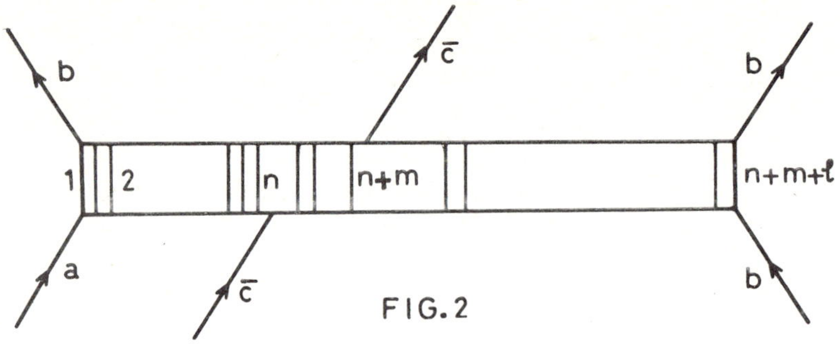

FIG.2

$$A_{n\ell m}(s,\vec{k}) = \Gamma_{\ell m}(\vec{k}) \frac{[\alpha \log s]^{n-1}}{(n-1)!} + \text{lower powers of } \log s \tag{24}$$

so that

$$A(s,\vec{k}) \equiv \sum_{n,\ell,m} A_{n\ell m} = \tilde{\Gamma}(\vec{k})\, s^{\alpha} \tag{25}$$

The "Regge intercept" α in this model will of course be a function of g, the coupling constant and μ, the mass of the field $\Phi(x)$. If they are adjusted to give α =1, then one will obtain the limiting fragmentation result. There are two levels of sophistication to this game. One is the leading log summation mentioned above where in each diagram with a given number of rungs only the leading term in log s is retained. The Mellin transform method dispenses with this approximation, and we found that both limiting fragmentation and pionization happen as $s \to \infty$, provided g and μ are adjusted to give a "Pomeron" intercept of α =1. More elaborate calculations of this type have been done, for instance, by Chang and Yan [9] where the family of diagrams summed is enlarged by using more complicated "rungs" than the ones in Fig. 2. It is also possible to sum such ladder like sequences using fancy renormalisation group equation methods. In this whole approach, apart from the approximation of summing only some subsequence of Feynman diagrams, such as cross channel ladders, it is pretty hard to get an exact sum for them. All one has been able to do is obtain the energy dependence of the inclusive distribution $f(\vec{k},E)$ as $E \to \infty$ and not the $\vec{k}$ distribution. The O(2,1) method also shares this deficiency.

Reggeon Couplings

Another area of inclusive reactions which few body physicists might look at profitably is that of Reggeon scattering cross sections and triple-Regge couplings. Let me first say what these are and how they arise (Finkelstein and Rajaraman [10]). Consider a given exclusive channel $a+b \to c + X_i$ contributing to the inclusive cross section $a+b \to c + X$. As $s \to \infty$ for fixed M and $\vec{k}_c$, one can make the usual Regge expansion for the amplitude $\langle cX_i | ab\rangle$

$$\langle c\, X_i | ab\rangle \simeq \beta_{bcR}(t)\, (s/M^2)^{\alpha(t)} \beta_{RaX_i}(t,M^2) \tag{26}$$

The inclusive cross section is given by

$$\omega \frac{d\sigma}{d\vec{k}} \sim \frac{1}{s} \sum_{X_i} |\langle c\, x_i | ab\rangle|^2$$

$$= \frac{1}{s} \beta^2_{bcR}(t) \left(s/M^2\right)^{2\alpha} \sum_{X_i} n_i \, |\beta_{RaX_i}(t, M^2)|^2 \tag{27}$$

where n_i is the multiplicity of particle c in the channel $|cX_i\rangle$. But

$$\sum_{X_i} n_i |\beta_{Rax_i}(t,M^2)|^2 \sim \sum_{X_i} |\langle X_i | aR\rangle|^2 \text{(symbolically)}$$

$$\sim M^2 \sigma_{aR}(M^2,t) \tag{28}$$

where σ_{aR} stands for the extrapolated Reggeon-particle "a" - total cross section (Fig. 3) at c.m. energy M for that pair and where I have omitted known constants of proportionality. The Reggeon is of course not a physical particle

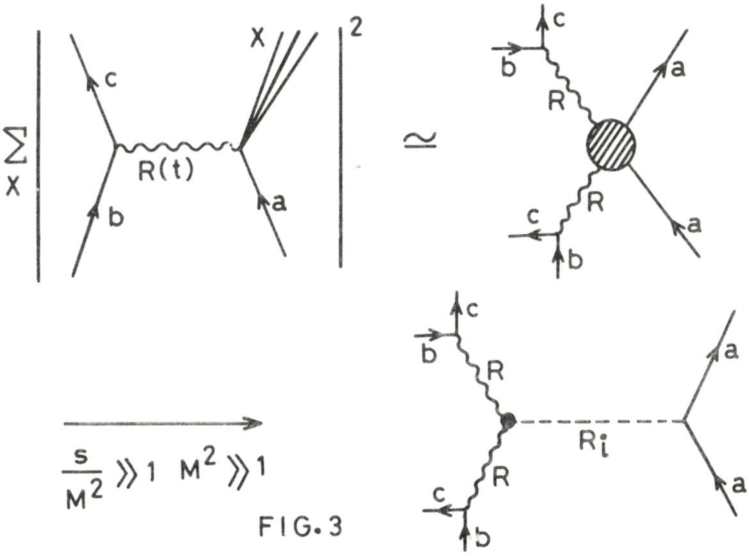

FIG. 3

except for these positive values of t for which $\alpha(t)$ is an integer. For other values of t, especially negative t as is the case here, the cross section $\sigma_{aR}(M^2,t)$ is a suitably analytically continued value in t and $\alpha(t)$. But the above equations allow us to "measure" it from the inclusive distributions for $s/M^2 \gg 1$. The other vertex factor $\beta_{bcR}(t)$ can be obtained from any two body process where b turns into c.

Another Reggeon coupling obtainable from inclusive distributions is the triple-Regge vertex, when both s/M^2 and M^2 are very large. Then $M^2\sigma_{aR}(M^2)$, which by optical theorem is related to the forward amplitude $f(aR \to aR)$, can further be expanded into contributions from leading Regge poles R_i in that process (because M, the c.m. energy in that process is large).

Thus $$M^2 \sigma_{aR}(M^2) \sim \sum_i c_i \beta_{aaR_i}(0)\left[(M^2)^{\alpha_{R_i}(0)}\right] \beta_{RRR_i}(t) \tag{29}$$

The function $\beta_{RRR_i}(t)$ which depends on the "mass" $\sqrt{t}$ of the Reggeon R is called the triple Regge vertex, and can be extracted from the inclusive distribution as outlined above.

What do few-body theorists have to do with all this? I feel that they might be able to try out models for extrapolating these Reggeon couplings and cross sections starting from physical values when the Reggeons are on the mass shell. Consider the vertex $\beta_{\rho\rho\rho}$ conneting three ρ mesons, for instance. Simple quark models treat the ρ as a J = 1 bound state of a $q\bar{q}$ pair using a simple potential between them such as a harmonic oscillator potential. The vertex $\beta_{\rho\rho\rho}(m_\rho^2)$ in this model would be the overlap integral

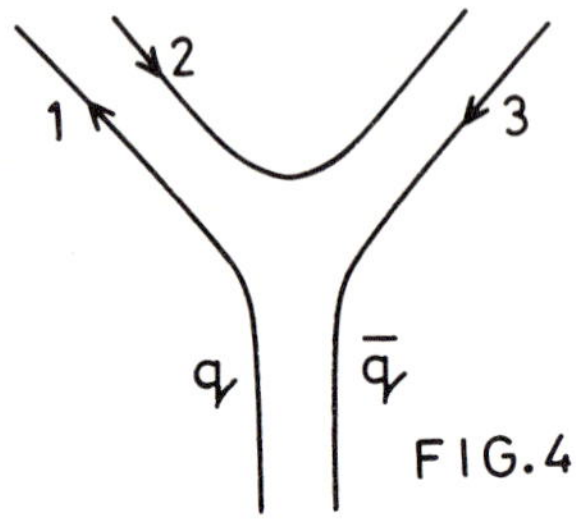

FIG.4

$$\int d\vec{r}_1 d\vec{r}_2 d\vec{r}_3\, \phi^*(r_{23})\phi^*(r_{12})\,\phi(r_{13})\, \exp\left\{i\vec{p}_1\cdot\vec{R}_{23} + i\vec{p}_3\cdot\vec{R}_{12} - i\vec{p}_2\cdot\vec{R}_{13}\right\} \tag{30}$$

where $\vec{r}_i$ are the coordinates of the ith quark and $\phi(r_{ij})$ is the wave function in terms of the $q\bar{q}$ pair (r_{ij}) and p_i are the momenta of the ρ mesons. When the appropriate J=1 wave function is used for the ρ meson, we get the physical vertex $\beta_{\rho\rho\rho}(m_\rho^2)$. To get the triple Regge vertex $\beta_{\rho\rho\rho}(t)$ where one ρ has zero mass and the other two a mass $\sqrt{t}$, and corresponding angular momenta $\alpha(t)$, one can continue to use (30) with suitably analytically continued wave functions. Given a potential theory-quark model, such analytical continuation is possible from the Schroedinger equation for the $q\bar{q}$ pair. I have given an admittedly crude idea. But so little work has been done by few body experts or quark modelists on the triple Regge vertex, that it might be useful if some of them work out an improved version of this approach. Similar work is clearly possible for the Reggeon-particle cross section σ_{aR} as well.

These then are some of the ways in which the field of inclusive reactions is intimately connected to the Few Body Problem.

It is a pleasure to thank Prof. A.N. Mitra for prompting me to think about this interesting question of relating inclusive reactions to few-body physics.

REFERENCES

1. R.P. Feynman, Phys. Rev.Letters 23 (1969) 1415.
2. J. Benecke, T.T. Chou, C.N. Yang and E. Yen, Phys. Rev. 188 (1969) 2159.
3. A.H. Mueller, Phys. Rev. D2 (1970) 2963.
4. J. Gillespie, "Final State Interactions," Holden-Day Inc.,San Francisco (1964).
5. B.H.J. McKellar and R. Rajaraman, Phys. Rev. C3 (1971) 1877.
6. M. Gell-mann, "Elementary Particles", Oppenheimer Memorial Lecture, Institute for Advanced Study Preprint (Oct. 1974).
7. For a review see R. Eden, P. Landshoff, D. Olive and J. Polkinghorne, "The Analytic s-Matrix" (Cambridge University Press,1966).
8. R. Rajaraman and A. Banerjee, Nuclear Physics B67 (1973),171 and Phys. Rev. D6 (1972) 3641.
9. S.J. Chang and T.M. Yan, Phys. Rev. D4 (1971) 537.
10. J. Finkelstein and R. Rajaraman, Phys. Rev. D5 (1972) 672 and Physics Letters 36B (1971) 459.
11. A.N. Mitra, "On the Formulation of Relativistic Hadron Couplings". Proceedings of the Workshop on Quarks and Hadron Structure, Erice (1975), Editor G. Morpurgo (to be published).

POLARIZATION PHENOMENA IN FEW-BODY SYSTEMS*

H. E. Conzett
Lawrence Berkeley Laboratory
University of California
Berkeley, California 94720

1. INTRODUCTION

Since the Quebec few-body conference [1] was held less than one and one-half years ago and the Zürich polarization symposium [2] took place only four months ago, it was anticipated that there would be rather few contributions on polarization phenomena at this conference. Therefore, this report includes, in addition, relevant material on the subject which either has been published since the Quebec conference or was reported at the Zürich symposium. My intention is to survey the recent developments in the study of polarization effects in the two-, three-, and several-nucleon systems. Since these measurements or calculations essentially have to do with cross sections for particular selected spin states, it is clear that these studies provide more detailed information and test more detailed calculations than is possible with the usual spin-averaged cross sections.

2. TWO-NUCLEON SYSTEMS

One of the nucleon-nucleon discussion sessions at the Quebec conference considered experiments for the next decade, with a stress on polarization experiments [3]. Included among the several suggestions for future experiments were: (1) a need for more accurate p - p cross section and polarization data below 20 MeV because of significant discrepancies among the existing data; (2) measurements of the neutron to proton polarization transfer (D_t) or the spin-correlation coefficient, C_{nn}, in n - p scattering at 50 MeV, in order to make a substantial improvement in the determination of ϵ_1, the $^3S_1$ - $^3D_1$ mixing parameter, since the available data left ϵ_1 undetermined between $-10°$ to $+3°$ at 50 MeV [4]; and (3) a general request for measurements above the inelastic threshold of total cross sections for various spin orientations of both beam and target, and for measurements of the analyzing powers and spin correlation coefficients, at the higher energies above the inelastic threshold. I am happy to report the significant progress that has already been made in these three areas at just the beginning of that decade.

Hutton et al. [5] have recently reported on measurements of the analyzing power in p-p scattering at 10 MeV. Their results as shown in Fig. 1. Note that the maximum (negative) value is 2×10^{-3} and the typical error is 10% of that, or 2×10^{-4}, so these are accurate data, with more than an order of magnitude better precision than had been attained previously. Traditionally, the low energy p-wave phase shifts have been obtained with phenomenological extrapolations from higher energies. The dot-dash curve in Fig. 1 is the analyzing power calculated from the phase shifts resulting from the 1 - 27.6 MeV analysis of Arndt et al. [6], in which the overall normalizations of the cross section data were kept fixed. That is clearly inconsistent with these data, whereas an analysis with the normalizations taken as adjustable parameters resulted in calculated analyzing powers that are in reasonable agreement

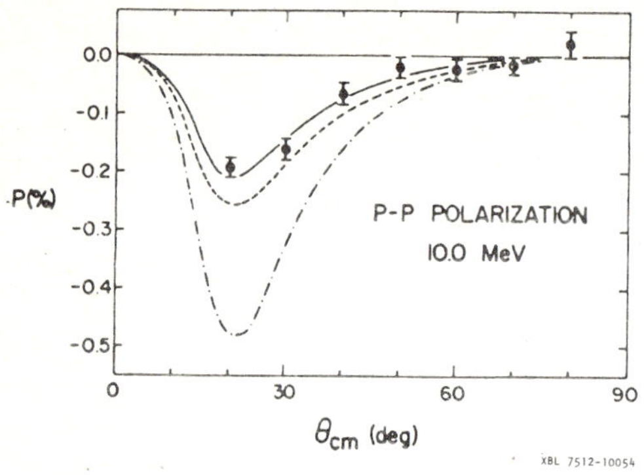

Fig. 1

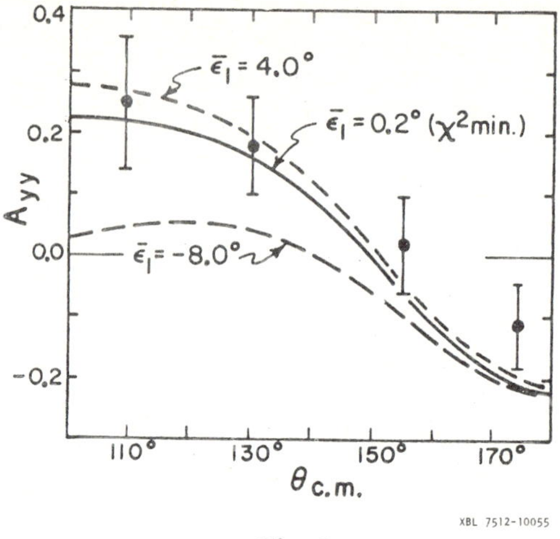

Fig. 2

with the data. The solid curve is the calculated analyzing power from a phase-shift analysis of these and nearby 9.9 MeV cross-section results. The s- and p-wave phase shifts determined in this analysis [5] are the first model independent determinations made below 25 MeV.

Proton-proton analyzing power data of similar accuracy at 16 MeV were reported at Zürich by Lovoi et al. [7], and the phase-shift prediction of MacGregor et al. [8] is in quite good agreement with them. Now that these two groups have shown that such accruacy is possible it seems clear that similar polarization measurements should be made at least at 25 and 50 MeV, the energies at which measurements of various p - p observables exist.

With respect to the n - p mixing parameter ϵ_1, Johnson et al. [9] reported at Zürich on their measurements at 50 MeV of the n - p spin correlation parameter A_{yy} (equal to C_{nn} for parity conserving or time-reversal invariant interactions). Their results are shown in Fig. 2, along with curves calculated from phase-shift solutions with the indicated values of ϵ_1. The $-8°$ value seems to be ruled out, but an additional overall normalization uncertainty of $\simeq$ 25% leaves the error on ϵ_1 considerably larger than the $\pm 1°$ envisioned by Binstock and Bryan. Their calculations, [4], in fact, with $\epsilon_1 = 2.8°$ gives A_{yy} (i.e. C_{nn}) = 0.38 at $\theta_{cm} = 120°$, which is a value 35% larger than that shown as the calculated result in Fig. 2. This discrepancy between the two calculations must certainly be resolved, but these measurements, requiring both a polarized neutron beam and polarized proton target, clearly demonstrate an impressive experimental achievement. Hopefully, this experiment will be further pursued in order to reduce the experimental errors and thus reduce the uncertainty in ϵ_1.

Krisch [10] reported on several unanticipated, thus interesting, results from p - p experiments at the Argonne ZGS (proton synchrotron) with transverse polarization of both target and beam at momenta up to 6 GeV/c. Total cross section measurements showed σ_{TOT} (spins antiparallel) to be 6 mb. larger than σ_{TOT} (spins parallel) at 2 GeV/c. This difference is more than 10% of the spin-averaged σ_{TOT}, and, as yet, no explanation of this large difference exists. Fig. 3 shows elastic p - p cross sections for various spin states, divided by the spin-averaged corss section. Note, for example, that $\sigma(\uparrow\uparrow)$ is about

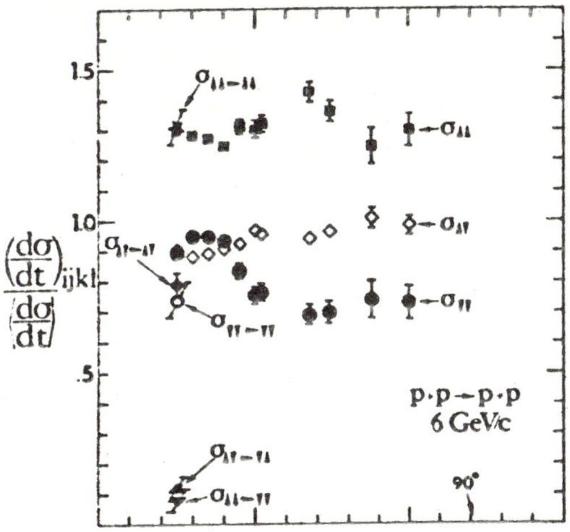

Fig. 3

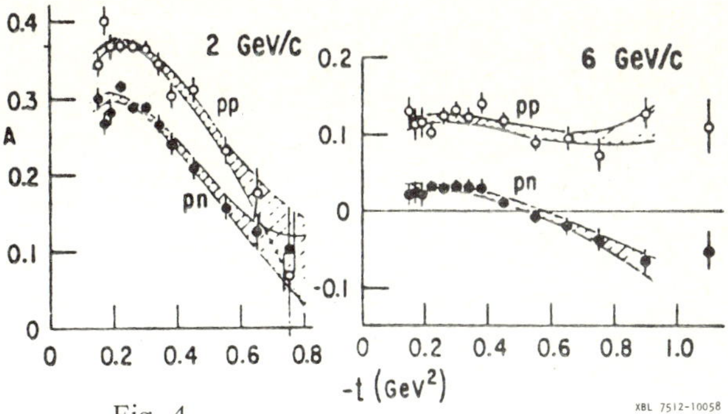

Fig. 4

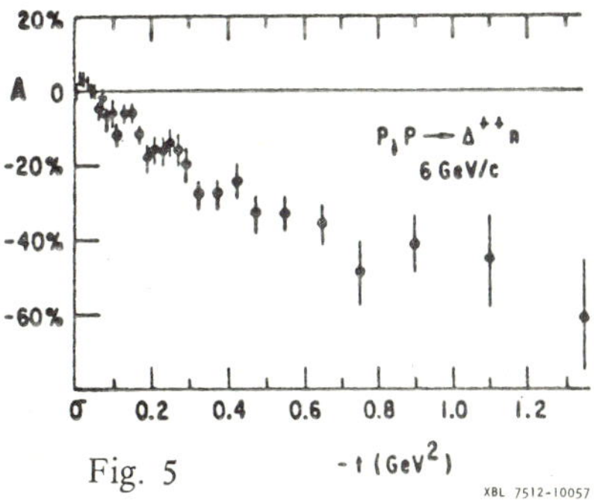

Fig. 5

twice $\sigma(\downarrow\downarrow)$ for the larger values of momentum transfer. Fig. 4 shows results of analyzing-power measurements in p-p and n-p elastic scattering at 2 and 6 GeV/c. Clearly the effects are substantial and they display a marked energy dependence. Finally, Fig. 5 shows analyzing-powers in the inelastic $p + p \rightarrow \Delta^{++} + n$ three-body final state reaction. Again the spin effects are large. There were no reported calculations of any of these observables, so the situation is reminiscent of that which existed with respect to elastic nucleon-deuteron scattering below 50 MeV several years ago. Some polarization measurements had been made, but the early exact three-body calculations were limited to s-wave N-N interactions, so they did not provide the observed polarizations. We can hope that the high energy theorists will provide calculations to reduce the present experimental-theoretical mismatch in as expeditious a manner as was accomplished by the lower energy three-body calculations.

3. THREE-NUCLEON SYSTEM

Since the three-nucleon polarization data and the relevant calculations were summarized at the Quebec conference, the work described here will serve as a natural extension of that report [11]. (See Fig. 1 of reference 11 for definitions of the spin-1 analyzing powers and of the polarization-transfer coefficients).

There is little new elastic N-d polarization data to discuss. Steinbock et al. [12] reported on their neutron analyzing power measurements in n-d scattering near 16 and 22 MeV, and their work confirmed that of Morris et al. [13] in finding good agreement with the p-d polarization data near those energies. There now remains only a single 35 MeV point at $\theta_{cm} = 120°$ for which the nucleon analyzing-powers in n-d and p-d scattering are reported to differ significantly [14]. All the other nucleon polarization data below 50 MeV agree, and thus support the principle of basic charge summetry of the nuclear interaction. On the theoretical side, Stolk and Tjon [15] have reported on their n-d elastic scattering calculations with the local Reid N-N potential, in contrast to the more generally used separable interactions. This is the first three-body calculation of polarization observables with local two-nucleon interactions. The s-wave contributions were calculated exactly, while the higher partial-wave contributions were calculated with a peturbative treatment. An example of their calculated polarizations, the neutron polarization at 14 MeV, is shown in Fig. 6. The indicated calculations included the following partial waves: dash ($^1S_0$, $^3S_1$ - $^3D_1$); dash-dot ($^1S_0$, $^3S_1$, $^1P_1$, $^3P_{0,1,2}$, no tensor); dash-dot-dot ($^1S_0$, $^3S_1$ - $^3D_1$, $^1P_1$, $^3P_{0,1,2}$); full curve adds $^1D_2$ and $^3D_2$ to the dash-dot-dot calculation. Note that the tensor interaction has a substantial effect (dash-dot to dash-dot-dot curve), as was found by Doleschall [16] in his exact calculation with separable potentials. Also, the effect was similar in depressing the back angle values and raising those at forward angles. The full-curve fit in Fig. 6 is not as good as that of Pieper's [17] analogous calculation with separable potentials, but at the higher energy of 46 MeV this local-potential calculation provides a better representation of the data than does the separable-potential model. So, there can be no claim yet of a better result from the local-potential calculation.

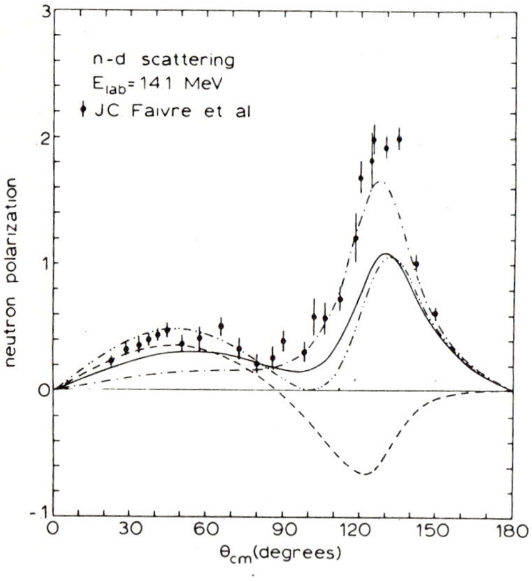

Fig. 6

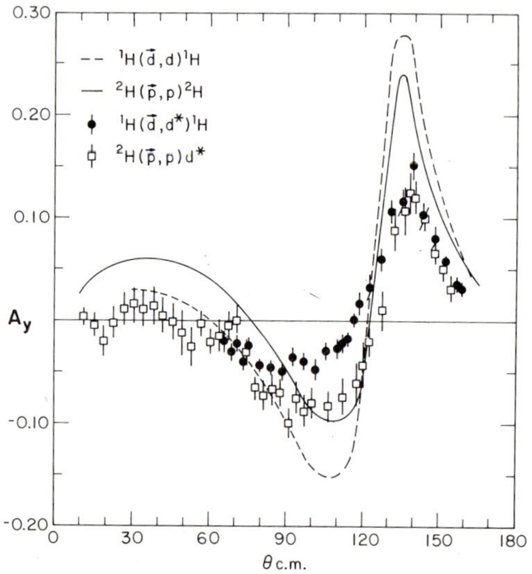

Fig. 7

Concerning polarization effects in the N-d breakup reaction, there have been recent noteworthy reports of both theoretical and experimental results. Rad et al. [18] have reported on measurements, at E_p = 22.7 MeV, of both the proton and deuteron (vector) analyzing powers in the p-d breakup transition to the np final-state interaction (FSI) region of the three-body spectrum. These data are shown in Fig. 7, where the smooth curves represent the elastic p-d proton and deuteron analyzing powers. The substantial values of these inelastic analyzing powers demonstrated a clear need for the exact N-d breakup calculations to include more than the simple s-wave N-N interactions that had been sufficient to fit the cross-section data. Such calculations are now underway. The first of these at E_n = 22.7 MeV, is reported in a contribution to this conference by Bruinsma and von Wageningen [19]. Their work shows that the tensor force affects the cross section in the region of the FSI peak but has little influence in the quasi-free scattering region. Their calculated deuteron tensor analyzing powers show very substantial values; an example is shown in Fig. 8. In particular, T_{20} exceeds 0.35 in the FSI region for the particular θ_1, θ_2 angle combination chosen here. These predictions of substantial tensor analyzing powers will certainly encourage experimentalists to measrue them. The calculated deuteron vector analyzing powers are less than 0.05 for this angle combination, so the situation seems to be comparable to that of the elastic N-d case, where the $^3S_1$ - $^3D_1$ tensor force was chiefly responsible for the observed tensor polarizations while the N-N p-wave interactions were the source of the vector polarizations. It is planned to next add the p-wave N-N interactions in these breakup calculations.

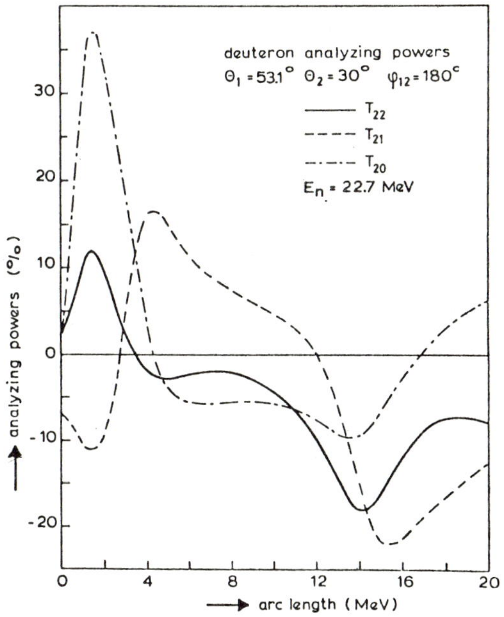

Fig. 8

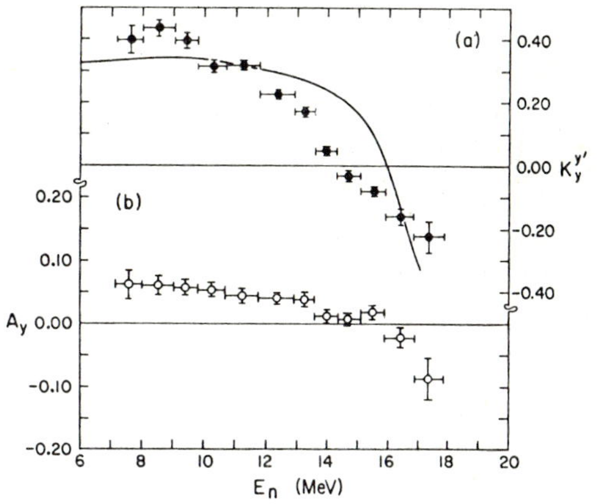

Fig. 9

Another polarization observable that has been measured recently in the p-d breakup reaction is the polarization transfer coefficient $K_y^{y'}(E_n)$ in the $D(\vec{p},\vec{n})2p$ reaction at E_p = 20.4 MeV and θ_n(lab) = 18°. These data are shown in Fig. 9, where the solid curve is the result from an exact three-body calculation limited to $^1S_0$ and $^3S_1$ separable N-N potentials [20]. Pieper [17] has shown that non-trivial values of $K_y^{y'}$ are obtained from such a calculation for N-d elastic scattering. Despite the qualitative agreement with the data in Fig. 9, the non-zero values of the analyzing power A_y, also shown in the figure, prove the necessity for other than s-wave interactions in the calculation.

There is now a real promise that the very fruitful interchange which occurred between theory and experiment with respect to polarization observables in elastic N-d scattering can now be repeated for the breakup reaction.

4. SEVERAL-NUCLEON SYSTEMS

There is such a wealth of polarization data on the mass 4, 5, and 6 systems that I will discuss here only selected examples of this work which seem to me to be particularly fruitful or unique. Most of these results were reported at the Zürich conference.

Dodder [21] described the Los Alamos group's program of multichannel R-matrix analysis of scattering and reaction data from few-nucleon systems. The aim of this program is to provide a simultaneous description of the experimental observables in all the reaction channels. As an example of the success of this work, Weitkamp [22] reported on measurements of three polarization-transfer coefficients in $\vec{p}$ - ^{3}He elastic scattering at three energies below 11 MeV. The R-matrix predictions of these coefficients, based on the previous analysis of several other observables in this mass-4 system, were in quite good agreement with the experimental data.

The Birmingham group [23] reported their measurements of analyzing powers in the elastic scattering of polarized ^{3}He from p, d, ^{3}He and ^{4}He at 32 MeV. These data are much more extensive and accurate than has been available from work using polarized ^{3}He targets, although the present limitation to a single energy restricts the utility of such data in an energy-dependent description of reactions involving these nuclei. However, there was no evidence of significant resonance effects in this energy region, and optical model analysis resulted in surprising good fits to these data, with spin-orbit potential strengths of 1 - 2 MeV. At the same time a Basel/Zürich group [24] reported on considerably improved and extended measurements in p + $^3\vec{\mathrm{He}}$ scattering between 2.3 and 8.8 MeV, using a polarized ^{3}He target. These ^{3}He analyzing-power data result in significantly better determinations of the p - ^{3}He phase shifts below 10 MeV.

The first measurements of polarizations in transitions to three-body final states of the mass-6 and mass-9 systems were discussed here by Slobodrian [25]. The final state proton polarizations were determined in the reactions ^{3}He(^{3}He, $\vec{p}$)pd and ^{6}Li(^{3}He, $\vec{p}$)$\alpha\alpha$. Fig. 10 displays their cross-section and polarization data for the first reaction, the (^{3}He, p) transition to the (unbound) ground state of ^{5}Li, at E(^{3}He) = 13.6 MeV. Significant proton polarization is seen. The solid curve which best fits the polarization data is a DWBA calculation for a deuteron-cluster transfer with J,L,S transfers of 2,1,1, respectively.

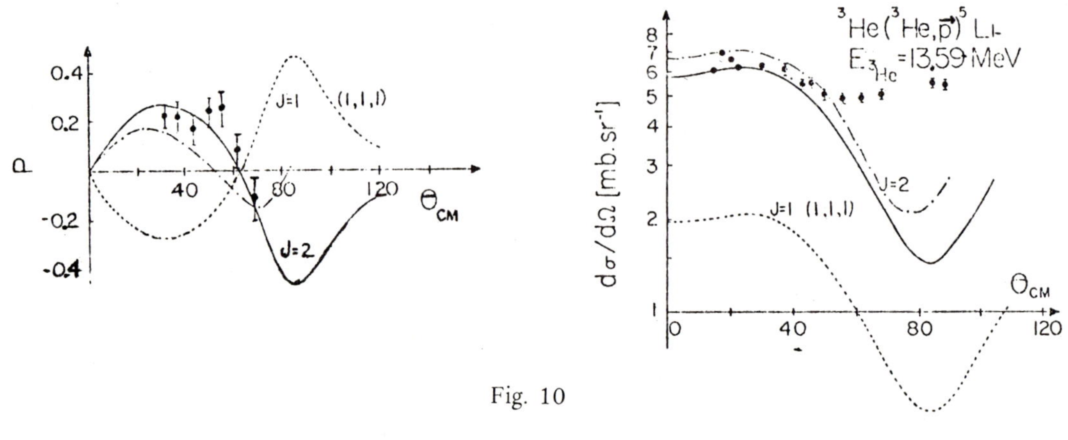

Fig. 10

Elastic scattering from ^{4}He has emerged as the principal means for determinations of the absolute polarizations of beams of polarized particles. Proof was given several years ago of the existence of points at which the analyzing power $A_y = 1.0$ in $\vec{p}$ - ^{4}He elastic scattering [26]. The establishment of $A_y = -1.0$ points in $\vec{t}$ - ^{4}He scattering was reported at Zürich [27]. Recent work has also been devoted to the search for similar points of maximum analyzing power in $\vec{d}$ - ^{4}He scattering. The more complicated spin-1 structure, as compared with spin-½, has led to some interesting developments. Analogous to the spin-½ case, the necessary and sufficient condition for the deuteron tensor analyzing-power component A_{yy} to be exactly unity has been established [28]. It is that $M_{11} + M_{1-1} = 0$, where M_{11} is a non-spin-flip element and M_{1-1} is a spin-flip element of the (3×3) d - ^{4}He M-matrix. From a phase-shift analysis of d - ^{4}He data, Grüebler et al. [29] showed that the equation $M_{11} + M_{1-1} = 0$ was satisfied at three points (E,θ). Fig. 11 illustrates the method for determination of one such point. Plotted are trajectories of the (complex) values of $M_{11} + M_{1-1}$ at several energies for different θ_{cm}. These were calculated from the known phase shifts. Because of continuity of these amplitudes in energy and angle, it is seen that at some point near E = 12.0 MeV, $\theta_{cm} = 55°$ $M_{11} + M_{1-1} = 0$, so that $A_{yy} = 1$. The experimental task was to find the point of relative maximum

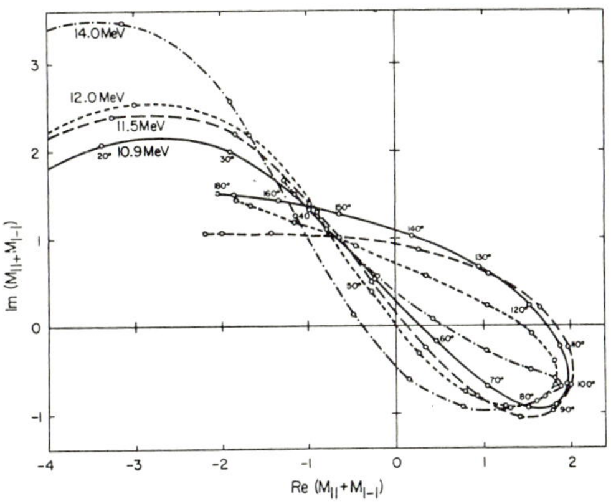

Fig. 11

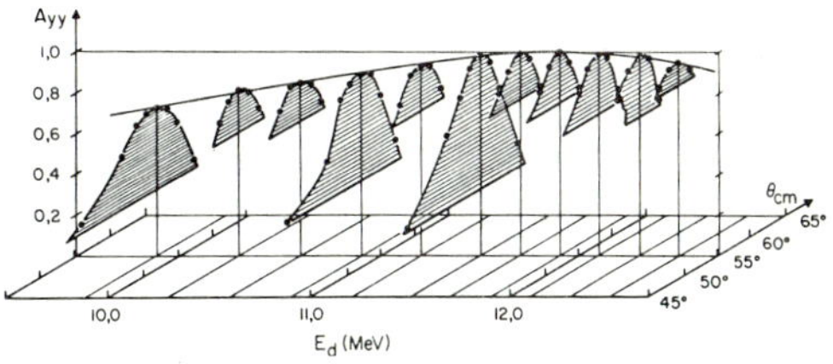

Fig. 12

A_{yy} in the region, to which one could then assign the value 1.0. Fig. 12 displays the data from that search in both E and θ_{cm}. The final determination was $A_{yy}(11.88\ \text{MeV}, 55°) = 1.0$ Grüebler et al. also examined the conditions imposed on the d - ^{4}He M-matrix in order that the vector analyzing power A_y reach the value ± 1.0. The conditions are so restrictive, requiring three equations among the M-matrix elements to be satisfied simultaneously, that at such a point <u>all</u> the analyzing-power components would be determined. Seiler et al. [30] then showed that this result followed from the basic condition that the density matrix is positive semi-definite, so that it is valid in reactions with the general spin structure $1 + a \rightarrow b + c$, where a, b, and c are arbitrary spins. The spin-1 analyzing power components are constrained to points within the cone shone in Fig. 13. It is easily seen that the points $A_y = \pm 1$ (points B or C) are quite unique, and at those points the other components are determined to be $A_{yy} = 1.0$, $A_{xx} = A_{zz} = -½$ (since $A_{xx} + A_{yy} + A_{zz} = 0$), and $A_{xz} = 0$. The a priori likelihood for an $A_y = \pm 1$ point in d - ^{4}He scattering is vanishingly small because of the requirement that three equations be satisfied simultaneously for a single (E, θ) point. In spite of that probability, an A_y value close to 1.0 in d - ^{4}He scattering was reported at Zürich by the Berkeley group [31]. Their measurements in the region of maximum A_y are shown in Fig. 14. Unlike the case of A_{yy}, it is not possible to prove that a point A_y = 1.0 has been reached. It is clear, however, that an $A_y \simeq$ 1.0 point exists, so it follows that the three equations among the M-matrix amplitudes are approximately satisfied. The particular importance of an A_y = 1.0 point is that here all of the analyzing-power components and, with a cross section measurement, all of the M-matrix amplitudes are determined.

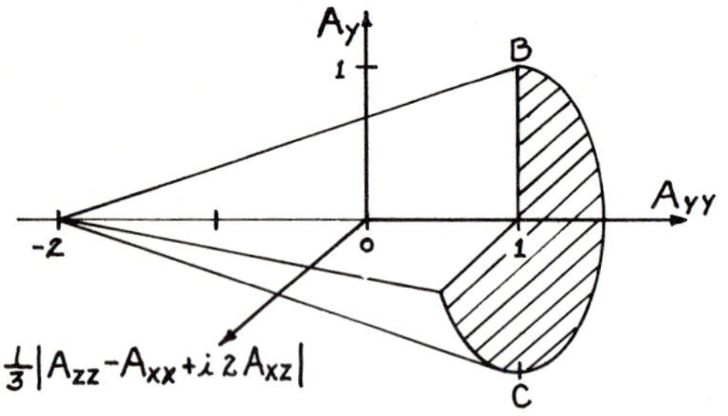

Fig. 13

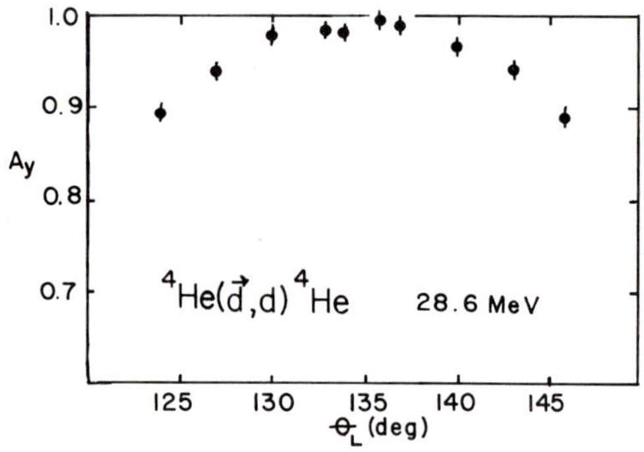

Fig. 14

5. POLARIZATION SYMMETRIES

There have been two rather recent developments that show the consequent effects on polarizations that follow from particular symmetries of the nuclear interaction. The first of these involves charge-symmetry or isospin conservation. Experiments on the ^{3}H(p,n)^{3}He and ^{7}Li(p,n)^{7}Be reactions [32] had shown that the neutron polarizations P_y and the proton analyzing powers A_y were essentially equal, with limited exceptions. This equality was seen to result from charge-symmetry [33], and it is an approximate equality because of the presence of the Coulomb interaction. However, it is in just those regions of significant difference between P_y and A_y that useful information can be derived on the structure of the nuclei studied [34]. Recent studies of P_y = A_y in (p,n) reactions on other nuclei were reported at Zürich. Fig. 15 shows the data of Lisowski et al. [35] on ^{9}Be. At 8.1 MeV the older P_y values (crosses) are quite different from those reported here (dots), and the latter are close to the measured A_y values (triangles). Final corrections to the A_y values from background subtractions are expected to improve the agreement. As is suggested by the comparison at 9.1 MeV, new measurements will be made there of P_y as a check on the old values (crosses).

The second development shows that for reactions with identical or charge-symmetric particles in the initial state, the angular-distribution symmetry of the analyzing powers can provide a clear signature of the reaction mechanism [36]. Consider, for example the ^{2}H($\vec{d}$, p)^{3}H reaction. There is, in general, no angular-distribution symmetry imposed on $A_y(\theta)$ by virtue of the identical initial-state deuterons, and indeed no symmetry was seen in such data below 20 MeV. However, if the reaction mechanism is a purely direct nucleon-transfer process it was seen that $A_y(\theta) = -A_y(\pi-\theta)$. Fig. 16 shows the two nucleon-transfer amplitudes which contribute. Since a and a' are deuterons the amplitudes for the transfers of

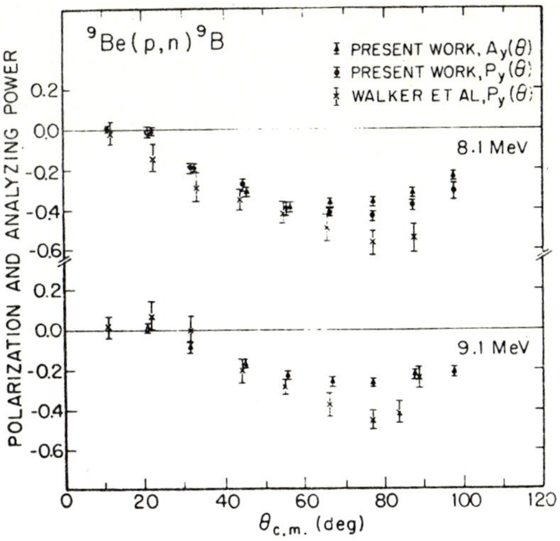

Fig. 15

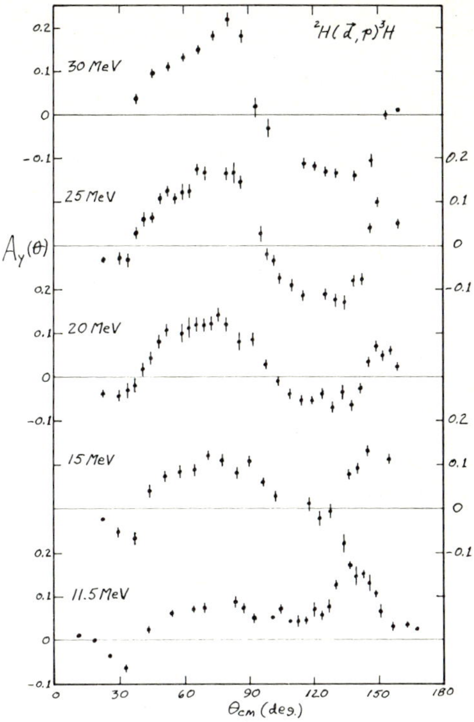

Fig. 17

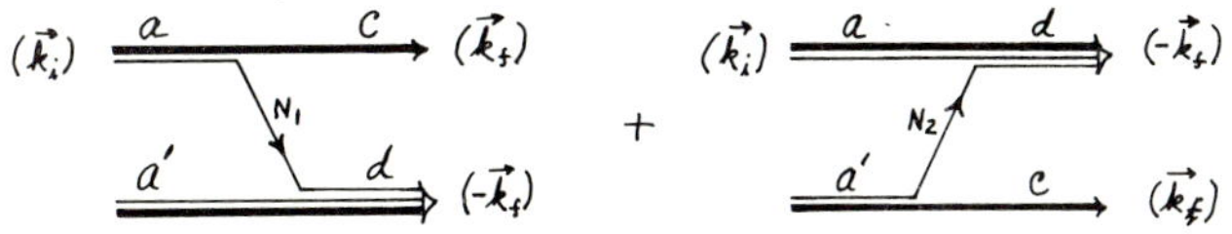

Fig. 16

neutrons N, and N_2 must be equal, and this equality imposes the symmetry on $A_y(\theta)$. Again, it is the deviations from the symmetry that provides the significant information. Fig. 17 shows $A_y(\theta)$ data for this reaction. The lower energy data show no symmetry, so it is clear that analysis in terms of ^{4}He intermediate states is required there. At the higher energies there is clear trend toward antisymmetry with respect to $\theta = \pi/2$, as is required of the direct nucleon-transfer process. Even there the antisymmetry is not exact, so this is evidence for contributions from the compound-nucleus mechanism at energies considerably above those for which excited states of ^{4}He are known to exist.

6. SUMMARY

Recent polarization studies in N-N scattering at and below 50 MeV have provided specific and significant improvements in the phase-shift parameters. High energy investigations with both polarized proton beams and targets have shown unexpectedly large spin effects, and this provides a challenge for theoretical effort to explain these results. Experimental and theoretical work on the three-nucleon problem continues to yield new and interesting results, with the emphasis now shifting to polarization studies in the breakup reaction. On-going work on several-nucleon systems continues to provide polarization data for general analyses, nuclear structure information, or specific resonance effects. Finally, the basic interaction symmetries continue to have unique and important consequences for polarization observables.

FOOTNOTE AND REFERENCES

*Work performed under the auspices of the U. S. Energy Research and Development Administration.

1. Few Body Problems in Nuclear and Particle Physics, eds. R. J. Slobodrian, B. Cujec, and K. Ramavataram (Les Presses de L'Université Laval, Quebec, 1975).

2. Proc. Fourth Int'l Symposium on Polarization Phenomena in Nuclear Reactions, Zürich, August 1975 (Birkhäuser Verlag, Basel), to be published.

3. J. Jovanovitch and J. Edgington, ref. 1, p. 170.

4. J. Binstock and R. Bryan, Phys. Rev. D 9 (1974) 2528.

5. J. D. Hutton, W. Haeberli, L. D. Knutson, and P. Signell, Phys. Rev. Lett. 35 (1975) 429, and ref. 2.

6. R. A. Arndt, R. H. Hackman, and L. D. Roper, Phys. Rev. C 9 (1974) 555.

7. P. A. Lovoi, G. G. Ohlsen, N. Jarmie, C. E. Moss, and D. M. Stupin, ref. 2.

8. M. H. MacGregor, R. A. Arndt, and R. M. Wright, Phys. Rev. 182 (1967) 1714.

9. S. W. Johnson, F. P. Brady, N. S. P. King, M. W. McNaughton, and P. Signell, ref. 2.. ..

10. A. D. Krisch, ref. 2; R. C. Fernow et al., Phys. Lett. 52B (1974) 243; W. de Boer et al., Phys. Rev. Lett. 34 (1975) 558.

11. H. E. Conzett, ref. 1, p. 566.

12. M. Steinbock, F. D. Brooks, and I. J. van Heerden, ref. 2.

13. C. L. Morris, R. Rotter, W. Dean, and S. T. Thornton, Phys. Rev. C 9 (1974) 1687.

14. J. Zamudio-Cristi, B. E. Bonner, F. P. Brady, J. A. Jungerman, and J. Wang, Phys. Rev. Lett. 31 (1973) 1009.

15. C. Stolk and J. A. Tjon, Phys. Rev. Lett. 35 (1975) 985.

16. P. Doleschall, Nucl. Phys. A220 (1974) 491.

17. S. C. Pieper, Nucl. Phys. A193 (1972) 529.

18. F. N. Rad, H. E. Conzett, R. Roy, and F. Seiler, Phys. Rev. Lett. 35 (1975) 1134.

19. J. Bruinsma and R. van Wageningen, 1975, contributed paper at this conference.

20. R. G. Graves, M. Jain, H. D. Knox, E. P. Chamberlin, and L. C. Northcliffe, Phys. Rev. Lett. 35 (1975) 917.

21. D. C. Dodder, ref. 2.

22. W. G. Weitkamp, W. Grüebler, V. König, P. A. Schmelzbach, R. Risler, and B. Jenny, ref. 2.

23. O. Karban, A. K. Basak, C. O. Blyth, J. B. A. England, J. M. Nelson, S. Roman, and G. G. Shute, ref. 2.

24. G. Szalosky, F. Seiler, W. Grüebler, and V. König, ref. 2.

25. R. J. Slobodrian, M. Irshad, R. Pigeon, C. Rioux, J. Asai, and S. Sen, paper at this conference.

26. G. R. Plattner and A. D. Bacher, Phys. Lett. 36B (1971) 211.

27. R. A. Hardekopf, G. G. Ohlsen, R. V. Poore, and N. Jarmie, ref. 2.

28. G. G. Ohlsen, P. A. Lovoi, G. C. Salzman, U. Meyer-Berkhout, C. K. Mitchell, and W. Grüebler, Phys. Rev. C 8 (1973) 1262.

29. W. Grüebler, P. A. Schmelzbach, V. König, R. Risler, B. Jenny, and D. Boerma, Nucl. Phys. A242 (1975) 285.

30. F. Seiler, F. N. Rad, and H. E. Conzett, ref. 2.

31. H. E. Conzett, F. Seiler, F. N. Rad, R. Roy, and R. M. Larimer, ref. 2.

32. R. C. Haight, J. J. Jarmer, J. E. Simmons, J. C. Martin, and T. R. Donoghue, Phys. Rev. Lett. 28 (1972) 1587; Phys. Rev. C 9 (1974) 1292; U. Rohrer and L. Brown, Nucl. Phys. A217 (1973) 525; Nucl. Phys. A221 (1974) 325.

33. H. E. Conzett, Phys. Lett. 51B (1974) 445; L. G. Arnold, ref. 2.

34. L. G. Arnold, R. G. Seyler, T. R. Donoghue, L. Brown, and U. Rohrer, Phys. Rev. Lett. 32 (1974) 310.

35. P. W. Lisowski, G. Mack, R. C. Byrd, W. Tornow, S. S. Skubic, R. L. Walter, and T. B. Clegg, ref. 2.

36. H. E. Conzett, ref. 2.

REMARKS ON POLARIZATION IN FEW-BODY REACTIONS

R.J. Slobodrian

Département de Physique, Laboratoire de Physique Nucléaire
Université Laval, Québec, G1K 7P4, Canada

This session consisted of papers which could be clearly differentiated on whether they were or they were not concerned with polarization effects. The following relates to the former. There were three invited papers authored W. Grüebler, S.S. Hanna, and L.C. Northcliffe and two contributed papers, by Ramavataram et al., and by Slobodrian et al. Hopefully quantity was compensated by quality. Figure 1 shows a qualitative drawing of the fields which are related and overlap :

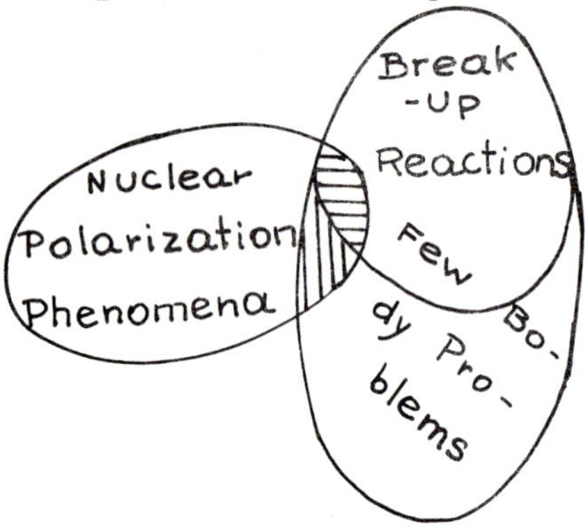

Fig. 1

Break-up reactions as a sub-set of our field overlaps with the field of polarization, which has its own Symposia.

H.E. Conzett's paper has detailed thoroughly both, the highlights of the last International Symposium held at Zürich and progress accomplished since. The purpose of these remarks is simply to sum up what we may have learned in this session as it took place, with some very lively discussions and comments.

We have learned from W. Grüebler of the very precise and careful work carried out at ETH-Zürich on studies with polarized deuterons. New concepts pervade now the few nucleon field, and the same path transited for nucleon-nucleon studies, where spin was initially considered an "unnecessary complication", is now in full progress here. A <u>complete experiment</u> is one where there is a full determination of the scattering matrix which in turn requires in general the measurement of a complete set of observables :

cross-sections polarizations - analysing powers	Easy to measure
polarization transfers correlation coefficients	Time consuming and more difficult

The number of observables increases rapidly in number for systems of higher spins. However, the question arises : can a measurement of an incomplete set of observables lead to unambiguous determination of the scattering matrix ? Grüebler answers with : yes, in certain cases. Particularly when determinations are made as a function of energy. For example, the results on d-α scattering between 0 and 17 MeV. This has been possible with cross section, vector and three tensor analysing powers. A significant proof of predictive powers of the subsequent phase shift analysis is given by the calculated σ_R (total reaction cross section) in good agreement with completely independent measurements, its agreement in yielding more easily visible the known resonances 2^+ and 1^+ of ^{6}Li and the possible prediction of a broad higher lying 1^+ resonance. Hence the method has a clear heuristic content. A careful comparison of the mirror reactions ^{2}H(d,p)^{3}H and ^{2}H(d,n)^{3}He has also been made. These reactions were measured simultaneously, hence there can be no question of relative normalizations. Grüebler has shown that most of the so called "discrepancy" and violation of charge symmetry in these two exit channels is taken into account by a simple correction of the energy scale for Coulomb effects. This is an important caveat in such matters, and the lesson is to refrain from statements unless Coulomb effects are taken into account. Clearly similar arguments can be made with respect to the determination of the T-matrix in reactions, a complete experiment ceases to be simply a kinematically complete experiment, unless other observables, related to the spin of the reaction products permits a determination of the transition matrix.

Non-conventional uses of polarized beams was the title given to a session at the recent Zürich Symposium, where the programme of S.S. Hanna at Stanford University on $(\vec{p},\gamma)$ reactions was discussed. The power of this new technique is now clear and I am afraid it will quickly cease to be a non-conventional method. It is relevant throughout the nuclear chart for stable targets. Its application to $^3H(\vec{p},\gamma)^4He$ reactions was the main theme here, and an impressive list of other few nucleon studies for the future was given. Significant progress was made and the determination of $\sigma_{E_2}(E)$ in the reaction mentioned above was shown in detail. Clearly an unambiguous disentangling of the multipole strengths in the region of the giant multipole resonances is now within reach.

Most impressive new results have been shown by L.C. Northcliffe, who has developped a presently unique facility at Texas A & M University, where a polarized proton beam with an average 75% polarization is available in intensities up to 1.6 μA ! With this intense beam precise neutron polarization experiments become feasible and this is the main theme of the facility, where there are helium neutron polarimeters available coupled to a four parameter on-line data acquisition system, and off-line analysis software permitting a considerable refining of results correcting for time delays, etc. In particular the study of the $^2H(\alpha,\vec{n})\alpha p$ reaction at 39.4 MeV has ruled out completely the old Q.F.S. calculations and favours a modified impulse approximation, although there is good room for theoretical improvement. Other studies are still in progress, but the importance of this programme for break-up studies is unquestionable. Perhaps direct $\vec{n}$ - n scattering will become feasible in a not too distant future.

It has become clear that even three nucleon studies, i.e. p+d and n+d → 3 nucleons are far from closed, both from the experimental and the theoretical side. The goals using polarized beams, targets or both are to measure a complete set of observables (or as many as possible). Similar remarks hold for four or more particle studies. On the theoretical side it is desirable that codes incorporate calculations with realistic interactions for the break-up channels with provisions for the calculations of analyzing powers, polarization transfers, etc. The subject of relativistic effects is also an important one, perhaps a crucial one. Spin dependent effects in hadronic interactions as in atomic physics or as the electromagnetic Lorentz force of classical physics, are perhaps also manifestations of relativistic effects in non-relativistic physics. A relativistic kinematics in the handling of experimental data or in dynamical calculations is far from satisfactory. We have learned from Franz Gross at this conference about the relevance in considering also the weak components of the wave function in relation to the dynamics of bound state problems. Undoubtedly such relevance exists also in the context of break-up reactions and polarization phenomena, dealing with scattering or unbound states. There should be at least estimates of the magnitude of errors committed when the dynamical calculation is non-covariant. Most probably such estimates will lead to the need of formulating a covariant calculation. Here it would be most appropriate if agreement is reached in what is a full covariant theory or calculation, as there exist at present several approaches. Perhaps it would be interesting to establish whether, in view of the extreme accelerations and decelerations of nucleons in the nuclear field, the context of special relativity is still adequate to treat correctly relativistic effects. We may be finally forced to drift into the unknown dangers of waters not yet sailed by nuclear theorists. Perhaps in some distant future a formulation will be found which will be relativistically correct and thus really exact.

Coming to the main theme of the conference : bridges, it has become apparent that a struggle still exists in few nucleon studies related to a clean separation of three body effects and off the energy shell effects. The hopes for a determination of the latter lie on good measure on measurements of bremsstrahlung due to processes depicted in fig. 2.

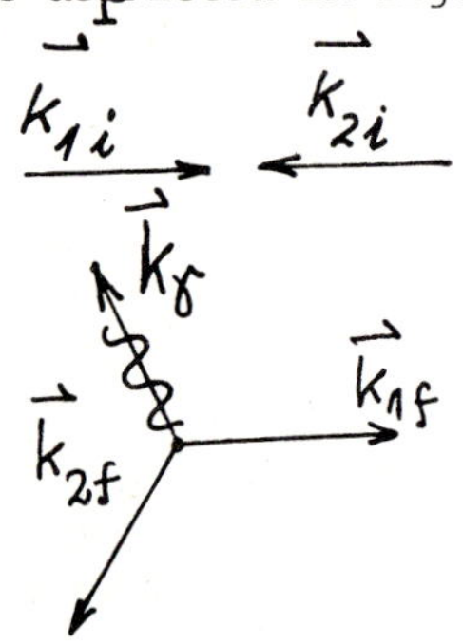

Fig. 2

The reaction is described by a T-matrix

$$T_{fi} = \langle \vec{k}_{1f}, \vec{k}_{2f}, \vec{k}_{\gamma} \; |V| \; \vec{k}_{1i}, \vec{k}_{2i} \rangle$$

Clearly experiments to-date are incomplete, as they simply measure a cross section for the process. It is questionable that such measurements can be conclusive, they never were, even for elastic scattering and even for the lowest energies assumptions were made to extrapolate the starting values of the p - and d - waves of the nuclear interaction.

Hence a bridge has to be built between polarization techniques and bremsstrahlung techniques to determine the T-matrix, performing complete experiments with polarized beams, polarized targets, or both. This should lead to an unambiguous determination of the two body off-shell interaction. Hence it will hopefully become possible to reach subsequently an unambiguous determination of three body forces in three body systems. For studies below 50 MeV the Texas A & M facility of Northcliffe is ideally suited. In particular, theoretical calculations agree still only qualitatively with the experimental cross section in that energy range, and hence the situation is still fluid even for this observable. At intermediate energies meson factories could contribute significantly to the solution of this problem if they increase the intensity of their polarized beams to levels comparable to those of the facility previously mentioned. In their energy range it is doubtful that real progress could be made without performing complete experiments in the sense indicated above.

This was a most timely, interesting and stimulating session, thanks to the enthousiasm of all participants.

BREAK-UP REACTIONS

Rapporteur's talk on papers submitted in session IV

H.G. Pugh

University of Maryland, College Park, Maryland 20742
(On leave of absence at National Science Foundation,
Washington, D. C. 20550)

This has turned out to be an excellent time to hold a conference. We have had time to digest the discussion from Quebec concerning the limitations on information available from breakup reactions; we have had an opportunity to reassess the existing data in the context of that discussion; we have had an opportunity to produce some very precise data which lead to a sharp confrontation between theory and experiment; finally the theorists have been at work in a direction which may help to resolve the problems that have arisen.

The papers submitted fall naturally under three headings:

- Three-nucleon Break-up Reactions
- Four-nucleon Break-up Reactions
- Three-and Four-Body Break-up of Systems with $A > 4$

I. Three- Nucleon Break-up Reaction

$$\underset{1}{p}+\underset{2}{d} \rightarrow \underset{3}{p}+\underset{4}{p}+\underset{5}{n}$$

I shall be concerned only with exclusive measurements, in which the momenta of all three final-state particles are determined. Historically the conferences in this series have marked milestones in the development of these studies.

Before UCLA: most measurements were in regions of phase space labelled Quasi-free Scattering (QFS) near $T_5=0$ or Final State Interaction (FSI) near T_{34} or T_{35} or $T_{45}=0$. Most analyses were in terms of Plane Wave Impulse Approximation (PWIA) for the QFS, or of Watson-Migdal model (WMM) for the FSI with perhaps some kind of crude distorted wave treatment.

In 1972 Ebenhoh's code became available for three-nucleon breakup based on the Faddeev formalism with S-wave separable potentials for the nucleon-nucleon interaction.

At UCLA: comparisons of data with results predicted by this code made it clear that for data up to 50 MeV all regions of phase space were qualitatively explained. This was an important milestone. Since then the QFS and FSI labels are still used for convenience in identifying regions of phase space but the corresponding PWIA and WMM theories are abandoned except for special purposes.

In 1973, Kloet and Tjon studied parameter sensitivity in the three-body calculation and in 1974 Brayshaw emphasized limits on the information available from three-body breakup measurements.

In QUEBEC, very close coordination between theory and experiment began, which has led to the present situation, where

At DELHI, very precise experimental information is available of sufficient quality that it becomes clear that the standard calculations are inadequate at all energies. It seems that we need:

better form factors
tensor forces (deuteron D-state)
higher partial waves
Coulomb corrections in intermediate states

Fortunately, a new calculation was presented at this conference which includes tensor forces and higher partial waves. This leads to the hope that by the next conference the theoretical situation will be in sufficiently good shape that a confrontation between experiment and theory will permit extraction of quantities of fundamental interest.

Papers Presented in This Session

(1) D. Miljanić, E. Andrade, and G. C. Phillips presented a new analysis of 1972 data at 12.5 MeV. They compared WMM for FSI and PWIA for QFS with the results of the Ebenhoh code (EC). They found:

that the EC fits the data within 20-30%;
that PWIA gives cross sections too high by an order of magnitude for QFS and also fails to give the correct ratio for pp and pn QFS;
that WMM does quite well in fitting the shape of the FSI but fails in detail, for reasons they discuss.

They conclude that use of EC makes obsolete the previous approaches which present no advantages at all when accuracy is of the essence.

(2) W.T.H. van Oers presented an invited paper which exemplified the change in approach since UCLA:

In 1973 the Texas A&M group found experimentally that there are regions of phase space where the single-scattering (Born) term in the amplitude tends to cancel the remainder or multiple-scattering part. Detailed calculations showed that in these regions the cross section is dominated by the $M(\frac{1}{2},0)$ amplitude - total spin $=\frac{1}{2}$ and nucleon-nucleon total spin=0. Jain, Rogers and Saylor suggested suitable kinematic conditions for experiments to exploit this feature.

van Oers made very precise measurements at 23.0 and 39.5 MeV under two sets of conditions:

I. θ_3=fixed; T_3=fixed; $T_{35}=T_{45}$=fixed; T_5=varied.

II. $\theta_3=\theta_4$; $T_3=T_4$; $T_{34}=T_{35}=T_{45}$=fixed; T_5=varied.

In both cases the choice of fixed quantities was made in a prior study using a detailed code.

Van Oers found important discrepancies near the minima, up to nearly 100% of the measured cross section. Away from the minima agreement is found to be within 20-30%.

(3) G.J.F. Bloomestijn, Y. Haitsma, R. Van Dantzig, and I. Slaus presented similar data taken at 15 MeV using the BOL 4π detector at IKO. They reached similar conclusions.

(4) N. Fujiwara, E. Hourany, H. Nakamura-Yokota, F. Reide, and T. Yuasa

studied the Lambert configuration at 156 MeV and compared the data with a multiple-scattering expansion by L'Huillier.

This kinematic configuration was described at Quebec by Lambert et al. who suggested that it might be particularly sensitive to off-shell or 3-body effects. It is a configuration in which the three final-state particles are collinear in the c.m. system and one is at rest. Clearly reaching this final state involves large momentum transfer to all particles.

The discrepancy between theory and experiment was found to be enormous; there was also very poor convergence of the multiple scattering series. This indicates that any short range information may be masked by multiple interactions. M.I. Haftel, E.L. Peterson and J.Wallace had already reached this conclusion in a preprint. They found that the Lambert configuration is not specially sensitive to off-shell effects, at least below 50MeV: fitting data in this region requires principally knowledge of on-shell parameters. In this preprint, Haftel et al. also found that the FSI region is the one most sensitive to off-shell effects.

(5) J.Doornbos, W. Krijgsman, and C.C. Jonker studied the pn FSI region at 20 MeV as a function of production angle. They compared with calculations including Coulomb corrections in the 3-body final state, and also analysed other data at 6.4 MeV and 16 MeV. They found that neither a Yamaguchi nor an exponential form factor could fit all the data to better than 20-30%.

(6) E.L. Peterson, M.I. Haftel, R.G. Allas, L.A. Beach, R.O. Bondelid, P.A. Treado and J.M. Lambert studied FSI at 45 MeV (off-shell sensitivity is expected to increase with energy). Their best fit, using an improved S-wave form factor, is good to about 20%. To do better they had to use an artificial value for the triton binding energy.

(7) R. Van Dantzig, B.J. Wielinga, G.J.F. Bloomestijn, and I.Slaus measured the total cross section for p-d breakup using the BOL 4π detector and integrating over the three-body breakup cross section, at an equivalent proton energy of 13.25 MeV. Their experiment gave 169 ± 36 mb to be compared with the theory of Kloet and Tjon which predicts 163 mb.

This is the only paper in which the agreement between experiment and theory is satisfactory. It may be because the total cross-section is the quantity for which the deficiencies in existing calculations are least important. On the other hand, the experimental uncertainty is still rather large: at the 20% level at which we have noticed that the theory fails for the large components of the differential cross section.

(8) B.Kuhn, H.Kumpf, J. Mosner, W.Neubert, and G.Schmidt presented a paper in which a wide variety of kinematic configurations were studied at 8.5 MeV. They tried a large variety of phase equivalent potentials, including Coulomb corrections in the final state. They obtained agreement to 20-30% overall, with better success in the QFS regions than in FSI.

(9) J. Bruinsma and R. van Wageningen presented breakup calculations including tensor forces and higher partial waves. This is an important step forward, since all the other papers presented used S-wave interactions only (except for L'Huillier's calculations at 156 MeV which used a multiple scattering formalism).

They find that the deuteron D-state makes up to a 25% difference in the FSI cross section but little difference in the QFS. They also predict values for vector and tensor polarization quantities which are known to need the higher partial waves.

It is very noticeable that since Quebec, experimenters no longer measure only the large cross-sections (particularly the QFS and FSI regions which the Ebenhoh and similar codes can fit to 20-30%). They now measure also small quantities, like the cross-section at the minima (Jain, Rogers, and Saylor kinematics) or at very large momentum transfers (Lambert kinematics or similar), for which the codes fail by up to 100%. There are now also data on polarization effects which sometimes the codes cannot predict even in principle. Thus the developments of Bruinsma and van Wageningen come just at the right time for a new level of sophistication.

II. Four-Nucleon Breakup Reactions

There were no exclusive measurements of breakup into four nucleons, but several of breakup into quasi-three-body final states.

(1) N. Fujiwara, E. Hourany, H. Nakamura-Yokota, F. Reide, V. Valkovic, and T. Yuasa studied the reaction:

$$p + {}^3He \rightarrow p + d + p \text{ at } 156 \text{ MeV}.$$

They observed p-d QFS and also a deep minimum corresponding to the p-d FSI.

(2) T. Yuasa, in an invited paper, described measurements of pairs of particles emitted in the reaction:

$$p + {}^3He \rightarrow p + p + p + n \text{ at } 156 \text{ MeV}.$$

She gave evidence for new 3-body final state interactions such as p-^{2}He (or excited states of the mass-three system) but the interpretation is preliminary and the analysis not yet definitive.

(3) S. Desremaux, A. Chisholm, P. Perrin and R. Bouchez studied n-^{3}H interactions at 14 MeV. They concentrated on measuring the proton spectrum to look for ^{3}n states. They are able to state an upper limit for the formation of such a state, but the sensitivity of this difficult experiment is not yet adequate.

(4) H. Kumpf suggested studying the reaction d+d $\rightarrow$ 2p + 2n in the regions of phase space where there should be an enhancement of the four-body breakup cross-section due to doubly-quasi-free scattering (or double spectator effect). In the c.m. system, for p-p scattering at 90° and two spectator neutrons, this can be illustrated as follows:

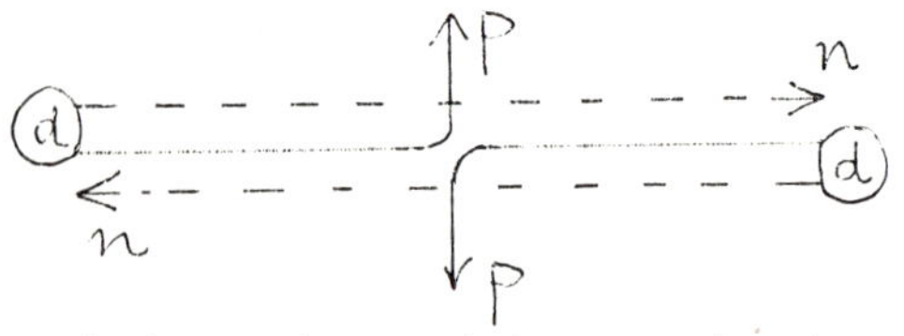

This process has already been observed in experiments at Maryland by N.S.Chant, B.Th. Leemann, C. C. Chang, N. Sarma and myself (not submitted to this conference). We bombarded deuterium with 80 MeV deuterons and measured protons emitted in coincidence at symmetric angles on either side of the beam direction. Our preliminary analysis with a plane-wave model indicates that we see the double spectator process, attenuated by about a factor of 10, but that we also see an additional enhancement due to a double final state interaction corresponding to the process d+d $\rightarrow d^* + d^*$.

While these conclusions are not definitive, it can definitely be stated that interesting 4-nucleon breakup results are becoming available and that theoretical attention to this process would be extremely valuable.

III. Three-and Four-Body Breakup of Systems With A > 4

Several papers considered systems with A > 4 but in none did the final state consist of all nucleons. Thus we are in the realm of phenomenology or spectroscopy. Two very recent reviews, by Roos and Chant, in the 2nd International Conference on Clustering Phenomena in Nuclei, concerning the extraction of spectroscopic information from quasi-free processes, are required reading for anyone working in this area.

(1) R. J. Slobodrian, R. Pigeon, M. Irshad, S. Sen and J. Asai presented new evidence for the reaction

$$^3He + {}^3He \rightarrow {}^2H + {}^4He + e^+ + \nu$$

which is of considerable astrophysical interest and seems to be much stronger than anticipated. The observed cross-section was 2.86 nb $MeV^{-1}sr^{-1}$ at 20°.

(2) D. Miljanić presented a criticism of Slobodrian's previous experiment on this reaction. However, the criticism does not seem to be adequately documented. Clearly more experiments would be useful.

(3) P.A. Treado, J.M. Lambert, R.A. Moyle, L.T. Myers, R.G. Allas, L.A. Beach, R. O. Bondelid, E. L. Peterson and I. Slaus presented detailed evidence for quasi-free scattering and quasi-free reactions in the $^2H + {}^3He$ reaction. Examples are:

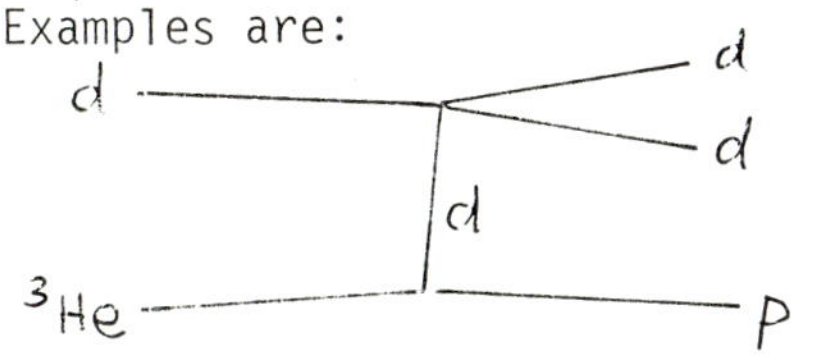

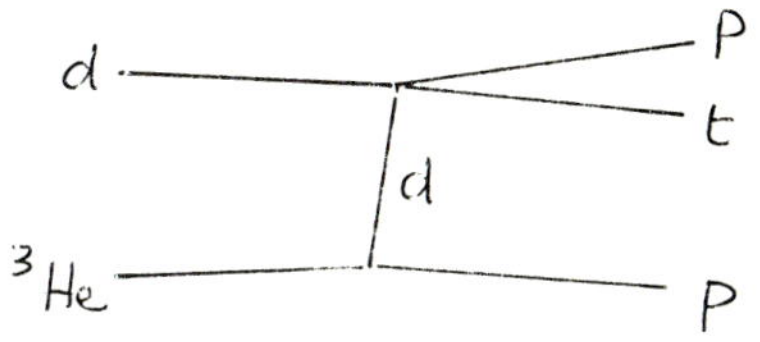

(4) The same authors, in the paper by R. G. Allas et al., presented similar evidence for the $^3He + {}^3He$ interaction.

(5) They also, in the paper by J. M. Lambert et al., give evidence for the double spectator process in the reaction $^3He + {}^3He \rightarrow d + d + p + p$, with both protons as spectators. This may be a cleaner example than the $d + d \rightarrow p+p + n+n$ reaction discussed previously because p-d final-state interactions are expected to be weak.

(6) R. Roy, R. J. Slobodrian and G. Goulard studied

$$^{10}B + d \rightarrow \alpha + \alpha + \alpha$$

and $^9Be + {}^3He \rightarrow \alpha + \alpha + \alpha$

to obtain spectroscopic information on states of ^{12}C.

IV. Summary

In the 3-nucleon breakup the data are now far ahead of theory: new codes are needed. It seems possible to include tensor forces and higher partial waves, which lead to important effects, but Coulomb interactions still present a problem.

In 4-nucleon breakup data are becoming available and interesting effects are being observed. Even simple calculations would be of great value in developing this area of study.

In A > 4 breakup new effects (QFR) are seen which are not present in 3,4 nucleon systems. However, the treatment must be phenomenological.

Finally, I would like to add my impression that not all of the difficulty in fitting data lies in the codes. I believe that the next conference will have to pay more attention to the two-nucleon problem and its connection to the three-nucleon problem.

DISCUSSION

Bleuler : I agree with the statement by Slaus that experiment is far ahead of theory. A full scale effort is needed: off-shell behaviour of nuclear forces from an enlarged meson-theory.

Slobodrian : I would like to point out that in three body break-up (or few body break-up) polarization observables may add an important handle for further improvement in the interpretation of experiment and, perhaps, the disentangling of three body and off-the-mass-shell effects.

Slaus : The totality of three nuclear scattering observables including polarization cannot disentangle off-the-energy-shell and three-nucleon forces. However, $n \geq 4$ - nucleon observables as well as the two-hadron lepton or photon interactions can - at least in principle - distinguish these two effects. Also, though break-up and polarization cannot distinguish three-body from off-E-shell effect, the combined influence is present and it is more pronounced for some observables in specific regions of phase space, and in general it increases as the incident energy increases.

Noyes : I wish to emphasize again, since the lesson still apparently has not been learned, that it has been rigorously proved that even if all two body hadronic observables (phase shifts and binding energies) and three body hadronic observables (elastic scattering and break-up including complete polarization analysis) are know it is still impossible to separate two-body off-shell effects from three-body forces. Of course, these can be disdinguished (if the theory is further articulated), by using a) Electromagnetic or weakly interacting probes, b) four or more body systems, or c) additional theoretical assumptions which are not empirically demonstrable.

Redish (first comment): While Noyes is correct that it is not possible to distinguish off-shell effects from 3-body force effects in hadronic 3-nucleon studies, I would like to point out that this arises from a very special situation. In the deuteron the single particle wave function and the pair-correlation function are essentially the same thing. In a heavier nucleus, however, a 3-body force responds to a pair correlation function, while an off-shell effect requires only a single-particle wave function. In principle, one should therefore be able to distinguish three-body and off-shell effects through a study of reactions with heavier nuclei.

Slaus : Though we should not abandon three nucleon problem because there is still a lot to do, let me emphasize again, that we should direct our attention to study $n \geq 4$ systems. These observables might reveal information on on-shell and off-shell NN interaction.

Levin : Necessity of deuteron D state has been known for several years in d + nucleus reactions. It is no surprise that it is also required in 3-body break-up. For small momentum transfer, D state effects are proportional to Q_d, but larger momentum transfer probe detailed shape.

Mitra (to Pugh): I was quite impressed with the assessment on the roles of tensor forces and form factors in the analysis of few-body break-up reactions. Especially interesting is the result of computation by Van Waganingen et al of the significant effect of tensor forces in bringing about agreement with experiment in 3-body break-up reactions. Your reference to form-factors in this context has wider ramifications. I am not referring particularly to the 3-body break-up reactions for which exact amplitudes are already calculable, but even for these, a precise knowledge of form factors is necessary for estimating the relative roles of different diagrams (particle exchanges, triangles, etc.) in the analysis of such reactions under appropriate kinematical conditions. For $n > 3$, especially four four-body reactions, form factors are now almost essential at the present state of the art (algebraic and computational) for the evaluation of these amplitudes, mostly via Feynman diagrams.

The main difficulty about the exact evaluation of form factors via overlap integrals is that the latter require nontrivial normalization with $n \geqslant 3$ wave functions, if one must evaluate these with realistic N-N input potentials only (without fresh empirical parametrization). On the other hand, with separable potentials and separable approximations to the kernels which appear in the process of successive reductions of three and four-body problems to effective two-body ones, a promising role can be played by the so-called spectator functions (which require only two-body normalizations), for the simulation of these form factors with much less tears. Some simple combinatorial arguments appear to simulate the exact form factors for n=3 within 10% accuracy over a wide range of momentum transfers of direct kinematical concern to present experiments.

Sandhas : An additional comment to Mitra's remark: his proposal to study few-body systems by an effective two-body formalism which intuitively introduces the essential exchange diagrams and vertex functions is fully supported by the pole-approximated four-body theory as described in my talk. Even the proposed choice of the normalization of the vertex-functions is justified in this case. To apply Mitra's proposal to $N \geqslant 4$, looks therefore very promising in particular if the experience about the shape of the vertex-functions, available now in the 4-body case, is taken into account.

A.K. Jain : First it is a comment on Pugh's remark that theory gives only 20-30% agreement with experiment. Our own experience in the case of QFS at least is that using parameter-free DWBA calculations one can predict the cross sections much more precisely. Secondly, in DWBA calculations the number of partial waves can be effectively increased to infinity by using the fact that only first few partial waves are distorted and larger partial waves are not influenced by optical model distortions due to the presence of large centrifugal barriers. Therefore beyond a certain ℓ one can use plane wave.

Pugh : I would like to emphasize that in making statements about the use of particular experiments and techniques to extract information it is no longer adequate to use arguments based on simple digrams or plane wave calculations. Multiple scattering effects have been shown to invalidate such choices in many instances now and complete calculations must be done even to design experiments whose fundamental information is to be extracted.

Noyes : We should recall from Conzett's Report at Laval that ϵ_1 and P_1 can be much more accurately measured in pd and nd than in free np scattering. Once the accurate pd and nd codes are available, experimentalists should make a concerted effort to measure these poorly known two-nucleon parameters.

QUARK PHYSICS

R. H. Dalitz
Department of Theoretical Physics
Oxford University

1. Introduction. The purpose of this paper is to give you some overall picture of the role of quarks in elementary particle theory today. Why should this concern a Conference on Few Body Problems? The reason is, I suppose, that the most fruitful picture of the so-called elementary particles today is that which envisages meson states as bound quark-antiquark systems and baryonic states as bound three-quark systems. However, in the older fields of quantum physics concerned with the few-nucleon systems, the few-atom systems and the few-electron systems, the problem now is how to make accurate and reliable calculations on the basis of known particle-particle forces, whereas for quark physics, we do not even know what properties should be assigned to the basic entities, much less what should be the nature of the forces between them. In this situation, the elaborate calculational methods which have been developed for the older fields are scarcely appropriate for quark physics, and so the similarities between these fields ceases rather soon after these opening remarks.

At present we are still feeling our way to determine what are the appropriate assumptions concerning the basis entities, even to know how many independent entities are necessary and what quantum numbers they should have, in order to account for the experimental information we have. I think that my present task is to draw you some kind of map to show the relationships between the various theoretical speculations now in course of development and between them and the kinds of empirical data which have been coming into the picture over the last few years.

2. Why Quarks? The unitary symmetry SU(3), in the octet version put forward by Gell-Mann [1], is one of the great facts of hadronic physics. The most striking evidence of this is provided by the symmetrical patterns obtained when the established particles and resonance states for a given spin-parity (JP) are represented by points on a plane with co-ordinates (I_3, Y). Here Y denotes the hypercharge, related with the strangeness s and the baryon number B by Y=(s+B), and I_3 denotes the 3-component of isospin, related with the charge Q by

$$Q = I_3 + \frac{1}{2}(s+B). \tag{2.1}$$

For the baryonic states (B=+1), singlet, octet and decuplet arrays are known, as illustrated in Fig.1. Complete patterns of these kinds are known for the Λ(1520) singlet (JP=3/2-); the baryon octet (JP=1/2+), and the low-lying decuplet (JP=3/2+). For the mesonic states (B=0), the patterns known are singlet and octet only, the two often occurring mixed together to form a nonet; these patterns are known completely for the pseudoscalar (JP=0-), vector (JP=1-) and tensor (JP=2+) mesons. There are really no well-established resonance or particle states with quantum numbers (I, I_3, Y) inconsistent with the quantum numbers allowed for the 1-, 8- or 10- patterns for baryons or for the 1- or 8- (or 1+8) patterns for mesons.

Sakata [2] realized rather early that the three basic additive conservation laws, those for charge Q, baryon number B and strangeness s, could readily be accounted for if there were three basis entities which carried these quantum numbers, and he proposed that this triplet T_i might consist of the three baryons (P, N, Λ), the meson states being formed by the binding of a T_i with a $\bar{T}_j$, where $\bar{T}_j$ denotes the antibaryon triplet ($\bar{P}$, $\bar{N}$, $\bar{\Lambda}$), and the other baryons (Σ, Ξ, Δ, Σ*, Ξ*, Ω, etc.) being formed from the binding of more complicated systems of these T_i and $\bar{T}_j$ entities. However, the discovery that the baryons actually formed an 8-pattern for SU(3) excluded this possibility, for there is no octet state with B=1 which can be formed from (n+1) of the T_i and n of the $\bar{T}_j$, when the T_i entities carry B=1. For the B=0 states, the mesons, the combination of T_i with $\bar{T}_j$ leads directly to the required 1- and 8- patterns, thus

$$3 \times \bar{3} = 8 + 1. \tag{2.2}$$

For the B≠0 states, the simplest octet which can be formed is a combination of three T_i, according to the successive reductions

$$3 \times (3 \times 3) = 3 \times (6 + \bar{3}) = (10 + 8) + (8 + 1), \tag{2.3}$$

but these states have baryon number B=+3.

Several years then passed before Gell-Mann [3] and Zweig [4] pointed out that this impasse could be relieved by supposing that the three basis entities were not (P, N, Λ) but were a unitary triplet t_i with fractional baryon number b. This allowed both reductions (2.2) and (2.3) to stand, with baryon number B=3b for the latter. The choice b=1/3 for the triplet t_i thus achieved a qualitative agreement with the data. Meson states with the simplest possible structure ($t_i\bar{t}_j$) were then limited to 1- and 8- patterns, as observed. Baryon states with B=+1 would then have the structure ($t_it_jt_k$), as the simplest possibility, and these would be limited to 1-, 8- and 10- patterns as observed, according to the reduction (2.3). Gell-Mann gave the name "quark" to these triplet entities, and the symbol q_i has become used for them, in place of t_i. This quark triplet consists of an isospin doublet (u, d) with zero strangeness, and an isospin singlet s with strangeness -1. The validity of the general relation (2.1) then required that the quarks should also have fractional charge, +2e/3 for the u quarks and -e/3 for the d and s quarks, where e denotes the proton charge.

Despite extensive searches, no evidence has yet been found for the existence of individual quarks with fractional charge. Nevertheless, their use in theoretical calculations concerning the hadronic (=strongly-interacting) particles has been immensely fruitful in the development of our basic concepts.

1. It led Gursey and Radicati [5] to propose an SU(6) symmetry for the hadrons and their interactions, in analogy with the SU(4) symmetry proposed and successfully developed by Wigner for the description of the properties of light nuclei. This SU(6) symmetry represented an invariance of the Hamiltonian for hadrons with respect to all unitary transformations in the product space $SU(3) \times SU(2)_\sigma$, where SU(3) acts in the space of unitary spin and $SU(2)_\sigma$ in the space of Pauli spin, just as the SU(4) symmetry acts in the product space $SU(2)_\tau \times SU(2)_\sigma$. With this SU(6) symmetry, connections were established between the pseudoscalar meson octet and the vector meson nonet, since these now belonged to the same SU(6) multiplet, a 35-dimensional representation of the SU(6) symmetry. Similarly, the baryon octet and the baryon decuplet now became joint members of a 56-dimensional SU(6) multiplet. Thus, the SU(6) symmetry was able to relate the properties of hadronic states with different spin-parity values; for example, it related the M1 matrix-element for the excitation $\gamma + P \to \Delta^+$ with the magnetic moment of the proton.

There are three SU(6) representations which can have relevance for the baryonic states, as long as these are limited to the qqq configurations. The three representations are those contained in the reduction

$$6 \times 6 \times 6 = 56_S + 2 \times 70_M + 20_A. \tag{2.4}$$

They may also be classified according to their permutation symmetry with respect to interchanges between the quark labels. The 56-representation is symmetric (symmetry S), the 20-representation is antisymmetric (A symmetry) and the 70-representation is two-dimensional, with mixed symmetry (symmetry M), as we have indicated in Eq.(2.4) by adding appropriate suffices.

2. It led naturally to the notion that the higher mesonic and baryonic resonances may correspond to the excitation of internal orbital motions within the $q\bar{q}$ and qqq systems, i.e. to a classification of resonance states in terms of SU(6) x O(3) supermultiplets. For mesons, the first-excited states would then be expected to correspond to $^1P_1$ and $^3P_{0,1,2}$ $q\bar{q}$ configurations, which implies that they should have positive parity with total spin limited to values J=0, 1 and 2, as is in fact observed experimentally in the mass range 1000-1450 MeV. Next would come the doubly-excited states, corresponding to $^1D_2$ and $^3D_{1,2,3}$ $q\bar{q}$ configurations; the mesonic states known for the mass region 1500-1700 MeV are indeed of negative parity and do have spin values J=1, 2 or 3. Similarly, assuming a shell-model structure for the qqq system, as proposed first by Greenberg [6], the first-excited baryonic supermultiplet would correspond to an internal $1s \rightarrow 1p$ excitation, giving internal orbital angular momentum L=1 with negative parity. The internal space wavefunction then necessarily has permutation symmetry M, and this can occur only together with an SU(6) wavefunction with the same symmetry M, namely the 70-representation, as is indicated in Eq.(2.4). This expectation that the first-excited baryon supermultiplet should have the character ($\underline{70}$; 1-) has now been well verified [7]. This supermultiplet includes nine SU(3) multiplets, all with negative parity, as follows:

$$(1)_{1/2},\ (1)_{3/2},\ (8)_{1/2},\ (8)_{3/2},\ (10)_{1/2},\ (10)_{3/2},\ (8)_{1/2},\ (8)_{3/2},\ (8)_{5/2}, \tag{2.5}$$

where we have used the notation $(\alpha)_J$ where α characterizes the SU(3) multiplet and J denotes its spin value. All of these SU(3) multiplets have now now been established to exist, and they are the only multiplets with negative parity below about 1950 MeV. Similarly, for the doubly-excited configurations, five baryon supermultiplets are predicted, all with positive parity and with the following structures:

$$(\underline{56},\ 2+),\ (\underline{56},\ 0+)^*,\ (\underline{70},\ 2+),\ (\underline{70},\ 0+),\ (\underline{20},\ 1+). \tag{2.6}$$

These two 56-dimensional supermultiplets are now well-established, all of their nucleonic states being known in some detail [7, 8]. There are also a few well-established resonance states which find natural interpretations [9] in terms of the two 70-supermultiplets listed in (2.6), although not sufficient to establish the existence of these supermultiplets.

As is now well known, the spectroscopy of mesonic and baryonic states is very rich. This fact alone already indicates that there must exist sub-units and sub-structure within the hadrons, in correspondence with the additional (internal) quantum numbers needed to describe their spectra of states. In view of the SU(3) symmetry, the simplest possibility for these sub-units is a quark triplet q = (u, d, s), and the evidence to date is that a simple quark sub-structure is quite sufficient to give hadronic spectra of a complexity quite comparable with that observed; the only question at present is whether or not the observed spectra may involve fewer supermultiplets than are predicted by this simplest of models. Almost all of the hadronic states established find a place within these SU(6) x O(3) supermultiplets, and have quantitative properties (such as partial rates for transitions to lower hadronic states, with emission of a pseudoscalar meson, a vector meson or a photon) which are in accord with their assignments.

The deviations from exact SU(3) symmetry are observed to be quite large; for example, the $\Omega(1672)$ and $\Delta(1236)$ states are both members of the same baryon decuplet. However, it appears that the primary SU(3)-breaking effect is just this, that $(\text{mass})^2$ for states in the same SU(3) multiplet increases almost linearly with the number of s quarks in the state. For the baryon octet, the mean square mass between the nucleon N(940) which has no s quarks and $\Xi(1315)$ which has two s quarks is at 1145 MeV; the deviations of $\Lambda(1115)$ and $\Sigma(1190)$ from this simple

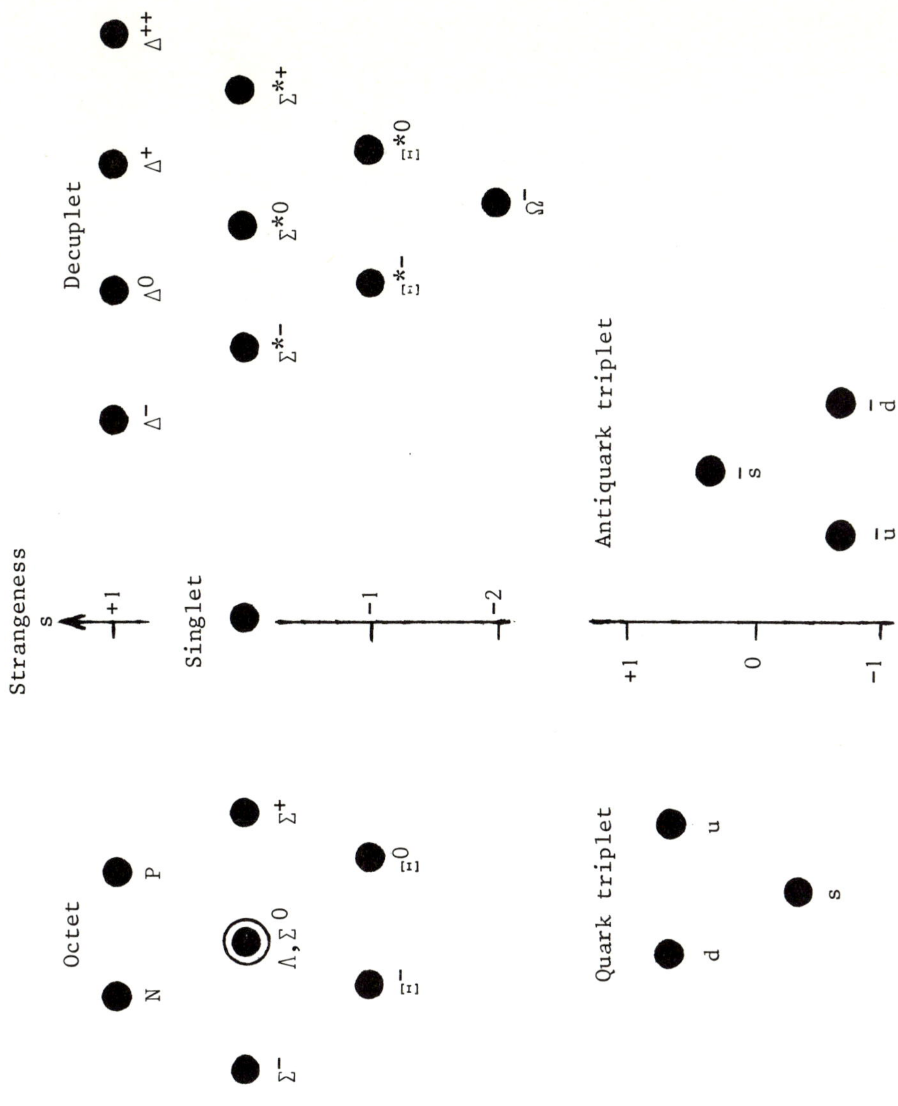

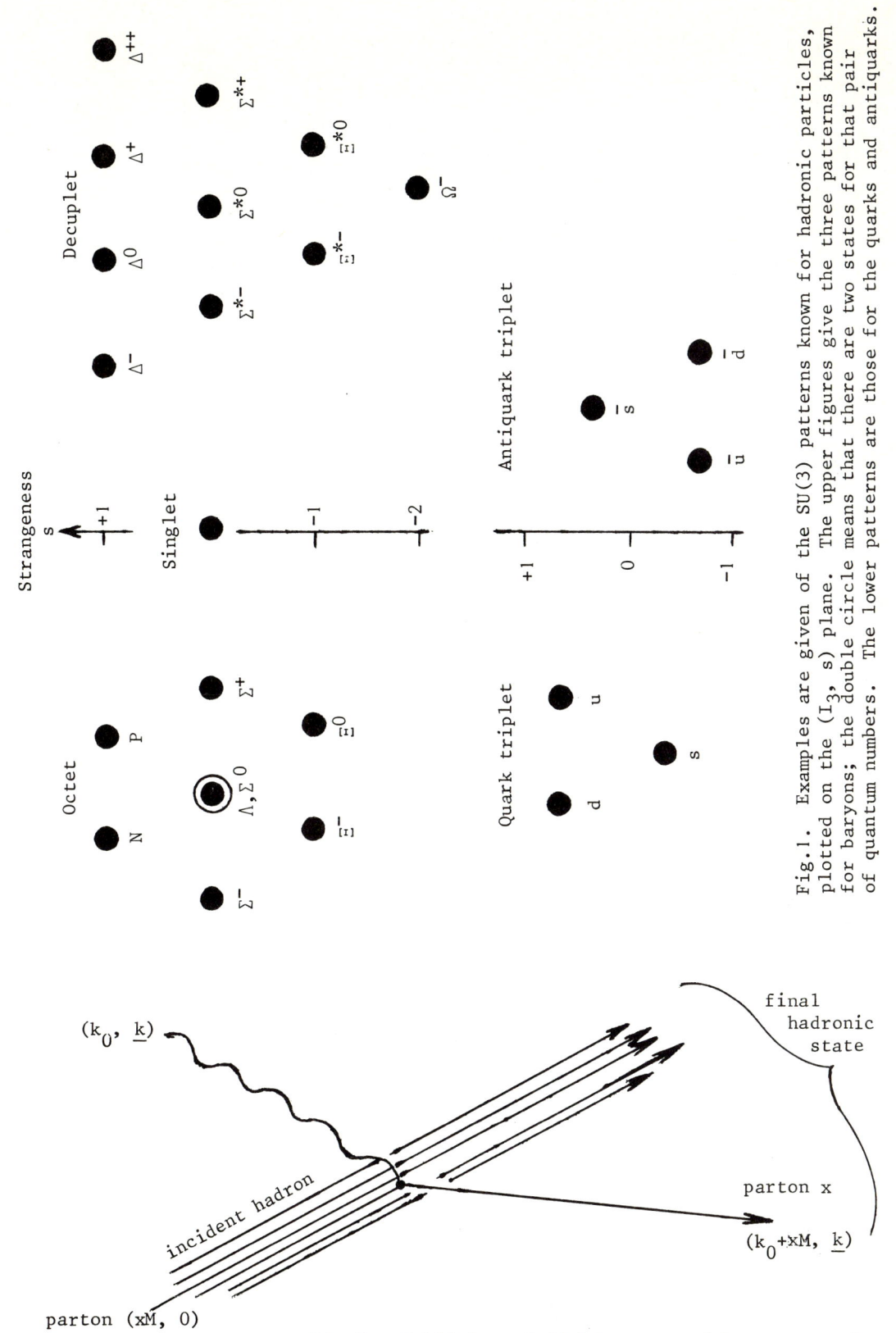

Fig.1. Examples are given of the SU(3) patterns known for hadronic particles, plotted on the (I_3, s) plane. The upper figures give the three patterns known for baryons; the double circle means that there are two states for that pair of quantum numbers. The lower patterns are those for the quarks and antiquarks.

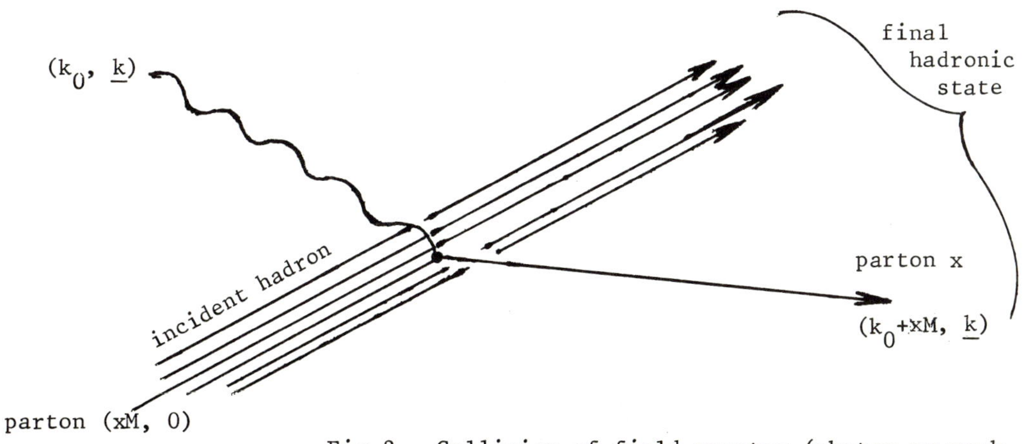

Fig.2. Collision of field quantum (photon or weak coupling) with parton x in the incident hadron

interpolation are substantially smaller (~15%) than their mass separation from N(940). For the vector meson nonet, the mean square mass between ρ(770) which has no s quarks and φ(1020) whose structure is essentially $(s\bar{s})$ is at 904 MeV; the deviation of $K^*(892)$ from this is less than 10% of its mass separation from ρ(770). This dominating linear s-dependence is generally attributed to a "mass difference" between the s quark and the (u, d) quarks, or less specifically as an additional "one-body contribution" to the energy of the state arising from each s or $\bar{s}$ quark in the state. The Λ(1115)-Σ(1190) mass difference shows that there do exist other SU(3)-breaking contributions to the energy, probably expressible as SU(3)-violating contributions to the q-q forces, but we shall not discuss these further here.

3. Deep Inelastic Lepton-Nucleon Interactions at High Energy. The second empirical strand in our story comes from the observation of "scaling" for the data on $e \to e'$ scattering and the $\nu \to \mu^-$ and $\bar{\nu} \to \mu^+$ reactions on proton and neutron targets, in the deep inelastic regime where the energy transfer k_0 and the momentum transfer $\underline{k}$ as seen in the laboratory frame are both large relative to the nucleon mass M. More precisely, for the ν and $\bar{\nu}$ reactions, the "scaling hypothesis" means that the differential cross-sections may be written in the following form (with + sign for ν, - sign for $\bar{\nu}$):

$$\frac{d^2\sigma^{\nu,\bar{\nu}}}{dxdy} = \frac{G^2ME}{\pi} \left(2xyF_1(x) + (1-y)F_2(x) - xy(1-\tfrac{1}{2}y)(F_1(x) \pm F_3(x))\right), \qquad (3.1)$$

where x and y are the scaling variables,

$$x = -k^2/(2p.k) = -k^2/(2Mk_0), \qquad y = k_0/E, \qquad (3.2)$$

and the three structure functions F_i are functions of x alone. In Eq.(3.1), the coefficient G is the beta-decay coupling constant, and in Eqs.(3.2), p denotes the energy-momentum four-vector for the target nucleon, given by (M, 0) in the laboratory frame, and E is the laboratory energy of the incident lepton.

Feynman [10] has given a physical picture for the interpretation of this scaling property. He supposed that any hadron is made up of "partons" which carry the fraction x of the hadron's mass (and therefore a fraction x of its energy and its momentum when the hadron is in motion), and that the primary lepton-hadron interaction at high momentum transfer consists of a scattering or reaction process involving one parton. The recoiling hadronic state observed results from secondary parton-parton interactions following this primary interaction, but we are not concerned here with the nature of the final hadrons but only with the total rate of the inelastic scattering or reaction process which transfers energy k_0 and momentum $\underline{k}$ from the lepton to the hadrons. The recoiling parton which has mass xM and was initially at rest, then carries off energy (k_0+xM) and momentum $\underline{k}$. These must satisfy the relation

$$(k_0 + xM)^2 - \underline{k}^2 = (xM)^2 \longrightarrow x = -k^2/2Mk_0 \qquad (3.3)$$

where k^2 denotes the quantity $(k_0^2-\underline{k}^2)$. Partons with a particular value of the parameter x can therefore be selected (in an infinite number of ways) by an appropriate choice of kinematics for the incident and outgoing leptons. The structure functions $F_i(x)$ are connected with properties appropriate to the partons x within the hadron considered, rather than with the properties of the probe particles, and they are therefore universal for processes involving this hadron.

Scaling was first established for deep-inelastic electron scattering [11, 12] and the statistics available are naturally far greater than for the neutrino processes. For the process (e, e'), the expression (3.1) holds, when the following changes are made:

(i) G^2 is replaced by the factor $(8\pi^2\alpha^2/(k^2)^2)$, where $\alpha = e^2/hc$,

(ii) the structure function F_3 is replaced by zero. In Eq.(3.1), the term F_3 occurs in consequence of parity non-conservation in the weak interactions, resulting from V-A interference. For the electromagnetic interaction, there is only the V coupling and therefore no such interference term occurs.

Further properties of the partons follow directly from these electron scattering data:

(a) spin $\frac{1}{2}$ for the partons implies $F_2(x) = 2xF_1(x)$, the Callan-Gross relation, and this is satisfied by the data, to better than 5% accuracy. Spin 0 partons would give $F_1 \equiv 0$, whereas spin 1 partons would lead to an additional q^2 dependence in expression (3.1), which would violate scaling. The data places a strong limit on the contributions which could be due to spin 0 or spin 1 partons.

(b) the partons must be point-like. If they had finite extension R, there would be an additional form factor $\phi^2(k^2R^2) \approx (1-k^2R^2/6)^{-2}$ in the expression (3.1). This factor does not have the scaling property and the data permits it only if R is sufficiently small. From the present data, one can probably conclude that $R < 0.05$ fm. There have been some positive indications of deviations from scaling reported recently from the comparison of deep-inelastic muon scattering at 56 and 150 GeV/c [13], but their interpretation is still considered controversial.

This point-like character of the partons was the most astonishing aspect of the early data on high-energy (e,e') inelastic scattering off protons, comparable in its nature and impact to the discovery by Rutherford of the point-like character of the charge distribution within atoms, indicated by the unexpectedly strong intensity he observed for α-particles back-scattered from atoms.

With SU(3) symmetry established, it is natural to identify the partons with some basic triplet entities t = (U, D, S), where (U, D) are an isospin doublet and S has strangeness -1. The probability distributions for these entities t within the proton are then denoted by U(x), D(x) and S(x), the probability distributions for their antiparticles $\bar{t}$ being denoted by $\bar{U}(x)$, $\bar{D}(x)$ and $\bar{S}(x)$. By charge symmetry, the probability distributions for the (U, D, S) entities within the neutron are then given by D(x), U(x) and S(x), respectively, those for their antiparticles $(\bar{U}, \bar{D}, \bar{S})$ being given by $\bar{D}(x)$, $\bar{U}(x)$ and $\bar{S}(x)$, respectively. The structure functions $F_i(x)$ can then be expressed simply in terms of these distributions. A particularly systematic discussion has been given by Barger and Phillips [14]. For example, for electron scattering we have

$$F_2^{eP}(x) = x\left\{Q_U^2(U(x) + \bar{U}(x)) + Q_D^2(D(x) + \bar{D}(x)) + Q_S^2(S(x) + \bar{S}(x))\right\}, \tag{3.4a}$$

$$F_2^{eN}(x) = x\left\{Q_D^2(U(x) + \bar{U}(x)) + Q_U^2(D(x) + \bar{D}(x)) + Q_S^2(S(x) + \bar{S}(x))\right\}, \tag{3.4b}$$

where Q_α denotes the charge of the corresponding triplet entity. For the νP and νN reactions, which involve only the parton transitions $D \to U$ and $\bar{U} \to \bar{D}$ (the $\Delta s=1$ transitions are neglected here because of the smallness of the Cabibbo angle θ_c), we have, similarly,

$$F_2^{\nu P} = 2x(D(x) + \bar{U}(x)), \tag{3.5a}$$

$$F_2^{\nu N} = 2x(U(x) + \bar{D}(x)), \tag{3.5b}$$

the latter following from the former by charge symmetry. We may also note the expressions for the interference term $F_3(x)$,

$$F_3^{\nu P} = 2(D(x) - \bar{U}(x)), \tag{3.6a}$$

$$F_3^{\nu N} = 2(U(x) - \bar{D}(x)). \tag{3.6b}$$

In principle, the analysis of all the possible processes will allow the determination of the six probability distributions involved. However, it is generally assumed that $S(x) = \bar{S}(x)$ since the nucleons have zero strangeness, although the proper expression of this fact is given by the integral relation

$$\int_0^1 (S(x) - \bar{S}(x))dx = 0. \tag{3.7}$$

We now wish to refer briefly to several sum rules which are related directly with parton properties. Since these sum rules involve integrals over all x, they are much better determined than any quantities which depend on details of the parton distributions.

(a) Expressions (3.6) lead to the following relation,

$$N(t) - N(\bar{t}) = \int_0^1 (U(x) + D(x) + S(x) - \bar{U}(x) - \bar{D}(x) - \bar{S}(x))dx$$

$$= \int_0^1 (D(x) - \bar{U}(x))dx + \int_0^1 (U(x) - \bar{D}(x))dx = \frac{1}{2}\int_0^1 (F_3^{\nu P} + F_3^{\nu N})dx, \tag{3.8}$$

where $N(t)$ and $N(\bar{t})$ denote the number of triplet entities and the number of anti-triplet entities, respectively, within the nucleons. The relation (3.7) has been used, since we are concerned with nucleon targets. The final expression from (3.8) may be evaluated directly from the ν and $\bar{\nu}$ interaction data [15] and has the value 3.2 ± 0.35. This tells us that the nucleons are made up of three triplet entities, together with some number of $t\bar{t}$ pairs. It is therefore consistent with the three-quark picture of the baryons built up from considerations of SU(3) symmetry and of hadron spectroscopy.

(b) Expressions (3.4) and (3.6) lead to the following expressions for quantities which can be deduced directly from the experimental data,

$$\int_0^1 (F_2^{eP} + F_2^{eN})dx = (Q_U^2 + Q_D^2)\int_0^1 (U(x) + D(x) + \bar{U}(x) + \bar{D}(x))xdx + $$

$$+ 2Q_S^2 \int_0^1 (S(x) + \bar{S}(x))xdx, \tag{3.9a}$$

$$\int_0^1 (F_2^{\nu P} + F_2^{\nu N})dx = 2\int_0^1 (U(x) + D(x) + \bar{U}(x) + \bar{D}(x))xdx, \tag{3.9b}$$

which have the ratio R, given by

$$R = \int_0^1 (F_2^{3P} + F_2^{eN})xdx/\int_0^1 (F_2^{\nu P} + F_2^{\nu N})xdx = \frac{1}{2}(Q_U^2 + Q_D^2) + Q_S^2\delta \tag{3.10}$$

where δ = (number of strange t)/(number of non-strange t) in the nucleon. It is already known [15], from the ratio 0.36 ± 0.02 of the total cross sections for ν

and $\bar{\nu}$ averaged over N and P, that the fraction of all antiparticle partons in neutrons and protons is at most 0.05 ± 0.02; assuming that $\bar{S}(x) = S(x)$ and that $\bar{S}(x)$ is not larger than $\bar{U}(x)$ and $\bar{D}(x)$, it is then reasonable to estimate $0 \leq \delta < 0.03$, and the δ term therefore gives a negligible contribution to (3.10). With the quark charges given in Table I, the predicted value for (3.10) is $R = 5/18 \approx 0.28$, irrespective of the precise value of δ in the allowed range, whereas the experimental value [15] is 0.29 ± 0.03. This constitutes a rather direct indication that the same fractional charges are to be assigned to the parton triplet as was proposed for the quarks by Gell-Mann and Zweig and therefore strengthens the case for the identification of the partons with quarks.

(c) A further relationship of importance concerns the total momentum of all the partons in the proton. With quark charge values for the parton triplet, we have from Eqs.(3.9-10),

$$\int_0^1 (U(x)+D(x)+S(x)+\bar{U}(x)+\bar{D}(x)+\bar{S}(X))x\,dx = 3\left(6\int_0^1 (F_2^{eP}+F_2^{eN})dx - \int_0^1 (F_2^{\nu P}+F_2^{\nu N})dx\right)/4, \quad (3.11a)$$

$$= 1 - \varepsilon_g. \quad (3.11b)$$

The left-hand side should equal +1, if the partons account for all the momentum of the proton. In fact, the experimental value for the right-hand side of (3.11a) is 0.46 ± 0.2, significantly less than unity, and we must conclude that there are other entities within the proton, which have neutral charge but carry momentum amounting to the fraction $\varepsilon_g = 0.54 \pm 0.2$ of the proton momentum. It is generally assumed that these neutral entities are vector gluons, whose role is to give rise to the binding forces between the quark-partons.

A more precise estimate for ε_g can be obtained using the electron scattering data alone. With quark charge values for the Q_α and the empirical value 16 for the left-hand side of Eq.(3.9a), we obtain from Eq.(3.9a) the result

$$\int_0^1 (U(x)+D(x)+S(x)+\bar{U}(x)+\bar{D}(x)+\bar{S}(x)x\,dx = \frac{9(1+\delta)}{5+2\delta}\int_0^1 (F_2^{eP}+F_2^{eN})dx = (0.54 \text{ to } 0.55)\pm 0.04, \quad (3.12)$$

as δ varies over its allowed range from 0 to 0.03, leading to the closer estimate $\varepsilon_g = 0.46 \pm 0.04$.

These results (a) and (b) above suggested that the nucleon could be considered as consisting of three "valence quarks", together with a sea of quark-antiquark pairs, as was proposed by Kuti and Weisskopf [17]. It is generally assumed that this sea has zero strangeness (so that $S(x) = \bar{S}(x)$) and zero isospin (so that $\bar{U}(x) = \bar{D}(x)$), and further that it is a unitary singlet, in which case we have

$$\bar{U}(x) = \bar{D}(x) = \bar{S}(x) = S(x). \quad (3.13)$$

With the assumptions (3.13), the electron scattering data has been fitted in terms of the three distribution functions $U(x)$, $D(x)$ and $\bar{U}(x)$ by McElhaney and Tuan [18] and by Barger and Phillips [14]. The functions obtained by the latter authors are plotted on Fig.3. The functions $xU(x)$ and $xD(x)$ plotted differ substantially in general magnitude, although having roughly similar forms, with broad maxima at $x \approx 0.2$. The function $x\bar{U}(x)$ characterizing the sea is quite different in form, being significant only for small x. These functions $U(x)$ and $D(x)$ include contributions to the quark distributions coming from the sea, of course. After transferring these contributions from $U(x)$ and $D(x)$ to the sea, we find that, in the integral for the total momentum carried by the quarks and antiquarks, the contribution from the sea is about 20% of the total given by (3.12), which was itself about 54% of the nucleon momentum.

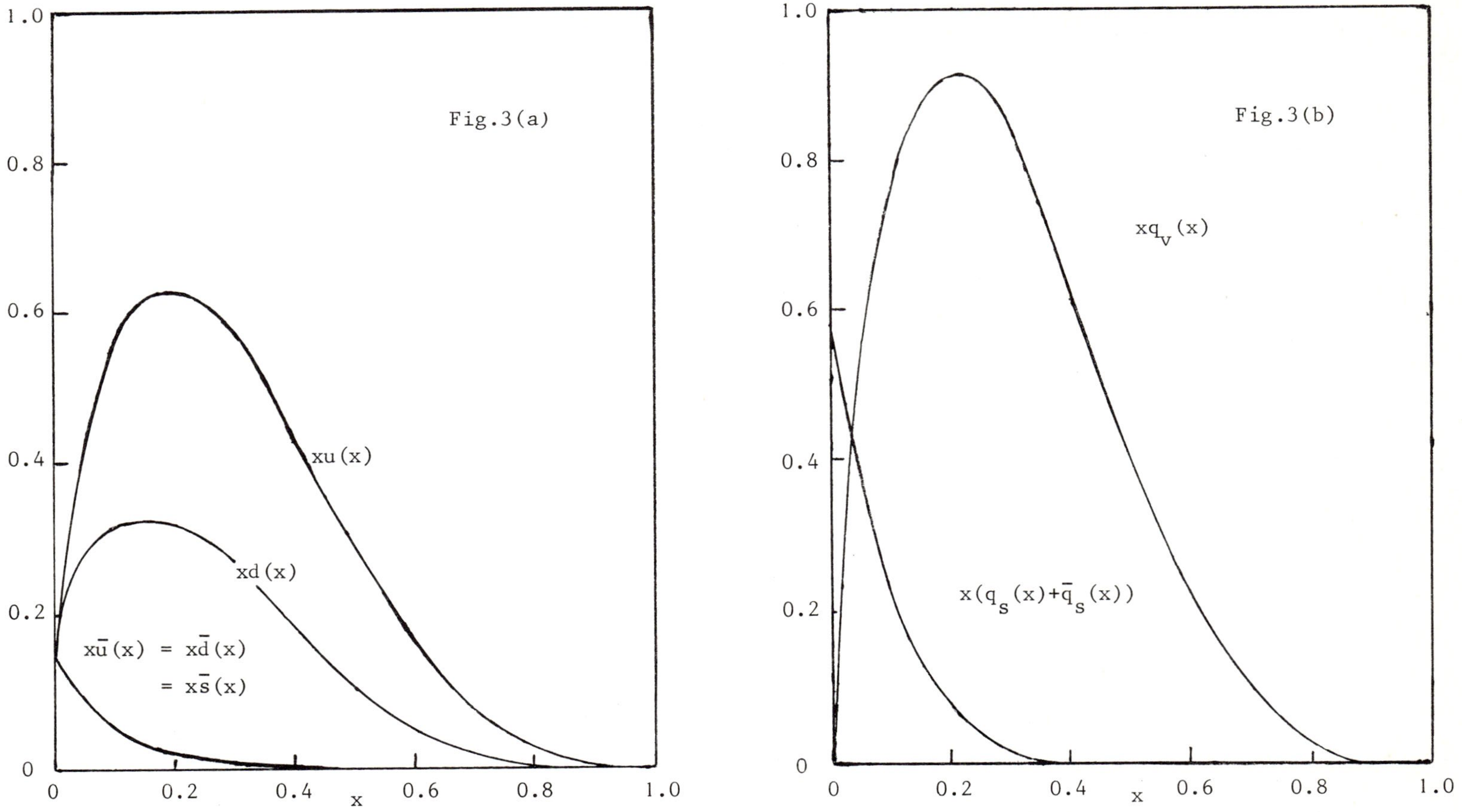

Fig.3. The parton distribution functions as parametrized by Barger and Phillips [14] are plotted as function of the scaling variable x, (a) for all the u and d quarks in the proton and for the antiquarks $\bar{u}$, $\bar{d}$ and $\bar{s}$ in the sea, and (b) separately for the valence quarks (distribution function $q_v(x)$) and for all the quarks $q_s(x)$ and antiquarks $\bar{q}_s(x)$ in the sea. Their derivation of these functions assumes that the sea is SU(3) singlet, so that $\bar{u}(x)=\bar{d}(x)=\bar{s}(x)=s(x)$.

At this point, the case for identifying the parton triplets with quarks is quite strong, and we shall henceforth assume that the t = (U, D, S) triplet is identical with the q = (u, d, s) triplet and we shall revert to the latter notation. We identify the valence quarks in the nucleon with the three quarks whose dynamics generate the spectrum of the baryons, and assume that the quark-antiquark sea is associated with the mechanisms which give rise to the binding forces between the valence quarks. However, the relationship between the (neutral) gluons and these quark-antiquark pairs from the sea is far from clear. Further, the quantitative relationship between the two models for the nucleons is still uncertain. The asymmetry between the u-quarks and the d-quarks in the valence quark distributions contrasts with the symmetrical space wavefunction found for the proton in Sec.2, on the basis of the quark shell-model for baryonic states. This suggests an appreciable SU(6)-breaking effect in the ground state qqq wavefunction, and this is certainly quite possible. We should emphasize that such an asymmetry can arise from an interference between the dominant 56 term in the nucleon wavefunction and any admixtures with 70 symmetry; an admixture of quite low intensity (≈5%, say) could well produce the effects observed. Some quantitative discussion of this question has been given by Alterelli et al. [19].

4. Colour. In Sec.2, we accepted without comment the experimental fact that the lowest baryonic supermultiplet belongs to the 56-dimensional representation of SU(6) symmetry. However, the 56 representation has permutation symmetry S. With considerable generality, the symmetry of a ground state space wavefunction is also expected to have S symmetry, and this is certainly the case for the wavefunction given by a three-quark shell model, since the ground state configuration is $(1s)^3$. Thus, the over-all wavefunction for the three-quark system, including space, spin and unitary spin factors, has S symmetry. This is inconsistent with the Spin-Statistics Theorem, which requires A symmetry for a system of spin-$\frac{1}{2}$ particles, and we must conclude that there are still further quantum numbers associated with the quark. Following Gell-Mann's proposal [20], this additional variable is now termed colour. In order to achieve its purpose of providing the required antisymmetric factor in the three-quark wavefunction, colour must have three and only three eigenstates for the quark, and these are generally referred to as red, blue and green. The required factor is then given by the determinant

$$\begin{vmatrix} q_r(1) & q_r(2) & q_r(3) \\ q_t(1) & q_t(2) & q_t(3) \\ q_g(1) & q_g(2) & q_g(3) \end{vmatrix} \tag{4.1}$$

This factor is both antisymmetric and unique. All three-quark states which have (4.1) as factor are referred to as "colour singlet". The symmetry group associated with colour might well have been a discrete group [21], but it appears more fruitful, as pointed out by Lipkin [22], to suppose the colour symmetry to be that of an SU(3)' group operating in the three-dimensional space spanned by (red, blue, green).

This SU(3)' scheme is based on three quark triplets with identical SU(3) quantum numbers, so that it is a particular form of the symmetry scheme SU(3) x SU(3)'. These three-quark triplets all have the usual fractional charge values of Gell-Mann and Zweig. The new feature introduced by Lipkin into this scheme is dynamical, namely the supposition that the leading terms in the qq and $q\bar{q}$ forces are generated by the exchange of SU(3)'-octet gluons. These colour gluons are all neutral in charge and couple with the quarks through the generators λ'_α of the SU(3)' group. As envisaged by De Rujula et al. [23], these gluons might well be the vector gauge fields associated with the colour symmetry SU(3)', but, although an attractive possibility, this is not essential for this scheme. One gluon exchange generates a potential between quarks i and j with the general form

$$U_{ij}(\underline{r}_{ij})\ \Sigma^{8}_{\alpha=1}\lambda'_{\alpha}(i)\lambda'_{\alpha}(j), \qquad (4.2)$$

the spin, unitary spin (SU(3)) and space dependence of the potential being included in the form U_{ij}. Lipkin pointed out that such forces would have the property that they strongly favour the SU(3)' singlet states for both $q\bar{q}$ and qqq configurations. The mean potential energy for a system of n(q) quarks and $n(\bar{q})$ antiquarks has roughly the following form:

$$\bar{V}(n, \bar{n}) = \frac{1}{2}\bar{U}(\mathcal{C} - \frac{4}{3}(n(q) + n(\bar{q})) \qquad (4.3)$$

where $\bar{U}$ denotes the mean value $\langle U_{ij}\rangle$ for all particle pairs in the system ($\bar{U}>0$ follows from the existence of baryons and mesons), and $\mathcal{C}$ is the eigenvalue of the SU(3)' Casimir operator for this system. The states of lowest mass occur for states with $\mathcal{C} = 0$ (i.e. colour singlets) and these have the low mass values observed only if $2\bar{U}/3 \approx M_q$. Thus, these remarks are directly appropriate only to models where the quarks are assumed to be very heavy.

With the quark dynamics given by the colour octet of gluons, as proposed by Lipkin, many things become clearer. It is convenient to speak in terms of the triality τ for a system of quarks and antiquarks, defined by

$$\tau = (n(q) - n(\bar{q}))\ (\text{modulo } 3). \qquad (4.4)$$

We may then consider the following questions:

1. Why the octet model? Note first that SU(3)'-singlet states necessarily have triality $\tau=0$. As mentioned above, with the interaction energy (4.2) and the expression (4.3), the strongly bound states are necessarily SU(3)'-singlet and therefore have zero triality. This explains why the observed hadronic states and their interactions follow the octet version of SU(3) symmetry, since this octet symmetry (more precisely, the symmetry of the factor group $SU(3)/Z_3$, where Z_3 denotes the centre of the SU(3) group), is concerned only with the states whose triality is zero.

2. Why are only (qqq) and ($q\bar{q}$) hadrons seen? We note first that, with the interaction form (4.2), the interaction of a quark or antiquark with an SU(3)'-singlet configuration of quarks and antiquarks is zero, in first approximation. This leads to the conclusion that a system with zero triality will generally be unstable with respect to the simplest SU(3)'-singlet systems which can be constructed from the quarks and antiquarks of this system, or at best weakly bound with respect to them. We shall discuss briefly two examples:

(a) The best-known example is the deuteron, which is a system of six quarks. This object is not a strongly bound hadron and prefers to spend most of its time as two SU(3)'-singlet (qqq) systems (= neutron + proton) with a relatively weak interaction between them; it happens that this six-quark system has mass just 2.2 MeV below the (N(qqq)+P(qqq)) threshold, but quite small changes in the interaction parameters would shift the mass above this threshold and allow complete break-up into (N+P).

(b) Another example is the controversial $Z_0^*(1780)$, for which the simplest possible quark structure would be ($qqqq\bar{q}$), which clearly has zero triality. According to the above remarks, this sytem must readily dissociate into the system (($q\bar{q}$)+(qqq)) and, of course, the Z_0^* mass is well above the lowest such threshold, namely the KN threshold. However, the orbital angular momentum between the two quark clumps K and N is L=1, so that there is a centrifugal barrier to hold them together; there may also be appreciable attraction between the two quark clumps due both to meson-exchange processes and to the strong inelastic processes which set in at the

K^*N threshold. The point to be made here is that there could be a resonance in the KN system not associated with a $(qqqq\bar{q})$ clump but resulting from secondary mechanisms, especially centrifugal barriers and the long-range forces arising from the exchange of light mesons between bound (SU(3)'-singlet) quark (and/or antiquark) clumps.

The conclusion is that multiquark states with triality $\tau\neq0$ will always lie high in mass, far beyond the hadronic mass values known at present; multiquark states with triality $\tau=0$ will readily break up into systems of $(q\bar{q})$ mesons and (qqq) baryons (or $(\bar{q}\bar{q}\bar{q})$ antibaryons) which will interact between themselves with forces relatively weak on the scale of the qq and $q\bar{q}$ binding forces. On the other hand, the (qqq) and $(q\bar{q})$ hadronic states are tightly bound and decay strongly only through a $q\bar{q}$ pair creation process. All of the hadronic states identified to date have properties consistent with these simple quark structures.

Nothing is known about the physical properties of the colour gluons which this proposal of Lipkin invokes. Possibly these are the neutral partons which are needed to account for the 46% of the nucleon momentum which cannot be accounted for by the quark partons.

One further consequence of the above picture has some interest. With the introduction of the colour variable, the Pauli principle for quarks operates in the normal way. For example, when two baryons come close together, so that their quark wavefunctions overlap strongly, the Pauli principle requires that some of the quarks be excited to higher orbitals, since there can only be three quarks in the (1s) orbit. Since this is energetically unfavourable, it corresponds phenomenologically to a short-range repulsive interaction between the two baryons. The situation is qualitatively the same as that of the effect of the Pauli principle for nucleons on the α-α interaction, discussed long ago by Herzenberg [24] and by others and more recently by Buck et al. [25], giving rise to effects which have been generally characterized by the introduction of a strongly-repulsive short-range term in the α-α potential. The appropriateness of this analogy for the interpretation of the hard core repulsion believed to exist in the nucleon-nucleon interaction was pointed out long ago [26] and it has been discussed again recently by Kukulin et al. [27] and by others. The nucleon-nucleon repulsion generated in this way will be finite in magnitude, with a range which will be determined by the size of the (qqq) structure within the nucleon and the range of the qq interaction. A crude estimate of the magnitude of this repulsive potential at $r_{12}=0$ can be made by considering the lowest state possible for six quarks moving in the potential used for the discussion of the (qqq) spectrum for baryons. In this state, three quarks must be excited to higher orbitals; since we are concerned with the interaction in a nucleon-nucleon S-state, the six quark state has parity + and the lowest configuration available for them is $(1s)^3(1p)^2$(1d and 2s). The energy required for this excitation is about 4(0.425) = 1.7 GeV, relative to the state of two baryons well-separated. It would therefore be reasonable to expect that the baryon-baryon interaction should have a soft-core repulsion of short-range, with a central value of about +2 GeV, but Buck et al. [25] have pointed out that there are other procedures possible for taking into account the effect of the Pauli principle in the interaction between complex systems, which may also prove advantageous for the calculation of baryon-baryon interaction effects.

5. Gauge Field Theories. On the theoretical side, great hopes have been aroused by the successes in the development of the mechanism of Spontaneous Symmetry Breaking. Ever since the work of Yang and Mills in 1954, the notion that there should exist a vector meson associated with each conserved quantum number, to serve as its gauge field (as the electromagnetic field serves as a gauge field for electric charge), has been very attractive. The difficulty was that all such gauge fields were predicted to have zero mass, as does the photon, the prototype gauge field. On the other hand, it was proved that such gauge field theories were renormalizable despite the vector character of the gauge fields, so that they were considered acceptable as theories, although they did not agree with experiment, of course.

The great steps forward were the recognition that spontaneous symmetry-breaking could occur in the majority of gauge field theories, and the proof by t'Hooft [28] that the renormalizability of the theory was then still maintained, as Weinberg [29] had speculated would be the case. A unification of the theories of electromagnetic and weak interactions then became possible, with the features:

(a) the gauge fields were initially a charge triplet W_i and a charge scalar S, all with zero mass. As a result of the spontaneous symmetry-breaking mechanism, the W_0 and S fields become mixed, yielding the photon field γ and a neutral field Z with mass. The photon mass remains zero because the law of conservation of charge holds exactly.

(b) the $W^\pm$ and Z particles which result are very massive. The simple theory given by Weinberg [29] leads to the estimates $m_W = (38/\sin\theta_W)$ GeV and $m_Z = (m_W/\cos\theta_W)$ GeV, where the Weinberg angle θ_W is believed to be about 0.5 rad.

(c) calculations with the resulting theory give finite answers for all physical quantities calculated, since the theory remains renormalizable after the spontaneous symmetry-breaking effects on its form.

This unified theory predicted that there should, in general, be neutral currents which couple with Z, the neutral carrier field for the weak interaction, as well as the well-known charged currents which couple with the charged carrier fields $W^\pm$. Independently, evidence was found in the neutrino experiments with Gargamelle [30] for (ν, ν') inelastic scattering off nuclei, a process which clearly requires the coupling of a neutral lepton current ($\bar{\nu}\gamma_\mu\nu$) with a neutral hadron current, and this evidence was soon supported by similar data obtained with the neutrino beams available with much higher energies at FNAL [31]. However, it was quickly realized that these neutral currents, ($\bar{e}\gamma_\mu(1+\gamma_5)e$ as well as ($\bar{\nu}\gamma_\mu\nu$)), did not couple appreciably with the well-known $\Delta s=\pm1$, $\Delta Q=0$ hadron currents. For example, the $\Delta Q=0$ decay process $K^+ \to \pi^+e^+e^-$ is weaker than the $\Delta Q=-1$ decay process $K^+ \to \pi^0e^+\nu$ by a factor of more than 10^{-5}; again, the $\Delta Q=0$ decay process $K_L \to \mu^+\mu^-$ is weaker than the $\Delta Q=-1$ decay process $K^+ \to \mu^+\nu$ by a factor of more than 10^{-8}. The absence of these neutral strangeness-changing weak interactions is therefore a rather striking effect.

It was pointed out by Glashow et al. [32] that this absence of $\Delta Q=0$, $\Delta s=\pm1$ weak interactions could be accounted for in an elegant way if a fourth quark c were introduced. Previously, it was known from experiment that the hadronic weak interaction currents were made up from the two fields (u, (d $\cos\theta_C$+s $\sin\theta_C$)), the charged current thus formed giving a good account of the $\Delta Q=\pm1$ weak interactions for both $\Delta s=0$ and $\Delta s=\pm1$. It was then natural to suppose that the $\Delta Q=0$ hadronic current would consist of the terms having the structures ($\bar{u}u$) and

$$((\bar{d}\cos\theta_C + \bar{s}\sin\theta_C)(d\cos\theta_C + s\sin\theta_C)), \tag{5.1}$$

where θ_C denotes the Cabibbo angle ($\theta_C=0.24\pm0.01$). This expression (5.1) includes a term ($\bar{d}s$) which gives rise to $\Delta s=\pm1$, $\Delta Q=0$ transitions. The suggestion of Glashow et al. was that a second set of weak currents should be formed from the two fields (c, (-d $\sin\theta_C$+s $\cos\theta_C$)). From these fields, following the same procedure, the $\Delta Q=0$ hadronic current consists of terms having the structures ($\bar{c}c$) and

$$((-\bar{d}\sin\theta_C + \bar{s}\cos\theta_C)(-d\sin\theta_C + s\cos\theta_C)). \tag{5.2}$$

The ($\bar{d}s$) term arising from the second set of fields is precisely opposite the ($\bar{d}s$) term from (5.1), thereby removing the possibility of any semi-leptonic weak interaction processes with $\Delta Q=0$ and $\Delta s=\pm1$.

Gauge theories are also of interest for the strong interactions, of course, as was realized by Yang and Mills from the beginning and emphasized by Sakurai [33] at a time when the large masses observed for the vector mesons appeared a major stumbling block to their identification as gauge particles. More recently, this idea has gained much interest again, especially since the demonstration that such gauge theories always lead to the property of Asymptotic Freedom, i.e. the property that the effective coupling constant becomes smaller and smaller the larger become the energy and momentum transfers involved in the interaction considered, or, equivalently, the closer the approach of the interacting quarks. This property may, in due course, allow us some quantitative understanding why the parton model has proved so successful to date, despite the fact that the partons are treated as if they were free, rather than as entities in a milieu of very strong interactions from the neighbouring partons. De Rujula et al. [23] have taken this property of Asymptotic freedom rather seriously and have argued that, in the region of close approach, the qq and $q\bar{q}$ forces will be dominated by one-gluon-exchange, since perturbation theory will be valid when the coupling constant reaches small values at separation distances where Asymptotic Freedom holds. With this view, the spin- and space-dependences of the potential U_{ij} in this region are prescribed, although the coupling constants are not. De Rujula et al. do not attempt to calculate from first principles the various expectation values $\langle U_{ij} \rangle$ which occur in the expressions for the energy of each state but prefer to use them as phenomenological parameters in the analysis of the data. Much concordance with the data is found - various effects have the sign predicted by their formulae and a reasonable magnitude - and it is clear that much further work along these lines still remains to be done, especially for the higher hadronic supermultiplets.

6. No Free Quarks? The successes of the quark-parton model contrast with the absence of any evidence for their existence in the free state, despite very considerable searches, both at accelerators and in Nature. With the evidence discussed above before us, the natural question is "Why are quarks not found?"

A number of speculative answers to this question have been given in the literature and are the subject of intensive research today. Collectively, these may be described as "quark confinement" theories. They differ greatly in the mechanism proposed as well as in detail.

The most ambitious speculation is that quark confinement may be a direct consequence of the gauge fields associated with the quantum number of colour. These fields are of long range, and the strength of the long range part of the field is directly proportional to the net colour of the source. This long range field is absent only for a quark system which has zero colour. In other words, the only hadronic systems which do not generate a long range colour gauge field are the SU(3)'-singlet hadrons, a class which includes all the hadrons we know. The hopes that the answer lies in this direction have been nourished by consideration of some simple (usually one space-dimensional) and calculable models. The most substantial attack along this line at present is the direct computational approach by Wilson [34] to the solution of a standard three-dimensional model with Lagrangian L, describing quarks (mass matrix M) and their octet of colour gauge fields (Lagrangian L(YM)) interacting through the gauge-covariant quark kinetic energy term ($\not{D}$),

$$L = L(YM) + i\bar{q}\not{D}q + \bar{q}Mq. \tag{6.1}$$

The procedure adopted appears bizarre, in that it replaces this problem by a well-defined lattice problem. Wilson proposes to solve the latter by established but lengthy procedures and then to treat the difference between two problems as a small perturbation. It appears to be only a matter of time before we know whether or not the picture which has been built up on the basis of simple models necessarily holds for this realistic three-dimensional model (6.1).

From the simple models which have been discussed, it has been conjectured that there will result a "quark confining potential" between two quarks (or for q and $\bar{q}$), which increases linearly with their separation r_{12}, at sufficiently large distances, and therefore prevents the two quarks from separating, no matter how much energy is put into the system. Of course, their mean square separation $\sqrt{(\bar{r}_{12}^2)}$

would then increase indefinitely with increasing internal energy, and it is far from clear how their fields will affect other quarks which may happen to be in the vicinity. In fact, this linear rise in the potential $U(r_{12})$ cannot continue indefinitely with increasing r_{12}, for this energy can be emitted from the system by particle pair-creation processes. At such distances, the effective potential U must have an imaginary part and its r_{12}-dependence will presumably also change.

Quite different approaches are being followed by groups at SLAC and MIT, these models being known generically as "bag models", although their initial philosophies are quite different. The SLAC model stems from normal field-theoretic concepts applied to some explicitly non-linear interaction Lagrangians. The "bag" represents a spatially extended coherent excitation of the vacuum, forming a closed surface, to which the quark has a very strong attraction. Inside the bag, the interactions between quarks are considered small compared with their attraction to the bag surface, and so the SLAC model has become known as the "shell" or "bubble" model. The work reported to date has been concerned mostly with the principles of the model and an excellent reivew of this has been given recently by Drell [35]. It has been found by Giles [36] that this bag is easily deformed, and this will complicate the detailed calculations needed for comparison with experiment.

The MIT bag model [37] for quark confinement has a somewhat more ad hoc basis. Essentially, they introduce a "volume tension" for the space within the bag, which contributes an energy proportional to the volume of the bag, and impose at its surface a boundary condition for the internal fields which ensures that there is no energy-momentum flow across the surface. The formalism is developed in such a way as to show that the content of this theory can be expressed in a completely Lorentz-invariant way. Inside the bag, there are quark and gluon fields satisfying the usual (covariant) equations of motion and interacting with each other, at least in principle. Since these fields are confined to the bag region, it is permissible for them to describe zero mass particles, but this is not a crucial matter; when the quark mass is taken small in this model, its precise value (finite or zero) is not important. In practice, the u and d quarks are taken to have zero mass, while the s quark is given a non-zero mass (of order 300 MeV) in order to give rise to the SU(3)-breaking effects observed. The internal motions are completely relativistic, so that there can be no SU(6) symmetry in this model. Nevertheless, with very few parameters involved, their discussion of the (qqq) ground state configuration $(1s_{1/2})^3$ gives a rather good account of the baryon octet and decuplet and their properties [38]. The real test will come when these calculations are extended to discuss the states we have attributed in Sec.2 to the SU(6) x O(3) supermultiplet (70, 1-). To date, all of these MIT calculations have been confined to spherically symmetric states; this allows the inclusion of the $p_{1/2}$ state, but not the $p_{3/2}$ state, in the quark configurations considered.

It is worth remarking that this bag model allows many more resonance states than the simple quark model. The gluon field is introduced explicitly and its excitation must also be considered; there is one class of mesonic states predicted for which there are no quarks or antiquarks inside the bag, only gluons. There are also mesonic states for which the bag contains two quarks and two antiquarks [39], and it has been argued that the broad (0+) mesonic states known with mass values about 1000 MeV may be of this class. Similarly, there are more baryonic states predicted than are given by the non-relativistic (qqq) model. There are many specific predictions made on the basis of the bag model and their experimental tests will hold much interest for the future.

7. ψ Meson Physics. The third empirical strand in our story comes from the data collected about the ψ meson family since the dramatic discovery of the J/ψ vector particle of mass 3095 MeV in November 1974 [40]. Two further vector particles, $\psi'=\psi(3684)$ and $\psi(\sim 4100)$, were reported from the SPEAR experiment at SLAC during 1975, and a further vector particle $\psi(4414)$ has become established quite recently 41 . The first two of these particles, J/ψ(3095) and ψ(3684) are notable for their astonishingly narrow decay widths, 0.07 MeV and 0.23 MeV, respectively, but

the upper states are broad, the widths being about 200 MeV for $\psi(4100)$ and 33 MeV for $\psi(4414)$.

The most immediate explanation of these new meson states was that they were related with the hypothetical fourth quark c, which had already been much discussed. In this view, J/ψ (3095) was the analogue of the vector meson $\phi(1020)$, the two mesons being the $^3S_1$ states for the $\bar{c}c$ and the $\bar{s}s$ systems, respectively. For the meson ψ', the fact that its major decay mode was a transition to $\psi(3.1)$ with the emission of an s-wave pion pair suggested that it had a similar structure to $\psi(3.1)$ and it was therefore assigned to the radially excited configuration 2^3S_1. The higher states $\psi(4100)$ and $\psi(4414)$ then receive a natural interpretation as the n=3 and n=4 states of the $^3S_1$ configuration. With the simple assumption that the $\bar{c}c$ potential has a linear radial dependence, as proposed on the basis of quark confinement models, this empirical sequence of mass values fit quite well the expectation [42-44] for the principal series of states n^3S_1, when the parameters are adjusted to fit the mass values for J/ψ and ψ'. This interpretation requires that the c quark should have a significantly larger mass than the s quark, by about 1 GeV, a larger mass difference than that between the s quark and the (u, d) quarks, various estimates of the latter giving values between 150 and 300 MeV.

The wide range of decay widths seen for the states of the ψ meson sequence requires comment. The widths for $\psi(4100)$ and $\psi(4414)$ are quite comparable with those for the well-known hadrons, when their decays are allowed through the strong interactions. This suggests that there exists some channel for hadronic particles with threshold mass between 3.7 and 4.1 GeV, so that decay into this channel is energetically allowed for $\psi(4100)$ but not for $\psi(3684)$. It is natural to suppose that this threshold has to do with new mesonic states involving one c or $\bar{c}$ quark, since the decay processes

$$(\bar{c}c) \rightarrow (\bar{c}q) + (\bar{q}c), \tag{7.1}$$

where q denotes the quark triplet (u, d, s), are certainly allowed through strong interactions, in general. The states $(\bar{q}c)$ have charm C=+1 and form an isospin doublet $D=((\bar{d}c), (\bar{u}c))$ with zero strangeness and an isospin singlet $S=(\bar{s}c)$ with strangeness s=+1. The lightest states for these C=+1 mesons will have $(\bar{q}c)$ configuration $^3S_1$ or $^1S_0$ and these will be distinguished by the appropriate suffix, V or PS, respectively. For the $^3S_1$ states, a plausible estimate of their masses may be made following the observation that $(\text{mass})^2$ for the established vector mesons is a linear function of the number of s and $\bar{s}$ quarks within them; thus we have for the masses,

$$m(\rho(770)^2 \approx m(\omega(783))^2, \tag{7.2a}$$

$$2m(K^*(892))^2 \approx m(\rho(770))^2 + m(\phi(1020))^2, \tag{7.2b}$$

which are satisfied within several percent. If we accept the equality for (7.2b), the mass 904 MeV would be predicted for the K^* mesons. These relationships (7.2) could be understood readily if the quark-antiquark interactions were independent of quark types and if the difference in $(\text{mass})^2$ between meson states were due to a difference in a "one-body operator" contribution from each quark or antiquark to the net $(\text{mass})^2$. If we accept this interpretation and follow the same prescription for the case of the c quark, then we obtain the estimates,

$$2m(D_V)^2 = m(\omega)^2 + m(J/\psi)^2 \qquad m(D_V) = 2257 \text{ MeV}, \tag{7.3a}$$

$$2m(S_V)^2 = m(\phi)^2 + m(J/\psi)^2 \qquad m(S_V) = 2304 \text{ MeV}. \tag{7.3b}$$

It is not obvious that this simple calculation should be correct, even if the

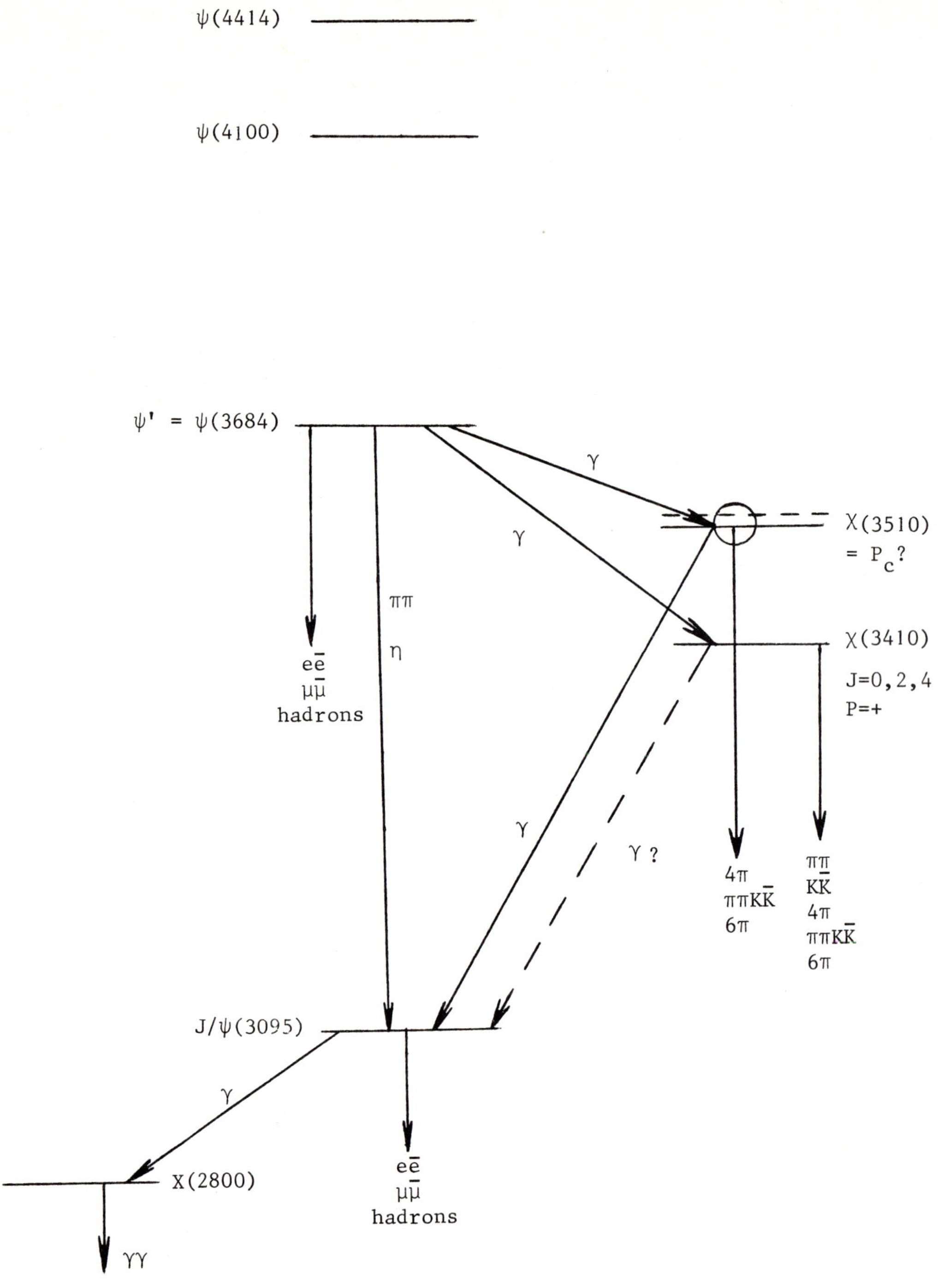

Fig.4. The present status of the experimental spectroscopy of the ψ meson family (JP=1- for the states labelled ψ). The situation at 3510 MeV is not clear. There are indications that some of the final hadron states have energy a little above this mass value. If there are two γ-rays from ψ' to the 3510 MeV region, they cannot be resolved. It is generally assumed that the state P_c reported from DESY is the same state as χ(3510). Undoubtedly, X(2880) has dominantly hadronic decay modes, but none of these has yet become established.

physical picture from which these relationships were obtained were to be valid. For example, the radial wavefunctions for the various $^3S_1$ states could be significantly modified by the effect of the widely differing values appropriate to the $(\text{mass})^2$ contribution from the individual quarks - the internal kinetic energy in each system will necessarily be a function of the appropriate quark $(\text{mass})^2$ values and the resulting differences may be expected to modify both the radial wavefunction $R_n(r)$ and the net mass for the state. Next we note the empirical relationship,

$$m(\pi(140))^2 - m(\rho(770))^2 \approx m(K(494))^2 - m(K^*(892))^2 = \delta \tag{7.4}$$

which suggests that, within the (35, L=0-) meson supermultiplet, the SU(6) symmetry of the $q\bar{q}$ forces is broken only by a spin-spin coupling independent of quark type (i.e. of SU(3)-singlet character). If we suppose that this property of the spin-spin interaction holds also for the $c\bar{s}$ and $c(\bar{d}, \bar{u})$ couplings, then we obtain similarly the relationships

$$m(D_{PS})^2 - m(D_V)^2 = m(S_{PS})^2 - m(S_V)^2 = \delta. \tag{7.5}$$

Using the right-hand expression in Eqs.(7.4) for δ, the estimates

$$m(D_{PS}) \approx 2130 \text{ MeV}, \qquad m(S_{PS}) \approx 2180 \text{ MeV}, \tag{7.6}$$

are then obtained. Of course, these are only rough estimates, but they are sufficient to indicate that it is plausible that the threshold for charmed particle pair production may lie at about 4 GeV and so account for the large widths observed for the ψ mesons $\psi(4100)$ and $\psi(4414)$. However, no evidence has yet been found in the final hadronic states studied in detail in the SPEAR experiments to demonstrate that the pair creation process $e^+e^- \rightarrow \bar{D}^+\bar{D}^-$ does actually occur, even at total c.m. energy 4100 MeV, and we now have the bizarre situation that there is a very strong case for the introduction of the charm quantum number C, at a time when we do not know empirically any hadron state which has non-zero value for this quantum number.

The extraordinarily narrow decay widths for J/ψ and ψ' are also a puzzle. They are reminiscent of the unusual narrowness of the $\phi(1020)$ decay width, $\Gamma_\phi = 4.2\pm0.2$ MeV. Of course, $\phi(1020)$ is above the $K\bar{K}$ threshold (mass 987 MeV for K^+K^-), and so the decays

$$(\bar{s}s) \qquad ((\bar{s}d) + (\bar{d}s)) \text{ and } ((\bar{s}u) + (\bar{u}s)) \tag{7.7}$$

are energetically possible. However, the decay $\phi \rightarrow \pi^+\pi^-\pi^0$, which involves an annihilation of the initial s and $\bar{s}$ quarks, has a partial decay width of only 0.65 ± 0.07 MeV. This has always been interpreted in terms of the Zweig rule [45] that all quark line graphs which involve annihilation of a quark and an antiquark which belong to the same incident or final particles should be omitted. Thus, if $\phi(1020)$ had exactly the structure $(\bar{s}s)$, and if its mass were below the threshold for the processes (7.7), its decay would be forbidden through strong interactions. The decays $\phi \rightarrow \pi^+\pi^-\pi^0$ actually observed are then attributed to a small admixture of the state $(\bar{u}u+\bar{d}d)$ in the physical $\phi(1020)$ state; this admixture can be calculated on the basis of SU(3)-breaking by quark "one-body operators" from the observed vector meson masses, and it was shown long ago by Glashow and Socolow [46] that the observed $\phi \rightarrow \pi^+\pi^-\pi^0$ decays fitted this interpretation quantitatively. The application of this Zweig rule to the question of hadronic decay modes for the J/ψ and ψ' mesons proceeds in complete analogy to the foregoing discussion for $\phi(1020)$ decay to pions. What has been so puzzling is to account for the high degree of $(\bar{c}c)$ purity which the J/ψ and ψ' states must have in order to account for such narrow hadronic decay widths for initial particles which are so heavy.

With this simple $(\bar{c}c)$ model for the ψ mesons, it was natural to ask where the P states of the system would lie in mass. This was discussed by Eichten et al. [44] and by Appelquist et al. [47] using potential forms suggested by the ideas of quark-confinement and of asymptotic freedom, with the conclusion that the three ($\bar{c}c$) states $^3P_{0,1,2}$ should lie at about 3450 MeV, the further state $^1P_1$ being not far from these in mass. It was predicted that all of the ^3P states could be reached by a γ-transition from the ψ' state, and that some of them might then make a γ-transition to J/ψ. These qualitative predictions have been borne out by SPEAR experiments [48] and by DORIS experiments at DESY (Hamburg) [49]. Two states, X(3410) and X(3510), have been established from observations on γ-emission from ψ'; the decay X(3510) $\to \gamma+J/\psi$ has been reported from both SPEAR and DESY, while DESY has also reported a few γ events attributed to X(3410) $\to \gamma+J/\psi$. The states X(3410) and X(3510) also undergo decay to hadronic final states and these hadronic data also give some preliminary indications that there may be a further state, X(3530). Finally, γ emission studies for the J/ψ meson also show indication of a decay process $J/\psi \to \gamma$+X(2800), where the state X is observed to decay subsequently to $\gamma+\gamma$, as well as to hadrons. It is rather plausible that X(2800) may be the lowest $^1S_0$ state (J^P=0−) for the ($\bar{c}c$) system, although its large separation from the $^3S_1$ state (=J/ψ) was rather unexpected. Decay to $\gamma\gamma$ is known to occur for all the other neutral pseudoscalar particles known, so that the observation $X\to\gamma\gamma$ is in accord with this assignment for X(2800).

It is clear that the new quantum number C associated with the quark c will also give rise to a rich spectrum of new baryon states. The patterns in Fig.1 will then become three-dimensional, the levels corresponding to baryonic states (qqq), (cqq), (ccq) and (ccc) as the quantum number C increases. The SU(3) multiplets become SU(4) multiplets; the generalization of the (1/2+) baryon octet is the SU(4) representation $\underset{\sim}{20}_M$, and that of the (3/2+) baryon decuplet is the $\underset{\sim}{20}_S$ representation. These representations consist of the following $(SU(3))_C$ multiplets,

$$\underset{\sim}{20}_S = (10)_0 + (6)_1 + (3)_2 + (1)_3, \quad \underset{\sim}{20}_M = (8)_0 + (6)_1 + (\bar{3})_1 + (3)_2, \qquad (7.8)$$

and they have been discussed in quantitative detail by De Rujula et al. [23] who have made predictions for the masses of all these sub-multiplets and discussed their decay modes. The three-dimensional patterns which generalize the baryon octet and decuplet patterns are depicted on Fig.5. The degree of symmetry which shows so clearly on these plots should not be over-estimated; the difference in mass value as we go from one C-layer to the next is very large, being of order 1500 MeV and corresponding to the large mass difference between J/ψ(3095) and ρ(770). However, it may well be that the cq and cc forces have relatively little spin-dependence. If this proves so, then the SU(6) symmetry we know already will be generalized to SU(8) symmetry, and the basic representations relevant for baryonic states will be $\underline{\underline{120}}_S$, $\underline{\underline{168}}_M$ and $\underline{\underline{56}}_A$, the permutation symmetry being indicated by the suffix. For given C, given s and given Q, the lightest baryonic state is expected to be semi-stable, decaying with lifetime of the general order of magnitude 10^{-13} to 10^{-15}sec. through weak interactions, either semi-leptonically or non-leptonically, to final hadrons.

Thus, a new field of baryonic and mesonic spectroscopy lies before us [50]. It is our expectation that the D and S spectroscopy will be most readily accessible through their pair production in e^+e^- storage ring experiments, despite the absence of evidence for this process in the present data. Some work on the charmed baryons B_C(s) may also prove possible in this way, through baryon-antibaryon pair creation. The study of these charmed particles appears an exceedingly difficult proposition for strong interaction experiments; the thresholds involved are very high and the cross sections correspondingly small, with an enormous number of competing processes from which these particular events are to be selected. Perhaps the most promising field for the study of the charmed baryons is that of neutrino-induced reactions, the reason being that the weak interactions violate almost all of the non-geometric selection rules we know in hadron physics, so that ΔC=+1 amplitudes will not be significantly less than those for other weak transitions. It is true,

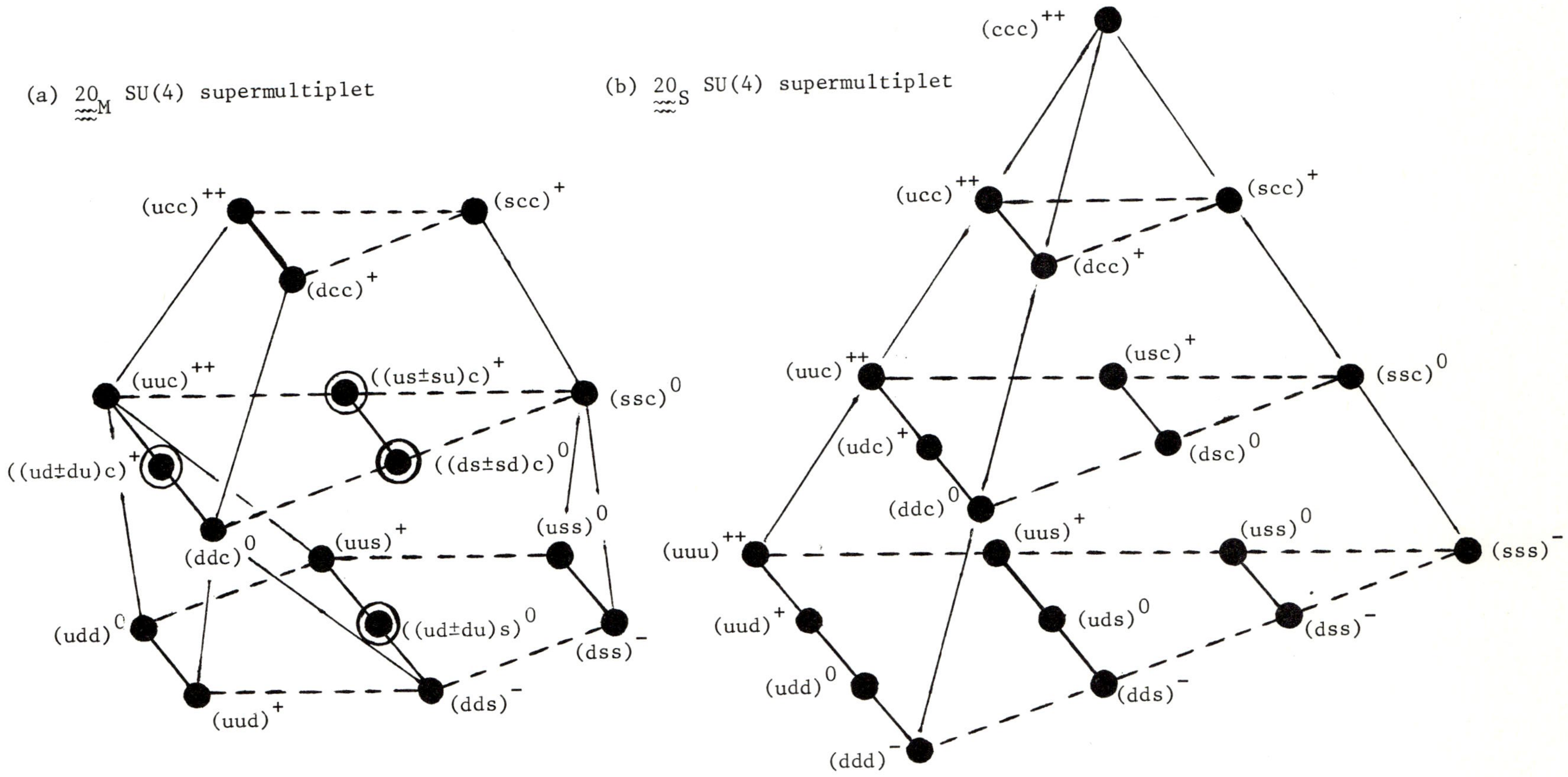

Fig.5. The states are given for the two SU(4) supermultiplets to which belong the low-lying baryons, spin-parity (1/2+) for the $\underset{\approx}{20}_M$ supermultiplet and spin-parity (3/2+) for the $\underset{\approx}{20}_S$ supermultiplet. The quark content and the net charge are both specified for each state; a double circle indicates that there are two configurations possible for that (Q, s, C). The vertical axis denotes charm C; the horizontal axes are isospin component I_3 and strangeness s. The bases of the two figures reproduce the octet and decuplet patterns displayed for C=0 baryons in Fig.1.

of course, that the neutrino-induced production of the $B_C(s)$ baryons with C=+1 will have a rather small cross section, but what matters is that these production cross sections will be comparable with the cross sections for the production of C=0 baryonic states, for comparable mass values. The individual events will, of course, still be quite complicated, because of the large energy release and the large number of decay sequences possible for charmed baryon decay.

It is interesting to mention here one neutrino-induced event [51] which could possibly be an example of charmed baryon production and decay, although this cannot be proved for an individual event, of course. The event is fitted as

$$\nu + p \to \mu^- + \Lambda + \pi^+ + \pi^+ + \pi^+ + \pi^- \tag{7.9}$$

with ($\Lambda\pi\pi\pi\pi$) mass 2426 MeV. It happens that this mass lies very close to the mass calculated by De Rujula et al. [23] for the C=+1, s=0, Q=+2 baryon belonging to the $S=\frac{3}{2}$, L=0+, 6-submultiplet of the ground state (120, L=0+) SU(8) x O(3) baryon supermultiplet. With this identification, reaction (7.9) would exemplify a $\Delta C=+1$, $\Delta s=0$, $\Delta Q=+1$ weak transition. De Rujula et al. had predicted that this particular baryon state would decay strongly by π^+ emission to the C=+1, s=0, Q=+1 member of the $S=\frac{1}{2}$, L=0+, $\bar{3}$-submultiplet of the same supermultiplet, the latter being the lightest baryon with these (C, s, Q) values. One of the three possible combinations of $(\Lambda\pi^+\pi^+\pi^-)$ gives a mass value in agreement with the value estimated by De Rujula et al. for the latter state. This state is expected to be semi-stable and to decay through the weak interactions. According to the weak interaction scheme of Glashow et al. [23], the weak current inducing the transitions $c\to s$ and $(u,d)\to(u,d)$ both have amplitude proportional to $\cos\theta_C$ (cf. paragraph above at Eqs. (5.1) and (5.2)) and the $\Delta C=-1$, $\Delta s=-1$, $\Delta Q=-1$ weak interaction resulting from them is expected to be as strong as the normal beta decay interaction $d\to u+e^-+\bar{\nu}$, and so Cazzoli et al. [51] have suggested that the $(\Lambda\pi^+\pi^-\pi^+)$ state results from the weak transition through this interaction

$$B^+(C=+1,\ s=0,\ S=1/2,\ L=0+,\ \bar{3}) \to \pi^+ + (\Lambda^* \text{ or } \Sigma^{*0}), \tag{7.10}$$

the final C=0, s=-1, Q=0 baryon then decaying through strong interactions to $(\Lambda\pi^+\pi^-)$. Obviously, this interpretation cannot be proved, on the basis of one event, but it does provide an interesting illustration of the kind of event which can be expected to occur in connection with the neutrino-excitation of charmed baryons.

In concluding this section, we must emphasize that charm is by no means the only possibility for the additional quantum number (or numbers) necessary to account for the data on ψ physics and related topics, although it is the simplest and most direct. Charm is certainly not an established quantum number, and by itself it cannot account for some of the most striking aspects of the data; dynamical assumptions like the Zweig rule are also necessary. Several other contenders have been strongly championed.

(i) The first to mention is the colour variable, no longer occurring as a mysterious and hidden part of the wavefunction, but as a dynamical variable. This is the possibility first put forward by Han and Nambu [52] in order to allow quarks to have integral charge. They invoked three non-equivalent triplets, (u_i, d_i, s_i) for i=1, 2, 3, where the charge states are (1, 0, 0), (1, 0, 0), and (0, -1, -1) in turn; we note that the average charges for u, d, s are then (2/3, -1/3, -1/3) so that the usual Gell-Mann Zweig results are obtained for any effect linear in the charge, when the states concerned are SU(3)'-singlet. However, the electromagnetic current is now not a colour singlet and can induce transitions in colour space. This SU(3) x SU(3)' scheme can come in forms [53] other than that of Han-Nambu. The implications of the Han-Nambu scheme for ψ physics have been explored especially by Feldman and Matthews [54] and by Marinescu and Stech [55].

(ii) The second is the paracharge scheme, developed rather far and in a critical spirit in a series of papers by Das et al. [56]. This is an SU(4) scheme in which the new additive quantum number is termed the paracharge Z, conserved exactly in all strong interactions. The electromagnetic current conserves paracharge except for an ad hoc Pauli moment term which has components giving $\Delta Z=\pm 1$, and it is this latter term which gives rise to J/ψ and ψ' production in e^+e^- annihilation.

These theories have much value, at the very least in presenting a different picture of the dynamics of ψ physics and causing the proponents of the conventional viewpoint to question how firmly their own assumptions are based on the empirical evidence; also, one of the unconventional theories may well turn out to be the correct theory, in the end.

8. Conclusion. The above sections may perhaps be seen as a picture of some confusion. A number of different areas of data are coming together, and theoretical ideas are being developed to explore in many different directions. In Fig.6, a map is provided, which may help you to find your way through this apparent confusion and to see the relationships which now appear to exist between the various areas and ideas. Models used in different areas are not necessarily consistent, but the links between these areas are not necessarily firm, so that apparent contradictions sometimes have to be let stand. Theoretical ideas and models which work well in some areas but fail in another area are not necessarily incorrect in the former areas; the relationship between these areas may not have been correctly understood. Our presentation has endeavoured to stress the possibility of a continuous thread connecting all these areas of data and of ideas, but it might well not be the correct thread.

For example, after all this talk, what is the mass of the quark? Well, some people favour:

(a) light quarks. Feynman has questioned whether the concept of mass for a quark-parton necessarily has any unique meaning, but, if pressed, he will say probably m=0. Others have argued for $m=M_P/3$, since this can give the nucleon magnetic moments correctly at once. The bag modellists are inclined to take m=0 for (u, d) quarks, although only as a convenience, and assign a finite mass of order 300 MeV for the s quark. Almost everybody will agree that the mass of the c quark must be quite large, at least of order 1.5 GeV, and new multiquark theories envisage further quarks with even higher mass values.

(b) heavy quarks. These are needed if the agreement of much data with SU(6) symmetry is not just a remarkable accident, since SU(6) symmetry does require Pauli spin to be an appropriate concept within hadronic states. This suggests that the quark motions within a hadron may be non-relativistic, and the non-relativistic quark models have contributed a great deal to our comprehension in hadron spectroscopy. Ultimately, however, the theory must be given in a covariant form, not only as a matter of principle, but also because the hadron decay processes generally involve the emission of highly relativistic particles. In a Lorentz-invariant context, calculations for heavy quark models have been pushed quite far by Joos and co-workers at DESY [57] using the Bethe-Salpeter equation and by Preparata [58] on a field-theoretic basis. Some valuable insights have come from these works, but what then is the relationship, if any, between partons and quarks?

How many quarks should there be? If we need a fourth quark c with a higher mass, why should the quark series stop after the fourth element? Many more have been proposed by various theoreticians, for various reasons, over recent months. On the side of experiment, Hom et al. [59] have just reported the observation of e^+e^- pairs with mass 5.97 GeV emitted from 400 GeV proton collisions with beryllium and attributed to the production and decay of a new meson Υ. The width measured is comparable with the resolution of the apparatus ($\sim$100 MeV at this mass) and the total cross section for production of the parent meson Υ is $(1.2\pm0.4)\times 10^{-35}/B_{e\bar{e}}$ c.m.^{2} per nucleon, where $B_{e\bar{e}}$ denotes the branching ratio for $\Upsilon\to e^+e^-$. From this we may conclude that the full width of this new meson Τ must again be quite small, on the hadronic energy scale. If its full width $\Gamma(T)$ were as much as 10 MeV, then a reasonable estimate for $B_{e\bar{e}}$ would be at most 10^{-3}, which would imply a production cross section for this massive meson Τ of order 10^{-32} cm^2/nucleon, a relatively high value. When data is available on the J/ψ production cross section in the same experiment, this argument will allow a rough lower limit to be placed on $B_{e\bar{e}}$, based on an educated guess for the mass dependence of the vector particle production cross section, and therefore an upper limit to be placed on $\Gamma(T)$, namely $\Gamma_{e\bar{e}}(T)/B_{e\bar{e}}$, a reasonable estimate of $\Gamma_{e\bar{e}}(T)$ being at most 10 keV. This meson Τ might well be a further state in the sequence $\omega = (\bar{u}u+\bar{d}d)/\sqrt{2}$, $\phi=\bar{s}s$, $J/\psi=\bar{c}c$, ... for some new and more massive quark; on the other hand, a SPEAR search for $e^+e^- \to T(5970)$, in steps of 1 MeV, has not yet detected evidence of this new meson.

We have invoked many quarks, at least twelve of them now, recalling that the (u, d, s, c) quarks each have three colour states. With so many basic objects to deal with, it is natural to ask whether the quarks themselves might not be composite, made up from some sub-quark entities. Discussions following this line of thought have been given recently by Greenberg [60], and also by Pati et al. [61]. The latter work is set in the wider framework of the Pati-Salam model of elementary particle interactions which considers four quarks with four colours, the fourth colour state being assigned to the four leptons (ν_e, e^-, μ^-, ν_μ), a rather radical proposal [62]. This model violates baryon number conservation, for the proton can decay into leptons. However, since the proton contains three hadronic quarks, its decay into leptons is not energetically possible unless all three of its hadronic quarks transform into leptons in the same transition. This involves a third order weak interaction, which gives a proton lifetime far longer than the lower limit (about 10^{20} years) provided by experiment today.

Great progress has been made over the past few years in opening up new experimental possibilities and new phenomena hitherto unsuspected. Our theoretical viewpoint has had to change greatly and areas of enquiry which previously seemed rather separate, such as hadron spectroscopy, electro- and photo-excitation processes, weak interaction processes and phenomena at asymptotic energies, now impinge closely one on the other in their implications. All the signs today are that we have a most interesting era ahead of us, in which the quark concept appears likely to play a central role, in the field of elementary particle physics.

References

1. M. Gell-Mann, "The Eightfold Way: A Theory of Strong Interaction Symmetry", Cal.Tech.rept.CTSL-20(1961); published in "The Eightfold Way", by M. Gell-Mann and Y. Ne'eman (W. A. Benjamin Inc., New York, 1964) p.11.

2. S. Sakata, Progr. Theoret. Phys. (Kyoto) 16 (1956) 686.

3. M. Gell-Mann, Phys. Letters 8 (1964) 214.

4. G. Zweig, "An SU_3 Model for Strong Interaction Symmetry and its Breaking", CERN Rept.No. 8182/TH401, January, 1964.

5. F. Gursey and L. A. Radicati, Phys.Rev. Letters 13 (1961) 173.

6. O. W. Greenberg, Phys.Rev.Letters 13 (1964) 598.

7. A. J. G. Hey, P. J. Litchfield and R. J. Cashmore, Nucl.Phys.B95 (1975) 516.

8. R. J. Cashmore, A. J. G. Hey and P. J. Litchfield, Nucl.Phys.B98 (1975) 237.

9. R. R. Horgan, Nucl.Phys.B71 (1974) 514.

10. R. P. Feynman, "Photon-Hadron Interactions" (W. A. Benjamin Inc., Reading,

Mass., 1972).

11. W. K. H. Panofsky, Proc.XIV Intl.Conf. on High Energy Physics (eds. J. Prentki and J. Steinberger, CERN, Geneva, 1968) p.23.

12. R. E. Taylor, Proc.1975 Intl.Symp. on Lepton and Photon Interactions at High Energies (ed. W. T. Kirk, SLAC, Stanford Univ., California, 1976) p.679.

13. L. W. Mo, Proc.1975 Intl.Symp. on Lepton and Photon Interactions at High Energies (ed. W. T. Kirk, SLAC, Stanford Univ., California, 1976) p.651.

14. V. Barger and R. J. N. Phillips, Nucl. Phys. B73 (1974) 269.

15. D. H. Perkins, Contemp. Phys. 16 (1975) 173.

16. G. Miller, E. Bloom, G. Buschhorn, D. Coward, H. De Staebler, J. Drees, C. Jordan, L. Mo, R. Taylor, J. Friedmann, G. Hartmann, H. Kendall and R. Verdier, Phys. Rev. D5 (1972) 528.

17. J. Kuti and V. F. Weisskopf, Phys. Rev. D4 (1971) 3418.

18. R. McElhaney and S. F. Tuan, Phys. Rev. D8 (1973) 2267.

19. G. Altarelli, N. Cabibbo, L. Maiani and R. Petronzio, Nucl.Phys.B69 (1974) 531.

20. M. Gell-Mann, in Elementary Particle Physics (ed. P. Urban, Springer-Verlag, Vienna, 1972) p.733.

21. R. H. Dalitz, in Hadron Interactions at Low Energies - Physics and Applications, Vol.1 (eds. D. Krupa and J. Pisut, VEDA, Slovak Acad. Sci., Bratislava, 1975), p.145.

22. H. J. Lipkin, Phys. Letters 45B (1973) 267.

23. A. De Rujula, H. Georgi and S. L. Glashow, Phys.Rev. D12 (1975) 147.

24. A. Herzenberg, Nuovo Cimento 1 (1955) 986 and 1008; Nucl.Phys. 3 (1955) 1.

25. B. Buck, H. Friedrich and C. Wheatley, "Justification of Local Potential Models for the Scattering of Complex Nuclei", Oxford preprint, 1976.

26. R. H. Dalitz, Proc.Oxford Intl.Conf. on Elementary Particles (ed. T. R. Walsh, Rutherford High Energy Lab., 1966), pp.178-9.

27. V. I. Kukulin, V. G. Neudatchen and Yu. F. Smirnov, Nucl.Phys.A245 (1975) 429. See also earlier references quoted there.

28. G. t'Hooft, Nucl.Phys. B33 (1971) 173 and B35 (1971) 167.

29. S. Weinberg, Phys. Rev. Letters 19 (1967) 1264.

30. F. J. Hasert et al., Phys. Letters 46B (1973)138; Nucl.Phys.B73 (1974) 1.

31. A. Benevenuti et al., Phys.Rev.Letters 32 (1974) 800; B. C. Barish et al., Phys. Rev. Letters 34 (1975) 538.

32. S. L. Glashow, J. Iliopoulos and L. Maiani, Phys. Rev. D2 (1970) 1285.

33. J. J. Sakurai, Ann. Phys. (N.Y.) 11 (1960) 1.

34. K. G. Wilson, "Quarks and Strings on a Lattice", Erice lecture notes, 1975,

preprint CLNS-321 (Cornell Univ., November, 1975).

35. S. D. Drell, "Quark Confinement Schemes in Field Theory", preprint SLAC-PUB-1683 (Stanford Univ., November, 1975).

36. R. C. Giles, "Semi-classical Dynamics of the 'SLAC Bag'", preprint SLAC-PUB-1682 (Stanford Univ., November, 1975).

37. A. Chodos, R. J. Jaffe, K. Johnson, C. B. Thorn and V. F. Weisskopf, Phys. Rev. D9 (1974) 3471.

38. A. Chodos, R. Jaffe, K. Johnson and C. B. Thorn, Phys.Rev.D10 (1974) 2599.

39. T. A. DeGrand and R. L. Jaffe, "Excited States of Confined Quarks", preprint CTP-529 (M.I.T., March, 1976).

40. Particle Data Group, Rev. Mod. Phys. 47 (1975) 535.

41. J. Siegrist et al., Phys. Rev. Letters 36 (1976) 700.

42. B. J. Harrington, S. Y. Park and A. Yildiz, Phys.Rev.Letters 34 (1975) 168.

43. E. P. Tryon, Phys. Rev. Letters 36 (1976) 455.

44. E. Eichten et al., Phys. Rev. Letters 34 (1975) 369.

45. G. Zweig, in "Symmetries in Elementary Particle Physics" (ed. A. Zichichi, Academic Press, New York, 1965), p.192.

46. S. L. Glashow and R. H. Socolow, Phys. Rev. Letters 15 (1965) 329.

47. T. Appelquist, A. De Rujula, H. E. Politzer and S. L. Glashow, Phys. Rev. Letters 34 (1975) 365.

48. G. J. Feldman, Proc.1975 Intl. Symp. on Lepton and Photon Interactions at High Energies (ed. W. T. Kirk, SLAC, Stanford Univ., California, 1976), p.39.

49. B. H. Wiik, ibid., p.69; H. Heintze, ibid., p.97.

50. R. H. Dalitz, "The Spectrum of Baryonic States", lecture at Workshop on Quark Models (Erice, Sicily, September 1975), to be published.

51. E. Cazzoli et al, Phys. Rev. Letters 34 (1975) 1125.

52. M. Y. Han and Y. Nambu, Phys. Rev. 139B (1965) 1006.

53. O. W. Greenberg and C. A. Nelson, Phys. Rev. Letters 20 (1968) 604; Phys. Rev. 179 (1969) 1354.

54. G. Feldman and P. T. Matthews, Phys.Rev.Letters 35 (1975) 344; Nuovo Cimento 31A (1976) 447; "Colour Symmetry and the ψ Particles", preprint ICTP/75/8, March, 1976.

55. N. Marinescu and B. Stech, "The New Mesons and Broken SU(3) x SU(3)-Colour", Univ. Heidelberg preprint (1976).

56. T. Das, P. P. Divakaran, L. K. Pandit and V. Singh, Phys.Rev.Letters 34 (1975) 770; Pramana 4 (1975) 105 and 5 (1975) 87; "Paracharge Phenomenology: Systematics of the New Hadrons", preprint TIFR/TH/76-9, submitted to Pramana.

57. M. Bohm, H. Joos and M. Krammer, "Dynamical Problems of the Relativistic Quark

Model", DESY Rept.73/20, May, 1973.

58. G. Preparata, Phys. Rev. D7 (1973) 2973.

59. D. C. Hom et al., "Observation of High Mass Dilepton Pairs in Hadron Collisions at 400 GeV", Phys. Rev. Letters, to be published (1976).

60. O. W. Greenberg, "Narrow Resonances above 3 GeV and Separate Localization of Ordinary and Colour SU(3)", Univ. Maryland Tech. Rept. 76-012, 1976.

61. J. C. Pati, A. Salam and J. Strathdee, Phys. Letters 59B (1975) 265.

62. J. C. Pati and A. Salam, Phys. Rev. D8 (1973) 1240 and D10 (1974) 275.

Table I. Quantum Numbers of the Quarks*†

Quark	u	d	s	c
Baryon Number B	+1/3	+1/3	+1/3	+1/3
Strangeness s	0	0	-1	0
Hypercharge Y=s+B	+1/3	+1/3	-2/3	+1/3
Charm C	0	0	0	+1
Isospin component I_3	+1/2	-1/2	0	0

* For the antiquarks ($\bar{u}$, $\bar{d}$, $\bar{s}$, $\bar{c}$) these additive quantum numbers are (-1) times the quantum numbers entered here for the (u, d, s, c) quarks, respectively.

† Each of these quarks has three colour states. The red, blue and green states for a given quark are identical in physical properties but are at the same time distinguishable, in principle (by their colour).

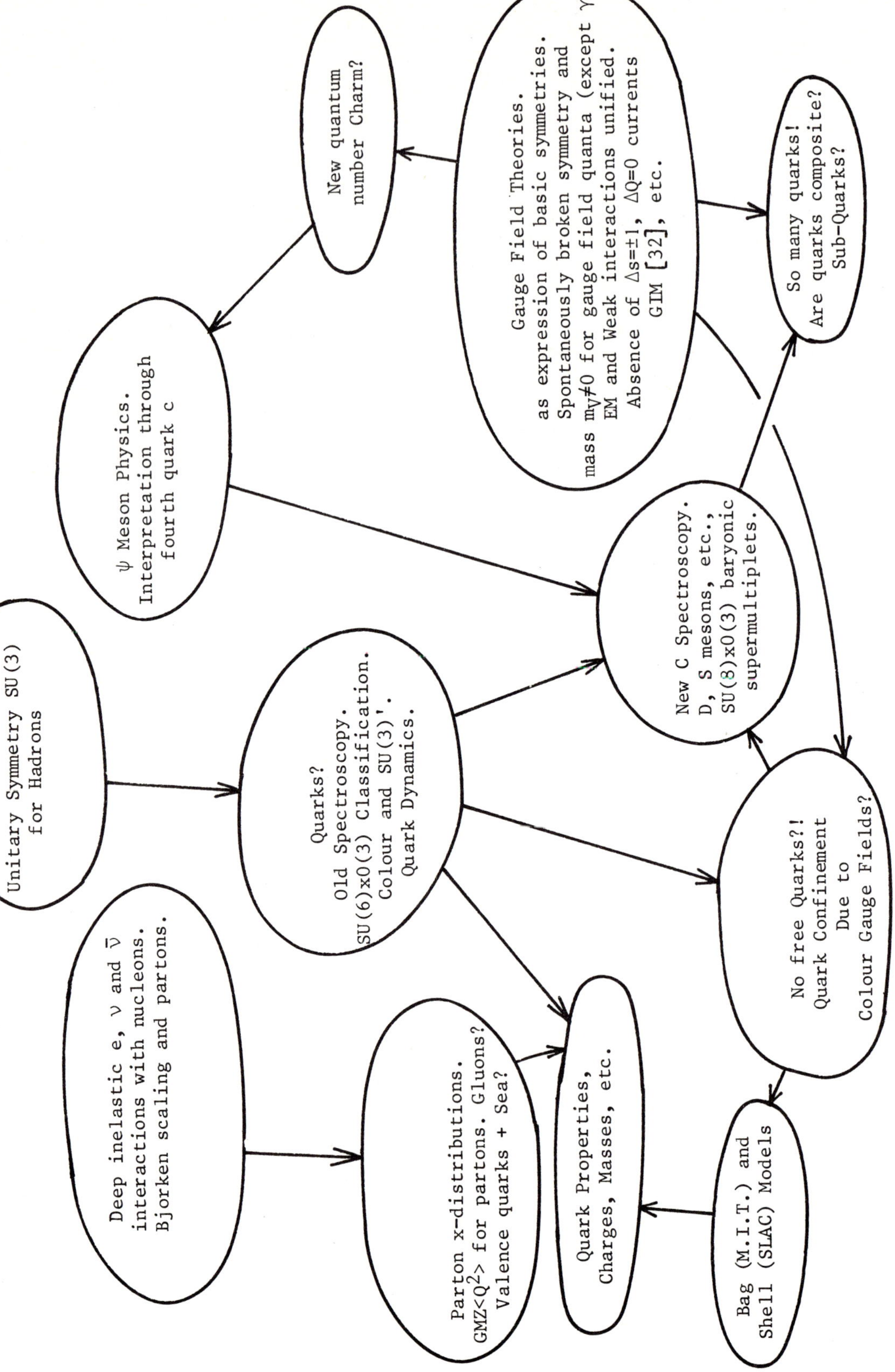
Unitary Symmetry SU(3) for Hadrons
Deep inelastic e, ν and $\bar{\nu}$ interactions with nucleons. Bjorken scaling and partons.
Quarks? Old Spectroscopy. SU(6)x0(3) Classification. Colour and SU(3)'. Quark Dynamics.
ψ Meson Physics. Interpretation through fourth quark c
New quantum number Charm?
Gauge Field Theories. as expression of basic symmetries. Spontaneously broken symmetry and mass $m_V \neq 0$ for gauge field quanta (except γ). EM and Weak interactions unified. Absence of Δs=±1, ΔQ=0 currents GIM [32], etc.
So many quarks! Are quarks composite? Sub-Quarks?
New C Spectroscopy. D, S mesons, etc., SU(8)x0(3) baryonic supermultiplets.
No free Quarks?! Quark Confinement Due to Colour Gauge Fields?
Parton x-distributions. GMZ<Q^2> for partons. Gluons? Valence quarks + Sea?
Quark Properties, Charges, Masses, etc.
Bag (M.I.T.) and Shell (SLAC) Models

Fig.6. MAP OF QUARK PHYSICS - end 1975

RESONANCES*

Gordon L. Shaw, Physics Department, University of California, Irvine, California 92717, U.S.A.

1. INTRODUCTION

There were not many papers presented in the Parallel Sessions dealing with resonances. Thus, in addition to summarizing these papers, I have been asked by the Organizing Committee to review some of the very exciting work on resonances in the past year following the discovery of the J/ψ. In this review I will point out a number of important few body aspects.

Section 2 will summarize work presented on generalized isobar models used to do resonance analysis. The important question of whether the A_1 is a true resonance is the focal point of this discussion. Relevant, new experiments and dynamical calculations lend support to the existence of the A_1. Section 3 will summarize papers concerning the $\Delta\Delta$ content of the deuteron. This seems to be a difficult question to answer quantitatively.

Section 4 will be devoted to review the many new developments in high energy spectroscopy starting with the remarkable simultaneous discovery of the very narrow meson resonance at 3.1 GeV at Brookhaven (the J) and at SLAC, (the ψ (3.1)). Most of the phenomena can be explained by the assumption of a new heavy c quark (possibly the charmed quark predicted by Glashow). I will single some problems in which few body systems play an important role.

In general, I will rely heavily on the excellent review of the quark model presented by the previous speaker, Professor Dalitz [1].

2. THE ISOBAR MODEL AND $J^P = 1^+$ MESONS

a) Resonance Analysis and Calculations

A great deal of effort has been devoted to the difficult problems of establishing resonances and their properties in three (or more) particle systems. It is usually assumed that these states are peripherally produced, and then decay via a series of quasi two body channels. There are a number of problems involved with this isobar model and we will illustrate most of them by discussing the A_1 produced in

$$\pi^- p \to \pi^- \pi^+ \pi^- p \ . \qquad (1)$$

The A_1 bump was found sometime ago in analyses of reaction (1) using the isobar model Fig. 1a. A complex fitting parameter $g_{\rho\pi}$ describes the scattering between the two pion system ρ (or ϵ) and the remaining pion. The phase of $g_{\rho\pi}$ studied with respect to a (nonresonant) reference amplitude, is found to be flat [2] rather than going through 90° (as for the A_2). Thus it was suggested that the A_1 is not a resonance but rather is an enhancement produced via the Deck mechanism [3] Fig. 1c.

Now as Aitchison [4] emphasized in his talk, the isobar diagram Fig. 1a is in principle incorrect since it fails for the three pion system to satisfy unitarity either in the two pion subchannels or in the three pion amplitude. Rescattering loop corrections of the type

* This work supported in part by the National Science Foundation

Fig. 1b should be added to the isobar amplitude Fig. 1a. A recent attempt to account for these unitarity effects via a "K matrix" formalism led to the same nonresonant behavior of the A_1 phase as well as a somewhat inferior fit [5]!

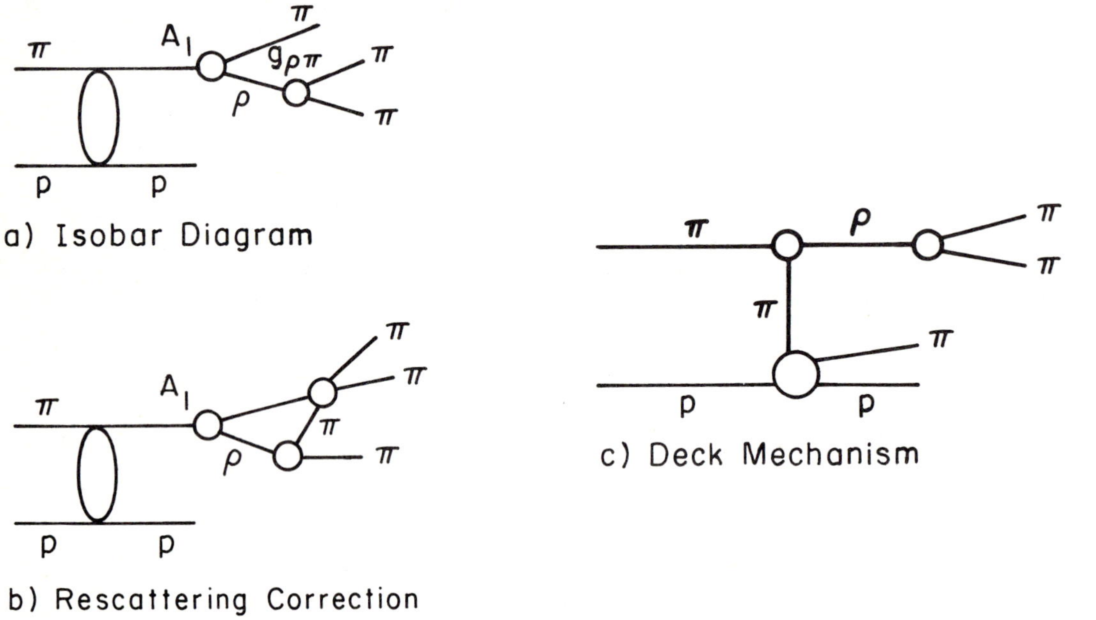

Figure 1. Diagrams for A_1 Production in Reaction (1).

Aitchison then stressed that some of the criticisms of the simple isobar model were overstated. In particular calculations such as the K-matrix approach introduced spurious singularities at the edge of the Dalitz plot [4,6]. The rapid variation of the amplitude introduced via the unitarity corrections is artificial!

Clearly, the experimentalist analyzing some 10^6 events cannot be solving three body integral equations in order to correctly include unitarity and analyticity. Thus, at present the familiar isobar model approach seems to be the safest way to analyze data. However, I am told by Aitchison [7] and by Brayshaw [8] that programs to put the three body dynamical equations in a form suitable for data analysis are underway and nearly completed. This should be extremely useful! I would note however that as one looks for resonances at 2 GeV and above, the dominant decay modes involve more than 3 particles. Thus it is of considerable importance for the four and five body equations to also be cast in simplified phenomenological forms for data analysis.

b) <u>Resolution of the A_1 Problem?</u>

There are two interesting papers that lend considerable support to the A_1 being a true resonance. This is quite important to the quark model description of meson spectroscopy. The quark model predicts 2 octets of axial vector mesons: An octet with $J^{PC} = 1^{++}$ in which the A_1 would reside and an octet with $J^{PC} = 1^{+-}$. Of these 2 octets only the B meson residing in the 1^{+-} multiplet has been positively identified.

A recent spark chamber experiment performed at SLAC [9] observed two K^* $J^P = 1^+$ resonances which would fit into these SU(3) octets as partners of the B and A_1. They looked at the reactions $K^{\pm}p \rightarrow (K^{\pm}\pi^+\pi^-)p$ and performed a detailed partial wave isobar analysis of 10^5 $(K\pi\pi)$ events. The parameters of these states (which both displayed appropriate Breit-Wigner behavior) are

Mass (MeV)	(MeV)	Decay (dominant)
~ 1300	~ 200	ρK
~ 1400	~ 160	$K^*(890)\pi$

Both of these resonances are substantially narrower than the A_1. Their existence puts the two SU(3) 1^+ octets on firm footing and thus lends (indirect) support to the A_1 being a resonance.

Then there is the important dynamical, relativistic three-body calculation of the A_1 presented by Brayshaw [10]. He finds an S matrix pole† on the second sheet of the complex energy plane at $M-i\Gamma/2 = (1160-90i)$ MeV but with no associated phase variation of the amplitude along the physical energy axis. He states that there is a crucial competition between $\rho\pi$ and $\epsilon\pi$ channels in the three pion system to yield this lack of phase variation (in anology to the phase behavior of highly inelastic two-body resonances). Brayshaw's results as well as his very interesting relativistic formulation with boundary conditions representing the short distance behavior should be studied in more detail.

3. THE $\Delta\Delta$ CONTENT OF THE DEUTERON

Papers presented by Igo [11] and Weber [12] dealt with the problem of isolating the Δ (1236) Δ (1236) content of the deuteron: (More generally, Weber discussed the possibility of NN^* configurations in the deuteron ground state, where N^* is a nucleon isobar.)

The idea is to produce Δ's in a collison process with a deuteron target. The key difficulty is to show that a Δ detected experimentally already existed performed in the deuteron rather than formed in the production process. Recent experiments

$$b + d \rightarrow b + \Delta + \Delta$$

for beams b of π's, K's, p's and $\bar{p}$'s have been performed [13]. Each experiment recorded a significant number of Δ (1236) events recoilling backward in the lab. One would suppose that these backward Δ's pre-existed in the deuteron. However, as Weber noted [12], surprisingly the πN angular distribution in the spectator Δ rest frame revealed considerable forward-backward asymmetry. This showed strong background contamination to pre-formed Δ's. Thus only an upper limit of $\leqslant 1\%$ total $\Delta\Delta$ probability is obtained from this data.

---

† Note that it is the S matrix pole which is the most model independent description of the position of a resonance. See, J. S. Ball, R. R. Campbell, P. S. Lee and G. L. Shaw, Phys. Rev. Letters 28, 1143 (1972).

Methods to suppress the background of nonresonant πN systems have been suggested which essentially rely on choosing appropriate momentum cuts or kinematical regions in looking at the data. These experiments lead $\Delta\Delta$ probabilities of $\leqslant .5\%$.

The observed $\Delta\Delta$ and NN^* probabilities rely on theoretical estimates for the reaction processes. Weber discusses the relevant calculations in some detail. Large uncertainties are introduced from use of the impulse approximation.

The nucleon and Δ (1236) are both members of the same SU(6) multiplet (56). Thus we might hope that looking at hypercharge Y = 2 states of 56 x 56, in which the deuteron would reside, would yield a prediction of the $\Delta\Delta$ content of the deuteron. Unfortunately the decomposition of 56 x 56 yields two representations 490 and 1050 in which the deuteron can fit [15]. In addition, experiments provide evidence [15] that the splitting between these two SU(6) representations is small compared with the SU(6) violating splitting between the SU(3) octet and decimal single baryon states in the 56. Thus due to expected mixing between the 490 and 1050, no SU(6) prediction on the $\Delta\Delta$ content of the deuteron can be made.

4. THE NEW PARTICLES

I will present a brief review of some of the exciting developments in particle physics following the remarkable discovery of the very narrow J/ψ in November, 1974. At Brookhaven, in the reaction

$$p + \mathrm{Be} \to e^+e^- + \ldots ,$$

the J particle was observed [16] as an enormous, narrow spike in the e^+e^- invariant distribution. Simultaneously, at Stanford, in the colliding e^+e^- beams at SPEAR, the same resonance, ψ, was observed [17] in the reactions

$$\begin{aligned} e^+e^- \to \text{"}\gamma\text{"} &\to e^+e^- , \\ &\to \mu^+\mu^- , \\ &\to \text{hadrons} . \end{aligned} \qquad (2)$$

Since then a family of related states denoted as ψ particles have been found in e^+e^- annihilation experiments at SPEAR and DORIS. (Excellent summaries are available in the 1975 Lepton-Photon Symposium Proceedings [18].) Figure 2, which shows some of the properties of these states reminds one of a typical nuclear physics level diagram. (Along these lines, it has been discussed at SPEAR that at some later time, it would be useful to fix the beam energies at the ψ (3095 MeV) and observe in great detail cascade decays into lower mass resonances, such as the A_1.)

It is now fairly generally accepted [18] that the ψ family of resonances is $c\bar{c}$ states where c is a new, heavy quark. Let us denote the usual u, d, s quarks by q so that, for example, the previously known mesons M are $q\bar{q}$ states. The narrowness of ψ (3095) and ψ (3684) is explained by the Zweig-Iizuka (Z-I) rule [19], discussed by Dalitz in the previous talk [1]: The Z-I rule greatly suppresses $\psi \to M\bar{M}$, and these two ψ's are below the threshold for the Z-I allowed decay $\psi \to D\bar{D}$ where D is a $c\bar{q}$ meson.

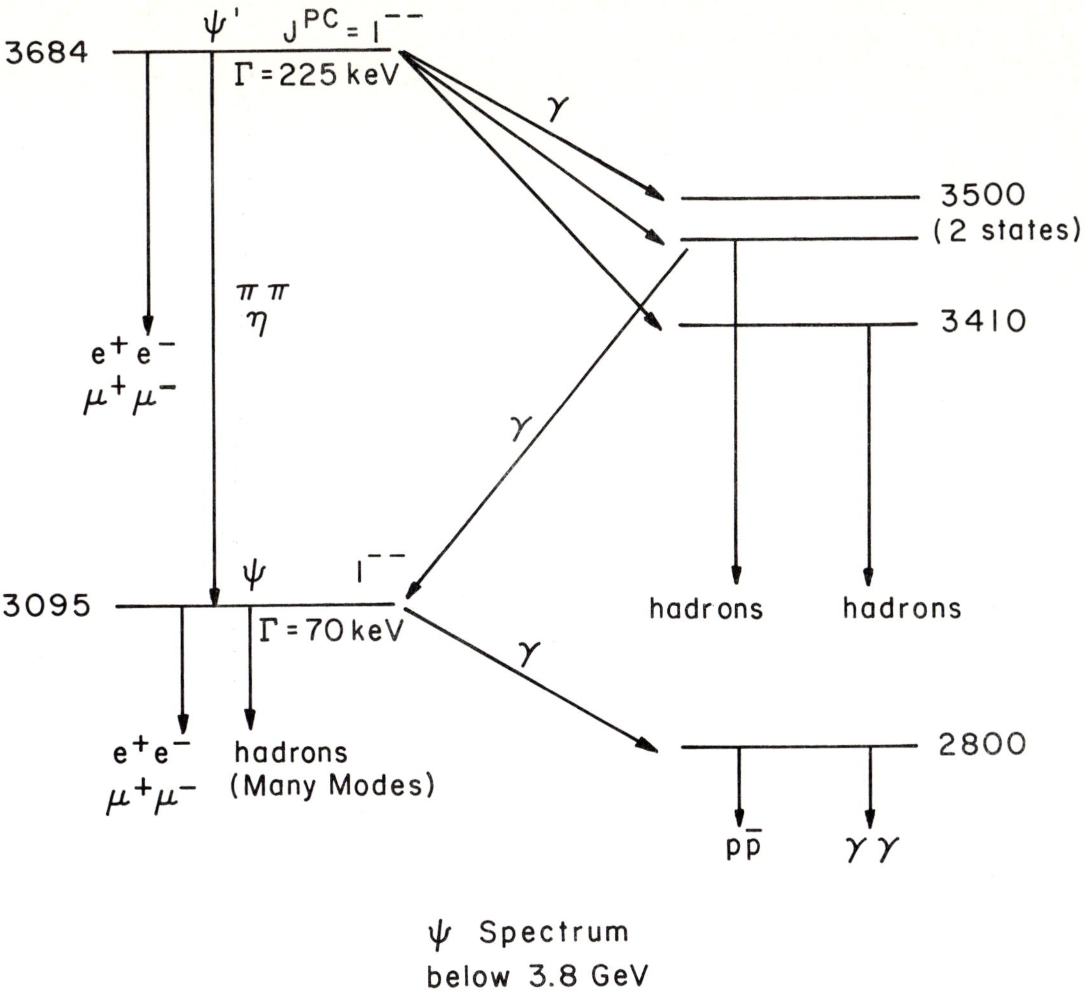

Figure 2. Some Properties of the Low Lying ψ States

One of the great excitements associated with the c quark is the good possibility that it is the "charmed" quark predicted by Glashow [20] in 1970 to explain the absence of strangeness changing neutral currents. As we now discuss, experiments [18] suggest the need for some new physics in addition to one new c quark!

The general features of $R \equiv \sigma(e^+e^- \rightarrow \text{hadrons})/\sigma(e^+e^- \rightarrow \mu^+\mu^-)$ determined at SPEAR are plotted in Figure 3.

In addition to the two spikes in R corresponding to ψ (3095) and ψ (3684), we note the two broad bumps at $E \sim 4.1$ and 4.4 GeV, and the two constant regions in R (with a transitional rise starting at ~ 3.6 GeV). Below 3.6 GeV R is roughly 2 whereas above 5 GeV R is roughly 5. The quark model predicts

$$R = \sum_i Q_i^2 \tag{3}$$

where Q_i is the charge of the i th quark and the sum goes over the types of quarks present in the hadrons being produced. Below 3.6, the u, d, s quarks contribute in Eq. (3) to R, each coming in three colors [1] so that in the usual Gell-Mann and Zweig fractionally charged model,

$$R = 3\left(\frac{4}{9} + \frac{1}{9} + \frac{1}{9}\right) = 2, \tag{4}$$

consistent with experiment.* It was thought that the rise in R is associated with the threshold for $D\bar{D}$ production. Adding a contribution from the c quark (charge 2/3) of $3 \times (4/9) = 1\ 1/3$ to the R value Eq. (4) yields the prediction of 3 1/3 above the "charmed" particle threshold. This is considerably below the experimental value of ~ 5.

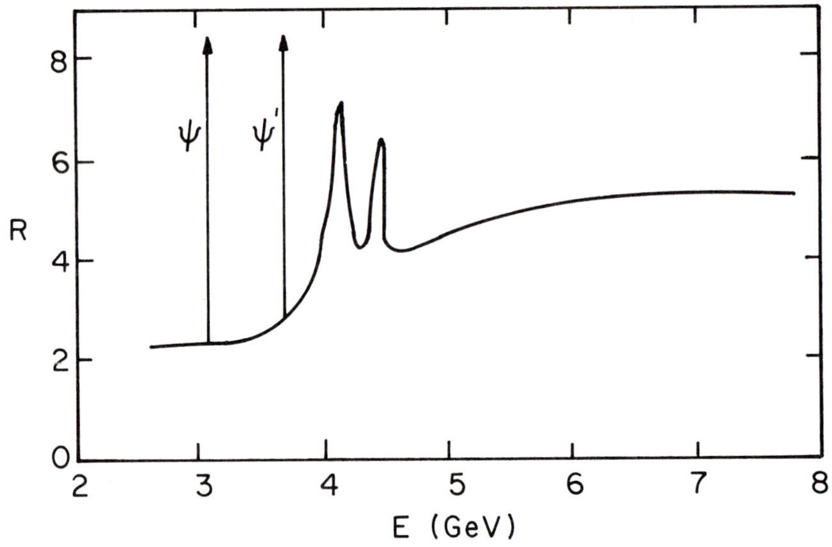

Figure 3. R Versus Total Center of Mass Energy E

Recently 64 events of the form,

$$e^+e^- \to e^\pm\mu^\mp + \geq 2 \text{ undetected particles}, \tag{5}$$

were found at SPEAR [21] for $E > 4$ GeV.

These events (5) have no conventional explanation, and it was suggested [21] that they represented pair production of new heavy (mass ~ 2 GeV) charged particles. The possibility of them being heavy leptons L was considerably favored [21] over a charged boson hypothesis. Note that in addition to purely leptonic decays of L yielding events (5), the dominant decays modes would involve hadrons and thus be counted in the measurement of R. Thus if a heavy lepton L has been found at SPEAR, one unit should be subtracted from R in Fig. 3 above the $L\bar{L}$ threshold (of ~ 4 GeV). This could explain much

* This agreement is an additional argument (to those mentioned by Dalitz [1] for the necessity of three colors.

of the descrepancy between the measured value of R and the charmed model prediction of 3 1/3. Other possible explanations for the large R value include excitation of Han-Numbu color† [22,23] as well as models with more quarks [18] (in addition to c, u, d, s). Direct searches for the charmed D mesons at SPEAR above 4 GeV have failed [18,24] to give positive evidence. (Higher statistics experiments are planned.) However, charmed particles may have been seen in ∂ scattering [25,26]. Some bubble chamber events [26] indicate that their decays involve K mesons as predicted [20].

Lets turn our attention to the bumps in R at $E \sim 4.1$ and 4.4 shown in Fig. 3. Recent experiments at SPEAR [24] show a great deal of structure within these bumps! Although the ψ family of narrow resonances (Fig. 2) below 3.8 GeV is adequately described as $c\bar{c}$ states, the existence of many closely spaced $J^{pc} = 1^{--}$ states between 3.9 and 4.4 GeV does not fit naturally into a scheme of radial excitations of $c\bar{c}$. Now examples of excitations of many closely spaced states of the same spin and parity abound in nuclear physics. For example, the excitations of the giant dipole resonance in light nuclei is described as the collective excitation of many particle-hole states of $J^p = 1^-$, differing only in the shell model orbitals of the various states. This gave us the initial motivation to suggest [27] that these new 1^{--} states are four quark composites, dynamically paired as, $(c\bar{q})$ $(\bar{c}q)$ which we denote as "new exotic" mesons E. If we limit q to be u and d quarks, there are eight (L=1) 1^{--} E states which couple to the photon (Eq. (1)), four with I=0 and four with I=1 (plus 32 other L=1 and L=0 nearly E states). We assume that the mass of D mesons [25,26] is such that $D\bar{D}$ pairs are not produced at SPEAR below 5 GeV. (The rise in R is then associated with non resonant E + M production.) Then the E's cannot fall apart into real $D\bar{D}$ pairs with a broad width but must rearrange the quarks in order to decay to ψ + M or E' + M. Such a rearrangement will suppress the widths yielding the relatively narrow (20 MeV) observed 1^{--} states. No such rearrangement is necessary for "ordinary exotic" e = $q\bar{q}\bar{q}q$ states. These e states should rapidly decay into $M\bar{M}$ with widths ~ 500 MeV [27]. The masses of the E's are about 1 GeV above the lowest ψ (Fig. 2). In analogy we expect the e's to be about 1 GeV above the M's, where their large width would make them difficult to observe.

5. CONCLUSION

As I mentioned at the beginning of this talk, there were relatively few papers at this Conference directly concerned with reasonances. I would guess that this will change, and at the next Few Body Conference considerably more discussion will be devoted to resonances. I will conclude by listing a few of the topics which I think may be of considerable interest: a) Casting the three and four body dynamical equations in a suitable form for resonance analysis. This is already being done for the three body system [7,8]. b) Dynamical three and four body relativistic calculations of resonances. Extending, e.g., Brayshaw's approach [10] would be worthwhile. In particular, one would like to see if several resonances occur in these calculations in a given partial wave as required by the data (and understood both in the context of the quark model and is daughter Regge trajectories). c) Use of nuclear and atomic physics concepts in the quark confinement problem. The MIT "bag model" [28,29] calculations are now concerned with quarks

† Quarks in the Han-Hambu model have integer charge.

contained in a "deformed bag", suggesting a unified nuclear model. The SLAC "bag model" [30] employs variational methods, and thus some of the variational methods, and thus some of the variational principles summarized here by Gerjuoy [31] and Spruch [32] might be of interest. d) We expect to learn more about the crucial questions of how many quarks and how many leptons exist. Also we hope to know more about a whole host of other very interesting, but speculative states such as "exotic" resonances (e.g., quark states $q\bar{q}\bar{q}q$ mentioned in Section 4), color gluons (which are thought to bind quarks), and magnetic monopoles.

REFERENCES

1. R. H. Dalitz, Invited Talk in These Proceedings.
2. G. Ascoli et al., Phys. Rev. D7 (1973) 669.
3. R. T. Deck, Phys. Rev. Lett. 13 (1964 169.
4. I. J. R. Aitchison, Invited Talk in These Proceedings; I. J. R. Aitchison and R. J. A. Golding, Phys. Lett 59B (1975) 288
5. G. Ascoli and H. W. Wyld, Phys. Rev. D12 (1975) 43.
6. R. Aaron et al., Phys. Rev. D12 (1975) 1984.
7. I. J. R. Aitchison, private communication.
8. D. D. Brayshaw, private communication.
9. G. W. Brandenburg et al., SLAC preprint (1975) unpublished.
10. D. D. Brayshaw, Phys. Rev. Lett. 36 (1976) 73.
11. G. J. Igo, Invited Talk in These Proceedings.
12. H. J. Weber, Invited Talk in These Proceedings.
13. Ref. 3-6 of Weber [12].
14. Ref. 8-11 of Weber [12].
15. F. J. Dyson and N. Xuong, Phys. Rev. Lett. 13 (1964) 815 and erratum Phys. Rev. Lett. 14 (1965) 339.
16. J. J. Aubert et al., Phys. Rev. Lett. 33 (1974) 1404.
17. J. E. Augustin et al., Phys. Rev. Lett. 33 (1974) 1406.
18. See, e.g., H. Harari in Proceedings of the 1975 International Conference on Lepton Interactions at High Energies, edt. by W. T. Kirk (SLAC, Stanford, California, 1975) p. 317; R. F. Schwitters, ibid, p. 5, G. Abrams, ibid, p. 25; G. J. Feldman, ibid, p. 39, B. H. Wiik, ibid, p. 69; F. J. Gilman, ibid, p. 131.
19. G. Zweig, CERN report N08419/TH412 (1964) unpublished; I. Iizuka, Suppl. to Prog. of Theor. Phys. 37 (1966) 21.
20. S. L. Glashow, J. Illiopolous and L. Maiani, Phys. Rev. D2 (1970) 1285.
21. M. L. Perl et al., Phys. Rev. Lett. 35 (1975) 1489.
22. Y. Nambu and M. Y. Han, Phys. Rev. D10 (1974) 674.
23. J. Pati and A. Salam, Phys. Rev. Lett. 36 (1976) 11.
24. G. Feldman, Invited Talk at the Irvine Conference on Quarks and the New Particles, Dec. 1975, unpublished.
25. A. Benvenuti et al., Phys. Rev. Lett. 35 (1975) 1203.
26. J. von Krogh, Invited Talk at the Irvine Conference, Dec. 1975, unpublished.
27. M. Bander, G. L. Shaw, P. Thomas and M. Meshkov, Phys. Rev. Letts. 36, to be published.
28. R. L. Jaffe and K. Johnson, MIT preprint 508, unpublished.
29. K. Johnson, private communication.
30. W. A. Bardeen et al., Phys. Rev. D11 (1975) 1094.
31. E. Gerjuoy, Invited Talk in These Proceedings.
32. L. Spruch, Invited Talk in These Proceedings.

MESON REACTIONS WITH FEW BODY SYSTEMS

I.R. AFNAN

School of Physical Sciences, Flinders University of South Australia,
Bedford Park, S.A. 5042, Australia.

I. INTRODUCTION - With the advent of meson facilities with high intensity pion beams, there is a renewed interest in the use of π-mesons as a probe of nuclear structure. Since the pion is the mediator of the nucleon-nucleon interaction, a study of nuclei with pion beams is in many ways similar to the study of atomic structure with photons. However, in the case of pions there are several added features not present in the atomic case: (i) the pion, unlike the photon, has a mass of $\sim$ 140MeV, thus pion absorption (production) gives rise to high momentum transfer. This in turn may be used to study the short range behavior of nuclear wave functions. (ii) Unlike electrons, nucleons have structure, and one may excite the nucleon in the nucleus. This allows one to study the pionic degrees of freedom of the nucleus, with the hope of understanding the importance of such degrees of freedom, and the role they play in nuclear structure. In other words does one need to include nucleon resonances in nuclear structure calculations, before one can evaluate such simple quantities as the binding energy and density distribution of nuclei. (iii) The pion with iso-spin 1 can excite certain degrees of freedom commonly not excited directly by other projectiles.

To make full use of these features of the pion one needs to resolve some basic problems (i) first, to understand the pion-nucleon interaction both in free space and in the nucleon media. In particular one needs to understand how one takes the pion-nucleon amplitude of the mass-shell (or off the energy-shell in a non-relativistic scheme). (ii) With the high momentum transfer involved in pion absorption (production) does the pion get absorbed on a single nucleon or several nucleons? Will one be able to knockout clusters of nucleons (e.g. α-particles). To answer these questions one needs to understand the mechanism for pion absorption or production. (iii) Since the nucleon-nucleon interaction is mediated by pions, one needs to include the pion degrees of freedom avoiding the problems of double counting and violation of the Pauli principle.

In the following I would like to readdress myself in more detail to the above problems, with specific examples, and recent attempts at answering some of these questions. In particular, I would like to discuss these problems within the framework of the papers presented at this conference. In Sec. II, I will discuss the pion-nucleon problem and how it manifests itself in pion-nucleus scattering. In Sec. III the possible mechanism for pion absorption (production) and attempts to relate these mechanisms will be discussed. Although in a complete theory one may not need to distinguish between the different mechanisms, it is vital at this stage to determine the important processes in order to understand the experimental data available, and as a guide to any new theory. In Sec. IV a model for the lightest nucleus (i.e. the deuteron), taking into consideration the above problems, is discussed with some interesting results.

II. PION-NUCLEUS SCATTERING - The amplitude for pion-nucleus scattering can be calculated in one of two ways: first, by calculating the multiple scattering series in which the pion is scattered off the individual nucleons. The success of this approach relies on the fact that the π-N interaction is weak as compared to the N-N interaction, and one need only calculate the first few terms in the multiple scattering series to see this. Furthermore, the final results can depend on the off-mass-shell behavior of the π-N amplitude. This was demonstrated in a contributed paper to this conference by M. Bleszynski *et al*[1] who calculated π-^{4}He amplitude to fourth order in the π-N interaction. They find that the π-^{4}He amplitude is very sensitive to the off-mass-shell behavior of the π-N amplitude.

The second approach to pion-nucleus scattering is through the intermediary of an optical potential, which is given in terms of the π-N t-matrix by the relation[2] (see Fig. 1)

$$\langle\vec{k}|U|\vec{k}\rangle = (A-1)\int d\vec{p}\,\phi^*(\vec{p}-\frac{A-1}{A}\vec{q})\langle\vec{k}',\vec{p}-\vec{q}-\frac{k}{A}|t^{\pi N}(w)|\vec{k},\vec{p}-\frac{\vec{k}}{A}\rangle\phi(p)$$

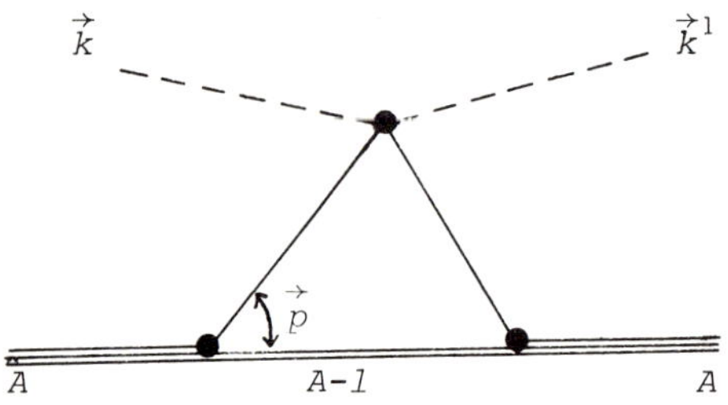

FIG. 1

with $\vec{q} = \vec{k}'-\vec{k}$. This potential which depends on the π-N t-matrix off the energy-shell, can be used in a two-body equation to give the π-A amplitude. A standard approximation[3] used in constructing the optical potential is to factor the π-N t-matrix out of the integral, and for the energy w use the free π-N energy. Recently, Landau and Thomas[2] have shown that at low energies (E_π = 50MeV), by choosing the three-body prescription for w and adjusting the binding energy of the nucleon to the core, one can improve the fit to the data for both $\pi^\pm$-^{4}He and π^+-^{12}C. This indicates that the π-A optical potential has uncertainties from both off-shell behavior of the π-N amplitude and the fact that one can not treat the π-N interaction in the nuclear medium to be the same as that in free space. In particular one needs to take into consideration the fact that the nucleon on which the pion is scattering, is bound. The results of Landau and Thomas indicate that the choice of energy w is important at low energies, and at present there is no precise definition of this energy.

For the interaction of pions with nucleons in nuclear medium, Scadron[4)] has suggested the use of current algebra in conjunction with PCAC to determine the off-mass shell π-N amplitude. This choice of π-N amplitude has been used to determine the contribution of three-body forces in nuclear matter[5)]. The main feature of such π-N amplitude is that it is consistent with the soft pion limit of Adler and Adler-Weisberger. In other words PCAC is effectively a constraint on the off-mass-shell behavior of the π-N amplitude. It should be interesting to compare the off-mass-shell behavior of the amplitude suggested by Scadron with that used in pion-nucleus scattering.

III. PION-ABSORPTION (PRODUCTION) IN NUCLEI - Since the announcement of the Upsala results for $(p,\pi^{\pm})$ reaction in nuclei, the interest in this reaction has increased dramatically. This was mainly due to the fact that π^+ production was considerably larger than the π^-, indicating the similarity of this reaction with (d,p) stripping. However, in this case the momentum transfer is much larger, raising the hope of learning about the short range behavior of the single particle wave function.

There have been two basic mechanism suggested for analysing the experimental data. The first is the analogue of stripping in which the proton emits a pion and is absorbed by the nucleus. In this model (Fig. 2) the neutron absorbed by the nucleus has a momentum transfer of ~ 600MeV/c for 185MeV incident proton.

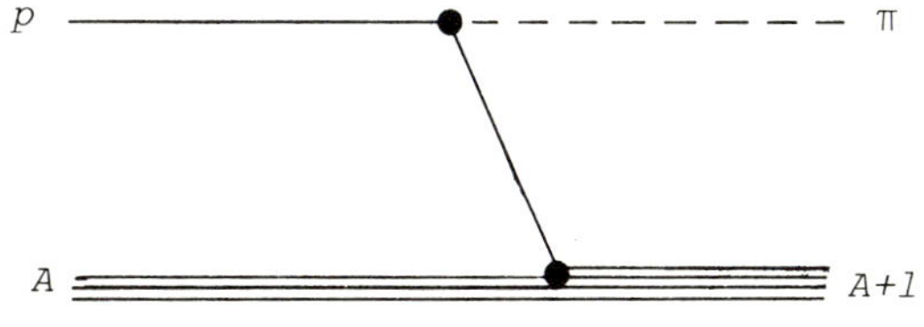

FIG. 2

Such high momentum transfer could give one information about the short range behavior of the single nucleon wave function. Note that in its simplest form, this mechanism gives zero cross section for π^- production. However, in the latest models based on the single nucleon mechanism (SNM) one uses a distorted wave in both initial and final states. Thus if the optical potential for

π+(A+1) is chosen to fit both elastic and double charge exchange scattering, one can get the π^- production cross section as well as the π^+ production cross section.

The second mechanism commonly used for this reaction is the two nucleon mechanism (TNM) originally suggested by Ruderman[6]. This is presented diagramatically in Fig. 3. Here, one writes the amplitude for pion production off nuclei in terms of the amplitude for $pp\to\pi^+d^*$, where d* stands for the deuteron or any two nucleon state. The momentum transfer in this case is shared by the two

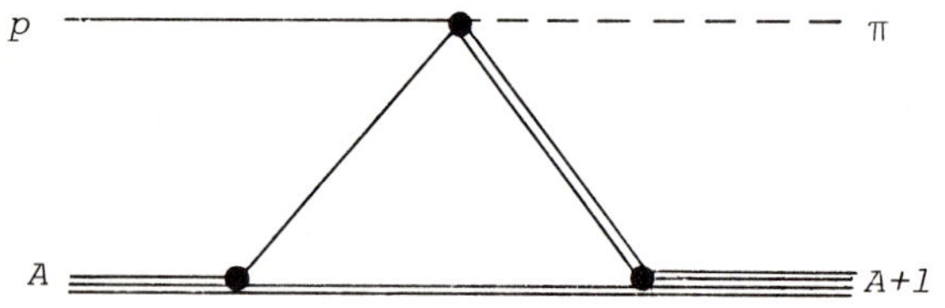

FIG. 3

nucleons and thus one has a smaller momentum transfer per nucleon than in SNM. This has led many to consider this mechanism more dominant than the SNM. There are several variations of this model which involve writing the amplitude for $pp\to\pi d^*$ in terms of the Mandelstam model[7] or by considering $pp\to\pi d^*$ to go through an intermediate NN* state[8].

In this conference, Bhasin[9] has suggested a third mechanism in which the pion emitted by the incident proton is scattered by the target nucleus (Fig. 4). This mechanism was applied to $d(p,\pi^+)t$ at E_p = 470MeV, with a resultant fit to the data both in the forward and backward angles.

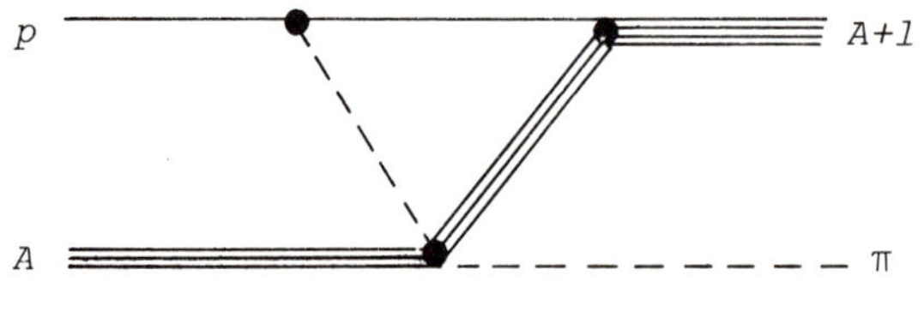

FIG. 4

Although all three mechanisms look distinct, there are many features that are common among the three. In the region of the Δ resonance, to get the energy dependence of the cross section one needs to include the resonance in one form or another into the amplitude. For SNM the Δ is present in the pion-nucleus optical potential, while for TNM the resonance is in the $pp \to \pi d^*$ amplitude and here one needs to calculate the full production amplitude. Finally, in the mechanism suggested by Bhasin (BM) for $d(p,\pi^+)t$, the resonance is in the π-d amplitude which is an input to the calculation. Thus all three mechanism suggested have incorporated the (3,3) resonance. The final fit to the energy dependence of the cross section is going to depend on how the different mechanisms shift and change the width of this resonance. This, in principle, can be checked by the experiment of Jones *et al*[10)] in progress at TRIUMF, where one can vary the energy of the proton beam.

If one examines these different mechanisms in some approximation one finds that they are related. Comparing the TNM with that proposed by Bhasin, one finds that they can be related through antisymmetry[8)], if both are dominated by Δ resonance. Thus in $d(p,\pi^+)t$ if the $pp \to \pi d^*$ in the TNM and $\pi d \to \pi d$ in the BM go through the Δ Figs. (5) and (6) then by exchanging 1 and 2 in the TNM one gets the BM result.

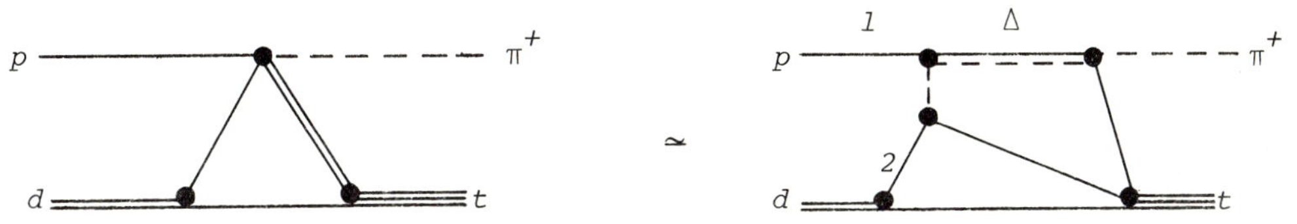

FIG. 5

FIG. 6

On the other hand if one takes the SNM with a pion optical potential in the final state, one represents the optical potential by the multiple scattering series (Fig. 7). At the same time if one replaces the π-d amplitude in BM by a multiple scattering series (Fig. 8), then the two series have terms that are related by changing the ordering of the interactions.

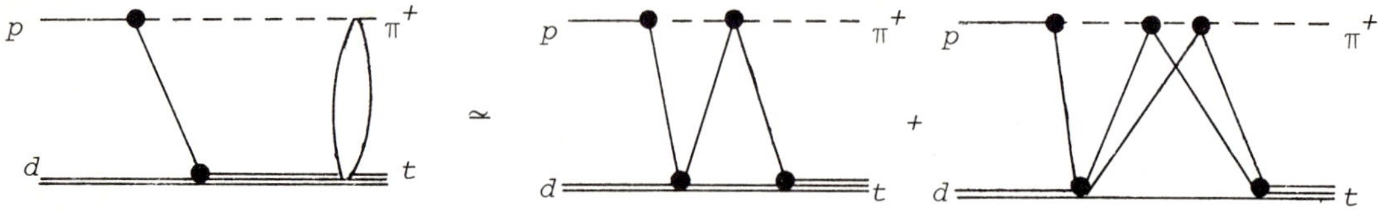

FIG. 7

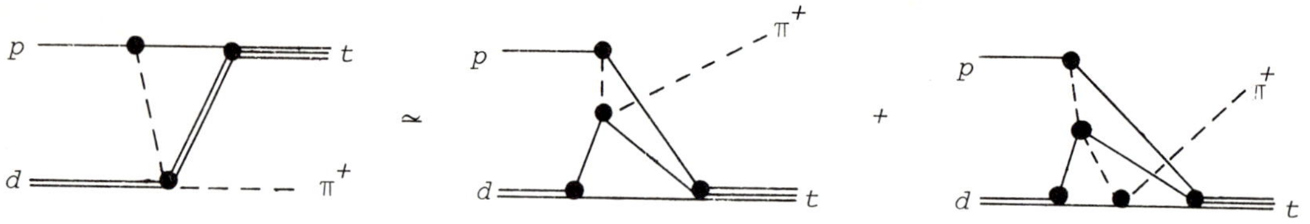

FIG. 8

One sees that in two ways these different mechanisms are related. In particular, one observes that the BM has some features of SNM, in that the outgoing pion multiple scatters off the other two nucleons, and on the other hand it is related to the TNM mechanism through the antisymmetry. It would be interesting to develop this model further for heavier nuclei in which case the π-d amplitude would be replaced by the π-A amplitude which one can get from the pion optical potential.

Finally, one turns to (π,NN) reaction where the pion is captured from a 1s orbit. In this case one can still consider SNM or TNM. However, the problem has the added complexity of a three-body final state. Here, Jain[11)] in a contributed paper to the conference has suggested that for s-wave pion capture one parametrizes the πd→NN amplitude in terms of two adjustable constants, and uses the standard optical potential for the final nucleons. This parametrization was applied to the calculation of (π^-,nn) amplitude on ^{6}Li and ^{12}C with reasonable agreement to the available data.

IV. DYNAMIC MODEL FOR THE πNN SYSTEM - In the above discussion of pion-nucleus optical potential I pointed to several approximations used to calculate pion-nucleus scattering. To test these approximations, and the degree of their validity, one needs now to consider the simplest pion-nucleus scattering where one can calculate the amplitude exactly in some model, and then test the different approximations. The simplest nucleus where that is possible is the deuteron. Here, pion-deuteron scattering can be treated as a three-body problem. The solution of the three-body equation will guarantee: (i) that two- and three-body unitarity are satisfied. (ii) An exact sum of the multiple scattering series including nucleon recoil are included. (iii) The correct energy for the π-N amplitude, which is an input to the three-body equations, and the treatment of the transformation for π-N to π-d center of mass are incorporated. Since the π-N amplitude has a pole in the P_{11} channel at the nucleon mass, one can incorporate the absorption channel into such a calculation. In this way one can treat the reactions

$$\begin{aligned} \pi + d &\to \pi + d \\ &\to \pi + N + N \\ &\to N + N \end{aligned}$$

and

$$\begin{aligned} N + N &\to N + N \\ &\to \pi + d \\ &\to \pi + N + N \end{aligned}$$

within the framework of the model. Such a scheme will allow one to examine the validity of the isobar model, and give one some insight into the dominant process for the TNM discussed in the last section. At the same time one can estimate the NN* component of the deuteron wave function, which has been a subject of considerable interest in recent years. Also, one can examine if the $^1D_2$ nucleon-nucleon partial wave has a resonance due to its strong coupling to the NΔ channel[12]. Such a resonance would also appear in the π-d channel, since all three channels are coupled in this scheme. Note, that in this model the Δ resonance is not treated as a particle, but would have the correct width.

The original idea for such a scheme was suggested by Lovelace[13], and has been implemented for the $\pi\pi$N system by Aaron, Amado and Young[14], and Aaron and Amado[15], with reasonable success. More recently Thomas and Afnan[16] have extended this scheme to the πNN system within the framework of non-relativistic Faddeev equations. They got results for N-N scattering, π-d scattering, and pion production in good agreement with experiments. For π-d elastic scattering there are now a number of both non-relativistic and relativistic calculations for the π-d scattering length[16,17] at pion energies of ~50MeV[18], and in the (3,3) resonance region[19]. In all these calculations the coupling to the N-N channel were excluded. Recently, Thomas[18] has calculated angular distribution for π-d elastic scattering at 47.7MeV. He finds an improved fit to the data over previous methods not based on three-body equations. He also examines the question of choice of energy in the π-N amplitude and how it effects the multiple scattering series and the total amplitude for π-d scattering. His results show that fixing the energy in the π-N amplitude leads to discrepancy as large as 50% at small angles with the corresponding correct three-body calculations. Thus for π-d scattering, a three-body approach gives good results below the (3,3) resonance, and it is hoped that with more experimental data one will be able to understand the limitation of some of the approximations in pion-nucleus scattering.

The results of phase shift analysis on π-^{4}He indicate that several partial waves appear to have resonance behavior[20]. This is based on Argand diagrams for these phase shifts. The question was raised by Hoenig and Rinat[21] if these results based on phase shift analysis do in fact prove the existence of resonances. To examine this question, they studied π-d scattering in Glauber theory, and found that this resonance behavior appears in more than one partial wave, and is a consequence of the (3,3) resonance in the π-N system. However, because of factorization in Glauber theory, their π-d amplitude does in fact have a pole. Examining this point more closely in the framework of three-body theory, Afnan and Thomas[22] have calculated the $J^{\pi} = 2^{+}$ partial wave for π-d scattering. They have calculated both the single scattering (s.s.) term Fig. 9, and the exact solution of the Faddeev equations, including only the (3,3)

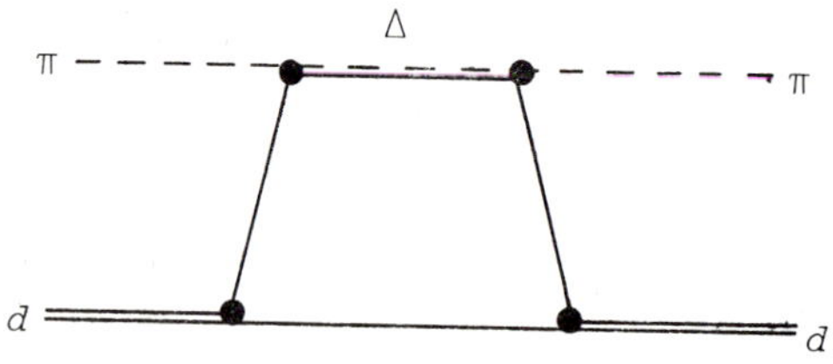

FIG. 9

resonant π-N amplitude, and S-wave deuteron. In the single scattering approximation the amplitude has a branch point due to the threshold for Δ production. However, in this case the amplitude has no pole. In Fig. 10, one presents the Argand diagram, together with the "speed"[23] which exhibits a Breit-Wigner form for both the s.s. and exact result. Although one cannot rule out the existence of a resonance pole for the exact amplitude, the s.s. result indicates that one can not establish the existence of resonance poles, in many body systems based only on Argand diagram and a Breit-Wigner peak in the speed. In fact, one needs to analytically continue the amplitude onto the second sheet to establish the resonance. This was demonstrated for the A_1 resonance by Brayshaw[24]. One notes from Fig. 10 that summing the multiple scattering series, narrows the peak and shifts it to lower energy. In the case where the pion is treated relativisticly the behavior of the single scattering amplitude remains the same, and the multiple scattering series is more convergent.

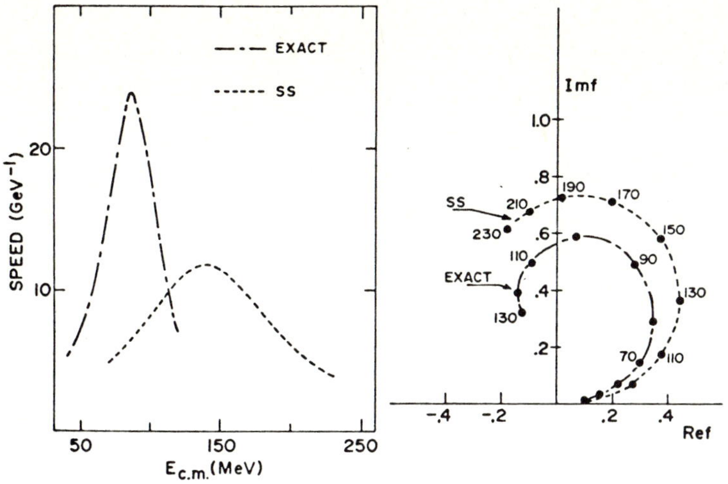

Figure 10.

A major improvement in treating the πNN system was presented at this conference by Brayshaw[25]. He presented results for N-N scattering based on a relativisticly covariant three-body boundary condition model[26]. By employing the boundary condition model he replaces the off-shell two-body amplitude by the on-shell amplitude obtained directly from experiment, plus an energy dependent boundary condition. In this way he avoids the problem of specifying the short range behavior of the two-body interaction to which pion production is sensitive. On the other hand the covariance of the theory removes the ambiguity in any non-relativistic reduction of the N↔Nπ vertex, which is present in most pion production calculations. By including the $^3S_1$ nucleon-nucleon interaction and the P_{11} and P_{33} pion-nucleon interaction he achieves good agreement with experiments in the $^1S_0$ phase shifts up to pion production threshold. He finds that the inclusion of the P_{33} pion-nucleon interaction gives rise to the appropriate repulsion. This effect of repulsion due to the inclusion of the P_{33} interaction is also present in the non-relativistic model of Thomas and Afnan[16], and is consistent with results of coupled channel calculations in which the NN channel is coupled to the NΔ channel.

For the inclusion of NN* component in the deuteron wave function, Weber[27] presented results based on a covariant three-body equations for the πNN system. He used the Aaron, Amado and Young[14] equations with the N* as a quasi-particle of (π-N) system. The main deviation from Aaron *et al* is that he employed the Wightman-Gardin definition of the relative momentum. This choice for the relative momentum first suggested by Namyslowski[28] for the three-body system guarantees the correct clustering property. He finds in this case that the NN* component is qualitatively similar to the nonrelativistic impulse approximation.

From the above discussion of the πNN system, one sees that one can get some valuable information not only for π-d elastic scattering, but also for N-N scattering and pion production. Considering the fact that the new meson facilities will be producing a considerable amount of high quality data some of which was presented at this conference, there is room for an extensive study of relativistic covariant three-body equations. Furthermore, an understanding of the πNN system will allow one to improve some of the approximations in heavier nuclei as seen from the results of Landau and Thomas[2].

The author would like to thank Professors H.J. Weber and V.S. Bhasin for many informative discussions during the conference.

REFERENCES:

1. M. Bleszynski, T. Joroszewicz and P. Osland, contributed to this conference.
2. R.H. Landau and A.W. Thomas, submitted to Phys. Letters.
3. R.H. Landau, S.C. Phatak and F. Tabakin, Ann. of Phys. 78 (1973) 299.
4. M.D. Scadron, invited talk at this conference.
5. B. McKeller, invited talk at this conference.
6. M. Ruderman, Phys. Rev. 81 (1952) 383.
7. A. Reitan, Nucl. Phys. B50 (1972) 166.
8. M.P. Locher and H.J. Weber, Nucl. Phys. B76 (1974) 400.
9. V.S. Bhasin, invited talk at this conference; V.S. Bhasin and I.M. Duck, Phys. Letters 46B (1973) 309.
10. G. Jones, University of British Columbia.
11. B.K. Jain, invited talk at this conference.
12. H. Suzuki, Prog. Theor. Phys. 54 (1975) 143.
13. Lovelace, Phys. Rev. 135 (1964) B1225; D.Z. Freedman, C. Lovelace and J.M. Namyslowski, Nuovo Cimento 43A (1966) 258.
14. R. Aaron, R.D. Amado, J.E. Young, Phys. Rev. 174 (1968) 2022.
15. R. Aaron, R.D. Amado, Phys. Rev. Lett. 27 (1971) 1316.
16. I.R. Afnan, A.W. Thomas in Proc. Int. Conf. on Few-Particle Problem in Nuclear Interaction, Los Angeles 1972 I. Slaus *et al* eds. (North Holland) p.861. A.W. Thomas and I.R. Afnan, Phys. Letters 45B (1973) 437; I.R. Afnan and A.W. Thomas, Phys. Rev. C10 (1974) 109.
17. N.M. Petrov and V.V. Peresypkin, Phys. Letters 44B (1973) 321; V.V. Peresypkin and N.M. Petrov, Nucl. Phys. A220 (1974) 277.
18. A.W. Thomas (preprint).
19. F. Myhrer and D.S. Koltun, Phys. Letters 46B (1973) 322; Nucl. Phys. B86 (1975) 441; V.B. Mandelzweig, H. Garcillazo and J.M. Eisenberg (preprint).
20. I.V. Falomkin *et al* Nuovo Cimento Lett. 5 (1972) 1125.
21. M. Hoenig and A.S. Rinat, Phys. Rev. C10 (1974) 2102.
22. I.R. Afnan and A.W. Thomas, Contribution in Sixth Int. Conf. on High Energy Physics and Nuclear Structure, Sante Fe, June 1975.
23. R.D. Tripp "Baryon Resonances" Internation School of Physics, Enrico Fermi Course 33 (Academic Press 1966).
24. D.D. Brayshaw SLAC-PUB-1677 (November 1975).
25. D.D. Brayshaw, invited talk to this conference.
26. D.D. Brayshaw, Phys. Rev. D11 (1975) 2583.
27. H.J. Weber, invited talk to this conference.
28. J.M. Namyslowski, Nuovo Cimento 57A (1968) 355.

SYMMETRIES*

Ernest M. Henley

Physics Department, University of Washington, Seattle, Washington 98195 U.S.A.

Nature is beautiful; part of its beauty resides in the slight imperfections which spoil perfect symmetry.

Physicists spend considerable effort in fathoming the symmetries of nature as well as in ferreting out the slight imperfections which often lead to a deeper understanding of physical phenomena. It is this endeavor in the subatomic world which is the topic of my talk. The subject is vast. I could not do it justice, even if time could be reversed at the end of my alloted period. Thus, I will restrict my presentation to three topics which are under active investigation. They are time reversal invariance, parity, and exotic (primarily second-class weak) currents. In view of the conference topic, I will concentrate on few body aspects of the problems, review recent progress, and finally indicate possible directions for future inquiries.

I. TIME REVERSAL INVARIANCE (TRI)

The primary interest in TRI at the present time continues to be our desire to understand the CP violation observed in the decay of the K^0. Although the present data is consistent with the superweak theory [1], milliweak ($\sim 10^{-3}$ × weak) theories [2] cannot be ruled out as yet. If the superweak theory is correct, then no effect should be seen in any other weak reactions than those involving the K^0. Probably the best and most decisive test of this theory is the measurement of the neutron dipole moment, which at present is known to be $\lesssim 3 \times 10^{-24}$ e-cm [3]. Since a milliweak model predicts $d_n \sim 10^{-23}$ e-cm [2],

$$|d_n| \sim e \times \frac{1}{M} \times \frac{GM^2}{4\pi} \times \frac{\alpha}{\pi}$$

$$\sim e \times 10^{-14}\ \text{cm} \times 10^{-6} \times 10^{-3} \sim 10^{-23}\ \text{e-cm} \quad , \tag{1}$$

(α is the fine structure constant, G is the weak coupling constant, and M is the mass of the nucleon) the present limit on the magnitude of the neutron electric dipole moment is close to ruling out many milliweak theories. However, for some of these theories, further inhibition factors occur so that d_n can even approach the superweak value of $d_n \sim 10^{-29}$ e-cm. What differentiates milliweak theories from the superweak one is that T-odd effects should appear in other weak processes in the former case. Interest in such tests has been rekindled by modern gauge theories of the weak interaction, and milliweak effects have been sought. The most precise test to date is a recent one carried out at Grenoble on the β decay of the neutron [5]. The T-odd correlation D in the decay rate R,

$$R \propto 1 + D \langle\vec{\sigma}_n\rangle \cdot \vec{p}_e \times \vec{p}_\nu \quad , \tag{2}$$

where $\langle\vec{\sigma}_n\rangle$ is the neutron polarization and $\vec{p}_e$ and $\vec{p}_\nu$ are the electron and neutrino momenta. The correlation coefficient D is a measure of the relative reality of the vector and axial vector coupling constants. The phase angle, ϕ_{V-A}, between these couplings was found to be [5]

$$\phi_{V-A} = 180.14 \pm 0.22^\circ \tag{3}$$

This result suggests that TRI is valid in weak and strong interactions to about 3 parts per 1000, i.e. the ratio of T-odd and T-even interactions is

$$\left|\frac{H_{T\text{-odd}}(\text{wk})}{H_{T\text{-even}}(\text{wk})}\right| \quad \text{and} \quad \left|\frac{H_{T\text{-odd}}(\text{hadr.})}{H_{T\text{-even}}(\text{hadr.})}\right| \lesssim 3 \times 10^{-3} . \qquad (4)$$

The limit set by tests of the same correlation in the β-decay of polarized ^{19}Ne is almost as good [6] and an improved experiment is being undertaken by Calaprice [7]. If this experiment indicates no violation of TRI at a level of $\sim 5 \times 10^{-4}$, then many milliweak theories will be in trouble. However, I remind you that β-decay between members of an isospin multiplet is a very insensitive test of TR violation due to second class currents [8].

Electromagnetic tests of TRI in nuclei [4,9], both at low and high energies, also have not revealed any effects to a level of $\sim 3 \times 10^{-3}$. Such tests attempt to discover TR violations due to either the electromagnetic interaction of hadrons or a millistrong ($\sim 10^{-3}$ × strong force) theory. However, these models are not totally ruled out, particularly if the TR-odd force is of short-range. Detailed balance experiments and polarization-asymmetry comparisons show no violation to about the level given by eq. (4). Earlier indications of a possible violation of TRI in the reaction $\gamma + n \leftrightarrow \pi^- + p$ close to the Δ(1236) resonance have now disappeared; an upper limit of 2° can be set on any T-odd phase [10]. Bryan [11] and colleagues have developed in great detail a model proposed by Sudarshan, in which TR-odd effects arise from A_1 exchange [11]. A comparison of polarization and asymmetry in high energy n-p scattering is suggested [10] for sensitive tests; as shown by Simonius [12], low energy nucleon-nucleon scattering is insensitive to TR violating effects because the partial wave amplitudes which contribute are $^3S_1 \leftrightarrow {}^3D_1$, $^1P_1 \leftrightarrow {}^3D_1$, $^1P_1 \leftrightarrow {}^3P_1$, $^1D_2 \leftrightarrow {}^3D_2$ and $^3P_2 \leftrightarrow {}^3F_2$. Only the latter is allowed in p-p scattering and all but the first phases are small in low energy n-p scattering. Simonius [12] has proposed a comparison of polarization and asymmetry in low energy p-^{3}He scattering to increase the sensitivity to the p-n interaction and yet have charged particles participate.

Since no T-odd effects have been observed in a reasonable variety of nuclear and electromagnetic tests of TRI to a few parts in 10^3, I have a prejudice against electromagnetic and millistrong theories, even though models can be concocted which give strong suppressions. An improvement by a factor of 5 in present tests would, it seems to me, make these theories untenable. However, at this level, various electromagnetic scattering corrections [4,8,13] begin to play a role and must be taken into account. It will not be possible to test milliweak theories in nuclear reactions or transitions. Improved and/or hindered β-transitions, as suggested by Barroso and Blin-Stoyle [14] can be used for such tests. An experiment which is also sensitive to T-violation due to second class currents, has been undertaken by Calaprice [7]. They measure the β-γ correlation proportional to $\langle \vec{J} \cdot \vec{p} \times \vec{k}\ \vec{J} \cdot \vec{k} \rangle$, where $\vec{J}$ is the nuclear spin, $\vec{p}$ is the momentum of the electron and $\vec{k}$ that of the photon, in the decay of oriented ^{56}Co. A preliminary result shows no asymmetry to about 1×10^{-3} [7]; it sets a limit on T-violation only if the ratio of Fermi to Gamow-Teller matrix elements is not zero. There is some controversy regarding this ratio [7]; if it is ~ 0.2, then the experiment of Calaprice et. al. sets a limit of roughly 5° on a T-odd phase [7].

II. PARITY

"In a really just cause the
weak conquer the strong."

(Sophocles, "Oedipus")

The interest in parity (P) stems in large part from the fact that measurements of parity violating (PV) effects may give us a better understanding of the weak forces between particles and of the general theory of weak interactions.

In contrast to TRI, violations of mirror asymmetry have been observed. However, progress in our understanding of parity violating effects has been slow. Theorists have not been sufficiently clever to explain the observations already made. Consequently experimental physicists have to hunt for tiny PV effects in light nuclei, where dynamical enhancements are rare, but, where the effects may be able to be explained. In light nuclei, definite PV signals have been observed only in ^{16}O and in the capture of very slow neutrons by protons [9].

New degrees of freedom have recently been injected by the spontaneously broken gauge models of the weak interaction and by the discoveries of neutral currents. But we do not yet know whether as suggested by Weinberg [15], these neutral currents give rise to PV effects or whether they are parity-conserving (e.g. pure vector) as suggested by De Rujula [16], by Fritzsch [16] and by others. Indeed, we do not even know whether the neutral current effects observed to be induced by muon neutrinos have counterparts in electron-neutrino and in charged lepton induced processes. Such effects are being sought.

What is the nature of the signals used to detect PV? Some of them are illustrated in Table I below.

a) Asymmetry:

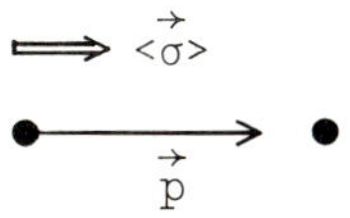

$$\underline{a} = \frac{\sigma_L - \sigma_R}{\sigma_L + \sigma_R} \propto \langle\vec{\sigma}\rangle \cdot \vec{p} \propto F_{wk}$$

b) Correlation:

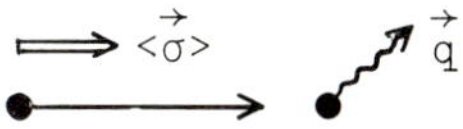

$$c = \frac{R_L - R_R}{R_L + R_R} \propto \langle\vec{\sigma}\rangle \cdot \vec{q} \propto F_{wk}$$

c) Circular Polarization:

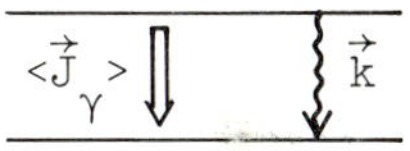

$$P_\gamma = \frac{R_L - R_L}{R_L + R_R} \propto \langle\vec{J}_\gamma\rangle \cdot \vec{k} \propto F_{wk}$$

d) Rate of Parity-Forbidden Transition:

$$i \overset{?}{\to} f \qquad R \propto |F_{wk}|^2$$

e) Optical Rotation:

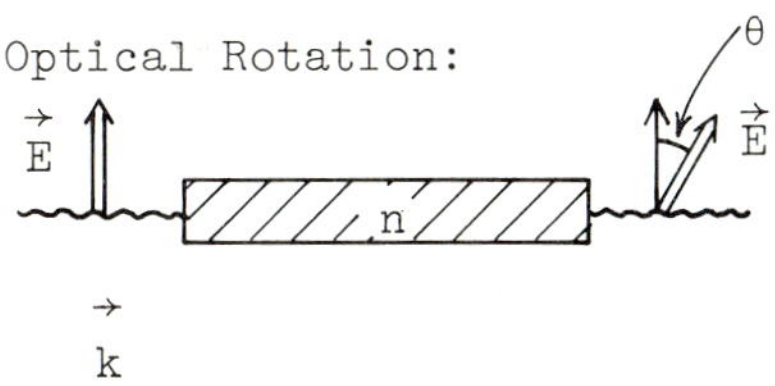

$$\theta = \tfrac{1}{2}\, k\, \mathrm{Re}(n_L - n_R) \propto F_{wk}$$

Table I. Parity-violating signals

The first four of these methods are familiar from subatomic physics tests. The last one is particularly useful in atomic searches of PV. It has been employed by a group at Seattle and I will refer to it anon.

A group of physicists from SLAC and Yale [17] are using a high energy (10-20 GeV) polarized electron beam to measure the asymmetry, a, in inelastic (inclusive) electron scattering from hydrogen. The reaction is:

$$e^- + p \rightarrow e^- + X \quad ,$$

where X is an unobserved hadronic state. The group hopes to measure the asymmetry to $\sim 1 \times 10^{-4}$ at large momentum transfer q^2. The sensitivity to the weak interaction is highest at large q^2 because of current conservation, which forbids PV at $q^2 = 0$ and because the interaction has a very short range. In the Weinberg model, the asymmetry, a, is predicted to be [18] (M = mass of proton)

$$\underline{a} \sim 10^{-5}\, q^2/M^2$$

in the deep inelastic region. The effect is approximately a factor of a $\approx 10^{-2}$ smaller if the neutral currents are parity-conserving [19].

Feinberg [20] has proposed electron scattering from nuclei to obtain the isospin dependence of the neutral current. He points out that a particularly favorable case is excitation of a 0^- state from a 0^+ ground state. An enhancement of ~ 100 occurs because the electromagnetic matrix element, with which the weak interaction one interferes, requires two photon exchange. This factor compensates the lower momentum transfers, q, accessible in electron nucleus scattering. As a particular example he cites the excitation of the isospin I=0, 0^- state at 10.95 MeV in ^{16}O and of the I=1, 0^- state at 12.8 MeV in the same nucleus. The Weinberg model predicts $\underline{a} \approx 10^{-3}$ for the latter state and 0 for the former if $\sin^2\theta_W \sim 0.35$. For elastic scattering on ^{16}O, the predicted asymmetry for 1 GeV electrons and $|q^2| \approx M^2/4 \approx 10\ fm^{-2}$ is considerably smaller, $\underline{a} \approx 3 \times 10^{-5}$, but the differential cross section is correspondingly larger, $d\sigma/d\Omega \approx 4 \times 10^{-32}\ cm^2/ster$.

At lower energies, correlations and asymmetry tests of P in muonic atoms have been proposed [20] and are being planned [21]. In light muonic atoms, the effective PV interaction consists of a nuclear spin-dependent term (from an axial nuclear coupling) and a similar one for the muon. In general, the latter one dominates (it is the only surviving term for spin zero nuclei) and can be expressed as a one-body δ-function potential (at the nucleus) in the non-relativistic limit

$$V_{PV} = \frac{G}{\sqrt{2}} \left[\frac{\vec{\sigma}\cdot\vec{p}}{2m_\mu} \delta^3(\vec{r}) + \delta^3(\vec{r}) \frac{\vec{\sigma}\cdot\vec{p}}{2m_\mu} \right] f_W \quad , \tag{5}$$

where G is the weak coupling constant, $GM^2 \approx 10^{-5}$, m_μ is the muon reduced mass, and f_W is a factor which depends on the particular neutral-current theory. For the Weinberg model, it is

$$f_W = \frac{1}{2}[N + (4\sin^2\theta_W - 1)Z], \tag{6}$$

where N, Z are the neutron and proton number and $\sin^2\theta_W \approx 0.35$. A large enhancement occurs for parity-mixing between the $2S_{1/2} - 2P_{1/2}$ states because they are almost degenerate in H-like atoms, as does a further one if one considers the $2S_{1/2} - 1S_{1/2}$ transition, where the P-allowed M1 mode is very slow, whereas the parity-admixed $2P_{1/2} - 1S_{1/2}$ E1 decay is fast. The case of ^6Li is shown in Fig. 1. The effect can be as large as a few %, but there are grave difficulties [20,21]:

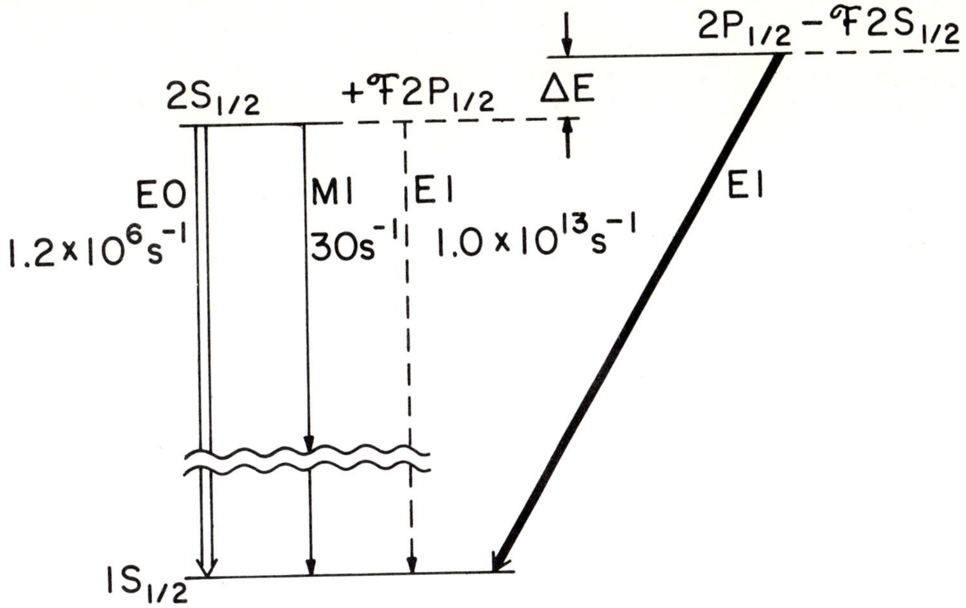

Fig. 1. Illustration of parity-mixing and decay rates for muonic ^{6}Li.

a) The population of the 2S state is small.
b) Stark-mixing depletes the $2S_{1/2}$ state.
c) Auger decay of the $2S_{1/2}$ state can be large.
d) Background effects from bremsstrahlung and from the 2γ decay mode of the 2S state can be sizable.
e) The 2S state has been observed only in ^{4}He by Zavattini [22] and colleagues.

For the above reasons Simons and others at CERN [21] and Fiorini [21] and colleagues have proposed PV searches involving somewhat higher atomic energy levels (e.g. 3P-3D) and in medium weight nuclei. Among light nuclei ^{4}He may be the best bet. We will probably have to wait until the time of the next conference to hear results.

Tests in ordinary atoms have also been suggested particularly in heavy atoms, where coherence enhances the PV effect [23]. At the University of Washington a group of physicists working with N. Fortson have searched for the optical rotation of plane polarized laser radiation passing through Bi vapor at a resonance [24]. The angle of rotation in one absorption length is predicted [25] to be $\approx 3 \times 10^{-7}$ by the Weinberg theory with $\sin^2\theta_W \approx 0.35$. Although this is not a light nucleus, I mention it because the group has seen no effect to $\sim 1 \times 10^{-6}$, which represents a factor of 1000 improvement over previous atomic PV searches. They hope to reach an accuracy of 1×10^{-7} in the near future.

In light atoms an interesting proposal has been put forward recently by Lewis [26] and Williams. They point out that by passing a metastable H beam through a magnetic field, the $2S_{1/2}$ and $2P_{1/2}$ levels can be made to cross, which leads to an enhanced (~ 20) PV mixing. Further, they propose rapid passage through a perpendicular electric field to induce the 2S decay or the use of a tunable laser to induce the 2S-3S transition. The difference of the absorption for L and R circularly polarized light which is expected to be $< 10^{-8}$ can be used, for instance, to measure the effect. The experiment is clearly a most difficult, but perhaps not impossible one. Since it is a basic lepton-hadron interaction which is being tested, it is a rewarding experiment.

As you can tell, there are many proposals, but no concrete results. This is not so in nuclear physics, where PV has been observed but the results are not understood.

Although neutral currents do lead to measurable consequences in nuclei, they cannot be sought here. The reason is that a hadronic charge-exchange can turn a charged current into an apparent neutral one and vice-versa. The nuclear effects of the neutral currents are model-dependent. In the Weinberg-Salam theories, a large ($\sim$ 20) enhancement of the isospin $\Delta I=1$ PV nuclear force occurs. It is the isospin character of the weak force between nucleons which I will use as a prop in the following discussion. Some salient features are illustrated in Table II.

ΔI	Lowest mass mesons exchanged	Anticipated theoretical property	Experimental evidence	Some possible experimental probes in light nuclei
0	ρ, ω, (2π)	Enhanced by $\Delta I=1/2$ rule (octet enhancement)	Required by PV ^{16}O decay	$^{16}O^* \to {}^{12}C+\alpha$: R $\vec{p}+d$: a $\vec{p}+p$: a $\vec{n}+p$: a
1	π, (2π)	Enhanced by neutral currents	?	$^{18}F^*(0^-) \to {}^{18}F(1^+)+\gamma$: P_γ $^{10}B^*(2^-) \to {}^{10}B^*(1^+)+\gamma$: P_γ $d+\alpha \to {}^6Li^*(0^+)$, $\to {}^6Li+\gamma$: R $\vec{n}+p \to d+\gamma$: a $\pi+\vec{p}$: a
2	ρ, (2π)	Suppressed by $\Delta I=1/2$ (rule {27} suppression)	Suggested by large P_γ in $n+p \to d+\gamma$	$n+p \to d+\gamma$: P_γ $\vec{p}+p$: a $\vec{p}+{}^3He$: a $\vec{p}+{}^3H$: a

Table II. Some properties of the PV internucleon force. The symbols R, a, P_γ are explained in Table I.

Let me use this table to summarize the present experimental situation and to suggest further tests in light nuclei. The most basic experiments are those which measure the PV force between nucleons. The experimental observation of a circular polarization in the capture of sub-thermal energy neutrons by protons is well known [27]

$$P_\gamma(\text{exp}) = -(1.3 \pm 0.45) \times 10^{-6} \quad .$$

Despite massive theoretical efforts [28,29] to understand this large number, success has not been achieved. There is now, I believe, general agreement that the theoretical expectation is

$$10^{-8} \lesssim P_\gamma\ (\text{th}) \approx 2 \times 10^{-8} \lesssim 10^{-7} \quad .$$

This result is relatively insensitive to neutral currents but is dependent on the short-range N-N force. The reason is that P_γ arises from ρ, ω and perhaps heavier meson exchanges. If octet enhancement results in suppression of the ΔI=2 part, as anticipated, then P_γ is very small, regardless of hadronic model; that is, the ΔI=2 part of the ρ-exchange parity-violating potential gives the dominant contribution to the theoretical value of P_γ, which it should be noted is opposite in sign to experiment. The large ΔI=2 matrix element has led McKellar [28,30] and Simonius [30] to propose that a large (∿ 50) enhancement of this piece of V_{PV} may be able to account for the Lobashov result without ruining agreement with other experiments. These other measurements are:

a) The PV forbidden α-decay of the 2^-(I=0) level in ^{16}O to $^{12}C+\alpha$, which depends only on the ΔI=0 component of the PV force and is in rough agreement with theory [4,9,28]. An improved calculation would be valuable here.

b) The asymmetry of photons, $\underline{a} = -(1.8 \pm 0.9) \times 10^{-4}$ emitted in the $1/2^- \rightarrow 1/2^+$ transition of ^{19}F [31]. The results here are sensitive to the ΔI=0 and ΔI=1 parts of V_{PV}. They are consistent with theory with or without neutral current enhancement of the ΔI=1 part of V_{PV} [32]. An improved measurement would be helpful.

c) The lack of asymmetry observed in the scattering of polarized protons by H at 15 MeV, $\underline{a} = -(1.7 \pm 3) \times 10^{-7}$ [33]. Theory, with ΔI=0 alone, gives $\underline{a} \sim 10^{-7}$ [34]. At 6 GeV, an asymmetry of order $\sim 10^{-5}$ is observed in a transmission experiment [33]; however, it appears to be explained in terms of Λ^0 production followed by weak (PV) decay to $p+\pi^-$. Again, a ΔI=0 theory predicts $\underline{a} \lesssim 10^{-7}$ [35]. These measurements are primarily sensitive to the ΔI=0 part of the PV force; the contribution of the ΔI=2 part is reduced by a Clebsch-Gordan coefficient of $(10)^{-1/2}$. If the errors of the experiment are reliable, the measurement nevertheless rules out the large enhancement of the ΔI=2 part of V_{PV} proposed by McKellar and Simonius to explain the Lobashov result. A further improvement in the experimental limit might net an observed asymmetry. Such an observation would go far towards unraveling the present puzzling situation.

Indeed, let me use Table II to indicate what can be done to help us unfold the detailed map of the nuclear weak interactions.

a) V_{PV} (ΔI=0):

Measurements of the asymmetry in p-p scattering should be carried out at other energies. Theory [34] predicts a peaking of $\underline{a}$ at about 50 MeV, $\underline{a} \approx 2 \times 10^{-7}$, and this energy is accessible. Medium energy tests at LAMPF can be helpful [35]. At very high energies, the use of high perpendicular momentum transfers may enhance the asymmetry in inclusive or total cross section measurements [35], since the weak interactions have a very short range. At almost all energies, the p-p scattering is sensitive primarily to the ΔI=0 part of V_{PV}. As a check of this hypothesis one can scatter protons from deuterons. Since the isospin of the deuteron is zero, there can be no contribution of $V_{PV}(\Delta I=2)$.

b) $V_{PV}(\Delta I=1)$:

The elusive isovector part of the PV internucleon force is of particular interest because the presence of neutral currents are expected to enhance it by $\sim$20 over the value predicted by the Cabibbo theory [36]. This enhancement may be helpful in explaining PV observations made in heavy nuclei.

The best place to seek this piece of the PV force is probably in the gamma decay of the 0^- level at 1.080 MeV to the ground state of ^{18}F [37]. Although this is an E1 transition, it is strongly hindered because both levels have I=0. There is a 0^+ level at 1.043 MeV in ^{18}F which has I=1; it is the admixing of this level which is expected to give the dominant contribution to the PV effect, e.g. P_γ. The experiment is being attempted by Adelberger [38] and colleagues. The circular polarization is predicted to be $\approx 5.7 \times 10^{-3}$ for the Weinberg model and 3.6×10^{-4} for the Cabibbo model [39]. A similar, but not as large, P_γ may occur in the decay of the 2^-(5.105 MeV) state to the 1^+(0.717 MeV) level in ^{10}B [37]. In ^{10}B there is an I=1, 2^+ state at 5.159 MeV, which should be the primary PV admixed component.

Another experiment which has been proposed and carried out is the $d(\alpha,{}^6Li)\gamma$ reaction through the 0^+(3.562 MeV) state of I=1 in ^6Li. However, the best limit obtained to date $\Gamma(3.562) < 8 \times 10^{-4}$ eV is still about 2 orders of magnitude larger than predicted by any theory [40].

A further experiment which is sensitive to $V_{PV}(\Delta I=1)$ is the asymmetry of photons emitted in the capture of very slow polarized neutrons by hydrogen. Until neutral currents were found, it was expected that $\underline{a} \sim 10^{-9}$, considerably smaller than P_γ from the capture of unpolarized neutrons. In the Weinberg model, the neutral currents push $\underline{a}$ up to $\sim 10^{-7}$, which is larger than P_γ and may be measurable. Fortunately Vignon, Wilson and colleagues paid no attention to theorists. They have undertaken a measurement of the asymmetry at the reactor of the Institut Laue-Langevin in Grenoble. The experiment is working at present; no effect has been seen to $\sim 10^{-6}$ [41].

Other tests of the $\Delta I=1$ PV force have been proposed [9,42]. Although there are no results to report at the present time on any measurement of $V_{PV}(\Delta I=1)$, this situation is likely to be different in the near future.

c) $V_{PV}(\Delta I=2)$:

The isotensor component of the PV force is expected to be small in a theory which enhances the octet part or suppresses the {27} contribution of the current-current interaction [9]. Such suppression is suggested by the $\Delta I=1/2$ rule observed to hold in strangeness changing non-leptonic decays. In contrast, an enhanced contribution has been called on to explain the results of Lobashov in the capture of unpolarized neutrons by hydrogen [30]. This important experiment clearly should be repeated; it is planned by Vignon and Wilson [34]. If the $\Delta I=2$ PV force is enhanced, then it should be detectable in the asymmetry of protons scattered by ^1H, ^3H, and ^3He.

There is presently no evidence for a large (e.g. $\gtrsim 10$) enhancement in p-p scattering.

What is one to conclude at the present time? It is my conviction that no amount of theoretical work will enlighten us. What is needed are a series of careful, definitive measurements. Because the experiments are so difficult, few of them have been carried out by more than one group. However, just because of their great difficulty, I believe that it is important that measured PV effects obtain independent confirmation. Of course, that is easy for a theorist to say! However, until further experimental data becomes available, I doubt that we can expect more understanding.

The tests which I believe to be most important are:

1) Determination of the PV nature of neutral currents (atomic, muonic atom, e^--p scattering, etc. tests).
2) Measurement of $V_{PV}(\Delta I=1)$ ($^{18}F^*$decay, $\vec{n}+p \rightarrow d+\gamma$).
3) Direct measurement of V_{PV} between nucleons ($\vec{p}+p$, $\vec{n}+p$ scattering).

III. "EXOTIC" CURRENTS

Since P. Sauer has presented to us a comprehensive talk on electromagnetic corrections to few-body hadronic processes, (and since time is not reversible in the real world) I will not comment on isospin symmetry. The major recent progress in this field and in the associated area of higher isospin components, e.g. tensor, of the electromagnetic current has been a conciliation of experimental results and a clarification of the theoretical implications. Improved searches for a $\Delta I=2$ electromagnetic current have failed to reveal such a component [10].

Let me turn then to the weak currents. Neutral currents are no longer considered exotic because they appear to be required by renormalizable theories. By contrast, second class axial currents are not required, but the evidence for them is on the increase again.

Historically, it was pointed out by Weinberg [43] that second class non-leptonic currents are not ruled out by theory, even though this behavior is opposite to that deduced from a generalization of weak leptonic currents and conservation of isospin for hadronic form factors.

These currents transform under G,

$$G = C\, e^{i\pi I_2} \quad , \tag{7}$$

as

$$GV_\mu^{II}G^{-1} = -V_\mu^{II} \ ,$$
$$GA_\mu^{II}G^{-1} = +A_\mu^{II} \ . \tag{8}$$

For the vector current, the most general form for free nucleons is:

$$\langle N(p')|V_\mu|N(p)\rangle = \bar{u}(p')(\gamma_\mu f_V^{\ I} + i\sigma_{\mu\nu}q^\nu f_W^{\ I} + q_\mu f_S^{\ II})u(p) \tag{9}$$

with $q = p'-p$. In Eq. (9), f_V and f_W are first-class form factors, whereas f_S is a second-class one. Current conservation (CVC) rules out a second-class current, i.e. $f_S=0$. Since the axial current is not conserved, even in theory, there is no such restriction for A_μ,

$$<N|A_\mu|N> = \bar{u}(\gamma_\mu f_A{}^I + i\sigma_{\mu\nu}q^\nu f_T{}^{II} + q_\mu f_P{}^I)\gamma^5 u \quad , \tag{10}$$

and the second-class current proportional to f_T need not be zero. Early evidence [44] for such a second class came from measurements of an asymmetry in the ratio, δ, of mirror β-decay rates in Gamow-Teller transitions,

$$\delta = (ft)^+/(ft)^- - 1 \quad . \tag{11}$$

Although an induced second-class current of order α or αWR, i.e. $f_T{}^{II} \approx \alpha/M$, with W the energy released and R the nuclear radius, might be expected due to Coulomb and radiative effects, the observed values of δ were more than an order of magnitude larger. The problem arises in the interpretation of the asymmetries. There are correction factors to the usual impulse approximation analysis due to binding energy, wave function, overlap, nuclear deformation, and other differences between the mirror pairs [44,45]. These corrections can account for a good part (or all) of the asymmetries observed in even A nuclei; it is more difficult to calculate the corrections for the larger values of δ found in odd A nuclei [44]. The observations clearly are sensitive to nuclear structure effects [44]. Observation in β-decays of $\Sigma^\pm$ and of the Λ^0, which are free of such effects, yield inconclusive results, but do not show a large second-class current contribution [44,45,46]. Since the burden of proof for a new finding rests on a definitive result, which cannot be accounted for by standard theory, one concludes that the comparison of mirror beta-decay rates does not provide such evidence.

For these reasons it was suggested that angular correlation tests for second-class currents in β-transitions between isospin pairs, e.g. $\alpha \to \beta + e^+ + \nu$ where α and β belong to the same isospin multiplet, should be carried out [8,47]. Such tests are much more model independent. Several tests are possible:

a) β-γ correlations in the decay of an unpolarized parent nucleus to an excited daughter.

b) β-α correlations in the decay of an unpolarized parent nucleus to a particle unstable daughter.

c) β-ν correlations in the decay of an unpolarized parent nucleus to a particle unstable daughter.

d) β-spin correlations in the decay of a polarized or aligned nucleus.

In mirror β-transitions, both Gamow-Teller and Fermi matrix elements generally contribute. The spin dependent correlation sensitive to $f_T{}^{II}$ thus also gets a contribution from the weak magnetism term, proportional to $f_W{}^I$; typically the correlation is proportional to $f_W{}^I \pm f_T{}^{II}$. Because of problems associated with the impulse approximation, the analysis is often carried out by treating the nuclei as particles [47]; the form factors corresponding to $f_W{}^I$ and $f_T{}^{II}$ are then usually denoted by b_I and d_{II}, respectively. Angular correlation tests thus require the secure knowledge that CVC is satisfied so that b can be deduced therefrom. Tests which measure the weak magnetism form factor b in mass 12 nuclei confirm CVC to approximately 20% [48]. No evidence against CVC has been found [44,48,49].

Recently several correlation tests have been carried out, two of which. although not in very light nuclei, do show a <u>positive</u> effect for second-class currents which remains to be explained.

The first of these tests, a β-α correlation measurement, was carried out by Garvey [49] and colleagues in the beta decay of ^8B and ^8Li to $^8Be^*(2^+)$. They find agreement with CVC and no evidence for a large second class currents.

The second measurement, by Sugimoto, Tanihata, and Göring uses polarized ^{12}B and ^{12}N.

They measure correlations of electrons and positrons emitted in the decays to the analog state in ^{12}C [50]. The difference in the correlation between the two parents, $\propto (b_I+d_{II})$, is found to be about twice that predicted by CVC for the weak magnetism term b_I. This is evidence for a second-class current contribution of the same order as the weak magnetism contribution. A similar result is obtained by Calaprice et. al. [51] from the correlation of positrons emitted in the analogue decay of polarized $^{19}Ne \rightarrow ^{19}F + e^+ + \nu$. If I understand the sign conventions, then the two results agree in sign as well as in magnitude.

The relationship of the measured asymmetries to basic nucleon currents are non-trivial, as I discussed earlier [52]. There are the usual corrections to the impulse approximation such as off-mass shell and nuclear structure effects. A simple impulse approximation implies $|f_T{}^{II}| \sim |f_W{}^I|$,

$$|f_T{}^{II}| \sim 2/M \sim |f_W{}^I| \quad , \tag{12}$$

which is large indeed and certainly larger than anticipated from an electromagnetic correction. It remains to be determined whether the observations can be ascribed to an electromagnetically induced nuclear structure effect such as that caused by differences among isospin multiplet members. One also must understand why the effect is not observed in the mass 8 nuclei.

Tests have also been proposed in muon capture and in ν scattering from nuclei [49], but these experiments are very difficult. Nevertheless, if the observations in the mass 12 and 19 nuclei cannot be explained by nuclear effects, then experiments in very light nuclei and especially in the decay of the neutron become crucial.

Clearly more work remains to be done before the dust settles on this interesting topic. There is thus no conclusion, and I shall close with an appropriate quote from Sir Winston Churchill:

> "Now this is not the end.
> It is not even the beginning of the end.
> But it is, perhaps, the end of the beginning."
>
> Winston Churchill

REFERENCES

*Supported in part by the U.S. Energy Research and Development Administration.

[1] L. Wolfenstein, Phys. Rev. Lett. 13 (1964) 180.

[2] For references, see L. Wolfenstein, Nucl. Phys. B77 (1974) 375.

[3] W.B. Dress, P. Perrin, P.D. Miller, T.M. Pendlebury, N.F. Ramsey, Annual Report Inst. Laue-Langevin, Grenoble (1974); W.B. Dress, P.D. Miller, and N.F. Ramsey, Phys. Rev. D7 (1973) 3147.

[4] For recent review, see e.g. A. Richter in Symposium on Interaction Studies in Nuclei, H. Jochim and B. Ziegler, eds. (North-Holland, Amsterdam, 1975) 191; F. Boehm in High Energy Physics and Nuclear Structure-1975, D.E. Nagle, A.S. Goldhaber, E.K. Hargrove, R.L. Burman, and B.G. Storms, eds. AIP Confer. Proc. No. 26, (Amer. Inst. Phys., New York, 1975) 488.

[5] R.I. Steinberg, P. Liaud, B. Vignon, and V.W. Hughes, Phys. Rev. Lett. 33 (1974) 41.

[6] F. Calaprice, E. Commins, and D.C. Girvin, Phys. Rev. D9 (1974) 519.

[7] F. Calaprice (private communication); F.P. Calaprice, S.J. Freedman, B. Osgood, and W. Tomlinson, Bull. Am. Phys. Soc. 20 (1975) 715.

[8] B.R. Holstein and S.B. Treiman, Phys. Rev. C3 (1971) 1921; C.W. Kim, Phys. Lett. 34B (1971) 383.

[9] E.M. Henley in High Energy and Nuclear Structure, G. Tibell, ed. (North-Holland, Amsterdam, 1974) 22; Ann. Rev. Nucl. Sci. 19 (1969) 367.

[10] J.C. Comiso, D.J. Blasberg, R.P. Haddock, B.M.K. Nefkens, P. Truoel, and L.J. Verhey, Phys. Rev. D12 (1975) 719.
[11] See R.A. Bryan, Phys. Rev. D10 (1974) 3854.
[12] M. Simonius, Phys. Lett. 58B (1975) 147.
[13] J.P. Hannon, Nucl. Phys. A177 (1971) 493.
[14] A. Barroso and R.J. Blin-Stoyle, Phys. Lett. 45B (1973) 178.
[15] S. Weinberg, Phys. Rev. Lett. 19 (1967) 1264; Phys. Rev. D5 (1972) 1412; A. Salam in Elementary Particle Physics, N. Svartholm, ed., (Almqvist and Wiksell, Stockholm, 1968) 367.
[16] A. De Rujula, H. Georgi, and S.L. Glashow, (to be published); H. Fritzsch, M. Gell-Mann, and P. Minkowski, Phys. Lett. 59B (1975) 256.
[17] V.W. Hughes (private communication); proposal No. 95 at SLAC.
[18] See e.g. S.M. Berman and J.R. Primack, Phys. Rev. D9 (1974) 2171 and D10 (1974) 3898; E. Derman, Phys. Rev. D7 (1973) 2755.
[19] E.M. Henley, A.H. Huffman, and D.U.L. Yu, Phys. Rev. D7 (1973) 943; L. Wolfenstein, Nucl. Phys. B72 (1974) 111.
[20] G. Feinberg (preprint) and in High Energy Physics and Nuclear Structure-1975, loc. cit., 468; J. Bernabeu, T.E.O. Ericson, and C. Jarlskog. Phys. Lett. 50B (1974) 467); G. Feinberg and M.Y. Chen, Phys. Rev. D10 (1974) 190 and D10 (1974) 3789.
[21] L.M. Simons, in Interaction Studies in Nuclei, A. Jochim and B. Ziegler, eds. (North-Holland, Amsterdam, 1975) 73; E. Fiorini, loc cit., 411; E.M. Henley, loc cit., 471; L.M. Simons, Helv. Phys. Acta. 48 (1975) 141.
[22] A. Bertin et. al., Nuovo Cimento 26B (1975) 433.
[23] See e.g., M.A. Bouchiat and C.C. Bouchiat, Phys. Lett. 48B (1974) 111; J. de Phys. 35 (1974) 899 and preprint; A.N. Moskalev, S.M. Bilenkii, N.A. Dadayan, and E. Kh. Khristova, Yad. Fiz. 21 (1975) 360 [transl. Sov. J. Nucl. Phys. 21 (1975) 189].
[24] D.C. Soreide, D.E. Roberts, E.G. Lindahl, L.L. Lewis, G.R. Apperson, and E.N. Fortson (submitted to Phys. Rev. Lett.); E.N. Fortson (private communication).
[25] M. Brimicombe, C.E. Loving, and P.G.H. Sandars (preprint); E.M. Henley in IX Interntl. Conf. on the Physics of Electronic and Atomic Collisions, Seattle, 1975 (to be published).
[26] R.R. Lewis and W.L. Williams, Phys. Lett. 59B (1975) 70.
[27] W.M. Lobashov, et al., Nucl. Phys. A197 (1972) 241.
[28] See e.g. P. Herczeg in High Energy Physics and Nuclear Structure, 1975, loc. cit., 504.
[29] H.J. Pirner and M.L. Rustgi, Nucl. Phys. A239 (1975) 427; G.N. Epstein, (preprint); B. Desplanques, Nucl. Phys. A242 (1975); K.R. Lassey and B.H.J. McKellar, Lettere Nuovo Cimento 13 (1975) 351 and Nucl. Phys. (to be published); B.A. Craver, E. Fishbach, Y.E. Kim, and A. Tubis (preprint); M. Gari and J. Schlitter, Phys. Lett. 59B (1975) 118; J.F. Donoghue (preprint).
[30] B.H.J. McKellar, Phys. Lett. (to be published) and Symposium on Interaction Studies in Nuclei, loc. cit.; 61; M. Simonius, loc. cit., 3.
[31] E.G. Adelberger, H.E. Swanson, M.D. Cooper, T.W. Tape, and T.A. Trainor, Phys. Rev. Lett. 34 (1975) 402.
[32] M. Gari, A.H. Huffman, J.B. McGrory, and R. Offerman, Phys. Rev. C11 (1975) 1485; M. Box, B.H.M. McKellar, P. Pick, and K. Lassey, Phys. Rev. C11 (1975) 1859.
[33] J.M. Potter, J.D. Bowman, C.F. Hwang, J.L. McKibben, R.E. Mischke, D.E. Nagle, P.G. Debrünner, H. Frauenfelder, and L.B. Sorensen, Phys. Rev. Lett. 33 (1974) 1307; D.E. Nagle in High Energy Physics and Nuclear Structure-1975, loc. cit., 497.
[34] V. Brown, E.M. Henley, and F.R. Krejs, Phys. Rev. C9 (1974) 935;M. Simonius, Phys. Lett. 41B (1972) 415; G.N. Epstein, Phys. Lett. 55B (1975) 249.
[35] E.M. Henley and F.R. Krejs, Phys. Rev. D11 (1975) 605; E. Fishbach and G.W. Look (preprint)
[36] See e.g. M.K. Gaillard and B.W. Lee, Phys. Rev. Lett. 33 (1973) 108; D. Bailin, A. Love, D.V. Nanopoulos, and G.G. Ross, Nucl. Phys. B59 (1973) 177; G. Alterelli and L. Maiani, Phys. Lett. 52B (1974) 351; M. Gari and J.H. Reid, Phys. Lett. 53B (1974) 237; J.F. Donoghue (preprint).

[37] E.M. Henley, Phys. Lett. 28B (1968) 1.
[38] E.G. Adelberger, C.A. Barnes, M. Lowry, R.E. Marrs, F.B. Morinigo, and H. Winkler (private communication).
[39] M. Gari, J.B. McGrory, and R. Offerman, Phys. Lett. 55B (1975) 277.
[40] J. Barrette, W. Del Blanco, P. Depommier, S. Kundu, N. Marquardt, and A. Richter, Nucl. Phys. A238 (1975) 176; E. Bellotti, E. Fiorini, P. Negri, A. Pullia, L. Zanotti, and I. Filosfo, Nuovo Cimento 29A (1975) 106.
[41] B. Vignon and R. Wilson (private communication).
[42] See e.g. E.M. Henley, T.E. Keliher, W.J. Pardee, and D.U.L. Yu, Phys. Rev. D9 (1974) 755; A. Andrási, B. Eman, J. Missimer, and D. Tadić, (preprint); S.S. Gershtein, V.N. Folomeshkin, and M. Yu. Khlopov, Yad. Fiz. 20 (1974) 737 [Transl. Sov. J. Nucl. Phys. 20 (1975) 395].
[43] S. Weinberg, Phys. Rev. 112 (1958) 1375.
[44] For a review, see D.H. Wilkinson in Few Particle Problems in the Nuclear Interaction, I. Slaus, S.A. Moszkowski, R.P. Haddock, and W.T.H. van Oers, eds. (North Holland, Amsterdam, 1972) 191; D.H. Wilkinson, Proc. Roy. Soc. (Edinburgh) A70 (1971/72) 307 and Phys. Lett. 48B (1974) 169.
[45] E.M. Henley, in Few Particle Problems in the Nuclear Interactions, loc. cit., 221; R.J. Blin-Stoyle, Fundamental Interactions and the Nucleus (North-Holland, Amsterdam, 1973), Chs. IV,V.
[46] I.V. Krive, Yad. Fiz. 20 (1974) 614 [Transl. Sov. J. Nucl. Phys. 20 (1975) 328].
[47] B. Holstein, Phys. Rev. C4 (1971) 740, 764; B. Holstein, Rev. Mod. Phys. 46 (1974) 789; C.W. Kim, Phys. Lett. 34B (1971) 383; T. Delorme and M. Rho, Nucl. Phys. B34 (1971) 317.
[48] C.S. Wu, Rev. Mod. Phys. Phys. 36 (1964) 619; C.S. Wu and S.A. Moszkowski, Beta Decay (Interscience, N.Y. 1966), Ch. 7; see also V.L. Telegdi and C.S. Wu in High Energy Physics and Nuclear Structure-1975, loc. cit., 452,453; H. Behrens and L. Szybisz, Z. Phys. A273 (1975) 177.
[49] R.E. Tribble and G.T. Garvey, Phys. Rev. Lett. 32 (1974) 314; A.M. Nathan, G.T. Garvey, P. Paul, and E.K. Warburton, Phys. Rev. Lett. 35 (1975) 1137.
[50] K. Sugimoto, I. Tanihata, and J. Göring, Phys. Rev. Lett. 34 (1975) 1533.
[51] F.P. Calaprice, S.J. Freedman, W.C. Mead, H.C. Vantine, Phys. Rev. Lett. 35 (1975) 1566.
[52] See e.g. M. Morita and I. Tanihata, Phys. Rev. Lett. 35 (1975) 26.

PHOTON AND LEPTON PROBES ON FEW BODY SYSTEMS

M.K. Sundaresan
Department of Physics, Carleton University, Ottawa, Ontario, Canada

This is a report on some dozen or more papers submitted to session VI of the conference. The papers can be broadly divided into two categories:(a) those involving photons (real or virtual), and, (b) those concerned with certain aspects of weak interactions. We present in summary form some of the high lights.

In an interesting contribution, Haftel has calculated the triton binding energy and the ^3He and ^3H charge form factors, both without and with exchange current corrections for various models of the internucleon force. One of the models includes a three nucleon (3N) force. This is probably the first time that such a calculation has been attempted which includes effects of 3N forces, meson exchange currents as well as possible off-shell effects. Haftel considers three models for the internucleon force: (1) Reid soft core (R) potential (2) de Tourreil-Sprung A super soft core potential (S), and, (3) the Graz potential (GRP). All these potentials are approximately equivalent upto 350 MeV in NN scattering except for ϵ_1, the $^3S_1$ - $^3D_1$ mixing coefficient. The two body parameters and the static deuteron properties as well as triton binding energy are given in Table 1 of this paper and reproduced here for convenience.

Table 1

Potential	a_s(fm) r_s(fm)	a_t(fm) r_t(fm)	E_d(MeV) E_T(MeV)	P_D(%) Q (fm^2)
Reid (R)	-17.1	5.39	2.2246	6.47
	2.80	1.72	6.34(7.0)	0.280
de Tourreil-	-17.3	5.50	2.2237	4.43
Sprung A (S)	2.84	1.85	7.02(7.64)	0.261
Graz (GRP)	-17.7	5.45	2.2248	3.69
	2.71	1.79	7.94(8.5)	0.313

In the fourth column of the table, the values not in parantheses are due to 3-channel calculations using $^1S_0$ and $^3S_1$+$^3D_1$ interactions. Values in parantheses are from more complete calculations of Gigneaux and Laverne, except in GRP where the result of a full Faddeev calculation is estimated.

An interesting fact mentioned by Haftel is that despite large variations in P_D in Table 1, all potentials give nearly the same deuteron electric form factor in agreement with experiment for $q^2 \lesssim 35$ fm^{-2}. The de Tourreil-Sprung potential (S) gives 0.7 MeV more binding for E_T than the Reid potential (R) because of lower P_D. The presence of supersoft core in (S) does not seem to play a significant role as the off-shell behaviours of (R) and (S) are nearly identical for momenta of interest. For GRP the lower P_D as well as the softer off shell behaviour accounts for the value of E_T which is even higher than for (S). Thus with (GRP) he has shown that it is possible to fit the experimental value of E_T with a two-body force without violating any two body constraints.

From GRP Haftel generates a potential GRPA with three-body forces from a three-body Hamiltonian with only two body forces i.e., $H = \sum_i T_i + \sum_{i<j} V_{ij}$, by using a two or three body unitary operator U and defining a transformed $\tilde{H} = UHU^\dagger$, where $\tilde{H}$ has the form $\tilde{H} = \sum_i T_i + \sum_{i<j} V_{ij} + V^{(3)}$. Here $V^{(3)}$ represents all terms that depend upon the coordinates of all three particles. Since U is unitary, E_T

is preserved, but $\Psi_T \to \tilde{\Psi}_T$ and $F(q^2) \to \tilde{F}(q^2)$. The starting two-body potential is GRP. Since GRPA also fits E_T, a test can be made as to whether both E_T and $F_{ch}(q^2)$ can be fitted if a three body force is included.

The results of his calculation for the trinucleon charge form factors with and without exchange currents are given in Figs. (2) and (3) of his contribution. It is interesting to note that in ^3He a much better fit can be obtained with exchange currents than without it. $F_{ch}(q^2)$ for ^3H is relatively insensitive to exchange currents, hence it provides a testing ground for 3N forces.

Next we consider the paper by Bleuler. In this paper Bleuler is concerned with the level scheme of F^{19} which plays an important role in experiments on weak parity violation. The level structure of F^{19} consists of parity doublets $(\frac{1}{2}\pm, \frac{3}{2}\pm, \frac{5}{2}\pm)$. On the basis of shell model one would expect levels with only positive parities in this region. To get a pair of levels with opposite parities and small energy separation between them, he introduces, within the framework of a sph-erical Hartree-Fock treatment, generalized single particle states which allow for an admixture of single particle states with the same angular momentum but opposite parities. Although the two-body forces are parity conserving, the question of whether the generalized Hartree-Fock solution might yield a lower total energy for (parity) admixed single particle states has been explored and answered in the affirmative.

Craver <u>et al</u> contributed papers to this conference (presented by Dr. Tubis), concerning the sensitivity of parity nonconservation effects in two nucleon systems due to different assumptions made on the short range behaviour of strong interactions. They have presented a calculation of α, the asymmetry parameter in the photon angular distribution with respect to initial neutron polarization in $n+p \to d+\gamma$. The calculation of α was done using a two nucleon strong interaction given by Reid soft core and several of its phase equivalents. Using Green's function techniques, calculations of quantities of interest are carried out which are exact with respect to strong interaction, but to first order in the weak interaction. If one splits α as $\alpha_0 + \alpha_1$ where α_0 and α_1 arise from isoscalar and isovector components of the parity violating Hamiltonian, they show that in the impulse approximation, the mixing of opposite parity states in the initial np states and the deuteron state, contribute only to α_1. Exchange currents make nonvanishing contributions to α_0. Knowledge of the values of α_0 and α_1 bear on the important question of a $\Delta S = 0, \Delta T = 1$ non-leptonic weak interaction.

McKellar <u>et al</u> have produced possible evidence for the enhancement of $\Delta S = 0$, $\Delta T = 2$ weak interactions in parity mixing in nuclei. They are concerned with the fact that a parity nonconserving nucleon-nucleon potential has been found which fits all available data on parity mixing in nuclei. These data are:

(1) α decay of 2^-, 8.88 MeV level in O^{16}
(in essential agreement with calculations)

(2) Parity mixing of 110 KeV γ-ray of F^{19}.
(in essential agreement for magnitude but not sign)

(3) Circular polarization of the γ in $n+p \to d+\gamma$ measured by Lobashov <u>et al</u>.
(this is in violent disagreement with standard calculations).

They note that the α decay experiment is sensitive to only the $\Delta T = 0$ term in the parity nonconserving (PNC) potential, while the F^{19} transition is sensitive to both the $\Delta T = 0$ and $\Delta T = 1$ parts of the PNC potential. A detailed analysis of the calculation concerned with the theory of the Lobashov experiment shows that the experiment is sensitive to $\Delta T = 0$ and $\Delta T = 2$ parts of the PNC potential.

The contribution of $\Delta T = 0$ part is too small compared to that of $\Delta T = 2$ part. An enhancement of $\Delta T = 0$ of about 10^3 is required to fix up the agreement with Lobashov experiment, but this would upset the agreement with the other two data. On the other hand, an enhancement by a factor of only 30 is required for the $\Delta T = 2$ part to obtain agreement for circular polarization without upsetting the agreement found in experiments on O^{16} and F^{19}. This enhancement of $\Delta T = 2$ potential is still compatible with the experimental limit on parity mixing in pp scattering measured at 15 MeV. If a repetition of this experiment lowers this limit, the enhanced $\Delta T = 2$ potential would be excluded. It is also possible to consider enhanced $\Delta T = 1$ potential to fix up the three experiments mentioned above. Their conclusions are: (1) Lobashov experiment should be repeated, (2) asymmetry measurement in pp scattering should be repeated to lower the limit, (3) asymmetry in np scattering should be measured-an asymmetry $\sim 10^{-6}$ is required for consistency with Lobashov experiment, and, (4) the calculations for O^{16} α decay should be redone.

Other papers submitted to this session involve calculations of electromagnetic charge form factor and coulomb energy for trinucleon systems using separable potentials, π^0 photoproduction, photo disintegration of the triton and α particle. There was also a paper on colour gluons and scaling in a unified gauge model of weak and electromagnetic interactions. As this last paper does not directly bear on interaction with few nucleon systems, we do not discuss this otherwise interesting paper.

For completeness, I would like to mention here two other items on which new information is available although not in the form of papers submitted to this conference. They are: (1) ratio of $n+p \to d+2\gamma$ to $n+p \to d+\gamma$, and (2) elastic e-d scattering and measurement of deuteron form factors for high q^2 ($8 \leqslant q^2 \leqslant 6$ $(GeV/c)^2$). We comment on these in turn.

Last year, Dress, Guet, Perrin and Miller (P.R.L. March 24, 1975), reported an experimental measurement of $n+p \to d+2\gamma$ process which was about 10^{-3} times the one photon raditive capture, $n+p \to d+\gamma$. This large value caused quite a stir in theoretical circles for it is quite difficult to understand. Blomquist and Ericson (Phys. Letters 57B (1975) 117) obtained a theoretical value for this ratio in the neighbourhood of 10^{-7}. This problem has recently been resolved. It has been established more or less conclusively that the 2γ's seen in the experiment are not really all from the process $n+p \to d+2\gamma$, but there is a large background of 2γ's arising from positron annihilations in flight. The positron annihilation also accounts for the energy distribution of the 2γ's.

The other item is a report of measurements by Arnold et al on elastic e-d scattering involving high momentum transfer (P.R.L. 35 (1975) 776), $0.8 < q^2 < 6$ $(GeV/c)^2$, using two high resolution spectrometers in coincidence. Their experimental results are in sharp disagreement with meson exchange effects expected to play a role here. It was expected qualitatively that in the impulse approximation for $q^2 \gtrsim 1$ $(GeV/c)^2$, the effect of the virtual photon striking one nucleon would be very small and that the cross section would be dominated by meson exchange effects or other processes which distributed the momentum equally between the two nucleons. Two different exchange current contributions, one by Chemtob, Moniz and Rho (P.R. C10 (1974) 334), and another by Blankenbecler and Gunion (P.R. D4 (1971) 718), have been compared with the data and one finds sharp disagreement. The 2γ contribution is also seen to be well below the experimental points and hence cannot be responsible for the disagreement. They have also compared the experimental data with potential model calculations with impulse approximation involving the Bethe-Reid soft core and Feshbach-Lomon models. There is a general agreement with experimental results within factors of 2 to 10. Relativistic corrections so far available are not reliable above $q^2 \sim 1$ $(GeV/c)^2$. Clearly, as was indicated by Prof. Gross at this Conference, more work has to be done here to see if one can fit the data with potential model and relativistic corrections. Also, isobar admixtures have been speculated as being present in the deuteron. These would provide high momentum components to the deuteron wave function and as a result would flatten out the corrections. However, the data shows no hint of flattering out, hence seriously questioning these speculations. An interesting feature of the data is that for high q^2, the fall off is like $(q^2)^{-5}$. This agrees with the prediction for the deuteron form factor based on dimensional scaling quark model which pictures the deuteron as a bound state of six quarks. To see this, imagine that a hadron is a bound state of n_H constituents, each carrying a finite fraction of the total hadron momentum. The amplitude for the scattering of a system of hadrons $AB \to CD$ is related to the scattering of their constituents integrated over possible momenta with the constraint that the total momentum of the constituents must equal the hadron momentum. If the total number of fields in initial and final states is n , the Feynman amplitude, M_n, has dimension of $(\text{length})^{n-4}$. If at large energy and momentum transfer, the only length scale is set by $(\sqrt{s})^{-1}$, then M_n $(\sqrt{s})^{4-n}$. If each of the n_H constituents of the hadron H carries a finite fraction of the momentum in the hadron rest frame, integration over possible momenta cannot produce a dependence on s. Therefore $M \sim M_n$ and $\frac{d\sigma}{dt}(AB \to CD) = \frac{1}{s^2}|M|^2 \sim s^{n-2} f(t/s)$. For ed scattering, $\frac{d\sigma}{dt} \sim t^{2-n} = \frac{1}{t^2}|F_d(t)|^2$. Thus $|F_d(t)|^2 \sim t^{4-n}$. Here $n = 2n_H + 2$, where the (+2) comes from the electron fields initially and finally, and n_H is the number of quarks (= 6 in the deuteron). Thus $F_d(t) \sim t^{1-n_H}$ and for $n_H = 6$, $F_d \sim t^{-5}$. If such dimensional scaling arguments are really correct it might be interesting to study e-^{4}He scattering at high q^2. (I am indebted for this remark to Dr. G. Shaw).

Now we turn to lepton interactions with few body systems. Recent interest has centered on the structure of weak neutral currents using muons and neutrinos as probes. Knowing the nuclear physics of few nucleon systems, one can use this knowledge to gain very interesting information on the detailed structure of weak neutral currents.

Exploration of neutral current effects can be done with the help of muonic atoms. The basic idea of this method has been described previously in papers by Bernabeu, Ericson and Jarlskog (Phys. Letters 50B (1974) 467) and by Feinberg and Chen (Phys. Rev. D10 (1974) 190), In very light muonic atoms such as atomic $2S_{1/2}$ and $2P_{1/2}$ levels are closely degenerate. Due to a combination of vacuum polarization and finite nucleon size effects, the $2S_{1/2}$ level is actually slightly (about 1 to 2 ev) below the $2P_{1/2}$ level rendering the $2S_{1/2}$ state metastable. When muons are stopped in helium or lithium and are captured in them, they rapidly cascade down through a series of Auger and radiative processes giving rise to characteristic muonic X-rays. A fraction of the muons end up in the $2S_{1/2}$ state and form metastable atoms. The modes of de-excitation of the metastable atom are: Auger process, 2γ process and 1γ process. The 1γ mode is M1 and is highly suppressed. The 2γ mode is the more probable mode. If in a muonic atom, in addition to the electromagnetic interaction, there exists parity nonconserving weak interaction (such as due to exchange of a neutral intermediate vector boson), the closely degenerate $2S_{1/2}$ and $2P_{1/2}$ levels get mixed, the net result of which is that the 1γ transition from the $2S_{1/2}$ level to the ground state of the atom should be strongly enhanced. One way of detecting the existence of a weak neutral current effect in lepton (muon)-hadron interaction is to look for the delayed enhanced 1γ decay from the $2S_{1/2}$ state of light muonic atoms. Such an enhancement can then be directly related to the parameter mixing the $2S_{1/2}$ and $2P_{1/2}$ levels and hence to the strength of the neutral current interaction. Other consequences of $2S_{1/2}$-$2P_{1/2}$ mixing due to weak interaction effects are:

(1) The emitted photon has a circular polarization

(2) If the initial lepton is polarized (as the muon is and retains most of it while cascading), there is an angular correlation $(\vec{\sigma} \cdot \vec{k})$ between the polari-

zation direction of the lepton ($\vec{\sigma}$) and the photon direction ($\vec{k}$).

(3) There also exists an angular correlation between the emitted photon and the decay electron from the muon in the ground state of the atom.

Although in principle the experiments suggested above sound quite simple, in practice they turn out to be quite difficult. In detecting the enhanced 1γ rate, one has the difficult problem of picking out the 1γ from among the background of γ's from the more probable 2γ mode. The measurement of various angular correlations is not easy either. A preliminary experiment looking for the metastable $2S_{1/2}$ state in light muonic atoms (μ-Li) and (μ-Be) has been tried by the Carleton-NRC-Virgnia collaboration at the SREL cyclotron in Newport News (Virginia). A major problem encountered has been the Auger de-excitation of the $2S_{1/2}$ state which seems to compete very strongly with the 2γ mode. The experiment has also been carried out by stopping muons in chemical compounds involving Li or Be, the hope being that the Auger process might be suppressed in these due to lack of availability of electrons for the Auger process to occur (long time required to refil the K-electron shell). Although chemical effects are seen in muonic X-rays, the Auger process still seems strong enough for the 2γ mode to proceed much less than the 1γ mode. Further experiments are planned especially with judicious choice of targets. So far the $2S_{1/2}$ state has been seen among light atoms only in muonic Helium. The best bet may very well be to do the experiment in muonic Helium.

Finally, I would like to spend a very few minutes reporting on some recent suggestions to use neutrino probes on few nucleon systems. The basic question one is concerned with here is the isospin structure of the weak neutral current, and, whether it is also (V,A) like the charged current or contains additional S,T,P pieces. Observation of elastic ν_μ p scattering and weak production of pions with neutrinos have been suggested as possible ways to detect isoscalar pieces of the weak neutral current. An important but difficult experiment is the study of elastic ν_μd scattering. Since <u>d</u> is an isoscalar object, this experiment will be particularly sensitive to isoscalar piece of the neutral current. It may be better to attempt to do the experiment at meson factories rather than at NAL because of suppression effects due to deuteron form factor at very high energies. Other experiments involving neutrino probes have been suggested such as giant dipole excitation etc., but these involve more than few nucleous in the nucleus and I have excluded them from discussion here.

DISCUSSION

Haftel : The total exchange current correction in ^3He is dominated by the pair current. If a πNN vertex function is used, this largely cancels the recoil term and the final result is close to that with the pair correction only.

Henley : The major so-called exchange current contribution, the pair-term, is in fact a direct result of gauge invariance and is not truly a relativistic effect.

Thakur : Apart from the off-shell effects, the pair term is also uncertain to the extent of whether one should use a vector form factor or an axial form factor.

Gross : I want to point out that the exchange currents referred to by Arnold and Chertok are of a different type from those which dominate the ^3He and ^3H form factor and $n+p \rightarrow d+\gamma$, so that they could be zero without contradicting the success of these other calculations.

Brayshaw : I agree with Namyslowki that many objections can be raised to the way that exchange current corrections are calculated. However, I want to stress the importance of EM probes of systems such as ^3He. I have rigorously shown that one cannot distinguish 2-and 3-body forces from three-body scattering. The only possibility is thus to probe the inner part of the wave function. I indicated one way to do this several years ago. In order to make this rigorous, of course, the question of exchange corrections must be clarified.

McKellar : With the advantage of hindsight one realises that the exchange current calculation of the deuteron form factor omitted meson-nuclear form factors which will certainly provide some reduction. Whether it can give a fit to the data, I don't know.

Sundaresan : It is difficult to say without detailed calculation.

Weber : I would like to make two comments.

1) Your remark and the corresponding implication in Arnold, Chertok, et al PRL (1975) (on e-d scattering at high q) that N* configurations in the deuteron ground state flatten out the elastic charge form factor, is not quite correct. In fact, for elastic scattering, N* configurations are almost as rigid as the np configuration. As exhibited in the N* review by H. Arenhovel and H.J. Weber, Springer Tracts 65 (1972) 58, the two structure functions A,B of elastic e-d scattering remain almost unchanged when the Δ is included - even at high momentum transfer q.

2) With respect to the work on the elastic form factor of ^3He, including meson exchange currents, I would like to emphasize the point mentioned earlier by F. Gross: If a recoil current contribution is included in conjunction with a non-retarded non-relativistic theory, it is largely spurious.

THE FEW CHARGED PARTICLE PROBLEM IN ATOMIC PHYSICS

E. Gerjuoy
Department of Physics, University of Pittsburgh
Pittsburgh, Pennsylvania 15260 USA

I. INTRODUCTION

My objective today is to survey recent progress in atomic physics problems involving a few charged particles (at most 4 or 5 say). Originally I was asked to speak on the "Few Electron Problem", but I assumed the sponsors of this Conference were intending that I keep around at least one positively charged particle, to prevent the electrons from all immediately cruising off to infinity. Nor am I going to confine myself to systems containing merely one positively charged particle. In modern atomic physics theory more and more attention is being paid to atom-atom or ion-atom collisions, in other words to problems involving two or more heavy centers of positive charge; to wholly ignore such problems would give a distorted view of present trends in atomic theory. Therefore I took the liberty of giving my talk its present title of "The Few Charged Particle Problem", so that you won't feel betrayed when I suddenly start talking about problems involving more than a single heavy particle. Similarly, I now will feel free to discuss reactions involving positrons, which in recent years have been enjoying an increasing theoretical vogue in the atomic physics area.

I have explained the title of my talk; next let me circumscribe its domain. In essence, I am going to be concerned solely with phenomena governed by the non-relativistic Schrodinger equation for particles interacting through purely static Coulomb fields. Here and there I may have a word to say about relativistic effects, or quantum electrodynamic corrections, but these will be just amplifying remarks, in circumstances where relativity or q.e.d. is needed to produce the last iota of agreement between theory and experiment. Similarly, unless otherwise stated I'm not going to worry about contributions to the Hamiltonian from spin-orbit coupling, hyperfine interactions, etc., which, if included explicitly in the interaction, would mean departures from the postulated purely Coulombic forces.

In other words, I am going to be concerned with recent progress in theoretical understanding of systems whose behavior can be well-described by the Hamiltonian

$$H = \sum_i \frac{-\hbar^2}{2m_i} \nabla^2 + \sum_{i>j} \frac{Z_i Z_j e^2}{r_{ij}} \tag{1}$$

where m and Ze denote the particle mass and charge respectively, where Z usually (but not always) equals -1, and where--for the purposes of the present talk--i is a relatively small number, $\lesssim 5$ say. Within this seemingly quite narrowly restricted domain can be found a surprisingly large number of atomic physics problems with practical as well as theoretical import. Moreover, even in the restricted atomic physics domain I have delineated, the theoretical methods which have been utilized are so varied, and have been applied so widely, that it is impossible to give a comprehensive survey in a talk of this length.

To get a reliable feeling for the scope of theoretical results on the few charged particle problem, one has to read the numerous recent reviews and tomes. Possibly the best single locus for wide-ranging discussions of atomic physics problems, including the few charged particle problem, are the meetings of ICPEAC, the International Conference on the Physics of Electronic and Atomic Collisions, which meets biannually, and whose Proceedings are published [1].

In the remainder of my time, therefore, I will describe a somewhat capriciously selected rather disconnected set of theoretical results, which you should regard as more illustrative than representative. Before doing so, however, I want to stress (probably unnecessarily) that the mere fact of restriction to the spin-independent Hamiltonian (1) does not mean I am wholly ignoring particle spin. Particle indistinguishability certainly cannot be ignored in atomic systems, and the proper treatment of particle indistinguishability requires the use of explicitly spin-dependent wave functions.

II. BOUND STATE CALCULATIONS

My introductory remarks now are concluded, and I turn to considerations of actual problems. I shall begin by discussing the status of bound state calculations in few charged particle systems. Very briefly, this status is as follows. Calculations of energy levels, where the Rayleigh-Ritz variational principle can be employed, are extremely accurate for one-center problems, are not quite so accurate for two-center problems, and on the whole are too cumbersome to be really useful for most three-center problems. The energies being discussed here are the bound state energies of the electrons for fixed locations of one, two or three massive centers of positive charge. The objective is to obtain these energies (to the required accuracy) ab initio, directly from the Hamiltonian (1) without any appeal to empirical information. As is well known, in the multi-center case these energies are the effective interaction potentials for the motions of the massive particles (namely the atomic nuclei) in the fields of the other massive particles and associated electron clouds (Born [6], Pauling [7]); once the electronic energies for fixed nuclear locations are known, computation of the actual bound state energies taking into account the nuclear motion is comparatively easy. The availability of this very useful so-called Born-Oppenheimer adiabatic approximation is a luxury denied to nuclear theorists however; it has been shown (Born [6]) that the smallness parameter measuring the validity of this two-stage method for finding the physical bound state energies is the fourth root of the ratio of the light and heavy masses.

A. Energy Levels of He and He-like Ions

To give you an idea of the present state of the art in atomic physics one-center few-electron problems, Table I shows results quoted by Accad [8] for the He isoelectronic sequence, i.e., for systems of two electrons in the field of a nucleus of positive charge $Z \geq 2$. Table I shows some of the P-state energies in wave numbers (1 ev $\cong$ 8000 cm^{-1}), in He, C V and Ne IX, as a function of the number of terms included in the series expansion of the trial wave function employed in the Rayleigh-Ritz. Evidently the calculations are accurate to quite a large number of significant figures. Table I illustrates the typical circumstance that to get the same accuracy an expansion in rather more terms is required as the states become more highly excited.

Table II shows that these Accad calculations are accurate enough to detect the Lamb shift in the He isoelectronic sequence. Note that in these two-electron systems we do not have (as in hydrogen) a pair of states which are initially degenerate but are split by the Lamb shift; in the He isoelectronic sequence the Lamb shift is not the entire energy splitting between the s and p levels involved, but instead shows up only as a very small correction to the normal energy splitting of s and p states in the non-Coulombic average effective field seen by each electron as it moves in the combined Coulomb fields of the nucleus and the other electron.

These Accad [8] He calculations probably are as elaborate as any which have been made; as many as 2300 terms were included in their expansions. Accad [8] uses a special technique developed by Pekeris [9] for the He isoelectronic sequence, involving an expansion in products of Laguerre polynomials in the sides of the triangle formed by the electrons and the nucleus. The particular expansion employed does not seem to be too critical in most one-center few electron problems however; for most purposes these problems appear to be within the powers of modern computers unless a particularly inappropriate expansion basis is used. In Li, for example, highly accurate energy level computations have been reported

Table I. P-State Energies in the He Isoelectronic Sequence Illustrating the Convergence as the Number of Terms in the Expansion is Increased

2^1P Energy E_{nr} (cm^{-1})

# Terms	He	C V	Ne IX
20	27166.016	678857.544	2206882.376
56	27176.090	678876.161	2206904.397
120	27176.640	678876.833	2206905.178
220	27176.683	678876.878	2206905.230
364	27176.688	678876.882	2206905.235

5^3P Energy E_{nr} (cm^{-1})

# Terms	He	C V	Ne IX
20	4473.342	110934.774	357980.567
56	4493.186	111004.909	358059.648
120	4505.607	111026.460	358082.758
220	4509.011	111029.548	358085.875
364	4509.695	111029.868	358086.176
560	4509.811	111029.902	358086.208

E_{nr} is calculated from the nonrelativistic Hamiltonian of Eq. (1), with the nuclear mass set equal to infinity.

$$1 \text{ ev} \cong 8000 \text{ cm}^{-1}$$

Table II. Theoretical and Experimental Wavelengths λ for some He-like Transitions

Transition		λ_{nr} (Å)	λ_{theor} (Å)	λ_{tot} (Å)	λ_{expt} (Å)
He I	2^1S-2^1P	20590.88	20586.51	20586.95	20586.92
Li II	2^1S-2^1P	9587.86	9583.43	9584.06	9584.04 ± 0.10
Be III	2^1S-2^1P	6149.1	6142.7	6143.5	6142.9
B IV	$2^3S-2^3P_0$	2834.39	2824.55	2825.25	2825.40
C V	$2^3S-2^3P_0$		2276.89	2277.84	2277.96
C V	$2^3S-2^3P_1$	2289.20	2277.49	2278.44	2278.63
C V	$2^3S-2^3P_2$		2270.52	2271.46	2271.59
N VI	$2^3S-2^3P_0$		1906.30	1907.53	1907.87
N VI	$2^3S-2^3P_1$	1921.12	1905.89	1907.12	1907.34
N VI	$2^3S-2^3P_2$		1895.49	1896.70	1896.32
O VII	$2^3S-2^3P_0$		1638.27	1639.82	1639.58 ± 0.08
O VII	$2^3S-2^3P_1$	1655.56	1636.52	1638.06	1637.96
O VII	$2^3S-2^3P_2$		1622.09	1623.60	1623.29

λ_{nr} = nonrelativistic value from nonrelativistic Hamiltonian of Eq. (1), nuclear mass = ∞.

λ_{theor} = value including relativistic and finite nuclear mass corrections.

λ_{tot} = λ_{theor} together with an estimate of the Lamb shift correction.

using configuration-interaction Hylleraas expansions (power series expansions in the inter-electron distances) (Sims [10]); using a variational formulation of Bethe-Goldstone theory (Nesbet[11]); and using Bethe-Goldstone theory in its conventional linked cluster perturbation expansion version (Lyons [12]).

B. Matrix Element Calculations

The wave functions obtained in the aforementioned He and Li calculations have been used to estimate various bound state matrix elements with good accuracy, including (among others) hyperfine interactions (Accad [8], Nesbet [11], Lyons [12]), radiative lifetimes of astrophysical interest (Onello [13], [14]), and atomic polarizabilities (Stevens [15]). Similar remarks pertain to other light elements in the first row of the periodic table (e.g., Stevens [15], Dehmer [16], Miller [17], Stewart [18], Nesbet [19]). For the most part these calculations merely have inserted--into standard expressions for the desired quantities--the best wave functions found in the energy computations. Consequently these calculations, of quantities other than the energy, tend to be less accurate than energy computations, for the well known reason that the customarily used Rayleigh-Ritz variational principle makes the error in the energy of second order when the error in the wave function is of first order. Use of these same wave functions in conventional expressions for the matrix elements, etc., leads to first order errors therefore.

In recent years, however, variational principles for matrix elements have been derived (Schwartz [41], Dalgarno [42], Delves [43]), and these derivations now have been generalized to the point that there exist routine procedures for constructing a variational principle for essentially any desired functional of bound state (or even continuum) eigenfunctions (Gerjuoy [44]). But these variational principles for quantities other than the energy always involve auxiliary trial functions (Gerjuoy [44]) in addition to the trial wave functions, and finding good trial auxiliary functions can be difficult. Probably for this reason, use of these variational formulations for quantities other than the energy has been comparatively limited.

Nonetheless, these variational formulations have been used, especially in the He isoelectronic sequence (Dalgarno [42]). Sahni [45] claims that with a comparatively simple Hartree-Fock approximation to the trial function the variational formulation yields expectation values of r^2, r and r^{-1} in He which are comparable with results from a conventional 2300 term Rayleigh-Ritz procedure. Very recently, it has been shown that the auxiliary trial functions (needed in the variational principle for quantities other than the energy) themselves can be estimated variationally (Gerjuoy [46]). Moreover, these variational techniques can be combined with formulas for bounds to yield so-called variational bounds, wherein trial parameters can be varied until the bound is as good as possible for the set of parameters employed (Blau [47]). Upper and lower bounds to oscillator strengths, and lower bounds to atomic polarizabilities, have been computed in the Be isoelectronic sequence (Sims [48], [49]). Anderson [50], using methods developed by Weinhold [51], has computed upper and lower bounds differing by only 0.25%, for the lifetime of the He-like Argon XVII 2^3S state. Shakeshaft [52], using the aforementioned variational principle for the auxiliary trial function, has computed the diamagnetic susceptibility of He to an accuracy of one part in 10^6 with a 20-parameter He trial function and a 50-parameter trial auxiliary function. Then, using these auxiliary trial functions in previously derived formulas (Blau [47]), upper and lower bounds on the He diamagnetic susceptibility differing by only one part in a thousand were obtained (Blau [69]), without any further variation to minimize the difference between the upper and lower bounds.

Despite the foregoing, all the evidence on the use of these new variational principles and formulas for bounds is not yet in. For most problems, it is not yet clear whether the desired accuracy is more readily obtained by working harder to improve the wave function in the Rayleigh-Ritz, or by using somewhat poorer wave functions in these more awkward variational principles and bound formulas. Investigations along these lines are needed, and not only in atomic physics; these variational principles and bound formulas have general applicability.

C. Two-Center and Three-Center Calculations

Two-center problems have much less symmetry than one-center problems, and three-center problems generally have no symmetry at all (other than reflection of the electrons in the plane containing the three centers). Consequently, three-center ab initio calculations are not often attempted, and rarely yield useful results. Everyone seems to agree that the theory is basically sound, i.e., that the same procedures that work in atomic physics one-center and two-center systems--namely use of the Born-Oppenheimer adiabatic approximation, judicious application of the Rayleigh-Ritz, etc.--should work in three-center systems; the difficulty is that in three-center systems, with so little symmetry, the number of terms needed in the expansion, and the amount of information that must be stored, overwhelm even modern computers.

Provided excessive accuracy is not required, calculation of the effective interaction potential in two-center problems, where there is symmetry for rotation of the electrons about the internuclear axis, is within the capability of modern computers using the methods discussed above. Indeed, two-center ab initio calculations of this sort appear to be the vogue right now. Nonetheless, for many applications, especially for applications to cross section prediction, the required accuracy of the two-center interatomic potentials can be hard to come by. An illustration of the care with which two-center potential curves often must be computed for reliable cross section prediction is provided by the annihilation of antihydrogen by hydrogen, a subject of recent theoretical interest (Morgan [32], Junker [33], Kolos [34]).

Consider a collision between neutral hydrogen and antihydrogen atoms. As the distance R between the proton and antiproton is reduced, the electron and positron become progressively more weakly bound to the heavy particles; indeed ultimately, when the proton and antiproton come very close to each other, the electron and positron each see a dipole, and a dipole moment less than 0.64 ea_0 cannot bind an electron or positron (Turner [35]). Thus when the heavy particles get sufficiently close to each other, the leptons can escape. Before the heavy particles get this close, however, it becomes energetically favorable for the leptons to form positronium, having a binding energy of half a Rydberg. The original binding energy of the hydrogen-antihydrogen pair was two Rydbergs. If the original center of mass kinetic energy was less than 3/2 Rydbergs therefore, the formation of positronium must leave the proton-antiproton pair in a bound state of protonium. This protonium bound state will be very highly excited, of course, because it will only have a few ev binding energy. It rapidly will cascade down to the ground state, however, and then annihilate; the positronium will similarly annihilate. In other words, for incident energies less than 3 Rydbergs in the laboratory system, about 40 ev, the cross section for hydrogen-antihydrogen annihilation is expected to be essentially identical to the cross section for rearrangement of hydrogen-antihydrogen into positronium and protonium

$$H(1s) + \bar{H}(1s) \rightarrow e^-e^+(1s) + p\bar{p}(\text{excited}) \qquad (2)$$

To compute the cross section for the rearrangement (2), our first step is the usual one--we calculate the H(1s)-$\bar{H}$(1s) interaction potential. This calculation was performed by Junker [33], with the result shown in Fig. 1. Using a 75-configuration trial function in the Rayleigh-Ritz, they found a peak at 3.089 Bohr radii. The peak height is very small; the difference between the energies at the maximum and minimum in Fig. 1 is only 0.0088 ev. Nevertheless, especially for lower incident energies and higher partial waves, this small peak acts as a potential barrier, preventing the nucleons from getting close enough for

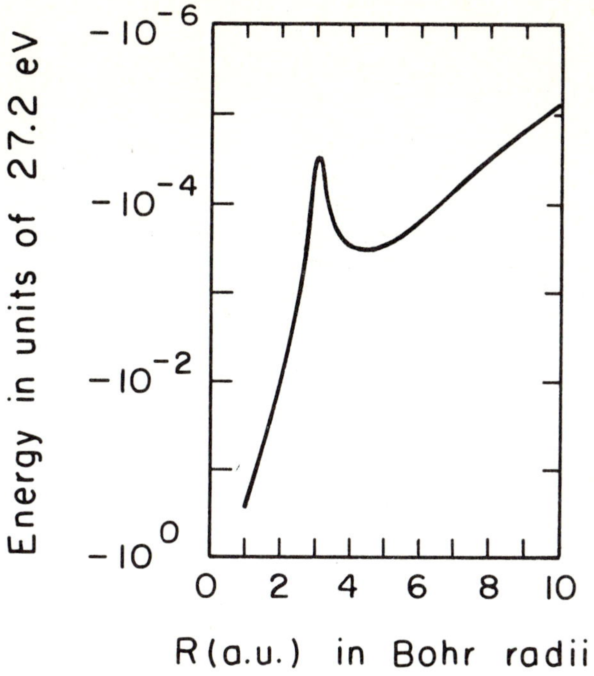

Fig. 1. Interaction Energy Between H(1s) and $\bar{H}$(1s)

positronium formation to occur, and therefore significantly reducing the rearrangement cross section below the value that might otherwise be expected.

Very recently, however, the H-$\bar{H}$ interaction potential has been recomputed by Kolos [34], using an expansion of about the same order as Junker [33], but with trial functions explicitly dependent on the positron-electron distance; in other words, Kolos used what might be termed a configuration-interaction Hyleraas expansion, rather than the more usual configuration-interaction expansion of Junker's [33], wherein the positron-electron distance did not explicitly appear. Kolos's [34] results are shown in Fig. 2, which also displays an earlier not ab initio estimate of the interaction made by Morgan [32]. The Kolos curve (labelled KMSW) shows but the barest trace of the peak found by Junker [33] (replotted as the curve JB in Fig. 2); because both calculations are variational, the Kolos curve, which lies below Junker's, should be more nearly correct. The disparities in the predicted rearrangement cross sections, resulting from the small deviations in the interatomic potential displayed in Fig. 2, are shown in Fig. 3 (Kolos [34]); it is seen that the cross section disparities can be an order of magnitude.

D. Excited States of H^-

This essentially concludes my detailed discussion of atomic physics few-particle bound state energy calculations. Before going on to continuum calculations, however, I do want to point out that even in one-center problems there remain some open atomic physics questions which are not wholly adequately answered by presently available theoretical and computational tools. For instance, recently there has been considerable interest in very highly singly excited, so-called Rydberg states of atoms and molecules. Such states are important in plasma diagnostics and energy balance, and their energies recently have become accessible to experimental measurement (Stebbings [36], Dehmer [37]). For these very highly

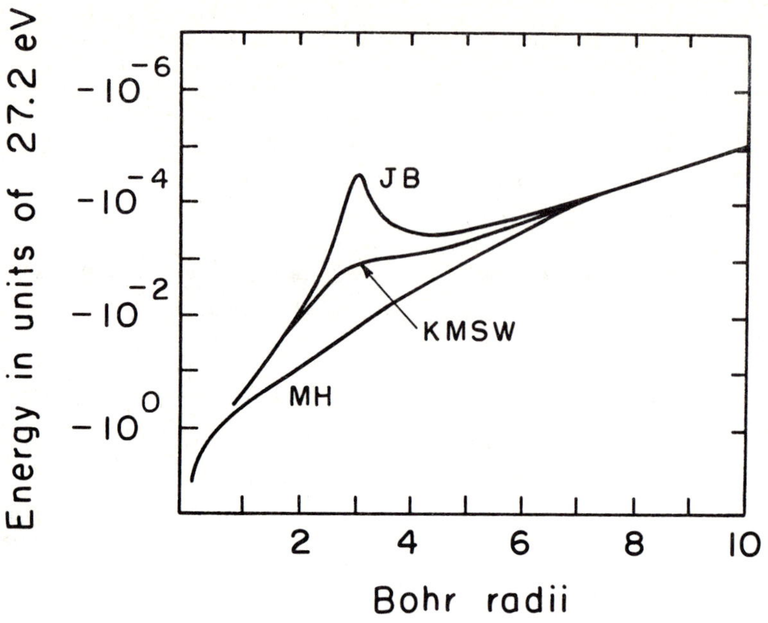

Fig. 2. Alternative Calculations of H-$\bar{H}$ Interaction Interatomic potential V as a function of the interbaryonic distance r_{ab}. Upper curve: Junker; middle curve: Kolos; lower curve: Morgan.

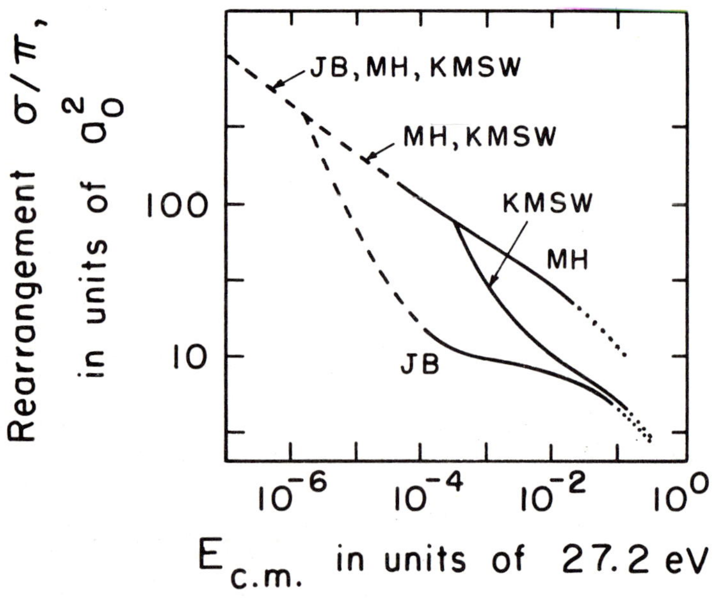

Fig. 3. Predicted H-$\bar{H}$ Rearrangement Cross Sections. MH-Morgan: Generally upper curve; KMSW-Kolos: Generally middle curve; JB-Junker: Generally lower curve. The more tentative portions of the curves are dashed.

excited Rydberg states the standard computational procedures which have been discussed are not suitable, but no really satisfactory alternative ab initio methods have been proposed.

Another open question of some theoretical interest is the existence of an excited bound state of H^-. To understand what's involved, consider Fig. 4 which shows the spectrum of two electrons in the field of an infinitely massive proton for two values of the strength λ of the electron-electron interaction. For $\lambda=0$, on the left side of Fig. 4, each electron is independently bound to the proton in a pure Coulomb field; the energy of the lowest $H^-(1s^2)$ state therefore lies one Rydberg below the ground state of hydrogen (the zero energy in Fig. 4). Since the second electron can have arbitrary positive energy at infinity, the continuum

$$\mathcal{H}(\lambda) = \frac{-\nabla_1^2}{2m} - \frac{\nabla_2^2}{2m} - \frac{e^2}{r_1} - \frac{e^2}{r_2} + \frac{\lambda e^2}{|\underset{\sim}{r}_1 - \underset{\sim}{r}_2|}$$

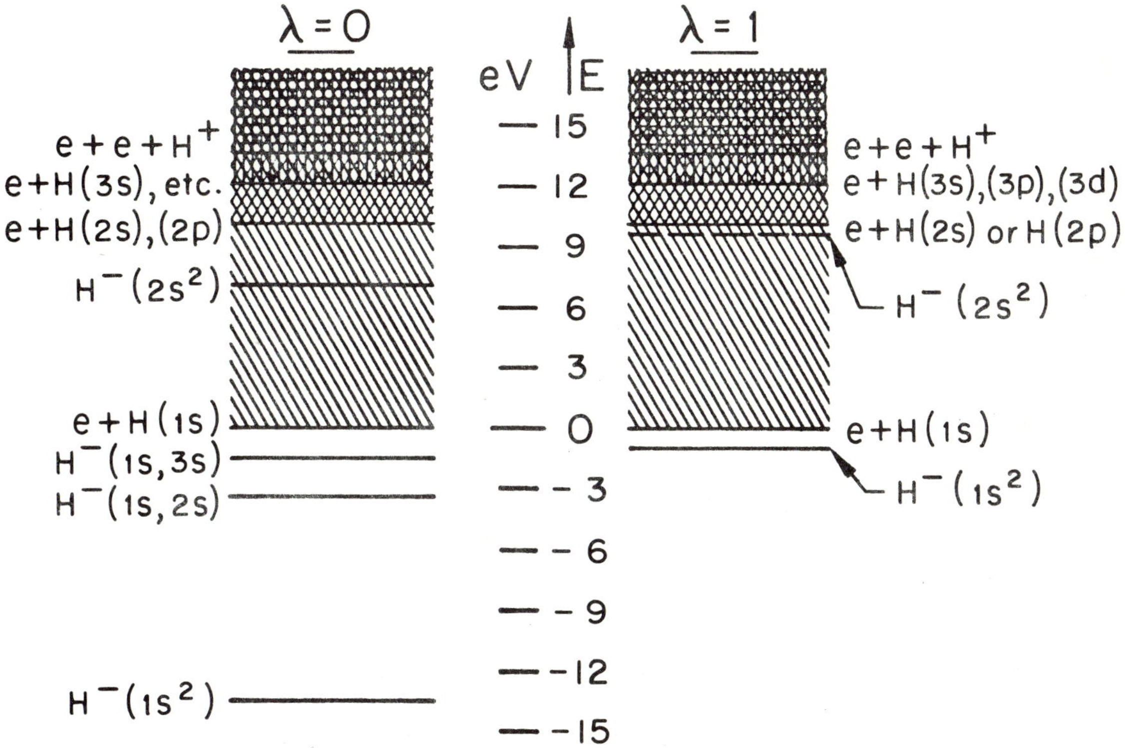

Fig. 4. Spectrum of the e^-e^-p System for Varying e^-e^- Interaction

starts at zero energy on the scale of Fig. 4; embedded in this continuum are an infinite number of bound states, however, which cannot decay by autoionization because there is no electron-electron interaction. In other words, the $H^-(2s^2)$ level shown, lying at 10.2 + 10.2 - 13.6 = 6.8 ev above the ground state of hydrogen, really has a quadratically integrable wave function, always neglecting coupling to the electromagnetic field, of course. New continua start at each excited state of atomic hydrogen, as shown by the various hatched areas in Fig. 4. Below the H(1s) continuum threshold lie an infinite number of bound states, each corresponding to having one electron in its ground H(1s) state and the other electron in an excited purely hydrogenic bound state.

All the above has been limited to the circumstance $\lambda = 0$. Now suppose the electron-electron interaction parameter is increased from zero to its proper physical value $\lambda = 1$. Just as soon as λ becomes different from zero, the aforementioned bound states lying in the continuum, such as the $H^-(2s^2)$ level shown, couple to the continuum through the now non-vanishing electron-electron interaction. Thus for $\lambda \neq 0$ these states are no longer quadratically integrable, and cannot be part of the real energy spectrum of the Hamiltonian $H(\lambda)$. Consequently as λ increases from $\lambda = 0$ the energies of these states move into the complex energy plane--the energies can't stay real and have to move somewhere as a function of λ. In other words, these states become the resonances in e^--H(1s) scattering, which I shall discuss in a moment. The known location of the $H^-(2s^2)$ resonance, at about 0.7 eV below the H(2s) continuum threshold, is shown on the right side of Fig. 4; the dashed line indicates that this level corresponds to a complex energy, i.e., is not a legitimate member of the spectrum of $H(\lambda=1)$.

The $\lambda = 1$ $H^-(2s^2)$ resonance lies above the $\lambda = 0$ $H^-(2s^2)$ bound state because increasing λ adds positive potential energy to the Hamiltonian. Correspondingly, as λ increases from 0, each of the infinite sequence of bound states lying below the H(1s) continuum threshold also increases in energy. As long as λ is less than unity, however, there must remain an infinite number of bound states below the H(1s) continuum threshold, because the outermost electron at long range from the H(1s) core is moving in an attractive r^{-1} field produced by an effective positive charge $(1-\lambda)e$. But at $\lambda = 1$, there is a sudden discontinuous transition to a field which no longer is r^{-1} at ∞, and which no longer supports an infinite number of bound states. The question is: How many bound states survive? To my knowledge, there is not yet a definite answer to this simple question. We know that the $\lambda=0$ -13.6 eV $H^-(1s^2)$ state moves up to -0.7 eV at $\lambda=1$, because this H^- state is observed and its energy is accurately predictable. But no one has been able to find any excited bound states of H^- lying below the H(1s) continuum threshold, although such bound states have been actively sought. On the other hand, I have not seen a proof that no such bound states exist, although rumors of the existence of such proofs have reached me.

However it is known (Drake [38]) that with the spin-independent Hamiltonian (1) there is a truly bound state of H^-, represented by a quadratically integrable wave function, lying in the continuum just below the H(2s) continuum threshold. This state remains bound for $\lambda = 1$, despite the remarks made earlier, because special selection rules prevent the coupling to the continuum which otherwise would be expected. Specifically, this H^- state is an even parity ^3P level, with nominal configuration assignment $2p^2$, lying about 0.0095 eV below the H(2s) continuum threshold, as found in a 50 term variational calculation. Because the Hamiltonian (1) preserves total orbital angular momentum, autoionization of this ^3P state to the ground H(1s) state of atomic hydrogen can occur only if the outgoing electron is in a p state; but this would mean the final two-electron state has odd parity, whereas the original ^3P level has even parity. Hence autoionization to H(1s) cannot occur; autoionization to excited states of atomic hydrogen is energetically impossible because this H^- ^3P state lies below the H(2s) continuum threshold. Therefore this $H^-(2p^2)^3P$ state does not couple to the continuum. A similar $He(2p^2)^3P$ bound state is embedded in the continuum (Drake [38]). Much the same type of argument shows that there is a $^4P_{5/2}$ bound state of He^- embedded in the continuum, even when the Hamiltonian (1) is modified to include the actual spin-orbit coupling (Holoien [39], Baranger [40]; this He^- $^4P_{5/2}$ state finds practical application in tandem Van de Graafs.

Another seemingly very elementary problem of binding, which actually turns out to be very difficult, is whether a positron can be bound to a hydrogen atom, in other words whether the positron analogue of H^- exists. We know the positron could be bound if it were sufficiently massive, because the molecular ion H_2^+ is stable. But what is the situation when the positron has its actual mass? Only very recently has it been shown that the three-body system pe^-e^+ just barely fails to be bound (Spruch [57]).

III. CONTINUUM CALCULATIONS

I now turn to atomic physics continuum problems. The variety of such problems, culminating typically in computation of a cross section, is so enormous that in a talk of this length I scarcely can even give you a feeling for the direction and content of current research. I already have given you some (very slight) indication of procedures in atom-atom collisions. In my remaining minutes I shall concentrate almost exclusively on the collisions of electrons with atomic hydrogen. This is the simplest and most fundamental atomic physics three-body continuum problem, corresponding to neutron-deutron collisions in nuclear physics. I begin with the resonances in e^--H scattering.

A. Resonances in e^--H(1s) Collisions

The possibility of observing true resonances in atomic collisions, describable by the Breit-Wigner formula, was first raised seriously in 1957 (Gerjuoy [2], Baranger [40]). Since then the study of such resonances, both theoretically and experimentally, has become a major subdivision of atomic physics. A review of the subject of atomic physics resonances has been given recently by Schulz [53], who first experimentally observed the resonances in the elastic scattering of electrons by atomic hydrogen. For the low lying resonances at any rate, it is fair to state that the theoretical and experimental positions of these resonances now are in very satisfactorily close agreement. Table III, from Bhatia [54], illustrates this assertion for the lowest resonance, which is associated with the ${}^1S_o(2s^2)$ complex energy eigenvalue of H^- discussed in connection with Fig. 4. The calculations listed in Table III all were originally performed assuming the proton is infinitely massive, but the results are accurate to so many significant figures that the finite mass corrections must be examined; the needed corrections

Table III. Theory of the Lowest e^--H(1s) Resonance

Method	Predicted Resonance E(ev)	Predicted Width (ev)
Close coupling	9.5603	0.0475
Kohn variational:		
Usual version	9.5574	0.0472
Modified (Nesbet)	9.571	0.0492
Modified (Chen)	9.5542	0.0411
Projection Operator	9.55724	
Stabilization	9.5572	0.0559
Complex Rotation	9.5572	0.0474
Experiment	9.558 ± 0.010	0.043 ± 0.006

have been incorporated into Table III. The widths of the resonances are less certain than their locations, both theoretically and experimentally (as can be inferred from Table III); nevertheless, the theoretical widths in Table III are no more than about 30% apart, and the data on this and other e^--H(1s) resonances (not listed in Table III) suggest this 30% figure is approximately the present discrepancy between theoretical and experimental widths of the lower lying resonances.

Conventional Calculations of Resonances

I want to briefly describe the various theoretical methods listed in Table III; as will be seen, several of these methods are very different from the others, a fact which makes the agreement between the various calculations quite remarkable. The close coupling method is the direct approach to solving the Schroedinger equation for the scattering problem, and the atomic physics procedures are basically identical with those employed in nuclear physics. The unsymmetrized two-electron wave function $\Psi(1,2)$ is expanded in hydrogenic eigenfunctions in one of the electrons (electron 2 say), with expansion coefficients which are unknown functions $F_n(r_1)$ in the coordinates of electron 1; $\Psi(1,2)$ then is suitably symmetrized in the coordinates of 1 and 2, and inserted into the Schrodinger equation. One gets an infinite set of coupled integro-differential equations in the unknown expansion coefficients F_n, which must be solved subject to the boundary condition that there was an incident wave only in the ground state channel. In practice, the expansion is truncated, and the resultant final set of equations is solved exactly. It is clear that one must include all the energetically open channels; the only question is how many closed channels must one include to get reasonably accurate results. Because of the rapidly decreasing spacing of the hydrogenic bound states as their energy increases, and because the total number of bound states is infinite, it is obvious that in e^--H(1s) scattering the close coupling method is most suitable at low energies, preferably below the 10.2 eV H(2s) excitation threshold. The most elaborate close coupling calculations have included the 1s, 2s, 2p, 3s, 3p, 3d states, plus so-called pseudostates, which are non-hydrogenic bound states included in the expansion to give some projection on the otherwise totally excluded hydrogenic bound states and continuum states lying above the included hydrogenic bound states.

The close coupling method can be given a variational formulation; the close coupling eigenphases of the scattering matrix approach their exact values monotonically as more and more closed channels are included in the expansion, provided all open channels are included. An alternative well known variational method is the Kohn, based on the stationary property of the scattering amplitude when properly expressed as a simple functional of the final scattering solution. Usually the trial wave function is written as a linear combination of some suitable basis set, with arbitrary coefficients. Making the functional stationary with respect to variation of these coefficients leads to a set of linear equations for the "best" set of these coefficients; the scattering phase shifts then are computed by inserting this "best" trial scattering solution into the variational principle. The Nesbet [55] modification treats the equations for the linear coefficients somewhat differently, in an attempt to avoid some spurious roots which can arise in the usual version of the Kohn variational procedure. The Chen modification (Chung [3]) reformulates the Kohn principle so as to separate the variations in the open and closed channels.

Less Conventional Resonance Calculations, Using Square Integrable Functions Only

The methods in Table III which I have described all involve explicit use of non-quadratically integrable functions at some stage of the calculation, as is to be expected; after all we are doing a scattering problem. However, a major feature of some very recent methods is the recognition that -- in atomic physics calculations at any rate -- accurate scattering predictions often are attainable using purely quadratically integrable functions.

The projection operator method of Table III is one way of avoiding non-quadratically integrable functions. The $H^-(2s^2)$ resonance lies below the H(2s) continuum threshold; therefore it couples to the continuum eigenfunctions only because the electron reaching infinity leaves behind a hydrogen atom in its ground state; were there no H(1s) state below $H^-(2s^2)$, the $H^-(2s^2)$ could not send an electron off to infinity, i.e., could not couple to the continuum. Therefore, in the projection operator method, one seeks the bound states of H^- in a subspace of the full Hilbert space, from which the H(1s) state has been projected out (Bhatia [54]). The desired $H^-(2s^2)$ energy level now can be obtained by direct application of the usual Rayleigh-Ritz variational procedure in the subspace. This $H^-(2s^2)$ eigenlevel in the subspace is not the physical resonance energy;

we know the $H^-(2s^2)$ eigenvalue in the full Hilbert space is complex (i.e., the resonance has a width) and there is a shift in the real part of the eigenvalue as one goes from the subspace to the full space. The level shift turns out to be small, however, so that a reasonably accurate estimate of its magnitude still suffices to give a very good answer for the location of the physical resonance; the level width is estimated similarly. The whole procedure is formally identical with the formalism introduced by Feshbach [58] in his unified theory of nuclear reactions, and in atomic physics the e^--H(1s) resonances of the $H^-(2s^2)$ type are known as Feshbach resonances. The projection operator method has very good accuracy only in electron scattering from hydrogen-like atoms or ions, however (e.g., in e-He^+(1s) scattering). In more complicated systems, as in the nuclear neutron-deuteron scattering problem, the projection operator removing the ground state of the target system is not known in simple closed form, and the construction of a Hilbert subspace exactly orthogonal to the ground state of the target is very difficult.

The stabilization method (Bhatia [59], Hazi [60]) applies the Rayleigh-Ritz to the $H^-(2s^2)$ eigenvalue problem without even attempting to introduce the aforementioned Hilbert subspace orthogonal to the hydrogen ground state. The stabilization method cannot be considered wholly well-founded therefore; on the other hand, it is capable of producing accurate results as Table III shows, and it can be applied to more complicated systems because now there is no need to have the exact projection operators. In the stabilization method one diagonalizes the Hamiltonian in some suitable quadratically integrable orthonormal basis, $u_1,\ldots,u_n$, and examines the eigenvalues in the vicinity of the expected resonance. As the number n of basis functions employed is increased, the computed eigenvalues being examined usually shift appreciably, as one expects because these computed eigenvalues at continuum energies can't correspond to actual bound states. In practice, however, some energy eigenvalues in the continuum are much more stable than other eigenvalues (i.e., shift much less than other eigenvalues), until the number n of basis functions used gets quite large; as n is increased farther, the previously stable eigenvalue now shifts as much as the others. The "stable" eigenvalues are identified with the resonances. If I understand what's going on (and I'm not sure I do, and I'm pretty sure no one understands it any better), the stabilization method is taking advantage of the fact that at a real energy close to a complex eigenvalue of narrow width, the exact solution to the Schroedinger equation is large in the vicinity of the potential and small outside; thus the exact solution near a resonance is essentially indistinguishable from a true quadratically integrable eigenfunction until the exact solution is expanded in eigenfunctions which extend out far enough to recognize that the wave function is oscillating at large distances instead of exponentially decreasing.

The rigorous definition of the complex resonant eigenvalues is that these eigenvalues correspond to poles of the scattering matrix, i.e., to solutions of the Schroedinger equation having the purely outgoing asymptotic form e^{ikr}/r in every channel, with no incoming wave in any channel. These resonance eigenfunctions can't be quadratically integrable however; if they were, they would be true eigenfunctions of the Hamiltonian, corresponding to purely real eigenvalues. In fact, the resonant k's have negative imaginary parts, and the wave function grows exponentially at infinity. Conversely, the lack of quadratic integrability means the resonant eigenvalues can't be determined via the usual Rayleigh-Ritz. In the complex rotation method, this difficulty is evaded by rotating the r-coordinates in complex coordinate space, i.e., one replaces every r by $re^{i\alpha}$. If α is sufficiently large, and it doesn't have to be very large for the k's encountered in practice, $e^{ikr} = e^{ikre^{i\alpha}}$ now will be exponentially decreasing at infinity despite the fact that k has an imaginary part, and the Rayleigh-Ritz now will be applicable. Of course, the replacement of r by $re^{i\alpha}$ amounts to introducing a new non-Hermitian Hamiltonian, and questions can be raised concerning the connection between the eigenvalues of the new and the original Hamiltonian. Nevertheless, the method does work, as Table III shows (Bardsley [61], Nuttall [62],

Doolen [63], Bain [56]).

B. Less Conventional Phase Shift Calculations

The more conventional methods I described earlier, namely the close coupling and the various versions of the Kohn variational, yield the phase shift at any incident energy. The resonances are those energies at which the phase shift increases by π as the incident electron energy is increased from below to above the resonance energy. Because the resonances are very narrow (the 0.04 $H^-(2s^2)$ width is much larger than most of the e^--H(1s) resonances), the conventional calculations will miss the resonances unless carried out with a very fine energy mesh. Use of a very fine energy mesh is quite expensive and time consuming however, especially if one is primarily interested in the cross sections at resonances. A virtue of the projection operator, stabilization and complex rotation methods is not only that they take advantage of square integrability properties, but that they lead directly to the resonance energies; on the other hand these latter methods cannot be used to find the scattering phase shift at an arbitrary assigned energy.

However, there are e^--H(1s) calculations I have not discussed which have yielded the phase shift at any given energy while also taking advantage of square integrability. I do not have time to more than mention these methods, which include extrapolation methods from complex energies (Nuttall [31]; approximating the Fredholm determinant in an L^2 basis (Heller [30], Reinhardt [29]; and the R-matrix method well known in nuclear physics (Burke [28]).

These last-mentioned methods, and some other square integrable techniques for scattering problems, have been reviewed recently by Bardsley [27]. One such interesting possibility is the so-called minimum variance method, which in any given basis defines the "best" wave function Ψ as the Ψ minimizing

$$S = \int dr \left| \omega(r)(H-E)\Psi(r) \right|^2 \tag{3}$$

where $\omega(r)$ is some suitably chosen weight function. Obviously S equals zero only when $\Psi(r)$ satisfies the Schrodinger equation everywhere. This method has been applied quite successfully to e^--H(1s) scattering (Read [25]). For potential scattering, but not for more complicated problems, Bardsley [26] has been able to show that the minimized S of Eq. (3) can yield upper and lower bounds on the phase shift.

The Faddeev equations also have been applied to e^--H(1s) scattering, taking advantage of the fact that the two-body T-matrix for Coulomb forces is known exactly in closed form. The results have been comparatively disappointing however (Chen [22], [23]); the Faddeev equations just are too hard to work with. Moreover there is the complication--stressed by Faddeev [24] when this Conference met at Birmingham in 1969--that because of the long range properties of the Coulomb force the Faddeev formalism may break down in e^--H(1s) collisions at energies above the ionization threshold, i.e., at incident kinetic energies above 13.6 eV.

Dispersion Relation in e^--H(1s) Scattering

With this discussion of the theory of e^--H(1s) scattering, especially near resonances, I largely must conclude my detailed remarks. Regrettably, there is no time for detailed examination of many other problems which are of considerable theoretical as well as practical interest to atomic physicists, and not a few of which would be of interest to this audience as well. However, I do want to at least mention that dispersion relations have been "derived" for e^--H and e^--He scattering (Gerjuoy [70]), and have been compared with experiment (Krall [71], McDowell [72], Byron [73]).I put "derived" in quotes because the original derivation recognized that in these multi-channel collisions there were some questions of mathematical rigor which had not been thoroughly examined. Assuming these questions can be satisfactorily resolved, the resultant dispersion relation in e^--H(1s) scattering is

$$Re(f - \frac{1}{2}g) = (f - \frac{1}{2}g)_{Born} + \frac{1}{4\pi^2} P \int_0^\infty dE' \frac{k'\sigma_t(E')}{E'-E} - \frac{1}{2} R \tag{4}$$

where f is the direct scattering amplitude and g the exchange amplitude; $E = \frac{\hbar^2 k^2}{2m}$; and σ_t is the total cross section, including all inelastic processes. The quantity R in (4) is a residue computed at the known $H^-(1s^2)$ bound state; if there were more than one bound H^- state below the H(1s) threshold, as discussed earlier, corresponding residues for such states would have to be added to the right side of (4).

In principle, therefore, Eq. (4) provides the possibility of experimentally deciding whether or not there are additional bound states of H^-, always assuming that the experiments can be made accurate enough and that Eq. (4) is valid. The experiments are getting to be very good, but unfortunately the latest indications are that the derivation of (4) is not valid. The trouble apparently arises in the exchange amplitude g, which seems to have singularities not anticipated nor readily taken into account in the derivation of (4) (Blum [74]). There is no obvious way to eliminate the exchange amplitude from (4), however, because the experimental cross section is the sum of the triplet and singlet scattering with respective weights 3/4 and 1/4. Indeed, it is because the singlet and triplet scattering satisfy separate optical theorems that (4) involves

$$Re(f - \frac{1}{2}g) = \frac{3}{4} Re(f - g) + \frac{1}{4} Re(f + g) \tag{5}$$

On the other hand, the dispersion relation for positron scattering from H(1s) does not involve exchange. The data on positron scattering are still coming in, but they indicate that the dispersion relation indeed is satisfied for positron scattering (Bransden [67]), i.e., that there is no trouble for the direct amplitude f. Note that the residue on the right side of (4) is absent from e^+-H(1s) scattering, because (as discussed above) the hydrogen atom can't bind a positron.

Glauber Theory in e^--H(1s) Scattering

I also want to mention that in the last few years there has been a frenzied application of Glauber theory to atomic collisions, both elastic and inelastic (Gerjuoy [21]). The initial and major work has been done on e^--H(1s) scattering, for excitation and ionization as well as for elastic scattering. At incident energies above about 50 eV the Glauber predictions are in reasonable agreement with experiment, and usually are considerably better than Born approximation. The Glauber predictions also are considerably better than close coupling calculations in this energy range above 50 eV, because the close coupling calculations become too difficult to carry through when there are so very many open channels, including the breakup (ionization) channel.

In fact, it is by now reasonably clear that in a sense the close coupling and the Glauber are complementary approximations. In the close coupling one expands the wave function in a finite basis set, but then the equations for the expansion coefficients are solved essentially exactly. On the other hand, the Glauber approximation for multi-channel scattering can be derived by inserting an eikonal approximation into the complete infinite set of exact close coupling equations (Byron [20], Gerjuoy [21]). In other words, it is understandable that the Glauber, which retains all the close coupling equations but solves them only approximately, should be preferable to close coupling at those higher energies where the close coupling calculations can't be carried through without ignoring many open channels. Fig. 5 shows how well the close coupling does at low energies; the better close coupling predictions go right through the absolute e^--H(1s) elastic scattering differential cross section measured by Williams [4].

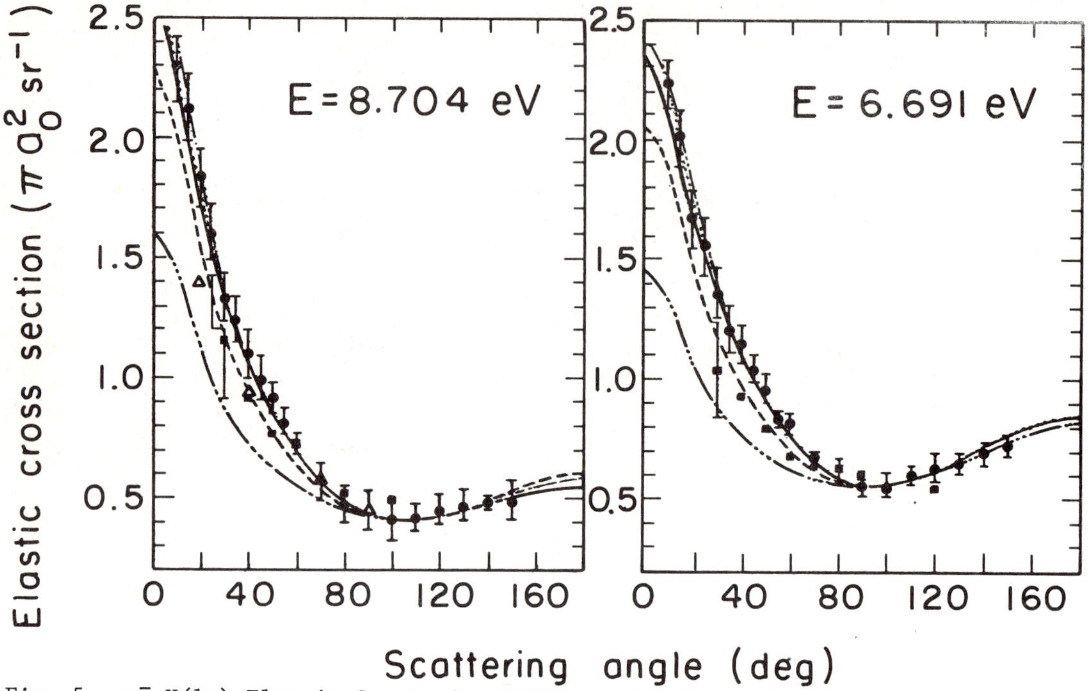

Fig. 5. e^--H(1s) Elastic Scattering Differential Cross Sections at Low Energies. The data points are absolute differential cross sections. The various theoretical curves include several close coupling calculations. The better (more states in expansion) close coupling predictions go right through the experimental points. These energies are much too low for Glauber to be useful.

Fig. 6 shows that at 200 ev, where close coupling calculations haven't even been attempted, the Glauber approximation quite accurately predicts the absolute differential cross sections for electrons scattered after exciting the hydrogen to the 2s or 2p level (Williams [5]). One obvious deficiency of the Glauber, however, is its prediction that positrons and electrons will have exactly the same cross section for scattering from the same target atom, including target atoms more complicated than atomic hydrogen. This prediction is not physically reasonable at intermediate incident electron energies, and seemingly is not borne out by the still rudimentary data.

A feature of the Glauber e^--H(1s) calculations worth stressing here is the fact that the resultant amplitude cannot be decomposed in the fashion common in nuclear and particle physics, namely into individual electron-electron and electron-proton Coulomb amplitudes, plus a double scattering correction term. This decomposition of the Glauber amplitude is not possible in e^--H(1s) collisions because for the individual long range Coulomb e^--e^- or e^--p potentials the Glauber phase integral, involving the integral of the potential over a straight line from $-\infty$ to ∞, diverges. The Glauber approximation is usable in e^--H(1s) collisions only because cancellation keeps finite the phase integral over the sum of the e^--e^- and e^--p potentials. On the other hand, the Glauber formula for the direct amplitude in e^--H(1s) scattering can be evaluated in closed form without any further approximation, for excitation as well as for elastic scattering (Gerjuoy [21]). This possibility is remarkable, because the Glauber formula for e^--H(1s) scattering is a complicated (at first sight intractable) five-dimensional integral. Apparently the possibility is a very special property of Coulomb forces; with Yukawa interactions the Glauber five-dimensional integral seems to be truly intractable. The advantage of the e$^-$-H(1s) closed form is that the Glauber predictions shown in Fig. 6 can be evaluated with scarcely more effort than is required to evaluate the Born approximation.

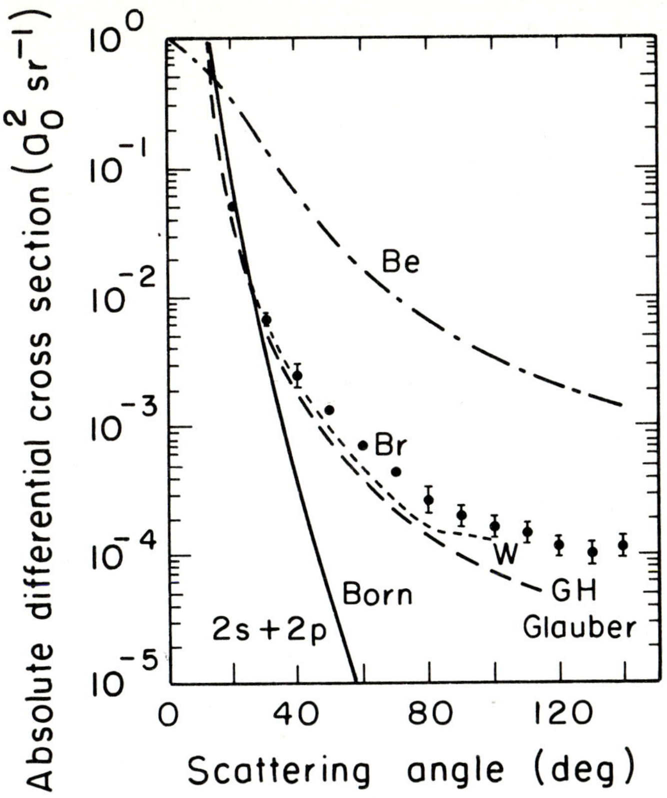

Fig. 6. e$^-$-H(1s) Inelastic Differential Cross Sections at 200 ev Incident Electron Energy. The data points are absolute differential cross sections for electrons following the excitation e$^-$+H(1s) → e$^-$+H(2s) or H(2p). The Glauber curve is labeled GH (Ghosh and Sil). The Born approximation (labeled 2s+2p) is far too small at wider angles. Close coupling calculations have not even been attempted at this energy.

e$^-$-H(1s) Ionization Near Threshold

In a Conference of this sort, it would be remiss of me to wholly ignore breakup reactions. Therefore I will briefly discuss the ionization of H(1s) by electrons. Theories of this ionization process almost universally ignore the profound complications resulting from the fact that the final state is composed of three charged particles, capable of influencing each other significantly even when all the particles are infinitely separated. The most widely used approximate treatment is the so-called Coulomb-Born, which computes the T-matrix using a plane wave for the incident electron and a Coulomb wave for the (usually) slower outgoing ionized electron. For incident electron energies above about 100 ev, the Coulomb-Born predictions of the total ionization cross section are quite good. Finer details of the ionization process--e.g., the ionized electron energy distribution, or the angular correlation of the outgoing electrons--are less accurately predicted by the Coulomb-Born, as might be expected because it is a relatively crude approximation after all. Better approximations, such as the distorted wave or Glauber approximations, are available however. In particular, there have been some elaborate Glauber calculations of e$^-$-H(1s) ionization recently (McGuire [64]), which claim to significantly improve the Coulomb-Born predictions. These Glauber calculations are very arduous, it should be noted; the analysis yielding Glauber e$^-$-H(1s) excitation amplitudes in closed form has not been generalized to ionization. Also I know of no serious Kohn variational calculations of e$^-$-H(1s) ionization.

The approximations under consideration in the preceding paragraph, notably the Coulomb-Born, Glauber and distorted wave approximations, all ignore the effect of the electron-electron repulsion on the asymptotic form of the final state wave function at infinity. This electron-electron interaction effect obviously is most important when the outgoing electrons are moving slowly; the success of the Coulomb-Born and the other approximations discussed suggests the electron-electron modification of the asymptotic form is relatively unimportant in e^--H(1s) at incident energies exceeding 100 eV. It is of interest to examine the e^--H(1s) ionization cross section near threshold therefore.

There are a number of competing theories of this threshold ionization cross section. For short range forces, the breakup cross section near threshold is proportional to $(E-E_t)^2$, where E_t is the threshold energy; this is a purely phase space result. Because the wave function in an attractive Coulomb field has an extra energy-dependent $1/\sqrt{k}$ normalization, the ionization cross section near threshold is modified (from the phase space result) to $(E-E_t)^{3/2}$ or $(E-E_t)$, depending respectively on whether one or both outgoing electrons are thought to be moving in the incompletely shielded 1/r long range field of the proton. Neither of these threshold cross section predictions have included the electron-electron interaction effects however. In fact, the only serious attempt to incorporate these effects is a purely classical calculation due to Wannier [65], which also reaches a result stemming largely from phase space considerations, but in this case of classical phase space. Wannier's result for the threshold behavior is

$$\sigma_{ion}(E) \sim (E-E_t)^{(-1+\sqrt{\frac{91}{3}})/4} \cong (E-E_t)^{1.127} \tag{6}$$

The almost unbelievable fact is that the data (McGowan [66]) indicate this is the threshold law for ionization by electrons. With the establishment of this experimental fact, quantum mechanical "derivations" of the Wannier result (6) have appeared (Rau [68]), but these derivations hardly are more than WKB approximations which beg the question of whether the postulated WKB form--which corresponds to Wannier's classical formulas--is a correct representation of the wave function near threshold. For this Conference the significance of the empirically verified formula (6) is that the Green's function for three charged particles--whose asymptotic form determines the energy dependence of the ionization cross section--must be a very complicated object indeed if it is to be capable of yielding quantum mechanically a power law like $(-1 + \sqrt{91/3})/4$.

REFERENCES

[1] Cf. e.g., J. S. Risley and R. Geballe, "Electronic and Atomic Collisions, Abstracts of Papers of the IXth International Conference on the Physics of Electronic and Atomic Collisions, July 1975" (Univ. of Washington 1975).
[2] E. Baranger and E. Gerjuoy, Phys. Rev. 106 (1957) 1182.
[3] K. T. Chung and J.C.Y. Chen, Phys. Rev. A6 (1972) 686.
[4] J. F. Williams, J. Phys. B:Atom. Mol. Phys. 8 (1975) 1683.
[5] J. F. Williams, J. Phys. B:Atom. Mol. Phys. 8 (1975) 1641.
[6] M. Born and J. R. Oppenheimer, Ann. d. Phys. 84 (1927) 457.
[7] L. Pauling and E. Bright Wilson "Introduction to Quantum Mechanics" (McGraw Hill 1935), 259ff.
[8] Y. Accad, C. L. Pekeris and B. Schiff, Phys. Rev. A4 (1971) 516.
[9] C. L. Pekeris, Phys. Rev. 112 (1958) 1649 and 127 (1962) 509.

[10] J. S. Sims and S. A. Hagstrom, Phys. Rev. A11 (1975) 418.
[11] R. K. Nesbet, Phys. Rev. A2 (1970) 661.
[12] J. D. Lyons, R. T. Pu and T. P. Das, Phys. Rev. 178 (1969) 103.
[13] J. S. Onello, L. Ford and A. Dalgarno, Phys. Rev. A10 (1974) 9.
[14] J. S. Onello and L. Ford, Phys. Rev. A11 (1975) 749.
[15] W. J. Stevens and F. P. Billingsley, Phys. Rev. A8 (1973) 2236.
[16] J. L. Dehmer, Mitio Inokuti and R. P. Saxon, Phys. Rev. A12 (1975) 102.
[17] J. H. Miller and H. P. Kelly, Phys. Rev. A3 (1971) 578 and A5 (1972) 516.
[18] R. F. Stewart, D. K. Watson and A. Dalgarno, J. Chem. Phys. 63, (1975) 3222.
[19] R. K. Nesbet, Phys. Rev. 175 (1968) 2.
[20] F. W. Byron, Phys. Rev. A4 (1971) 1907.
[21] E. Gerjuoy and B. K. Thomas, Rep. Prog. Phys. 37 (1974) 1345.
[22] J.C.Y. Chen, Computer Phys. Comm. 6 (1974) 336.
[23] J.C.Y. Chen, in T.R. Govers and F. J. de Heer, (see ref. 31) 232.
[24] L. D. Faddeev, in "The Three-body Problem" (North-Holland 1970) 154.
[25] F. H. Read and J. R. Soto-Montiel, J. Phys. B:Atom. Mol. Phys. 6 (1973) L15.
[26] J. N. Bardsley, E. Gerjuoy and C. V. Sukumar, Phys. Rev. A6 (1972) 1813.
[27] J. N. Bardsley, "Theory of Low-Energy Electron-Atom Collisions and Related Processes", invited talk ICPEAC IX July 1975 (see ref. 1), to be published in the Proceedings of that Conference.
[28] P. G. Burke, Comments Atom. Mol. Phys. 3 (1972) 31.
[29] W. P. Reinhardt, D. W. Oxtoby and T. N. Rescigno, Phys. Rev. Lett. 28 (1972) 401.
[30] E. J. Heller, T. N. Rescigno and W. P. Reinhardt, Phys. Rev. A8 (1973) 2946.
[31] J. Nuttall, in T. R. Govers and F. J. de Heer, "Physics of Electronic and Atomic Collisions, VII ICPEAC 1971" (North Holland 1972) 265.
[32] D. L. Morgan and V. Hughes, Phys. Rev. D2 (1970) 1389.
[33] B. R. Junker and J. N. Bardsley, Phys. Rev. Lett. 28 (1972) 1227.
[34] W. Kolos, D. L. Morgan, D. M. Schrader and L. Wolniewicz, Phys. Rev. A11 (1975) 1792.
[35] J. E. Turner, V. E. Anderson and K. Fox, Phys. Rev. 174 (1968) 81.
[36] R. F. Stebbings, Bull. Am. Phys. Soc. 20 (1975) 1453.
[37] P. M. Dehmer and W. A. Chupka, Bull. Am. Phys. Soc. 20 (1975) 1457. See also the other Abstracts on p. 1458.
[38] G. W. F. Drake, Phys. Rev. Lett. 24 (1970) 126.
[39] E. Holoien and J. Midtdal, Proc. Phys. Soc. (London) A68(1955) 815.
[40] E. Baranger and E. Gerjuoy, Proc. Phys. Soc. (London) 72 (1958) 326.
[41] C. Schwartz. Ann. Phys. (N.Y.) 2 (1959) 156 and 170.
[42] A. Dalgarno and A. L. Stewart, Proc. Roy. Soc. (London) A257 (1960) 534.
[43] L. M. Delves, Nucl. Phys. 41 (1963) 497 and 45 (1963) 313.
[44] E. Gerjuoy, A.R.P. Rau and L. Spruch, Phys. Rev. A8 (1973) 662.
[45] V. Sahni and J. B. Krieger, Phys. REv. A8 (1973) 65.
[46] E. Gerjuoy, A.R.P. Rau, L. Rosenberg and Larry Spruch, Phys.Rev. A9 (1974) 108.
[47] R. Blau, A.R.P. Rau and L. Spruch, Phys. Rev. A8 (1973) 119.
[48] J. S. Sims and R. C. Whitten, Phys. Rev. A8 (1973) 2220.
[49] J. S. Sims and J. R. Rumble, Phys. Rev. A8 (1973) 2231.
[50] M. T. Anderson and F. Weinhold, Phys. Rev. A11 (1975) 442.
[51] F. Weinhold, Phys. Rev. Lett. 25 (1970) 907,J. Chem. Phys. 54 (1971) 1874.
[52] R. Shakeshaft, L. Rosenberg and L. Spruch, to be published.
[53] G. Schulz, Rev. Mod. Phys. 45 (1973) 378 and 423.
[54] A. K. Bhatia and A. Temkin, Phys. Rev. A11 (1975) 2018.
[55] R. K. Nesbet and J. D. Lyons, Phys. Rev. A4 (1971) 1812.
[56] R. A. Bain, J. N. Bardsley, B. R. Junker and C. V. Sukumar, J. Phys. B: Atom.Mol. Phys. 7 (1974) 2189.
[57] L. Spruch, private communication.
[58] H. Feshbach, Ann. Phys. (N.Y.) 5 (1958) 357; 19 (1962) 287.
[59] A. K. Bhatia, Phys. Rev. A6 (1972) 120.
[60] A. U. Hazi and H. S. Taylor, Phys. Rev. A1 (1970) 1109.
[61] J. N. Bardsley and B. R. Junker, J. Phys. B:Atom. Mol. Phys. 5 (1972) L178.

[62] J. Nuttall, Bull. Am. Phys. Soc. 17 (1972) 598.
[63] G. Doolen, J. Nuttall and R. W. Stagat, Phys. Rev. A10 (1974) 1012.
[64] J. McGuire, M. B. Hidalgo, G. D. Doolen and J. Nuttall, Phys. Rev. A7 (1973) 973.
[65] G. H. Wannier, Phys. Rev. 90 (1953) 817.
[66] J. W. McGowan and E. M. Clarke, Phys. Rev. 167 (1968) 43.
[67] B. H. Bransden and P. K. Hutt, J. Phys. B:Atom. Mol. Phys. 8 (1975) 603.
[68] A.R.P. Rau, Phys. Rev. A4 (1971) 207.
[69] R. Blau, to be published.
[70] E. Gerjuoy and N. A. Krall, Phys. Rev. 119 (1960) 705; 127 (1962) 2105.
[71] N. A. Krall and E. Gerjuoy, Phys. Rev. 120 (1960) 143.
[72] M.R.C. McDowell, Comments on Atom. Mol. Phys. 4 (1974) 147.
[73] F. W. Byron, F. J. de Heer and C. J. Joachain, Phys. Rev. Lett. 35 (1975) 1147.
[74] K. Blum and P. G. Burke, to be published.

A REPORT ON SOME FEW-BODY PROBLEMS IN ATOMIC PHYSICS*

Larry Spruch
Physics Department, New York University, New York, New York 10003, U.S.A.

The material to be covered will include some topics discussed in the parallel session: high-energy scattering from a two-center potential, channel coupling array theory, and the binding energy of $e^-+e^-+e^+$. It will also include some subjects of a slightly more exotic character: positron production in energetic heavy-ion collisions, charge exchange at high incident energy, and hydrogen atoms in intense magnetic fields.

"DIVING": POSITRON PRODUCTION IN ENERGETIC HEAVY-ION COLLISIONS

I can still remember the sense of awe induced on learning, a long time ago, that solutions of the one-particle Dirac equation for an electron in the field of a point nucleus of charge Z break down when $\alpha Z = (Z/137)$ exceeds 1. Thus, for example, the energy of the 1s ground state, $E_{gd} = \mu c^2 (1 - \alpha^2 Z^2)^{1/2}$, cannot go below zero; E_{gd} is imaginary for $\alpha Z > 1$, signifying that the Hamiltonian is no longer self-adjoint. [Using the uncertainty principle and the fact that the kinetic energy T approaches pc for large kinetic energies, one can give a heuristic argument which suggests that any one-particle relativistic equation will become meaningless for Z sufficiently large. Thus, let the (relatively smooth) ground-state wave function have the dimension R. Then as Z increases T will approach $k_1\hbar c/R$ while $V = -Ze^2/r$ will approach $-k_2Ze^2/R$, where k_1 and k_2 are dimensionless constants of order unity. As αZ approaches $k \equiv k_1/k_2$, the energy E = T+V approaches zero. For $\alpha Z > k$, the ground state solution of the relativistic one-particle equation ceases to exist. (Detailed analysis gives k=1 for the Dirac equation as noted above, and k = 1/2 for the Klein-Gordon equation.) For the Schroedinger equation, on the other hand, $T = p^2/2\mu$ is at least of order $\hbar^2/(2\mu R^2)$ and dominates $V = -Ze^2/r$, which will be of order $-Ze^2/R$, for all Z for any R; only (attractive) potentials that behave as $1/r^2$ can dominate T. Indeed, on rewriting the Dirac or Klein-Gordon equations with a Coulomb potential as Schroedinger equations the effective V contains an attractive $(1/r^2)$ term.]

Often when the one-particle Dirac equation breaks down one can ascribe the difficulty to its neglect of the possibility of pair production. In the point Coulomb case, however, the true source of the difficulty is not the one-particle approximation but simply the singular nature of V(r), with the uncertainty principle consequences noted above. If one studies the one-particle Dirac equation for the potential generated by a heavy (finite) nucleus, E_{gd} reaches zero at a value of Z necessarily somewhat greater than 137 -- it happens at $Z=Z_0 \approx 150$ -- and, without any breakdown, continues below zero. Difficulties first arise at the critical value $Z_{crit} \simeq 170$, for which $E_{gd} = -\mu c^2$, and the electron dives into the negative-energy sea. At this stage, one has a bound state embedded in a continuum (the negative energy sea), a problem that can be treated (Fano [1]) in atomic and nuclear physics. One expects an empty K-shell bound state that dives into the negative energy sea to be filled by a negative energy electron, with the concomitant appearance of a positron. The filled K-shell state cannot be a true bound state, since it is embedded in the continuum, but there will be a polarization of the vacuum which will give a charge distribution that simulates that of a K-shell electron. The K-shell bound state energy will have a width, as will therefore the energy of the emitted positron.

---

* Work supported by the U.S. Office of Naval Research under contract No. N000 14-67-A-0467-0007, and by the U.S. National Science Foundation under Grant No. MPS-7500131.

A number of authors (Fulcher [2]) have pointed out that it is extremely useful to reinterpret the above comments. They redefine the vacuum as the state in which all negative energy solutions of the Dirac equation are occupied; these solutions include bound states as well as the continuum below $-\mu c^2$. (Without this reinterpretation, the energy for a system with $Z \gtrsim Z_0$ and with one or two electrons in the K-shell would lie below the energy of the vacuum). For $Z_{crit} > Z > Z_0$, and for a system with one K-shell orbit occupied by an "electron" and one unoccupied, the "electron" then represents a positive energy positron with energy less than μc^2, that is, a bound positron. As Z approaches Z_{crit} the energy of the positron approaches μc^2 and it is liberated. (See figure).

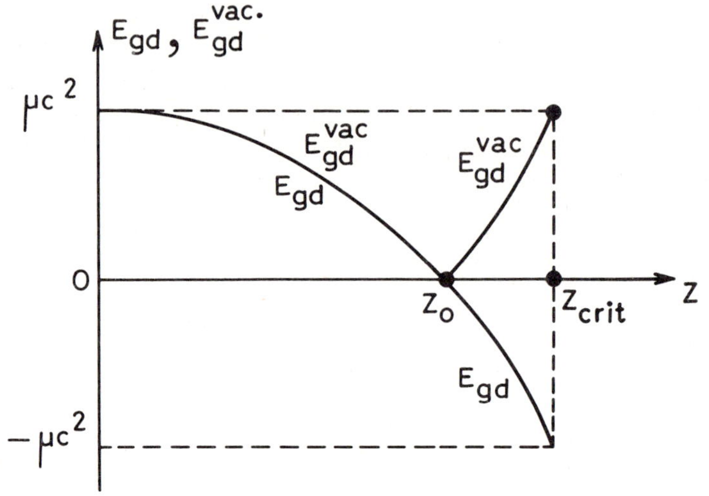

Figure: Energy in the field of a physical nucleus for one of the two 1s states occupied. E_{gd} is the energy defined by the one-particle Dirac theory. E_{gd}^{vac} is the energy measured with respect to the vacuum, defined as having all negative energy levels, bound states and continuum, occupied; both 1s states are therefore unfilled for $Z < Z_0$, and both are filled for $Z_0 < Z < Z_{crit}$. For $Z < Z_{crit}$, we then have, with Σ a step function,

$$E_{gd}^{vac} = E_{gd} - 2E_{gd}\Sigma(Z-Z_0) = |E_{gd}|.$$

The domain $\alpha Z > 1$ is an extremely interesting one for quantum electrodynamics, and many field theorists have indeed been looking at the problem. Interest in the subject has been very much stimulated in recent years by the prospect of achieving what amounts to nuclei of very large Z. We cannot now hope to produce real nuclei with very large Z, but we can hope to accelerate ionized heavy atoms, with charges $Z_1 \gg 1$, and have them strike targets composed of atoms with charges $Z_2 \gg 1$, and cause the nuclei to reach a separation less than the dimension of the ground state of an electron in the field of nuclei of finite size of charges Z_1 and Z_2 at that separation, for a time appreciably greater than the orbiting time of an electron in the field of such a nuclear pair. If, for example, a Uranium ion could be accelerated to 700 MeV in the laboratory frame -- this corresponds to only about 3 MeV per nucleon -- and if the U ion were to impinge on a U atom, the two nuclei would have a distance of closest approach of 35 fermis. At this distance, with the total charge (184) necessarily greater than Z_{crit}, an electron has an energy $-\mu c^2$ and "atomic" positron production, as it is sometimes called, becomes possible. "Molecular" positron production might be a more appropriate name. Invoking the adiabatic approximation, E_{gd} is calculated variationally using the one-particle Dirac equation with a two-center force; the electron is in a $1s\sigma$ state.

Atomic positron production cannot take place in the fashion just described unless, in the course of the uranium atom - uranium ion collision, a K-shell vacancy is produced. It is extremely difficult to estimate the probability $P_{1s\sigma}(0)$ of producing such a vacancy. At 1600 MeV in the lab frame, estimates of the cross section for atomic positron production range from about 10^{-29} to 10^{-26} cm^2. While atomic positron production has not yet been achieved, a great deal is being learned about the energetic heavy-ion collision process from studies of molecular

X-rays produced during such collisions. These studies should make possible reasonably accurate estimates of $P_{1s\sigma}(0)$ since some of the X-rays originate in transitions of electrons from occupied higher molecular energy levels to $1s\sigma$ vacancies produced during the collision.

Apart from their interest with respect to estimates of $P_{1s\sigma}(0)$, the X-rays emitted by the superheavy molecules temporarily formed in heavy-ion collisions are of fundamental interest since they determine the effects of quantum electrodynamics for strong fields, with $\alpha(Z_1+Z_2)>1$. For example, for $Z_1+Z_2\sim150$, the vacuum polarization contribution to the $^2P_{3/2}$ - $1s_{1/2}$ transition will be about 5 KeV and will exceed the self-energy contribution.

The only purpose of my remarks on atomic positron production and the spectroscopy of superheavy molecules is to generate further interest; the difficult aspects of the problem have scarcely been touched upon in this discussion. There is now a considerable body of literature on the subject; it can easily be traced back from any of a number of recent articles (Müller [3]). We note, incidentally, that the best sources for review articles on atomic physics include the volumes of proceedings of ICPEAC (International Conference on the Physics of Electronic and Atomic Collisions), nine thus far, and of ICAP (International Conference on Atomic Physics), four thus far, as well as Comments on Atomic and Molecular Physics, and Advances in Atomic and Molecular Physics.

HIGH-ENERGY SCATTERING FROM A TWO-CENTER POTENTIAL

In the analysis of "diving", one must study the two-center bound state problem within the context of the Dirac theory to determine E_{gd}. This leads naturally to the question of scattering by a two-center potential in Dirac theory, a topic addressed by Drs. Chandel and Mukherjee in a paper submitted to this conference. The authors consider a fixed orientation of the centers, and work with spheroidal coordinates. As opposed to the Schroedinger case, the problem is not separable. Proceeding in the usual fashion, they therefore rewrite the Dirac equation as

$$L\Psi = (V^2 + P^2 + 2\varepsilon V + V^2)\Psi = i(\alpha.\nabla V)\Psi ,$$

where ε is the energy and $P^2 = \varepsilon^2 - 1$. With Ψ_0 defined by $L\Psi_0 = 0$ and appropriate boundary conditions, they use a Sommerfeld-Maue approximation to approximate Ψ by $\Psi = \Psi_0 + \Psi_1$, where Ψ_1 can be expressed in terms of Ψ_0. They neglect V^2, which will presumably be dominated by εV for ε sufficiently large, and restrict themselves to potentials of the form

$$2\varepsilon V(\underset{\sim}{r}) = U(\xi)/(\xi^2 - \eta^2),$$

where $\underset{\sim}{r}$, $\underset{\sim}{r}_a$, and $\underset{\sim}{r}_b$ are measured with respect to the origin at the midpoint of the two centers and with respect to the centers, which are at aseparation R, and where $\xi = (r_a + r_b)/R$ and $\eta = (r_a - r_b)/R$. $V(\underset{\sim}{r})$ will be of the prescribed form if the centers are point nuclei, and for a number of other cases. The problem is then separable, and Ψ_0 can be expressed in terms of spheroidal phase shifts $\delta_{m\ell}$ defined by one-dimensional radial equations. The $\delta_{m\ell}$ are determined numerically for small values of m and ℓ and are approximated by Coulomb phase shifts for larger values of ℓ. To obtain the observed cross-section, one must average over all orientations of the nuclei.

Their analysis seems to be the first study of such a scattering problem. A number of open questions remain. These include the question of the advantages and disadvantages of this approach with respect to

i) A multiple scattering approach, in which one adds the scattering amplitudes from each center -- each of these could be obtained with very high precision since the Dirac problem with one spherically symmetric center of force is separable

-- and, further, accounts for say one additional scattering from the other center. This will surely be the appropriate approach at sufficiently high energies.

ii) an eikonal approximation

iii) a variational approximation.

As a separate question, would there be any purpose in using a Padé approximation based on the calculated "lower" phase shifts for estimating the "higher" phase shifts, rather than switching to Coulomb phase shifts?

CHANNEL COUPLING ARRAY THEORY

The two-body problem can today be thought of as "solvable". That is not quite the case for many three-body problems and it is anything but the case for almost all four-body problems. New formal and computational approaches are therefore to be welcomed.

Considerable formal progress was made with the introduction of the Faddeev equations. In these equations, the full wavefunction is "partitioned" into components, related to the pair of particles which interact first; for other than atomic physics with its Coulomb potentials, the equations can be extremely helpful in studies of existence and convergence questions, but computationally they have perhaps not quite lived up to their original promise.

Some results using different but still interesting sets of partitions were reported on in two mini-invited paper at this conference, by Levin and by Kouri. The channel coupling array approach can be applied to bound state and to continuum problems, and can be formulated as coupled differential or integral equations; further, as already noted, many different partitions are possible. To begin, we consider the determination of the energy E of the spatially symmetric ground state of a pair of electrons, numbered 1 and 2, in the field of a proton. We study the coupled equations for the channel components $\psi_1(r_1,r_2)$ and $\psi_2(r_1,r_2)$,

$$(H_1-E)\psi_1 + V_2\psi_2 = 0$$

$$(H_2-E)\psi_2 + V_1\psi_1 = 0,$$

where

$$H_1 = K_1 + (K_2 - e^2/r_2), \qquad V_1 = (-e^2/r_1) + (e^2/r_{12}),$$

$$H_2 = K_2 + (K_1 - e^2/r_1), \qquad V_2 = (-e^2/r_2) + (e^2/r_{12}),$$

and where the K's are kinetic energy operators. Note that the full Hamiltonian can be written as

$$H = H_1 + V_1 = H_2 + V_2 .$$

Adding the coupled equations immediately gives

$$(H - E)(\psi_1 + \psi_2) = 0 .$$

With the identification

$$\Psi = \psi_1 + \psi_2$$

it follows that the coupled equations reduce to the Schrodinger equation, for both bound state and scattering problems. (It does not follow from the above discussion that the solutions ψ_1 and ψ_2 are unique, nor even everywhere finite.)

The channel coupling array theory, in integral equation form, had been applied previously to low energy electron-hydrogen atom scattering, with reasonable success. The conference report by Levin contains some results for the same scattering problem obtained with the coupled differential equations above; the results were neither much better nor much worse than those obtained by more standard approaches. The report also contains some bound state calculations; the bound state area is one to which the theory has only recently been applied. The first bound state result is the binding energy of H^-. In my opinion, the result is not very impressive. On the other hand, the results of the determination of the energy E(R) of an electron in the field of two protons fixed at a separation R, and of the equilibrium position, are quite impressive. These results are obtained from the coupled equations given above, with H_1, H_2, V_1 and V_2 redefined. The details are in Levin's report.

Not many results have been obtained thus far, and one cannot yet pinpoint the problems for which the approach can be expected to be particularly useful. While waiting for further results, it may be worthwhile to look at ψ_1 and ψ_2 in slightly more detail. Levin and his co-workers interpret ψ_1 and ψ_2 for scattering and bound state problems in terms of the asymptotic behavior in different channels. With the understanding that I have had no time to consider this particular point, let me consider a slightly different viewpoint for bound state problems. Considering again two electrons in the field of a proton, in the spatially symmetric ground state, we begin by noting that $\psi_1 = \psi_2 = \frac{1}{2}\Psi$ does not represent a solution of the coupled equations. Rather, since $\Psi(1,2) = \Psi(2,1)$, it follows that ψ_1 and ψ_2 can be written as

$$\psi_1 = \frac{1}{2}\Psi + \Omega ,$$
$$\psi_2 = \frac{1}{2}\Psi - \Omega ,$$

where $\Omega(1,2) = -\Omega(2,1)$. In assessing the merits of the approach, it should be brone in mind that it requires the estimation not only of Ψ but of Ω. Only if the additional effort involved in estimating Ω is more than compensated for by the accuracy achieved can the approach be considered useful. Hopefully, the Ω might somehow build in some of the more relevant dynamics. On inserting the expressions for ψ_1 and ψ_2 in terms of Ψ and Ω into the coupled equations, and subtracting one from the other, Ω is found to satisfy

$$\left[K_1 + K_2 - (e^2/r_{12}) - E\right]\Omega(1,2) = \left[(e^2/r_2) - (e^2/r_1)\right]\Psi .$$

This equation may possibly help in the interpretation of the physical significance of $\Omega(1,2)$.

Note too that the coupled equations can be cast in matrix form, $(\underset{\sim}{H} - E\underset{\sim}{1})\vec{\Psi} = 0$, where the column vector $\vec{\Psi}$ has elements ψ_1 and ψ_2, and where the matrix $\underset{\sim}{H}$ is not Hermitean. This need not be a source of real difficulty, but one must formulate the calculational procedure appropriately if the variational principles for the energy and for scattering parameters are to be preserved.

Apart from its possible numerical use, the theory offers an alternative approach in the formal introduction of exchange effects. Tobocoman recently used the approach, in conjunction with the Feshbach projection operator technique, to obtain exchange effects in the theory of radiative decay.

Many aspects of scattering theory are being reexamined within the context of the channel coupling array theory. In a paper at the conference, Kouri and Goldflam derived a new form for the optical potential $\tilde{U}_\alpha$ for a model problem in which only two arrangements, α and β, are present. Thus, the leading term in $\tilde{U}_\alpha$ is given by $PV_\alpha G_\beta^+ V_\beta P$, where G_β is a Green's function and P is a projection operator onto the elastic scattering state; this is as opposed to the leading term $PV_\alpha P$ in Feshbach's approach. The new form of the leading term thus has more built into it; in particular, it is energy dependent and implies the opening of the second channel at the appropriate energy. Many numerical calculations will have to be made before one will know the advantages which can be extracted by utilizing the freedom available in the channel coupling array approach, but the approach does allow large numbers of (equivalent) starting points, and different starting points suggest and allow different approximations with different resultant accuracy. Thus, for example, the above form $\tilde{U}_\alpha$ gives a connection between channels α and β, a connection more difficult to obtain in the Feshbach approach.

CHARGE EXCHANGE AT HIGH INCIDENT ENERGY

The high energy limit of the proton-hydrogen atom charge exchange process in the forward direction was first studied some 50 years ago (Thomas [4]). It is now known that at a sufficiently high incident velocity the two-collision contribution to the reaction $p+H \rightarrow H+p$ dominates over the one-collision contribution. That is an astonishing result. The argument that leads to that result is fairly complicated and no attempt will be made to reproduce it, but a few comments may be helpful. Charge transfer (or capture) can be expected to take place only if the electron emerges from the target with a velocity comparable to the high velocity $\vec{v}$ of the incident proton. Crudely speaking, in a simple high-energy collision between the (heavy) incident proton and the (light) target electron, the velocities will generally be comparable after the collision only if they are at least roughly comparable before the collision; in any event, the electron velocity before the collision will tend to be very high. For this to be the case, we must invoke the initial velocity distribution of the electron in the target. The probability of finding a high velocity component for an electron in the 1s state of hydrogen is very small, and cuts down the one-collision contribution to the cross-section; using the (quantum mechanical) first Born approximation, Brinkman and Kramers [5] showed that the contribution went as v^{-12}. The two-collision contribution is reduced, relative to the one-collision contribution, by the need of a second interaction, but in the two-collision process there is the saving feature that the target electron need not have a velocity close to $\vec{v}$. Rather, in Rutherford scattering with the incident (undeflected) proton, the electron, which can now be taken to be at rest initially, recoils toward the target proton where it undergoes a second Rutherford scattering and emerges with a velocity close to that of the proton. Using semi-classical non-relativistic arguments, Thomas [4] showed that the two-collision contribution to the cross-section behaves as v^{-11}, and will therefore dominate over the one-collision for v sufficiently large.

Correspondingly, in a non-relativistic quantum treatment of the process (Drisko [6]), the second Born term dominates over the first Born term. (These results are more than intriguing; they can throw light on the high-energy behavior of the Green's function.) It has also been shown that the second Born term dominates over any higher order Born term (Dettmann [7]), but it is by no means clear that the Born series converges. In the impact parameter approximation, and for short-ranged interactions, it has been shown (Shakeshaft [8]) that the asymptotic form of the cross-section is indeed determined by the second Born term. (In the impact parameter approach, the protons are treated classically, with their orbits taken to be straight line motion at constant velocity. The potential seen by the electron is then time-dependent, and the amplitude for charge transfer is characterized by a transition amplitude. The proof that the second Born term dominates is then based on the use of variational upper and lower bounds for transition amplitudes (Spruch [9]) which are applicable to almost any time-dependent process.)

Though of considerable intrinsic interest nevertheless, the above discussion is not of practical interest for two quite different reasons. Firstly, the second Born term does not dominate the first until the incident velocity is in the relativistic domain, but the analysis presumes the validity of non-relativistic theory. Secondly, the second Born term for radiationless charge exchange is itself dominated by the radiative charge exchange process, $p+H \rightarrow H+p+h\nu$ (Schnopper [10]; Briggs [11]).

It may seem surprising that the radiative charge exchange process can dominate over the radiationless charge exchange process; if a process can proceed without radiation (as is the case for charge exchange) the corresponding process with radiation will normally have a cross-section smaller by roughly a factor of α. We can understand the dominance of the radiative process in the present instance as follows (Shakeshaft [12]). In the radiative process, the target proton to which the electron is initially attached generates a momentum distribution of the electron which gives a Compton profile to the cross-section, but has almost no effect on the total cross-section; for an incident velocity large compared to the characteristic internal velocity of the electron, αc, the radiative process can therefore be thought of as the radiative capture of a free electron by the incident proton. For the radiationless process, on the other hand, the target nucleus plays an essential role, since a free electron cannot undergo radiationless capture. The radiationless process is therefore rather more complicated than the radiative one, and has a smaller cross-section.

Experimental data on the radiationless process was recently obtained (Schnopper [10]), and the process was analyzed theoretically (Briggs [11]). Since the photon cannot carry off much momentum, radiative capture of the effectively free electron will contain as a matrix element the Fourier transform $\phi_f(-m\vec{v}/\hbar)$ of the final hydrogenic bound state, where $-\vec{v}$ is the incident velocity. As noted above, the high velocity component of a hydrogenic state is small, and the radiative capture cross-section will therefore be small. However, putting in the density of states, it goes as v^{-5}, and though it also has a factor c^{-3}, it will therefore ultimately dominate; it dominates, in fact, when the incident proton has an energy of 9 MeV.

We should at least note that there are processes other than charge exchange for which the first Born term does not give the asymptotic form of the scattering amplitude. These include, for example, excitation of the 2s state in the large momentum transfer limit (Joachain [13]).

HYDROGEN ATOMS IN INTENSE MAGNETIC FIELDS

The subject of atoms in weak magnetic fields has played a significant role in the history of physics, but magnetic fields (B-fields) which are so strong that they cannot be treated as perturbations of the Coulomb interactions present represent a new, exotic and fascinating subject. The maximum steady $\vec{B}$-field presently attainable, of the order of 10^5 Gauss, can represent such a $\vec{B}$-field in some solids, where a large dielectric constant and an effective electron mass different from the mass of the free electron can cut down electric relative to magnetic forces. The $\vec{B}$-fields in the atmospheres of some white dwarfs, and of neutron stars, of the order of 10^7 and 10^{12} Gauss, respectively, also represent such $\vec{B}$-fields. Indeed, for the latter case, that is, for a uniform $\vec{B}$-field which we take to be parallel to the z-axis, and which is of the order of 10^{12}G, the motion of the electron (s) of a light atom in the x-y (or, equivalently, $\rho-\phi$) plane is effectively determined by the $\vec{B}$-field. The angular momentum projection along the z-axis, a constant of the motion, will be $m\hbar$, with $m = \ldots -1,0,1, \ldots$. Landau showed a long time ago that the ground state is infinitely degenerate, with $m = 0,-1,-2, \ldots$, and with the quantum number for motion along ρ equal to zero. The ground state energy is $\hbar^2/(\mu\hat{\rho}^2)$, where μ is the electron mass and where $\hat{\rho}^2 = 2\hbar c/(eB)$, and the probability as a function of ρ is sharply peaked about the value $\hat{\rho}_m \equiv (|m|+1/2)^{1/2}\hat{\rho}$. With a_0 the Bohr radius, and with $B \simeq 10^{12}$ G, one finds $\hat{\rho} \simeq a_0/20$, and a lowest excitation energy of order 20 KeV; excitation by the weak atomic Coulomb fields can therefore be ignored. Motion of the electrons parallel to the z-axis is determined by the Coulomb interactions.

We choose our zero energy reference level to be that of the free electron in the (intense) $\vec{B}$-field, and seek the energy E_m and normalized wave function of a hydrogen atom in the $\vec{B}$-field. (For hydrogen-like ions, we need merely replace e^2 by Ze^2 and a_0 by a_0/Z, but we must have the e in $\hat{\rho}$ untouched.)

For $\vec{B}$ large enough, the three-dimensional problem reduces to the one-dimensional problem defined by

$$[T_z + V_m(z) - E_m]\,\Psi_m(z) = 0 ,$$

where $V_m(z)$ is the weighted average of $(-e^2/r)$ with respect to the Landau probability density for motion in the $\rho - \phi$ plane. A very good approximation for $V_m(z)$, especially for large $|m|$, is

$$V_m(z) = -e^2/(\hat{\rho}_m^2 + z^2)^{1/2} .$$

E_m cannot be determined analytically.

A rough approximation $V_m(z)$ is provided by

$$\tilde{V}_m(z) \equiv -e^2/(\hat{\rho}_m + |z|) .$$

(Both $V_m(z)$ and $\tilde{V}_m(z)$ represent regularized one-dimensional Coulomb potentials, potentials which are not singular at the origin but which differ very little from the one-dimensional Coulomb potential for $|z|$ of order a_0.) For $B \sim \infty$, it is known (Loudon [14]) that the associated ground state energy E_m, for the given m, is

$$\tilde{E}_m \sim E_{\text{deep } m} \equiv -(e^2/2a_0)4\ln^2(a_0/\hat{\rho}_m).$$

The corresponding normalized ground state wave function, $\tilde{\Psi}_m(z)$, is

$$\tilde{\Psi}_m(z) = (\tilde{K}_m)^{1/2} \exp(-\tilde{K}_m|z|),\quad \tilde{K}_m \equiv (2/a_0)\ln(a_0/\hat{\rho}_m) .$$

The energy lies much deeper than $(-e^2/2a_0)$ because the electron occupies roughly a cylindrical region of radius $\hat{\rho}_m$ and length a_0, spending thereby much more of its time near the origin than does the electron in the laboratory hydrogen atom. (It does not "cost" any kinetic energy to confine the electron to $\rho \lesssim \hat{\rho}_m$; the large kinetic energy associated with that confinement defines the zero energy reference level.)

The result for $\tilde{E}_m$ was obtained by solving for $\tilde{\Psi}_m(z)$ as an infinite power series and then letting $B \sim \infty$. A much prettier derivation can be given (Herrick [15]). Since $V(z)$ represents a regularized one-dimensional Coulomb interaction for which the (symmetric) ground state wavefunction $\Psi(z)$ is expected to have dimension ka_0, where k is of order unity, we can approximate V(z) by a delta-function and thereby reduce the problem to a trivial one. Thus, we can write

$$\int_{-\infty}^{\infty}\Psi^2(z)V(z)dz \simeq 2\int_0^{ka_0}\Psi^2(z)V(z)dz$$

$$\simeq 2\,\Psi^2(0)\int_0^{ka_0}V(z)dz = -2C\int_{-\infty}^{\infty}\Psi^2(z)\,\delta(z)dz,$$

where we wrote $\Psi^2(z) = \Psi^2(0) + \{\Psi^2(z) - \Psi^2(0)\}$ and used the fact that the contribution involving $\{\ \}$ is of lower order since $\{\ \}$ vanishes at the origin and thereby eliminates the near-singularity of V(z). Note that

$$C \equiv -\int_0^{ka_0} V(z)dz$$

is positive. Thus, for $B \sim \infty$, we can approximate V(z) by $-2C\delta(z)$. But for

$$[T_z - 2C\delta(z) - E]\psi(z) = 0$$

we readily find, with $K \equiv 2\mu C/\hbar^2$,

$$E = -(2\mu/\hbar^2)C^2, \quad \psi(z) = K^{1/2}\exp(-K|z|).$$

For V(z) equal to $V_m(z)$ or $\tilde{V}_m(z)$, one immediately finds the leading term for both Em and $\tilde{E}_m$ to be $E_{deep,m}$. (k appears only as a factor in the argument of a log term and can be dropped.) $\psi_m(z)$ and $\tilde{\psi}_m(z)$ also follow immediately.

The true value of E_m for small m and for $B \simeq 10^{12}$ G is somewhat under 200 eV: $E_{deep,m}$ is off by the order of 100%. The result $E_{deep,m}$ is therefore largely academic, since it becomes valid only for B so large that the electron should be described relativistically. The approach might sometimes be useful nevertheless; one could use $\psi(z)$, but with K as a variational parameter, as a trial function in a Rayleigh-Ritz calculation. Further, the binding energy of H^- is known to be about 15 eV; the two electrons in H^- partially shield one another from the proton, and the validity of the approximation

$$[T_{z1} + T_{z2} + 2C\{\delta(z_1-z_2) - \delta(z_1) - \delta(z_2)\} - E]\,\psi(z_1,z_2) = 0$$

should begin at a lower value of B than the corresponding approximation for H. The solution $\psi(z_1,z_2)$, modified to include a variational parameter, could serve as a useful trial function. A nicer example is that of an electron in the field of two protons at a separation R. One should easily be able to determine the energy E(R) of the electron, and the equilibrium separation R_{eq}, analytically. An extension of the above approach that could be used to extract the dominant energy dependence for problems in three and higher dimensions would be extremely useful, but it isn't clear how to proceed.

We note without proof that, as opposed to the laboratory situation, the binding energy of two H atoms in an intense $\vec{B}$-field far exceeds the sum of the binding energies of two isolated H-atoms in intense $\vec{B}$-fields (Ruderman [16]). (It is not then surprising that one can obtain long tightly bound chains of H-atoms, and indeed of other atoms, and that, in turn, the chains form tightly bound fibres, but since this would take us out of the few electron domain we will not mention it.)

K-HARMONICS AND THE GROUND STATE ENERGY OF $e^-+e^-+e^+$

The most commonly used technique for the estimation of the ground state energy E of a system is the Rayleigh-Ritz method. One introduces a trial wave function Ψ_t and determines the energy estimate E_t, where

$$E \approx E_t \equiv (\Psi_t, H\Psi_t)/(\Psi_t, \Psi_t)$$

and where H is the full Hamiltonian. We consider two forms for Ψ_t:

$$i)\quad \Psi_t = \sum_{n=1}^{M} c_n \phi_n$$

$$ii)\quad \Psi_t = \sum_{n=1}^{2} x_n(x_i)\, u_n(x_j)$$

In the first form, the ϕ_n are known orthonormal functions, while the constants c_n are to be determined. By demanding that E_t be stationary with respect to variations of the c_n, the problem can be reduced to a set of linear equations for the c_n, equations which can be readily solved.

In the second form, the u_n are known orthonormal functions of some of the internal variables, and the χ_n are unknown functions of the remaining internal variables. By demanding that E_t be stationary with respect to arbitrary variations of the χ_n, one obtains a set of coupled differential equations that define the χ_n and E.

It is by no means clear that the second approach is the better one. True, one determines not simply constants but entire functions, but at the price of solving differential equations. The relative merits of the two approaches can only be judged after one has gained experience by performing a number of calculations.

In an interesting paper submitted to the conference, Roychoudhury, Sural and Roy have estimated the ground state energy E of $e^-+e^-+e^+$ using a version of the second approach. Separating out the centres of mass motion, one remains with six internal coordinates. These can be taken to be a length ρ and five angles, denoted collectively by Ω_ρ. The authors choose the X_i to be the single variable ρ, the X_j to be the Ω_ρ, and the $u_n(\Omega_\rho)$ to be the K harmonic functions used previously (Simonov [17]) in some nuclear physics problems. Introducing a Q by Q matrix $\underset{\sim}{H}$ with elements

$$H_{mn} \equiv (u_m, Hu_n),$$

where the integration is over the Ω_ρ, but not over the ρ, and introducing a vector $\vec{\chi}$ with elements χ_n, we find

$$(H - E\underset{\sim}{1})\vec{\chi} = 0,$$

where $\underset{\sim}{1}$ is the Q by Q unit matrix. Writing H = K+V, where K is a kinetic energy operator, where V is the sum of three coulomb interactions, and where K and V can be obtained explicitly, we can rewrite the coupled differential equations as

$$(\underset{\sim}{K} - E\underset{\sim}{1})\underset{\sim}{\chi} = -\underset{\sim}{V}\underset{\sim}{\chi}.$$

The Green's function matrix G(E) defined by

$$(K - E\underset{\sim}{1})\underset{\sim}{G} = -\underset{\sim}{1}$$

can also be obtained explicitly, and we remain with

$$\vec{\chi} = \underset{\sim}{G}\underset{\sim}{V}\vec{\chi}.$$

These coupled integral equations are solved numerically for the χ_n and for E.

One uses only the "lowest" of the $u_n(\Omega_\rho)$. For Q = 2 and Q = 4, the energy in atomic units is found to be -0.256 and -0.259, respectively. The best available estimate based on a Hylleraas type (that is, one that is a function of the interparticle coordinates) trial function is -0.262. It would be useful to have an estimate of the effort involved in the different approaches, in terms of machine time, for example.

The K-harmonic approach has the disadvantage that convergence in an expansion in angle variables is normally relatively slow composed to an expansion in interparticle coordinates. For four-particle systems, on the other hand, the analytic work for an interparticle coordinate expansion would probably be very much more difficult than for a K-harmonic expansion.

REFERENCES

1. U. Fano, Phys. Rev. 124 (1961) 1866.
2. L. Fulcher and A. Klein, Phys. Rev. D8 (1973) 2455;
   J. Rafelski, B. Müller, and W. Greiner, Nucl. Phys. B68 (1974) 685.
3. See, for example, B. Müller, R.K. Smith and W. Greiner, Atomic Physics 4 (Plenum Press, 1975), edited by zu Putlitz, Weber, and Winnacker,p.209, and W. Meyerhof, Comments Atom . Mol. Phys. 5 (1975) 33.
4. L.H. Thomas, Proc. R. Soc., A114 (1967) 461. See alsl R.S. Mapleton, Theory of Charge Exchange (Wiley, New York, 1972), p.200.
5. H.C. Brinkmann and H.A. Kramers, Proc. Acad. Sci., Amsterdam, 33 (1930) 973.
6. R.M. Drisko, Thesis (Carnegie Institute of Technology, 1955) (Unpublished).
7. K. Dettmann and G. Liebfried, Z. Phys. 218 (1969) 1;
   K. Dettmann, Springer Tracts Mod. Phys. 58 (1971) 119.
8. R. Shakeshaft and L. Spruch, Phys. Rev. A8 (1973) 206.
9. L. Spruch, in Lectures in Theoretical Physics - Atomic Collisions, edited by S. Geltman, K.T. Mahanthappa, and W.E. Brittin (Gordon and Breach, New York, 1969) Vol. XIC, p. 77.
10. H.W. Schnopper et al, Phys. Rev. Lett. 29 (1972) 898;
    P. Kienle et al,Phys. Rev. Lett. 31 (1973) 1099.
11. J.S. Briggs and K. Dettmann, Phys. Rev. Lett 33 (1974) 1123.
12. R. Shakeshaft and L. Spruch, to be published.
13. See, for example, C.J. Joachain and C. Quigg, Rev. Mod. Phys. 46 (1974) 279; and F.W. Byron, Jr., Atomic Physics 4 (Plenum Press, 1975), edited by zu Putlitz, Weber, and Winnacker, p. 337.
14. R. Loudon, Amer. J. of Phys. 27 (1959) 649 and L.K. Haines and D.R. Robert, Amer. J. of Phys. 37 (1969) 1145. For the hydrogen atom, see, for example, Y. Yafet, R.W. Keyes, and E.N. Adams, J. Phys. Chem. Solids 1 (1956) 137 and R. Cohen, J. Lodequai and M, Ruderman, Phys. Rev. Letter 25 (1970) 467.
15. D.R. Herrick and F.H. Stillinger, Phys. Rev. All (1975) 42 and J. Math. Phys. 16 (1975) 1047; D. Maison, Unpublished.
16. M. Ruderman, Ann. Rev. Astr. and Ap. 10 (1972) 427.
17. Y.A. Simonov, Sov. J. Nucl. Phys. 7 (1966) 722.

NUCLEAR REACTIONS AND SCATTERING -- FEW-BODY ASPECTS[†]

F. S. Levin[*]
Physics Department
Brown University
Providence, Rhode Island 02912

1. INTRODUCTION

Compared to the few-nucleon problem, where detailed and accurate numerical calculations based on various representations of the nucleon-nucleon interaction can be carried out, the field of direct nuclear reaction theory may appear to be primitive and is empirical. Generally the theoretical analyses are two-body in character, and at best, the many-body nature of nuclear reactions is brought into the analysis (if at all) by coupling together different two-body channels: three-body, four-body, etc., channels are never explicitly considered. In terms of this drastic truncation in the description, direct reaction theory can only be considered as overwhelmingly successful for elastic and inelastic scattering, knockout reactions, and single and double particle transfer reactions. The theory is less good for multi-nucleon transfer.

The reduction of an exact many-body transition amplitude to an approximate form such as is used in analyses of direct reactions involves various assumptions, both theoretical and technical. Furthermore, some of the reactions noted above can be considered, at least in the context of our orientation towards a many-body description, as three-body systems or three-body processes. Thus, the attempt to use three-body methods, or their many-body extensions, as a means for justifying approximations and understanding the physics of direct reactions is an obvious step, one actually begun nearly 15 years ago. The successes of direct reaction theory make this a compelling procedure to undertake.

There are two overlapping categories into which research efforts can be placed. One is the use of three-body models, either as testing grounds for approximations, or as models sufficiently analogous to the many-body case that some of the physics of the latter can be deduced from the former. Naturally, these two aspects can overlap. The second category is the use of three-body methods or their many-body generalizations as applied to realistic nuclear reactions. An example is the development of connected-kernel equations to use as a basis for going beyond the distorted wave Born approximation and including multistep processes.

Excluding the actual three-nucleon problem, there have been many studies which fall into one or both of the above categories, some of which include work on deuteron stripping and elastic scattering [1-3], knock-out processes [4], resonance reactions [5], and the optical model [6]. My own opinion is that much less understanding of direct reaction theory has resulted than is possible, and that considerable progress is yet to be made. Suggestions for this are given later.

The purpose of session VIII of this conference was to present and discuss recent work in the area of few-body aspects of nuclear reactions and scattering. In particular it was hoped that all the papers for this session would fall into one of the above two categories. This hope was not realized. Accordingly, only two of the contributed papers (those of Mukherjee, and Mukherjee and Roy) were selected for presentation along with the invited papers. For some of these latter papers, their relation to the title of the session was marginal at best.

The task of rapporteur combines that of a guide and critic. In the present case, it seemed desirable to provide a map of some of the area through which I will be acting as a guide. To this end, the next part of my presentation is an idiosyncratic summary of some aspects of direct reaction theory, into which the papers of session VIII can then be fitted, sometimes as round pegs in triangular holes. In addition, a few recent papers outside the scope of this session but inside the scope of the few nucleon problem and thus, presumably, of interest to few-body physicists, are also mentioned. No attempt is made to be complete, either in reviewing the direct reaction literature (indeed, much is omitted) or in discussing the papers of session VIII.

2. SELECTED REVIEW OF DIRECT REACTION THEORY

Most of the experiments of the direct reaction type have been carried out at lab energies of less than 50 MeV per particle in the projectile; the Q values are usually in the range $|Q| \lesssim 20$ MeV. Typical experiments are (p,p'), (α,α'), (d,p), (p,n), (p,2p), (^{3}He,p), ($\bar{p}$,α), (^{15}N,^{14}N), etc. Inelastic scattering and single particle transfer are the best understood, and in general angular distributions are better fitted than polarizations.

Apart from the (p,2p) case direct reactions involve only two-body initial and final channels; symbolically, we write

$$a + A \rightarrow B + b \quad , \tag{1}$$

with A and B the target and residual nuclei, and a the projectile. For a transfer reaction, a may be considered to be formed of b + y, so that B = (A + y), the parentheses indicating a bound state.

In terms of cm relative motion wave vectors $\vec{k}_i$ and $\vec{k}_f$, the reaction is

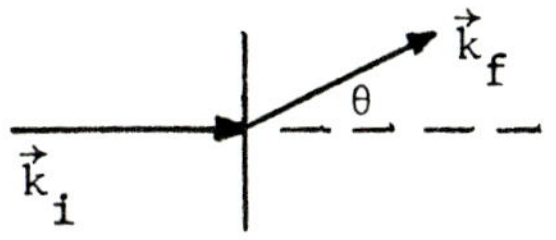

where θ is the cm scattering angle. The exact transition amplitude A for (1) is

$$A = \langle \vec{k}_f \Phi_b \Phi_B | V_{bB} | \Psi_{\vec{k}_i} \rangle \quad , \tag{2}$$

where Φ_j is the bound state of particle j, V_{bB} is the full interaction between b and B, and $\Psi_{\vec{k}_i}$ is full Schrödinger wave function generated by a incident on A.

Since $\Psi_{\vec{k}_i}$ is the full many-body wave function it is at present beyond calculation. Hence, A can only be evaluated approximately. The major thrust of all direct reaction theories is to try to determine an approximation to $\Psi_{\vec{k}_i}$ (after transforming the $\langle \vec{k}_f | V_{bB}$ term in (2)), such that the important contributions to the matrix element A are also given by the approximation. Since Φ_b and V_{bB} are both short-range, the important contributions to A come from the region of configuration space near to B. Because of this, the asymptotic portion of $\Psi_{\vec{k}_i}$ is not important, in particular that region of configuration space corresponding to breakup states (either of b or B).

The procedures for approximating $\Psi_{\vec{k}_i}$ are empirical. That is, one introduces a set of approximations that lead from (2) to an amplitude of the desired form, but no real justification for most of the approximations used has ever been given apart from physical arguments and success. The most widely used, simple, approximate amplitude is that of the distorted wave (DW) approximation [7,8];

other simple approximations for (d,p) reactions which do not employ deuteron optical potentials have been used [9], but a detailed account of them will not be given. In the DW approximation for reaction (1), A is replaced by A^{DWB}:

$$A \rightarrow A^{DWB} = \langle X_f \Phi_b \Phi_B | \hat{V}_f | \Phi_A \Phi_a X_i \rangle \quad , \tag{3}$$

where $\hat{V}_f$ is a portion of V_{bB} and the X's are states in one-body ("optical") absorptive potentials representing average effects of the many-body a + A and b + B interactions. The B in DWB denotes "Born" and refers to the fact that V_f occurs only once in A^{DWB}.

The physical argument underlying (3) is that a direct reaction such as (1) involves very few nuclear degrees of freedom, essentially proceeding by directly attaching y to the ground state of A to form B in its final state. Schematically A^{DWB} describes (1) by

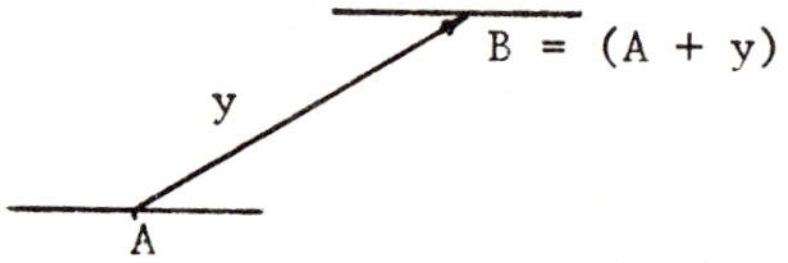

All competing channel processes are subsumed in the absorptive part of the potentials generating the X's, which in the standard DWBA yield asymptotically the elastic a + A or b + B phase shifts at the energies of interest.

Deuteron stripping is the best known example of a direct reaction. In standard DWBA, the average d + A and p + B interactions are the elastic (one-body) optical potentials generating d + A and p + B elastic scattering phase shifts. Almost all attempts to improve DWBA within the framework of the one-step amplitude typified by (3) have been based on replacements of the deuteron elastic optical model wave function, thus altering the model. One of these, the adiabatic model [10], is discussed below.

A key feature in the DWBA analysis of (d,p) reactions follows from the (usual) choice $\hat{V}_f = V_{np}$, where V_{np} is the deuteron binding interaction. With this choice the nuclear overlap in (3) can be immediately evaluated to give $\langle \Phi_B | V_{np} | \Phi_A \rangle = S^{\frac{1}{2}} \langle \varphi_n | V_{np}$, where φ_n is taken to be the single particle neutron state and the spectroscopic factor S is a measure of the strength with which the ground state of A, Φ_A, is present in Φ_B. Alternately, S is a measure of the degree to which Φ_B is a single-particle state. All nuclear structure information is contained in S. Stripping angular distributions based on DWBA are of the form $d\sigma/d\Omega = S\sigma(\theta)$, so that the magnitude of the experimental cross section determines S, if $\sigma(\theta)$ can be calculated accurately enough.

The DWBA for deuteron stripping has been very successful. The overall features of experimental angular distributions have been well fitted by theory, and quite often values of S are extracted that are believed to be reliable to 20% or better. Empirically, DWBA works best for single particle transfer reactions, and becomes less reliable as the number of transferred particles in y increases. A popularly held and theoretically attractive notion is that multistep processes, in which successive transfers occur, are important competing processes. A simple reaction having a multistep channel that contributes coherently with the one-step process is (^{3}He,p). The one-step process describes direct attachment of a deuteron onto the target A forming the final nucleus B = (A + d). The reaction may also occur in two steps, with A first gaining a proton, the intermediate light projectile being a deuteron; a neutron is subsequently stripped from the deuteron onto the intermediate nucleus C = (A + p) to form B = (C + n), with the final proton emerging. Theoretical means for including the effects of multistep processes exist [11,12]; however, the procedures suffer from a lack of firm theoretical foundation [13]. A similar remark holds as well for the theoretical procedure for including effects of excitation of the initial or final nuclei. Three-particle final states can also be treated using DW methods, as for example in the (p,2p) reaction. Here, the amplitude is given by a matrix element similar to (3), except that $\hat{V}_f$ is replaced by two-body t matrices. A recent discussion of the validity of this approach to (p,2p) reactions, within the framework of the DW method, is given by Redish and Young [4]. An interesting and successful application of this approach has been to the study of atomic wave functions using (e,2e) reactions by McCarthy and collaborators [15].

It will be noticed that deuteron-nucleus processes play a special role in the DW method, being the best "understood" theoretically and the best fitted experimentally. Only this case will be considered herein, partly due to space limitations and partly because 3-body methods have mainly been applied to this case.

An intermediate approximation between (3) and (2) for the (d,p) case is

$$A \simeq M = \langle X_f \Phi_p \Phi_B | V_{np} | \Phi_A \psi_d \rangle \quad , \tag{4}$$

where the replacement $\hat{V}_f = V_{np}$ has been made, Φ_p is the final proton spinor (for simplicity, spin-orbit forces are neglected in the potential generating X_f), and the approximation $\Psi_{\vec{k}_i} \simeq \Phi_A \psi_d$ has been used in the matrix element. Here, ψ_d is the state describing motion of the deuteron in the presence of the target nucleus A.

It should be clear that ψ_d is a true three-body wave function, and that with approximation (4), the opportunity to include three-body effects is manifest. One may well argue that for a direct reaction, only the ground state of the target is important (we assume no strong inelastic scattering in contrast for example to the case for collective 2^+ or 3^- excited-target states), but it is far from obvious that in ψ_d, only the deuteron elastic portion $\Phi_a X_i$ which occurs in (3) is important. In particular, even though the breakup or stripping parts of ψ_d may be unimportant asymptotically, they could and probably do alter the short-range behavior of ψ_d away from $\Phi_d X_i$ (Φ_d is the deuteron ground state wave function). Equation (4) has been the basis for various new approaches to the one-step direct reaction amplitude, some of which are noted below. One consequence has been a departure from the standard input of the elastic deuteron distorted wave state X_i (as occurs in the DW matrix element), thus providing concrete examples of the remark that $A \simeq A^{DWB}$ and approximate ways of evaluating A^{DWB} are intimately connected.

The two alterations to Eq. (3) that we consider in detail are inclusion of the deuteron D-state in A^{DWB} (finite-range effect) and the adiabatic approximation to (4), which leads to a form of Eq. (3) in which X_i is no longer the "elastic" distorted wave. A few other attempts at improving A^{DWB} will also be commented on but not in detail.

For the (d,p) case, $\hat{V}_f = V_{np}$ allows the $\langle \Phi_B | \hat{V}_f | \Phi_A \rangle$ overlap to be evaluated without recourse to a detailed knowledge of the nuclear wave functions, leading to

$$A^{DWB} = S^{1/2} \langle X_f \varphi_n | V_{np} | \Phi_d X_i \rangle \quad . \tag{5}$$

Suppressing the magnetic quantum number, Φ_d can be written in the usual way as a linear combination of S and D states:

$$r\Phi_d = u(r) Y_{011} + w(r) Y_{211} \quad , \tag{6}$$

where Y_{LSJ} is the usual vector spherical harmonic and u and w are the radial S and D state wave functions. Since $w(0) = 0$, then use of the zero-range approximation in (5) means that the D state w makes no contribution to A^{DWB}. Because the zero-range approximation greatly reduces the complexity of the matrix element in (5), it was widely used in evaluating A^{DWB} [7,8], with apparently little consequence for angular distribution analyses. Systematic investigations of the effects of w were initiated by Johnson [16] and Johnson and Santos [17], who showed that some improvement to certain angular distribution data could be achieved by retaining w in finite-range calculations. However, spectacular improvements were obtained to fits of the data on angular distributions of the tensor polarizations $T_{2q}(\theta)$, as discussed in detail for example by Johnson et al. [18], and in many other works as well. This is a direct manifestation of the D state, since calculations of the vector polarization $T_{11}(\theta)$ show relatively little sensitivity to w. Instead, $T_{11}(\theta)$ is sensitive to the presence of spin-orbit interactions in the optical potentials.

These results on $T_{20}(\theta)$ and $T_{00}(\theta)$ lead to a remarkable possibility, originally communicated to me by R.C. Johnson: studies of these tensor polarizations (or tensor analyzing powers) provide a nuclear reaction laboratory for probing the structure of the deuteron! If the reaction theory analysis is sufficiently reliable, then new or at least confirmatory information on deuteron properties can be obtained from the polarization data. For example, the normalization constant for w is roughly proportional to the deuteron quadrupole moment Q_d, while at low momentum transfer (not too large values of k_i and k_f), w(k) varies linearly with k, $w(k) \propto Q_d k$, so that the D state contribution to A^{DWB} is proportional to Q_d and, roughly, an integral involving only linear values of momentum transfer. But at larger values of momentum transfer one departs from the linear dependence of w on k into new regions. Not only are these interesting regions to probe (in momentum space), but for some models of the nucleon-nucleon interaction, the resulting D state may be larger than the S state at these values of k [19], thus giving rise to amplitudes whose properties will surely be different than those for the low momentum case. This will be true for angular distributions. Such possibilities should be an additional stimulus to new work seeking to understand the validity of the distorted wave approach.

Of presumably equal interest to participants at this conference is the extremely interesting, logical extension of these ideas to the triton. Knutsen et al. [20] have recently reported on measurements and analysis of tensor analyzing powers observed in (d,t) experiments using the University of Wisconsin polarized deuteron beam. In this case, T_{2q} is sensitive to the triton D state, and the results are consistent with present theoretical estimates of the strength of the D state. Just as with the (d,p) reaction though, one has the hope of using the (d,t) reaction at higher energies to study the higher k (non-linear) portion of the fourier transform of the triton D state. The prospect of obtaining new information about the deuteron and the triton, rather than about nuclear structure, through nuclear reactions, is an exciting development in a field mostly known for its taxonomic aspects (i.e., the use of direct reactions to extract nuclear structure information).

From these remarks we next turn to recent developments based on Eq. (4). Two similar approaches which abandon the use of deuteron optical potentials in favor of a momentum space description of individual neutron and proton motion (weighted with the fourier transform of Φ_d) have been proposed and discussed by Pearson and collaborators [9] and by Butler and collaborators [9]. The latter approach (referred to as the BHMM method) is of special interest here because it has been studied in a three-body model by McKellar, results of which are included in his invited paper for session VIII, which we comment on below. A note worthy result of the BHMM method is the replacement of the spectroscopic amplitude $S^{\frac{1}{2}}$ occurring in Eq. (5) by the factor $[S^{\frac{1}{2}}/(1 - S)]$. Extraction of S from experimental data using the BHMM method generally leads to smaller values than with DWBA. In particular, the BHMM method must be re-examined when pure single-particle states (S = 1) are considered, since the factor $(1 - S) \to 0$ for $S \to 1$. However, the BHMM matrix element, when calculated exactly and not using the BHMM approximations also goes to zero in such a way that the amplitude is non-zero and finite [9].

Although BHMM provides reasonable fits to angular distributions, it has not been widely used. One possible reason for this, apart from the predominance of DWBA codes in the hands or machines of nuclear reaction experimentalists, is the unsettled question of the missing $f_{7/2}$ single particle strength for the $^{40}Ca(d,p)$ ^{41}Ca reaction: BHMM predict roughly 20-40% of the strength to be in the continuum, although no significant concentration of $f_{7/2}$ strength there has ever been observed. It is also assumed, of course, rightly or wrongly, that with so much strength to be accounted for in the continuum, a significant portion of it would show up as a pronounced bump in $^{40}Ca(n,n)^{41}Ca$, rather than be spread out so widely as to be essentially unobservable. We return to this point below in our discussion of McKellar's paper.

Of much greater use than the methods noted above is the adiabatic model approximation to ψ_d of Eq. (4), introduced by Johnson and Soper [10]. The basic idea is to replace the three-body state ψ_d by the two-body expression $\Phi_d F$, where, unlike X_i, F is explicitly constructed so as to contain contributions from the breakup continuum. It is here that the contrast between the methods familiar to physicists active in the few-body problem and those used by nuclear reaction theorists is greatest. The procedure followed not only in the adiabatic model but also be workers seeking to improve it [21] is to expand ψ_d in a complete set of deuteron states:

$$\psi_d = \Phi_d X_i + \int d^3q \; \Phi_{\vec{q}} X_{\vec{q}} \quad , \tag{7}$$

where $\Phi_{\vec{q}}$ is an interacting n-p continuum state and $X_{\vec{q}}$ is a scattering coefficient. In the adiabatic model, it is assumed that only those $X_{\vec{q}}$ corresponding to low relative energies are important; one eventually [10] is lead to an equation for F in $\psi_d \simeq \Phi_d F$ (F depends only on the CM coordinate $2\vec{R} = \vec{r}_n + \vec{r}_p$) containing a local, absorptive potential which significantly differs from the usual elastic optical potential. Thus, in the Watanabe model for elastic deuteron scattering [10], the potential is given by $\langle\Phi_d|U_n + U_p|\Phi_d\rangle$ with U_i the optical potential for particle i. In the adiabatic model, F is generated by a potential proportional to $\langle V_{np}|U_n + U_p|\Phi_d\rangle$, and F is approximated by $\langle V_{np}|\psi_d\rangle$, where the implied integrations are over relative n-p coordinates. This latter potential contains breakup effects.

The adiabatic-model approximations finesse the complications arising from the presence of the integral term in (7). Thus, an inherently difficult starting point, and one scarcely used in the three-body problem, is never explicitly examined, the idea again being to improve an approximation to ψ_d in the region where $V_{np} \neq 0$, rather than to calculate breakup amplitudes as well as the two-body amplitudes as in conventional three-body methods. We note here that in a three-body, S-wave interaction model, replacement of ψ_d by $\Phi_d F$ is <u>exact</u> in the (d,p) matrix element [1]. Because of this, one can compare the effects of the approximation $\psi_d \simeq \Phi_d X_i$, where X_i is the exact elastic scattering function for the model, with $\psi_d \to \Phi_d F$; i.e., compare an "exact" DWBA with an exact two-body replacement for ψ_d. This was done by Bouldin and Levin, who examined amplitudes, angular distributions, and also compared the radial dependence of F and X_i for various energies [1]. The conclusions were that while DWBA was a good approximation globally, the peak cross sections were such that $(d\sigma/d\Omega)_{DW} < (d\sigma/d\Omega)_{exact}$, and hence in many-body calculations one should use adiabatic rather than DW.

The adiabatic model has led to significant improvements in fits to angular distribution data and has also provided an argument for using non-elastic optical potentials. Subsequent investigations using discretized approximations to the integral term in (7) have attempted to justify the adiabatic model [21]. The consensus is that there are important effects which would invalidate the model were there not cancellations. Hence, the adiabatic model seems to work by accident. An alternate approach to understanding the model has been proposed by Johnson and Tandy [22], though detailed calculations remain to be done.

This brief review of the distorted wave method has been explored in detail for the best known case, the (d,p) reaction. We have tried to emphasize the motives for using the DW type of approximation: to approximate, as well as is feasible,in terms of two-body states, the many-body wave function in the spatial region which makes the most significant contribution to the matrix element and not to take into account the correct asymptotic behavior. So far, the method has been successful beyond understanding or reasonable expectations, at least for certain kinds of reactions and measurements as noted earlier. To date, and including the contributions to session VIII, three-body models and methods have not led to much fundamental understanding of these successes. They remain largely empirical. The model problems that have been solved are indeed physically interesting but do not yet bridge the gap to justification of the many-body approximations. Some possibilities for doing this are noted in the next section where brief comments on some of the papers of session VIII are given. There are other possibilities as well, but these involve attempted solutions of the three-body problem with interactions that are both local and absorptive. So far there has been no rush to undertake such calculations.

3. A BRIEF GUIDE TO SALIENT FEATURES OF PAPERS PRESENTED IN SESSION VIII

My purpose, where possible, is to place the papers of VIII into the context of the preceding remarks. Not all papers are discussed below. For example, the work of Komarov, though presented in VIII and commented on in the oral version of my rapporteur's discussion, has been published with the papers of session II, and at the advice of the editors, my remarks on this paper have been deleted from this written version.

Three of the papers in VIII are related to the adiabatic model. MUKHERJEE and ROY present an alternative to this model (though one seemingly not too dissimilar), and find good fits to data. The agreement is no worse than with the adiabatic model, and in some cases is better. However, calculations of breakup are not in satisfactory agreement with experiment. No physical arguments are given indicating how their method differs from the adiabatic or why their method should be successful. MUKHERJEE uses the coupled (d,d) and (d,p) equations based on the non-orthogonal basis expansion pioneered by Rawitscher [23]; the agreement with experiment is good, although (as in the past) no justification for the validity of the approximations is attempted. Also discussed is stripping to the continuum (a special form of breakup) via DWBA. Neither of these papers deals with the subject of session VIII, although they are interesting because they present alternatives to the adiabatic model, and this model could be investigated using a three-body approach (see the reference to application of the method of Tandy et al. below). The third paper, however, that of NAMYSLOWSKI, is germane. He investigates a three-body model of heavy-ion processes (2 heavy cores and a single nucleon) with local, non-absorptive potentials, which in the realistic cases of interest will be coulomb potentials. For these cases, the simplification provided by separable potentials is not available, and so an extremely interesting although mathematically non-trivial approximation scheme is used. First breakup effects are included using an adiabatic model. Second, the equations of this model are linearized. Third, an eikonal correction is introduced, the advantage here being lack of partial waves. As an alternate scheme to DWBA for heavy-ion reactions, and one that includes some breakup effects without partial-waving, the method, if the calculations are not prohibitively complicated, would be a welcome one.

Three other papers deal directly with the topic of session VIII. MCKELLAR discusses three main ideas. First is a method for treating the wave functions of an energy dependent potential. Second is a separable potential model in which the target has excited states. It is here that he argues, that for this model, there is an unexpectedly large portion of the spectroscopic strength in the continuum. This conclusion violates intuition but supports the BHMM ^{40}Ca result discussed in section 2. Since the argument is based on analytic approximations, it would seem to be well worth verifying numerically. (This is of course one important use to which models can be put.) The third topic is a study of (d,p) reactions for a three-body model based on the preceding excited-core separable potential. Neither DWBA nor BHMM are found to agree with the exact angular distribution at angles >50° for 15 MeV $\leq$ E $\leq$ 45 MeV for this model. Furthermore, S_{BHMM} is energy dependent for E < 5 MeV, just as in the many-body case [8], while for E > 5 MeV, $S_{DWB} \approx S_{BHMM} > S_{exact}$; the relationship $S_{DWB} > S_{exact}$ is consistent with the findings of Bouldin and Levin [1] discussed above that $(d\sigma/d\Omega)_{DWB} < (d\sigma/d\Omega)_{exact}$. It will be interesting to learn whether McKellar's speculation that two-step processes can improve the preceding approximate results is verified or not.

A completely different three-body model is discussed by TANDY, REDISH, and BOLLE. They begin by extending the Watson multiple scattering theory to include identical particle effects through use of two approximations: they ignore all many-body intermediate states and they assume only a single scattering as the input to the integral equation for the many-body transition operator [6]. Keeping only one active nucleon among the A (identical) target particles, they find a pair of coupled integral equations for elastic scattering and pickup that includes both the correct normalization and an interesting (and heretofore neglected) overcounting factor. The knockout (breakup) amplitude is obtainable from that for pickup. This model provides a new and possibly very useful mechanism for exploring quantitatively the role played by various effects such as deuteron breakup or multistep contributions in, e.g., deuteron-induced and other direct reactions, and numerical results will be of considerable interest when they become available.

A more conventional three-body model is used by LEVIN to test a new unitary approximation method developed by Kouri, Levin and Sandhas [24] based on the channel coupling array theory [25]. The model used is that of Mitra [26] and approximate deuteron elastic and stripping calculations are compared with the exact calculations of Bouldin and Levin for this model [1]. Although the approximate (d,d) cross sections are not in good agreement with the exact ones, the exact and approximate higher energy (11.2 and 15.12 MeV) (d,p) results are. An implication of this is that the unitary-type method for including multistep processes introduced in the third portion of this paper may be a reasonable procedure to follow. It is based on connected kernel equations, and, once DWBA-type matrix elements are evaluated,involves solving only algebraic equations for the various amplitudes. It will be interesting to see the results of calculations based on this approach, at least because of the great simplicity achieved in obtaining approximate amplitudes.

A new method for introducing a "crude" approximation to a complicated many-body problem, which method can then be systematically improved, is discussed in the paper of SCHMID and ZIEGELMANN. They argue that Pauli principle exchange effects and the presence of bound states are the most important crude (non-detailed) aspects of composite particle scattering. Their method is that of orthogonality scattering, based on the idea that the Pauli principle implies certain states are occupied, and therefore that scattering states must be orthogonal to them. There is some freedom available in how the orthogonality constraint is constructed, and they propose to exploit this in achieving fits to data. Calculations of the quartet neutron-deuteron phase shifts are presented which are in good agreement with empirical ones, and more complex systems are being studied. The method appears promising.

For the three-body problem, breakup can be treated exactly (in principle and approximately in practice for local interactions), so that rescattering effects, fsi, etc., can be determined. Prior to the extensive theoretical work on the three-body problem, a qualitative method for describing certain three-body final states was in use, viz., that of sequential decay [27]. This method is used by HEISS to study three-body final states in α + d and ^{3}He + ^{3}He reactions. Both

$d\sigma/d\Omega$ and $P(\theta)$ were calculated; spin was found to be very important. The $\alpha + d$ results are better than the $^3He + {}^3He$ ones, and it will be interesting to see if the latter discrepancy can be removed. As a possible empirical approach to three-body final states in composite particle scattering, the method is also a promising one.

The paper of VANZANI is mainly a review of new and old work on many-body scattering, with emphasis on showing how his new connected kernel equations either reduce to other equations or can be put into form suggestive for future calculations. The most interesting aspect of his paper is a reference in section 2 to calculations of neutron and proton scattering from light nuclei using single term separable potentials in each partial wave; good fits to experimental phase shifts are obtained. It will be interesting to see if this method can be extended to other targets, and to composite particle projectiles. The other applications of his many-body equations remain for the future.

Finally, there is a broad review by ALI of aspects of the interaction between various light nuclear systems, such as nucleon $+\alpha$, $\alpha + \alpha$, α + nucleus, etc. The main theoretical areas discussed are the use of the resonating group method, the use of separable potentials and the folding model optical potential. This review seems reasonably up to date and is recommended even for non-specialists who may be interested in this area of nuclear reactions.

REFERENCES

† Rapporteur's report for session VIII.

* Work supported in part by the U.S. E.R.D.A.

[1] D.P. Bouldin and F.S. Levin, Phys. Letts. 42B(1972)167 and references cited therein.
[2] B.H.J. McKellar, J. Phys. G1(1975)180, and references cited therein.
[3] I. Lovas, Ann. Phys. 89(1975)96.
[4] S.K. Young and E.F. Redish, Phys. Rev. C10(1974)498, and references cited.
[5] I. Lovas and E. Denes, Phys. Rev. C7(1973)937 and references cited therein, and K. Schaefer, Nucl. Phys. A140(1970)9 and references cited therein.
[6] P.C. Tandy, E.F. Redish and D. Bolle, Phys. Rev. Letts. 35(1975)921; D.J. Ernst and R.M. Thaler, ibid., 36(1976)222.
[7. N. Austern, Direct Nuclear Reaction Theories, (Wiley, New York, 1970).
[8] F.S. Levin in Reaction Dynamics by F.S. Levin and H. Feshbach, (Gordon and Breach, New York, 1973).
[9] See (8) for references to the Pearson and BHMM approximations.
[10] R.C. Johnson and P.J.R. Soper, Phys. Rev. C1(1970)976; J.D. Harvey and R.C. Johnson, Phys. Rev. C3(1970)636, J. Phys. A7(1974)2017.
[11] Eg., T. Udagawa, H.H. Wolter, and W.R. Coker, Phys. Rev. Letts. 31(1973)1507.
[12] R.S. MacKintosh, Nucl. Phys. A230(1974)195.
[13] D. Robson, Phys. Rev. C7(1973)1; N. Austern and C.M. Vincent, Phys. Rev. C10 (1974)2623.
[14] T. Tamura, Phys. Rpts. 14(1974)59.
[15] Eg., I.E. McCarthy, J. Phys. B6(1973)2358; and Annual Progress Reports, Physics Discipline, Flinders University.
[16] R.C. Johnson, Nucl. Phys. A90(1967)289.
[17] R.C. Johnson and F.D. Santos, Parts. and Nuclei 2(1971)285.
[18] R.C. Johnson et al., Nucl. Phys. A208(1973)221.
[19] F.D. Santos, Ph.D. thesis, University of Surrey, 1968.
[20] L.D. Knutsen et al., Phys. Rev. Letts. 35(1975)1570.
[21] J.P. Farrell, C.M. Vincent, and N. Austern, preprint (1975), and references cited therein; G. Rawitscher, Phys. Rev. C11(1975)1152.
[22] R.C. Johnson and P.C. Tandy, Nucl. Phys. A235(1974)56.
[23] G. Rawitscher, Phys. Rev. 163(1967)1223.
[24] D.J. Kouri, F.S. Levin, and W. Sandhas, submitted to Phys. Rev.
[25] Eg., D.J. Kouri and F.S. Levin, Nucl. Phys. A253(1975)395.
[26] A.N. Mitra, Phys. Rev. 139(1965)B1472.
[27] Eg., Rev. Mod. Phys. 37(1965); F.S. Levin, Ann. Phys. 46(1968)41.

PART 4

POSTDEADLINE PAPERS

MULTIPLE SCATTERING APPROACH TO THE π ^{4}He SCATTERING AT INTERMEDIATE ENERGIES

M. Błeszyński, T. Jaroszewicz
Institute of Nuclear Physics, Kraków, Poland
P. Osland
International Centre for Theoretical Physics, Trieste, Italy

The elastic scattering of medium energy poins by ^{4}He is a case where all the standard approximations seem to break down. Standard Glauber theory which works reasonably well at higher energies does also provide a reasonable description of ^{16}O and ^{12}C scattering at these energies [1] but it fails for ^{4}He. One could ascribe this to the following

(i) The noneikonal corrections to the Glauber model are needed.

(ii) Since the ^{4}He is a relatively small and strongly bound nucleus, the off-mass-shell dependence of the elementary π N amplitude should be important.

(iii) π ^{4}He amplitude could be sensitive to the form of the ^{4}He ground state density for small separations between the nucleons.

In particular, (ii) and (iii) are closely connected, since the off-shell behaviour of the π N t-matrix elements is relevant only in the region of overlap of the interactions of the incident pion with the target nucleons. Therefore, for the density of ^{4}He with repulsive correlations at small internuclear distances the differences between the π ^{4}He amplitudes calculated with various off-shell prescriptions should be smaller than for the simple gaussian wave function.

In order to investigate quantitavely the above-mentioned problems we have calculated π ^{4}He scattering amplitude using the complete multiple scattering formula up to the terms of the fourth order

$$T=\sum_{i=1}^{4} t_i+\sum_{i\neq k\neq 1}^{4} t_i G_o t_k+\sum_{i\neq j\neq k=1}^{4} t_i G_o t_j G_o t_k+\sum_{i\neq j\neq k\neq l=1}^{4} t_i G_o t_j G_o t_k G_o t_l \qquad (1)$$

where t_i are the elementary π N t-matrices for the scattering on i-th nucleon and G_o is the free wave pion propagator. Formula (1) has been recently applied to the scattering of protons from ^{4}He in [2] . In the π ^{4}He Breit frame we can represent formula (1) by the sum of diagramms of the form:

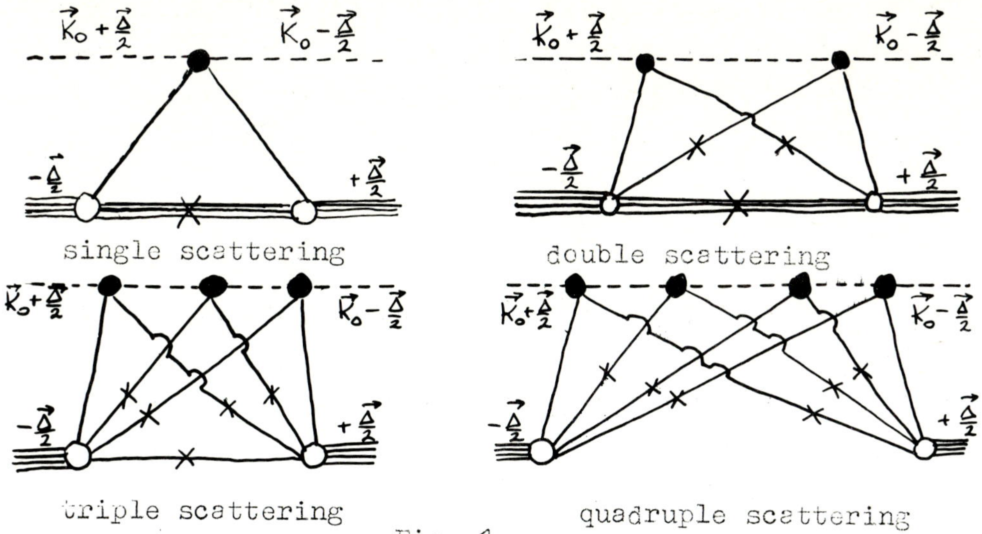

Fig. 1

where crosses denote putting the (spectator) particles on their mass shell and $\vec{K}_0=\frac{1}{2}(\vec{k}_i+\vec{k}_f)$, $\vec{\Delta}=\vec{k}_i-\vec{k}_f$, $\vec{k}_i$ and $\vec{k}_f$ are the initial and final pion momenta.

The elementary πN t-matix elements have been parametrized in the form

$$\langle \vec{p}'|t|\vec{p}\rangle = a + b(|\vec{p}|,|\vec{p}'|)\vec{p}\cdot\vec{p}' + i\vec{\sigma}\cdot(\vec{p}\times\vec{p})\,c(|\vec{p}|,|\vec{p}'|). \quad (2)$$

The on-shell values of the parameters b,c have been calculated from the CERN phase-shift data for the πN C.M. energies corresponding to the πN vertices in the above diagramms and then transformed to the πN Breit frames. In order to take into account the Fermi motion we have averaged a, b and c with the nucleon single particle densities in the momentum space representation.

The ^4He ground state density was expressed as a linear combination of gaussians with parameters which are consistent with the ^4He formfactor measured in the electron scattering experiments.

We have studied several prescriptions for the off-shell dependence of πN t-matrix elements. From our preliminary analysis it turns out that $\pi\,^4He$ amplitude is very sensitive to the off-shell continuation especially for large angle scattering. This is due to the fact that the principal value parts of propagator G_0 in the multiple scattering terms for p-wave scattering are in general comparable with and/or even bigger than the delta parts.

It can be concluded that a study of $\pi\,^4$He scattering at intermediate energies can provide useful information on the off-mass-shell dependence of the π N scattering amplitude.

References

[1] K. Bjornenak, J. Finjord, P. Osland and A. Reitan, Nucl. Phys. B27, (1971) 598.

[2] M. Błeszyński and T. Jaroszewicz, Phys. Lett. B56, (1975) 247.

THREE NUCLEON POTENTIAL IN NUCLEAR MATTER

S.A. Coon*, B.R. Barrett+, M.A. Scadron+, D.W.E. Blatt† and B.H.J. McKellar**

*Institut de Physique, Université de Liège, Belgium and Institut für Theoretische Physik, Technischen Universität, Hannover, W. Germany.

+Department of Physics, University of Arizona, Tucson, U.S.A.

†School of Computing Science, N.S.W.I.T., Sydney, Australia.

**School of Physics, University of Melbourne, Australia, and Service de Physique Théorique, CEN-Saclay, BP n° 2, 91190 Gif-sur-Yvette, France.

In this paper we present a new calculation of the two-pion exchange three nucleon potential for use in nuclear matter calculations. The new features of our potential compared to the older potential of Fujita and Miyazawa [1], which has been used in most previous calculations [2] in nuclear matter, are:

i) The potential is based on a πN scattering amplitude which is constructed to satisfy the constraints imposed by current algebra and the most recent πN scattering data [3].

ii) There is a significant s wave πN amplitude, although the well known cancellation at the Adler point is imposed.

Our method is an extension of that of Coon, Barrett and Scadron [4], who because they were using the approximate methods of Brown and Green [5] to evaluate the contribution of the three-body force to the energy of nuclear matter, were able to work directly with the πN amplitude in the forward direction where several simplifications occur.

We wish to construct a potential in nuclear matter where we eventually will average over the spins and isospins of the nucleons in computing the energy. We perform this average over the spin and isospin of the "middle" nucleon immediately, denoting it by $\langle \ldots \rangle$.

The resulting πN amplitude which we require is $T = -\langle u_f F^{(+)} u_i \rangle$, in the rest frame of the incident nucleon, using the notation of ref. 4 (any undefined symbols in the following are defined there).

We write $F^{(+)}(\nu, t, q^2, q'^2)$, (which we need off pion mass shell) as a pole term plus background

$$F^{(+)}(\nu, t, q^2, q'^2) = \frac{g^2}{m} K(q^2) K(q'^2) \frac{\nu_B^2}{\nu_B^2 - \nu^2} + \bar{F}^{(+)}(\nu, t, q^2, q'^2),$$

where $K(q^2)$ is the πN form factor, and $\bar{F}^{(+)}$ the "background" term. $\bar{F}^{(+)}$ is required to satisfy the current algebra constraints when one pion is soft and the other is on the mass shell $q^2 = \mu^2$ (the Adler point), and when both pions are soft.

This enables us to determine the coefficients in the expansion

$$\bar{F}^{(+)}(0, t, q^2, q'^2) \Big|_{q_0 = q_0' = 0} = \alpha + \beta \underset{\sim}{q} . \underset{\sim}{q}' + \gamma(\underset{\sim}{q}^2 + \underset{\sim}{q}'^2) + \ldots$$

in terms of the πN_σ term, σ, the π decay constant f_π, and on shell quantities. The next step is to make a similar expansion of the pole term, with the forward propagating Born term T_{FPBT} subtracted to remove the iterated OPEP contribution. We then obtain the final equation for $A = T - T_{FPBT}$

$$A(0, t, q^2, q'^2) \Big|_{q_o = q'_o = 0} = K(q^2)\, K(q'^2)\, \{a + b\, \underset{\sim}{q}.\underset{\sim}{q}' + c(\underset{\sim}{q}'^2 + \underset{\sim}{q}^2)\ \ldots\}$$

The parameters a, b, c are found to be:

$$a = \frac{\sigma}{f_\pi^2} \qquad = 1.14\ \mu^{-1}$$

$$b = -\frac{2\sigma}{f_\pi^2}\left(\frac{1}{\mu^2} + \frac{1}{4m^2}\right) + \frac{2}{\mu^2}\bar{F}^{(+)}(0, \mu^2, \mu^2, \mu^2) \qquad = -2.55\ \mu^{-3}$$

$$c = +\frac{\sigma}{f_\pi^2}\left(\frac{1}{\mu^2} + \frac{1}{4m^2}\right) - \frac{g^2}{4m^3} + K'(0)\frac{\sigma}{f_\pi^2} \qquad = 1.09\ \mu^{-3}.$$

Where the numerical values use the results of ref. 3, and the derivative of $K(q^2)$ is obtained from $K'(0) = \mu^{-2}\{1 - mg_A/(f_\pi g)\}$ which follows from the Goldberger Treiman relation.

The resulting potential, for use in nuclear matter calculations is expressible as an s wave term V_s coming from $a + c(\vec{q}^2 + \vec{q}'^2)$ and a p wave term V_p arising from $b\,\vec{q}.\vec{q}'$. The p wave potential is given by eq. (2.1) of ref. 2, with the identification

$$C_p = -\frac{1}{8}\frac{g^2}{(4\pi)^2}\frac{b\mu^6}{m^2}$$

The s wave potential is

$$V_s = \frac{g^2\mu^2}{4(4\pi)^2 m^2}(\tau_1.\tau_3)(\sigma_1.\nabla_1)(\sigma_3.\nabla_3)\Big\{(a-2\mu^2 c)\, Z_1(r_{12})\, Z_1(r_{23}) + c\left[Z_o(r_{12})\, Z_1(r_{23}) + Z_1(r_{12})\, Z_o(r_{23})\right]\Big\}$$

where

$$Z_n(r) = \frac{4\pi}{\mu}\int \frac{d\underset{\sim}{q}}{(2\pi)^3}\, e^{i\underset{\sim}{q}.\underset{\sim}{r}}\, \frac{H(\underset{\sim}{q}^2)}{(\underset{\sim}{q}^2+\mu^2)^n}$$

where $H(q^2) = K^2(-q^2)\,K_1(-q^2)$, and $K_1(q^2)$ is the pion propagator "form factor" defined in terms of the propagator $D(q^2)$ by $D(q^2) \propto K_1(q^2)\,(q^2 - \mu^2)^{-1}$.
It is hoped that we will be able to present results of a calculation of the contribution of this potential to the binding energy of nuclear matter.

REFERENCES

[1] J. Fujita and H. Miyazawa, Prog. Theor. Phys. 17, 260 (1957).
[2] D.W.E. Blatt and B.H.J. McKellar, Phys. Rev. C11, 614 (1975), and references therein.
[3] M.D. Scadron and L.R. Thebaud, Phys. Rev. D9, 1544 (1974).
[4] S.A. Coon, M.D. Scadron and B.R. Barrett, Nucl. Phys. A242, 467 (1975).
[5] G.E. Brown and A.M. Green, Nucl. Phys. A137, 1 (1969).

VARIATIONAL AND EXACT STUDIES OF THE BOUND STATE OF THREE HELIUM ATOMS*

T. K. LIM

Department of Physics and Atmospheric Science

Drexel University, Philadelphia, Pa. 19104, U.S.A.

INTRODUCTION

It is only just recently that the study of small helium molecules has received deserved attention [1,2]. In spite of its importance in the elucidation of clustering phenomena in the liquid state, in astrophysical and high-pressure applications, in the quantum theory of virial coefficients, and in the choice of the best from rival interatomic potentials, the structure of $(^{3,4}He)_n$ systems has been an untouched subject.

(i) What is the minimum number of atoms of $(^{3,4}He)$ for which a molecular bound state exists?

(ii) Can the theoretical discussion of bound clusters of rare-gas atoms be based on phenomenological, pairwise, interatomic potentials?

(iii) Do molecular bound states affect the virial and transport coefficients of the rare gases?

A decade ago, pioneers in few-nucleon physics such as Primakoff[3], Blatt[4], Reiner[5], Schmid and Tang[6] wrestled briefly with these questions, then left the problem to use their skills elsewhere because "more accurate data on gaseous helium at low temperatures are needed badly"[4].

Lately, this lament has been heard; experimentalists in molecular and low-temperature physics have begun to respond and there is now information that

(i) the trimer $(^4He)_3$ exists[7],

(ii) elastic scattering cross sections for ^4He-^4He collisions can be fitted well by a simple phenomenological potential[8], and

(iii) the ^3He-^3He interaction is slightly more attractive than that for ^4He-^4He[9].

The situation is promising for few-body theoreticians to test their arsenal of techniques on this problem.

*Supported in part by the National Science Foundation, USA

In this paper, I report preliminary results from an extensive and ongoing theoretical study of (^{4}He) trimers. If the helium atoms can be regarded as separate entities interacting through two-body, central potentials, then the (^{4}He)$_3$ molecular system constitutes the simplest of three-body systems, viz. three non-relativistic, identical, spinless entities interacting pairwise through a central, local potential.

It is obvious the sophisticated and powerful battery of three-nucleon techniques, embodied in present-day variational and exact, Faddeev methods, can be brought to bear on this physically-interesting problem. One should expect subtle but important differences in the treatment of trimer and trinucleon systems. These should be analogous to those which occur between liquid-helium and nuclear-matter calculations [10] and are a reflection of the differing well-shapes and strengths of the basic two-body interaction (see Fig. 1). Thus, the nuclear force

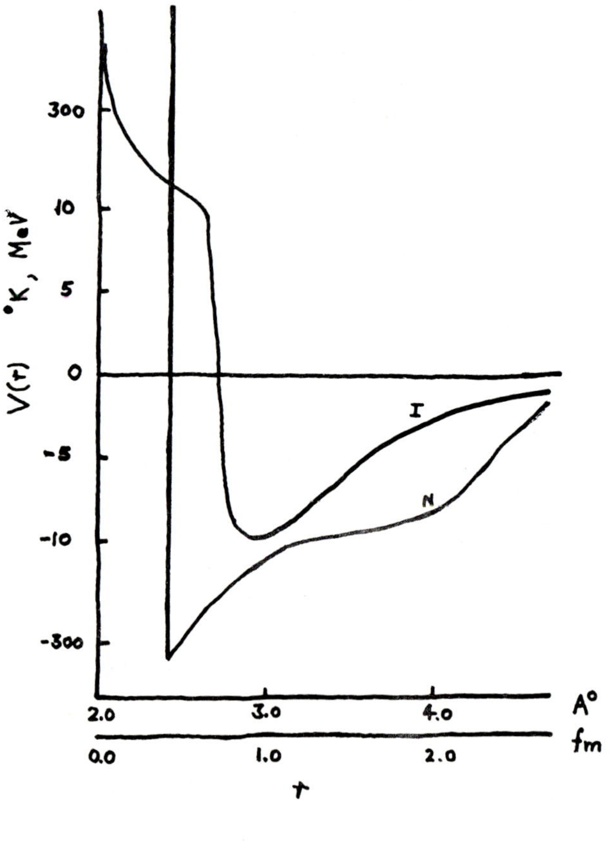

Fig. 1 Representative nuclear and interatomic potentials.

supports a bound two-body system; the interatomic potential for helium does not.

THE INTERATOMIC POTENTIAL OF ^{4}HE

An important ingredient in any work on three-body systems is the nature of the interparticle forces working within them. It has been shown from various analyses [11] that potentials between pairs of ^{4}He atoms are adequate substitutes for the dynamic structure of the interacting electronic systems. The traditional strategy in developing effective working potentials has been to determine their parameters from phenomenological fits to virial coefficients; transport coefficients; crystal properties; high-resolution, elastic, differential cross sections; and from crossed-molecular-beam experiments. The range of such few-parameter potentials, some Fourier-transformable some not, include those of Farrar and Lee [8], Bennewitz [9], Beck [11], and the Lennard-Jones (6,12) type [6]. As of now, no one potential has been granted favored status. I have completed trimer calculations using the Beck and two hard-core square-well potentials, one of which is phase-equivalent to the Lennard-Jones (6,12) used by Schmid [6,12].

NUMERICAL PROCEDURES AND RESULTS

The variational method is represented in this work by the equivalent two-body method (ETBM). The ETBM is particularly effective for three-boson systems [13] and, for our limited purpose of determining the existence of ^{4}He trimers, serves us well. I refrain from a full discussion of the numerical aspects of the ETBM since most of these are well-known. Suffice it to mention the outstanding features of the numerical analysis as applied to the trimer.

(i) The product function f(r) input as trial function in the iteration procedure is the ground state for relative motion in the two-body potential V(r) with reduced mass Mm/2:

$$\frac{d^2}{dr^2}\left[r f(r)\right] + \frac{2Mm}{\hbar^2}\left[E - V(r)\right]\left[r f(r)\right] = 0$$

M is a variational parameter and has optimum values ranging from 1.0 to 1.03 depending on the potential.

(ii) The Numerov method is used for the solution of the integro-differential equation which arises in the ETBM. The accuracy of this method is insufficient at large r so f(r) is replaced at the end of each iteration, by its asymptotic behavior in this region, viz. $\exp(-\beta r)/\sqrt{r}$, where β is determined such that continuity in f and f' are maintained at a variable matching radius.

(iii) Differentiations are executed with the aid of a six-point numerical procedure in which an interval of length 6h is used. I used h = 0.08 Å. Integrations are performed using Simpson's rule with the same step length.

(iv) For the Beck potential, V(r) is replaced by ∞ for $r \leq 2$ Å, to avoid difficulties with numerical quadrature. Since the trimer is a loosely-bound system, I have taken the integration range out to 80 Å.

The Faddeev method has been used by Kruger [14] to study the trimer system. However, Kruger's trimer exhibits the spectacular collapse phenomenon first observed by Osborn [15]. This experience demonstrates that the repulsive cores of the ^{4}He-^{4}He potentials can present formidable computational difficulties. The system of three ^{4}He atoms interacting through Herzfeld potentials is a tractable problem for Faddeev techniques; hence the choice of these potentials for the present study.

To be concise, I refrain once again from a detailed discussion of the numerical aspects of the method. The main features are:

(i) The one-term unitary pole expansion of Harms [16] is used to expand the local potential, which is taken to be effective only in S states. The form factor is obtained by invoking the approach of Kowalski and Feldman [17,18].

(ii) The energy variable s in the UPA is taken to be zero since V(r) is just unable to support a bound dimer.

(iii) The integral equation for the trimer is solved by direct matrix inversion using the discretization afforded by quadrature.

The results obtained with the ETBM and the UPA are displayed on Table I and in Fig. 2. A bound trimer is found in every case. The agreement between the ETBM and the UPA for SQA3 [4] is excellent. Since the ETBM yields the largest binding energy for trial functions of

TABLE I. The binding energy and r.m.s. radius of $(^4He)_3$.

Potential	Kin. energy (°K)	Pot. energy (°K)	Bind. energy (°K)	Radius (Å)	Faddeev UPA (°K)
Beck [11]	0.829	-0.943	-0.114	26.6	-
SQA3 [4]	1.517	-1.590	-0.073	-	-0.070
Burkhardt [12]	-	-	-	-	-0.023

product type and is superior to almost all other variational methods for three-boson systems, I conjecture that $(^4He)_3$ is bound by close to 0.15°K. I used the wave function from the final iteration in the Beck calculation to evaluate $\langle r_{ij}^2 \rangle^{1/2}$, the root-mean-square interparticle separation; I found this to be 26.6 Å, indicating that the cluster of three ^{4}He atoms has considerable size. This is also intimated by the gradual drop-off in f(r) (see Fig. 2).

It is evident from the results of this work that the proven methods of few-nucleon physics are adequate for the study of weakly-bound trimers of rare-gas atoms. The success of Faddeev theory reported here has encouraged me to look beyond the Herzfeld potential to other realistic potentials. Because the Faddeev formulation is tailor-made for the

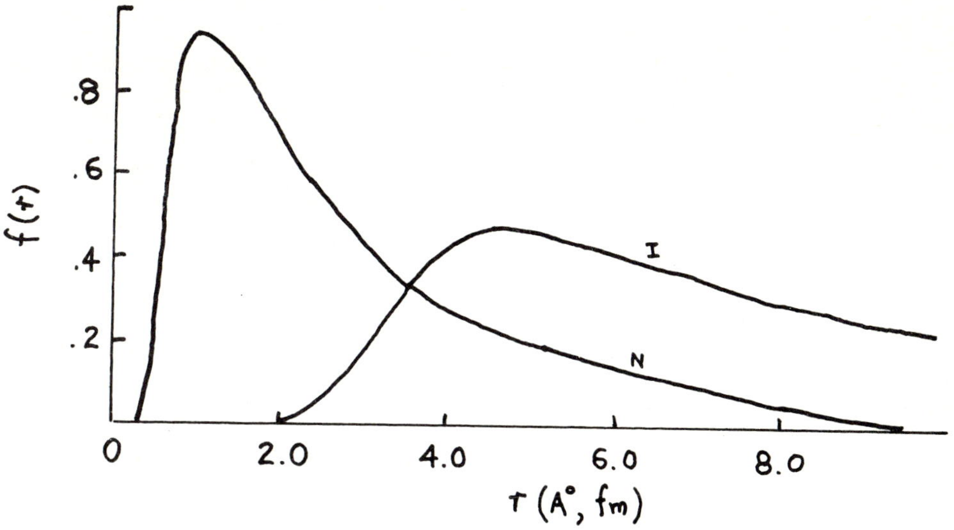

Fig. 2 Plot of nuclear and interatomic functions.

description of scattering processes, it may have application in reaction kinetics, a subject of great interest to chemists [19].

REFERENCES

1 L. W. Bruch and I. J. McGee, J. Chem. Phys. 52 (1970) 5884.
2 T. K. Lim and M. A. Zuniga, J. Chem. Phys. 63 (1975) 2245.
3 N. Bernardes and H. Primakoff, J. Chem. Phys. 30 (1959) 691.
4 J. M. Blatt, J. N. Lyness and S. Y. Larsen, Phys. Rev. 131 (1963) 2131.
5 A. S. Reiner, Phys. Rev. 151 (1966) 170.
6 E. W. Schmid, J. Schwager, Y. C. Tang and R. C. Herndon, Physica (Hague) 31 (1965) 1143.
7 A. P. J. van Deursen and J. Reuss, J. Chem. Phys. 63 (1975) 4559.
8 J. M. Farrar and Y. T. Lee, J. Chem. Phys. 56 (1972) 5801.
9 H. G. Bennewitz, et al., Z. Phys. 253 (1972) 435.
10 H. B. Ghassib and J. M. Irvine, J. Low Temp. Phys. 18 (1975) 201.
11 D. E. Beck, Mol. Phys. 14 (1968) 311; 15 (1968) 332.
12 T. W. Burkhardt, Ann. Phys. 47 (1968) 516.
13 H. Fiedeldey, et al., Nucl. Phys. A113 (1968) 543.
14 H. Kruger, Phys. Rev. Lett. 28 (1972) 71.
15 T. Osborn, Stanford LINAC Rep. No. 79 (1967).
16 E. Harms, Phys. Rev. C1 (1970) 1667.
17 K. L. Kowalski and D. Feldman, J. Math. Phys. 2 (1961) 499; 4 (1963) 507.
18 V. S. Mathur, A. V. Lagu and C. Maheshwari, Nucl. Phys. A178 (1972) 365.
19 A. Kuppermann, to be published.

STUDIES OF THE $^{2}H(p,2p)n$ REACTION FOR CONSTANT NN RELATIVE ENERGIES*

W.T.H. van Oers

Cyclotron Laboratory, Department of Physics,
University of Manitoba, Winnipeg, Manitoba, R3T 2N2

Numerous experiments have been performed to investigate the disintegration of the deuteron by nucleons. Most of these experiments have been concerned with regions in phase space where two-body effects prevail as in quasi-free scattering processes and final-state interaction processes. In kinematically complete experiments this implies that coplanar geometries were chosen. The ultimate aim of all these experiments is a detailed comparison with the predictions of Faddeev type three-body calculations which until now employ S-wave separable or local NN potentials. Even relaxing the condition that comparisons should only be made with predictions calculated using NN potentials which are phase equivalent (i.e., potentials which reproduce equally well the NN on-energy-shell experimental information), the observed differences are not much larger than the experimental uncertainties. Consequently, relatively little has been learned from such comparisons.

In order to find a region where the differences between calculations with different potential models become really significant, Kloet and Tjon [1] performed a four-dimensional phase space search mapping the differences of the predictions for two local S-wave NN potentials. It was found that the greatest differences occurred in the regions of cross-section minima where the breakup amplitude for the state with total spin S = 1/2 and spin for the pair of identical nucleons $S_{NN} = 0$ (denoted as M(1/2,0) is the only non-zero amplitude. The occurrence of such minima is due to a delicate cancellation of the single-scattering or Born part with the multiple-scattering part or remainder of this breakup amplitude. The value of the cross-section in these regions depends then in a subtle way on the dynamics of the three-body scattering as expressed in the Faddeev equations.

In the present study two procedures were chosen to investigate the interference effects. In the first procedure the variation of the breakup cross-section was studied for fixed values of the final-state NN relative energies and a fixed value of the momentum of one of the emerging protons. By choosing, as suggested by Jain et al. [2], particular equal relative energies between the two protons and the neutron one ensures that the cross-section is dominated by the M(1/2,0) amplitude. For an incident energy of 39.5 MeV three-body calculations were made of loci at various angles of the fixed proton momentum and as function of the energy of the unobserved neutron using the two local potential sets MT13 and MT14 of Malfliet and Tjon [3] and the separable YY potential used by Jain and Doolen [4]. The MT13 and MT14 potential sets differ in that the second one does not possess a repulsive core in the $^{3}S_{1}$ NN state. Both potential sets have a repulsive core in the $^{1}S_{0}$ NN state. It was decided to perform a breakup cross-section measurement along the locus defined by $\theta_3 = 27.1°$. Here, at the position of the minimum in the cross-section, there is a difference of a factor of two between the MT13 and MT14 potential set predictions, yet the width of this sensitive region was not so narrow that it might be washed out by finite experimental resolution effects. In the notation adopted 1 is the incident proton, 2 is the target deuteron, 3 and 4 are the observed protons, and 5 is the unobserved neutron.

In the second procedure, the variation of the breakup cross-section was studied along loci which are defined by the requirements of equal values for the final-state NN relative energies and of symmetry with respect to the incident beam direction for the momenta of the two emerging protons. In this kinematical situation only the crucial M(1/2,0) doublet amplitude contributes. Again strong destructive interference minima are observed with the MT13 potential set predictions at least a factor two smaller than the MT14 potential set predictions at the location of the minimum. Two incident proton energies were selected :

23.0 and 39.5 MeV. The lower energy was chosen because earlier work on n-d scattering [5] indicated that for incident energies up to 22 MeV the S-wave potential model predictions for the differential elastic and total cross-sections were in good agreement with the available data. Now it is known that both the NN tensor force and P-wave and D-wave interactions must be included to obtain acceptable agreement with the experimental n-d vector and tensor polarizations [6– 9]. It is thus of interest to see how well the S-wave potential model calculations reproduce the breakup cross-sections at 23.0 MeV. Furthermore if genuine three-body forces are of importance, one expects to find at loci with equal NN relative energies the greatest sensitivity. At the higher energy even greater sensitivity to the NN potential sets was found. At the most sensitive point, the minimum in the cross-section, the MT14 potential set and the YY separable potential predictions were factors of three and four larger than the MT13 potential set prediction, respectively. However, it should be borne in mind that the potentials considered are not phase equivalent and thus the predicted cross-sections will in part reflect the difference in the on-energy-shell behaviour of the T-matrices calculated from these potentials. Nevertheless an experimental test of the above three NN interaction models appeared to be warranted by the greater sensitivity at 39.5 MeV.

The experiment was performed at the University of Manitoba cyclotron facility. Momentum analyzed proton beams were incident on a hemispherical gas cell containing deuterium gas of 99.5 % purity at a pressure of approximately two atmospheres. The two emerging protons from the $^2H(p,2p)n$ reaction were detected in two counter telescopes, one of which could be moved in a horizontal plane through the incident beam direction while the second one could be moved to practically any position in the upper hemisphere. Each counter telescope consisted of a 200 μm thick surface barrier ΔE detector for timing and particle identification, a 5 mm thick lithium drifted silicon E detector for energy determination of the breakup protons, and a 500 μm thick VETO detector for rejection of elastically scattered protons. Data were accumulated in a computer as two 64 × 64 channel T_3 versus T_4 arrays, one for real plus accidental coincidence events and one for accidental coincidence events. The data were also recorded event by event on magnetic tape for later off-line analysis. The differential cross-sections $d\sigma/(dT_3 d\Omega_3 d\Omega_4)$ were obtained using the measured gas scattering geometry, the target-gas pressure and temperature, and a Faraday cup integration of the incident beam current. Corrections to the data were made for the effects of dead time and pile up rejection. Further experimental details are given in Ref. [10].

In the first procedure energy sharing spectra were measured at angle pairs which correspond to seven locations along the θ_3 = 27.1° locus [11]. The locus was further defined by T_3 = 17.87 MeV and $T_{35} = T_{45}$ = 14.37 MeV. In each spectrum the value of the cross-section at T_3 = 17.87 MeV was extracted by fitting a polynomial curve to approximately fourteen data points in this region. The values obtained in this manner were plotted as a function of neutron energy T_5 (Fig. 1). The error bars indicate one standard deviation due to counting statistics. The uncertainty in the absolute normalization resulting from systematic errors is estimated to be within ± 7 %.

In the figure, the data at large values of T_5 provide better agreement with the MT13 potential set prediction. The MT13 potential is the most realistic of the three S-wave NN potentials tested as it provides the better fit to the $^1S_0$ and $^3S_1$ NN phases up to 300 MeV and also gives a triton binding energy of 8.3 MeV, close to the experimental value of 8.48 MeV. In the region of the cross-section minimum the experimental data are a factor of three larger than the predictions for the MT13 potential set. At this incident energy there are two possible reasons for the discrepancy. In the first place only NN interacting pairs with zero orbital angular momentum were considered in the calculation of the scattering amplitudes. In the second place the deuteron wave function was taken to be a pure $^3S_1$ bound state wave function as opposed to a more realistic one which is an admixture of $^3S_1$ and $^3D_1$ states coupled by a tensor force. It should be noted that all currently available breakup calculations do not consider NN interactions with $\ell > 0$. With the help of a plane wave impulse approximation calculation one finds that a deuteron D-state probability of 8 % may fill the interference minimum to a degree in agreement with experiment [11].

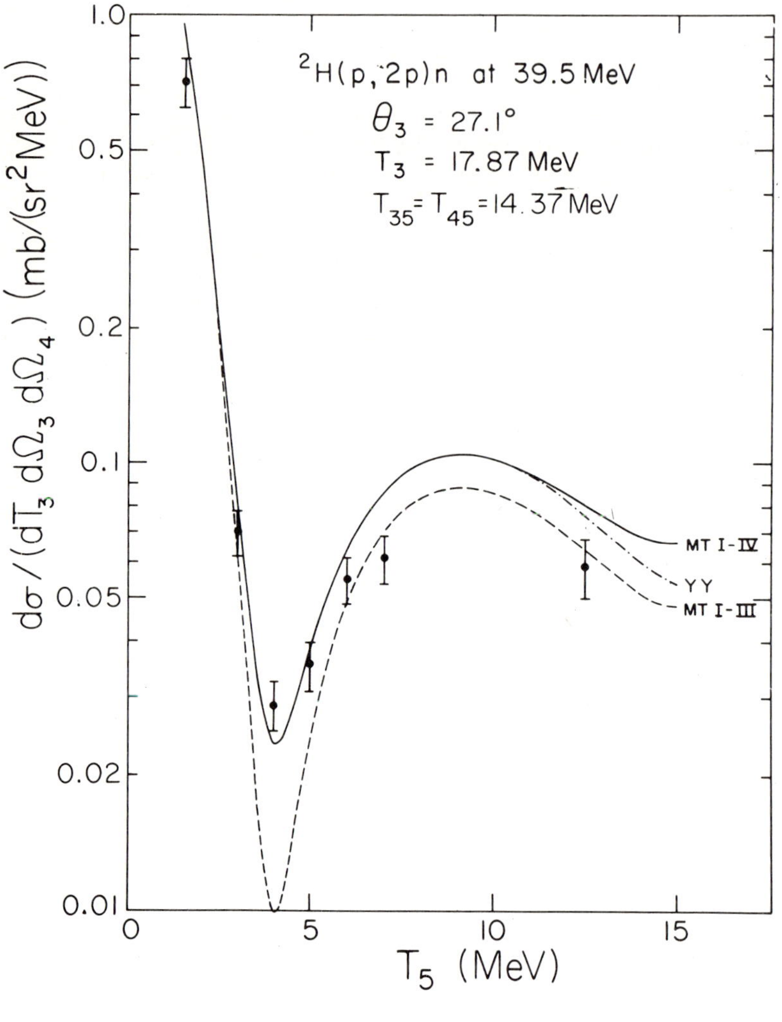

Fig. 1

In the second procedure at 23.0 MeV, energy sharing spectra were measured which correspond to eight locations along the locus of constant NN relative energies. The locus is defined by $T_{34} = T_{45} = T_{35} = 6.55$ MeV and furthermore by $\theta_3 = \theta_4$ and $T_3 = T_4$. The reproducibility of the data was checked by repeating the experimental measurement corresponding to $T_5 = 0.24$ MeV with excellent results. Differential cross-sections were extracted from the eight energy sharing spectra by smoothing the data using a polynomial in the region of the specified value of T_3. The values obtained in this manner were plotted as function of neutron energy (see Fig. 2). The failure of the MT13 potential set to predict the breakup cross-section, in particular at the interference minimum, lends support to the argument given above that it is essential to take account of the D-state of the deuteron in further calculations. Doleschall [8] for instance concluded that N-d elastic scattering polarization observables at 22.7 MeV exhibit significant sensitivity to the NN tensor force but are insensitive to the behaviour of the on-energy-shell NN T-matrix at higher energies, i.e. 100 to 400 MeV. Hence it is not likely that the difference between the present p-d breakup data and the MT13 and MT14 potential set predictions would have been significantly reduced had these potentials produced a better fit to the $^1S_0$ and $^3S_1$ phases. In the multiple scattering region (large T_5 energy values) the theoretical calculations agree with the experimental results within one standard deviation.

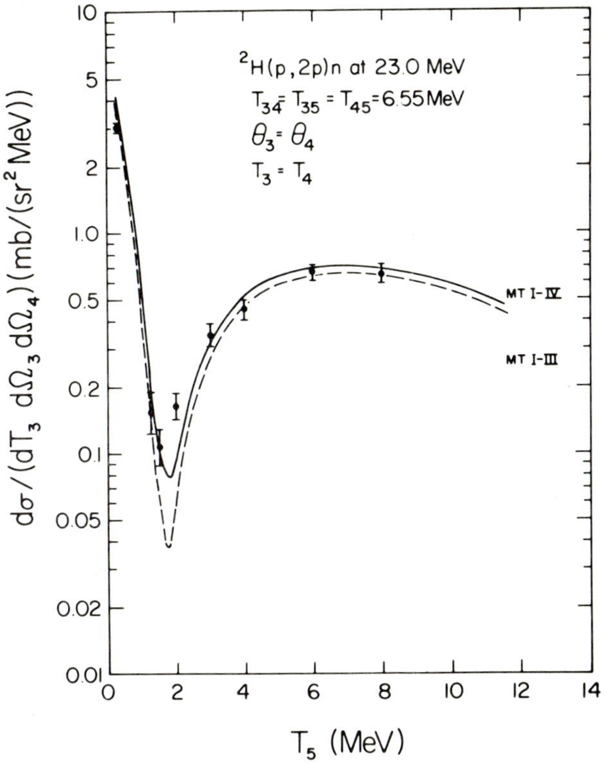

Fig. 2

Finally at 39.5 MeV energy sharing spectra were measured again corresponding to eight locations along the locus of constant NN relative energies. The locus is defined by $T_{34} = T_{45} = T_{35} = 12.05$ MeV and furthermore by $\theta_3 = \theta_4$ and $T_3 = T_4$. Fig. 3 shows four of the energy sharing spectra measured. In these spectra the error bars indicate the statistical uncertainty after background subtraction. Also shown are the MT13 and MT14 potential set predictions. The breakup cross-section as a function of T_5 is given in Fig. 4. To determine the influence of finite geometry effects, some of the points along this locus, including the one at the interference minimum, were remeasured with considerably smaller apertures. In each case the two data points agreed well within their respective errors. At the coplanar locations of all three loci measured, the theoretical curves require a renormalization of ~ 0.8 to obtain agreement with the experimental data. This is in agreement with the reanalysis of the p + d breakup data of Klein et al [12] at 16 MeV by Bruinsma et al. [13] using separable potentials.

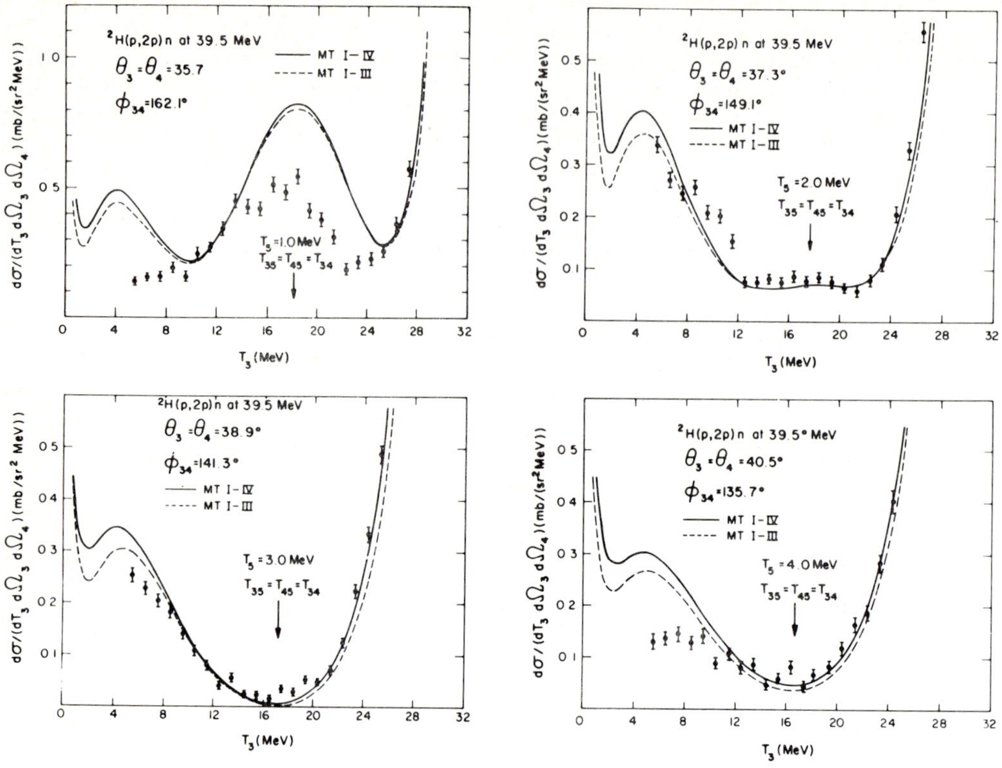

²H(p,2p)n at 39.5 MeV
θ3 = θ4 = 35.7
φ34 = 162.1°
MT I–IV
MT I–III
T5 = 1.0 MeV
T35 = T45 = T34
dσ/(dT3 dΩ3 dΩ4) (mb/(sr² MeV))
T3 (MeV)
²H(p,2p)n at 39.5 MeV
θ3 = θ4 = 37.3°
φ34 = 149.1°
MT I–IV
MT I–III
T5 = 2.0 MeV
T35 = T45 = T34
²H(p,2p)n at 39.5 MeV
θ3 = θ4 = 38.9°
φ34 = 141.3°
MT I–IV
MT I–III
T5 = 3.0 MeV
T35 = T45 = T34
²H(p,2p)n at 39.5° MeV
θ3 = θ4 = 40.5°
φ34 = 135.7°
MT I–IV
MT I–III
T5 = 4.0 MeV
T35 = T45 = T34

Fig. 3

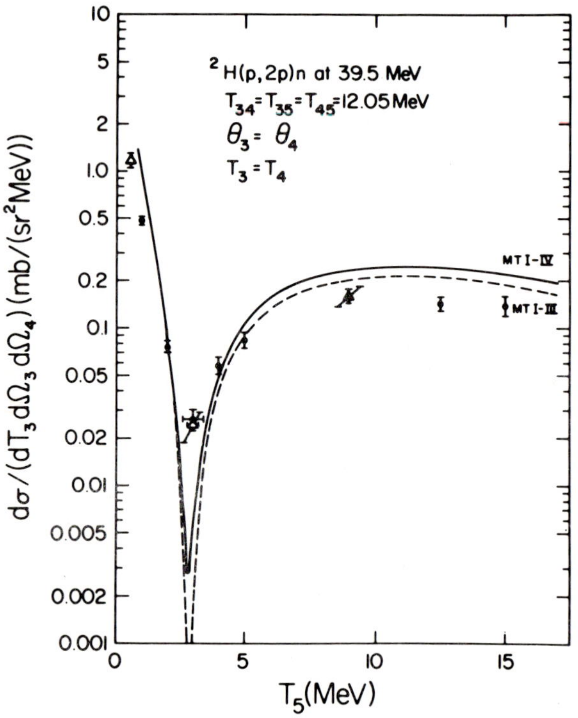

²H(p,2p)n at 39.5 MeV
T34 = T35 = T45 = 12.05 MeV
θ3 = θ4
T3 = T4
MT I–IV
MT I–III
dσ/(dT3 dΩ3 dΩ4) (mb/(sr²MeV))
T5(MeV)

Fig. 4

At 23.0 MeV Petersen et al. [14] found discrepancies as large as 30 % between theory and experiment with respect to the height of the quasi-free scattering peaks. An improvement in the form of the NN interaction potential will be necessary to resolve the present difficulty at low T_5 values. With regard to the discrepancies at the location of the interference minimum, the same comments can be made as above. For large T_5 energy values the data give closer agreement with the preferred MT13 local potential set prediction. Note that in this region all data points are within two standard deviations of the theoretical predictions. Unfortunately there exists an intrinsic difficulty in separating out the effects due to the off-energy-shell behaviour of the two nucleon T-matrices and those due to genuine three-body forces. Both effects should be strongest in the kinematical situations chosen in the present experiment. However no conclusions can be drawn regarding the latter points before improvements in the calculations have been made by incorporating a NN tensor force which couples the $^3S_1$ and $^3D_1$ states of the deuteron with a deuteron D-state probability of 6 - 8 % and other NN interactions characterized by orbital angular momenta $\ell > 0$. Furthermore the two-nucleon T-matrix elements containing off-energy-shell variation should maintain the same on-energy-shell behaviour.

REFERENCES

[1] W.M. Kloet and J.A. Tjon, Nucl. Phys. A210 (1973) 380.
[2] M. Jain, J.G. Rogers and D.P. Saylor, Phys. Rev. Lett. 31 (1973) 838.
[3] R.A. Malfliet and J.A. Tjon, Nucl. Phys. A127 (1969) 161.
[4] M. Jain and G.D. Doolen, Phys. Rev. C8 (1973) 124.
[5] W.M. Kloet and J.A. Tjon, Ann. Phys. (N.Y.) 79 (1973) 407.
[6] J.C. Aarons and I.H. Sloan, Nucl. Phys. A182 (1972) 369.
[7] S.C. Pieper, Nucl. Phys. A193 (1972) 529.
[8] P. Doleschall, Nucl. Phys. A220 (1974) 491.
[9] C. Stolk and J.A. Tjon, Phys. Rev. Lett. 35 (1975) 985.
[10] A.M. McDonald, Ph. D. Thesis, University of Manitoba (1976), unpublished.
[11] A.M. McDonald, D.I. Bonbright, W.T.H. van Oers, J.W. Watson, J.C. Rogers, J.M. Cameron, J. Soukup, W.M. Kloet and J.A. Tjon, Phys. Rev. Lett. 34 (1975) 488.
[12] H. Klein, H. Eichner, H.J. Helten, H. Kretzer, K. Prescher, H. Stehle and W.W. Wohlfarth, Nucl. Phys. A199 (1973) 169.
[13] J. Bruinsma, W. Ebenhöh, J.H. Stuivenberg and R. van Wageningen, Nucl. Phys. A228 (1974) 52.
[14] E.L. Petersen, M.J. Haftel, R.G. Allas, L.A. Beach, R.O. Bondelid, P.A. Treado, J.M. Lambert, M. Jain and J.M. Wallace, Phys. Rev. C9 (1974) 508.

* Work supported in part by the Atomic Energy Control Board of Canada.

NONLOCALITY IN (HELIUM-HELIUM) INTERATOMIC POTENTIALS

Y.S.T.Rao
Department of Physics, Calicut University
Calicut 673 635, India

INTRODUCTION

Interatomic potentials are usually taken to be purely local. With this assumption, the potential is obtained through the use of Born-Oppenheimer approximation. This usually leads to a long range attractive Van der Waal's potential and a short range steep repulsion. When such potentials are used in the study of many body systems the results are not very satisfactory. For instance in the Brueckner theoretic calculations[1] of the ground state of Liquid Helium-3, one obtains only 0.10 K per particle for the binding energy at a saturation density of 0.0053 particles per Angstrom cube(corresponding to the Fermi momentum k_f =0.54 A^{-1}). The corresponding observed values are 2.5 K at 0.0166 A^{-3} (and k_f =0.79 A^{-1}).

The low density and poor binding resulting from these calculations is due to the large radius of the repulsive core. Within the framework of local potentials, it is not possible to reduce the core radius. Though it is possible that higher order corrections in Brueckner theory might improve the results, it is worth studying the consequences of replacing part of the local repulsion by a momentum dependent repulsion.

EFFECT OF NONLOCALITY

If we assume that we do not know anything about the internal structure of Helium atoms, then the interaction between two atoms is to be deduced only from the phaseshifts when one atom is scattered by another. However it is known that the set of phase-shifts determine the potential uniquely only if we assume that the potential is local. If we allow nonlocal momentum dependent potentials, phaseshifts do not determine the potential uniquely anymore. The uncertainity in the potential, when the phaseshifts are specified is referred to as the off shell uncertainity. It is known from the experience with nuclear matter and with finite nuclei[2], that different phaseshift-equivalent potentials yield different results. By a suitable choice of offshell continuation the binding energy of nuclear matter can be changed by as much as 10 Mev per particle.

In the same spirit, nonlocal potentials that are phaseequivalent to the local Helium-Helium interatomic potential of Frost and Musulin[4] were generated and were used in the lowest order Brueckner calculations of Liquid Helium-3. As in the case of nuclear matter, here also there is a linear relation between the binding energy per particle and the hardness of the potential as measured by the wound integral K. As the potential is made softer and softer, the binding energy and the equilibrium density increase. It can be seen from the table below that a very soft potential, say with K = 0.2 K_L (where K_L is the wound integral for the local potential), gives a remarkable improvement in the eesults[3]. This potential corresponds to a repulsive core at about c = 1.6 A instead of being at the usual value of 2.6 A. The additional repulsion is reflected through the momentum dependence.

K/K_T k_f=	0.76	0.80	0.84
0.0	2.93	3.62	4.59
0.2	1.62	1.73	1.74
0.4	0.70	0.52	0.30
0.6	0.12	-0.33	-0.86
0.8	-0.51	-0.96	-1.55
1.0(local)	-0.93	-1.44	-1.95

Table 1: Binding energy per particle (in ^{o}K) as a function of Fermi momentum k_f (in A^{-1}) and average wound integral K, for nonlocal phaseshift equivalent potentials.

HOW TO GENERATE NONLOCALITY

To clarify the situation, we now give an empirical method to find out if there is nonlocality in the interatomic interaction. Consider the scattering of a Helium atom by another below the inelastic threshold (a few eV). If the phaseshifts obtained by the exact calculation of the six-body-system(two nuclei and four electrons) disagree with the phaseshifts obtained from the local potential given by the Born-Oppenheimer approximation, then we have the evidence of nonlocality. However such calculations are not practicable and no calculations on such lines exist.

A more useful approach would be through the use of the generator coordinate method(5). Instead of constraining the nuclei at a fixed distance of R, as is done in the conventional calculations, we could use R as the generator coordinate. Consider the more general wavefunction Ψ (x, r_1, r_2, r_3, r_4) where x is the internuclear separation and r_i are the electronic coordinates. We restrict only to such class of wavefunctions that give identical expectation values for the internuclear separation $\langle x \rangle$ =R, and standard deviation σ of the same. Within such a class, we minimise the expectation value of the total Hamiltonian by proper choice of dependence on r_i . This expectation value thus obtained, V(R,σ) is our nonlocal analogue of the local potential, to which it reduces as tends to zero. From σ dependence of V(R,σ), one can estimate the nonlocality. For better results full generator coordinate method is to be used.

REFERENCES

1. I.Ramarao and Y.S.T.Rao Pramana 5 (1975) 227.
   K.A.Brueckner and J.A.Gammel: Phys.Rev.109(1958)1040.
2. M.I.Haftel and F.Tabakin Phys.Rev. C3(1971)921.
3. Y.S.T.Rao and I.Ramarao: Pramana (to be published.)
4. A.A.Frost and B.Musulin: J.Chem.Phys. 46(1954)378.
5. D.L.Hill and J.A.Wheeler: Phys.Rev. 89(1953)1102.
6. B.Giraud and J.Le Tourneux: Nucl.Phys.A197(1972)410.

THREE BODY PROBLEMS WITH LOCAL POTENTIALS IN COORDINATE SPACE

T.Sasakawa and T.Sawada[+]

Department of Physics, Tohoku University, Sendai, Japan

Department of Applied Mathematics, Faculty of Engineering Science, Osaka University, Toyonaka, Japan[+]

The three body problem is formulated in the coordinate space by the method proposed by Sasakawa several years ago[1]. By this method, one can treat a local nucleon-nucleon potential exactly without separable approximation as usually done in the momentum space method. The scattering amplitude is expressed in a manner that satisfies the unitarity of the S-matrix. One can treat bound states and resonances, if any, as the poles of the amplitude. The denominator and the numerator of the amplitude are determinants, whose dimensionalities are equal to the number of components of the asymptotic waves. Each element of this determinent and the numerator is calculated by solving the Schrödinger-Faddeev differential equation. Each component of the asymptotic wave serves as the boundary condition of each differential equation.

If we denote by $|u\rangle$ the wave function of deuteron times the spherical Bessel function, the Faddeev equation reads

$$|\psi\rangle = |u\rangle + GV\Lambda\mathcal{J}|\psi\rangle, \qquad (1)$$

where $\mathcal{J}$ denotes the sum of permutation operaters $P_{123} + P_{132}$, and Λ the Racah coefficient for spin-isospin functions. $|\psi\rangle$ and $|u\rangle$ are 3×1 matrices, while G and V are 3×3 matrices. Let $|\varphi\rangle$ be the wave function of deuteron and $|\varphi\rangle\langle\varphi|$ the projection operator on the deuteron state. Let g be the two-body Green function, which vanishes when the spectator particle is removed far from the deuteron. We denote by $|w\rangle$ the product of the deuteron wave function and the spherical Hankel function. Then Eq.(1) reads

$$|\psi\rangle = |u\rangle - |w\rangle T^{(e)} + |\varphi\rangle g\langle\varphi|V\Lambda\mathcal{J}|\psi\rangle + G(1 - P)V\Lambda\mathcal{J}|\psi\rangle, \qquad (2)$$

where $T^{(e)}$ denotes the elastic amplitude defined by

$$T^{(e)} = \langle u|V\Lambda\mathcal{J}|\psi\rangle . \qquad (3)$$

We denote by G^o the Green function in the three-body free space with the boundary condition that vanishes at large hyperradius. The Green function G^o is real. We define a real matrix $\mathcal{R}$ by

$$\mathcal{R} = V + VG^o\mathcal{R} \qquad . \qquad (4)$$

We denote by H_n the Hankel function of integral order, divided by R^2 (R ; the hyperradius), times the K - harmonics.
Then we can show that

$$G = G' - (1 + G^o\mathcal{R})\sum_n |H_n\rangle\langle F_n| (1 + VG). \qquad (5)$$

Using Eq.(5) into Eq.(2), we obtain

$$|\psi\rangle = |u\rangle - |w\rangle T^{(e)} - (1 + G^o\mathcal{R})\sum_n |H_n\rangle T_n^{(B)} + \{G^o\mathcal{R} - G'PV + |\varphi\rangle g\langle\varphi|V\}\Lambda\mathcal{J}|\psi\rangle, \qquad (6)$$

where

$$G' = G^o + G^oVG'$$

and $T^{(B)}$ is the breakup amplitude defined by

$$T_n^{(B)} = \langle F_n| (1 + VG)(1 - P)V\Lambda\mathcal{J}|\psi\rangle \qquad . \qquad (7)$$

If we eliminate $|\psi\rangle$ from Eqs.(3),(6) and (7), we obtain on-shell algebraic equations for $T^{(e)}$ and $T_n^{(B)}$.

In practice, the operator $\mathcal{G}$ involved in Eq.(3) makes the integrand singular. Therefore, we must modify the above formalism slightly. Let $f(\rho)$ be a short range arbitrary function, subjects to a condition

$$-\frac{2M}{\hbar^2} k_c \int_c^{\infty} j_\ell(k_c\rho) f(\rho)\rho^2 d\rho = 1 \qquad . \tag{8}$$

Then, the function $\bar{h}_\ell(\rho)$, which is the solution of the equation

$$\left[\frac{d^2}{d\rho^2} + \frac{2}{\rho}\frac{d}{d\rho} - \frac{\ell(\ell+1)}{\rho^2} + k_c^2\right]\bar{h}_\ell(\rho) = \frac{2M}{\hbar^2} f(\rho) \tag{9}$$

is regular but behaves asymptotically as $ih_\ell^{(1)}(k_o\rho)$. Denoting

$$|h_\ell\rangle \equiv i\, h_\ell^{(1)}(k_c\rho) \quad , \quad |\bar{h}_\ell\rangle \equiv \bar{h}_\ell(\rho) \quad , \tag{10}$$

we introduce a two-body Green function $\bar{g}$, which is regular near the origin and vanishes at large ρ , by

$$\bar{g} = g - |h_\ell\rangle\langle j_\ell| + |\bar{h}_\ell\rangle\langle j_\ell| \quad . \tag{11}$$

Defining $|\bar{w}\rangle$ by the product of $|\bar{h}_\ell\rangle$ and $|\varphi\rangle$, we rewrite Eq.(6), in place of $|w\rangle$ by $|\bar{w}\rangle$ and g by $\bar{g}$. Defining the wave matrix Ω by

$$\Omega = \frac{1}{1 - \{G^o\mathcal{R} - G'PV + |\varphi\rangle g\langle\varphi|\, V\}\Lambda\, \mathcal{G}} \quad , \tag{12}$$

we eliminate $|\psi\rangle$ from Eqs.(3),(7) and Eq.(6) so modified. Then we obtain the on-shell algebraic equations

$$T^{(e)} = \langle u|V\Lambda\mathcal{G}\Omega|u\rangle - \langle u|V\Lambda\mathcal{G}\Omega|\bar{w}\rangle T^{(e)} - \sum_n \langle u|V\Lambda\mathcal{G}\Omega(1 + G^o\mathcal{R})|H_n\rangle T_n^{(B)}$$

and

$$T_n^{(B)} = \langle F_n|\bar{\mathcal{R}}\Lambda\mathcal{G}\Omega|u\rangle - \langle F_n|\bar{\mathcal{R}}\Lambda\mathcal{G}\Omega|\bar{w}\rangle T^{(e)} - \sum_{n'} \langle F_n|\bar{\mathcal{R}}\Lambda\mathcal{G}\Omega(1 + G^o\mathcal{R})|H_{n'}\rangle T_{n'}^{(B)}$$

$$- \sum_{n'}\langle F_n|\mathcal{R}|H_{n'}\rangle T_{n'}^{(B)} \quad ,$$

$$\bar{\mathcal{R}} \equiv \mathcal{R} - G_0^{-1}G'PV.$$

$\Omega|u\rangle$, $\Omega|\bar{w}\rangle$ and $\Omega|H_n\rangle$ are obtained numerically by solving the original differential equations taking the boundary conditions at large distances as $|u\rangle$, $|\bar{w}\rangle$ and $|H_n\rangle$, respectively.

1) T.Sasakawa, Supplement, Prog. Theor. Phys.27(1963),1.
T.Sasakawa, Treatment of Three-Body Problems in Coordinate Space, International Symposium on Mathematical Problems in Theoretical Physics (Springer, 1975) pp.491-496.

CONTRIBUTOR INDEX

VII International Conference

on "Few Body Problems in Nuclear and Particle Physics"

December 29, 1975 - January 3, 1976

List of Registrants

Afnan, I.R.
School of Physical Sciences,
Flinders University
Bedford Park, S.A. 5042, Australia

Afzal, S.A.
Atomic Energy Centre,
P.O. Box 164, Dacca-2
Bangladesh

Ahmad, M.
Department of Physics
Kashmir University,
Srinagar-190006

Aitchison, I.J.R.
Department of Theoretical Physics
University of Oxford, 12 Parks Road
Oxford OX1-3PQ, England

Akaishi, Y.
Department of Physics,
Hokkaido University, Kita 10
Nishi 8, Sapporo, Japan

Ali, S.
Atomic Energy Centre, P.O.Box 164,
Dacca-2, Bangladesh

Allas, R.G.
Naval Research Laboratory
Code 6611, Washington, D.C. 20375
U.S.A.

Alt, E.O.
Institut für Physik,
Universität Mainz, D65 Mainz,
Jakob-Welderweg 11,
West-Germany

Auluck, F.C.
Department of Physics and Atrophysics
University of Delhi
Delhi-110007.

Bakker, B.L.G.
Natuurkundig Laboratorium der Vrije
Universiteit, De Boelelaan 1081,
Amsterdam-Buitenveldert,
The Netherlands

Banerjee, D.K.
Physics Department
University College of Science,
92 Acharya P.C. Road, Calcutta-700009

Banerjee, M.
Department of Physics and Astronomy
University of Maryland
College Park, Maryland 20742, U.S.A.

Bansal, R.K.
Department of Physics
Panjab University,
Chandigarh-160014

Bencze, G.
Central Research Institute for Physics
H-1525, Budapest, P.O. Box 49,
Hungary

Bhamathi, G.
Department of Theoretical Physics
University of Madras
Madras-600025

Bhasin, V.S.
Department of Physics and Astrophysics
University of Delhi, Delhi-110007

Bhatia, R.P.
Regional College
Ajmer-305004

Bhowmik, B.
Department of Physics and Astrophysics
University of Delhi
Delhi-110007

Bleszynski, M.
Institute of Nuclear Physics
Radzikovskiego 152
31-342 Krakow, Poland

Bleuler, K.
Institut für Theoretische Kernphysik
D-5300 Bonn Nassallee 14-16
West-Germany

Bondelid, R.O.
Naval Research Laboratory
Code 6610, Washington,
D.C. 20375, U.S.A.

Bosco, B.
Instituto di Fisica Teorica
Universita di Firenze
50125 Firenze, Italy

Bouchez, B.
Institut des Sciences Nucléaires
Universite de Grenoble
Cedex No. 257-38, Grenoble, France

Brayshaw, D.D.
SLAC, Box 4349
Stanford University
Stanford, CA 94305, U.S.A.

Cameron, J.M.
Nuclear Research Centre
The University of Alberta
Edmonton, Alberta, Canada

Chandler, C.
Department of Physics and Astronomy
University of New Mexico
New Mexico 87131, U.S.A.

Chatterji, D.
Department of Physics
Visva-Bharati University
Santiniketan-732135, West-Bengal

Chavda, L.K.
Physics Department
I.I.T., Bombay-400076

Conzett , H.E.
Lawrence Berkeley Lab.
University of California
Berkeley, CA-94720, U.S.A.

Dalitz, R.
Department of Theoretical Physics
Oxford University, 12 Parks Road,
Oxford OX1-3PQ , England

Deka, A.K.
Physics Department
Dibrugarh University
Dibrugarh-786004

Divatia, A.S.
VEC Project, BARC
I-AF, Bidhar Nagar
Calcutta-700064

Doornbos, J.
Natuurkundig Laboratorium
Vrye Universiteit, Nat-Lab VV,
de Boelelaan 1081, Amsterdam
The Netherlands

Englefield, M.J.
Department of Mathematics
Monash University
Clayton, VIC 3168, Australia

Fabre de la Ripelle, M.
Division de Physique Théorique
Institut de Physique Nucléaire
I.P.N. 91406 Orsay, France

Fang, K.K.
Department of Physics
Hensselaer Polytechnic Institute
Troy, N.Y.-12181, U.S.A.

Ganguly, N.K.
VEC Shed, BARC
Trombay, Bombay-400085

Gautam, V.
Theoretical Physics Department
Indian Association for the Cultivation
of Science, Calcutta-700032

Gerjuoy, E.
Department of Physics,
Pittsburgh University,
Pittsburgh, Pennsylvania-15213,U.S.A.

Glantz, L.
Tandem Accelerator Laboratory
Box 533 S-751 21 Uppsala, Sweden

Gross, F.
College of William and Mary,
Department of Physics
Williamsburg, Va 23185, USA

Gruebler, W.
Laboratorium für Kernphysik
Eidg. Technische Hochschule
Hönggerberg CH-8049 Zurich, Switzerland

Gupta, V.K.
Department of Physics and Astrophysics
University of Delhi,
Delhi-110007.

Hackenbroich, H.H.
Institut für Theoret. Physik der
Universitat Köln, 5 Köln
Universitätsstr 14, West-Germany.

Haftel, M.I.
Naval Research Laboratory
Washington, D.C. 20375, U.S.A.

Hanna, S.S.
Department of Physics
Stanford University
Stanford, California 94305, U.S.A.

Hans, H.S.
Department of Physics,
Panjab University
Chandigarh-160014

Heiss, P.
Institut für Theoret. Physik
der Universitat Köln, 5 Köln
Universitätsstr 14, West-Germany

Henley, E.M.
Physics Department, FM-15
University of Washington
Seattle, Wa, 98195, U.S.A.

Holinde, K.
Institut für Theoretische Kernphysik
D-53 Bonn, Nubollee 14-16,
West-Germany

Horgan, R.
Department of Physics
Oxford University, 12 Parks Road
Oxford, England.

Igo, G.
Department of Physics
University of California
405 Hilgard Avenue
Los Angeles, Calif. 90024,U,S.A.

Iyengar, P.K.
BARC, Bombay-400085

Jacob, H.
Department of Physics and Astrophysics
University of Delhi
Delhi-110007

Jain, A.K.
Van de Graaff Laboratory
BARC, Bombay-400 085

Jain, B.K.
Nuclear Physics Division
BARC, Bombay-400 085

Jain, M.
Texas A and M University and L.A.S.L.
PDORW MS=458, L.A.S.L. Los Alamos,
N.M.-87545, U.S.A.

Kabir, P.K.
Department of Physics,
University of Virginia
Charlottesville, Virginia 22901, U.S.A.

Karlsson, B.R.
Institute of Theoretical Physics
Fack, S-40220 Goteborg, Sweden

Katyal, D.L.
St. Stephen's College
University of Delhi, Delhi-110007

Kaushal, R.S.
Ramjas College,
University of Delhi, Delhi-110007

Kim, Y.E.
Department of Physics
Purdue University, W. Lafayette,
Indiana 47907, U.S.A.

Kok, L.P.
Institute for Theoretical Physics
University of Groningen, P.O.Box 800,
Groningen, The Netherlands

Komarov, V.
Institute of Nuclear Physics,
Moscow State University
Moscow, U.S.S.R.

Kothari, D.S.
Department of Physis and Astrophysics
University of Delhi
Delhi-110007

Kothari, L.S.
Department of Physics and Astrophysics
University of Delhi
Delhi-110007

Kouri, D.
Department of Physics
University of Houston,
Houston, Texas 77004, U.S.A.

Kuhn, B.
Zentralinstitut für Kernforschung der
Akademie der Wissenschaften der DDR
Rossendorf, East-Germany

Kumpf, H.
Zentralinstitut für Kernforschung der
Akademie der Wissenschaften der DDR
Rossendorf, East-Germany

Kundu, D.N.
Saha Institute of Nuclear Physics
92, Acharya P.C. Road
Calcutta-700009

Lagu, A.V.
Physics Department
Baranas Hindu University
Varanasi-221005

Lambert, J.M.
Department of Physics
University of Georgetown
37th and O Streets, N.W. Wash.,
D.C. 20007, U.S.A.

Levin, F.S.
Physics Department, Brown University
Providence, Rhode Island 02912, U.S.A.

Lim, T.K.
Department of Physics and Atmospheric
Sciences, Drexel University,
Philadelphia, Pennsylvania-19104,U.S.A.

Lodhi, M.A.K.
Department of Physics
Texas Tech University
P.O. Vox 4180, Lubbock, Texas 79409
U.S.A.

Lovas, I.
Central Research Institute for Physics
H-1525 Budapest 114,
P.O. Box 49, Hungary

Maheshwari, C.
Physics Department
Banaras Hindu University
Varanasi-221005

Majumdar, C.K.
Department of Physics
Calcutta University
92 Acharya P.C. Road, Calcutta-700009

Mantri, A.N.
Physics Department
Banaras Hindu University
Varanasi-221005

Mathur, V.S.
Department of Physics
Banaras Hindu University
Varanasi-221005

McCarthy, I.E.
Department of Physics
The Flinders University of South Australia
Bedford Park, SA 5042, Australia.

McKellar, B.H.J.
School of Physics
University of Melbourne,
Parkville ViC 3052, Australia

Mehdi, S.S.
Department of Physics and Astrophysics
University of Delhi,
Delhi-110007

Mehta, G.K.
Department of Physics
I.I.T., Kanpur 208016

Mehta, M.K.
Van de Graaff Lab,
Nuclear Physics Division
BARC, Bombay-400085

Menon, M.G.K.
Department of Electronics,
Vigyan Bhawan, Annexe-II,
New Delhi-110001

Mitra, A.N.
Department of Physics and Astrophysics
University of Delhi
Delhi-110007

Monga, S.K.
Hans Raj College,
University of Delhi
Delhi-110007

Moravcsik, M.
Institute of Theoretical Science
University of Oregon
Eugene, Oregon-97403, U.S.A.

Morioka, S.
Department of Nuclear Engineering
Osaka University, Suita,
Osaka, Japan

Mosconi, B.
Instituto di Fisica Teorica
Universita di Firenze,
50125 Firenze, Italy

Mukherjee, Saila
Department of Physics
H.P. University
Simla

Mukherjee, Shankar
Physics Department
Banaras Hindu University
Varanasi-221005

Mukherjee, Suprokash
Saha Institute of Nuclear Physics
92 Acharya P.C. Road, Calcutta-700009

Nagchaudhuri, B.D.
J.N.U., New Delhi-110057

Namyslowski, J.
Institute of Theoretical Physics
Warsaw University, 00-681,
Warsaw, ul. Hoza 69, Poland

Narsimhan, V.L.
Department of Physics,
I.I.T., Powai, Bombay-400076

Northcliffe, L.
M.S. 458 Los Alamos Scientific Lab.,
Los Alamos, New Mexico-87544
U.S.A.

Noyes, H.P.
Stanford Linear Accelerator Center
Stanford, Calif.-94305, U.S.A.

Oryu, S.
Department of Physics,
Faculty of Science and Technology
Science University of Tokyo
Noda, Chiba 278, Japan

Pancholi, S.C.
Department of Physics and Astrophysics
University of Delhi
Delh-110007

Pandya, S.P.
Physical Research Laboratory
Navrangpura, Ahmedabad-380 009

Parashar, D.N.
Department of Physics
A.R.S.D. College
New Delhi-110021

Petersen, E.L.
Naval Research Laboratory
Code 6611, Washington, D.C. 20375
U.S.A.

Piragino, G.
Instituto di Fisica Dell'Universita di
Torino, Corso M. D'Azeglio, 46 - 10125
Torino, Italy

Pisent, G.
Instituto di Fisica Dell'Universita
Via Marzolo 8, 35100 Padova, Italy

Prakash, Y.
Physics Department
Jammu University
Jammu-180001

Pugh, H.G.
Department of Physics and Astronomy
University of Maryland, College Park,
Maryland-20742, U.S.A.

Rajaraman, R.
Department of Physics and Astrophysics
University of Delhi
Delhi-110007

Rajasekaran, G.
Tata Institute of Fundamental Research
Homi Bhabha Road
Bombay-400005

Ramachandran, R.
Department of Physics
I.I.T.,Kanpur-208016

Ramanna, R.
BARC, Bombay-40085

Rao, Y.S.T.
Department of Physics
Calicut University
Calicut-673635

Redish, E.
Department of Physics
University of Maryland
College Park, Maryland 20742
U.S.A.

Richardson, J.R.
Triumf
University of British Columbia
Vancouver,B.C. Canada

Rodney, W.
Nationa. Science Foundation
1800 G Street, N.W.,
Washington, D.C. 20550, U.S.A.

Saha, A.
Bose Institute
93/1, Acharya P.C. Road
Calcutta-700009

Sandhas, W.
Physikalisches Institut,
Universität Bonn
11-13, Endenicher Allee
D-53 Bonn, West-Germany

Sarma, N.
Nuclear Physics Division
BARC, Bombay-400085

Sasakawa, T.
Department of Physics,
Tohoku University
Katahiracho, 980-Sendai,Japan

Sauer, P.U.
Theoretical Physics
Technical University
Appelstrasse 1,3000 Hannover
West-Germany

Scadron, M.
Physics Department,
University of Arizona
Tucson, Arizona-85721, U.S.A.

Schmid, E.
Institut Fuer Theoretische Physik
74 Tuebingen, Auf der Morgenstelle 14
West-Germaby.

Schwager, J.
Institut f Theoret. Physik
Universtat Tubingen
Morgenstelle, D-74 Tubingen
West Germany

Shah, M.S.
Department of Physics
Faculty of Science, M.S. University
Baroda-390002

Sharma, J.S.
Ramjas College,
University of Delhi
Delhi-110007

Sharma, V.K.
Dharma Samaj College
Aligarh-202001

Shaw, G.
Department of Physics
University of California
Irvine, California-92664, U.S.A.

Singh, P.P.
Physics Department
Indiana University,
Swain West, Bloomington, IND.,47401,U.S.A.

Singh, V.
Tata Institute of Fundamental Research
Homi Bhabha Road, Bombay-400005

Slaus, I.
Institute "R. Bošković"
POB 1016, Bijenička 54,
41001 Zagreb, Yugoslavia

Sloan, I.H.
School of Mathematics,
University of N.S.W.
Kensington, N.S.W.-2033, Australia

Slobodrian, R.J.
Département de Physique
Université Laval
G1K 7P4 Quebec P.Q. Canada

Sood, P.C.
Nuclear Research Laboratory,
Banaras Hindu University
Varanasi-221005

Spruch, L.
Department of Physics
New York University
New York, New York-10003, U.S.A.

Srinivasa Rao, K.
Matscience, Adyar,
Madras-600020

Srivastava, B.K.
Physics Department
I.I.T., Kharagpur-2
West Bengal

Srivastava, M.K.
Department of Physics
University of Roorkee
Roorkee-247667

Sundaresan, M.K.
Department of Physics,
Carleton University
Ottawa, Canada

Tamagaki, R.
Department of Physics
Kyoto University
Sakyo-ku, Kyoto, Japan

Tandy, P.
Department of Physics
University of Maryland
College Park, Maryland 20742,U.S.A.

Tjon, J.A.
Institute for Theoretical Physics
Sorbonne laan 4, Utrecht
The Netherlands

Tubis, A.
Department of Physics
Purdue University
W. Lafayette, IN-47907, U.S.A.

van Haeringen, H.
Department of Physics
Free University
de Boelelaan 1081, Amsterdam,
The Netherlands

van Oers, W.T.H.
Department of Physics
University of Manitoba
Winnipeg, Manitaba, Canada R3T 2N2

van Wageningen, R.
Natuurkundig Laboratorium der Vrije
Universiteit, De Boelelaan 1081,
Amsterdam-Buitenveldert,
The Netherlands

Vanzani, V.
Istituto di Fisica "G.Galilei"
via Marzolo 8, 35100-Padova,
Italy

Verma, A.R.
National Physical Laboratory
Hillside Road, New Delhi-110012

Vinh Mau, R.
Laboratoire de Physique Theorique
Universite de Paris VI, Tour 16, 4,
Place Jussieu-75230 Paris Cedex
05-France

Warke, C.S.
Tata Institute of Fundamental Research
Homi Bhabha Road
Bombay-400005

Weber, H.J.
Department of Physics
University of Virginia
Charlottesville, Va 22901, U.S.A.

Wolfe, D.M.
Department of Physics and Astronomy,
University of New Mexico
Albuquerque, 87131, U.S.A.

Yuasa, T.
Institut de Physique Nucleairé
B.P. No. 1, 91406 Orsay,
France

Zankel, H.
Institut Für Theoretische Physik der
Universität Graz, Universitätsplatz 5,
A-8010 Graz, Austria

Zingl, H.
Institut für Theoretische Physik der
Universität Graz, Universitatsplatz 5,
A-8010 Graz, Austria

List of Observers

Anand, J.D.
Department of Physics & Astrophysics
University of Delhi
Delhi-110007

Ansari, A.
Institute of Physics
A/105 Saheed Nagar
Bhubaneswar-751007

Bondopadhyay, D.
Department of Physics & Astrophysics
University of Delhi
Delhi-110007

Garg, V.P.
Department of Physics
Panjabi University
Patiala-147002

Govil, I.M.
Physics Department
Panjab University
Chandigarh-160014

Goyal, D.P.
Department of Physics & Astrophysics
University of Delhi
Delhi-110007

Goyal, R.K.
Department of Physics
Panjabi University
Patiala-147002

Jayaraman, J.
Department of Physics
Calicut University
Calicut-673635

Nair, S.C.K.
Physics Department
Calicut University
Calicut-673635

Naqvi, J.
Department of Physics
Aligarh Muslim University
Aligarh, U.P.

Praharaj, C.R.
Institute of Physics
A/105 Saheed Nagar
Bhabaneswar-751007

Rizvi, S.H.
Physics Section
Engineering College
Aligarh Muslim University
Aligarh, U.P.

Sharma, D.K.
Department of Physics
Science College
Patna University, Patna-800005

Sharma, M.L.
K.G.K. College,
Moradabad-244001

Sharma, S.D.
Physics Department
Panjabi University
Patiala-147002

Shivpuri, R.K.
Department of Physics & Astrophysics
University of Delhi
Delhi-110007

Sirohi, A.P.S.
K.G.K. College
Moradabad-244001

Thakur, J.N.
Department of Physics
Science College, Patna University
Patna-800005

Zahid, M.
Physics Department
Jamia Millia
Jamianagar, New Delhi-110025